THE

# PRACTICAL

# MODEL CALCULATOR,

FOR THE

ENGINEER, MECHANIC, MACHINIST,
MANUFACTURER OF ENGINE-WORK, NAVAL ARCHITECT,
MINER, AND MILLWRIGHT.

---

BY


OLIVER BYRNE,

CIVIL, MILITARY, AND MECHANICAL ENGINEER.

*Compiler and Editor of the "Dictionary of Machines, Mechanics, Engine-work, and Engineering;" Author of "The Companion for Machinists, Mechanics, and Engineers;" Author and Inventor of a New Science, termed "The Calculus of Form," a substitute for the differential and Integral Calculus; "The Elements of Euclid by Colours," and numerous other Mathematical and Mechanical Works. Surveyor-General of the English Settlements in the Falkland Isles. Professor of Mathematics, College of Civil Engineers, London.*


---


PHILADELPHIA:
PUBLISHED BY HENRY CAREY BAIRD.
(SUCCESSOR TO E. L. CAREY.)
SOUTH-EAST CORNER MARKET AND FIFTH STREETS.
1852.

STEREOTYPED BY L. JOHNSON AND CO.
PHILADELPHIA.
PRINTED BY T. K. AND P. G. COLLINS.

THE

# PRACTICAL MODEL CALCULATOR.

## WEIGHTS AND MEASURES.

### THE UNIT OF LENGTH.

THE YARD.—If a pendulum vibrating seconds in vacuo, in Philadelphia, be divided into 2509 equal parts, 2310 of such equal parts is the length of the standard yard; the measures are taken on brass rods at the temperature of 32° Fahrenheit. This yard will not be in error the ten-millionth part of an inch.

2310 : 2509 as 1· to 1·086142 nearly.

### THE UNIT OF WEIGHT.

*The Pound*, avoirdupois, is 27·7015 cubic inches of distilled water, weighed in air, at the temperature of maximum density, 39°·82; the barometer at 30 inches.

### THE LIQUID UNIT.

*The Gallon*, 231 cubic inches, contains 8·3388822 pounds avoirdupois, equal 58372·1754 grains troy of distilled water, at 39°·82 Fah.; the barometer at 30 inches.

### UNIT OF DRY CAPACITY.

*The Bushel* contains 2150·42 cubic inches, 77·627412 pounds avoirdupois, 543391·89 grains of distilled water, at the temperature of maximum density; the barometer at 30 inches.

The French unit of length or distance is the metre, and is the ten-millionth of the quadrant of the globe, measured from the equator to the pole.

The French *Metre* = 3·2808992 English *feet* linear measure = 39·3707904 *inches*.

| *For Multiples the following Greek words are used:* | | *For Divisors the following Latin words are used:* | |
|---|---|---|---|
| *Deca* for ............ | 10 times. | *Deci* for the | 10*th* part. |
| *Hecto* — ............ | 100 times. | *Centi* — | 100*th* part. |
| *Kilo* — ............ | 1000 times. | *Milli* — | 1000*th* part. |
| *Myria* — ............ | 10000 times. | Thus a *Kilometre* = | 1000 metres. |

$$Millimetre = \frac{\text{metre}}{1000}$$

The square *Deca Metre*, called the *Are*, is the element of land measure in France, which = 1076·42996 *square feet* English.

The *Stere* is a cubic metre = 35·316582 *cubic feet* English.

The *Litre* for liquid measure is a cubic decimetre = 1·76077 *imperial pints* English, at the temperature of melting ice; a *litre* of distilled water weighs 15434 *grains troy.*

The unit of weight is the *gramme:* it is the weight of a cubic centimetre of distilled water, or of a millilitre, and therefore equal to 15·434 *grains troy.*

The kilogramme is the weight of a cubic decimetre of distilled water, at the temperature of maximum density, 4° centigrade.

The pound troy contains 5760 grains.

The pound avoirdupois contains 7000 grains.

The English imperial gallon contains 277·274 cubic inches; and the English corn bushel contains eight such gallons, or 2218·192 cubic inches.

APOTHECARIES' WEIGHT.

| | | | | |
|---|---|---|---|---|
| Grains | | | marked | gr. |
| 20 Grains | make | 1 Scruple | — | sc. or ℈ |
| 3 Scruples | — | 1 Dram | — | dr. or ʒ |
| 8 Drams | — | 1 Ounce | — | oz. or ℥ |
| 12 Ounces | — | 1 Pound | — | lb. or ℔. |

| gr. | sc. | dr. | oz. | lb. |
|---|---|---|---|---|
| 20 = | 1 | | | |
| 60 = | 3 = | 1 | | |
| 480 = | 24 = | 8 = | 1 | |
| 5760 = | 288 = | 96 = | 12 = | 1 |

This is the same as troy weight, only having some different divisions. Apothecaries make use of this weight in compounding their medicines; but they buy and sell their drugs by avoirdupois weight.

AVOIRDUPOIS WEIGHT.

| | | | | |
|---|---|---|---|---|
| Drams | | | marked | dr. |
| 16 Drams | make | 1 Ounce | — | oz. |
| 16 Ounces | — | 1 Pound | — | lb. |
| 28 Pounds | — | 1 Quarter | — | qr. |
| 4 Quarters | — | 1 Hundred Weight | — | cwt. |
| 20 Hundred Weight | — | 1 Ton | — | ton. |

| dr. | oz. | lb. | qr. | cwt. | ton. |
|---|---|---|---|---|---|
| 16 = | 1 | | | | |
| 256 = | 16 = | 1 | | | |
| 7168 = | 448 = | 28 = | 1 | | |
| 28672 = | 1792 = | 112 = | 4 = | 1 | |
| 573440 = | 35840 = | 2240 = | 80 = | 20 = | 1 |

By this weight are weighed all things of a coarse or drossy nature, as Corn, Bread, Butter, Cheese, Flesh, Grocery Wares, and some Liquids; also all Metals except Silver and Gold.

| | | Oz. | Dwt. | Gr. | |
|---|---|---|---|---|---|
| *Note,* that 1 lb. avoirdupois | = | 14 | 11 | 15½ | troy. |
| 1 oz. — | = | 0 | 18 | 5½ | — |
| 1 dr. — | = | 0 | 1 | 3½ | — |

### TROY WEIGHT.

| | | | | | | | |
|---|---|---|---|---|---|---|---|
| Grains.....................marked Gr. | Gr. | | Dwt. | | | | |
| 24 Grains make 1 Pennyweight Dwt. | 24 | = | 1 | | Oz. | | |
| 20 Pennyweights 1 Ounce Oz. | 480 | = | 20 | = | 1 | | Lb. |
| 12 Ounces 1 Pound Lb. | 5760 | = | 240 | = | 12 | = | 1 |

By this weight are weighed Gold, Silver, and Jewels.

### LONG MEASURE.

| | | | | |
|---|---|---|---|---|
| 3 | Barley-corns............make | 1 Inch..............marked | | In. |
| 12 | Inches...................... — | 1 Foot.............. | — | Ft. |
| 3 | Feet........................ — | 1 Yard.............. | — | Yd. |
| 6 | Feet........................ — | 1 Fathom.......... | — | Fth. |
| 5 | Yards and a half....... — | 1 Pole or Rod..... | — | Pl. |
| 40 | Poles...................... — | 1 Furlong......... | — | Fur. |
| 8 | Furlongs.................. — | 1 Mile.............. | — | Mile. |
| 3 | Miles....................... — | 1 League.......... | — | Lea. |
| $69\frac{1}{6}$ | Miles nearly............ — | 1 Degree........... | — | Deg. or ° |

| In. | | Ft. | | | | | | | | |
|---|---|---|---|---|---|---|---|---|---|---|
| 12 | = | 1 | | Yd. | | | | | | |
| 36 | = | 3 | = | 1 | | Pl. | | | | |
| 198 | = | $16\frac{1}{2}$ | = | $5\frac{1}{2}$ | = | 1 | | Fur. | | |
| 7920 | = | 660 | = | 220 | = | 40 | = | 1 | | Mile. |
| 63360 | = | 5280 | = | 1760 | = | 320 | = | 8 | = | 1 |

### CLOTH MEASURE.

| | | | |
|---|---|---|---|
| 2 Inches and a quarter....make | 1 Nail...................marked | | Nl. |
| 4 Nails....................... — | 1 Quarter of a Yard.. | — | Qr. |
| 3 Quarters................... — | 1 Ell Flemish.......... | — | E F. |
| 4 Quarters................... — | 1 Yard.................. | — | Yd. |
| 5 Quarters................... — | 1 Ell English........... | — | E E. |
| 4 Qrs. $1\frac{1}{5}$ Inch............. — | 1 Ell Scotch ........... | — | E S. |

### SQUARE MEASURE.

| | | | | |
|---|---|---|---|---|
| 144 | Square Inches........make | 1 Sq. Foot..............marked | | Ft. |
| 9 | Square Feet.......... — | 1 Sq. Yard............ | — | Yd. |
| $30\frac{1}{4}$ | Square Yards........ — | 1 Sq. Pole............. | — | Pole. |
| 40 | Square Poles......... — | 1 Rood.................. | — | Rd. |
| 4 | Roods................. — | 1 Acre.................. | — | Acr. |

| Sq. Inc. | | Sq. Ft. | | | | | | | | |
|---|---|---|---|---|---|---|---|---|---|---|
| 144 | = | 1 | | Sq. Yd. | | | | | | |
| 1296 | = | 9 | = | 1 | | Sq. Pl. | | | | |
| 39204 | = | $272\frac{1}{4}$ | = | $30\frac{1}{4}$ | = | 1 | | Rd. | | |
| 1568160 | = | 10890 | = | 1210 | = | 40 | = | 1 | | Acr. |
| 6272640 | = | 43560 | = | 4840 | = | 160 | = | 4 | = | 1 |

When three dimensions are concerned, namely, length, breadth, and depth or thickness, it is called cubic or solid measure, which is used to measure Timber, Stone, &c.

The cubic or solid Foot, which is 12 inches in length, and breadth, and thickness, contains 1728 cubic or solid inches, and 27 solid feet make one solid yard.

### DRY, OR CORN MEASURE.

| | | | | |
|---|---|---|---|---|
| 2 Pints | make | 1 Quart | marked | Qt. |
| 2 Quarts | — | 1 Pottle | — | Pot. |
| 2 Pottles | — | 1 Gallon | — | Gal. |
| 2 Gallons | — | 1 Peck | — | Pec. |
| 4 Pecks | — | 1 Bushel | — | Bu. |
| 8 Bushels | — | 1 Quarter | — | Qr. |
| 5 Quarters | — | 1 Weigh or Load | — | Wey. |
| 2 Weys | — | 1 Last | — | Last. |

Pts. Gal.
8 = 1 Pec.
16 = 2 = 1 Bu.
64 = 8 = 4 = 1 Qr.
512 = 64 = 32 = 8 = 1 Wey.
2560 = 320 = 160 = 40 = 5 = 1 Last.
5120 = 640 = 320 = 80 = 10 = 2 = 1

### WINE MEASURE.

| | | | | |
|---|---|---|---|---|
| 2 Pints | make | 1 Quart | marked | Qt. |
| 2 Quarts | — | 1 Gallon | — | Gal. |
| 42 Gallons | — | 1 Tierce | — | Tier. |
| 63 Gallons or 1½ Tier. | — | 1 Hogshead | — | Hhd. |
| 2 Tierces | — | 1 Puncheon | — | Pun. |
| 2 Hogsheads | — | 1 Pipe or Butt | — | Pi. |
| 2 Pipes | — | 1 Tun | — | Tun. |

Pts. Qts.
2 = 1 Gal.
8 = 4 = 1 Tier.
336 = 168 = 42 = 1 Hhd.
504 = 252 = 63 = 1½ = 1 Pun.
672 = 336 = 84 = 2 = 1½ = 1 Pi.
1008 = 504 = 126 = 3 = 2 = 1½ = 1 Tun.
2016 = 1008 = 252 = 6 = 4 = 3 = 2 = 1.

### ALE AND BEER MEASURE.

| | | | | |
|---|---|---|---|---|
| 2 Pints | make | 1 Quart | marked | Qt. |
| 4 Quarts | — | 1 Gallon | — | Gal. |
| 36 Gallons | — | 1 Barrel | — | Bar. |
| 1 Barrel and a half | — | 1 Hogshead | — | Hhd. |
| 2 Barrels | — | 1 Puncheon | — | Pun. |
| 2 Hogsheads | — | 1 Butt | — | Butt. |
| 2 Butts | — | 1 Tun | — | Tun. |

Pts. Qt.
2 = 1 Gal.
8 = 4 = 1 Bar.
288 = 144 = 36 = 1 Hhd.
432 = 216 = 54 = 1½ = 1 Butt.
864 = 432 = 108 = 3 = 2 = 1

## OF TIME.

| | | | | |
|---|---|---|---|---|
| 60 Seconds | make | 1 Minute | marked | M. or |
| 60 Minutes | — | 1 Hour | — | Hr. |
| 24 Hours | — | 1 Day | — | Day. |
| 7 Days | — | 1 Week | — | Wk. |
| 4 Weeks | — | 1 Month | — | Mo. |
| 13 Months, 1 Day, 6 Hours, or 365 Days, 6 Hours. | — | 1 Julian Year | — | Yr. |

| Sec. | Min. | | | | |
|---|---|---|---|---|---|
| 60 = | 1 | Hr. | | | |
| 3600 = | 60 = | 1 | Day. | | |
| 86400 = | 1440 = | 24 = | 1 | Wk. | |
| 604800 = | 10080 = | 168 = | 7 | = 1 | Mo. |
| 2419200 = | 40320 = | 672 = | 28 | = 4 = | 1 |
| 31557600 = | 525960 = | 8766 = | 365¼ | = 1 Year. | |

| | Wk. | Da. | Hr. | | Mo. | Da. | Hr. | |
|---|---|---|---|---|---|---|---|---|
| Or | 52 | 1 | 6 | = | 13 | 1 | 6 | = 1 Julian Year. |

| | Da. | Hr. | M. | Sec. | |
|---|---|---|---|---|---|
| But | 365 | 5 | 48 | 48 | = 1 Solar Year. |

The time of rotation of the earth on its axis is called a *sidereal day*, for the following reason: If a permanent object be placed on the surface of the earth, always retaining the same position, it may be so located as to be posited in the same plane with the observer and some selected fixed star at the same instant of time; although this coincidence may be but momentary, still this coincidence continually recurs, and the interval elapsed between two consecutive coincidences has always throughout all ages appeared the same.

It is this interval that is called a sidereal day.

The sidereal day increased in a certain ratio, and called the *mean solar day*, has been adopted as the standard of time.

Thus, 366·256365160 *sidereal days* = 366·256365160 − 1 or 365·256365160 *mean solar days*, whence sidereal day : mean solar day : : 365·256365160 : 366·256365160 : : 0·997269672 : 1 or as 1 : 1·002737803, when 23 *hours*, 56 *minutes* 4·0996608 *sec.* of mean solar time = 1 *sidereal day;* and 24 *hours*, 3 *minutes*, 56·5461797 *sec.* of sidereal time = 1 *mean solar day.*

The *true solar day* is the interval between two successive coincidences of the sun with a fixed object on the earth's surface, bringing the sun, the fixed object, and the observer in the same plane.

This interval is variable, but is susceptible of a maximum and minimum, and oscillates about that mean period which is called a mean solar day.

*Apparent* or *true time* is that which is denoted by the sun-dial, from the apparent motion of the sun in its diurnal revolution, and differs several minutes in certain parts of the ecliptic from the mean time, or that shown by the clock. The difference is called the equation of time, and is set down in the almanac, in order to ascertain the true time.

# ARITHMETIC.

Arithmetic is the art or science of numbering; being that branch of Mathematics which treats of the nature and properties of numbers. When it treats of whole numbers, it is called *Common Arithmetic;* but when of broken numbers, or parts of numbers, it is called *Fractions.*

*Unity*, or a *Unit*, is that by which every thing is called one; being the beginning of number; as one man, one ball, one gun.

*Number* is either simply one, or a compound of several units; as one man, three men, ten men.

An *Integer* or *Whole Number*, is some certain precise quantity of units; as one, three, ten. These are so called as distinguished from *Fractions*, which are broken numbers, or parts of numbers; as one-half, two-thirds, or three-fourths.

## NOTATION AND NUMERATION.

Notation, or Numeration, teaches to denote or express any proposed number, either by words or characters; or to read and write down any sum or number.

The numbers in Arithmetic are expressed by the following ten digits, or Arabic numeral figures, which were introduced into Europe by the Moors about eight or nine hundred years since: viz. 1 one, 2 two, 3 three, 4 four, 5 five, 6 six, 7 seven, 8 eight, 9 nine, 0 cipher or nothing. These characters or figures were formerly all called by the general name of *Ciphers;* whence it came to pass that the art of Arithmetic was then often called *Ciphering.* Also, the first nine are called *Significant Figures*, as distinguished from the cipher, which is quite insignificant of itself.

Besides this value of those figures, they have also another, which depends upon the place they stand in when joined together; as in the following Table:

| &c. | Hundreds of Millions. | Tens of Millions. | Millions. | Hundreds of Thousands. | Tens of Thousands. | Thousands. | Hundreds. | Tens. | Units. |
|---|---|---|---|---|---|---|---|---|---|
| &c. | 9 | 8 | 7 | 6 | 5 | 4 | 3 | 2 | 1 |
| | | 9 | 8 | 7 | 6 | 5 | 4 | 3 | 2 |
| | | | 9 | 8 | 7 | 6 | 5 | 4 | 3 |
| | | | | 9 | 8 | 7 | 6 | 5 | 4 |
| | | | | | 9 | 8 | 7 | 6 | 5 |
| | | | | | | 9 | 8 | 7 | 6 |
| | | | | | | | 9 | 8 | 7 |
| | | | | | | | | 9 | 8 |
| | | | | | | | | | 9 |

Here any figure in the first place, reckoning from right to left, denotes only its own simple value; but that in the second place denotes ten times its simple value; and that in the third place a hundred times its simple value; and so on; the value of any figure, in each successive place, being always ten times its former value.

Thus, in the number 1796, the 6 in the first place denotes only six units, or simply six; 9 in the second place signifies nine tens, or ninety; 7 in the third place, seven hundred; and the 1 in the fourth place, one thousand; so that the whole number is read thus—one thousand seven hundred and ninety-six.

As to the cipher 0, it stands for nothing of itself, but being joined on the right-hand side to other figures, it increases their value in the same tenfold proportion: thus, 5 signifies only five; but 50 denotes 5 tens, or fifty; and 500 is five hundred; and so on.

For the more easily reading of large numbers, they are divided into periods and half-periods, each half-period consisting of three figures; the name of the first period being units; of the second, millions; of the third, millions of millions, or bi-millions, contracted to billions; of the fourth, millions of millions of millions, or tri-millions, contracted to trillions; and so on. Also, the first part of any period is so many units of it, and the latter part so many thousands.

The following Table contains a summary of the whole doctrine:

| Periods. | Quadrill.; | | Trillions; | | Billions; | | Millions; | | Units. | |
|---|---|---|---|---|---|---|---|---|---|---|
| Half-per. | th. | un. | th. | un. | th. | un. | th. | un. | th. | un. |
| Figures. | 123, | 456; | 789, | 098; | 765, | 432; | 101, | 234; | 567, | 890. |

Numeration is the reading of any number in words that is proposed or set down in figures.

Notation is the setting down in figures any number proposed in words.

### OF THE ROMAN NOTATION.

The Romans, like several other nations, expressed their numbers by certain letters of the alphabet. The Romans only used seven numeral letters, being the seven following capitals: viz. I for *one*; V for *five*; X for *ten*; L for *fifty*; C for a *hundred*; D for *five hundred*; M for a *thousand*. The other numbers they expressed by various repetitions and combinations of these, after the following manner:

| | |
|---|---|
| 1 = I. | |
| 2 = II. | As often as any character is repeated, so many times is its value repeated. |
| 3 = III. | |
| 4 = IIII. or IV. | A less character before a greater diminishes its value. |
| 5 = V. | |
| 6 = VI. | A less character after a greater increases its value. |
| 7 = VII. | |
| 8 = VIII. | |
| 9 = IX. | |
| 10 = X. | |
| 50 = L. | |
| 100 = C. | |
| 500 = D or IↃ. | For every Ↄ annexed, this becomes ten times as many. |
| 1000 = M or CIↃ. | For every C and Ↄ, placed one at each end, it becomes ten times as much. |
| 2000 = MM. | |
| 5000 = $\overline{\text{V}}$ or IↃↃ. | A bar over any number increases it 1000 fold. |
| 6000 = $\overline{\text{VI}}$. | |
| 10000 = $\overline{\text{X}}$ or CCIↃↃ. | |
| 50000 = $\overline{\text{L}}$ or IↃↃↃ. | |
| 60000 = $\overline{\text{LX}}$. | |
| 100000 = $\overline{\text{C}}$ or CCCIↃↃↃ. | |
| 1000000 = $\overline{\text{M}}$ or CCCCIↃↃↃↃ. | |
| 2000000 = $\overline{\text{MM}}$. | |
| &c. &c. | |

## EXPLANATION OF CERTAIN CHARACTERS.

There are various characters or marks used in Arithmetic and Algebra, to denote several of the operations and propositions; the chief of which are as follow:

| | |
|---|---|
| + signifies *plus*, or addition. | : :: : ....... proportion. |
| − .......... *minus*, or subtraction. | = .......... equality. |
| × .......... multiplication. | $\sqrt{\ }$ .......... square root. |
| ÷ .......... division. | $\sqrt[3]{\ }$ .......... cube root, &c. |

Thus,

5 + 3, denotes that 3 is to be added to 5 = 8.
6 − 2, denotes that 2 is to be taken from 6 = 4.
7 × 3, denotes that 7 is to be multiplied by 3 = 21.
8 ÷ 4, denotes that 8 is to be divided by 4 = 2.
2 : 3 :: 4 : 6, shows that 2 is to 3 as 4 is to 6, and thus, 2×6=3×4.
6 + 4 = 10, shows that the sum of 6 and 4 is equal to 10.
$\sqrt{3}$, or $3^{\frac{1}{2}}$, denotes the square root of the number 3 = 1·7320508.
$\sqrt[3]{5}$, or $5^{\frac{1}{3}}$, denotes the cube root of the number 5 = 1·709976.
$7^2$, denotes that the number 7 is to be squared = 49.
$8^3$, denotes that the number 8 is to be cubed = 512.
&c.

## RULE OF THREE.

THE RULE OF THREE teaches how to find a fourth proportional to three numbers given. Whence it is also sometimes called the Rule of Proportion. It is called the Rule of Three, because three terms or numbers are given to find the fourth; and because of its great and extensive usefulness, it is often called the Golden Rule.

This Rule is usually considered as of two kinds, namely, Direct and Inverse.

The Rule of Three Direct is that in which more requires more, or less requires less. As in this: if 3 men dig 21 yards of trench in a certain time, how much will 6 men dig in the same time? Here more requires more, that is, 6 men, which are more than 3 men, will also perform more work in the same time. Or when it is thus: if 6 men dig 42 yards, how much will 3 men dig in the same time? Here, then, less requires less, or 3 men will perform proportionally less work than 6 men in the same time. In both these cases, then, the Rule, or the Proportion, is Direct; and the stating must be

thus, As 3 : 21 : : 6 : 42,
or thus, As 6 : 42 : : 3 : 21.

But, the Rule of Three Inverse is when more requires less, or less requires more. As in this: if 3 men dig a certain quantity of trench in 14 hours, in how many hours will 6 men dig the like quantity? Here it is evident that 6 men, being more than 3, will perform an equal quantity of work in less time, or fewer hours. Or thus: if 6 men perform a certain quantity of work in 7 hours, in how many hours will 3 men perform the same? Here less requires more, for 3 men will take more hours than 6 to perform the same work. In both these cases, then, the Rule, or the Proportion, is Inverse; and the stating must be

thus, As 6 : 14 : : 3 : 7,
or thus, As 3 : 7 : : 6 : 14.

And in all these statings the fourth term is found, by multiplying the 2d and 3d terms together, and dividing the product by the 1st term.

Of the three given numbers, two of them contain the supposition, and the third a demand. And for stating and working questions of these kinds observe the following general Rule:

RULE.—State the question by setting down in a straight line the three given numbers, in the following manner, viz. so that the 2d term be that number of supposition which is of the same kind that the answer or 4th term is to be; making the other number of supposition the 1st term, and the demanding number the 3d term, when the question is in direct proportion; but contrariwise, the other number of supposition the third term, and the demanding number the 1st term, when the question has inverse proportion.

Then, in both cases, multiply the 2d and 3d terms together, and divide the product by the first, which will give the answer, or 4th term sought, of the same denomination as the second term.

*Note*, If the first and third terms consist of different denominations, reduce them both to the same; and if the second term be a compound number, it is mostly convenient to reduce it to the lowest denomination mentioned. If, after division, there be any remainder, reduce it to the next lower denomination, and divide by the same divisor as before, and the quotient will be of this last denomination. Proceed in the same manner with all the remainders, till they be reduced to the lowest denomination which the second term admits of, and the several quotients taken together will be the answer required.

*Note* also, The reason for the foregoing Rules will appear when we come to treat of the nature of Proportions. Sometimes also two or more statings are necessary, which may always be known from the nature of the question.

An engineer having raised 100 yards of a certain work in 24 days with 5 men, how many men must he employ to finish a like quantity of work in 15 days?

```
     da. men.   da. men.
As 15 : 5 : : 24 : 8 Ans.
               5
         15 ) 120 ( 8 Answer.
              120
```

## COMPOUND PROPORTION.

COMPOUND PROPORTION teaches how to resolve such questions as require two or more statings by Simple Proportion; and that, whether they be Direct or Inverse.

In these questions, there is always given an odd number of terms, either five, or seven, or nine, &c. These are distinguished into terms of supposition and terms of demand, there being always one term more of the former than of the latter, which is of the same kind with the answer sought.

RULE.—Set down in the middle place that term of supposition which is of the same kind with the answer sought. Take one of the other terms of supposition, and one of the demanding terms which is of the same kind with it; then place one of them for a first term, and the other for a third, according to the directions given in the Rule of Three. Do the same with another term of supposition, and its corresponding demanding term; and so on if there be more terms of each kind; setting the numbers under each other which fall all on the left-hand side of the middle term, and the same for the others on the right-hand side. Then to work.

*By several Operations.*—Take the two upper terms and the middle term, in the same order as they stand, for the first Rule of Three question to be worked, whence will be found a fourth term. Then take this fourth number, so found, for the middle term of a second Rule of Three question, and the next two under terms in the general stating, in the same order as they stand, finding a fourth term from them; and so on, as far as there are any numbers in the general stating, making always the fourth number resulting from each simple stating to be the second term of the next following one. So shall the last resulting number be the answer to the question.

*By one Operation.*—Multiply together all the terms standing under each other, on the left-hand side of the middle term; and, in like manner, multiply together all those on the right-hand side of it. Then multiply the middle term by the latter product, and divide the result by the former product, so shall the quotient be the answer sought.

How many men can complete a trench of 135 yards long in 8 days, when 16 men can dig 54 yards in 6 days?

*General stating.*

```
yds.  54 : 16 men :: 135 yds.
days   8               6 days
     ---             ---
     432             810
                      16
                    ----
                    4860
                    810
               432 ) 12960 ( 30 Ans. by one operation.
                     1296
                     -----
                         0
```

*The same by two operations.*

```
1st.                              2d.
As 54 : 16 :: 135 : 40      As 8 : 40 :: 6 : 30
                16                       6
               ---                  8 ) 240 ( 30 Ans.
               810                      24
              135                       --
         54 ) 2160 (40                   0
              216
              ----
                 0
```

## OF COMMON FRACTIONS.

A FRACTION, or broken number, is an expression of a part, or some parts, of something considered as a whole.

It is denoted by two numbers, placed one below the other, with a line between them:

thus, $\dfrac{3 \text{ numerator}}{4 \text{ denominator}}$ } which is named three-fourths.

The Denominator, or number placed below the line, shows how many equal parts the whole quantity is divided into; and represents the Divisor in Division. And the Numerator, or number set above the line, shows how many of those parts are expressed by the Fraction; being the remainder after division. Also, both these numbers are, in general, named the Terms of the Fractions.

Fractions are either Proper, Improper, Simple, Compound, or Mixed.

A Proper Fraction is when the numerator is less than the denominator; as $\frac{1}{2}$, or $\frac{2}{3}$, or $\frac{3}{4}$, &c.

An Improper Fraction is when the numerator is equal to, or exceeds, the denominator; as $\frac{3}{3}$, or $\frac{5}{4}$, or $\frac{7}{5}$, &c.

A Simple Fraction is a single expression denoting any number of parts of the integer; as $\frac{3}{8}$, or $\frac{3}{2}$.

A Compound Fraction is the fraction of a fraction, or several fractions connected with the word *of* between them; as $\frac{1}{2}$ of $\frac{2}{3}$, or $\frac{3}{5}$ of $\frac{5}{6}$ of 3, &c.

A Mixed Number is composed of a whole number and a fraction together; as $3\frac{1}{4}$, or $12\frac{5}{6}$, &c.

A whole or integer number may be expressed like a fraction, by writing 1 below it, as a denominator; so 3 is $\frac{3}{1}$, or 4 is $\frac{4}{1}$, &c.

A fraction denotes division; and its value is equal to the quotient obtained by dividing the numerator by the denominator; so $\frac{12}{4}$ is equal to 3, and $\frac{20}{5}$ is equal to 4.

Hence, then, if the numerator be less than the denominator, the value of the fraction is less than 1. If the numerator be the same as the denominator, the fraction is just equal to 1. And if the numerator be greater than the denominator, the fraction is greater than 1.

## REDUCTION OF FRACTIONS.

Reduction of Fractions is the bringing them out of one form or denomination into another, commonly to prepare them for the operations of Addition, Subtraction, &c., of which there are several cases.

*To find the greatest common measure of two or more numbers.*

The Common Measure of two or more numbers is that number which will divide them both without a remainder: so 3 is a common measure of 18 and 24; the quotient of the former being 6, and of the latter 8. And the greatest number that will do this, is the greatest common measure: so 6 is the greatest common measure of 18 and 24; the quotient of the former being 3, and of the latter 4, which will not both divide farther.

Rule.—If there be two numbers only, divide the greater by the less; then divide the divisor by the remainder; and so on, dividing always the last divisor by the last remainder, till nothing remains; then shall the last divisor of all be the greatest common measure sought.

When there are more than two numbers, find the greatest common measure of two of them, as before; then do the same for that common measure and another of the numbers; and so on, through all the numbers; then will the greatest common measure last found be the answer.

If it happen that the common measure thus found is 1, then the numbers are said to be incommensurable, or to have no common measure.

To find the greatest common measure of 1998, 918, and 522.

```
918 ) 1998 ( 2                So 54 is the greatest common measure
      1836                             of 1998 and 918.
      ----
       162 ) 918 ( 5          Hence 54 ) 522 ( 9
             810                         486
             ---                         ---
             108 ) 162 ( 1                36 ) 54 ( 1
                   108                         36
                   ---                         --
                    54 ) 108 ( 2               18 ) 36 ( 2
                         108                        36
                         ---                        --
```

So that 18 is the answer required.

*To abbreviate or reduce fractions to their lowest terms.*

Rule.—Divide the terms of the given fraction by any number that will divide them without a remainder; then divide these quotients again in the same manner; and so on, till it appears that there is no number greater than 1 which will divide them: then the fraction will be in its lowest terms.

Or, divide both the terms of the fraction by their greatest common measure, and the quotients will be the terms of the fraction required, of the same value as at first.

That dividing both the terms of the fraction by the same number, whatever it be, will give another fraction equal to the former, is evident. And when those divisions are performed as often as can be done, or when the common divisor is the greatest possible, the terms of the resulting fraction must be the least possible.

1. Any number ending with an even number, or a cipher, is divisible, or can be divided by 2.

2. Any number ending with 5, or 0, is divisible by 5.

3. If the right-hand place of any number be 0, the whole is divisible by 10; if there be 2 ciphers, it is divisible by 100; if 3 ciphers, by 1000; and so on, which is only cutting off those ciphers.

4. If the two right-hand figures of any number be divisible by 4, the whole is divisible by 4. And if the three right-hand figures be divisible by 8, the whole is divisible by 8; and so on.

5. If the sum of the digits in any number be divisible by 3, or by 9, the whole is divisible by 3, or by 9.

6. If the right-hand digit be even, and the sum of all the digits be divisible by 6, then the whole will be divisible by 6.

7. A number is divisible by 11 when the sum of the 1st, 3d, 5th, &c., or of all the odd places, is equal to the sum of the 2d, 4th, 6th, &c., or of all the even places of digits.

8. If a number cannot be divided by some quantity less than the square of the same, that number is a prime, or cannot be divided by any number whatever.

9. All prime numbers, except 2 and 5, have either 1, 3, 7, or 9, in the place of units; and all other numbers are composite, or can be divided.

10. When numbers, with a sign of addition or subtraction between them, are to be divided by any number, then each of those numbers must be divided by it. Thus, $\frac{10+8-4}{2}=5+4-2=7$.

11. But if the numbers have the sign of multiplication between them, only one of them must be divided. Thus, $\frac{10\times 8\times 3}{6\times 2}=$ $\frac{10\times 4\times 3}{6\times 1}=\frac{10\times 4\times 1}{2\times 1}=\frac{10\times 2\times 1}{1\times 1}=\frac{20}{1}=20$.

Reduce $\frac{144}{240}$ to its least terms.

$\frac{144}{240}=\frac{72}{120}=\frac{36}{60}=\frac{18}{30}=\frac{9}{15}=\frac{3}{5}$, the answer.

Or thus:

```
144 ) 240 ( 1
      144
       96 ) 144 ( 1
             96
             48 ) 96 ( 2
                  96
```

Therefore 48 is the greatest common measure, and 48 ) $\frac{144}{240}=\frac{3}{5}$ the answer, the same as before.

*To reduce a mixed number to its equivalent improper fraction.*

RULE.—Multiply the whole number by the denominator of the fraction, and add the numerator to the product; then set that sum above the denominator for the fraction required.

Reduce $23\frac{2}{5}$ to a fraction.

```
 23
  5
115
  2
117
  5
```

Or, $\frac{(23\times 5)+2}{5}=\frac{117}{5}$.

*To reduce an improper fraction to its equivalent whole or mixed number.*

RULE.—Divide the numerator by the denominator, and the quotient will be the whole or mixed number sought.

Reduce $\frac{12}{3}$ to its equivalent number.

Here $\frac{12}{3}$ or $12\div 3=4$.

Reduce $\frac{15}{7}$ to its equivalent number.

Here $\frac{15}{7}$ or $15\div 7=2\frac{1}{7}$.

Reduce $\frac{749}{17}$ to its equivalent number.

```
Thus, 17 ) 749 ( 44 1/17
           68
           69
           68
            1
```

So that $\frac{749}{17}=44\frac{1}{17}$

*To reduce a whole number to an equivalent fraction, having a given denominator.*

RULE.—Multiply the whole number by the given denominator, then set the product over the said denominator, and it will form the fraction required.

Reduce 9 to a fraction whose denominator shall be 7.

Here $9 \times 7 = 63$, then $\frac{63}{7}$ is the answer.

For $\frac{63}{7} = 63 \div 7 = 9$, the proof.

*To reduce a compound fraction to an equivalent simple one.*

RULE.—Multiply all the numerators together for a numerator, and all the denominators together for the denominator, and they will form the simple fraction sought.

When part of the compound fraction is a whole or mixed number, it must first be reduced to a fraction by one of the former cases.

And, when it can be done, any two terms of the fraction may be divided by the same number, and the quotients used instead of them. Or, when there are terms that are common, they may be omitted.

Reduce $\frac{1}{2}$ of $\frac{2}{3}$ of $\frac{3}{4}$ to a simple fraction.

$$\text{Here } \frac{1 \times 2 \times 3}{2 \times 3 \times 4} = \frac{6}{24} = \frac{1}{4}.$$

Or, $\frac{1 \times 2 \times 3}{2 \times 3 \times 4} = \frac{1}{4}$, by omitting the twos and threes.

Reduce $\frac{2}{3}$ of $\frac{3}{5}$ of $\frac{10}{11}$ to a simple fraction.

$$\text{Here } \frac{2 \times 3 \times 10}{3 \times 5 \times 11} = \frac{60}{165} = \frac{12}{33} = \frac{4}{11}.$$

Or, $\frac{2 \times 3 \times 10}{3 \times 5 \times 11} = \frac{4}{11}$, the same as before.

*To reduce fractions of different denominators to equivalent fractions, having a common denominator.*

RULE.—Multiply each numerator into all the denominators except its own for the new numerators; and multiply all the denominators together for a common denominator.

It is evident, that in this and several other operations, when any of the proposed quantities are integers, or mixed numbers, or compound fractions, they must be reduced, by their proper rules, to the form of simple fractions.

Reduce $\frac{1}{2}$, $\frac{2}{3}$, and $\frac{3}{4}$ to a common denominator.

$1 \times 3 \times 4 = 12$ the new numerator for $\frac{1}{2}$.
$2 \times 2 \times 4 = 16$........................ for $\frac{2}{3}$.
$3 \times 2 \times 3 = 18$........................ for $\frac{3}{4}$.
$2 \times 3 \times 4 = 24$ the common denominator.

Therefore, the equivalent fractions are $\frac{12}{24}$, $\frac{16}{24}$, and $\frac{18}{24}$.

Or, the whole operation of multiplying may be very well performed mentally, and only set down the results and given fractions thus: $\frac{1}{2}, \frac{2}{3}, \frac{3}{4} = \frac{12}{24}, \frac{16}{24}, \frac{18}{24} = \frac{6}{12}, \frac{8}{12}, \frac{9}{12}$, by abbreviation.

When the denominators of two given fractions have a common measure, let them be divided by it; then multiply the terms of each given fraction by the quotient arising from the other's denominator.

When the less denominator of two fractions exactly divides the greater, multiply the terms of that which hath the less denominator by the quotient.

When more than two fractions are proposed, it is sometimes convenient first to reduce two of them to a common denominator, then these and a third; and so on, till they be all reduced to their least common denominator.

*To find the value of a fraction in parts of the integer.*

Rule.—Multiply the integer by the numerator, and divide the product by the denominator, by Compound Multiplication and Division, if the integer be a compound quantity.

Or, if it be a single integer, multiply the numerator by the parts in the next inferior denomination, and divide the product by the denominator. Then, if any thing remains, multiply it by the parts in the next inferior denomination, and divide by the denominator as before; and so on, as far as necessary; so shall the quotients, placed in order, be the value of the fraction required.

What is the value of $\frac{3}{5}$ of a pound troy? 7 oz. 4 dwts.
What is the value of $\frac{5}{16}$ of a cwt.? 1 qr. 7 lb.
What is the value of $\frac{5}{8}$ of an acre? 2 ro. 20 po.
What is the value of $\frac{3}{10}$ of a day? 7 hrs. 12 min.

*To reduce a fraction from one denomination to another.*

Rule.—Consider how many of the less denomination make one of the greater; then multiply the numerator by that number, if the reduction be to a less name, or the denominator, if to a greater.

Reduce $\frac{2}{7}$ of a cwt. to the fraction of a pound.

$$\tfrac{2}{7} \times \tfrac{4}{1} \times \tfrac{28}{1} = \tfrac{32}{1}.$$

ADDITION OF FRACTIONS.

*To add fractions together that have a common denominator.*

Rule.—Add all the numerators together, and place the sum over the common denominator, and that will be the sum of the fractions required.

If the fractions proposed have not a common denominator, they must be reduced to one. Also, compound fractions must be reduced to simple ones, and mixed numbers to improper fractions; also, fractions of different denominations to those of the same denomination.

To add $\frac{3}{5}$ and $\frac{4}{5}$ together. Here $\frac{3}{5} + \frac{4}{5} = \frac{7}{5} = 1\frac{2}{5}$.
To add $\frac{3}{5}$ and $\frac{5}{6}$ together. $\frac{3}{5} + \frac{5}{6} = \frac{18}{30} + \frac{25}{30} = \frac{43}{30} = 1\frac{13}{30}$.
To add $\frac{5}{8}$ and $7\frac{1}{2}$ and $\frac{1}{3}$ of $\frac{3}{4}$ together.

$$\tfrac{5}{8} + 7\tfrac{1}{2} + \tfrac{1}{3} \text{ of } \tfrac{3}{4} = \tfrac{5}{8} + \tfrac{15}{2} + \tfrac{1}{4} = \tfrac{5}{8} + \tfrac{60}{8} + \tfrac{2}{8} = \tfrac{67}{8} = 8\tfrac{3}{8}.$$

## SUBTRACTION OF FRACTIONS.

RULE.—Prepare the fractions the same as for Addition; then subtract the one numerator from the other, and set the remainder over the common denominator, for the difference of the fractions sought.

To find the difference between $\frac{5}{6}$ and $\frac{1}{6}$.

Here $\frac{5}{6} - \frac{1}{6} = \frac{4}{6} = \frac{2}{3}$.

To find the difference between $\frac{3}{4}$ and $\frac{5}{7}$.

$\frac{3}{4} - \frac{5}{7} = \frac{21}{28} - \frac{20}{28} = \frac{1}{28}$.

## MULTIPLICATION OF FRACTIONS.

MULTIPLICATION of any thing by a fraction implies the taking some part or parts of the thing; it may therefore be truly expressed by a compound fraction; which is resolved by multiplying together the numerators and the denominators.

RULE.—Reduce mixed numbers, if there be any, to equivalent fractions; then multiply all the numerators together for a numerator, and all the denominators together for a denominator, which will give the product required.

Required the product of $\frac{3}{4}$ and $\frac{2}{9}$.

Here $\frac{3}{4} \times \frac{2}{9} = \frac{6}{36} = \frac{1}{6}$.

Or, $\frac{3}{4} \times \frac{2}{9} = \frac{1}{2} \times \frac{1}{3} = \frac{1}{6}$.

Required the continued product of $\frac{2}{3}$, $3\frac{1}{4}$, 5, and $\frac{3}{4}$ of $\frac{3}{5}$.

$$\text{Here } \frac{2}{3} \times \frac{13}{4} \times \frac{5}{1} \times \frac{3}{4} \times \frac{3}{5} = \frac{13 \times 3}{4 \times 2} = \frac{39}{8} = 4\frac{7}{8}.$$

## DIVISION OF FRACTIONS.

RULE.—Prepare the fractions as before in Multiplication; then divide the numerator by the numerator, and the denominator by the denominator, if they will exactly divide; but if not, then invert the terms of the divisor, and multiply the dividend by it, as in Multiplication.

Divide $\frac{25}{9}$ by $\frac{5}{3}$.

Here $\frac{25}{9} \div \frac{5}{3} = \frac{5}{3} = 1\frac{2}{3}$, by the first method.

Divide $\frac{5}{9}$ by $\frac{2}{15}$.

Here $\frac{5}{9} \div \frac{2}{15} = \frac{5}{9} \times \frac{15}{2} = \frac{5}{3} \times \frac{5}{2} = \frac{25}{6} = 4\frac{1}{6}$, by the latter.

## RULE OF THREE IN FRACTIONS.

RULE.—Make the necessary preparations as before directed; then multiply continually together the second and third terms, and the first with its terms inverted as in Division, for the answer. This is only multiplying the second and third terms together, and dividing the product by the first, as in the Rule of Three in whole numbers.

If $\frac{3}{8}$ of a yard of velvet cost $\frac{2}{5}$ of a dollar, what will $\frac{5}{16}$ of a yard cost?

$$\text{Here } \frac{3}{8} : \frac{2}{5} :: \frac{5}{16} : \frac{8}{3} \times \frac{2}{5} \times \frac{5}{16} = \tfrac{1}{3} \text{ of a dollar.}$$

## DECIMAL FRACTIONS.

A Decimal Fraction is that which has for its denominator a unit (1) with as many ciphers annexed as the numerator has places; and it is usually expressed by setting down the numerator only, with a point before it on the left hand. Thus, $\frac{5}{10}$ is ·5, and $\frac{25}{100}$ is ·25, and $\frac{75}{1000}$ is ·075, and $\frac{124}{100000}$ is ·00124; where ciphers are prefixed to make up as many places as are in the numerator, when there is a deficiency of figures.

A mixed number is made up of a whole number with some decimal fraction, the one being separated from the other by a point. Thus, 3·25 is the same as $3\frac{25}{100}$, or $\frac{325}{100}$.

Ciphers on the right hand of decimals make no alteration in their value; for ·5, or ·50, or ·500, are decimals having all the same value, being each $= \frac{5}{10}$ or $\frac{1}{2}$. But if they are placed on the left hand, they decrease the value in a tenfold proportion. Thus, ·5 is $\frac{5}{10}$ or 5 tenths, but ·05 is only $\frac{5}{100}$ or 5 hundreths, and ·005 is but $\frac{5}{1000}$ or 5 thousandths.

The first place of decimals, counted from the left hand towards the right, is called the place of primes, or 10ths; the second is the place of seconds, or 100ths; the third is the place of thirds, or 1000ths; and so on. For, in decimals, as well as in whole numbers, the values of the places increase towards the left hand, and decrease towards the right, both in the same tenfold proportion; as in the following Scale or Table of Notation:

| millions. | hundred thousands. | ten thousands. | thousands. | hundreds. | tens. | units. | tenth parts. | hundredth parts. | thousandths parts. | ten thousandths parts. | hundred thousandth parts. | millionth parts. |
|---|---|---|---|---|---|---|---|---|---|---|---|---|
| 3 | 3 | 3 | 3 | 3 | 3 | 3 | 3 | 3 | 3 | 3 | 3 | 3 |

### ADDITION OF DECIMALS.

Rule.—Set the numbers under each other according to the value of their places, like as in whole numbers; in which state the decimal separating points will stand all exactly under each other. Then, beginning at the right hand, add up all the columns of number as in integers, and point off as many places for decimals as are in the greatest number of decimal places in any of the lines that are added; or, place the point directly below all the other points.

To add together 29·0146, and 3146·5, and 2109, and 62417, and 14·16.

```
  29·0146
3146·5
2109·
    ·62417
  14·16
---------
5299·29877, the sum.
```

The sum of 376·25 + 86·125 + 637·4725 + 6·5 + 41·02 + 358·865 = 1506.2325.

The sum of 3·5 + 47·25 + 2.0073 + 927·01 + 1·5 = 981.2673.

The sum of 276 + 54·321 + 112 + 0.65 + 12·5 + ·0463 = 455·5173.

## SUBTRACTION OF DECIMALS.

RULE.—Place the numbers under each other according to the value of their places, as in the last rule. Then, beginning at the right hand, subtract as in whole numbers, and point off the decimals as in Addition.

To find the difference between 91.73 and 2.138.

```
91·73
 2·138
89·592 the difference.
```

The difference between 1·9185 and 2·73 = 0·8115.

The difference between 214·81 and 4·90142 = 209·90858.

The difference between 2714 and ·916 = 2713·084.

## MULTIPLICATION OF DECIMALS.

RULE.—Place the factors, and multiply them together the same as if they were whole numbers. Then point off in the product just as many places of decimals as there are decimals in both the factors. But if there be not so many figures in the product, then supply the defect by prefixing ciphers.

```
Multiply ·321096
by         ·2465
         1605480
        1926576
       1284384
       642192.
   ·0791501640 the product.
```

Multiply 79·347 by 23·15, and we have 1836·88305.

Multiply ·63478 by ·8204, and we have ·520773512.

Multiply ·385746 by ·00464, and we have ·00178986144.

### CONTRACTION I.

*To multiply decimals by* 1 *with any number of ciphers, as* 10, *or* 100, *or* 1000, *&c.*

This is done by only removing the decimal point so many places farther to the right hand as there are ciphers in the multiplier: and subjoining ciphers if need be.

The product of 51·3 and 1000 is 51300.

The product of 2·714 and 100 is 271·4.

The product of ·916 and 1000 is 916.

The product of 21·31 and 10000 is 213100.

### CONTRACTION II.

*To contract the operation, so as to retain only as many decimals in the product as may be thought necessary, when the product would naturally contain several more places.*

Set the units' place of the multiplier under that figure of the multiplicand whose place is the same as is to be retained for the last in the product; and dispose of the rest of the figures in the inverted or contrary order to what they are usually placed in. Then, in multiplying, reject all the figures that are more to the right than each multiplying figure; and set down the products, so that their right hand figures may fall in a column straight below each other; but observing to increase the first figure of every line with what would arise from the figures omitted, in this manner, namely, 1 from 5 to 14, 2 from 15 to 24, 3 from 25 to 34, &c.; and the sum of all the lines will be the product as required, commonly to the nearest unit in the last figure.

To multiply 27·14986 by 92·41035, so as to retain only four places of decimals in the product.

| Contracted way. | Common way. |
|---:|---:|
| 27·14986 | 27·14986 |
| 53014·29 | 92·41035 |
| 24434874 | 13\|574930 |
| 542997 | 81\|44958 |
| 108599 | 2714\|986 |
| 2715 | 108599\|44 |
| 81 | 542997\|2 |
| 14 | 24434874\| |
| 2508·9280 | 2508·9280\|650510 |

DIVISION OF DECIMALS.

Rule.—Divide as in whole numbers; and point off in the quotient as many places for decimals, as the decimal places in the dividend exceed those in the divisor.

When the places of the quotient are not so many as the rule requires, let the defect be supplied by prefixing ciphers.

When there happens to be a remainder after the division; or when the decimal places in the divisor are more than those in the dividend; then ciphers may be annexed to the dividend, and the quotient carried on as far as required.

| | |
|---|---|
| 179) ·48624097 ( ·00271643 | ·2685) 27·00000 ( 100·55865 |
| 1282 | 15000 |
| 294 | 15750 |
| 1150 | 23250 |
| 769 | 17700 |
| 537 | 15900 |
| 000 | 24750 |

Divide 234·70525 by 64·25. 3·653.

Divide 14 by ·7854. 17·825.

Divide 2175·68 by 100. 21·7568.

Divide ·8727587 by ·162. 5·38739.

CONTRACTION I.

When the divisor is an integer, with any number of ciphers annexed; cut off those ciphers, and remove the decimal point in the dividend as many places farther to the left as there are ciphers cut off, prefixing ciphers if need be; then proceed as before.

Divide 45·5 by 2100.

```
21·00)·455(·0216, &c.
        35
        140
         14
```

CONTRACTION II.

Hence, if the divisor be 1 with ciphers, as 10, or 100, or 1000, &c.; then the quotient will be found by merely moving the decimal point in the dividend so many places farther to the left as the divisor has ciphers; prefixing ciphers if need be.

So, 217·3 ÷ 100 = 2·173, and 419 ÷ 10 = 41·9.
And 5·16 ÷ 100 = ·0516, and ·21 ÷ 1000 = ·00021.

CONTRACTION III.

When there are many figures in the divisor; or only a certain number of decimals are necessary to be retained in the quotient, then take only as many figures of the divisor as will be equal to the number of figures, both integers and decimals, to be in the quotient, and find how many times they may be contained in the first figures of the dividend, as usual.

Let each remainder be a new dividend; and for every such dividend, leave out one figure more on the right hand side of the divisor; remembering to carry for the increase of the figures cut off, as in the 2d contraction in Multiplication.

When there are not so many figures in the divisor as are required to be in the quotient, begin the operation with all the figures, and continue it as usual till the number of figures in the divisor be equal to those remaining to be found in the quotient, after which begin the contraction.

Divide 2508·92806 by 92·41035, so as to have only four decimals in the quotient, in which case the quotient will contain six figures.

| Contracted. | Common way. |
|---|---|
| 92·4103,5)2508·928,06(27·1498 | 92·4103,5)2508·928,06(27·1498 |
| 660721 | 66072106 |
| 13849 | 13848610 |
| 4608 | 46075750 |
| 912 | 91116100 |
| 80 | 79467850 |
| 6 | 5539570 |

## REDUCTION OF DECIMALS.

*To reduce a common fraction to its equivalent decimal.*

Rule.—Divide the numerator by the denominator as in Division of Decimals, annexing ciphers to the numerator as far as necessary; so shall the quotient be the decimal required.

Reduce $\frac{7}{24}$ to a decimal.

$$24 = 4 \times 6. \quad \text{Then } 4\,)\,7\cdot \quad 6\,)\,1\cdot750000 \quad \cdot291666, \&c.$$

$\frac{3}{8}$ reduced to a decimal, is ·375.
$\frac{1}{25}$ reduced to a decimal, is ·04.
$\frac{3}{192}$ reduced to a decimal, is ·015625.
$\frac{275}{3842}$ reduced to a decimal, is ·071577, &c.

CASE II.

*To find the value of a decimal in terms of the inferior denominations.*

RULE.—Multiply the decimal by the number of parts in the next lower denomination; and cut off as many places for a remainder, to the right hand, as there are places in the given decimal.

Multiply that remainder by the parts in the next lower denomination again, cutting off for another remainder as before.

Proceed in the same manner through all the parts of the integer; then the several denominations, separated on the left hand, will make up the value required.

What is the value of ·0125 lb. troy:— 3 dwts.
What is the value of ·4694 lb. troy:— 5 oz. 12 dwt. 15·744 gr.
What is the value of ·625 cwt.:— 2 qr. 14 lb.
What is the value of ·009943 miles:— 17 yd. 1 ft. 5·98848 in.
What is the value of ·6875 yd.:— 2 qr. 3 nls.
What is the value of ·3375 ac.:— 1 rd. 14 poles.
What is the value of ·2083 hhd. of wine:— 13·1229 gal.

CASE III.

*To reduce integers or decimals to equivalent decimals of higher denominations.*

RULE.—Divide by the number of parts in the next higher denomination; continuing the operation to as many higher denominations as may be necessary, the same as in Reduction Ascending of whole numbers.

Reduce 1 dwt. to the decimal of a pound troy.

| | |
|---|---|
| 20 | 1 dwt. |
| 12 | 0·05 oz. |
| | 0·004166, &c. lb. |

Reduce 7 dr. to the decimal of a pound avoird.:— ·02734375 lb.
Reduce 2·15 lb. to the decimal of a cwt.:— ·019196 cwt.
Reduce 24 yards to the decimal of a mile:— ·013636, &c. miles.
Reduce ·056 poles to the decimal of an acre:— ·00035 ac.
Reduce 1·2 pints of wine to the decimal of a hhd.:— ·00238 hhd.
Reduce 14 minutes to the decimal of a day:— ·009722, &c. da.
Reduce ·21 pints to the decimal of a peck:— ·013125 pec.

*When there are several numbers, to be reduced all to the decimal of the highest.*

Set the given numbers directly under each other, for dividends, proceeding orderly from the lowest denomination to the highest.

Opposite to each dividend, on the left hand, set such a number for a divisor as will bring it to the next higher name; drawing a perpendicular line between all the divisors and dividends.

Begin at the uppermost, and perform all the divisions; only observing to set the quotient of each division, as decimal parts, on the right hand of the dividend next below it; so shall the last quotient be the decimal required.

Reduce 5 oz. 12 dwts. 16 gr. to lbs.:— ·46944, &c. lb.

### RULE OF THREE IN DECIMALS.

RULE.—Prepare the terms by reducing the vulgar fractions to decimals, any compound numbers either to decimals of the higher denominations, or to integers of the lower, also the first and third terms to the same name: then multiply and divide as in whole numbers.

Any of the convenient examples in the Rule of Three or Rule of Five in Integers, or Common Fractions, may be taken as proper examples to the same rules in Decimals.—The following example, which is the first in Common Fractions, is wrought here to show the method.

If $\frac{3}{8}$ of a yard of velvet cost $\frac{2}{5}$ of a dollar, what will $\frac{5}{16}$ yd. cost?

yd. \$ yd. \$

$\frac{3}{8}$ = ·375 ·375 : ·4 :: ·3125 : ·333, &c.

·4

$\frac{2}{5}$ = ·4 ·375 ) ·12500 ( ·333333, $33\frac{1}{3}$ cts.

1250

125

$\frac{5}{16}$ = ·3125.

## DUODECIMALS.

DUODECIMALS, or CROSS MULTIPLICATION, is a rule made use of by workmen and artificers, in computing the contents of their works.

Dimensions are usually taken in feet, inches, and quarters; any parts smaller than these being neglected as of no consequence. And the same in multiplying them together, or casting up the contents.

RULE.—Set down the two dimensions, to be multiplied together, one under the other, so that feet stand under feet, inches under inches, &c.

Multiply each term in the multiplicand, beginning at the lowest, by the feet in the multiplier, and set the result of each straight under its corresponding term, observing to carry 1 for every 12, from the inches to the feet.

In like manner, multiply all the multiplicand by the inches and parts of the multiplier, and set the result of each term one place removed to the right hand of those in the multiplicand; omitting, however, what is below parts of inches, only carrying to these the proper number of units from the lowest denomination.

Or, instead of multiplying by the inches, take such parts of the multiplicand as these are of a foot.

Then add the two lines together, after the manner of Compound Addition, carrying 1 to the feet for 12 inches, when these come to so many.

| Multiply | 4 f. | 7 inc. | | Multiply | 14 f. | 9 inc. |
|---|---|---|---|---|---|---|
| by | 6 | 4 | | by | 4 | 6 |
| | 27 | 6 | | | 59 | 0 |
| | 1 | $6\frac{1}{3}$ | | | 7 | $4\frac{1}{2}$ |
| | 29 | $0\frac{1}{3}$ | | | 66 | $4\frac{1}{2}$ |

## INVOLUTION.

INVOLUTION is the raising of Powers from any given number, as a root.

A Power is a quantity produced by multiplying any given number, called the Root, a certain number of times continually by itself. Thus,

$2 = 2$ is the root, or first power of 2.
$2 \times 2 = 4$ is the 2d power, or square of 2.
$2 \times 2 \times 2 = 8$ is the 3d power, or cube of 2.
$2 \times 2 \times 2 \times 2 = 16$ is the 4th power of 2, &c.

And in this manner may be calculated the following Table of the first nine powers of the first nine numbers.

TABLE OF THE FIRST NINE POWERS OF NUMBERS.

| 1st | 2d. | 3d. | 4th. | 5th. | 6th. | 7th. | 8th. | 9th. |
|---|---|---|---|---|---|---|---|---|
| 1 | 1 | 1 | 1 | 1 | 1 | 1 | 1 | 1 |
| 2 | 4 | 8 | 16 | 32 | 64 | 128 | 256 | 512 |
| 3 | 9 | 27 | 81 | 243 | 729 | 2187 | 6561 | 19683 |
| 4 | 16 | 64 | 256 | 1024 | 4096 | 16384 | 65536 | 262144 |
| 5 | 25 | 125 | 625 | 3125 | 15625 | 78125 | 390625 | 1953125 |
| 6 | 36 | 216 | 1296 | 7776 | 46656 | 279936 | 1679616 | 10077696 |
| 7 | 49 | 343 | 2401 | 16807 | 117649 | 823543 | 5764801 | 40353607 |
| 8 | 64 | 512 | 4096 | 32768 | 262144 | 2097152 | 16777216 | 134217728 |
| 9 | 81 | 729 | 6561 | 59049 | 531441 | 4782969 | 43046721 | 387420489 |

The Index or Exponent of a Power is the number denoting the height or degree of that power; and it is 1 more than the number of multiplications used in producing the same. So 1 is the index or exponent of the 1st power or root, 2 of the 2d power or square, 3 of the 3d power or cube, 4 of the 4th power, and so on.

Powers, that are to be raised, are usually denoted by placing the index above the root or first power.

So $2^2 = 4$, is the 2d power of 2.
$2^3 = 8$, is the 3d power of 2.
$2^4 = 16$, is the 4th power of 2.
$540^4$, is the 4th power of $540 = 85030560000$.

When two or more powers are multiplied together, their product will be that power whose index is the sum of the exponents of the factors or powers multiplied. Or, the multiplication of the powers answers to the addition of the indices. Thus, in the following powers of 2.

| | 1st. | 2d. | 3d. | 4th. | 5th. | 6th. | 7th. | 8th. | 9th. | 10th. |
|---|---|---|---|---|---|---|---|---|---|---|
| | 2 | 4 | 8 | 16 | 32 | 64 | 128 | 256 | 512 | 1024 |
| or, | $2^1$ | $2^2$ | $2^3$ | $2^4$ | $2^5$ | $2^6$ | $2^7$ | $2^8$ | $2^9$ | $2^{10}$ |

Here, $4 \times 4 = 16$, and $2 + 2 = 4$ its index;
and $8 \times 16 = 128$, and $3 + 4 = 7$ its index;
also $16 \times 64 = 1024$, and $4 + 6 = 10$ its index.

The 2d power of 45 is 2025.
The square of 4·16 is 17·3056.
The 3d power of 3·5 is 42·875.
The 5th power of ·029 is ·000000020511149.
The square of $\frac{2}{3}$ is $\frac{4}{9}$.
The 3d power of $\frac{5}{9}$ is $\frac{125}{729}$.
The 4th power of $\frac{3}{4}$ is $\frac{81}{256}$.

## EVOLUTION.

EVOLUTION, or the reverse of Involution, is the extracting or finding the roots of any given powers.

The root of any number, or power, is such a number as, being multiplied into itself a certain number of times, will produce that power. Thus, 2 is the square root or 2d root of 4, because $2^2 = 2 \times 2 = 4$; and 3 is the cube root or 3d root of 27, because $3^3 = 3 \times 3 \times 3 = 27$.

Any power of a given number or root may be found exactly, namely, by multiplying the number continually into itself. But there are many numbers of which a proposed root can never be exactly found. Yet, by means of decimals we may approximate or approach towards the root to any degree of exactness.

These roots, which only approximate, are called Surd roots; but those which can be found quite exact, are called Rational roots. Thus, the square root of 3 is a surd root; but the square root of 4 is a rational root, being equal to 2: also, the cube root of 8 is rational, being equal to 2; but the cube root of 9 is surd, or irrational.

Roots are sometimes denoted by writing the character $\surd$ before the power, with the index of the root against it. Thus, the third root of 20 is expressed by $\sqrt[3]{20}$; and the square root or 2d root of it is $\sqrt{20}$, the index 2 being always omitted when the square root is designed.

When the power is expressed by several numbers, with the sign + or − between them, a line is drawn from the top of the sign over all the parts of it; thus, the third root of 45 − 12 is $\sqrt[3]{45 - 12}$, or thus, $\sqrt[3]{}(45 - 12)$, enclosing the numbers in parentheses.

But all roots are now often designed like powers, with fractional indices: thus, the square root of 8 is $8^{\frac{1}{2}}$, the cube root of 25 is $25^{\frac{1}{3}}$, and the 4th root of 45 — 18 is $\overline{45 - 18}|^{\frac{1}{4}}$, or, $(45 - 18)^{\frac{1}{4}}$.

## TO EXTRACT THE SQUARE ROOT.

RULE.—Divide the given number into periods of two figures each, by setting a point over the place of units, another over the place of hundreds, and so on, over every second figure, both to the left hand in integers, and to the right in decimals.

Find the greatest square in the first period on the left hand, and set its root on the right hand of the given number, after the manner of a quotient figure in Division.

Subtract the square thus found from the said period, and to the remainder annex the two figures of the next following period for a dividend.

Double the root above mentioned for a divisor, and find how often it is contained in the said dividend, exclusive of its right-hand figure; and set that quotient figure both in the quotient and divisor.

Multiply the whole augmented divisor by this last quotient figure, and subtract the product from the said dividend, bringing down to the next period of the given number, for a new dividend.

Repeat the same process over again, namely, find another new divisor, by doubling all the figures now found in the root; from which, and the last dividend, find the next figure of the root as before, and so on through all the periods to the last.

The best way of doubling the root to form the new divisor is by adding the last figure always to the last divisor, as appears in the following examples. Also, after the figures belonging to the given number are all exhausted, the operation may be continued into decimals at pleasure, by adding any number of periods of ciphers, two in each period.

To find the square root of 29506624.

$2\dot{9}5\dot{0}6\dot{6}2\dot{4}$ ( 5432 the root.

| | |
|---|---|
| | 29506624 |
| | 25 |
| 104 | 450 |
| 4 | 416 |
| 1083 | 3466 |
| 3 | 3249 |
| 10862 | 21724 |
| 2 | 21724 |

*When the root is to be extracted to many places of figures, the work may be considerably shortened, thus:*

Having proceeded in the extraction after the common method till there be found half the required number of figures in the root, or one figure more; then, for the rest, divide the last remainder by its corresponding divisor, after the manner of the third contraction in Division of Decimals; thus,

To find the root of 2 to nine places of figures.

```
            2(1·4142
            1
        ----------
        24 | 100
         4 |  96
        281 | 400
          1 | 281
        ------------
        2824 | 11900
           4 | 11296
        --------------
        28282 | 60400
            2 | 56564
        ----------------
        28284)  3836(1356
                1008
                 160
                  19
                   2
      1·41421356 the root required.
```

The square root of ·000729 is ·027.
The square root of 3 is 1·732050.
The square root of 5 is 2·236068.
The square root of 6 is 2·449489.

### RULES FOR THE SQUARE ROOTS OF COMMON FRACTIONS AND MIXED NUMBERS.

First, prepare all common fractions by reducing them to their least terms, both for this and all other roots. Then,

1. Take the root of the numerator and of the denominator for the respective terms of the root required. And this is the best way if the denominator be a complete power; but if it be not, then,

2. Multiply the numerator and denominator together; take the root of the product: this root being made the numerator to the denominator of the given fraction, or made the denominator to the numerator of it, will form the fractional root required.

$$\text{That is, } \sqrt{\frac{a}{b}} = \frac{\sqrt{a}}{\sqrt{b}} = \frac{\sqrt{ab}}{b} = \frac{a}{\sqrt{ab}}.$$

And this rule will serve whether the root be finite or infinite.

3. Or reduce the common fraction to a decimal, and extract its root.

4. Mixed numbers may be either reduced to improper fractions, and extracted by the first or second rule; or the common fraction may be reduced to a decimal, then joined to the integer, and the root of the whole extracted.

The root of $\frac{25}{36}$ is $\frac{5}{6}$.
The root of $\frac{9}{49}$ is $\frac{3}{7}$.
The root of $\frac{9}{12}$ is 0·866025.
The root of $\frac{5}{12}$ is 0·645497.
The root of $17\frac{3}{8}$ is 4·168333.

By means of the square root, also, may readily be found the 4th root, or the 8th root, or the 16th root, &c.; that is, the root of any power whose index is some power of the number 2; namely, by extracting so often the square root as is denoted by that power of 2; that is, two extractions for the 4th root, three for the 8th root, and so on.

So, to find the 4th root of the number 21035·8, extract the square root twice as follows:

```
          21035·8000              (145·037237 (12·0431407, the 4th root.
          1                        1
 24  |  110               22  |  45
  4  |   96                2  |  44
 285 |  1435            2404  |  10372
   5 |  1425               4  |   9616
29003 | 108000          24083 |   75637
    6 |  87009              6 |   72249
         20991 (7237             3388 (1407
           687                    980
           107                     17
            20
```

TO EXTRACT THE CUBE ROOT.

1. DIVIDE the page into three columns (I), (II), (III), in order, from left to right, so that the breadth of the columns may increase in the same order. In column (III) write the given number, and divide it into periods of three figures each, by putting a point over the place of units, and also over every third figure, from thence to the left in whole numbers, and to the right in decimals.

2. Find the nearest less cube number to the first or left-hand period; set its root in column (III), separating it from the right of the given number by a curve line, and also in column (I); then multiply the number in (I) by the root figure, thus giving the square of the first root figure, and write the result in (II); multiply the number in (II) by the root figure, thus giving the cube of the first root figure, and write the result below the first or left-hand period in (III); subtract it therefrom, and annex the next period to the remainder for a dividend.

3. In (I) write the root figure below the former, and multiply the sum of these by the root figure; place the product in (II), and add the two numbers together for a trial divisor. Again, write the root figure in (I), and add it to the former sum.

4. With the number in (II) as a trial divisor of the dividend, omitting the two figures to the right of it, find the next figure of the root, and annex it to the former, and also to the number in (I). Multiply the number now in (I) by the new figure of the root, and write the product as it arises in (II), but extended two places of figures more to the right, and the sum of these two numbers will be the corrected divisor; then multiply the corrected divisor by the

last root figure, placing the product as it arises below the dividend; subtract it therefrom, annex another period, and proceed precisely as described in (3), for correcting the columns (I) and (II). Then with the new trial divisor in (II), and the new dividend in (III), proceed as before.

When the trial divisor is not contained in the dividend, after two figures are omitted on the right, the next root figure is 0, and therefore one cipher must be annexed to the number in (I); two ciphers to the number in (II); and another period to the dividend in (III).

When the root is interminable, we may contract the work very considerably, after obtaining a few figures in the decimal part of the root, if we omit to annex another period to the remainder in (III); cut off one figure from the right of (II), and two figures from (I), which will evidently have the effect of cutting off *three* figures from each column; and then work with the numbers on the left, as in contracted multiplication and division of decimals.

Find the cube root of 21035·8 to ten places of decimals.

| (I) | (II) | (III) |
|---|---|---|
| 2 | 4 | 21035·8 ( 27·60491055944 |
| 2 | 8 | 8 |
| 4 | 12 . . | 13035 |
| 2 | 4 6 9 | 11683 |
| 67 | 1 6 6 9 | 1352800 |
| 7 | 5 1 8 | 1341576 |
| 74 | 2 1 8 7 . . | 11224 . . . . . . |
| 7 | 4 8 9 6 | 9142444864 |
| 8 16 | 2 2 3 5 9 6 | 2081555136 |
| 6 | 4 9 3 2 | 2057415281 |
| 8 22 | 2 2 8 5 2 8 . . . . | 24139855 |
| 6 | 3 3 1 2 1 6 | 22860923 |
| 8 28 04 | 2 2 8 5 6 1 1 2 1 6 | 1278932 |
| 4 | 3 3 1 2 3 2 | 1143046 |
| 8 28 08 | 2 2 8 5 9 4 2 4 4\|8 | 135886 |
| 4 | 7 4 5 3\|1 | 114305 |
| \|·8\|28\|12 | 2 2 8 6 0 1 6 9 7\|9 | 21581 |
| | 7 4 5 3\|1 | 20575 |
| | 2 2 8 6 0 9 1 5\|1 | 1006 |
| | 8\|3 | 914 |
| | 2 2 8 6 0 9 2 3\|4 | 92 |
| | 8\|3 | 91 |
| | 2\|2\|8\|6\|0\|9\|3\|2 | 1 |

Required the cube roots of the following numbers:—

| | |
|---|---|
| 48228544, 46656, and 15069223. | 364, 36, and 247. |
| 64481·201, and 28991029248. | 40·1, and 3072. |
| 12821119155125, and ·000076765625. | 23405, and ·0425. |
| $\frac{13824}{42875}$, and 16. | $\frac{24}{25}$, and 2·519842. |
| $91\frac{1}{8}$, and $7\frac{6}{7}$. | 4·5, and 1·98802366. |

### TO EXTRACT ANY ROOT WHATEVER.

LET N be the given power or number, $n$ the index of the power, A the assumed power, $r$ its root, R the required root of N.

Then, as the sum of $n + 1$ times A and $n - 1$ times N, is to the sum of $n + 1$ times N and $n - 1$ times A, so is the assumed root $r$, to the required root R.

Or, as half the said sum of $n + 1$ times A and $n - 1$ times N, is to the difference between the given and assumed powers, so is the assumed root $r$, to the difference between the true and assumed roots; which difference, added or subtracted, as the case requires, gives the true root nearly.

That is, $(n + 1) . \mathrm{A} + (n - 1) . \mathrm{N} : (n + 1) . \mathrm{N} + (n - 1) \cdot \mathrm{A} :: r : \mathrm{R}$.

Or, $(n + 1) . \frac{1}{2}\mathrm{A} + (n - 1) . \frac{1}{2}\mathrm{N} : \mathrm{A} \backsim \mathrm{N} :: r : \mathrm{R} \backsim r$.

And the operation may be repeated as often as we please, by using always the last found root for the assumed root, and its $n$th power for the assumed power A.

*To extract the 5th root of* 21035·8.

Here it appears that the 5th root is between 7·3 and 7·4. Taking 7·3, its 5th power is 20730·71593. Hence then we have,

$\mathrm{N} = 21035{\cdot}8$; $r = 7{\cdot}3$; $n = 5$; $\frac{1}{2} . (n + 1) = 3$; $\frac{1}{2}.(n - 1) = 2$.
A = 20730·716
N − A = 305·084

A = 20730·716 N = 21035·8
3 2
3 A = 62192.148 42071·6
2 N = 42071·6

As 104263·7 : 305·084 :: 7·3 : ·0213605
7·3
915252
2135588
104263·7 ) 2227·1132 ( ·0213605, the difference.
14184 7·3 = $r$ add
3758
630 7·321360 = R, the root, true to
5 the last figure.

The 6th root of 21035.8 is 5·254037.
The 6th root of 2 is 1·122462.
The 7th root of 21035·8 is 4·145392.
The 7th root of 2 is 1·104089.
The 9th root of 2 is 1·080059.

## OF RATIOS, PROPORTIONS, AND PROGRESSIONS.

NUMBERS are compared to each other in two different ways: the one comparison considers the difference of the two numbers, and is named Arithmetical Relation, and the difference sometimes Arithmetical Ratio: the other considers their quotient, and is called

Geometrical Relation, and the quotient the Geometrical Ratio. So, of these two numbers 6 and 3, the difference or arithmetical ratio is $6 - 3$ or 3; but the geometrical ratio is $\frac{6}{3}$ or 2.

There must be two numbers to form a comparison: the number which is compared, being placed first, is called the Antecedent; and that to which it is compared the Consequent. So, in the two numbers above, 6 is the antecedent, and 3 is the consequent.

If two or more couplets of numbers have equal ratios, or equal differences, the equality is named Proportion, and the terms of the ratios Proportionals. So, the two couplets, 4, 2 and 8, 6 are arithmetical proportionals, because $4 - 2 = 8 - 6 = 2$; and the two couplets 4, 2 and 6, 3 are geometrical proportionals, because $\frac{4}{2} = \frac{6}{3} = 2$, the same ratio.

To denote numbers as being geometrically proportional, a colon is set between the terms of each couplet to denote their ratio; and a double colon, or else a mark of equality between the couplets or ratios. So, the four proportionals, 4, 2, 6, 3, are set thus, $4 : 2 :: 6 : 3$, which means that 4 is to 2 as 6 is to 3; or thus, $4 : 2 = 6 : 3$; or thus, $\frac{4}{2} = \frac{6}{3}$, both which mean that the ratio of 4 to 2 is equal to the ratio of 6 to 3.

Proportion is distinguished into Continued and Discontinued. When the difference or ratio of the consequent of one couplet and the antecedent of the next couplet is not the same as the common difference or ratio of the couplets, the proportion is discontinued. So, 4, 2, 8, 6 are in discontinued arithmetical proportion, because $4 - 2 = 8 - 6 = 2$, whereas, $2 - 8 = -6$; and 4, 2, 6, 3 are in discontinued geometrical proportion, because $\frac{4}{2} = \frac{6}{3} = 2$, but $\frac{2}{6} = \frac{1}{3}$, which is not the same.

But when the difference or ratio of every two succeeding terms is the same quantity, the proportion is said to be continued, and the numbers themselves a series of continued proportionals, or a progression. So, 2, 4, 6, 8 form an arithmetical progression, because $4 - 2 = 6 - 4 = 8 - 6 = 2$, all the same common difference; and 2, 4, 8, 16, a geometrical progression, because $\frac{4}{2} = \frac{8}{4} = \frac{16}{8} = 2$, all the same ratio.

When the following terms of a Progression exceed each other, it is called an Ascending Progression or Series; but if the terms decrease, it is a Descending one.

So, 0, 1, 2, 3, 4, &c., is an ascending arithmetical progression,
but 9, 7, 5, 3, 1, &c., is a descending arithmetical progression:
Also, 1, 2, 4, 8, 16, &c., is an ascending geometrical progression,
and 16, 8, 4, 2, 1, &c., is a descending geometrical progression.

## ARITHMETICAL PROPORTION AND PROGRESSION.

THE first and last terms of a Progression are called the Extremes; and the other terms lying between them, the Means.

The most useful part of arithmetical proportions is contained in the following theorems:

THEOREM 1.—If four quantities be in arithmetical proportion, the sum of the two extremes will be equal to the sum of the two means.

Thus, of the four 2, 4, 6, 8, here $2 + 8 = 4 + 6 = 10$.

THEOREM 2.—In any continued arithmetical progression, the sum of the two extremes is equal to the sum of any two means that are equally distant from them, or equal to double the middle term when there is an uneven number of terms.

Thus, in the terms 1, 3, 5, it is $1 + 5 = 3 + 3 = 6$.

And in the series 2, 4, 6, 8, 10, 12, 14, it is $2 + 14 = 4 + 12 = 6 + 10 = 8 + 8 = 16$.

THEOREM 3.—The difference between the extreme terms of an arithmetical progression, is equal to the common difference of the series multiplied by one less than the number of the terms.

So, of the ten terms, 2, 4, 6, 8, 10, 12, 14, 16, 18, 20, the common difference is 2, and one less than the number of terms 9; then the difference of the extremes is $20 - 2 = 18$, and $2 \times 9 = 18$ also.

Consequently, the greatest term is equal to the least term added to the product of the common difference multiplied by 1 less than the number of terms.

THEOREM 4.—The sum of all the terms of any arithmetical progression is equal to the sum of the two extremes multiplied by the number of terms, and divided by 2; or the sum of the two extremes multiplied by the number of the terms gives double the sum of all the terms in the series.

This is made evident by setting the terms of the series in an inverted order under the same series in a direct order, and adding the corresponding terms together in that order. Thus,

| | | | | | | | | |
|---|---|---|---|---|---|---|---|---|
| in the series, | 1, | 3, | 5, | 7, | 9, | 11, | 13, | 15; |
| ...... inverted, | 15, | 13, | 11, | 9, | 7, | 5, | 3, | 1; |
| the sums are, | 16 + | 16 + | 16 + | 16 + | 16 + | 16 + | 16 + | 16, |

which must be double the sum of the single series, and is equal to the sum of the extremes repeated so often as are the number of the terms.

From these theorems may readily be found any one of these five parts; the two extremes, the number of terms, the common difference, and the sum of all the terms, when any three of them are given, as in the following Problems:

### PROBLEM I.

*Given the extremes and the number of terms, to find the sum of all the terms.*

RULE.—Add the extremes together, multiply the sum by the number of terms, and divide by 2.

The extremes being 3 and 19, and the number of terms 9; required the sum of the terms?

$$\begin{array}{r} 19 \\ 3 \\ \hline 22 \\ 9 \\ \hline 2\,)\,198 \\ \hline 99 \end{array} = \text{the sum.}$$

Or, $\frac{19 + 3}{2} \times 9 = \frac{22}{2} \times 9 = 11 \times 9 = 99.$

The strokes a clock strikes in one whole revolution of the index, or in 12 hours, is 78.

PROBLEM II.

*Given the extremes, and the number of terms; to find the common difference.*

RULE.—Subtract the less extreme from the greater, and divide the remainder by 1 less than the number of terms, for the common difference.

The extremes being 3 and 19, and the number of terms 9; required the common difference?

$$\begin{array}{r} 19 \\ 3 \\ \hline 8\,)\,16 \\ \hline 2 \end{array} \qquad \text{Or, } \frac{19-3}{9-1} = \frac{16}{8} = 2.$$

If the extremes be 10 and 70, and the number of terms 21; what is the common difference, and the sum of the series?

The com. diff. is 3, and the sum is 840.

PROBLEM III.

*Given one of the extremes, the common difference, and the number of terms; to find the other extreme, and the sum of the series.*

RULE.—Multiply the common difference by 1 less than the number of terms, and the product will be the difference of the extremes: therefore add the product to the less extreme, to give the greater; or subtract it from the greater, to give the less.

Given the least term 3, the common difference 2, of an arithmetical series of 9 terms; to find the greatest term, and the sum of the series?

$$\begin{array}{rl} 2 & \\ 8 & \\ \hline 16 & \\ 3 & \\ \hline 19 & \text{the greatest term.} \\ 3 & \text{the least.} \\ \hline 22 & \text{sum.} \\ 9 & \text{number of terms.} \\ \hline 2\,)\,198 & \\ \hline 99 & \text{the sum of the series.} \end{array}$$

If the greatest term be 70, the common difference 3, and the number of terms 21; what is the least term and the sum of the series?

The least term is 10, and the sum is 840.

PROBLEM IV.

*To find an arithmetical mean proportional between two given terms.*

RULE.—Add the two given extremes or terms together, and take half their sum for the arithmetical mean required. Or, subtract the less extreme from the greater, and half the remainder will be the common difference; which, being added to the less extreme, or subtracted from the greater, will give the mean required.

To find an arithmetical mean between the two numbers 4 and 14.

| Here, 14 | Or, 14 | Or, 14 |
|---|---|---|
| 4 | 4 | 5 |
| 2 ) 18 | 2 ) 10 | 9 |
| 9 | 5 the com. dif. | |
| | 4 the less extreme. | |
| | 9 | |

So that 9 is the mean required by both methods.

### PROBLEM V.

*To find two arithmetical means between two given extremes.*

RULE.—Subtract the less extreme from the greater, and divide the difference by 3, so will the quotient be the common difference; which, being continually added to the less extreme, or taken from the greater, gives the means.

To find two arithmetical means between 2 and 8.

Here 8
2
3 ) 6
com. dif. 2

Then $2 + 2 = 4$ the one mean,
and $4 + 2 = 6$ the other mean.

### PROBLEM VI.

*To find any number of arithmetical means between two given terms or extremes.*

RULE.—Subtract the less extreme from the greater, and divide the difference by 1 more than the number of means required to be found, which will give the common difference; then this being added continually to the least term, or subtracted from the greatest, will give the mean terms required.

To find five arithmetical means between 2 and 14.

Here 14
2
6 ) 12
com. dif. 2

Then, by adding this com. dif. continually, the means are found, 4, 6, 8, 10, 12.

## GEOMETRICAL PROPORTION AND PROGRESSION.

THE most useful part of Geometrical Proportion is contained in the following theorems:

THEOREM 1.—If four quantities be in geometrical proportion, the product of the two extremes will be equal to the product of the two means.

Thus, in the four 2, 4, 3, 6 it is $2 \times 6 = 3 \times 4 = 12$.

And hence, if the product of the two means be divided by one of the extremes, the quotient will give the other extreme. So, of the above numbers, the product of the means $12 \div 2 = 6$ the one extreme, and $12 \div 6 = 2$ the other extreme; and this is the foundation and reason of the practice in the Rule of Three.

THEOREM 2.—In any continued geometrical progression, the product of the two extremes is equal to the product of any two means that are equally distant from them, or equal to the square of the middle term when there is an uneven number of terms.

Thus, in the terms 2, 4, 8, it is $2 \times 8 = 4 \times 4 = 16$.
And in the series 2, 4, 8, 16, 32, 64, 128,
it is $2 \times 128 = 4 \times 64 = 8 \times 32 = 16 \times 16 = 256$.

THEOREM 3.—The quotient of the extreme terms of a geometrical progression is equal to the common ratio of the series raised to the power denoted by one less than the number of the terms.

So, of the ten terms 2, 4, 8, 16, 32, 64, 128, 256, 512, 1024, the common ratio is 2, one less than the number of terms 9; then the quotient of the extremes is $\frac{1024}{2} = 512$, and $2^9 = 512$ also.

Consequently, the greatest term is equal to the least term multiplied by the said power of the ratio whose index is one less than the number of terms.

THEOREM 4.—The sum of all the terms of any geometrical progression is found by adding the greatest term to the difference of the extremes divided by one less than the ratio.

So, the sum 2, 4, 8, 16, 32, 64, 128, 256, 512, 1024, (whose ratio is 2,) is $1024 + \frac{1024 - 2}{2 - 1} = 1024 + 1022 = 2046$.

The foregoing, and several other properties of geometrical proportion, are demonstrated more at large in Byrne's Doctrine of Proportion. A few examples may here be added to the theorems just delivered, with some problems concerning mean proportionals.

The least of ten terms in geometrical progression being 1, and the ratio 2, what is the greatest term, and the sum of all the terms?

The greatest term is 512, and the sum 1023.

PROBLEM I.

*To find one geometrical mean proportional between any two numbers.*

RULE.—Multiply the two numbers together, and extract the square root of the product, which will give the mean proportional sought.

Or, divide the greater term by the less, and extract the square root of the quotient, which will give the common ratio of the three terms: then multiply the less term by the ratio, or divide the greater term by it, either of these will give the middle term required.

To find a geometrical mean between the two numbers 3 and 12.

| *First way.* | *Second way.* |
|---|---|
| 12 | 3 ) 12 ( 4, its root, is 2, the ratio. |
| 3 | |
| 36 ( 6 the mean. | Then, $3 \times 2 = 6$ the mean. |
| 36 | Or, $12 \div 2 = 6$ also. |

PROBLEM II.

*To find two geometrical mean proportionals between any two numbers.*

RULE.—Divide the greater number by the less, and extract the cube root of the quotient, which will give the common ratio of the terms. Then multiply the least given term by the ratio for the first mean, and this mean again by the ratio for the second mean; or, divide the greater of the two given terms by the ratio for the greater mean, and divide this again by the ratio for the less mean.

To find two geometrical mean proportionals between 3 and 24.

Here, 3 ) 24 ( 8, its cube root, 2 is the ratio.
Then, $3 \times 2 = 6$, and $6 \times 2 = 12$, the two means.
Or, $24 \div 2 = 12$, and $12 \div 2 = 6$, the same.
That is, the two means between 3 and 24, are 6 and 12.

PROBLEM III.

*To find any number of geometrical mean proportionals between two numbers.*

RULE.—Divide the greater number by the less, and extract such root of the quotient whose index is one more than the number of means required, that is, the 2d root for 1 mean, the 3d root for 2 means, the 4th root for 3 means, and so on; and that root will be the common ratio of all the terms. Then with the ratio multiply continually from the first term, or divide continually from the last or greatest term.

To find four geometrical mean proportionals between 3 and 96.

Here, 3 ) 96 ( 32, the 5th root of which is 2, the ratio.
Then, $3 \times 2 = 6$, and $6 \times 2 = 12$, and $12 \times 2 = 24$, and $24 \times 2 = 48$.
Or, $96 \div 2 = 48$, and $48 \div 2 = 24$, and $24 \div 2 = 12$, and $12 \div 2 = 6$.
That is, 6, 12, 24, 48 are the four means between 3 and 96.

OF MUSICAL PROPORTION.

THERE is also a third kind of proportion, called Musical, which, being but of little or no common use, a very short account of it may here suffice.

Musical proportion is when, of three numbers, the first has the same proportion to the third, as the difference between the first and second has to the difference between the second and third.

As in these three, 6, 8, 12;
where, $6 : 12 :: 8 - 6 : 12 - 8$,
that is, $6 : 12 :: 2 : 4$.

When four numbers are in Musical Proportion; then the first has the same proportion to the fourth, as the difference between the first and second has to the difference between the third and fourth.

As in these, 6, 8, 12, 18;
where, $6 : 18 :: 8 - 6 : 18 - 12$,
that is, $6 : 18 :: 2 : 6$.

When numbers are in Musical Progression, their reciprocals are in Arithmetical Progression; and the converse, that is, when numbers are in Arithmetical Progression, their reciprocals are in Musical Progression.

So, in these Musicals 6, 8, 12, their reciprocals $\frac{1}{6}$, $\frac{1}{8}$, $\frac{1}{12}$, are in arithmetical progression; for $\frac{1}{6} + \frac{1}{12} = \frac{3}{12} = \frac{1}{4}$; and $\frac{1}{8} + \frac{1}{8} = \frac{2}{8} = \frac{1}{4}$; that is, the sum of the extremes is equal to double the mean, which is the property of arithmeticals.

## FELLOWSHIP, OR PARTNERSHIP.

FELLOWSHIP is a rule by which any sum or quantity may be divided into any number of parts, which shall be in any given proportion to one another.

By this rule are adjusted the gains, or losses, or charges of partners in company; or the effects of bankrupts, or legacies in case of a deficiency of assets or effects; or the shares of prizes, or the numbers of men to form certain detachments; or the division of waste lands among a number of proprietors.

Fellowship is either Single or Double. It is Single, when the shares or portions are to be proportional each to one single given number only; as when the stocks of partners are all employed for the same time: and Double, when each portion is to be proportional to two or more numbers; as when the stocks of partners are employed for different times.

### SINGLE FELLOWSHIP.

GENERAL RULE.—Add together the numbers that denote the proportion of the shares. Then,

As the sum of the said proportional numbers
Is to the whole sum to be parted or divided,
So is each several proportional number
To the corresponding share or part.

Or, As the whole stock is to the whole gain or loss,
So is each man's particular stock to his particular share of the gain or loss.

*To prove the work.*—Add all the shares or parts together, and the sum will be equal to the whole number to be shared, when the work is right.

To divide the number 240 into three such parts, as shall be in proportion to each other as the three numbers, 1, 2, and 3.

Here 1 + 2 + 3 = 6 the sum of the proportional numbers.

Then, as 6 : 240 :: 1 : 40 the 1st part,
and, as 6 : 240 :: 2 : 80 the 2d part,
also as 6 : 240 :: 3 : 120 the 3d part.
Sum of all 240, the proof.

Three persons, A, B, C, freighted a ship with 340 tuns of wine; of which, A loaded 110 tuns, B 97, and C the rest: in a storm, the seamen were obliged to throw overboard 85 tuns; how much must each person sustain of the loss?

Here, 110 + 97 = 207 tuns, loaded by A and B;
theref., 340 − 207 = 133 tuns, loaded by C.
hence, as 340 : 85 :: 110
or, as 4 : 1 :: 110 : $27\frac{1}{2}$ tuns = A's loss;
and, as 4 : 1 :: 97 : $24\frac{1}{4}$ tuns = B's loss;
also, as 4 : 1 :: 133 : $33\frac{1}{4}$ tuns = C's loss.
Sum 85 tuns, the proof.

### DOUBLE FELLOWSHIP.

Double Fellowship, as has been said, is concerned in cases in which the stocks of partners are employed or continued for different times.

Rule.—Multiply each person's stock by the time of its continuance; then divide the quantity, as in Single Fellowship, into shares in proportion to these products, by saying:

As the total sum of all the said products
Is to the whole gain or loss, or quantity to be parted,
So is each particular product
To the corresponding share of the gain or loss.

## SIMPLE INTEREST.

Interest is the premium or sum allowed for the loan, or forbearance of money.

The money lent, or forborne, is called the Principal.

The sum of the principal and its interest, added together, is called the Amount.

Interest is allowed at so much per cent. per annum, which premium per cent. per annum, or interest of a $100 for a year, is called the Rate of Interest. So,

When interest is at 3 per cent. the rate is 3;
........................ 4 per cent. ............. 4;
........................ 5 per cent. ............. 5;
........................ 6 per cent. ............. 6.

Interest is of two sorts: Simple and Compound.

Simple Interest is that which is allowed for the principal lent or forborne only, for the whole time of forbearance.

As the interest of any sum, for any time, is directly proportional to the principal sum, and also to the time of continuance; hence arises the following general rule of calculation.

General Rule.—As $100 is to the rate of interest, so is any given principal to its interest for one year. And again,

As one year is to any given time, so is the interest for a year just found to the interest of the given sum for that time.

*Otherwise.*—Take the interest of one dollar for a year, which, multiply by the given principal, and this product again by the time of loan or forbearance, in years and parts, for the interest of the proposed sum for that time.

When there are certain parts or years in the time, as quarters, or months, or days, they may be worked for either by taking the aliquot, or like parts of the interest of a year, or by the Rule of Three, in the usual way. Also, to divide by 100, is done by only pointing off two figures for decimals.

### COMPOUND INTEREST.

Compound Interest, called also Interest upon Interest, is that which arises from the principal and interest, taken together, as it becomes due at the end of each stated time of payment.

Rules.—1. Find the amount of the given principal, for the time of the first payment, by Simple Interest. Then consider this amount as a new principal for the second payment, whose amount calculate as before; and so on, through all the payments to the last, always accounting the last amount as a new principal for the next payment. The reason of which is evident from the definition of Compound Interest. Or else,

2. Find the amount of one dollar for the time of the first payment, and raise or involve it to the power whose index is denoted by the number of payments. Then that power multiplied by the given principal will produce the whole amount. From which the said principal being subtracted, leaves the Compound Interest of the same; as is evident from the first rule.

## POSITION.

Position is a method of performing certain questions which cannot be resolved by the common direct rules. It is sometimes called False Position, or False Supposition, because it makes a supposition of false numbers to work with, the same as if they were the true ones, and by their means discovers the true numbers sought. It is sometimes also called Trial and Error, because it proceeds by *trials* of false numbers, and thence finds out the true ones by a comparison of the *errors*.

Position is either Single or Double.

### SINGLE POSITION.

Single Position is that by which a question is resolved by means of one supposition only.

Questions which have their results proportional to their suppositions belong to Single Position; such as those which require the multiplication or division of the number sought by any proposed number; or, when it is to be increased or diminished by itself, or any parts of itself, a certain proposed number of times.

Rule.—Take or assume any number for that required, and perform the same operations with it as are described or performed in the question.

Then say, as the result of the said operation is to the position or number assumed, so is the result in the question to the number sought.

A person, after spending $\frac{1}{3}$ and $\frac{1}{4}$ of his money, has yet remaining $60, what had he at first?

| Suppose he had at first $120 | | Proof. | |
|---|---|---|---|
| Now $\frac{1}{3}$ of 120 is | 40 | $\frac{1}{3}$ of 144 is | 48 |
| $\frac{1}{4}$ of it is | 30 | $\frac{1}{4}$ of 144 is | 36 |
| their sum is | 70 | their sum | 84 |
| which taken from | 120 | taken from | 144 |
| leaves | 50 | leaves | 60 as per question. |

Then, 50 : 120 :: 60 : 144.

What number is that, which multiplied by 7, and the product divided by 6, the quotient may be 14? 12.

## PERMUTATIONS AND COMBINATIONS.

The *Permutations* of any number of quantities signify the changes which these quantities may undergo with respect to their order.

Thus, if we take the quantities *a*, *b*, *c*; then, *a b c*, *a c b*, *b a c*, *b c a*, *c a b*, *c b a*, are the permutations of these three quantities taken *all together*; *a b*, *a c*, *b a*, *b c*, *c a*, *c b*, are the permutations of these quantities taken *two* and *two*; *a*, *b*, *c*, are the permutation of these quantities taken singly, or *one* and *one*, &c.

The number of the permutations of the eight letters, *a*, *b*, *c*, *d*, *e*, *f*, *g*, *h*, is 40320; becomes,

$$1 . 2 . 3 . 4 . 5 . 6 . 7 . 8 = 40320.$$

## DOUBLE POSITION.

Double Position is the method of resolving certain questions by means of two suppositions of false numbers.

To the Double Rule of Position belong such questions as have their results not proportional to their positions: such are those, in which the numbers sought, or their parts, or their multiples, are increased or diminished by some given absolute number, which is no known part of the number sought.

Take or assume any two convenient numbers, and proceed with each of them separately, according to the conditions of the question, as in Single Position; and find how much each result is different from the result mentioned in the question, noting also whether the results are too great or too little.

Then multiply each of the said errors by the contrary supposition, namely, the first position by the second error, and the second position by the first error.

If the errors are alike, divide the difference of the products by the difference of the errors, and the quotient will be the answer.

But if the errors are unlike, divide the sum of the products by the sum of the errors, for the answer.

The errors are said to be alike, when they are either both too great, or both too little; and unlike, when one is too great and the other too little.

What number is that, which, being multiplied by 6, the product increased by 18, and the sum divided by 9, the quotient shall be 20.

Suppose the two numbers, 18 and 30. Then

| First position. | | Second position. | | Proof. |
|---|---|---|---|---|
| 18 | | 30 | | 27 |
| 6 | mult. | 6 | | 6 |
| 108 | | 180 | | 162 |
| 18 | add. | 18 | | 18 |
| 9) 126 | | 9) 198 | | 9) 180 |
| 14 | results. | 22 | | 20 |
| 20 | true res. | 20 | | |
| + 6 | errors unlike. | — 2 | | |
| 2d pos. 30 | mult. | 18 | 1st pos. | |
| Errors { 2 180 | | 36 | | |
| { 6 36 | | | | |
| Sum 8) 216 | sum of products. | | | |
| 27 | answer sought. | | | |

Find, by trial, two numbers, as near the true number as possible, and operate with them as in the question; marking the errors which arise from each of them.

Multiply the difference of the two numbers, found by trial, by the least error, and divide the product by the difference of the errors, when they are alike, but by their sum when they are unlike.

Add the quotient, last found, to the number belonging to the least error, when that number is too little, but subtract it when too great, and the result will give the true quantity sought.

## MENSURATION OF SUPERFICIES.

THE *area* of any figure is the measure of its surface, or the space contained within the bounds of that surface, without any regard to thickness.

A square whose side is one inch, one foot, or one yard, &c. is called the *measuring unit*, and the area or content of any figure is computed by the number of those squares contained in that figure.

*To find the area of a parallelogram; whether it be a square, a rectangle, a rhombus, or a rhomboides.*—Multiply the length by the perpendicular height, and the product will be the area.

The perpendicular height of the parallelogram is equal to the area divided by the base.

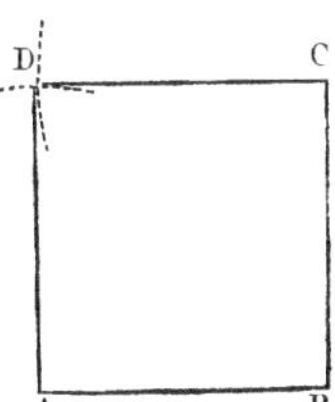


Required the area of the square ABCD whose side is 5 feet 9 inches.

*Here* 5 *ft.* 9 *in.* = 5·75 : *and* $\overline{5{\cdot}75}|^2$ = 5·75 × 5·75 = 33·0625 *feet* = 33 *fe.* 0 *in.* 9 *pa.* = *area required.*

Required the area of the rectangle ABCD, whose length AB is 13·75 chains, and breadth BC 9·5 chains.

*Here* $13·75 \times 9·5 = 130·625$; *and* $\frac{130·625}{10} = 13·0625$ *ac.* $=$ 13 *ac.* 0 *ro.* 10 *po.* = *area required.*

Required the area of the rhombus ABCD, whose length AB is 12 feet 6 inches, and its height DE 9 feet 3 inches.

*Here* 12 *fe.* 6 *in.* = 12·5, *and* 9 *fe.* 3 *in.* = 9·25.

*Whence*, $12·5 \times 9·25 = 115·625$ *fe.* = 115 *fe.* 7 *in.* 6 *pa.* = *area required.*

What is the area of the rhomboides ABCD, whose length AB is 10·52 chains, and height DE 7·63 chains.

*Here* $10·52 \times 7·63 = 80·2676$; *and* $\frac{80·2676}{10} = 8·02676$ *acres* = 8 *ac.* 0 *ro.* 4 *po. area required.*

*To find the area of a triangle.*—Multiply the base by the perpendicular height, and half the product will be the area.

The perpendicular height of the triangle is equal to twice the area divided by the base.

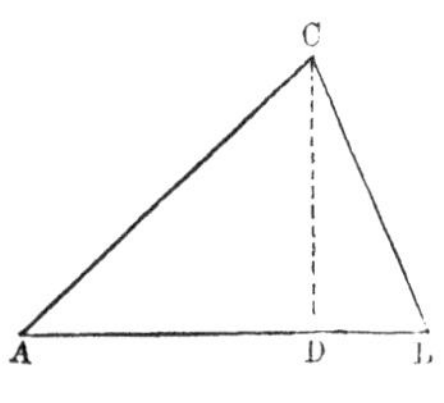


Required the area of the triangle ABC, whose base AB is 10 feet 9 inches, and height DC 7 feet 3 inches.

*Here* 10 *fe.* 9 *in.* = 10·75, *and* 7 *fe.* 3 *in.* = 7·25.

*Whence*, $10·75 \times 7·25 = 77·9375$, *and* $\frac{77·9375}{2} = 38·96875$ *feet* = 38 *fe.* 11 *in.* $7\frac{1}{2}$ *pa.* = *area required.*

*To find the area of a triangle whose three sides only are given.*—From half the sum of the three sides subtract each side severally.

Multiply the half sum and the three remainders continually together, and the square root of the product will be the area required.

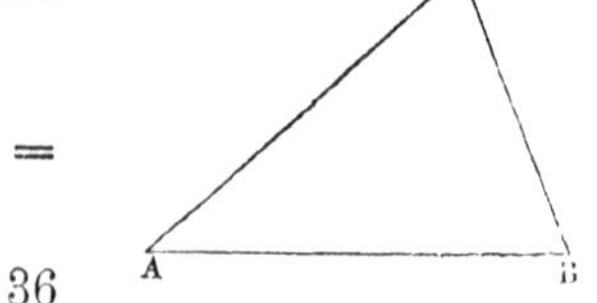


Required the area of the triangle ABC, whose three sides BC, CA, and AB are 24, 36, and 48 chains respectively.

*Here* $\frac{24 + 36 + 48}{2} = \frac{108}{2} = 54$ = $\frac{1}{2}$ *sum of the sides.*

Also, $54 - 24 = 30$ *first diff.*; $54 - 36 = 18$ *second diff.*; *and* $54 - 48 = 6$ *third diff.*

*Whence*, $\sqrt{54 \times 30 \times 18 \times 6} = \sqrt{174960} = 418{\cdot}282$ = *area required.*

*Any two sides of a right angled triangle being given to find the third side.*—When the two legs are given to find the hypothenuse, add the square of one of the legs to the square of the other, and the square root of the sum will be equal to the hypothenuse.

*When the hypothenuse and one of the legs are given to find the other leg.*—From the square of the hypothenuse take the square of the given leg, and the square root of the remainder will be equal to the other leg.

In the right angled triangle ABC, the base AB is 56, and the perpendicular BC 33, what is the hypothenuse?

*Here* $56^2 + 33^2 = 3136 + 1089 = 4225$, *and* $\sqrt{4225} = 65$ = *hypothenuse* AC.

If the hypothenuse AC be 53, and the base AB 45, what is the perpendicular BC?

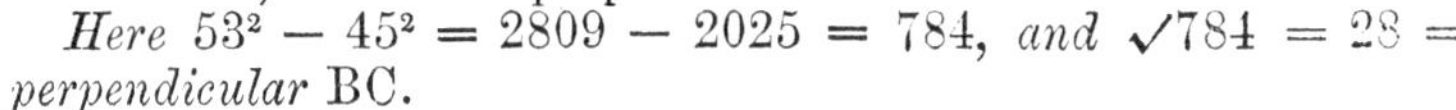

*Here* $53^2 - 45^2 = 2809 - 2025 = 784$, *and* $\sqrt{784} = 28$ = *perpendicular* BC.

*To find the area of a trapezium.*—Multiply the diagonal by the sum of the two perpendiculars falling upon it from the opposite angles, and half the product will be the area.

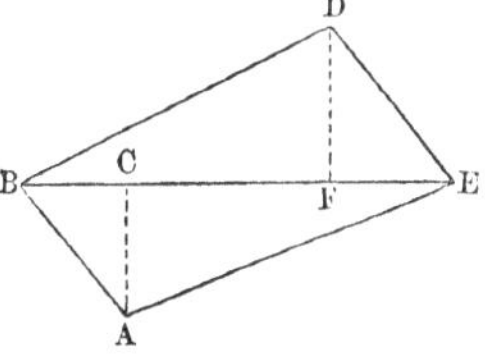


Required the area of the trapezium BAED, whose diagonal BE is 84, the perpendicular AC 21, and DF 28.

*Here* $\overline{28 + 21} \times 84 = 49 \times 84 = 4116$,

*and* $\frac{4116}{2} = 2058$ *the area required.*

*To find the area of a trapezoid, or a quadrangle, two of whose opposite sides are parallel.*—Multiply the sum of the parallel sides by the perpendicular distance between them, and half the product will be the area.

Required the area of the trapezoid ABCD, whose sides AB and DC are 321·51 and 214·24, and perpendicular DE 171·16.

*Here* $321{\cdot}51 + 214{\cdot}24 = 535{\cdot}75$ = *sum of the parallel sides* AB, DC.

*Whence*, $535{\cdot}75 \times 171{\cdot}16$ (*the perp.* DE) = $91698{\cdot}9700$, *and* $\frac{91698{\cdot}9700}{2} = 45849{\cdot}485$ *the area required.*

*To find the area of a regular polygon.*—Multiply half the perimeter of the figure by the perpendicular falling from its centre upon one of the sides, and the product will be the area.

The perimeter of any figure is the sum of all its sides.

Required the area of the regular pentagon ABCDE, whose side AB, or BC, &c., is 25 feet, and the perpendicular OP 17·2 feet.

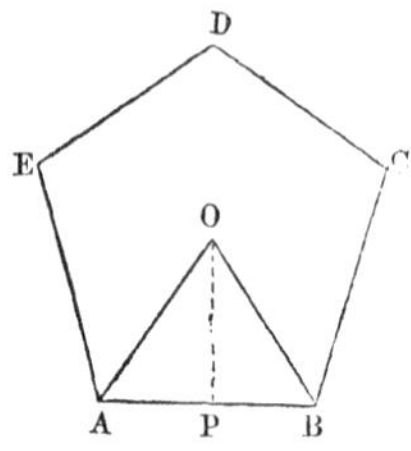


*Here* $\frac{25 \times 5}{2} = 62{\cdot}5 =$ *half perimeter, and* $62{\cdot}5 \times 17{\cdot}2 = 1075$ *square feet = area required.*

*To find the area of a regular polygon, when the side only is given.*—Multiply the square of the side of the polygon by the number standing opposite to its name in the following table, and the product will be the area.

| No. of sides. | Names. | Multipliers. | No. of sides. | Names. | Multipliers. |
|---|---|---|---|---|---|
| 3 | Trigon or equil. △ | 0·433013 | 8 | Octagon | 4·828427 |
| 4 | Tetragon or square | 1·000000 | 9 | Nonagon | 6·181824 |
| 5 | Pentagon | 1·720477 | 10 | Decagon | 7·694209 |
| 6 | Hexagon | 2·598076 | 11 | Undecagon | 9·365640 |
| 7 | Heptagon | 3·633912 | 12 | Duodecagon | 11·196152 |

The angle OBP, together with its tangent, for any polygon of not more than 12 sides, is shown in the following table:

| No. of sides. | Names. | Angle OBP. | Tangents. |
|---|---|---|---|
| 3 | Trigon | 30° | $\cdot 57735 = \frac{1}{3}\sqrt{3}$ |
| 4 | Tetragon | 45° | $1{\cdot}00000 = 1 \times 1$ |
| 5 | Pentagon | 54° | $1{\cdot}37638 = \sqrt{1 + \frac{2}{5}\sqrt{5}}$ |
| 6 | Hexagon | 60° | $1{\cdot}73205 = \sqrt{3}$ |
| 7 | Heptagon | $64°\frac{2}{7}$ | 2·07652 |
| 8 | Octagon | $67°\frac{1}{2}$ | $2{\cdot}41421 = 1 + \sqrt{2}$ |
| 9 | Nonagon | 70° | 2·74747 |
| 10 | Decagon | 72° | $3{\cdot}07768 = \sqrt{5 + 2\sqrt{5}}$ |
| 11 | Undecagon | $73°\frac{7}{11}$ | 3·40568 |
| 12 | Duodecagon | 75° | $3{\cdot}73205 = 2 + \sqrt{3}$ |

Required the area of a pentagon whose side is 15.

The number opposite pentagon in the table is 1·720477.

*Hence* $1{\cdot}720477 \times 15^2 = 1{\cdot}720477 \times 225 = 387{\cdot}107325 =$ *area required.*

*The diameter of a circle being given to find the circumference, or the circumference being given to find the diameter.*—Multiply the diameter by 3·1416, and the product will be the circumference, or

Divide the circumference by 3·1416, and the quotient will be the diameter.

As 7 is to 22, so is the diameter to the circumference; or as 22 is to 7, so is the circumference to the diameter.

As 113 is to 355, so is the diameter to the circumference; or, as 352 is to 115, so is the circumference to the diameter.

If the diameter of a circle be 17, what is the circumference?

*Here* $3{\cdot}1416 \times 17 = 53{\cdot}4072 =$ *circumference.*

If the circumference of a circle be 354, what is the diameter?

$$\textit{Here}\ \frac{354{\cdot}000}{3{\cdot}1416} = 112{\cdot}681 = \textit{diameter}.$$

*To find the length of any arc of a circle.*—When the chord of the arc and the versed sine of half the arc are given:

To 15 times the square of the chord, add 33 times the square of the versed sine, and reserve the number.

To the square of the chord, add 4 times the square of the versed sine, and the square root of the sum will be *twice the chord of half the arc.*

Multiply twice the chord of half the arc by 10 times the square of the versed sine, divide the product by the reserved number, and add the quotient to twice the chord of half the arc: the sum will be the length of the arc very nearly.

*When the chord of the arc, and the chord of half the arc are given.*—From the square of the chord of half the arc subtract the square of half the chord of the arc, the remainder will be the square of the versed sine: then proceed as above.

When the diameter and the versed sine of half the arc are given:

From 60 times the diameter subtract 27 times the versed sine, and reserve the number.

Multiply the diameter by the versed sine, and the square root of the product will be the *chord of half the arc.*

Multiply twice the chord of half the arc by 10 times the versed sine, divide the product by the reserved number, and add the quotient to twice the chord of half the arc; the sum will be the length of the arc very nearly.

When the diameter and chord of the arc are given, the versed sine may be found thus: From the square of the diameter subtract the square of the chord, and extract the square root of the remainder. Subtract this root from the diameter, and half the remainder will give the versed sine of half the arc.

The square of the chord of half the arc being divided by the diameter will give the versed sine, or being divided by the versed sine will give the diameter.

The length of the arc may also be found by multiplying together the number of degrees it contains, the radius and the number ·01745329.

Or, as 180 is to the number of degrees in the arc, so is 3·1416 times the radius, to the length of the arc.

Or, as 3 is to the number of degrees in the arc, so is ·05236 times the radius to the length of the arc.

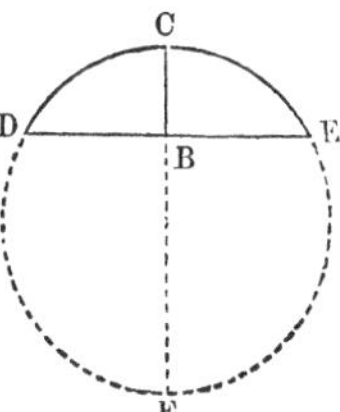


If the chord DE be 48, and the versed sine CB 18, what is the length of the arc?

*Here* $48^2 \times 15 = 34560$

$18^2 \times 33 = 10692$

45252 *reserved number.*

$48^2 = 2304 =$ *the square of the chord.*

$18^2 \times 4 = 1296 = 4$ *times the square of the versed sine.*

$\sqrt{3600} = 60 =$ *twice the chord of half the arc.*

*Now* $\frac{60 \times 18^2 \times 10}{45252} = \frac{194400}{45252} = 4{\cdot}2959$, *which added to twice the chord of half the arc gives* $64{\cdot}2959 =$ *the length of the arc.*

$$\begin{array}{r} 50 \times 60 = 3000 \\ 18 \times 27 = \phantom{0}486 \\ \hline 2514 \end{array}$$ *reserved number.*

$AC = \sqrt{50 \times 18} = 30 =$ *the chord of half the arc.*

$\frac{\overline{30 \times 2 \times 18 \times 10}}{2514} = \frac{10800}{2514} = 4{\cdot}2959$, *which added to twice the chord of half the arc gives* $64{\cdot}2959 =$ *the length of the arc.*

*To find the area of a circle.*—Multiply half the circumference by half the diameter, and the product will be the area.

Or take $\frac{1}{4}$ of the product of the whole circumference and diameter.

What is the area of a circle whose diameter is 42, and circumference 131·946?

$$\begin{array}{r} 2\,)\,131{\cdot}946 \\ \hline 65{\cdot}973 = \tfrac{1}{2}\ \textit{circumference.} \\ 21 = \tfrac{1}{2}\ \textit{diameter.} \\ \hline 65973 \\ 131946\phantom{0} \\ \hline 1385{\cdot}433 = \textit{area required.} \end{array}$$

What is the area of a circle whose diameter is 10 feet 6 inches, and circumference 31 feet 6 inches?

| *fe.* | *in.* | |
|---|---|---|
| 15 | 9 | $= 15{\cdot}75 = \frac{1}{2}$ *circumference.* |
| 5 | 3 | $= 5{\cdot}25 = \frac{1}{2}$ *diameter.* |

$$\begin{array}{r} 7875 \\ 3150\phantom{0} \\ 7875\phantom{00} \\ \hline 82{\cdot}6875 \\ 12 \\ \hline 8{\cdot}2500 \end{array}$$

82 *feet* 8 *inches.*

Multiply the square of the diameter by ·7854, and the product will be the area; or,

Multiply the square of the circumference by ·07958, and the product will be the area.

The following table will also show most of the useful problems relating to the circle and its equal or inscribed square.

Diameter × ·8862 = side of an equal square.
Circumf. × ·2821 = side of an equal square.
Diameter × ·7071 = side of the inscribed square.

Circumf. × ·2251 = side of the inscribed square.
Area × ·6366 = side of the inscribed square.
Side of a square × 1·4142 = diam. of its circums. circle.
Side of a square × 4·443 = circumf. of its circums. circle.
Side of a square × 1·128 = diameter of an equal circle.
Side of a square × 3·545 = circumf. of an equal circle.
What is the area of a circle whose diameter is 5?

$$\begin{array}{r} 7854 \\ 25 \\ \hline 39270 \\ 15708\phantom{0} \\ \hline 19{\cdot}6350 \end{array}$$

25 = *square of the diameter.*

19·6350 = *the answer.*

*To find the area of a sector, or that part of a circle which is bounded by any two radii and their included arc.*—Find the length of the arc, then multiply the radius, or half the diameter, by the length of the arc of the sector, and half the product will be the area.

If the diameter or radius is not given, add the square of half the chord of the arc, to the square of the versed sine of half the arc; this sum being divided by the versed sine, will give the diameter.

The radius AB is 40, and the chord BC of the whole arc 50, required the area of the sector.

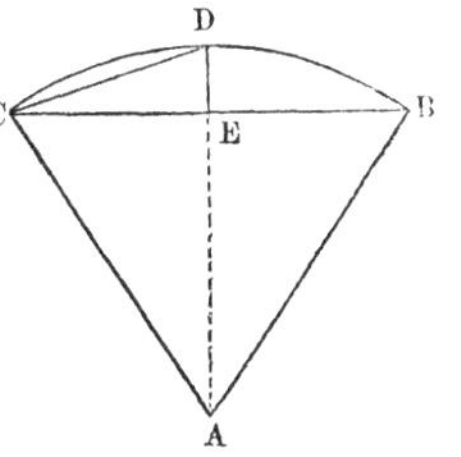


$\dfrac{80 - \sqrt{80^2 - 50^2}}{2} = 8{\cdot}7750 =$ *the versed sine of half the arc.*

$\overline{80 \times 60} - \overline{8{\cdot}7750 \times 27} = 4563{\cdot}0750 =$ *the reserved number.*

$2 \times \sqrt{8{\cdot}7750 \times 80} = 52{\cdot}9906 =$ *twice the chord of half the arc.*

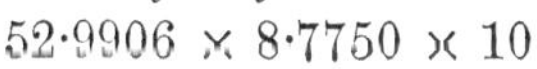

$\dfrac{52{\cdot}9906 \times 8{\cdot}7750 \times 10}{4563{\cdot}0750} = 1{\cdot}0190,$ *which added to twice the chord of half the arc gives* 54·0096 *the length of the arc.*

And $\dfrac{54{\cdot}0096 \times 40}{2} = 1080{\cdot}1920 =$ *area of the sector required.*

As 360 is to the degrees in the arc of a sector, so is the area of the whole circle, whose radius is equal to that of the sector, to the area of the sector required.

For a semicircle, a quadrant, &c. take one half, one quarter, &c. of the whole area.

The radius of a sector of a circle is 20, and the degrees in its arc 22; what is the area of the sector?

*Here the diameter is* 40.

*Hence, the area of the circle* = $40^2 \times {\cdot}7854 = 1600 \times {\cdot}7854 =$ 1256·64.

*Now,* 360° : 22° : : 1256·64 : 76·7947 = *area of the sector.*

*To find the area of a segment of a circle.*—Find the area of the sector, having the same arc with the segment, by the last problem.

Find the area of the triangle formed by the chord of the segment, and the radii of the sector.

Then the sum, or difference, of these areas, according as the segment is greater or less than a semicircle, will be the area required.

The difference between the versed sine and radius, multiplied by half the chord of the arc, will give the area of the triangle.

The radius OB is 10, and the chord AC 10; what is the area of the segment ABC?

$CD = \frac{AC^2}{CE} = \frac{100}{20} = 5 =$ *the versed sine of half the arc.*

$\overline{20 \times 60} - \overline{5 \times 27} = 1065 =$ *the reserved number.*

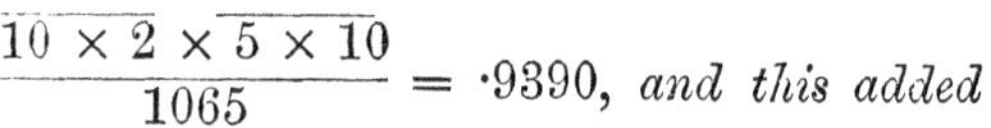

$\frac{\overline{10 \times 2} \times \overline{5 \times 10}}{1065} = {\cdot}9390,$ *and this added to twice the chord of half the arc gives* $20{\cdot}9390 =$ *the length of the arc.*

$\frac{20{\cdot}9390 \times 10}{2} = 104{\cdot}6950 =$ *area of the sector* OACB.

$OD = OC = CD = 5$ *the perpendicular height of the triangle.*

$AD = \sqrt{AO^2 - OD^2} = \sqrt{75} = 8{\cdot}6603 = \frac{1}{2}$ *the chord of the arc.*

$8{\cdot}6603 \times 5 = 43{\cdot}3015 =$ *the area of the triangle* AOB.

$104{\cdot}6950 - 43{\cdot}3015 = 61{\cdot}3935 =$ *area of the segment required;* it being in this case less than a semicircle.

Divide the height, or versed sine, by the diameter, and find the quotient in the table of versed sines.

Multiply the number on the right hand of the versed sine by the square of the diameter, and the product will be the area.

When the quotient arising from the versed sine divided by the diameter, has a remainder or fraction after the third place of decimals; having taken the area answering to the first three figures, subtract it from the next following area, multiply the remainder by the said fraction, and add the product to the first area, then the sum will be the area for the whole quotient.

If the chord of a circular segment be 40, its versed sine 10, and the diameter of the circle 50, what is the area?

$$
\begin{array}{r l}
5{\cdot}0\,)\,1{\cdot}0 & \\
{\cdot}2 & = \textit{tabular versed sine.} \\
{\cdot}111823 & = \textit{tabular segment.} \\
2500 & = \textit{square of 50.} \\
\hline
55911500 & \\
223646\phantom{00} & \\
\hline
279{\cdot}557500 & = \textit{area required.}
\end{array}
$$

*To find the area of a circular zone, or the space included between any two parallel chords and their intercepted arcs.*—From the greater chord subtract half the difference between the two, multiply the remainder by the said half difference, divide the product by the breadth of the zone, and add the quotient to the breadth. To the square of this number add the square of the less chord, and the square root of the sum will be the diameter of the circle.

Now, having the diameter EG, and the two chords AB and DC, find the areas of the segments ABEA, and DCED, the difference of which will be the area of the zone required.

The difference of the tabular segments multiplied by the square of the circle's diameter will give the area of the zone.

When the larger segment AEB is greater than a semicircle, find the areas of the segments AGB, and DCE, and subtract their sum from the area of the whole circle: the remainder will be the area of the zone.

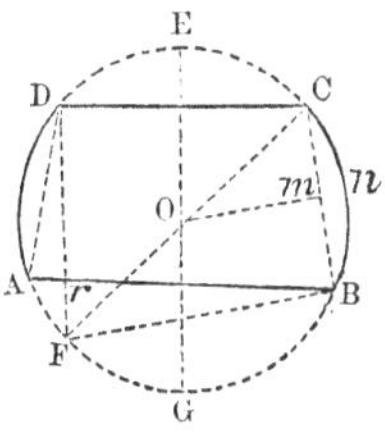

The greater chord AB is 20, the less DC 15, and their distance D*r* $17\frac{1}{2}$: required the area of the zone ABCD.

$\frac{20 - 15}{2} = 2{\cdot}5 = \frac{1}{2}$ = *the difference between the chords.*

$17{\cdot}5 + \frac{(20 - 2{\cdot}5) \times 2{\cdot}5}{17{\cdot}5} = 17{\cdot}5 + 2{\cdot}5 = 20$ = DF.

*And* $\sqrt{20^2 + 15^2} = \sqrt{625} = 25$ = *the diameter of the circle.*

*The segment* AEB *being greater than a semicircle, we find the versed sine of* DCE = 2·5, *and that of* AGB = 5.

*Hence* $\frac{2{\cdot}5}{25} = {\cdot}100$ = *tabular versed sine of* DEC.

*And* $\frac{5}{25} = {\cdot}200$ = *tabular versed sine of* AGB.

*Now* $\cdot040875 \times 25^2$ = *area of seg.* DEC = 25·546875
*And* $\cdot111823 \times 25^2$ = *area of seg.* AGB = 69·889375
*sum* 95·436250
$\cdot7854 \times 25^2$ = *area of the whole circle,* = 490·87500
*Difference* = *area of the zone* ABCD = 395·43875

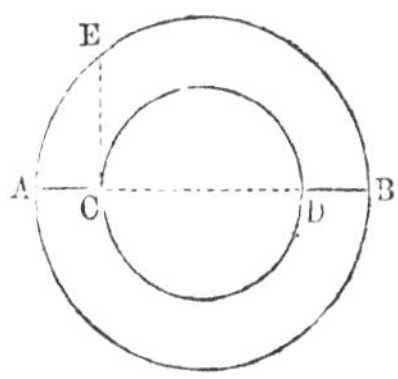

*To find the area of a circular ring, or the space included between the circumference of two concentric circles.*—The difference between the areas of the two circles will be the area of the ring.

Or, multiply the sum of diameters by their difference, and this product again by ·7854, and it will give the area required.

The diameters AB and CD are 20 and 15: required the area of

the circular ring, or the space included between the circumferences of those circles.

*Here* $\overline{AB + CD} \times \overline{AB - CD} = 35 \times 5 = 175$, *and* $175 \times \cdot 7854 = 137 \cdot 4450$ = *area of the ring required.*

*To find the areas of lunes, or the spaces between the intersecting arcs of two eccentric circles.*—Find the areas of the two segments from which the lune is formed, and their difference will be the area required.

The following property is one of the most curious:

If ABC be a right angled triangle, and semicircles be described on the three sides as diameters, then will the said triangle be equal to the two lunes D and F taken together.

For the semicircles described on AC and BC = the one described on AB, from each take the segments cut off by AC and BC, then will the lunes AFCE and BDCG = the triangle ACB.

The length of the chord AB is 40, the height DC 10, and DE 4: required the area of the lune ACBEA.

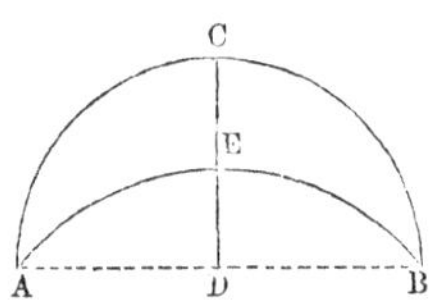


*The diameter of the circle of which* ACB

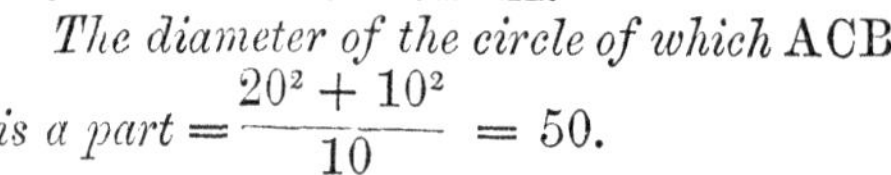

*is a part* $= \frac{20^2 + 10^2}{10} = 50.$

*And the diameter of the circle of which* AEB *is a part* $= \frac{20^2 + 4^2}{4} = 104.$

*Now having the diameter and versed sines, we find,*

*The area of seg.* ACB $= \cdot 111823 \times 50^2 = 279 \cdot 5575$
*And area of seg.* AEB $= \cdot 009955 \times 104^2 = 107 \cdot 6733$
*Their difference is the area of the lune* AEBCA *required,* $= 171 \cdot 8842$

*To find the area of an irregular polygon, or a figure of any number of sides.*—Divide the figure into triangles and trapeziums, and find the area of each separately.

Add these areas together, and the sum will be equal to the area of the whole polygon.

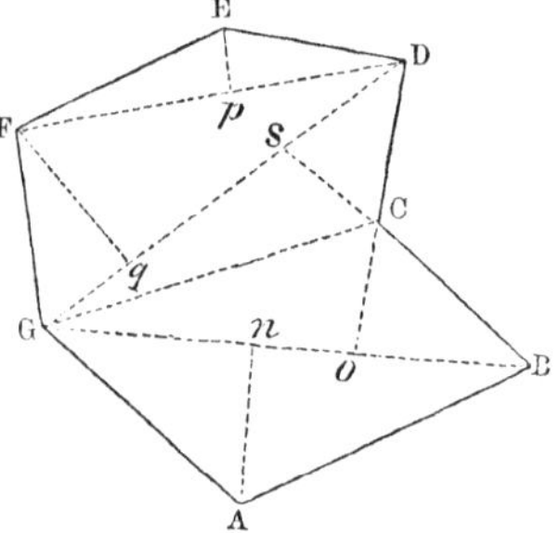


Required the area of the irregular figure ABCDEFGA, the following lines being given:

GB = 30·5 A$n$ = 11·2, CO = 6
GD = 29 F$q$ = 11 C$s$ = 6·6
FD = 24·8 E$p$ = 4 ......

*Here* $\frac{An + Co}{2} \times GB = \frac{11 \cdot 2 + 6}{2} \times 30 \cdot 5 + 8 \cdot 6 \times 30 \cdot 5 = 262 \cdot 3$ = *area of the trapezium* ABCG.

*And* $\frac{Fq + Cs}{2} \times GD = \frac{11 + 6{\cdot}6}{2} \times 29 = 8{\cdot}8 \times 29 = 255{\cdot}2 =$ *area of the trapezium* GCDF.

*Also,* $\frac{FD \times Ep}{2} = \frac{24{\cdot}8 \times 4}{2} = \frac{99{\cdot}2}{2} = 49{\cdot}6 =$ *area of the triangle* FDE.

*Whence* 262·3 + 255·2 + 49·6 = 567·1 = *area of the whole figure required.*

DECIMAL APPROXIMATIONS FOR FACILITATING CALCULATIONS IN MENSURATION.

| | | | |
|---|---|---|---|
| Lineal feet multiplied by | ·00019 | = | miles. |
| — yards | ·000568 | = | — |
| Square inches — | ·007 | = | square feet. |
| — yards — | ·0002067 | = | acres. |
| Circular inches — | ·00546 | = | square feet. |
| Cylindrical inches — | ·0004546 | = | cubic feet. |
| — feet — | ·02909 | = | cubic yards. |
| Cubic inches — | ·00058 | = | cubic feet. |
| — feet — | ·03704 | = | cubic yards. |
| — — — | 6·232 | = | imperial gallons. |
| — inches — | ·003607 | = | — |
| Cylindrical feet — | 4·895 | = | — |
| — inches — | ·002832 | = | — |
| Cubic inches — | ·263 | = | lbs. avs. of cast iron. |
| — — | ·281 | = | — wrought do. |
| — — | ·283 | = | — steel. |
| — — | ·3225 | = | — copper. |
| — — | ·3037 | = | — brass. |
| — — | ·26 | = | — zinc. |
| — — | ·4103 | = | — lead. |
| — — | ·2636 | = | — tin. |
| — — | ·4908 | = | — mercury. |
| Cylindrical inches — | ·2065 | = | — cast iron. |
| — — | ·2168 | = | — wrought iron. |
| — — | ·2223 | = | — steel. |
| — — | ·2533 | = | — copper. |
| — — | ·2385 | = | — brass. |
| — — | ·2042 | = | — zinc. |
| — — | ·3223 | = | — lead. |
| — — | ·207 | = | — tin. |
| — — | ·3854 | = | — mercury. |
| Avoirdupois lbs. — | ·009 | = | cwts. |
| — — | ·00045 | = | tons. |
| 183·346 circular inches | | = | 1 square foot. |
| 2200 cylindrical inches | | = | 1 cubic foot. |
| French metres × 3·281 | | = | feet. |
| — kilogrammes × 2·205 | | = | avoirdupois lb. |
| — grammes × ·002205 | | = | avoirdupois lbs. |

Diameter of a sphere × ·806 = dimensions of equal cube.
Diameter of a sphere × ·6667 = length of equal cylinder.
Lineal inches × ·0000158 = miles.
A French cubic foot = 2093·47 cubic inches.
Imperial gallons × ·7977 = New York gallons.

The average quantity of water that falls in rain and snow at Philadelphia is 36 inches.

At West Point the variation of the magnetic needle, Nov. 16th, 1839, was 7° 58′ 27″ West, and the dip 73° 26′ 28″.

## DECIMAL EQUIVALENTS TO FRACTIONAL PARTS OF LINEAL MEASURES.

**One inch, the integer or whole number.**

| Decimal | are equal to | Decimal | are equal to | Decimal | are equal to |
|---|---|---|---|---|---|
| ·96875 | $\frac{7}{8} + \frac{3}{32}$ | ·625 | $\frac{5}{8}$ | ·28125 | $\frac{1}{4} + \frac{1}{32}$ |
| ·9375 | $\frac{7}{8} + \frac{1}{16}$ | ·59375 | $\frac{1}{2} + \frac{3}{32}$ | ·25 | $\frac{1}{4}$ |
| ·90625 | $\frac{7}{8} + \frac{1}{32}$ | ·5625 | $\frac{1}{2} + \frac{1}{16}$ | ·21875 | $\frac{1}{8} + \frac{3}{32}$ |
| ·875 | $\frac{7}{8}$ | ·53125 | $\frac{1}{2} + \frac{1}{32}$ | ·1875 | $\frac{1}{8} + \frac{1}{16}$ |
| ·84375 | $\frac{3}{4} + \frac{3}{32}$ | ·5 | $\frac{1}{2}$ | ·15625 | $\frac{1}{8} + \frac{1}{32}$ |
| ·8125 | $\frac{3}{4} + \frac{1}{16}$ | ·46875 | $\frac{3}{8} + \frac{3}{32}$ | ·125 | $\frac{1}{8}$ |
| ·78125 | $\frac{3}{4} + \frac{1}{32}$ | ·4375 | $\frac{3}{8} + \frac{1}{16}$ | ·09375 | $\frac{3}{32}$ |
| ·75 | $\frac{3}{4}$ | ·40625 | $\frac{3}{8} + \frac{1}{32}$ | ·0625 | $\frac{1}{16}$ |
| ·71875 | $\frac{5}{8} + \frac{3}{32}$ | ·375 | $\frac{3}{8}$ | ·03125 | $\frac{1}{32}$ |
| ·6875 | $\frac{5}{8} + \frac{1}{16}$ | ·34375 | $\frac{1}{4} + \frac{3}{32}$ | | |
| ·65625 | $\frac{5}{8} + \frac{1}{32}$ | ·3125 | $\frac{1}{4} + \frac{1}{16}$ | | |

**One foot, or 12 inches, the integer.**

| Decimal | are equal to | Decimal | are equal to | Decimal | are equal to |
|---|---|---|---|---|---|
| ·9166 | 11 inches. | ·4166 | 5 inches. | ·0625 | $\frac{3}{4}$ of in. |
| ·6338 | 10 — | ·3333 | 4 — | ·05208 | $\frac{5}{8}$ — |
| ·75 | 9 — | ·25 | 3 — | ·04166 | $\frac{1}{2}$ — |
| ·6666 | 8 — | ·1666 | 2 — | ·03125 | $\frac{3}{8}$ — |
| ·5833 | 7 — | ·0833 | 1 — | ·02083 | $\frac{1}{4}$ — |
| ·5 | 6 — | ·07291 | $\frac{7}{8}$ — | ·01041 | $\frac{1}{8}$ — |

**One yard, or 36 inches, the integer.**

| Decimal | are equal to | Decimal | are equal to | Decimal | are equal to |
|---|---|---|---|---|---|
| ·9722 | 35 inches. | ·6389 | 23 inches. | ·3055 | 11 inches. |
| ·9444 | 34 — | ·6111 | 22 — | ·2778 | 10 — |
| ·9167 | 33 — | ·5833 | 21 — | ·25 | 9 — |
| ·8889 | 32 — | ·5556 | 20 — | ·2222 | 8 — |
| ·8611 | 31 — | ·5278 | 19 — | ·1944 | 7 — |
| ·8333 | 30 — | ·5 | 18 — | ·1667 | 6 — |
| ·8056 | 29 — | ·4722 | 17 — | ·1389 | 5 — |
| ·7778 | 28 — | ·4444 | 16 — | ·1111 | 4 — |
| ·75 | 27 — | ·4167 | 15 — | ·0833 | 3 — |
| ·7222 | 26 — | ·3889 | 14 — | ·0555 | 2 — |
| ·6944 | 25 — | ·3611 | 13 — | ·0278 | 1 — |
| ·6667 | 24 — | ·3333 | 12 — | | |

*Table containing the Circumferences, Squares, Cubes, and Areas of Circles, from 1 to 100, advancing by a tenth.*

| Diam. | Circum. | Square. | Cube. | Area. | Diam. | Circum. | Square. | Cube. | Area. |
|---|---|---|---|---|---|---|---|---|---|
| 1 | 3·1416 | 1 | 1 | ·7854 | 9 | 28·2744 | 81 | 729 | 63·6174 |
| ·1 | 3·4557 | 1·21 | 1·331 | ·9503 | ·1 | 28·5885 | 82·81 | 753·571 | 65·0389 |
| ·2 | 3·7699 | 1·44 | 1·728 | 1·1309 | ·2 | 28·9027 | 84·64 | 778·688 | 66·4762 |
| ·3 | 4·0840 | 1·69 | 2·197 | 1·3273 | ·3 | 29·2168 | 86·49 | 804·357 | 67·9292 |
| ·4 | 4·3982 | 1·96 | 2·744 | 1·5393 | ·4 | 29·5310 | 88·36 | 830·584 | 69·3979 |
| ·5 | 4·7124 | 2·25 | 3·375 | 1·7671 | ·5 | 29·8452 | 90·25 | 857·375 | 70·8823 |
| ·6 | 5·0265 | 2·56 | 4·096 | 2·0106 | ·6 | 30·1593 | 92·16 | 884·736 | 72·3824 |
| ·7 | 5·3407 | 2·89 | 4·913 | 2·2698 | ·7 | 30·4735 | 94·09 | 912·673 | 73·8982 |
| ·8 | 5·6548 | 3·24 | 5·832 | 2·5446 | ·8 | 30·7876 | 96·04 | 941·192 | 75·4298 |
| ·9 | 5·9690 | 3·61 | 6·859 | 2·8352 | ·9 | 31·1018 | 98·01 | 970·299 | 76·9770 |
| 2 | 6·2832 | 4 | 8 | 3·1416 | 10 | 31·4160 | 100 | 1000 | 78·5400 |
| ·1 | 6·5973 | 4·41 | 9·261 | 3·4636 | ·1 | 31·7301 | 102·01 | 1030·301 | 80·1186 |
| ·2 | 6·9115 | 4·84 | 10·648 | 3·8013 | ·2 | 32·0443 | 104·04 | 1061·208 | 81·7130 |
| ·3 | 7·2256 | 5·29 | 12·167 | 4·1547 | ·3 | 32·3580 | 106·09 | 1092·727 | 83·3230 |
| ·4 | 7·5398 | 5·76 | 13·824 | 4·5239 | ·4 | 32·6726 | 108·16 | 1124·864 | 84·9488 |
| ·5 | 7·8540 | 6·25 | 15·625 | 4·9087 | ·5 | 32·9868 | 110·25 | 1157·625 | 86·5903 |
| ·6 | 8·1681 | 6·76 | 17·576 | 5·3093 | ·6 | 33·3009 | 112·36 | 1191·016 | 88·2475 |
| ·7 | 8·4823 | 7·29 | 19·683 | 5·7255 | ·7 | 33·6151 | 114·49 | 1225·043 | 89·9204 |
| ·8 | 8·7964 | 7·84 | 21·952 | 6·1575 | ·8 | 33·9292 | 116·64 | 1259·712 | 91·6090 |
| ·9 | 9·1106 | 8·41 | 24·389 | 6·6052 | ·9 | 34·2434 | 118·81 | 1295·029 | 93·3133 |
| 3 | 9·4248 | 9 | 27· | 7·0686 | 11 | 34·5576 | 121 | 1331 | 95·0334 |
| ·1 | 9·7389 | 9·61 | 29·791 | 7·5476 | ·1 | 34·8717 | 123·21 | 1367·631 | 96·7691 |
| ·2 | 10·0531 | 10·24 | 32·768 | 8·0424 | ·2 | 35·1859 | 125·44 | 1404·928 | 98·5205 |
| ·3 | 10·3672 | 10·89 | 35·937 | 8·5530 | ·3 | 35·5010 | 127·69 | 1442·897 | 100·2877 |
| ·4 | 10·6814 | 11·56 | 39·304 | 9·0792 | ·4 | 35·8142 | 129·96 | 1481·544 | 102·0705 |
| ·5 | 10·9956 | 12·25 | 42·875 | 9·6211 | ·5 | 36·1284 | 132·25 | 1520·875 | 103·8691 |
| ·6 | 11·3097 | 12·96 | 46·656 | 10·1787 | ·6 | 36·4425 | 134·56 | 1560·896 | 105·6834 |
| ·7 | 11·6239 | 13·69 | 50·653 | 10·7521 | ·7 | 36·7567 | 136·89 | 1601·613 | 107·5134 |
| ·8 | 11·9380 | 14·44 | 54·872 | 11·3411 | ·8 | 37·0708 | 139·24 | 1643·032 | 109·3590 |
| ·9 | 12·2522 | 15·21 | 59·319 | 11·9459 | ·9 | 37·3840 | 141·61 | 1685·159 | 111·2204 |
| 4 | 12·5664 | 16 | 64 | 12·5664 | 12 | 37·6992 | 144 | 1728 | 113·0976 |
| ·1 | 12·8805 | 16·81 | 68·921 | 13·2025 | ·1 | 38·0133 | 146·41 | 1771·561 | 114·9904 |
| ·2 | 13·1947 | 17·64 | 74·088 | 13·8544 | ·2 | 38·3275 | 148·84 | 1815·848 | 116·8989 |
| ·3 | 13·5088 | 18·49 | 79·507 | 14·5220 | ·3 | 38·6416 | 151·29 | 1860·867 | 118·8231 |
| ·4 | 13·8230 | 19·36 | 85·184 | 15·2053 | ·4 | 38·9558 | 153·76 | 1906·624 | 120·7631 |
| ·5 | 14·1372 | 20·25 | 91·125 | 15·9043 | ·5 | 39·2700 | 156·25 | 1953·125 | 122·7187 |
| ·6 | 14·4513 | 21·16 | 97·336 | 16·6190 | ·6 | 39·5841 | 158.76 | 2000·376 | 124·6901 |
| ·7 | 14·7655 | 22·09 | 103·823 | 17·3494 | ·7 | 39·8983 | 161·29 | 2048·383 | 126·6771 |
| ·8 | 15·0796 | 23·04 | 110·592 | 18·0956 | ·8 | 40·2124 | 163·84 | 2097·152 | 128·6799 |
| ·9 | 15·3938 | 24·01 | 117·649 | 18·8574 | ·9 | 40·5266 | 166·41 | 2146·689 | 130·6984 |
| 5 | 15·7080 | 25 | 125 | 19·6350 | 13 | 40·8408 | 169 | 2197 | 132·7326 |
| ·1 | 16·0221 | 26·01 | 132·651 | 20·4282 | ·1 | 41·1549 | 171·61 | 2248·091 | 134·7824 |
| ·2 | 16·3363 | 27·04 | 140·608 | 21·2372 | ·2 | 41·4691 | 174·24 | 2299·968 | 136·8480 |
| ·3 | 16·6504 | 28·09 | 148·877 | 22·0618 | ·3 | 41·7832 | 176·89 | 2352·637 | 138·9294 |
| ·4 | 16·9646 | 29·16 | 157·464 | 22·9022 | ·4 | 42·0974 | 179·56 | 2406·104 | 141·0264 |
| ·5 | 17·2788 | 30·25 | 166·375 | 23·7583 | ·5 | 42·4116 | 182·25 | 2460·375 | 143·1391 |
| ·6 | 17·5929 | 31·36 | 175·616 | 24·6301 | ·6 | 42·7257 | 184·96 | 2515·456 | 145·2675 |
| ·7 | 17·9071 | 32·49 | 185·193 | 25·5176 | ·7 | 43·0399 | 187·69 | 2571·353 | 147·4117 |
| ·8 | 18·2212 | 33·64 | 195·112 | 26·4208 | ·8 | 43·3540 | 190·44 | 2628·072 | 149·5715 |
| ·9 | 18·5354 | 34·81 | 205·379 | 27·3397 | ·9 | 43·6682 | 193·21 | 2685·619 | 151·7471 |
| 6 | 18·8496 | 36 | 216 | 28·2744 | 14 | 43·9824 | 196 | 2744 | 153·9384 |
| ·1 | 19·1007 | 37·21 | 226·981 | 29·2247 | ·1 | 44·2965 | 198·81 | 2803·221 | 156·1453 |
| ·2 | 19·4779 | 38·44 | 238·328 | 30·1907 | ·2 | 44·6107 | 201·64 | 2863·288 | 158·3680 |
| ·3 | 19·7920 | 39·69 | 250·047 | 31·1725 | ·3 | 44·9248 | 204·49 | 2924·207 | 160·6064 |
| ·4 | 20·1062 | 40·96 | 262·144 | 32·1699 | ·4 | 45·2390 | 207·36 | 2985·984 | 162·8605 |
| ·5 | 20·4204 | 42·25 | 274·625 | 33·1831 | ·5 | 45·5532 | 210·25 | 3048·625 | 165·1303 |
| ·6 | 20·7345 | 43·56 | 287·496 | 34·2120 | ·6 | 45·8073 | 213·16 | 3112·136 | 167·4158 |
| ·7 | 21·0487 | 44·89 | 300·763 | 35·2566 | ·7 | 46·1815 | 216·09 | 3176·523 | 169·7179 |
| ·8 | 21·3628 | 46·24 | 314·432 | 36·3168 | ·8 | 46·4956 | 219·04 | 3241·792 | 172·0340 |
| ·9 | 21·6770 | 47·61 | 328·509 | 37·3928 | ·9 | 46·8098 | 222·01 | 3307·949 | 174·3666 |
| 7 | 21·9912 | 49 | 343 | 38·4846 | 15 | 47·1240 | 225 | 3375 | 176·7150 |
| ·1 | 22·3053 | 50·41 | 357·911 | 39·5920 | ·1 | 47·4381 | 228·01 | 3442·951 | 179·0790 |
| ·2 | 22·6195 | 51·84 | 373·248 | 40·7151 | ·2 | 47·7523 | 231·04 | 3511·808 | 181·4588 |
| ·3 | 22·9336 | 53·29 | 389·017 | 41·8539 | ·3 | 48·0664 | 234·09 | 3581·577 | 183·8542 |
| ·4 | 23·2478 | 54·76 | 405·224 | 43·0085 | ·4 | 48·3806 | 237·16 | 3652·264 | 186·2654 |
| ·5 | 23·5620 | 56·25 | 421·875 | 44·1787 | ·5 | 48·6948 | 240·25 | 3723·875 | 188·6923 |
| ·6 | 23·8761 | 57·76 | 438·976 | 45·3647 | ·6 | 49·0089 | 243·36 | 3796·416 | 191·1349 |
| ·7 | 24·1903 | 59·29 | 456·533 | 46·5663 | ·7 | 49·3231 | 246·49 | 3869·893 | 193·5932 |
| ·8 | 24·5044 | 60·84 | 474·552 | 47·7837 | ·8 | 49·6372 | 249·64 | 3944·312 | 196·0672 |
| ·9 | 24·8186 | 62·41 | 493·039 | 49·0168 | ·9 | 49·9514 | 252·81 | 4019·679 | 198·5569 |
| 8 | 25·1328 | 64 | 512 | 50·2656 | 16 | 50·2656 | 256 | 4096 | 201·0624 |
| ·1 | 25·4469 | 65·61 | 531·441 | 51·5300 | ·1 | 50·5797 | 259·21 | 4173·281 | 203·5835 |
| ·2 | 25·7611 | 67·24 | 551·368 | 52·8102 | ·2 | 50·8939 | 262·44 | 4251·528 | 206·1209 |
| ·3 | 26·0752 | 68·89 | 571·787 | 54·1062 | ·3 | 51·2080 | 265·69 | 4330·747 | 208·6723 |
| ·4 | 26·3894 | 70·56 | 592·704 | 55·4178 | ·4 | 51·5224 | 268·96 | 4410·944 | 211·1411 |
| ·5 | 26·7036 | 72·25 | 614·125 | 56·7451 | ·5 | 51·8364 | 272·25 | 4492·125 | 213·8251 |
| ·6 | 27·0177 | 73·96 | 636·056 | 58·0881 | ·6 | 52·1505 | 275·56 | 4574·296 | 216·4248 |
| ·7 | 27·3319 | 75·69 | 658·503 | 59·4469 | ·7 | 52·4647 | 278·89 | 4657·463 | 219·0402 |
| ·8 | 27·6460 | 77·44 | 681·472 | 60·8213 | ·8 | 52·7788 | 282·24 | 4741·632 | 221·6712 |
| ·9 | 27·9602 | 79·21 | 704·969 | 62·2115 | ·9 | 53·0930 | 285·61 | 4826·809 | 224·3180 |

| Diam. | Circum. | Square. | Cube. | Area. | Diam. | Circum. | Square. | Cube. | Area. |
|---|---|---|---|---|---|---|---|---|---|
| 17 | 53·4072 | 289 | 4913 | 226·9806 | 25 | 78·5400 | 625 | 15625 | 490·8750 |
| ·1 | 53·7213 | 292·41 | 5000·211 | 229·6588 | ·1 | 78·8541 | 630·01 | 15813·251 | 494·3098 |
| ·2 | 54·0355 | 295·84 | 5088·448 | 232·3527 | ·2 | 79·1683 | 635·04 | 16003·008 | 498·7604 |
| ·3 | 54·3496 | 299·29 | 5177·717 | 235·0623 | ·3 | 79·4824 | 640·09 | 16194·277 | 502·7266 |
| ·4 | 54·6038 | 302·76 | 5268·024 | 237·7877 | ·4 | 79·7966 | 645·16 | 16387·064 | 506·7086 |
| ·5 | 54·9780 | 306·25 | 5359·375 | 240·5287 | ·5 | 80·8108 | 650·25 | 16581·375 | 510·7063 |
| ·6 | 55·2921 | 309·76 | 5451·776 | 243·2855 | ·6 | 80·4249 | 655·36 | 16777·216 | 514·7196 |
| ·7 | 55·6063 | 313·29 | 5545·233 | 246·0579 | ·7 | 80·7391 | 660·49 | 16974·593 | 518·7488 |
| ·8 | 55·9204 | 316·84 | 5639·752 | 248·8461 | ·8 | 81·0532 | 665·64 | 17173·512 | 522·7936 |
| ·9 | 56·2346 | 320·41 | 5735·339 | 251·6500 | ·9 | 81·3674 | 670·81 | 17373·979 | 526·8541 |
| 18 | 56·5488 | 324 | 5832 | 254·4696 | 26 | 81·6816 | 676 | 17576 | 530·9304 |
| ·1 | 56·8629 | 327·61 | 5929·741 | 257·3048 | ·1 | 81·9976 | 681·21 | 17779·581 | 535·0223 |
| ·2 | 57·1771 | 331·24 | 6028·568 | 260·1558 | ·2 | 82·3099 | 686·44 | 17984·728 | 539·1299 |
| ·3 | 57·4912 | 334·89 | 6128·487 | 263·0226 | ·3 | 82·6240 | 691·69 | 18191·447 | 543·2533 |
| ·4 | 57·8054 | 338·56 | 6229·504 | 265.9050 | ·4 | 82·9382 | 696·96 | 18399·744 | 547·3923 |
| ·5 | 58·1196 | 342·25 | 6331·625 | 268·8031 | ·5 | 83·2524 | 702·25 | 18609·625 | 551·5471 |
| ·6 | 58·4337 | 345·96 | 6434·856 | 271·7169 | ·6 | 83·5665 | 707·56 | 18821·096 | 555·7176 |
| ·7 | 58·7479 | 349·69 | 6539·203 | 274·6465 | ·7 | 83·8807 | 712·89 | 19034·163 | 559·9038 |
| ·8 | 59·0620 | 353·44 | 6644·672 | 277·5917 | ·8 | 84·1948 | 718·24 | 19248·832 | 564·1056 |
| 9 | 59·3762 | 357·21 | 6751·269 | 280·5527 | ·9 | 84·5090 | 723·61 | 19465·109 | 568·3232 |
| 19 | 59·6904 | 361 | 6859 | 283·5294 | 27 | 84·8232 | 729 | 19683 | 572·5566 |
| ·1 | 60·0045 | 364·81 | 6967·871 | 286·5217 | ·1 | 85·1373 | 734·41 | 19902·511 | 576·8056 |
| ·2 | 60·3187 | 368·64 | 7077·888 | 289·5298 | ·2 | 85·4515 | 739·84 | 20123·648 | 581·0703 |
| ·3 | 60·6328 | 372.49 | 7189·057 | 292·5536 | ·3 | 85·7656 | 745·29 | 20346·417 | 585·3507 |
| ·4 | 60·9470 | 376.36 | 7301·384 | 295·5931 | ·4 | 86·0798 | 750·76 | 20570·824 | 589·6469 |
| ·5 | 61·2612 | 380·25 | 7414·875 | 298·6483 | ·5 | 86·3940 | 756·25 | 20796·875 | 593·9587 |
| ·6 | 61·5753 | 384·16 | 7529·536 | 301·7192 | ·6 | 86·7081 | 761·76 | 21024·576 | 598·2863 |
| ·7 | 61·8895 | 388·09 | 7645·373 | 304·8060 | ·7 | 87·0223 | 767·29 | 21253·933 | 602·6295 |
| ·8 | 62·2036 | 392·04 | 7762·392 | 307·9082 | ·8 | 87·3364 | 772·84 | 21484·952 | 606·9885 |
| ·9 | 62·5178 | 396·01 | 7880·599 | 311·0252 | ·9 | 87·6506 | 778·41 | 21717·639 | 611·3632 |
| 20 | 62·8320 | 400 | 8000 | 314·1600 | 28 | 87·9648 | 784 | 21952 | 615·7536 |
| ·1 | 63·1461 | 404·01 | 8120·601 | 317·3094 | ·1 | 88·2789 | 789·61 | 22188·041 | 620·1596 |
| ·2 | 63·4603 | 408·04 | 8242·408 | 320·4746 | ·2 | 88·5931 | 795·24 | 22425·768 | 624·5814 |
| ·3 | 63·7744 | 412·09 | 8365·427 | 323·6554 | ·3 | 88·9072 | 800·89 | 22665·187 | 629·0190 |
| ·4 | 64·0886 | 416·16 | 8489·664 | 326·8520 | ·4 | 89·2214 | 806·56 | 22906·304 | 633·4722 |
| ·5 | 64·4028 | 420·25 | 8615·125 | 330·0643 | ·5 | 89·5356 | 812·25 | 23149·125 | 637·9411 |
| ·6 | 64·7161 | 424·36 | 8741·816 | 333·2923 | ·6 | 89·8497 | 817·96 | 23393·656 | 642·4257 |
| ·7 | 65·0311 | 428·49 | 8869·743 | 336·5360 | ·7 | 90·1639 | 823·69 | 23639·903 | 646·9261 |
| ·8 | 65·3452 | 432·64 | 8998·912 | 339·7954 | ·8 | 90·4780 | 829·44 | 23887·872 | 651·4421 |
| ·9 | 65·6594 | 436·81 | 9129·329 | 343·0705 | ·9 | 90·7922 | 835·21 | 24137·569 | 655·9739 |
| 21 | 65·9736 | 441 | 9261 | 346·3614 | 29 | 91·1064 | 841 | 24389 | 660·5214 |
| ·1 | 66·2870 | 445·21 | 9393·931 | 349·6679 | ·1 | 91·4205 | 846·81 | 24642·171 | 665·0845 |
| ·2 | 66·6012 | 449·44 | 9528·128 | 352·9901 | ·2 | 91·7347 | 852·64 | 24897·088 | 669·6634 |
| ·3 | 66·7916 | 453·69 | 9663·597 | 356·3281 | ·3 | 92·0488 | 858·49 | 25153·757 | 674·2580 |
| ·4 | 67·2930 | 457·96 | 9800·344 | 359·6817 | ·4 | 92·3630 | 864·36 | 25412·184 | 678·8683 |
| ·5 | 67·5444 | 462·25 | 9938·375 | 363·0511 | ·5 | 92·6772 | 870·25 | 25672·375 | 683·4943 |
| ·6 | 67·8585 | 466·56 | 10077·696 | 366·4362 | ·6 | 92·9913 | 876·16 | 25934·336 | 688·1360 |
| ·7 | 68·1727 | 470·89 | 10218·313 | 369·8370 | ·7 | 93·3055 | 882·09 | 26198·073 | 692·7934 |
| ·8 | 68·4868 | 475·24 | 10360·232 | 373·2534 | ·8 | 93·6196 | 888·04 | 26463·592 | 697·4666 |
| ·9 | 68·8010 | 479·61 | 10503·459 | 376·6856 | ·9 | 93·9338 | 894·01 | 26730·899 | 702·1554 |
| 22 | 69·1152 | 484 | 10648 | 380·1336 | 30 | 94·2480 | 900 | 27000 | 706·8600 |
| ·1 | 69·4293 | 488·41 | 10793·861 | 383·5972 | ·1 | 94·5621 | 906·01 | 27270·901 | 711·5802 |
| ·2 | 69·7435 | 492·84 | 10941·048 | 387·0765 | ·2 | 94·8763 | 912·04 | 27543·608 | 716·3162 |
| ·3 | 70·0576 | 497·29 | 11089·567 | 390·5751 | ·3 | 95·1904 | 918·09 | 27818·127 | 721·0678 |
| ·4 | 70·3718 | 501·76 | 11239·424 | 394·0823 | ·4 | 95·5046 | 924·16 | 28094·464 | 725·8352 |
| ·5 | 70·6860 | 506·25 | 11390·625 | 397·6087 | ·5 | 95·8188 | 930·25 | 28372·625 | 730·6183 |
| ·6 | 71·0001 | 510·76 | 11543·176 | 401·1509 | ·6 | 96·1329 | 936·36 | 28652·616 | 735·4171 |
| ·7 | 71·3143 | 515·29 | 11697·083 | 404·7087 | ·7 | 96·4471 | 942·49 | 28934·443 | 740·2316 |
| ·8 | 71·6284 | 519·84 | 11852·352 | 408·2823 | ·8 | 96·7612 | 948·64 | 29218·112 | 745·0618 |
| ·9 | 71·9426 | 524·41 | 12008·989 | 411·8716 | ·9 | 97·0754 | 954·81 | 29503·629 | 749·9077 |
| 23 | 72·2568 | 529 | 12167 | 415·4766 | 31 | 97·3896 | 961 | 29791 | 754·7694 |
| ·1 | 72·5709 | 533·61 | 12326·391 | 419·0972 | ·1 | 97·7037 | 967·21 | 30080·231 | 759·6467 |
| ·2 | 72·8851 | 538·24 | 12487·168 | 422·7336 | ·2 | 98·0179 | 973·44 | 30371·328 | 764·5397 |
| ·3 | 73·1992 | 542·89 | 12649·337 | 426·3858 | ·3 | 98·3320 | 979·69 | 30664·297 | 769·4485 |
| ·4 | 73·5134 | 547·56 | 12812·904 | 430·0536 | ·4 | 98·6452 | 985·96 | 30959·144 | 774·3729 |
| ·5 | 73·8276 | 552·25 | 12977·875 | 433·7371 | ·5 | 98·9604 | 992·25 | 31255·875 | 779·3131 |
| ·6 | 74·1417 | 556·96 | 13144·256 | 437·4363 | ·6 | 99·2745 | 998·56 | 31554·496 | 784·2689 |
| ·7 | 74·4559 | 561·69 | 13312·053 | 441·1511 | ·7 | 99·5887 | 1004·89 | 31855·013 | 789·2406 |
| ·8 | 74·7680 | 566·44 | 13481·272 | 444·8819 | ·8 | 99·9028 | 1011·24 | 32157·432 | 794·2278 |
| ·9 | 75·0882 | 571·21 | 13651·919 | 448·6283 | ·9 | 100·2170 | 1017·61 | 32461·759 | 799·2308 |
| 24 | 75·3984 | 576 | 13824 | 452·3904 | 32 | 100·5312 | 1024 | 32768 | 804·2496 |
| ·1 | 75·7125 | 580·81 | 13997·541 | 456·1681 | ·1 | 100·8453 | 1030·41 | 33076·161 | 809·2840 |
| ·2 | 76·0267 | 585·64 | 14172·488 | 459·9616 | ·2 | 101·1595 | 1036·84 | 38386·248 | 814·3341 |
| ·3 | 76·3408 | 590·49 | 14348·907 | 463·7708 | ·3 | 101·4736 | 1043·29 | 33698·267 | 819·3999 |
| ·4 | 76·6523 | 595·36 | 14526·784 | 467·5957 | ·4 | 101·7478 | 1049·76 | 34012·224 | 824·4815 |
| ·5 | 76·9692 | 600·25 | 14706·125 | 471·4363 | ·5 | 102·1020 | 1056·25 | 34328·125 | 829·5787 |
| ·6 | 77·2833 | 605·16 | 14886·936 | 475·2926 | ·6 | 102·4161 | 1062·76 | 34645·976 | 834·6917 |
| ·7 | 77·5975 | 610·09 | 15069·223 | 479·1646 | ·7 | 102·7303 | 1069·29 | 34965·783 | 839·8203 |
| ·8 | 77·9116 | 615·04 | 15252·992 | 483·0524 | ·8 | 103·0444 | 1075·84 | 35287·552 | 844·9647 |
| ·9 | 78·2258 | 620·01 | 15438·249 | 486·9558 | ·9 | 103·3586 | 1082·41 | 35611·289 | 850·1248 |

| Diam. | Circum. | Square. | Cube. | Area. | Diam. | Circum. | Square. | Cube. | Area. |
|---|---|---|---|---|---|---|---|---|---|
| 33 | 103·6728 | 1089 | 35937 | 855·3006 | 41 | 128·8056 | 1681 | 68921 | 1320·2574 |
| ·1 | 103·9869 | 1095·61 | 36264·691 | 860·4920 | ·1 | 129·1197 | 1689·21 | 69426·531 | 1326·7055 |
| ·2 | 104·3011 | 1102·24 | 36594·368 | 865·6992 | ·2 | 129·4323 | 1697·44 | 69934·528 | 1333·1693 |
| ·3 | 104·6151 | 1108·89 | 36926·037 | 870·9222 | ·3 | 129·7480 | 1705·69 | 70444·997 | 1339·6489 |
| ·4 | 104·9294 | 1115·56 | 37259·704 | 876·1608 | ·4 | 130·0622 | 1713·96 | 70957·944 | 1346·1441 |
| ·5 | 105·2436 | 1122·25 | 37595·375 | 881·4151 | ·5 | 130·3764 | 1722·25 | 71473·375 | 1352·6551 |
| ·6 | 105·5577 | 1128·96 | 37933·056 | 886·6851 | ·6 | 130·6905 | 1730·56 | 71991·296 | 1359·1818 |
| ·7 | 105·8719 | 1135·69 | 38272·753 | 891·9709 | ·7 | 131·0047 | 1738·89 | 72511·713 | 1365·7242 |
| ·8 | 106·1860 | 1142·44 | 38614·472 | 897·2723 | ·8 | 131·3188 | 1747·24 | 73034·632 | 1372·2822 |
| ·9 | 106·5002 | 1149·21 | 38958·219 | 902·5895 | ·9 | 131·6320 | 1755·61 | 73560·059 | 1378·8560 |
| 34 | 106·8144 | 1156 | 39304 | 907·9224 | 42 | 131·9472 | 1764 | 74088 | 1385·4456 |
| ·1 | 107·1285 | 1162·81 | 39651·821 | 913·2709 | ·1 | 132·2613 | 1772·41 | 74618·461 | 1392·0508 |
| ·2 | 107·4272 | 1169·64 | 40001·688 | 918·6352 | ·2 | 132·5755 | 1780·84 | 75151·448 | 1398·6717 |
| ·3 | 107·7568 | 1176·49 | 40353·607 | 924·0115 | ·3 | 132·8896 | 1789·29 | 75686·967 | 1405·3083 |
| ·4 | 108·0710 | 1183·36 | 40707·584 | 929·4109 | ·4 | 133·2038 | 1797·76 | 76225·024 | 1411·9607 |
| ·5 | 108·3852 | 1190·25 | 41063·625 | 934·8223 | ·5 | 133·5180 | 1806·25 | 76765·625 | 1418·6287 |
| ·6 | 108·6993 | 1197·16 | 41421·736 | 940·2494 | ·6 | 133·8321 | 1814·76 | 77308·776 | 1425·3125 |
| ·7 | 109·0352 | 1204·09 | 41781·923 | 945·6922 | ·7 | 134·1463 | 1823·29 | 77854·483 | 1432·0119 |
| ·8 | 109·3076 | 1211·04 | 42144·192 | 951·1508 | ·8 | 134·4604 | 1831·84 | 78402·752 | 1438·7271 |
| ·9 | 109·6418 | 1218·01 | 42508·549 | 956·6250 | ·9 | 134·7746 | 1840·41 | 78958·589 | 1445·4580 |
| 35 | 109·9560 | 1225 | 42875 | 962·1150 | 43 | 135·0888 | 1849 | 79507 | 1452·2046 |
| ·1 | 110·2701 | 1232·01 | 43243·551 | 967·6206 | ·1 | 135·4029 | 1857·61 | 80062·991 | 1458·9668 |
| ·2 | 110·5843 | 1239·04 | 43614·208 | 973·1420 | ·2 | 135·7171 | 1866·24 | 80621·568 | 1465·7448 |
| ·3 | 110·8984 | 1246·09 | 43986·977 | 978·6790 | ·3 | 136·0332 | 1874·89 | 81182·737 | 1472·5385 |
| ·4 | 111·2126 | 1253·16 | 44361·864 | 984·2318 | ·4 | 136·3454 | 1883·56 | 81746·504 | 1479·3480 |
| ·5 | 111·5268 | 1260·25 | 44738·875 | 989·8003 | ·5 | 136·6596 | 1892·25 | 82312·875 | 1486·1731 |
| ·6 | 111·8409 | 1267·36 | 45118·016 | 995·3845 | ·6 | 136·9737 | 1900·96 | 82881·856 | 1493·0139 |
| ·7 | 112·1551 | 1274·49 | 45499·293 | 1000·9843 | ·7 | 137·2879 | 1909·69 | 83453·453 | 1499·8705 |
| ·8 | 112·4692 | 1281·64 | 45882·712 | 1006·6000 | ·8 | 137·6020 | 1918·44 | 84027·672 | 1506·7427 |
| ·9 | 112·7834 | 1288·81 | 46268·279 | 1012·2313 | ·9 | 137·9162 | 1927·21 | 84604·519 | 1513·6287 |
| 36 | 113·0976 | 1296 | 46656 | 1017·8784 | 44 | 138·2304 | 1936 | 85184 | 1520·5344 |
| ·1 | 113·4117 | 1303·21 | 47045·831 | 1023·5411 | ·1 | 138·5445 | 1944·81 | 85766·121 | 1527·4537 |
| ·2 | 113·7259 | 1310·44 | 47437·928 | 1029·2195 | ·2 | 138·8587 | 1953·64 | 86350·888 | 1534·3888 |
| ·3 | 114·0400 | 1317·69 | 47832·147 | 1034·9131 | ·3 | 139·1728 | 1962·49 | 86938·307 | 1541·3396 |
| ·4 | 114·3542 | 1324·96 | 48228·544 | 1040·6235 | ·4 | 139·4870 | 1971·36 | 87528·384 | 1548·3061 |
| ·5 | 114·6684 | 1332·25 | 48627·125 | 1046·3491 | ·5 | 139·8012 | 1980·25 | 88121·125 | 1555·2883 |
| ·6 | 114·9825 | 1339·56 | 49027·896 | 1052·0904 | ·6 | 140·1153 | 1989·16 | 88716·536 | 1562·2862 |
| ·7 | 115·2967 | 1346·89 | 49430·863 | 1057·8474 | ·7 | 140·4295 | 1998·09 | 89314·623 | 1569·2998 |
| ·8 | 115·6108 | 1354·24 | 49836·032 | 1063·6200 | ·8 | 140·7436 | 2007·04 | 89915·392 | 1576·3292 |
| ·9 | 115·9250 | 1361·61 | 50243·409 | 1069·4084 | ·9 | 141·0578 | 2016·01 | 90518·849 | 1583·3742 |
| 37 | 116·2392 | 1369 | 50653 | 1075·2126 | 45 | 141·3720 | 2025 | 91125 | 1590·4350 |
| ·1 | 116·5533 | 1376·41 | 51064·811 | 1081·0324 | ·1 | 141·6861 | 2034·01 | 91733·851 | 1597·5114 |
| ·2 | 116·8675 | 1383·84 | 51478·848 | 1086·8679 | ·2 | 142·0003 | 2043·04 | 92345·408 | 1604·6036 |
| ·3 | 117·1816 | 1391·29 | 51895·117 | 1092·7191 | ·3 | 142·3144 | 2052·09 | 92959·677 | 1611·7114 |
| ·4 | 117·4958 | 1398·76 | 52313·624 | 1098·5862 | ·4 | 142·6286 | 2061·16 | 93576·664 | 1618·8350 |
| ·5 | 117·8100 | 1406·25 | 52734·375 | 1104·4687 | ·5 | 142·9428 | 2070·25 | 94196·375 | 1625·9743 |
| ·6 | 118·1241 | 1413·76 | 53157·376 | 1110·3671 | ·6 | 143·2569 | 2079·36 | 94818·816 | 1633·1293 |
| ·7 | 118·4383 | 1421·29 | 53582·633 | 1116·2811 | ·7 | 143·5711 | 2088·49 | 95443·993 | 1640·3020 |
| ·8 | 118·7524 | 1428·84 | 54010·152 | 1122·2109 | ·8 | 143·8852 | 2097·64 | 96071·912 | 1647·4864 |
| ·9 | 119·0666 | 1436·41 | 54439·939 | 1128·1564 | ·9 | 144·1994 | 2106·81 | 96702·579 | 1654·6885 |
| 38 | 119·3808 | 1444 | 54872 | 1134·1176 | 46 | 144·5136 | 2116 | 97336 | 1661·9064 |
| ·1 | 119·6949 | 1451·61 | 55306·341 | 1140·0946 | ·1 | 144·8277 | 2125·21 | 97972·181 | 1669·1399 |
| ·2 | 120·0091 | 1459·24 | 55742·968 | 1146·0870 | ·2 | 145·1419 | 2134·44 | 98611·128 | 1676·3891 |
| ·3 | 120·3232 | 1466·89 | 56181·887 | 1152·0954 | ·3 | 145·4560 | 2143·69 | 99252·847 | 1683·6541 |
| ·4 | 120·6374 | 1474·56 | 56623·104 | 1158·1194 | ·4 | 145·7702 | 2152·96 | 99897·344 | 1690·9347 |
| ·5 | 120·9516 | 1482·25 | 57066·625 | 1164·1591 | ·5 | 146·0844 | 2162·25 | 100544·625 | 1698·2311 |
| ·6 | 121·2657 | 1489·96 | 57512·456 | 1170·2145 | ·6 | 146·3985 | 2171·56 | 101194·696 | 1705·5432 |
| ·7 | 121·5799 | 1497·69 | 57960·603 | 1176·2857 | ·7 | 146·7127 | 2180·89 | 101847·563 | 1712·8710 |
| ·8 | 121·8940 | 1505·44 | 58411·072 | 1182·3725 | ·8 | 147·0268 | 2190·24 | 102503·232 | 1720·2144 |
| ·9 | 122·2082 | 1513·21 | 58863·869 | 1188·4651 | ·9 | 147·3410 | 2199·61 | 103161·709 | 1727·5736 |
| 39 | 122·5224 | 1521 | 59319 | 1294·5394 | 47 | 147·6552 | 2209 | 103823 | 1734·9486 |
| ·1 | 122·8365 | 1528·81 | 59776·471 | 1200·7273 | ·1 | 147·9693 | 2218·41 | 104487·111 | 1742·3392 |
| ·2 | 123·1507 | 1536·64 | 60236·288 | 1206·8770 | ·2 | 148·2835 | 2227·84 | 105154·048 | 1749·7455 |
| ·3 | 123·4648 | 1544·49 | 60698·457 | 1213·0424 | ·3 | 148·5976 | 2237·29 | 105823·817 | 1757·1675 |
| ·4 | 123·7790 | 1552·36 | 61162·984 | 1219·2243 | ·4 | 148·9118 | 2246·76 | 106496·424 | 1764·6045 |
| ·5 | 124·0932 | 1560·25 | 61629·875 | 1225·4203 | ·5 | 149·2260 | 2256·25 | 107171·875 | 1772·0587 |
| ·6 | 124·4073 | 1568·16 | 62099·136 | 1231·6328 | ·6 | 149·5361 | 2265·76 | 107850·176 | 1779·5279 |
| ·7 | 124·7215 | 1576·09 | 62570·773 | 1237·8610 | ·7 | 149·8543 | 2275·29 | 108531·333 | 1787·0127 |
| ·8 | 125·0356 | 1584·04 | 63044·792 | 1244·1210 | ·8 | 150·1684 | 2284·84 | 109215·352 | 1794·5133 |
| ·9 | 125·3498 | 1592·01 | 63521·199 | 1250·3646 | ·9 | 150·4826 | 2294·41 | 109902·239 | 1802·0296 |
| 40 | 125·6640 | 1600 | 64000 | 1256·6400 | 48 | 150·7968 | 2304 | 110592 | 1809·5616 |
| ·1 | 125·9781 | 1608·01 | 64481·201 | 1262·9310 | ·1 | 151·1109 | 2313·61 | 111284·641 | 1817·1092 |
| ·2 | 126·2923 | 1616·04 | 64964·808 | 1269·2388 | ·2 | 151·4251 | 2323·24 | 111980·168 | 1824·6726 |
| ·3 | 126·6064 | 1624·09 | 65450·827 | 1275·5602 | ·3 | 151·7392 | 2332·89 | 112678·587 | 1832·2518 |
| ·4 | 126·9206 | 1632·16 | 65939·264 | 1281·8984 | ·4 | 152·0534 | 2342·56 | 113379·904 | 1839·8466 |
| ·5 | 127·2348 | 1640·25 | 66430·125 | 1288·2523 | ·5 | 152·3676 | 2352·25 | 114084·125 | 1847·4571 |
| ·6 | 127·5489 | 1648·36 | 66923·416 | 1294·6219 | ·6 | 152·6817 | 2361·96 | 114791·256 | 1855·0833 |
| ·7 | 127·8631 | 1656·49 | 67419·143 | 1301·0071 | ·7 | 152·9959 | 2371·69 | 115501·303 | 1862·7253 |
| ·8 | 128·1772 | 1664·64 | 67917·312 | 1307·4082 | ·8 | 153·3100 | 2381·44 | 116214·272 | 1870·3829 |
| ·9 | 128·4914 | 1672·81 | 68417·929 | 1313·8249 | ·9 | 153·6242 | 2391·21 | 116930·169 | 1878·0563 |

| Diam. | Circum. | Square. | Cube. | Area. | Diam. | Circum. | Square. | Cube. | Area. |
|---|---|---|---|---|---|---|---|---|---|
| 49 | 153·9384 | 2401 | 117649 | 1885·7454 | 57 | 179·0712 | 3249 | 185193 | 2551·7646 |
| ·1 | 154·2525 | 2410·81 | 118370·771 | 1893·4501 | ·1 | 179·3853 | 3260·41 | 186169·411 | 2560·7260 |
| ·2 | 154·5667 | 2420·64 | 119095·488 | 1901·1706 | ·2 | 179·6995 | 3271·84 | 187149·248 | 2569·7031 |
| ·3 | 154·8808 | 2430·49 | 119823·157 | 1908·9068 | ·3 | 180·0136 | 3283·29 | 188132·517 | 2578·6959 |
| ·4 | 155·1950 | 2440·36 | 120553·784 | 1916·6587 | ·4 | 180·3278 | 3294·76 | 189119·224 | 2587·7045 |
| ·5 | 155·5092 | 2450·25 | 121287·375 | 1924·4263 | ·5 | 180·6420 | 3306·25 | 190109·375 | 2596·7287 |
| ·6 | 155·8233 | 2460·16 | 122023·936 | 1932·2096 | ·6 | 180·9561 | 3317·76 | 191102·976 | 2605·7687 |
| ·7 | 156·1375 | 2470·09 | 122763·473 | 1940·0086 | ·7 | 181·2803 | 3329·29 | 192100·033 | 2614·8243 |
| ·8 | 156·4516 | 2480·04 | 123505·992 | 1947·8234 | ·8 | 181·5844 | 3340·84 | 193100·552 | 2623·8957 |
| ·9 | 156·7558 | 2490·01 | 124251·499 | 1955·6538 | ·9 | 181·8986 | 3352·41 | 194104·539 | 2632·9828 |
| 50 | 157·0800 | 2500 | 125000 | 1963·5000 | 58 | 182·2128 | 3364 | 195112 | 2642·0856 |
| ·1 | 157·3941 | 2510·01 | 125751·501 | 1971·3618 | ·1 | 182·5269 | 3375·61 | 196122·941 | 2651·2046 |
| ·2 | 157·7083 | 2520·04 | 126506·008 | 1979·2394 | ·2 | 182·8411 | 3387·24 | 197137·368 | 2660·3382 |
| ·3 | 158·0224 | 2530·09 | 127263·527 | 1987·1326 | ·3 | 183·1552 | 3398·89 | 198155·287 | 2669·4882 |
| ·4 | 158·3366 | 2540·16 | 128024·064 | 1995·0416 | ·4 | 183·4694 | 3410·56 | 199176·704 | 2678·6538 |
| ·5 | 158·6508 | 2550·25 | 128787·625 | 2002·9663 | ·5 | 183·7836 | 3422·25 | 200201·625 | 2687·8351 |
| ·6 | 158·9649 | 2560·36 | 129554·216 | 2010·9067 | ·6 | 184·0977 | 3433·96 | 201230·056 | 2697·0321 |
| ·7 | 159·2791 | 2570·49 | 130323·843 | 2018·8628 | ·7 | 184·4119 | 3445·69 | 202262·003 | 2706·2449 |
| ·8 | 159·5932 | 2580·64 | 131096·512 | 2026·8346 | ·8 | 184·7260 | 3457·44 | 203297·472 | 2715·4733 |
| ·9 | 159·9074 | 2590·81 | 131872·229 | 2034·8770 | ·9 | 185·0402 | 3469·21 | 204336·469 | 2724·7175 |
| 51 | 160·2216 | 2601 | 132651 | 2042·8254 | 59 | 185·3544 | 3481 | 205379 | 2733·9774 |
| ·1 | 160·5357 | 2611·21 | 133432·831 | 2050·8443 | ·1 | 185·6685 | 3492·81 | 206425·071 | 2743·2529 |
| ·2 | 160·8499 | 2621·44 | 134217·728 | 2058·8784 | ·2 | 185·9827 | 3504·64 | 207474·688 | 2752·5442 |
| ·3 | 161·1640 | 2631·69 | 135005·697 | 2066·9293 | ·3 | 186·2696 | 3516·49 | 208527·857 | 2761·8512 |
| ·4 | 161·4782 | 2641·96 | 135796·744 | 2074·9953 | ·4 | 186·6110 | 3528·36 | 209584·584 | 2771·1739 |
| ·5 | 161·7924 | 2652·25 | 136590·875 | 2083·0771 | ·5 | 186·9252 | 3540·25 | 210644·875 | 2780·5123 |
| ·6 | 162·1065 | 2662·56 | 137388·096 | 2091·1746 | ·6 | 187·2393 | 3552·16 | 211708·736 | 2789·8664 |
| ·7 | 162·4207 | 2672·89 | 138188·413 | 2099·2878 | ·7 | 187·5535 | 3564·09 | 212776·173 | 2799·2362 |
| ·8 | 162·7348 | 2683·24 | 138991·832 | 2107·4166 | ·8 | 187·8676 | 3576·04 | 213847·192 | 2808·6218 |
| ·9 | 163·0490 | 2693·61 | 139798·359 | 2115·5612 | ·9 | 188·1818 | 3588·01 | 214921·799 | 2818·0230 |
| 52 | 163·3632 | 2704 | 140608 | 2123·7216 | 60 | 188·4960 | 3600 | 216000 | 2827·4400 |
| ·1 | 163·6773 | 2714·41 | 141420·761 | 2131·8976 | ·1 | 188·8101 | 3612·01 | 217081·801 | 2836·8726 |
| ·2 | 163·9935 | 2724·84 | 142236·648 | 2140·0893 | ·2 | 189·1243 | 3624·04 | 218167·208 | 2846·3210 |
| ·3 | 164·3056 | 2735·29 | 143055·667 | 2148·2967 | ·3 | 189·4384 | 3636·09 | 219256·227 | 2855·7850 |
| ·4 | 164·6198 | 2745·76 | 143877·824 | 2156·5199 | ·4 | 189·7526 | 3648·16 | 220348·864 | 2865·2648 |
| ·5 | 164·9340 | 2756·25 | 144703·125 | 2164·7587 | ·5 | 190·0668 | 3660·25 | 221445·125 | 2874·7603 |
| ·6 | 165·2481 | 2766·76 | 145531·576 | 2173·0133 | ·6 | 190·3809 | 3672·36 | 222545·016 | 2884·2615 |
| ·7 | 165·5623 | 2777·29 | 146363·183 | 2181·2835 | ·7 | 190·6951 | 3684·49 | 223648·543 | 2893·7984 |
| ·8 | 165·8764 | 2787·84 | 147197·952 | 2189·5695 | ·8 | 191·0092 | 3696·64 | 224755·712 | 2903·3410 |
| ·9 | 166·1906 | 2798·41 | 148035·889 | 2197·8712 | ·9 | 191·3234 | 3708·81 | 225866·529 | 2912·8993 |
| 53 | 166·5048 | 2809 | 148877 | 2206·1886 | 61 | 191·6376 | 3721 | 226981 | 2922·4734 |
| ·1 | 166·8189 | 2819·61 | 149721·291 | 2214·5216 | ·1 | 191·9517 | 3733·21 | 228099·131 | 2932·0631 |
| ·2 | 167·1331 | 2830·24 | 150568·768 | 2222·8704 | ·2 | 192·2659 | 3745·44 | 229220·928 | 2941·6685 |
| ·3 | 167·4472 | 2940·89 | 151419·437 | 2231·2350 | ·3 | 192·5800 | 3757·69 | 230346·397 | 2951·2897 |
| ·4 | 167·7614 | 2851·56 | 152273·304 | 2239·6152 | ·4 | 192·8942 | 3769·96 | 231475·544 | 2960·9265 |
| ·5 | 168·0756 | 2862·25 | 153130·375 | 2248·0111 | ·5 | 193·2084 | 3782·25 | 232608·375 | 2970·5791 |
| ·6 | 168·3897 | 2872·96 | 153990·656 | 2256·4227 | ·6 | 193·5225 | 3794·56 | 233744·896 | 2980·2474 |
| ·7 | 168·7049 | 2883·69 | 154854·153 | 2264·8701 | ·7 | 193·8367 | 3806·89 | 234885·113 | 2989·9314 |
| ·8 | 169·0180 | 2894·44 | 155720·872 | 2273·2931 | ·8 | 194·1508 | 3819·24 | 236029·032 | 2999·6300 |
| ·9 | 169·3322 | 2905·21 | 156590·819 | 2281·7519 | ·9 | 194·4650 | 3831·61 | 237176·659 | 3009·3464 |
| 54 | 169·6464 | 2916 | 157464 | 2290·2264 | 62 | 194·7792 | 3844 | 238328 | 3019·0776 |
| ·1 | 169·9605 | 2926·81 | 158340·421 | 2298·7165 | ·1 | 195·0933 | 3856·41 | 239483·061 | 3028·8244 |
| ·2 | 170·2747 | 2937·64 | 159220·088 | 2307·2224 | ·2 | 195·4075 | 3868·84 | 240641·848 | 3038·5809 |
| ·3 | 170·5888 | 2948·49 | 160103·007 | 2315·7440 | ·3 | 195·7216 | 3881·29 | 241804·367 | 3048·3651 |
| ·4 | 170·9030 | 2959·36 | 160989·184 | 2324·2813 | ·4 | 196·0358 | 3893·76 | 242970·624 | 3058·1591 |
| ·5 | 171·2172 | 2970·25 | 161878·625 | 2332·8343 | ·5 | 196·3500 | 3906·25 | 244140·625 | 3067·9687 |
| ·6 | 171·5313 | 2981·16 | 162771·336 | 2341·4030 | ·6 | 196·6641 | 3918·76 | 245314·376 | 3077·7941 |
| ·7 | 171·8455 | 2992·09 | 163667·323 | 2349·9874 | ·7 | 196·9783 | 3931·29 | 246491·883 | 3087·6341 |
| ·8 | 172·1596 | 3003·04 | 164566·592 | 2358·5876 | ·8 | 197·2924 | 3943·84 | 247673·152 | 3097·4919 |
| ·9 | 172·4738 | 3014·01 | 165469·149 | 2367·2034 | ·9 | 197·6066 | 3956·41 | 248858·189 | 3107·3644 |
| 55 | 172·7880 | 3025 | 166375 | 2375·8350 | 63 | 197·9208 | 3969 | 250047 | 3117·2526 |
| ·1 | 173·1021 | 3036·01 | 167284·151 | 2384·4822 | ·1 | 198·2349 | 3981·61 | 251239·591 | 3127·1564 |
| ·2 | 173·4163 | 3047·04 | 168196·608 | 2393·1452 | ·2 | 198·5491 | 3994·24 | 252435·968 | 3137·0758 |
| ·3 | 173·7304 | 3058·09 | 169112·377 | 2401·8238 | ·3 | 198·8632 | 4006·89 | 253636·137 | 3147·0114 |
| ·4 | 174·0446 | 3069·16 | 170031·464 | 2410·5182 | ·4 | 199·1774 | 4019·56 | 254840·104 | 3156·9664 |
| ·5 | 174·3588 | 3080·25 | 170953·875 | 2419·2283 | ·5 | 199·4916 | 4032·25 | 256047·875 | 3166·9291 |
| ·6 | 174·6729 | 3091·36 | 171879·616 | 2427·9541 | ·6 | 199·8057 | 4044·96 | 257259·456 | 3176·9115 |
| ·7 | 174·9771 | 3102·49 | 172808·693 | 2436·6956 | ·7 | 200·1199 | 4057·69 | 258474·853 | 3186·9097 |
| ·8 | 175·3092 | 3113·64 | 173741·112 | 2445·4528 | ·8 | 200·4340 | 4070·44 | 259694·072 | 3196·9235 |
| ·9 | 175·6154 | 3124·81 | 174676·879 | 2454·2257 | ·9 | 200·7482 | 4083·21 | 260917·119 | 3206·9531 |
| 56 | 175·9296 | 3136 | 175616 | 2463·0144 | 64 | 201·0624 | 4096 | 262144 | 3216·9984 |
| ·1 | 176·2437 | 3147·21 | 176558·481 | 2471·8187 | ·1 | 201·3765 | 4108·81 | 263374·721 | 3227·0593 |
| ·2 | 176·5579 | 3158·44 | 177504·328 | 2480·6387 | ·2 | 201·6907 | 4121·64 | 264609·288 | 3237·1360 |
| ·3 | 176·8720 | 3169·69 | 178453·547 | 2489·4745 | ·3 | 202·0048 | 4134·49 | 265847·707 | 3247·2284 |
| ·4 | 177·1862 | 3180·96 | 179406·144 | 2498·3259 | ·4 | 202·3190 | 4147·36 | 267089·984 | 3257·3365 |
| ·5 | 177·5004 | 3192·25 | 180362·125 | 2507·1931 | ·5 | 202·6332 | 4160·25 | 268336·125 | 3267·4603 |
| ·6 | 177·8145 | 3203·56 | 181321·496 | 2516·0760 | ·6 | 202·9473 | 4173·16 | 269586·136 | 3277·5998 |
| ·7 | 178·1287 | 3214·89 | 182284·263 | 2524·9736 | ·7 | 203·2615 | 4186·09 | 270840·023 | 3287·7550 |
| ·8 | 178·4428 | 3226·24 | 183250·432 | 2533·8888 | ·8 | 203·5756 | 4199·04 | 272097·792 | 3297·9260 |
| ·9 | 178·7570 | 3237·61 | 184220·009 | 2542·8188 | ·9 | 203·8898 | 4212·01 | 273359·449 | 3308·1126 |

| Diam. | Circum. | Square. | Cube. | Area. | Diam. | Circum. | Square. | Cube. | Area. |
|---|---|---|---|---|---|---|---|---|---|
| 65 | 204·2040 | 4225 | 274625 | 3318·3150 | 73 | 229·3368 | 5329 | 389017 | 4185·3966 |
| ·1 | 204·5181 | 4238·01 | 275894·451 | 3328·5340 | ·1 | 229·6509 | 5343·61 | 390617·891 | 4196·8712 |
| ·2 | 204·8323 | 4251·04 | 277167·808 | 3338·7668 | ·2 | 229·9651 | 5358·24 | 392223·168 | 4208·3614 |
| ·3 | 205·1464 | 4264·09 | 278445·077 | 3349·0162 | ·3 | 230·2792 | 5372·89 | 393832·837 | 4219·8678 |
| ·4 | 205·4606 | 4277·16 | 279726·264 | 3359·2814 | ·4 | 230·5934 | 5387·56 | 395446·904 | 4231·3896 |
| ·5 | 205·7748 | 4290·25 | 281011·375 | 3369·5623 | ·5 | 230·9076 | 5402·25 | 397065·375 | 4242·9271 |
| ·6 | 206·0889 | 4303·36 | 282300·416 | 3379·8589 | ·6 | 231·2217 | 5416·96 | 398688·256 | 4254·4803 |
| ·7 | 206·4031 | 4316·49 | 283593·393 | 3390·1712 | ·7 | 231·5359 | 5431·69 | 400315·553 | 4266·0493 |
| ·8 | 206·7172 | 4329·64 | 284890·312 | 3400·4992 | ·8 | 231·8500 | 5446·44 | 401947·272 | 4277·6339 |
| ·9 | 207·0314 | 4342·81 | 286191·179 | 3410·8429 | ·9 | 232·1642 | 5461·21 | 403583·419 | 4289·2343 |
| 66 | 207·3456 | 4356 | 287496 | 3421·2024 | 74 | 232·4784 | 5476 | 405224 | 4300·8504 |
| ·1 | 207·6597 | 4369·21 | 288804·781 | 3431·5775 | ·1 | 232·7925 | 5490·81 | 406869·021 | 4312·4821 |
| ·2 | 207·9739 | 4382·44 | 290117·528 | 3441·9633 | ·2 | 233·1067 | 5505·64 | 408518·488 | 4324·1296 |
| ·3 | 208·2880 | 4395·69 | 291434·247 | 3452·3749 | ·3 | 233·4208 | 5520·49 | 410172·407 | 4335·7928 |
| ·4 | 208·6022 | 4408·96 | 292754·944 | 3462·7971 | ·4 | 233·7350 | 5535·36 | 411830·784 | 4347·4717 |
| ·5 | 208·9164 | 4422·25 | 294079·625 | 3473·2351 | ·5 | 234·0492 | 5550·25 | 413493·625 | 4359·1663 |
| ·6 | 209·2305 | 4435·56 | 295408·296 | 3483·6888 | ·6 | 234·3633 | 5565·16 | 415160·936 | 4370·8766 |
| ·7 | 209·5447 | 4448·89 | 296740·963 | 3494·1640 | ·7 | 234·6775 | 5580·09 | 416832·723 | 4382·6026 |
| ·8 | 209·8588 | 4462·24 | 298077·632 | 3504·6432 | ·8 | 234·9916 | 5595·04 | 418508·992 | 4394·3448 |
| ·9 | 210·1730 | 4475·61 | 299418·309 | 3515·1430 | ·9 | 235·3058 | 5610·01 | 420189·749 | 4406·1018 |
| 67 | 210·4872 | 4489 | 300763 | 3525·6606 | 75 | 235·6200 | 5625 | 421875 | 4417·8750 |
| ·1 | 210·8013 | 4502·41 | 302111·711 | 3536·1928 | ·1 | 235·9341 | 5640·01 | 423564·751 | 4429·6638 |
| ·2 | 211·1155 | 4515·84 | 303464·448 | 3546·7407 | ·2 | 236·2483 | 5655·04 | 425259·008 | 4441·4684 |
| ·3 | 211·4296 | 4529·29 | 304821·217 | 3557·3043 | ·3 | 236·5624 | 5670·09 | 426957·777 | 4453·2886 |
| ·4 | 211·7438 | 4542·76 | 306182·024 | 3567·8837 | ·4 | 236·8766 | 5685·16 | 428661·064 | 4465·1246 |
| ·5 | 212·0580 | 4556·25 | 307546·875 | 3578·4787 | ·5 | 237·1908 | 5700·25 | 430368·875 | 4476·9763 |
| ·6 | 212·3721 | 4569·76 | 308915·776 | 3589·0895 | ·6 | 237·5049 | 5715·36 | 432081·216 | 4488·8437 |
| ·7 | 212·6863 | 4583·29 | 310288·733 | 3599·7159 | ·7 | 237·8191 | 5730·49 | 433798·093 | 4500·7268 |
| ·8 | 213·0004 | 4596·84 | 311665·752 | 3610·3581 | ·8 | 238·1332 | 5745·64 | 435519·512 | 4512·6256 |
| ·9 | 213·3146 | 4610·41 | 313046·839 | 3621·0160 | ·9 | 238·4474 | 5760·81 | 437245·479 | 4524·5401 |
| 68 | 213·6288 | 4624 | 314432 | 3631·6896 | 76 | 238·7616 | 5776 | 438976 | 4536·4704 |
| ·1 | 213·9429 | 4637·61 | 315821·241 | 3642·3788 | ·1 | 239·0757 | 5791·21 | 440711·081 | 4548·4163 |
| ·2 | 214·2571 | 4651·24 | 317214·568 | 3653·0838 | ·2 | 239·3899 | 5806·44 | 442450·728 | 4560·3787 |
| ·3 | 214·5712 | 4664·89 | 318611·987 | 3663·8040 | ·3 | 239·7040 | 5821·69 | 444194·947 | 4572·3553 |
| ·4 | 214·8854 | 4678·56 | 320013·504 | 3674·5410 | ·4 | 240·0182 | 5836·96 | 445943·744 | 4584·3583 |
| ·5 | 215·1996 | 4692·25 | 321419·125 | 3685·2931 | ·5 | 240·3324 | 5852·25 | 447697·125 | 4596·3571 |
| ·6 | 215·5137 | 4705·96 | 322828·856 | 3696·0060 | ·6 | 240·6465 | 5867·56 | 449455·096 | 4608·3816 |
| ·7 | 215·8279 | 4719·69 | 324242·703 | 3706·8445 | ·7 | 240·9607 | 5882·89 | 451217·663 | 4620·4218 |
| ·8 | 216·1420 | 4733·44 | 325660·672 | 3717·6437 | ·8 | 241·2748 | 5898·24 | 452984·832 | 4632·4776 |
| ·9 | 216·4562 | 4747·21 | 327082·769 | 3728·4587 | ·9 | 241·5987 | 5913·61 | 454756·609 | 4644·5492 |
| 69 | 216·7704 | 4761 | 328509 | 3739·2894 | 77 | 241·9032 | 5929 | 456533 | 4656·6366 |
| ·1 | 217·0845 | 4774·81 | 329939·371 | 3750·1357 | ·1 | 242·2173 | 5944·41 | 458314·011 | 4668·7396 |
| ·2 | 217·3987 | 4788·64 | 331373·888 | 3760·9978 | ·2 | 242·5315 | 5959·84 | 460099·648 | 4680·8583 |
| ·3 | 217·7128 | 4802·49 | 332812·557 | 3771·8756 | ·3 | 242·8456 | 5975·29 | 461889·917 | 4692·9927 |
| ·4 | 218·0270 | 4816·36 | 334255·384 | 3782·7691 | ·4 | 243·1598 | 5990·76 | 463684·824 | 4705·1429 |
| ·5 | 218·3412 | 4830·25 | 335702·375 | 3793·6783 | ·5 | 243·4740 | 6006·25 | 465484·375 | 4717·3087 |
| ·6 | 218·6553 | 4844·16 | 337153·536 | 3804·6032 | ·6 | 243·7881 | 6021·76 | 467288·576 | 4729·4903 |
| ·7 | 218·9695 | 4858·09 | 338608·873 | 3815·5438 | ·7 | 244·1023 | 6037·29 | 469097·433 | 4741·6875 |
| ·8 | 219·2836 | 4872·04 | 340068·392 | 3826·5002 | ·8 | 244·4164 | 6052·84 | 470910·952 | 4753·9005 |
| ·9 | 219·5978 | 4886·01 | 341532·099 | 3837·4722 | ·9 | 244·7306 | 6068·41 | 472729·139 | 4766·1292 |
| 70 | 219·9120 | 4900 | 343000 | 3848·4600 | 78 | 245·0448 | 6084 | 474552 | 4778·3736 |
| ·1 | 220·2261 | 4914·01 | 344472·101 | 3859·4952 | ·1 | 245·3589 | 6099·61 | 476379·541 | 4790·6336 |
| ·2 | 220·5403 | 4928·04 | 345948·408 | 3870·4826 | ·2 | 245·6731 | 6115·24 | 478211·768 | 4802·9094 |
| ·3 | 220·8544 | 4942·09 | 347428·927 | 3881·5174 | ·3 | 245·9872 | 6130·89 | 480048·687 | 4815·2010 |
| ·4 | 221·1686 | 4956·16 | 348913·664 | 3892·5600 | ·4 | 240·3014 | 6146·56 | 481890·304 | 4827·5082 |
| ·5 | 221·4828 | 4970·25 | 350402·625 | 3903·6343 | ·5 | 246·6156 | 6162·25 | 483736·625 | 4839·8311 |
| ·6 | 221·7969 | 4984·36 | 351895·816 | 3914·7163 | ·6 | 246·9297 | 6177·96 | 485587·656 | 4852·1697 |
| ·7 | 222·1111 | 4998·49 | 353393·243 | 3925·8140 | ·7 | 247·2439 | 6193·69 | 487443·403 | 4864·5241 |
| ·8 | 222·4252 | 5012·64 | 354894·912 | 3036·9274 | ·8 | 247·5480 | 6209·44 | 489303·872 | 4876·8973 |
| ·9 | 222·7394 | 5026·81 | 356400·829 | 3948·0565 | ·9 | 247·8722 | 6225·21 | 491169·069 | 4889·2799 |
| 71 | 223·0536 | 5041 | 357911 | 3959·2014 | 79 | 248·1864 | 6241 | 493039 | 4901·6814 |
| ·1 | 223·3677 | 5055·21 | 359425·431 | 3970·3619 | ·1 | 248·5005 | 6256·81 | 494913·671 | 4914·0985 |
| ·2 | 223·6819 | 5069·44 | 360944·128 | 3981·5381 | ·2 | 248·8147 | 6272·64 | 496793·088 | 4926·5314 |
| ·3 | 223·9960 | 5083·69 | 362467·097 | 3992·7301 | ·3 | 249·1288 | 6288·49 | 498677·257 | 4938·9820 |
| ·4 | 224·3102 | 5097·96 | 363994·344 | 4003·9373 | ·4 | 249·4430 | 6304·36 | 500566·184 | 4951·4443 |
| ·5 | 224·6244 | 5112·25 | 365525·875 | 4015·1611 | ·5 | 249·7572 | 6320·25 | 502459·875 | 4963·9243 |
| ·6 | 224·9385 | 5126·56 | 367061·696 | 4026·4002 | ·6 | 250·0713 | 6336·16 | 504358·336 | 4976·4840 |
| ·7 | 225·2527 | 5140·89 | 368601·813 | 4037·6550 | ·7 | 250·3855 | 6352·09 | 506261·573 | 4988·9314 |
| ·8 | 225·5668 | 5155·24 | 370146·232 | 4048·9254 | ·8 | 250·6996 | 6368·04 | 508169·592 | 5001·4586 |
| ·9 | 225·8810 | 5169·61 | 371694·959 | 4060·2116 | ·9 | 251·0138 | 6384·01 | 510082·399 | 5014·0014 |
| 72 | 226·1952 | 5184 | 373248 | 4071·5136 | 80 | 251·3280 | 6400 | 512000 | 5026·5600 |
| ·1 | 226·5093 | 5198·41 | 374805·361 | 4082·8332 | ·1 | 251·6421 | 6416·01 | 513922·401 | 5039·1342 |
| ·2 | 226·8235 | 5212·84 | 376367·048 | 4094·1645 | ·2 | 251·9563 | 6432·04 | 515849·608 | 5051·7242 |
| ·3 | 227·1376 | 5227·29 | 377933·067 | 4105·5125 | ·3 | 252·2704 | 6448·09 | 517781·627 | 5064·3298 |
| ·4 | 227·4518 | 5241·76 | 379503·424 | 4116·8793 | ·4 | 252·5846 | 6464·16 | 519718·464 | 5076·9552 |
| ·5 | 227·7660 | 5256·25 | 381078·125 | 4128·2587 | ·5 | 252·8988 | 6480·25 | 521660·125 | 5089·5883 |
| ·6 | 228·0801 | 5270·76 | 382657·176 | 4139·6524 | ·6 | 253·2129 | 6496·36 | 523606·616 | 5102·2411 |
| ·7 | 228·3943 | 5285·29 | 384240·583 | 4151·0667 | ·7 | 253·5271 | 6512·49 | 525557·943 | 5114·9096 |
| ·8 | 228·7084 | 5299·84 | 385828·352 | 4162·4943 | ·8 | 253·8412 | 6528·64 | 527514·112 | 5127·5938 |
| ·9 | 229·0226 | 5314·41 | 387420·489 | 4173·9376 | ·9 | 254·1554 | 6544·81 | 529475·129 | 5140·2937 |

| Diam. | Circum. | Square. | Cube. | Area. | Diam. | Circum. | Square. | Cube. | Area. |
|---|---|---|---|---|---|---|---|---|---|
| 81 | 254·4696 | 6561 | 531441 | 5153·0094 | 89 | 279·6024 | 7921 | 704969 | 6221·1534 |
| ·1 | 254·7837 | 6577·21 | 533411·731 | 5165·7407 | ·1 | 279·9165 | 7938·81 | 707347·971 | 6235·1413 |
| ·2 | 255·0979 | 6593·44 | 535387·328 | 5178·4877 | ·2 | 280·2307 | 7956·64 | 709732·288 | 6249·1450 |
| ·3 | 255·4120 | 6609·69 | 537367·797 | 5191·2505 | ·3 | 280·5448 | 7974·49 | 712121·957 | 6263·1644 |
| ·4 | 255·7262 | 6625·96 | 539353·144 | 5204·0285 | ·4 | 280·8590 | 7992·36 | 714516·984 | 6277·1995 |
| ·5 | 256·0404 | 6642·25 | 541343·375 | 5216·8231 | ·5 | 281·1732 | 8010·25 | 716917·375 | 6291·2035 |
| ·6 | 256·3545 | 6658·56 | 543338·496 | 5229·6330 | ·6 | 281·4873 | 8028·16 | 719323·136 | 6305·3168 |
| ·7 | 256·6687 | 6674·89 | 545338·513 | 5242·4586 | ·7 | 281·8825 | 8046·09 | 721734·273 | 6319·3990 |
| ·8 | 256·9828 | 6691·24 | 547343·432 | 5255·2998 | ·8 | 282·1156 | 8064·04 | 724150·792 | 6333·4970 |
| ·9 | 257·2970 | 6707·61 | 549353·259 | 5268·1568 | ·9 | 282·4298 | 8082·01 | 726572·699 | 6347·6813 |
| 82 | 257·6112 | 6724 | 551368 | 5281·0296 | 90 | 282·7440 | 8100 | 729000 | 6361·7400 |
| ·1 | 257·9253 | 6740·41 | 553387·661 | 5293·9180 | ·1 | 283·0581 | 8118·01 | 731432·701 | 6375·8850 |
| ·2 | 258·2395 | 6756·84 | 555412·248 | 5306·8221 | ·2 | 283·3723 | 8136·04 | 733870·808 | 6390·0458 |
| ·3 | 258·5536 | 6773·29 | 557441·767 | 5319·7439 | ·3 | 283·6864 | 8154·09 | 736314·327 | 6404·2222 |
| ·4 | 258·8646 | 6789·76 | 559476·224 | 5332·6775 | ·4 | 284·0006 | 8172·16 | 738763·264 | 6418·4144 |
| ·5 | 259·1820 | 6806·25 | 561515·625 | 5345·6287 | ·5 | 584·3148 | 8190·25 | 741217·625 | 6432·6223 |
| ·6 | 259·4961 | 6822·76 | 563559·976 | 5358·5957 | ·6 | 284·6289 | 8208·36 | 743677·416 | 6446·8474 |
| ·7 | 259·8103 | 6839·29 | 565609·283 | 5371·5983 | ·7 | 284·9431 | 8226·49 | 746142·643 | 6461·0852 |
| ·8 | 260·1244 | 6855·84 | 567663·552 | 5384·5762 | ·8 | 285·2572 | 8244·64 | 748613·312 | 6475·3402 |
| ·9 | 260·4386 | 6872·41 | 569722·789 | 5397·5908 | ·9 | 285·5714 | 8262·81 | 751089·429 | 6489·6109 |
| 83 | 260·7528 | 6889 | 571787 | 5410·6206 | 91 | 285·8856 | 8281 | 753571 | 6503·8974 |
| ·1 | 261·0669 | 6905·61 | 573856·191 | 5423·6660 | ·1 | 286·1997 | 8299·21 | 756058·031 | 6518·1995 |
| ·2 | 261·3811 | 6922·24 | 575930·368 | 5436·7272 | ·2 | 286·5139 | 8317·44 | 758550·528 | 6532·5173 |
| ·3 | 261·6952 | 6938·89 | 578009·537 | 5449·8042 | ·3 | 286·8290 | 8335·69 | 761048·497 | 6546·8909 |
| ·4 | 262·0094 | 6955·56 | 580093·704 | 5462·8968 | ·4 | 287·1422 | 8353·96 | 763551·944 | 6561·2081 |
| ·5 | 262·3236 | 6972·25 | 582182·875 | 5476·0051 | ·5 | 287·4564 | 8372·25 | 766060·875 | 6575·5651 |
| ·6 | 262·6376 | 6988·96 | 584277·056 | 5489·1291 | ·6 | 287·7705 | 8390·56 | 768575·296 | 6589·9458 |
| ·7 | 262·9519 | 7005·69 | 586376·253 | 5502·2689 | ·7 | 288·0847 | 8408·89 | 771095·213 | 6604·3222 |
| ·8 | 263·2640 | 7022·44 | 588480·472 | 5515·4243 | ·8 | 288·3988 | 8427·24 | 773620·632 | 6618·7542 |
| ·9 | 263·5802 | 7039·21 | 590589·719 | 5528·5958 | ·9 | 288·7130 | 8445·61 | 776151·559 | 6633·1820 |
| 84 | 263·8944 | 7056 | 592704 | 5541·7824 | 92 | 289·0272 | 8464 | 778688 | 6647·6256 |
| ·1 | 264·2085 | 7072·81 | 594823·321 | 5554·9849 | ·1 | 289·3413 | 8482·41 | 781229·961 | 6662·0848 |
| ·2 | 264·5227 | 7089·64 | 596947·688 | 5568·2032 | ·2 | 289·6555 | 8500·84 | 783777·448 | 6676·5597 |
| ·3 | 264·8368 | 7106·49 | 599077·107 | 5581·4372 | ·3 | 289·9696 | 8519·29 | 786330·467 | 6691·0161 |
| ·4 | 265·1510 | 7123·36 | 601211·584 | 5594·6869 | ·4 | 290·2838 | 8537·76 | 788889·024 | 6705·5567 |
| ·5 | 265·4652 | 7140·25 | 603351·125 | 5607·9523 | ·5 | 290·5980 | 8556·25 | 791453·125 | 6720·0787 |
| ·6 | 265·7793 | 7157·16 | 605495·736 | 5621·2334 | ·6 | 290·9121 | 8574·76 | 794022·776 | 6734·6165 |
| ·7 | 266·0935 | 7174·09 | 607645·423 | 5634·5682 | ·7 | 291·2263 | 8593·29 | 796597·983 | 6749·1699 |
| ·8 | 266·4076 | 7191·04 | 609800·192 | 5647·8428 | ·8 | 291·5404 | 8611·84 | 799178·752 | 6763·7391 |
| ·9 | 266·7218 | 7208·01 | 611960·049 | 5661·1710 | ·9 | 291·8546 | 8630·41 | 801765·089 | 6778·3240 |
| 85 | 267·0360 | 7225 | 614125 | 5674·5150 | 93 | 292·1688 | 8649 | 804357 | 6792·9246 |
| ·1 | 267·3501 | 7242·01 | 616295·051 | 5687·8746 | ·1 | 292·4829 | 8667·61 | 806954·491 | 6807·5408 |
| ·2 | 267·6643 | 7259·04 | 618470·208 | 5701·2500 | ·2 | 292·7971 | 8686·24 | 809557·568 | 6822·1730 |
| ·3 | 267·9784 | 7276·09 | 620650·477 | 5714·6410 | ·3 | 293·1112 | 8704·89 | 812166·237 | 6836·8206 |
| ·4 | 268·2926 | 7293·16 | 622835·864 | 5728·0478 | ·4 | 293·4254 | 8723·56 | 814780·504 | 6851·4840 |
| ·5 | 268·6068 | 7310·25 | 625026·375 | 5741·4703 | ·5 | 293·7396 | 8742·25 | 817400·375 | 6866·1631 |
| ·6 | 268·9209 | 7327·36 | 627222·016 | 5754·9085 | ·6 | 294·0537 | 8760·96 | 820025·856 | 6880·8579 |
| ·7 | 269·2351 | 7344·49 | 629422·793 | 5768·3624 | ·7 | 294·3679 | 8779·69 | 822656·953 | 6895·5685 |
| ·8 | 269·5492 | 7361·64 | 631628·712 | 5781·8320 | ·8 | 294·6820 | 8798·44 | 825293·672 | 6910·2947 |
| ·9 | 269·8634 | 7378·81 | 633839·779 | 5795·3173 | ·9 | 294·9962 | 8817·21 | 827936·019 | 6925·0367 |
| 86 | 270·1776 | 7396 | 636056 | 5808·8184 | 94 | 295·3104 | 8836 | 830584 | 6939·7944 |
| ·1 | 270·4917 | 7413·21 | 638277·381 | 5822·3351 | ·1 | 295·6245 | 8854·81 | 833237·621 | 6954·5677 |
| ·2 | 270·8059 | 7430·44 | 640503·928 | 5835·8675 | ·2 | 295·9387 | 8873·64 | 835896·888 | 6969·3568 |
| ·3 | 271·1200 | 7447·69 | 642735·647 | 5849·4157 | ·3 | 296·2436 | 8892·49 | 838561·807 | 6984·1614 |
| ·4 | 271·4342 | 7464·96 | 644972·544 | 5862·9795 | ·4 | 296·5670 | 8911·36 | 841232·384 | 6998·9821 |
| ·5 | 271·7484 | 7482·25 | 647214·625 | 5876·5591 | ·5 | 296·8812 | 8930·25 | 843908·625 | 7013·8183 |
| ·6 | 272·0665 | 7499·56 | 649461·896 | 5890·1541 | ·6 | 297·1953 | 8949·16 | 846590·536 | 7028·6702 |
| ·7 | 272·3767 | 7516·89 | 651714·363 | 5903·7654 | ·7 | 297·5095 | 8968·09 | 849278·123 | 7043·5025 |
| ·8 | 272·6908 | 7534·24 | 653972·032 | 5917·3920 | ·8 | 297·8236 | 8987·04 | 851971·392 | 7058·4180 |
| ·9 | 273·0050 | 7551·61 | 656234·909 | 5931·0344 | ·9 | 298·1378 | 9006·01 | 854670·349 | 7073·3202 |
| 87 | 273·3192 | 7569 | 658503 | 5944·6926 | 95 | 298·4520 | 9025 | 857375 | 7088·2350 |
| ·1 | 273·6333 | 7586·41 | 660776·311 | 5958·3644 | ·1 | 298·7661 | 9044·01 | 860085·351 | 7103·1654 |
| ·2 | 273·9875 | 7603·84 | 663054·848 | 5972·0559 | ·2 | 299·0723 | 9063·04 | 862801·408 | 7118·1116 |
| ·3 | 274·2616 | 7621·29 | 665338·617 | 5985·7691 | ·3 | 299·3944 | 9082·09 | 865523·177 | 7133·0734 |
| ·4 | 274·5758 | 7638·76 | 667627·624 | 5999·4821 | ·4 | 299·7086 | 9101·16 | 868250·664 | 7148·0510 |
| ·5 | 274·8900 | 7656·25 | 669921·875 | 6013·2187 | ·5 | 300·0228 | 9120·25 | 870983·875 | 7163·0443 |
| ·6 | 275·2041 | 7673·76 | 672221·376 | 6026·9711 | ·6 | 300·3369 | 9139·36 | 873722·816 | 7178·0533 |
| ·7 | 275·5183 | 7691·29 | 674526·133 | 6040·7391 | ·7 | 300·6511 | 9158·49 | 876467·493 | 7193·0780 |
| ·8 | 275·8324 | 7708·84 | 676836·152 | 6054·5149 | ·8 | 300·9652 | 9177·64 | 879217·912 | 7208·1184 |
| ·9 | 276·1466 | 7726·41 | 679151·439 | 6068·3224 | ·9 | 301·2794 | 9196·81 | 881974·079 | 7223·1745 |
| 88 | 276·4608 | 7744 | 681472 | 6082·1376 | 96 | 301·5936 | 9216 | 884736 | 7238·2464 |
| ·1 | 276·7749 | 7761·61 | 683797·841 | 6095·9684 | ·1 | 301·9077 | 9235·21 | 887503·681 | 7253·3339 |
| ·2 | 277·0891 | 7779·24 | 686128·968 | 6109·8150 | ·2 | 302·2219 | 9254·44 | 890277·128 | 7268·4371 |
| ·3 | 277·4032 | 7796·89 | 688465·387 | 6123·6774 | ·3 | 302·5360 | 9273·69 | 893056·347 | 7283·5561 |
| ·4 | 277·7174 | 7814·56 | 690807·104 | 6137·5554 | ·4 | 302·8502 | 9292·96 | 895841·344 | 7298·6907 |
| ·5 | 278·0316 | 7832·25 | 693154·122 | 6151·4491 | ·5 | 303·1644 | 9312·25 | 898632·125 | 7313·8411 |
| ·6 | 278·3457 | 7849·96 | 695506·456 | 6165·3585 | ·6 | 303·4785 | 9331·56 | 901428·696 | 7329·0072 |
| ·7 | 278·6599 | 7867·69 | 697864·103 | 6179·2837 | ·7 | 303·7927 | 9350·89 | 904231·063 | 7344·1890 |
| ·8 | 278·9750 | 7885·44 | 700227·072 | 6193·2245 | ·8 | 304·1068 | 9370·24 | 907039·232 | 7359·3864 |
| ·9 | 279·2882 | 7903·21 | 702595·369 | 6207·1811 | ·9 | 304·4210 | 9389·61 | 909853·209 | 7374·5996 |

| Diam. | Circum. | Square. | Cube. | Area. | Diam. | Circum. | Square. | Cube. | Area. |
|---|---|---|---|---|---|---|---|---|---|
| 97 | 304·7352 | 9409 | 912673 | 7389·8286 | ·6 | 309·7617 | 9721·96 | 958585·256 | 7635·6273 |
| ·1 | 305·0493 | 9428·41 | 915498·611 | 7405·0732 | ·7 | 310·0759 | 9741·69 | 961504·803 | 7651·1933 |
| ·2 | 305·3635 | 9447·84 | 918330·048 | 7420·3335 | ·8 | 310·3960 | 9761·44 | 964430·272 | 7666·6349 |
| ·3 | 305·6776 | 9467·29 | 921167·317 | 7435·6095 | ·9 | 310·7042 | 9781·21 | 967361·669 | 7682·1623 |
| ·4 | 305·9918 | 9486·76 | 924010·424 | 7450·9013 | 99 | 311·0184 | 9801 | 970299 | 7697·7054 |
| ·5 | 306·3060 | 9506·25 | 926859·375 | 7466·2087 | ·1 | 311·3325 | 9820·81 | 973242·271 | 7713·2641 |
| ·6 | 306·6201 | 9525·76 | 929714·176 | 7481·5319 | ·2 | 311·6467 | 9840·64 | 976191·488 | 7728·8386 |
| ·7 | 306·9363 | 9545·29 | 932574·833 | 7496·8707 | ·3 | 311·9608 | 9860·49 | 979146·657 | 7744·4288 |
| ·8 | 307·2484 | 9564·84 | 935441·352 | 7512·2253 | ·4 | 312·2750 | 9880·36 | 982107·784 | 7760·0347 |
| ·9 | 307·5626 | 9584·41 | 938313·739 | 7527·5956 | ·5 | 812·5892 | 9900·25 | 985074·875 | 7775·6563 |
| 98 | 307·8768 | 9604 | 941192 | 7542·9816 | ·6 | 312·9033 | 9920·16 | 988047·936 | 7791·2936 |
| ·1 | 308·1909 | 9623·61 | 944076·141 | 7558·3832 | ·7 | 313·2175 | 9940·09 | 991026·973 | 7806·9466 |
| ·2 | 308·5051 | 9643·24 | 946966·168 | 7573·8006 | ·8 | 313·5116 | 9960·04 | 994011·992 | 7822·6154 |
| ·3 | 308·8192 | 9662·89 | 949862·087 | 7589·2338 | ·9 | 313·8458 | 9980·01 | 997002·999 | 7838·2998 |
| ·4 | 309·1334 | 9682·56 | 952763·904 | 7604·6826 | 100 | 314·1600 | 10000 | 1000000 | 7854·0000 |
| ·5 | 309·4476 | 9702·25 | 955671·625 | 7620·1471 | | | | | |

## A TABLE *of the Length of Circular Arcs, radius being unity.*

| Degree. | Length. | Degree. | Length. | Min. | Length. | Sec. | Length. |
|---|---|---|---|---|---|---|---|
| 1 | 0·0174533 | 60 | 1·0471976 | 1 | 0·0002909 | 1 | 0·000048 |
| 2 | 0·0349066 | 70 | 1·2217305 | 2 | 0·0005818 | 2 | 0·000097 |
| 3 | 0·0523599 | 80 | 1·3962634 | 3 | 0·0008727 | 3 | 0·0000145 |
| 4 | 0·0698132 | 90 | 1·5707963 | 4 | 0·0011636 | 4 | 0·0000194 |
| 5 | 0·0872665 | 100 | 1·7453293 | 5 | 0·0014544 | 5 | 0·0000242 |
| 6 | 0·1047198 | 120 | 2·0943951 | 6 | 0·0017453 | 6 | 0·0000291 |
| 7 | 0·1221730 | 150 | 2·6179939 | 7 | 0·0020362 | 7 | 0·0000339 |
| 8 | 0·1396263 | 180 | 3·1415927 | 8 | 0·0023271 | 8 | 0·0000388 |
| 9 | 0·1570796 | 210 | 3·6651914 | 9 | 0·0026180 | 9 | 0·0000436 |
| 10 | 0·1745329 | 240 | 4·1887902 | 10 | 0·0029089 | 10 | 0·0000485 |
| 20 | 0·3490659 | 270 | 4·7123890 | 20 | 0·0058178 | 20 | 0·0000970 |
| 30 | 0·5235988 | 300 | 5·2359878 | 30 | 0·0087266 | 30 | 0·0001454 |
| 40 | 0·6981317 | 330 | 5·7595865 | 40 | 0·0116355 | 40 | 0·0001939 |
| 50 | 0·8726646 | 360 | 6·2831853 | 50 | 0·0145444 | 50 | 0·0002424 |

Required the length of a circular arc of 37° 42′ 58″?

30° = 0·5235988
7° = 0·1221730
40′ = 0·0116355
2′ = 0·0020368
50″ = 0·0002424
8″ = 0·0000388

The length 0·6582703 required in terms of the radius.

1207° Fahrenheit = 1° of Wedgewood's pyrometer. Iron melts at about 166° Wedgewood; 200362° Fahrenheit.

Sound passes in air at a velocity of 1142 feet a second, and in water at a velocity of 4700 feet.

Freezing water gives out 140° of heat, and may be cooled as low as 20°. All solids absorb heat when becoming a fluid, and the quantity of heat that renders a substance fluid is termed its caloric of fluidity, or latent heat. Fluids in vacuo boil with 124° less heat, than when under the pressure of the atmosphere.

AREAS *of the Segments and Zones of a Circle of which the* DIAMETER *is Unity, and supposed to be divided into* 1000 *equal parts.*

| Height. | Area of Segment. | Area of Zone. | Height. | Area of Segment. | Area of Zone. | Height. | Area of Segment. | Area of Zone. |
|---|---|---|---|---|---|---|---|---|
| ·001 | ·000042 | ·001000 | ·051 | ·015119 | ·050912 | ·101 | ·041476 | ·100309 |
| ·002 | ·000119 | ·002000 | ·052 | ·015561 | ·051906 | ·102 | ·042080 | ·101288 |
| ·003 | ·000219 | ·003000 | ·053 | ·016007 | ·052901 | ·103 | ·052687 | ·102267 |
| ·004 | ·000337 | ·004000 | ·054 | ·016457 | ·053895 | ·104 | ·043296 | ·103246 |
| ·005 | ·000470 | ·005000 | ·055 | ·016911 | ·054890 | ·105 | ·043908 | ·104223 |
| ·006 | ·000618 | ·006000 | ·056 | ·017369 | ·055883 | ·106 | ·044522 | ·105201 |
| ·007 | ·000779 | ·007000 | ·057 | ·017831 | ·056877 | ·107 | ·045139 | ·106178 |
| ·008 | ·000951 | ·008000 | ·058 | ·018296 | ·057870 | ·108 | ·045759 | ·107155 |
| ·009 | ·001135 | ·009000 | ·059 | ·018766 | ·058863 | ·109 | ·046381 | ·108131 |
| ·010 | ·001329 | ·010000 | ·060 | ·019239 | ·059856 | ·110 | ·047005 | ·109107 |
| ·011 | ·001533 | ·011000 | ·061 | ·019716 | ·060849 | ·111 | ·047632 | ·110082 |
| ·012 | ·001746 | ·011999 | ·062 | ·020196 | ·061841 | ·112 | ·048262 | ·111057 |
| ·013 | ·001968 | ·012999 | ·063 | ·020680 | ·062833 | ·113 | ·048894 | ·112031 |
| ·014 | ·002199 | ·013998 | ·064 | ·021168 | ·063825 | ·114 | ·049528 | ·113004 |
| ·015 | ·002438 | ·014998 | ·065 | ·021659 | ·064817 | ·115 | ·050165 | ·113978 |
| ·016 | ·002685 | ·015997 | ·066 | ·022154 | ·065807 | ·116 | ·050804 | ·114951 |
| ·017 | ·002940 | ·016997 | ·067 | ·022652 | ·066799 | ·117 | ·051446 | ·115924 |
| ·018 | ·003202 | ·017996 | ·068 | ·023154 | ·067790 | ·118 | ·052090 | ·116896 |
| ·019 | ·003471 | ·018996 | ·069 | ·023659 | ·068782 | ·119 | ·052736 | ·117867 |
| ·020 | ·003748 | ·019995 | ·070 | ·024168 | ·069771 | ·120 | ·053385 | ·118838 |
| ·021 | ·004031 | ·020994 | ·071 | ·024680 | ·070761 | ·121 | ·054036 | ·119809 |
| ·022 | ·004322 | ·021993 | ·072 | ·025195 | ·071751 | ·122 | ·054689 | ·120779 |
| ·023 | ·004618 | ·022992 | ·073 | ·025714 | ·072740 | ·123 | ·055345 | ·121748 |
| ·024 | ·004921 | 023991 | ·074 | ·026236 | ·073729 | ·124 | ·056003 | ·122717 |
| ·025 | ·005230 | ·024990 | ·075 | ·026761 | ·074718 | ·125 | ·056663 | ·123686 |
| ·026 | ·005546 | ·025989 | ·076 | ·027289 | ·075707 | ·126 | ·057326 | ·124654 |
| ·027 | ·005867 | ·026987 | ·077 | ·027821 | ·076695 | ·127 | ·057991 | ·125621 |
| ·028 | ·006194 | ·027986 | ·078 | ·028356 | ·077683 | ·128 | ·058658 | ·126588 |
| ·029 | ·006527 | ·028984 | ·079 | ·028894 | ·078670 | ·129 | ·059327 | ·127555 |
| ·030 | ·006865 | ·029982 | ·080 | ·029435 | ·079658 | ·130 | ·059999 | ·128521 |
| ·031 | ·007209 | ·030980 | ·081 | ·029979 | ·080645 | ·131 | ·060672 | ·129486 |
| ·032 | ·007558 | ·031978 | ·082 | ·030526 | ·081631 | ·132 | ·061348 | ·130451 |
| ·033 | ·007913 | ·032976 | ·083 | ·031076 | ·082618 | ·133 | ·062026 | ·131415 |
| ·034 | ·008273 | ·033974 | ·084 | ·031629 | ·083604 | ·134 | ·062707 | ·132379 |
| ·035 | ·008638 | ·034972 | ·085 | ·032186 | ·084589 | ·135 | ·063389 | ·133342 |
| ·036 | ·009008 | ·035969 | ·086 | ·032745 | ·085574 | ·136 | ·064074 | ·134304 |
| ·037 | ·009383 | ·036967 | ·087 | ·033307 | ·086559 | ·137 | ·064760 | ·135266 |
| ·038 | ·009763 | ·037965 | ·088 | ·033872 | ·087544 | ·138 | ·065449 | ·136228 |
| ·039 | ·010148 | ·038962 | ·089 | ·034441 | ·088528 | ·139 | ·066140 | ·137189 |
| ·040 | ·010537 | ·039958 | ·090 | ·035011 | ·089512 | ·140 | ·066833 | ·138149 |
| ·041 | ·010931 | ·040954 | ·091 | ·035585 | ·090496 | ·141 | ·067528 | ·139109 |
| ·042 | ·011330 | ·041951 | ·092 | ·036162 | ·091479 | ·142 | ·068225 | ·140068 |
| ·043 | ·011734 | ·042947 | ·093 | ·036741 | ·092461 | ·143 | ·068924 | ·141026 |
| ·044 | ·012142 | ·043944 | ·094 | ·037323 | ·093444 | ·144 | ·069625 | ·141984 |
| ·045 | ·012554 | ·044940 | ·095 | ·037909 | ·094426 | ·145 | ·070328 | ·142942 |
| ·046 | ·012971 | ·045935 | ·096 | ·038496 | ·095407 | ·146 | ·071033 | ·143898 |
| ·047 | ·013392 | ·046931 | ·097 | .039087 | ·096388 | ·147 | ·071741 | ·144854 |
| ·048 | ·013818 | ·047927 | ·098 | .039680 | ·097369 | ·148 | ·072450 | ·145810 |
| ·049 | ·014247 | ·048922 | ·099 | .040276 | ·098350 | ·149 | ·073161 | ·146765 |
| ·050 | ·014681 | ·049917 | ·100 | .040875 | ·099330 | ·150 | ·073874 | ·147719 |

| Height. | Area of Seg. | Area of Zone. | Height. | Area of Seg. | Area of Zone. | Height. | Area of Seg. | Area of Zone. |
|---|---|---|---|---|---|---|---|---|
| ·151 | ·074589 | ·148674 | ·206 | ·116650 | ·200915 | ·261 | ·163140 | ·248608 |
| ·152 | ·075306 | ·149625 | ·207 | ·117460 | ·200924 | ·262 | ·164019 | ·249461 |
| ·153 | ·076026 | ·150578 | ·208 | ·118271 | ·201835 | ·263 | ·164899 | ·250212 |
| ·154 | ·076747 | ·151530 | ·209 | ·119083 | ·202744 | ·264 | ·165780 | ·251162 |
| ·155 | ·077469 | ·152481 | ·210 | ·119897 | ·203652 | ·265 | ·166663 | ·252011 |
| ·156 | ·078194 | ·153431 | ·211 | ·120712 | ·204559 | ·266 | ·167546 | ·252851 |
| ·157 | ·078921 | ·154381 | ·212 | ·121529 | ·205465 | ·267 | ·168430 | ·253704 |
| ·158 | ·079649 | ·155330 | ·213 | ·122347 | ·206370 | ·268 | ·169315 | ·254549 |
| ·159 | ·080380 | ·156278 | ·214 | ·123167 | ·207274 | ·269 | ·170202 | ·255392 |
| ·160 | ·081112 | ·157226 | ·215 | ·123988 | ·208178 | ·270 | ·171080 | ·256235 |
| ·161 | ·081846 | ·158173 | ·216 | ·124810 | ·209080 | ·271 | ·171978 | ·257075 |
| ·162 | ·082582 | ·159119 | ·217 | ·125634 | ·209981 | ·272 | ·172867 | ·257915 |
| ·163 | ·083320 | ·160065 | ·218 | ·126459 | ·210882 | ·273 | ·173758 | ·258754 |
| ·164 | ·084059 | ·161010 | ·219 | ·127285 | ·211782 | ·274 | ·174649 | ·259591 |
| ·165 | ·084801 | ·161954 | ·220 | ·128113 | ·212680 | ·275 | ·175542 | ·260427 |
| ·166 | ·085544 | ·162898 | ·221 | ·128942 | ·213577 | ·276 | ·176435 | ·261261 |
| ·167 | ·086289 | ·163841 | ·222 | ·129773 | ·214474 | ·277 | ·177330 | ·262094 |
| ·168 | ·087036 | ·165784 | ·223 | ·130605 | ·215369 | ·278 | ·178225 | ·262926 |
| ·169 | ·087785 | ·165725 | ·224 | ·131438 | ·216264 | ·279 | ·179122 | ·263757 |
| ·170 | ·088535 | ·166666 | ·225 | ·132272 | ·217157 | ·280 | ·180019 | ·264586 |
| ·171 | ·089287 | ·167606 | ·226 | ·133108 | ·218050 | ·281 | ·180918 | ·265414 |
| ·172 | ·090041 | ·168549 | ·227 | ·133945 | ·218941 | ·282 | ·181817 | ·266240 |
| ·173 | ·090797 | ·160484 | ·228 | ·134784 | ·219832 | ·283 | ·182718 | ·267065 |
| ·174 | ·091554 | ·170422 | ·229 | ·135624 | ·220721 | ·284 | ·183619 | ·267889 |
| ·175 | ·092313 | ·171359 | ·230 | ·136465 | ·221610 | ·285 | ·184521 | ·268711 |
| ·176 | ·093074 | ·172295 | ·231 | ·137307 | ·222497 | ·286 | ·185425 | ·269532 |
| ·177 | ·093836 | ·173231 | ·232 | ·138150 | ·223354 | ·287 | ·186329 | ·270352 |
| ·178 | ·094601 | ·174166 | ·233 | ·138995 | ·224269 | ·288 | ·187234 | ·271170 |
| ·179 | ·095366 | ·175100 | ·234 | ·139841 | ·225153 | ·289 | ·188140 | ·271987 |
| ·180 | ·096134 | ·176033 | ·235 | ·140688 | ·226036 | ·290 | ·189047 | ·272802 |
| ·181 | ·096903 | ·176966 | ·236 | ·141537 | ·226919 | ·291 | ·189955 | ·273616 |
| ·182 | ·097674 | ·177897 | ·237 | ·142387 | ·227800 | ·292 | ·190864 | ·274428 |
| ·183 | ·098447 | ·178828 | ·238 | ·143238 | ·228680 | ·293 | ·191775 | ·275239 |
| ·184 | ·099221 | ·179759 | ·239 | ·144091 | ·229559 | ·294 | ·192684 | ·276049 |
| ·185 | ·099997 | ·180688 | ·240 | ·144944 | ·230439 | ·295 | ·193596 | ·276857 |
| ·186 | ·100774 | ·181617 | ·241 | ·145799 | ·231313 | ·296 | ·194509 | ·277664 |
| ·187 | ·101553 | ·182545 | ·242 | ·146655 | ·232189 | ·297 | ·195422 | ·278469 |
| ·188 | ·102334 | ·183472 | ·243 | ·147512 | ·233063 | ·298 | ·196337 | ·279273 |
| ·189 | ·103116 | ·184398 | ·244 | ·148371 | ·233937 | ·299 | ·197252 | ·280075 |
| ·190 | ·103900 | ·185323 | ·245 | ·149230 | ·234809 | ·300 | ·198168 | ·280876 |
| ·191 | ·104685 | ·186248 | ·246 | ·150091 | ·235680 | ·301 | ·199085 | ·281675 |
| ·192 | ·105472 | ·187172 | ·257 | ·150953 | ·236550 | ·302 | ·200003 | ·282473 |
| ·193 | ·106261 | ·188094 | ·248 | ·151816 | ·237419 | ·303 | ·200922 | ·283269 |
| ·194 | ·107051 | ·189016 | ·249 | ·152680 | ·238287 | ·304 | ·201841 | ·284063 |
| ·195 | ·107842 | ·189938 | ·250 | ·153546 | ·239153 | ·305 | ·202761 | ·284857 |
| ·196 | ·108636 | ·190858 | ·251 | ·154412 | ·240019 | ·306 | ·203683 | ·285648 |
| ·197 | ·109430 | ·191777 | ·252 | ·155280 | ·240883 | ·307 | ·204605 | ·286438 |
| ·198 | ·110226 | ·192696 | ·253 | ·156149 | ·241746 | ·308 | ·205527 | ·287227 |
| ·199 | ·111024 | ·193614 | ·254 | ·157019 | ·242608 | ·309 | ·206451 | ·288014 |
| ·200 | ·111823 | ·194531 | ·255 | ·157890 | ·243469 | ·310 | ·207376 | ·288799 |
| ·201 | ·112624 | ·195447 | ·256 | ·158762 | ·244328 | ·311 | ·208301 | ·289583 |
| ·202 | ·113426 | ·196362 | ·257 | ·159636 | ·245187 | ·312 | ·209227 | ·290365 |
| ·203 | ·114230 | ·197277 | ·258 | ·160510 | ·246044 | ·313 | ·210154 | ·291146 |
| ·204 | ·115035 | ·198190 | ·259 | ·161386 | ·246900 | ·314 | ·211082 | ·291925 |
| ·205 | ·115842 | ·199103 | ·260 | ·162263 | ·247755 | ·315 | ·212011 | ·292702 |

| Height. | Area of Seg. | Area of Zone. | Height. | Area of Seg. | Area of Zone. | Height. | Area of Seg. | Area of Zone. |
|---|---|---|---|---|---|---|---|---|
| ·316 | ·212940 | ·293478 | ·371 | ·265144 | ·333372 | ·426 | ·318970 | ·366463 |
| ·317 | ·213871 | ·294252 | ·372 | ·266111 | ·334041 | ·427 | ·319959 | ·366985 |
| ·318 | ·214802 | ·295025 | ·373 | ·267078 | ·334708 | ·428 | ·320948 | ·367504 |
| ·319 | ·215733 | ·295796 | ·374 | ·268045 | ·335373 | ·429 | ·321938 | ·368019 |
| ·320 | ·216666 | ·296565 | ·375 | ·269013 | ·336036 | ·430 | ·322928 | ·368531 |
| ·321 | ·217599 | ·297333 | ·376 | ·269982 | ·336696 | ·431 | ·323918 | ·369040 |
| ·322 | ·218533 | ·298098 | ·377 | ·270951 | ·337354 | ·432 | ·324909 | ·369545 |
| ·323 | ·219468 | ·298863 | ·378 | ·271920 | ·338010 | ·433 | ·325900 | ·370047 |
| ·324 | ·220404 | ·299625 | ·379 | ·272890 | ·338663 | ·434 | ·326892 | ·370545 |
| ·325 | ·221340 | ·300386 | ·380 | ·273861 | ·339314 | ·435 | ·327882 | ·371040 |
| ·326 | ·222277 | ·301145 | ·381 | ·274832 | ·339963 | ·436 | ·328874 | ·371531 |
| ·327 | ·223215 | ·301902 | ·382 | ·275803 | ·340609 | ·437 | ·329866 | ·372019 |
| ·328 | ·224154 | ·302658 | ·383 | ·276775 | ·341253 | ·438 | ·330858 | ·372503 |
| ·329 | ·225093 | ·303412 | ·384 | ·277748 | ·341895 | ·439 | ·331850 | ·372983 |
| ·330 | ·226033 | ·304164 | ·385 | ·278721 | ·342534 | ·440 | ·332843 | ·373460 |
| ·331 | ·226974 | ·304914 | ·386 | ·279694 | ·343171 | ·441 | ·333836 | ·373933 |
| ·332 | ·227915 | ·305663 | ·387 | ·280668 | ·343805 | ·442 | ·334829 | ·374403 |
| ·333 | ·228858 | ·306410 | ·388 | ·281642 | ·344437 | ·443 | ·335822 | ·374868 |
| ·334 | ·229801 | ·307155 | ·389 | ·282617 | ·345067 | ·444 | ·336816 | ·375330 |
| ·335 | ·230745 | ·307898 | ·390 | ·283592 | ·345694 | ·445 | ·337810 | ·375788 |
| ·336 | ·231689 | ·308640 | ·391 | ·284568 | ·346318 | ·446 | ·338804 | ·376242 |
| ·337 | ·232634 | ·309379 | ·392 | ·285544 | ·346940 | ·447 | ·339798 | ·376692 |
| ·338 | ·233580 | ·310117 | ·393 | ·286521 | ·347560 | ·448 | ·340793 | ·377138 |
| ·339 | ·234526 | ·310853 | ·394 | ·287498 | ·348177 | ·449 | ·341787 | ·377580 |
| ·340 | ·235473 | ·311588 | ·395 | ·288476 | ·348791 | ·450 | ·342782 | ·378018 |
| ·341 | ·236421 | ·312319 | ·396 | ·289453 | ·349403 | ·451 | ·343777 | ·378452 |
| ·342 | ·237369 | ·313050 | ·397 | ·290432 | ·350012 | ·452 | ·344772 | ·378881 |
| ·343 | ·238318 | ·313778 | ·398 | ·291411 | ·350619 | ·453 | ·345768 | ·379307 |
| ·344 | ·239268 | ·314505 | ·399 | ·292390 | ·351223 | ·454 | ·346764 | ·379728 |
| ·345 | ·240218 | ·315230 | ·400 | ·293369 | ·351824 | ·455 | ·347759 | ·380145 |
| ·346 | ·241169 | ·315952 | ·401 | ·294349 | ·352423 | ·456 | ·348755 | ·380557 |
| ·347 | ·242121 | ·316673 | ·402 | ·295330 | ·353019 | ·457 | ·349752 | ·380965 |
| ·348 | ·243074 | ·317393 | ·403 | ·296311 | ·353612 | ·458 | ·350748 | ·381369 |
| ·349 | ·244026 | ·318110 | ·404 | ·297292 | ·354202 | ·459 | ·351745 | ·381768 |
| ·350 | ·244980 | ·318825 | ·405 | ·298273 | ·354790 | ·460 | ·352742 | ·382162 |
| ·351 | ·245934 | ·319538 | ·406 | ·299255 | ·355376 | ·461 | ·353739 | ·382551 |
| ·352 | ·246889 | ·320249 | ·407 | ·300238 | ·355958 | ·462 | ·354736 | ·382936 |
| ·353 | ·247845 | ·320958 | ·408 | ·301220 | ·356537 | ·463 | ·355732 | ·383316 |
| ·354 | ·248801 | ·321666 | ·409 | ·302203 | ·357114 | ·464 | ·356730 | ·383691 |
| ·355 | ·249757 | ·322371 | ·410 | ·303187 | ·357688 | ·465 | ·357727 | ·384061 |
| ·356 | ·250715 | ·323075 | ·411 | ·304171 | ·358258 | ·466 | ·358725 | ·384426 |
| ·357 | ·251673 | ·323775 | ·412 | ·305155 | ·358827 | ·467 | ·359723 | ·384786 |
| ·358 | ·252631 | ·324474 | ·413 | ·306140 | ·359392 | ·468 | ·360721 | ·385144 |
| ·359 | ·253590 | ·325171 | ·414 | ·307125 | ·359954 | ·469 | ·361719 | ·385490 |
| ·360 | ·254550 | ·325866 | ·415 | ·308110 | ·360513 | ·470 | ·362717 | ·385834 |
| ·361 | ·255510 | ·326559 | ·416 | ·309095 | ·361070 | ·471 | ·363715 | ·386172 |
| ·362 | ·256471 | ·327250 | ·417 | ·310081 | ·361623 | ·472 | ·364713 | ·386505 |
| ·363 | ·257433 | ·327939 | ·418 | ·311068 | ·362173 | ·473 | ·365712 | ·386832 |
| ·364 | ·258395 | ·328625 | ·419 | ·312054 | ·362720 | ·474 | ·366710 | ·387153 |
| ·365 | ·259357 | ·329310 | ·420 | ·313041 | ·363264 | ·475 | ·367709 | ·387469 |
| ·366 | ·260320 | ·329992 | ·421 | ·314029 | ·363805 | ·476 | ·368708 | ·387778 |
| ·367 | ·261284 | ·330673 | ·422 | ·315016 | ·364343 | ·477 | ·369707 | ·388081 |
| ·368 | ·262248 | ·331351 | ·423 | ·316004 | ·364878 | ·478 | ·370706 | ·388377 |
| ·369 | ·263213 | ·332027 | ·424 | ·316992 | ·365410 | ·479 | ·371704 | ·388669 |
| ·370 | ·264178 | ·332700 | ·425 | ·317981 | ·365939 | ·480 | ·372704 | ·388951 |

| Height. | Area of Seg. | Area of Zone. | Height. | Area of Seg. | Area of Zone. | Height. | Area of Seg. | Area of Zone. |
|---|---|---|---|---|---|---|---|---|
| ·481 | ·373703 | ·389228 | ·491 | ·383699 | ·391564 | ·496 | ·388699 | ·392362 |
| ·482 | ·374702 | ·389497 | ·492 | ·384699 | ·391748 | ·497 | ·389699 | ·392480 |
| ·483 | ·375702 | ·389759 | ·493 | ·385699 | ·391920 | ·498 | ·390699 | ·392580 |
| ·484 | ·376702 | ·390014 | ·494 | ·386699 | ·392081 | ·499 | ·391699 | ·392657 |
| ·435 | ·377701 | ·390261 | ·495 | ·387699 | ·392229 | ·500 | ·392699 | ·392699 |
| ·486 | ·378701 | ·390500 | | | | | | |
| ·487 | ·379700 | ·390730 | | | | | | |
| ·488 | ·380700 | ·390953 | | | | | | |
| ·489 | ·381699 | ·391166 | | | | | | |
| ·490 | ·382699 | ·391370 | | | | | | |

*To find the area of a segment of a circle.*

RULE.—Divide the height, or versed sine, by the diameter of the circle, and find the quotient in the column of heights.

Then take out the corresponding area, in the column of areas, and multiply it by the square of the diameter; this will give the area of the segment.

Required the area of a segment of a circle, whose height is $3\frac{1}{4}$ feet, and the diameter of the circle 50 feet.

$3\frac{1}{4} = 3{\cdot}25$; and $3{\cdot}25 \div 50 = {\cdot}065$.

·065, by the Table, = ·021659; and $\cdot021659 \times 50^2 = 54{\cdot}147500$, the area required.

*To find the area of a circular zone.*

RULE 1.—When the zone is less than a semi-circle, divide the height by the longest chord, and seek the quotient in the column of heights. Take out the corresponding area, in the next column on the right hand, and multiply it by the square of the longest chord.

Required the area of a zone whose longest chord is 50, and height 15.

$15 \div 50 = {\cdot}300$; and ·300, by the Table, = ·280876.

Hence $\cdot280876 \times 50^2 = 702{\cdot}19$, the area of the zone.

RULE 2.—When the zone is greater than a semi-circle, take the height on each side of the diameter of the circle.

Required the area of a zone, the diameter of the circle being 50, and the height of the zone on each side of the line which passes through the diameter of the circle 20 and 15 respectively.

20 : 50 — ·400; ·400, by the Table, = ·351824; and $\cdot351824 \times 50^2 = 879{\cdot}56$.

$15 \div 50 = {\cdot}300$; ·300, by the Table, =·280876; and $\cdot280876 \times 50^2 = 702{\cdot}19$. Hence $879{\cdot}56 + 702{\cdot}19 = 1581{\cdot}75$.

*Approximating rule to find the area of a segment of a circle.*

RULE.—Multiply the chord of the segment by the versed sine, divide the product by 3, and multiply the remainder by 2.

Cube the height, or versed sine, find how often twice the length of the chord is contained in it, and add the quotient to the former product; this will give the area of the segment very nearly.

Required the area of the segment of a circle, the chord being 12, and the versed sine 2.

$$12 \times 2 = 24;\ \frac{24}{3} = 8;\ \text{and } 8 \times 2 = 16.$$

$$2^3 \div 24 = {\cdot}3333.$$

Hence $16 + {\cdot}3333 = 16{\cdot}3333$, the area of the segment very nearly.

| Height of Arc. | Length of Arc. | Height of Arc. | Length of Arc. | Height of Arc. | Length of Arc. | Height of Arc. | Length of Arc. | Height of Arc. | Length of Arc. |
|---|---|---|---|---|---|---|---|---|---|
| ·100 | 1·02645 | ·181 | 1·08519 | ·261 | 1·17275 | ·341 | 1·28583 | ·421 | 1·42041 |
| ·101 | 1·02698 | ·182 | 1·08611 | ·262 | 1·17401 | ·342 | 1·28739 | ·422 | 1·42222 |
| ·102 | 1·02752 | ·183 | 1·08704 | ·263 | 1·17527 | ·343 | 1·28895 | ·423 | 1·42402 |
| ·103 | 1·02806 | ·184 | 1·08797 | ·264 | 1·17655 | ·344 | 1·29052 | ·424 | 1·42583 |
| ·104 | 1·02860 | ·185 | 1·08890 | ·265 | 1·17784 | ·345 | 1·29209 | ·425 | 1·42764 |
| ·105 | 1·02914 | ·186 | 1·08984 | ·266 | 1·17912 | ·346 | 1·29366 | ·426 | 1·42945 |
| ·106 | 1·02970 | ·187 | 1·09079 | ·267 | 1·18040 | ·347 | 1·29523 | ·427 | 1·43127 |
| ·107 | 1·03026 | ·188 | 1·09174 | ·268 | 1·18162 | ·348 | 1·29681 | ·428 | 1·43309 |
| ·108 | 1·03082 | ·189 | 1·09269 | ·269 | 1·18294 | ·349 | 1·29839 | ·429 | 1·43491 |
| ·109 | 1·03139 | ·190 | 1·09365 | ·270 | 1·18428 | ·350 | 1·29997 | ·430 | 1·43673 |
| ·110 | 1·03196 | ·191 | 1·09461 | ·271 | 1·18557 | ·351 | 1·30156 | ·431 | 1·43856 |
| ·111 | 1·03254 | ·192 | 1·09557 | ·272 | 1·18688 | ·352 | 1·30315 | ·432 | 1·44039 |
| ·112 | 1·03312 | ·193 | 1·09654 | ·273 | 1·18819 | ·353 | 1·30474 | ·433 | 1·44222 |
| ·113 | 1·03371 | ·194 | 1·09752 | ·274 | 1·18969 | ·354 | 1·30634 | ·434 | 1·44405 |
| ·114 | 1·03430 | ·195 | 1·09850 | ·275 | 1·19082 | ·355 | 1·30794 | ·435 | 1·44589 |
| ·115 | 1·03490 | ·196 | 1·09949 | ·276 | 1·19214 | ·356 | 1·30954 | ·436 | 1·44773 |
| ·116 | 1·03551 | ·197 | 1·10048 | ·277 | 1·19345 | ·357 | 1·31115 | ·437 | 1·44957 |
| ·117 | 1·03611 | ·198 | 1·10147 | ·278 | 1·19477 | ·358 | 1·31276 | ·438 | 1·45142 |
| ·118 | 1·03672 | ·199 | 1·10247 | ·279 | 1·19610 | ·359 | 1·31437 | ·439 | 1·45327 |
| ·119 | 1·03734 | ·200 | 1·10348 | ·280 | 1·19743 | ·360 | 1·31599 | ·440 | 1·45512 |
| ·120 | 1·03797 | ·201 | 1·10447 | ·281 | 1·19887 | ·361 | 1·31761 | ·441 | 1·45697 |
| ·121 | 1·03860 | ·202 | 1·10548 | ·282 | 1·20011 | ·362 | 1·31923 | ·442 | 1·45883 |
| ·122 | 1·03923 | ·203 | 1·10650 | ·283 | 1·20146 | ·363 | 1·32086 | ·443 | 1·46069 |
| ·123 | 1·03987 | ·204 | 1·10752 | ·284 | 1·20282 | ·364 | 1·32249 | ·444 | 1·46255 |
| ·124 | 1·04051 | ·205 | 1·10855 | ·285 | 1·20419 | ·365 | 1·32413 | ·445 | 1·46441 |
| ·125 | 1·04116 | ·206 | 1·10958 | ·286 | 1·20558 | ·366 | 1·32577 | ·446 | 1·46628 |
| ·126 | 1·04181 | ·207 | 1·11062 | ·287 | 1·20696 | ·367 | 1·32741 | ·447 | 1·46815 |
| ·127 | 1·04247 | ·208 | 1·11165 | ·288 | 1·20828 | ·368 | 1·32905 | ·448 | 1·47002 |
| ·128 | 1·04313 | ·209 | 1·11269 | ·289 | 1·20967 | ·369 | 1·33069 | ·449 | 1·47189 |
| ·129 | 1·04380 | ·210 | 1·11374 | ·290 | 1·21202 | ·370 | 1·33234 | ·450 | 1·47377 |
| ·130 | 1·04447 | ·211 | 1·11479 | ·291 | 1·21239 | ·371 | 1·33399 | ·451 | 1·47565 |
| ·131 | 1·04515 | ·212 | 1·11584 | ·292 | 1·21381 | ·372 | 1·33564 | ·452 | 1·47753 |
| ·132 | 1·04584 | ·213 | 1·11692 | ·293 | 1·21520 | ·373 | 1·33730 | ·453 | 1·47942 |
| ·133 | 1·04652 | ·214 | 1·11796 | ·294 | 1·21658 | ·374 | 1·33896 | ·454 | 1·48131 |
| ·134 | 1·04722 | ·215 | 1·11904 | ·295 | 1·21794 | ·375 | 1·34063 | ·455 | 1·48320 |
| ·135 | 1·04792 | ·216 | 1·12011 | ·296 | 1·21926 | ·376 | 1·34229 | ·456 | 1·48509 |
| ·136 | 1·04862 | ·217 | 1·12118 | ·297 | 1·22061 | ·377 | 1·34396 | ·457 | 1·48699 |
| ·137 | 1·04932 | ·218 | 1·12225 | ·298 | 1·22203 | ·378 | 1·34563 | ·458 | 1·48889 |
| ·138 | 1·05003 | ·219 | 1·12334 | ·299 | 1·22347 | ·379 | 1·34731 | ·459 | 1·49079 |
| ·139 | 1·05075 | ·220 | 1·12445 | ·300 | 1·22495 | ·380 | 1·34899 | ·460 | 1·49269 |
| ·140 | 1·05147 | ·221 | 1·12556 | ·301 | 1·22635 | ·381 | 1·35068 | ·461 | 1·49460 |
| ·141 | 1·05220 | ·222 | 1·12663 | ·302 | 1·22776 | ·382 | 1·35237 | ·462 | 1·49651 |
| ·142 | 1·05293 | ·223 | 1·12774 | ·303 | 1·22918 | ·383 | 1·35406 | ·463 | 1·49842 |
| ·143 | 1·05367 | ·224 | 1·12885 | ·304 | 1·23061 | ·384 | 1·35575 | ·464 | 1·50033 |
| ·144 | 1·05441 | ·225 | 1·12997 | ·305 | 1·23205 | ·385 | 1·35744 | ·465 | 1·50224 |
| ·145 | 1·05516 | ·226 | 1·13108 | ·306 | 1·23349 | ·386 | 1·35914 | ·466 | 1·50416 |
| ·146 | 1·05591 | ·227 | 1·13219 | ·307 | 1·23494 | ·387 | 1·36084 | ·467 | 1·50608 |
| ·147 | 1·05667 | ·228 | 1·13331 | ·308 | 1·23636 | ·388 | 1·36254 | ·468 | 1·50800 |
| ·148 | 1·05743 | ·229 | 1·13444 | ·309 | 1·23780 | ·389 | 1·36425 | ·469 | 1·50992 |
| ·149 | 1·05819 | ·230 | 1·13557 | ·310 | 1·23925 | ·390 | 1·36596 | ·470 | 1·51185 |
| ·150 | 1·05896 | ·231 | 1·13671 | ·311 | 1·24070 | ·391 | 1·36767 | ·471 | 1·51378 |
| ·151 | 1·05973 | ·232 | 1·13786 | ·312 | 1·24216 | ·392 | 1·36939 | ·472 | 1·51571 |
| ·152 | 1·06051 | ·233 | 1·13903 | ·313 | 1·24360 | ·393 | 1·37111 | ·473 | 1·51764 |
| ·153 | 1·06130 | ·234 | 1·14020 | ·314 | 1·24506 | ·394 | 1·37283 | ·474 | 1·51958 |
| ·154 | 1·06209 | ·235 | 1·14136 | ·315 | 1·24654 | ·395 | 1·37455 | ·475 | 1·52152 |
| ·155 | 1·06288 | ·236 | 1·14247 | ·316 | 1·24801 | ·396 | 1·37628 | ·476 | 1·52346 |
| ·156 | 1·06368 | ·237 | 1·14363 | ·317 | 1·24946 | ·397 | 1·37801 | ·477 | 1·52541 |
| ·157 | 1·06449 | ·238 | 1·14480 | ·318 | 1·25095 | ·398 | 1·37974 | ·478 | 1·52736 |
| ·158 | 1·06530 | ·239 | 1·14597 | ·319 | 1·25243 | ·399 | 1·38148 | ·479 | 1·52931 |
| ·159 | 1·06611 | ·240 | 1·14714 | ·320 | 1·25391 | ·400 | 1·38322 | ·480 | 1·53126 |
| ·160 | 1·06693 | ·241 | 1·14831 | ·321 | 1·25539 | ·401 | 1·38496 | ·481 | 1·53322 |
| ·161 | 1·06775 | ·242 | 1·14949 | ·322 | 1·25686 | ·402 | 1·38671 | ·482 | 1·53518 |
| ·162 | 1·06858 | ·243 | 1·15067 | ·323 | 1·25836 | ·403 | 1·38846 | ·483 | 1·53714 |
| ·163 | 1·06941 | ·244 | 1·15186 | ·324 | 1·25987 | ·404 | 1·39021 | ·484 | 1·53910 |
| ·164 | 1·07025 | ·245 | 1·15308 | ·325 | 1·26137 | ·405 | 1·39196 | ·485 | 1·54106 |
| ·165 | 1·07109 | ·246 | 1·15429 | ·326 | 1·26286 | ·406 | 1·39372 | ·486 | 1·54302 |
| ·166 | 1·07194 | ·247 | 1·15549 | ·327 | 1·26437 | ·407 | 1·39548 | ·487 | 1·54499 |
| ·167 | 1·07279 | ·248 | 1·15670 | ·328 | 1·26588 | ·408 | 1·39724 | ·488 | 1·54696 |
| ·168 | 1·07365 | ·249 | 1·15791 | ·329 | 1·26740 | ·409 | 1·39900 | ·489 | 1·54893 |
| ·169 | 1·07451 | ·250 | 1·15912 | ·330 | 1·26892 | ·410 | 1·40077 | ·490 | 1·55090 |
| ·170 | 1·07537 | ·251 | 1·16033 | ·331 | 1·27044 | ·411 | 1·40254 | ·491 | 1·55288 |
| ·171 | 1·07624 | ·252 | 1·16157 | ·332 | 1·27196 | ·412 | 1·40432 | ·492 | 1·55486 |
| ·172 | 1·07711 | ·253 | 1·16279 | ·333 | 1·27349 | ·413 | 1·40610 | ·493 | 1·55685 |
| ·173 | 1·07799 | ·254 | 1·16402 | ·334 | 1·27502 | ·414 | 1·40788 | ·494 | 1·55854 |
| ·174 | 1·07888 | ·255 | 1·16526 | ·335 | 1·27656 | ·415 | 1·40966 | ·495 | 1·56083 |
| ·175 | 1·07977 | ·256 | 1·16649 | ·336 | 1·27810 | ·416 | 1·41145 | ·496 | 1·56282 |
| ·176 | 1·08066 | ·257 | 1·16774 | ·337 | 1·27864 | ·417 | 1·41324 | ·497 | 1·56481 |
| ·177 | 1·08156 | ·258 | 1·16899 | ·338 | 1·28118 | ·418 | 1·41503 | ·498 | 1·56680 |
| ·178 | 1·08246 | ·259 | 1·17024 | ·339 | 1·28273 | ·419 | 1·41682 | ·499 | 1·56879 |
| ·179 | 1·08337 | ·260 | 1·17150 | ·340 | 1·28428 | ·420 | 1·41861 | ·500 | 1·57079 |
| ·180 | 1·08428 | | | | | | | | |

# PROPORTIONS OF THE LENGTHS OF SEMI-ELLIPTIC ARCS.

| Height of Arc. | Length of Arc. | Height of Arc. | Length of Arc. | Height of Arc. | Length of Arc. | Height of Arc. | Length of Arc. | Height of Arc. | Length of Arc. |
|---|---|---|---|---|---|---|---|---|---|
| ·100 | 1·04162 | ·157 | 1·10113 | ·214 | 1·66678 | ·271 | 1·23835 | ·328 | 1.31472 |
| ·101 | 1·04262 | ·158 | 1·10224 | ·215 | 1·16799 | ·272 | 1·23966 | ·329 | 1·31610 |
| ·102 | 1·04362 | ·159 | 1·10335 | ·216 | 1·16920 | ·273 | 1·24097 | ·330 | 1·31748 |
| ·103 | 1·04462 | ·160 | 1·10447 | ·217 | 1·17041 | ·274 | 1·24228 | ·331 | 1·31886 |
| ·104 | 1·04562 | ·161 | 1·10560 | ·218 | 1·17163 | ·275 | 1·24359 | ·332 | 1·32024 |
| ·105 | 1.04662 | ·162 | 1·10672 | ·219 | 1·17285 | ·276 | 1·24480 | ·333 | 1·32162 |
| ·106 | 1·04762 | ·163 | 1·10784 | ·220 | 1·17407 | ·277 | 1·24612 | ·334 | 1·32300 |
| ·107 | 1·04862 | ·164 | 1·10896 | ·221 | 1·17529 | ·278 | 1·24744 | ·335 | 1·32438 |
| ·108 | 1·04962 | ·165 | 1·11008 | ·222 | 1·17651 | ·279 | 1·24876 | ·336 | 1·32576 |
| ·109 | 1·05063 | ·166 | 1·11120 | ·223 | 1·17774 | ·280 | 1·25010 | ·337 | 1·32715 |
| ·110 | 1·05164 | ·167 | 1·11232 | ·224 | 1·17897 | ·281 | 1·25142 | ·338 | 1·32854 |
| ·111 | 1·05265 | ·168 | 1·11344 | ·225 | 1·18020 | ·282 | 1·25274 | ·339 | 1·32993 |
| ·112 | 1·05366 | ·169 | 1·11456 | ·226 | 1·18143 | ·283 | 1·25406 | ·340 | 1·33132 |
| ·113 | 1·05467 | ·170 | 1·11569 | ·227 | 1·18266 | ·284 | 1·25538 | ·341 | 1·33272 |
| ·114 | 1·05568 | ·171 | 1·11682 | ·228 | 1·18390 | ·285 | 1·25670 | ·342 | 1·33412 |
| ·115 | 1·05669 | ·172 | 1·11795 | ·229 | 1·18514 | ·286 | 1·25803 | ·343 | 1·33552 |
| ·116 | 1·05770 | ·173 | 1·11908 | ·230 | 1·18638 | ·287 | 1·25936 | ·344 | 1·33692 |
| ·117 | 1·05872 | ·174 | 1·12021 | ·231 | 1·18762 | ·288 | 1·26069 | ·345 | 1·33833 |
| ·118 | 1·05974 | ·175 | 1·12134 | ·232 | 1·18886 | ·289 | 1·26202 | ·346 | 1·33974 |
| ·119 | 1·06076 | ·176 | 1·12247 | ·233 | 1·19010 | ·290 | 1·26335 | ·347 | 1·34115 |
| ·120 | 1·06178 | ·177 | 1·12360 | ·234 | 1·19134 | ·291 | 1·26468 | ·348 | 1·34256 |
| ·121 | 1·06280 | ·178 | 1·12473 | ·235 | 1·19258 | ·292 | 1·26601 | ·349 | 1·34397 |
| ·122 | 1·06382 | ·179 | 1·12586 | ·236 | 1·19382 | ·293 | 1·26734 | ·350 | 1·34539 |
| ·123 | 1·06484 | ·180 | 1·12699 | ·237 | 1·19506 | ·294 | 1·26867 | ·851 | 1·34681 |
| ·124 | 1·06586 | ·181 | 1·12813 | ·238 | 1·19630 | ·295 | 1·27000 | ·352 | 1·34823 |
| ·125 | 1·06689 | ·182 | 1·12927 | ·239 | 1·19755 | ·296 | 1·27133 | ·353 | 1·34965 |
| ·126 | 1·06792 | ·183 | 1·13041 | ·240 | 1·19880 | ·297 | 1·27267 | ·354 | 1·35108 |
| ·127 | 1·06895 | ·184 | 1·13155 | ·241 | 1·20005 | ·298 | 1·27401 | ·355 | 1·35251 |
| ·128 | 1·06998 | ·185 | 1·13269 | ·242 | 1·20130 | ·299 | 1·27535 | ·356 | 1·35394 |
| ·129 | 1·07001 | ·186 | 1·13383 | ·243 | 1·20255 | ·300 | 1·27669 | ·357 | 1·35537 |
| ·130 | 1·07204 | ·187 | 1·13497 | ·244 | 1·20380 | ·301 | 1·27803 | ·358 | 1·35680 |
| ·131 | 1·07308 | ·188 | 1·13611 | ·245 | 1·20506 | ·302 | 1·27937 | ·359 | 1·35823 |
| ·132 | 1·07412 | ·189 | 1·13726 | ·246 | 1·20632 | ·303 | 1·28071 | ·360 | 1·35967 |
| ·133 | 1·07516 | ·190 | 1·13841 | ·247 | 1·20758 | ·304 | 1·28205 | ·361 | 1·36111 |
| ·134 | 1·07621 | ·191 | 1·13956 | ·248 | 1·20884 | ·305 | 1·28339 | ·362 | 1·36255 |
| ·135 | 1·07726 | ·192 | 1·14071 | ·249 | 1·21010 | ·306 | 1·28474 | ·363 | 1·36399 |
| ·136 | 1·07831 | ·193 | 1·14186 | ·250 | 1·21136 | ·307 | 1·28609 | ·364 | 1·36543 |
| ·137 | 1·07937 | ·194 | 1·14301 | ·251 | 1·21263 | ·308 | 1·28744 | ·365 | 1·36688 |
| ·138 | 1·08043 | ·195 | 1·14416 | ·252 | 1·21390 | ·309 | 1·28879 | ·366 | 1·36833 |
| ·139 | 1·08149 | ·196 | 1·14531 | ·253 | 1·21517 | ·310 | 1·29014 | ·367 | 1·36978 |
| ·140 | 1·08255 | ·197 | 1·14646 | ·254 | 1.21644 | ·311 | 1·29149 | ·368 | 1·37123 |
| ·141 | 1·08362 | ·198 | 1·14762 | ·255 | 1·21772 | ·312 | 1·29285 | ·369 | 1·37268 |
| ·142 | 1·08469 | ·199 | 1·14888 | ·256 | 1·21900 | ·313 | 1·29421 | ·370 | 1·37414 |
| ·143 | 1·08576 | ·200 | 1·15014 | ·257 | 1·22028 | ·314 | 2·29557 | ·371 | 1·37662 |
| ·144 | 1·08684 | ·201 | 1·15131 | ·258 | 1·22156 | ·315 | 1·29603 | ·372 | 1·37708 |
| ·145 | 1·08792 | ·202 | 1·15248 | ·259 | 1·22284 | ·316 | 1·29829 | ·373 | 1·37854 |
| ·146 | 1·08901 | ·203 | 1·15366 | ·260 | 1·22412 | ·317 | 1·29965 | ·374 | 1·38000 |
| ·147 | 1·09010 | ·204 | 1·15484 | ·261 | 1·22541 | ·318 | 1·30102 | ·375 | 1·38146 |
| ·148 | 1·09119 | ·205 | 1·15602 | ·262 | 1·22670 | ·319 | 1·30239 | ·376 | 1·38292 |
| ·149 | 1·09228 | ·206 | 1·15720 | ·263 | 1·22799 | ·320 | 1·30376 | ·377 | 1·38439 |
| ·150 | 1·09330 | ·207 | 1·15838 | ·264 | 1·22928 | ·321 | 1·30513 | ·378 | 1·38585 |
| ·151 | 1·09448 | ·208 | 1·15957 | ·265 | 1·23057 | ·322 | 1·30650 | ·379 | 1·38732 |
| ·152 | 1·09558 | ·209 | 1·16076 | ·266 | 1·23186 | ·323 | 1·30787 | ·380 | 1·38879 |
| ·153 | 1·09669 | ·210 | 1·16196 | ·267 | 1·23315 | ·324 | 1·30924 | ·381 | 1·39024 |
| ·154 | 1·09780 | ·211 | 1·16316 | ·268 | 1·23445 | ·325 | 1·31061 | ·382 | 1·39169 |
| ·155 | 1·09891 | ·212 | 1·16436 | ·269 | 1·23575 | ·326 | 1·31198 | ·383 | 1·39314 |
| ·156 | 1·10002 | ·213 | 1·16557 | ·270 | 1·23705 | ·327 | 1·31335 | ·384 | 1·39459 |

| Height of Arc. | Length of Arc. | Height of Arc. | Length of Arc. | Height of Arc. | Length of Arc. | Height of Arc. | Length of Arc. | Height of Arc. | Length of Arc. |
|---|---|---|---|---|---|---|---|---|---|
| ·385 | 1·39605 | ·447 | 1·48850 | ·509 | 1·58474 | ·571 | 1·68195 | ·633 | 1·78172 |
| ·386 | 1·39751 | ·448 | 1·49003 | ·510 | 1·58629 | ·572 | 1·68354 | ·634 | 1·78335 |
| ·387 | 1·39897 | ·449 | 1·49157 | ·511 | 1·58784 | ·573 | 1·68513 | ·635 | 1·78498 |
| ·388 | 1·40043 | ·450 | 1·49311 | ·512 | 1·58940 | ·574 | 1·68672 | ·636 | 1·78660 |
| ·389 | 1·40189 | ·451 | 1·49465 | ·513 | 1·59096 | ·575 | 1·68831 | ·637 | 1·78823 |
| ·390 | 1·40335 | ·452 | 1·49618 | ·514 | 1·59252 | ·576 | 1·68990 | ·638 | 1·78986 |
| ·391 | 1·40481 | ·453 | 1·49771 | ·515 | 1·59408 | ·577 | 1·69149 | ·639 | 1·79149 |
| ·392 | 1·40627 | ·454 | 1·49924 | ·516 | 1·59564 | ·578 | 1·69308 | ·640 | 1·79312 |
| ·393 | 1·40773 | ·455 | 1·50077 | ·517 | 1·59720 | ·579 | 1·69467 | ·641 | 1·79475 |
| ·394 | 1·40919 | ·456 | 1·50230 | ·518 | 1·59876 | ·580 | 1·69626 | ·642 | 1·79638 |
| ·395 | 1·41065 | ·457 | 1·50383 | ·519 | 1·60032 | ·581 | 1·69785 | ·643 | 1·79801 |
| ·396 | 1·41211 | ·458 | 1·50536 | ·520 | 1·60188 | ·582 | 1·69945 | ·644 | 1·79964 |
| ·397 | 1·41357 | ·459 | 1·50689 | ·521 | 1·60344 | ·583 | 1·70105 | ·645 | 1·80127 |
| ·398 | 1·41504 | ·460 | 1·50842 | ·522 | 1·60500 | ·584 | 1·70264 | ·646 | 1·80290 |
| ·399 | 1·41651 | ·461 | 1·50996 | ·523 | 1·60656 | ·585 | 1·70424 | ·647 | 1·80454 |
| ·400 | 1·41798 | ·462 | 1·51150 | ·524 | 1·60812 | ·586 | 1·70584 | ·648 | 1·80617 |
| ·401 | 1·41945 | ·463 | 1·51304 | ·525 | 1·60968 | ·587 | 1·70745 | ·649 | 1·80780 |
| ·402 | 1·42092 | ·464 | 1·51458 | ·526 | 1·61124 | ·588 | 1·70905 | ·650 | 1·80943 |
| ·403 | 1·42239 | ·465 | 1·51612 | ·527 | 1·61280 | ·589 | 1·71065 | ·651 | 1·81107 |
| ·404 | 1·42386 | ·466 | 1·51766 | ·528 | 1·61436 | ·590 | 1·71225 | ·652 | 1·81271 |
| ·405 | 1·42533 | ·467 | 1·51920 | ·529 | 1·61592 | ·591 | 1·71286 | ·653 | 1·81435 |
| ·406 | 1·42681 | ·468 | 1·52074 | ·530 | 1·61748 | ·592 | 1·71546 | ·654 | 1·81599 |
| ·407 | 1·42829 | ·469 | 1·52229 | ·531 | 1·61904 | ·593 | 1·71707 | ·655 | 1·81763 |
| ·408 | 1·42977 | ·470 | 1·52384 | ·532 | 1·62060 | ·594 | 1·71868 | ·656 | 1·81928 |
| ·409 | 1·43125 | ·471 | 1·52539 | ·533 | 1·62216 | ·595 | 1·72029 | ·657 | 1·82091 |
| ·410 | 1·43273 | ·472 | 1·52691 | ·534 | 1·62372 | ·596 | 1·72190 | ·658 | 1·82255 |
| ·411 | 1·43421 | ·473 | 1·52849 | ·535 | 1·62528 | ·597 | 1·72350 | ·659 | 1·82419 |
| ·412 | 1·43569 | ·474 | 1·53004 | ·536 | 1·62684 | ·598 | 1·72511 | ·660 | 1·82583 |
| ·413 | 1·43718 | ·475 | 1·53159 | ·537 | 1·62840 | ·599 | 1·72672 | ·661 | 1·82747 |
| ·414 | 1·43867 | ·476 | 1·53314 | ·538 | 1·62996 | ·600 | 1·72833 | ·662 | 1·82911 |
| ·415 | 1·44016 | ·477 | 1·53469 | ·539 | 1·63152 | ·601 | 1·72994 | ·663 | 1·83075 |
| ·416 | 1·44165 | ·478 | 1·53625 | ·540 | 1·63309 | ·602 | 1·73155 | ·664 | 1·83240 |
| ·417 | 1·44314 | ·479 | 1·53781 | ·541 | 1·63465 | ·603 | 1·73316 | ·665 | 1·83404 |
| ·418 | 1·44463 | ·480 | 1·53937 | ·542 | 1·63623 | ·604 | 1·73477 | ·666 | 1·83568 |
| ·419 | 1·44613 | ·481 | 1·54093 | ·543 | 1·63780 | ·605 | 1·73638 | ·667 | 1·83733 |
| ·420 | 1·44763 | ·482 | 1·54249 | ·544 | 1·63937 | ·606 | 1·73799 | ·668 | 1·83897 |
| ·421 | 1·44913 | ·483 | 1·54405 | ·545 | 1·64094 | ·607 | 1·73960 | ·669 | 1·84061 |
| ·422 | 1·45064 | ·484 | 1·54561 | ·546 | 1·64251 | ·608 | 1·74121 | ·670 | 1·84226 |
| ·423 | 1·45214 | ·485 | 1·54718 | ·547 | 1·64408 | ·609 | 1·74283 | ·671 | 1·84391 |
| ·424 | 1·45364 | ·486 | 1·54875 | ·548 | 1·64565 | ·610 | 1·74444 | ·672 | 1·84556 |
| ·425 | 1·45515 | ·487 | 1·55032 | ·549 | 1·64722 | ·611 | 1·74605 | ·673 | 1·84720 |
| ·426 | 1·45665 | ·488 | 1·55189 | ·550 | 1·64879 | ·612 | 1·74767 | ·674 | 1·84885 |
| ·427 | 1·45815 | ·489 | 1·55346 | ·551 | 1·65036 | ·613 | 1·74929 | ·675 | 1·85050 |
| ·428 | 1·45966 | ·490 | 1·55503 | ·552 | 1·65193 | ·614 | 1·75091 | ·676 | 1·85215 |
| ·429 | 1·46167 | ·491 | 1·55660 | ·553 | 1·65350 | ·615 | 1·75252 | ·677 | 1·85379 |
| ·430 | 1·46268 | ·492 | 1·55817 | ·554 | 1·65507 | ·616 | 1·75414 | ·678 | 1·85544 |
| ·431 | 1·46419 | ·493 | 1·55974 | ·555 | 1·65665 | ·617 | 1·75576 | ·679 | 1·85709 |
| ·432 | 1·46570 | ·494 | 1·56131 | ·556 | 1·65823 | ·618 | 1·75738 | ·680 | 1·85874 |
| ·433 | 1·46721 | ·495 | 1·56289 | ·557 | 1·65981 | ·619 | 1·75900 | ·681 | 1·86039 |
| ·434 | 1·46872 | ·496 | 1·56447 | ·558 | 1·66139 | ·620 | 1·76062 | ·682 | 1·86205 |
| ·435 | 1·47023 | ·497 | 1·56605 | ·559 | 1·66297 | ·621 | 1·76224 | ·683 | 1·86370 |
| ·436 | 1·47174 | ·498 | 1·56763 | ·560 | 1·66455 | ·622 | 1·76386 | ·684 | 1·86535 |
| ·437 | 1·47326 | ·499 | 1·56921 | ·561 | 1·66613 | ·623 | 1·76548 | ·685 | 1·86700 |
| ·438 | 1·47478 | ·500 | 1·57089 | ·562 | 1·66771 | ·624 | 1·76710 | ·686 | 1·86866 |
| ·439 | 1·47630 | ·501 | 1·57234 | ·563 | 1·66929 | ·625 | 1·76872 | ·687 | 1·87031 |
| ·440 | 1·47782 | ·502 | 1·57389 | ·564 | 1·67087 | ·626 | 1·77034 | ·688 | 1·87196 |
| ·441 | 1·47934 | ·503 | 1·57544 | ·565 | 1·67245 | ·627 | 1·77197 | ·689 | 1·87362 |
| ·442 | 1·48086 | ·504 | 1·57699 | ·566 | 1·67403 | ·628 | 1·77359 | ·690 | 1·87527 |
| ·443 | 1·48238 | ·505 | 1·57854 | ·567 | 1·67561 | ·629 | 1·77521 | ·691 | 1·87693 |
| ·444 | 1·48391 | ·506 | 1·58009 | ·568 | 1·67719 | ·630 | 1·77684 | ·692 | 1·87859 |
| ·445 | 1·48544 | ·507 | 1·58164 | ·569 | 1·67877 | ·631 | 1·77847 | ·693 | 1·88024 |
| ·446 | 1·48697 | 508 | 1·58319 | ·570 | 1·68036 | ·632 | 1·78009 | ·694 | 1·88190 |

| Height of Arc. | Length of Arc. | Height of Arc. | Length of Arc. | Height of Arc. | Length of Arc. | Height of Arc. | Length of Arc. | Height of Arc. | Length of Arc. |
|---|---|---|---|---|---|---|---|---|---|
| ·695 | 1·88356 | ·757 | 1·98794 | ·818 | 2·09360 | ·879 | 2·20292 | ·940 | 2·31479 |
| ·696 | 1·88522 | ·758 | 1·98964 | ·819 | 2·09536 | ·880 | 2·20474 | ·941 | 2·31666 |
| ·697 | 1·88688 | ·759 | 1·99134 | ·820 | 2·09712 | ·881 | 2·20656 | ·942 | 2·31852 |
| ·698 | 1·88854 | ·760 | 1·99305 | ·821 | 2·09888 | ·882 | 2·20839 | ·943 | 2·32038 |
| ·699 | 1·89020 | ·761 | 1·99476 | ·822 | 2·10065 | ·883 | 2·21022 | ·944 | 2·32224 |
| ·700 | 1·89186 | ·762 | 1·99647 | ·823 | 2·10242 | ·884 | 2·21205 | ·945 | 2·32411 |
| ·701 | 1·89352 | ·763 | 1·99818 | ·824 | 2·10419 | ·885 | 2·21388 | ·946 | 2·32598 |
| ·702 | 1·89519 | ·764 | 1·99989 | ·825 | 2·10596 | ·886 | 2·21571 | ·947 | 2·32785 |
| ·703 | 1·89685 | ·765 | 2·00160 | ·826 | 2·10773 | ·887 | 2·21754 | ·948 | 2·32972 |
| ·704 | 1·89851 | ·766 | 2·00331 | ·827 | 2·10950 | ·888 | 2·21937 | ·949 | 2·33160 |
| ·705 | 1·90017 | ·767 | 2·00502 | ·828 | 2·11127 | ·889 | 2·22120 | ·950 | 2·33348 |
| ·706 | 1·90184 | ·768 | 2·00673 | ·829 | 2·11304 | ·890 | 2·22303 | ·951 | 2·33537 |
| ·707 | 1·90350 | ·769 | 2·00844 | ·830 | 2·11481 | ·891 | 2·22486 | ·952 | 2·33726 |
| ·708 | 1·90517 | ·770 | 2·01016 | ·831 | 2·11659 | ·892 | 2·22670 | ·953 | 2·33915 |
| ·709 | 1·90684 | ·771 | 2·01187 | ·832 | 2·11837 | ·893 | 2·22854 | ·954 | 2·34104 |
| ·710 | 1·90852 | ·772 | 2·01359 | ·833 | 2·12015 | ·894 | 2·23038 | ·955 | 2·34293 |
| ·711 | 1·91019 | ·773 | 2·01531 | ·834 | 2·12193 | ·895 | 2·23222 | ·956 | 2·34483 |
| ·712 | 1·91187 | ·774 | 2·01702 | ·835 | 2·12371 | ·896 | 2·23406 | ·957 | 2·34673 |
| ·713 | 1·91355 | ·775 | 2·01874 | ·836 | 2·12549 | ·897 | 2·23590 | ·958 | 2·34862 |
| ·714 | 1·91523 | ·776 | 2·02045 | ·837 | 2·12727 | ·898 | 2·23774 | ·959 | 2·35051 |
| ·715 | 1·91691 | ·777 | 2·02217 | ·838 | 2·12905 | ·899 | 2·23958 | ·960 | 2·35241 |
| ·716 | 1·91859 | ·778 | 2·02389 | ·839 | 2·13083 | ·900 | 2·24142 | ·961 | 2·35431 |
| ·717 | 1·92027 | ·779 | 2·02561 | ·840 | 2·13261 | ·901 | 2·24325 | ·962 | 2·35621 |
| ·718 | 1·92195 | ·780 | 2·02733 | ·841 | 2·13439 | ·902 | 2·24508 | ·963 | 2·35810 |
| ·719 | 1·92363 | ·781 | 2·02907 | ·842 | 2·13618 | ·903 | 2·24691 | ·964 | 2·36000 |
| ·720 | 1·92531 | ·782 | 2·03080 | ·843 | 2·13797 | ·904 | 2.24874 | ·965 | 2·36191 |
| ·721 | 1·92700 | ·783 | 2·03252 | ·844 | 2·13976 | ·905 | 2·25057 | ·966 | 2·36381 |
| ·722 | 1·92868 | ·784 | 2·03425 | ·845 | 2·14155 | ·906 | 2·25240 | ·967 | 2·36571 |
| ·723 | 1·93036 | ·785 | 2·03598 | ·846 | 2·14334 | ·907 | 2·25423 | ·968 | 2·36762 |
| ·724 | 1·93204 | ·786 | 2·03771 | ·847 | 2·14513 | ·908 | 2·25606 | ·969 | 2·36952 |
| ·725 | 1·93373 | ·787 | 2·03944 | ·848 | 2·14692 | ·909 | 2·25789 | ·970 | 2·37143 |
| ·726 | 1·93541 | ·788 | 2·04117 | ·849 | 2·14871 | ·910 | 2·25972 | ·971 | 2·37334 |
| ·727 | 1·93710 | ·789 | 2·04290 | ·850 | 2·15050 | ·911 | 2·26155 | ·972 | 2·37525 |
| ·728 | 1·93878 | ·790 | 2·04462 | ·851 | 2·15229 | ·912 | 2·26338 | ·973 | 2·37716 |
| ·729 | 1·94046 | ·791 | 2·04635 | ·852 | 2·15409 | ·913 | 2·26521 | ·974 | 2·37908 |
| ·730 | 1·94215 | ·792 | 2·04809 | ·853 | 2·15589 | ·914 | 2·26704 | ·975 | 2·38100 |
| ·731 | 1·94383 | ·793 | 2·04983 | ·854 | 2·15770 | ·915 | 2·26888 | ·976 | 2·38291 |
| ·732 | 1·94552 | ·794 | 2·05157 | ·855 | 2·15950 | ·916 | 2·27071 | ·977 | 2·38482 |
| ·733 | 1·94721 | ·795 | 2·05331 | ·856 | 2·16130 | ·917 | 2·27254 | ·978 | 2·38673 |
| ·734 | 1·94890 | ·796 | 2·05505 | ·857 | 2·16309 | ·918 | 2·27437 | ·979 | 2·38864 |
| ·735 | 1 95050 | ·797 | 2·05679 | ·858 | 2·16489 | ·919 | 2·27620 | ·980 | 2·39055 |
| ·736 | 1·95228 | ·798 | 2·05853 | ·859 | 2·16668 | ·920 | 2·27803 | ·981 | 2·39247 |
| ·737 | 1·95397 | ·799 | 2·06027 | ·860 | 2·16848 | ·921 | 2·27987 | ·982 | 2·39439 |
| ·738 | 1·95566 | ·800 | 2·06202 | ·861 | 2·17028 | ·922 | 2·28170 | ·983 | 2·39631 |
| ·739 | 1·95735 | ·801 | 2·06377 | ·862 | 2·17209 | ·923 | 2·28354 | ·984 | 2·39823 |
| ·740 | 1·95994 | ·802 | 2·06552 | ·863 | 2·17389 | ·924 | 2·28537 | ·985 | 2·40016 |
| ·741 | 1·96074 | ·803 | 2·06727 | ·864 | 2·17570 | ·925 | 2·28720 | ·986 | 2·40208 |
| ·742 | 1·96244 | ·804 | 2·06901 | ·865 | 2·17751 | ·926 | 2·28903 | ·987 | 2·40400 |
| ·743 | 1·96414 | ·805 | 2·07076 | ·866 | 2·17932 | ·927 | 2·29086 | ·988 | 2·40592 |
| ·744 | 1·96583 | ·806 | 2·07251 | ·867 | 2·18113 | ·928 | 2·29270 | ·989 | 2·40784 |
| ·745 | 1·96753 | ·807 | 2·07427 | ·868 | 2·18294 | ·929 | 2·29453 | ·990 | 2·40976 |
| ·746 | 1·96923 | ·808 | 2·07602 | ·869 | 2·18475 | ·930 | 2·29636 | ·991 | 2·41169 |
| ·747 | 1·97093 | ·809 | 2·07777 | ·870 | 2·18656 | ·931 | 2·29820 | ·992 | 2·41362 |
| ·748 | 1·97262 | ·810 | 2·07953 | ·871 | 2·18837 | ·932 | 2·30004 | ·993 | 2·41556 |
| ·749 | 1·97432 | ·811 | 2·08128 | ·872 | 2·19018 | ·933 | 2·30188 | ·994 | 2·41749 |
| ·750 | 1·97602 | ·812 | 2·08304 | ·873 | 2·19200 | ·934 | 2·30373 | ·995 | 2·41943 |
| ·751 | 1·97772 | ·813 | 2·08480 | ·874 | 2·19382 | ·935 | 2·30557 | ·996 | 2·42136 |
| ·752 | 1·97943 | ·814 | 2·08656 | ·875 | 2·19564 | ·936 | 2·30741 | ·997 | 2·42329 |
| ·753 | 1·98113 | ·815 | 2·08832 | ·876 | 2·19746 | ·937 | 2·30926 | ·998 | 2·42522 |
| ·754 | 1·98283 | ·816 | 2·09008 | ·877 | 2·19928 | ·938 | 2·31111 | ·999 | 2·42715 |
| ·755 | 1·98453 | ·817 | 2·09198 | ·878 | 2·20110 | ·939 | 2·31295 | ·1000 | 2·42908 |
| ·756 | 1·98623 | | | | | | | | |

*To find the length of an arc of a circle, or the curve of a right semi-ellipse.*

Rule.—Divide the height by the base, and the quotient will be the height of an arc of which the base is unity. Seek, in the Table of Circular or of Semi-elliptical arcs, as the case may be, for a number corresponding to this quotient, and take the length of the arc from the next right-hand column. Multiply the number thus taken out by the base of the arc, and the product will be the length of the arc or curve required.

In a Bridge, suppose the profiles of the arches are the arcs of circles; the span of the middle arch is 240 feet and the height 24 feet; required the length of the arc.

$24 \div 240 = \cdot 100$; and ·100, by the Table, is 1·02645.

Hence $1\cdot 02645 \times 24 = 246\cdot 34800$ feet, the length required.

The profiles of the arches of a Bridge are all equal and similar semi-ellipses; the span of each is 120 feet, and the rise 18 feet; required the length of the curve.

$28 \div 120 = \cdot 233$; and ·233 by the Table, is 1·19010.

Hence $1\cdot 19010 \times 120 = 142\cdot 81200$ feet, the length required.

In this example there is, in the division of 28 by 120, a remainder of 40, or one-third part of the divisor; consequently, the answer, 142·81200, is rather less than the truth. But this difference, in even so large an arch, is little more than half an inch; therefore, except where extreme accuracy is required, it is not worth computing.

These Tables are equally useful in estimating works which may be carried into practice, and the quantity of work to be executed from drawings to a scale.

As the Tables do not afford the means of finding the lengths of the curves of elliptical arcs which are less than half of the entire figure, the following geometrical method is given to supply the defect.

Let the curve, of which the length is required to be found, be ABC.

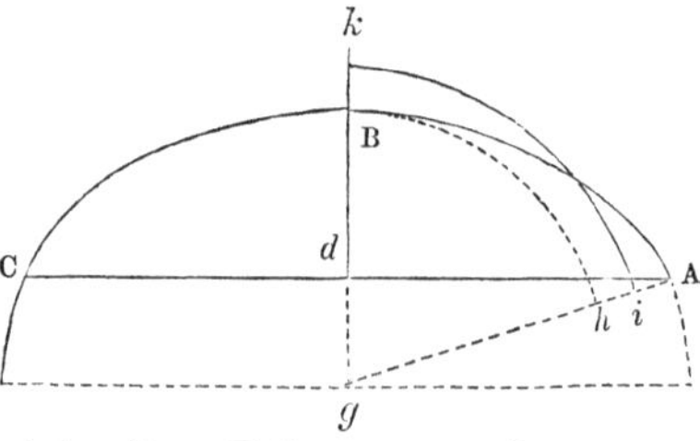


Produce the height line B$d$ to meet the centre of the curve in $g$. Draw the right line A$g$, and from the centre $g$, with the distance $g$B describe an arc B$h$, meeting A$g$ in $h$. Bisect A$h$ in $i$, and from the centre $g$ with the radius $gi$ describe the arc $ik$, meeting $d$B produced to $k$; then $ik$ is half the arc ABC.

A TABLE *of the Reciprocals of Numbers; or the* DECIMAL FRACTIONS *corresponding to* VULGAR FRACTIONS *of which the Numerator is unity or* 1.

[In the following Tables, the Decimal fractions are Reciprocals of the Denominators of those opposite to them; and their product is = unity.

To find the Decimal corresponding to a fraction having a higher Numerator than 1, multiply the Decimal opposite to the given Denominator, by the given Numerator. Thus, the Decimal corresponding to $\frac{1}{64}$ being ·015625, the Decimal to $\frac{15}{64}$ will be ·015625 × 15 = ·234375.]

| Fraction or Numb. | Decimal or Reciprocal. | Fraction or Numb. | Decimal or Reciprocal. | Fraction or Numb. | Decimal or Reciprocal. |
|---|---|---|---|---|---|
| 1/2 | ·5 | 1/47 | ·0212766 | 1/92 | ·010869565 |
| 1/3 | ·333333333 | 1/48 | ·020833333 | 1/93 | ·010752688 |
| 1/4 | ·25 | 1/49 | ·020408163 | 1/94 | ·010638298 |
| 1/5 | ·2 | 1/50 | ·02 | 1/95 | ·010526316 |
| 1/6 | ·166666667 | 1/51 | ·019607843 | 1/96 | ·010416667 |
| 1/7 | ·142857143 | 1/52 | ·019230769 | 1/97 | ·010309278 |
| 1/8 | ·125 | 1/53 | ·018867925 | 1/98 | ·010204082 |
| 1/9 | ·111111111 | 1/54 | ·018518519 | 1/99 | ·01010101 |
| 1/10 | ·1 | 1/55 | ·018181818 | 1/100 | ·01 |
| 1/11 | ·090909091 | 1/56 | ·017857143 | 1/101 | ·00990099 |
| 1/12 | ·083333333 | 1/57 | ·01754386 | 1/102 | ·009803922 |
| 1/13 | ·076923077 | 1/58 | ·017241379 | 1/103 | ·009708738 |
| 1/14 | ·071428571 | 1/59 | ·016949153 | 1/104 | ·009615385 |
| 1/15 | ·066666667 | 1/60 | ·016666667 | 1/105 | ·00952381 |
| 1/16 | ·0625 | 1/61 | ·016393443 | 1/106 | ·009433962 |
| 1/17 | ·058823529 | 1/62 | ·016129032 | 1/107 | ·009345794 |
| 1/18 | ·055555556 | 1/63 | ·015873016 | 1/108 | ·009259259 |
| 1/19 | ·052631579 | 1/64 | ·015625 | 1/109 | ·009174312 |
| 1/20 | ·05 | 1/65 | ·015384615 | 1/110 | ·009090909 |
| 1/21 | ·047619048 | 1/66 | ·015151515 | 1/111 | ·009009009 |
| 1/22 | ·045454545 | 1/67 | ·014925373 | 1/112 | ·008928571 |
| 1/23 | ·043478261 | 1/68 | ·014705882 | 1/113 | ·008849558 |
| 1/24 | ·041666667 | 1/69 | ·014492754 | 1/114 | ·00877193 |
| 1/25 | ·04 | 1/70 | ·014285714 | 1/115 | ·008695652 |
| 1/26 | 038461538 | 1/71 | ·014084517 | 1/116 | ·00862069 |
| 1/27 | ·037037037 | 1/72 | ·013888889 | 1/117 | ·008547009 |
| 1/28 | ·035714286 | 1/73 | ·01369863 | 1/118 | ·008474576 |
| 1/29 | ·034482759 | 1/74 | ·013513514 | 1/119 | ·008403361 |
| 1/30 | ·033333333 | 1/75 | ·013333333 | 1/120 | ·008333333 |
| 1/31 | ·032258065 | 1/76 | ·013157895 | 1/121 | ·008264463 |
| 1/32 | ·03125 | 1/77 | ·012987013 | 1/122 | ·008196721 |
| 1/33 | ·030303030 | 1/78 | ·012820513 | 1/123 | ·008130081 |
| 1/34 | ·029411765 | 1/79 | ·012658228 | 1/124 | ·008064516 |
| 1/35 | ·028571429 | 1/80 | ·0125 | 1/125 | ·008 |
| 1/36 | ·027777778 | 1/81 | ·012345679 | 1/126 | ·007936508 |
| 1/37 | ·027027027 | 1/82 | ·012195122 | 1/127 | ·007874016 |
| 1/38 | ·026315789 | 1/83 | ·012048193 | 1/128 | ·0078125 |
| 1/39 | ·025641026 | 1/84 | ·011904762 | 1/129 | ·007751938 |
| 1/40 | ·025 | 1/85 | ·011764706 | 1/130 | ·007692308 |
| 1/41 | ·024390244 | 1/86 | ·011627907 | 1/131 | ·007633588 |
| 1/42 | ·023809524 | 1/87 | ·011494253 | 1/132 | ·007575758 |
| 1/43 | ·023255814 | 1/88 | ·011363636 | 1/133 | ·007518797 |
| 1/44 | ·022727273 | 1/89 | ·011235955 | 1/134 | ·007462687 |
| 1/45 | ·022222222 | 1/90 | ·011111111 | 1/135 | ·007407407 |
| 1/46 | ·02173913 | 1/91 | ·010989011 | 1/136 | ·007352941 |

| Fraction or Numb. | Decimal or Reciprocal. | Fraction or Numb. | Decimal or Reciprocal. | Fraction or Numb. | Decimal or Reciprocal. |
|---|---|---|---|---|---|
| 1/137 | ·00729927 | 1/198 | ·005050505 | 1/259 | ·003861004 |
| 1/138 | ·007246377 | 1/199 | ·005025126 | 1/260 | ·003846154 |
| 1/139 | ·007194245 | 1/200 | ·005 | 1/261 | ·003831418 |
| 1/140 | ·007142857 | 1/201 | ·004975124 | 1/262 | ·003816794 |
| 1/141 | ·007092199 | 1/202 | ·004950495 | 1/263 | ·003802281 |
| 1/142 | ·007042254 | 1/203 | ·004926108 | 1/264 | ·003787879 |
| 1/143 | ·006993007 | 1/204 | ·004901961 | 1/265 | ·003773585 |
| 1/144 | ·006944444 | 1/205 | ·004878049 | 1/266 | ·003759398 |
| 1/145 | ·006896552 | 1/206 | ·004854369 | 1/267 | ·003745318 |
| 1/146 | ·006849315 | 1/207 | ·004830918 | 1/268 | ·003731343 |
| 1/147 | ·006802721 | 1/208 | ·004807692 | 1/269 | ·003717472 |
| 1/148 | ·006756757 | 1/209 | ·004784689 | 1/270 | ·003703704 |
| 1/149 | ·006711409 | 1/210 | ·004761905 | 1/271 | ·003690037 |
| 1/150 | ·006666667 | 1/211 | ·004739336 | 1/272 | ·003676471 |
| 1/151 | ·006622517 | 1/212 | ·004716981 | 1/273 | ·003663004 |
| 1/152 | ·006578947 | 1/213 | ·004694836 | 1/274 | ·003649635 |
| 1/153 | ·006535948 | 1/214 | ·004672897 | 1/275 | ·003636364 |
| 1/154 | ·006493506 | 1/215 | ·004651163 | 1/276 | ·003623188 |
| 1/155 | ·006451613 | 1/216 | ·00462963 | 1/277 | ·003610108 |
| 1/156 | ·006410256 | 1/217 | ·004608295 | 1/278 | ·003597122 |
| 1/157 | ·006369427 | 1/218 | ·004587156 | 1/279 | ·003584229 |
| 1/158 | ·006329114 | 1/219 | ·00456621 | 1/280 | ·003571429 |
| 1/159 | ·006289308 | 1/220 | ·004545455 | 1/281 | ·003558719 |
| 1/160 | ·00625 | 1/221 | ·004524887 | 1/282 | ·003546099 |
| 1/161 | ·00621118 | 1/222 | ·004504505 | 1/283 | ·003533569 |
| 1/162 | ·00617284 | 1/223 | ·004484305 | 1/284 | ·003522127 |
| 1/163 | ·006134969 | 1/224 | ·004464286 | 1/285 | ·003508772 |
| 1/164 | ·006097561 | 1/225 | ·004444444 | 1/286 | ·003496503 |
| 1/165 | ·006060606 | 1/226 | ·004424779 | 1/287 | ·003484321 |
| 1/166 | ·006024096 | 1/227 | ·004405286 | 1/288 | ·003472222 |
| 1/167 | ·005988024 | 1/228 | ·004385965 | 1/289 | ·003460208 |
| 1/168 | ·005952381 | 1/229 | ·004366812 | 1/290 | ·003448276 |
| 1/169 | ·00591716 | 1/230 | ·004347826 | 1/291 | ·003436426 |
| 1/170 | ·005882353 | 1/231 | ·004329004 | 1/292 | ·003424658 |
| 1/171 | ·005847953 | 1/232 | ·004310345 | 1/293 | ·003412969 |
| 1/172 | ·005813953 | 1/233 | ·004291845 | 1/294 | ·003401361 |
| 1/173 | ·005780347 | 1/234 | ·004273504 | 1/295 | ·003389831 |
| 1/174 | ·005747126 | 1/235 | ·004255319 | 1/296 | ·003378378 |
| 1/175 | ·005714286 | 1/236 | ·004237288 | 1/297 | ·003367003 |
| 1/176 | ·005681818 | 1/237 | ·004219409 | 1/298 | ·003355705 |
| 1/177 | ·005649718 | 1/238 | ·004201681 | 1/299 | ·003344482 |
| 1/178 | ·005617978 | 1/239 | ·0041841 | 1/300 | ·003333333 |
| 1/179 | ·005586592 | 1/240 | ·004166667 | 1/301 | ·003322259 |
| 1/180 | ·005555556 | 1/241 | ·004149378 | 1/302 | ·003311258 |
| 1/181 | ·005524862 | 1/242 | ·004132231 | 1/303 | ·00330133 |
| 1/182 | ·005494505 | 1/243 | ·004115226 | 1/304 | ·003289474 |
| 1/183 | ·005464481 | 1/244 | ·004098361 | 1/305 | ·003278689 |
| 1/184 | ·005434783 | 1/245 | ·004081633 | 1/306 | ·003267974 |
| 1/185 | ·005405405 | 1/246 | ·004065041 | 1/307 | ·003257329 |
| 1/186 | ·005376344 | 1/247 | ·004048583 | 1/308 | ·003246753 |
| 1/187 | ·005347594 | 1/248 | ·004032258 | 1/309 | ·003236246 |
| 1/188 | ·005319149 | 1/249 | ·004016064 | 1/310 | ·003225806 |
| 1/189 | ·005291005 | 1/250 | ·004 | 1/311 | ·003215434 |
| 1/190 | ·005263158 | 1/251 | ·003984064 | 1/312 | ·003205128 |
| 1/191 | ·005235602 | 1/252 | ·003968254 | 1/313 | ·003194888 |
| 1/192 | ·005208333 | 1/253 | ·003952569 | 1/314 | ·003184713 |
| 1/193 | ·005181347 | 1/254 | ·003937008 | 1/315 | ·003174603 |
| 1/194 | ·005154639 | 1/255 | ·003921569 | 1/316 | ·003164557 |
| 1/195 | ·005128205 | 1/256 | ·00390625 | 1/317 | ·003154574 |
| 1/196 | ·005102041 | 1/257 | ·003891051 | 1/318 | ·003144654 |
| 1/197 | ·005076142 | 1/258 | ·003875969 | 1/319 | ·003134796 |

| Fraction or Numb. | Decimal or Reciprocal. | Fraction or Numb. | Decimal or Reciprocal. | Fraction or Numb. | Decimal or Reciprocal. |
|---|---|---|---|---|---|
| 1/320 | ·003125 | 1/381 | ·002624672 | 1/442 | ·002262443 |
| 1/321 | ·003115265 | 1/382 | ·002617801 | 1/443 | ·002257336 |
| 1/322 | ·00310559 | 1/383 | ·002610966 | 1/444 | ·002252252 |
| 1/323 | ·003095975 | 1/384 | ·002604167 | 1/445 | ·002247191 |
| 1/324 | ·00308642 | 1/385 | ·002597403 | 1/446 | ·002242152 |
| 1/325 | ·003076923 | 1/386 | ·002590674 | 1/447 | ·002237136 |
| 1/326 | ·003067485 | 1/387 | ·002583979 | 1/448 | ·002232143 |
| 1/327 | ·003058104 | 1/388 | ·00257732 | 1/449 | ·002227171 |
| 1/328 | ·00304878 | 1/389 | ·002570694 | 1/450 | ·002222222 |
| 1/329 | ·003039514 | 1/390 | ·002564103 | 1/451 | ·002217295 |
| 1/330 | ·003030303 | 1/391 | ·002557545 | 1/452 | ·002212389 |
| 1/331 | ·003021148 | 1/392 | ·00255102 | 1/453 | ·002207506 |
| 1/332 | ·003012048 | 1/393 | ·002544529 | 1/454 | ·002202643 |
| 1/333 | ·003003003 | 1/394 | ·002538071 | 1/455 | ·002197802 |
| 1/334 | ·002994012 | 1/395 | ·002531646 | 1/456 | ·002192982 |
| 1/335 | ·002985075 | 1/396 | ·002525253 | 1/457 | ·002188184 |
| 1/336 | ·00297619 | 1/397 | ·002518892 | 1/458 | ·002183406 |
| 1/337 | ·002967359 | 1/398 | ·002512563 | 1/459 | ·002178649 |
| 1/338 | ·00295858 | 1/399 | ·002506266 | 1/460 | ·002173913 |
| 1/339 | ·002949853 | 1/400 | ·0025 | 1/461 | ·002169197 |
| 1/340 | ·002941176 | 1/401 | ·002493766 | 1/462 | ·002164502 |
| 1/341 | ·002932551 | 1/402 | ·002487562 | 1/463 | ·002159827 |
| 1/342 | ·002923977 | 1/403 | ·00248139 | 1/464 | ·002155172 |
| 1/343 | ·002915452 | 1/404 | ·002475248 | 1/465 | ·002150538 |
| 1/344 | ·002906977 | 1/405 | ·002469136 | 1/466 | ·002145923 |
| 1/345 | ·002898551 | 1/406 | ·002463054 | 1/467 | ·002141328 |
| 1/346 | ·002890173 | 1/407 | ·002457002 | 1/468 | ·002136752 |
| 1/347 | ·002881844 | 1/408 | ·00245098 | 1/469 | ·002132196 |
| 1/348 | ·002873563 | 1/409 | ·002444988 | 1/470 | ·00212766 |
| 1/349 | ·00286533 | 1/410 | ·002439024 | 1/471 | ·002123142 |
| 1/350 | ·002857143 | 1/411 | ·00243309 | 1/472 | ·002118644 |
| 1/351 | ·002849003 | 1/412 | ·002427184 | 1/473 | ·002114165 |
| 1/352 | ·002840909 | 1/413 | ·002421308 | 1/474 | ·002109705 |
| 1/353 | ·002832861 | 1/414 | ·002415459 | 1/475 | ·002105263 |
| 1/354 | ·002824859 | 1/415 | ·002409639 | 1/476 | ·00210084 |
| 1/355 | ·002816901 | 1/416 | ·002406846 | 1/477 | ·002096486 |
| 1/356 | ·002808989 | 1/417 | ·002398082 | 1/478 | ·00209205 |
| 1/357 | ·00280112 | 1/418 | ·002392344 | 1/479 | ·002087683 |
| 1/358 | ·002793296 | 1/419 | ·002386635 | 1/480 | ·002083333 |
| 1/359 | ·002785515 | 1/420 | ·002380952 | 1/481 | ·002079002 |
| 1/360 | ·002777778 | 1/421 | ·002375297 | 1/482 | ·002074689 |
| 1/361 | ·002770083 | 1/422 | ·002369668 | 1/483 | ·002070393 |
| 1/362 | ·002762431 | 1/423 | ·002364066 | 1/484 | ·002066116 |
| 1/363 | ·002754821 | 1/424 | ·002358491 | 1/485 | ·002061856 |
| 1/364 | ·002747235 | 1/425 | ·002352941 | 1/486 | ·002057613 |
| 1/365 | ·002739726 | 1/426 | ·002347418 | 1/487 | ·002053388 |
| 1/366 | ·00273224 | 1/427 | ·00234192 | 1/488 | ·00204918 |
| 1/367 | ·002724796 | 1/428 | ·002336449 | 1/489 | ·00204499 |
| 1/368 | ·002717391 | 1/429 | ·002331002 | 1/490 | ·002040816 |
| 1/369 | ·002710027 | 1/430 | ·002325581 | 1/491 | ·00203666 |
| 1/370 | ·002702703 | 1/431 | ·002320186 | 1/492 | ·00203252 |
| 1/371 | ·002695418 | 1/432 | ·002314815 | 1/493 | ·002028398 |
| 1/372 | ·002688172 | 1/433 | ·002309469 | 1/494 | ·002024291 |
| 1/373 | ·002680965 | 1/434 | ·002304147 | 1/495 | ·002020202 |
| 1/374 | ·002673797 | 1/435 | ·002298851 | 1/496 | ·002016129 |
| 1/375 | ·002666667 | 1/436 | ·002293578 | 1/497 | ·002012072 |
| 1/376 | ·002659574 | 1/437 | ·00228833 | 1/498 | ·002008032 |
| 1/377 | ·00265252 | 1/438 | ·002283105 | 1/499 | ·002004008 |
| 1/378 | ·002645503 | 1/439 | ·002277904 | 1/500 | ·002 |
| 1/379 | ·002638521 | 1/440 | ·002272727 | 1/501 | ·001996008 |
| 1/380 | ·002631579 | 1/441 | ·002267574 | 1/502 | ·001992032 |

| Fraction or Numb. | Decimal or Reciprocal. | Fraction or Numb. | Decimal or Reciprocal. | Fraction or Numb. | Decimal or Reciprocal. |
|---|---|---|---|---|---|
| 1/503 | ·001988072 | 1/564 | ·00177305 | 1/625 | ·0016 |
| 1/504 | ·001984127 | 1/565 | ·001769912 | 1/626 | ·001597444 |
| 1/505 | ·001980198 | 1/566 | ·001766784 | 1/627 | ·001594896 |
| 1/506 | ·001976285 | 1/567 | ·001763668 | 1/628 | ·001592357 |
| 1/507 | ·001972387 | 1/568 | ·001760563 | 1/629 | ·001589825 |
| 1/508 | ·001968504 | 1/569 | ·001757469 | 1/630 | ·001587302 |
| 1/509 | ·001964637 | 1/570 | ·001754386 | 1/631 | ·001584786 |
| 1/510 | ·001960784 | 1/571 | ·001751313 | 1/632 | ·001582278 |
| 1/511 | ·001956947 | 1/572 | ·001748252 | 1/633 | ·001579779 |
| 1/512 | ·001953125 | 1/573 | ·001745201 | 1/634 | ·001577287 |
| 1/513 | ·001949318 | 1/574 | ·00174216 | 1/635 | ·001574803 |
| 1/514 | ·001945525 | 1/575 | ·00173913 | 1/636 | ·001572327 |
| 1/515 | ·001941748 | 1/576 | ·001736111 | 1/637 | ·001569859 |
| 1/516 | ·001937984 | 1/577 | ·001733102 | 1/638 | ·001567398 |
| 1/517 | ·001934236 | 1/578 | ·001730104 | 1/639 | ·001564945 |
| 1/518 | ·001930502 | 1/579 | ·001727116 | 1/640 | ·0015625 |
| 1/519 | ·001926782 | 1/580 | ·001724138 | 1/641 | ·001560062 |
| 1/520 | ·001923077 | 1/581 | ·00172117 | 1/642 | ·001557632 |
| 1/521 | ·001919386 | 1/582 | ·001718213 | 1/643 | ·00155521 |
| 1/522 | ·001915709 | 1/583 | ·001715266 | 1/644 | ·001552795 |
| 1/523 | ·001912046 | 1/584 | ·001712329 | 1/645 | ·001550388 |
| 1/524 | ·001908397 | 1/585 | ·001709402 | 1/646 | ·001547988 |
| 1/525 | ·001904762 | 1/586 | ·001706485 | 1/647 | ·001545595 |
| 1/526 | ·001901141 | 1/587 | ·001703578 | 1/648 | ·00154321 |
| 1/527 | ·001897533 | 1/588 | ·00170068 | 1/649 | ·001540832 |
| 1/528 | ·001893939 | 1/589 | ·001697793 | 1/650 | ·001538462 |
| 1/529 | ·001890359 | 1/590 | ·001694915 | 1/651 | ·001536098 |
| 1/530 | ·001886792 | 1/591 | ·001692047 | 1/652 | ·001533742 |
| 1/531 | ·001883239 | 1/592 | ·001689189 | 1/653 | ·001531394 |
| 1/532 | ·001879699 | 1/593 | ·001686341 | 1/654 | ·001529052 |
| 1/533 | ·001876173 | 1/594 | ·001683502 | 1/655 | ·001526718 |
| 1/534 | ·001872659 | 1/595 | ·001680672 | 1/656 | ·00152439 |
| 1/535 | ·001869159 | 1/596 | ·001677852 | 1/657 | ·00152207 |
| 1/536 | ·001865672 | 1/597 | ·001675042 | 1/658 | ·001519751 |
| 1/537 | ·001862197 | 1/598 | ·001672241 | 1/659 | ·001517451 |
| 1/538 | ·001858736 | 1/599 | ·001669449 | 1/660 | ·001515152 |
| 1/539 | ·001855288 | 1/600 | ·001666667 | 1/661 | ·001512859 |
| 1/540 | ·001851852 | 1/601 | ·001663894 | 1/662 | ·001510574 |
| 1/541 | ·001848429 | 1/602 | ·00166113 | 1/663 | ·001508296 |
| 1/542 | ·001845018 | 1/603 | ·001658375 | 1/664 | ·001506024 |
| 1/543 | ·001841621 | 1/604 | ·001655629 | 1/665 | ·001503759 |
| 1/544 | ·001838235 | 1/605 | ·001652893 | 1/666 | ·001501502 |
| 1/545 | ·001834862 | 1/606 | ·001650165 | 1/667 | ·00149925 |
| 1/546 | ·001831502 | 1/607 | ·001647446 | 1/668 | ·001497006 |
| 1/547 | ·001828154 | 1/608 | ·001644737 | 1/669 | ·001494768 |
| 1/548 | ·001824818 | 1/609 | ·001642036 | 1/670 | ·001492537 |
| 1/549 | ·001821494 | 1/610 | ·001639344 | 1/671 | ·001490313 |
| 1/550 | ·001818182 | 1/611 | ·001636661 | 1/672 | ·001488095 |
| 1/551 | ·001814882 | 1/612 | ·001633987 | 1/673 | ·001485884 |
| 1/552 | ·001811594 | 1/613 | ·001631321 | 1/674 | ·00148368 |
| 1/553 | ·001808318 | 1/614 | ·001628664 | 1/675 | ·001481481 |
| 1/554 | ·001805054 | 1/615 | ·001626016 | 1/676 | ·00147929 |
| 1/555 | ·001801802 | 1/616 | ·001623377 | 1/677 | ·001477105 |
| 1/556 | ·001798561 | 1/617 | ·001620746 | 1/678 | ·001474926 |
| 1/557 | ·001795332 | 1/618 | ·001618123 | 1/679 | ·001472754 |
| 1/558 | ·001792115 | 1/619 | ·001615509 | 1/680 | ·001470588 |
| 1/559 | ·001788909 | 1/620 | ·001612903 | 1/681 | ·001468429 |
| 1/560 | ·001785714 | 1/621 | ·001610306 | 1/682 | ·001466276 |
| 1/561 | ·001782531 | 1/622 | ·001607717 | 1/683 | ·001464129 |
| 1/562 | ·001779359 | 1/623 | ·001605136 | 1/684 | ·001461988 |
| 1/563 | ·001776199 | 1/624 | ·001602564 | 1/685 | ·001459854 |

| Fraction or Numb. | Decimal or Reciprocal. | Fraction or Numb. | Decimal or Reciprocal. | Fraction or Numb. | Decimal or Reciprocal. |
|---|---|---|---|---|---|
| 1/686 | ·001457726 | 1/747 | ·001338688 | 1/808 | ·001237624 |
| 1/687 | ·001455604 | 1/748 | ·001336898 | 1/809 | ·001236094 |
| 1/688 | ·001453488 | 1/749 | ·001335113 | 1/810 | ·001234568 |
| 1/689 | ·001451379 | 1/750 | ·001333333 | 1/811 | ·001233046 |
| 1/690 | ·001449275 | 1/751 | ·001331558 | 1/812 | ·001231527 |
| 1/691 | ·001447178 | 1/752 | ·001329787 | 1/813 | ·001230012 |
| 1/692 | ·001445087 | 1/753 | ·001328021 | 1/814 | ·001228501 |
| 1/693 | ·001443001 | 1/754 | ·00132626 | 1/815 | ·001226994 |
| 1/694 | ·001440922 | 1/755 | ·001324503 | 1/816 | ·001225499 |
| 1/695 | ·001438849 | 1/756 | ·001322751 | 1/817 | ·00122399 |
| 1/696 | ·001436782 | 1/757 | ·001321004 | 1/818 | ·001222494 |
| 1/697 | ·00143472 | 1/758 | ·001319261 | 1/819 | ·001221001 |
| 1/698 | ·001432665 | 1/759 | ·001317523 | 1/820 | ·001219512 |
| 1/699 | ·001430615 | 1/760 | ·001315789 | 1/821 | ·001218027 |
| 1/700 | ·001428571 | 1/761 | ·00131406 | 1/822 | ·001216545 |
| 1/701 | ·001426534 | 1/762 | ·001312336 | 1/823 | ·001215067 |
| 1/702 | ·001424501 | 1/763 | ·001310616 | 1/824 | ·001213592 |
| 1/703 | ·001422475 | 1/764 | ·001308901 | 1/825 | ·001212121 |
| 1/704 | ·001420455 | 1/765 | ·00130719 | 1/826 | ·001210654 |
| 1/705 | ·00141844 | 1/766 | ·001305483 | 1/827 | ·00120919 |
| 1/706 | ·001416431 | 1/767 | ·001303781 | 1/828 | ·001207729 |
| 1/707 | ·001414427 | 1/768 | ·001302083 | 1/829 | ·001206273 |
| 1/708 | ·001412429 | 1/769 | ·00130039 | 1/830 | ·001204819 |
| 1/709 | ·001410437 | 1/770 | ·001298701 | 1/831 | ·001203369 |
| 1/710 | ·001408451 | 1/771 | ·001297017 | 1/832 | ·001201923 |
| 1/711 | ·00140647 | 1/772 | ·001295337 | 1/833 | ·00120048 |
| 1/712 | ·001404494 | 1/773 | ·001293661 | 1/834 | ·001199041 |
| 1/713 | ·001402525 | 1/774 | ·00129199 | 1/835 | ·001197605 |
| 1/714 | ·00140056 | 1/775 | ·001290323 | 1/836 | ·001196172 |
| 1/715 | ·001398601 | 1/776 | ·00128866 | 1/837 | ·001194743 |
| 1/716 | ·001396648 | 1/777 | ·001287001 | 1/838 | ·001193317 |
| 1/717 | ·0013947 | 1/778 | ·001285347 | 1/839 | ·001191895 |
| 1/718 | ·001392758 | 1/779 | ·001283697 | 1/840 | ·001190476 |
| 1/719 | ·001390821 | 1/780 | ·001282051 | 1/841 | ·001189061 |
| 1/720 | ·001388889 | 1/781 | ·00128041 | 1/842 | ·001187648 |
| 1/721 | ·001386963 | 1/782 | ·001278772 | 1/843 | ·00118624 |
| 1/722 | ·001385042 | 1/783 | ·001277139 | 1/844 | ·001184834 |
| 1/723 | ·001383126 | 1/784 | ·00127551 | 1/845 | ·001183432 |
| 1/724 | ·001381215 | 1/785 | ·001273885 | 1/846 | ·001182033 |
| 1/725 | ·00137931 | 1/786 | ·001272265 | 1/847 | ·001180638 |
| 1/726 | ·00137741 | 1/787 | ·001270648 | 1/848 | ·001179245 |
| 1/727 | ·001375516 | 1/788 | ·001269036 | 1/849 | ·001177856 |
| 1/728 | ·001373626 | 1/789 | ·001267427 | 1/850 | ·001176471 |
| 1/729 | ·001371742 | 1/790 | ·001265823 | 1/851 | ·001175088 |
| 1/730 | ·001369863 | 1/791 | ·001264223 | 1/952 | ·001173709 |
| 1/731 | ·001367989 | 1/792 | ·001262626 | 1/853 | ·001172333 |
| 1/732 | ·00136612 | 1/793 | ·001261034 | 1/854 | ·00117096 |
| 1/733 | ·001364256 | 1/794 | ·001259446 | 1/855 | ·001169591 |
| 1/734 | ·001362398 | 1/795 | ·001257862 | 1/856 | ·001168224 |
| 1/735 | ·001360544 | 1/796 | ·001256281 | 1/857 | ·001166861 |
| 1/736 | ·001358696 | 1/797 | ·001254705 | 1/858 | ·001165501 |
| 1/737 | ·001356852 | 1/798 | ·001253133 | 1/859 | ·001164144 |
| 1/738 | ·001355014 | 1/799 | ·001251364 | 1/860 | ·001162791 |
| 1/739 | ·00135318 | 1/800 | ·00125 | 1/861 | ·00116144 |
| 1/740 | ·001351351 | 1/801 | ·001248439 | 1/862 | ·001160093 |
| 1/741 | ·001349528 | 1/802 | ·001246883 | 1/863 | ·001158749 |
| 1/742 | ·001347709 | 1/803 | ·00124533 | 1/864 | ·001157407 |
| 1/743 | ·001345895 | 1/804 | ·001243781 | 1/865 | ·001156069 |
| 1/744 | ·001344086 | 1/805 | ·001242236 | 1/866 | ·001154734 |
| 1/745 | ·001342282 | 1/806 | ·001240695 | 1/867 | ·001153403 |
| 1/746 | ·001340483 | 1/807 | ·001239157 | 1/868 | ·001152074 |

| Fraction or Numb. | Decimal or Reciprocal. | Fraction or Numb. | Decimal or Reciprocal. | Fraction or Numb. | Decimal or Reciprocal. |
|---|---|---|---|---|---|
| 1/869 | ·001150748 | 1/913 | ·00109529 | 1/957 | ·001044932 |
| 1/870 | ·001149425 | 1/914 | ·001094092 | 1/958 | ·001043841 |
| 1/871 | ·001148106 | 1/915 | ·001092896 | 1/959 | ·001042753 |
| 1/872 | ·001146789 | 1/916 | ·001091703 | 1/960 | ·001041667 |
| 1/873 | ·001145475 | 1/917 | ·001090513 | 1/961 | ·001040583 |
| 1/874 | ·001144165 | 1/918 | ·001089325 | 1/962 | ·001039501 |
| 1/875 | ·001142857 | 1/919 | ·001088139 | 1/963 | ·001038422 |
| 1/876 | ·001141553 | 1/920 | ·001086957 | 1/964 | ·001037344 |
| 1/877 | ·001140251 | 1/921 | ·001085776 | 1/965 | ·001036269 |
| 1/878 | ·001138952 | 1/922 | ·001084599 | 1/966 | ·001035197 |
| 1/879 | ·001137656 | 1/923 | ·001083423 | 1/967 | ·001034126 |
| 1/880 | ·001136364 | 1/924 | ·001082251 | 1/968 | ·001033058 |
| 1/881 | ·001135074 | 1/925 | ·001081081 | 1/969 | ·001031992 |
| 1/882 | ·001133787 | 1/926 | ·001079914 | 1/970 | ·001030928 |
| 1/883 | ·001132503 | 1/927 | ·001078749 | 1/971 | ·001029866 |
| 1/884 | ·001131222 | 1/928 | ·001077586 | 1/972 | ·001028807 |
| 1/885 | ·001129944 | 1/929 | ·001076426 | 1/973 | ·001027749 |
| 1/886 | ·001128668 | 1/930 | ·001075269 | 1/974 | ·001026694 |
| 1/887 | ·001127396 | 1/931 | ·001074114 | 1/975 | ·001025641 |
| 1/888 | ·001126126 | 1/932 | ·001072961 | 1/976 | ·00102459 |
| 1/889 | ·001124859 | 1/933 | ·001071811 | 1/977 | ·001023541 |
| 1/890 | ·001123596 | 1/934 | ·001070664 | 1/978 | ·001022495 |
| 1/891 | ·001122334 | 1/935 | ·001069519 | 1/979 | ·00102145 |
| 1/892 | ·001121076 | 1/936 | ·001068376 | 1/980 | ·001020408 |
| 1/893 | ·001119821 | 1/937 | ·001067236 | 1/981 | ·001019168 |
| 1/894 | ·001118568 | 1/938 | ·001066098 | 1/982 | ·00101833 |
| 1/895 | ·001117818 | 1/939 | ·001064963 | 1/983 | ·001017294 |
| 1/896 | ·001116071 | 1/940 | ·00106383 | 1/984 | ·00101626 |
| 1/897 | ·001114827 | 1/941 | ·001062699 | 1/985 | ·001015228 |
| 1/898 | ·001113586 | 1/942 | ·001061571 | 1/986 | ·001014199 |
| 1/899 | ·001112347 | 1/943 | ·001060445 | 1/987 | ·001013171 |
| 1/900 | ·001111111 | 1/944 | ·001059322 | 1/988 | ·001012146 |
| 1/901 | ·001109878 | 1/945 | ·001058201 | 1/989 | ·001011122 |
| 1/902 | ·001108647 | 1/946 | ·001057082 | 1/990 | ·001010101 |
| 1/903 | ·00110742 | 1/947 | ·001055966 | 1/991 | ·001009082 |
| 1/904 | ·001106195 | 1/948 | ·001054852 | 1/992 | ·001008065 |
| 1/905 | ·001104972 | 1/949 | ·001053741 | 1/993 | ·001007049 |
| 1/906 | ·001103753 | 1/950 | ·001052632 | 1/994 | ·001006036 |
| 1/907 | ·001102536 | 1/951 | ·001051525 | 1/995 | ·001005025 |
| 1/908 | ·001101322 | 1/952 | ·00105042 | 1/996 | ·001004016 |
| 1/909 | ·00110011 | 1/953 | ·001049318 | 1/997 | ·001003009 |
| 1/910 | ·001098901 | 1/954 | ·001048218 | 1/998 | ·001002004 |
| 1/911 | ·001091695 | 1/955 | ·00104712 | 1/999 | ·001001001 |
| 1/912 | ·001096491 | 1/956 | ·001046025 | 1/1000 | ·001 |

Divide 80000 by 971.

By the above Table we find that 1 divided by 971 gives ·001029866, and ·001029866 × 80000 = 82·38928.

What is the sum of $\frac{5}{883}$ and $\frac{2}{953}$?

$$5 \times \frac{1}{883} = \cdot 001132503 \times 5 = \cdot 005662515$$

$$2 \times \frac{1}{953} = \cdot 001049318 \times 2 = \cdot 002098636$$

$$\therefore \frac{5}{883} + \frac{2}{953} = \cdot 007761141$$

## WEIGHTS AND VALUES IN DECIMAL PARTS.

**TROY WEIGHT.** — Dec. parts of a lb.

| Ozs. | Decimals. |
|---|---|
| 11 | ·916666 |
| 10 | ·833333 |
| 9 | ·75 |
| 8 | ·666666 |
| 7 | ·583333 |
| 6 | ·5 |
| 5 | ·416666 |
| 4 | ·333333 |
| 3 | ·25 |
| 2 | ·166666 |
| 1 | ·083333 |
| **Dwts.** | **Decimals.** |
| 19 | ·079166 |
| 18 | ·075 |
| 17 | ·070833 |
| 16 | ·066666 |
| 15 | ·0625 |
| 14 | ·058333 |
| 13 | ·054166 |
| 12 | ·05 |
| 11 | ·045833 |
| 10 | ·041666 |
| 9 | ·0375 |
| 8 | ·033333 |
| 7 | ·029166 |
| 6 | ·025 |
| 5 | ·020833 |
| 4 | ·016666 |
| 3 | ·0125 |
| 2 | ·008333 |
| 1 | ·004166 |
| **Grs.** | **Decimals.** |
| 15 | ·002604 |
| 14 | ·002430 |
| 13 | ·002257 |
| 12 | ·002083 |
| 11 | ·001910 |
| 10 | ·001736 |
| 9 | ·001562 |
| 8 | ·001389 |
| 7 | ·001215 |
| 6 | ·001042 |
| 5 | ·000868 |
| 4 | ·000694 |
| 3 | ·000521 |
| 2 | ·000347 |
| 1 | ·000173 |

**AVOIRDUPOIS WEIGHT.** — Dec. parts of a cwt.

| Qrs. | Decimals. |
|---|---|
| 3 | ·75 |
| 2 | ·5 |
| 1 | ·25 |
| **lbs.** | **Decimals.** |
| 27 | ·241071 |
| 26 | ·232142 |
| 25 | ·223214 |
| 24 | ·214286 |
| 23 | ·205357 |
| 22 | ·196428 |
| 21 | ·187500 |
| 20 | ·178572 |
| 19 | ·169643 |
| 18 | ·160714 |
| 17 | ·151785 |
| 16 | ·142856 |
| 15 | ·133928 |
| 14 | ·125 |
| 13 | ·116071 |
| 12 | ·107143 |
| 11 | ·098214 |
| 10 | ·089286 |
| 9 | ·080357 |
| 8 | ·071428 |
| 7 | ·0625 |
| 6 | ·053571 |
| 5 | ·044643 |
| 4 | ·035714 |
| 3 | ·026786 |
| 2 | ·017857 |
| 1 | ·008928 |
| **Ozs.** | **Decimals.** |
| 15 | ·008370 |
| 14 | ·007812 |
| 13 | ·007254 |
| 12 | ·006696 |
| 11 | ·006138 |
| 10 | ·005580 |
| 9 | ·005022 |
| 8 | ·004464 |
| 7 | ·003906 |
| 6 | ·003348 |
| 5 | ·002790 |
| 4 | ·002232 |
| 3 | ·001674 |
| 2 | ·001116 |
| 1 | ·000558 |

**AVOIRDUPOIS WEIGHT.** — Dec. parts of a lb.

| Ozs. | Decimals. |
|---|---|
| 15 | ·9375 |
| 14 | ·875 |
| 13 | ·8125 |
| 12 | ·75 |
| 11 | ·6875 |
| 10 | ·625 |
| 9 | ·5625 |
| 8 | ·5 |
| 7 | ·4375 |
| 6 | ·375 |
| 5 | ·3125 |
| 4 | ·25 |
| 3 | ·1875 |
| 2 | ·125 |
| 1 | ·0625 |
| **Drs.** | **Decimals.** |
| 15 | ·058593 |
| 14 | ·054686 |
| 13 | ·050780 |
| 12 | ·046874 |
| 11 | ·042968 |
| 10 | ·039062 |
| 9 | ·035156 |
| 8 | ·03125 |
| 7 | ·027343 |
| 6 | ·023437 |
| 5 | ·019531 |
| 4 | ·015625 |
| 3 | ·011718 |
| 2 | ·007812 |
| 1 | ·003906 |

**LONG MEASURE.** — Dec. parts of a foot.

| Ins. | Decimals. |
|---|---|
| 11 | ·916666 |
| 10 | ·833333 |
| 9 | ·75 |
| 8 | ·666666 |
| 7 | ·583333 |
| 6 | ·5 |
| 5 | ·416666 |
| 4 | ·333333 |
| 3 | ·25 |
| 2 | ·166666 |
| 1 | ·083333 |

*To find the solidity of a cube, the height of one of its sides being given.*—Multiply the side of the cube by itself, and that product again by the side, and it will give the solidity required.

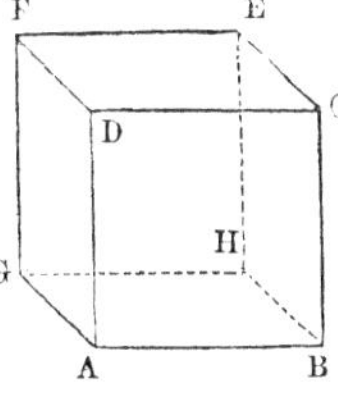


The side AB, or BC, of the cube ABCDFGHE, is 25·5: what is the solidity?

*Here* $AB^3 = (22·5)|^3 = 25·5 \times 25·5 \times 25·5 = 25·5 \times 650·25 = 16581·375$, *content of the cube.*

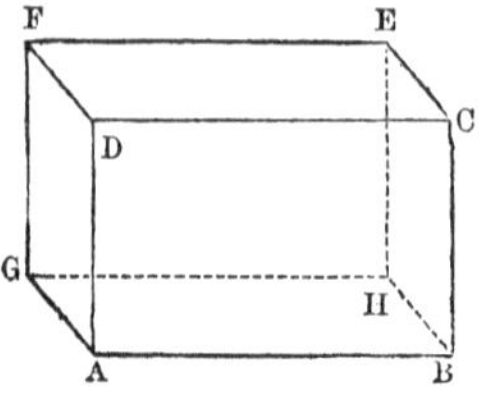


*To find the solidity of a parallelopipedon.*—Multiply the length by the breadth, and that product again by the depth or altitude, and it will give the solidity required.

Required the solidity of a parallelopipedon ABCDFEHG, whose length AB is 8 feet, its breadth FD $4\frac{1}{2}$ feet, and the depth or altitude AD $6\frac{3}{4}$ feet?

*Here* AB $\times$ AD $\times$ FD $= 8 \times 6{\cdot}75 \times 4.5 = 54 \times 4{\cdot}5 = 243$ *solid feet, the contents of the parallelopipedon.*

*To find the solidity of a prism.*—Multiply the area of the base into the perpendicular height of the prism, and the product will be the solidity.

What is the solidity of the triangular prism ABCFED, whose length AB is 10 feet, and either of the equal sides, BC, CD, or DB, of one of its equilateral ends BCD, $2\frac{1}{2}$ feet?

*Here* $\frac{1}{4} \times 2{\cdot}5^2 \times \sqrt{3} = \frac{1}{4} \times 6{\cdot}25 \times \sqrt{3} = 1{\cdot}5625 \times \sqrt{3} = 1{\cdot}5625 \times 1{\cdot}732 = 2{\cdot}70625 =$ *area of the base* BCD.

*Or,* $\dfrac{2{\cdot}5 + 2{\cdot}5 + 2{\cdot}5}{2} = \dfrac{7{\cdot}5}{2} = 3{\cdot}75 = \frac{1}{2}$ *sum of the sides,* BC, CD, DB, *of the triangle* CDB.

*And* $3{\cdot}75 - 2{\cdot}5 = 1{\cdot}25$, $\therefore$ $1{\cdot}25$, $1{\cdot}25$ *and* $1{\cdot}25 = 3$ *differences.*

*Whence* $\sqrt{3{\cdot}75 \times 1{\cdot}25 \times 1{\cdot}25 \times 1{\cdot}25} = \sqrt{3{\cdot}75 \times 1{\cdot}25^3} = \sqrt{7{\cdot}32421875} = 2{\cdot}7063 =$ *area of the base as before,*

*And* $2{\cdot}7063 \times 10 = 27{\cdot}063$ *solid feet, the content of the prism required.*

*To find the convex surface of a cylinder.*—Multiply the periphery or circumference of the base, by the height of the cylinder, and the product will be the convex surface.

What is the convex surface of the right cylinder ABCD, whose length BC is 20 feet, and the diameter of its base AB 2 feet?

*Here* $3{\cdot}1416 \times 2 = 6{\cdot}2832 =$ *periphery of the base* AB.

*And* $6{\cdot}2832 \times 20 = 125{\cdot}6640$ *square feet, the convexity required.*

*To find the solidity of a cylinder.*—Multiply the area of the base by the perpendicular height of the cylinder, and the product will be the solidity.

What is the solidity of the cylinder ABCD, the diameter of whose base AB is 30 inches, and the height BC 50 inches.

*Here* $\cdot7854 \times 30^2 = \cdot7854 \times 900 = 706{\cdot}86 =$ *area of the base* AB.

*And* $706{\cdot}86 \times 50 = 35343$ *cubic inches; or* $\dfrac{35343}{1728} = 20{\cdot}4531$ *solid feet.*

The four following cases contain all the rules for finding the superficies and solidities of *cylindrical ungulas.*

*When the section is parallel to the axis of the cylinder.*

RULE.—Multiply the length of the arc line of the base by the height of the cylinder, and the product will be the *curve surface.*

Multiply the area of the base by the height of the cylinder, and the product will be the *solidity.*

*When the section passes obliquely through the opposite sides of the cylinder.*

RULE.—Multiply the circumference of the base of the cylinder by half the sum of the greatest and least lengths of the ungula, and the product will be the *curve surface.*

Multiply the area of the base of the cylinder by half the sum of the greatest and least lengths of the ungula, and the product will be the *solidity.*

*When the section passes through the base of the cylinder, and one of its sides.*

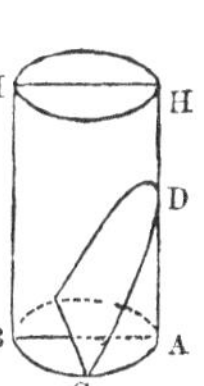


RULE.—Multiply the sine of half the arc of the base by the diameter of the cylinder, and from this product subtract the product of the arc and cosine.

Multiply the difference thus found, by the quotient of the height divided by the versed sine, and the product will be the *curve surface.*

From $\frac{2}{3}$ of the cube of the right sine of half the arc of the base, subtract the product of the area of the base and the cosine of the said half arc.

Multiply the difference, thus found, by the quotient arising from the height divided by the versed sine, and the product will be the *solidity.*

*When the section passes obliquely through both ends of the cylinder.*

RULE.—Conceive the section to be continued, till it meets the side of the cylinder produced; then say, as the difference of the versed sines of half the arcs of the two ends of the ungula is to the versed sine of half the arc of the less end, so is the height of the cylinder to the part of the side produced.

Find the surface of each of the ungulas, thus formed, and their difference will be the *surface.*

In like manner find the solidities of each of the ungulas, and their difference will be the *solidity.*

*To find the convex surface of a right cone.*—Multiply the circumference of the base by the slant height, or the length of the side of the cone, and half the product will be the surface required.

The diameter of the base AB is 3 feet, and the slant height AC or BC 15 feet; required the convex surface of the cone ACB.

*Here* $3{\cdot}1416 \times 3 = 9{\cdot}4248$ = *circumference of the base* AB.

*And* $\frac{9{\cdot}4248 \times 15}{2} = \frac{141{\cdot}3720}{2} = 70{\cdot}686$ *square feet, the convex surface required.*

*To find the convex surface of the frustum of a right cone.*—Multiply the sum of the perimeters of the two ends, by the slant height of the frustum, and half the product will be the surface required.

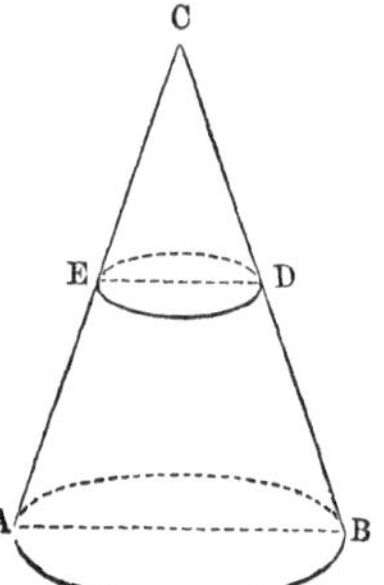


In the frustum ABDE, the circumferences of the two ends AB and DE are 22·5 and 15·75 respectively, and the slant height BD is 26; what is the convex surface?

*Here* $\frac{(22{\cdot}5 + 15{\cdot}75) \times 26}{2} = \overline{22{\cdot}5 + 15{\cdot}75} \times 13 = 38{\cdot}25 \times 13 = 497{\cdot}25$ = *convex surface.*

*To find the solidity of a cone or pyramid.*—Multiply the area of the base by one-third of the perpendicular height of the cone or pyramid, and the product will be the solidity.

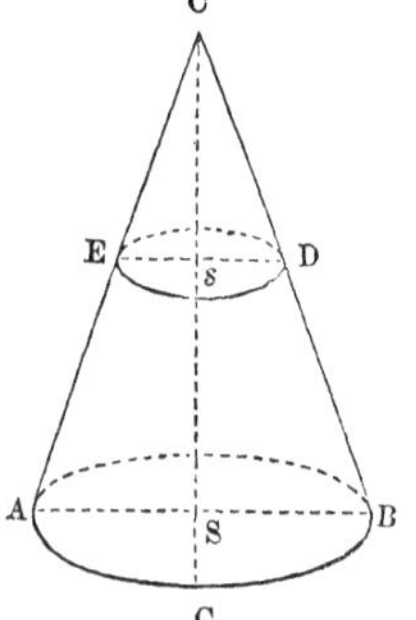


Required the solidity of the cone ACB, whose diameter AB is 20, and its perpendicular height CS 24.

*Here* $\cdot 7854 \times 20^2 = \cdot 7854 \times 400 = 314{\cdot}16$ = *area of the base* AB.

*And* $314{\cdot}16 \times \frac{24}{3} = 314{\cdot}16 \times 8 = 2513{\cdot}28$ = *solidity required.*

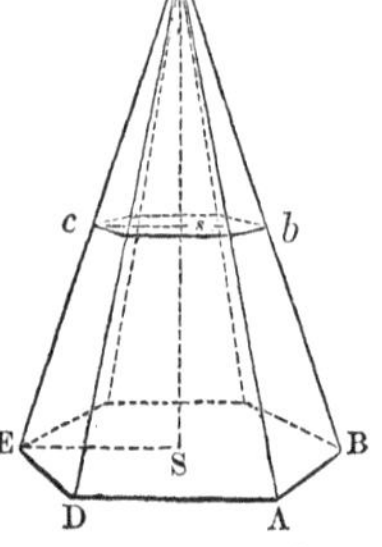


Required the solidity of the hexagonal pyramid ECBD, each of the equal sides of its base being 40, and the perpendicular height CS 60.

*Here* 2·598076 (*multiplier when the side is* 1) $\times 40^2 = 2{\cdot}598076 \times 1600 = 4156{\cdot}9216$ = *area of the base.*

*And* $4156{\cdot}9216 \times \frac{60}{3} = 4156{\cdot}9216 \times 20 = 83138{\cdot}432$ *solidity.*

*To find the solidity of a frustum of a cone or pyramid.*—For the frustum of a cone, the diameters or circumferences of the two ends, and the height being given.

Add together the square of the diameter of the greater end, the square of the diameter of the less end, and the product of the two diameters; multiply the sum by ·7854, and the product by the height; $\frac{1}{3}$ of the last product will be the solidity. Or,

Add together the square of the circumference of the greater end, the square of the circumference of the less end, and the product of the two circumferences; multiply the sum by ·07958, and the product by the height; $\frac{1}{3}$ of the last product will be the solidity.

*For the frustum of a pyramid whose sides are regular polygons.*—Add together the square of a side of the greater end, the square of a side of the less end, and the product of these two sides; multiply the sum by the proper number in the Table of Superficies, and the product by the height; $\frac{1}{3}$ of the last product will be the solidity.

*When the ends of the pyramids are not regular polygons.*—Add together the areas of the two ends and the square root of their product; multiply the sum by the height, and $\frac{1}{3}$ of the product will be the solidity.

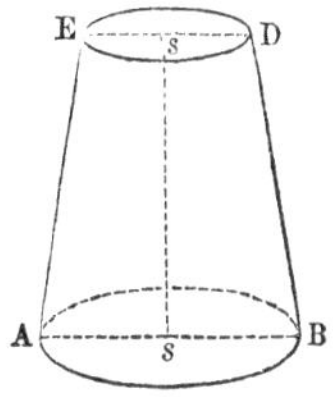

What is the solidity of the frustum of the cone EABD, the diameter of whose greater end AB is 5 feet, that of the less end ED, 3 feet, and the perpendicular height S*s*, 9 feet?

$$\frac{(5^2 + 3^2 + \overline{5 \times 3}) \times \cdot7854 \times 9}{3} = \frac{346\cdot3614}{3} =$$

115·4538 *solid feet, the content of the frustum.*

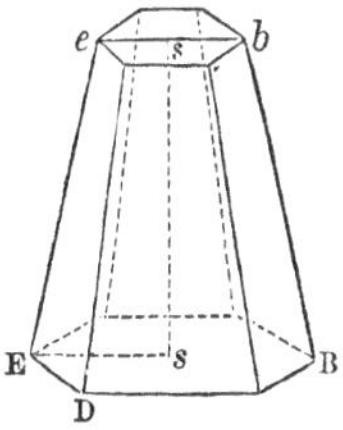

What is the solidity of the frustum *e*EDB*b* of a hexagonal pyramid, the side ED of whose greater end is 4 feet, that *eb* of the less end 3 feet, and the height S*s*, 9 feet?

$$\frac{(4^2 + 3^2 + \overline{4 \times 3}) \times 2\cdot598076 \times 9}{3} = \frac{865\cdot159308}{3}$$

$= 288\cdot386436$ *solid feet, the solidity required.*

The following cases contain all the rules for finding the superficies and solidities of conical ungulas.

*When the section passes through the opposite extremities of the ends of the frustum.*

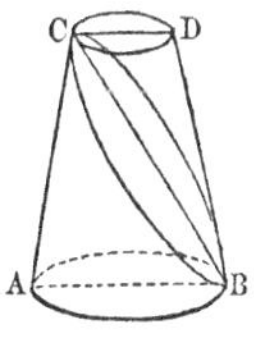

Let D = AB the diameter of the greater end; $d$ = CD, the diameter of the less end; $h$ = perpendicular height of the frustum, and $n$ = ·7854.

Then $\dfrac{d^2 - d\sqrt{\mathrm{D}d}}{\mathrm{D} - d} \times \dfrac{n\mathrm{D}h}{3}$ = solidity of the greater elliptic ungula ADB.

$\dfrac{\mathrm{D}\sqrt{\mathrm{D}d} - d^2}{\mathrm{D} - d} \times \dfrac{ndh}{3}$ = solidity of the less ungula ACD.

$\dfrac{(\mathrm{D}^{\frac{3}{2}} - d^{\frac{3}{2}})^2}{\mathrm{D} - d} \times \dfrac{nh}{3}$ = difference of these hoofs.

And $\dfrac{n}{\mathrm{D} - d}\sqrt{4h^2 + (\mathrm{D} - d^2)} \times (\mathrm{D}^2 - \dfrac{\mathrm{D} + d}{2}\sqrt{\mathrm{D}d}$ = curve surface of ADB.

*When the section cuts off parts of the base, and makes the angle* D$r$B *less than the angle* CAB.

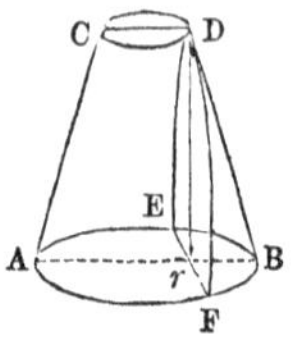

Let S = tabular segment, whose versed sine is B$r$÷D; $s$ — tab. seg. whose versed sine is $\overline{\text{B}r - (\text{D} - d)} \div d$, and the other letters as above.

The $(\text{S} \times \text{D}^3 - s \times d^3 \times \frac{\text{B}r}{\text{B}r - \overline{\text{D} - d}} \sqrt{\frac{\text{B}r}{\text{B}r - \overline{\text{D} - d}}} \times \frac{\frac{1}{2}h}{\text{D} - d}$ = solidity of the elliptic hoof EFBD.

And $\frac{1}{\text{D} - d} \sqrt{4h^2 + (\text{D} - d)^2} \times (\text{seg. FBE} - \frac{d^2}{\text{D}^2} \times \frac{\frac{1}{2} \times (\text{D} + d) - \text{A}r}{d - \text{A}r} \times \sqrt{\frac{\text{B}r}{d - \text{A}r}} \times$ seg. of the circle AB, whose height is $\text{D} \times \frac{d - \text{A}r}{d})$ = convex surface of EFBD.

*When the section is parallel to one of the sides of the frustum.*

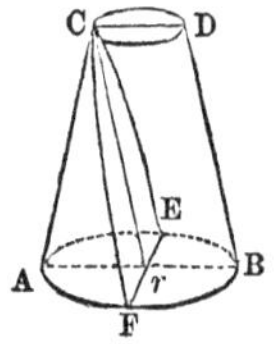

Let A = area of the base FBE, and the other letters as before.

Then $(\frac{\text{A} \times \text{D}}{\text{D} - d} - \frac{4}{3}d \sqrt{(\text{B} - d) \times d}) \times \frac{1}{3}h$ = solidity of the parabolic hoof EFBD.

And $\frac{1}{\text{D} - d} \sqrt{4h^2 \times (\text{D} - d)^2} \times (\text{seg. FBE} - \frac{2}{3}\overline{\text{D} - d} \times \sqrt{d \times \overline{\text{D} - d}})$ = convex surface of EFBD.

*When the section cuts off part of the base, and makes the angle* D$r$B *greater than the angle* CAB.

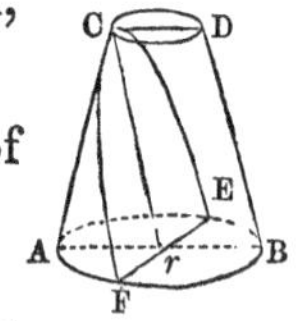

Let the area of the hyperbolic section EDF = A, and the area of the circular seg. EBF = $a$.

Then $\frac{\frac{1}{3}h}{\text{D} - h} \times (a \times \text{D} - \text{A} \times \frac{d \times \text{E}r}{\text{C}r})$ = solidity of the hyperbolic ungula EFBD.

And $\frac{1}{\text{D} - d} \times \sqrt{4h^2 + (\text{D} - d)^2} \times (\text{cir. seg. EBF} - \frac{d^2}{\text{D}^2} \times \frac{\text{B}r - \frac{1}{2}(\text{D} - d)}{\text{B}r - \overline{\text{D} - d}} \sqrt{\frac{\text{B}r}{\text{B}r - \overline{d - \text{D}}}}$ = curve surface of EFBD.

The transverse diameter of the hyp. seg. $= \frac{d \times \text{C}r}{\text{D} - d - \text{B}r}$ and the conjugate $= d \sqrt{\frac{\text{B}r}{\text{D} - d - \text{B}r}}$, from which its area may be found by the former rules.

*To find the solidity of a cuneus or wedge.*—Add twice the length of the base to the length of the edge, and reserve the number.

Multiply the height of the wedge by the breadth of the base, and this product by the reserved number; $\frac{1}{6}$ of the last product will be the solidity.

How many solid feet are there in a wedge, whose base is 5 feet 4 inches long, and 9 inches broad, the length of the edge being 3 feet 6 inches, and the perpendicular height 2 feet 4 inches?

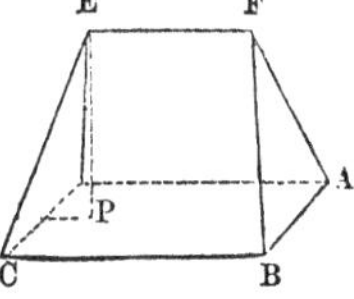


*Here* $\frac{(64 \times 2 + 42) \times 28 \times 9}{6} = \frac{(128 + 42) \times 28 \times 9}{6} = \frac{170 \times 28 \times 9}{6} = \frac{170 \times 28 \times 3}{2} = 170 \times 14 \times 3 = 7140$ *solid inches.*

*And* $7140 \div 1728 = 4{\cdot}1319$ *solid feet, the content.*

*To find the solidity of a prismoid.*—To the sum of the areas of the two ends add four times the area of a section parallel to and equally distant from both ends, and this last sum multiplied by $\frac{1}{6}$ of the height will give the solidity.

The length of the middle rectangle is equal to half the sum of the lengths of the rectangles of the two ends, and its breadth equal to half the sum of the breadths of those rectangles.

What is the solidity of a rectangle prismoid, the length and breadth of one end being 14 and 12 inches, and the corresponding sides of the other 6 and 4 inches, and the perpendicular $30\frac{1}{2}$ feet.

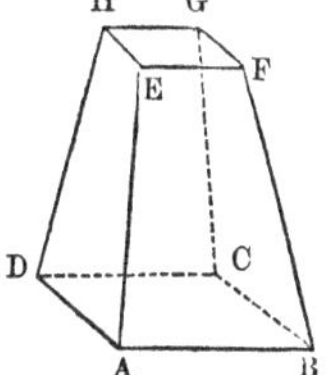


*Here* $14 \times 12 + \overline{6 \times 4} = 168 + 24 = 192 =$ *sum of the area of the two ends.*

*Also* $\frac{14 + 6}{2} = \frac{20}{2} = 10 =$ *length of the middle rectangle.*

*And* $\frac{12 + 4}{2} = \frac{16}{2} = 8 =$ *breadth of the middle rectangle.*

*Whence* $10 \times 8 \times 4 = 80 \times 4 = 320 = 4$ *times the area of the middle rectangle.*

*Or* $(320 + 192) \times \frac{366}{6} = 512 \times 61 = 31232$ *solid inches.*

*And* $31232 \div 1728 = 18{\cdot}074$ *solid feet, the content.*

*To find the convex surface of a sphere.*—Multiply the diameter of the sphere by its circumference, and the product will be the convex superficies required.

The curve surface of any zone or segment will also be found by multiplying its height by the whole circumference of the sphere.

What is the convex superficies of a globe BCG whose diameter BG is 17 inches?

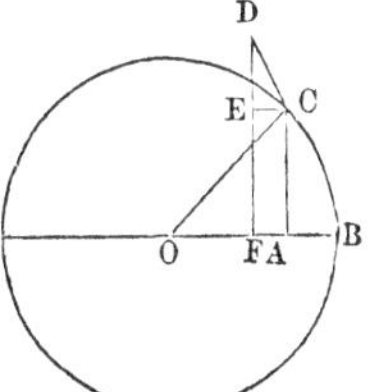


*Here* $3{\cdot}1416 \times 17 \times 17 = 53{\cdot}4072 \times 17 = 907{\cdot}9224$ *square inches.*

*And* $907{\cdot}9224 \div 144 = 6{\cdot}305$ *square feet.*

*To find the solidity of a sphere or globe.*—Multiply the cube of the diameter by ·5236, and the product will be the solidity.

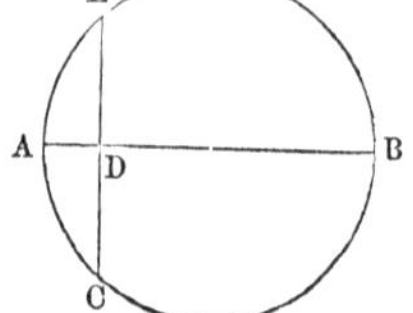


What is the solidity of the sphere AEBC, whose diameter AB is 17 inches?

*Here* $17^3 \times \cdot5236 = 17 \times 17 \times 17 \times \cdot5236 = 289 \times 17 \times 5236 = 4913 \times \cdot5236 = 2572\cdot4468$ *solid inches.*

*And* $2572\cdot4468 \div 1728 = 1\cdot48868$ *solid feet.*

*To find the solidity of the segment of a sphere.*—To three times the square of the radius of its base add the square of its height, and this sum multiplied by the height, and the product again by ·5236, will give the solidity. Or,

From three times the diameter of the sphere subtract twice the height of the segment, multiply by the square of the height, and that product by ·5236; the last product will be the solidity.

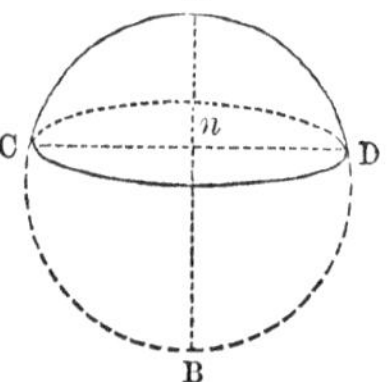


The radius C*n* of the base of the segment CAD is 7 inches, and the height A*n* 4 inches; what is the solidity?

*Here* $(7^2 \times 3 + 4^2) \times 4 \times \cdot5236 = (49 \times 3 + 4^2) \times 4 \times \cdot5236 = (147 + 4^2) \times 4 \times \cdot5236 = (147 + 16) \times 4 \times \cdot5236 = 163 \times 4 \times \cdot5236 = 652 \times \cdot5236 = 341\cdot3872$ *solid inches.*

*To find the solidity of a frustum or zone of a sphere.*—To the sum of the squares of the radii of the two ends, add one-third of the square of their distance, or of the breadth of the zone, and this sum multiplied by the said breadth, and the product again by 1·5708, will give the solidity.

What is the solid content of the zone ABCD, whose greater diameter AB is 20 inches, the less diameter CD 15 inches, and the distance *nm* of the two ends 10 inches?

*Here* $(10^2 + 7\cdot5^2 + \frac{10^2}{3}) \times 10 \times 1\cdot5708 = (100 + 56\cdot25 + 33\cdot33) \times 10 \times 1\cdot5708 = 189\cdot58 \times 10 \times 1\cdot5708 = 1895\cdot8 \times 1\cdot5708 = 2977\cdot92264$ *solid inches.*

*To find the solidity of a spheroid.*—Multiply the square of the revolving axe by the fixed axe, and this product again by ·5236, and it will give the solidity required.

$\cdot5236$ is $= \frac{1}{6}$ of $3\cdot1416$.

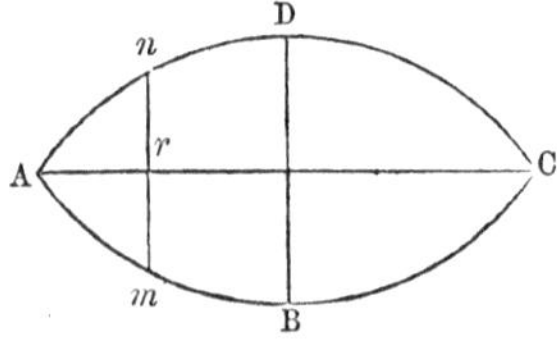


In the prolate spheroid ABCD, the transverse, or fixed axe AC is 90, and the conjugate or revolving axe DB is 70; what is the solidity?

*Here* $DB^2 \times AC \times \cdot5236 = 70^2 \times 90 \times \cdot5236 = 4900 \times 90 \times \cdot5236 = 441000 \times \cdot5236 = 230907\cdot6 =$ *solidity required.*

*To find the content of the middle frustum of a spheroid, its length, the middle diameter, and that of either of the ends, being given, when the ends are circular or parallel to the revolving axis.*—To twice the square of the middle diameter add the square of the diameter of either of the ends, and this sum multiplied by the length of the frustum, and the product again by ·2618, will give the solidity.

Where $\cdot2618 = \frac{1}{12}$ of 3·1416.

In the middle frustum of a spheroid EFGH, the middle diameter DB is 50 inches, and that of either of the ends EF or GH is 40 inches, and its length *nm* 18 inches; what is its solidity?

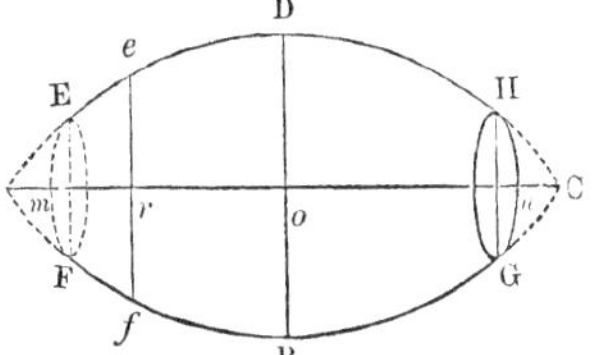


*Here* $(50^2 \times 2 + 40^2) \times 18 \times \cdot2618 = (2500 \times 2 + 1600) \times 18 \times \cdot2618 = (5000 + 1600) \times 18 \times \cdot2618 = 6600 \times 18 \times \cdot2618 = 118800 \times \cdot2613 = 31101\cdot84$ *cubic inches.*

*When the ends are elliptical or perpendicular to the revolving axis.*—Multiply twice the transverse diameter of the middle section by its conjugate diameter, and to this product add the product of the transverse and conjugate diameters of either of the ends.

Multiply the sum thus found by the distance of the ends or the height of the frustum, and the product again by ·2618, and it will give the solidity required.

In the middle frustum ABCD of an oblate spheroid, the diameters of the middle section EF are 50 and 30, those of the end AD 40 and 24, and its height *ne* 18; what is the solidity?

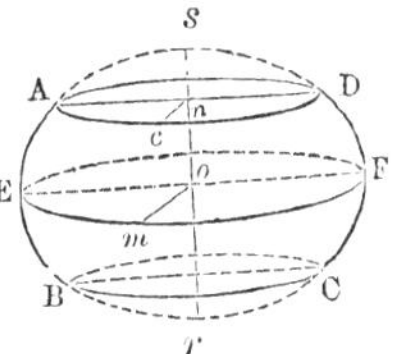


*Here* $(50 \times 2 \times 30 + \overline{40 \times 24}) \times 18 \times \cdot2618 = (3000 + 960) \times 18 \times \cdot2618 = 3960 \times 18 \times \cdot2618 = 71280 \times \cdot2618 = 18661\cdot104 =$ *the solidity.*

*To find the solidity of the segment of a spheroid, when the base is parallel to the revolving axis.*—Divide the square of the revolving axis by the square of the fixed axe, and multiply the quotient by the difference between three times the fixed axe and twice the height of the segment.

Multiply the product thus found by the square of the height of the segment, and this product again by ·5236, and it will give the solidity required.

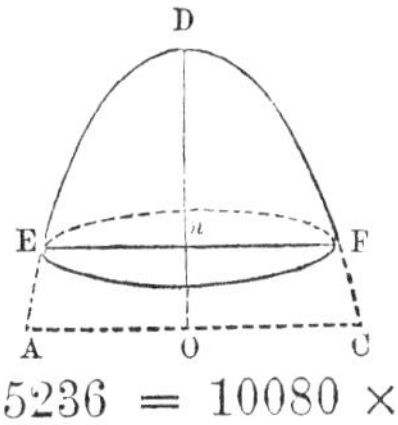

In the prolate spheroid DEFD, the transverse axis 2 DO is 100, the conjugate AC 60, and the height D*n* of the segment EDF 10; what is the solidity?

*Here* $(\frac{60^2}{100^2} \times \overline{300 - 20}) \times 10^2 \times \cdot5236 = \cdot36 \times 280 \times 10^2 \times \cdot5236 = 100\cdot80 \times 100 \times \cdot5236 = 10080 \times \cdot5236 = 5277\cdot888 =$ *the solidity.*

*When the base is perpendicular to the revolving axis.*—Divide the fixed axe by the revolving axe, and multiply the quotient by the difference between three times the revolving axe and twice the height of the segment.

Multiply the product thus found by the square of the height of the segment, and this product again by ·5236, and it will give the solidity required.

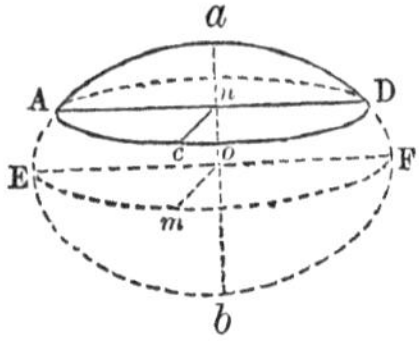


In the prolate spheroid $aEbF$, the transverse axe EF is 100, the conjugate $ab$ 60, and the height $an$ of the segment $a$AD 12; what is the solidity?

*Here* 156 (= *diff. of* $3ab$ *and* $2an$) $\times 1\frac{2}{3}$ (= EF $\div ab$ $\times$ 144 (= *square of* $an$) $\times$ ·5236 $= \frac{156 \times 5}{3} \times 144 \times \cdot5236 = 52 \times 5 \times 144 \times \cdot5236 = 260 \times 144 \times \cdot5236 = 37440 \times \cdot5236 = 19603\cdot584$ = *the solidity.*

*To find the solidity of a parabolic conoid.*—Multiply the area of the base by half the altitude, and the product will be the content.

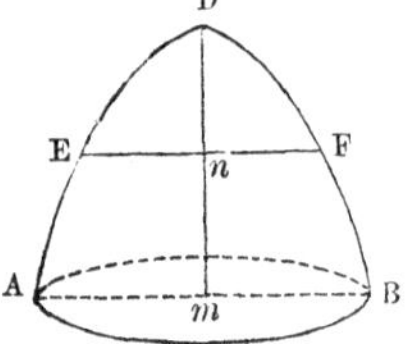


What is the solidity of the paraboloid ADB, whose height D$m$ is 84, and the diameter BA of its circular base 48?

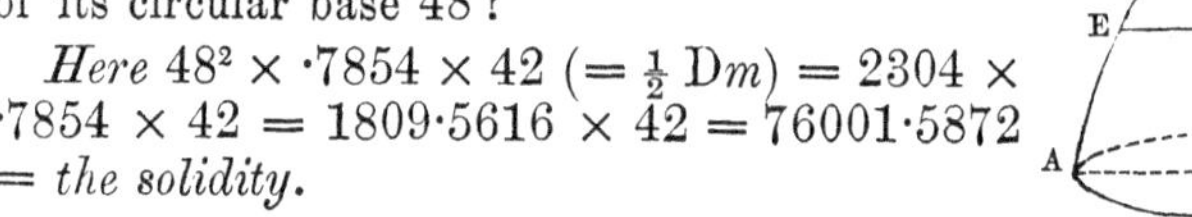

*Here* $48^2 \times \cdot7854 \times 42$ (= $\frac{1}{2}$ D$m$) $= 2304 \times \cdot7854 \times 42 = 1809\cdot5616 \times 42 = 76001\cdot5872$ = *the solidity.*

*To find the solidity of the frustum of a paraboloid, when its ends are perpendicular to the axe of the solid.*—Multiply the sum of the squares of the diameters of the two ends by the height of the frustum, and the product again by ·3927, and it will give the solidity.

Required the solidity of the parabolic frustum ABC$d$, the diameter AB of the greater end being 58, that of the less end $dc$ 30, and the height $no$ 18.

*Here* $(58^2 + 30^2) \times 18 \times \cdot3927 = (3364 + 900) \times 18 \times \cdot3927 = 4264 \times 18 \times \cdot3927 = 76752 \times \cdot3927 = 30140\cdot5104$ = *the solidity.*

*To find the solidity of an hyperboloid.*—To the square of the radius of the base add the square of the middle diameter between the base and the vertex, and this sum multiplied by the altitude, and the product again by ·5236 will give the solidity.

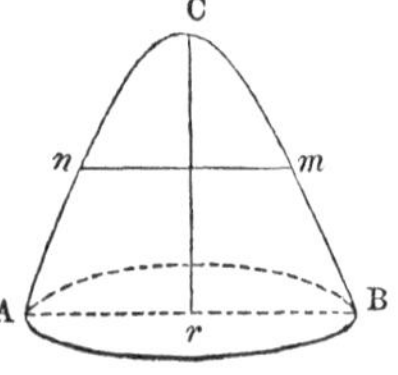


In the hyperboloid ACB, the altitude C$r$ is 10, the radius A$r$ of the base 12, and the middle diameter $nm$ 15·8745; what is the solidity?

*Here* $\overline{15\cdot8745^2 + 12^2} \times 10 \times \cdot5236 = \overline{251\cdot99975 + 144} \times 10 \times \cdot5236 = 395\cdot99975 \times 10 \times \cdot5236 = 3959\cdot9975 \times \cdot5236 = 2073\cdot454691$ = *the solidity.*

*To find the solidity of the frustum of an hyperbolic conoid.*—Add together the squares of the greatest and least semi-diameters, and the square of the whole diameter in the middle; then this sum being multiplied by the altitude, and the product again by ·5236, will give the solidity.

In the hyperbolic frustum ADCB, the length *rs* is 20, the diameter AB of the greater end 32, that DC of the less end 24, and the middle diameter *nm* 28·1708; required the solidity.

*Here* $(16^2 + 12^2 + 28{\cdot}1708^2) \times 20 \times {\cdot}52359 = (256 + 144 + 793{\cdot}5939) \times 20 \times {\cdot}52359 = 1193{\cdot}5939 \times 20 \times {\cdot}52359 = 23871{\cdot}878 \times {\cdot}52359 = 12499{\cdot}07660202 =$ *solidity.*

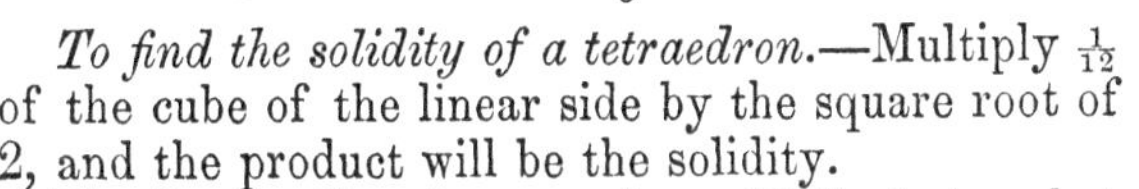

*To find the solidity of a tetraedron.*—Multiply $\frac{1}{12}$ of the cube of the linear side by the square root of 2, and the product will be the solidity.

The linear side of a tetraedron ABC*n* is 4; what is the solidity?

$$\frac{4^3}{12} \times \sqrt{2} = \frac{4 \times 4 \times 4}{12} \times \sqrt{2} = \frac{4 \times 4}{3} \times \sqrt{2} = \frac{16}{3} \times \sqrt{2} = \frac{16}{3} \times 1{\cdot}414 = \frac{22{\cdot}624}{3} = 7{\cdot}5413 = \textit{solidity.}$$

*To find the solidity of an octaedron.*—Multiply $\frac{1}{3}$ of the cube of the linear side by the square root of 2, and the product will be the solidity.

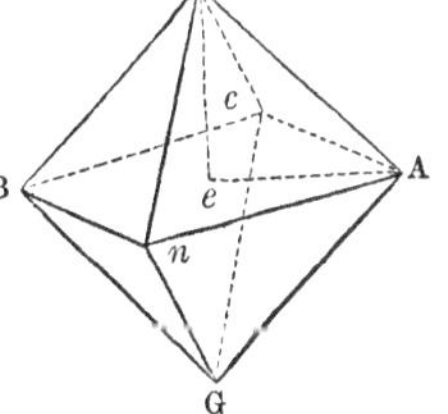


What is the solidity of the octaedron BGAD, whose linear side is 4?

$$\frac{4^3}{3} \times \sqrt{2} = \frac{64}{3} \times \sqrt{2} = 21{\cdot}333, \times \sqrt{2} = 21{\cdot}333 \times 1{\cdot}414 = 30{\cdot}16486 = \textit{solidity.}$$

*To find the solidity of a dodecaedron.*—To 21 times the square root of 5 add 47, and divide the sum by 40: then the square root of the quotient being multiplied by five times the cube of the linear side will give the solidity.

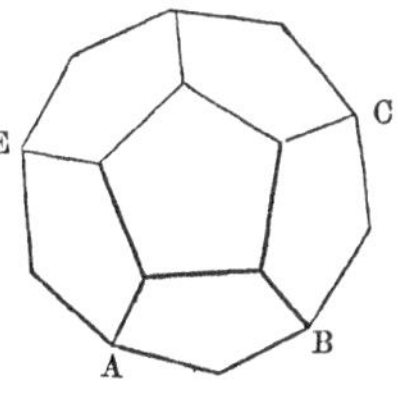


The linear side of the dodecaedron ABCDE is 3; what is the solidity?

$$\sqrt{\frac{21\sqrt{5}+47}{40}} \times 27 \times 5 = \sqrt{\frac{21 \times 2{\cdot}23606+47}{40}} \times 27 \times 5 = \sqrt{\frac{46{\cdot}95726+47}{40}} \times 135 = 206{\cdot}901$$

*solidity.*

*To find the solidity of an icosaedron.*—To three times the square root of 5 add 7, and divide the sum by 2; then the square root of this quotient being multiplied by $\frac{5}{6}$ of the cube of the linear side will give the solidity.

That is $\frac{5}{6}$ $S^3 \times \sqrt{(\frac{7 + 3\sqrt{5}}{2})}$ = solidity when S is = to the linear side.

The linear side of the icosaedron ABCDEF is 3; what is the solidity?

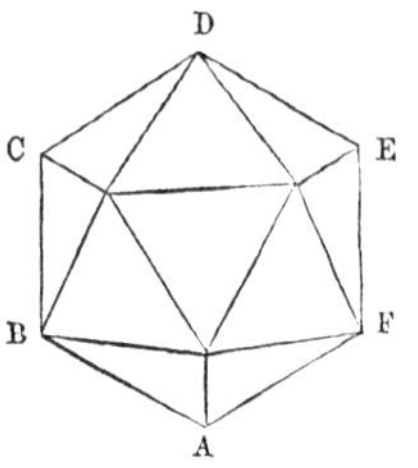

$$\sqrt{\frac{3\sqrt{5} + 7}{2} \times \frac{5 \times 3^2}{6}} = \sqrt{\frac{3 \times 2{\cdot}23606 + 7}{2} \times \frac{5 \times 27}{6}} = \sqrt{\frac{6{\cdot}70818 + 7}{2} \times \frac{5 \times 9}{2}} = \sqrt{\frac{13{\cdot}70818}{2} \times \frac{45}{2}} = \sqrt{6{\cdot}85409} \times 22{\cdot}5 = 2{\cdot}61803$$

$\times$ 22·5 = 58·9056 = *solidity.*

The superficies and solidity of any of the five regular bodies may be found as follows:

RULE 1. Multiply the tabular area by the square of the linear edge, and the product will be the superficies.

2. Multiply the tabular solidity by the cube of the linear edge, and the product will be the solidity.

*Surfaces and Solidities of the Regular Bodies.*

| No. of Sides. | Names. | Surfaces. | Solidities. |
|---|---|---|---|
| 4 | Tetraedron | 1.73205 | 0.11785 |
| 6 | Hexaedron | 6.00000 | 1.00000 |
| 8 | Octaedron | 3.46410 | 0.47140 |
| 12 | Dodecaedron | 20.64578 | 7.66312 |
| 20 | Icosaedron | 8.66025 | 2.18169 |

*To find the convex superficies of a cylindric ring.*—To the thickness of the ring add the inner diameter, and this sum being multiplied by the thickness, and the product again by 9.8696, will give the superficies.

The thickness of A*c* of a cylindric ring is 3 inches, and the inner diameter *cd* 12 inches; what is the convex superficies?

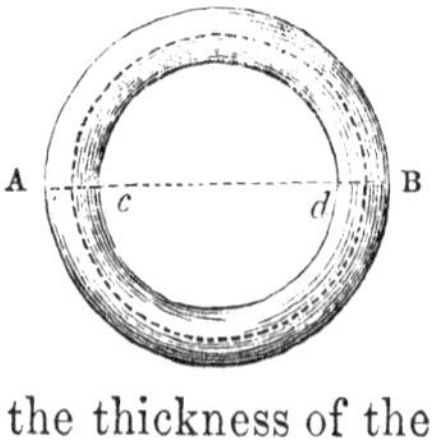

$\overline{12 + 3} \times 3 \times 9{\cdot}8696 = 15 \times 3 \times 9{\cdot}8696 = 45 \times 9{\cdot}8696 = 444{\cdot}132$ = *superficies.*

*To find the solidity of a cylindric ring.*—To the thickness of the ring add the inner diameter, and this sum being multiplied by the square of half the thickness, and the product again by 9·8696, will give the solidity.

What is the solidity of an anchor ring, whose inner diameter is 8 inches, and thickness in metal 3 inches?

$\overline{8 + 3} \times \left|\frac{3}{2}\right|^2 \times 9{\cdot}8696 = 11 \times 1{\cdot}5^2 \times 9{\cdot}8693 = 11 \times 2{\cdot}25 \times 9{\cdot}8696 = 24{\cdot}75 \times 9{\cdot}8696 = 244{\cdot}2726 =$ *solidity*.

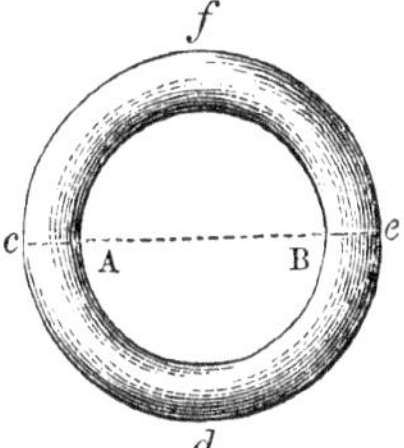


The inner diameter AB of the cylindric ring *cdef* equals 18 feet, and the sectional diameter *c*A or B*e* equals 9 inches; required the convex surface and solidity of the ring.

18 feet $\times$ 12 = 216 *inches, and* $\overline{216 + 9} \times 9 \times 9{\cdot}8696 = 19985{\cdot}94$ *square inches.*

$\overline{216 + 9} \times 9^2 \times 2{\cdot}4674 = 44968{\cdot}365$ *cubic inches.*

In the formation of a hoop or ring of wrought iron, it is found in practice that in bending the iron, the side or edge which forms the interior diameter of the hoop is upset or shortened, while at the same time the exterior diameter is drawn or lengthened; therefore, the proper diameter by which to determine the length of the iron in an unbent state, is the distance from centre to centre of the iron of which the hoop is composed: *hence the rule to determine the length of the iron.* If it is the interior diameter of the hoop that is given, add the thickness of the iron; but if the exterior diameter, subtract from the given diameter the thickness of the iron, multiply the sum or remainder by 3·1416, and the product is the length of the iron, in equal terms of unity.

Supposing the interior diameter of a hoop to be 32 inches, and the thickness of the iron $1\frac{1}{4}$, what must be the proper length of the iron, independent of any allowance for shutting?

$$\overline{32 + 1{\cdot}25} = 33{\cdot}25 \times 3{\cdot}1416 = 104{\cdot}458 \textit{ inches.}$$

But the same is obtained simply by inspection in the Table of Circumferences.

Thus, 33·25 = 2 *feet* $9\frac{1}{4}$ *in.*, opposite to which is 8 feet $8\frac{1}{2}$ inches.

Again, let it be required to form a hoop of iron $\frac{7}{8}$ inch in thickness, and $16\frac{1}{2}$ inches outside diameter.

$$16{\cdot}5 - {\cdot}875 = 15{\cdot}625, \text{ or 1 foot } 3\tfrac{5}{8} \text{ inches;}$$

opposite to which, in the Table of Circumferences, is 4 feet 1 inch, independent of any allowance for shutting.

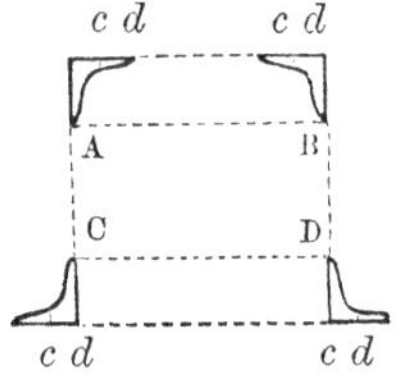


The length for angle iron, of which to form a ring of a given diameter, varies according to the strength of the iron at the root; and the rule is, for a ring with the flange outside, *add* to its required interior diameter, twice the extreme strength of the iron at the root; or, for a ring with the flange inside, *subtract* twice the extreme strength; and the sum or remainder is the diameter by which to determine the length of the angle iron. Thus, suppose two angle iron rings similar to the following be required, the exterior diameter AB, and interior diameter CD, each to be 1 foot $10\frac{1}{2}$ inches, and the extreme strength of the iron at the root *cd*, *cd*, &c, $\frac{7}{8}$ of an inch;

twice $\frac{7}{8} = 1\frac{3}{4}$, and 1 ft. $10\frac{1}{2}$ in. + $1\frac{3}{4}$ = 2 ft. $\frac{1}{4}$ in., opposite to which, in the Table of Circumferences, is 6 ft. $4\frac{1}{4}$ in., the length of the iron for CD; and 1 ft. $10\frac{1}{2}$ in. − $1\frac{3}{4}$ = 1 ft. $8\frac{3}{4}$ in., opposite to which is 5 ft. $5\frac{1}{4}$ in., the length of the iron for AB.

But observe, as before, that the necessary allowance for shutting must be added to the length of the iron, in addition to the length as expressed by the Table.

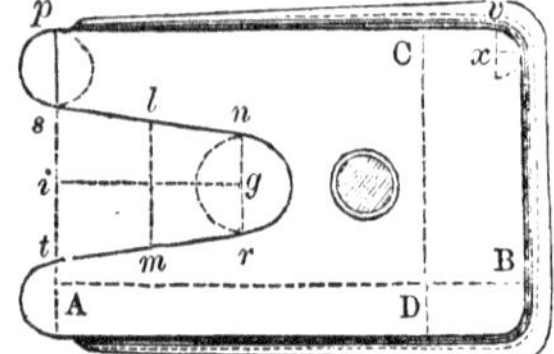


Required the capacity in gallons of a locomotive engine tender tank, 2 feet 8 inches in depth, and its superficial dimensions the following, with reference to the annexed plan:

| | | | | | |
|---|---|---|---|---|---|
| Length, or dist. between | A and B | = 10 ft. | $2\frac{3}{4}$ in. | or, | 122·75 in. |
| Breadth " | C and D | = 6 | $7\frac{1}{2}$ | | 79·5 |
| Length " | $i$ and $g$ | = 3 | $10\frac{3}{4}$ | | 46·75 |
| Mean breadth of coke-space or | $lm$ | = 3 | $1\frac{1}{4}$ | | 37·25 |
| Diameter of circle | $rn$ | = 2 | $8\frac{1}{4}$ | | 32·25 |
| " " | $ps$ | = 1 | $6\frac{1}{2}$ | | 18·5 |
| Radius of back corners | $vx$ | = | 4 | | 4 |

Then, 122·75 × 79·5 = 9758·525 square inches, as a rectangle.
And $18{\cdot}5^2 \times {\cdot}7854$ = 268·8 " " area of circle formed by the two ends.

Total 10027·325 " " from which deduct the area of the coke-space, and the difference of area between the semicircle formed by the two back corners, and that of a rectangle of equal length and breadth;

Then 46·75 × 37·25 = 1731·4375 area of $r$, $n$, $s$, $t$, in sq. ins.

$$\frac{32{\cdot}25^2 \times {\cdot}7854}{2} = 408{\cdot}4$$ area of half the circle $rn$.

Radius of back corners = 4 inches;
consequently $8^2 \times {\cdot}7854 = 25{\cdot}13$, the semicircle's area; and $8 \times 4 = 32 - 25{\cdot}13 = 6{\cdot}87$ inches taken off by rounding the corners.

Hence, $\overline{1731{\cdot}4375 + 408{\cdot}4 + 6{\cdot}87} = 2146{\cdot}707$, and
10027·235 − 2146·707 = 7880·618 square inches, or whole area in plan,
7880·618 × 32 the depth = 252179·776 cubic inches, and 252179·776 divided by 231 gives 1091·6873 the content in gallons.

TABLES *by which to facilitate the Mensuration of Timber.*

## 1. Flat or Board Measure.

| Breadth in inches. | Area of a lineal foot. | Breadth in inches. | Area of a lineal foot. | Breadth in inches. | Area of a lineal foot. |
|---|---|---|---|---|---|
| $\frac{1}{4}$ | ·0208 | 4 | ·3334 | 8 | ·6667 |
| $\frac{1}{2}$ | ·0417 | $4\frac{1}{4}$ | ·3542 | $8\frac{1}{4}$ | ·6875 |
| $\frac{3}{4}$ | ·0625 | $4\frac{1}{2}$ | ·375 | $8\frac{1}{2}$ | ·7084 |
| 1 | ·0834 | $4\frac{3}{4}$ | ·3958 | $8\frac{3}{4}$ | ·7292 |
| $1\frac{1}{4}$ | ·1042 | 5 | ·4167 | 9 | ·75 |
| $1\frac{1}{2}$ | ·125 | $5\frac{1}{4}$ | ·4375 | $9\frac{1}{4}$ | ·7708 |
| $1\frac{3}{4}$ | ·1459 | $5\frac{1}{2}$ | ·4583 | $9\frac{1}{2}$ | ·7917 |
| 2 | ·1667 | $5\frac{3}{4}$ | ·4792 | $9\frac{3}{4}$ | ·8125 |
| $2\frac{1}{4}$ | ·1875 | 6 | ·5 | 10 | ·8334 |
| $2\frac{1}{2}$ | ·2084 | $6\frac{1}{4}$ | ·5208 | $10\frac{1}{4}$ | ·8542 |
| $2\frac{3}{4}$ | ·2292 | $6\frac{1}{2}$ | ·5416 | $10\frac{1}{2}$ | ·875 |
| 3 | ·25 | $6\frac{3}{4}$ | ·5625 | $10\frac{3}{4}$ | ·8959 |
| $3\frac{1}{4}$ | ·2708 | 7 | ·5833 | 11 | ·9167 |
| $3\frac{1}{2}$ | ·2916 | $7\frac{1}{4}$ | ·6042 | $11\frac{1}{4}$ | ·9375 |
| $3\frac{3}{4}$ | ·3125 | $7\frac{1}{2}$ | ·625 | $11\frac{1}{2}$ | ·9583 |
| | | $7\frac{3}{4}$ | ·6458 | $11\frac{3}{4}$ | ·9792 |

*Application and Use of the Table.*

Required the number of square feet in a board or plank $16\frac{1}{2}$ feet in length and $9\frac{3}{4}$ inches in breadth.

Opposite $9\frac{3}{4}$ is $·8125 \times 16·5 = 13·4$ square feet.

A board 1 foot $2\frac{3}{4}$ inches in breadth, and 21 feet in length; what is its superficial content in square feet?

Opposite $2\frac{3}{4}$ is ·2292, to which add the 1 foot; then $1·2292 \times 21 = 25·8$ square feet.

In a board $15\frac{1}{2}$ inches at one end, 9 inches at the other, and $14\frac{1}{2}$ feet in length, how many square feet?

$$\frac{15·5 + 9}{2} = 12\tfrac{1}{4}, \text{ or } 1·0208; \text{ and } 1·0208 \times 14·5 = 14·8 \text{ sq. ft.}$$

The solidity of round or unsquared timber may be found with much more accuracy by the succeeding Rule:—Multiply the square of one-fifth of the mean girth by twice the length, and the product will be the solidity, very near the truth.

A piece of timber is 30 feet long, and the mean girth is 128 inches, what is the solidity?

$$\frac{128}{5} = 25·6.$$

$$\text{Then } \frac{25·6^2 \times 60}{144} = 273·06 \text{ cubic feet.}$$

This is nearer the truth than if one-fourth the girth be employed.

## 2. Cubic or Solid Measure.

| Mean ¼ girt in inches. | Cubic feet in each lineal foot. | Mean ¼ girt in inches. | Cubic feet in each lineal foot. | Mean ¼ girt in inches. | Cubic feet in each lineal foot. | Mean ¼ girt in inches. | Cubic feet in each lineal foot. |
|---|---|---|---|---|---|---|---|
| 6 | ·25 | 12 | 1 | 18 | 2·25 | 24 | 4 |
| 6¼ | ·272 | 12¼ | 1·042 | 18¼ | 2·313 | 24¼ | 4·084 |
| 6½ | ·294 | 12½ | 1·085 | 18½ | 2·376 | 24½ | 4·168 |
| 6¾ | ·317 | 12¾ | 1·129 | 18¾ | 2·442 | 24¾ | 4·254 |
| 7 | ·340 | 13 | 1·174 | 19 | 2·506 | 25 | 4·34 |
| 7¼ | ·364 | 13¼ | 1·219 | 19¼ | 2·574 | 25¼ | 4·428 |
| 7½ | ·39 | 13½ | 1·265 | 19½ | 2·64 | 25½ | 4·516 |
| 7¾ | ·417 | 13¾ | 1·313 | 19¾ | 2·709 | 25¾ | 4·605 |
| 8 | ·444 | 14 | 1·361 | 20 | 2·777 | 26 | 4·694 |
| 8¼ | ·472 | 14¼ | 1·41 | 20¼ | 2·898 | 26¼ | 4·785 |
| 8½ | ·501 | 14½ | 1·46 | 20½ | 2·917 | 26½ | 4·876 |
| 8¾ | ·531 | 14¾ | 1·511 | 20¾ | 2·99 | 26¾ | 4·969 |
| 9 | ·562 | 15 | 1·562 | 21 | 3·062 | 27 | 5·062 |
| 9¼ | ·594 | 15¼ | 1·615 | 21¼ | 3·136 | 27¼ | 5·158 |
| 9½ | ·626 | 15½ | 1·668 | 21½ | 3·209 | 27½ | 5·252 |
| 9¾ | ·659 | 15¾ | 1·772 | 21¾ | 3·285 | 27¾ | 5·348 |
| 10 | ·694 | 16 | 1·777 | 22 | 3·362 | 28 | 5·444 |
| 10¼ | ·73 | 16¼ | 1·833 | 22¼ | 3·438 | 28¼ | 5·542 |
| 10½ | ·766 | 16½ | 1·89 | 22½ | 3·516 | 28½ | 5·64 |
| 10¾ | ·803 | 16¾ | 1·948 | 22¾ | 3·598 | 28¾ | 5·74 |
| 11 | ·84 | 17 | 2·006 | 23 | 3·673 | 29 | 5·84 |
| 11¼ | ·878 | 17¼ | 2·066 | 23¼ | 3·754 | 29¼ | 5·941 |
| 11½ | ·918 | 17½ | 2·126 | 23½ | 3·835 | 29½ | 6·044 |
| 11¾ | ·959 | 17¾ | 2·187 | 23¾ | 3·917 | 29¾ | 6·146 |

In the cubic estimation of timber, custom has established the rule of ¼, the mean girt being the side of the square considered as the cross sectional dimensions; hence, multiply the number of cubic feet by lineal foot as in the Table of Cubic Measure opposite the ¼ girt, and the product is the solidity of the given dimensions in cubic feet.

Suppose the mean ¼ girt of a tree 21¼ inches, and its length 16 feet, what are its contents in cubic feet?

$$3{\cdot}136 \times 16 = 50{\cdot}176 \text{ cubic feet.}$$

Battens, Deals, and Planks are each similar in their various lengths, but differing in their widths and thicknesses, and hence their principal distinction: thus, a batten is 7 inches by 2½, a deal 9 by 3, and a plank 11 by 3, these being what are termed the standard dimensions, by which they are bought and sold, the length of each being taken at 12 feet; therefore, in estimating for the proper value of any quantity, nothing more is required than their lineal dimensions, by which to ascertain the number of times 12 feet, there are in the given whole.

Suppose I wish to purchase the following:

|  |  |  |
|---|---|---|
| 7 of | 6 feet | $6 \times 7 = 42$ feet |
| 5 | 14 | $14 \times 5 = 70$ |
| 11 | 19 | $19 \times 11 = 209$ |
| and 6 | 21 | $21 \times 6 = 126$ |

12 ) 447 ) 37·25 standard deals.

TABLE *showing the number of Lineal Feet of Scantling of various dimensions, which are equal to a Cubic Foot.*

| | Inches. | | Ft. | In. |
|---|---|---|---|---|
| 2 inches by | 2 | require in length | 36 | 0 |
| | $2\frac{1}{2}$ | | 28 | 9 |
| | 3 | | 24 | 0 |
| | $3\frac{1}{2}$ | | 20 | 7 |
| | 4 | | 18 | 0 |
| | $4\frac{1}{2}$ | | 16 | 0 |
| | 5 | | 14 | 5 |
| | $5\frac{1}{2}$ | | 13 | 1 |
| | 6 | | 12 | 0 |
| | $6\frac{1}{2}$ | | 11 | 1 |
| | 7 | | 10 | 3 |
| | $7\frac{1}{2}$ | | 9 | 7 |
| | 8 | | 9 | 0 |
| | $8\frac{1}{2}$ | | 8 | 6 |
| | 9 | | 8 | 0 |
| | $9\frac{1}{2}$ | | 7 | 7 |
| | 10 | | 7 | 3 |
| | $10\frac{1}{2}$ | | 6 | 10 |
| | 11 | | 6 | 6 |
| | $11\frac{1}{2}$ | | 6 | 4 |
| | 12 | | 6 | 0 |
| 3 inches by | 3 | | 16 | 0 |
| | $3\frac{1}{2}$ | | 13 | 8 |
| | 4 | | 12 | 0 |
| | $4\frac{1}{2}$ | | 10 | 8 |
| | 5 | | 9 | 7 |
| | $5\frac{1}{2}$ | | 9 | 0 |
| | 6 | | 8 | 0 |
| | $6\frac{1}{2}$ | | 7 | 4 |
| | 7 | | 6 | 10 |
| | $7\frac{1}{2}$ | | 6 | 4 |
| | 8 | | 6 | 0 |
| | $8\frac{1}{2}$ | | 5 | 8 |
| | 9 | | 5 | 4 |
| | $9\frac{1}{2}$ | | 5 | 0 |
| | 10 | | 4 | 10 |
| | $10\frac{1}{2}$ | | 4 | 6 |
| | 11 | | 4 | 4 |
| | $11\frac{1}{2}$ | | 4 | 2 |
| | 12 | | 4 | 0 |
| 4 inches by | 4 | require in length | 9 | 0 |
| | $4\frac{1}{2}$ | | 8 | 0 |
| | 5 | | 7 | 2 |
| | $5\frac{1}{2}$ | | 6 | 6 |
| | 6 | | 6 | 0 |
| | $6\frac{1}{2}$ | | 5 | 6 |
| | 7 | | 5 | 1 |
| | $7\frac{1}{2}$ | | 4 | 9 |
| | 8 | | 4 | 6 |
| | $8\frac{1}{2}$ | | 4 | 3 |
| | 9 | | 4 | 0 |
| | $9\frac{1}{2}$ | | 3 | 9 |
| | 10 | | 3 | 7 |
| | $10\frac{1}{2}$ | | 3 | 5 |
| | 11 | | 3 | 3 |
| | $11\frac{1}{2}$ | | 3 | 2 |
| | 12 | | 3 | 0 |
| 5 inches by | 5 | | 5 | 9 |
| | $5\frac{1}{2}$ | | 5 | 3 |
| | 6 | | 4 | 10 |
| | $6\frac{1}{2}$ | | 4 | 5 |
| | 7 | | 4 | 1 |
| | $7\frac{1}{2}$ | | 3 | 10 |
| | 8 | | 3 | 7 |
| | $8\frac{1}{2}$ | | 3 | 5 |
| | 9 | | 3 | 2 |
| | $9\frac{1}{2}$ | | 3 | 0 |
| | 10 | | 2 | 10 |
| | $10\frac{1}{2}$ | | 2 | 9 |
| | 11 | | 2 | 8 |
| | $11\frac{1}{2}$ | | 2 | 6 |
| | 12 | | 2 | 4 |
| 6 inches by | 6 | | 4 | 0 |
| | $6\frac{1}{2}$ | | 3 | 8 |
| | 7 | | 3 | 5 |
| | $7\frac{1}{2}$ | | 3 | 2 |
| | 8 | | 3 | 0 |
| | $8\frac{1}{2}$ | | 2 | 10 |
| | 9 | | 2 | 8 |
| 6 inches by | $9\frac{1}{2}$ | require in length | 2 | 6 |
| | 10 | | 2 | 5 |
| | $10\frac{1}{2}$ | | 2 | 3 |
| | 11 | | 2 | 2 |
| | $11\frac{1}{2}$ | | 2 | 1 |
| | 12 | | 2 | 0 |
| 7 inches by | 7 | | 2 | 11 |
| | $7\frac{1}{2}$ | | 2 | 9 |
| | 8 | | 2 | 6 |
| | $8\frac{1}{2}$ | | 2 | 5 |
| | 9 | | 2 | 3 |
| | $9\frac{1}{2}$ | | 2 | 2 |
| | 10 | | 2 | 1 |
| | $10\frac{1}{2}$ | | 1 | 11 |
| | 11 | | 1 | 10 |
| | $11\frac{1}{2}$ | | 1 | 9 |
| | 12 | | 1 | 8 |
| 8 inches by | 8 | | 2 | 3 |
| | $8\frac{1}{2}$ | | 2 | 1 |
| | 9 | | 2 | 0 |
| | $9\frac{1}{2}$ | | 1 | 10 |
| | 10 | | 1 | 9 |
| | $10\frac{1}{2}$ | | 1 | 8 |
| | 11 | | 1 | 7 |
| | $11\frac{1}{2}$ | | 1 | 7 |
| | 12 | | 1 | 6 |
| 9 inches by | 9 | | 1 | 9 |
| | $9\frac{1}{2}$ | | 1 | 8 |
| | 10 | | 1 | 7 |
| | $10\frac{1}{2}$ | | 1 | 6 |
| | 11 | | 1 | 5 |
| | $11\frac{1}{2}$ | | 1 | 4 |
| | 12 | | 1 | 4 |
| 10 in. by | 10 | | 1 | 5 |
| | $10\frac{1}{2}$ | | 1 | 4 |
| | 11 | | 1 | 4 |
| | $11\frac{1}{2}$ | | 1 | 3 |

Hewn and sawed timber are measured by the cubic foot. The *unit* of board measure is a superficial foot one inch thick.

*To measure round timber.*—Multiply the length in feet by the square of $\frac{1}{4}$ of the mean girth in inches, and the product divided by 144 gives the content in cubic feet.

The $\frac{1}{4}$ girths of a piece of timber, taken at five points, equally distant from each other, are 24, 28, 33, 35, and 40 inches; the length 30 feet, what is the content?

$$\frac{24 + 28 + 33 + 35 + 40}{5} = 32.$$

$$\text{Then } \frac{32^2 \times 30}{144} = 213\tfrac{1}{3} \text{ cubic feet.}$$

TABLE *containing the Superficies and Solid Content of Spheres, from 1 to 12, and advancing by a tenth.*

| Diam. | Superficies. | Solidity. | Diam. | Superficies. | Solidity. | Diam. | Superficies. | Solidity. |
|---|---|---|---|---|---|---|---|---|
| 1·0 | 3·1416 | ·5236 | 4·7 | 69·3979 | 54·3617 | 8·4 | 221·6712 | 310·3398 |
| ·1 | 3·8013 | ·6969 | ·8 | 72·3824 | 57·9059 | ·5 | 226·9806 | 321·5558 |
| ·2 | 4·5239 | ·9047 | ·9 | 75·4298 | 61·6010 | ·6 | 232·3527 | 333·0389 |
| ·3 | 5·3093 | 1·1503 | 5·0 | 78·5400 | 65·4500 | ·7 | 237·7877 | 344·7921 |
| ·4 | 6·1575 | 1·4367 | ·1 | 81·7130 | 69·4560 | ·8 | 243·2855 | 356·8187 |
| ·5 | 7·0686 | 1·7671 | ·2 | 84·9488 | 73·6223 | ·9 | 248·8461 | 369·1217 |
| ·6 | 8·0424 | 2·1446 | ·3 | 88·2475 | 77·9519 | 9·0 | 254·4696 | 381·7044 |
| ·7 | 9·0792 | 2·5724 | ·4 | 91·6090 | 82·4481 | ·1 | 260·1558 | 394·5697 |
| ·8 | 10·1787 | 3·0536 | ·5 | 95·0334 | 87·1139 | ·2 | 265·9130 | 407·7210 |
| ·9 | 11·3411 | 3·5913 | ·6 | 98·5205 | 91·9525 | ·3 | 271·7169 | 421·1613 |
| 2·0 | 12·5664 | 4·1888 | ·7 | 102·0705 | 96·9670 | ·4 | 277·5917 | 434·8937 |
| ·1 | 13·8544 | 4·8490 | ·8 | 105·6834 | 102·1606 | ·5 | 283·5294 | 448·9215 |
| ·2 | 15·2053 | 5·5752 | ·9 | 109·3590 | 107·5364 | ·6 | 289·5298 | 463·2477 |
| ·3 | 16·6190 | 6·3706 | 6·0 | 113·0976 | 113·0976 | ·7 | 295·5931 | 477·7755 |
| ·4 | 18·0956 | 7·2382 | ·1 | 116·8989 | 118·8472 | ·8 | 301·7192 | 492·8081 |
| ·5 | 19·6350 | 8·1812 | ·2 | 120·7631 | 124·7885 | ·9 | 307·9082 | 508·0485 |
| ·6 | 21·2372 | 9·2027 | ·3 | 124·6901 | 130·9246 | 10·0 | 314·1600 | 523·6000 |
| ·7 | 22·9022 | 10·3060 | ·4 | 128·6799 | 137·2585 | ·1 | 320·4746 | 539·4656 |
| ·8 | 24·6300 | 11·4940 | ·5 | 132·7326 | 143·7936 | ·2 | 326·8520 | 555·6485 |
| ·9 | 26·4208 | 12·7700 | ·6 | 136·8480 | 150·5329 | ·3 | 333·2923 | 572·1518 |
| 3.0 | 28·2744 | 14·1372 | ·7 | 141·0264 | 157·4795 | ·4 | 339·7954 | 588·9784 |
| ·1 | 30·1907 | 15·5985 | ·8 | 145·2675 | 164·6365 | ·5 | 346·3614 | 606·1324 |
| ·2 | 32·1699 | 17·1573 | ·9 | 149·5715 | 172·0073 | ·6 | 352·9901 | 623·6159 |
| ·3 | 34·2120 | 18·8166 | 7·0 | 153·9384 | 179·5948 | ·7 | 359·6817 | 641·4325 |
| ·4 | 36·3168 | 20·5795 | ·1 | 158·3680 | 187·4021 | ·8 | 366·4362 | 659·5852 |
| ·5 | 38·4846 | 22·4493 | ·2 | 162·8605 | 195·4326 | ·9 | 373·2534 | 678·0771 |
| ·6 | 40·7151 | 24·4290 | ·3 | 167·4158 | 203·6893 | 11·0 | 380·1336 | 696·9116 |
| ·7 | 43·0085 | 26·5219 | ·4 | 172·0340 | 212·1752 | ·1 | 387·0765 | 716·0915 |
| ·8 | 45·3647 | 28·7309 | ·5 | 176·7150 | 220·8937 | ·2 | 394·0823 | 735·6200 |
| ·9 | 47·7837 | 31·0594 | ·6 | 181·4588 | 229·8478 | ·3 | 401·1509 | 755·5008 |
| 4·0 | 50·2656 | 33·5104 | ·7 | 186·2654 | 239·0511 | ·4 | 408·2823 | 775·7364 |
| ·1 | 52·8102 | 36·0870 | ·8 | 191·1349 | 248·4754 | ·5 | 415·4766 | 796·3301 |
| ·2 | 55·4178 | 38·7924 | ·9 | 196·0672 | 258·1552 | ·6 | 422·7336 | 817·2851 |
| ·3 | 58·0881 | 41·6298 | 8·0 | 201·0624 | 268·0832 | ·7 | 430·0536 | 838·6045 |
| ·4 | 60·8213 | 44·6023 | ·1 | 206·1203 | 278·2625 | ·8 | 437·4363 | 860·2915 |
| ·5 | 63·6174 | 47·7130 | ·2 | 211·2411 | 288·6962 | ·9 | 444·8819 | 882·3492 |
| ·6 | 66·4782 | 50·9651 | ·3 | 216·4248 | 299·3876 | 12·0 | 452·3904 | 904·7808 |

## *To reduce Solid Inches into Solid Feet.*

1728 Solid Inches to one Solid Foot.

| Feet. | Inches. | Feet. | Inches. | Feet. | Inches. | Feet. | Inches. | Feet. | Inches. | Feet. | Inches. |
|---|---|---|---|---|---|---|---|---|---|---|---|
| 1 | =1728 | 18 | =31104 | 35 | =60480 | 52 | =88956 | 69 | =119232 | 85 | =146880 |
| 2 | 3456 | 19 | 32832 | 36 | 62208 | 53 | 91584 | 70 | 120960 | 86 | 148608 |
| 3 | 5184 | 20 | 34560 | 37 | 63936 | 54 | 93312 | 71 | 122688 | 87 | 150336 |
| 4 | 6912 | 21 | 36288 | 38 | 65664 | 55 | 95040 | 72 | 124416 | 88 | 152064 |
| 5 | 8640 | 22 | 38016 | 39 | 67392 | 56 | 96768 | 73 | 126144 | 89 | 153792 |
| 6 | 10368 | 23 | 39744 | 40 | 69120 | 57 | 98496 | 74 | 127872 | 90 | 155520 |
| 7 | 12096 | 24 | 41472 | 41 | 70848 | 58 | 100224 | 75 | 129600 | 91 | 157248 |
| 8 | 13824 | 25 | 43200 | 42 | 72576 | 59 | 101952 | 76 | 131328 | 92 | 158976 |
| 9 | 15552 | 26 | 44928 | 43 | 74304 | 60 | 103680 | 77 | 133056 | 93 | 160704 |
| 10 | 17280 | 27 | 46656 | 44 | 76032 | 61 | 105408 | 78 | 134784 | 94 | 162432 |
| 11 | 19008 | 28 | 48384 | 45 | 77760 | 62 | 107136 | 79 | 136512 | 95 | 164160 |
| 12 | 20736 | 29 | 50112 | 46 | 79488 | 63 | 108864 | 80 | 138240 | 96 | 165888 |
| 13 | 22464 | 30 | 51840 | 47 | 81216 | 64 | 110592 | 81 | 139968 | 97 | 167616 |
| 14 | 24192 | 31 | 53568 | 48 | 82944 | 65 | 112320 | 82 | 141696 | 98 | 169344 |
| 15 | 25920 | 32 | 55296 | 49 | 84672 | 66 | 114048 | 83 | 143424 | 99 | 171072 |
| 16 | 27648 | 33 | 57024 | 50 | 86400 | 67 | 115776 | 84 | 145152 | 100 | 172800 |
| 17 | 29376 | 34 | 58752 | 51 | 88128 | 68 | 117504 | | | | |

## CUTTINGS AND EMBANKMENTS.

THE angle of repose upon railways, or that incline on which a carriage would rest in whatever situation it was placed, is said to be at 1 in 280, or nearly 19 feet per mile; at any greater rise than this, the force of gravity overcomes the horizontal traction, and carriages will not rest, or remain quiescent upon the line, but will of themselves run down the line with accelerated velocity. The angle of practical effect is variously stated, ranging from 1 in 75 to 1 in 330.

The width of land required for a railway must vary with the depth of the cuttings and length of embankments, together with the slopes necessary to be given to suit the various materials of which the cuttings are composed: thus, rock will generally stand when the sides are vertical; chalk varies from $\frac{1}{6}$ to 1, to 1 to 1; gravel $1\frac{1}{2}$ to 1; coal $1\frac{1}{2}$ to 1; clay 1 to 1, &c.; but where land can be obtained at a reasonable rate, it is always well to be on the safe side.

The following Table is calculated for the purpose of ascertaining the extent of any cutting in cubic yards, for 1 chain, 22 yards, or 66 feet in length, the slopes or angles of the sides being those which are most in general practice, and formation level equal 30 feet.

*Slopes* 1 *to* 1.

| Depth of cutting in feet. | Half width at top in feet. | Content in cubic yards per chain. | Content of 1 perpendicular ft. in breadth. | Content of 3 perpendicular ft. in breadth. | Content of 6 perpendicular ft. in breadth. | Depth of cutting in feet. | Half width at top in feet. | Content in cubic yards per chain. | Content of 1 perpendicular ft. in breadth. | Content of 3 perpendicular ft. in breadth. | Content of 6 perpendicular ft. in breadth. |
|---|---|---|---|---|---|---|---|---|---|---|---|
| 1 | 16 | 75·78 | 2·44 | 7·33 | 14·67 | 26 | 41 | 3599·11 | 63·55 | 190·67 | 381·33 |
| 2 | 17 | 156·42 | 4·89 | 14·67 | 29·33 | 27 | 42 | 3762·00 | 65·99 | 198·00 | 396·00 |
| 3 | 18 | 242·00 | 7·33 | 22·00 | 44·00 | 28 | 43 | 3969·78 | 68·43 | 205·33 | 410·67 |
| 4 | 19 | 332·44 | 9·78 | 29·33 | 58·67 | 29 | 44 | 4182·44 | 70·88 | 212·67 | 425·33 |
| 5 | 20 | 427·78 | 12·22 | 36·67 | 73·33 | 30 | 45 | 4400·00 | 73·32 | 220·00 | 440·00 |
| 6 | 21 | 528·00 | 14·67 | 44·00 | 88·00 | 31 | 46 | 4622·44 | 75·77 | 227·33 | 454·67 |
| 7 | 22 | 633·11 | 17·11 | 51·33 | 102·67 | 32 | 47 | 4849·78 | 78·22 | 234·67 | 469·33 |
| 8 | 23 | 743·11 | 19·56 | 58·67 | 117·33 | 33 | 48 | 5082·00 | 80·67 | 242·00 | 484·00 |
| 9 | 24 | 858·00 | 22·00 | 66·00 | 132·00 | 34 | 49 | 5319·11 | 83·11 | 249·33 | 498·67 |
| 10 | 25 | 977·78 | 24·44 | 73·33 | 146·67 | 35 | 50 | 5561·11 | 85·55 | 256·67 | 513·33 |
| 11 | 26 | 1102·44 | 26·89 | 80·67 | 161·33 | 36 | 51 | 5808·00 | 88·00 | 264·00 | 528·00 |
| 12 | 27 | 1232·00 | 29·33 | 88·00 | 176·00 | 37 | 52 | 6059·78 | 90·44 | 271·33 | 542·67 |
| 13 | 28 | 1366·44 | 31·78 | 95·33 | 190·67 | 38 | 53 | 6316·44 | 92·39 | 278·67 | 557·33 |
| 14 | 29 | 1505·78 | 34·22 | 102·67 | 205·33 | 39 | 54 | 6578·00 | 95·33 | 286·00 | 572·00 |
| 15 | 30 | 1650·00 | 36·66 | 110·00 | 220·00 | 40 | 55 | 6844·44 | 97·77 | 293·33 | 586·67 |
| 16 | 31 | 1799·11 | 39·11 | 117·33 | 234·67 | 41 | 56 | 7115·78 | 100·22 | 300·67 | 601·33 |
| 17 | 32 | 1953·11 | 41·55 | 124·67 | 249·33 | 42 | 57 | 7392·00 | 102·66 | 308·00 | 616·00 |
| 18 | 33 | 2112·00 | 43·99 | 132·00 | 264·00 | 43 | 58 | 7673·11 | 105·11 | 315·33 | 630·67 |
| 19 | 34 | 2275·78 | 46·44 | 139·33 | 278·67 | 44 | 59 | 7959·11 | 107·55 | 322·67 | 645·33 |
| 20 | 35 | 2444·44 | 48·89 | 146·67 | 293·33 | 45 | 60 | 8250·00 | 109·99 | 330·00 | 660·00 |
| 21 | 36 | 2618·00 | 51·33 | 154·00 | 308·00 | 46 | 61 | 8545·78 | 112·44 | 337·33 | 674·67 |
| 22 | 37 | 2796·44 | 53·77 | 161·33 | 322·67 | 47 | 62 | 8846·44 | 114·88 | 344·67 | 689·33 |
| 23 | 38 | 2979·78 | 56·21 | 168·67 | 337·33 | 48 | 63 | 9152·00 | 117·33 | 352·00 | 704·00 |
| 24 | 39 | 3168·00 | 58·66 | 176·00 | 352·00 | 49 | 64 | 9462·44 | 119·77 | 359·33 | 718·67 |
| 25 | 40 | 3361·11 | 61·10 | 183·33 | 366·67 | 50 | 65 | 9777·78 | 122·21 | 366·67 | 733·33 |

## *Slopes* 1½ *to* 1.

| Depth of cutting in feet. | Half width at top in feet. | Content in cubic yards per chain. | Content of 1 perpendicular ft. in breadth. | Content of 3 perpendicular ft. in breadth. | Content of 6 perpendicular ft. in breadth. | Depth of cutting in feet. | Half width at top in feet. | Content in cubic yards per chain. | Content of 1 perpendicular ft. in breadth. | Content of 3 perpendicular ft. in breadth. | Content of 6 perpendicular ft. in breadth. |
|---|---|---|---|---|---|---|---|---|---|---|---|
| 1 | 16½ | 77·00 | 2·44 | 7·33 | 14·67 | 26 | 54 | 4385·33 | 63·55 | 190·67 | 381·33 |
| 2 | 18 | 161·33 | 4·89 | 14·67 | 29·33 | 27 | 55½ | 4653·00 | 65·99 | 198·00 | 396·00 |
| 3 | 19½ | 253·00 | 7·33 | 22·00 | 44·00 | 28 | 57 | 4928·00 | 68·43 | 205·33 | 410·67 |
| 4 | 21 | 352·00 | 9·78 | 29·33 | 58·67 | 29 | 58½ | 5210·33 | 70·88 | 212·67 | 425·33 |
| 5 | 22½ | 453·33 | 12·22 | 36·67 | 73·33 | 30 | 60 | 5500·00 | 73·32 | 220·00 | 440·00 |
| 6 | 24 | 572·00 | 14·67 | 44·00 | 88·00 | 31 | 61½ | 5797·00 | 75·77 | 227·33 | 454·67 |
| 7 | 25½ | 693·00 | 17·11 | 51·33 | 102·67 | 32 | 63 | 6101·33 | 78·22 | 234·67 | 469·33 |
| 8 | 27 | 821·33 | 19·56 | 58·67 | 117·33 | 33 | 64½ | 6413·00 | 80·67 | 242·00 | 484·00 |
| 9 | 28½ | 957·00 | 22·00 | 66·00 | 132·00 | 34 | 66 | 6732·00 | 83·11 | 249·33 | 498·67 |
| 10 | 30 | 1100·00 | 24·44 | 73·33 | 146·67 | 35 | 67½ | 7058·33 | 85·55 | 256·67 | 513·33 |
| 11 | 31½ | 1250·33 | 26·89 | 80·67 | 161·33 | 36 | 69 | 7392·00 | 88·00 | 264·00 | 528·00 |
| 12 | 33 | 1408·00 | 29·33 | 88·00 | 176·00 | 37 | 70½ | 7733·00 | 90·44 | 271·33 | 542·67 |
| 13 | 34½ | 1573·00 | 31·78 | 95·33 | 190·67 | 38 | 72 | 8081·33 | 92·39 | 278·67 | 557·33 |
| 14 | 36 | 1745·33 | 34·22 | 102·67 | 205·33 | 39 | 73½ | 8437·00 | 95·33 | 286·00 | 572·00 |
| 15 | 37½ | 1925·00 | 36·66 | 110·00 | 220·00 | 40 | 75 | 8800·00 | 97·77 | 293·33 | 586·67 |
| 16 | 39 | 2112·00 | 39·11 | 117·33 | 234·67 | 41 | 76½ | 9170·33 | 100·22 | 300·67 | 601·33 |
| 17 | 40½ | 2306·33 | 41·55 | 124·67 | 249·33 | 42 | 78 | 9548·00 | 102·66 | 308·00 | 616·00 |
| 18 | 42 | 2508·00 | 43·99 | 132·00 | 264·00 | 43 | 79½ | 9933·00 | 105·11 | 315·33 | 630·67 |
| 19 | 43½ | 2717·00 | 46·44 | 139·33 | 278·67 | 44 | 81 | 10325·33 | 107·55 | 322·67 | 645·33 |
| 20 | 45 | 2933·33 | 48·89 | 146·67 | 293·33 | 45 | 82½ | 10725·00 | 109·99 | 330·00 | 660·00 |
| 21 | 46½ | 3157·00 | 51·33 | 154·00 | 308·00 | 46 | 84 | 11132·00 | 112·44 | 337·33 | 674·67 |
| 22 | 48 | 3388·00 | 53·77 | 161·33 | 322·67 | 47 | 85½ | 11546·33 | 114·88 | 344·67 | 689·33 |
| 23 | 49½ | 3626·33 | 56·21 | 168·67 | 337·33 | 48 | 87 | 11968·00 | 117·33 | 352·00 | 704·00 |
| 24 | 51 | 3872·00 | 58·66 | 176·00 | 352·00 | 49 | 88½ | 12397·00 | 119·77 | 359·33 | 718·67 |
| 25 | 52½ | 4125·00 | 61·10 | 183·33 | 366·67 | 50 | 90 | 12833·33 | 122·21 | 366·67 | 733·33 |

## *Slopes* 2 *to* 1.

| Depth of cutting in feet. | Half width at top in feet. | Content in cubic yards per chain. | Content of 1 perpendicular ft. in breadth. | Content of 3 perpendicular ft. in breadth. | Content of 6 perpendicular ft. in breadth. | Depth of cutting in feet. | Half width at top in feet. | Content in cubic yards per chain. | Content of 1 perpendicular ft. in breadth. | Content of 3 perpendicular ft. in breadth. | Content of 6 perpendicular ft. in breadth. |
|---|---|---|---|---|---|---|---|---|---|---|---|
| 1 | 17 | 78·22 | 2·44 | 7·33 | 14·67 | 26 | 67 | 5211·55 | 63·55 | 190·67 | 381·33 |
| 2 | 19 | 166·22 | 4·89 | 14·67 | 29·33 | 27 | 69 | 5544·00 | 65·99 | 198·00 | 396·00 |
| 3 | 21 | 264·00 | 7·33 | 22·00 | 44·00 | 28 | 71 | 5886·22 | 68·43 | 205·33 | 410·67 |
| 4 | 23 | 371·55 | 9·78 | 29·33 | 58·67 | 29 | 73 | 6238·22 | 70·88 | 212·67 | 425·33 |
| 5 | 25 | 488·89 | 12·22 | 36·67 | 73·33 | 30 | 75 | 6600·00 | 73·32 | 220·00 | 440·00 |
| 6 | 27 | 616·00 | 14·67 | 44·00 | 88·00 | 31 | 77 | 6971·55 | 75·77 | 227·33 | 454·67 |
| 7 | 29 | 752·89 | 17·11 | 51·33 | 102·67 | 32 | 79 | 7352·89 | 78·22 | 234·67 | 469·33 |
| 8 | 31 | 899·55 | 19·56 | 58·67 | 117·33 | 33 | 81 | 7744·00 | 80·67 | 242·00 | 484·00 |
| 9 | 33 | 1056·00 | 22·00 | 66·00 | 132·00 | 34 | 83 | 8144·89 | 83·11 | 249·33 | 498·67 |
| 10 | 35 | 1222·22 | 24·44 | 73·33 | 146·67 | 35 | 85 | 8555·55 | 85·55 | 256·67 | 513·33 |
| 11 | 37 | 1398·22 | 26·89 | 80·67 | 161·33 | 36 | 87 | 8976·00 | 88·00 | 264·00 | 528·00 |
| 12 | 39 | 1584·00 | 29·33 | 88·00 | 176·00 | 37 | 89 | 9406·22 | 90·44 | 271·33 | 542·67 |
| 13 | 41 | 1779·55 | 31·78 | 95·33 | 190 67 | 38 | 91 | 9846·22 | 92·39 | 278·67 | 557·33 |
| 14 | 43 | 1984·89 | 34·22 | 102·67 | 205·33 | 39 | 93 | 10296·00 | 95·33 | 286·00 | 572·00 |
| 15 | 45 | 2200·00 | 36·66 | 110·00 | 220·00 | 40 | 95 | 10755·55 | 97·77 | 293·33 | 586·67 |
| 16 | 47 | 2424·89 | 39·11 | 117·33 | 234·67 | 41 | 97 | 11224·89 | 100·22 | 300·67 | 601·33 |
| 17 | 49 | 2659·55 | 41·55 | 124·67 | 249·33 | 42 | 99 | 11704·00 | 102·66 | 308·00 | 616·00 |
| 18 | 51 | 2904·00 | 43·99 | 132·00 | 264·00 | 43 | 101 | 12192·89 | 105·11 | 315·33 | 630·67 |
| 19 | 53 | 3158·22 | 46·44 | 139·33 | 278·67 | 44 | 103 | 12691·55 | 107·55 | 322·67 | 645·33 |
| 20 | 55 | 34·2222 | 48·89 | 146·67 | 293·33 | 45 | 105 | 13200·00 | 109·99 | 330·00 | 660·00 |
| 21 | 57 | 3696·00 | 51·33 | 154·00 | 308·00 | 46 | 107 | 13718·22 | 112·44 | 337·33 | 674·67 |
| 22 | 59 | 3979·55 | 53·77 | 161·33 | 322·67 | 47 | 109 | 14246·22 | 114·88 | 344·67 | 689·33 |
| 23 | 61 | 4272·89 | 56·21 | 168·67 | 337·33 | 48 | 111 | 14784·00 | 117·33 | 352·00 | 704·00 |
| 24 | 63 | 4576·00 | 58·66 | 176·00 | 352·00 | 49 | 113 | 15331·55 | 119·77 | 359·33 | 718·67 |
| 25 | 65 | 4888·89 | 61·10 | 183·33 | 366·67 | 50 | 115 | 15888·89 | 122·21 | 366·67 | 733·33 |

By the fourth, fifth, and sixth columns in each table, the number of cubic yards is easily ascertained at any other width of formation level above or below 30 feet, having the same slopes as by the tables, thus:—

Suppose an excavation of 40 feet in depth, and 33 feet in width at formation level, whose slopes or sides are at an angle of 2 to 1, required the extent of excavation in cubic yards:

$$10755{\cdot}55 + 293{\cdot}33 = 11048{\cdot}88 \text{ cubic yards.}$$

The number of cubic yards in any other excavation may be ascertained by the following simple rule:

To the width at formation level in feet, add the horizontal length of the side of the triangle formed by the slope, multiply the sum by the depth of the cutting, or excavation, and by the length, also in feet; divide the product by 27, and the quotient is the content in cubic yards.

Suppose a cutting of any length, and of which take 1 chain, its depth being $14\frac{1}{2}$ feet, width at the bottom 28 feet, and whose sides have a slope of $1\frac{1}{4}$ to 1, required the content in cubic yards:

$$14{\cdot}5 \times 1{\cdot}25 = \overline{18{\cdot}125 + 28} \times 14 = 645{\cdot}75 \times 66 = \frac{42619{\cdot}5}{27} = 1578{\cdot}5 \text{ cubic yards.}$$

$$\frac{l}{6}\left\{(b + rh')\,h' + (b + rh)\,h + 4\left[b + r\frac{h + h'}{2}\right]\frac{h + h'}{2}\right\}$$

gives the content of any cutting. In words, this formula will be:— To the area of each end, add four times the middle area; the sum multiplied by the length and divided by 6 gives the content. The breadth at the bottom of cutting $= b$; the perpendicular depth of cutting at the higher end $= h$; the perpendicular depths of cutting at the lower end $= h'$; $l$, the length of the solid; and $rh'$ the ratio of the perpendicular height of the slope to the horizontal base, multiplied by the height $h'$. $rh$, the ratio $r$, of the perpendicular height of the slope, to the horizontal base, multiplied by the height $h$.

Let $b = 30$; $h = 50$; $h' = 20$; $l = 84$ feet; and 2 to 5 or $\frac{2}{5}$ the ratio of the perpendicular height of the slope to the horizontal base:

$$\frac{84}{6}\left\{(30 + \tfrac{2}{5} \times 20)\,20 + (30 + \tfrac{2}{5} \times 50)\,50 + 4\left[30 + \tfrac{2}{5}\frac{50 + 20}{2}\right]\frac{50 + 20}{2}\right\} = 14\left\{38 \times 20 + 50 \times 50 + 4 \times 44 \times 35\right\} = 131880$$

cubic feet. $\frac{131880}{27} = 4884{\cdot}44$ cubic yards.

This rule is one of the most useful in the mensuration of solids, it will give the content of any irregular solid very nearly, whether it be bounded by right lines or not.

TABLE *of Squares, Cubes, Square and Cube Roots of Numbers.*

| Number. | Squares. | Cubes. | Square Roots. | Cube Roots. | Reciprocals. |
|---|---|---|---|---|---|
| 1 | 1 | 1 | 1·0000000 | 1·0000000 | ·100000000 |
| 2 | 4 | 8 | 1·4142136 | 1·2599210 | ·500000000 |
| 3 | 9 | 27 | 1·7320508 | 1·4422496 | ·333333333 |
| 4 | 16 | 64 | 2·0000000 | 1·5874011 | ·250000000 |
| 5 | 25 | 125 | 2·2360680 | 1·7099759 | ·200000000 |
| 6 | 36 | 216 | 2·4494897 | 1·8171206 | ·166666667 |
| 7 | 49 | 343 | 2·6457513 | 1·9129312 | ·142857143 |
| 8 | 64 | 512 | 2·8284271 | 2·0000000 | ·125000000 |
| 9 | 81 | 729 | 3·0000000 | 2·0800837 | ·111111111 |
| 10 | 100 | 1000 | 3·1622777 | 2·1544347 | ·100000000 |
| 11 | 121 | 1331 | 3·3166248 | 2·2239801 | ·090909091 |
| 12 | 144 | 1728 | 3·4641016 | 2·2894286 | ·083333333 |
| 13 | 169 | 2197 | 3·6055513 | 2·3513347 | ·076923077 |
| 14 | 196 | 2744 | 3·7416574 | 2·4101422 | ·071428571 |
| 15 | 225 | 3375 | 3·8729833 | 2·4662121 | ·066666667 |
| 16 | 256 | 4096 | 4·0000000 | 2·5198421 | ·062500000 |
| 17 | 289 | 4913 | 4·1231056 | 2·5712816 | ·058823529 |
| 18 | 324 | 5832 | 4·2426407 | 2·6207414 | ·055555556 |
| 19 | 361 | 6859 | 4·3588989 | 2·6684016 | ·052631579 |
| 20 | 400 | 8000 | 4·4721360 | 2·7144177 | ·050000000 |
| 21 | 441 | 9261 | 4·5825757 | 2·7589243 | ·047619048 |
| 22 | 484 | 10648 | 4·6904158 | 2·8020393 | ·045454545 |
| 23 | 529 | 12167 | 4·7958315 | 2·8438670 | ·043478261 |
| 24 | 576 | 13824 | 4·8989795 | 2·8844991 | ·041666667 |
| 25 | 625 | 15625 | 5·0000000 | 2·9240177 | ·040000000 |
| 26 | 676 | 17576 | 5·0990195 | 2·9624960 | ·038461538 |
| 27 | 729 | 19683 | 5·1961524 | 3·0000000 | ·037037037 |
| 28 | 784 | 21952 | 5·2915026 | 3·0365889 | ·035714286 |
| 29 | 841 | 24389 | 5·3851648 | 3·0723168 | ·034482759 |
| 30 | 900 | 27000 | 5·4772256 | 3·1072325 | ·033333333 |
| 31 | 961 | 29791 | 5·5677644 | 3·1413806 | ·032258065 |
| 32 | 1024 | 32768 | 5·6568542 | 3·1748021 | ·031250000 |
| 33 | 1089 | 35937 | 5·7445626 | 3·2075343 | ·030303030 |
| 34 | 1156 | 39304 | 5·8309519 | 3·2396118 | ·029411765 |
| 35 | 1225 | 42875 | 5·9160798 | 3·2710663 | ·028571429 |
| 36 | 1296 | 46656 | 6·0000000 | 3·3019272 | ·027777778 |
| 37 | 1369 | 50653 | 6·0827625 | 3·3322218 | ·027027027 |
| 38 | 1444 | 54872 | 6·1644140 | 3·3619754 | ·026315789 |
| 39 | 1521 | 59319 | 6·2449980 | 3·3912114 | ·025641026 |
| 40 | 1600 | 64000 | 6·3245553 | 3·4199519 | ·025000000 |
| 41 | 1681 | 68921 | 6·4031242 | 3·4482172 | ·024390244 |
| 42 | 1764 | 74088 | 6·4807407 | 3·4760266 | ·023809524 |
| 43 | 1849 | 79507 | 6·5574385 | 3·5033981 | ·023255814 |
| 44 | 1936 | 85184 | 6·6332496 | 3·5303483 | ·022727273 |
| 45 | 2025 | 91125 | 6·7082039 | 3·5568933 | ·022222222 |
| 46 | 2116 | 97336 | 6·7823300 | 3·5830479 | ·021739130 |
| 47 | 2209 | 103823 | 6·8556546 | 3·6088261 | ·021276600 |
| 48 | 2304 | 110592 | 6·9282032 | 3·6342411 | ·020833333 |
| 49 | 2401 | 117649 | 7·0000000 | 3·6593057 | ·020408163 |
| 50 | 2500 | 125000 | 7·0710678 | 3·6840314 | ·020000000 |
| 51 | 2601 | 132651 | 7·1414284 | 3·7084298 | ·019607843 |
| 52 | 2704 | 140608 | 7·2111026 | 3·7325111 | ·019230769 |
| 53 | 2809 | 148877 | 7·2801099 | 3·7562858 | ·018867925 |
| 54 | 2916 | 157464 | 7·3484692 | 3·7797631 | ·018518519 |
| 55 | 3025 | 166375 | 7·4161985 | 3·8029525 | ·018181818 |
| 56 | 3136 | 175616 | 7·4833148 | 3·8258624 | ·017857143 |
| 57 | 3249 | 185193 | 7·5498344 | 3·8485011 | ·017543860 |

| Number. | Squares. | Cubes. | Square Roots. | Cube Roots. | Reciprocals. |
|---|---|---|---|---|---|
| 58 | 3364 | 195112 | 7·6157731 | 3·8708766 | ·017241379 |
| 59 | 3481 | 205379 | 7·6811457 | 3·8929965 | ·016949153 |
| 60 | 3600 | 216000 | 7·7459667 | 3·9148676 | ·016666667 |
| 61 | 3721 | 226981 | 7·8102497 | 3·9304972 | ·016393443 |
| 62 | 3844 | 238328 | 7·8740079 | 3·9578915 | ·016129032 |
| 63 | 3969 | 250047 | 7·9372539 | 3·9790571 | ·015873016 |
| 64 | 4096 | 262144 | 8·0000000 | 4·0000000 | ·015625000 |
| 65 | 4225 | 274625 | 8·0622577 | 4·0207256 | ·015384615 |
| 66 | 4356 | 287496 | 8·1240384 | 4·0412401 | ·015151515 |
| 67 | 4489 | 300763 | 8·1853528 | 4·0615480 | ·014925373 |
| 68 | 4624 | 314432 | 8·2462113 | 4·0816551 | ·014705882 |
| 69 | 4761 | 328509 | 8·3066239 | 4·1015661 | ·014492754 |
| 70 | 4900 | 343000 | 8·3666003 | 4·1212853 | ·014285714 |
| 71 | 5041 | 357911 | 8·4261498 | 4·1408178 | ·014084517 |
| 72 | 5184 | 373248 | 8·4852814 | 4·1601676 | ·013888889 |
| 73 | 5329 | 389017 | 8·5440037 | 4·1793390 | ·013698630 |
| 74 | 5476 | 405224 | 8·6023253 | 4·1983364 | ·013513514 |
| 75 | 5625 | 421875 | 8·6602540 | 4·2171633 | ·013333333 |
| 76 | 5776 | 438976 | 8·7177979 | 4·2358236 | ·013157895 |
| 77 | 5929 | 456533 | 8·7749644 | 4·2543210 | ·012987013 |
| 78 | 6084 | 474552 | 8·8317609 | 4·2726586 | ·012820513 |
| 79 | 6241 | 493039 | 8·8881944 | 4·2908404 | ·012658228 |
| 80 | 6400 | 512000 | 8·9442719 | 4·3088695 | ·012500000 |
| 81 | 6561 | 531441 | 9·0000000 | 4·3267487 | ·012345679 |
| 82 | 6724 | 551368 | 9·0553851 | 4·3444815 | ·012195122 |
| 83 | 6889 | 571787 | 9·1104336 | 4·3620707 | ·012048193 |
| 84 | 7056 | 592704 | 9·1651514 | 4·3795191 | ·011904762 |
| 85 | 7225 | 614125 | 9·2195445 | 4·3968296 | ·011764706 |
| 86 | 7396 | 636056 | 9·2736185 | 4·4140049 | ·011627907 |
| 87 | 7569 | 658503 | 9·3273791 | 4·4310476 | ·011494253 |
| 88 | 7744 | 681472 | 9·3808315 | 4·4470692 | ·011363636 |
| 89 | 7921 | 704969 | 9·4339811 | 4·4647451 | ·011235955 |
| 90 | 8100 | 729000 | 9·4868330 | 4·4814047 | ·011111111 |
| 91 | 8281 | 753571 | 9·5393920 | 4·4979414 | ·010989011 |
| 92 | 8464 | 778688 | 9·5916630 | 4·5143574 | ·010869565 |
| 93 | 8649 | 804357 | 9·6436508 | 4·5306549 | ·010752688 |
| 94 | 8836 | 830584 | 9·6953597 | 4·5468359 | ·010638298 |
| 95 | 9025 | 857374 | 9·7467943 | 4·5629026 | ·010526316 |
| 96 | 9216 | 884736 | 9·7979590 | 4·5788570 | ·010416667 |
| 97 | 9409 | 912673 | 9·8488578 | 4·5947009 | ·010309278 |
| 98 | 9604 | 941192 | 9·8994949 | 4·6104363 | ·010204082 |
| 99 | 9801 | 970299 | 9·9498744 | 4·6260650 | ·010101010 |
| 100 | 10000 | 1000000 | 10·0000000 | 4·6415888 | ·010000000 |
| 101 | 10201 | 1030301 | 10·0498756 | 4·6570095 | ·009900990 |
| 102 | 10404 | 1061208 | 10·0995049 | 4·6723287 | ·009803922 |
| 103 | 10609 | 1092727 | 10·1488916 | 4·6875482 | ·009708738 |
| 104 | 10816 | 1124864 | 10·1980390 | 4·7026694 | ·009615385 |
| 105 | 11025 | 1157625 | 10·2469508 | 4·7176940 | ·009523810 |
| 106 | 11236 | 1191016 | 10·2956301 | 4·7326235 | ·009433962 |
| 107 | 11449 | 1225043 | 10·3440804 | 4·7474594 | ·009345794 |
| 108 | 11664 | 1259712 | 10·3923048 | 4·7622032 | ·009259259 |
| 109 | 11881 | 1295029 | 10·4403065 | 4·7768562 | ·009174312 |
| 110 | 12100 | 1331000 | 10·4880885 | 4·7914199 | ·009090909 |
| 111 | 12321 | 1367631 | 10·5356538 | 4·8058995 | ·009009009 |
| 112 | 12544 | 1404928 | 10·5830052 | 4·8202845 | ·008928571 |
| 113 | 12769 | 1442897 | 10·6301458 | 4·8345881 | ·008849558 |
| 114 | 12996 | 1481544 | 10·6770783 | 4·8488076 | ·008771930 |
| 115 | 13225 | 1520875 | 10·7238053 | 4·8629442 | ·008695652 |
| 116 | 13456 | 1560896 | 10·7703296 | 4·8769990 | ·008620690 |
| 117 | 13689 | 1601613 | 10·8166538 | 4·8909732 | ·008547009 |
| 118 | 13924 | 1643032 | 10·8627805 | 4·9048681 | ·008474576 |
| 119 | 14161 | 1685159 | 10·9087121 | 4·9186847 | ·008403361 |

| Number. | Squares. | Cubes. | Square Roots. | Cube Roots. | Reciprocals. |
|---|---|---|---|---|---|
| 120 | 14400 | 1728000 | 10·9544512 | 4·9324242 | ·008333333 |
| 121 | 14641 | 1771561 | 11·0000000 | 4·9460874 | ·008264463 |
| 122 | 14884 | 1815848 | 11·0453610 | 4·9596757 | ·008196721 |
| 123 | 15129 | 1860867 | 11·0905365 | 4·9731898 | ·008130081 |
| 124 | 15376 | 1906624 | 11·1355287 | 4·9866310 | ·008064516 |
| 125 | 15625 | 1953125 | 11·1803399 | 5·0000000 | ·008000000 |
| 126 | 15876 | 2000376 | 11·2249722 | 5·0132979 | ·007936508 |
| 127 | 16129 | 2048383 | 11·2694277 | 5·0265257 | ·007874016 |
| 128 | 16384 | 2097152 | 11·3137085 | 5·0396842 | ·007812500 |
| 129 | 16641 | 2146689 | 11·3578167 | 5·0527743 | ·007751938 |
| 130 | 16900 | 2197000 | 11·4017543 | 5·0657970 | ·007692308 |
| 131 | 17161 | 2248091 | 11·4455231 | 5·0787531 | ·007633588 |
| 132 | 17424 | 2299968 | 11·4891253 | 5·0916434 | ·007575758 |
| 133 | 17689 | 2352637 | 11·5325626 | 5·1044687 | ·007518797 |
| 134 | 17956 | 2406104 | 11·5758369 | 5·1172299 | ·007462687 |
| 135 | 18225 | 2460375 | 11·6189500 | 5·1299278 | ·007407407 |
| 136 | 18496 | 2515456 | 11·6619038 | 5·1425632 | ·007352941 |
| 137 | 18769 | 2571353 | 11·7046999 | 5·1551367 | ·007299270 |
| 138 | 19044 | 2628072 | 11·7473444 | 5·1676493 | ·007246377 |
| 139 | 19321 | 2685619 | 11·7898261 | 5·1801015 | ·007194245 |
| 140 | 19600 | 2744000 | 11·8321596 | 5·1924941 | ·007142857 |
| 141 | 19881 | 2803221 | 11·8743421 | 5·2048279 | ·007092199 |
| 142 | 20164 | 2863288 | 11·9163753 | 5·2171034 | ·007042254 |
| 143 | 20449 | 2924207 | 11·9582607 | 5·2293215 | ·006993007 |
| 144 | 20736 | 2985984 | 12·0000000 | 5·2414828 | ·006944444 |
| 145 | 21025 | 3048625 | 12·0415946 | 5·2535879 | ·006896552 |
| 146 | 21316 | 3112136 | 12·0830460 | 5·2656374 | ·006849315 |
| 147 | 21609 | 3176523 | 12·1243557 | 5·2776321 | ·006802721 |
| 148 | 21904 | 3241792 | 12·1655251 | 5·2895725 | ·006756757 |
| 149 | 22201 | 3307949 | 12·2065556 | 5·3014592 | ·006711409 |
| 150 | 22500 | 3375000 | 12·2474487 | 5·3132928 | ·006666667 |
| 151 | 22801 | 3442951 | 12·2882057 | 5·3250740 | ·006622517 |
| 152 | 23104 | 3511008 | 12·3288280 | 5·3368033 | ·006578947 |
| 153 | 23409 | 3581577 | 12·3693169 | 5·3484812 | ·006535948 |
| 154 | 23716 | 3652264 | 12·4096736 | 5·3601084 | ·006493506 |
| 155 | 24025 | 3723875 | 12·4498996 | 5·3716854 | ·006451613 |
| 156 | 24336 | 3796416 | 12·4899960 | 5·3832126 | ·006410256 |
| 157 | 24649 | 3869893 | 12·5299641 | 5·3946907 | ·006369427 |
| 158 | 24964 | 3944312 | 12·5698051 | 5·4061202 | ·006329114 |
| 159 | 25281 | 4019679 | 12·6095202 | 5·4175015 | ·006289308 |
| 160 | 25600 | 4096000 | 12·6491106 | 5·4288352 | ·006250000 |
| 161 | 25921 | 4173281 | 12·6885775 | 5·4401218 | ·006211180 |
| 162 | 26244 | 4251528 | 12·7279221 | 5·4513618 | ·006172840 |
| 163 | 26569 | 4330747 | 12·7671453 | 5·4625556 | ·006134969 |
| 164 | 26896 | 4410944 | 12·8062485 | 5·4737037 | ·006097561 |
| 165 | 27225 | 4492125 | 12·8452326 | 5·4848066 | ·006060606 |
| 166 | 27556 | 4574296 | 12·8840987 | 5·4958647 | ·006024096 |
| 167 | 27889 | 4657463 | 12·9228480 | 5·5068784 | ·005988024 |
| 168 | 28224 | 4741632 | 12·9614814 | 5·5178484 | ·005952381 |
| 169 | 28561 | 4826809 | 13·0000000 | 5·5287748 | ·005917160 |
| 170 | 28900 | 4913000 | 13·0384048 | 5·5396583 | ·005882353 |
| 171 | 29241 | 5000211 | 13·0766968 | 5·5504991 | ·005847953 |
| 172 | 29584 | 5088448 | 13·1148770 | 5·5612978 | ·005813953 |
| 173 | 29929 | 5177717 | 13·1529464 | 5·5720546 | ·005780347 |
| 174 | 30276 | 5268024 | 13·1909060 | 5·5827702 | ·005747126 |
| 175 | 30625 | 5359375 | 13·2287566 | 5·5934447 | ·005714286 |
| 176 | 30976 | 5451776 | 13·2664992 | 5·6040787 | ·005681818 |
| 177 | 31329 | 5545233 | 13·3041347 | 5·6146724 | ·005649718 |
| 178 | 31684 | 5639752 | 13·3416641 | 5·6252263 | ·005617978 |
| 179 | 32041 | 5735339 | 13·3790882 | 5·6357408 | ·005586592 |
| 180 | 32400 | 5832000 | 13·4164079 | 5·6462162 | ·005555556 |
| 181 | 32761 | 5929741 | 13·4536240 | 5·6566528 | ·005524862 |

| Number. | Squares. | Cubes. | Square Roots. | Cube Roots. | Reciprocals. |
|---|---|---|---|---|---|
| 182 | 33124 | 6028568 | 13·4907376 | 5·6670511 | ·005494505 |
| 183 | 33489 | 6128487 | 13·5277493 | 5·6774114 | ·005464481 |
| 184 | 33856 | 6229504 | 13·5646600 | 5·6877340 | ·005434783 |
| 185 | 34225 | 6331625 | 13·6014705 | 5·6980192 | ·005405405 |
| 186 | 34596 | 6434856 | 13·6381817 | 5·7082675 | ·005376344 |
| 187 | 34969 | 6539203 | 13·6747943 | 5·7184791 | ·005347594 |
| 188 | 35344 | 6644672 | 13·7113092 | 5·7286543 | ·005319149 |
| 189 | 35721 | 6751269 | 13·7477271 | 5·7387936 | ·005291005 |
| 190 | 36100 | 6859000 | 13·7840488 | 5·7488971 | ·005263158 |
| 191 | 36481 | 6967871 | 13·8202750 | 5·7589652 | ·005235602 |
| 192 | 36864 | 7077888 | 13·8564065 | 5·7689982 | ·005208333 |
| 193 | 37249 | 7189517 | 13·8924400 | 5·7789966 | ·005181347 |
| 194 | 37636 | 7301384 | 13·9283883 | 5·7889604 | ·005154639 |
| 195 | 38025 | 7414875 | 13·9642400 | 5·7988900 | ·005128205 |
| 196 | 38416 | 7529536 | 14·0000000 | 5·8087857 | ·005102041 |
| 197 | 38809 | 7645373 | 14·0356688 | 5·8186479 | ·005076142 |
| 198 | 39204 | 7762392 | 14·0712473 | 5·8284867 | ·005050505 |
| 199 | 39601 | 7880599 | 14·1067360 | 5·8382725 | ·005025126 |
| 200 | 40000 | 8000000 | 14·1421356 | 5·8480355 | ·005000000 |
| 201 | 40401 | 8120601 | 14·1774469 | 5·8577660 | ·004975124 |
| 202 | 40804 | 8242408 | 14·2126704 | 5·8674673 | ·004950495 |
| 203 | 41209 | 8365427 | 14·2478068 | 5·8771307 | ·004926108 |
| 204 | 41616 | 8489664 | 14·2828569 | 5·8867653 | ·004901961 |
| 205 | 42025 | 8615125 | 14·3178211 | 5·8963685 | ·004878049 |
| 206 | 42436 | 8741816 | 14·3527001 | 5·9059406 | ·004854369 |
| 207 | 42849 | 8869743 | 14·3874946 | 5·9154817 | ·004830918 |
| 208 | 43264 | 8998912 | 14·4222051 | 5·9249921 | ·004807692 |
| 209 | 43681 | 9129329 | 14·4568323 | 5·9344721 | ·004784689 |
| 210 | 44100 | 9261000 | 14·4913767 | 5·9439220 | ·004761905 |
| 211 | 44521 | 9393931 | 14·5258390 | 5·9533418 | ·004739336 |
| 212 | 44944 | 9528128 | 14·5602198 | 5·9627320 | ·004716981 |
| 213 | 45369 | 9663597 | 14·5945195 | 5·9720926 | ·004694836 |
| 214 | 45796 | 9800344 | 14·6287388 | 5·9814240 | ·004672897 |
| 215 | 46225 | 9938375 | 14·6628783 | 5·9907264 | ·004651163 |
| 216 | 46656 | 10077696 | 14·6969385 | 6·0000000 | ·004629630 |
| 217 | 47089 | 10218313 | 14·7309199 | 6·0092450 | ·004608295 |
| 218 | 47524 | 10360232 | 14·7648231 | 6·0184617 | ·004587156 |
| 219 | 47961 | 10503459 | 14·7986486 | 6·0276502 | ·004566210 |
| 220 | 48400 | 10648000 | 14·8323970 | 6·0368107 | ·004545455 |
| 221 | 48841 | 10793861 | 14·8660687 | 6·0459435 | ·004524887 |
| 222 | 49284 | 10941048 | 14·8996644 | 6·0550489 | ·004504505 |
| 223 | 49729 | 11089567 | 14·9331845 | 6·0641270 | ·004484305 |
| 224 | 50176 | 11239424 | 14·9666295 | 6·0731779 | ·004464286 |
| 225 | 50625 | 11390625 | 15·0000000 | 6·0824020 | ·004444444 |
| 226 | 51076 | 11543176 | 15·0332964 | 6·0991994 | ·004424779 |
| 227 | 51529 | 11697083 | 15·0665192 | 6·1001702 | ·004405286 |
| 228 | 51984 | 11852352 | 15·0996689 | 6·1091147 | ·004385965 |
| 229 | 52441 | 12008989 | 15·1327460 | 6·1180332 | ·004366812 |
| 230 | 52900 | 12167000 | 15·1657509 | 6·1269257 | ·004347826 |
| 231 | 53361 | 12326391 | 15·1986842 | 6·1357924 | ·004329004 |
| 232 | 53824 | 12487168 | 15·2315462 | 6·1446337 | ·004310345 |
| 233 | 54289 | 12649337 | 15·2643375 | 6·1534495 | ·004291845 |
| 234 | 54756 | 12812904 | 15·2970585 | 6·1622401 | ·004273504 |
| 235 | 55225 | 12977875 | 15·3297097 | 6·1710058 | ·004255319 |
| 236 | 55696 | 13144256 | 15·3622915 | 6·1797466 | ·004237288 |
| 237 | 56169 | 13312053 | 15·3948043 | 6·1884628 | ·004219409 |
| 238 | 56644 | 13481272 | 15·4272486 | 6·1971544 | ·004201681 |
| 239 | 57121 | 13651919 | 15·4596248 | 6·2058218 | ·004184100 |
| 240 | 57600 | 13824000 | 15·4919334 | 6·2144650 | ·004166667 |
| 241 | 58081 | 13997521 | 15·5241747 | 6·2230843 | ·004149378 |
| 242 | 58564 | 14172488 | 15·5563492 | 6·2316797 | ·004132231 |
| 243 | 59049 | 14348907 | 15·5884573 | 6·2402515 | ·004115226 |

| Number. | Squares. | Cubes. | Square Roots. | Cube Roots. | Reciprocals. |
|---|---|---|---|---|---|
| 244 | 59536 | 14526784 | 15·6204994 | 6·2487998 | ·004098361 |
| 245 | 60025 | 14706125 | 15·6524758 | 6·2573248 | ·004081633 |
| 246 | 60516 | 14886936 | 15·6843871 | 6·2658266 | ·004065041 |
| 247 | 61009 | 15069223 | 15·7162336 | 6·2743054 | ·004048583 |
| 248 | 61504 | 15252992 | 15·7480157 | 6·2827613 | ·004032258 |
| 249 | 62001 | 15438249 | 15·7797338 | 6·2911946 | ·004016064 |
| 250 | 62500 | 15625000 | 15·8113883 | 6·2996053 | ·004000000 |
| 251 | 63001 | 15813251 | 15·8429795 | 6·3079935 | ·003984064 |
| 252 | 63504 | 16003008 | 15·8745079 | 6·3163596 | ·003968254 |
| 253 | 64009 | 16194277 | 15·9059737 | 6·3247035 | ·003952569 |
| 254 | 64516 | 16387064 | 15·9373775 | 6·3330256 | ·003937008 |
| 255 | 65025 | 16581375 | 15·9687194 | 6·3413257 | ·003921569 |
| 256 | 65536 | 16777216 | 16·0000000 | 6·3496042 | ·003906250 |
| 257 | 66049 | 16974593 | 16·0312195 | 6·3578611 | ·003891051 |
| 258 | 66564 | 17173512 | 16·0623784 | 6·3660968 | ·003875969 |
| 259 | 67081 | 17373979 | 16·0934769 | 6·3743111 | ·003861004 |
| 260 | 67600 | 17576000 | 16·1245155 | 6·3825043 | ·003846154 |
| 261 | 68121 | 17779581 | 16·1554944 | 6·3906765 | ·003831418 |
| 262 | 68644 | 17984728 | 16·1864141 | 6·3988279 | ·003816794 |
| 263 | 69169 | 18191447 | 16·2172747 | 6·4069585 | ·003802281 |
| 264 | 69696 | 18399744 | 16·2480768 | 6·4150687 | ·003787879 |
| 265 | 70225 | 18609625 | 16·2788206 | 6·4231583 | ·003773585 |
| 266 | 70756 | 18821096 | 16·3095064 | 6·4312276 | ·003759398 |
| 267 | 71289 | 19034163 | 16·3401346 | 6·4392767 | ·003745318 |
| 268 | 71824 | 19248832 | 16·3707055 | 6·4473057 | ·003731343 |
| 269 | 72361 | 19465109 | 16·4012195 | 6·4553148 | ·003717472 |
| 270 | 72900 | 19683000 | 16·4316767 | 6·4633041 | ·003703704 |
| 271 | 73441 | 19902511 | 16·4620776 | 6·4712736 | ·003690037 |
| 272 | 73984 | 20123648 | 16·4924225 | 6·4792236 | ·003676471 |
| 273 | 74529 | 20346417 | 16·5227116 | 6·4871541 | ·003663004 |
| 274 | 75076 | 20570824 | 16·5529454 | 6·4950653 | ·003649635 |
| 275 | 75625 | 20796875 | 16·5831240 | 6·5029572 | ·003636364 |
| 276 | 76176 | 21024576 | 16·6132477 | 6·5108300 | ·003623188 |
| 277 | 76729 | 21253933 | 16·6433170 | 6·5186839 | ·003610108 |
| 278 | 77284 | 21484952 | 16·6783320 | 6·5265189 | ·003597122 |
| 279 | 77841 | 21717639 | 16·7032931 | 6·5343351 | ·003584229 |
| 280 | 78400 | 21952000 | 16·7332005 | 6·5421326 | ·003571429 |
| 281 | 78961 | 22188041 | 16·7630546 | 6·5499116 | ·003558719 |
| 282 | 79524 | 22425768 | 16·7928556 | 6·5576722 | ·003546099 |
| 283 | 80089 | 22665187 | 16·8226038 | 6·5654144 | ·003533569 |
| 284 | 80656 | 22906304 | 16·8522995 | 6·5731385 | ·003522127 |
| 285 | 81225 | 23149125 | 16·8819430 | 6·5808443 | ·003508772 |
| 286 | 81796 | 23393656 | 16·9115345 | 6·5885323 | ·003496503 |
| 287 | 82369 | 23639903 | 16·9410743 | 6·5962023 | ·003484321 |
| 288 | 82944 | 23887872 | 16·9705627 | 6·6038545 | ·003472222 |
| 289 | 83521 | 24137569 | 17·0000000 | 6·6114890 | ·003460208 |
| 290 | 84100 | 24389000 | 17·0293864 | 6·6191060 | ·003448276 |
| 291 | 84681 | 24642171 | 17·0587221 | 6·6267054 | ·003436426 |
| 292 | 85264 | 24897088 | 17·0880075 | 6·6342874 | ·003424658 |
| 293 | 85849 | 25153757 | 17·1172428 | 6·6418522 | ·003412969 |
| 294 | 86436 | 25412184 | 17·1464282 | 6·6493998 | ·003401361 |
| 295 | 87025 | 25672375 | 17·1755640 | 6·6569302 | ·003389831 |
| 296 | 87616 | 25934836 | 17·2046505 | 6·6644437 | ·003378378 |
| 297 | 88209 | 26198073 | 17·2336879 | 6·6719403 | ·003367003 |
| 298 | 88804 | 26463592 | 17·2626765 | 6·6794200 | ·003355705 |
| 299 | 89401 | 26730899 | 17·2916165 | 6·6868831 | ·003344482 |
| 300 | 90000 | 27000000 | 17·3205081 | 6·6943295 | ·003333333 |
| 301 | 90601 | 27270901 | 17·3493516 | 6·7017593 | ·003322259 |
| 302 | 91204 | 27543608 | 17·3781472 | 6·7091729 | ·003311258 |
| 303 | 91809 | 27818127 | 17·4068952 | 6·7165700 | ·003301330 |
| 304 | 92416 | 28094464 | 17·4355958 | 6·7239508 | ·003289474 |
| 305 | 93025 | 28372625 | 17·4642492 | 6·7313155 | ·003278689 |

| Number. | Squares. | Cubes. | Square Roots. | Cube Roots. | Reciprocals. |
|---|---|---|---|---|---|
| 306 | 93636 | 28652616 | 17·4928557 | 6·7386641 | ·003267974 |
| 307 | 94249 | 28934443 | 17·5214155 | 6·7459967 | ·003257329 |
| 308 | 94864 | 29218112 | 17·5499288 | 6·7533134 | ·003246753 |
| 309 | 95481 | 29503609 | 17·5783958 | 6·7606143 | ·003236246 |
| 310 | 96100 | 29791000 | 17·6068169 | 6·7678995 | ·003225806 |
| 311 | 96721 | 30080231 | 17·6351921 | 6·7751690 | ·003215434 |
| 312 | 97344 | 30371328 | 17·6635217 | 6·7824229 | ·003205128 |
| 313 | 97969 | 30664297 | 17·6918060 | 6·7896613 | ·003194888 |
| 314 | 98596 | 30959144 | 17·7200451 | 6·7968844 | ·003184713 |
| 315 | 99225 | 31255875 | 17·7482393 | 6·8040921 | ·003174603 |
| 316 | 99856 | 31554496 | 17·7763888 | 6·8112847 | ·003164557 |
| 317 | 100489 | 31855013 | 17·8044938 | 6·8184620 | ·003154574 |
| 318 | 101124 | 32157432 | 17·8325545 | 6·8256242 | ·003144654 |
| 319 | 101761 | 32461759 | 17·8605711 | 6·8327714 | ·003134796 |
| 320 | 102400 | 32768000 | 17·8885438 | 6·8399037 | ·003125000 |
| 321 | 103041 | 33076161 | 17·9164729 | 6·8470213 | ·003115265 |
| 322 | 103684 | 33386248 | 17·9443584 | 6·8541240 | ·003105590 |
| 323 | 104329 | 33698267 | 17·9722008 | 6·8612120 | ·003095975 |
| 324 | 104976 | 34012224 | 18·0000000 | 6·8682855 | ·003086420 |
| 325 | 105625 | 34328125 | 18·0277564 | 6·8753433 | ·003076923 |
| 326 | 106276 | 34645976 | 18·0554701 | 6·8823888 | ·003067485 |
| 327 | 106929 | 34965783 | 18·0831413 | 6·8894188 | ·003058104 |
| 328 | 107584 | 35287552 | 18·1107703 | 6·8964345 | ·003048780 |
| 329 | 108241 | 35611289 | 18·1383571 | 6·9034359 | ·003039514 |
| 330 | 108900 | 35937000 | 18·1659021 | 6·9104232 | ·003030303 |
| 331 | 109561 | 36264691 | 18·1934054 | 6·9173964 | ·003021148 |
| 332 | 110224 | 36594368 | 18·2208672 | 6·9243556 | ·003012048 |
| 333 | 110889 | 36926037 | 18·2482876 | 6·9313088 | ·003003003 |
| 334 | 111556 | 37259704 | 18·2756669 | 6·9382321 | ·002994012 |
| 335 | 112225 | 37595375 | 18·3030052 | 6·9451496 | ·002985075 |
| 336 | 112896 | 37933056 | 18·3303028 | 6·9520533 | ·002976190 |
| 337 | 113569 | 38272753 | 18·3575598 | 6·9589434 | ·002967359 |
| 338 | 114244 | 38614472 | 18·3847763 | 6·9658198 | ·002958580 |
| 339 | 114921 | 38958219 | 18·4119526 | 6·9726826 | ·002949853 |
| 340 | 115600 | 39304000 | 18·4390889 | 6·9795321 | ·002941176 |
| 341 | 116281 | 39651821 | 18·4661853 | 6·9863681 | ·002932551 |
| 342 | 116964 | 40001688 | 18·4932420 | 6·9931906 | ·002923977 |
| 343 | 117649 | 40353607 | 18·5202592 | 7·0000000 | ·002915452 |
| 344 | 118336 | 40707584 | 18·5472370 | 7·0067962 | ·002906977 |
| 345 | 119025 | 41063625 | 18·5741756 | 7·0135791 | ·002898551 |
| 346 | 119716 | 41421736 | 18·6010752 | 7·0203490 | ·002890173 |
| 347 | 120409 | 41781923 | 18·6279360 | 7·0271058 | ·002881844 |
| 348 | 121104 | 42144192 | 18·6547581 | 7·0338497 | ·002873563 |
| 349 | 121801 | 42508549 | 18·6815417 | 7·0405860 | ·002865330 |
| 350 | 122500 | 42875000 | 18·7082869 | 7·0472987 | ·002857143 |
| 351 | 123201 | 43243551 | 18·7349940 | 7·0540041 | ·002849003 |
| 352 | 123904 | 43614208 | 18·7616630 | 7·0606967 | ·002840909 |
| 353 | 124609 | 43986977 | 18·7882942 | 7·0673767 | ·002832861 |
| 354 | 125316 | 44361864 | 18·8148877 | 7·0740440 | ·002824859 |
| 355 | 126025 | 44738875 | 18·8414437 | 7·0806988 | ·002816901 |
| 356 | 126736 | 45118016 | 18·8679623 | 7·0873411 | ·002808989 |
| 357 | 127449 | 45499293 | 18·8944436 | 7·0939709 | ·002801120 |
| 358 | 128164 | 45882712 | 18·9208879 | 7·1005885 | ·002793296 |
| 359 | 128881 | 46268279 | 18·9472953 | 7·1071937 | ·002785515 |
| 360 | 129600 | 46656000 | 18·9736660 | 7·1137866 | ·002777778 |
| 361 | 130321 | 47045831 | 19·0000000 | 7·1203674 | ·002770083 |
| 362 | 131044 | 47437928 | 19·0262976 | 7·1269360 | ·002762431 |
| 363 | 131769 | 47832147 | 19·0525589 | 7·1334925 | ·002754821 |
| 364 | 132496 | 48228544 | 19·0787840 | 7·1400370 | ·002747253 |
| 365 | 133225 | 48627125 | 19·1049732 | 7·1465695 | ·002739726 |
| 366 | 133956 | 49027896 | 19·1311265 | 7·1530901 | ·002732240 |
| 367 | 134689 | 49430863 | 19·1572441 | 7·1595988 | ·002724796 |

| Number. | Squares. | Cubes. | Square Roots. | Cube Roots. | Reciprocals. |
|---|---|---|---|---|---|
| 368 | 135424 | 49836032 | 19·1833261 | 7·1660957 | ·002717391 |
| 369 | 136161 | 50243409 | 19·2093727 | 7·1725809 | ·002710027 |
| 370 | 136900 | 50653000 | 19·2353841 | 7·1790544 | ·002702703 |
| 371 | 137641 | 51064811 | 19·2613603 | 7·1855162 | ·002695418 |
| 372 | 138384 | 51478848 | 19·2873015 | 7·1919663 | ·002688172 |
| 373 | 139129 | 51895117 | 19·3132079 | 7·1984050 | ·002680965 |
| 374 | 139876 | 52313624 | 19·3390796 | 7·2048322 | ·002673797 |
| 375 | 140625 | 52734375 | 19·3649167 | 7·2112479 | ·002666667 |
| 376 | 141376 | 53157376 | 19·3907194 | 7·2176522 | ·002659574 |
| 377 | 142129 | 53582633 | 19·4164878 | 7·2240450 | ·002652520 |
| 378 | 142884 | 54010152 | 19·4422221 | 7·2304268 | ·002645503 |
| 379 | 143641 | 54439939 | 19·4679223 | 7·2367972 | ·002638521 |
| 380 | 144400 | 54872000 | 19·4935887 | 7·2431565 | ·002631579 |
| 381 | 145161 | 55306341 | 19·5192213 | 7·2495045 | ·002624672 |
| 382 | 145924 | 55742968 | 19·5448203 | 7·2558415 | ·002617801 |
| 383 | 146689 | 56181887 | 19·5703858 | 7·2621675 | ·002610966 |
| 384 | 147456 | 56623104 | 19·5959179 | 7·2684824 | ·002604167 |
| 385 | 148225 | 57066625 | 19·6214169 | 7·2747864 | ·002597403 |
| 386 | 148996 | 57512456 | 19·6468827 | 7·2810794 | ·002590674 |
| 387 | 149769 | 57960603 | 19·6723156 | 7·2873617 | ·002583979 |
| 388 | 150544 | 58411072 | 19·6977156 | 7·2936330 | ·002577320 |
| 389 | 151321 | 58863869 | 19·7230829 | 7·2998936 | ·002570694 |
| 390 | 152100 | 59319000 | 19·7484177 | 7·3061436 | ·002564103 |
| 391 | 152881 | 59776471 | 19·7737199 | 7·3123828 | ·002557545 |
| 392 | 153664 | 60236288 | 19·7989899 | 7·3186114 | ·002551020 |
| 393 | 154449 | 60698457 | 19·8242276 | 7·3248295 | ·002544529 |
| 394 | 155236 | 61162984 | 19·8494332 | 7·3310369 | ·002538071 |
| 395 | 156025 | 61629875 | 19·8746069 | 7·3372339 | ·002531646 |
| 396 | 156816 | 62099136 | 19·8997487 | 7·3434205 | ·002525253 |
| 397 | 157609 | 62570773 | 19·9248588 | 7·3495966 | ·002518892 |
| 398 | 158404 | 63044792 | 19·9499373 | 7·3557624 | ·002512563 |
| 399 | 159201 | 63521199 | 19·9749844 | 7·3619178 | ·002506266 |
| 400 | 160000 | 64000000 | 20·0000000 | 7·3680630 | ·002500009 |
| 401 | 160801 | 64481201 | 20·0249844 | 7·3741979 | ·002493766 |
| 402 | 161604 | 64964808 | 20·0499377 | 7·3803227 | ·002487562 |
| 403 | 162409 | 65450827 | 20·0748599 | 7·3864373 | ·002481390 |
| 404 | 163216 | 65939264 | 20·0997512 | 7·3925418 | ·002475248 |
| 405 | 164025 | 66430125 | 20·1246118 | 7·3986363 | ·002469136 |
| 406 | 164836 | 66923416 | 20·1494417 | 7·4047206 | ·002463054 |
| 407 | 165649 | 67419143 | 20·1742410 | 7·4107950 | ·002457002 |
| 408 | 166464 | 67917312 | 20·1990099 | 7·4168595 | ·002450980 |
| 409 | 167281 | 68417929 | 20·2237484 | 7·4229142 | ·002444988 |
| 410 | 168100 | 68921000 | 20·2484567 | 7·4289589 | ·002439024 |
| 411 | 168921 | 69426531 | 20·2731349 | 7·4349938 | ·002433090 |
| 412 | 169744 | 69934528 | 20·2977831 | 7·4410189 | ·002427184 |
| 413 | 170569 | 70444997 | 20·3224014 | 7·4470343 | ·002421308 |
| 414 | 171396 | 70957944 | 20·3469899 | 7·4530399 | ·002415459 |
| 415 | 172225 | 71473375 | 20·3715488 | 7·4590359 | ·002409639 |
| 416 | 173056 | 71991296 | 20·3960781 | 7·4650223 | ·002406846 |
| 417 | 173889 | 72511713 | 20·4205779 | 7·4709991 | ·002398082 |
| 418 | 174724 | 73034632 | 20·4450483 | 7·4769664 | ·002392344 |
| 419 | 175561 | 73560059 | 20·4694895 | 7·4829242 | ·002386635 |
| 420 | 176400 | 74088000 | 20·4939015 | 7·4888724 | ·002380952 |
| 421 | 177241 | 74618461 | 20·5182845 | 7·4948113 | ·002375297 |
| 422 | 178084 | 75151448 | 20·5426386 | 7·5007406 | ·002369668 |
| 423 | 178929 | 75686967 | 20·5669638 | 7·5066607 | ·002364066 |
| 424 | 179776 | 76225024 | 20·5912603 | 7·5125715 | ·002358491 |
| 425 | 180625 | 76765625 | 20·6155281 | 7·5184730 | ·002352941 |
| 426 | 181476 | 77308776 | 20·6397674 | 7·5243652 | ·002347418 |
| 427 | 182329 | 77854483 | 20·6639783 | 7·5302482 | ·002341920 |
| 428 | 183184 | 78402752 | 20·6881609 | 7·5361221 | ·002336449 |
| 429 | 184041 | 78953589 | 20·7123152 | 7·5419867 | ·002331002 |

| Number. | Squares. | Cubes. | Square Roots. | Cube Roots. | Reciprocals. |
|---|---|---|---|---|---|
| 430 | 184900 | 79507000 | 20·7364414 | 7·5478423 | ·002325581 |
| 431 | 185761 | 80062991 | 20·7605395 | 7·5536888 | ·002320186 |
| 432 | 186624 | 80621568 | 20·7846097 | 7·5595263 | ·002314815 |
| 433 | 187489 | 81182737 | 20·8086520 | 7·5653548 | ·002309469 |
| 434 | 188356 | 81746504 | 20·8326667 | 7·5711743 | ·002304147 |
| 435 | 189225 | 82312875 | 20·8566536 | 7·5769849 | ·002298851 |
| 436 | 190096 | 82881856 | 20·8806130 | 7·5827865 | ·002293578 |
| 437 | 190969 | 83453453 | 20·9045450 | 7·5885793 | ·002288330 |
| 438 | 191844 | 84027672 | 20·9284495 | 7·5943633 | ·002283105 |
| 439 | 192721 | 84604519 | 20·9523268 | 7·6001385 | ·002277904 |
| 440 | 193600 | 85184000 | 20·9761770 | 7·6059049 | ·002272727 |
| 441 | 194481 | 85766121 | 21·0000000 | 7·6116626 | ·002267574 |
| 442 | 195364 | 86350888 | 21·0237960 | 7·6174116 | ·002262443 |
| 443 | 196249 | 86938307 | 21·0475652 | 7·6231519 | ·002257336 |
| 444 | 197136 | 87528384 | 21·0713075 | 7·6288837 | ·002252252 |
| 445 | 198025 | 88121125 | 21·0950231 | 7·6346067 | ·002247191 |
| 446 | 198916 | 88716536 | 21·1187121 | 7·6403213 | ·002242152 |
| 447 | 199809 | 89314623 | 21·1423745 | 7·6460272 | ·002237136 |
| 448 | 200704 | 89915392 | 21·1660105 | 7·6517247 | ·002232143 |
| 449 | 201601 | 90518849 | 21·1896201 | 7·6574138 | ·002227171 |
| 450 | 202500 | 91125000 | 21·2132034 | 7·6630943 | ·002222222 |
| 451 | 203401 | 91733851 | 21·2367606 | 7·6687665 | ·002217295 |
| 452 | 204304 | 92345408 | 21·2602916 | 7·6744303 | ·002212389 |
| 453 | 205209 | 92959677 | 21·2837967 | 7·6800857 | ·002207506 |
| 454 | 206116 | 93576664 | 21·3072758 | 7·6857328 | ·002202643 |
| 455 | 207025 | 94196375 | 21·3307290 | 7·6913717 | ·002197802 |
| 456 | 207936 | 94818816 | 21·3541565 | 7·6970023 | ·002192982 |
| 457 | 208849 | 95443993 | 21·3775583 | 7·7026246 | ·002188184 |
| 458 | 209764 | 96071912 | 21·4009346 | 7·7082388 | ·002183406 |
| 459 | 210681 | 96702579 | 21·4242853 | 7·7188448 | ·002178649 |
| 460 | 211600 | 97336000 | 21·4476106 | 7·7194426 | ·002173913 |
| 461 | 212521 | 97972181 | 21·4709106 | 7·7250325 | ·002169197 |
| 462 | 213444 | 98611128 | 21·4941853 | 7·7306141 | ·002164502 |
| 463 | 214369 | 99252847 | 21·5174348 | 7·7361877 | ·002159827 |
| 464 | 215296 | 99897344 | 21·5406592 | 7·7417532 | ·002155172 |
| 465 | 216225 | 100544625 | 21·5638587 | 7·7473109 | ·002150538 |
| 466 | 217156 | 101194696 | 21·5870331 | 7·7528606 | ·002145923 |
| 467 | 218089 | 101847563 | 21·6101828 | 7·7584023 | ·002141328 |
| 468 | 219024 | 102503232 | 21·6333077 | 7·7639361 | ·002136752 |
| 469 | 219961 | 103161709 | 21·6564078 | 7·7694620 | ·002132196 |
| 470 | 220900 | 103823000 | 21·6794834 | 7·7749801 | ·002127660 |
| 471 | 221841 | 104487111 | 21·7025344 | 7·7804904 | ·002123142 |
| 472 | 222784 | 105154048 | 21·7255610 | 7·7859928 | ·002118644 |
| 473 | 223729 | 105828817 | 21·7485632 | 7·7914875 | ·002114165 |
| 474 | 224676 | 106496424 | 21·7715411 | 7·7969745 | ·002109705 |
| 475 | 225625 | 107171875 | 21·7944947 | 7·8024538 | ·002105263 |
| 476 | 226576 | 107850176 | 21·8174242 | 7·8079254 | ·002100840 |
| 477 | 227529 | 108531333 | 21·8403297 | 7·8133892 | ·002096486 |
| 478 | 228484 | 109215352 | 21·8632111 | 7·8188456 | ·002092050 |
| 479 | 229441 | 109902239 | 21·8860686 | 7·8242942 | ·002087683 |
| 480 | 230400 | 110592000 | 21·9089023 | 7·8297353 | ·002083333 |
| 481 | 231361 | 111284641 | 21·9317122 | 7·8351688 | ·002079002 |
| 482 | 232324 | 111980168 | 21·9544984 | 7·8405949 | ·002074689 |
| 483 | 233289 | 112678587 | 21·9772610 | 7·8460134 | ·002070393 |
| 484 | 234256 | 113379904 | 22·0000000 | 7·8514244 | ·002066116 |
| 485 | 235225 | 114084125 | 22·0227155 | 7·8568281 | ·002061856 |
| 486 | 236196 | 114791256 | 22·0454077 | 7·8622242 | ·002057613 |
| 487 | 237169 | 115501303 | 22·0680765 | 7·8676130 | ·002053388 |
| 488 | 238144 | 116214272 | 22·0907220 | 7·8729944 | ·002049180 |
| 489 | 239121 | 116930169 | 22·1133444 | 7·8783684 | ·002044990 |
| 490 | 240100 | 117649000 | 22·1359436 | 7·8837352 | ·002040816 |
| 491 | 241081 | 118370771 | 22·1585198 | 7·8890946 | ·002036660 |

| Number. | Squares. | Cubes. | Square Roots. | Cube Roots. | Reciprocals. |
|---|---|---|---|---|---|
| 492 | 242064 | 119095488 | 22·1810730 | 7·8944468 | ·002032520 |
| 493 | 243049 | 119823157 | 22·2036033 | 7·8997917 | ·002028398 |
| 494 | 244036 | 120553784 | 22·2261108 | 7·9051294 | ·002024291 |
| 495 | 245025 | 121287375 | 22·2485955 | 7·9104599 | ·002020202 |
| 496 | 246016 | 122023936 | 22·2710575 | 7·9157832 | ·002016129 |
| 497 | 247009 | 122763473 | 22·2934968 | 7·9210994 | ·002012072 |
| 498 | 248004 | 123505992 | 22·3159136 | 7·9264085 | ·002008032 |
| 499 | 249001 | 124251499 | 22·3383079 | 7·9317104 | ·002004008 |
| 500 | 250000 | 125000000 | 22·3606798 | 7·9370053 | ·002000000 |
| 501 | 251001 | 125751501 | 22·3830293 | 7·9422931 | ·001996008 |
| 502 | 252004 | 126506008 | 22·4053565 | 7·9475739 | ·001992032 |
| 503 | 253009 | 127263527 | 22·4276615 | 7·9528477 | ·001988072 |
| 504 | 254016 | 128024064 | 22·4499443 | 7·9581144 | ·001984127 |
| 505 | 255025 | 128787625 | 22·4722051 | 7·9633743 | ·001980198 |
| 506 | 256036 | 129554216 | 22·4944438 | 7·9686271 | ·001976285 |
| 507 | 257049 | 130323843 | 22·5166605 | 7·9738731 | ·001972387 |
| 508 | 258064 | 131096512 | 22·5388553 | 7·9791122 | ·001968504 |
| 509 | 259081 | 131872229 | 22·5610283 | 7·9843444 | ·001964637 |
| 510 | 260100 | 132651000 | 22·5831796 | 7·9895697 | ·001960784 |
| 511 | 261121 | 133432831 | 22·6053091 | 7·9947883 | ·001956947 |
| 512 | 262144 | 134217728 | 22·6274170 | 8·0000000 | ·001953125 |
| 513 | 263169 | 135005697 | 22·6495033 | 8·0052049 | ·001949318 |
| 514 | 264196 | 135796744 | 22·6715681 | 8·0104032 | ·001945525 |
| 515 | 265225 | 136590875 | 22·6936114 | 8·0155946 | ·001941748 |
| 516 | 266256 | 137388096 | 22·7156334 | 8·0207794 | ·001937984 |
| 517 | 267289 | 138188413 | 22·7376341 | 8·0259574 | ·001934236 |
| 518 | 268324 | 138991832 | 22·7596134 | 8·0311287 | ·001930502 |
| 519 | 269361 | 139798359 | 22·7815715 | 8·0362935 | ·001926782 |
| 520 | 270400 | 140608000 | 22·8035085 | 8·0414515 | ·001923077 |
| 521 | 271411 | 141420761 | 22·8254244 | 8·0466030 | ·001919386 |
| 522 | 272484 | 142236648 | 22·8473193 | 8·0517479 | ·001915709 |
| 523 | 273529 | 143055667 | 22·8691933 | 8·0568862 | ·001912046 |
| 524 | 274576 | 143877824 | 22·8910463 | 8·0620180 | ·001908397 |
| 525 | 275625 | 144703125 | 22·9128785 | 8·0671432 | ·001904762 |
| 526 | 276676 | 145531576 | 22·9346899 | 8·0722620 | ·001901141 |
| 527 | 277729 | 146363183 | 22·9564806 | 8·0773743 | ·001897533 |
| 528 | 278784 | 147197952 | 22·9782506 | 8·0824800 | ·001893939 |
| 529 | 279841 | 148035889 | 23·0000000 | 8·0875794 | ·001890359 |
| 530 | 280900 | 148877001 | 23·0217289 | 8·0926723 | ·001886792 |
| 531 | 281961 | 149721291 | 23·0434372 | 8·0977589 | ·001883239 |
| 532 | 283024 | 150568768 | 23·0651252 | 8·1028390 | ·001879699 |
| 533 | 284089 | 151419437 | 23·0867928 | 8·1079128 | ·001876173 |
| 534 | 285156 | 152273304 | 23·1084400 | 8·1129803 | ·001872659 |
| 535 | 286225 | 153130375 | 23·1300670 | 8·1180414 | ·001869159 |
| 536 | 287296 | 153990656 | 23·1516738 | 8·1230962 | ·001865672 |
| 537 | 288369 | 154854153 | 23·1732605 | 8·1281447 | ·001862197 |
| 538 | 289444 | 155720872 | 23·1948270 | 8·1331870 | ·001858736 |
| 539 | 290521 | 156590819 | 23·2163735 | 8·1382230 | ·001855288 |
| 540 | 291600 | 157464000 | 23·2379001 | 8·1432529 | ·001851852 |
| 541 | 292681 | 158340421 | 23·2594067 | 8·1482765 | ·001848429 |
| 542 | 293764 | 159220088 | 23·2808935 | 8·1532939 | ·001845018 |
| 543 | 294849 | 160103007 | 23·3023604 | 8·1583051 | ·001841621 |
| 544 | 295936 | 160989184 | 23·3238076 | 8·1633102 | ·001838235 |
| 545 | 297025 | 161878625 | 23·3452351 | 8·1683092 | ·001834862 |
| 546 | 298116 | 162771336 | 23·3666429 | 8·1733020 | ·001831502 |
| 547 | 299209 | 163667323 | 23·3880311 | 8·1782888 | ·001828154 |
| 548 | 300304 | 164566592 | 23·4093998 | 8·1832695 | ·001824818 |
| 549 | 301401 | 165469149 | 23·4307490 | 8·1882441 | ·001821494 |
| 550 | 302500 | 166375000 | 23·4520788 | 8·1932127 | ·001818182 |
| 551 | 303601 | 167284151 | 23·4733892 | 8·1981753 | ·001814882 |
| 552 | 304704 | 168196608 | 23·4946802 | 8·2031319 | ·001811594 |
| 553 | 305809 | 169112377 | 23·5159520 | 8·2080825 | ·001808318 |

| Number. | Squares. | Cubes. | Square Roots. | Cube Roots. | Reciprocals. |
|---|---|---|---|---|---|
| 554 | 306916 | 170031464 | 23·5372046 | 8·2130271 | ·001805054 |
| 555 | 308025 | 170953875 | 23·5584380 | 8·2179657 | ·001801802 |
| 556 | 309136 | 171879616 | 23·5796522 | 8·2228985 | ·001798561 |
| 557 | 310249 | 172808693 | 23·6008474 | 8·2278254 | ·001795332 |
| 558 | 311364 | 173741112 | 23·6220236 | 8·2327463 | ·001792115 |
| 559 | 312481 | 174676879 | 23·6431808 | 8·2376614 | ·001788909 |
| 560 | 313600 | 175616000 | 23·6643191 | 8·2425706 | ·001785714 |
| 561 | 314721 | 176558481 | 23·6854386 | 8·2474740 | ·001782531 |
| 562 | 315844 | 177504328 | 23·7065392 | 8·2523715 | ·001779359 |
| 563 | 316969 | 178453547 | 23·7276210 | 8·2572635 | ·001776199 |
| 564 | 318096 | 179406144 | 23·7486842 | 8·2621492 | ·001773050 |
| 565 | 319225 | 180362125 | 23·7697286 | 8·2670294 | ·001769912 |
| 566 | 320356 | 181321496 | 23·7907545 | 8·2719039 | ·001766784 |
| 567 | 321489 | 182284263 | 23·8117618 | 8·2767726 | ·001763668 |
| 568 | 322624 | 183250432 | 23·8327506 | 8·2816255 | ·001760563 |
| 569 | 323761 | 184220009 | 23·8537209 | 8·2864928 | ·001757469 |
| 570 | 324900 | 185193000 | 23·8746728 | 8·2913444 | ·001754386 |
| 571 | 326041 | 186169411 | 23·8956063 | 8·2961903 | ·001751313 |
| 572 | 327184 | 187149248 | 23·9165215 | 8·3010304 | ·001748252 |
| 573 | 328329 | 188132517 | 23·9374184 | 8·3058651 | ·001745201 |
| 574 | 329476 | 189119224 | 23·9582971 | 8·3106941 | ·001742160 |
| 575 | 330625 | 190109375 | 23·9791576 | 8·3155175 | ·001739130 |
| 576 | 331776 | 191102976 | 24·0000000 | 8·3203353 | ·001736111 |
| 577 | 332927 | 192100033 | 24·0208243 | 8·3251475 | ·001733102 |
| 578 | 334084 | 193100552 | 24·0416306 | 8·3299542 | ·001730104 |
| 579 | 335241 | 194104539 | 24·0624188 | 8·3347553 | ·001727116 |
| 580 | 336400 | 195112000 | 24·0831891 | 8·3395509 | ·001724138 |
| 581 | 337561 | 196122941 | 24·1039416 | 8·3443410 | ·001721170 |
| 582 | 338724 | 197137368 | 24·1246762 | 8·3491256 | ·001718213 |
| 583 | 339889 | 198155287 | 24·1453929 | 8·3539047 | ·001715266 |
| 584 | 341056 | 199176704 | 24·1660919 | 8·3586784 | ·001712329 |
| 585 | 342225 | 200201625 | 24·1867732 | 8·3634466 | ·001709402 |
| 586 | 343396 | 201230056 | 24·2074369 | 8·3682095 | ·001706485 |
| 587 | 344569 | 202262003 | 24·2280829 | 8·3729668 | ·001703578 |
| 588 | 345744 | 203297472 | 24·2487113 | 8·3777188 | ·001700680 |
| 589 | 346921 | 204336469 | 24·2693222 | 8·3824653 | ·001697793 |
| 590 | 348100 | 205379000 | 24·2899156 | 8·3872065 | ·001694915 |
| 591 | 349281 | 206425071 | 24·3104996 | 8·3919428 | ·001692047 |
| 592 | 350464 | 207474688 | 24·3310501 | 8·3966729 | ·001689189 |
| 593 | 351649 | 208527857 | 24·3515913 | 8·4013981 | ·001686341 |
| 594 | 352836 | 209584584 | 24·3721152 | 8·4061180 | ·001683502 |
| 595 | 354025 | 210644875 | 24·3926218 | 8·4108326 | ·001680672 |
| 596 | 355216 | 211708736 | 24·4131112 | 8·4155419 | ·001677852 |
| 597 | 356409 | 212776173 | 24·4335834 | 8·4202460 | ·001675042 |
| 598 | 357604 | 213847192 | 24·4540385 | 8·4249448 | ·001672241 |
| 599 | 358801 | 214921799 | 24·4744765 | 8·4296383 | ·001669449 |
| 600 | 360000 | 216000000 | 24·4948974 | 8·4343267 | ·001666667 |
| 601 | 361201 | 217081801 | 24·5153013 | 8·4390098 | ·001663894 |
| 602 | 362404 | 218167208 | 24·5356883 | 8·4436877 | ·001661130 |
| 603 | 363609 | 219256227 | 24·5560583 | 8·4483605 | ·001658375 |
| 604 | 364816 | 220348864 | 24·5764115 | 8·4530281 | ·001655629 |
| 605 | 366025 | 221445125 | 24·5967478 | 8·4576906 | ·001652893 |
| 606 | 367236 | 222545016 | 24·6170673 | 8·4623479 | ·001650165 |
| 607 | 368449 | 223648543 | 24·6373700 | 8·4670001 | ·001647446 |
| 608 | 369664 | 224755712 | 24·6576560 | 8·4716471 | ·001644737 |
| 609 | 370881 | 225866529 | 24·6779254 | 8·4762892 | ·001642036 |
| 610 | 372100 | 226981000 | 24·6981781 | 8·4809261 | ·001639344 |
| 611 | 373321 | 228099131 | 24·7184142 | 8·4855579 | ·001636661 |
| 612 | 374544 | 229220928 | 24·7386338 | 8·4901848 | ·001633987 |
| 613 | 375769 | 230346397 | 24·7588368 | 8·4948065 | ·001631321 |
| 614 | 376996 | 231475544 | 24·7790234 | 8·4994233 | ·001628664 |
| 615 | 378225 | 232608375 | 24·7991935 | 8·5040350 | ·001626016 |

| Number. | Squares. | Cubes. | Square Roots. | Cube Roots. | Reciprocals. |
|---|---|---|---|---|---|
| 616 | 379456 | 233744896 | 24·8193473 | 8·5086417 | ·001623377 |
| 617 | 380689 | 234885113 | 24·8394847 | 8·5132435 | ·001620746 |
| 618 | 381924 | 236029032 | 24·8596058 | 8·5178403 | ·001618123 |
| 619 | 383161 | 237176659 | 24·8797106 | 8·5224331 | ·001615509 |
| 620 | 384400 | 238328000 | 24·8997992 | 8·5270189 | ·001612903 |
| 621 | 385641 | 239483061 | 24·9198716 | 8·5316009 | ·001610306 |
| 622 | 386884 | 240641848 | 24·9399278 | 8·5361780 | ·001607717 |
| 623 | 388129 | 241804367 | 24·9599679 | 8·5407501 | ·001605136 |
| 624 | 389376 | 242970624 | 24·9799920 | 8·5453173 | ·001602564 |
| 625 | 390625 | 244140625 | 25·0000000 | 8·5498797 | ·001600000 |
| 626 | 391876 | 245134376 | 25·0199920 | 8·5544372 | ·001597444 |
| 627 | 393129 | 246491883 | 25·0399681 | 8·5589899 | ·001594896 |
| 628 | 394384 | 247673152 | 25·0599282 | 8·5635377 | ·001592357 |
| 629 | 395641 | 248858189 | 25·0798724 | 8·5680807 | ·001589825 |
| 630 | 396900 | 250047000 | 25·0998008 | 8·5726189 | ·001587302 |
| 631 | 398161 | 251239591 | 25·1197134 | 8·5771523 | ·001584786 |
| 632 | 399424 | 252435968 | 25·1396102 | 8·5816809 | ·001582278 |
| 633 | 400689 | 253636137 | 25·1594913 | 8·5862247 | ·001579779 |
| 634 | 401956 | 254840104 | 25·1793566 | 8·5907238 | ·001577287 |
| 635 | 403225 | 256047875 | 25·1992063 | 8·5952380 | ·001574803 |
| 636 | 404496 | 257259456 | 25·2190404 | 8·5997476 | ·001572327 |
| 637 | 405769 | 258474853 | 25·2388589 | 8·6042525 | ·001569859 |
| 638 | 407044 | 259694072 | 25·2586619 | 8·6087526 | ·001567398 |
| 639 | 408321 | 260917119 | 25·2784493 | 8·6132480 | ·001564945 |
| 640 | 409600 | 262144000 | 25·2982213 | 8·6177388 | ·001562500 |
| 641 | 410881 | 263374721 | 25·3179778 | 8·6222248 | ·001560062 |
| 642 | 412164 | 264609288 | 25·3377189 | 8·6267063 | ·001557632 |
| 643 | 413449 | 265847707 | 23·3574447 | 8·6311830 | ·001555210 |
| 644 | 414736 | 267089984 | 25·3771551 | 8·6356551 | ·001552795 |
| 645 | 416125 | 268336125 | 25·3968502 | 8·6401226 | ·001550388 |
| 646 | 417316 | 269585136 | 25·4165302 | 8·6445855 | ·001547988 |
| 647 | 418609 | 270840023 | 25·4361947 | 8·6490437 | ·001545595 |
| 648 | 419904 | 272097792 | 25·4558441 | 8·6534974 | ·001543210 |
| 649 | 421201 | 273359449 | 25·4754784 | 8·6579465 | ·001540832 |
| 650 | 422500 | 274625000 | 25·4950976 | 8·6623911 | ·001538462 |
| 651 | 423801 | 275894451 | 25·5147013 | 8·6668310 | ·001536098 |
| 652 | 425104 | 277167808 | 25·5342907 | 8·6712665 | ·001533742 |
| 653 | 426409 | 278445077 | 25·5538647 | 8·6756974 | ·001531394 |
| 654 | 427716 | 279726264 | 25·5734237 | 8·6801237 | ·001529052 |
| 655 | 429025 | 281011375 | 25·5929678 | 8·6845456 | ·001526718 |
| 656 | 430336 | 282300416 | 25·6124969 | 8·6889630 | ·001524390 |
| 657 | 431639 | 283593393 | 25·6320112 | 8·6933759 | ·001522070 |
| 658 | 432964 | 284890312 | 25·6515107 | 8·6977843 | ·001519751 |
| 659 | 434281 | 286191179 | 25·6709953 | 8·7021882 | ·001517451 |
| 660 | 435600 | 287496000 | 25·6904652 | 8·7065877 | ·001515152 |
| 661 | 436921 | 288804781 | 25·7099203 | 8·7109827 | ·001512859 |
| 662 | 438244 | 290117528 | 25·7293607 | 8·7153734 | ·001510574 |
| 663 | 439569 | 291434247 | 25·7487864 | 8·7197596 | ·001508296 |
| 664 | 440896 | 292754944 | 25·7681975 | 8·7241414 | ·001506024 |
| 665 | 442225 | 294079625 | 25·7875939 | 8·7285187 | ·001503759 |
| 666 | 443556 | 295408296 | 25·8069758 | 8·7328918 | ·001501502 |
| 667 | 444899 | 296740963 | 25·8263431 | 8·7372604 | ·001499250 |
| 668 | 446224 | 298077632 | 25·8456960 | 8·7416246 | ·001497006 |
| 669 | 447561 | 299418309 | 25·8650343 | 8·7459846 | ·001494768 |
| 670 | 448900 | 300763000 | 25·8843582 | 8·7503401 | ·001492537 |
| 671 | 450241 | 302111711 | 25·9036677 | 8·7546913 | ·001490313 |
| 672 | 451584 | 303464448 | 25·9229628 | 8·7590383 | ·001488095 |
| 673 | 452929 | 304821217 | 25·9422435 | 8·7633809 | ·001485884 |
| 674 | 454276 | 306182024 | 25·9615100 | 8·7677192 | ·001483680 |
| 675 | 455625 | 307546875 | 25·9807621 | 8·7720532 | ·001481481 |
| 676 | 456976 | 308915776 | 26·0000000 | 8·7763830 | ·001479290 |
| 677 | 458329 | 310288733 | 26·0192237 | 8·7807084 | ·001477105 |

| Number. | Squares. | Cubes. | Square Roots. | Cube Roots. | Reciprocals. |
|---|---|---|---|---|---|
| 678 | 459684 | 311665752 | 26·0384331 | 8·7850296 | ·001474926 |
| 679 | 461041 | 313046839 | 26·0576284 | 8·7893466 | ·001472754 |
| 680 | 462400 | 314432000 | 26·0768096 | 8·7936593 | ·001470588 |
| 681 | 463761 | 315821241 | 26·0959767 | 8·7979679 | ·001468429 |
| 682 | 465124 | 317214568 | 26·1151297 | 8·8022721 | ·001466276 |
| 683 | 466489 | 318611987 | 26·1342687 | 8·8065722 | ·001464129 |
| 684 | 467856 | 320013504 | 26·1533937 | 8·8108681 | ·001461988 |
| 685 | 469225 | 321419125 | 26·1725047 | 8·8151598 | ·001459854 |
| 686 | 470596 | 322828856 | 26·1916017 | 8·8194474 | ·001457726 |
| 687 | 471969 | 324242703 | 26·2106848 | 8·8237307 | ·001455604 |
| 688 | 473344 | 325660672 | 26·2297541 | 8·8280099 | ·001453488 |
| 689 | 474721 | 327082769 | 26·2488095 | 8·8322850 | ·001451379 |
| 690 | 476100 | 328509000 | 26·2678511 | 8·8365559 | ·001449275 |
| 691 | 477481 | 329939371 | 26·2868789 | 8·8408227 | ·001447178 |
| 692 | 478864 | 331373888 | 26·3058929 | 8·8450854 | ·001445087 |
| 693 | 480249 | 332812557 | 26·3248932 | 8·8493440 | ·001443001 |
| 694 | 481636 | 334255384 | 26·3438797 | 8·8535985 | ·001440922 |
| 695 | 483025 | 335702375 | 26·3628527 | 8·8578489 | ·001438849 |
| 696 | 484416 | 337153536 | 26·3818119 | 8·8620952 | ·001436782 |
| 697 | 485809 | 338608873 | 26·4007576 | 8·8663375 | ·001434720 |
| 698 | 487204 | 340068392 | 26·4196896 | 8·8705757 | ·001432665 |
| 699 | 488601 | 341532099 | 26·4386081 | 8·8748099 | ·001430615 |
| 700 | 490000 | 343000000 | 26·4575131 | 8·8790400 | ·001428571 |
| 701 | 491401 | 344472101 | 26·4764046 | 8·8832661 | ·001426534 |
| 702 | 492804 | 345948408 | 26·4952826 | 8·8874882 | ·001424501 |
| 703 | 494209 | 347428927 | 26·5141472 | 8·8917063 | ·001422475 |
| 704 | 495616 | 348913664 | 26·5329983 | 8·8959204 | ·001420455 |
| 705 | 497025 | 350402625 | 26·5518361 | 8·9001304 | ·001418440 |
| 706 | 498436 | 351895816 | 26·5706605 | 8·9043366 | ·001416431 |
| 707 | 499849 | 353393243 | 26·5894716 | 8·9085387 | ·001414427 |
| 708 | 501264 | 354894912 | 26·6082694 | 8·9127369 | ·001412429 |
| 709 | 502681 | 356400829 | 26·6270539 | 8·9169311 | ·001410437 |
| 710 | 504100 | 357911000 | 26·6458252 | 8·9211214 | ·001408451 |
| 711 | 505521 | 359425431 | 26·6645833 | 8·9253078 | ·001406470 |
| 712 | 506944 | 360944128 | 26·6833281 | 8·9294902 | ·001404494 |
| 713 | 508369 | 362467097 | 26·7020598 | 8·9336687 | ·001402525 |
| 714 | 509796 | 363994344 | 26·7207784 | 8·9378433 | ·001400560 |
| 715 | 511225 | 365525875 | 26·7394839 | 8·9420140 | ·001398601 |
| 716 | 512656 | 367061696 | 26·7581763 | 8·9461809 | ·001396648 |
| 717 | 514089 | 368601813 | 26·7768557 | 8·9503438 | ·001394700 |
| 718 | 515524 | 370146232 | 26·7955220 | 8·9545029 | ·001392758 |
| 719 | 516961 | 371694959 | 26·8141754 | 8·9586581 | ·001390821 |
| 720 | 518400 | 373248000 | 26·8328157 | 8·9628095 | ·001388889 |
| 721 | 519841 | 374805361 | 26·8514432 | 8·9669570 | ·001386963 |
| 722 | 521284 | 376367048 | 26·8700577 | 8·9711007 | ·001385042 |
| 723 | 522729 | 377933067 | 26·8886593 | 8·9752406 | ·001383126 |
| 724 | 524176 | 379503424 | 26·9072481 | 8·9793766 | ·001381215 |
| 725 | 525625 | 381078125 | 26·9258240 | 8·9835089 | ·001379310 |
| 726 | 527076 | 382657176 | 26·9443872 | 8·9876373 | ·001377410 |
| 727 | 528529 | 384240583 | 26·9629375 | 8·9917620 | ·001375516 |
| 728 | 529984 | 385828352 | 26·9814751 | 8·9958899 | ·001373626 |
| 729 | 531441 | 387420489 | 27·0000000 | 9·0000000 | ·001371742 |
| 730 | 532900 | 389017000 | 27·0185122 | 9·0041134 | ·001369863 |
| 731 | 534361 | 390617891 | 27·0370117 | 9·0082229 | ·001367989 |
| 732 | 535824 | 392223168 | 27·0554985 | 9·0123288 | ·001366120 |
| 733 | 537289 | 393832837 | 27·0739727 | 9·0164309 | ·001364256 |
| 734 | 538756 | 395446904 | 27·0924344 | 9·0205293 | ·001362398 |
| 735 | 540225 | 397065375 | 27·1108834 | 9·0246239 | ·001360544 |
| 736 | 541696 | 398688256 | 27·1293199 | 9·0287149 | ·001358696 |
| 737 | 543169 | 400315553 | 27·1477149 | 9·0328021 | ·001356852 |
| 738 | 544644 | 401947272 | 27·1661554 | 9·0368857 | ·001355014 |
| 739 | 546121 | 403583419 | 27·1845544 | 9·0409655 | ·001353180 |

| Number. | Squares. | Cubes. | Square Roots. | Cube Roots. | Reciprocals. |
|---|---|---|---|---|---|
| 740 | 547600 | 405224000 | 27·2029140 | 9·0450419 | ·001351351 |
| 741 | 549801 | 406869021 | 27·2213152 | 9·0491142 | ·001349528 |
| 742 | 550564 | 408518488 | 27·2396769 | 9·0531831 | ·001347709 |
| 743 | 552049 | 410172407 | 27·2580263 | 9·0572482 | ·001345895 |
| 744 | 553536 | 411830784 | 27·2763634 | 9·0613098 | ·001344086 |
| 745 | 555025 | 413493625 | 27·2946881 | 9·0653677 | ·001342282 |
| 746 | 556516 | 415160936 | 27·3130006 | 9·0694220 | ·001340483 |
| 747 | 558009 | 416832723 | 27·3313007 | 9·0734726 | ·001338688 |
| 748 | 559504 | 418508992 | 27·3495887 | 9·0775197 | ·001336898 |
| 749 | 561001 | 420189749 | 27·3678644 | 9·0815631 | ·001335113 |
| 750 | 562500 | 421875000 | 27·3861279 | 9·0856030 | ·001333333 |
| 751 | 564001 | 423564751 | 27·4043792 | 9·0896352 | ·001331558 |
| 752 | 565504 | 425259008 | 27·4226184 | 9·0936719 | ·001329787 |
| 753 | 567009 | 426957777 | 27·4408455 | 9·0977010 | ·001328021 |
| 754 | 568516 | 428661064 | 27·4590604 | 9·1017265 | ·001326260 |
| 755 | 570025 | 430368875 | 27·4772633 | 9·1057485 | ·001324503 |
| 756 | 571536 | 432081216 | 27·4954542 | 9·1097669 | ·001322751 |
| 757 | 573049 | 433798093 | 27·5136330 | 9·1137818 | ·001321004 |
| 758 | 574564 | 435519512 | 27·5317998 | 9·1177931 | ·001319261 |
| 759 | 576081 | 437245479 | 27·5499546 | 9·1218010 | ·001317523 |
| 760 | 577600 | 438976000 | 27·5680975 | 9·1258053 | ·001315789 |
| 761 | 579121 | 440711081 | 27·5862284 | 9·1298061 | ·001314060 |
| 762 | 580644 | 442450728 | 27·6043475 | 9·1338034 | ·001312336 |
| 763 | 582169 | 444194947 | 27·6224546 | 9·1377971 | ·001310616 |
| 764 | 583696 | 445943744 | 27·6405499 | 9·1417874 | ·001308901 |
| 765 | 585225 | 447697125 | 27·6586334 | 9·1457742 | ·001307190 |
| 766 | 586756 | 449455096 | 27·6767050 | 9·1497576 | ·001305483 |
| 767 | 588289 | 451217663 | 27·6947648 | 9·1537375 | ·001303781 |
| 768 | 589824 | 452984832 | 27·7128129 | 9·1577139 | ·001302083 |
| 769 | 591361 | 454756609 | 27·7308492 | 9·1616869 | ·001300390 |
| 770 | 592900 | 456533000 | 27·7488739 | 9·1656565 | ·001298701 |
| 771 | 594441 | 458314011 | 27·7668868 | 9·1696225 | ·001297017 |
| 772 | 595984 | 460099648 | 27·7848880 | 9·1735852 | ·001295337 |
| 773 | 597529 | 461889917 | 27·8028775 | 9·1775445 | ·001293661 |
| 774 | 599076 | 463684824 | 27·8208555 | 9·1815003 | ·001291990 |
| 775 | 600625 | 465484375 | 27·8388218 | 9·1854527 | ·001290323 |
| 776 | 602176 | 467288576 | 27·8567766 | 9·1894018 | ·001288660 |
| 777 | 603729 | 469097433 | 27·8747197 | 9·1933474 | ·001287001 |
| 778 | 605284 | 470910952 | 27·8926514 | 9·1972897 | ·001285347 |
| 779 | 606841 | 472729139 | 27·9105715 | 9·2012286 | ·001283697 |
| 780 | 608400 | 474552000 | 27·9284801 | 9·2051641 | ·001282051 |
| 781 | 609961 | 476379541 | 27·9463772 | 9·2090962 | ·001280410 |
| 782 | 611524 | 478211768 | 27·9642629 | 9·2130250 | ·001278772 |
| 783 | 613089 | 480048687 | 27·9821372 | 9·2169505 | ·001277139 |
| 784 | 614656 | 481890304 | 28·0000000 | 9·2208726 | ·001275510 |
| 785 | 616225 | 483736625 | 28·0178515 | 9·2247914 | ·001273885 |
| 786 | 617796 | 485587656 | 28·0356915 | 9·2287068 | ·001272265 |
| 787 | 619369 | 487443403 | 28·0535203 | 9·2326189 | ·001270648 |
| 788 | 620944 | 489303872 | 28·0713377 | 9·2365277 | ·001269036 |
| 789 | 622521 | 491169069 | 28·0891438 | 9·2404333 | ·001267427 |
| 790 | 624100 | 493039000 | 28·1069386 | 9·2443355 | ·001265823 |
| 791 | 625681 | 494913671 | 28·1247222 | 9·2482344 | ·001264223 |
| 792 | 627624 | 496793088 | 28·1424946 | 8·2521300 | ·001262626 |
| 793 | 628849 | 498677257 | 28·1602557 | 9·2560224 | ·001261034 |
| 794 | 630436 | 500566184 | 28·1780056 | 9·2599114 | ·001259446 |
| 795 | 632025 | 502459875 | 28·1957444 | 9·2637973 | ·001257862 |
| 796 | 633616 | 504358336 | 28·2134720 | 9·2676798 | ·001256281 |
| 797 | 635209 | 506261573 | 28·2311884 | 9·2715592 | ·001254705 |
| 798 | 636804 | 508169592 | 28·2488938 | 9·2754352 | ·001253133 |
| 799 | 638401 | 510082399 | 28·2665881 | 9·2793081 | ·001251364 |
| 800 | 640000 | 512000000 | 28·2842712 | 9·2831777 | ·001250000 |
| 801 | 641601 | 513922401 | 28·3019434 | 9·2870444 | ·001248439 |

| Number. | Squares. | Cubes. | Square Roots. | Cube Roots. | Reciprocals. |
|---|---|---|---|---|---|
| 802 | 643204 | 515849608 | 28·3196045 | 9·2909072 | ·001246883 |
| 803 | 644809 | 517781627 | 28·3372546 | 9·2947671 | ·001245330 |
| 804 | 646416 | 519718464 | 28·3548938 | 9·2986239 | ·001243781 |
| 805 | 648025 | 521660125 | 28·3725219 | 9·3024775 | ·001242236 |
| 806 | 649636 | 523606616 | 28·3901391 | 9·3063278 | ·001240695 |
| 807 | 651249 | 525557943 | 28·4077454 | 9·3101750 | ·001239157 |
| 808 | 652864 | 527514112 | 28·4253408 | 9·3140190 | ·001237624 |
| 809 | 654481 | 529475129 | 28·4429253 | 9·3178599 | ·001236094 |
| 810 | 656100 | 531441000 | 28·4604989 | 9·3216975 | ·001234568 |
| 811 | 657721 | 533411731 | 28·4780617 | 9·3255320 | ·001233046 |
| 812 | 659344 | 535387328 | 28·4956137 | 9·3293634 | ·001231527 |
| 813 | 660969 | 537367797 | 28·5131549 | 9·3331916 | ·001230012 |
| 814 | 662596 | 539353144 | 28·5306852 | 9·3370167 | ·001228501 |
| 815 | 664225 | 541343375 | 28·5482048 | 9·3408386 | ·001226994 |
| 816 | 665856 | 543338496 | 28·5657137 | 9·3446575 | ·001225499 |
| 817 | 667489 | 545338513 | 28·5832119 | 9·3484731 | ·001223990 |
| 818 | 669124 | 547343432 | 28·6006993 | 9·3522857 | ·001222494 |
| 819 | 670761 | 549353259 | 28·6181760 | 9·3560952 | ·001221001 |
| 820 | 672400 | 551368000 | 28·6356421 | 9·3599016 | ·001219512 |
| 821 | 674041 | 553387661 | 28·6530976 | 9·3637049 | ·001218027 |
| 822 | 675684 | 555412248 | 28·6705424 | 9·3675051 | ·001216545 |
| 823 | 677329 | 557441767 | 28·6879716 | 9·3713022 | ·001215067 |
| 824 | 678976 | 559476224 | 28·7054002 | 9·3750963 | ·001213592 |
| 825 | 680625 | 561515625 | 28·7228132 | 9·3788873 | ·001212121 |
| 826 | 682276 | 563559976 | 28·7402157 | 9·3826752 | ·001210654 |
| 827 | 683929 | 565609283 | 28·7576077 | 9·3864600 | ·001209190 |
| 828 | 685584 | 567663552 | 28·7749891 | 9·3902419 | ·001207729 |
| 829 | 687241 | 569722789 | 28·7923601 | 9·3940206 | ·001206273 |
| 830 | 688900 | 571787000 | 28·8097206 | 9·3977964 | ·001204819 |
| 831 | 690561 | 573856191 | 28·8270706 | 9·4015691 | ·001203369 |
| 832 | 692224 | 575930368 | 28·8444102 | 9·4053387 | ·001201923 |
| 833 | 693889 | 578009537 | 28·8617394 | 9·4091054 | ·001200480 |
| 834 | 695556 | 580093704 | 28·8790582 | 9·4128690 | ·001199041 |
| 835 | 697225 | 582182875 | 28·8963666 | 9·4166297 | ·001197605 |
| 836 | 698896 | 584277056 | 28·9136646 | 9·4203873 | ·001196172 |
| 837 | 700569 | 586376253 | 28·9309523 | 9·4241420 | ·001194743 |
| 838 | 702244 | 588480472 | 28·9482297 | 9·4278936 | ·001193317 |
| 839 | 703921 | 590589719 | 28·9654967 | 9·4316423 | ·001191895 |
| 840 | 705600 | 592704000 | 28·9827535 | 9·4353800 | ·001190476 |
| 841 | 707281 | 594823321 | 29·0000000 | 9·4391307 | ·001189061 |
| 842 | 708964 | 596947688 | 29·0172363 | 9·4428704 | ·001187648 |
| 843 | 710649 | 590077107 | 29·0344623 | 9·4466072 | ·001186240 |
| 844 | 712336 | 601211584 | 29·0516781 | 9·4503410 | ·001184834 |
| 845 | 714025 | 603351125 | 29·0688837 | 9·4540719 | ·001183432 |
| 846 | 715716 | 605495736 | 29·0860791 | 9·4577999 | ·001182033 |
| 847 | 717409 | 607645423 | 29·1032644 | 9·4615249 | ·001180638 |
| 848 | 719104 | 609800192 | 29·1204396 | 9·4652470 | ·001179245 |
| 849 | 720801 | 611960049 | 29·1376046 | 9·4689661 | ·001177856 |
| 850 | 722500 | 614125000 | 29·1547595 | 9·4726824 | ·001176471 |
| 851 | 724201 | 616295051 | 29·1719043 | 9·4763957 | ·001175088 |
| 852 | 725904 | 618470208 | 29·1890390 | 9·4801061 | ·001173709 |
| 853 | 727609 | 620650477 | 29·2061637 | 9·4838136 | ·001172333 |
| 854 | 729316 | 622835864 | 29·2232784 | 9·4875182 | ·001170960 |
| 855 | 731025 | 625026375 | 29·2403830 | 9·4912200 | ·001169591 |
| 856 | 732736 | 627222016 | 29·2574777 | 9·4949188 | ·001168224 |
| 857 | 734449 | 629422793 | 29·2745623 | 9·4986147 | ·001166861 |
| 858 | 736164 | 631628712 | 29·2916370 | 9·5023078 | ·001165501 |
| 859 | 737881 | 633839779 | 29·3087018 | 9·5059980 | ·001164144 |
| 860 | 739600 | 636056000 | 29·3257566 | 9·5096854 | ·001162791 |
| 861 | 741321 | 638277381 | 29·3428015 | 9·5133699 | ·001161440 |
| 862 | 743044 | 640503928 | 29·3598365 | 9·5170515 | ·001160093 |
| 863 | 744769 | 642735647 | 29·3768616 | 9·5207303 | ·001158749 |

| Number. | Squares. | Cubes. | Square Roots. | Cube Roots. | Reciprocals. |
|---|---|---|---|---|---|
| 864 | 746496 | 644972544 | 29·3938769 | 9·5244063 | ·001157407 |
| 865 | 748225 | 647214625 | 29·4108823 | 9·5280794 | ·001156069 |
| 866 | 749956 | 649461896 | 29·4278779 | 9·5317497 | ·001154734 |
| 867 | 751689 | 651714363 | 29·4448637 | 9·5354172 | ·001153403 |
| 868 | 753424 | 653972032 | 29·4618397 | 9·5390818 | ·001152074 |
| 869 | 755161 | 656234909 | 29·4788059 | 9·5427437 | ·001150748 |
| 870 | 756900 | 658503000 | 29·4957624 | 9·5464027 | ·001149425 |
| 871 | 758641 | 660776311 | 29·5127091 | 9·5500589 | ·001148106 |
| 872 | 760384 | 663054848 | 29·5296461 | 9·5537123 | ·001146789 |
| 873 | 762129 | 665338617 | 29·5465734 | 9·5573630 | ·001145475 |
| 874 | 763876 | 667627624 | 29·5634910 | 9·5610108 | ·001144165 |
| 875 | 765625 | 669921875 | 29·5803989 | 9·5646559 | ·001142857 |
| 876 | 767376 | 672221376 | 29·5972972 | 9·5682782 | ·001141553 |
| 877 | 769129 | 674526133 | 29·6141858 | 9·5719377 | ·001140251 |
| 878 | 770884 | 676836152 | 29·6310648 | 9·5755745 | ·001138952 |
| 879 | 772641 | 679151439 | 29·6479342 | 9·5792085 | ·001137656 |
| 880 | 774400 | 681472000 | 29·6647939 | 9·5828397 | ·001136364 |
| 881 | 776161 | 683797841 | 29·6816442 | 9·5864682 | ·001135074 |
| 882 | 777924 | 686128968 | 29·6984848 | 9·5900937 | ·001133787 |
| 883 | 779689 | 688465387 | 29·7153159 | 9·5937169 | ·001132503 |
| 884 | 781456 | 690807104 | 29·7321375 | 9·5973373 | ·001131222 |
| 885 | 783225 | 693154125 | 29·7489496 | 9·6009548 | ·001129944 |
| 886 | 784996 | 695506456 | 29·7657521 | 9·6045696 | ·001128668 |
| 887 | 786769 | 697864103 | 29·7825452 | 9·6081817 | ·001127396 |
| 888 | 788544 | 700227072 | 29·7993289 | 9·6117911 | ·001126126 |
| 889 | 790321 | 702595369 | 29·8161030 | 9·6153977 | ·001124859 |
| 890 | 792100 | 704969000 | 29·8328678 | 9·6190017 | ·001123596 |
| 891 | 793881 | 707347971 | 29·8496231 | 9·6226030 | ·001122334 |
| 892 | 795664 | 707932288 | 29·8663690 | 9·6262016 | ·001121076 |
| 893 | 797449 | 712121957 | 29·8831056 | 9·6297975 | ·001119821 |
| 894 | 799236 | 714516984 | 29·8998328 | 9·6333907 | ·001118568 |
| 895 | 801025 | 716917375 | 29·9165506 | 9·6369812 | ·001117818 |
| 896 | 802816 | 719323136 | 29·9332591 | 9·6405690 | ·001116071 |
| 897 | 804609 | 721734273 | 29·9499583 | 9·6441542 | ·001114827 |
| 898 | 806404 | 724150792 | 29·9666481 | 9·6477367 | ·001113586 |
| 899 | 808201 | 726572699 | 29·9833287 | 9·6513166 | ·001112347 |
| 900 | 810000 | 729000000 | 30·0000000 | 9·6548938 | ·001111111 |
| 901 | 811801 | 731432701 | 30·0166621 | 9·6584684 | ·001109878 |
| 902 | 813604 | 733870808 | 30·0333148 | 9·6620403 | ·001108647 |
| 903 | 815409 | 736314327 | 30·0499584 | 9·6656096 | ·001107420 |
| 904 | 817216 | 738763264 | 30·0665928 | 9·6691762 | ·001106195 |
| 905 | 819025 | 741217625 | 30·0832179 | 9·6727403 | ·001104972 |
| 906 | 820836 | 743677416 | 30·0998339 | 9·6763017 | ·001103753 |
| 907 | 822649 | 746142643 | 30·1164407 | 9·6798604 | ·001102536 |
| 908 | 824464 | 748613312 | 30·1330383 | 9·6834166 | ·001101322 |
| 909 | 826281 | 751089429 | 30·1496269 | 9·6869701 | ·001100110 |
| 910 | 828100 | 753571000 | 30·1662063 | 9·6905211 | ·001098901 |
| 911 | 829921 | 756058031 | 30·1827765 | 9·6940694 | ·001097695 |
| 912 | 831744 | 758550825 | 30·1993377 | 9·6976151 | ·001096491 |
| 913 | 833569 | 761048497 | 30·2158899 | 9·7011583 | ·001095290 |
| 914 | 835396 | 763551944 | 30·2324329 | 9·7046989 | ·001094092 |
| 915 | 837225 | 766060875 | 30·2489669 | 9·7082369 | ·001092896 |
| 916 | 839056 | 768575296 | 30·2654919 | 9·7117723 | ·001091703 |
| 917 | 840889 | 771095213 | 30·2820079 | 9·7153051 | ·001090513 |
| 918 | 842724 | 773620632 | 30·2985148 | 9·7188354 | ·001089325 |
| 919 | 844561 | 776151559 | 30·3150128 | 9·7223631 | ·001088139 |
| 920 | 846400 | 778688000 | 30·3315018 | 9·7258883 | ·001086957 |
| 921 | 848241 | 781229961 | 30·3479818 | 9·7294109 | ·001085776 |
| 922 | 850084 | 783777448 | 30·3644529 | 9·7329309 | ·001084599 |
| 923 | 851929 | 786330467 | 30·3809151 | 9·7364484 | ·001083423 |
| 924 | 853776 | 788889024 | 30·3973683 | 9·7399634 | ·001082251 |
| 925 | 855625 | 791453125 | 30·4138127 | 9·7434758 | ·001081081 |

| Number. | Squares. | Cubes. | Square Roots. | Cube Roots. | Reciprocals. |
|---|---|---|---|---|---|
| 926 | 857476 | 794022776 | 30·4302481 | 9·7469857 | ·001079914 |
| 927 | 859329 | 796597983 | 30·4466747 | 9·7504930 | ·001078749 |
| 928 | 861184 | 799178752 | 30·4630924 | 9·7539979 | ·001077586 |
| 929 | 863041 | 801765089 | 30·4795013 | 9·7575002 | ·001076426 |
| 930 | 864900 | 804357000 | 30·4959014 | 9·7610001 | ·001075269 |
| 931 | 866761 | 806954491 | 30·5122926 | 9·7644974 | ·001074114 |
| 932 | 868624 | 809557568 | 30·5286750 | 9·7679922 | ·001072961 |
| 933 | 870489 | 812166237 | 30·5450487 | 9·7714845 | ·001071811 |
| 934 | 872356 | 814780504 | 30·5614136 | 9·7749743 | ·001070664 |
| 935 | 874225 | 817400375 | 30·5777697 | 9·7784616 | ·001069519 |
| 936 | 876096 | 820025856 | 30·5941171 | 9·7829466 | ·001068376 |
| 937 | 877969 | 822656953 | 30·6104557 | 9·7854288 | ·001067236 |
| 938 | 879844 | 825293672 | 30·6267857 | 9·7889087 | ·001066098 |
| 939 | 881721 | 827936019 | 30·6431069 | 9·7923861 | ·001064963 |
| 940 | 883600 | 830584000 | 30·6594194 | 9·7958611 | ·001063830 |
| 941 | 885481 | 833237621 | 30·6757233 | 9·7993336 | ·001062699 |
| 942 | 887364 | 835896888 | 30·6920185 | 9·8028036 | ·001061571 |
| 943 | 889249 | 838561807 | 30·7083051 | 9·8062711 | ·001060445 |
| 944 | 891136 | 841232384 | 30·7245830 | 9·8097362 | ·001059322 |
| 945 | 893025 | 843908625 | 30·7408523 | 9·8131989 | ·001058201 |
| 946 | 894916 | 846590536 | 30·7571130 | 9·8166591 | ·001057082 |
| 947 | 896808 | 849278123 | 30·7733651 | 9·8201169 | ·001055966 |
| 948 | 898704 | 851971392 | 30·7896086 | 9·8235723 | ·001054852 |
| 949 | 900601 | 854670349 | 30·8058436 | 9·8270252 | ·001053741 |
| 950 | 902500 | 857375000 | 30·8220700 | 9·8304757 | ·001052632 |
| 951 | 904401 | 860085351 | 30·8382879 | 9·8339238 | ·001051525 |
| 952 | 906304 | 862801408 | 30·8544972 | 9·8373695 | ·001050420 |
| 953 | 908209 | 865523177 | 30·8706981 | 9·8408127 | ·001049318 |
| 954 | 910116 | 868250664 | 30·8868904 | 9·8442536 | ·001048218 |
| 955 | 912025 | 870983875 | 30·9030743 | 9·8476920 | ·001047120 |
| 956 | 913936 | 873722816 | 30·9192477 | 9·8511280 | ·001046025 |
| 957 | 915849 | 876467493 | 30·9354166 | 9·8545617 | ·001044932 |
| 958 | 917764 | 879217912 | 30·9515751 | 9·8579929 | ·001043841 |
| 959 | 919681 | 881974079 | 30·9677251 | 9·8614218 | ·001042753 |
| 960 | 921600 | 884736000 | 30·9838668 | 9·8648483 | ·001041667 |
| 961 | 923521 | 887503681 | 31·0000000 | 9·8682724 | ·001040583 |
| 962 | 925444 | 890277128 | 31·0161248 | 9·8716941 | ·001039501 |
| 963 | 927369 | 893056347 | 31·0322413 | 9·8751135 | ·001038422 |
| 964 | 929296 | 895841344 | 31·0483494 | 9·8785305 | ·001037344 |
| 965 | 931225 | 898632125 | 31·0644491 | 9·8819451 | ·001036269 |
| 966 | 933156 | 901428696 | 31·0805405 | 9·8853574 | ·001035197 |
| 967 | 935089 | 904231063 | 31·0966236 | 9·8887673 | ·001034126 |
| 968 | 937024 | 907039232 | 31·1126984 | 9·8921749 | ·001033058 |
| 969 | 938961 | 909853209 | 31·1287648 | 9·8955801 | ·001031992 |
| 970 | 940900 | 912673000 | 31·1448230 | 9·8989830 | ·001030928 |
| 971 | 942841 | 915498611 | 31·1608729 | 9·9023835 | ·001029866 |
| 972 | 944784 | 918330048 | 31·1769145 | 9·9057817 | ·001028807 |
| 973 | 946729 | 921167317 | 31·1929479 | 9·9091776 | ·001027749 |
| 974 | 948676 | 924010424 | 31·2089731 | 9·9125712 | ·001026694 |
| 975 | 950625 | 926859375 | 31·2249900 | 9·9159624 | ·001025641 |
| 976 | 952576 | 929714176 | 31·2409987 | 9·9193513 | ·001024590 |
| 977 | 954529 | 932574833 | 31·2569992 | 9·9227379 | ·001023541 |
| 978 | 956484 | 935441352 | 31·2729915 | 9·9261222 | ·001022495 |
| 979 | 958441 | 938313739 | 31·2889757 | 9·9295042 | ·001021450 |
| 980 | 960400 | 941192000 | 31·3049517 | 9·9328839 | ·001020408 |
| 981 | 962361 | 944076141 | 31·3209195 | 9·9362613 | ·001019168 |
| 982 | 964324 | 946966168 | 31·3368792 | 9·9396363 | ·001018330 |
| 983 | 966289 | 949862087 | 31·3528308 | 9·9430092 | ·001017294 |
| 984 | 968256 | 952763904 | 31·3687743 | 9·9463797 | ·001016260 |
| 985 | 970225 | 955671625 | 31·3847097 | 9·9497479 | ·001015228 |
| 986 | 972196 | 958585256 | 31·4006369 | 9·9531138 | ·001014199 |
| 987 | 974169 | 961504803 | 31·4165561 | 9·9564775 | ·001013171 |

| Number. | Squares. | Cubes. | Square Roots. | Cube Roots. | Reciprocals. |
|---|---|---|---|---|---|
| 988 | 976144 | 964430272 | 31·4324673 | 9·9598389 | ·001012146 |
| 989 | 978121 | 967361669 | 31·4483704 | 9·9631981 | ·001011122 |
| 990 | 980100 | 970299000 | 31·4642654 | 9·9665549 | ·001010101 |
| 991 | 982081 | 973242271 | 31·4801525 | 9·9699055 | ·001009082 |
| 992 | 984064 | 976191488 | 31·4960315 | 9·9732619 | ·001008065 |
| 993 | 986049 | 979146657 | 31·5119025 | 9·9766120 | ·001007049 |
| 994 | 988036 | 982107784 | 31·5277655 | 9·9799599 | ·001006036 |
| 995 | 990025 | 985074875 | 31·5436206 | 9·9833055 | ·001005025 |
| 996 | 992016 | 988047936 | 31·5594677 | 9·9866488 | ·001004016 |
| 997 | 994009 | 991026973 | 31·5753068 | 9·9899900 | ·001003009 |
| 998 | 996004 | 994011992 | 31·5911380 | 9·9933289 | ·001002004 |
| 999 | 998001 | 997002999 | 31·6069613 | 9·9966656 | ·001001001 |
| 1000 | 1000000 | 1000000000 | 31·6227766 | 10·0000000 | ·001000000 |
| 1001 | 1000201 | 1003003001 | 31·6385840 | 10·0033222 | ·0009990010 |
| 1002 | 1004004 | 1006012008 | 31·6543866 | 10·0066622 | ·0009980040 |
| 1003 | 1006009 | 1009027027 | 31·6701752 | 10·0099899 | ·0009970090 |
| 1004 | 1008016 | 1012048064 | 31·6859590 | 10·0133155 | ·0009960159 |
| 1005 | 1010025 | 1015075125 | 31·7017349 | 10·0166389 | ·0009950249 |
| 1006 | 1010036 | 1018108216 | 31·7175030 | 10·0199601 | ·0009940358 |
| 1007 | 1014049 | 1021147343 | 31·7332633 | 10·0232791 | ·0009930487 |
| 1008 | 1016064 | 1024192512 | 31·7490157 | 10·0265958 | ·0009920635 |
| 1009 | 1018081 | 1027243729 | 31·7647603 | 10·0299104 | ·0009910803 |
| 1010 | 1020100 | 1030301000 | 31·7804972 | 10·0332228 | ·0009900990 |
| 1011 | 1020121 | 1033364331 | 31·7962262 | 10·0365330 | ·0009891197 |
| 1012 | 1024144 | 1036433728 | 31·8119474 | 10·0398410 | ·0009881423 |
| 1013 | 1026169 | 1039509197 | 31·8276609 | 10·0431469 | ·0009871668 |
| 1014 | 1028196 | 1042590744 | 31·8433666 | 10·0464506 | ·0009861933 |
| 1015 | 1030225 | 1045678375 | 31·8590646 | 10·0497521 | ·0009852217 |
| 1016 | 1032256 | 1048772096 | 31·8747549 | 10·0530514 | ·0009842520 |
| 1017 | 1034289 | 1051871913 | 31·8904374 | 10·0563485 | ·0009832842 |
| 1018 | 1036324 | 1054977832 | 31·9061123 | 10·0596435 | ·0009823183 |
| 1019 | 1038361 | 1058089859 | 31·9217794 | 10·0629364 | ·0009813543 |
| 1020 | 1040400 | 1061208000 | 31·9374388 | 10·0662271 | ·0009803922 |
| 1021 | 1042441 | 1064332261 | 31·9530906 | 10·0695156 | ·0009794319 |
| 1022 | 1044484 | 1067462648 | 31·9687347 | 10·0728020 | ·0009784736 |
| 1023 | 1046529 | 1070599167 | 31·9843712 | 10·0760863 | ·0009775171 |
| 1024 | 1048576 | 1073741824 | 32·0000000 | 10·0793684 | ·0009765625 |
| 1025 | 1050625 | 1076890625 | 32·0156212 | 10·0826484 | ·0009756098 |
| 1026 | 1052676 | 1080045576 | 32·0312348 | 10·0859262 | ·0009746589 |
| 1027 | 1054729 | 1083206683 | 32·0468407 | 10·0892019 | ·0009737098 |
| 1028 | 1056784 | 1086373952 | 32·0624391 | 10·0924755 | ·0009727626 |
| 1029 | 1058841 | 1089547389 | 32·0780298 | 10·0957469 | ·0009718173 |
| 1030 | 1060900 | 1092727000 | 32·0936131 | 10·0990163 | ·0009708738 |
| 1031 | 1062961 | 1095912791 | 32·1091887 | 10·1022835 | ·0009699321 |
| 1032 | 1065024 | 1099104768 | 32·1247568 | 10·1055487 | ·0009689922 |
| 1033 | 1067089 | 1102302937 | 32·1403173 | 10·1088117 | ·0009680542 |
| 1034 | 1069156 | 1105507304 | 32·1558704 | 10·1120726 | ·0009671180 |
| 1035 | 1071225 | 1108717875 | 32·1714159 | 10·1153314 | ·0009661836 |
| 1036 | 1073296 | 1111934656 | 32·1869539 | 10·1185882 | ·0009652510 |
| 1037 | 1075369 | 1115157653 | 32·2024844 | 10·1218428 | ·0009643202 |
| 1038 | 1077444 | 1118386872 | 32·2180074 | 10·1250953 | ·0009633911 |
| 1039 | 1079521 | 1121622319 | 32·2335229 | 10·1283457 | ·0009624639 |
| 1040 | 1081600 | 1124864000 | 32·2490310 | 10·1315941 | ·0009615385 |
| 1041 | 1083681 | 1128111921 | 32·2645316 | 10·1348403 | ·0009606148 |
| 1042 | 1085764 | 1131366088 | 32·2800248 | 10·1380845 | ·0009596929 |
| 1043 | 1087849 | 1134626507 | 32·2955105 | 10·1413266 | ·0009587738 |
| 1044 | 1089936 | 1137893184 | 32·3109888 | 10·1445667 | ·0009578544 |
| 1045 | 1092025 | 1141166125 | 32·3264598 | 10·1478047 | ·0009569378 |
| 1046 | 1094116 | 1144445336 | 32·3419233 | 10·1510406 | ·0009560229 |
| 1047 | 1096209 | 1147730823 | 32·3573794 | 10·1542744 | ·0009551098 |
| 1048 | 1098304 | 1151022592 | 32·3728281 | 10·1575062 | ·0009541985 |
| 1049 | 1100401 | 1154320649 | 32·3882695 | 10·1607359 | ·0009532888 |

| Number. | Squares. | Cubes. | Square Roots. | Cube Roots. | Reciprocals. |
|---|---|---|---|---|---|
| 1050 | 1102500 | 1157625000 | 32·4037035 | 10·1639636 | ·0009523810 |
| 1051 | 1104601 | 1160935651 | 32·4191301 | 10·1671893 | ·0009514748 |
| 1052 | 1106704 | 1164252608 | 32·4345495 | 10·1704129 | ·0009505703 |
| 1053 | 1108809 | 1167575877 | 32·4499615 | 10·1736344 | ·0009496676 |
| 1054 | 1110916 | 1170905464 | 32·4653662 | 10·1768539 | ·0009487666 |
| 1055 | 1113125 | 1174241375 | 32·4807635 | 10·1800714 | ·0009478673 |
| 1056 | 1115136 | 1177583616 | 32·4961536 | 10·1832868 | ·0009469697 |
| 1057 | 1117249 | 1180932193 | 32·5115364 | 10·1865002 | ·0009460738 |
| 1058 | 1119364 | 1184287112 | 32·5269119 | 10·1897116 | ·0009451796 |
| 1059 | 1121481 | 1187648379 | 32·5422802 | 10·1929209 | ·0009442871 |
| 1060 | 1123600 | 1191016000 | 32·5576412 | 10·1961283 | ·0009433962 |
| 1061 | 1125721 | 1194389981 | 32·5729949 | 10·1993336 | ·0009425071 |
| 1062 | 1127844 | 1197770328 | 32·5883415 | 10·2025369 | ·0009416196 |
| 1063 | 1129969 | 1201157047 | 32·6035807 | 10·2057382 | ·0009407338 |
| 1064 | 1132096 | 1204550144 | 32·6190129 | 10·2089375 | ·0009398496 |
| 1065 | 1134225 | 1207949625 | 32·6343377 | 10·2121347 | ·0009389671 |
| 1066 | 1136356 | 1211355496 | 32·6496554 | 10·2153300 | ·0009380863 |
| 1067 | 1138489 | 1214767763 | 32·6649659 | 10·2185233 | ·0009372071 |
| 1068 | 1140624 | 1218186432 | 32·6802693 | 10·2217146 | ·0009363296 |
| 1069 | 1142761 | 1221611509 | 32·6955654 | 10·2249039 | ·0009354537 |
| 1070 | 1144900 | 1225043000 | 32·7108544 | 10·2280912 | ·0009345794 |
| 1071 | 1147041 | 1228480911 | 32·7261363 | 10·2312766 | ·0009337068 |
| 1072 | 1149184 | 1231925248 | 32·7414111 | 10·2344599 | ·0009328358 |
| 1073 | 1151329 | 1235376017 | 32·7566787 | 10·2376413 | ·0009319664 |
| 1074 | 1153476 | 1238833224 | 32·7719392 | 10·2408207 | ·0009310987 |
| 1075 | 1155625 | 1242296875 | 32·7871926 | 10·2439981 | ·0009302326 |
| 1076 | 1157776 | 1245766976 | 32·8024398 | 10·2471735 | ·0009293680 |
| 1077 | 1159929 | 1249243533 | 32·8176782 | 10·2503470 | ·0009285051 |
| 1078 | 1162084 | 1252726552 | 32·8329103 | 10·2535186 | ·0009276438 |
| 1079 | 1164241 | 1256216039 | 32·8481354 | 10·2566881 | ·0009267841 |
| 1080 | 1166400 | 1259712000 | 32·8633535 | 10·2598557 | ·0009259259 |
| 1081 | 1168561 | 1263214441 | 32·8785644 | 10·2630213 | ·0009250694 |
| 1082 | 1170724 | 1266723368 | 32·8937684 | 10·2661850 | ·0009242144 |
| 1083 | 1172889 | 1270238787 | 32·9089653 | 10·2693467 | ·0009233610 |
| 1084 | 1175056 | 1273760704 | 32·9241553 | 10·2725065 | ·0009225092 |
| 1085 | 1177225 | 1277289125 | 32·9393382 | 10·2756644 | ·0009216590 |
| 1086 | 1179396 | 1280824056 | 32·9545141 | 10·2788203 | ·0009208103 |
| 1087 | 1181569 | 1284365503 | 32·9696830 | 10·2819743 | ·0009199632 |
| 1088 | 1183744 | 1287913472 | 32·9848450 | 10·2851264 | ·0009191176 |
| 1089 | 1185921 | 1291467969 | 33·0000000 | 10·2882765 | ·0009182736 |
| 1090 | 1188100 | 1295029000 | 33·0151480 | 10·2914247 | ·0009174312 |
| 1091 | 1190281 | 1298596571 | 33·0302891 | 10·2945709 | ·0009165903 |
| 1092 | 1192464 | 1302170688 | 33·0454233 | 19·2977153 | ·0009157509 |
| 1093 | 1194649 | 1305751357 | 33·0605505 | 10·3008577 | ·0009149131 |
| 1094 | 1196836 | 1309338584 | 33·0756708 | 10·3039982 | ·0009140768 |
| 1095 | 1199025 | 1312932375 | 33·0907842 | 10·3071368 | ·0009132420 |
| 1096 | 1201216 | 1316532736 | 33·1058907 | 10·3102735 | ·0009124008 |
| 1097 | 1203409 | 1320139673 | 33·1209903 | 10·3134083 | ·0009115770 |
| 1098 | 1205604 | 1323753192 | 33·1360830 | 10·3165411 | ·0009107468 |
| 1099 | 1207801 | 1327373299 | 33·1511689 | 10·3196721 | ·0009099181 |
| 1100 | 1210000 | 1331000000 | 33·1662479 | 10·3228012 | ·0009090909 |
| 1101 | 1212201 | 1334633301 | 33·1813200 | 10·3259284 | ·0009082652 |
| 1102 | 1214404 | 1338273208 | 33·1963853 | 10·3290537 | ·0009074410 |
| 1103 | 1216609 | 1341919727 | 33·2114438 | 10·3321770 | ·0009066183 |
| 1104 | 1218816 | 1345572864 | 33·2266955 | 10·3352985 | ·0009057971 |
| 1105 | 1221025 | 1349232625 | 33·2415403 | 10·3384181 | ·0009049774 |
| 1106 | 1223236 | 1352899016 | 33·2565783 | 10·3415358 | ·0009041591 |
| 1107 | 1225449 | 1356572043 | 33·2716095 | 10·3446517 | ·0009033424 |
| 1108 | 1227664 | 1360251712 | 33·2866339 | 10·3477657 | ·0009025271 |
| 1109 | 1229881 | 1363938029 | 33·3016516 | 10·3508778 | ·0009017133 |
| 1110 | 1232100 | 1367631000 | 33·3166625 | 10·3539880 | ·0009009009 |
| 1111 | 1234321 | 1371330631 | 33·3316666 | 10·3570964 | ·0009000900 |

| Number. | Squares. | Cubes. | Square Roots. | Cube Roots. | Reciprocals. |
|---|---|---|---|---|---|
| 1112 | 1236544 | 1375036928 | 33·3466640 | 10·3602029 | ·0008992806 |
| 1113 | 1238769 | 1378749897 | 33·3616546 | 10·3633076 | ·0008984726 |
| 1114 | 1240996 | 1382469544 | 33·3766385 | 10·3664103 | ·0008976661 |
| 1115 | 1243225 | 1386195875 | 33·3916157 | 10·3695113 | ·0008968610 |
| 1116 | 1245456 | 1389928896 | 33·4065862 | 10·3726103 | ·0008960753 |
| 1117 | 1247689 | 1393668613 | 33·4215499 | 10·3757076 | ·0008952551 |
| 1118 | 1249924 | 1397415032 | 33·4365070 | 10·3788030 | ·0008944544 |
| 1119 | 1252161 | 1401168159 | 33·4514573 | 10·3818965 | ·0008936550 |
| 1120 | 1254400 | 1404928000 | 33·4664011 | 10·3849882 | ·0008928571 |
| 1121 | 1256641 | 1408694561 | 33·4813381 | 10·3880781 | ·0008960607 |
| 1122 | 1258884 | 1412467848 | 33·4962684 | 10·3911661 | ·0008912656 |
| 1123 | 1261129 | 1416247867 | 33·5111921 | 10·3942527 | ·0008904720 |
| 1124 | 1263376 | 1420034624 | 33·5261092 | 10·3973366 | ·0008896797 |
| 1125 | 1265625 | 1423828125 | 33·5410196 | 10·4004192 | ·0008888889 |
| 1126 | 1267876 | 1427628376 | 33·5559234 | 10·4034999 | ·0008880995 |
| 1127 | 1270129 | 1431435383 | 33·5708206 | 10·4065787 | ·0008873114 |
| 1128 | 1272384 | 1435249152 | 33·5857112 | 10·4096557 | ·0008865248 |
| 1129 | 1274641 | 1439069689 | 33·6005952 | 10·4127310 | ·0008857396 |
| 1130 | 1276900 | 1442897000 | 33·6154726 | 10·4158044 | ·0008849558 |
| 1131 | 1279161 | 1446731091 | 33·6303434 | 10·4188760 | ·0008841733 |
| 1132 | 1281424 | 1450571968 | 33·6452077 | 10·4219458 | ·0008833922 |
| 1133 | 1283689 | 1454419637 | 33·6600653 | 10·4250138 | ·0008826125 |
| 1134 | 1285956 | 1458274104 | 33·6749165 | 10·4280800 | ·0008818342 |
| 1135 | 1288225 | 1462135375 | 33·6897610 | 10·4311443 | ·0008810573 |
| 1136 | 1290496 | 1466003456 | 33·7045991 | 10·4342069 | ·0008802817 |
| 1137 | 1292769 | 1469878353 | 33·7174306 | 10·4372677 | ·0008795075 |
| 1138 | 1295044 | 1473760072 | 33·7340556 | 10·4403677 | ·0008787346 |
| 1139 | 1297321 | 1477648619 | 33·7490741 | 10·4433839 | ·0008779631 |
| 1140 | 1299600 | 1481544000 | 33·7638860 | 10·4464393 | ·0008771930 |
| 1141 | 1301881 | 1485446221 | 33·7786915 | 10·4494929 | ·0008764242 |
| 1142 | 1304164 | 1489355288 | 33·7934905 | 10·4525448 | ·0008756567 |
| 1143 | 1306449 | 1493271207 | 33·8082830 | 10·4555948 | ·0008748906 |
| 1144 | 1308736 | 1497193984 | 33·8230691 | 10·4586431 | ·0008741259 |
| 1145 | 1311025 | 1501123625 | 33·8378486 | 10·4616896 | ·0008733624 |
| 1146 | 1313316 | 1505060136 | 33·8526218 | 10·4647343 | ·0008726003 |
| 1147 | 1315609 | 1509003523 | 33·8673884 | 10·4677773 | ·0008718396 |
| 1148 | 1317904 | 1512953792 | 33·8821487 | 10·4708158 | ·0008710801 |
| 1149 | 1320201 | 1516910949 | 33·8969025 | 10·4738579 | ·0008703220 |
| 1150 | 1322500 | 1520875000 | 33·9116499 | 10·4768955 | ·0008695652 |
| 1151 | 1324801 | 1524845951 | 33·9263909 | 10·4799314 | ·0008688097 |
| 1152 | 1327104 | 1528823808 | 33·9411255 | 10·4829656 | ·0008680556 |
| 1153 | 1329409 | 1532808577 | 33·9558537 | 10·4859980 | ·0008673027 |
| 1154 | 1331716 | 1536800264 | 33·9705755 | 10·4890286 | ·0008665511 |
| 1155 | 1334025 | 1540798875 | 33·9852910 | 10·4920575 | ·0008658009 |
| 1156 | 1336336 | 1544804416 | 34·0000000 | 10·4950847 | ·0008650519 |
| 1157 | 1338649 | 1548816893 | 34·0147027 | 10·4981101 | ·0008643042 |
| 1158 | 1340964 | 1552836312 | 34·0293990 | 10·5011337 | ·0008635579 |
| 1159 | 1343281 | 1556862679 | 34·0440890 | 10·5041556 | ·0008628128 |
| 1160 | 1345600 | 1560896000 | 34·0587727 | 10·5071757 | ·0008620690 |
| 1161 | 1347921 | 1564936281 | 34·0734501 | 10·5101942 | ·0008613244 |
| 1162 | 1350244 | 1568983528 | 34·0881211 | 10·5132109 | ·0008605852 |
| 1163 | 1352569 | 1573037749 | 34·0127858 | 10·5162259 | ·0008598452 |
| 1164 | 1354896 | 1577098944 | 34·1174442 | 10·5192391 | ·0008591065 |
| 1165 | 1357225 | 1581167125 | 34·1320963 | 10·5222506 | ·0008583691 |
| 1166 | 1359556 | 1585242296 | 34·1467422 | 10·5252604 | ·0008576329 |
| 1167 | 1361889 | 1589324463 | 34·1613817 | 10·5282685 | ·0008568980 |
| 1168 | 1364224 | 1593413632 | 34·1760150 | 10·5312749 | ·0008561644 |
| 1169 | 1366561 | 1597509809 | 34·1906420 | 10·5342795 | ·0008554320 |
| 1170 | 1368900 | 1601613000 | 34·2052627 | 10·5372825 | ·0008547009 |
| 1171 | 1371241 | 1605723211 | 34·2198773 | 10·5402837 | ·0008539710 |
| 1172 | 1373584 | 1609840448 | 34·2344855 | 10·5432832 | ·0008532423 |
| 1173 | 1375929 | 1613964717 | 34·2490875 | 10·5462810 | ·0008525149 |

| Number. | Squares. | Cubes. | Square Roots. | Cube Roots. | Reciprocals. |
|---|---|---|---|---|---|
| 1174 | 1378276 | 1618096024 | 34·2636834 | 10·5492771 | ·0008517888 |
| 1175 | 1380625 | 1622234375 | 34·2782730 | 10·5522715 | ·0008510638 |
| 1176 | 1382976 | 1626379776 | 34·2928564 | 10·5552642 | ·0008503401 |
| 1177 | 1385329 | 1630532233 | 34·3074336 | 10·5582552 | ·0008496177 |
| 1178 | 1387684 | 1634691752 | 34·3220046 | 10·5612445 | ·0008488964 |
| 1179 | 1390041 | 1638858339 | 34·3365694 | 10·5642322 | ·0008481764 |
| 1180 | 1392400 | 1643032000 | 34·3511281 | 10·5672181 | ·0008471576 |
| 1181 | 1394761 | 1647212741 | 34·3656805 | 10·5702024 | ·0008467401 |
| 1182 | 1397124 | 1651400568 | 34·3802268 | 10·5731849 | ·0008460237 |
| 1183 | 1399489 | 1655595487 | 34·3947670 | 10·5761658 | ·0008453085 |
| 1184 | 1401856 | 1659797504 | 34·4093011 | 10·5791449 | ·0008445946 |
| 1185 | 1404225 | 1664006625 | 34·4238289 | 10·5821225 | ·0008438819 |
| 1186 | 1406596 | 1668222856 | 34·4383507 | 10·5850983 | ·0008431703 |
| 1187 | 1408969 | 1672446203 | 34·4528663 | 10·5880725 | ·0008424600 |
| 1188 | 1411344 | 1676676672 | 34·4673759 | 10·5910450 | ·0008417508 |
| 1189 | 1413721 | 1680914629 | 34·4818793 | 10·5940158 | ·0008410429 |
| 1190 | 1416100 | 1685159000 | 34·4963766 | 10·5969850 | ·0008403361 |
| 1191 | 1418481 | 1689410871 | 34·5108678 | 10·5999525 | ·0008396306 |
| 1192 | 1420864 | 1693669888 | 34·5253530 | 10·6029184 | ·0008389262 |
| 1193 | 1423249 | 1697936057 | 34·5398321 | 10·6058826 | ·0008382320 |
| 1194 | 1425636 | 1702209384 | 34·5543051 | 10·6088451 | ·0008375209 |
| 1195 | 1428025 | 1706489875 | 34·5687720 | 10·6118060 | ·0008368201 |
| 1196 | 1430416 | 1710777536 | 34·5832329 | 10·6147652 | ·0008361204 |
| 1197 | 1432809 | 1715072373 | 34·5976879 | 10·6177228 | ·0008354219 |
| 1198 | 1435204 | 1719374392 | 34·6121366 | 10·6206788 | ·0008347245 |
| 1199 | 1437601 | 1723683599 | 34·6265794 | 10·6236331 | ·0008340284 |
| 1200 | 1440000 | 1728000000 | 34·6410162 | 10·6265857 | ·0008333333 |
| 1201 | 1442401 | 1732323601 | 34·6554469 | 10·6295367 | ·0008326395 |
| 1202 | 1444804 | 1736654408 | 34·6698716 | 10·6324860 | ·0008319468 |
| 1203 | 1447209 | 1740992427 | 34·6842904 | 10·6354338 | ·0008312552 |
| 1204 | 1449616 | 1745337664 | 34·6987031 | 10·6383799 | ·0008305648 |
| 1205 | 1452025 | 1749690125 | 34·7131099 | 10·6413244 | ·0008298755 |
| 1206 | 1454436 | 1754049816 | 34·7275107 | 10·6442672 | ·0008291874 |
| 1207 | 1456849 | 1758416743 | 34·7419055 | 10·6472085 | ·0008285004 |
| 1208 | 1459264 | 1762790912 | 34·7562944 | 10·6501480 | ·0008278146 |
| 1209 | 1461681 | 1767172329 | 34·7706773 | 10·6530860 | ·0008271299 |
| 1210 | 1464100 | 1771561000 | 34·7850543 | 10·6560223 | ·0008264463 |
| 1211 | 1466521 | 1775956931 | 34·7994253 | 10·6589570 | ·0008257638 |
| 1212 | 1468944 | 1780360128 | 34·8137904 | 10·6618902 | ·0008250825 |
| 1213 | 1471369 | 1784770597 | 34·8281495 | 10·6648217 | ·0008244023 |
| 1214 | 1473796 | 1789188344 | 34·8425028 | 10·6677516 | ·0008237232 |
| 1215 | 1476225 | 1793613375 | 34·8568501 | 10·6706799 | ·0008230453 |
| 1216 | 1478656 | 1798045696 | 34·8711915 | 10·6736066 | ·0008223684 |
| 1217 | 1481089 | 1802485313 | 34·8855271 | 10·6765317 | ·0008216927 |
| 1218 | 1483524 | 1806932232 | 34·8998567 | 10·6794552 | ·0008210181 |
| 1219 | 1485961 | 1811386459 | 34·9141805 | 10·6823771 | ·0008203445 |
| 1220 | 1488400 | 1815848000 | 34·9284984 | 10·6852973 | ·0008196721 |
| 1221 | 1490841 | 1820316861 | 34·9428104 | 10·6882160 | ·0008190008 |
| 1222 | 1493284 | 1824793048 | 34·9571166 | 10·6911331 | ·0008183306 |
| 1223 | 1495729 | 1829276567 | 34·9714169 | 10·6940486 | ·0008176615 |
| 1224 | 1498176 | 1833764247 | 34·9857114 | 10·6969625 | ·0008169935 |
| 1225 | 1500625 | 1838265625 | 35·0000000 | 10·6998748 | ·0008163265 |
| 1226 | 1503276 | 1842771176 | 35·0142828 | 10·7027855 | ·0008156607 |
| 1227 | 1505529 | 1847284083 | 35·0285598 | 10·7056947 | ·0008149959 |
| 1228 | 1507984 | 1851804352 | 35·0428309 | 10·7086023 | ·0008143322 |
| 1229 | 1510441 | 1856331989 | 35·0570963 | 10·7115083 | ·0008136696 |
| 1230 | 1512900 | 1860867000 | 35·0713558 | 10·7144127 | ·0008130081 |
| 1231 | 1515361 | 1865409391 | 35·0856096 | 10·7173155 | ·0008123477 |
| 1232 | 1517824 | 1869959168 | 35·0998575 | 10·7202168 | ·0008116883 |
| 1233 | 1520289 | 1874516337 | 35·1140997 | 10·7231165 | ·0008110300 |
| 1234 | 1522756 | 1879080904 | 35·1283361 | 10·7260146 | ·0008103728 |
| 1235 | 1525225 | 1883652875 | 35·1425568 | 10·7289112 | ·0008097166 |

| Number. | Squares. | Cubes. | Square Roots. | Cube Roots. | Reciprocals. |
|---|---|---|---|---|---|
| 1236 | 1527696 | 1888232256 | 35·1567917 | 10·7318062 | ·0008090615 |
| 1237 | 1530169 | 1892819053 | 35·1710108 | 10·7346997 | ·0008084074 |
| 1238 | 1532644 | 1897413272 | 35·1852242 | 10·7375916 | ·0008077544 |
| 1239 | 1535121 | 1902014919 | 35·1994318 | 10·7404819 | ·0008071025 |
| 1240 | 1537600 | 1906624000 | 35·2136337 | 10·7433707 | ·0008064516 |
| 1241 | 1540081 | 1911240521 | 35·2278299 | 10·7462579 | ·0008058018 |
| 1242 | 1542564 | 1915864488 | 35·2420204 | 10·7491436 | ·0008051530 |
| 1243 | 1545049 | 1920495907 | 35·2562051 | 10·7520277 | ·0008045052 |
| 1244 | 1547536 | 1925134784 | 35·2703842 | 10·7549103 | ·0008038585 |
| 1245 | 1550025 | 1929781125 | 35·2845575 | 10·7577913 | ·0008032129 |
| 1246 | 1552521 | 1934434936 | 35·2987252 | 10·7606708 | ·0008025682 |
| 1247 | 1555009 | 1939096223 | 35·3128872 | 10·7635488 | ·0008019246 |
| 1248 | 1557504 | 1943764992 | 35·3270435 | 10·7664252 | ·0008012821 |
| 1249 | 1560001 | 1948441249 | 35·3411941 | 10·7693001 | ·0008006405 |
| 1250 | 1562500 | 1953125000 | 35·3553391 | 10·7721735 | ·0008000000 |
| 1251 | 1565001 | 1957816251 | 35·3694784 | 10·7750453 | ·0007993605 |
| 1252 | 1567504 | 1962515008 | 35·3836120 | 10·7779156 | ·0007987220 |
| 1253 | 1570009 | 1967221277 | 35·3977400 | 10·7807843 | ·0007980846 |
| 1254 | 1572516 | 1971935064 | 35·4118624 | 10·7836516 | ·0007974482 |
| 1255 | 1575025 | 1976656375 | 35·4259792 | 10·7865173 | ·0007968127 |
| 1256 | 1577536 | 1981385216 | 35·4400903 | 10·7893815 | ·0007961783 |
| 1257 | 1580049 | 1986121593 | 35·4541958 | 10·7922441 | ·0007955449 |
| 1258 | 1582564 | 1990865512 | 35·4682957 | 10·7951053 | ·0007949126 |
| 1259 | 1585081 | 1995616979 | 35·4823900 | 10·7979649 | ·0007942812 |
| 1260 | 1587600 | 2000376000 | 35·4964787 | 10·8008230 | ·0007936508 |
| 1261 | 1590121 | 2005142581 | 35·5105618 | 10·8036797 | ·0007930214 |
| 1262 | 1592644 | 2009916728 | 35·5246393 | 10·8065348 | ·0007923930 |
| 1263 | 1595166 | 2014698447 | 35·5387113 | 10·8093884 | ·0007917656 |
| 1264 | 1597696 | 2019487744 | 35·5527777 | 10·8122404 | ·0007911392 |
| 1265 | 1600225 | 2024284625 | 35·5668385 | 10·8150909 | ·0007905138 |
| 1266 | 1602756 | 2029089096 | 35·5808937 | 10·8179400 | ·0007898894 |
| 1267 | 1605289 | 2033901163 | 35·5949434 | 10·8207876 | ·0007892660 |
| 1268 | 1607824 | 2038720832 | 35·6089876 | 10·8236336 | ·0007886435 |
| 1269 | 1610361 | 2043548109 | 35·6230262 | 10·8264782 | ·0007880221 |
| 1270 | 1612900 | 2048383000 | 35·6370593 | 10·8293213 | ·0007874016 |
| 1271 | 1615441 | 2053225511 | 35·6510869 | 10·8321629 | ·0007867821 |
| 1272 | 1617984 | 2058075648 | 35·6651090 | 10·8350030 | ·0007861635 |
| 1273 | 1620529 | 2062933417 | 35·6791255 | 10·8378416 | ·0007855460 |
| 1274 | 1623076 | 2067798824 | 35·6931366 | 10·8406788 | ·0007849294 |
| 1275 | 1625625 | 2072671875 | 35·7071421 | 10·8435144 | ·0007843137 |
| 1276 | 1628176 | 2077552576 | 35·7211422 | 10·8463485 | ·0007836991 |
| 1277 | 1630729 | 2082440933 | 35·7351367 | 10·8491812 | ·0007830854 |
| 1278 | 1633284 | 2087336952 | 35·7491258 | 10·8520125 | ·0007824726 |
| 1279 | 1635841 | 2092240639 | 35·7631095 | 10·8548422 | ·0007818608 |
| 1280 | 1638400 | 2097152000 | 35·7770876 | 10·8576704 | ·0007812500 |
| 1281 | 1640961 | 2102071841 | 35·7910603 | 10·8604972 | ·0007806401 |
| 1282 | 1643524 | 2106997768 | 35·8050276 | 10·8633225 | ·0007800312 |
| 1283 | 1646089 | 2111932187 | 35·8189894 | 10·8661454 | ·0007794232 |
| 1284 | 1648656 | 2116874304 | 35·8329457 | 10·8689687 | ·0007788162 |
| 1285 | 1651225 | 2121824125 | 35·8468966 | 10·8717897 | ·0007782101 |
| 1286 | 1653796 | 2126781656 | 35·8608421 | 10·8746091 | ·0007776050 |
| 1287 | 1656369 | 2131746903 | 35·8747822 | 10·8774271 | ·0007770008 |
| 1288 | 1658944 | 2136719872 | 35·8887169 | 10·8802436 | ·0007763975 |
| 1289 | 1661521 | 2141700569 | 35·9026461 | 10·8830587 | ·0007757952 |
| 1290 | 1664100 | 2146689000 | 35·9165699 | 10·8858723 | ·0007751938 |
| 1291 | 1666681 | 2151685171 | 35·9304884 | 10·8886845 | ·0007745933 |
| 1292 | 1669264 | 2156689088 | 35·9444015 | 10·8914952 | ·0007739938 |
| 1293 | 1671849 | 2161700757 | 35·9583092 | 10·8943044 | ·0007733952 |
| 1294 | 1674436 | 2166720184 | 35·9722115 | 10·8971123 | ·0007727975 |
| 1295 | 1677025 | 2171747375 | 35·9861084 | 10·8999186 | ·0007722008 |
| 1296 | 1679616 | 2176782336 | 36·0000000 | 10·9027235 | ·0007716049 |
| 1297 | 1682209 | 2181825073 | 36·0138862 | 10·9055269 | ·0007710100 |

| Number. | Squares. | Cubes. | Square Roots. | Cube Roots. | Reciprocals. |
|---|---|---|---|---|---|
| 1298 | 1684804 | 2186875592 | 36·0277671 | 10·9083290 | ·0007704160 |
| 1299 | 1687401 | 2191933899 | 36·0416426 | 10·9111296 | ·0007698229 |
| 1300 | 1690000 | 2197000000 | 36·0555128 | 10·9139287 | ·0007692308 |
| 1301 | 1692601 | 2202073901 | 36·0693776 | 10·9167265 | ·0007686395 |
| 1302 | 1695204 | 2207155608 | 36·0832371 | 10·9195228 | ·0007680492 |
| 1303 | 1697809 | 2212245127 | 36·0970913 | 10·9223177 | ·0007674579 |
| 1304 | 1700416 | 2217342464 | 36·1109402 | 10·9251111 | ·0007668712 |
| 1305 | 1703025 | 2222447625 | 36·1247837 | 10·9279031 | ·0007662835 |
| 1306 | 1705636 | 2227560616 | 36·1386220 | 10·9306937 | ·0007656968 |
| 1307 | 1708249 | 2232681443 | 36·1524550 | 10·9334829 | ·0007651109 |
| 1308 | 1710864 | 2237810112 | 36·1662826 | 10·9362706 | ·0007645260 |
| 1309 | 1713481 | 2242946629 | 36·1801050 | 10·9390569 | ·0007639419 |
| 1310 | 1716100 | 2248091000 | 36·1939221 | 10·9418418 | ·0007633588 |
| 1311 | 1718721 | 2253243231 | 36·2077340 | 10·9446253 | ·0007627765 |
| 1312 | 1721344 | 2258403328 | 36·2215406 | 10·9475074 | ·0007621951 |
| 1313 | 1723969 | 2263571297 | 36·2353419 | 10·9501880 | ·0007616446 |
| 1314 | 1726596 | 2268747144 | 36·2491379 | 10·9529673 | ·0007610350 |
| 1315 | 1729225 | 2273930875 | 36·2626287 | 10·9557451 | ·0007604563 |
| 1316 | 1731856 | 2279122496 | 36·2767143 | 10·9585215 | ·0007598784 |
| 1317 | 1734489 | 2284322013 | 36·2904246 | 10·9612965 | ·0007593014 |
| 1318 | 1737124 | 2289529432 | 36·3042697 | 10·9640701 | ·0007587253 |
| 1319 | 1739761 | 2294744759 | 36·3180396 | 10·9668423 | ·0007581501 |
| 1320 | 1742400 | 2299968000 | 36·3318042 | 10·9696131 | ·0007575758 |
| 1321 | 1745041 | 2305199161 | 36·3455637 | 10·9723825 | ·0007570023 |
| 1322 | 1747684 | 2310438248 | 36·3593179 | 10·9751505 | ·0007564297 |
| 1323 | 1750329 | 2315685267 | 36·3730670 | 10·9779171 | ·0007558579 |
| 1324 | 1752976 | 2320940224 | 36·3868108 | 10·9806823 | ·0007552870 |
| 1325 | 1755625 | 2326203125 | 36·4005494 | 10·9834462 | ·0007547170 |
| 1326 | 1758276 | 2331473976 | 36·4142829 | 10·9862086 | ·0007541478 |
| 1327 | 1760929 | 2336752783 | 36·4280112 | 10·9889696 | ·0007535795 |
| 1328 | 1763584 | 2342039552 | 36·4417343 | 10·9917293 | ·0007530120 |
| 1329 | 1766241 | 2347334289 | 36·4554523 | 10·9944876 | ·0007524454 |
| 1330 | 1768900 | 2352637000 | 36·4691650 | 10·9972445 | ·0007518797 |
| 1331 | 1771561 | 2357947691 | 36·4828727 | 11·0000000 | ·0007513148 |
| 1332 | 1774224 | 2363266368 | 36·4965752 | 11·0027541 | ·0007507508 |
| 1333 | 1776889 | 2368593037 | 36·5102725 | 11·0055069 | ·0007501875 |
| 1334 | 1779556 | 2373927704 | 36·5239647 | 11·0082583 | ·0007496252 |
| 1335 | 1782225 | 2379270375 | 36·5376518 | 11·0110082 | ·0007490637 |
| 1336 | 1784896 | 2384621056 | 36·5513388 | 11·0137569 | ·0007485030 |
| 1337 | 1787569 | 2389979753 | 36·5650106 | 11·0165041 | ·0007479432 |
| 1338 | 1790244 | 2395346472 | 36·5786823 | 11·0192500 | ·0007473842 |
| 1339 | 1792921 | 2400721219 | 36·5923489 | 11·0219915 | ·0007468260 |
| 1340 | 1795600 | 2406104000 | 36·6060104 | 11·0247377 | ·0007462687 |
| 1341 | 1798281 | 2411494821 | 36·6196668 | 11·0274795 | ·0007457122 |
| 1342 | 1800964 | 2416893688 | 36·6333181 | 11·0302199 | ·0007451565 |
| 1343 | 1803649 | 2422300607 | 36·6469144 | 11·0329590 | ·0007446016 |
| 1344 | 1806336 | 2427715584 | 36·6606056 | 11·0356967 | ·0007440476 |
| 1345 | 1809025 | 2433138625 | 36·6742416 | 11·0384330 | ·0007434944 |
| 1346 | 1811716 | 2438569736 | 36·6878726 | 11·0411680 | ·0007429421 |
| 1347 | 1814409 | 2444008923 | 36·7014986 | 11·0439017 | ·0007423905 |
| 1348 | 1817104 | 2449456192 | 36·7151195 | 11·0466339 | ·0007418398 |
| 1349 | 1819801 | 2454911549 | 36·7287353 | 11·0493649 | ·0007412898 |
| 1350 | 1822500 | 2460375000 | 36·7423461 | 11·0520945 | ·0007407407 |
| 1351 | 1825201 | 2465846551 | 36·7559519 | 11·0548227 | ·0007401924 |
| 1352 | 1827904 | 2471326208 | 36·7695526 | 11·0575497 | ·0007396450 |
| 1353 | 1830609 | 2476813977 | 36·7831483 | 11·0602752 | ·0007390983 |
| 1354 | 1833316 | 2482309864 | 36·7967390 | 11·0629994 | ·0007385524 |
| 1355 | 1836025 | 2487813875 | 36·8103246 | 11·0657222 | ·0007380074 |
| 1356 | 1838736 | 2493326016 | 36·8239053 | 11·0684437 | ·0007374631 |
| 1357 | 1841449 | 2498846293 | 36·8374809 | 11·0711639 | ·0007369197 |
| 1358 | 1844164 | 2504374712 | 36·8510515 | 11·0738828 | ·0007363770 |
| 1359 | 1846881 | 2509911279 | 36·8646172 | 11·0766003 | ·0007358352 |

| Number. | Squares. | Cubes. | Square Roots. | Cube Roots. | Reciprocals. |
|---|---|---|---|---|---|
| 1360 | 1849600 | 2515456000 | 36·8781778 | 11·0793165 | ·0007352941 |
| 1361 | 1852321 | 2521008881 | 36·8917335 | 11·0820314 | ·0007347539 |
| 1362 | 1855044 | 2526569928 | 36·9052842 | 11·0847449 | ·0007342144 |
| 1363 | 1857769 | 2532139147 | 36·9188299 | 11·0874571 | ·0007336757 |
| 1364 | 1860496 | 2537716544 | 36·9323706 | 11·0901679 | ·0007331378 |
| 1365 | 1863225 | 2543302125 | 36·9459064 | 11·0928775 | ·0007326007 |
| 1366 | 1865956 | 2548895896 | 36·9594372 | 11·0955857 | ·0007320644 |
| 1367 | 1868689 | 2554497863 | 36·9729631 | 11·0982926 | ·0007315289 |
| 1368 | 1871424 | 2560108032 | 36·9864840 | 11·1009982 | ·0007309942 |
| 1369 | 1874161 | 2565726409 | 37·0000000 | 11·1037025 | ·0007304602 |
| 1370 | 1876900 | 2571353000 | 37·0135110 | 11·1064054 | ·0007299270 |
| 1371 | 1879641 | 2576987811 | 37·0270172 | 11·1091070 | ·0007293946 |
| 1372 | 1882384 | 2582630848 | 37·0405184 | 11·1118073 | ·0007288630 |
| 1373 | 1885129 | 2588282117 | 37·0540146 | 11·1145064 | ·0007283321 |
| 1374 | 1887876 | 2593941624 | 37·0675060 | 11·1172041 | ·0007278020 |
| 1375 | 1890625 | 2599609375 | 37·0899924 | 11·1199004 | ·0007272727 |
| 1376 | 1893376 | 2605285376 | 37·0944740 | 11·1225955 | ·0007267442 |
| 1377 | 1896129 | 2610969633 | 37·1079506 | 11·1252893 | ·0007262164 |
| 1378 | 1898884 | 2616662152 | 37·1214224 | 11·1279817 | ·0007256894 |
| 1379 | 1901641 | 2622362939 | 37·1348893 | 11·1306729 | ·0007251632 |
| 1380 | 1904400 | 2628072000 | 37·1483512 | 11·1333628 | ·0007246377 |
| 1381 | 1907161 | 2633789341 | 37·1618084 | 11·1360514 | ·0007241130 |
| 1382 | 1909924 | 2639514968 | 37·1752606 | 11·1387386 | ·0007235890 |
| 1383 | 1912689 | 2645248887 | 37·1887079 | 11·1414246 | ·0007230658 |
| 1384 | 1915456 | 2650991104 | 37·2021505 | 11·1441093 | ·0007225434 |
| 1385 | 1918225 | 2656741625 | 37·2155881 | 11·1467926 | ·0007220217 |
| 1386 | 1920996 | 2662500456 | 37·2290209 | 11·1494747 | ·0007215007 |
| 1387 | 1923769 | 2668267603 | 37·2424489 | 11·1521555 | ·0007209805 |
| 1388 | 1926544 | 2674043072 | 37·2558720 | 11·1548350 | ·0007204611 |
| 1389 | 1929321 | 2679826869 | 37·2692903 | 11·1575133 | ·0007199424 |
| 1390 | 1932100 | 2685619000 | 37·2827037 | 11·1601903 | ·0007194245 |
| 1391 | 1934881 | 2691419471 | 37·2961124 | 11·1628659 | ·0007189073 |
| 1392 | 1937664 | 2697228288 | 37·3095162 | 11·1655403 | ·0007183908 |
| 1393 | 1940449 | 2703045457 | 37·3229152 | 11·1682134 | ·0007178751 |
| 1394 | 1943236 | 2708870984 | 37·3363094 | 11·1708852 | ·0007173601 |
| 1395 | 1946025 | 2714704875 | 37·3496988 | 11·1735558 | ·0007168459 |
| 1396 | 1948816 | 2720547136 | 37·3630834 | 11·1762250 | ·0007163324 |
| 1397 | 1951609 | 2726397773 | 37·3764632 | 11·1788930 | ·0007158196 |
| 1398 | 1954404 | 2732256792 | 37·3898382 | 11·1815598 | ·0007153076 |
| 1399 | 1957201 | 2738124199 | 37·4032084 | 11·1842252 | ·0007147963 |
| 1400 | 1960000 | 2744000000 | 37·4165738 | 11·1868894 | ·0007142857 |
| 1401 | 1962801 | 2749884201 | 37·4299345 | 11·1895523 | ·0007137759 |
| 1402 | 1965604 | 2755776808 | 37·4432904 | 11·1922139 | ·0007132668 |
| 1403 | 1968409 | 2761677827 | 37·4566416 | 11·1948743 | ·0007127584 |
| 1404 | 1971216 | 2767587264 | 37·4699880 | 11·1975334 | ·0007122507 |
| 1405 | 1974025 | 2773505123 | 37·4833296 | 11·2001913 | ·0007117438 |
| 1406 | 1976836 | 2779431416 | 37·4966665 | 11·2028479 | ·0007112376 |
| 1407 | 1979649 | 2785366143 | 37·5099987 | 11·2055032 | ·0007107321 |
| 1408 | 1982464 | 2791309312 | 37·5233261 | 11·2081573 | ·0007102273 |
| 1409 | 1985281 | 2797260929 | 37·5366487 | 11·2108101 | ·0007097232 |
| 1410 | 1988100 | 2803221000 | 37·5499667 | 11·2134617 | ·0007092199 |
| 1411 | 1990921 | 2809189531 | 37·5632799 | 11·2161120 | ·0007087172 |
| 1412 | 1993744 | 2815166528 | 37·5765885 | 11·2187611 | ·0007082153 |
| 1413 | 1996569 | 2821151997 | 37·5898922 | 11·2214089 | ·0007077141 |
| 1414 | 1999396 | 2827145944 | 37·6031913 | 11·2240054 | ·0007072136 |
| 1415 | 2002225 | 2833148375 | 37·6164857 | 11·2267007 | ·0007067138 |
| 1416 | 2005056 | 2839159296 | 37·6297754 | 11·2293448 | ·0007062147 |
| 1417 | 2007889 | 2845178713 | 37·6430604 | 11·2319876 | ·0007057163 |
| 1418 | 2010724 | 2851206632 | 37·6563407 | 11·2346292 | ·0007052186 |
| 1419 | 2013561 | 2857243059 | 37·6696164 | 11·2372696 | ·0007047216 |
| 1420 | 2016400 | 2863288000 | 37·6828874 | 11·2399087 | ·0007042254 |
| 1421 | 2019241 | 2869341461 | 37·6961536 | 11·2425465 | ·0007037298 |

| Number. | Squares. | Cubes. | Square Roots. | Cube Roots. | Reciprocals. |
|---|---|---|---|---|---|
| 1422 | 2022084 | 2875403448 | 37·7094153 | 11·2451831 | ·0007032349 |
| 1423 | 2024929 | 2881473967 | 37·7226722 | 11·2478185 | ·0007027407 |
| 1424 | 2027776 | 2887553024 | 37·7359245 | 11·2504527 | ·0007022472 |
| 1425 | 2030625 | 2893640625 | 37·7491722 | 11·2530856 | ·0007017544 |
| 1426 | 2033476 | 2899736776 | 37·7624152 | 11·2557173 | ·0007012623 |
| 1427 | 2036329 | 2905841483 | 37·7756535 | 11·2583478 | ·0007007708 |
| 1428 | 2039184 | 2911954752 | 37·7888873 | 11·2609770 | ·0007002801 |
| 1429 | 2042041 | 2918076589 | 37·8021163 | 11·2636050 | ·0006997901 |
| 1430 | 2044900 | 2924207000 | 37·8153408 | 11·2662318 | ·0006993007 |
| 1431 | 2047761 | 2930345991 | 37·8285606 | 11·2688573 | ·0006988120 |
| 1432 | 2050624 | 2936493568 | 37·8417759 | 11·2714816 | ·0006983240 |
| 1433 | 2053489 | 2942649737 | 37·8549864 | 11·2741047 | ·0006978367 |
| 1434 | 2056356 | 2948814504 | 37·8681924 | 11·2767266 | ·0006973501 |
| 1435 | 2059225 | 2954987875 | 37·8813938 | 11·2793472 | ·0006968641 |
| 1436 | 2062096 | 2961169856 | 37·8945906 | 11·2819666 | ·0006963788 |
| 1437 | 2064969 | 2967360453 | 37·9077828 | 11·2845849 | ·0006958942 |
| 1438 | 2067844 | 2973559672 | 37·9209704 | 11·2872019 | ·0006954103 |
| 1439 | 2070721 | 2979767519 | 37·9341538 | 11·2898177 | ·0006949270 |
| 1440 | 2073600 | 2985984000 | 37·9473319 | 11·2924323 | ·0006944444 |
| 1441 | 2076481 | 2992209121 | 37·9605058 | 11·2950457 | ·0006939625 |
| 1442 | 2079364 | 3098442888 | 37·9736751 | 11·2976579 | ·0006934813 |
| 1443 | 2082249 | 3004685307 | 37·9868398 | 11·3002688 | ·0006930007 |
| 1444 | 2085136 | 3010936384 | 38·0000000 | 11·3028786 | ·0006925208 |
| 1445 | 2088025 | 3017196125 | 38·0131556 | 11·3054871 | ·0006920415 |
| 1446 | 2080916 | 3023464536 | 38·0263067 | 11·3080945 | ·0006915629 |
| 1447 | 2093809 | 3029741623 | 38·0394532 | 11·3107006 | ·0006910850 |
| 1448 | 2096704 | 3036027392 | 38·0525952 | 11·3133056 | ·0006906078 |
| 1449 | 2099601 | 3042321849 | 38·0657326 | 11·3159094 | ·0006901312 |
| 1450 | 2102500 | 3048625000 | 38·0788655 | 11·3185119 | ·0006896552 |
| 1451 | 2105401 | 3054936851 | 38·0919939 | 11·3211132 | ·0006891799 |
| 1452 | 2108304 | 3061257408 | 38·1051178 | 11·3237134 | ·0006887052 |
| 1453 | 2111209 | 3067586777 | 38·1182371 | 11·3263124 | ·0006882312 |
| 1454 | 2114116 | 3073924664 | 38·1313519 | 11·3289102 | ·0006877579 |
| 1455 | 2117025 | 3080271375 | 38·1444622 | 11·3315067 | ·0006872852 |
| 1456 | 2119936 | 3086626816 | 38·1575681 | 11·3341022 | ·0006868132 |
| 1457 | 2122849 | 3092990993 | 38·1706693 | 11·3366964 | ·0006863412 |
| 1458 | 2125764 | 3099363912 | 38·1837662 | 11·3392894 | ·0006858711 |
| 1459 | 2128681 | 3105745579 | 38·1968585 | 11·3418813 | ·0006854010 |
| 1460 | 2131600 | 3112136000 | 38·2099463 | 11·3444719 | ·0006849315 |
| 1461 | 2134521 | 3118535181 | 38·2230297 | 11·3470614 | ·0006844627 |
| 1462 | 2137444 | 3124943128 | 38·2361085 | 11·3496497 | ·0006839945 |
| 1463 | 2140369 | 3131359847 | 38·2491829 | 11·3522368 | ·0006835270 |
| 1464 | 2143296 | 3137785344 | 38·2622529 | 11·3548227 | ·0006830601 |
| 1465 | 2146225 | 3144219625 | 38·2753184 | 11·3574075 | ·0006825939 |
| 1466 | 2149156 | 3150662696 | 38·2883794 | 11·3599911 | ·0006821282 |
| 1467 | 2152089 | 3157114563 | 38·3014360 | 11·3625735 | ·0006816633 |
| 1468 | 2155024 | 3163575232 | 38·3144881 | 11·3651547 | ·0006811989 |
| 1469 | 2157961 | 3170044709 | 38·3275358 | 11·3677347 | ·0006807352 |
| 1470 | 2160900 | 3176523000 | 38·3405790 | 11·3703136 | ·0006802721 |
| 1471 | 2163841 | 3183010111 | 38·3536178 | 11·3728914 | ·0006798097 |
| 1472 | 2166784 | 3189506048 | 38·3666522 | 11·3754679 | ·0006793478 |
| 1473 | 2169729 | 3196010817 | 38·3796821 | 11·3780433 | ·0006788866 |
| 1474 | 2172676 | 3202524424 | 38·3927076 | 11·3806175 | ·0006784261 |
| 1475 | 2175625 | 3209046875 | 38·4057287 | 11·3831906 | ·0006779661 |
| 1476 | 2178576 | 3215578176 | 38·4187454 | 11·3857625 | ·0006775068 |
| 1477 | 2181529 | 3222118333 | 38·4317577 | 11·3883332 | ·0006770481 |
| 1478 | 2184484 | 3228667352 | 38·4447656 | 11·3909028 | ·0006765900 |
| 1479 | 2187441 | 3235225239 | 38·4577691 | 11·3934712 | ·0006761325 |
| 1480 | 2190400 | 3241792000 | 38·4707681 | 11·3960384 | ·0006756757 |
| 1481 | 2193361 | 3248367641 | 38·4837627 | 11·3986045 | ·0006752194 |
| 1482 | 2196324 | 3254952168 | 38·4967530 | 11·4011695 | ·0006747638 |
| 1483 | 2199289 | 3261545587 | 38·5097390 | 11·4037332 | ·0006743088 |

| Number. | Squares. | Cubes. | Square Roots. | Cube Roots. | Reciprocals. |
|---|---|---|---|---|---|
| 1484 | 2202256 | 3268147904 | 38·5227206 | 11·4062959 | ·0006738544 |
| 1485 | 2205225 | 3274759125 | 38·5356977 | 11·4088574 | ·0006734007 |
| 1486 | 2208196 | 3281379256 | 38·5486705 | 11·4114177 | ·0006729474 |
| 1487 | 2211169 | 3288008303 | 38·5616389 | 11·4139769 | ·0006724950 |
| 1488 | 2214144 | 3294646272 | 38·5746030 | 11·4165349 | ·0006720430 |
| 1489 | 2217121 | 3301293169 | 38·5875627 | 11·4190918 | ·0006715917 |
| 1490 | 2220100 | 3307949000 | 38·6005181 | 11·4206476 | ·0006711409 |
| 1491 | 2223081 | 3314613771 | 38·6134691 | 11·4242022 | ·0006706908 |
| 1492 | 2226004 | 3321287488 | 38·6264158 | 11·4267556 | ·0006702413 |
| 1493 | 2229049 | 3227970157 | 38·6393582 | 11·4293079 | ·0006697924 |
| 1494 | 2232036 | 3334661784 | 38·6522962 | 11·4318591 | ·0006693440 |
| 1495 | 2235025 | 3341362375 | 38·6652299 | 11·4344092 | ·0006688963 |
| 1496 | 2238016 | 3348071936 | 38·6781593 | 11·4369581 | ·0006684492 |
| 1497 | 2241009 | 3354790473 | 38·6910843 | 11·4395059 | ·0006680027 |
| 1498 | 2244004 | 3361517992 | 38·7040050 | 11·4420525 | ·0006675567 |
| 1499 | 2247001 | 3368254499 | 38·7169214 | 11·4445980 | ·0006671114 |
| 1500 | 2250000 | 3375000000 | 38·7298335 | 11·4471424 | ·0006666667 |
| 1501 | 2253001 | 3381754501 | 38·7427412 | 11·4496857 | ·0006662225 |
| 1502 | 2256004 | 3388518008 | 38·7556447 | 11·4522278 | ·0006657790 |
| 1503 | 2259009 | 3395290527 | 38·7685439 | 11·4547688 | ·0006553360 |
| 1504 | 2262016 | 3402072064 | 38·7814389 | 11·4573087 | ·0006648936 |
| 1505 | 2265025 | 3408862625 | 38·7943294 | 11·4598476 | ·0006644518 |
| 1506 | 2268036 | 3415662216 | 38·8072158 | 11·4623850 | ·0006640106 |
| 1507 | 2271049 | 3422470843 | 38·8200978 | 11·4649215 | ·0006635700 |
| 1508 | 2274064 | 3429288512 | 38·8329757 | 11·4674568 | ·0006631300 |
| 1509 | 2277081 | 3436115229 | 38·8458491 | 11·4699911 | ·0006626905 |
| 1510 | 2280100 | 3442951000 | 38·8587184 | 11·4725242 | ·0006622517 |
| 1511 | 2283121 | 3449795831 | 38·8715834 | 11·4750562 | ·0006618134 |
| 1512 | 2286144 | 3456649728 | 38·8844442 | 11·4775871 | ·0006613757 |
| 1513 | 2289169 | 3463512697 | 38·8973006 | 11·4801169 | ·0006609385 |
| 1514 | 2292196 | 3470384744 | 38·9101529 | 11·4826455 | ·0006605020 |
| 1515 | 2295225 | 3477265875 | 38·9230009 | 11·4851731 | ·0006600660 |
| 1516 | 2298256 | 3484156096 | 38·9358447 | 11·4876995 | ·0006596306 |
| 1517 | 2301289 | 3491055413 | 38·9486841 | 11·4902249 | ·0006591958 |
| 1518 | 2304324 | 3597963832 | 38·9615194 | 11·4927491 | ·0006587615 |
| 1519 | 2307361 | 3504881359 | 38·9743505 | 11·4952722 | ·0006583278 |
| 1520 | 2310400 | 3511808000 | 38·9871774 | 11·4977942 | ·0006578947 |
| 1521 | 2313441 | 3518743761 | 39·0000000 | 11·5003151 | ·0006574622 |
| 1522 | 2316484 | 3525688648 | 39·0128184 | 11·5028348 | ·0006570302 |
| 1523 | 2319529 | 3532642667 | 39·0256326 | 11·5053535 | ·0006565988 |
| 1524 | 2322576 | 3539605824 | 39·0384426 | 11·5078711 | ·0006561680 |
| 1525 | 2325625 | 3546578125 | 39·0512483 | 11·5103876 | ·0006557377 |
| 1526 | 2328676 | 3553559576 | 39·0640499 | 11·5129030 | ·0006553080 |
| 1527 | 2331729 | 3567549552 | 39·0768473 | 11·5154173 | ·0006548788 |
| 1528 | 2334784 | 3560558183 | 39·0896406 | 11·5179305 | ·0006544503 |
| 1529 | 2337841 | 3574558889 | 39·1024296 | 11·5204425 | ·0006540222 |
| 1530 | 2340900 | 3581577000 | 39·1152144 | 11·5229535 | ·0006535948 |
| 1531 | 2343961 | 3588604291 | 39·1279951 | 11·5254634 | ·0006531679 |
| 1532 | 2347024 | 3595640768 | 39·1407716 | 11·5279722 | ·0006527415 |
| 1533 | 2350089 | 3602686437 | 39·1535439 | 11·5304799 | ·0006523157 |
| 1534 | 2353156 | 3609741304 | 39·1663120 | 11·5329865 | ·0006518905 |
| 1535 | 2356225 | 3616805375 | 39·1790760 | 11·5354920 | ·0000514658 |
| 1536 | 2359256 | 3623878656 | 39·1918359 | 11·5379965 | ·0006510417 |
| 1537 | 2362369 | 3630961153 | 39·2045915 | 11·5404998 | ·0006506181 |
| 1538 | 2365444 | 3638052872 | 39·2173431 | 11·5430021 | ·0006501951 |
| 1539 | 2368521 | 3645153819 | 39·2300905 | 11·5455033 | ·0006497726 |
| 1540 | 2371600 | 3652264000 | 39·2428337 | 11·5480034 | ·0006493506 |
| 1541 | 2374681 | 3659383421 | 39·2555728 | 11·5505025 | ·0006489293 |
| 1542 | 2377764 | 3666512088 | 39·2683078 | 11·5530004 | ·0006485084 |
| 1543 | 2380849 | 3673650007 | 39·2810387 | 11·5554972 | ·0006480881 |
| 1544 | 2383936 | 3680797184 | 39·2937654 | 11·5579931 | ·0006476684 |
| 1545 | 2387025 | 3687953625 | 39·3064880 | 11·5604878 | ·0006472492 |

| Number. | Squares. | Cubes. | Square Roots. | Cube Roots. | Reciprocals. |
|---|---|---|---|---|---|
| 1546 | 2390116 | 3695119336 | 39·3192065 | 11·5629815 | ·0006468305 |
| 1547 | 2393209 | 3702294323 | 39·3319208 | 11·5654740 | ·0006464124 |
| 1548 | 2396304 | 3709478592 | 39·3446311 | 11·5679655 | ·0006459948 |
| 1549 | 2399401 | 3716672149 | 39·3573373 | 11·5704559 | ·0006455778 |
| 1550 | 2402500 | 3723875000 | 39·3700394 | 11·5729453 | ·0006451613 |
| 1551 | 2405601 | 3731087151 | 39·3827373 | 11·5754336 | ·0006447453 |
| 1552 | 2408704 | 3738308608 | 39·3954312 | 11·5779208 | ·0006443299 |
| 1553 | 2411809 | 3745539377 | 39·4081210 | 11·5804069 | ·0006439150 |
| 1554 | 2414916 | 3752779464 | 39·4208067 | 11·5828919 | ·0006435006 |
| 1555 | 2418025 | 3760028875 | 39·4334883 | 11·5853759 | ·0006430868 |
| 1556 | 2421136 | 3767287616 | 39·4461658 | 11·5878588 | ·0006426735 |
| 1557 | 2424249 | 3774555693 | 39·4588393 | 11·5903407 | ·0006422608 |
| 1558 | 2427364 | 3781833112 | 39·4715087 | 11·5928215 | ·0006418485 |
| 1559 | 2430481 | 3789119879 | 39·4841740 | 11·5953013 | ·0006414368 |
| 1560 | 2433600 | 3796416000 | 39·4968353 | 11·5977799 | ·0006410256 |
| 1561 | 2436721 | 3803721481 | 39·5094925 | 11·6002576 | ·0006406150 |
| 1562 | 2439844 | 3811036328 | 39·5221457 | 11·6027342 | ·0006402049 |
| 1563 | 2442969 | 3818360547 | 39·5347948 | 11·6052097 | ·0006397953 |
| 1564 | 2446096 | 3825641444 | 39·5474399 | 11·6076841 | ·0006393862 |
| 1565 | 2449225 | 3833037125 | 39·5600809 | 11·6101575 | ·0006389776 |
| 1566 | 2452356 | 3840389496 | 39·5727179 | 11·6126299 | ·0006385696 |
| 1567 | 2455489 | 3847751263 | 39·5853508 | 11·6151012 | ·0006381621 |
| 1568 | 2458624 | 3855123432 | 39·5979797 | 11·6175715 | ·0006377551 |
| 1569 | 2461761 | 3862503009 | 39·6106046 | 11·6200407 | ·0006373486 |
| 1570 | 2464900 | 3869883000 | 39·6232255 | 11·6225088 | ·0006369427 |
| 1571 | 2468041 | 3877292411 | 39·6358424 | 11·6249759 | ·0006365372 |
| 1572 | 2471184 | 3884701248 | 39·6484552 | 11·6274420 | ·0006361323 |
| 1573 | 2474329 | 3892119157 | 39·6610640 | 11·6299070 | ·0006357279 |
| 1574 | 2477476 | 3899547224 | 39·6736688 | 11·6323710 | ·0006353240 |
| 1575 | 2480625 | 3906984375 | 39·6862696 | 11·6348339 | ·0006349206 |
| 1576 | 2483776 | 3914430976 | 39·6988665 | 11·6372957 | ·0006345178 |
| 1577 | 2486929 | 3921887033 | 39·7114593 | 11·6397566 | ·0006341154 |
| 1578 | 2490084 | 3929352552 | 39·7240481 | 11·6422164 | ·0006337136 |
| 1579 | 2493241 | 3936827539 | 39·7366329 | 11·6446751 | ·0006333122 |
| 1580 | 2496400 | 3944312000 | 39·7492138 | 11·6471329 | ·0006329114 |
| 1581 | 2499561 | 3951805941 | 39·7617907 | 11·6495895 | ·0006325111 |
| 1582 | 2502724 | 3959309368 | 39·7743636 | 11·6520452 | ·0006321113 |
| 1583 | 2505889 | 3966822287 | 39·7869325 | 11·6544998 | ·0006317119 |
| 1584 | 2509056 | 3974344704 | 39·7994976 | 11·6569534 | ·0006313131 |
| 1585 | 2512225 | 3981876625 | 39·8120585 | 11·6594059 | ·0006309148 |
| 1586 | 2515396 | 3989418056 | 39·8246155 | 11·6618574 | ·0006305170 |
| 1587 | 2518569 | 3996969003 | 39·8371686 | 11·6643079 | ·0006301197 |
| 1588 | 2521744 | 4004529472 | 39·8497177 | 11·6667574 | ·0006297229 |
| 1589 | 2524921 | 4012099469 | 39·8622628 | 11·6692058 | ·0006293266 |
| 1590 | 2528100 | 4014679000 | 39·8748040 | 11·6716532 | ·0006289308 |
| 1591 | 2531281 | 4027268071 | 39·8873413 | 11·6740996 | ·0006285355 |
| 1592 | 2534464 | 4034866688 | 39·8998747 | 11·6765449 | ·0006281407 |
| 1593 | 2537649 | 4042474857 | 39·9124041 | 11·6789892 | ·0006277464 |
| 1594 | 2540836 | 4050092584 | 39·9249295 | 11·6814325 | ·0006273526 |
| 1595 | 2544025 | 4057719875 | 39·9374511 | 11·6838748 | ·0006269592 |
| 1596 | 2547216 | 4065356736 | 39·9499687 | 11·6863161 | ·0006265664 |
| 1597 | 2550409 | 4073003173 | 39·9624824 | 11·6887563 | ·0006261741 |
| 1598 | 2553604 | 4080659192 | 39·9749922 | 11·6911955 | ·0006257822 |
| 1599 | 2556801 | 4088324799 | 39·9874980 | 11·6936337 | ·0006253909 |
| 1600 | 2560000 | 4096000000 | 40·0000000 | 11·6960709 | ·0006250000 |

*To find the square or cube root of a number consisting of integers and decimals.*

RULE.—Multiply the difference between the root of the integer part of the given number, and the root of the next higher integer number, by the decimal part of the given number, and add the product to the root of the given integer number; the sum is the root required.

Required the square root of 20·321.

Square root of 21 = 4·5825
Do. 20 = 4·4721

·1104 × ·321 + 4·4721 = 4·5075384, the square root required.

Required the cube root of 16·42.

Cube root of 17 = 2·5712
Do. 16 = 2·5198

·0514 × ·42 + 2·5198 = 2·541388, the cube root required.

*To find the squares of numbers in arithmetical progression; or, to extend the foregoing table of squares.*

RULE.—Find, in the usual way, the squares of the first two numbers, and subtract the less from the greater. Set down the square of the larger number, in a separate column, and add to it the difference already found, with the addition of 2, as a constant quantity; the product will be the square of the next following number.

The square of 1500 .................= 2250000.........2250000
The square of 1499 .................= 2247001

Difference...... 2999 + 2 = 3001

The square of 1501.........................................2253001
Difference...... 3001 + 2 = 3003

The square of 1502.........................................2256004

*To find the square of a greater number than is contained in the table.*

RULE 1.—If the number required to be squared exceed by 2, 3, 4, or any other number of times, any number contained in the table, let the square affixed to the number in the table be multiplied by the square of 2, 3, or 4, &c., and the product will be the answer sought.

Required the square of 2595.

2595 is three times greater than 865; and the square of 865, by the table, is 748225.

Then, $748225 \times 3^2 = 6734025$.

RULE 2.—If the number required to be squared be an odd number, and do not exceed twice the amount of any number contained in the table, find the two numbers nearest to each other, which, added together, make that sum; then the sum of the squares of these two numbers, by the table, multiplied by 2, will exceed the square required by 1.

Required the square of 1865.

The two nearest numbers (932 + 933) = 1865.

Then, by table $(932^2 = 868624) + (933^2 = 870489) = 1739113 \times 2 = 3478226 - 1 = 3478225$.

*To find the cube of a greater number than is contained in the table.*

RULE.—Proceed, as in squares, to find how many times the number required to be cubed exceeds the number contained in the table. Multiply the cube of that number by the cube of as many times as the number sought exceeds the number in the table, and the product will be the answer required.

Required the cube of 3984.

3984 is 4 times greater than 996; and the cube of 996, by the table, is 988047936.

$$\text{Then, } 988047936 \times 4^3 = 63235067904.$$

*To find the square or cube root of a higher number than is in the table.*

RULE.—Refer to the table, and seek in the column of squares or cubes the number nearest to that number whose root is sought, and the number from which that square or cube is derived will be the answer required, when decimals are not of importance.

Required the square root of 542869.

In the Table of Squares, the nearest number is 543169; and the number from which that square has been obtained is 737.

$$\text{Therefore, } \sqrt{542869} = 737 \text{ nearly.}$$

*To find more nearly the cube root of a higher number than is in the table.*

RULE.—Ascertain, by the table, the nearest cube number to the number given, and call it the assumed cube.

Multiply the assumed cube, and the given number, respectively, by 2; to the product of the assumed cube add the given number, and to the product of the given number add the assumed cube.

Then, by proportion, as the sum of the assumed cube is to the sum of the given number, so is the root of the assumed cube to the root of the given number.

Required the cube root of 412568555.

By the table, the nearest number is 411830784, and its cube root is 744.

$$\text{Therefore, } 411830784 \times 2 + 412568555 = 1236230123.$$

$$\text{And, } 412568555 \times 2 + 411830784 = 1236967894.$$

Hence, as 1236230123 : 1236967894 : : 744 : 744·369, very nearly.

*To find the square or cube root of a number containing decimals.*

RULE.—Subtract the square root or cube root of the *integer* of the given number from the root of the next higher number, and multiply the difference by the decimal part. The product, added to the root of the integer of the given number will be the answer required.

Required the square root of 321·62.

$\sqrt{321} = 17·9164729$, and $\sqrt{322} = 17·9443584$; the difference $(·0278855) \times ·62 + 17·9164729 = 17·9337619$.

*To obtain the square root or cube root of a number containing decimals, by inspection.*

RULE.—The square or cube root of a number containing decimals may be found at once by inspection of the tables, by taking the figures cut off in the *number*, by the decimal point, in *pairs* if for the square root, and in *triads* if for the cube root. The following example will show the results obtained, by simple inspection of the tables, from the figures 234, and from the numbers formed by the addition of the decimal point or of ciphers.

| Number. | Square Root. | Cube Root. |
|---|---|---|
| ·00234 | ·0483735465* | ·132761439† |
| ·0234 | ·152970585 | ·284‡ |
| ·2340 | ·483735465 | ·61622401 |
| 2·34 | 1·52970585 | 1·32761439 |
| 23·40 | 4·83735465 | 2·860 |
| 234 | 15·2970585 | 6·1622401 |
| 2340 | 48·3735465 | 13·2761439 |
| 23400 | 152·970585 | 28·60 |

*To find the cubes of numbers in arithmetical progression, or to extend the preceding table of cubes.*

RULE.—Find the cubes of the first two numbers, and subtract the less from the greater. Then, multiply the least of the two numbers cubed by 6, add the product, with the addition of 6 as a constant quantity, to the difference; and thus, adding 6 each time to the sum last added, form a first series of differences.

To form a second series of differences, bring down, in a separate column, the cube of the highest of the above numbers, and add the difference to it. The amount will be the cube of the next general number.

Required the cubes of 1501, 1502, and 1503.

| *First series of differences.* | | | *Second series of differences.* | | |
|---|---|---|---|---|---|
| By Tab. 1500 = | 3375000000 | | Then, | 3375000000 | Cube of 1500 |
| 1499 = | 3368254499 | | Diff. for 1500 = | 6754501 | |
| | 6745501 | difference. | | 3381754501 | Cube of 1501 |
| 1499 × 6 + 6 = | 9000 | | Diff. for 1501 = | 6763507 | |
| | 6754501 | diff. of 1500 | | 3388518008 | Cube of 1502 |
| 9000 + 6 = | 9006 | | Diff. for 1502 = | 6772519 | |
| | 6763507 | diff. of 1501 | | 3395290527 | Cube of 1503 |
| 9006 + 6 = | 9012 | | | &c., &c. | |
| | 6772519 | diff. of 1502 | | | |
| | &c., &c. | | | | |

* Derived from ·002340 by means of 2340.

† Derived from ·002340 by means of 2340.

‡ The nearest result by simple inspection is obtained for ·023 by 23. But four places correct can always be obtained by looking in the table of cubes for the nearest triad or triads, in this instance for 23400; the cube beginning with the figures 23393 is that of 2860, whence ·2860 is true to the last place, and is afterwards substituted.

## TABLE *of the Fourth and Fifth Powers of Numbers.*

| Number. | 4th Power. | 5th Power. | Number. | 4th Power. | 5th Power. |
|---|---|---|---|---|---|
| 1 | 1 | 1 | 76 | 33362176 | 2535525376 |
| 2 | 16 | 32 | 77 | 35153041 | 2706784157 |
| 3 | 81 | 243 | 78 | 37015056 | 2887174368 |
| 4 | 256 | 1024 | 79 | 38950081 | 3077056399 |
| 5 | 625 | 3125 | 80 | 40960000 | 3276800000 |
| 6 | 1296 | 7776 | 81 | 43046721 | 3486784401 |
| 7 | 2401 | 16807 | 82 | 45212176 | 3707398432 |
| 8 | 4096 | 32768 | 83 | 47458321 | 3939040643 |
| 9 | 6561 | 59049 | 84 | 49787136 | 4182119424 |
| 10 | 10000 | 100000 | 85 | 52200625 | 4437053125 |
| 11 | 14641 | 161051 | 86 | 54708016 | 4704270176 |
| 12 | 20736 | 248832 | 87 | 57289761 | 4984209207 |
| 13 | 28561 | 371293 | 88 | 59969536 | 5277319168 |
| 14 | 38416 | 537824 | 89 | 62742241 | 5584059449 |
| 15 | 50625 | 759375 | 90 | 65610000 | 5904900000 |
| 16 | 65536 | 1048576 | 91 | 68574961 | 6240321451 |
| 17 | 83521 | 1419857 | 92 | 71639296 | 6590815232 |
| 18 | 104976 | 1889568 | 93 | 74805201 | 6596883693 |
| 19 | 130321 | 2476099 | 94 | 78074896 | 7339040224 |
| 20 | 160000 | 3200000 | 95 | 81450625 | 7737809375 |
| 21 | 194481 | 4084101 | 96 | 84934656 | 8153726976 |
| 22 | 234256 | 5153632 | 97 | 88529281 | 8587340257 |
| 23 | 279841 | 6436343 | 98 | 92236816 | 9039207968 |
| 24 | 331776 | 7962624 | 99 | 96059601 | 9509900499 |
| 25 | 390625 | 9765625 | 100 | 100000000 | 10000000000 |
| 26 | 456976 | 11881376 | 101 | 104060401 | 10510100501 |
| 27 | 531441 | 14348907 | 102 | 108243216 | 11040808032 |
| 28 | 614656 | 17210368 | 103 | 112550881 | 11592740743 |
| 29 | 707281 | 20511149 | 104 | 116985856 | 12166529024 |
| 30 | 810000 | 24300000 | 105 | 121550625 | 12762815625 |
| 31 | 923521 | 28629151 | 106 | 126247696 | 13382255776 |
| 32 | 1048576 | 33554432 | 107 | 131079601 | 14025517307 |
| 33 | 1185921 | 39135393 | 108 | 136048896 | 14693280768 |
| 34 | 1336336 | 45435424 | 109 | 141158161 | 15386239549 |
| 35 | 1500625 | 52521875 | 110 | 146410000 | 16105100000 |
| 36 | 1679616 | 60466176 | 111 | 151807041 | 16850581551 |
| 37 | 1874161 | 69343957 | 112 | 157351936 | 17623416832 |
| 38 | 2085136 | 79235168 | 113 | 163047361 | 18424351793 |
| 39 | 2313441 | 90224199 | 114 | 168896016 | 19254145824 |
| 40 | 2560000 | 102400000 | 115 | 174900625 | 20113571875 |
| 41 | 2825761 | 115856201 | 116 | 181063936 | 21003416576 |
| 42 | 3111696 | 130691232 | 117 | 187388721 | 21924480357 |
| 43 | 3418801 | 147008443 | 118 | 193877776 | 22877577568 |
| 44 | 3748096 | 164916224 | 119 | 200533921 | 23863536599 |
| 45 | 4100625 | 184528125 | 120 | 207360000 | 24883200000 |
| 46 | 4477456 | 205962976 | 121 | 214358881 | 25937424601 |
| 47 | 4879681 | 229345007 | 122 | 221533456 | 27027081632 |
| 48 | 5308416 | 254803968 | 123 | 228886641 | 28153056843 |
| 49 | 5764801 | 282475249 | 124 | 236421376 | 29316250624 |
| 50 | 6250000 | 312500000 | 125 | 244140625 | 30517578125 |
| 51 | 6765201 | 345025251 | 126 | 252047376 | 31757969376 |
| 52 | 7311616 | 380204032 | 127 | 260144641 | 33038369407 |
| 53 | 7890481 | 418195493 | 128 | 268435456 | 34359738368 |
| 54 | 8503056 | 459165024 | 129 | 276922881 | 35723051649 |
| 55 | 9150625 | 503284375 | 130 | 285610000 | 37129300000 |
| 56 | 9834496 | 550731776 | 131 | 294499921 | 38579489651 |
| 57 | 10556001 | 601692057 | 132 | 303595776 | 40074642432 |
| 58 | 11316496 | 656356768 | 133 | 312900721 | 41615795893 |
| 59 | 12117361 | 714924299 | 134 | 322417936 | 43204003424 |
| 60 | 12960000 | 777600000 | 135 | 332150625 | 44840334375 |
| 61 | 13845841 | 844596301 | 136 | 342102016 | 46525874176 |
| 62 | 14776336 | 916132832 | 137 | 352275361 | 48261724457 |
| 63 | 15752961 | 992436543 | 138 | 362673936 | 50049003168 |
| 64 | 16777216 | 1073741824 | 139 | 373301041 | 51888844699 |
| 65 | 17850625 | 1160290625 | 140 | 384160000 | 53782400000 |
| 66 | 18974736 | 1252332576 | 141 | 395254161 | 55730836701 |
| 67 | 20151121 | 1350125107 | 142 | 406586896 | 57735339232 |
| 68 | 21381376 | 1453933568 | 143 | 418161601 | 59797108943 |
| 69 | 22667121 | 1564031349 | 144 | 429981696 | 61917364224 |
| 70 | 24010000 | 1680700000 | 145 | 442050625 | 64097340625 |
| 71 | 25411681 | 1804229351 | 146 | 454371856 | 66338290976 |
| 72 | 26873856 | 1934917632 | 147 | 466948881 | 68641485507 |
| 73 | 28398241 | 2073071593 | 148 | 479785216 | 71008211968 |
| 74 | 29986576 | 2219006624 | 149 | 492884401 | 73439775749 |
| 75 | 31640625 | 2373046875 | 150 | 506250000 | 75937500000 |

## TABLE *of Hyperbolic Logarithms.*

| N. | Logarithm. | N. | Logarithm. | N. | Logarithm. | N. | Logarithm. |
|---|---|---|---|---|---|---|---|
| 1·01 | ·0099503 | 1·58 | ·4574248 | 2·15 | ·7654678 | 2·72 | 1·0006318 |
| 1·02 | ·0198026 | 1·59 | ·4637340 | 2·16 | ·7701082 | 2·73 | 1·0043015 |
| 1·03 | ·0295588 | 1·60 | ·4700036 | 2·17 | ·7747271 | 2·74 | 1·0079579 |
| 1·04 | ·0392207 | 1·61 | ·4762341 | 2·18 | ·7793248 | 2·75 | 1·0116008 |
| 1·05 | ·0487902 | 1·62 | ·4824261 | 2·19 | ·7839015 | 2·76 | 1·0152306 |
| 1·06 | ·0582689 | 1·63 | ·4885800 | 2·20 | ·7884573 | 2·77 | 1·0188473 |
| 1·07 | ·0676586 | 1·64 | ·4946962 | 2·21 | ·7929925 | 2·78 | 1·0224509 |
| 1·08 | ·0769610 | 1·65 | ·5007752 | 2·22 | ·7975071 | 2·79 | 1·0260415 |
| 1·09 | ·0861777 | 1·66 | ·5068175 | 2·23 | ·8020015 | 2·80 | 1·0296194 |
| 1·10 | ·0953102 | 1·67 | ·5128236 | 2·24 | ·8064758 | 2·81 | 1·0331844 |
| 1·11 | ·1043600 | 1·68 | ·5187937 | 2·25 | ·8109302 | 2·82 | 1·0367368 |
| 1·12 | ·1133287 | 1·69 | ·5247285 | 2·26 | ·8153648 | 2·83 | 1·0402766 |
| 1·13 | ·1222176 | 1·70 | ·5306282 | 2·27 | ·8197798 | 2·84 | 1·0438040 |
| 1·14 | ·1310283 | 1·71 | ·5364933 | 2·28 | ·8241754 | 2·85 | 1·0473189 |
| 1·15 | ·1397619 | 1·72 | ·5423242 | 2·29 | ·8285518 | 2·86 | 1·0508216 |
| 1·16 | ·1484200 | 1·73 | ·5481214 | 2·30 | ·8329091 | 2·87 | 1·0543120 |
| 1·17 | ·1570037 | 1·74 | ·5538851 | 2·31 | ·8372475 | 2·88 | 1·0577902 |
| 1·18 | ·1655144 | 1·75 | ·5596157 | 2·32 | ·8415671 | 2·89 | 1·0612564 |
| 1·19 | ·1739533 | 1·76 | ·5653138 | 2·33 | ·8458682 | 2·90 | 1·0647107 |
| 1·20 | ·1823215 | 1·77 | ·5709795 | 2·34 | ·8501509 | 2·91 | 1·0681530 |
| 1·21 | ·1906203 | 1·78 | ·5766133 | 2·35 | ·8544153 | 2·92 | 1·0715836 |
| 1·22 | ·1988508 | 1·79 | ·5822156 | 2·36 | ·8586616 | 2·93 | 1·0750024 |
| 1·23 | ·2070141 | 1·80 | ·5877866 | 2·37 | ·8628899 | 2·94 | 1·0784095 |
| 1·24 | ·2151113 | 1·81 | ·5933268 | 2·38 | ·8671004 | 2·95 | 1·0818051 |
| 1·25 | ·2231435 | 1·82 | ·5988365 | 2·39 | ·8712933 | 2·96 | 1·0851892 |
| 1·26 | ·2311117 | 1·83 | ·6043159 | 2·40 | ·8754687 | 2·97 | 1·0885619 |
| 1·27 | ·2390169 | 1·84 | ·6097655 | 2·41 | ·8796267 | 2·98 | 1·0919233 |
| 1·28 | ·2468600 | 1·85 | ·6151856 | 2·42 | ·8837675 | 2·99 | 1·0952733 |
| 1·29 | ·2546422 | 1·86 | ·6205764 | 2·43 | ·8878912 | 3·00 | 1·0986123 |
| 1·30 | ·2623642 | 1·87 | ·6259384 | 2·44 | ·8919980 | 3·01 | 1·1019400 |
| 1·31 | ·2700271 | 1·88 | ·6312717 | 2·45 | ·8960880 | 3·02 | 1·1052568 |
| 1·32 | ·2776317 | 1·89 | ·6365768 | 2·46 | ·9001613 | 3·03 | 1·1085626 |
| 1·33 | ·2851789 | 1·90 | ·6418538 | 2·47 | ·9042181 | 3·04 | 1·1118575 |
| 1·34 | ·2926696 | 1·91 | ·6471032 | 2·48 | ·9082585 | 3·05 | 1·1151415 |
| 1·35 | ·3001045 | 1·92 | ·6523251 | 2·49 | ·9122826 | 3·06 | 1·1184149 |
| 1·36 | ·3074846 | 1·93 | ·6575200 | 2·50 | ·9162907 | 3·07 | 1·1216775 |
| 1·37 | ·3148107 | 1·94 | ·6626879 | 2·51 | ·9202827 | 3·08 | 1·1249295 |
| 1·38 | ·3220834 | 1·95 | ·6678293 | 2·52 | ·9242589 | 3·09 | 1·1281710 |
| 1·39 | ·3293037 | 1·96 | ·6729444 | 2·53 | ·9282193 | 3·10 | 1·1314021 |
| 1·40 | ·3364722 | 1·97 | ·6780335 | 2·54 | ·9321640 | 3·11 | 1·1346227 |
| 1·41 | ·3435897 | 1·98 | ·6830968 | 2·55 | ·9360933 | 3·12 | 1·1378330 |
| 1·42 | ·3506568 | 1·99 | ·6881346 | 2·56 | ·9400072 | 3·13 | 1·1410330 |
| 1·43 | ·3576744 | 2·00 | ·6931472 | 2·57 | ·9439058 | 3·14 | 1·1442227 |
| 1·44 | ·3646431 | 2·01 | ·6981347 | 2·58 | ·9477893 | 3·15 | 1·1474024 |
| 1·45 | ·3715635 | 2·02 | ·7030974 | 2·59 | ·9516578 | 3·16 | 1·1505720 |
| 1·46 | ·3784364 | 2·03 | ·7080357 | 2·60 | ·9555114 | 3·17 | 1·1537315 |
| 1·47 | ·3852624 | 2·04 | ·7129497 | 2·61 | ·9593502 | 3·18 | 1·1568811 |
| 1·48 | ·3920420 | 2·05 | ·7178397 | 2·62 | ·9631743 | 3·19 | 1·1600209 |
| 1·49 | ·3987761 | 2·06 | ·7227059 | 2·63 | ·9669838 | 3·20 | 1·1631508 |
| 1·50 | ·4054651 | 2·07 | ·7275485 | 2·64 | ·9707789 | 3·21 | 1·1662709 |
| 1·51 | ·4121096 | 2·08 | ·7323678 | 2·65 | ·9745596 | 3·22 | 1·1693813 |
| 1·52 | ·4187103 | 2·09 | ·7371640 | 2·66 | ·9783261 | 3·23 | 1·1724821 |
| 1·53 | ·4252677 | 2·10 | ·7419373 | 2·67 | ·9820784 | 3·24 | 1·1755733 |
| 1·54 | ·4317824 | 2·11 | ·7466879 | 2·68 | ·9858167 | 3·25 | 1·1786549 |
| 1·55 | ·4382549 | 2·12 | ·7514160 | 2·69 | ·9895411 | 3·26 | 1·1817271 |
| 1·56 | ·4446858 | 2·13 | ·7561219 | 2·70 | ·9932517 | 3·27 | 1·1847899 |
| 1·57 | ·4510756 | 2·14 | ·7608058 | 2·71 | ·9969486 | 3·28 | 1·1878434 |

| N. | Logarithm. | N. | Logarithm. | N. | Logarithm. | N. | Logarithm. |
|---|---|---|---|---|---|---|---|
| 3·29 | 1·1908875 | 3·91 | 1·3635373 | 4·53 | 1·5107219 | 5·15 | 1·6389967 |
| 3·30 | 1·1939224 | 3·92 | 1·3660916 | 4·54 | 1·5129269 | 5·16 | 1·6409365 |
| 3·31 | 1·1969481 | 3·93 | 1·3686394 | 4·55 | 1·5151272 | 5·17 | 1·6428726 |
| 3·32 | 1·1999647 | 3·94 | 1·3711807 | 4·56 | 1·5173226 | 5·18 | 1·6448050 |
| 3·33 | 1·2029722 | 3·95 | 1·3737156 | 4·57 | 1·5195132 | 5·19 | 1·6467336 |
| 3·34 | 1·2059707 | 3·96 | 1·3762440 | 4·58 | 1·5216990 | 5·20 | 1·6486586 |
| 3·35 | 1·2089603 | 3·97 | 1·3787661 | 4·59 | 1·5238800 | 5·21 | 1·6505798 |
| 3·36 | 1·2119409 | 3·98 | 1·3812818 | 4·60 | 1·5260563 | 5·22 | 1·6524974 |
| 3·37 | 1·2149127 | 3·99 | 1·3837912 | 4·61 | 1·5282278 | 5·23 | 1·6544112 |
| 3·38 | 1·2178757 | 4·00 | 1·3862943 | 4·62 | 1·5303947 | 5·24 | 1·6563214 |
| 3·39 | 1·2208299 | 4·01 | 1·3887912 | 4·63 | 1·5325568 | 5·25 | 1·6582280 |
| 3·40 | 1·2237754 | 4·02 | 1·3912818 | 4·64 | 1·5347143 | 5·26 | 1·6601310 |
| 3·41 | 1·2267122 | 4·03 | 1·3937663 | 4·65 | 1·5368672 | 5·27 | 1·6620303 |
| 3·42 | 1·2296405 | 4·04 | 1·3962446 | 4·66 | 1·5390154 | 5·28 | 1·6639260 |
| 3·43 | 1·2325605 | 4·05 | 1·3987168 | 4·67 | 1·5411590 | 5·29 | 1·6658182 |
| 3·44 | 1·2354714 | 4·06 | 1·4011829 | 4·68 | 1·5432981 | 5·30 | 1·6677068 |
| 3·45 | 1·2383742 | 4·07 | 1·4036429 | 4·69 | 1·5454325 | 5·31 | 1·6695918 |
| 3·46 | 1·2412685 | 4·08 | 1·4060969 | 4·70 | 1·5475625 | 5·32 | 1·6714733 |
| 3·47 | 1·2441545 | 4·09 | 1·4085449 | 4·71 | 1·5496879 | 5·33 | 1·6733512 |
| 3·48 | 1·2470322 | 4·10 | 1·4109869 | 4·72 | 1·5518087 | 5·34 | 1·6752256 |
| 3·49 | 1·2499017 | 4·11 | 1·4134230 | 4·73 | 1·5539252 | 5·35 | 1·6770965 |
| 3·50 | 1·2527629 | 4·12 | 1·4158531 | 4·74 | 1·5560371 | 5·36 | 1·6789639 |
| 3·51 | 1·2556160 | 4·13 | 1·4182774 | 4·75 | 1·5581446 | 5·37 | 1·6808278 |
| 3·52 | 1·2584609 | 4·14 | 1·4206957 | 4·76 | 1·5602476 | 5·38 | 1·6826882 |
| 3·53 | 1·2612978 | 4·15 | 1·4231083 | 4·77 | 1·5623462 | 5·39 | 1·6845453 |
| 3·54 | 1·2641266 | 4·16 | 1·4255150 | 4·78 | 1·5644405 | 5·40 | 1·6863989 |
| 3·55 | 1·2669475 | 4·17 | 1·4279160 | 4·79 | 1·5665304 | 5·41 | 1·6882491 |
| 3·56 | 1·2697605 | 4·18 | 1·4303112 | 4·80 | 1·5686159 | 5·42 | 1·6900958 |
| 3·57 | 1·2725655 | 4·19 | 1·4327007 | 4·81 | 1·5706971 | 5·43 | 1·6919391 |
| 3·58 | 1·2753627 | 4·20 | 1·4350845 | 4·82 | 1·5727739 | 5·44 | 1·6937790 |
| 3·59 | 1·2781521 | 4·21 | 1·4374626 | 4·83 | 1·5748464 | 5·45 | 1·6956155 |
| 3·60 | 1·2809338 | 4·22 | 1·4398351 | 4·84 | 1·5769147 | 5·46 | 1·6974487 |
| 3·61 | 1·2837077 | 4·23 | 1·4422020 | 4·85 | 1·5789787 | 5·47 | 1·6992786 |
| 3·62 | 1·2864740 | 4·24 | 1·4445632 | 4·86 | 1·5810384 | 5·48 | 1·7011051 |
| 3·63 | 1·2892326 | 4·25 | 1·4469189 | 4·87 | 1·5830939 | 5·49 | 1·7029282 |
| 3·64 | 1·2919836 | 4·26 | 1·4492691 | 4·88 | 1·5851452 | 5·50 | 1·7047481 |
| 3·65 | 1·2947271 | 4·27 | 1·4516138 | 4·89 | 1·5871923 | 5·51 | 1·7065646 |
| 3·66 | 1·2974631 | 4·28 | 1·4539530 | 4·90 | 1·5892352 | 5·52 | 1·7083778 |
| 3·67 | 1·3001916 | 4·29 | 1·4562867 | 4·91 | 1·5912739 | 5·53 | 1·7101878 |
| 3·68 | 1·3029127 | 4·30 | 1·4586149 | 4·92 | 1·5933085 | 5·54 | 1.7119944 |
| 3·69 | 1·3056264 | 4·31 | 1·4609379 | 4·93 | 1·5953389 | 5·55 | 1·7137979 |
| 3·70 | 1·3083328 | 4·32 | 1·4632553 | 4·94 | 1·5973653 | 5·56 | 1·7155981 |
| 3·71 | 1·3110318 | 4·33 | 1·4655675 | 4·95 | 1·5993875 | 5·57 | 1·7173950 |
| 3·72 | 1·3137236 | 4·34 | 1·4678743 | 4·96 | 1·6014057 | 5·58 | 1·7191887 |
| 3·73 | 1·3164082 | 4·35 | 1·4701758 | 4·97 | 1·6034198 | 5·59 | 1·7209792 |
| 3·74 | 1·3190856 | 4·36 | 1·4724720 | 4·98 | 1·6054298 | 5·60 | 1·7227666 |
| 3·75 | 1·3217558 | 4·37 | 1·4747630 | 4·99 | 1·6074358 | 5·61 | 1·7245507 |
| 3·76 | 1·3244189 | 4·38 | 1·4770487 | 5·00 | 1·6094379 | 5·62 | 1·7263316 |
| 3·77 | 1·3270749 | 4·39 | 1·4793292 | 5·01 | 1·6114359 | 5·63 | 1·7281094 |
| 3·78 | 1·3297240 | 4·40 | 1·4816045 | 5·02 | 1·6134300 | 5·64 | 1·7298840 |
| 3·79 | 1·3323660 | 4·41 | 1·4838746 | 5·03 | 1·6154200 | 5·65 | 1·7316555 |
| 3·80 | 1·3350010 | 4·42 | 1·4861396 | 5·04 | 1·6174060 | 5·66 | 1·7334238 |
| 3·81 | 1·3376291 | 4·43 | 1·4883995 | 5·05 | 1·6193882 | 5·67 | 1·7351891 |
| 3·82 | 1·3402504 | 4·44 | 1·4906543 | 5·06 | 1·6213664 | 5·68 | 1·7369512 |
| 3·83 | 1·3428648 | 4·45 | 1·4929040 | 5·07 | 1·6233408 | 5·69 | 1·7387102 |
| 3·84 | 1·3454723 | 4·46 | 1·4951487 | 5·08 | 1·6253112 | 5·70 | 1·7404661 |
| 3·85 | 1·3480731 | 4·47 | 1·4973883 | 5·09 | 1·6272778 | 5·71 | 1·7422189 |
| 3·86 | 1·3506671 | 4·48 | 1·4996230 | 5·10 | 1·6292405 | 5·72 | 1·7439687 |
| 3·87 | 1·3532544 | 4·49 | 1·5018527 | 5·11 | 1·6311994 | 5·73 | 1·7457155 |
| 3·88 | 1·3558351 | 4·50 | 1·5040774 | 5·12 | 1·6331544 | 5·74 | 1·7474591 |
| 3·89 | 1·3584091 | 4·51 | 1·5062971 | 5·13 | 1·6351056 | 5·75 | 1·7491998 |
| 3·90 | 1·3609765 | 4·52 | 1·5085119 | 5·14 | 1·6370530 | 5·76 | 1·7509374 |

| N. | Logarithm. | N. | Logarithm. | N. | Logarithm. | N. | Logarithm. |
|---|---|---|---|---|---|---|---|
| 5·77 | 1·7526720 | 6·39 | 1·8547342 | 7·01 | 1·9473376 | 7·63 | 2·0320878 |
| 5·78 | 1·7544036 | 6·40 | 1·8562979 | 7·02 | 1·9487632 | 7·64 | 2·0333976 |
| 5·79 | 1·7561323 | 6·41 | 1·8578592 | 7·03 | 1·9501866 | 7·65 | 2·0347056 |
| 5·80 | 1·7578579 | 6·42 | 1·8594181 | 7·04 | 1·9516080 | 7·66 | 2·0360119 |
| 5·81 | 1·7595805 | 6·43 | 1·8609745 | 7·05 | 1·9530275 | 7·67 | 2·0373166 |
| 5·82 | 1·7613002 | 6·44 | 1·8625285 | 7·06 | 1·9544449 | 7·68 | 2·0386195 |
| 5·83 | 1·7630170 | 6·45 | 1·8640801 | 7·07 | 1·9558604 | 7·69 | 2·0399207 |
| 5·84 | 1·7647308 | 6·46 | 1·8656293 | 7·08 | 1·9572739 | 7·70 | 2·0412203 |
| 5·85 | 1·7664416 | 6·47 | 1·8671761 | 7·09 | 1·9586853 | 7·71 | 2·0425181 |
| 5·86 | 1·7681496 | 6·48 | 1·8687205 | 7·10 | 1·9600947 | 7·72 | 2·0438143 |
| 5·87 | 1·7698546 | 6·49 | 1·8702625 | 7·11 | 1·9615022 | 7·73 | 2·0451088 |
| 5·88 | 1·7715567 | 6·50 | 1·8718021 | 7·12 | 1·9629077 | 7·74 | 2·0464016 |
| 5·89 | 1·7732559 | 6·51 | 1·8733394 | 7·13 | 1·9643112 | 7·75 | 2·0476928 |
| 5·90 | 1·7749523 | 6·52 | 1·8748743 | 7·14 | 1·9657127 | 7·76 | 2·0489823 |
| 5·91 | 1·7766458 | 6·53 | 1·8764069 | 7·15 | 1·9671123 | 7·77 | 2·0502701 |
| 5·92 | 1·7783364 | 6·54 | 1·8779371 | 7·16 | 1·9685099 | 7·78 | 2·0515563 |
| 5·93 | 1·7800242 | 6·55 | 1·8794650 | 7·17 | 1·9699056 | 7·79 | 2·0528408 |
| 5·94 | 1·7817091 | 6·56 | 1·8809906 | 7·18 | 1·9712993 | 7·80 | 2·0541237 |
| 5·95 | 1·7833912 | 6·57 | 1·8825138 | 7·19 | 1·9726911 | 7·81 | 2·0554049 |
| 5·96 | 1·7850704 | 6·58 | 1·8840347 | 7·20 | 1·9740810 | 7·82 | 2·0566845 |
| 5·97 | 1·7867469 | 6·59 | 1·8855533 | 7·21 | 1·9754689 | 7·83 | 2·0579624 |
| 5·98 | 1·7884205 | 6·60 | 1·8870696 | 7·22 | 1·9768549 | 7·84 | 2·0592388 |
| 5·99 | 1·7900914 | 6·61 | 1·8885837 | 7·23 | 1·9782390 | 7·85 | 2·0605135 |
| 6·00 | 1·7917594 | 6·62 | 1·8900954 | 7·24 | 1·9796212 | 7·86 | 2·0617866 |
| 6·01 | 1·7934247 | 6·63 | 1·8916048 | 7·25 | 1·9810014 | 7·87 | 2·0630580 |
| 6·02 | 1·7950872 | 6·64 | 1·8931119 | 7·26 | 1·9823798 | 7·88 | 2·0643278 |
| 6·03 | 1·7967470 | 6·65 | 1·8946168 | 7·27 | 1·9837562 | 7·89 | 2·0655961 |
| 6·04 | 1·7984040 | 6·66 | 1·8961194 | 7·28 | 1·9851308 | 7·90 | 2·0668627 |
| 6·05 | 1·8000582 | 6·67 | 1·8976198 | 7·29 | 1·9865035 | 7·91 | 2·0681277 |
| 6·06 | 1·8017098 | 6·68 | 1·8991179 | 7·30 | 1·9878743 | 7·92 | 2·0693911 |
| 6·07 | 1·8033586 | 6·69 | 1·9006138 | 7·31 | 1·9892432 | 7·93 | 2·0706530 |
| 6·08 | 1·8050047 | 6·70 | 1·9021075 | 7·32 | 1·9906103 | 7·94 | 2·0719132 |
| 6·09 | 1·8066481 | 6·71 | 1·9035989 | 7·33 | 1·9919754 | 7·95 | 2·0731719 |
| 6·10 | 1·8082887 | 6·72 | 1·9050881 | 7·34 | 1·9933387 | 7·96 | 2·0744290 |
| 6·11 | 1·8099267 | 6·73 | 1·9065751 | 7·35 | 1·9947002 | 7·97 | 2·0756845 |
| 6·12 | 1·8115621 | 6·74 | 1·9080600 | 7·36 | 1·9960599 | 7·98 | 2·0769384 |
| 6·13 | 1·8131947 | 6·75 | 1·9095425 | 7·37 | 1·9974177 | 7·99 | 2·0781907 |
| 6·14 | 1·8148247 | 6·76 | 1·9110228 | 7·38 | 1·9987736 | 8·00 | 2·0794415 |
| 6·15 | 1·8164520 | 6·77 | 1·9125011 | 7·39 | 2·0001278 | 8·01 | 2·0806907 |
| 6·16 | 1·8180767 | 6·78 | 1·9139771 | 7·40 | 2·0014800 | 8·02 | 2·0819384 |
| 6·17 | 1·8196988 | 6·79 | 1·9154509 | 7·41 | 2·0028305 | 8·03 | 2·0831845 |
| 6·18 | 1·8213182 | 6·80 | 1·9169226 | 7·42 | 2·0041790 | 8·04 | 2·0844290 |
| 6·19 | 1·8229351 | 6·81 | 1·9183921 | 7·43 | 2·0055258 | 8·05 | 2·0856720 |
| 6·20 | 1·8245493 | 6·82 | 1·9198594 | 7·44 | 2·0068708 | 8·06 | 2·0869135 |
| 6·21 | 1·8261608 | 6·83 | 1·9213247 | 7·45 | 2·0082140 | 8·07 | 2·0881534 |
| 6·22 | 1·8277699 | 6·84 | 1·9227877 | 7·46 | 2·0095553 | 8·08 | 2·0893918 |
| 6·23 | 1·8293763 | 6·85 | 1·9242486 | 7·47 | 2·0108949 | 8·09 | 2·0906287 |
| 6·24 | 1·8309801 | 6·86 | 1·9257074 | 7·48 | 2·0122327 | 8·10 | 2·0918640 |
| 6·25 | 1·8325814 | 6·87 | 1·9271641 | 7·49 | 2·0135687 | 8·11 | 2·0930984 |
| 6·26 | 1·8341801 | 6·88 | 1·9286186 | 7·50 | 2·0149030 | 8·12 | 2·0943306 |
| 6·27 | 1·8357763 | 6·89 | 1·9300710 | 7·51 | 2·0162354 | 8·13 | 2·0955613 |
| 6·28 | 1·8373699 | 6·90 | 1·9315214 | 7·52 | 2·0175661 | 8·14 | 2·0967905 |
| 6·29 | 1·8389610 | 6·91 | 1·9329696 | 7·53 | 2·0188950 | 8·15 | 2·0980182 |
| 6·30 | 1·8405496 | 6·92 | 1·9344157 | 7·54 | 2·0202221 | 8·16 | 2·0992444 |
| 6·31 | 1·8421356 | 6·93 | 1·9358598 | 7·55 | 2·0215475 | 8·17 | 2·1004691 |
| 6·32 | 1·8437191 | 6·94 | 1·9373017 | 7·56 | 2·0228711 | 8·18 | 2·1016923 |
| 6·33 | 1·8453002 | 6·95 | 1·9387416 | 7·57 | 2·0241929 | 8·19 | 2·1029140 |
| 6·34 | 1·8468787 | 6·96 | 1·9401794 | 7·58 | 2·0255131 | 8·20 | 2·1041341 |
| 6·35 | 1·8484547 | 6·97 | 1·9416152 | 7·59 | 2·0268315 | 8·21 | 2·1053529 |
| 6·36 | 1·8500283 | 6·98 | 1·9430489 | 7·60 | 2·0281482 | 8·22 | 2·1065702 |
| 6·37 | 1·8515994 | 6·99 | 1·9444805 | 7·61 | 2·0294631 | 8·23 | 2·1077861 |
| 6·38 | 1·8531680 | 7·00 | 1·9459101 | 7·62 | 2·0307763 | 8·24 | 2·1089998 |

| N. | Logarithm. | N. | Logarithm. | N. | Logarithm. | N. | Logarithm. |
|---|---|---|---|---|---|---|---|
| 8·25 | 2·1102128 | 8·69 | 2·1621729 | 9·13 | 2·2115656 | 9·57 | 2·2586332 |
| 8·26 | 2·1114243 | 8·70 | 2·1633230 | 9·14 | 2·2126603 | 9·58 | 2·2596776 |
| 8·27 | 2·1126343 | 8·71 | 2·1644718 | 9·15 | 2·2137538 | 9·59 | 2·2607209 |
| 8·28 | 2·1138428 | 8·72 | 2·1656192 | 9·16 | 2·2148461 | 9·60 | 2·2617631 |
| 8·29 | 2·1150499 | 8·73 | 2·1667653 | 9·17 | 2·2159372 | 9·61 | 2·2628042 |
| 8·30 | 2·1162555 | 8·74 | 2·1679101 | 9·18 | 2·2170272 | 9·62 | 2·2638442 |
| 8·31 | 2·1174596 | 8·75 | 2·1690536 | 9·19 | 2·2181160 | 9·63 | 2·2648832 |
| 8·32 | 2·1186622 | 8·76 | 2·1701959 | 9·20 | 2·2192034 | 9·64 | 2·2659211 |
| 8·33 | 2·1198634 | 8·77 | 2·1713367 | 9·21 | 2·2202898 | 9·65 | 2·2669579 |
| 8·34 | 2·1210632 | 8·78 | 2·1724763 | 9·22 | 2·2213750 | 9·66 | 2·2679936 |
| 8·35 | 2·1222615 | 8·79 | 2·1736146 | 9·23 | 2·2224590 | 9·67 | 2·2690282 |
| 8·36 | 2·1234584 | 8·80 | 2·1747517 | 9·24 | 2·2235418 | 9·68 | 2·2700618 |
| 8·37 | 2·1246539 | 8·81 | 2·1758874 | 9·25 | 2·2246235 | 9·69 | 2·2710944 |
| 8·38 | 2·1258479 | 8·82 | 2·1770218 | 9·26 | 2·2257040 | 9·70 | 2·2721258 |
| 8·39 | 2·1270405 | 8·83 | 2·1781550 | 9·27 | 2·2267833 | 9·71 | 2·2731562 |
| 8·40 | 2·1282317 | 8·84 | 2·1792868 | 9·28 | 2·2278615 | 9·72 | 2·2741856 |
| 8·41 | 2·1294214 | 8·85 | 2·1804174 | 9·29 | 2·2289385 | 9·73 | 2·2752138 |
| 8·42 | 2·1306098 | 8·86 | 2·1815467 | 9·30 | 2·2300144 | 9·74 | 2·2762411 |
| 8·43 | 2·1317967 | 8·87 | 2·1826747 | 9·31 | 2·2310890 | 9·75 | 2·2772673 |
| 8·44 | 2·1329822 | 8·88 | 2·1838015 | 9·32 | 2·2321626 | 9·76 | 2·2782924 |
| 8·45 | 2·1341664 | 8·89 | 2·1849270 | 9·33 | 2·2332350 | 9·77 | 2·2793165 |
| 8·46 | 2·1353491 | 8·90 | 2·1860512 | 9·34 | 2·2343062 | 9·78 | 2·2803395 |
| 8·47 | 2·1365304 | 8·91 | 2·1871742 | 9·35 | 2·2353763 | 9·79 | 2·2813614 |
| 8·48 | 2·1377104 | 8·92 | 2·1882959 | 9·36 | 2·2364452 | 9·80 | 2·2823823 |
| 8·49 | 2·1388889 | 8·93 | 2·1894163 | 9·37 | 2·2375130 | 9·81 | 2·2834022 |
| 8·50 | 2·1400661 | 8·94 | 2·1905355 | 9·38 | 2·2385797 | 9·82 | 2·2844211 |
| 8·51 | 2·1412419 | 8·95 | 2·1916535 | 9·39 | 2·2396452 | 9·83 | 2·2854389 |
| 8·52 | 2·1424163 | 8·96 | 2·1927702 | 9·40 | 2·2407096 | 9·84 | 2·2864556 |
| 8·53 | 2·1435893 | 8·97 | 2·1938856 | 9·41 | 2·2417729 | 9·85 | 2·2874714 |
| 8·54 | 2·1447609 | 8·98 | 2·1949998 | 9·42 | 2·2428350 | 9·86 | 2·2884861 |
| 8·55 | 2·1459312 | 8·99 | 2·1961128 | 9·43 | 2·2438960 | 9·87 | 2·2894998 |
| 8·56 | 2·1471001 | 9·00 | 2·1972245 | 9·44 | 2·2449559 | 9·88 | 2·2905124 |
| 8·57 | 2·1482676 | 9·01 | 2·1983350 | 9·45 | 2·2460147 | 9·89 | 2·2915241 |
| 8·58 | 2·1494339 | 9·02 | 2·1994443 | 9·46 | 2·2470723 | 9·90 | 2·2925347 |
| 8·59 | 2·1505987 | 9·03 | 2·2005523 | 9·47 | 2·2481288 | 9·91 | 2·2635443 |
| 8·60 | 2·1517622 | 9·04 | 2·2016591 | 9·48 | 2·2491843 | 9·92 | 2·2945529 |
| 8·61 | 2·1529243 | 9·05 | 2·2027647 | 9·49 | 2·2502386 | 9·93 | 2·2955604 |
| 8·62 | 2·1540851 | 9·06 | 2·2038691 | 9·50 | 2·2512917 | 9·94 | 2·2965670 |
| 8·63 | 2·1552445 | 9·07 | 2·2049722 | 9·51 | 2·2523438 | 9·95 | 2·2975725 |
| 8·64 | 2·1564026 | 9·08 | 2·2060741 | 9·52 | 2·2533948 | 9·96 | 2·2985770 |
| 8·65 | 2·1575593 | 9·09 | 2·2071748 | 9·53 | 2·2544446 | 9·97 | 2·2995806 |
| 8·66 | 2·1587147 | 9·10 | 2·2082744 | 9·54 | 2·2554934 | 9·98 | 2·3005831 |
| 8·67 | 2·1598687 | 9·11 | 2·2093727 | 9·55 | 2·2565411 | 9·99 | 2·3015846 |
| 8·68 | 2·1610215 | 9·12 | 2·2104697 | 9·56 | 2·2575877 | 10·00 | 2·3025851 |

Logarithms were invented by Juste Byrge, a Frenchman, and not by Napier. See "Biographie Universelle," "The Calculus of Form," article 822, and "The Practical, Short, and Direct Method of Calculating the Logarithm of any given Number and the Number corresponding to any given Logarithm," discovered by Oliver Byrne, the author of the present work. Juste Byrge also invented the proportional compasses, and was a profound astronomer and mathematician. The common Logarithm of a number multiplied by 2·302585052994 gives the hyperbolic Logarithm of that number. The common Logarithm of 2·22 is ·346353 ∴ 2·302585 × ·346353 = ·7975071 the hyperbolic Logarithm. The application of Logarithms to the calculations of the Engineer will be treated of hereafter.

## COMBINATIONS OF ALGEBRAIC QUANTITIES.

THE following practical examples will serve to illustrate the method of combining or representing numbers or quantities algebraically; the chief object of which is, to help the memory with respect to the use of the *signs* and *letters*, or *symbols*.

Let $a = 6$, $b = 4$, $c = 3$, $d = 2$, $e = 1$, and $f = 0$.

Then will, (1) $2a + b = 12 + 4 = 16$.

(2) $ab + 2c - d = 24 + 6 - 2 = 28$.

(3) $a^2 - b^2 + e + f = 36 - 16 + 1 + 0 = 21$.

(4) $b^2 \times (a - b) = 16 \times (6 - 4) = 16 \times 2 = 32$.

(5) $3abc - 7de = 216 - 14 = 202$.

(6) $2(a - b)(5c - 2d) = (12 - 8) \times (15 - 4) = 44$.

(7) $\dfrac{c^2 - e^2}{d + f} \times (a - c) = \dfrac{9 - 1}{2 + 0} \times (6 - 3) = 4 \times 3 = 12$.

(8) $\surd(a^2 - 2b^2) + d - f = \surd(36 - 32) + 2 - 0 = 4$.

(9) $3ab - (a - b - c + d) = 72 - 1 = 71$.

(10) $3ab - (a - b - c - d) = 72 + 3 = 75$.

(11) $\dfrac{\surd 2abc}{\surd(ab - 4d)} \times (c + d) = \dfrac{\surd 144}{\surd(24 - 8)} \times (3 + 2) = 15$.

In solving the following questions, the letters $a$, $b$, $c$, &c. are supposed to have the same values as before, namely, 6, 4, 3, &c.; but any other values might have been assigned to them; therefore, do not suppose that $a$ must necessarily be 6, nor that $b$ must be 4, for the letter $a$ may be put for any known *quantity*, *number*, or *magnitude* whatever; thus $a$ may represent 10 *miles*, or 50 *pounds*, or any *number* or *quantity*, or it may represent 1 *globe*, or 2 *cubic feet*, &c.; the same may be said of $b$, or any other letter.

(1) $a + b - c = 7$.

(2) $3bc - d + e = 35$.

(3) $2a^2 + c^2 - d + f = 79$.

(4) $\dfrac{a^2}{b} \times (b - c + d) = 27$.

(5) $5c^2d - a^2 + 4de = 62$.

(6) $4(a^2 - b^2)(c - e) = 160$.

(7) $\dfrac{a^2 - b^2}{c + d} \times (d^2 + c^2) = 52$.

(8) $\surd(2a^2 + 2d^2) + bc - f = 20$.

(9) $4a^2b - (c^2 - d - e) = 570$.

(10) $\dfrac{\surd 4a^2}{\surd(10d^2 - 4cd)} \times \dfrac{a}{d} - c^2 = 0$.

In the use of algebraic symbols, $3\sqrt[3]{4a - b}$ signifies the same thing as $3(4a - b)^{\frac{1}{3}}$.

$4(c + d)^{\frac{1}{2}}(a + b)^{\frac{1}{3}}$, or $4 \times \overline{c + d}^{\frac{1}{2}} \times \overline{a + b}^{\frac{1}{3}}$, signifies the same thing as $4\sqrt{c + d} \cdot \sqrt[3]{a + b}$.

# THE STEAM ENGINE.

THE particular example which we shall select is that of an engine having 8 feet stroke and 64 inch cylinder.

The breadth of the web of the crank at the paddle centre is the breadth which the web would have if it were continued to the paddle centre. Suppose that we wished to know the breadth of the web of crank of an engine whose stroke is 8 feet and diameter of cylinder 64 inches. The proper breadth of the web of crank at paddle centre would in this case be about 18 inches.

*To find the breadth of crank at paddle centre.*—Multiply the square of the length of the crank in inches by 1·561, and then multiply the square of the diameter of cylinder in inches by ·1235; multiply the square root of the sum of these products by the square of the diameter of the cylinder in inches; divide the product by 45; finally extract the cube root of the quotient. The result is the breadth of the web of crank at paddle centre.

Thus, to apply this rule to the particular example which we have selected, we have

48 = length of crank in inches.
48
———
2304
1·561 = constant multiplier.
———
3596·5
505·8 found below.
———
4102·3

64 = diameter of cylinder.
64
———
4096
1235 = constant multiplier.
———
505·8

and $\sqrt{4102·3}$ = 64·05 nearly.
4096 = square of the diameter of the cylinder.
———
45) 262348·5
———
5829·97

and $\sqrt[3]{5829·97}$ = 18 nearly.

Suppose that we wished the proper thickness of the large eye of crank for an engine whose stroke is 8 feet and diameter of cylinder 64 inches. The proper thickness for the large eye of crank is 5·77 inches.

RULE.—*To find the thickness of large eye of crank.*—Multiply the square of the length of the crank in inches by 1·561, and then multiply the square of the diameter of the cylinder in inches by ·1235; multiply the sum of these products by the square of the diameter of the cylinder in inches; afterwards, divide the product by 1828·28; divide this quotient by the length of the crank in inches; finally extract the cube root of the quotient. The result is the proper thickness of the large eye of crank in inches.

Thus, to apply this rule to the particular example which we have selected, we have

```
          48 = length of crank in inches.
          48
        ----
        2304
       1·561 constant multiplier.
      ------
      3596·5
       505·8
      ------
      4102·3
         64 = diameter of cylinder in inches.
         64
       ----
       4096
       1235 = constant multiplier.
      -----
      505·8

             4102·3
               4096 = square of diameter.
         ----------
      48) 16803020·8
         -----------
1828·28) 350062·94
         ---------
            191·47
```

and $\sqrt[3]{191·47} = 5·77$ nearly.

The proper thickness of the web of crank at paddle shaft centre is the thickness which the web ought to have if continued to centre of the shaft. Suppose that it were required to find the proper thickness of web of crank at shaft centre for an engine whose stroke is 8 feet and diameter of cylinder 64 inches. The proper thickness of the web at shaft centre in this case would be 8·97 inches.

RULE.—*To find the thickness of the web of crank at paddle shaft centre.*—Multiply the square of the length of crank in inches by 1·561, and then multiply the square of the diameter in inches by ·1235; multiply the square root of the sum of these products by the square of the diameter of the cylinder in inches; divide this quotient by 360; finally extract the cube root of the quotient. The result is the thickness of the web of crank at paddle shaft centre in inches.

Thus, to apply the rule to the particular example which we have selected, we have

48 = length of crank in inches.
48
2304
1·561 = constant multiplier.
3596·5
505·8
4102·3
64 = diameter of cylinder.
64
4096
1235 = constant multiplier.
505·8

And $\sqrt{4102 \cdot 3} = 64 \cdot 05$ nearly.
4096 = square of diameter.
360) 262348·5
728·75

And $\sqrt[3]{782 \cdot 75} = 9$ nearly.

Suppose that it were required to find the proper diameter for the paddle shaft journal of an engine whose stroke is 8 feet and diameter of cylinder 64 inches. The proper diameter of the paddle shaft journal in this case is 14·06 inches.

RULE.—*To find the diameter of the paddle shaft journal.*—Multiply the square of the diameter of cylinder in inches by the length of the crank in inches; extract the cube root of the product; finally multiply the result by ·242. The final product is the diameter of the paddle shaft journal in inches.

Thus, to apply this rule to the particular example which we have before selected, we have

64 = diameter of cylinder in inches.
64
4096
48 = length of crank in inches.
196608

and $\sqrt[3]{196608} = 58 \cdot 148$
but $58 \cdot 148 \times \cdot 242 = 14 \cdot 07$ inches.

Suppose it were required to find the proper length of the paddle shaft journal for an engine whose stroke is 8 feet, and diameter of cylinder 64 inches. The proper length of the paddle shaft journal would be, in this case, 17·59 inches.

The following rule serves for engines of all sizes:

RULE.—*To find the length of the paddle shaft journal.*—Multiply the square of the diameter of the cylinder in inches by the length of the crank in inches; extract the cube root of the quotient; multiply the result by ·303. The product is the length of the paddle shaft journal in inches. (The length of the paddle shaft journal is $1\frac{1}{4}$ times the diameter.)

To apply this rule to the example which we have selected, we have

$$\begin{array}{rl} 64 & = \text{diameter of cylinder in inches.} \\ 64 & \\ \hline 4096 & \\ 48 & = \text{length of crank in inches.} \\ \hline 196608 & \end{array}$$

$$\text{and } \sqrt[3]{196608} = 58 \cdot 148$$

$$\therefore \text{length of journal} = 58 \cdot 148 \times \cdot 303 = 17 \cdot 60 \text{ inches.}$$

We shall now calculate the proper dimensions of some of those parts which do not depend upon the length of the stroke. Suppose it were required to find the proper dimensions of the respective parts of a marine engine the diameter of whose cylinder is 64 inches.

Diameter of crank-pin journal = 90·9 inches, or about 9 inches.

Length of crank-pin journal = 10·18 inches, or nearly $10\frac{1}{5}$ inches.

Breadth of the eye of cross-head = 2·64 inches, or between $2\frac{1}{2}$ and $2\frac{3}{4}$ inches.

Depth of the eye of cross-head = 18·37 inches, or very nearly $18\frac{1}{2}$ inches.

Diameter of the journal of cross-head = 5·5 inches, or $5\frac{1}{2}$ inches.

Length of journal of cross-head = 6·19 inches, or very nearly $6\frac{1}{5}$ inches.

Thickness of the web of cross-head at middle = 4·6 inches, or somewhat more than $4\frac{1}{2}$ inches.

Breadth of web of cross-head at middle = 17·15 inches, or between $17\frac{1}{10}$ and $17\frac{1}{5}$ inches.

Thickness of web of cross-head at journal = 3·93 inches, or very nearly 4 inches.

Breadth of web of cross-head at journal = 6·46 inches, or nearly $6\frac{1}{2}$ inches.

Diameter of piston rod = 6·4 inches, or $6\frac{2}{5}$ inches.

Length of part of piston rod in piston = 12·8 inches, or $12\frac{4}{5}$ inches.

Major diameter of part of piston rod in cross-head = 06·8 inches, or nearly $6\frac{1}{10}$ inches.

Minor diameter of part of piston rod in cross-head = 5·76 inches, or $5\frac{3}{4}$ inches.

Major diameter of part of piston rod in piston = 8·96 inches, or nearly 9 inches.

Minor 'diameter of part of piston rod in piston = 7·36 inches, or between $7\frac{1}{4}$ and $7\frac{1}{2}$ inches.

Depth of gibs and cutter through cross-head = 6·72 inches, or very nearly $6\frac{3}{4}$ inches.

Thickness of gibs and cutter through cross-head = 1·35 inches, or between $1\frac{1}{4}$ and $1\frac{1}{2}$ inches.

Depth of cutter through piston = 5·45 inches, or nearly $5\frac{1}{2}$ inches.

Thickness of cutter through piston = 2·24 inches, or nearly $2\frac{1}{4}$ inches.

Diameter of connecting rod at ends = 6·08 inches, or nearly $6\frac{1}{10}$ inches.

Major diameter of part of connecting rod in cross-tail = 6·27 inches, or about $6\frac{1}{4}$ inches.

Minor diameter of part of connecting rod in cross-tail = 5·76 inches, or nearly $5\frac{3}{4}$ inches.

Breadth of butt = 9·98 inches, or very nearly 10 inches.

Thickness of butt = 8 inches.

Mean thickness of strap at cutter = 2·75 inches, or $2\frac{3}{4}$ inches.

Mean thickness of strap above cutter = 2·06 inches, or somewhat more than 2 inches.

Distance of cutter from end of strap = 3·08 inches, or very nearly $3\frac{1}{10}$ inches.

Breadth of gibs and cutter through cross-tail = 6·73 inches, or very nearly $6\frac{3}{4}$ inches.

Breadth of gibs and cutter through butt = 7·04 inches, or somewhat more than 7 inches.

Thickness of gibs and cutter through butt = 1·84 inches, or between $1\frac{3}{4}$ and 2 inches.

These results are calculated from the following rules, which give correct results for all sizes of engines.

Rule 1. *To find the diameter of crank-pin journal.*—Multiply the diameter of the cylinder in inches by ·142. The result is the diameter of crank-pin journal in inches.

Rule 2. *To find the length of crank-pin journal.*—Multiply the diameter of the cylinder in inches by ·16. The product is the length of the crank-pin journal in inches.

Rule 3. *To find the breadth of the eye of cross-head.*—Multiply the diameter of the cylinder in inches by ·041. The product is the breadth of the eye in inches.

Rule 4. *To find the depth of the eye of cross-head.*—Multiply the diameter of the cylinder in inches by ·286. The product is the depth of the eye of cross-head in inches.

Rule 5. *To find the diameter of the journal of cross-head.*—Multiply the diameter of the cylinder in inches by ·086. The product is the diameter of the journal in inches.

Rule 6. *To find the length of the journal of cross-head.*—Multiply the diameter of the cylinder in inches by ·097. The product is the length of the journal in inches.

Rule 7. *To find the thickness of the web of cross-head at middle.*—Multiply the diameter of the cylinder in inches by ·072. The product is the thickness of the web of cross-head at middle in inches.

Rule 8. *To find the breadth of web of cross-head at middle.*—Multiply the diameter of the cylinder in inches by ·268. The product is the breadth of the web of cross-head at middle in inches.

RULE 9. *To find the thickness of the web of cross-head at journal.*—Multiply the diameter of the cylinder in inches by ·061. The product is the thickness of the web of cross-head at journal in inches.

RULE 10. *To find the breadth of web of cross-head at journal.*—Multiply the diameter of the cylinder in inches by ·101. The product is the breadth of the web of cross-head at journal in inches.

RULE 11. *To find the diameter of the piston rod.*—Divide the diameter of the cylinder in inches by 10. The quotient is the diameter of the piston rod in inches.

RULE 12. *To find the length of the part of the piston rod in the piston.*—Divide the diameter of the cylinder in inches by 5. The quotient is the length of the part of the piston rod in the piston in inches.

RULE 13. *To find the major diameter of the part of piston rod in cross-head.*—Multiply the diameter of the cylinder in inches by ·095. The product is the major diameter of the part of piston rod in cross-head in inches.

RULE 14. *To find the minor diameter of the part of piston rod in cross-head.*—Multiply the diameter of the cylinder in inches by ·09. The product is the minor diameter of the part of piston rod in cross-head in inches.

RULE 15. *To find the major diameter of the part of piston rod in piston.*—Multiply the diameter of the cylinder in inches by ·14. The product is the major diameter of the part of piston rod in piston in inches.

RULE 16. *To find the minor diameter of the part of piston rod in piston.*—Multiply the diameter of the cylinder in inches by ·115. The product is the minor diameter of the part of piston rod in piston.

RULE 17. *To find the depth of gibs and cutter through cross-head.*—Multiply the diameter of the cylinder in inches by ·105. The product is the depth of the gibs and cutter through cross-head.

RULE 18. *To find the thickness of the gibs and cutter through cross-head.*—Multiply the diameter of the cylinder in inches by ·021. The product is the thickness of the gibs and cutter through cross-head.

RULE 19. *To find the depth of cutter through piston.*—Multiply the diameter of the cylinder in inches by ·085. The product is the depth of the cutter through piston in inches.

RULE 20. *To find the thickness of cutter through piston.*—Multiply the diameter of the cylinder in inches by ·035. The product is the thickness of cutter through piston in inches.

RULE 21. *To find the diameter of connecting rod at ends.*—Multiply the diameter of the cylinder in inches by ·095. The product is the diameter of the connecting rod at ends in inches.

RULE 22. *To find the major diameter of the part of connecting rod in cross-tail.*—Multiply the diameter of the cylinder in inches by ·098. The product is the major diameter of the part of connecting rod in cross-tail.

RULE 23. *To find the minor diameter of the part of connecting rod in cross-tail.*—Multiply the diameter of the cylinder in inches by ·09. The product is the minor diameter of the part of connecting rod in cross-tail in inches.

RULE 24. *To find the breadth of butt.*—Multiply the diameter of the cylinder in inches by ·156. The product is the breadth of the butt in inches.

RULE 25. *To find the thickness of the butt.*—Divide the diameter of the cylinder in inches by 8. The quotient is the thickness of the butt in inches.

RULE 26. *To find the mean thickness of the strap at cutter.*—Multiply the diameter of the cylinder in inches by ·043. The product is the mean thickness of the strap at cutter.

RULE 27. *To find the mean thickness of the strap above cutter.*—Multiply the diameter of the cylinder in inches by ·032. The product is the mean thickness of the strap above cutter.

RULE 28. *To find the distance of cutter from end of strap.*—Multiply the diameter of the cylinder in inches by ·048. The product is the distance of cutter from end of strap in inches.

RULE 29. *To find the breadth of the gibs and cutter through cross-tail.*—Multiply the diameter of the cylinder in inches by ·105. The product is the breadth of the gibs and cutter through cross-tail.

RULE 30. *To find the breadth of the gibs and cutter through butt.*—Multiply the diameter of the cylinder in inches by ·11. The product is the breadth of the gibs and cutter through butt in inches.

RULE 31. *To find the thickness of the gibs and cutter through butt.*—Multiply the diameter of the cylinder in inches by ·029. The product is the thickness of the gibs and cutter through butt in inches.

To find other parts of the engine which do not depend upon the stroke. Suppose it were required to find the thickness of the small eye of crank for an engine the diameter of whose cylinder is 64 inches. According to the rule, the proper thickness of the small eye of crank is 4·04 inches. Again, suppose it were required to find the length of the small eye of crank. Hence, according to the rule, the proper length of the small eye of crank is 11·94 inches. Again, supposing it were required to find the proper thickness of the web of crank at pin centre; that is to say, the thickness which it would have if continued to the pin centre. According to the rule, the proper thickness for the web of crank at pin centre is 7·04 inches. Again, suppose it were required to find the breadth of the web of crank at pin centre; that is to say, the breadth which it would have if it were continued to the pin centre. Hence, according to the rule, the proper breadth of the web of crank at pin centre is 10·24 inches.

These results are calculated from the following rules, which give the proper dimensions for engines of all sizes:

RULE 1. *To find the breadth of the small eye of crank.*—Multiply the diameter of the cylinder in inches by ·063. The product is the proper breadth of the small eye of crank in inches.

RULE 2. *To find the length of the small eye of crank.*—Multiply the diameter of the cylinder in inches by ·187. The product is the proper length of the small eye of crank in inches.

RULE 3. *To find the thickness of the web of crank at pin centre.*—Multiply the diameter of the cylinder in inches by ·11. The product is the proper thickness of the web of crank at pin centre in inches.

RULE 4. *To find the breadth of the web of crank at pin centre.*—Multiply the diameter of the cylinder in inches by ·16. The product is the proper breadth of crank at pin centre in inches.

To illustrate the use of the succeeding rules, let us take the particular example of an engine of 8 feet stroke and 64-inch cylinder, and let us suppose that the length of the connecting rod is 12 feet, and the side rod 10 feet. We find by a previous rule that the diameter of the connecting rod at ends is 6·08, and the ratio between the diameters at middle and ends of a connecting rod, whose length is 12 feet, is 1·504. Hence, the proper diameter at middle of the connecting rod = 6·08 × 1·504 inches = 9·144 inches. And again, we find the diameter of cylinder side rods at ends, for the particular engine which we have selected, is 4·10, and the ratio between the diameters at middle and ends of cylinder side rods, whose lengths are 10 feet, is 1·42. Hence, according to the rules, the proper diameter of the cylinder side rods at middle is equal to 4·1 × 1·42 inches = 5·82 inches.

To find some of those parts of the engine which do not depend upon the stroke. Suppose we take the particular example of an engine the diameter of whose cylinder is 64 inches. We find from the following rules that

Diameter of cylinder side rods at ends = 4·1 inches, or $4\frac{1}{10}$ inches.

Breadth of butt = 4·93 inches, or very nearly 5 inches.

Thickness of butt = 3·9 inches, or $3\frac{9}{10}$ inches.

Mean thickness of strap at cutter = 2·06 inches, or a little more than 2 inches.

Mean thickness of strap below cutter = 1·47 inches, or very nearly $1\frac{1}{2}$ inches.

Depths of gibs and cutter = 5·12 inches, or a little more than $5\frac{1}{10}$ inches.

Thickness of gibs and cutter = 1·03 inches, or a little more than 1 inch.

Diameter of main centre journal = 11·71 inches, or very nearly $11\frac{3}{4}$ inches.

Length of main centre journal = 17·6 inches, or $17\frac{3}{5}$ inches.

Depth of eye round end studs of lever = 4·75 inches, or $4\frac{3}{4}$ inches.

Thickness of eye round end studs of lever = 3·33 inches, or $3\frac{1}{3}$ inches.

Diameter of end studs of lever = 4·48 inches, or very nearly $4\frac{1}{2}$ inches.

Length of end studs of lever = 4·86 inches, or between $4\frac{3}{4}$ and 5 inches.

Diameter of air-pump studs = 2·91 inches, or nearly 3 inches.

Length of air-pump studs = 3·16 inches, or nearly $3\frac{1}{5}$ inches.

These results were obtained from the following rules, which will be found to give the proper dimensions for all sizes of engines.

RULE 1. *To find the diameter of cylinder side rods at ends.*—Multiply the diameter of the cylinder in inches by ·065. The product is the diameter of the cylinder side rods at ends in inches.

RULE 2. *To find the breadth of butt in inches.*—Multiply the diameter of the cylinder in inches by ·077. The product is the breadth of the butt in inches.

RULE 3. *To find the thickness of the butt.*—Multiply the diameter of the cylinder in inches by ·061. The product is the thickness of the butt in inches.

RULE 4. *To find the mean thickness of strap at cutter.*—Multiply the diameter of the cylinder in inches by ·032. The product is the mean thickness of the strap at cutter.

RULE 5. *To find the mean thickness of strap below cutter.*—Multiply the diameter of the cylinder in inches by ·023. The product is the mean thickness of strap below cutter in inches.

RULE 6. *To find the depth of gibs and cutter.*—Multiply the diameter of the cylinder in inches by ·08. The product is the depth of the gibs and cutter in inches.

RULE 7. *To find the thickness of gibs and cutter.*—Multiply the diameter of the cylinder in inches by ·016. The product is the thickness of gibs and cutter in inches.

RULE 8. *To find the diameter of the main centre journal.*—Multiply the diameter of the cylinder in inches by ·183. The product is the diameter of the main centre journal in inches.

RULE 9. *To find the length of the main centre journal.*—Multiply the diameter of the cylinder in inches by ·275. The product is the diameter of the cylinder in inches.

RULE 10. *To find the depth of eye round end studs of lever.*—Multiply the diameter of the cylinder in inches by ·074. The product is the depth of the eye round end studs of lever in inches.

RULE 11. *To find the thickness of eye round end studs of lever.*—Multiply the diameter of the cylinder in inches by ·052. The product is the thickness of eye round end studs of lever in inches.

RULE 12. *To find the diameter of the end studs of lever.*—Multiply the diameter of the cylinder in inches by ·07. The product is the diameter of the end studs of lever in inches.

RULE 13. *To find the length of the end studs of lever.*—Multiply the diameter of the cylinder in inches by ·076. The product is the length of the end studs of lever in inches.

RULE 14. *To find the diameter of the air-pump studs.*—Multiply the diameter of the cylinder in inches by ·045. The product is the diameter of the air-pump studs in inches.

RULE 15. *To find the length of the air-pump studs.*—Multiply the diameter of the cylinder in inches by ·049. The product is the length of the air-pump studs in inches.

The next rule gives the proper depth in inches across the centre of the side lever, when, as is generally the case, the side lever is of cast iron. It will be observed that the depth is made to depend upon the diameter of the cylinder and the length of the lever, and not at all upon the length of the stroke, except indeed in so far as the length of the lever may depend upon the length of the stroke. Suppose it were required to find the proper depth across the centre of a side lever whose length is 20 feet, and the diameter of the cylinder 64 inches. According to the rule, the proper depth across the centre would be 39·26 inches.

The following rule will give the proper dimensions for any size of engine:

RULE.—*To find the depth across the centre of the side lever.*—Multiply the length of the side lever in feet by ·7423; extract the cube root of the product, and reserve the result for a multiplier. Then square the diameter of the cylinder in inches; extract the cube root of the result. The product of the final result and the reserved multiplier is the depth of the side lever in inches across the centre.

Thus, to apply this rule to the particular example which we have selected, we have

20 = length of side lever in feet.
·7423 = constant multiplier.

14·846

and $\sqrt[3]{14·846}$ = 2·458 nearly.

64 = diameter of cylinder in inches.
64

4096

and $\sqrt[3]{4096}$ = 16

Hence depth at centre = 16 × 2·458 inches = 39·33 inches, or between 39¼ and 39½ inches.

---

The next set of rules give the dimensions of several of the parts of the air-pump machinery which depend upon the diameter of the cylinder only. To illustrate the use of these rules, let us take the particular example of an engine the diameter of whose cylinder is 64 inches. We find from the succeeding rules successively,

Diameter of air-pump = 38·4 inches, or 38⅜ inches.

Thickness of the eye of air-pump cross-head = 1·58 inches, or a little more than $1\frac{1}{2}$ inches.

Depth of eye of air-pump cross-head = 11·01, or about 11 inches.

Diameter of end journals of air-pump cross-head = 3·29 inches, or somewhat more than $3\frac{1}{4}$ inches.

Length of end journals of air-pump cross-head = 3·7 inches, or $3\frac{7}{10}$ inches.

Thickness of the web of air-pump cross-head at middle = 2·76 inches, or a little more than $2\frac{3}{4}$ inches.

Depth of web of air-pump cross-head at middle = 10·29 inches, or somewhat more than $10\frac{1}{4}$ inches.

Thickness of web of air-pump cross-head at journal = 2·35 inches, or about $2\frac{3}{8}$ inches.

Depth of web of air-pump cross-head at journal = 3·89 inches, or about $3\frac{7}{8}$ inches.

Diameter of air-pump piston rod when made of copper = 4·27 inches, or about $4\frac{1}{4}$ inches.

Depth of gibs and cutter through air-pump cross-head = 4·04 inches, or a little more than 4 inches.

Thickness of gibs and cutter through air-pump cross-head = ·81 inches, or about $\frac{7}{8}$ inch.

Depth of cutter through piston = 3·27 inches, or somewhat more than $3\frac{1}{4}$ inches.

Thickness of cutter through piston = 1·34 inches, or about $1\frac{3}{8}$ inches.

These results were obtained from the following rules, and give the proper dimensions for all sizes of engines:

RULE 1. *To find the diameter of the air-pump.*—Multiply the diameter of the cylinder in inches by ·6. The product is the diameter of air-pump in inches.

RULE 2. *To find the thickness of the eye of air-pump cross-head.*—Multiply the diameter of the cylinder in inches by ·025. The product is the thickness of the eye of air-pump cross-head in inches.

RULE 3. *To find the depth of eye of air-pump cross-head.*—Multiply the diameter of the cylinder in inches by ·171. The product is the depth of the eye of air-pump cross-head in inches.

RULE 4. *To find the diameter of the journals of air-pump cross-head.*—Multiply the diameter of the cylinder in inches by ·051. The product is the diameter of the end journals.

RULE 5. *To find the length of the end journals for air-pump cross-head.*—Multiply the diameter of the cylinder in inches by 058. The product is the length of the air-pump cross-head journals in inches.

RULE 6. *To find the thickness of the web of air-pump cross-head at middle.*—Multiply the diameter of the cylinder in inches by ·043. The product is the thickness at middle of the web of air-pump cross-head in inches.

RULE 7. *To find the depth at middle of the web of air-pump cross-head.*—Multiply the diameter of the cylinder in inches by ·161.

The product is the depth at middle of air-pump cross-head in inches.

RULE 8. *To find the thickness of the web of air-pump cross-head at journals.*—Multiply the diameter of the cylinder in inches by ·037. The product is the thickness of the web of air-pump cross-head at journals in inches.

RULE 9. *To find the depth of the air-pump cross-head web at journals.*—Multiply the diameter of the cylinder in inches by ·061. The product is the depth at journals of the web of air-pump cross-head.

RULE 10. *To find the diameter of the air-pump piston rod when of copper.*—Multiply the diameter of the cylinder in inches by ·067. The product is the diameter of the air-pump piston rod, when of copper, in inches.

RULE 11. *To find the depth of gibs and cutter through air-pump cross-head.*—Multiply the diameter of the cylinder in inches by ·063. The product is the depth of the gibs and cutter through air-pump cross-head in inches.

RULE 12. *To find the thickness of the gibs and cutter through air-pump cross-head.*—Multiply the diameter of the cylinder in inches by ·013. The product is the thickness of the gibs and cutter in inches.

RULE 13. *To find the depth of cutter through piston.*—Multiply the diameter of the cylinder in inches by ·051. The product is the depth of the cutter through piston in inches.

RULE 14. *To find the thickness of cutter through air-pump piston.*—Multiply the diameter of the cylinder in inches by ·021. The product is the thickness of the cutter through air-pump piston.

---

The next seven rules give the dimensions of the remaining parts of the engine which do not depend upon the stroke. To exemplify their use, suppose it were required to find the corresponding dimensions for an engine the diameter of whose cylinder is 64 inches. According to the rule, the proper diameter of the air-pump side rod would be 2·48 inches. Hence, according to the rule, the proper breadth of butt is 2·95 inches. According to the rule, the proper thickness of butt is 2·35 inches. According to the rule, the mean thickness of strap at cutter ought to be 1·24 inches. Hence, according to the rule, the mean thickness of strap below cutter is ·91 inch. According to the rule, the proper depth for the gibs and cutter is 2·94 inches. According to the rule, the proper thickness of the gibs and cutter is ·63 inches.

The following rules give the correct dimensions for all sizes of engines:

RULE 1. *To find the diameter of air-pump side rod at ends.*—Multiply the diameter of the cylinder in inches by ·039. The product is the diameter of the air-pump side rod at ends in inches.

RULE 2. *To find the breadth of butt for air-pump.*—Multiply the diameter of the cylinder in inches by ·046. The product is the breadth of butt in inches.

RULE 3. *To find the thickness of butt for air-pump.*—Multiply the diameter of the cylinder in inches by ·037. The product is the thickness of butt for air-pump in inches.

RULE 4. *To find the mean thickness of strap at cutter.*—Multiply the diameter of the cylinder in inches by ·019. The product is the mean thickness of strap at cutter for air-pump in inches.

RULE 5. *To find the mean thickness of strap below cutter.*—Multiply the diameter of the cylinder in inches by 0·14. The product is the mean thickness of strap below cutter in inches.

RULE 6. *To find the depth of gibs and cutter for air-pump.*—Multiply the diameter of the cylinder in inches by 0·48. The product is the depth of gibs and cutter for air-pump in inches.

RULE 7. *To find the thickness of gibs and cutter for air-pump.*—Divide the diameter of the cylinder in inches by 100. The quotient is the proper thickness of the gibs and cutter for air-pump in inches.

With regard to other dimensions made to depend upon the nominal horse power of the engine:—Suppose that we take the particular example of an engine whose stroke is 8 feet, and diameter of cylinder 64 inches. We find that the nominal horse power of this engine is nearly 175. Hence we have successively,

Diameter of valve shaft at journal in inches = 4·85, or between 4¾ and 5 inches.

Diameter of parallel motion shaft at journal in inches = 3·91, or very nearly 4 inches.

Diameter of valve rod in inches = 2·44, or about 2⅜ inches.

Diameter of radius rod at smallest part in inches = 1·97, or very nearly 2 inches.

Area of eccentric rod, at smallest part, in square inches = 8·37, or about 8⅜ square inches.

Sectional area of eccentric hoop in square inches = 8·75, or 8¾ square inches.

Diameter of eccentric pin in inches = 2·24, or 2¼ inches.

Breadth of valve lever for eccentric pin at eye in inches = 5·7, or very nearly 5¾ inches.

Thickness of valve lever for eccentric pin at eye in inches = 3.

Breadth of parallel motion crank at eye = 4·2 inches, or very nearly 4¼ inches.

Thickness of parallel motion crank at eye = 1·76 inches, or about 1¾ inches.

---

To find the area in square inches of each steam port. Suppose it were required to find the area of each steam port for an engine whose stroke is 8 feet, and diameter of cylinder 64 inches. According to the rule, the area of each steam port would be 202·26 square inches.

With regard to the rule, we may remark that the area of the

steam port ought to depend principally upon the cubical content of the cylinder, which again depends entirely upon the product of the square of the diameter of the cylinder and the length of the stroke of the engine. It is well known, however, that the quantity of steam admitted by a small hole does not bear so great a proportion to the quantity admitted by a larger one, as the area of the one does to the area of the other; and a certain allowance ought to be made for this. In the absence of correct theoretical information on this point, we have attempted to make a proper allowance by supplying a constant; but of course this plan ought only to be regarded as an approximation. Our rule is as follows:

Rule.—*To find the area of each steam port.*—Multiply the square of the diameter of the cylinder in inches by the length of the stroke in feet; multiply this product by 11; divide the last product by 1800; and, finally, to the quotient add 8. The result is the area of each steam port in square inches.

To show the use of this rule, we shall apply it to a particular example. We shall apply it to an engine whose stroke is 6 feet, and diameter of cylinder 30 inches. Then, according to the rule, we have

30 = diameter of the cylinder in inches.
30
900 = square of diameter.
6 = length of stroke in feet.
5400
11
59400 ÷ 1800 = 33
8 = constant to be added.
41 = area of steam port in square inches.

When the length of the opening of steam port is from any circumstance found, the corresponding depth in inches may be found, by dividing the number corresponding to the particular engine, by the given length in inches: conversely, the length may be found, when for some reason or other the depth is fixed, by dividing the number corresponding to the particular engine, by the given depth in inches: the quotient is the length in inches.

The next rule is useful for determining the diameter of the steam pipe branching off to any particular engine. Suppose it were required to find the diameter of the branch steam pipe for an engine whose stroke is 8 feet, and diameter of cylinder 64 inches. According to the rule, the proper diameter of the steam pipe would be 13·16 inches.

The following rule will be found to give the proper diameter of steam pipe for all sizes of engines.

Rule.—*To find the diameter of branch steam pipe.*—Multiply together the square of the diameter of the cylinder in inches, the length of the stroke in feet, and ·00498; to the product add 10·2, and extract the square root of the sum. The result is the diameter of the steam pipe in inches.

To exemplify the use of this rule we shall take an engine whose stroke is 8 feet, and diameter of cylinder 64 inches. In this case we have as follows:—

64 = diameter of cylinder in inches.
64
4096 = square of diameter.
8 = length of stroke in feet.
32768
·00498 = constant multiplier.
163·18
10·2 = constant to be added.
173·38
and $\sqrt{173 \cdot 38} = 13 \cdot 16$.

To find the diameter of the pipes connected with the engine. They are made to depend upon the nominal horse power of the engine. Suppose it were required to apply this rule to determine the size of the pipes for two marine engines, whose strokes are each 8 feet, and diameters of cylinder each 64 inches. We find the nominal horse power of each of these engines to be 174·3. Hence, according to the rules, we have in succession,

Diameter of waste water pipe = 15·87 inches, or between 15¾ and 16 inches.

Area of foot-valve passage = 323 square inches.

Area of injection pipe = 14·88 square inches.

If the injection pipe be cylindrical, then by referring to the table of areas of circles, we see that its diameter would be about 4⅜ inches.

Diameter of feed pipe = 4·12 inches, or between 4 and 4¼ inches.

Diameter of waste steam pipe = 12·17 inches, or nearly 12¼ inches.

Diameter of safety valve,

When one is used = 14·05 inches.
When two are used = 9·94 inches.
When three are used = 8·12 inches.
When four are used = 7·04 inches.

These results were obtained from the following rules, which will give the correct dimensions for all sizes of engines.

Rule 1. *To find the diameter of waste water pipe.*—Multiply the square root of the nominal horse power of the engine by 1·2. The product is the diameter of the waste water pipe in inches.

Rule 2. *To find the area of foot-valve passage.*—Multiply the nominal horse power of the engine by 9; divide the product by 5; add 8 to the quotient. The sum is the area of foot-valve passage in square inches.

RULE 3. *To find the area of injection pipe.*—Multiply the nominal horse power of the engine by ·069; to the product add 2·81. The sum is the area of the injection pipe in square inches.

RULE 4. *To find the diameter of feed pipe.*—Multiply the nominal horse power of the engine by ·04; to the product add 3; extract the square root of the sum. The result is the diameter of the feed pipe in inches.

RULE 5. *To find the diameter of waste steam pipe.*—Multiply the collective nominal horse power of the engines by ·375; to the product add 16·875; extract the square root of the sum. The final result is the diameter of the waste steam pipe in inches.

RULE 6. *To find the diameter of the safety valve when only one is used.*—To one-half the collective nominal horse power of the engines add 22·5; extract the square root of the sum. The result is the diameter of the safety valve when only one is used.

RULE 7. *To find the diameter of the safety valve when two are used.*—Multiply the collective nominal horse power of the engines by ·25; to the product add 11·25; extract the square root of the sum. The result is the diameter of the safety valve when two are used.

RULE 8. *To find the diameter of the safety valve when three are used.*—To one-sixth of the collective nominal horse power of the engines add 7·5; extract the square root of the sum. The result is the diameter of the safety valve where three are used.

RULE 9. *To find the diameter of the safety valve when four are used.*—Multiply the collective nominal horse power of the engines by ·125; to the product add 5·625; extract the square root of the sum. The result is the diameter of the safety valve when four are used.

Another rule for safety valves, and a preferable one for low pressures, is to allow ·8 of a circular inch of area per nominal horse power.

The next rule is for determining the depth across the web of the main beam of a land engine. Suppose we wished to find the proper depth at the centre of the main beam of a land engine whose main beam is 16 feet long, and diameter of cylinder 64 inches. According to the rule, the proper depth of the web across the centre is 46·17 inches. This rule gives correct dimensions for all sizes of engines.

RULE.—*To find the depth of the web at the centre of the main beam of a land engine.*—Multiply together the square of the diameter of the cylinder in inches, half the length of the main beam in feet, and the number 3; extract the cube root of the product. The result is the proper depth of the web of the main beam across the centre in inches, when the main beam is constructed of cast iron.

To illustrate this rule we shall take the particular example of an engine whose main beam is 20 feet long, and the diameter of the cylinder 64 inches. In this case we have

64 = diameter of cylinder in inches.<br>
64<br>
4096 = square of the diameter.<br>
10 = $\frac{1}{2}$ length of main beam in feet.<br>
40960<br>
3 = constant multiplier.<br>
122880

| | | |
|---|---|---|
| 0 | 0 | 122880 ( 49·714 = $\sqrt[3]{122880}$ |
| 4 | 16 | 64 |
| 4 | 16 | 58880 |
| 4 | 32 | 53649 |
| 8 | 4800 | 5231 |
| 4 | 1161 | 5112 |
| 120 | 5961 | 119 |
| 9 | 1242 | 74 |
| 129 | 7203 | 35 |
| 9 | 10 | |
| 138 | 730 | |
| 9 | 10 | |
| 147 | 741 | |

To find the depth of the main beam across the ends. Suppose it were required to find the depth at ends of a cast-iron main beam whose length is 20 feet, when the diameter of the cylinder is 64 inches. The proper depth will be 19·89 inches.

The following rule gives the proper dimensions for all sizes of engines.

RULE.—*To find the depth of main beam at ends.*—Multiply together the square of the diameter of the cylinder in inches, half the length of the main beam in feet, and the number ·192; extract the cube root of the product. The result is the depth in inches of the main beam at ends, when of cast iron.

To illustrate this rule, let us apply it to the particular example of an engine whose main beam is 20 feet long, and the diameter of the cylinder 64 inches. In this case we have as follows:

64 = diameter of cylinder in inches.<br>
64<br>
4096 = square of diameter of cylinder.<br>
10 = $\frac{1}{2}$ length of main beam in feet.<br>
40960<br>
·192 = constant multiplier.<br>
7864·32

| 0 | 0 | 7864·32 ( 19·89 = $\sqrt[3]{7864 \cdot 32}$ |
|---|---|---|
| 1 | 1 | 1 |
| 1 | 1 | 6864 |
| 1 | 2 | 5859 |
| 2 | 300 | 1005 |
| 1 | 351 | 898 |
| 30 | 651 | 107 |
| 9 | 432 | |
| 39 | 1083 | |
| 9 | 4 | |
| 48 | 112 | |
| 9 | 4 | |
| 57 | 116 | |

so that, according to the rule, the depth at ends is nearly 20 inches.

To find the dimensions of the feed-pump in cubic inches. Suppose we take the particular example of an engine whose stroke is 8 feet, and diameter of cylinder 64 inches. The proper content of the feed-pump would be 1093·36 cubic inches. Suppose, now, that the cold-water pump was suspended from the main beam at a fourth of the distance between the centre and the end, so that its stroke would be 2 feet, or 24 inches. In this case the area of the pump would be equal to 1093·36 ÷ 24 = 45·556 square inches; so that we conclude that the diameter is between $7\frac{1}{2}$ and $7\frac{3}{4}$ inches. Conversely, suppose that it was wished to find the stroke of the pump when the diameter was 5 inches. We find the area of the pump to be 19·635 square inches; so that the stroke of the feed-pump must be equal to 1093·36 ÷ 19·635 = 55·69 inches, or very nearly $55\frac{3}{4}$ inches.

This rule will be found to give correct dimensions for all sizes of engines:

RULE.—*To find the content of the feed-pump.*—Multiply the square of the diameter of the cylinder in inches by the length of the stroke in feet; divide the product by 30. The quotient is the content of the feed-pump in cubic inches.

Thus, for an engine whose stroke is 6 feet, and diameter of cylinder 50 inches, we have,

```
         50 = diameter of cylinder.
         50
       ----
       2500 = square of the diameter of the cylinder.
          6 = length of stroke in feet.
   30)15000
      -----
        500 = content of feed-pump in cubic inches.
```

To determine the content of the cold-water pump in cubic feet. To illustrate this, suppose we take the particular example of an engine whose stroke is 8 feet, and diameter of cylinder 64 inches. Suppose, now, the stroke of the pump to be 5 feet, then the area equal to 7·45 ÷ 5 = 1·49 square feet = 214·56 square inches; we see that the diameter of the pump is about 16½ inches. Again, suppose that the diameter of the cold-water pump was 20 inches, and that it was required to find the length of its stroke. The area of the pump is 314·16 square inches, or 314·16 ÷ 144 = 2·18 square feet; so that the stroke of the pump is equal to 7·45 ÷ 2·18 = 3·42 feet.

The content is calculated from the following rule, which will be found to give correct dimensions for all sizes of engines:

RULE.—*To find the content of the cold-water pump.*—Multiply the square of the diameter of the cylinder in inches by the length of the stroke in feet; divide the product by 4400. The quotient is the content of the cold-water pump in cubic feet.

To explain this rule, we shall take the particular example of an engine whose stroke is 5½ feet, and diameter of cylinder 60 inches. In this case we have in succession,

$$
\begin{array}{rl}
60 & = \text{diameter of cylinder in inches.} \\
60 & \\
\hline
3600 & = \text{square of the diameter of cylinder.} \\
5\tfrac{1}{2} & = \text{length of stroke in feet.} \\
\hline
4400)\,19800 & \\
\hline
4{\cdot}5 & = \text{content of cold water pump in cubic feet.}
\end{array}
$$

To determine the proper thickness of the large eye of crank for fly-wheel shaft when the crank is of cast iron. The crank is sometimes cast on the shaft, and of course the thickness of the large eye is not then so great as when the crank is only keyed on the shaft, or rather there is then no large eye at all. To illustrate the use of this rule, we shall apply it to the particular example of an engine whose stroke is 8 feet, and diameter of cylinder 64 inches. Hence, according to the rule, the proper thickness of the large eye of crank when of cast iron is 8·07 inches. For a marine engine of 8 feet stroke and 64 inch cylinder, the thickness of the large eye of crank is about 5¾ inches. The difference is thus about 2¼ inches, which is an allowance for the inferiority of cast iron to malleable iron.

The following rule will be found to give correct dimensions for all sizes of engines:

RULE.—*To find the thickness of the large eye of crank for fly-wheel shaft when of cast iron.*—Multiply the square of the length of the crank in inches by 1·561, and then multiply the square of the diameter of the cylinder in inches by ·1235; multiply the sum of these products by the square of the diameter of cylinder in inches; divide this product by 666·283; divide this quotient by the length of the crank in inches; finally extract the cube root of the quotient.

The result is the proper thickness of the large eye of crank for fly-wheel shaft in inches, when of cast iron.

As this rule is rather complicated, we shall show its application to the particular example already selected.

48 = length of crank in inches.

48

2304 = square of length of crank in inches.

1·561 = constant multiplier.

3596·5

64 = diameter of cylinder in inches.

64

4096 = square of the diameter of cylinder.

·1235 = constant multiplier.

505·8

3596·5

4102·3 = sum of products.

4096 = square of the diameter of cylinder.

666·283) 16803020·8

length of crank=48) 25219·045

525·397

and $\sqrt[3]{525·397} = 8·07$ nearly.

To find the breadth of the web of crank at the centre of the fly-wheel shaft, that is to say, the breadth which it would have if it were continued to the centre of the fly-wheel shaft. Suppose it were required to find the breadth of the crank at the centre of the fly-wheel shaft for an engine whose stroke is 8 feet, and diameter of cylinder 64 inches. According to the rule, the proper breadth is 22·49 inches. According to a former rule, the breadth of the web of a cast iron crank of an engine whose stroke is 8 feet, and diameter of cylinder 64 inches, is about 18 inches. The difference between these two is about 4½ inches; which is not too great an allowance for the inferiority of the cast iron.

The following rule will be found to give correct dimensions for all sizes of engines:

Rule.—*To find the breadth of the web of crank at fly-wheel shaft, when of cast iron.*—Multiply the square of the length of the crank in inches by 1·561, and then multiply the square of the diameter of the cylinder in inches by ·1235; multiply the square root of the sum of these products by the square of the diameter of the cylinder in inches; divide the product by 23·04, and finally extract the cube root of the quotient. The final result is the breadth of the crank at the centre of the fly-wheel shaft, when the crank is of cast iron.

As this rule is rather complicated, we shall illustrate it by showing its application to the particular example of an engine whose stroke is 8 feet, and diameter of cylinder 64 inches.

64 = diameter of cylinder in inches.
64
4096 = square of the diameter of cylinder.
·1235 = constant multiplier.
505·8

48 = length of crank in inches.
48
2304 = square of the length of crank.
1·561 = constant multiplier.
3596·5
505·8
4102·3 = sum of products.
$\sqrt{4102 \cdot 3} = 64 \cdot 05$ nearly.
4096 = square of the diameter of [cylinder.
constant divisor = 23·04) 262348·5
11386·66 nearly.
and $\sqrt[3]{11386 \cdot 66} = 22 \cdot 49$.

To determine the thickness of the web of crank at the centre of the fly-wheel shaft; that is to say, the thickness which it would have if it were continued so far. Suppose it were required to find the thickness of web of crank at the centre of fly-wheel shaft of an engine whose stroke is 8 feet, and diameter of cylinder 64 inches. According to the rule, the proper thickness would be 11·26 inches. The proper thickness of web at centre of paddle shaft for a marine engine whose stroke is 8 feet, and diameter of cylinder 64 inches, is nearly 9 inches. The difference between the two thicknesses is about 2¼ inches, which is not too great an allowance for the inferiority of cast iron to malleable iron.

The following rule will be found to give correct dimensions for all sizes of engines:

RULE.—*To find the thickness of the web of crank at centre of fly-wheel shaft, when of cast iron.*—Multiply the square of the length of the crank in inches by 1·561, and then multiply the square of the diameter of the cylinder in inches by ·1235; multiply the square root of the sum of these products by the square of the diameter of the cylinder in inches; divide this product by 184·32; finally extract the cube root of the quotient. The result is the thickness of the web of crank at the centre of the fly-wheel shaft when of cast iron, in inches.

As this rule is rather complicated, we shall illustrate it by applying it to the particular engine which we have already selected.

$$
\begin{array}{rl}
48 & = \text{length of crank in inches.} \\
48 & \\
\hline
2304 & = \text{square of length of crank.} \\
1{\cdot}561 & = \text{constant multiplier.} \\
\hline
3596{\cdot}5 & \\
 & \\
64 & = \text{diameter of cylinder in inches.} \\
64 & \\
\hline
4096 & = \text{square of the diameter of cylinder.} \\
1235 & = \text{constant multiplier.} \\
\hline
505{\cdot}8 & \\
3596{\cdot}5 & \\
\hline
4102{\cdot}3 & = \text{sum of products.}
\end{array}
$$

and $\sqrt{4102{\cdot}3} = 64{\cdot}05$ nearly.

$4096$ = square of diameter.

Constant divisor $= 184{\cdot}32)\overline{262348{\cdot}5}$

$1423{\cdot}33$

and $\sqrt[3]{1423{\cdot}33} = 11{\cdot}24$

To find the proper diameter of the fly-wheel shaft at its smallest part, when, as is usually the case, it is of cast iron. Suppose it were required to find the diameter of the fly-wheel shaft for an engine whose stroke is 8 feet, and diameter of cylinder 64 inches. According to the rule, the diameter would be 17·59 inches. It is obvious enough that the fly-wheel shaft stands in much the same relation to the land engine, as the paddle shaft does to the marine engine. According to a former rule, the diameter of the paddle shaft journal of a marine engine whose stroke is 8 feet, and diameter of cylinder 64 inches, is about 14 inches. The difference betwixt the diameter of the paddle shaft for the marine engine, and the diameter of the fly-wheel shaft for the corresponding land engine is about $3\frac{1}{2}$ inches. This will be found to be a very proper allowance for the different circumstances connected with the land engine.

The following rule will be found to give correct dimensions for all sizes of engines.

RULE.—*To find the diameter of the fly-wheel shaft at smallest part, when it is of cast iron.*—Multiply the square of the diameter of the cylinder in inches by the length of the crank in inches; extract the cube root of the product; finally multiply the result by ·3025. The result is the diameter of the fly-wheel shaft at smallest part in inches.

We shall illustrate this rule by applying it to the particular engine which we have already selected.

64 = diameter of cylinder in inches.
64

4096 = square of the diameter.
48 = length of crank in inches.

196608

| | | |
|---|---|---|
| 0 | 0 | 196608 (58·15 = $\sqrt[3]{196608}$ |
| 5 | 25 | 125 |
| 5 | 25 | 71608 |
| 5 | 50 | 70112 |
| 10 | 7500 | 1496 |
| 5 | 1264 | 1011 |
| 150 | 8764 | 485 |
| 8 | 1328 | |
| 158 | 10092 | |
| 8 | 2 | |
| 166 | 1011 | |
| 8 | 2 | |
| 174 | 1013 | |

and 58·15 × ·3025 = 1759
which agrees with the number given by a former rule.

To determine the sectional area of the fly-wheel rim when of cast iron. Suppose it were required to find the sectional area of the rim of a fly-wheel for an engine whose stroke is 8 feet, and diameter of cylinder 64 inches, the diameter of the fly-wheel itself being 30 feet. According to the rule, the sectional area of the rim in square inches = 146·4 × ·813 = 119·02. We may remark that this calculation has been made on the supposition that the fly-wheel is so connected with the engine, as to make exactly one revolution for each double stroke of the piston. If the fly-wheel is so connected with the engine as to make more than one revolution for each double stroke, then the rim does not need to be so heavy as we make it. If, on the contrary, the fly-wheel does not make a complete revolution for each double stroke of the engine, then it ought to be heavier than this rule makes it.

Rule.—*To find the sectional area of the rim of the fly-wheel when of cast iron.*—Multiply together the square of the diameter of the cylinder in inches, the square of the length of the stroke in feet, the cube root of the length of the stroke in feet, and 6·125; divide the final product by the cube of the diameter of the fly-wheel in feet. The quotient is the sectional area of the rim of fly-wheel in square inches, provided it is of cast iron.

As this rule is rather complicated, we shall endeavour to illustrate it by showing its application to a particular engine. We shall apply the rule to determine the sectional area of the rim of fly-wheel for an engine whose stroke is 8 feet, diameter of cylinder 50 inches; the diameter of the fly-wheel being 20 feet. For this engine we have as follows:

$$\begin{array}{rl} 2500 & = \text{square of diameter of cylinder.} \\ 64 & = \text{square of the length of stroke.} \\ \hline 160000 & \\ 2 & = \text{cube root of the length of stroke.} \\ \hline 320000 & \\ 6 \cdot 125 & = \text{constant multiplier.} \\ \hline 1960000 & \end{array}$$

therefore sectional area in square inches $= 1960000 \div 20^3 = 1960000 \div 8000 = 1960 \div 8 = 245$.

In the following formulas we denote the diameter of the cylinder in inches by D, the length of the crank in inches by R, the length of the stroke in feet, and the nominal horse power of the engine by H.P.

### MARINE ENGINES.—DIMENSIONS OF SEVERAL OF THE PARTS OF THE SIDE LEVER.

Depth of eye round end studs of lever $= \cdot 074 \times D$.
Thickness of eye round end studs of lever $= \cdot 052 \times D$.
Diameter of end studs, in inches $= \cdot 07 \times D$.
Length of end studs, in inches $= \cdot 076 \times D$.
Diameter of air-pump studs, in inches $= \cdot 045 \times D$.
Length of air-pump studs, in inches $= \cdot 049 \times D$.

Depth of cast iron side lever across centre, in inches $= D^{\frac{2}{3}} \times \{\cdot 7423 \times \text{length of lever in feet}\}^{\frac{1}{3}}$.

### MARINE ENGINE.—DIMENSIONS OF SEVERAL PARTS OF AIR-PUMP CROSS-HEAD.

Diameter of air-pump, in inches $= \cdot 6 \times D$.
Thickness of eye for air-pump rod, in inches $= \cdot 025 \times D$.
Depth of eye for air-pump rod, in inches $= \cdot 171 \times D$.
Diameter of end journals, in inches $= \cdot 051 \times D$.
Length of end journals, in inches $= \cdot 058 \times D$.
Thickness of web at middle, in inches $= \cdot 043 \times D$.
Depth of web at middle, in inches $= \cdot 161 \times D$.
Thickness of web at journal $= \cdot 037 \times D$.
Depth of web at journal $= \cdot 061 \times D$.

### MARINE ENGINE.—DIMENSIONS OF THE PARTS OF AIR-PUMP PISTON-ROD.

Diameter of air-pump piston-rod, when of copper, in inches $= \cdot 067 \times D$.

Depth of gibs and cutter through cross-head, in inches $= \cdot 063 \times D$.

Thickness of gibs and cutter through cross-head, in inches = 013 × D.

Depth of cutter through piston, in inches = ·051 × D.

Thickness of cutter through piston, in inches = ·021 × D.

### MARINE ENGINE.—DIMENSIONS OF THE REMAINING PARTS OF THE AIR-PUMP MACHINERY.

Diameter of air-pump side rods at ends, in inches = ·039 × D.

Breadth of butt, in inches = ·046 × D.

Thickness of butt, in inches = ·037 × D.

Mean thickness of strap at cutter, in inches = ·019 × D.

Mean thickness of strap below cutter, in inches = ·014 × D.

Depth of gibs and cutter, in inches = ·048 × D.

Thickness of gibs and cutter in inches = D ÷ 100.

### MARINE AND LAND ENGINES.—AREA OF STEAM PORTS.

Area of each steam port, in square inches = $11 \times l \times D^2 \div 1800 + 8$.

### MARINE AND LAND ENGINES.—DIMENSIONS OF BRANCH STEAM PIPES.

Diameter of each branch steam pipe = $\sqrt{\cdot 00498 \times l \times D^2 \times 10{\cdot}2}$.

### MARINE ENGINE.—DIMENSIONS OF SEVERAL OF THE PIPES CONNECTED WITH THE ENGINE.

Diameter of waste water pipe, in inches = $1{\cdot}2 \times \sqrt{\text{H.P.}}$

Area of foot-valve passage, in square inches = 1·8 × H.P. + 8.

Area of injection pipe, in square inches = ·069 × H.P. + 2·81.

Diameter of feed pipe, in inches = $\sqrt{\cdot 04 \times \text{H.P.} + 3}$.

Diameter of waste steam pipe in inches = $\sqrt{\cdot 375 \times \text{H.P.} + 16{\cdot}875}$.

### MARINE AND LAND ENGINES.—DIMENSIONS OF SAFETY-VALVES.

Diam. of safety-valve, when one only is used = $\sqrt{\cdot 5 \times \text{H.P.} + 22{\cdot}5}$.

Diam. of safety-valve, when two are used = $\sqrt{\cdot 25 \times \text{H.P.} + 11{\cdot}25}$.

Diam. of safety-valve, when three are used = $\sqrt{\cdot 167 \times \text{H.P.} + 7{\cdot}5}$.

Diam. of safety-valve, when four are used = $\sqrt{\cdot 125 \times \text{H.P.} + 5{\cdot}625}$.

### LAND ENGINE.—DIMENSIONS OF MAIN BEAM.

Depth of web of main beam across centre =

$$\sqrt[3]{3 \times D^2 \times \text{half length of main beam in feet.}}$$

Depth of main beam at ends =

$$\sqrt[3]{\cdot 192 \times D^2 \times \text{half length of main beam, in feet.}}$$

### LAND AND MARINE ENGINES.—CONTENT OF FEED-PUMP.

Content of feed-pump, in cubic inches = $D^2 \times l \div 30$.

### LAND ENGINES.—CONTENT OF COLD WATER PUMP.

Content of cold water pump, in cubic feet = $D^2 \times l \div 4400$.

LAND ENGINES.—DIMENSIONS OF CRANK.

Thickness of large eye of crank, in inches =

$$\sqrt[3]{D^2 \times (1{\cdot}561 \times R^2 + {\cdot}1235\ D^2) \div (R \times 666{\cdot}283)}.$$

Breadth of web of crank at fly-wheel shaft centre, in inches =

$$\sqrt[3]{D^2 \times \sqrt{(1{\cdot}561 \times R^2 + {\cdot}1235 \times D^2)} \div 23{\cdot}04}.$$

Thickness of web of crank at fly-wheel shaft centre, in inches =

$$\sqrt[3]{D^2 \times \sqrt{(1{\cdot}561 \times R^2 + {\cdot}1235 \times D^2)} \div 184{\cdot}32}.$$

LAND ENGINES.—DIMENSIONS OF FLY-WHEEL SHAFT.

Diameter of fly-wheel shaft, when of cast iron $= 3025 \times \sqrt[3]{R \times D^2}$.

---

## DIMENSIONS OF PARTS OF LOCOMOTIVES.

DIAMETER OF CYLINDER.

In locomotive engines, the diameter of the cylinder varies less than either the land or the marine engine. In few of the locomotive engines at present in use is the diameter of the cylinder greater than 16 inches, or less than 12 inches. The length of the stroke of nearly all the locomotive engines at present in use is 18 inches, and there are always two cylinders, which are generally connected to cranks upon the axle, standing at right angles with one another.

AREA OF INDUCTION PORTS.

RULE.—*To find the size of the steam ports for the locomotive engine.*—Multiply the square of the diameter of the cylinder by ·068. The product is the proper size of the steam ports in square inches.

Required the proper size of the steam ports of a locomotive engine whose diameter is 15 inches. Here, according to the rule, size of steam ports $= {\cdot}068 \times 15 \times 15 = {\cdot}068 \times 225 = 15{\cdot}3$ square inches, or between $15\frac{1}{4}$ and $15\frac{1}{2}$ square inches.

After having determined the area of the ports, we may easily find the depth when the length is given, or, conversely, the length when the depth is given. Thus, suppose we knew that the length was 8 inches, then we find that the depth should be $15{\cdot}3 \div 8 = 1{\cdot}9125$ inches, or nearly 2 inches; or suppose we knew the depth was 2 inches, then we would find that the length was $15{\cdot}3 \div 2 = 7{\cdot}65$ inches, or nearly $7\frac{3}{4}$ inches.

AREA OF EDUCTION PORTS.

The proper area for the eduction ports may be found from the following rule.

RULE.—*To find the area of the eduction ports.*—Multiply the square of the diameter of the cylinder in inches by ·128. The product is the area of the eduction ports in square inches.

Required the area of the eduction ports of a locomotive engine, when the diameter of the cylinders is 13 inches. In this example we have, according to the rule,

Area of eduction port = ·128 × 13² = ·128 × 169 = 21·632 inches, or between 21½ and 21¾ square inches.

### BREADTH OF BRIDGE BETWEEN PORTS.

The breadth of the bridges between the eduction port and the induction ports is usually between ¾ inch and 1 inch.

### DIAMETER OF BOILER.

It is obvious that the diameter of the boiler may vary very considerably; but it is limited chiefly by considerations of strength; and 3 feet are found a convenient diameter. Rules for the strength of boilers will be given hereafter.

RULE.—*To find the inside diameter of the boiler.*—Multiply the diameter of the cylinder in inches by 3·11. The product is the inside diameter of the boiler in inches.

Required the inside diameter of the boiler for a locomotive engine, the diameter of the cylinders being 15 inches.

In this example we have, according to the rule,

Inside diameter of boiler = 15 × 3·11 = 46·65 inches,

or about 3 feet 10⅝ inches.

### LENGTH OF BOILER.

The length of the boiler is usually in practice between 8 feet and 8½ feet.

### DIAMETER OF STEAM DOME, INSIDE.

It is obvious that the diameter of the steam dome may be varied considerably, according to circumstances; but the first indication is to make it large enough. It is usual, however, in practice, to proportion the diameter of the steam dome to the diameter of the cylinder; and there appears to be no great objection to this. The following rule will be found to give the diameter of the dome usually adopted in practice.

RULE.—*To find the diameter of the steam dome.*—Multiply the diameter of the cylinder in inches by 1·43. The product is the diameter of the dome in inches.

Required the diameter of the steam dome for a locomotive engine whose diameter of cylinders is 13 inches. In this example we have, according to the rule,

Diameter of steam dome = 1·43 × 13 = 18·59 inches,

or about 18½ inches.

### HEIGHT OF STEAM DOME.

The height of the steam dome may vary. Judging from practice, it appears that a uniform height of 2½ feet would answer very well.

### DIAMETER OF SAFETY-VALVE.

In practice the diameter of the safety-valve varies considerably. The following rule gives the diameter of the safety-valve usually adopted in practice.

RULE.—*To find the diameter of the safety-valve.*—Divide the diameter of the cylinder in inches by 4. The quotient is the diameter of the safety-valve in inches.

Required the diameter of the safety-valves for the boiler of a locomotive engine, the diameter of the cylinder being 13 inches. Here, according to the rule, diameter of safety-valve $= 13 \div 4 = 3\frac{1}{4}$ inches. A larger size, however, is preferable, as being less likely to stick.

### DIAMETER OF VALVE SPINDLE.

The following rule will be found to give the correct diameter of the valve spindle. It is entirely founded on practice.

RULE.—*To find the diameter of the valve spindle.*—Multiply the diameter of the cylinder in inches by ·076. The product is the proper diameter of the valve spindle.

Required the diameter of the valve spindle for a locomotive engine whose cylinders' diameters are 13 inches.

In this example we have, according to the rule, diameter of valve spindle $= 13 \times {\cdot}076 = {\cdot}988$ inches, or very nearly 1 inch.

### DIAMETER OF CHIMNEY.

It is usual in practice to make the diameter of the chimney equal to the diameter of the cylinder. Thus a locomotive engine whose cylinders' diameters are 15 inches would have the inside diameter of the chimney also 15 inches, or thereabouts. This rule has, at least, the merit of simplicity.

### AREA OF FIRE-GRATE.

The following rule determines the area of the fire-grate usually given in practice. We may remark, that the area of the fire-grate in practice follows a more certain rule than any other part of the engine appears to do; but it is in all cases much too small, and occasions a great loss of power by the urging of the blast it renders necessary, and a rapid deterioration of the furnace plates from excessive heat. There is no good reason why the furnace should not be nearly as long as the boiler: it would then resemble the furnace of a marine boiler, and be as manageable.

RULE.—*To find the area of the fire-grate.*—Multiply the diameter of the cylinder in inches by ·77. The product is the area of the fire-grate in superficial feet.

Required the area of the fire-grate of a locomotive engine, the diameters of the cylinders being 15 inches.

In this example we have, according to the rule,

$$\text{Area of fire-grate} = {\cdot}77 \times 15 = 11{\cdot}55 \text{ square feet,}$$

or about $11\frac{1}{2}$ square feet. Though this rule, however, represents the usual practice, the area of the fire-grate should not be contingent upon the size of the cylinder, but upon the quantity of steam to be raised.

### AREA OF HEATING SURFACE.

In the construction of a locomotive engine, one great object is to obtain a boiler which will produce a sufficient quantity of steam with as little bulk and weight as possible. This object is admirably accomplished in the construction of the boiler of the locomotive engine. This little barrel of tubes generates more steam in an hour than was formerly raised from a boiler and fire occupying a considerable house. This favourable result is obtained simply by exposing the water to a greater amount of heating surface.

In the usual construction of the locomotive boiler, it is obvious that we can only consider four of the six faces of the inside fire-box as effective heating surface; viz. the crown of the box, and the three perpendicular sides. The circumferences of the tubes are also effective heating surface; so that the whole effective heating surface of a locomotive boiler may be considered to be the four faces of the inside fire-box, plus the sum of the surfaces of the tubes. Understanding this to be the effective heating surface, the following rule determines the average amount of heating surface usually given in practice.

RULE.—*To find the effective heating surface.*—Multiply the square of the diameter of the cylinder in inches by 5; divide the product by 2. The quotient is the area of the effective heating surface in square feet.

Required the effective heating surface of the boiler of a locomotive engine, the diameters of the cylinders being 15 inches.

In this example we have, according to the rule,

Effective heating surface $= 15^2 \times 5 \div 2 = 225 \times 5 \div 2 = 1125 \div 2 = 562\frac{1}{2}$ square feet.

According to the rule which we have given for the fire-grate, the area of the fire-grate for this boiler would be about $11\frac{1}{2}$ square feet. We may suppose, therefore, the area of the crown of the box to be 12 square feet. The area of the three perpendicular sides of the inside fire-box is usually three times the area of the crown; so that the effective heating surface of the fire-box is 48 square feet. Hence the heating surface of the tubes $= 526{\cdot}5 - 48 = 478{\cdot}5$ square feet. The inside diameters of the tubes are generally about $1\frac{3}{4}$ inches; and therefore the circumference of a section of these tubes, according to the table, is 5·4978 inches. Hence, supposing the tube to be $8\frac{1}{2}$ feet long, the surface of one $= 5{\cdot}4978 \times 8\frac{1}{2} \div 12 = {\cdot}45815 \times 8\frac{1}{2} = 3{\cdot}8943$ square feet; and, therefore, the number of tubes $= 478{\cdot}5 \div 3{\cdot}8943 = 123$ nearly. The amount of heating surface, however, like that of grate surface, is properly a function of the quantity of steam to be raised, and the proportions of both, given hereafter, will be found to answer well for boilers of every description.

### AREA OF WATER-LEVEL.

This, of course, varies with the different circumstances of the boiler. The average area may be found from the following rule.

RULE.—*To find the area of the water-level.*—Multiply the diameter of the cylinder in inches by 2·08. The product is the area of the water-level in square feet.

Required the area of the water-level for a locomotive engine, whose cylinders' diameters are 14 inches.

In this case we have, according to the rule,

Area of water-level = $14 \times 2{\cdot}08 = 29{\cdot}12$ square feet.

### CUBICAL CONTENT OF WATER IN BOILER.

This, of course, varies not only in different boilers, but also in the same boiler at different times. The following rule is supposed to give the average quantity of water in the boiler.

RULE.—*To find the cubical content of the water in the boiler.*—Multiply the square of the diameter of the cylinder in inches by 9: divide the product by 40. The quotient is the cubical content of the water in the boiler in cubic feet.

Required the average cubical content of the water in the boiler of a locomotive engine, the diameters of the cylinders being 14 inches. In this example we have, according to the rule,

Cubical content of water = $9 \times 14^2 \div 40 = 44{\cdot}1$ cubic feet.

### CONTENT OF FEED-PUMP.

In the locomotive engine, the feed-pump is generally attached to the cross-head, and consequently it has the same stroke as the piston. As we have mentioned before, the stroke of the locomotive engine is generally in practice 18 inches. Hence, assuming the stroke of the feed-pump to be constantly 18 inches, it only remains for us to determine the diameter of the ram. It may be found from the following rule.

RULE.—*To find the diameter of the feed-pump ram.*—Multiply the square of the diameter of the cylinder in inches by ·011. The product is the diameter of the ram in inches.

Required the diameter of the ram for the feed-pump for a locomotive engine whose diameter of cylinder is 14 inches. In this example we have, according to the rule,

Diameter of ram = $\cdot 011 \times 14^2 = \cdot 011 \times 196 = 2{\cdot}156$ inches, or between 2 and $2\frac{1}{4}$ inches.

### CUBICAL CONTENT OF STEAM ROOM.

The quantity of steam in the boiler varies not only for different boilers, but even for the same boiler in different circumstances. But when the locomotive is in motion, there is usually a certain proportion of the boiler filled with the steam. Including the dome and the steam pipe, the content of the steam room will be found usually to be somewhat less than the cubical content of the water.

But as it is desirable that it should be increased, we give the following rule.

RULE.—*To find the cubical content of the steam room.*—Multiply the square of the diameter of the cylinder in inches by 9; divide the product by 40. The quotient is the cubical content of the steam room in cubic feet.

Required the cubical content of the steam room in a locomotive boiler, the diameters of the cylinders being 12 inches.

In this example we have, according to the rule,

Cubical content of steam room $= 9 \times 12^2 \div 40 = 9 \times 144 \div 40 =$ 32·4 cubic feet.

### CUBICAL CONTENT OF INSIDE FIRE-BOX ABOVE FIRE-BARS.

The following rule determines the cubical content of fire-box usually given in practice.

RULE.—*To find the cubical content of inside fire-box above fire-bars.*—Divide the square of the diameter of the cylinder in inches by 4. The quotient is the content of the inside fire-box above fire-bars in cubic feet.

Required the content of inside fire-box above fire-bars in a locomotive engine, when the diameters of the cylinders are each 15 inches.

In this example we have, according to the rule,

Content of inside fire-box above fire-bars $= 15^2 \div 4 = 225 \div 4 =$ $56\frac{1}{4}$ cubic feet.

### THICKNESS OF THE PLATES OF BOILER.

In general, the thickness of the plates of the locomotive boiler is $\frac{3}{8}$ inch. In some cases, however, the thickness is only $\frac{5}{16}$ inch.

### INSIDE DIAMETER OF STEAM PIPE.

The diameter usually given to the steam pipe of the locomotive engine may be found from the following rule.

RULE.—*To find the diameter of the steam pipe of the locomotive engine.*—Multiply the square of the diameter of the cylinder in inches by ·03. The product is the diameter of the steam pipe in inches.

Required the diameter of the steam pipe of a locomotive engine, the diameter of the cylinder being 13 inches. Here, according to the rule, diameter of steam pipe $= ·03 \times 13^2 = ·03 \times 169 = 5·07$ inches; or a very little more than 5 inches. The steam pipe is usually made too small in engines intended for high speeds.

### DIAMETER OF BRANCH STEAM PIPES.

The following rule gives the usual diameter of the branch steam pipe for locomotive engines.

RULE.—*To find the diameter of the branch steam pipe for the locomotive engine.*—Multiply the square of the diameter of the cylinder in inches by ·021. The product is the diameter of the branch steam pipe for the locomotive engine in inches.

Required the diameter of the branch steam pipes for a locomotive engine, when the cylinder's diameter is 15 inches. Here, according to the rule, diameter of branch pipe $= \cdot 021 \times 15^2 = \cdot 021 \times 225 = 4\cdot 725$ inches, or about $4\frac{3}{4}$ inches.

### DIAMETER OF TOP OF BLAST PIPE.

The diameter of the top of the blast pipe may be found from the following rule.

RULE.—*To find the diameter of the top of the blast pipe.*—Multiply the square of the diameter of the cylinder in inches by 0·17. The product is the diameter of the top of the blast pipe in inches.

The diameter of a locomotive engine is 13 inches; required the diameter of the blast pipe at top. Here, according to the rule, diameter of blast pipe at top $= \cdot 017 \times 13^2 = \cdot 017 \times 169 = 2\cdot 873$ inches, or between $2\frac{3}{4}$ and 3 inches; but the orifice of the blast pipe should always be made as large as the demands of the blast will permit.

### DIAMETER OF FEED PIPES.

There appear to be no theoretical considerations which would lead us to determine exactly the proper size of the feed pipes. Judging from practice, however, the following rule will be found to give the proper dimensions.

RULE.—*To find the diameter of the feed pipes.*—Multiply the diameter of the cylinder in inches by ·141. The product is the proper diameter of the feed pipes.

Required the diameter of the feed pipes for a locomotive engine, the diameter of the cylinder being 15 inches.

In this example we have, according to the rule,

Diameter of feed-pipe $= 15 \times \cdot 141 = 2\cdot 115$ inches,

or between 2 and $2\frac{1}{4}$ inches.

### DIAMETER OF PISTON ROD.

The diameter of the piston rod for the locomotive engine is usually about one-seventh the diameter of the cylinder. Making practice our guide, therefore, we have the following rule.

RULE.—*To find the diameter of the piston rod for the locomotive engine.*—Divide the diameter of the cylinder in inches by 7. The quotient is the diameter of the piston rod in inches.

The diameter of the cylinder of a locomotive engine is 15 inches; required the diameter of the piston rod. Here, according to the rule, diameter of piston rod $= 15 \div 7 = 2\frac{1}{7}$ inches.

### THICKNESS OF PISTON.

The thickness of the piston in locomotive engines is usually about two-sevenths of the diameter of the cylinder. Making practice our guide, therefore, we have the following rule.

RULE.—*To find the thickness of the piston in the locomotive engine.*—Multiply the diameter of the cylinder in inches by 2; divide the product by 7. The quotient is the thickness of the piston in inches.

The diameter of the cylinder of a locomotive engine is 14 inches; required the thickness of the piston. Here, according to the rule, thickness of piston $= 2 \times 14 \div 7 = 4$ inches.

### DIAMETER OF CONNECTING RODS AT MIDDLE.

The following rule gives the diameter of the connecting rod at middle. The rule, we may remark, is entirely founded on practice.

RULE.—*To find the diameter of the connecting rod at middle of the locomotive engine.*—Multiply the diameter of the cylinder in inches by ·21. The product is the diameter of the connecting rod at middle in inches.

Required the diameter of the connecting rods at middle for a locomotive engine, the diameter of the cylinders being twelve inches.

For this example we have, according to the rule,

Diameter of connecting rods at middle $= 12 \times \cdot 21 = 2 \cdot 52$ inches, or $2\frac{1}{2}$ inches.

### DIAMETER OF BALL ON CROSS-HEAD SPINDLE.

The diameter of the ball on the cross-head spindle may be found from the following rule.

RULE.—*To find the diameter of the ball on cross-head spindle of a locomotive engine.*—Multiply the diameter of the cylinder in inches by ·23. The product is the diameter of the ball on the cross-head spindle.

Required the diameter of the ball on the cross-head spindle of a locomotive engine, when the diameter of the cylinder is 15 inches. Here, according to the rule,

Diameter of ball $= \cdot 23 \times 15 = 3 \cdot 45$ inches, or nearly $3\frac{1}{2}$ inches.

### DIAMETER OF THE INSIDE BEARINGS OF THE CRANK AXLE.

It is obvious that the inside bearings of the crank axle of the locomotive engine correspond to the paddle-shaft journal of the marine engine, and to the fly-wheel shaft journal of the land-engine. We may conclude, therefore, that the proper diameter of these bearings ought to depend jointly upon the length of the stroke and the diameter of the cylinder. In the locomotive engine the stroke is usually 18 inches, so that we may consider that the diameter of the bearing depends solely upon the diameter of the cylinder. The following rule will give the diameter of the inside bearing.

RULE.—*To find the diameter of the inside bearing for the locomotive engine.*—Extract the cube root of the square of the diameter of the cylinder in inches; multiply the result by ·96. The product is the proper diameter of the inside bearing of the crank axle for the locomotive engine.

Required the diameter of the inside bearing of the crank axle for a locomotive engine whose cylinders are of 13-inch diameters. In this example we have, according to the rule,

13 = diameter of cylinder in inches.
13
169 = square of the diameter of cylinder.

| | | |
|---|---|---|
| 0 | 0 | 169(5·5289 = $\sqrt[3]{169}$ |
| 5 | 25 | 125 |
| 5 | 25 | 44000 |
| 5 | 50 | 41375 |
| 10 | 7500 | 2625 |
| 5 | 775 | 1820 |
| 150 | 8275 | 805 |
| 5 | 800 | 726 |
| 155 | 9075 | 79 |
| 5 | 3 | |
| 160 | 910 | |
| 5 | 3 | |
| 165 | 913 | |

and diameter of bearing = 5·5289 × ·96 = 5·31 inches nearly; or between $5\frac{1}{4}$ and $5\frac{1}{2}$ inches.

### DIAMETER OF THE OUTSIDE BEARINGS OF THE CRANK AXLE.

The crank axle, in addition to resting upon the inside bearings, is sometimes also made to rest partly upon outside bearings. These outside bearings are added only for the sake of steadiness, and they do not need to be so strong as the inside bearings. The proper size of the diameter of these bearings may be found from the following rule.

RULE.—*To find the diameter of outside bearings for the locomotive engine.*—Multiply the square of the diameters of the cylinders in inches by ·396; extract the cube root of the product. The result is the diameter of the outside bearings in inches.

Required the proper diameter of the outside bearings for a locomotive engine, the diameter of its cylinders being 15 inches.

In this example we have, according to the rule,

15 = diameter of cylinders in inches.
15
225 = square of diameter of cylinder.
·396 = constant multiplier.
89·1

| | | |
|---|---|---|
| 0 | 0 | 89·1(4·466 = $\sqrt[3]{89·1}$ |
| 4 | 16 | 64 |
| 4 | 16 | 25100 |
| 4 | 32 | 21184 |
| 8 | 4800 | 3916 |
| 4 | 496 | 3528 |
| 120 | 5296 | 388 |
| 4 | 512 | 358 |
| 124 | 5808 | |
| 4 | 8 | |
| 128 | 588 | |
| 4 | 8 | |
| 132 | 596 | |

Hence diameter of outside bearing = 4·466 inches, or very nearly $4\frac{1}{2}$ inches.

### DIAMETER OF PLAIN PART OF CRANK AXLE.

It is usual to make the plain part of crank axle of the same sectional area as the inside bearings. Hence, to determine the sectional area of the plain part when it is cylindrical, we have the following rule.

RULE.—*To determine the diameter of the plain part of crank axle for the locomotive engine.*—Extract the cube root of the square of the diameter of the cylinder in inches; multiply the result by ·96. The product is the proper diameter of the plain part of the crank axle of the locomotive engine in inches.

Required the diameter of the plain part of the crank axle for the locomotive engine, whose cylinders' diameters are 14 inches. In this example we have, according to the rule,

| | |
|---|---|
| 14 | = diameter of cylinder in inches. |
| 14 | |
| 196 | = square of the diameter of cylinder. |

| | | |
|---|---|---|
| 0 | ·0 | 196(5·808 = $\sqrt[3]{196}$ |
| 5 | 25 | 125 |
| 5 | 25 | 71·000 |
| 5 | 50 | 70·112 |
| 10 | 7500 | ·888 |
| 5 | 1264 | |
| 150 | 8764 | |
| 8 | 1328 | |
| 158 | 10092 | |
| 8 | | |
| 166 | | |
| 8 | | |
| 174 | | |

Hence the plain part of crank axle = 5·808 × ·96 = 5·58 nearly, or a little more than 5½ inches.

### DIAMETER OF CRANK PIN.

The following rule gives the proper diameter of the crank pin. It is obvious that the crank pin of the locomotive engine is not altogether analogous to the crank pin of the marine or land engine, and, like them, ought to depend upon the diameter of the cylinder, as it is usually formed out of the solid axle.

RULE.—*To find the diameter of the crank pin for the locomotive engine.*—Multiply the diameter of the cylinder in inches by ·404. The product is the diametor of the crank pin in inches.

Required the diameter of the crank pin of a locomotive engine whose cylinders' diameters are 15 inches.

In this example we have, according to the rule,

Diameter of crank pin = 15 × ·404 = 6·06 inches, or about 6 inches.

### LENGTH OF CRANK PIN.

The length of the crank pin usually given in practice may be found from the following rule.

RULE.—*To find the length of the crank pin.*—Multiply the diameter of the cylinder in inches by ·233. The product is the length of the crank pins in inches.

Required the length of the crank pins for a locomotive engine with a diameter of cylinder of 13 inches.

In this example we have, according to the rule,

Length of crank pin = 13 × ·233 = 3·029 inches,

or about 3 inches. The part of the crank axle answering to the crank pin is usually rounded very much at the corners, both to give additional strength, and to prevent side play.

These then are the chief dimensions of locomotive engines according to the practice most generally followed. The establishment of express trains and the general exigencies of steam locomotion are daily introducing innovations, the effect of which is to make the engines of greater size and power: but it cannot be said that a plan of locomotive engine has yet been contrived that is free from grave objections. The most material of these defects is the necessity that yet exists of expending a large proportion of the power in the production of a draft; and this evil is traceable to the inadequate area of the fire-grate, which makes an enormous rush of air through the fire necessary to accomplish the combustion of the fuel requisite for the production of the steam. To gain a sufficient area of fire-grate, an entirely new arrangement of engine must be adopted: the furnace must be greatly lengthened, and perhaps it may be found that short upright tubes, or the very ingenious arrangement of Mr. Dimpfell, of Philadelphia, may be introduced with advantage. Upright tubes have been found to be more effectual in raising steam than horizontal tubes; but the tube plate in the case of upright tubes would be more liable to burn.

We here give the preceding rules in formulas, in the belief that those well acquainted with algebraic symbols prefer to have a rule expressed as a formula, as they can thus see at once the different operations to be performed. In the following formulas we denote the diameter of the cylinder in inches by D.

LOCOMOTIVE ENGINE.—PARTS OF THE CYLINDER.

Area of induction ports, in square inches = $\cdot068 \times D^2$.
Area of eduction ports, in square inches = $\cdot128 \times D^2$.
Breadth of bridge between ports between $\frac{3}{4}$ inch and 1 inch.

LOCOMOTIVE ENGINE.—PARTS OF BOILER.

Diameter of boiler, in inches = $3\cdot11 \times D$.
Length of boiler between 8 feet and 12 feet.
Diameter of steam dome, inside, in inches = $1\cdot43 \times D$.
Height of steam dome = $2\frac{1}{2}$ feet.
Diameter of safety valve, in inches = $D \div 4$.
Diameter of valve spindle, in inches = $\cdot076 \times D$.
Diameter of chimney, in inches = D.
Area of fire-grate, in square feet = $\cdot77 \times D$.
Area of heating surface, in square feet = $5 \times D^2 \div 2$.
Area of water level, in square feet = $2\cdot08 \times D$.
Cubical content of water in boiler, in cubic feet = $9 \times D^2 \div 40$.
Diameter of feed-pump ram, in inches = $\cdot011 \times D^2$.
Cubical content of steam room, in cubic feet = $9 \times D^2 \div 40$.
Cubical content of inside fire-box above fire bars, in cubic feet = $D^2 \div 4$.
Thickness of the plates of boiler = $\frac{3}{8}$ inch.

LOCOMOTIVE ENGINE.—DIMENSIONS OF SEVERAL PIPES.

Inside diameter of steam pipe, in inches = $\cdot03 \times D^2$.
Inside diameter of branch steam pipe, in inches = $\cdot021 \times D^2$.
Inside diameter of the top of blast pipe = $\cdot017 \times D^2$.
Inside diameter of the feed pipes = $\cdot141 \times D$.

LOCOMOTIVE ENGINE.—DIMENSIONS OF SEVERAL MOVING PARTS.

Diameter of piston rod, in inches = $D \div 7$.
Thickness of piston, in inches = $2\ D \div 7$.
Diameter of connecting rods at middle, in inches = $\cdot21 \times D$.
Diameter of the ball on cross-head spindle, in inches = $\cdot23 \times D$.
Diameter of the inside bearings of the crank axle, in inches = $96 \times \sqrt[3]{D^2}$.
Diameter of the plain part of crank axle, in inches = $\cdot96 \times \sqrt[3]{D^2}$.
Diameter of the outside bearings of the crank axle, in inches = $\sqrt[3]{\cdot396 \times D^2}$.
Diameter of crank pin, in inches = $\cdot404 \times D$.
Length of crank pin, in inches = $\cdot233 \times D$.

TABLE *of the Pressure of Steam, in Inches of Mercury, at different Temperatures.*

| Temperature, Fahrenheit. | Dalton. | Ure. | Young. | Ivory. | Tredgold. | Southern. | Robison. | Watt. |
|---|---|---|---|---|---|---|---|---|
| 0° | 0·08 | ... | ... | ... | ... | ... | ... | ... |
| 10 | 0·12 | ... | ... | ... | ... | ... | ... | ... |
| 20 | 0·17 | ... | 0·11 | ... | ... | ... | ... | ... |
| 32 | 0·26 | 0·20 | 0·18 | ... | 0·17 | 0·16 | 0·00 | ... |
| 40 | 0·34 | 0·25 | 0·20 | ... | 0·24 | 0·22 | 0·10 | ... |
| 50 | 0·49 | 0·36 | 0·36 | 0·36 | 0·37 | 0·33 | 0·20 | ... |
| 60 | 0·65 | 0·52 | 0·53 | ... | 0·55 | 0·48 | 0·35 | ... |
| 70 | 0·87 | 0·73 | 0·75 | 0·73 | 0·78 | 0·68 | 0·55 | 0·77 |
| 80 | 1·16 | 1·01 | 1·05 | ... | 1·11 | 0·95 | 0·82 | ... |
| 90 | 1·59 | 1·36 | 1·44 | 1·36 | 1·53 | 1·34 | 1·18 | ... |
| 100 | 2·12 | 1·86 | 1·95 | ... | 2·08 | 1·84 | 1·60 | 1·55 |
| 110 | 2·79 | 2·45 | 2·62 | 2·46 | 2·79 | 2·56 | 2·25 | ... |
| 120 | 3·63 | 3·30 | 3·46 | ... | 3·68 | 3·46 | 3·00 | ... |
| 130 | 4·71 | 4·37 | 4·54 | 4·41 | 4·81 | 4·43 | 3·95 | ... |
| 140 | 6·05 | 5·78 | 5·88 | ... | 6·21 | 5·75 | 5·15 | 5·14 |
| 150 | 7·73 | 7·53 | 7·55 | 7·42 | 7·94 | 7·46 | 6·72 | ... |
| 160 | 9·79 | 9·60 | 9·62 | ... | 10·05 | 9·52 | 8·65 | 8·92 |
| 170 | 12·31 | 12·05 | 12·14 | 12·05 | 12·60 | 12·14 | 11·05 | 11·37 |
| 180 | 15·38 | 15·16 | 15·23 | ... | 15·67 | 15·20 | 14·05 | 12·73 |
| 190 | 18·98 | 19·00 | 18·96 | 18·93 | 19·00 | ... | 17·85 | 19·00 |
| 200 | 23·51 | 23·60 | 23·44 | ... | 23·71 | ... | 22·65 | ... |
| 210 | 28·82 | 28·88 | 28·81 | 28·81 | 28·86 | ... | 28·62 | ... |
| 212 | 30·00 | 30·00 | 30·00 | 30·00 | 30·00 | 30·00 | 30·00 | 29·40 |
| 220 | 35·18 | 35·54 | 35·19 | ... | 34·92 | ... | 35·8 | 33·65 |
| 230 | 44·60 | 43·10 | 42·47 | 42·63 | 42·00 | ... | 44·5 | 40 |
| 240 | 53·45 | 51·70 | 51·66 | ... | 50·24 | ... | 54·9 | 49·0 |

TABLE *of the Temperature of Steam at different Pressures in Atmospheres.*

| Pressure in Atmospheres. | French Academy. | Dr. Ure. | Young. | Ivory. | Tredgold. | Southern. | Robison. | Watt. | Franklin Institute. |
|---|---|---|---|---|---|---|---|---|---|
| 1st At. | 212·0° | 212° | 212° | 212° | 212° | ... | 212° | 212° | 212° |
| 2d At. | 250·5 | 250·0 | 240·3 | 249 | 250 | 250·3 | ... | 252·5 | 250·0 |
| 3d At. | 275·2 | 275·0 | 271 | ... | 274 | ... | 267 | ... | 275·2 |
| 4th At. | 293·7 | 291·5 | 288 | 290 | 294 | 293·4 | ... | ... | 291·5 |
| 5th At. | 308·8 | 304·5 | 302 | ... | 309 | ... | ... | ... | 304·5 |
| 6th At. | 320·4 | 315·5 | ... | ... | 322 | ... | ... | ... | 315·5 |
| 7th At. | 331·7 | 325·5 | ... | ... | ... | ... | ... | ... | 326·5 |
| 8th At. | 342·0 | 336·0 | ... | 337 | 342 | 343·6 | ... | ... | 336·0 |
| 9th At. | 350·0 | 345·0 | ... | ... | ... | ... | ... | ... | 345·0 |
| 10th At. | 358·9 | ... | ... | ... | ... | ... | ... | ... | 352·5 |
| 11th At. | 366·8 | ... | ... | ... | ... | ... | ... | ... | ... |
| 12th At. | 374·0 | ... | ... | ... | 372 | ... | ... | ... | ... |
| 13th At. | 380·6 | ... | ... | ... | ... | ... | ... | ... | ... |
| 14th At. | 386·9 | ... | ... | ... | ... | ... | ... | ... | ... |
| 15th At. | 392·8 | ... | ... | ... | ... | ... | ... | ... | 383·8 |
| 16th At. | 398·5 | ... | ... | ... | ... | ... | ... | ... | ... |
| 17th At. | 403·8 | ... | ... | ... | ... | ... | ... | ... | ... |
| 18th At. | 408·9 | ... | ... | ... | ... | ... | ... | ... | ... |
| 19th At. | 413·9 | ... | ... | ... | ... | ... | ... | ... | ... |
| 20th At. | 418·5 | ... | ... | ... | 414 | ... | ... | ... | 405 |
| 30th At. | 457·2 | ... | ... | ... | ... | ... | ... | ... | ... |
| 40th At. | 466·6 | ... | ... | ... | ... | ... | ... | ... | ... |
| 50th At. | 510·6 | ... | ... | ... | ... | ... | ... | ... | ... |

TABLE *of the Expansion of Air by Heat.*

| Fahren. | | Fahren. | | Fahren. | |
|---|---|---|---|---|---|
| 32 | 1000 | 61 | 1069 | 90 | 1132 |
| 33 | 1002 | 62 | 1071 | 91 | 1134 |
| 34 | 1004 | 63 | 1073 | 92 | 1136 |
| 35 | 1007 | 64 | 1075 | 93 | 1138 |
| 36 | 1009 | 65 | 1077 | 94 | 1140 |
| 37 | 1012 | 66 | 1030 | 95 | 1142 |
| 38 | 1015 | 67 | 1080 | 96 | 1144 |
| 39 | 1018 | 68 | 1034 | 97 | 1146 |
| 40 | 1021 | 69 | 1087 | 98 | 1148 |
| 41 | 1023 | 70 | 1089 | 99 | 1150 |
| 42 | 1025 | 71 | 1091 | 100 | 1152 |
| 43 | 1027 | 72 | 1093 | 110 | 1173 |
| 44 | 1030 | 73 | 1095 | 120 | 1194 |
| 45 | 1032 | 74 | 1097 | 130 | 1215 |
| 46 | 1034 | 75 | 1099 | 140 | 1235 |
| 47 | 1036 | 76 | 1101 | 150 | 1255 |
| 48 | 1038 | 77 | 1104 | 160 | 1275 |
| 49 | 1040 | 78 | 1106 | 170 | 1295 |
| 50 | 1043 | 79 | 1108 | 180 | 1315 |
| 51 | 1045 | 80 | 1110 | 190 | 1334 |
| 52 | 1047 | 81 | 1112 | 200 | 1364 |
| 53 | 1050 | 82 | 1114 | 210 | 1372 |
| 54 | 1052 | 83 | 1116 | 212 | 1376 |
| 55 | 1055 | 84 | 1118 | 302 | 1558 |
| 56 | 1057 | 85 | 1121 | 392 | 1739 |
| 57 | 1059 | 86 | 1123 | 482 | 1919 |
| 58 | 1062 | 87 | 1125 | 572 | 2098 |
| 59 | 1064 | 88 | 1128 | 680 | 2312 |
| 60 | 1066 | 89 | 1130 | | |

## STRENGTH OF MATERIALS.

The chief materials, of which it is necessary to record the strength in this place, are cast and malleable iron; and many experiments have been made at different times upon each of these substances, though not with any very close correspondence. The following is a summary of them:—

| Materials. | C | S | E | M |
|---|---|---|---|---|
| Iron, cast { from | 16300 | 8100 | 69120000 | 5530000 |
| { to | 36000 | | | |
| —— Malleable | 60000 | 9000 | 91440000 | 6770000 |
| —— Wire | 80000 | | | |

The first column of figures, marked C, contains the mean strength of cohesion on an inch section of the material; the second, marked S, the constant for transverse strains; the third, marked E, the constant for deflections; and the fourth, marked M, the modulus of elasticity. The introduction of the hot blast iron brought with it the impression that it was less strong than that previously in use, and the experiments which had previously been confided in as giving results near enough the truth, for all practical purposes, were no longer considered to be applicable to the new state of things. New experiments were therefore made. The following Table gives, we have no doubt, results as nearly correct as can be required or attained:—

## RESULTS OF EXPERIMENTS ON THE STRENGTH AND OTHER PROPERTIES OF CAST IRON.

In the following Table each bar is reduced to exactly one inch square; and the transverse strength, which may be taken as a criterion of the value of each Iron, is obtained from a mean between the experiments upon it;—first on bars 4 ft. 6 in. between the supports; and next on those of half the length, or 2 ft. 3 in. between the supports. All the other results are deduced from the 4 ft. 6 in. bars. In all cases the weights were laid on the middle of the bar.

| Name of Iron. | Specific Gravity. | Modulus of elasticity in lbs. per sq. inch, or stiffness. | Breaking weight in lbs. of bars 4 ft. 6 in. between supports. | Breaking weight in lbs. of bars 2 ft. 3 in. reduced to 4 ft. 6in. between supports. | Mean breaking weight in lbs. (S.) | Ultimate deflection of 4 ft. 6 in. bars, in parts of an inch. | Power of the 4 ft. 6 in. bars to resist impact. | Colour. |
|---|---|---|---|---|---|---|---|---|
| Dickerson's, Newark, N. J······ | 7·030 | 18470000 | 510 | 532 | 600 | 1·530 | 991 | Gray |
| Ponkey, No. 3. Cold Blast ······ | 7·122 | 17211000 | 567 | 595 | 581 | 1·747 | 992 | Whitish gray |
| Devon, No. 3. Hot Blast* ······ | 7·251 | 22473650 | 537 | — | 537 | 1·09 | 589 | White |
| Oldberry, No. 3. Hot Blast······ | 7·300 | 22733400 | 543 | 517 | 530 | 1·005 | 549 | White |
| Pattison, N. J. Hot Blast*······ | 7·056 | 17873100 | 520 | 534 | 527 | 1·365 | 710 | Whitish gray |
| Beaufort, No. 3. Hot Blast······ | 7·069 | 16802000 | 505 | 529 | 517 | 1·599 | 807 | Dullish gray |
| Pennsylvanian ················ | 7·8 | 15379500 | 500 | 515 | 502 | 1·815 | 889 | Dark gray |
| Bute, No. 1. Cold Blast ········ | 7·066 | 15163000 | 495 | 487 | 491 | 1·764 | 872 | Bluish gray |
| Wind Mill End, No. 2. Cold Blast | 7·071 | 16490000 | 483 | 495 | 489 | 1·581 | 765 | Dark gray |
| Old Park, No. 2. Cold Blast ···· | 7·049 | 14607000 | 441 | 529 | 485 | 1·621 | 718 | Gray |
| Beaufort, No. 2. Hot Blast······ | 7·108 | 16301000 | 478 | 470 | 474 | 1·512 | 729 | Dull gray |
| Low Moor, No. 2. Cold Blast···· | 7·055 | 14509500 | 462 | 483 | 472 | 1·852 | 855 | Dark gray |
| Buffery, No. 1. Cold Blast* ···· | 7·079 | 15381200 | 463 | — | 463 | 1·55 | 721 | Gray |
| Brimbo, No. 2. Cold Blast ······ | 7·017 | 14911666 | 466 | 453 | 459 | 1·748 | 815 | Light gray |
| Apedale, No. 2. Hot Blast ······ | 7·017 | 14852000 | 457 | 455 | 456 | 1·730 | 791 | Light gray |
| Oldberry, No. 2. Cold Blast ···· | 7·059 | 14307500 | 453 | 457 | 455 | 1·811 | 822 | Dark gray |
| Pentwyn, No. 2················ | 7·038 | 15193000 | 438 | 473 | 455 | 1·484 | 650 | Bluish gray |
| Maesteg, No. 2 ················ | 7·038 | 13959500 | 453 | 455 | 454 | 1·957 | 886 | Dark gray |
| Muirkirk, No. 1. Cold Blast*···· | 7·113 | 14003550 | 443 | 464 | 453 | 1·734 | 770 | Bright gray |
| Adelphi, No. 2. Cold Blast······ | 7·080 | 13815500 | 441 | 457 | 449 | 1·759 | 777 | Light gray |
| Blania, No. 3. Cold Blast ······· | 7·159 | 14281466 | 433 | 464 | 448 | 1·726 | 747 | Bright gray |
| Devon, No. 3. Cold Blast* ······ | 7·285 | 22907700 | 448 | — | 448 | ·790 | 353 | Light gray |
| Gartsherrie, No. 3. Hot Blast · | 7·017 | 13894000 | 427 | 467 | 447 | 1·557 | 998 | Light gray |
| Frood, No. 2. Cold Blast········ | 7·031 | 13112666 | 460 | 434 | 447 | 1·825 | 841 | Light gray |
| Lane End, No. 2. ·············· | 7·028 | 15787666 | 444 | — | 444 | 1·414 | 629 | Dark gray |
| Carron, No. 3. Cold Blast*······ | 7·094 | 16246966 | 444 | 443 | 443 | 1·336 | 593 | Gray |
| Dundyvan, No. 3. Cold Blast···· | 7·087 | 16534000 | 456 | 430 | 443 | 1·469 | 674 | Dull gray |
| Maesteg (Marked Red)·········· | 7·038 | 13971500 | 440 | 444 | 442 | 1·887 | 830 | Bluish gray |
| Corbyns Hall, No. 2············ | 7·007 | 13845866 | 430 | 454 | 442 | 1.687 | 727 | Gray |
| Pontypool, No. 2··············· | 7·080 | 13136500 | 439 | 441 | 440 | 1·857 | 816 | Dull blue |
| Wallbrook, No. 3 ·············· | 6·979 | 15394766 | 432 | 449 | 440 | 1·443 | 625 | Light gray |
| Milton, No. 3. Hot Blast········ | 7·051 | 15852500 | 427 | 449 | 438 | 1·368 | 585 | Gray |
| Buffery, No. 1. Hot Blast*······ | 6·998 | 13730500 | 436 | — | 436 | 1·64 | 721 | Dull gray |
| Level, No. 1. Hot Blast········· | 7·080 | 15452500 | 461 | 403 | 432 | 1·516 | 699 | Light gray |
| Pant, No. 2···················· | 6·975 | 15280900 | 408 | 455 | 431 | 1·251 | 511 | Light gray |
| Level, No. 2. Hot Blast········· | 7·031 | 15241000 | 419 | 439 | 429 | 1·358 | 570 | Dull gray |
| W. S. S., No. 2················ | 7·041 | 14953333 | 413 | 446 | 429 | 1·339 | 554 | Light gray |
| Eagle Foundry, No. 2. Hot Blast | 7·038 | 14211000 | 408 | 446 | 427 | 1·512 | 618 | Bluish gray |
| Elsicar, No. 2. Cold Blast······· | 6·928 | 12586500 | 446 | 408 | 427 | 2·224 | 992 | Gray |
| Varteg, No. 2. Hot Blast ······· | 7·007 | 15012000 | 422 | 430 | 426 | 1·450 | 621 | Gray |
| Coltham, No. 1. Hot Blast······ | 7·128 | 15510066 | 464 | 385 | 424 | 1·532 | 716 | Whitish gray |
| Carroll, No. 2. Cold Blast ······ | 7·069 | 17036000 | 430 | 408 | 419 | 1·231 | 530 | Gray |
| Muirkirk, No. 1. Hot Blast*····· | 6·953 | 13294400 | 417 | 419 | 418 | 1·570 | 656 | Bluish gray |
| Bierley, No. 2················· | 7·185 | 16156133 | 404 | 432 | 418 | 1·222 | 494 | Dark gray |
| Coed-Talon, No. 2. Hot Blast* ·· | 6·969 | 14322500 | 409 | 424 | 416 | 1·882 | 771 | Bright gray |
| Coed-Talon, No. 2. Cold Blast*·· | 6·955 | 14304000 | 408 | 418 | 413 | 1·470 | 600 | Gray |
| Monkland, No. 2. Hot Blast ···· | 6·916 | 12259500 | 402 | 404 | 403 | 1·762 | 709 | Bluish gray |
| Ley's Works, No. 1. Hot Blast·· | 6·957 | 11539333 | 392 | — | 392 | 1·890 | 742 | Bluish gray |
| Milton, No. 1. Hot Blast ······· | 6·976 | 11974500 | 353 | 386 | 369 | 1·525 | 538 | Gray |
| Plaskynaston, No. 2. Hot Blast · | 6·916 | 13341633 | 378 | 337 | 357 | 1·366 | 517 | Light gray |

The irons with asterisks are taken from Experiments on Hot and Cold Blast Iron.

RULE.—To find from the above Table the breaking weight in rectangular bars, generally. Calling $b$ and $d$ the breadth and depth in inches, and $l$ the distance between the supports, in feet, and putting 4·5 for 4 ft. 6 in., we have $\frac{4{\cdot}5 \times b\, d^2\, S}{l}$ = breaking weight in lbs.,—the value of S being taken from the above Table.

*For example:*—What weight would be necessary to break a bar of Low Moor Iron, 2 inches broad, 3 inches deep, and 6 feet between the supports? According to the rule given above, we have $b = 2$ inches, $d = 3$ inches, $l = 6$ feet, $S = 472$ from the Table. Then $\frac{4{\cdot}5 \times b\, d^2\, S}{l} = \frac{4{\cdot}5 \times 2 \times 3^2 \times 472}{6} = 6372$ lbs., the breaking weight.

TABLE *of the Cohesive Power of Bodies whose Cross Sectional Areas equal one Square Inch.*

| METALS. | Cohesive Power in lbs. |
|---|---|
| Swedish bar iron | 65,000 |
| Russian do | 59,470 |
| English do | 56,000 |
| Cast steel | 134,256 |
| Blistered do | 133,152 |
| Shear do | 127,632 |
| Wrought copper | 33,892 |
| Hard gun-metal | 36,368 |
| Cast copper | 19,072 |
| Yellow brass, cast | 17,968 |
| Cast iron | 17,628 |
| Tin, cast | 4,736 |
| Bismuth, cast | 3,250 |
| Lead, cast | 1,824 |
| Elastic power or direct tension of wrought iron, medium quality | 22,400 |

NOTE.—A bar of iron is extended ·000096, or nearly one ten-thousandth part of its length, for every ton of direct strain per square inch of sectional area.

## CENTRE OF GRAVITY.

*The centre of gravity* of a body is that point within it which continually endeavours to gain the lowest possible situation; or it is that point on which the body, being freely suspended, will remain at rest in all positions. The centre of gravity of a body does not always exist within the matter of which the body is composed, there being bodies of such forms as to preclude the possibility of this being the case, but it must either be surrounded by the constituent matter, or so placed that the particles shall be symmetrically situated, with respect to a vertical line in which the position of the centre occurs. Thus, the centre of gravity of a ring is not in the substance of the ring itself, but, if the ring be uniform, it will be in the axis of its circumscribing cylinder; and if the ring varies in form or density, it will be situated nearest to those parts where the weight or density is greatest. Varying the position of a body will not cause any change in the situation of the centre of gravity; for any change of position the body undergoes will only have the effect of altering the directions of the sustaining forces, which will still preserve their parallelism. When a body is suspended by any other point than its centre of gravity, it will not rest unless that centre be in the same vertical line with the point of suspension; for, in every other position, the force which is intended to insure the equilibrium will not directly oppose the resultant of gravity upon the particles of the body, and of course the equilibrium will not obtain; the directions of the forces of gravity upon the constituent particles are all parallel to one another and perpendicular to the horizon. If a heavy body be sustained by two or more forces, their lines of direction must meet either at the centre of gravity, or in the vertical line in which it occurs.

A body cannot descend or fall downwards, unless it be in such a position that by its motion the centre of gravity descends. If a body stands on a plane, and a line be drawn perpendicular to the horizon, and if this perpendicular line fall within the base of the body, it will be supported without falling; but if the perpendicular falls without the base of the body, it will overset. For when the perpendicular falls within the base, the body cannot be moved at all without raising the centre of gravity; but when the perpendicular falls without the base towards any side, if the body be moved towards that side, the centre of gravity will descend, and consequently the body will overset in that direction. If a perpendicular to the horizon from the centre of gravity fall upon the extremity of the base, the body may continue to stand, but the least force that can be applied will cause it to overset in that direction; and the nearer the perpendicular is to any side the easier the body will be made to fall on that side, but the nearer the perpendicular is to the middle of the base the firmer the body will stand. If the centre of gravity of a body be supported, the whole body is supported, and the place of the centre of gravity must be considered as the place of the body, and it is always in a line which is perpendicular to the horizon.

In any two bodies, the common centre of gravity divides the line that joins their individual centres into two parts that are to one another reciprocally as the magnitudes of the bodies. The products of the bodies multiplied by their respective distances from the common centre of gravity are equal. If a weight be laid upon any point of an inflexible lever which is supported at the ends, the pressure on each point of the support will be inversely as the respective distances from the point where the weight is applied. In a system of three bodies, if a line be drawn from the centre of gravity of any one of them to the common centre of the other two, then the common centre of all the three bodies divides the line into two parts that are to each other reciprocally as the magnitude of the body from which the line is drawn to the sum of the magnitudes of the other two; and, consequently, the single body multiplied by its distance from the common centre of gravity is equal to the sum of the other bodies multiplied by the distance of their common centre from the common centre of the system.

If there be taken any point in the straight line or lever joining the centres of gravity of two bodies, the sum of the two products of each body multiplied by its distance from that point is equal to the product of the sum of the bodies multiplied by the distance of their common centre of gravity from the same point. The two bodies have, therefore, the same tendency to turn the lever about the assumed point, as if they were both placed in their common centre of gravity. Or, if the line with the bodies moves about the assumed point, the sum of the momenta is equal to the momentum of the sum of the bodies placed at their common centre of gravity. The same property holds with respect to any number of bodies whatever, and also when the bodies are not placed in the line, but in perpendiculars to it passing through the bodies. If any plane pass through the assumed point, perpendicular to the line in which it subsists, then the distance of the common centre of gravity of all the bodies from that plain is equal to the sum of all the momenta divided by the sum of all the bodies. We may here specify the positions of the centre of gravity in several figures of very frequent occurrence.

In a straight line, or in a straight bar or rod of uniform figure and density, the position of the centre of gravity is at the middle of its length. In the plane of a triangle the centre of gravity is situated in the straight line drawn from any one of the angles to the middle of the opposite side, and at two-thirds of this line distant from the angle where it originates, or one-third distant from the base. In the surface of a trapezium the centre of gravity is in the intersections of the straight lines that join the centres of the opposite triangles made by the two diagonals. The centre of gravity of the surface of a parallelogram is at the intersection of the diagonals, or at the intersection of the two lines which bisect the figure from its opposite sides. In any regular polygon the centre of gravity is at the same point as the centre of magnitude. In a circular arc the position of the centre of gravity is distant from the centre of the circle by the measure of a fourth proportional to the arc, radius, and chord. In a semicircular arc the position of the centre of gravity is distant from the centre by the measure of a third proportional to the arc of the quadrant and the radius. In the sector of a circle the position of the centre of gravity is distant from the centre of the circle by a fourth proportional to three times the arc of the sector, the chord of the arc, and the diameter of the circle. In a circular segment, the position of the centre of gravity is distant from the centre of the circle by a space which is equal to the cube or third power of the chord divided by twelve times the area of the segment. In a semicircle the position of the centre of gravity is distant from the centre of the circle by a space which is equal to four times the radius divided by the constant number $3{\cdot}1416 \times 3 = 9{\cdot}4248$. In a parabola the position of the centre of gravity is distant from the vertex by three-fifths of the axis. In a semi-parabola the position of the centre of gravity is at the intersection of the co-ordinates, one of which is parallel to the base, and distant from it by two-fifths of the axis, and the other parallel to the axis, but distant from it by three-eighths of the semi-base.

The centres of gravity of the surface of a cylinder, a cone, and conic frustum, are respectively at the same distances from the origin as are the centres of gravity of the parallelogram, the triangle, and the trapezoid, which are sections passing along the axes of the respective solids. The centre of gravity of the surface of a spheric segment is at the middle of the versed sine or height. The centre of gravity of the convex surface of a spherical zone is at the middle of that portion of the axis of the sphere intercepted by its two bases. In prisms and cylinders the position of the centre of gravity is at the middle of the straight line that joins the centres of gravity of their opposite ends. In pyramids and cones the centre of gravity is in the straight line that joins the vertex with the centre of gravity of the base, and at three-fourths of its length from the vertex, and one-fourth from the base. In a semisphere, or semispheroid, the position of the centre of gravity is distant from the centre by three-eighths of the radius. In a parabolic conoid the position of the centre of gravity is distant from the base by one-third of the axis, or two-thirds of the axis distant from the vertex. There are several other bodies and figures of which the position of the centre of gravity is known; but as the position in those cases cannot be defined without algebra, we omit them.

## CENTRIPETAL AND CENTRIFUGAL FORCES.

*Central forces* are of two kinds, *centripetal* and *centrifugal*. *Centripetal force* is that force by which a body is attracted or impelled towards a certain fixed point as a centre, and that point towards which the body is urged is called the *centre of attraction* or the *centre of force*. *Centrifugal force* is that force by which a body endeavours to recede from the centre of attraction, and from which it would actually fly off in the direction of a tangent if it were not prevented by the action of the centripetal force. These two forces are therefore antagonistic; the action of the one being directly opposed to that of the other. It is on the joint action of these two forces that all curvilinear motion depends. Circular motion is that affection of curvilinear motion where the body is constrained to move in the circumference of a circle: if it continues to move so as to describe the entire circle, it is denominated *rotatory motion*, and the body is said to revolve in a circular orbit, the centre of which is called the centre of motion. In all circular motions the deflection or deviation from the rectilinear course is constantly the same at every point of the orbit, in which case the centripetal and centrifugal forces are equal to one another. In circular orbits the centripetal forces, by which equal bodies placed at equal distances from the centres of force are attracted or drawn towards those centres, are proportional to the quantities of matter in the central bodies. This is manifest, for since all attraction takes place towards some particular body, every particle in the attracting body must produce its individual effect; consequently, a body containing twice the quantity of matter will exert twice the attractive energy, and a body containing thrice the quantity of matter will operate with thrice the attractive force, and so on according to the quantity of matter in the attracting body.

Any body, whether large or small, when placed at the same distance from the centre of force, is attracted or drawn through equal spaces in the same time by the action of the central body. This is obvious from the consideration that although a body two or three times greater is urged with two or three times greater an attractive force, yet there is two or three times the quantity of matter to be moved; and, as we have shown elsewhere, the velocity generated in a given time is directly proportional to the force by which it is generated, and inversely as the quantity of matter in the moving or attracted body. But the force which in the present instance is the weight of the body is proportional to the quantity of matter which it contains; consequently, the velocity generated is directly and inversely proportional to the quantity of matter in the attracted body, and is, therefore, a given or a constant quantity. Hence, the centripetal force, or force towards the centre of the circular orbit, is not measured by the magnitude of the revolving body, but only by the space which it describes or passes over in a given time. When a body revolves in a circular orbit, and is retained in it by means of a centripetal force directed to the centre, the actual velocity of the revolving body at every point of its revolution is equal to that which it would acquire by falling perpendicularly with the same uniform force through one-fourth of the diameter, or one-half the radius of its orbit; and this velocity is the same as would be acquired by a second body in falling through half the radius, whilst the first body, in revolving in its orbit, describes a portion of the circumference which is equal in length to half the diameter of the circle. Consequently, if a body revolves uniformly in the circumference of a circle by means of a given centripetal force, the portion of the circumference which it describes in any time is a mean proportional between the diameter of the circle and the space which the body would descend perpendicularly in the same time, and with the same given force continued uniformly.

The *periodic time*, in the doctrine of central forces, is the time occupied by a body in performing a complete revolution round the centre, when that body is constrained to move in the circumference by means of a centripetal force directed to that point; and when the body revolves in a circular orbit, the periodic time, or the time of performing a complete revolution, is expressed by the term $\pi t \sqrt{\frac{d}{s}}$, and the velocity or space passed over in the time $t$ will be $\sqrt{d s}$; in which expressions $d$ denotes the diameter of the circular orbit described by the revolving body, $s$ the space descended in any time by a body falling perpendicularly downwards with the same uniform force, $t$ the time of descending through the space, $s$ and $\pi$ the circumference of a circle whose diameter is unity. If several bodies revolving in circles round the same or different centres be retained in their orbits by the action of centripetal forces directed to those points, the periodic times will be directly as the square roots of the radii or distances of the revolving bodies, and inversely as the square roots of the centripetal forces, or, what is the same thing, the squares of the periodic times are directly as the radii, and inversely as the centripetal forces.

## CENTRE OF GYRATION.

The *centre of gyration* is that point in which, if all the constituent particles, or all the matter contained in a revolving body, or system of bodies, were concentrated, the same angular velocity would be generated in the same time by a given force acting at any place as would be generated by the same force acting similarly on the body or system itself according to its formation.

The *angular motion* of a body, or system of bodies, is the motion of a line connecting any point with the centre or axis of motion, and is the same in all parts of the same revolving system.

In different unconnected bodies, each revolving about a centre, the angular velocity is directly proportional to the absolute velocity, and inversely as the distance from the centre of motion; so that, if the absolute velocities of the revolving bodies be proportional to their radii or distances, the angular velocities will be equal. If the axis of motion passes through the centre of gravity, then is this centre called the principal centre of gyration.

The distance of the centre of gyration from the point of suspension, or the axis of motion in any body or system of bodies, is a geometrical mean between the centres of gravity and oscillation from the same point or axis; consequently, having found the distances of these centres in any proposed case, the square root of their product will give the distance of the centre of gyration. If any part of a system be conceived to be collected in the centre of gyration of that particular part, the centre of gyration of the whole system will continue the same as before; for the same force that moved this part of the system before along with the rest will move it now without any change; and consequently, if each part of the system be collected into its own particular centre, the common centre of the whole system will continue the same. If a circle be described about the centre of gravity of any system, and the axis of rotation be made to pass through any point of the circumference, the distance of the centre of gyration from that point will always be the same.

If the periphery of a circle revolve about an axis passing through the centre, and at right angles to its plane, it is the same thing as if all the matter were collected into any one point in the periphery. And moreover, the plane of a circle or a disk containing twice the quantity of matter as the said periphery, and having the same diameter, will in an equal time acquire the same angular velocity. If the matter of a revolving body were actually to be placed in the centre of gyration, it ought either to be arranged in the circumference, or in two points of the circumference diametrically opposite to each other, and equally distant from the centre of motion, for by this means the centre of motion will coincide with the centre of gravity, and the body will revolve without any lateral force on any side. These are the chief properties connected with the centre of gyration, and the following are a few of the cases in which its position has been ascertained.

In a right line, or a cylinder of very small diameter revolving about one of its extremities, the distance of the centre of gyration from the centre of motion is equal to the length of the revolving line or cylinder multiplied by the square root of $\frac{1}{3}$. In the plane of a circle, or a cylinder revolving about the axis, it is equal to the radius multiplied by the square root of $\frac{1}{2}$. In the circumference of a circle revolving about the diameter it is equal to the radius multiplied by the square root of $\frac{1}{2}$. In the plane of a circle revolving about the diameter it is equal to one-half the radius. In a thin circular ring revolving about one of its diameters as an axis it is equal to the radius multiplied by the square root of $\frac{1}{2}$. In a solid globe revolving about the diameter it is equal to the radius multiplied by the square root of $\frac{2}{5}$. In the surface of a sphere revolving about the diameter it is equal to the radius multiplied by the square root of $\frac{2}{3}$. In a right cone revolving about the axis it is equal to the radius of the base multiplied by the square root of $\frac{3}{10}$. In all these cases the distance is estimated from the centre of the axis of motion. We shall have occasion to illustrate these principles when we come to treat of fly-wheels in the construction of the different parts of steam engines.

When bodies revolving in the circumferences of different circles are retained in their orbits by centripetal forces directed to the centres, the periodic times of revolution are directly proportional to the distances or radii of the circles, and inversely as the velocities of motion; and the periodic times, under like circumstances, are directly as the velocities of motion, and inversely as the centripetal forces. If the times of revolution are equal, the velocities and centripetal forces are directly as the distances or radii of the circles. If the centripetal forces are equal, the squares of the times of revolution and the squares of the velocities are as the distances or radii of the circles. If the times of revolution are as the radii of the circles, the velocities will be equal, and the centripetal forces reciprocally as the radii.

If several bodies revolve in circular orbits round the same or different centres, the velocities are directly as the distances or radii, and inversely as the times of revolution. The velocities are directly as the centripetal forces and the times of revolution. The squares of the velocities are proportional to the centripetal forces, and the distances or radii of the circles. When the velocities are equal, the times of revolution are proportional to the radii of the circles in which the bodies revolve, and the radii of the circles are inversely as the centripetal forces. If the velocities be proportional to the distances or radii of the circles, the centripetal forces will be in the same ratio, and the times of revolution will be equal.

If several bodies revolve in circular orbits about the same or different centres, the centripetal forces are proportional to the distances or radii of the circles directly, and inversely as the squares of the times of revolution. The centripetal forces are directly proportional to the velocities, and inversely as the times of revolution. The centripetal forces are directly as the squares of the velocities, and inversely as the distances or radii of the circles. When the centripetal forces are equal, the velocities are proportional to the times of revolution, and the distances as the squares of the times or as the squares of the velocities. When the central forces are proportional to the distances or radii of the circles, the times of revolution are equal. If several bodies revolve in circular orbits about the same or different centres, the radii of the circles are directly proportional to the centripetal forces, and the squares of the periodic times. The distances or radii of the circles are directly as the velocities and periodic times. The distances or radii of the circles are directly as the squares of the velocities, and reciprocally as the centripetal forces. If the distances are equal, the centripetal forces are directly as the squares of the velocities, and reciprocally as the squares of the times of revolution; the velocities also are reciprocally as the times of revolution. The converse of these principles and properties are equally true; and all that has been here stated in regard to centripetal forces is similarly true of centrifugal forces, they being equal and contrary to each other.

The quantities of matter in all attracting bodies, having other bodies revolving about them in circular orbits, are proportional to the cubes of the distances directly, and to the squares of the times of revolution reciprocally. The attractive force of a body is directly proportional to the quantity of matter, and inversely as the square of the distance. If the centripetal force of a body revolving in a circular orbit be proportional to the distance from the centre, a body let fall from the upper extremity of the vertical diameter will reach the centre in the same time that the revolving body describes one-fourth part of the orbit. The velocity of the descending body at any point of the diameter is proportional to the ordinate of the circle at that point; and the time of falling through any portion of the diameter is proportional to the arc of the circumference whose versed sine is the space fallen through. All the times of falling from any altitudes whatever to the centre of the orbit will be equal; for these times are equal to one-fourth of the periodic times, and these times, under the specified conditions, are equal. The velocity of the descending body at the centre of the circular orbit is equal to the velocity of the revolving body.

These are the chief principles that we need consider regarding the motion of bodies in circular orbits; and from them we are led to the consideration of bodies suspended on a centre, and made to revolve in a circle beneath the suspending point, so that when the body describes the circumference of a circle, the string or wire by which it is suspended describes the surface of a cone. A body thus revolving is called a *conical pendulum*, and this species of pendulum, or, as it is usually termed, the *governor*, is of great importance in mechanical arrangements, being employed to regulate the movements of steam engines, water-wheels, and other mechanism. As we shall have occasion to show the construction and use of this instrument when treating of the parts and proportions of engines, we need not do more at present than state the principles on which its action depends. We must, however, previously say a few words on the properties of the simple pendulum, or that which, being suspended from a centre, is made to vibrate from side to side in the same vertical plane.

### PENDULUMS.

If a pendulum vibrates in a small circular arc, the time of performing one vibration is to the time occupied by a heavy body in falling perpendicularly through half the length of the pendulum as the circumference of a circle is to its diameter. All vibrations of the same pendulum made in very small circular arcs, are made in very nearly the same time. The space described by a falling body in the time of one vibration is to half the length of the pendulum as the square of the circumference of a circle is to the square of the diameter. The lengths of two pendulums which by vibrating describe similar circular arcs are to each other as the squares of the times of vibration. The times of pendulums vibrating in small circular arcs are as the square roots of the lengths of the pendulums. The velocity of a pendulum at the lowest point of its path is proportional to the chord of the arc through which it descends to acquire that velocity. Pendulums of the same length vibrate in the same time, whatever the weights may be. From which we infer, that all bodies near the earth's surface, whether they be heavy or light, will fall through equal spaces in equal times, the resistance of the air not being considered.

The lengths of pendulums vibrating in the same time in different positions of the earth's surface are as the forces of gravity in those positions. The times wherein pendulums of the same length will vibrate by different forces of gravity are inversely as the square roots of the forces. The lengths of pendulums vibrating in different places are as the forces of gravity at those places and the squares of the times of vibration. The times in which pendulums of any length perform their vibrations are directly as the square roots of their lengths, and inversely as the square roots of the gravitating forces. The forces of gravity at different places on the earth's surface are directly as the lengths of the pendulums, and inversely as the squares of the times of vibration. These are the chief properties of a simple pendulum vibrating in a vertical plane, and the principal problems that arise in connection with it are the following, viz.:

*To find the length of a pendulum that shall make any number of vibrations in a given time;* and secondly, *having given the length of a pendulum, to find the number of vibrations it will make in any time given.*—These are problems of very easy solution, and the rules for resolving them are simply as follow:—For the first, the rule is, multiply the square of the number of seconds in the given time by the constant number 39·1015, and divide the product by the square of the number of vibrations, for the length of the pendulum in inches. For the second, it is, multiply the square of the number of seconds in the given time by the constant number 39·1393, divide the product by the given length of the pendulum in inches, and extract the square root of the quotient for the number of vibrations sought. The number 39·1015 is the length of a pendulum in inches, that vibrates seconds, or sixty times in a minute, in the latitude of Philadelphia.

Suppose a pendulum is found to make 35 vibrations in a minute; what is the distance from the centre of suspension to the centre of oscillation?

Here, by the rule, the number of seconds in the given time is 60; hence we get $60 \times 60 \times 39{\cdot}1015 = 140765{\cdot}4$, which, being divided by $35 \times 35 = 1225$, gives $140765{\cdot}4 \div 1225 = 114{\cdot}9105$ inches for the length required.

The length of a pendulum between the centre of suspension and the centre of oscillation is 64 inches; what number of vibrations will it make in 60 seconds?

By the rule we have $60 \times 60 \times 39{\cdot}1015 = 140765{\cdot}4$, which, being divided by 64, gives $140765{\cdot}4 \div 64 = 2199{\cdot}46$, and the square root of this is $2199{\cdot}46 = 46{\cdot}9$, number of vibrations sought. When the given time is a minute, or 60 seconds, as in the two examples proposed above, the product of the constant number 39·1015 by the square of the time, or 140765·4, is itself a constant quantity, which, being kept in mind, will in some measure facilitate the process of calculation in all similar cases. We now return to the consideration of the conical pendulum, or that in which the ball revolves about a vertical axis in the circumference of a circular plane which is parallel to the horizon.

## CONICAL PENDULUM.

If a pendulum be suspended from the upper extremity of a vertical axis, and be made to revolve about that axis by a conical motion, which constrains the revolving body to move in the circumference of a circle whose plane is parallel to the horizon, then the time in which the pendulum performs a revolution about the axis can easily be found.

Let CD be the pendulum in question, suspended from C, the upper extremity of the vertical axis CD, and let the ball or body B, by revolving about the said axis, describe the circle BE AH, the plane of which is parallel to the horizon; it is proposed to assign the time of description, or the time in which the body B performs a revolution about the axis CD, at the distance BD.

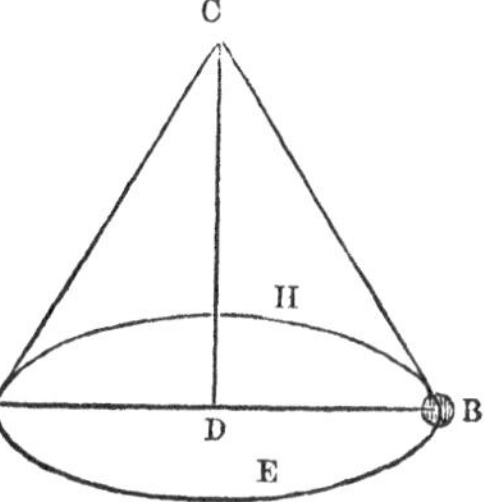


Conceive the axis CD to denote the weight of the revolving body, or its force in the direction of gravity; then, by the Composition and Resolution of Forces, CB will denote the force or tension of the string or wire that retains the revolving body in the direction CB, and BD the force tending to the centre of the plane of revolution at D. But, by the general laws of motion and forces previously laid down, if the time be given, the space described will be directly proportional to the force; but, by the laws of gravity, the space fallen perpendicularly from rest, in one second of time, is $g = 16\frac{1}{12}$ feet; consequently we have CD : BD : : $16\frac{1}{12} : \frac{16\frac{1}{12}.BD}{CD}$, the space described towards D by the force in BD in one second. Consequently, by the laws of centripetal forces, the periodic time, or the time of the body revolving in the circle BEAH, is expressed by the term $\pi\sqrt{\frac{2\cdot CD}{16\frac{1}{12}}}$, where $\pi = 3\cdot1416$, the circumference of a circle whose diameter is unity; or putting $t$ to denote the time, and expressing the height CD in feet, we get $t = 0\cdot2832 \sqrt{\frac{CD}{12 \times 32\frac{1}{6}}}$, or, by reducing the expression to its simplest form, it becomes $t = 0\cdot31986\sqrt{CD}$, where CD must be estimated in inches, and $t$ in seconds. Here we have obtained an expression of great simplicity, and the practical rule for reducing it may be expressed in words as follows:

Rule.—Multiply the square root of the height, or the distance between the point of suspension and the centre of the plane of revolution, in inches, by the constant fraction 0·31986, and the product will be the time of revolution in seconds.

In what time will a conical pendulum revolve about its vertical axis, supposing the distance between the point of suspension and the centre of the plane of revolution to be 39·1393 inches, which is the length of a simple pendulum that vibrates seconds in latitude 51° 30′?

The square root of 39·1393 is 6·2561; consequently, by the rule, we have, $6{\cdot}2561 \times 0{\cdot}31986 = 2{\cdot}0011$ seconds for the time of revolution sought. It consequently revolves 30 times in a minute, as it ought to do by the theory of the simple pendulum.

By reversing the process, the height of the cone, or the distance between the point of suspension and the centre of the plane of revolution, corresponding to any given time, can easily be ascertained; for we have only to divide the number of seconds in the given time by the constant decimal 0·31986, and the square of the quotient will be the required height in inches. Thus, suppose it were required to find the height of a conical pendulum that would revolve 30 times in a minute. Here the time of revolution is 2 seconds for $60 \div 30 = 2$; therefore, by division, it is $2 \div 0{\cdot}31986 = 6{\cdot}2527$, which, being squared, gives $6{\cdot}2527 = 39{\cdot}0961$ inches, or the length of a simple pendulum that vibrates seconds very nearly. In all conical pendulums the times of revolution, or the periodic times, are proportional to the square roots of the heights of the cones. This is manifest, for in the foregoing equation of the periodic time the numbers 6·2832 and 386, or $12 \times 32\frac{1}{6}$, are constant quantities, consequently $t$ varies as $\sqrt{CD}$.

If the heights of the cones, or the distances between the points of suspension and the centres of the planes of revolution, be the same, the periodic times, or the times of revolution, will be the same, whatever may be the radii of the circles described by the re-

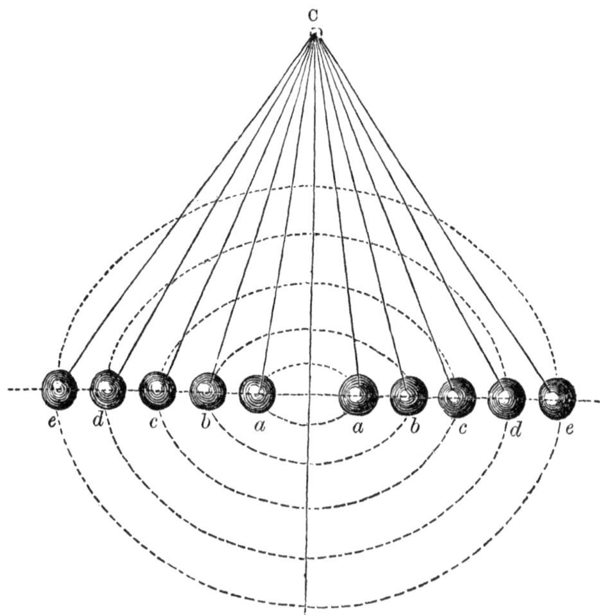


volving bodies. This will be clearly understood by contemplating the subjoined diagram, where all the pendulums $Ca$, $Cb$, $Cc$, $Cd$, and $Ce$, having the common axis CD, will revolve in the same time; and if they are all in the same vertical plane when first put in motion, they will continue to revolve in that plane, whatever be the velocity, so long as the common axis or height of the cone remains the same. This will become manifest, if we conceive an inflexible bar or rod of iron to pass through the centres of all the balls as well as the common axis, for then the bar and the several balls must all revolve in the same time; but if any one of them should be allowed to rise higher, its velocity would be increased; and if it descends, the velocity will be decreased.

Half the periodic time of a conical pendulum is equal to the time of vibration of a simple pendulum, the length of which is equal to the axis or height of the cone; that is, the simple pendulum makes two oscillations or vibrations from side to side, or it arrives at the same point from which it departed, in the same time that the conical pendulum revolves about its axis. The space descended by a falling body in the time of one revolution of the conical pendulum is equal to $3{\cdot}1416^2$ multiplied by twice the height or axis of the cone. The periodic time, or the time of one revolution is equal to the product of $3{\cdot}1416 \sqrt{2}$ multiplied by the time of falling through the height of the cone. The weight of a conical pendulum, when revolving in the circumference of a circle, bears the same proportion to the centrifugal force, or its tendency to fly off in a straight line, as the axis or height of the cone bears to the radius of the plane of revolution; consequently, when the height of the cone is equal to the radius of its base, the centripetal or centrifugal force is equal to the power of gravity.

These are the principles on which the action of the conical pendulum depends; but as we shall hereafter have occasion to consider it more at large, we need not say more respecting it in this place. Before dismissing the subject, however, it may be proper to put the reader in possession of the rules for calculating the position of the centre of oscillation in vibrating bodies, in a few cases where it has been determined, these being the cases that are of the most frequent occurrence in practice.

The *centre of oscillation* in a vibrating body is that point in the line of suspension, in which, if all the matter of the system were collected, any force applied there would generate the same angular motion in a given time as the same force applied at the centre of gravity. The centres of oscillation for several figures of very frequent use, suspended from their vertices and vibrating flatwise, are as follow :—

In a right line, or parallelogram, or a cylinder of very small diameter, the centre of oscillation is at two-thirds of the length from the point of suspension. In an isosceles triangle the centre of oscillation is at three-fourths of the altitude. In a circle it is five-fourths of the radius. In the common parabola it is five-sevenths of its altitude. In a parabola of any order it is $\left(\frac{2n + 1}{3n + 1}\right) \times$ altitude, where $n$ denotes the order of the figure.

In bodies vibrating laterally, or in their own plane, the centres of oscillation are situated as follows; namely, in a circle the centre of oscillation is at three-fourths of the diameter; in a rectangle, suspended at one of its angles, it is at two-thirds of the diagonal; in a parabola, suspended by the vertex, it is five-sevenths of the axis, increased by one-third of the parameter; in a parabola, suspended by the middle of its base, it is four-sevenths of the axis, increased by half the parameter; in the sector of a circle it is three times the arc of the sector multiplied by the radius, and divided by four times the chord; in a right cone it is four-fifths of the axis or height, increased by the quotient that arises when the square of the radius of the base is divided by five times the height; in a globe or sphere it is the radius of the sphere, plus the length of the thread by which it is suspended, plus the quotient that arises when twice the square of the radius is divided by five times the sum of the radius and the length of the suspending thread. In all these cases the distance is estimated from the point of suspension, and since the centres of oscillation and percussion are in one and the same point, whatever has been said of the one is equally true of the other.

## THE TEMPERATURE AND ELASTIC FORCE OF STEAM.

In estimating the mechanical action of steam, the intensity of its elastic force must be referred to some known standard measure, such as the pressure which it exerts against a square inch of the surface that contains it, usually reckoned by so many pounds avoirdupois upon the square inch. The intensity of the elastic force is also estimated by the inches in height of a vertical column of mercury, whose weight is equal to the pressure exerted by the steam on a surface equal to the base of the mercurial column. It may also be estimated by the height of a vertical column of water measured in feet; or generally, the elastic force of any fluid may be compared with that of atmospheric air when in its usual state of temperature and density; this is equal to a column of mercury 30 inches or $2\frac{1}{2}$ feet in height.

When the temperature of steam is increased, respect being had to its density, the elastic force, or the effort to separate the parts of the containing vessel and occupy a larger space, is also increased; and when the temperature is diminished, a corresponding and proportionate diminution takes place in the intensity of the emancipating effort or elastic power. It consequently follows that there must be some law or principle connecting the temperature of steam with its elastic force; and an intimate acquaintance with this law, in so far as it is known, must be of the greatest importance in all our researches respecting the theory and the mechanical operations of the *steam engine.*

*To find a theorem, by means of which it may be ascertained when a general law exists, and to determine what that law is, in cases where it is known to obtain.*—Suppose, for example, that it is required to assign the nature of the law that subsists between the temperature of steam and its elastic force, on the supposition that the elasticity is proportional to some power of the temperature, and unaffected by any other constant or co-efficient, except the exponent by which the law is indicated. Let E and $e$ be any two values of the elasticity, and T, $t$, the corresponding temperatures deducted from observation. It is proposed to ascertain the powers of T and $t$, to which E and $e$ are respectively proportional. Let $n$ denote the index or exponent of the required power; then by the conditions of the problem admitting that a law exists, we get, $T^n : t^n :: E : e$; but by the principles of proportion, it is $\frac{t^n}{T^n} = \frac{e}{E}$; and if this be expressed logarithmically, it is $n \times \log. \frac{t}{T} = \log. \frac{e}{E}$, and by reducing the equation in respect of $n$, it finally becomes

$$n = \frac{\log. e - \log. E}{\log. t - \log. T}.$$

The theorem that we have here obtained is in its form sufficiently simple for practical application; it is of frequent occurrence in physical science, but especially so in inquiries respecting the motion of bodies moving in air and other resisting media; and it is even applicable to the determination of the planetary motions themselves. The process indicated by it in the case that we have chosen, is simply, *To divide the difference of the logarithms of the elasticities by the difference of the logarithms of the corresponding temperatures, and the quotient will express that power of the temperature to which the elasticity is proportional.*

Take as an example the following data:—In two experiments it was found that when the temperature of steam was 250·3 and 343·6 degrees of Fahrenheit's scale, the corresponding elastic forces were 59·6 and 238·4 inches of the mercurial column respectively. From these data it is required to determine the law which connects the temperature with the elastic force on the supposition that a law does actually exist under the specified conditions. The process by the rule is as follows:

| | | |
|---|---|---|
| Greater temperature, 343·6 | log. | 2·5352941 |
| Lesser temperature, 250·3 | log. | 2·3984608 |
| Remainder | = | 0·1368333 |
| Greater elastic force, 238·4 | log. | 2·3773063 |
| Lesser elastic force, 59·6 | log. | 1·7752463 |
| Remainder | = | 0·6020600 |

Let the second of these remainders be divided by the first, as directed in the rule, and we get $n = 6020600 \div 1368333 = 4{\cdot}3998$, the exponent sought. Consequently, by taking the nearest unit, for the sake of simplicity, we shall have, according to this result, the following analogy, viz.:

$$T^{4\cdot4} : t^{4\cdot4} :: E : e;$$

that is, the elasticities are proportional to the 4·4 power of the temperatures very nearly.

Now this law is rigorously correct, as applied to the particular cases that furnished it; for if the two temperatures and one elasticity be given, the other elasticity will be found as indicated by the above analogy; or if the two elasticities and one temperature be given, the other temperature will be found by a similar process. It by no means follows, however, that the principle is general, nor could we venture to affirm that the exponent here obtained will accurately represent the result of any other experiments than those from which it is deduced, whether the temperature be higher or lower than that of boiling water; but this we learn from it, that the index which represents the law of elasticity is of a very high order, and that the general equation, whatever its form may be, must involve other conditions than those which we have assumed in the foregoing investigation. The theorem, however, is valuable to practical men, not only as being applicable to numerous other branches of mechanical inquiry, but as leading directly to the methods by which some of the best rules have been obtained for calculating the elasticity of steam, when in contact with the liquid from which it is generated.

We now proceed to apply our formula to the determination of a general law, or such as will nearly represent the class of experiments on which it rests; and for this purpose we must first assign the limits, and then inquire under what conditions the limitations take place, for by these limitations we must in a great measure be guided in determining the ultimate form of the equation which represents the law of elasticity.

The limits of elasticity will be readily assigned from the following considerations, viz.: In the first place, it is obvious that steam cannot exist when the cohesive attraction of the particles is of greater intensity than the repulsive energy of the caloric or matter of heat interposed between them; for in this case, the change from an elastic fluid to a solid may take place without passing through the intermediate stage of liquidity: hence we infer that there must be a temperature at which the elastic force is nothing, and this temperature, whatever may be its value, corresponds to the lower limit of elasticity. The higher limit will be discovered by similar considerations, for it must take place when the density of steam is the same as that of water, which therefore depends on the *modulus of elasticity* of water. The modulus of elasticity of any substance is the measure of its elastic force; that of water at 60° of temperature is 22,100 atmospheres. Thus, for instance, suppose a given quantity of water to be confined in a close vessel which it exactly fills, and let it be exposed to a high degree of temperature, then it is obvious that in this state no steam would be produced, and the force which is exerted to separate the parts of the vessel is simply the expansive force of compressed water; we therefore have the following proportion. As the expanded volume of water is to the quantity of expansion, so is the modulus of elasticity of water to the elastic force of steam of the same density as water.

Having therefore assigned the limits beyond which the elastic force of steam cannot reach, we shall now proceed to apply the principle of our formula to the determination of the general law which connects the temperature with the elastic force; and for this purpose, in addition to the notation which we have already laid down, let $c$ denote some constant quantity that affects the elasticity, and $d$ the temperature at which the elasticity vanishes; then since this temperature must be applied subtractively, we have from the foregoing principle, $c\,\mathrm{E} = (\mathrm{T} - \delta)^n$, and $c\,e = (t - \delta)^n$. From either of these equations, therefore, the constant quantity $c$ can be determined in terms of the rest when they are known; thus we have $c = \frac{(\mathrm{T} - \delta)^n}{\mathrm{E}}$, and $c = \frac{(t - \delta)^n}{e}$, and by comparing these two independent values of $c$, the value of $n$ becomes known; for $\frac{(\mathrm{T} - \delta)^n}{\mathrm{E}} = \frac{(t - \delta)^n}{e}$, and consequently

$$n = \frac{\log. e - \log. \mathrm{E}}{\log. (t - \delta) - \log. (\mathrm{T} - \delta).} \quad . \; . \; . \; . \; (\mathrm{A}).$$

In this equation the value of the symbol $\delta$ is unknown; in order therefore to determine it, we must have another independent expression for the value of $n$; and in order to this, let the elasticities E and $e$ become E′ and $e'$ respectively; while the corresponding temperatures T and $t$ assume the values T′ and $t'$; then by a similar process to the above, we get $\frac{(\mathrm{T}' - \delta)^n}{\mathrm{E}'} = \frac{(t' - \delta)}{e'}$, and

$$n = \frac{\log. e' - \log. \mathrm{E}'}{\log. (t' - \delta) - \log. (\mathrm{T}' - \delta)} \quad . \; . \; . \; (\mathrm{B}).$$

Let the equations (A) and (B) be compared with each other, and we shall then have an expression involving only the unknown quantity $\delta$, for it must be understood that the several temperatures with their corresponding elasticities are to be deduced from experiment; and in consequence, the law that we derive from them must be strictly empirical; thus we have

$$\frac{\log. e - \log. \mathrm{E}.}{\log. (t - \delta) - \log. (\mathrm{T} - \delta)} = \frac{\log. e' - \log. \mathrm{E}}{\log. (t' - \delta) - \log. (\mathrm{T}' - \delta)} \quad . \; . \; (\mathrm{C}).$$

We have no direct method of reducing expressions of this sort, and the usual process is therefore by approximation, or by the rule of trial and error, and it is in this way that the value of the quantity $\delta$ must be found; and for the purpose of performing the reduction, we shall select experiments performed with great care, and may consequently be considered as representing the law of elasticity with very great nicety.

T = 212·0 Fahrenheit E = 29·8 inches of mercury.
$t$ = 250·3 $e$ = 59·6
T′ = 293·4 E′ = 119·2
$t'$ = 343·6 $e'$ = 238·4

Therefore, by substituting these numbers in equation (C), and making a few trials, we find that $\delta = -50°$, and substituting this in either of the equations (A) or (B), we get $n = 5·08$; and finally, by substituting these values of $\delta$ and $n$ in either of the expressions for the constant quantity $c$, we get $c = 64674730000$, the 5·08 root of which is 134·27 very nearly; hence we have

$$F = \left\{\frac{t + 50}{134·27}\right\}^{5·08} \quad . \quad . \quad . \quad . \text{ (D)}.$$

Where the symbol F denotes generally the elastic force of the steam in inches of mercury, and $t$ the corresponding temperature in degrees of Fahrenheit's thermometer, the logarithm of the denominator of the fraction is 2·1279717, which may be used as a constant in calculating the elastic force corresponding to any given temperature. We have thus discovered a rule of a very simple form; it errs in defect; but this might have been remedied by assuming two points near one extremity of the range of experiment, and two points near the other extremity; and by substituting the observed numbers in equation (C), different constants and a more correct exponent would accordingly have been obtained. Mr. Southern has, by pursuing a method somewhat analogous to that which is here described, found his experiments to be very nearly represented by

$$F = \left\{\frac{t + 51·3}{135·767}\right\}^{5·13}$$

But even here the formula errs in defect, for he has found it necessary to correct it by adding the arbitrary decimal 0·1; and thus modified, it becomes

$$F = \left\{\frac{t + 51.3}{135·767}\right\}^{5·13} + 0·1. \quad . \quad . \quad . \quad . \text{ (E)}.$$

Our own formula may also be corrected by the application of some arbitrary constant of greater magnitude; but as our motive for tracing the steps of investigation in the foregoing case was to exemplify the method of determining the law of elasticity, our end is answered; for we consider it a very unsatisfactory thing merely to be put in possession of a formula purporting to be applicable to some particular purpose, without at the same time being put in possession of the method by which that formula was obtained, and the principles on which it rests. Having thus exhibited the principles and the method of reduction, the reader will have greater confidence as regards the consistency of the processes that he may be called upon to perform. The operation implied by equation (E) may be expressed in words as follows:—

RULE.—To the given temperature in degrees of Fahrenheit's thermometer add 51·3 degrees and divide the sum by 135·767; to the 5·13 power of the quotient add the constant fraction $\frac{1}{10}$, and the sum will be the elastic force in inches of mercury.

The process here described is that which is performed by the rules of common arithmetic; but since the index is affected by a fraction, it is difficult to perform in that way: we must therefore have recourse to logarithms as the only means of avoiding the difficulty. The rule adapted to these numbers is as follows:—

RULE FOR LOGARITHMS.—To the given temperature in degrees of Fahrenheit's thermometer add 51·3 degrees; then, from the logarithm of the sum subtract 2·1327940 or the logarithm of 135·767, the denominator of the fraction; multiply the remainder by the index 5·13, and to the natural number answering to the sum add the constant fraction $\frac{1}{10}$; the sum will be the elastic force in inches of mercury.

If the temperature of steam be 250·3 degrees as indicated by Fahrenheit's thermometer, what is the corresponding elastic force in inches of mercury?

By the rule it is 250·3 + 51·3 = 301·6 log. 2·4794313
constant den. = 135·767 log. 2·1327940 subtract

remainder = 0·3466373
31·5 inverted

17331865
346637
103991

natural number 60·013 log. 1·7782493

If this be increased by $\frac{1}{10}$, we get 60·113 inches of mercury for the elastic force of steam at 250·3 degrees of Fahrenheit.

By simply reversing the process or transposing equation (E), the temperature corresponding to any given elastic force can easily be found; the transformed expression is as follows, viz.:

$$t = 135{\cdot}767\,(F - 0{\cdot}1)^{\frac{1}{5.13}} - 51{\cdot}3 \quad . \quad . \quad . \quad . \quad (F).$$

Since, in consequence of the complicated index, the process of calculation cannot easily be performed by common arithmetic, it is needless to give a rule for reducing the equation in that way; we shall therefore at once give the rule for performing the process by logarithms.

RULE.—From the given elastic force in inches of mercury, subtract the constant fraction 0·1; divide the logarithm of the remainder by 5·13, and to the quotient add the logarithm 2·1327940; find the natural number answering to the sum of the logarithms, and from the number thus found subtract the constant 51·3, and the remainder will be the temperature sought.

Supposing the elastic force of steam or the vapour of water to be equivalent to the weight of a vertical column of mercury, the height of which is 238·4 inches; what is the corresponding temperature in degrees of Fahrenheit's thermometer?

Here, by proceeding as directed in the rule, we have 238·4 − 0·1 =

238·3, and dividing the logarithm of this remainder by the constant exponent 5·13, we get

| | | | |
|---|---|---|---|
| log. 238·3 ÷ 5·13 | = | 2·3771240 ÷ 5·13 = 0·4633770 | |
| constant co-efficient | = 135·767 | - - log. 2·1327940 | add |
| natural number | = 394·61 | - - - log. 2·5961710 | sum |
| constant temperature | = 51·3 | subtract | |
| required temperature | = 343·31 | degrees of Fahrenheit's thermometer. | |

The temperature by observation is 343·6 degrees, giving a difference of only 0·29 of a degree in defect. For low temperature or low pressure steam, that is, steam not exceeding the simple pressure of the atmosphere, M. Pambour gives

$$p = 0{\cdot}04948 + \left(\frac{t + 51{\cdot}3}{155{\cdot}7256}\right)^{5{\cdot}13} \quad . \quad . \quad . \quad (G).$$

In which equation the symbol $p$ denotes the pressure in pounds avoirdupois per square inch, and $t$ the temperature in degrees of Fahrenheit's thermometer. When this expression is reduced in reference to temperature, it is

$$t = 155{\cdot}7256\,(p - 0{\cdot}04948)^{\frac{1}{5{\cdot}13}} - 51{\cdot}3 \quad . \quad . \quad . \quad . \quad (H).$$

The formula of Tredgold is well known. The equation, in its original form, is

$$177 f^{\frac{1}{6}} = t + 100 \quad . \quad . \quad . \quad . \quad (I):$$

where $f$ denotes the elastic force of steam in inches of mercury, and $t$ the temperature in degrees of Fahrenheit's thermometer. The same formula, as modified and corrected by M. Millet, becomes

$$179{\cdot}0773 f^{\frac{1}{6}} = t + 103 \quad . \quad . \quad . \quad . \quad (K).$$

Dr. Young of Dublin constructed a formula which was adapted to the experiments of his countryman Dr. Dalton: it assumed a form sufficiently simple and elegant; it is thus expressed—

$$f = (1 + 0.0029\,t)^7 \quad . \quad . \quad . \quad . \quad (L):$$

where the symbol $f$ denotes the elastic force of steam expressed in atmospheres of 30 inches of mercury, and $t$ the temperature in degrees estimated above 212 of Fahrenheit. This formula is not applicable in practice, especially in high temperatures, as it deviates very widely and rapidly from the results of observation: it is chiefly remarkable as being made the basis of a numerous class of theorems somewhat varied, but of a more correct and satisfactory character. The Commission of the French Academy represented their experiments by means of a formula constructed on the same principles: it is thus expressed—

$$f = (1 + 0{\cdot}7153\,t)^5 \quad . \quad . \quad . \quad . \quad (M):$$

where $f$ denotes the elastic force of the steam expressed in atmospheres of 0·76 metres or 29·922 inches of mercury, and $t$ the temperature estimated above 100 degrees of the centigrade thermometer; but when the same formula is so transformed as to be expressed in the usual terms adopted in practice, it is

$$p = (0{\cdot}2679 + 0{\cdot}0067585\,t)^5 \quad . \quad . \quad . \quad . \quad (N):$$

where $p$ is the pressure in pounds per square inch, and $t$ the temperature in degrees of Fahrenheit's scale, estimated above 212 or simple atmospheric pressure.

The committee of the Franklin Institute adopted the exponent 6, and found it necessary to change the constant 0·0029 into 0·00333; thus modified, they represented their experiments by the equation

$$p = (0{\cdot}460467 + 0{\cdot}00521478\,t)^6 \quad . \quad . \quad . \quad . \quad (O).$$

By combining Dr. Dalton's experiments with the mean between those of the French Academy and the Franklin Institute, we obtain the following equations, the one being applicable for temperatures below 212 degrees, and the other for temperatures above that point as far as 50 atmospheres. Thus, for low pressure steam, that is, for steam of less temperature than 212, it is

$$f = \left(\frac{t + 175}{387}\right)^{7{\cdot}71307} \quad . \quad . \quad . \quad . \quad (P):$$

and for steam above the temperature of 212, it is

$$f = \left(\frac{t + 121}{333}\right)^{6{\cdot}42} \quad . \quad . \quad . \quad . \quad (Q).$$

In consequence therefore of the high and imposing authority from which these formulas are deduced, we shall adopt them in all our subsequent calculations relative to the steam engine; and in order to render their application easy and familiar, we shall translate them into rules in words at length, and illustrate them by the resolution of appropriate numerical examples; and for the sake of a systematic arrangement, we think proper to branch the subject into a series of problems, as follows:

*The temperature of steam being given in degrees of Fahrenheit's thermometer, to find the corresponding elastic force in inches of mercury.*—The problem, as here propounded, is resolved by one or other of the last two equations, and the process indicated by the arrangement is thus expressed:—

Rule.—To the given temperature expressed in degrees of Fahrenheit's thermometer, add the constant temperature 175; find the logarithm answering to the sum, from which subtract the constant 2·587711; multiply the remainder by the index 7·71307, and the product will be the logarithm of the elastic force in atmospheres of 30 inches of mercury when the given temperature is less than 212 degrees. But when the temperature is greater than 212, increase it by 121; then, from the logarithm of the temperature thus increased, subtract the constant logarithm 2·522444, multiply the remainder by the exponent 6·42, and the product will be the logarithm of the elastic force in atmospheres of 30 inches of mercury; which being multiplied by 30 will give the force in inches, or if multiplied by 14·76 the result will be expressed in pounds avoirdupois per square inch.

When steam is generated under a temperature of 187 degrees of Fahrenheit's thermometer, what is its corresponding elastic force in atmospheres of 30 inches of mercury?

In this example, the given temperature is less than 212 degrees: it will therefore be resolved by the first clause of the preceding rule, in which the additive constant is 175; hence we get

$$\begin{array}{rl} 187 + 175 = 362\ldots\text{log.} & 2{\cdot}558709 \\ \text{Constant divisor} = 387\ldots\text{log.} & \underline{2{\cdot}587711} \text{ subtract} \\ & 9{\cdot}970998 \times 7{\cdot}71307 = 9{\cdot}773393 \end{array}$$

And the corresponding natural number is 0·5934 atmospheres, or 17·802 inches of mercury, the elastic force required, or if expressed in pounds per square inch, it is 0·5934 × 14·76 = 8·76 lbs. very nearly. If the temperature be 250 degrees of Fahrenheit, the process is as follows:

$$\begin{array}{rl} 250 + 121 = 371\ldots\text{log.} & 2{\cdot}569374 \\ \text{Constant divisor} = 333\ldots\text{log.} & \underline{2{\cdot}522444} \text{ subtract} \\ & 0{\cdot}046930 \times 6{\cdot}42 = 0{\cdot}301291 \end{array}$$

And the corresponding natural number is 2·0012 atmospheres, or 60·036 inches of mercury, and in pounds per square inch it is 2·0012 × 14·76 = 29·54 lbs. very nearly.

It is sometimes convenient to express the results in inches of mercury, without a previous determination in atmospheres, and for this purpose the rule is simply as follows:

Rule.—Multiply the given temperature in degrees of Fahrenheit's thermometer by the constant coefficient 1·5542, and to the product add the constant number 271·985; then from the logarithm of the sum subtract the constant logarithm 2·587711, and multiply the remainder by the exponent 7·71307; the natural number answering to the product, considered as a logarithm, will give the elastic force in inches of mercury. This answers to the case when the temperature is less than 212 degrees; but when it is above that point proceed as follows:

Multiply the given temperature in degrees of Fahrenheit's thermometer by the constant coefficient 1·69856, and to the product add the constant number 205·526; then from the logarithm of the sum subtract the constant logarithm 2·522444, and multiply the remainder by the exponent 6·42; the natural number answering to the product considered as a logarithm, will give the elastic force in inches of mercury. Take, for example, the temperatures as assumed above, and the process, according to the rule, is as follows:

187 × 1·5542 = 290·6354
Constant = 271·985 add

Sum = 562·6204...log. 2·750216
Constant = 387..........log. 2·587711 subtract

0·162505 × 7·71307 = 1·253408

And the natural number answering to this logarithm is 17·923 inches of mercury. By the preceding calculation the result is 17·802; the slight difference arises from the introduction of the decimal constants, which in consequence of not terminating at the proper place are taken to the nearest unit in the last figure, but the process is equally true notwithstanding. For the higher temperature, we get

250 × 1·69856 = 424·640
Constant = 205·526 add

Sum = 630.166......log. 2·799456
Constant = 333............log. 2·522444 subtract

0·277011 × 6·42 = 1·778410

And the natural number answering to this logarithm is 60·036 inches of mercury, agreeing exactly with the result obtained as above.

It is moreover sometimes convenient to express the force of the steam in pounds per square inch, without a previous determination in atmospheres or inches of mercury; and when the equations are modified for that purpose, they supply us with the following process, viz.:

Multiply the given temperature by the constant coefficient 1·41666, and to the product add the constant number 247·9155; then, from the logarithm of the sum subtract the constant logarithm 2·587711, and multiply the remainder by the index 7·71307; the natural number answering to the product will give the pressure in pounds per square inch, when the temperature is less than 212 degrees; but for all greater temperatures the process is as follows:

Multiply the given temperature by the constant coefficient 1·5209, and to the product add the constant number 184·0289; then, from the logarithm of the sum subtract the constant logarithm 2·522444, and multiply the remainder by the exponent 6·42; the natural or common number answering to the product, will express the force of the steam in pounds per square inch. If any of these results be multiplied by the decimal 0·7854, the product will be the corresponding pressure in pounds per circular inch. Taking, therefore, the temperatures previously employed, the operation is as follows:

187 × 1·41666 = 264·9155
Constant = 247·9155 add

Sum = 512.8310.log. 2·709974
Constant = 387........log. 2·587711 subtract

0·122263 × 7·71307 = 0·942656

And the number answering to this logarithm is 8·763 lbs. per square inch, and 8·763 × 0·7854 = 6·8824 lbs. per circular inch, the proportion in the two cases being as 1 to 0·7554. Again, for the higher temperature, it is

250 × 1·5209 = 380·2250
Constant = 184·0289 add

Sum = 564·2539......log. 2·751475
Constant = 333............log. 2·522444 subtract

0·229031 × 6·42 = 1·470279

And the number answering to this logarithm is 29·568 lbs. per square inch, or 29568 × 0·7854 = 23·2226 lbs. per circular inch.

We have now to reverse the process, and determine the temperature corresponding to any given power of the steam, and for this purpose we must so transpose the formulas (P) and (Q), as to express the temperature in terms of the elastic force, combined with given constant numbers; but as it is probable that many of our readers would prefer to see the theorems from which the rules are deduced, we here subjoin them.

For the lower temperature, or that which does not exceed the temperature of boiling water, we get

$$t = 249f^{\frac{1}{7.71307}} - 175 \quad . \; . \; . \; . \quad (R).$$

Where $t$ denotes the temperature in degrees of Fahrenheit's thermometer, and $f$ the elastic force in inches of mercury, less than 30 inches, or one atmosphere; but when the elastic force is greater than one atmosphere, the formula for the corresponding temperature is as follows:

$$t = 196f^{\frac{1}{6.42}} - 121 \quad . \; . \; . \; . \quad (S).$$

In the construction of these formulas, we have, for the sake of simplicity, omitted the fractions that obtain in the coefficient of $f$; for since they are very small, the omission will not produce an error of any consequence; indeed, no error will arise on this account, as we retain the correct logarithms, a circumstance that enables the computer to ascertain the true value of the coefficients whenever it is necessary so to do; but in all cases of actual practice, the results derived from the integral coefficients will be quite sufficient. The rule supplied by the equations (R) and (S) is thus expressed:

When the elastic force is less than the pressure of the atmosphere, that is, less than 30 inches of the mercurial column,—

Rule.—Divide the logarithm of the given elastic force in inches of mercury, by the constant index 7·71307, and to the quotient add the constant logarithm 2·396204; then from the common or natural number answering to the sum, subtract the constant temperature 175 degrees, and the remainder will be the temperature sought in degrees of Fahrenheit's thermometer. But when the elastic force exceeds 30 inches, or one atmosphere, the following rule applies:

Divide the logarithm of the given elastic force in inches of mercury by the constant index 6·42, and to the quotient add the constant logarithm 2·292363: then, from the natural number answering to the sum subtract the constant temperature 121 degrees, and the remainder will be the temperature sought. Similar rules might be constructed for determining the temperature, when the pressure in pounds per square inch is given; but since this is a less useful case of the problem, we have thought proper to omit it. We therefore proceed to exemplify the above rules, and for this purpose we shall suppose the pressure in the two cases to be equivalent to the weight of 19 and 60 inches of mercury respectively. The operations will therefore be as follows:

Log. 19 ÷ 7·71307 = 1·278754 ÷ 7·71307 = 0·165791<br>
Constant coefficient = 249....................log. 2·396204 add

Natural number = 364·75................log. 2·561994<br>
Constant temperature = 175 subtract

Required temperature = 189·75 degrees of Fahrenheit's scale.

For the higher elastic force the operation is as follows:

Log. 60 ÷ 6·42 = 1·778151 ÷ 6·42 = 0·276969<br>
Constant coefficient = 196 ...............log. 2·292363 add

Natural number = 370·97............log. 2·569332<br>
Constant temperature = 121 subtract

Required temperature = 249·97 degrees of Fahrenheit's scale.

All the preceding results, as computed by our rules, agree as nearly with observation as can be desired: but they have all been obtained on the supposition that the steam is in contact with the liquid from which it is generated; and in this case it is evident that the steam must always attain an elastic force corresponding to the temperature; and in accordance to any increase of pressure, supposing the temperature to remain the same, a quantity of it corresponding to the degree of compression must simply be condensed into water, and in consequence will leave the diminished space occupied by steam of the original degree of tension; or otherwise to express it, if the temperature and pressure invariably correspond with each other, it is impossible to increase the density and elasticity of the steam except by increasing the temperature at the same time; and, contrariwise, the temperature cannot be increased without at the same time increasing the elasticity and density. This being admitted, it is obvious that under these circumstances the steam must always maintain its maximum of pressure and density: but if it be separated from the liquid that produces it, and if its temperature in this case be increased, it will be found not to possess a higher degree of elasticity than a volume of atmospheric air similarly confined, and heated to the same temperature. Under this new condition, the state of maximum density and elasticity ceases; for it is obvious that since no water is present, there cannot be any more steam generated by an increase of temperature; and consequently the force of the steam is only that which confines it to its original bulk, and is measured by the effort which it exerts to expand itself. Our next object, therefore, is to inquire what is the law of elasticity of steam under the conditions that we have here specified.

The specific gravity of steam, its density, and the volume which it occupies at different temperatures, have been determined by experiment with very great precision; and it has also been ascertained that the expansion of vapour by means of heat is regulated by the same laws as the expansion of the other gases, viz. that all gases expand from unity to 1·375 in bulk by 180 degrees of temperature; and again, that steam obeys the law discovered by Boyle and Mariotte, contracting in volume in proportion to the degree of pressure which it sustains. We have therefore to inquire what space a given quantity of water converted into steam will occupy at a given pressure; and from thence we can ascertain the specific gravity, density, and volume at all other pressures.

When a gas or vapour is submitted to a constant pressure, the quantity which it expands by a given rise of temperature is calculated by the following theorem,

$$v' = v\left(\frac{t' + 459}{t + 459}\right)\text{.............(T)}$$

where $t$ and $t'$ are the temperatures, and $v$, $v'$ the corresponding volumes before and after expansion; hence this rule.

Rule.—To each of the temperatures before and after expansion, add the constant experimental number 459; divide the greater sum by the lesser, and multiply the quotient by the volume at the lower temperature, and the product will give the expanded volume.

If the volume of steam at the temperature of 212 degrees of Fahrenheit be 1711 times the bulk of the water that produces it, what will be its volume at the temperature of 250·3 degrees, supposing the pressure to be the same in both cases?

Here, by the rule, we have $212 + 459 = 671$, and $250{\cdot}3 + 459 = 709{\cdot}3$; consequently, by dividing the greater by the lesser, and multiplying by the given volume, we get $\frac{709{\cdot}3}{671} \times 1711 = 1808{\cdot}66$ for the volume at the temperature of 250·3 degrees.

Again, if the elastic force at the lower temperature and the corresponding volume be given, the elastic force at the higher temperature can readily be found; for it is simply as the volume the vapour occupies at the lower temperature is to the volume at the higher temperature, or what it would become by expansion, so is the elastic force given to that required.

If the volume which steam occupies under any given pressure and temperature be given, the volume which it will occupy under any proposed pressure can readily be found by reversing the preceding process, or by referring to chemical tables containing the specific gravity of the gases compared with air as unity at the same pressure and temperature. Now, air at the mean state of the atmosphere has a specific gravity of $1\frac{2}{9}$ as compared with water at 1000; and the bulks are inversely as the specific gravities, according to the general laws of the properties of matter previously announced; hence it follows that air is 818 times the bulk of an equal weight of water, for $1000 \div 1\frac{2}{9} = 818{\cdot}18$. But, by the experiments of Dr. Dalton, it has been found that steam of the same pressure and temperature has a specific gravity of ·625 compared with air as unity; consequently, we have only to divide the number 818·18 by ·625, and the quotient will give the proportion of volume of the vapour to one of the liquid from which it is generated; thus we get $818{\cdot}18 \div {\cdot}625 = 1309$; that is, the volume of steam at 60 degrees of Fahrenheit, its force being 30 inches of mercury, is 1309 times the volume of an equal weight of water; hence it follows, from equation (T), that when the temperature increases to $t'$, the volume becomes

$$v' = 1309 \times \left(\frac{459 + t'}{459 + 60}\right) = 2{\cdot}524(459 + t');$$

and from this expression, the volume corresponding to any specified elastic force $f$, and temperature $t'$, may easily be found; for it is inversely as the compressing force: that is,

$$f : 30 : : 2{\cdot}525(459 + t') : v';$$

consequently, by working out the analogy, we get

$$v = \frac{75{\cdot}67(459 + t')}{f} \quad . \; . \; . \; . \; (\mathrm{U}).$$

By this theorem is found the volume of steam as compared with that of the water producing it, when under a pressure corresponding to the temperature. The rule in words is as follows:

RULE.—Calculate the elastic force in inches of mercury by the rule already given for that purpose, and reserve it for a divisor. To the given temperature add the constant number 459, and multiply the sum by 75·67; then divide the product by the reserved divisor, and the quotient will give the volume sought.

When the temperature of steam is 250·3 degrees of Fahrenheit's thermometer, what is the volume, compared with that of water?

The temperature being greater than 212 degrees, the force is calculated by the rule to equation (Q), and the process is as follows:

$250{\cdot}3 + 121 = 371{\cdot}3$ log. 2·5697249
Constant divisor $= 333$ log. 2·5224442 subtract
0·0472807×6·42=0·3035421
Atmosphere = 30 inches of mercury log. 1·4771213 add
Elastic force = 60·348 log. 1·7806634 } sub.
Again it is,
$459 + 250{\cdot}3 = 709{\cdot}3$ log. 2·8508300 } add
Constant coefficient $= 75{\cdot}67$ log. 1·8789237 } 4·7297537 } sub.
Volume = 889·39 times that of water, log. 2·9490903 remainder.

Thus we have given the method of calculating the elastic force of steam when the temperature is given either in atmospheres or inches of mercury, and also in pounds or the square or circular inch: we have also reversed the process, and determined the temperature corresponding to any given elastic force. We have, moreover, shown how to find the volume corresponding to different temperatures, when the pressure is constant; and, finally, we have calculated the volume, when under a pressure due to the elastic force. These are the chief subjects of calculation as regards the properties of steam; and we earnestly advise our readers to render themselves familiar with the several operations. The calculations as regards the motion of steam in the parts of an engine to produce power, will be considered in another part of the present treatise.

The equation (U), we may add, can be exhibited in a different form involving only the temperature and known quantities; for since the expressions (P) and (Q) represent the elastic force in terms of the temperature, according as it is under or above 212 degrees of Fahrenheit, we have only to substitute those values of the elastic force when reduced to inches of mercury, instead of the symbol $f$ in equation (U), and we obtain, when the temperature is less than 212 degrees,

$$\text{Vol.} = 75{\cdot}67(\text{tem.} + 459) \div ({\cdot}004016 \times \text{tem.} + {\cdot}702807)^{7{\cdot}71307} \quad \text{(V).}$$

and when the temperature exceeds 212 degrees, the expression becomes

$$\text{Vol.} = 75{\cdot}67(\text{tem.} + 459) \div {\cdot}005101 \times \text{tem.} + {\cdot}617195)^{6{\cdot}42} \quad \text{(W.)}$$

These expressions are simple in their form, and easily reduced; but, in pursuance of the plan we have adopted, it becomes necessary to express the manner of their reduction in words at length, as follows:

RULE.—When the given temperature is under 212 degrees, multiply the temperature in degrees of Fahrenheit's thermometer by the constant fraction ·004016, and to the product add the constant increment ·702807; multiply the logarithm of the sum by the index 7·71307, and find the natural or common number answering to the product, which reserve for a divisor. To the temperature add the constant number 459, and multiply the sum by the coefficient 75·67 for a dividend; divide the latter result by the former, and the quotient will express the volume of steam when that of water is unity.

Again, when the given temperature is greater than 212 degrees, multiply it by the fraction ·005101, and to the product add the constant increment ·617195; multiply the logarithm of the sum by the index 6·42, and reserve the natural number answering to the product for a divisor; find the dividend as directed above, which, being divided by the divisor, will give the volume of steam when that of the water is unity.

How many cubic feet of steam will be supplied by one cubic foot of water, under the respective temperatures of 187 and 293·4 degrees of Fahrenheit's thermometer?

Here, by the rule, we have

$187 \times 0{\cdot}004016 = 0{\cdot}750992$
Constant increment $= 0{\cdot}702807$

Sum $= 1{\cdot}453799 \log.{\cdot}1625043 \times 7{\cdot}71307 = 1{\cdot}2534069$

and the number answering to this logarithm is 17·92284, the divisor. But $187 + 459 = 646$, and $646 \times 75{\cdot}67 = 48882{\cdot}82$, the dividend; hence, by division, we get $48882{\cdot}82 \div 17{\cdot}92284 = 2727{\cdot}4$ cubic feet of steam from one cubic foot of water.

Again, for the higher temperature, it is

$293{\cdot}4 \times 0{\cdot}005101 = 1{\cdot}496633$
Constant increment $= 0{\cdot}617195$

Sum $= 2{\cdot}113828 \log.\ 0{\cdot}3250696 \times 642 = 2{\cdot}0869468$;

and the number answering to this logarithm is 122·165, the divisor. But $293{\cdot}4 + 459 = 752{\cdot}4$, and $752{\cdot}4 \times 75{\cdot}67 = 56934{\cdot}108$, the dividend; therefore, by division, we get $56934{\cdot}108 \div 122{\cdot}165 = 466{\cdot}04$ cubic feet of steam from one cubic foot of water.

The preceding is a very simple process for calculating the volume which the steam of a cubic foot of water will occupy when under a pressure due to a given temperature and elastic force; and since a knowledge of this particular is of the utmost importance in calculations connected with the steam engine, it is presumed that our readers will find it to their advantage to render themselves familiar with the method of obtaining it. The above example includes both cases of the problem, a circumstance which gives to the operation, considered as a whole, a somewhat formidable appearance: but it would be difficult to conceive a case in actual practice where the application of both the formulas will be required at one and the same time; the entire process must therefore be considered as embracing only one of the cases above exemplified; and consequently it can be performed with the greatest facility by every person who is acquainted with the use of logarithms; and those unacquainted with the application of logarithms ought to make themselves masters of that very simple mode of computation.

Another thing which it is necessary sometimes to discover in reasoning on the properties of steam as referred to its action in a steam engine, is the weight of a cubic foot, or any other quantity of it, expressed in grains, corresponding to a given temperature and pressure. Now, it has been ascertained by experiment, that when the temperature of steam is 60 degrees of Fahrenheit, and the pressure equal to 30 inches of mercury, the weight of a cubic foot in grains is 329·4; but the weight is directly proportional to the elastic force, for the elastic force is proportional to the density: consequently, if $f$ denote any other elastic force, and $w$ the weight in grains corresponding thereto, then we have

$$30 : f :: 329{\cdot}4 : w = 10{\cdot}98 f,$$

the weight of a cubic foot of vapour at the force $f$, and temperature 60 degrees of Fahrenheit. Let $t$ denote the temperature at the force $f$; then by equation (T), we have $v = \frac{459 + t}{459 + 60} = \frac{459 + t}{519}$, the volume at the temperature $t$, supposing the volume at 60 degrees to be unity; that is, one cubic foot. Now, since the densities are inversely proportional to the spaces which the vapour occupies, we have $\frac{(459 + t)}{519} : 1 :: w : w' = \frac{519w}{459 + t}$; but by the preceding analogy, the value of $w$ is $10{\cdot}98f$; therefore, by substitution, we get

$$w' = \frac{5698{\cdot}62f}{459 + t} \quad . \quad . \quad . \quad . \quad (X).$$

This equation expresses the weight in grains of a cubic foot of steam at the temperature $t$ and force $f$; and if we substitute the value of $f$, from equations (P) and (Q), reduced to inches of mercury, and modified for the two cases of temperature below and above 212 degrees of Fahrenheit, we shall obtain, in the first case,

$$w' = (0{\cdot}012324 \times \text{temp.} + 2{\cdot}155611)^{7{\cdot}71307} \div (\text{temp.} + 459) \ldots . (Y)$$

and for the second case, where the temperature exceeds 212, it is

$$w' = (0{\cdot}01962 \times \text{temp.} + 2{\cdot}37374)^{6{\cdot}42} \div (\text{temp.} + 459) \ldots (Z)$$

These two equations, like those marked (V) and (W) are sufficiently simple in their form, and offer but little difficulty in their application. The rule for their reduction when expressed in words at length, is as follows:

RULE.—When the temperature is less than 212 degrees, multiply the given temperature, in degrees of Fahrenheit's thermometer, by the fraction 0·012324, and to the product add the constant increment 2·155611; then multiply the logarithm of the sum by the index 7·71307, and from the product subtract the logarithm of the temperature, increased by 459; the natural number answering to the remainder will be the weight of a cubic foot in grains.

Again, when the temperature exceeds 212, multiply it by the fraction 0·01962, and to the product add the constant increment 2·37374; then multiply the logarithm of the sum by the index 6·42, and from the product subtract the logarithm of the temperature increased by 459; the natural number answering to the remainder will be the weight of a cubic foot in grains.

Supposing the temperatures to be as in the preceding example, what will be the weight of a cubic foot in grains for the two cases?

Here, by the rule, we have

187 × 0·012324 = 2·304588
Constant increment = 2·155611

Sum = 4·460199 log. 0·6493542 × 7·71307 = 5·0085143
187 + 459 = 646 . . . . . . log. 2·8102325, subtract

Natural number = 157·863 grains per cubic foot log. 2·1982818

For the higher temperature, it is

| | | | |
|---|---|---|---|
| 293·4 × 0·01962 = | 5·756508 | | |
| Constant increment = | 2·373740 | | |
| Sum = | 8·130248 | log. 0·9101038 × 6·42 = | 5·8428664 |
| 293·4 + 459 = | 752·4 | . . . . . log. | 2·8764488, subtract |
| Natural number = | 925·59 grains per cubic foot | . log. | 2·9664176 |

Here again the operation resolves both cases of the problem; but in practice only one of them can be required.

### THE MOTION OF ELASTIC FLUIDS.

The next subject that claims our attention is the velocity with which elastic fluids or vapours move in pipes or confined passages. It is a well-known fact in the doctrine of pneumatics, that the motion of free elastic fluids depends upon the temperature and pressure of the atmosphere; and, consequently, when an elastic fluid is confined in a close vessel, it must be similarly circumstanced with regard to temperature and pressure as it would be in an atmosphere competent to exert the same pressure upon it. The simplest and most convenient way of estimating the motion of an elastic fluid is to assign the height of a column of uniform density, capable of producing the same pressure as that which the fluid sustains in its state of confinement; for under the pressure of such a column, the velocity into a perfect vacuum will be the same as that acquired by a heavy body in falling through the height of the homogeneous column, a proper allowance being made for the contraction at the aperture or orifice through which the fluid flows.

When a passage is opened between two vessels containing fluids of different densities, the fluid of greatest density rushes out of the vessel that contains it, into the one containing the rarer fluid, and the velocity of influx at the first instant of the motion is equal to that which a heavy body acquires in falling through a certain height, and that height is equal to the difference of two uniform columns of the fluid of greatest density, competent to produce the pressures under which the fluids are originally confined; and the velocity of motion at any other instant is proportional to the square root of the difference between the heights of the uniform columns producing the pressures at that instant. Hence we infer that the velocity of motion continually decreases,—the density of the fluids in the two vessels approaching nearer and nearer to an equality, and after a certain time an equilibrium obtains, and the velocity of motion ceases.

It is abundantly confirmed by observation and experiment, that oblique action produces very nearly the same effect in the motion of elastic fluids through apertures as it does in the case of water; and it has moreover been ascertained that eddies take place under similar circumstances, and these eddies must of course have a tendency to retard the motion: it therefore becomes necessary, in all the calculations of practice, to make some allowance for the retardation that takes place in passing the orifice; and this end is most conveniently answered by modifying the constant coefficient according to the nature of the aperture through which the motion is made. Numerous experiments have been made to ascertain the effect of contraction in orifices of different forms and under different conditions, and amongst those which have proved the most successful in this respect, we may mention the experiments of Du Buat and Eytelwein, the latter of whom has supplied us with a series of coefficients, which, although not exclusively applicable to the case of the steam engine, yet, on account of their extensive utility, we take the liberty to transcribe. They are as follow:—

1. For the velocity of motion that would result from the direct unretarded action of the column of the fluid that produces it, we have ........ $3\ V = \sqrt{579h}$
2. For an orifice or tube in the form of the contracted vein ........ $10\ V = \sqrt{6084h}$
3. For wide openings having the sill on a level with the bottom of the reservoir ...
4. For sluices with walls in a line with the orifice ........ } $10\ V = \sqrt{5929h}$ (Nos. 3–5)
5. For bridges with pointed piers ........
6. For narrow openings having the sill on a level with the bottom of the reservoir ...
7. For small openings in a sluice with side walls ........ } $10\ V = \sqrt{4761h}$ (Nos. 6–9)
8. For abrupt projections ........
9. For bridges with square piers ........
10. For openings in sluices without side walls $10\ V = \sqrt{2601h}$
11. For openings or orifices in a thin plate ..... $V = \sqrt{25h}$
12. For a straight tube from 2 to 3 diameters in length projecting outwards ........ $10\ V = \sqrt{4225}$
13. For a tube from 2 to 3 diameters in length projecting inwards ........ $10\ V = \sqrt{2976{\cdot}25h}$

It is necessary to observe, that in all these equations V is the velocity of motion in feet per second, and $h$ the height of the column producing it, estimated also in feet. Nos. 1, 2, 11, 12, and 13 are those which more particularly apply to the usual passages for the steam in a steam engine; but since all the others meet their application in the every-day practice of the civil engineer, we have thought it useful to supply them.

## MOTION OF STEAM IN AN ENGINE.

We have already stated that the best method of estimating the motion of an elastic fluid, such as steam or the vapour of water, is to assign the height of a uniform column of that fluid capable of producing the pressure: the determination of this column is therefore the leading step of the inquiry; and since the elastic force of steam is usually reckoned in inches of mercury, 30 inches being equal to the pressure of the atmosphere, the subject presents but little difficulty; for we have already seen that the height of a column of water of the temperature of 60 degrees, balancing a column of 30 inches of mercury, is 34·023 feet; the corresponding column of steam must therefore be as its relative bulk and elastic force; hence we have $30 : 34{\cdot}023 : f v : h = 1{\cdot}1341 f v$, where $f$ is the elastic force of the steam in inches of mercury, $v$ the corresponding volume or bulk when that of water is unity, and $h$ the height of a uniform column of the fluid capable of producing the pressure due to the elastic force; consequently, in the case of a direct unretarded action, the velocity into a perfect vacuum, according to No. 1 of the preceding class of formulas, is $V = 8{\cdot}542 \sqrt{f v}$; but for the best form of pipes, or a conical tube in form of the contracted vein, the velocity into a vacuum, according to No. 2, becomes $V = 8{\cdot}307 \sqrt{f v}$; and for pipes of the usual construction, No. 12 gives $V = 6{\cdot}922 \sqrt{f v}$; No. 13 gives $V = 5{\cdot}804 \sqrt{f v}$; and in the case of a simple orifice in a thin plate, we get from No. 11 $V = 5{\cdot}322 \sqrt{f v}$. The consideration of all these equations may occasionally be required, but our researches will at present be limited to that arising from No. 12, as being the best adapted for general practice; and for the purpose of shortening the investigation, we shall take no further notice of the case in which the temperature of the steam is below 212 degrees of Fahrenheit; for the expression which indicates the velocity into a vacuum being independent of the elastic force, a separate consideration for the two cases is here unnecessary.

It has been shown in the equation marked (U), that the volume of steam which is generated from an unit of water, is $v = \frac{75{\cdot}67 \text{ (temp. + 459)}}{f}$; let this value of $v$ be substituted for it in the equation $V = 6{\cdot}922 \sqrt{f v}$, and we obtain for the velocity into a vacuum for the usual form of steam passages, as follows, viz.:

$$V = 60{\cdot}2143 \sqrt{(\text{temp.} + 459)}.$$

This is a very neat and simple expression, and the object determined by it is a very important one: it therefore merits the reader's utmost attention, especially if he is desirous of becoming familiar with the calculations in reference to the motion of steam. The rule which the equation supplies, when expressed in words at length, is as follows:—

Rule.—To the temperature of the steam, in degrees of Fahrenheit's thermometer, add the constant number or increment 459, and multiply the square root of the sum by 60·2143; the product will be the velocity with which the steam rushes into a vacuum in feet per second.

With what velocity will steam of 293·4 degrees of Fahrenheit's thermometer rush into a vacuum when under a pressure due to the elastic force corresponding to the given temperature.

| | | |
|---|---|---|
| By the rule it is 293·4 + 459 = 752·4 | ½ log. 1·4382244 | |
| Constant coefficient = 60·2143 | log. 1·7797018 | add |
| Velocity into a vacuum in feet per second = 1651·68 | log. 3·2179262 | |

This is the velocity into a perfect vacuum, when the motion is made through a straight pipe of uniform diameter; but when the pipe is alternately enlarged and contracted, the velocity must necessarily be reduced in proportion to the nature of the contraction; and it is further manifest, that every bend and angle in a pipe will be attended with a correspondent diminution in the velocity of motion: it therefore behoves us, in the actual construction of steam passages, to avoid these causes of loss as much as possible; and where they cannot be avoided altogether, such forms should be adopted as will produce the smallest possible retarding effect. In cases where the forms are limited by the situation and conditions of construction, such corrections should be applied as the circumstances of the case demand; and the amount of these corrections must be estimated according to the nature of the obstructions themselves. For each right-angled bend, the diminution of velocity is usually set down as being about one-tenth of its unobstructed value; but whether this conclusion be correct or not, it is at least certain that the obstruction in the case of a right-angled bend is much greater than in that of a gradually curved one. It is a very common thing, especially in steam vessels, for the main steam pipe to send off branches at right angles to each cylinder, and it is easy to see that a great diminution in the velocity of the steam must take place here. In the expansion valve chest a further obstruction must be met with, probably to the extent of reducing the velocity of the steam two-tenths of its whole amount.

These proportional corrections are not to be taken as the results of experiments that have been performed for the purpose of determining the effect of the above causes of retardation: we have no experiments of this sort on which reliance can be placed; and, in consequence, such elements can only be inferred from a comparison of the principles that regulate the motion of other fluids under similar circumstances: they will, however, greatly assist the engineer in arriving at an approximate estimate of the diminution that takes place in the velocity in passing any number of obstructions, when the precise nature of those obstructions can be ascertained. In the generality of practical cases, if the constant coefficient 60·2143 be reduced in the ratio of 650 to 450, the resulting constant 41·6868 may be employed without introducing an error of any consequence.

## OF THE ASCENT OF SMOKE AND HEATED AIR IN CHIMNEYS.

The subject of chimney flues, with the ascent of smoke and heated air, is another case of the motion of elastic fluids, in which, by a change of temperature, an atmospheric column assumes a different density from another, where no such alteration of temperature occurs. The proper construction of chimneys is a matter of very great importance to the practical engineer, for in a close fireplace, designed for the generation of steam, there must be a considerable draught to accomplish the intended purpose, and this depends upon the three following particulars, viz.:

1. The height of the chimney from the throat to the top.
2. The area of the transverse section.
3. The temperature at which the smoke and heated air are allowed to enter it.

The formula for determining the power of the chimney may be investigated in the following manner:

Put $h$ = the height in feet from the place where the flue enters to the top of the chimney,
$b$ = the number of cubic feet of air of atmospheric density that the chimney must discharge per hour,
$a$ = the area of the aperture in square inches through which $b$ cubic feet of air must pass when expanded by a change of temperature,
$v$ = the velocity of ascent in feet per second,
$t'$ = the temperature of the external air, and
$t$ = the temperature of the air to be discharged by the chimney.

Now the force producing the motion in this case is manifestly the difference between the weight of a column of the atmospheric air and another of the air discharged by the chimney: and when the temperature of the atmospheric air is at 52 degrees of Fahrenheit's thermometer, this difference will be indicated by the term $h\left(\frac{t'-t}{t'+459}\right)$; the velocity of ascent will therefore be

$v = \sqrt{64\frac{2}{5}\, h \left\{\frac{t'-t}{t'+459}\right\}}$ feet per second, and the quantity of air discharged per second will therefore be, $a\sqrt{64\frac{2}{5}\left\{\frac{t'-t}{t'+459}\right\}}$, supposing that there is no contraction in the stream of air; but it is found by experiment, that in all cases the contraction that takes place diminishes the quantity discharged, by about three-eighths of the whole; consequently, the quantity discharged per hour in cubic feet becomes

$$b = 125\cdot69\, a\sqrt{\frac{h\,(t'-t)}{t'+459}}.$$

This would be the quantity discharged, provided there were no increase of volume in consequence of the change of temperature; but air expands from $b$ to $\frac{b\,(t'+459)}{t+459}$ for $t'-t$ degrees of temperature, as has been shown elsewhere; consequently, by comparison, we have

$$\frac{b\,(t'+459)}{t+459} = 125\cdot69\, a\sqrt{\frac{h\,(t'-t)}{t+459}}.$$

From this equation, therefore, any one of the quantities which it involves can be found, when the others are given: it however supposes that there is no other cause of diminution but the contraction at the aperture; but this can seldom if ever be the case; for eddies, loss of heat, obstructions, and change of direction in the chimney, will diminish the velocity, and consequently a larger area will be required to suffer the heated air to pass. A sufficient allowance for these causes of retardation will be made, if we change the coefficient 125·69 to 100; and in this case the equation for the area of section becomes

$$a = b \sqrt{(t' + 459)^3} \div 100\,(t + 459) \sqrt{h\,(t' - t)}.$$

And if we take the mean temperature of the air of the atmosphere at 52 degrees of Fahrenheit, and make an allowance of 16 degrees for the difference of density between atmospheric air and coal smoke, our equation will ultimately assume the form

$$a = b \sqrt{(t' + 459)^3} \div 51100 \sqrt{h\,(t' - t - 16)}.$$

It has been found by experiment that 200 cubic feet of air of atmospheric density are required for the complete combustion of one pound of coal, and the consumption of ten pounds of coal per hour is usually reckoned equivalent to one horse power: it therefore appears that 2000 cubic feet of air per hour must pass through the fire for each horse power of the engine. This is a large allowance, but it is the safest plan to calculate in excess in the first instance; for the chimney may afterwards be convenient, even if considerably larger than is necessary. The rule for reducing the equation is as follows:—

RULE.—Multiply the number of horse power of the engine by the $\frac{3}{2}$ power of the temperature at which the air enters the chimney, increased by 459; then divide the product by 25·55 times the square root of the height of the chimney in feet, multiplied by the difference of temperature, less 16 degrees, and the quotient will be the area of the chimney in square inches.

Suppose the height of the chimney for a 40-horse engine to be 70 feet, what should be its area when the difference between the temperature at which the air enters the flue, and that of the atmosphere is 250 degrees?

Here, by the rule, we have,

250 + 52 = 302, the temperature at which the air enters [the flue.
Constant increment = 459

Sum = 761 .......................... log. 2·8813847
3

2)8·6441541

4·3220770
Number of horse power = 40 .......................... log. 1·6020600

5·9241370

| | | |
|---|---|---|
| | | 5·9241370 |
| 250 − 16 = 234 . . . . | log. 2·3692159 | |
| height = 70 feet . . | log. 1·8450980 | |
| | 2)4·2143139 | |
| | 2·1071569 | |
| Constant = 25·55 . . | log. 1.4073909 . . . | 3·5145478 |

Hence the area of the chimney in square inches is 256·79, log. 2·4095892; and in this way may the area be calculated for any other case; but particular care must be taken to have the data accurately determined before the calculation is begun. In the above example the particulars are merely assumed; but even that is sufficient to show the process of calculation, which is more immediately the object of the present inquiry. It is right, however, to add, that recent experiments have greatly shaken the doctrine that it is beneficial to make chimneys small at the top, though such is the way in which they are, nevertheless, still constructed, and our rules must have reference to the present practice. It appears, however, that it would be the best way to make chimneys expand as they ascend, after the manner of a trumpet, with its mouth turned downwards: but these experiments require further confirmation.

The method of calculation adopted above is founded on the principle of correcting the temperature for the difference between the specific gravity of atmospheric air and that of coal-smoke, the one being unity and the other 1·05; there is, however, another method, somewhat more elegant and legitimate, by employing the specific gravity of coal-smoke itself: the investigation is rather tedious and prolix, but the resulting formula is by no means difficult; and since both methods give the same result when properly calculated, we make no further apology for presenting our readers with another rule for obtaining the same object. The formula is as follows:

$$a = \frac{b\,(t' + 459)}{2757{\cdot}5}\sqrt{\frac{1}{h\,(t' - 77{\cdot}55)}}$$

where $a$ is the area of the transverse section of the chimney in square inches, $b$ the quantity of atmospheric air required for combustion of the coal in cubic feet per hour, $h$ the height of the chimney in feet, and $t'$ the temperature at which the air enters the flue after passing through the fire. The rule for performing this process is thus expressed:

Rule.—From the temperature at which the air enters the chimney, subtract the constant decrement 77·55; multiply the remainder by the height of the chimney in feet, divide unity by the product, and extract the square root of the quotient. To the temperature of the heated air, add the constant number 459; multiply the sum by the number of cubic feet required for combustion per hour, and divide the product by the number 2757·5; then multiply the quotient by the square root found as above, and the product will be the number of square inches in the transverse section of the chimney.

Suppose a mass of fuel in a state of combustion to require 5000 cubic feet of air per hour, what must be the size of the chimney when its height is 100 feet, the temperature at which the heated air enters the chimney being 200 degrees of Fahrenheit's thermometer?

By the rule we have 200−77·55=122·45 . . log. 2·0879588
Height of the chimney=100 . . . . log. 2·0000000

4·0879588

2)5·9120412

7·9560206

200+459=659 . . . log. 2·8188854
5000 . . . log. 3·6989700 } add 3·0773399
2757·5 ar. co. log. 6·5594845

1·0333605 10·798 in.

This appears to be a very small flue for the quantity of air that passes through it per hour; but it must be observed that we have assumed a great height for the shaft, which has the effect of creating a very powerful draught, thereby drawing off the heated air with great rapidity.

The advantage of a high flue is so very great, that the reader may be desirous of knowing to what height a chimney of a given base may be carried with safety, in cases where it is inconvenient to secure it with lateral stays; and, as an approximate rule for this purpose is not difficult of investigation, we think proper to supply it here.

When the chimney is equally wide throughout its whole height, the formula is

$$s = h\sqrt{\frac{156}{12000 - \frac{1}{3}\,h\,w}};$$

but when the side of the base is double the size of the top, the equation becomes

$$s = h\sqrt{\frac{104}{12000 - 0{\cdot}42\,h\,w}};$$

where $s$ is the side of the base in feet, $h$ the height, and $m$ the weight of one cubic foot of the material. When the chimney stalk is not square, but longer on the one side than the other, $s$ must be the least dimension. The proportion of solid wall to a given base, as sanctioned by experience, is about two-thirds of its area, consequently $w$ ought to be two-thirds of the weight of a cubic foot of brickwork. Now, a cubic foot of dried brickwork is, on an average, 114 lbs.; consequently $w = 76$ lbs.; and if this be substituted in the foregoing equations, we get for a chimney of equal size throughout,

$$s = h\sqrt{\frac{156}{1200 - 25\,h}};$$

and when the chimney tapers to one-half the size at top, it is

$$s - h\sqrt{\frac{104}{12000 - 32\,h}};$$

where it may be remarked that 12000 lbs. is the cohesive force of one square foot of mortar; and in the investigation of the formulas we have assumed the greatest force of the wind on a square foot of surface at 52 lbs. These equations are too simple in their form to require elucidation from us; we therefore leave the reduction as an exercise to the reader, who it is presumed will find no difficulty in resolving the several cases that may arise in the course of his practice.

$$v = \sqrt{\frac{2\,g\,\mathrm{H}\,a\,t\,\mathrm{D}}{\mathrm{D} + 2\,g\,\mathrm{K}\,(\mathrm{L} + \mathrm{H}}},$$

is the expression given by M. Péclet for the velocity of smoke in a chimney. $v$, the velocity; $t$, the temperature, whose maximum value is about 300° centigrade; $g = 32\frac{1}{6}$ feet; D, the diameter of the chimney; H, the height; L, the length of horizontal flues, supposing them formed into a cylinder of the same diameter as that of the chimney. K = ·0127 for brick, = ·005 for sheet-iron, and = ·0025 for cast-iron chimneys. $a$ = ·00365.

Let L=60; H=150; D=5; K=·005; $2g=64\frac{1}{3}$; $t$=300°; $a$=·00365. Then $v = \sqrt{\frac{2g\,\mathrm{H}\,a\,t\,\mathrm{D}}{\mathrm{D}+2\,g\,\mathrm{K}(\mathrm{H}+\mathrm{L})}} = 26{\cdot}986$ feet.

A cubic foot of water raised into steam is reckoned equivalent to a horse power, and to generate the steam with sufficient rapidity, an allowance of one square foot of fire-bars, and one square yard of effective heating surface, are very commonly made in practice, at least in land engines. These proportions, however, greatly vary in different cases; and in some of the best marine engine boilers, where the area of fire-grate is restricted by the breadth of the vessel, and the impossibility of firing long furnaces effectually at sea, half a square foot of fire-grate per horse power is a very common proportion. Ten cubic feet of water in the boiler per horse power, and ten cubic feet of steam room per horse power, have been assigned as the average proportion of these elements; but the fact is, no general rule can be formed upon the subject, for the proportions which would be suitable for a wagon boiler would be inapplicable to a tubular boiler, whether marine or locomotive; and good examples will in such cases be found a safer guide than rules which must often give a false result. A capacity of three cubic feet per horse power is a common enough proportion of furnace-room, and it is a good plan to make the furnaces of a considerable width, as they can then be fired more effectually, and do not produce so much smoke as if they are made narrow. As regards the question of draft, there is a great difference of opinion among engineers upon the subject, some preferring a very slow draft and others a rapid one. It is obvious that the question of draft is virtually that of the area of fire-grate, or of the quantity of fuel consumed upon a given area of grate surface, and the weight of fuel burned on a foot of fire-grate per hour varies in different cases in practice from $3\frac{1}{2}$ to 80 lbs. Upon the quickness of the draft again hinges the question of the proper thickness of the stratum of incandescent fuel upon the grate; for if the draft be very strong, and the fire at the same time be thin, a great deal of uncombined oxygen will escape up through the fire, and a needless refrigeration of the contents of the flues will be thereby occasioned; whereas, if the fire be thick, and the draft be sluggish, much of the useful effect of the coal will be lost by the formation of carbonic oxide. The length of the circuit made by the smoke varies in almost every boiler, and the same may be said of the area of the flue in its cross section, through which the smoke has to pass. As an average, about one-fifth of the area of fire-grate for the area of the flue behind the bridge, diminished to half that amount for the area of the chimney, has been given as a good proportion, but the examples which we have given, and the average flue area of the boilers which we shall describe, may be taken as a safer guide than any such loose statements. When the flue is too long, or its sectional area is insufficient, the draft becomes insufficient to furnish the requisite quantity of steam; whereas if the flue be too short or too large in its area, a large quantity of the heat escapes up the chimney, and a deposition of soot in the flues also takes place. This last fault is one of material consequence in the case of tubular boilers consuming bituminous coal, though indeed the evil might be remedied by blocking some of the tubes up. The area of water-level is about 5 feet per horse power in land boilers. In many cases, however, it is much less; but it is always desirable to make the area of the water-level as large as possible, as, when it is contracted, not only is the water-level subject to sudden and dangerous fluctuations, but water is almost sure to be carried into the cylinder with the steam, in consequence of the violent agitation of the water, caused by the ascent of a large volume of steam through a small superficies. It would be an improvement in boilers, we think, to place over each furnace an inverted vessel immerged in the water, which might catch the steam in its ascent, and deliver it quietly by a pipe rising above the water-level. The water-level would thus be preserved from any inconvenient agitation, and the weight of water within the boiler would be diminished at the same time that the original depth of water over the furnaces was preserved. It would also be an improvement to make the sides of the furnaces of marine boilers sloping, instead of vertical, as is the common practice, for the steam could then ascend freely at the instant of its formation, instead of being entangled among the rivets and landings of the plates, and superinducing an overheating of the plates by preventing a free access of the water to the metal.

We have, in the following table, collected a few of the principal results of experiments made on steam boilers.

TABLE I.

| NATURE OF THE BOILERS USED. | Mean of Huel Towan, and United Mines boilers, in Cornwall. | Boiler, at Warwick. | Mean of 8 experiments at the Albion Mills, Clithero, Preston, and New River Water Company. | Atmospheric Engine, at Long Benton, 1772. | Mean of 11 of M. de Pambour's experiments. | Cornish boiler at the East London Water Works, 1839. | Another boiler at the East London Water Works, 1839. |
|---|---|---|---|---|---|---|---|
| | Cylindrical with internal flue. | Wagon. | Wagon. | Circular or Hay-stack. | Locomotive. | Cylindrical with internal flue. | Wagon with internal flue. |
| Total area of heated surface in square feet | 962 | 152 | 342·8 | 459 | 334·6 | 798 | 588 |
| Length of circuit made by the heat in feet | 155 | 50 66 | 72·5 | 52·8 | 7·0 | 83·1 | 78 |
| Area of fire grates in square feet | 23·66 | 23·33 | 26·09 | 35·10 | 7·03 | 14·25 | 37·26 |
| Weight of fuel burned on each square foot of grate, per hour, in lbs. | 3·46 | 4·00 | 10·75 | 20·34 | 79·33 | 46·82 | 13·31 |
| Cub. ft. of water evaporated from initial temperature by 112 lbs. of fuel | 18·87 | 16·44 | 13·91 | 14·11 | 11·14 | | |
| Cubic feet of water evaporated per hour from initial temperature | 13·81 | 13·79 | 34·40 | 90·7 | 55·18 | | |
| Square feet of heated surface for each cubic foot of water evaporated per hour | 69·58 | 11·00 | 9·96 | 5·06 | 6·06 | 17·17 | |
| Square feet of heated surface for each square foot of grate | 40·65 | 6·51 | 13·13 | 13·08 | 47·59 | 56·0 | 15·78 |
| Pressure of steam above the atmosphere in lbs. | 42·2 | 2·5 | 3·68 | 1·5 | 50 | 15·45 | |

The economical effects of expansion will be found to be very clearly exhibited in the next table. The duties are recorded in the fifth line from the top, and the degree of expansion in the bottom line. It will be observed, that the order in which the different engines stand in respect of superiority of duty is the same as in respect of amount of expansion. The Holmbush engine has a duty of 140,484,848 lbs. raised 1 foot by 1 cwt. of coals, and the steam acts expansively over ·83 of the whole stroke; while the water-works' Cornish engine has only a duty of 105,664,118 lbs., and expands the steam over only ·687 of the whole stroke. Again, comparing the second and last engines together, the Albion Mills engine has a duty of 25,756,752 lbs., and no expansive action. The water-works' engine, again, acts expansively over one-half of its stroke, and has an increased duty of 46,602,333 lbs. Other causes, of course, may influence these comparisons, especially the last, where one engine is a double-acting rotative engine, and the other a single-acting pumping one; but there can be no doubt that the expansive action in the latter is the *principal* cause of its more economical performance.

The heating surface per horse power allowed by some engineers is about 9 square feet in wagon boilers, reckoning the total surface as effective surface, if the boilers be of a considerable size; but in the case of small boilers, the proportion is larger. The total

# TABLE II.

| | Atmospheric Engine, Long Benton, Northumberland, date 1772. | Non-expansive rotative condensing Engine, Albion Mills, London, date 1786. | Holmbush, Cornish, condensing Engine, single acting for pumping water. Steam acts expansively after the first sixth of the stroke. 1836. | Noncondensing double-acting Engine, nonexpansive, Congleton, Cheshire. 1823. | Cornish Engine, East London Water Works. | Pumping Engine at East London Water Works. |
|---|---|---|---|---|---|---|
| Diameter of cylinder in inches.................................. | 52 | 34 | 50 | 13 | $79\frac{3}{4}$ | $59\frac{5}{8}$ |
| Length of stroke in feet.......................................... | 7 | 8 | 9·1 | 4 | 10 | 7·91 |
| Number of strokes per minute................................ | 12 | 16 | 4·63 | 27·5 | 7 | 11·5 |
| Pressure on the piston, above or below the atmosphere in lbs., per square inch...................................... | . . . . | Estimated at —2·5 | +30 | +20 | +5·17 | +2·15 |
| Weight in lbs. raised one foot by 112 lbs. of coals......... | 12,600,000 | 25,756,752 | 140,484,848 | 12,418,560 | 105,664,118 | 46,602,333 |
| Do. do. by one pound of water, as steam | 14,280 | 28,489 | 119,097 | 15,840 | 110,716 | 53,369 |
| Effective power of the engine at time of experiment in horse power.................................................. | 40·5 | 50·0 | 26·48 | 12·0 | . . . . | |
| Efficiency of the steam, its efficiency in the Albion Mills being unity.............................................. | ·501 | 1·000 | 4·180 | ·556 | 3·89 | 1·87 |
| Efficiency of the fuel, its efficiency in the Albion Mills being unity................................................... | ·480 | 1·000 | 5·454 | ·482 | 4·1 | 1·81 |
| Distance of the piston from the end of its stroke when the steam is cut off in parts of the length of stroke. | 0 | 0 | ·833 | 0 | ·687 | ·5 |

heating surface of a two horse power wagon boiler is, according to Fitzgerald's proportions, 30 square feet, or 15 ft. per horse power; whereas, in the case of a 45 horse power boiler the total heating surface is 438 square feet, or 9·6 ft. per horse power. The capacity of steam room is 8¾ cubic feet per horse power, in the two horse power boiler, and 5¾ cubic feet in the 20 horse power boiler; and in the larger class of boilers, such as those suitable for 30 and 45 horse power engines, the capacity of the steam room does not fall below this amount, and indeed is nearer 6 than 5¾ cubic feet per horse power. The content of water is 18½ cubic feet per horse power in the two horse power boiler, and 15 cubic feet per horse power in the 20 horse. power boiler. In marine boilers about the same proportions obtain in most particulars. The original boilers of one or two large steamers were proportioned with about half a square foot of fire grate per horse power, and 10 square feet of flue and furnace surface, reckoning the total amount as effective; but in the boilers of other vessels a somewhat smaller proportion of heating surface was adopted. In some cases we have found that, in their marine flue boilers, 9 square feet of flue and furnace surface are requisite to boil off a cubic foot of water per hour, which is the proportion that obtains in some land boilers; but inasmuch as in modern engines the nominal considerably exceeds the actual power, they allow 11 square feet of heating surface per nominal horse power in their marine boilers, and they reckon, as effective heating surface, the tops of the flues, and the whole of the sides of the flues, but not the bottoms. They have been in the habit of allowing for the capacity of the steam space in marine boilers 16 times the content of the cylinder; but as there are two cylinders, this is equivalent to 8 times the content of both cylinders, which is the proportion commonly followed in land engines, and which agrees very nearly with the proportion of between 5 and 6 cubic feet of steam room per horse power. Taking, for example, an engine with 23 inches diameter of cylinder and 4 feet stroke, which will be 18·4 horse power—the area of the cylinder will be 415·476 square inches, which, multiplied by 48, the number of inches in the stroke, will give 19942·848 for the capacity of the cylinder in cubic inches; 8 times this is 159542·784 cubic inches, or 92·3 cubic feet; 92·3 divided by 18·4 is rather more than 5 cubic feet per horse power. There is less necessity, however, that the steam space should be large when the flow of steam from the boiler is very uniform, as it will be where there are two engines attached to the boiler at right angles with one another, or where the engines work at a great speed, as in the case of locomotive engines. A high steam chest too, by rendering boiling over into the steam pipes, or priming as it is called, more difficult, obviates the necessity for so large a steam space; and the use of steam of a high pressure, worked expansively, has the same operation; so that in modern marine boilers, of the tubular construction, where the whole of these modifying circumstances exist, there is no necessity for so large a proportion of steam room as 5 or 6 cubic feet per horse power, and about half that amount more nearly represents the general practice. Many allow 0·64 of a square foot per nominal horse power of grate bars in their marine boilers, and a good effect arises from this proportion; but sometimes so large an area of fire grate cannot be conveniently got, and the proportion of half a square foot per horse power seems to answer very well in engines working with some expansion, and is now very widely adopted. With this allowance, there will be about 22 square feet of heating surface per square foot of fire grate; and if the consumption of fuel be taken at 6 lbs. per nominal horse power per hour, there will be 12 lbs. of coal consumed per hour on each square foot of grate. The flues of all flue boilers diminish in their calorimeter as they approach the chimney; some very satisfactory boilers have been made by allowing a proportion of 0·6 of a square foot of fire grate per nominal horse power, and making the sectional area of the flue at the largest part $\frac{1}{7}$th of the area of fire grate, and the smallest part, where it enters the chimney, $\frac{1}{11}$th of the area of the fire grate; but in some of the boilers proportioned on this plan the maximum sectional area is only $\frac{1}{7\cdot5}$ or $\frac{1}{8\cdot5}$, according to the purposes of the boiler. These proportions are retained whether the boiler is flue or tubular, and from 14 to 16 square feet of tube surface is allowed per nominal horse power; but such boilers, although they may give abundance of steam, are generally, perhaps needlessly, bulky.

We shall therefore conclude our remarks upon the subject by introducing a table of the comparative evaporative power of different kinds of coal, which will prove useful, by affording data for the comparison of experiments upon different boilers when different kinds of coal are used.

TABLE *of the Comparative Evaporative Power of different kinds of Coal.*

| No. | Description of Coals. | Water evaporated per lb. of Coals. |
|---|---|---|
| | | *Lbs.* |
| 1 | The best Welsh | 9·493 |
| 2 | Anthracite American | 9·14 |
| 3 | The best small Pittsburgh | 8·526 |
| 4 | Average small Newcastle | 8·074 |
| 5 | Pennsylvanian | 10·45 |
| 6 | Coke from Gas-works | 7·908 |
| 7 | Coke and Newcastle, small, $\frac{1}{2}$ and $\frac{1}{2}$ | 7·897 |
| 8 | Welsh and Newcastle, mixed $\frac{1}{2}$ and $\frac{1}{2}$ | 7·865 |
| 9 | Derbyshire and small Newcastle, $\frac{1}{2}$ and $\frac{1}{2}$ | 7·710 |
| 10 | Average large Newcastle | 7·658 |
| 11 | Derbyshire | 6·772 |
| 12 | Blythe Main, Northumberland | 6·600 |

*Strength of boilers.*—The extension of the expansive method of employing steam to boilers of every denomination, and the gradual introduction in connection therewith of a higher pressure than formerly, makes the question of the strength of boilers one of great and increasing importance. This topic was very successfully elucidated, a few years ago, by a committee of the Franklin Institute, Philadelphia, and we shall here recapitulate a few of the more important of the conclusions at which they arrived. Iron boiler plate was found to increase in tenacity as its temperature was raised, until it reached a temperature of 550° above the freezing point, at which point its tenacity began to diminish. The following table exhibits the cohesive strength at different temperatures.

At 32° to 80° the tenacity was = 56,000 lbs., or 1-7th below its maximum.
At 570° —— = 66,500 lbs., the maximum.
At 720° —— = 55,000 lbs., the same nearly as at 32°.
At 1050° —— = 32,000 lbs., nearly ½ of the maximum.
At 1240° —— = 22,000 lbs., nearly ⅓ of the maximum.
At 1317° —— = 9,000 lbs., nearly 1-7th of the maximum.
At 3000° iron becomes fluid.

The difference in strength between strips of iron cut in the direction of the fibre, and strips cut across the grain, was found to be about 6 per cent. in favour of the former. Repeated piling and welding was found to increase the tenacity and closeness of the iron, but welding together different kinds of iron was found to give an unfavourable result; riveting plates was found to occasion a diminution in their strength, to the extent of about one-third. The accidental overheating of a boiler was found to reduce its strength from 65,000 lbs. to 45,000 lbs. per square inch. Taking into account all these contingencies, it appears expedient to limit the tensile force upon boilers in actual use to about 3000 lbs. per square inch of iron.

Copper follows a different law, and appears to diminish in strength by every addition of heat, reckoning from the freezing point. The square of the diminution of strength seems to keep pace with the cube of the temperature, as appears by the following table:—

TABLE *showing the Diminution of Strength of* COPPER *Boiler Plates by additions to the Temperature, the Cohesion at* 32° *being* 32,800 *lbs. per Square Inch.*

| No. | Temperature above 32°. | Diminution of Strength. | No. | Temperature above 32°. | Diminution of Strength. |
|---|---|---|---|---|---|
| 1 | 90° | 0·0175 | 9 | 660° | 0·3425 |
| 2 | 180 | 0·0540 | 10 | 769 | 0·4398 |
| 3 | 270 | 0·0926 | 11 | 812 | 0·4944 |
| 4 | 360 | 0·1513 | 12 | 880 | 0·5581 |
| 5 | 450 | 0·2046 | 13 | 984 | 0·6691 |
| 6 | 460 | 0·2133 | 14 | 1000 | 0·6741 |
| 7 | 513 | 0·2446 | 15 | 1200 | 0·8861 |
| 8 | 529 | 0·2558 | 16 | 1300 | 1·0000 |

In the case of iron, the following are the results when tabulated after a similar fashion.

TABLE *of Experiments on* IRON *Boiler Plate at High Temperature; the Mean Maximum Tenacity being at* 550° = 65,000 *lbs. per Square Inch.*

| Temperature observed. | Diminution of Tenacity observed. | Temperature observed. | Diminution of Tenacity observed. |
|---|---|---|---|
| 550° | 0·0000 | 824° | 0·2010 |
| 570 | 0·0869 | 932 | 0·3324 |
| 596 | 0·0899 | 947 | 0·3593 |
| 600 | 0·0964 | 1030 | 0·4478 |
| 630 | 0·1047 | 1111 | 0·5514 |
| 562 | 0·1155 | 1155 | 0·6000 |
| 722 | 0·1436 | 1159 | 0·6011 |
| 732 | 0·1491 | 1187 | 0·6352 |
| 734 | 0·1535 | 1237 | 0·6622 |
| 766 | 0·1589 | 1245 | 0·6715 |
| 770 | 0·1627 | 1317 | 0·7001 |

The application of stays to marine boilers, especially in those parts of the water spaces which lie in the wake of the furnace bars, has given engineers much trouble; the $\frac{3}{8}$ plate, of which ordinary boilers are composed, is hardly thick enough to retain a stay with security by merely tapping the plate, whereas, if the stay be riveted, the head of the rivet will in all probability be soon burnt away. The best practice appears to be to run the stays used for the water spaces in this situation, in a line somewhat beneath the level of the bars, so that they may be shielded as much as possible from the fire, while those which are required above the level of the bars should be kept as nearly as possible towards the crown of the furnace, so as to be removed from the immediate contact of the fire. Screw bolts with a fine thread tapped into the plate, and with a thin head upon the one side, and a thin nut made of a piece of boiler plate on the other, appear to be the best description of stay that has yet been contrived. The stays between the sides of the boiler shell, or the bottom of the boiler and the top, present little difficulty in their application, and the chief thing that is to be attended to is to take care that there be plenty of them; but we may here remark that we think it an indispensable thing, when there is any high pressure of steam to be employed, that the furnace crown be stayed to the top of the boiler. This, it will be observed, is done in the boilers of the Tagus and Infernal; and we know of no better specimen of staying than is afforded by those boilers.

### AREA OF STEAM PASSAGES.

RULE.—To the temperature of steam in the boiler add the constant increment 459; multiply the sum by 11025; and extract the square root of the product. Multiply the length of stroke by the number of strokes per minute; divide the product by the square root just found; and multiply the square root of the quotient by the diameter of the cylinder; the product will be the diameter of the steam passages.

Let it be required to determine the diameter of the steam passages in an engine of which the diameter of the cylinder is 48 inches, the length of stroke $4\frac{1}{2}$ feet, and the number of strokes per minute 26, supposing the temperature under which the steam is generated to be 250 degrees of Fahrenheit's thermometer.

Here by the rule we get $\sqrt{11025(250 + 459)} = 2795{\cdot}84$; the number of strokes is 26, and the length of stroke $4\frac{1}{2}$ feet; hence it is $\delta = d\sqrt{\frac{117}{2795{\cdot}84}} = 0{\cdot}20456d = 0{\cdot}20456 \times 48 = 9{\cdot}819$ inches; so that the diameter of the steam passages is a little more than one-fifth of the diameter of the cylinder. The same rule will answer for high and low pressure engines, and also for the passages into the condenser.

LOSS OF FORCE BY THE DECREASE OF TEMPERATURE IN THE STEAM PIPES.

RULE.—From the temperature of the surface of the steam pipes subtract the temperature of the external air; multiply the remainder by the length of the pipes in feet, and again by the constant number or coefficient 1·68; then divide the product by the diameter of the pipe in inches drawn into the velocity of the steam in feet per second, and the quotient will express the diminution of temperature in degrees of Fahrenheit's thermometer.

Let the length of the steam pipe be 16 feet and its diameter 5 inches, and suppose the velocity of the steam to be about 95 feet per second, what will be the diminution of temperature, on the supposition that the steam is at 250° and the external air at 60° of Fahrenheit?

Here, by the note to the above rule, the temperature of the surface of the steam pipe is $250 - 250 \times 0{\cdot}05 = 237{\cdot}5$; hence we get

$$t'' = \frac{1{\cdot}68 \times 16(237{\cdot}5 - 60)}{5 \times 95} = 10{\cdot}044 \text{ degrees.}$$

If we examine the manner of the composition of the above equation, it will be perceived that, since the diameter of the pipe and the velocity of motion enter as divisors, the loss of heat will be less as these factors are greater; but, on the other hand, the loss of heat will be greater in proportion to the length of pipe and the temperature of the steam. Since the steam is reduced from a higher to a lower temperature during its passage through the steam pipes, it must be attended with a corresponding diminution in the elastic force; it therefore becomes necessary to ascertain to what extent the force is reduced, in consequence of the loss of heat that takes place in passing along the pipes. This is an inquiry of some importance to the manufacturers of steam engines, as it serves to guard them against a very common mistake into which they are liable to fall, especially in reference to steamboat engines, where it is usual to cause the pipe to pass round the cylinder, instead of carrying it in the shortest direction from the boiler, in order to decrease the quantity of surface exposed to the cooling effect of the atmosphere.

RULE.—From the temperature of the surface of the steam pipe subtract the temperature of the external air; multiply the remainder by the length of the pipe in feet, and again by the constant fractional coefficient 0·00168; divide the product by the diameter of the pipe in inches drawn into the velocity of steam in feet per second, and subtract the quotient from unity; then multiply the difference thus obtained by the elastic force corresponding to the temperature of steam in the boiler, and the product will be the elastic force of the steam as reduced by cooling in passing through the pipes.

Let the dimensions of the pipe, the temperature of the steam, and its velocity through the passages, be the same as in the preceding example, what will be the quantity of reduction in the elastic force occasioned by the effect of cooling in traversing the steam pipe?

Since the elastic force of the steam in the boiler enters the equation from which the above rule is deduced, it becomes necessary in the first place to calculate its value; and this is to be done by a rule already given, which answers to the case in which the temperature is greater than 212°; thus we have

$250 \times 1·69856 = 424·640$
Constant number $= 205·526$ add

Sum $= 630·166$ ...... log. 2·79945
Constant divisor $= 333$............ log. 2·522444 subtract

$0·277011 \times 6·42 = 1·778410$,

which is the logarithm of 60·036 inches of mercury.

Again, we have $250 - 0·05 \times 250 = 237·5$; consequently, by multiplying as directed in the rule, we get $237·5 \times 0·00168 \times 16 = 6·384$, which being divided by $95 \times 5 = 475$, gives 0·01344; and by taking this from unity and multiplying the remainder by the elastic force as calculated above, the value of the reduced elastic force becomes

$$f' = 60·036\,(1 - 0·01344) = 59·229 \text{ inches of mercury.}$$

The loss of force is therefore $60·036 - 59·229 = 0·807$ inches of mercury, which amounts to $\frac{1}{75}$th part of the entire elastic force of the steam in the boiler as generated under the given temperature, being a quantity of sufficient importance to claim the attention of our engineers.

### FEED WATER.

The quantity of water required to supply the waste occasioned by evaporation from a boiler, or, as it is technically termed, the "feed water" required by a boiler working with any given pressure, is easily determinable. For, since the relative volumes of water and steam at any given pressure are known, it becomes necessary merely to restore the quantity of water by the feed pump equivalent to that abstracted in the form of steam, which the known relation of the density to the pressure of the steam renders of easy accomplishment. In practice, however, it is necessary that the feed pump should be able to supply a much larger quantity of water than what theory prescribes, as a great waste of water sometimes occurs from leakage or priming, and it is necessary to provide against such contingencies. The feed pump is usually made of such dimensions as to be capable of supplying 3½ times the water that the boiler will evaporate, and in low pressure engines, where the cylinder is double acting and the feed pump single acting, this proportion will be maintained by making the pump a 240th of the capacity of the cylinder. In low pressure engines the pressure in the boiler may be taken at 5 lbs. above the pressure of the atmosphere, or 20 lbs. in all; and as high pressure steam is merely low pressure steam compressed into a smaller compass, the size of the feed pump relatively to the size of the cylinder must obviously vary in the direct proportion of the pressure. If, then, the feed pump be 1-240th of the capacity of the cylinder when the total pressure of the steam is 20 lbs., it must be 1-120th of the capacity of the cylinder when the total pressure of the steam is 40 lbs., or 25 lbs. above the atmosphere. This law of variation is expressed by the following rule, which gives the capacity of feed pump proper for all pressures:—Multiply the capacity of the cylinder in cubic inches by the *total* pressure of the steam in lbs. per square inch, or the pressure in lbs. per square inch on the safety valve, plus 15, and divide the product by 4800; the quotient is the capacity of the feed pump in cubic inches, when the feed pump is single acting and the engine double acting. If the feed pump be double acting, or the engine single acting, the capacity of the pump must be just one-half what is given by this rule.

### CONDENSING WATER.

It was found that the most beneficial temperature of the hot well was 100 degrees. If, therefore, the temperature of the steam be 212°, and the latent heat 1000°, then 1212° may be taken to represent the heat contained in the steam, or 1112° if we deduct the temperature of the hot well. If the temperature of the injection water be 50°, then 50 degrees of cold are available for the abstraction of heat, and as the total quantity of heat to be abstracted is that requisite to raise the quantity of water in the steam 1112 degrees, or 1112 times that quantity, one degree, it would raise one-fiftieth of this, or 22·24 times the quantity of water in the steam, 50 degrees. A cubic inch of water, therefore, raised into steam, will require 22·24 cubic inches of water at 50 degrees for its condensation, and will form therewith 23·24 cubic inches of hot water at 100 degrees. It has been a practice to allow about a wine pint (28·9 cubic inches) of injection water for every cubic inch of water evaporated from the boiler. The usual capacity for the cold water pump is $\frac{1}{48}$th of the capacity of the cylinder, which allows some water to run to waste. As a maximum effect is obtained when the temperature of the hot well is about 100°, it will not be advisable to reduce it below that temperature in practice. With the superior vacuum due to a temperature of 70° or 80° the admission of so much cold water into the condenser becomes necessary,—and which has afterwards to be pumped out in opposition to the pressure of the atmosphere,—so that the gain in the vacuum does not equal the loss of power occasioned by the additional load upon the pump, and there is, therefore, a clear loss by the reduction of the temperature below 100°, if such reduction be caused by the admission of an additional quantity of water. If the reduction of temperature, however, be caused by the use of colder water, there is a gain produced by it, though the gain will within certain limits be greater, if advantage be taken of the lowness of the temperature to diminish the quantity of injection.

---

## SAFETY VALVES.

RULE.—Add 459 to the temperature of the steam in degrees of Fahrenheit; divide the sum by the product of the elastic force of the steam in inches of mercury, into its excess above the weight of the atmosphere in inches of mercury; multiply the square root of the quotient by ·0653; multiply this product by the number of cubic feet per hour of water evaporated, and this last product is the theoretical area of the orifice of the safety valve in square inches.

To apply this to an example—which, however, it must be remembered, will give a result much too small for practice.

Required the least area of a safety valve of a boiler suited for a 250 horse power engine, working with steam 6 lbs. more than the atmosphere on the square inch.

In this case the total pressure is equal to 21 lbs. per square inch; and as in round numbers one pound of pressure is equal to about two inches of mercury, it follows that $f = 42$ inches of mercury.

It will be necessary to calculate $t$ from formula (S) already given. The operation is as follows:—

$$\log. 42 \div 6{\cdot}42 = 1{\cdot}623249 \div 6{\cdot}42 = 0{\cdot}252842$$

$$\text{constant co-efficient} = 196 \qquad 2{\cdot}292363$$

$$\overline{2{\cdot}545205}$$

$$\text{natural number} = 350{\cdot}92$$

$$\text{constant temperature} = 121$$

$$t = \overline{229{\cdot}92}$$

$$\text{therefore } \sqrt{\frac{459 + t}{f\,(f - 30)}} = \sqrt{\frac{459 + 229{\cdot}92}{42 \times 12}}$$

$$= \sqrt{\frac{688{\cdot}92}{50{\cdot}4}} = \sqrt{1{\cdot}3669} = 1{\cdot}168;$$

$$\text{therefore } x = {\cdot}0653 \times 1{\cdot}168 \times \text{N} = {\cdot}0757\ \text{N}.$$

We have stated in a former part of this work that a cubic foot of water evaporated per hour is equivalent to one horse power; therefore in this case N = 250 and $x$ = 18·925 sq. in.

As another example. Required the proper area of the safety valve of a boiler suited to an engine of 500 horse power, when it is wished that the steam should never acquire an elastic force greater than 60 lbs. on the square inch above the atmosphere.

In this case the whole elastic force of the steam is 75 lbs.; and as 1 pound corresponds in round numbers to 2 inches of mercury, it follows that $f = 150$. It will be necessary to calculate the temperature corresponding to this force. The operation is as follows:—

Log. 150 ÷ 6·42 = 2·176091 ÷ 6·42 = ·338955
constant co-efficient = 196 log. 2·292363 add
natural number = 427·876 2·631318
constant temperature = 121
required temperature 306·876 degrees of Fahrenheit's scale

$$\text{therefore } \frac{459 + t}{f(f - 30)} = \frac{459 + 306 \cdot 876}{150(150 - 30)} = \frac{765 \cdot 876}{150 \times 120} = \frac{765 \cdot 896}{18000}$$

$$= \cdot 043549 \text{; therefore } \sqrt{\frac{459 + t}{f(f - 30)}} = \sqrt{\cdot 042549} = \cdot 20628.$$

Hence the required area = ·0653 × ·20628 × 500 = ·01347 × 500 = 6·735 square inches.

If the area of the safety valve of a boiler suited for an engine of 500 horse power be required, when it is wished the steam should never acquire a greater temperature than 300°, it will be necessary to calculate the elastic force corresponding to this temperature; and by formula for this purpose, the required area = ·0653 × ·231 × 500 = ·0151 × 500 = 7·55 square inches. It will be perceived from these examples that the greater the elasticity and the higher the corresponding temperature the less is the area of the safety valve. This is just as might have been expected, for then the steam can escape with increased velocity. We may repeat that the results we have arrived at are much less than those used in practice. For the sake of safety, the orifices of the safety valve are intentionally made much larger than what theory requires; usually $\frac{8}{10}$ of a square inch per horse power is the ordinary proportion allowed in the case of low pressure engines.

## THE SLIDE VALVE.

The four following practical rules are applicable alike to short slide and long D valves.

RULE I.—*To find how much cover must be given on the steam side in order to cut the steam off at any given part of the stroke.*—From the length of the stroke of the piston, subtract the length of that part of the stroke that is to be made before the steam is cut off. Divide the remainder by the length of the stroke of the piston, and extract the square root of the quotient. Multiply the square root thus found by half the length of the stroke of the valve, and from the product take half the lead, and the remainder will be the cover required.

RULE II.—*To find at what part of the stroke any given amount of cover on the steam side will cut off the steam.*—Add the cover on the steam side to the lead; divide the sum by half the length of stroke of the valve. In a table of natural sines find the arc whose sine is equal to the quotient thus obtained. To this arc add 90°, and from the sum of these two arcs subtract the arc whose cosine is equal to the cover on the steam side divided by half the stroke of the valve. Find the cosine of the remaining arc, add 1 to it, and multiply the sum by half the stroke of the piston, and the product is the length of that part of the stroke that will be made by the piston before the steam is cut off.

RULE III.—*To find how much before the end of the stroke, the exhaustion of the steam in front of the piston will be cut off.*—To the cover on the steam side add the lead, and divide the sum by half the length of the stroke of the valve. Find the arc whose sine is equal to the quotient, and add 90° to it. Divide the cover on the exhausting side by half the stroke of the valve, and find the arc whose cosine is equal to the quotient. Subtract this arc from the one last obtained, and find the cosine of the remainder. Subtract this cosine from 2, and multiply the remainder by half the stroke of the piston. The product is the distance of the piston from the end of its stroke when the exhaustion is cut off.

RULE IV.—*To find how far the piston is from the end of its stroke, when the steam that is propelling it by expansion is allowed to escape to the condenser.*—To the cover on the steam side add the lead, divide the sum by half the stroke of the valve, and find the arc whose sine is equal to the quotient. Find the arc whose cosine is equal to the cover on the exhausting side, divided by half the stroke of the valve. Add these two arcs together, and subtract 90°. Find the cosine of the residue, subtract it from 1, and multiply the remainder by half the stroke of the piston. The product is the distance of the piston from the end of its stroke, when the steam that is propelling it is allowed to escape to the condenser. In using these rules, all the dimensions are to be taken in inches, and the answers will be found in inches also.

From an examination of the formulas we have given on this subject, it will be perceived (supposing that there is no lead) that the part of the stroke where the steam is cut off, is determined by the proportion which the cover on the steam side bears to the length of the stroke of the valve: so that in all cases where the cover bears the same proportion to the length of the stroke of the valve, the steam will be cut off at the same part of the stroke of the piston.

In the first line, accordingly, of Table I., will be found eight different parts of the stroke of the piston designated; and directly below each, in the second line, is given the quantity of cover requisite to cause the steam to be cut off at that particular part of the stroke. The different sizes of the cover are given in the second line, in decimal parts of the length of the stroke of the valve; so that, to get the quantity of cover corresponding to any of the given degrees of expansion, it is only necessary to take the decimal in the second line, which stands under the fraction in the first, that marks the degree of expansion, and multiply that decimal by the length you intend to make the stroke of the valve. Thus, suppose you have an engine in which you wish to have the steam cut off when the piston is a quarter of the length of its stroke from the end of it, look in the table, and you will find in the third column from the left, $\frac{1}{4}$. Directly under that, in the second line, you have the decimal ·250. Suppose that you think 18 inches will be a convenient length for the stroke of the valve, multiply the decimal ·250 by 18, which gives $4\frac{1}{2}$. Hence we learn that with an 18 inch stroke for the valve, $4\frac{1}{2}$ inches of cover on the steam side will cause the steam to be cut off when the piston has still a quarter of its stroke to perform.

Half the stroke of the valve must always be *at least* equal to the cover on the steam side added to the breadth of the port. By the "breadth" of the port, we mean its dimension in the direction of the valve's motion; in short, its perpendicular depth when the cylinder is upright. The words "cover" and "lap" are synonymous. Consequently, as the cover, in this case, must be $4\frac{1}{2}$ inches, and as half the stroke of the valve is 9 inches, the breadth of the port cannot be more than $(9 - 4\frac{1}{2} = 4\frac{1}{2})$ $4\frac{1}{2}$ inches. If this breadth of port is not enough, we must increase the stroke of the valve; by which means we shall get both the cover and the breadth of the port proportionally increased. Thus, if we make the length of valve stroke 20 inches, we shall have for the cover $·250 \times 20 = 5$ inches, and for the breadth of the port $10 - 5 = 5$ inches.

TABLE I.

| Distance of the piston from the termination of its stroke, when the steam is cut off, in parts of the length of its stroke. | $\frac{8}{24}$ or $\frac{1}{3}$ | $\frac{7}{24}$ | $\frac{6}{24}$ or $\frac{1}{4}$ | $\frac{5}{24}$ | $\frac{4}{24}$ or $\frac{1}{6}$ | $\frac{3}{24}$ or $\frac{1}{8}$ | $\frac{2}{24}$ or $\frac{1}{12}$ | $\frac{1}{24}$ |
|---|---|---|---|---|---|---|---|---|
| Cover on the steam side of the valve, in decimal parts of the length of its stroke. | ·289 | ·270 | ·250 | ·228 | ·204 | ·177 | ·144 | ·102 |

This table, as we have already intimated, is computed on the supposition that the valve is to have no lead; but, if it is to have lead, all that is necessary is to subtract half the proposed lead from the cover found from the table, and the remainder will be the proper quantity of cover to give to the valve. Suppose that, in the last example, the valve was to have $\frac{1}{4}$ inch of lead, we would subtract $\frac{1}{8}$ inch from the 5 inches found for the cover by the table: that would leave $4\frac{7}{8}$ inches for the quantity of cover that the valve ought to have.

### TABLE II.

| Length of the stroke of the valve. Inches. | Cover required on the steam side of the valve to cut the steam off at any of the under-noted parts of the stroke. | | | | | | | |
|---|---|---|---|---|---|---|---|---|
| | $\frac{1}{3}$ | $\frac{7}{24}$ | $\frac{1}{4}$ | $\frac{5}{24}$ | $\frac{1}{6}$ | $\frac{1}{8}$ | $\frac{1}{12}$ | $\frac{1}{24}$ |
| 24 | 6·94 | 6·48 | 6·00 | 5·47 | 4·90 | 4·25 | 3·47 | 2·45 |
| 23½ | 6·79 | 6·34 | 5·88 | 5·36 | 4·79 | 4·16 | 3·39 | 2·39 |
| 23 | 6·65 | 6·21 | 5·75 | 5·24 | 4·69 | 4·07 | 3·32 | 2·34 |
| 22½ | 6·50 | 6·07 | 5·62 | 5·13 | 4·59 | 3·98 | 3·25 | 2·29 |
| 22 | 6·36 | 5·94 | 5·50 | 5·02 | 4·49 | 3·89 | 3·13 | 2·24 |
| 21½ | 6·21 | 5·80 | 5·38 | 4·90 | 4·39 | 3·80 | 3·10 | 2·19 |
| 21 | 6·07 | 5·67 | 5·25 | 4·79 | 4·28 | 3·72 | 3·03 | 2·14 |
| 20½ | 5·92 | 5·53 | 5·12 | 4·67 | 4·18 | 3·63 | 2·96 | 2·09 |
| 20 | 5·78 | 5·40 | 5·00 | 4·56 | 4·08 | 3·54 | 2·89 | 2·04 |
| 19½ | 5·64 | 5·26 | 4·87 | 4·45 | 3·98 | 3·45 | 2·82 | 1·99 |
| 19 | 5·49 | 5·13 | 4·75 | 4·33 | 3·88 | 3·36 | 2·74 | 1·94 |
| 18½ | 5·34 | 4·99 | 4·62 | 4·22 | 3·77 | 3·27 | 2·67 | 1·88 |
| 18 | 5·20 | 4·86 | 4·50 | 4·10 | 3·67 | 3·19 | 2·60 | 1·83 |
| 17½ | 5·06 | 4·72 | 4·37 | 3·99 | 3·57 | 3·10 | 2·53 | 1·78 |
| 17 | 4·91 | 4·59 | 4·25 | 3·88 | 3·47 | 3·01 | 2·45 | 1·73 |
| 16½ | 4·77 | 4·45 | 4·12 | 3·76 | 3·36 | 2·92 | 2·38 | 1·68 |
| 16 | 4·62 | 4·32 | 4·00 | 3·65 | 3·26 | 2·83 | 2·31 | 1·63 |
| 15½ | 4·48 | 4·18 | 3·87 | 3·53 | 3·16 | 2·74 | 2·24 | 1·58 |
| 15 | 4·33 | 4·05 | 3·75 | 3·42 | 3·06 | 2·65 | 2·16 | 1·53 |
| 14½ | 4·19 | 3·91 | 3·62 | 3·31 | 2·96 | 2·57 | 2·09 | 1·48 |
| 14 | 4·05 | 3·78 | 3·50 | 3·19 | 2·86 | 2·48 | 2·02 | 1·43 |
| 13½ | 3·90 | 3·64 | 3·37 | 3·08 | 2·75 | 2·39 | 1·95 | 1·37 |
| 13 | 3·76 | 3·51 | 3·25 | 2·96 | 2·65 | 2·30 | 1·88 | 1·32 |
| 12½ | 3·61 | 3·37 | 3·12 | 2·85 | 2·55 | 2·21 | 1·80 | 1·27 |
| 12 | 3·47 | 3·24 | 3·00 | 2·74 | 2·45 | 2·12 | 1·73 | 1·22 |
| 11½ | 3·32 | 3·10 | 2·87 | 2·62 | 2·35 | 2·03 | 1·66 | 1·17 |
| 11 | 3·18 | 2·97 | 2·75 | 2·51 | 2·24 | 1·95 | 1·58 | 1·12 |
| 10½ | 3·03 | 2·83 | 2·62 | 2·39 | 2·14 | 1·86 | 1·51 | 1·07 |
| 10 | 2·89 | 2·70 | 2·50 | 2·28 | 2·04 | 1·77 | 1·44 | 1·02 |
| 9½ | 2·65 | 2·56 | 2·37 | 2·17 | 1·93 | 1·68 | 1·32 | ·96 |
| 9 | 2·60 | 2·43 | 2·25 | 2·05 | 1·84 | 1·59 | 1·30 | ·92 |
| 8½ | 2·46 | 2·29 | 2·12 | 1·94 | 1·73 | 1·50 | 1·23 | ·86 |
| 8 | 2·31 | 2·16 | 2·00 | 1·82 | 1·63 | 1·42 | 1·15 | ·81 |
| 7½ | 2·16 | 2·02 | 1·87 | 1·71 | 1·53 | 1·33 | 1·08 | ·76 |
| 7 | 2·02 | 1·89 | 1·75 | 1·60 | 1·43 | 1·24 | 1·01 | ·71 |
| 6½ | 1·88 | 1·75 | 1·62 | 1·48 | 1·32 | 1·15 | ·94 | ·66 |
| 6 | 1·73 | 1·62 | 1·50 | 1·37 | 1·22 | 1·06 | ·86 | ·61 |
| 5½ | 1·58 | 1·48 | 1·37 | 1·25 | 1·12 | ·97 | ·79 | ·56 |
| 5 | 1·44 | 1·35 | 1·25 | 1·14 | 1·02 | ·88 | ·72 | ·51 |
| 4½ | 1·30 | 1·21 | 1·12 | 1·03 | ·92 | ·80 | ·65 | ·46 |
| 4 | 1·16 | 1·08 | 1·00 | ·91 | ·82 | ·71 | ·58 | ·41 |
| 3½ | 1·01 | ·94 | ·87 | ·80 | ·71 | ·62 | ·50 | ·35 |
| 3 | ·86 | ·81 | ·75 | ·68 | ·61 | ·53 | ·44 | ·30 |

Table II. is an extension of Table I. for the purpose of obviating, in most cases, the necessity of even the very small degree of trouble required in multiplying the stroke of the valve by one of the decimals in Table I. The first line of Table II. consists, as in Table I., of eight fractions, indicating the various parts of the stroke at which the steam may be cut off. The first column on the left hand consists of various numbers that represent the different lengths that may be given to the stroke of the valve, diminishing, by half-inches, from 24 inches to 3 inches. Suppose that you wish the steam cut off at any of the eight parts of the stroke indicated in the first line of the table, (say at $\frac{1}{6}$ from the end of the stroke,) you find $\frac{1}{6}$ at the top of the sixth column from the left. Look for the proposed length of stroke of the valve (say 17 inches) in the first column on the left. From 17, in that column, run along the line towards the right, and in the sixth column, and directly under the $\frac{1}{6}$ at the top, you will find 3·47, which is the cover required to cause the steam to be cut off at $\frac{1}{6}$ from the end of the stroke, if the valve has no lead. If you wish to give it lead, (say $\frac{1}{4}$ inch,) subtract the half of that, or $\frac{1}{8} = \cdot 125$ inch from 3·47, and you will have $3\cdot 47 - \cdot 125 = 3\cdot 345$ inches, the quantity of cover that the valve should have.

To find the greatest breadth that we can give to the port in this case, we have, as before, half the length of stroke, $8\frac{1}{2} - 3\cdot 345 = 5\cdot 155$ inches, which is the greatest breadth we can give to the port with this length of stroke. It is scarcely necessary to observe that it is not at all essential that the port should be so broad as this; indeed, where great length of stroke in the valve is not inconvenient, it is always an advantage to make it travel farther than is just necessary to make the port full open; because, when it travels farther, both the exhausting and steam ports are more quickly opened, so as to allow greater freedom of motion to the steam.

The manner of using this table is so simple, that we need not trouble the reader with more examples. We pass on, therefore, to explain the use of Table III.

Suppose that the piston of a steam engine is making its downward stroke, that the steam is entering the upper part of the cylinder by the upper steam-port, and escaping from below the piston by the lower exhausting-port; then, if (as is generally the case) the slide valve has some cover on the steam side, the upper port will be closed before the piston gets to the bottom of the stroke, and the steam above then acts expansively, while the communication between the bottom of the cylinder and the condenser still continues open, to allow any vapour from the condensed water in the cylinder, or any leakage past the piston, to escape into the condenser; but, before the piston gets to the bottom of the cylinder, this passage to the condenser will also be cut off by the valve closing the lower port. Soon after the lower port is thus closed, the upper port will be opened towards the condenser, so as to allow the steam that has been acting expansively to escape. Thus, before the piston has completed its stroke, the propelling power is removed from behind it, and a resisting power is opposed before it, arising from the vapour in the cylinder, which has no longer any passage open to the condenser. It is evident, that if there is no cover on the exhausting side of the valve, the exhausting port before the piston will be closed, and the one behind it opened, at the same time; but, if there is any cover on the exhausting side, the port before the piston will be closed before that behind it is opened; and the interval between the closing of the one and the opening of the other will depend on the quantity of cover on the exhausting side of the valve. Again, the position of the piston in the cylinder, when these ports are closed and opened respectively, will depend on the quantity of cover that the valve has on the steam side. If the cover is large enough to cut the steam off when the piston is yet a considerable distance from the end of its stroke, these ports will be closed and opened at a proportionably early part of the stroke; and when it is attempted to obtain great expansion by the slide-valve alone, without an expansion-valve, considerable loss of power is incurred from this cause.

Table III. is intended to show the parts of the stroke where, under any given arrangement of slide valve, these ports close and open respectively, so that thereby the engineer may be able to estimate how much of the efficiency of the engine he loses, while he is trying to add to the power of the steam by increasing the expansion in this manner. In the table, there are eight double columns, and at the heads of these columns are eight fractions, as before, representing so many different parts of the stroke at which the steam may be supposed to be cut off.

In the left-hand single column in each double one, are four decimals, which represent the distance of the piston (in terms of the length of its stroke) from the end of its stroke when the exhausting-port before it is opened, corresponding with the degree of expansion indicated by the fraction at the top of the double column and the cover on the exhausting side opposite to these decimals respectively in the left-hand column. The right-hand single column in each double one contains also each four decimals, which show in the same way at what part of the stroke the exhausting-port behind the piston is opened. A few examples will, perhaps, explain this best.

Suppose we have an engine in which the slide valve is made to cut the steam off when the piston is 1-3d from the end of its stroke, and that the cover on the exhausting side of the valve is 1-8th of the whole length of its stroke. Let the stroke of the piston be 6 feet, or 72 inches. We wish to know when the exhausting-port before the piston will be closed, and when the one behind it will be opened. At the top of the left-hand double column, the given degree of expansion (1-3d) is marked, and in the extreme left column we have at the top the given amount of cover (1-8th). Opposite the 1-8th, in the first double column, we have ·178 and ·033, which decimals, multiplied respectively by 72, the length of the stroke, will give the required positions of the piston: thus 72×·178=12·8 inches = distance of the piston from the end of the stroke when the exhausting-port *before* the piston is shut; and 72 × ·033 = 2·38 inches = distance of the piston from the end of its stroke when the exhausting-port behind it is opened.

# TABLE III.

| Steam cut off at | | Cover on the exhausting side of the valve in parts of the length of its stroke. 1-8th | 1-16th | 1-32d | 0 |
|---|---|---|---|---|---|
| 1-3d from the end of the stroke. | Distance of the piston from the end of its stroke, when the exhausting-port before it is shut (in parts of the stroke). | ·178 | ·130 | ·113 | ·092 |
| | Distance of the piston from the end of its stroke, when the exhausting-port behind it is opened (in parts of the stroke). | ·033 | ·060 | ·073 | ·092 |
| 7-24ths from the end of the stroke. | Distance of the piston from the end of its stroke, when the exhausting-port before it is shut (in parts of the stroke). | ·161 | ·118 | ·101 | ·082 |
| | Distance of the piston from the end of its stroke, when the exhausting-port behind it is opened (in parts of the stroke). | ·026 | ·052 | ·066 | ·082 |
| 1-4th from the end of the stroke. | Distance of the piston from the end of its stroke, when the exhausting-port before it is shut (in parts of the stroke). | ·143 | ·100 | ·085 | ·067 |
| | Distance of the piston from the end of its stroke, when the exhausting-port behind it is opened (in parts of the stroke). | ·019 | ·040 | ·051 | ·067 |
| 5-24ths from the end of the stroke. | Distance of the piston from the end of its stroke, when the exhausting-port before it is shut (in parts of the stroke). | ·126 | ·085 | ·069 | ·055 |
| | Distance of the piston from the end of its stroke, when the exhausting-port behind it is opened (in parts of the stroke). | ·012 | ·030 | ·042 | ·055 |
| 1-6th from the end of the stroke. | Distance of the piston from the end of its stroke, when the exhausting-port before it is shut (in parts of the stroke). | ·109 | ·071 | ·053 | ·043 |
| | Distance of the piston from the end of its stroke, when the exhausting-port behind it is opened (in parts of the stroke). | ·008 | ·022 | ·033 | ·043 |
| 1-8th from the end of the stroke. | Distance of the piston from the end of its stroke, when the exhausting-port before it is shut (in parts of the stroke). | ·093 | ·058 | ·043 | ·033 |
| | Distance of the piston from the end of its stroke, when the exhausting-port behind it is opened (in parts of the stroke). | ·004 | ·015 | ·023 | ·033 |
| 1-12th from the end of the stroke. | Distance of the piston from the end of its stroke, when the exhausting-port before it is shut (in parts of the stroke.) | ·074 | ·043 | ·033 | ·022 |
| | Distance of the piston from the end of its stroke, when the exhausting-port behind it is opened (in parts of the stroke). | ·001 | ·008 | ·013 | ·022 |
| 1-24th from the end of the stroke. | Distance of the piston from the end of its stroke, when the exhausting-port before it is shut (in parts of the stroke). | ·053 | ·027 | ·024 | ·011 |
| | Distance of the piston from the end of its stroke, when the exhausting-port behind it is opened (in parts of the stroke). | ·001 | ·002 | ·004 | ·011 |

To take another example. Let the stroke of the valve be 16 inches, the cover on the exhausting side $\frac{1}{2}$ inch, the cover on the steam side $3\frac{1}{4}$ inches, the length of the stroke of the piston 60 inches. It is required to ascertain all the particulars of the working of this valve. The cover on the exhausting side is evidently $\frac{1}{32}$ of the length of the valve stroke. Again, looking at 16 in the left-hand column of Table II., we find in the same horizontal line 3·26, or very nearly $3\frac{1}{4}$ under $\frac{1}{6}$ at the head of the column, thus showing that the steam will be cut off at $\frac{1}{6}$ from the end of the stroke. Again, under $\frac{1}{6}$ at the head of the fifth double column from the left in Table III., and in a horizontal line with $\frac{1}{32}$ in the left-hand column, we have ·053 and ·033. Hence, $·053 \times 60 = 3·18$ inches = distance of the piston from the end of its stroke when the exhausting-port before it is shut, and $·033 \times 60 = 1·98$ inches = distance of the piston from the end of its stroke when the exhausting-port behind it is opened. If in this valve the cover on the exhausting side were increased (say to 2 inches, or $\frac{1}{8}$ of the stroke,) the effect would be to make the port before the valve be shut sooner in the proportion of ·109 to ·053, and the port behind it later in the proportion of ·008 to ·033 (see Table III.) Whereas, if the cover on the exhausting side were removed entirely, the port before the piston would be shut and that behind it opened at the same time, and (see bottom of fifth double column, Table III.) the distance of the piston from the end of its stroke at that time would be $·043 \times 60 = 2·58$ inches.

An inspection of Table III. shows us the effect of increasing the expansion by the slide-valve in augmenting the loss of power occasioned by the imperfect action of the eduction passages. Referring to the bottom line of the table, we see that the eduction passage before the piston is closed, and that behind it opened, (thus destroying the whole moving power of the engine,) when the piston is ·092 from the end of its stroke, the steam being cut off at $\frac{1}{3}$ from the end. Whereas, if the steam is only cut off at $\frac{1}{24}$ from the end of the stroke, the moving power is not withdrawn till only ·011 of the stroke remains uncompleted. It will also be observed that increasing the cover on the exhausting side has the effect of retaining the action of the steam longer *behind* the piston, but it at the same time causes the eduction-port *before* it to be closed sooner.

A very cursory examination of the action of the slide valve is sufficient to show that the cover on the steam side should always be greater than on the exhausting side. If they are equal, the steam would be admitted on one side of the piston at the same time that it was allowed to escape from the other; but universal experience has shown that when this is the case, a very considerable part of the power of the engine is destroyed by the resistance opposed to the piston, by the exhausting steam not getting away to the condenser with sufficient rapidity. Hence we see the necessity of the cover on the exhausting side being always less than the cover on the steam side; and the difference should be the greater the higher the velocity of the piston is intended to be, because the quicker the piston moves the passage for the waste steam requires to be the larger, so as to admit of its getting away to the condenser with as great rapidity as possible. In locomotive or other engines, where it is not wished to expand the steam in the cylinder at all, the slide valve is sometimes made with very little cover on the steam side: and in these circumstances, in order to get a sufficient difference between the cover on the steam and exhausting sides of the valve, it may be necessary not only to take away all the cover on the exhausting side, but to take off still more, so as to make both exhausting passages be in some degree open, when the valve is at the middle of its stroke. This, accordingly, is sometimes done in such circumstances as we have described; but, when there is even a small degree of cover on the steam side, this plan of taking *more than all* the cover off the exhausting side ought never to be resorted to, as it can serve no good purpose, and will materially increase an evil we have already explained, viz. the opening of the exhausting-port behind the piston before the stroke is nearly completed. The tables apply equally to the common short slide three-ported valves and to the long D valves.

In fig. 1 is exhibited a common arrangement of the valves in lo-

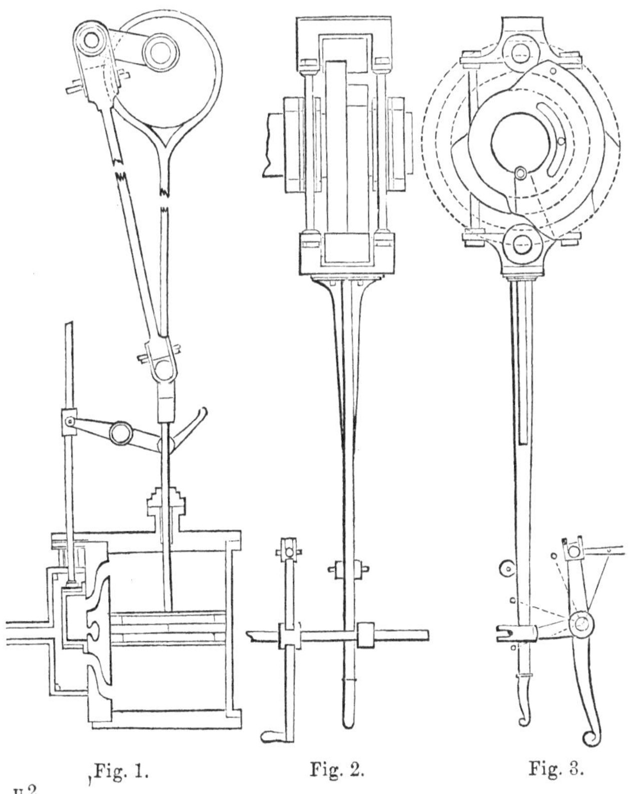

Fig. 1. Fig. 2. Fig. 3.

comotive engines, and in figs. 2 and 3 is shown an arrangement for working valves by a shifting cam, by which the amount of expansion may be varied. This particular arrangement, however, is antiquated, and is now but little used.

The extent to which expansion can be carried beneficially by means of lap upon the valve is about one-third of the stroke; that is, the valve may be made with so much lap, that the steam will be cut off when one-third of the stroke has been performed, leaving the residue to be accomplished by the agency of the expanding steam; but if more lap be put on than answers to this amount of expansion, a very distorted action of the valve will be produced, which will impair the efficiency of the engine. If a further amount of expansion than this is wanted, it may be accomplished by wire-drawing the steam, or by so contracting the steam passage, that the pressure within the cylinder must decline when the speed of the piston is accelerated, as it is about the middle of the stroke. Thus, for example, if the valve be so made as to shut off the steam by the time two-thirds of the stroke have been performed, and the steam be at the same time throttled in the steam pipe, the full pressure of the steam within the cylinder cannot be maintained except near the beginning of the stroke where the piston travels slowly; for as the speed of the piston increases, the pressure necessarily subsides, until the piston approaches the other end of the cylinder, where the pressure would rise again but that the operation of the lap on the valve by this time has had the effect of closing the communication between the cylinder and steam pipe, so as to prevent more steam from entering. By throttling the steam, therefore, in the manner here indicated, the amount of expansion due to the lap may be doubled, so that an engine with lap enough upon the valve to cut off the steam at two-thirds of the stroke, may, by the aid of wire-drawing, be virtually rendered capable of cutting off the steam at one-third of the stroke. The usual manner of cutting off the steam, however, is by means of a separate valve, termed an expansion valve; but such a device appears to be hardly necessary in many engines. In the Cornish engines, where the steam is cut off in some cases at one-twelfth of the stroke, a separate valve for the admission of steam, other than that which permits its escape, is of course indispensable; but in common rotative engines, which may realize expansive efficacy by throttling, a separate expansive valve does not appear to be required. In all engines there is a point beyond which expansion cannot be carried with advantage, as the resistance to be surmounted by the engine will then become equal to the impelling power; but in engines working with a high pressure of steam that point is not so speedily attained.

In high pressure, as contrasted with condensing engines, there is always the loss of the vacuum, which will generally amount to 12 or 13 lbs. on the square inch, and in high pressure engines there is a benefit arising from the use of a very high pressure over a pressure of a moderate account. In all high pressure engines, there is a diminution in the power caused by the counteracting pressure of the atmosphere on the educting side of the piston ; for the force of the piston in its descent would obviously be greater, if there was a vacuum beneath it ; and the counteracting pressure of the atmosphere is relatively less when the steam used is of a very high pressure. It is clear, that if we bring down the pressure of the steam in a high pressure engine to the pressure of the atmosphere, it will not exert any power at all, whatever quantity of steam may be expended, and if the pressure be brought nearly as low as that of the atmosphere, the engine will exert only a very small amount of power ; whereas, if a very high pressure be employed, the pressure of the atmosphere will become relatively as small in counteracting the impelling pressure, as the attenuated vapour in the condenser of a condensing engine is in resisting the lower pressure which is there employed. Setting aside loss from friction, and supposing the vacuum to be a perfect one, there would be no benefit arising from the use of steam of a high pressure in condensing engines, for the same weight of steam used without expansion, or with the same measure of expansion, would produce at every pressure the same amount of mechanical power. A piston with a square foot of area, and a stroke of three feet with a pressure of one atmosphere, would obviously lift the same weight through the same distance, as a cylinder with half a square foot of area, a stroke of three feet, and a pressure of two atmospheres. In the one case, we have three cubic feet of steam of the pressure of one atmosphere, and in the other case $1\frac{1}{2}$ cubic feet of the pressure of two atmospheres. But there is the same weight of steam, or the same quantity of heat and water in it, in both cases ; so that it appears a given weight of steam would, under such circumstances, produce a definite amount of power, without reference to the pressure. In the case of ordinary engines, however, these conditions do not exactly apply ; the vacuum is not a perfect one, and the pressure of the resisting vapour becomes relatively greater as the pressure of the steam is diminished ; the friction also becomes greater from the necessity of employing larger cylinders, so that even in the case of condensing engines, there is a benefit arising from the use of steam of a considerable pressure. Expansion cannot be carried beneficially to any great extent, unless the initial pressure be considerable ; for if steam of a low pressure were used, the ultimate tension would be reduced to a point so nearly approaching that of the vapour in the condenser, that the difference would not suffice to overcome the friction of the piston ; and a loss of power would be occasioned by carrying expansion to such an extent. In some of the Cornish engines, the steam is cut off at one-twelfth of the stroke ; but there would be a loss arising from carrying the expansion so far, instead of a gain, unless the pressure of the steam were considerable. It is clear, that in the case of engines which carry expansion very far, a very perfect vacuum in the condenser is more important than it is in other cases. Nothing can be easier than to compute the ultimate pressure of expanded steam, so as to see at what point expansion ceases to be productive of benefit; for as the pressure of expanded steam is inversely as the space occupied, the terminal pressure when the expansion is twelve times is just one-twelfth of what it was at first, and so on, in all other projections. The total pressure should be taken as the initial pressure—not the pressure on the safety valve, but that pressure plus the pressure of the atmosphere.

In high pressure engines, working at from 70 to 90 lbs. on the square inch, as in the case of locomotives, the efficiency of a given quantity of water raised into steam may be considered to be about the same as in condensing engines. If the pressure of steam in a high pressure engine be 120 lbs., or 125 lbs. above the atmosphere, then the resistance occasioned by the atmosphere will cause a loss of $\frac{1}{8}$th of the power. If the pressure of the steam in a low pressure engine be 16 lbs. on the square inch, or 11 lbs. above the atmosphere, and the tension of the vapour in the condenser be equivalent to 4 inches of mercury, or 2 lbs. of pressure on the square inch, then the resistance occasioned by this rare vapour will also cause a loss of $\frac{1}{8}$th of the power. A high pressure engine, therefore, with a pressure of 105 lbs. above the atmosphere, works with only the same loss from resistance to the piston, as a low pressure engine with a pressure of 1 lb. above the atmosphere, and with these proportions the power produced by a given weight of steam will be the same, whether the engine be high pressure or condensing.

## SPHEROIDAL CONDITION OF WATER IN BOILERS.

Some of the more prominent causes of boiler explosions have been already enumerated; but explosions have in some cases been attributed to the spheroidal condition of the water in the boiler, consequent upon the flues becoming red-hot from a deficiency of water, the accumulation of scale, or otherwise. The attachment of scale, from its imperfect conducting power, will cause the iron to be unduly heated; and if the scale be accidentally detached, a partial explosion may occur in consequence. It is found, that a sudden disengagement of steam does not immediately follow the contact of water with the hot metal, for water thrown upon red-hot iron is not immediately converted into steam, but assumes the spheroidal form and rolls about in globules over the surface. These globules, however high the temperature of the metal may be on which they are placed, never rise above the temperature of 205°, and give off but very little steam; but if the temperature of the metal be lowered, the water ceases to retain the spheroidal form, and comes into intimate contact with the metal, whereby a rapid disengagement of steam takes place. If water be poured into a very hot copper flask, the flask may be corked up, as there will be scarce any steam produced so long as the high temperature is maintained; but so soon as the temperature is suffered to fall below 350° or 400°, the spheroidal condition being no longer maintainable, steam is generated with rapidity, and the cork will be projected from the mouth of the flask with great force. In a boiler, no doubt, where there is a considerable head of water, the repellant action of the spheroidal globules will be more effectually counteracted than in the small vessels employed in experimental researches. But it is doubtful whether in all boilers there may not be something of the spheroidal action perpetually in operation, and leading to effects at present mysterious or inexplicable.

One of the most singular phenomena attending the spheroidal condition is, that the vapour arising from a spheroid is of a far higher temperature than the spheroid itself. Thus, if a thermometer be held in the atmosphere of vapour which surrounds a spheroid of water, the mercury, instead of standing at 205°, as would be the case if it had been immersed in the spheroid, will rise to a point determinable by the temperature of the vessel in which the spheroid exists. In the case of a spheroid, for example, existing within a crucible raised to a temperature of 400°, the thermometer, if held in the vapour, will rise to that point; and if the crucible be made red-hot, the thermometer will be burst, from the boiling point of mercury having been exceeded. A part of this effect may, indeed, be traced to direct radiation, yet it appears indisputable, from the experiments which have been made, that the vapour of a liquid spheroid is much hotter than the spheroid itself.

## EXPANSION.

At page 131 we have given a table of hyperbolic or Byrgean logarithms, for the purpose of facilitating computations upon this subject.

Let the pressure of the steam in the boiler be expressed by unity, and let $x$ represent the space through which the piston has moved whilst urged by the expanding steam. The density will then be $\frac{1}{1 + x}$, and, assuming that the densities and elasticities are proportionate, $\frac{d\,x}{1 + x}$ will be the differential of the efficiency, and the efficiency itself will be the integral of this, or, in other words, the hyperbolic logarithm of the denominator; wherefore the efficiency of the whole stroke will be $1 + \log. (1 + x)$.

Supposing the pressure of the atmosphere to be 15 lbs., $15 + 35 = 50$ lbs., and if the steam be cut off at $\frac{1}{4}$th of the stroke, it will be expanded into four times its original volume; so that at the termination of the stroke, its pressure will be $50 \div 4 = 12{\cdot}2$ lbs., or 2·8 lbs. less than the atmospheric pressure.

When the steam is cut off at one-fourth, it is evident that $x = 3$. In such case the efficiency is

$$1 + \log. (1 + 3), \text{ or } 1 + \log. 4.$$

The hyperbolic logarithm of 4 is 1·386294, so that the efficiency of the steam becomes 2·386294; that is, by cutting off the steam at $\frac{1}{4}$, more than twice the effect is produced with the same consumption of fuel; in other words, one-half of the fuel is saved.

This result may thus be expressed in words:—Divide the length of the stroke through which the steam expands by the length of stroke performed with the full pressure, which last portion call 1; the hyperbolic logarithm of the quotient is the increase of efficiency due to expansion. We introduce on the following page more detailed tables, to facilitate the computation of the power of an engine working expansively, or rather to supersede the necessity of entering into a computation at all in each particular case.

The first column in each of the following tables contains the initial pressure of the steam in pounds, and the remaining columns contain the mean pressure of steam throughout the stroke, with the different degrees of expansion indicated at the top of the columns, and which express the portion of the stroke during which the steam acts expansively. Thus, for example, if steam be admitted to the cylinder at a pressure of 3 pounds per square inch, and be cut off within $\frac{1}{8}$th of the end of the stroke, the mean pressure during the whole stroke will be 2·96 pounds per square inch. In like manner, if steam at the pressure of 3 pounds per square inch were cut off after the piston had gone through $\frac{1}{8}$th of the stroke, leaving the steam to expand through the remaining $\frac{7}{8}$th, the mean pressure during the whole stroke would be 1·164 pounds per square inch.

### FRICTION.

The friction of iron sliding upon brass, which has been oiled and then wiped dry, so that no film of oil is interposed, is about $\frac{1}{11}$ of the pressure; but in machines in actual operation, where there is a film of oil between the rubbing surfaces, the fraction is only about one-third of this amount, or $\frac{1}{33}$d of the weight. The tractive resistance of locomotives at low speeds, which is entirely made up of friction, is in some cases $\frac{1}{500}$th of the weight; but on the average about $\frac{1}{300}$th of the load, which nearly agrees with my former statement. If the total friction be $\frac{1}{300}$th of the load, and the rolling friction be $\frac{1}{1000}$th of the load, then the friction of attrition must be $\frac{1}{429}$th of the load; and if the diameter of the wheels be 36 in., and the diameter of the axles be 3 in., which are common proportions, the friction of attrition must be increased in the proportion of 36 to 3, or 12 times, to represent the friction of the rubbing surface when moving with the velocity of the carriage. $\frac{12}{429}$ths are about $\frac{1}{35}$th of the load, which does not differ much from the proportion of $\frac{1}{33}$d, as previously stated. While this, however, is the average result, the friction is a good deal less in some cases. Engineers, in some experiments upon the friction, found the friction to amount to less than $\frac{1}{40}$th of the weight; and in some experiments upon the friction of locomotive axles, it was found that by ample lubrication the friction might be made as little as $\frac{1}{60}$th of the weight, and the traction, with the ordinary size of wheels, would in such a case be about $\frac{1}{500}$th of the weight. The function of lubricating substances is to prevent the rubbing surfaces from coming into contact, whereby abrasion would be produced, and unguents are effectual in this

## EXPANDED STEAM.—MEAN PRESSURE AT DIFFERENT DENSITIES AND RATE OF EXPANSION.

*The column headed 0 contains the initial pressure in lbs., and the remaining columns contain the mean pressure in lbs., with different grades of expansion.*

**Expansion by Eighths.**

| 0 | 1/8 | 2/8 | 3/8 | 4/8 | 5/8 | 6/8 | 7/8 |
|---|---|---|---|---|---|---|---|
| 3 | 2·96 | 2·89 | 2·75 | 2·53 | 2·22 | 1·789 | 1·154 |
| 4 | 3·95 | 3·85 | 3·67 | 3·38 | 2·96 | 2·386 | 1·539 |
| 5 | 4·948 | 4·818 | 4·593 | 4·232 | 3·708 | 2·982 | 1·924 |
| 6 | 5·937 | 5·782 | 5·512 | 5·079 | 4·450 | 3·579 | 2·309 |
| 7 | 6·927 | 6·746 | 6·431 | 5·925 | 5·241 | 4·175 | 2·694 |
| 8 | 7·917 | 7·710 | 7·350 | 6·772 | 5·934 | 4·772 | 3·079 |
| 9 | 8·906 | 8·673 | 8·268 | 7·618 | 6·675 | 5·368 | 3·463 |
| 10 | 9·896 | 9·637 | 9·187 | 8·465 | 7·417 | 5·965 | 3·848 |
| 11 | 10·885 | 10·601 | 10·106 | 9·311 | 8·159 | 6·561 | 4·233 |
| 12 | 11·875 | 11·565 | 10·925 | 10·158 | 8·901 | 7·158 | 4·618 |
| 13 | 12·865 | 12·528 | 11·943 | 11·004 | 9·642 | 7·754 | 5·003 |
| 14 | 13·854 | 13·492 | 12·862 | 11·851 | 10·384 | 8·531 | 5·388 |
| 15 | 14·844 | 14·456 | 13·781 | 12·697 | 11·126 | 8·947 | 5·773 |
| 16 | 15·834 | 15·420 | 14·700 | 13·544 | 11·868 | 9·544 | 6·158 |
| 17 | 16·823 | 16·383 | 15·618 | 14·390 | 12·609 | 10·140 | 6·542 |
| 18 | 17·813 | 17·347 | 16·537 | 15·237 | 13·351 | 10·737 | 6·927 |
| 19 | 18·702 | 18·311 | 17·448 | 16·803 | 14·093 | 11·333 | 7·312 |
| 20 | 19·792 | 19·275 | 18·375 | 16·930 | 14·835 | 11·930 | 7·697 |
| 25 | 24·740 | 24·093 | 22·968 | 21·162 | 18·543 | 14·912 | 9·621 |
| 30 | 29·688 | 28·912 | 27·562 | 25·395 | 22·252 | 17·895 | 11·546 |
| 35 | 34·636 | 33·731 | 33·156 | 29·627 | 25·961 | 20·877 | 13·470 |
| 40 | 39·585 | 38·550 | 36·750 | 33·860 | 29·670 | 23·860 | 15·395 |
| 45 | 44·533 | 43·368 | 41·343 | 38·092 | 33·378 | 26·842 | 17·319 |
| 50 | 49·481 | 48·187 | 45·937 | 42·325 | 37·067 | 29·825 | 19·243 |

**Expansion by Tenths.**

| 0 | 1/10 | 2/10 | 3/10 | 4/10 | 5/10 | 6/10 | 7/10 | 8/10 | 9/10 |
|---|---|---|---|---|---|---|---|---|---|
| 3 | 2·980 | 2·930 | 2·830 | 2·710 | 2·539 | 2·299 | 1·981 | 1·668 | 0·990 |
| 4 | 3·974 | 3·913 | 3·780 | 3·614 | 3·386 | 3·065 | 2·642 | 2·087 | 1·320 |
| 5 | 4·968 | 4·892 | 4·725 | 4·518 | 4·232 | 3·832 | 3·303 | 2·609 | 1·651 |
| 6 | 5·961 | 5·870 | 5·670 | 5·421 | 5·079 | 4·598 | 3·963 | 3·130 | 1·981 |
| 7 | 6·955 | 6·848 | 6·615 | 6·325 | 5·925 | 5·364 | 4·624 | 3·652 | 2·311 |
| 8 | 7·948 | 7·827 | 7·560 | 7·228 | 6·772 | 6·131 | 5·284 | 4·174 | 2·641 |
| 9 | 8·942 | 8·805 | 8·505 | 8·132 | 7·618 | 6·897 | 5·945 | 4·696 | 2·971 |
| 10 | 9·936 | 9·784 | 9·450 | 9·036 | 8·465 | 7·664 | 6·606 | 5·218 | 3·302 |
| 11 | 10·929 | 10·762 | 10·395 | 9·939 | 9·311 | 8·430 | 7·266 | 5·739 | 3·632 |
| 12 | 11·923 | 11·740 | 11·340 | 10·843 | 10·158 | 9·196 | 7·927 | 6·261 | 3·962 |
| 13 | 12·856 | 12·719 | 12·285 | 11·746 | 10·994 | 9·963 | 8·587 | 6·783 | 4·292 |
| 14 | 13·910 | 13·967 | 13·230 | 12·650 | 11·851 | 10·729 | 9·248 | 7·305 | 4·622 |
| 15 | 14·904 | 14·676 | 14·175 | 13·554 | 12·697 | 11·496 | 9·909 | 7·827 | 4·953 |
| 16 | 15·897 | 15·654 | 15·120 | 14·457 | 13·544 | 12·262 | 10·569 | 8·348 | 5·283 |
| 17 | 16·891 | 16·632 | 16·065 | 15·361 | 14·051 | 13·028 | 11·230 | 8·870 | 5·613 |
| 18 | 17·884 | 17·611 | 17·010 | 16·264 | 15·237 | 13·795 | 11·890 | 9·392 | 5·944 |
| 19 | 18·878 | 18·589 | 17·955 | 17·168 | 16·083 | 14·561 | 12·551 | 9·914 | 6·273 |
| 20 | 19·872 | 19·568 | 18·900 | 18·072 | 16·930 | 15·328 | 13·212 | 10·436 | 6·600 |
| 25 | 24·840 | 24·460 | 23·625 | 22·590 | 21·162 | 19·160 | 16·515 | 13·040 | 8·255 |
| 30 | 29·808 | 29·352 | 28·350 | 27·108 | 25·395 | 22·992 | 19·818 | 15·654 | 9·906 |
| 35 | 34·776 | 34·244 | 33·075 | 31·626 | 29·627 | 26·824 | 23·121 | 18·263 | 11·557 |
| 40 | 39·744 | 39·136 | 37·800 | 36·144 | 33·860 | 30·656 | 26·224 | 20·872 | 13·208 |
| 45 | 44·912 | 44·028 | 42·525 | 40·662 | 38·092 | 34·888 | 29·727 | 23·481 | 14·859 |
| 50 | 49·680 | 48·920 | 47·250 | 45·180 | 42·325 | 38·320 | 33·030 | 26·090 | 16·510 |

respect in the proportion of their viscidity; but if the viscidity of the unguent be greater than what suffices to keep the surfaces asunder, an additional resistance will be occasioned; and the nature of the unguent selected should always have reference, therefore, to the size of the rubbing surfaces, or to the pressure per square inch upon them. With oil, the friction appears to be a minimum when the pressure on the surface of a bearing is about 90 lbs. per square inch: the friction from too small a surface increases twice as rapidly as the friction from too large a surface; added to which, the bearing, when the surface is too small, wears rapidly away. For all sorts of machinery, the oil of Patrick Sarsfield Devlan, of Reading, Pa., is the best.

## HORSE POWER.

A horse power is an amount of mechanical force capable of raising 33,000 lbs. one foot high in a minute. The average force exerted by the strongest horses, amounting to 33,000 lbs., raised one foot high in the minute, was adopted, and has since been retained. The efficacy of engines of a given size, however, has been so much increased, that the dimensions answerable to a horse power then, will raise much more than 33,000 lbs. one foot high in the minute now; so that an *actual* horse power, and a *nominal* horse power are no longer convertible terms. In some engines every nominal horse power will raise 52,000 lbs. one foot high in the minute, in others 60,000 lbs., and in others 66,000 lbs.; so that an actual and nominal horse power are no longer comparable quantities,—the one being a unit of dimension, and the other a unit of force. The actual horse power of an engine is ascertained by an instrument called an indicator; but the nominal power is ascertained by a reference to the dimensions of the cylinder, and may be computed by the following rule:—Multiply the square of the diameter of the cylinder in inches by the velocity of the piston in feet per minute, and divide the product by 6,000; the quotient is the number of nominal horses power. In using this rule, however, it is necessary to adopt the speed of piston which varies with the length of the stroke. The speed of piston with a two feet stroke is, according to this system, 160 per minute; with a 2 ft. 6 in. stroke, 170; 3 ft., 180; 3 ft., 6 in., 189; 4 ft., 200; 5 ft., 215; 6 ft., 228; 7 ft., 245; 8 ft., 256 ft.

By ascertaining the ratio in which the velocity of the piston increases with the length of the stroke, the element of velocity may be cast out altogether; and this for most purposes is the most convenient method of procedure. To ascertain the nominal power by this method, multiply the square of the diameter of the cylinder in inches by the cube root of the stroke in feet, and divide the product by 47; the quotient is the number of nominal horses power of the engine. This rule supposes a uniform effective pressure upon the piston of 7 lbs. per square inch; the effective pressure upon the piston of 4 horse power engines of some of the best makers has been estimated at 6·8 lbs. per square inch, and the pressure increased slightly with the power, and became 6·94 lbs. per square inch in engines of 100 horse power; but it appears to be more convenient to take a uniform pressure of 7 lbs. for all powers. Small engines, indeed, are somewhat less effective in proportion than large ones; but the difference can be made up by slightly increasing the pressure in the boiler; and small boilers will bear such an increase without inconvenience.

Nominal power, it is clear, cannot be transformed into actual power, for the nominal horse power expresses the size of an engine, and the actual horse power the number of times 33,000 lbs. it will lift one foot high in a minute. To find the number of times 33,000 lbs. or 528 cubic feet of water, an engine will raise one foot high in a minute,—or, in other words, the actual power,—we first find the pressure in the cylinder by means of the indicator, from which we deduct a pound and a half of pressure for friction, the loss of power in working the air pump, &c.; multiply the area of the piston in square inches by this residual pressure, and by the motion of the piston, in feet per minute, and divide by 33,000; the quotient is the actual number of horse power. The same result is attained by squaring the diameter of the cylinder, multiplying by the pressure per square inch, as shown by the indicator, less a pound and a half, and by the motion of the piston in feet, and dividing by 42,017. The quantity thus arrived at, will, in the case of nearly all modern engines, be very different from that obtained by multiplying the square of the diameter of the cylinder by the cube root of the stroke, and dividing by 47, which expresses the nominal power; and the actual and nominal power must by no means be confounded, as they are totally different things. The duty of an engine is the work done in relation to the fuel consumed, and in ordinary mill or marine engines it can only be ascertained by the indicator, as the load upon such engines is variable, and cannot readily be determined; but in the case of engines for pumping water, where the load is constant, the number of strokes performed by the engine represents the duty; and a mechanism to register the number of strokes made by the engine in a given time, is a sufficient test of the engine's performance.

In high pressure engines the actual power is readily ascertained by the indicator, by the same process by which the actual power of low pressure engines is ascertained. The friction of a locomotive engine when unloaded, is found by experiment to be about 1 lb. per square inch on the surface of the pistons, and the additional friction caused by any additional resistance is estimated at about ·14 of that resistance; but it will be a sufficiently near approximation to the power consumed by friction in high pressure engines, if we make a deduction of a pound and a half from the pressure on that account, as in the case of low pressure engines. High pressure engines, it is true, have no air pump to work; but the deduction of a pound and a half of pressure is relatively a much smaller one where the pressure is high than where it does not much exceed the pressure of the atmosphere. The rule, therefore, for the actual horse power of a high pressure engine will stand thus:—Square the diameter of the cylinder in inches, multiply by the pressure of the steam in the cylinder per square inch, less 1½ lbs., and by the speed of the piston in feet per minute, and divide by 42,017; the quotient is the actual horse power. The nominal horse power of a high pressure engine has never been defined; but it should obviously hold the same relation to the actual power as that which obtains in the case of condensing engines, so that an engine of a given nominal horse power may be capable of performing the same work, whether high pressure or condensing. This relation is maintained in the following rule, which expresses the nominal horse power of high pressure engines:—Multiply the square of the diameter of the cylinder in inches by the pressure on the piston in pounds per square inch, and by the speed of the piston in feet per minute, and divide the product by 120,000; the quotient is the power of the engine in nominal horses power. If the pressure upon the piston be 80 lbs. per square inch, the operation may be abbreviated by multiplying the square of the diameter of the cylinder by the speed of the piston, and dividing by 1,500, which will give the same result. This rule for nominal horse power, however, is not representative of the dimensions of the cylinder; but a rule for the nominal horse power of high pressure engines which shall discard altogether the element of velocity, is easily constructed; and, as different pressures are used in different engines, the pressure must become an element in the computation. The rule for the nominal power will therefore stand thus:—Multiply the square of the diameter of the cylinder in inches by the pressure on the piston in pounds per square inch, and the cube root of the stroke in feet, and divide the product by 940; the quotient is the power of the engine in nominal horse power, the engine working at the ordinary speed of 128 times the cube root of the stroke.

A summary of the results arrived at by these rules is given in the following tables, which, for the convenience of reference, we introduce.

---

## PARALLEL MOTION.

RULE I.—*In such a combination of two levers as is represented in Figs. 1 and 2, page 245, to find the length of radius bar required for any given length of lever C G, and proportion of parts of the link, G E and F E, so as to make the point E move in a perpendicular line.*—Multiply the length of G C by the length of the segment G E, and divide the product by the length of the segment F E. The quotient is the length of the radius bar.

RULE II.—(*Fig. 2, page 245.*) *The length of the radius bar and of C G being given, to find the length of the segment (F E) of the link next the radius bar.*—Multiply the length of C G by the

TABLE *of Nominal Horse Power of Low Pressure Engines.*

| Diameter of Cylinder in Inches. | LENGTH OF STROKE IN FEET. | | | | | | | | | | | |
|---|---|---|---|---|---|---|---|---|---|---|---|---|
| | 1 | 1½ | 2 | 2½ | 3 | 3½ | 4 | 4½ | 5 | 5½ | 6 | 7 |
| 4 | ·34 | ·39 | ·43 | ·46 | ·49 | ·52 | ·54 | ·56 | ·58 | ·60 | ·62 | ·65 |
| 5 | ·53 | ·61 | ·67 | ·72 | ·76 | ·81 | ·84 | ·88 | ·91 | ·94 | ·96 | 1·02 |
| 6 | ·76 | ·87 | ·96 | 1·04 | 1·10 | 1·16 | 1·22 | 1·26 | 1·31 | 1·35 | 1·39 | 1·47 |
| 7 | 1·04 | 1·19 | 1·31 | 1·41 | 1·50 | 1·58 | 1·65 | 1·72 | 1·78 | 1·84 | 1·89 | 1·99 |
| 8 | 1·36 | 1·56 | 1·72 | 1·85 | 1·96 | 2·07 | 2·16 | 2·25 | 2·33 | 2·40 | 2·47 | 2·60 |
| 9 | 1·72 | 1·97 | 2·17 | 2·34 | 2·49 | 2·62 | 2·74 | 2·84 | 2·95 | 3·04 | 3·13 | 3·30 |
| 10 | 2·13 | 2·44 | 2·68 | 2·89 | 3·07 | 3·23 | 3·38 | 3·51 | 3·64 | 3·76 | 3·87 | 4·07 |
| 11 | 2·57 | 2·95 | 3·24 | 3·49 | 3·77 | 3·91 | 4·15 | 4·25 | 4·40 | 4·54 | 4·68 | 4·92 |
| 12 | 3·06 | 3·51 | 3·86 | 4·16 | 4·42 | 4·65 | 4·86 | 5·06 | 5·24 | 5·41 | 5·57 | 5·86 |
| 13 | 3·60 | 4·12 | 4·53 | 4·88 | 5·19 | 5·46 | 5·64 | 5·94 | 6·15 | 6·35 | 6·53 | 6·88 |
| 14 | 4·17 | 4·77 | 5·25 | 5·66 | 6·01 | 6·33 | 6·62 | 6·88 | 7·13 | 7·36 | 7·58 | 7·98 |
| 15 | 4·77 | 5·48 | 6·03 | 6·50 | 6·90 | 7·27 | 7·60 | 7·90 | 8·19 | 8·45 | 8·70 | 9·16 |
| 16 | 5·45 | 6·23 | 6·86 | 7·39 | 7·86 | 8·27 | 8·65 | 8·99 | 9·31 | 9·61 | 9·90 | 10·42 |
| 17 | 6·15 | 7·04 | 7·75 | 8·35 | 8·86 | 9·34 | 9·76 | 10·15 | 10·52 | 10·85 | 11·17 | 11·76 |
| 18 | 6·89 | 7·89 | 8·68 | 9·36 | 9·94 | 10·47 | 10·94 | 11·38 | 11·79 | 12·17 | 12·53 | 13·19 |
| 19 | 7·68 | 8·79 | 9·68 | 10·42 | 11·17 | 11·66 | 12·19 | 12·68 | 13·13 | 13·56 | 13·96 | 14·69 |
| 20 | 8·51 | 9·74 | 10·72 | 11·55 | 12·27 | 12·92 | 13·51 | 14·05 | 14·55 | 15·02 | 15·46 | 16·28 |
| 22 | 10·30 | 11·79 | 12·97 | 13·98 | 14·85 | 15·63 | 16·62 | 17·30 | 17·65 | 18·18 | 18·71 | 19·70 |
| 24 | 12·26 | 14·03 | 15·44 | 16·63 | 17·67 | 18·61 | 19·45 | 20·23 | 20·95 | 31·63 | 22·27 | 23·44 |
| 26 | 14·39 | 16·46 | 18·12 | 19·52 | 20·75 | 21·84 | 22·56 | 23·75 | 24·6 | 25·39 | 26·14 | 27·51 |
| 28 | 16·68 | 19·09 | 21·02 | 22·64 | 24·06 | 25·33 | 26·48 | 27·54 | 28·52 | 29·44 | 30·31 | 31·90 |
| 30 | 19·15 | 21·92 | 24·13 | 25·99 | 27·62 | 29·07 | 30·40 | 31·61 | 32·74 | 33·80 | 34·80 | 36·63 |
| 32 | 21·79 | 24·96 | 27·51 | 29·57 | 31·42 | 33·08 | 34·59 | 35·97 | 37·26 | 38·46 | 39·59 | 41·68 |
| 34 | 24·60 | 28·16 | 30·99 | 33·39 | 35·44 | 37·34 | 39·04 | 40·60 | 42·06 | 43·41 | 44·69 | 47·05 |
| 36 | 27·57 | 31·56 | 34·74 | 37·42 | 39·77 | 41·87 | 43·77 | 45·52 | 47·15 | 48·67 | 50·11 | 52·75 |
| 38 | 30·72 | 35·17 | 38·71 | 41·69 | 44·66 | 46·64 | 48·77 | 50·72 | 52·54 | 54·23 | 55·83 | 58·78 |
| 40 | 34·04 | 38·97 | 42·89 | 46·20 | 49·10 | 51·69 | 54·04 | 56·20 | 58·21 | 60·09 | 61·86 | 65·12 |
| 42 | 37·53 | 42·96 | 47·29 | 50·94 | 54·13 | 56·98 | 59·58 | 61·96 | 64·18 | 66·25 | 68·21 | 71·78 |
| 44 | 41·19 | 47·15 | 51·90 | 55·91 | 59·38 | 62·54 | 66·46 | 68·00 | 70·44 | 72·71 | 74·85 | 78·79 |
| 46 | 45·02 | 51·54 | 56·72 | 61·10 | 64·88 | 68·19 | 71·43 | 74·33 | 76·69 | 79·47 | 81·81 | 86·12 |
| 48 | 49·02 | 56·11 | 61·76 | 66·53 | 70·70 | 74·42 | 77·82 | 80·94 | 83·83 | 86·53 | 89·08 | 93·78 |
| 50 | 53·19 | 60·89 | 67·02 | 72·19 | 76·71 | 80·76 | 84·44 | 87·82 | 90·96 | 93·89 | 96·65 | 101·7 |
| 52 | 57·55 | 65·86 | 72·48 | 78·08 | 83·00 | 87·35 | 90·25 | 94·98 | 98·40 | 101·55 | 104·5 | 110·0 |
| 54 | 62·04 | 71·02 | 78·17 | 84·20 | 89·48 | 94·20 | 98·49 | 102·4 | 106·1 | 109·5 | 112·7 | 118·7 |
| 56 | 66·72 | 76·38 | 84·07 | 90·55 | 96·23 | 101·30 | 105·9 | 110·1 | 114·1 | 117·8 | 121·2 | 127·6 |
| 58 | 71·58 | 81·93 | 90·18 | 97·14 | 103·2 | 108·6 | 113·6 | 118·2 | 122·4 | 126·3 | 129·2 | 136·7 |
| 60 | 76·60 | 87·68 | 96·50 | 103·9 | 110·4 | 116·3 | 121·6 | 126·4 | 131·0 | 135·2 | 139·2 | 146·5 |
| 62 | 81·79 | 93·62 | 103·04 | 111·0 | 117·96 | 124·18 | 129·81 | 135·03 | 139·86 | 144·37 | 148·6 | 156·7 |
| 64 | 87·15 | 99·84 | 110·0 | 118·3 | 125·7 | 132·3 | 138·3 | 143·9 | 149·0 | 153·82 | 158·4 | 166·7 |
| 66 | 92·68 | 106·1 | 116·8 | 125·8 | 133·6 | 140·7 | 147·3 | 153·0 | 158·5 | 163·6 | 168·4 | 177·3 |
| 68 | 98·40 | 112·6 | 123·9 | 133·6 | 141·8 | 149·4 | 156·2 | 162·4 | 168·2 | 173·6 | 178·8 | 188·2 |
| 70 | 104·26 | 119·3 | 131·3 | 141·5 | 150·4 | 158·3 | 165·5 | 172·1 | 178·2 | 184·0 | 189·4 | 199·4 |
| 72 | 110·30 | 126·2 | 139·0 | 149·7 | 159·1 | 167·4 | 175·1 | 182·1 | 188·6 | 194·7 | 200·4 | 211·0 |
| 74 | 116·5 | 133·4 | 146·8 | 158·1 | 167·9 | 176·7 | 185·4 | 192·4 | 199·2 | 205·7 | 211·6 | 223·4 |
| 76 | 122·9 | 140·7 | 154·8 | 166·8 | 178·6 | 186·6 | 195·0 | 202·9 | 210·1 | 216·9 | 223·3 | 235·1 |
| 78 | 129·4 | 148·2 | 163·1 | 175·6 | 186·7 | 196·5 | 205·4 | 212·1 | 221·4 | 228·5 | 235·2 | 247·6 |
| 80 | 136·2 | 155·8 | 171·6 | 184·8 | 196·4 | 206·7 | 216·1 | 224·8 | 232·8 | 240·4 | 247·4 | 260·5 |
| 82 | 143·0 | 163·8 | 180·2 | 194·2 | 206·2 | 217·3 | 226·9 | 237·8 | 244·6 | 252·5 | 260·0 | 273·8 |
| 84 | 150·1 | 171·8 | 189·1 | 203·8 | 216·5 | 227·9 | 238·3 | 247·8 | 256·7 | 265·0 | 272·8 | 287·1 |
| 86 | 157·4 | 180·1 | 198·2 | 213·6 | 227·0 | 237·8 | 247·4 | 258·2 | 269·1 | 277·8 | 286·0 | 301·0 |
| 88 | 164·8 | 188·6 | 207·6 | 223·6 | 237·5 | 250·2 | 261·6 | 272·0 | 281·7 | 290·8 | 299·4 | 315·2 |
| 90 | 172·3 | 197·3 | 217·1 | 233·9 | 248·6 | 261·7 | 273·6 | 284·5 | 291·7 | 304·2 | 313·2 | 329·7 |

length of the link G F, and divide the product by the sum of the lengths of the radius bar and of C G. The quotient is the length required.

RULE III.—(*Figs.* 3 *and* 4, *pages* 246 *and* 247.) *To find the length of the radius bar* (*F H*), *the length of C G being given.*—Square the length of C G, and divide it by the length of D G. The quotient is the length required.

RULE IV.—(*Figs.* 3 *and* 4, *pages* 246 *and* 247.) *To find the length of the radius bar, the horizontal distance of its centre* (*H*) *from the main centre being given.*—To this given horizontal distance, add half the versed sine (D N) of the arc described by the end of beam (D). Square this sum. Take the same sum, and add to it the length of

TABLE *of Nominal Horse Power of High Pressure Engines.*

| Diameter of Cylinder in Inches. | Length of Stroke in Feet. | | | | | | | | | | | |
|---|---|---|---|---|---|---|---|---|---|---|---|---|
| | 1 | 1½ | 2 | 2½ | 3 | 3½ | 4 | 4½ | 5 | 5½ | 6 | 7 |
| 2 | ·25 | ·29 | ·32 | ·35 | ·37 | ·38 | ·40 | ·42 | ·44 | ·45 | ·46 | ·49 |
| 2½ | ·39 | ·45 | ·50 | ·54 | ·57 | ·60 | ·63 | ·66 | ·68 | ·70 | ·72 | ·76 |
| 3 | ·57 | ·65 | ·72 | ·78 | ·83 | ·87 | ·91 | ·95 | ·98 | 1·01 | 1·04 | 1·10 |
| 3½ | ·78 | ·89 | ·98 | 1·06 | 1·13 | 1·19 | 1·24 | 1·29 | 1·34 | 1·38 | 1·42 | 1·49 |
| 4 | 1·02 | 1·17 | 1·29 | 1·38 | 1·47 | 1·56 | 1·62 | 1·68 | 1·74 | 1·80 | 1·86 | 1·95 |
| 4½ | 1·29 | 1·48 | 1·63 | 1·75 | 1·86 | 1·96 | 2·05 | 2·13 | 2·21 | 2·28 | 2·35 | 2·47 |
| 5 | 1·59 | 1·83 | 2·01 | 2·16 | 2·28 | 2·43 | 2·52 | 2·64 | 2·73 | 2·82 | 2·88 | 3·06 |
| 5½ | 1·93 | 2·21 | 2·43 | 2·62 | 2·78 | 2·93 | 3·12 | 3·18 | 3·50 | 3·42 | 3·51 | 3·69 |
| 6 | 2·28 | 2·61 | 2·88 | 3·12 | 3·30 | 3·48 | 3·66 | 3·78 | 3·93 | 4·05 | 4·17 | 4·41 |
| 6½ | 2·69 | 3·09 | 3·39 | 3·66 | 3·90 | 4·08 | 4·23 | 4·44 | 4·62 | 4·77 | 4·89 | 5·16 |
| 7 | 3·12 | 3·57 | 3·93 | 4·23 | 4·50 | 4·74 | 4·95 | 5·16 | 5·34 | 5·52 | 5·67 | 5·97 |
| 7½ | 3·60 | 4·11 | 4·53 | 4·86 | 5·19 | 5·46 | 5·70 | 5·94 | 6·15 | 6·33 | 6·51 | 6·87 |
| 8 | 4·08 | 4·68 | 5·16 | 5·55 | 5·88 | 6·21 | 6·48 | 6·75 | 6·99 | 7·20 | 7·41 | 7·80 |
| 8½ | 4·62 | 5·28 | 5·82 | 6·27 | 6·63 | 6·99 | 7·32 | 7·62 | 7·89 | 8·13 | 8·37 | 8·82 |
| 9 | 5·16 | 5·91 | 6·51 | 7·02 | 7·47 | 7·86 | 8·22 | 8·52 | 8·85 | 9·12 | 9·39 | 9·90 |
| 9½ | 5·76 | 6·60 | 7·26 | 7·80 | 8·37 | 8·76 | 9·15 | 9·51 | 9·84 | 10·17 | 10·47 | 10·01 |
| 10 | 6·39 | 7·32 | 8·04 | 8·67 | 9·21 | 9·69 | 10·14 | 10·53 | 10·92 | 11·28 | 11·61 | 12·21 |
| 10½ | 7·05 | 8·04 | 8·88 | 9·54 | 10·14 | 10·68 | 11·16 | 11·61 | 12·03 | 12·42 | 12·78 | 13·47 |
| 11 | 7·71 | 8·85 | 9·72 | 10·47 | 11·31 | 11·73 | 12·45 | 12·75 | 13·20 | 13·62 | 14·04 | 14·76 |
| 11½ | 8·43 | 9·66 | 10·62 | 11·46 | 12·15 | 12·78 | 13·80 | 13·92 | 14·61 | 14·91 | 15·33 | 16·14 |
| 12 | 9·18 | 10·53 | 11·58 | 12·41 | 13·26 | 13·95 | 14·58 | 15·18 | 15·72 | 16·23 | 16·71 | 17·58 |
| 12½ | 9·96 | 11·40 | 12·57 | 13·53 | 14·37 | 15·15 | 15·84 | 16·47 | 17·04 | 17·58 | 18·12 | 19·08 |
| 13 | 10·80 | 12·36 | 13·59 | 14·64 | 15·57 | 16·38 | 16·92 | 17·82 | 18·45 | 19·05 | 19·59 | 21·64 |
| 13½ | 11·64 | 13·32 | 14·64 | 15·78 | 16·77 | 17·67 | 18·48 | 19·20 | 19·89 | 20·52 | 21·15 | 22·26 |
| 14 | 12·51 | 14·31 | 15·75 | 16·98 | 18·03 | 18·99 | 19·86 | 20·64 | 21·39 | 22·08 | 22·74 | 23·94 |
| 14½ | 13·41 | 15·36 | 16·92 | 18·21 | 19·35 | 20·37 | 21·30 | 22·14 | 22·95 | 23·70 | 24·39 | 25·62 |
| 15 | 14·31 | 16·44 | 18·09 | 19·50 | 20·70 | 21·81 | 22·80 | 23·70 | 24·57 | 25·35 | 26·10 | 27·48 |
| 16 | 16·35 | 18·69 | 20·58 | 22·17 | 23·58 | 24·81 | 25·95 | 26·97 | 27·93 | 28·83 | 29·70 | 31·26 |
| 17 | 18·45 | 21·12 | 23·25 | 25·05 | 26·58 | 28·02 | 29·28 | 30·45 | 31·56 | 32·55 | 33·57 | 35·28 |
| 18 | 20·67 | 23·67 | 26·04 | 28·08 | 29·82 | 31·41 | 32·82 | 34·14 | 35·37 | 36·51 | 37·59 | 39·57 |
| 19 | 23·04 | 26·37 | 29·04 | 31·26 | 33·51 | 34·98 | 36·57 | 38·04 | 39·39 | 40·68 | 41·88 | 44·07 |
| 20 | 25·53 | 29·22 | 32·16 | 34·65 | 36·81 | 38·76 | 40·53 | 42·15 | 43·65 | 45·06 | 46·38 | 48·84 |
| 22 | 30·90 | 35·37 | 38·91 | 41·94 | 44·55 | 46·89 | 49·86 | 51·90 | 52·95 | 54·54 | 56·13 | 59·10 |
| 24 | 36·78 | 42·09 | 46·32 | 49·89 | 53·01 | 55·83 | 58·35 | 60·69 | 62·85 | 64·89 | 66·81 | 70·32 |
| 26 | 43·17 | 49·38 | 54·36 | 58·56 | 62·25 | 65·52 | 67·68 | 71·25 | 73·80 | 76·17 | 78·42 | 82·53 |
| 28 | 50·04 | 57·27 | 63·06 | 67·92 | 72·18 | 75·99 | 79·44 | 82·62 | 85·56 | 88·32 | 90·93 | 95·70 |
| 30 | 57·45 | 65·76 | 72·39 | 77·97 | 82·86 | 87·21 | 91·20 | 94·83 | 98·22 | 101·40 | 104·4 | 109·9 |
| 32 | 65·37 | 74·88 | 82·53 | 88·71 | 94·26 | 99·24 | 103·7 | 107·9 | 111·8 | 115·4 | 118·7 | 125·0 |
| 34 | 73·80 | 84·48 | 92·9 | 100·22 | 106·3 | 112·0 | 117·1 | 121·8 | 126·2 | 130·2 | 134·0 | 141·1 |
| 36 | 82·71 | 94·68 | 104·2 | 112·2 | 119·3 | 125·6 | 131·3 | 136·5 | 141·4 | 146·0 | 150·3 | 158·2 |
| 38 | 92·16 | 105·5 | 116·1 | 125·0 | 134·0 | 136·9 | 146·3 | 152·1 | 157·6 | 162·7 | 167·5 | 176·3 |
| 40 | 102·1 | 116·9 | 129·6 | 128·6 | 147·3 | 155·1 | 162·1 | 168·6 | 174·6 | 180·2 | 185·6 | 195·3 |
| 42 | 112·6 | 128·9 | 141·8 | 152·8 | 162·4 | 170·9 | 178·7 | 185·9 | 192·5 | 198·7 | 204·6 | 215·3 |
| 44 | 123·5 | 141·4 | 155·7 | 167·7 | 178·1 | 187·6 | 199·4 | 204·0 | 211·3 | 218·1 | 224·5 | 236·3 |
| 46 | 135·0 | 154·6 | 170·1 | 183·3 | 194·6 | 204·6 | 214·3 | 223·0 | 230·0 | 238·4 | 245·4 | 258·3 |
| 48 | 147·0 | 168·3 | 185·3 | 199·6 | 212·1 | 223·2 | 233·4 | 242·8 | 251·5 | 259·6 | 267·2 | 281·3 |
| 50 | 159·6 | 182·6 | 201·0 | 216·5 | 230·1 | 242·3 | 253·3 | 263·4 | 272·9 | 281·6 | 289·9 | 305·1 |
| 52 | 172·6 | 197·6 | 217·4 | 234·2 | 249·0 | 262·0 | 270·7 | 284·9 | 295·2 | 304·6 | 313·5 | 330·0 |
| 54 | 186·1 | 213·0 | 234·5 | 252·6 | 268·4 | 282·6 | 295·4 | 307·2 | 318·3 | 328·5 | 338·1 | 356·1 |
| 56 | 200·1 | 229·1 | 252·2 | 271·6 | 288·7 | 303·9 | 317·7 | 330·3 | 342·3 | 353·4 | 363·6 | 382·8 |
| 58 | 214·7 | 245·8 | 270·5 | 291·4 | 309·6 | 325·8 | 340·8 | 354·6 | 367·2 | 378·9 | 389·7 | 410·1 |
| 60 | 229·8 | 263·0 | 289·5 | 311·7 | 331·2 | 348·9 | 364·8 | 379·2 | 393·0 | 405·6 | 417·6 | 439·5 |

the beam (C D). Divide the square previously found by this last sum, and the quotient is the length sought.

RULE V.—(*Figs.* 5 *and* 6, *pages* 247, 248.)—*To find the length of the radius bar, C G and P Q being given.*—Square C G, and multiply the square by the length of the side rod (P D): call this product A. Multiply Q D by the length of the side lever (C D). From this product subtract the product of D P into C G, and divide A by the remainder. The quotient is the length required.

RULE VI.—(*Figs.* 5 *and* 6, *pages* 247, 248.) *To find the length of the radius bar; P Q, and the horizontal distance of the centre H of the radius bar from the main centre being given.*—To the given horizontal distance add half the versed sine (D N) of the arc described

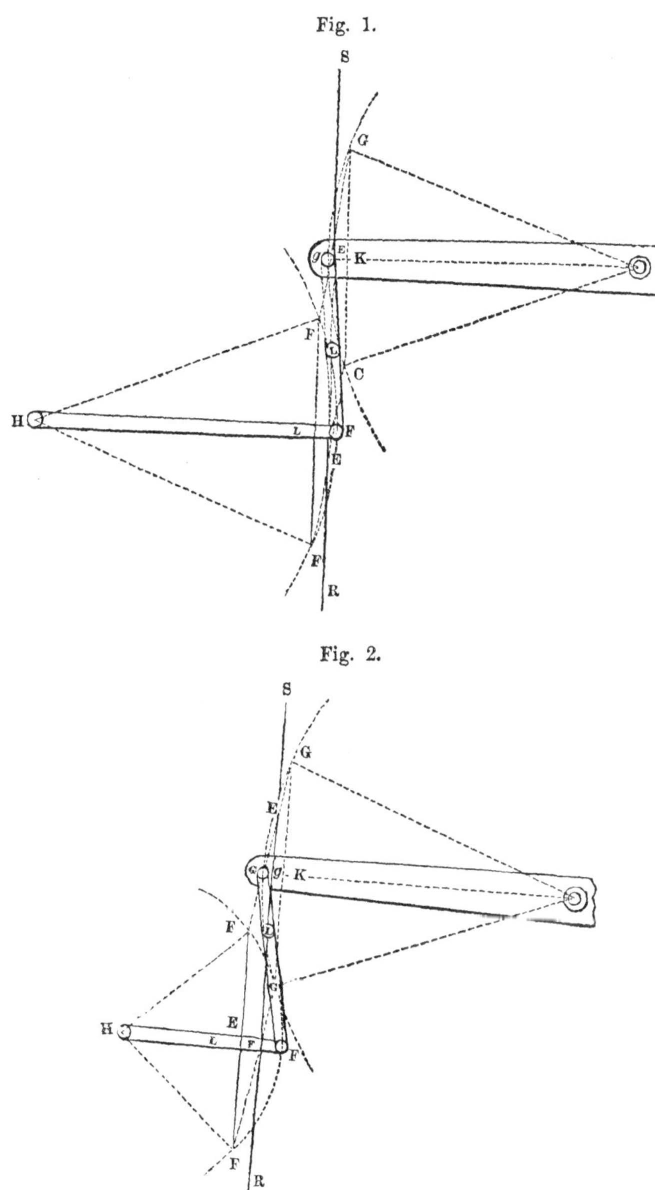

Fig. 1.

Fig. 2.

by the extremity (D) of the side lever. Square this sum and multiply the square by the length of the side rod (P D). Call this product A. Take the same horizontal distance as before added to the same half versed sine (D N), and multiply the sum by the length of the side rod (P D): to the product add the product of the length of

Fig. 3.

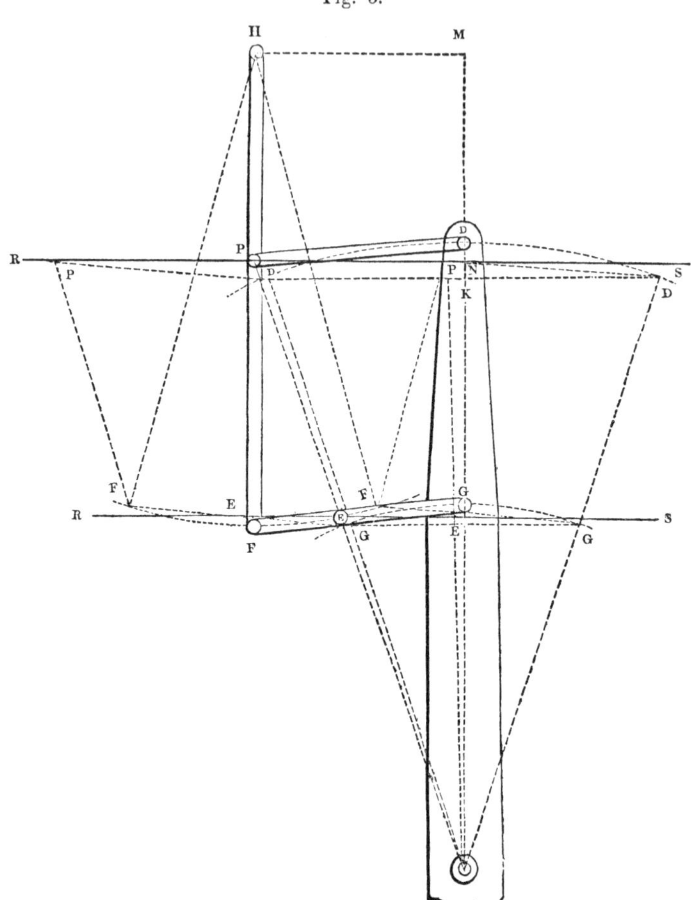


the side lever C D into the length of Q D, and divide A by the sum. The quotient will be the length required.

When the centre H of the radius has its position determined, rules 4 and 6 will always give the length of the radius bar F H. To get the length of C G, it will only be necessary to draw through the point F a line parallel to the side rod D P, and the point where that line cuts D C will be the position of the pin G.

In using these formulas and rules, the dimensions must all be taken in the same measure; that is, either all in feet, or all in inches; and when great accuracy is required, the corrections given in Table (A) must be added to or subtracted from the calculated length of the radius bar, according as it is less or greater than the length of C G, the part of the beam that works it.

1. Rule 4.—Let the horizontal distance (M C) of the centre (H)

Fig. 4.

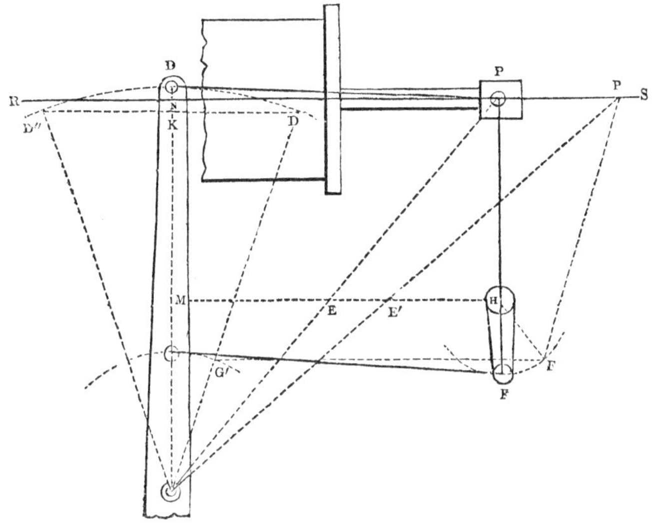

Fig. 5.

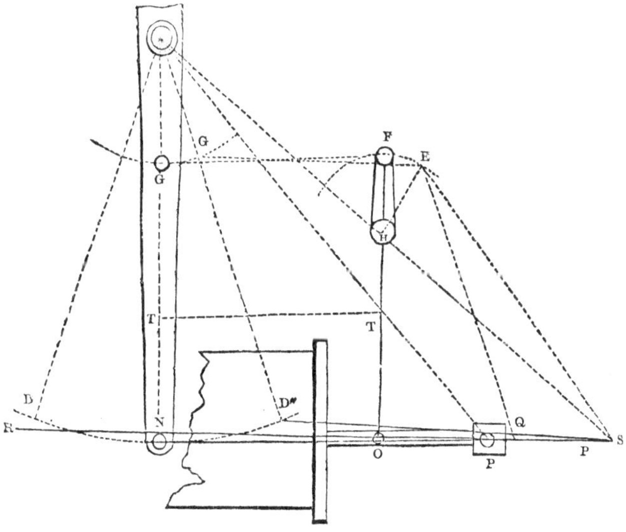


of the radius bar from the main centre be equal to 51 inches; the half versed sine D N = 3 inches, and D C = 126 inches; then by the rule we will have

$$\frac{(51 + 3)^2}{51 + 3 + 126} = \frac{(54)^2}{180} = \frac{2916}{180} = 16{\cdot}2 \text{ inches},$$

which is the required length of the radius bar (F H).

Fig. 6.

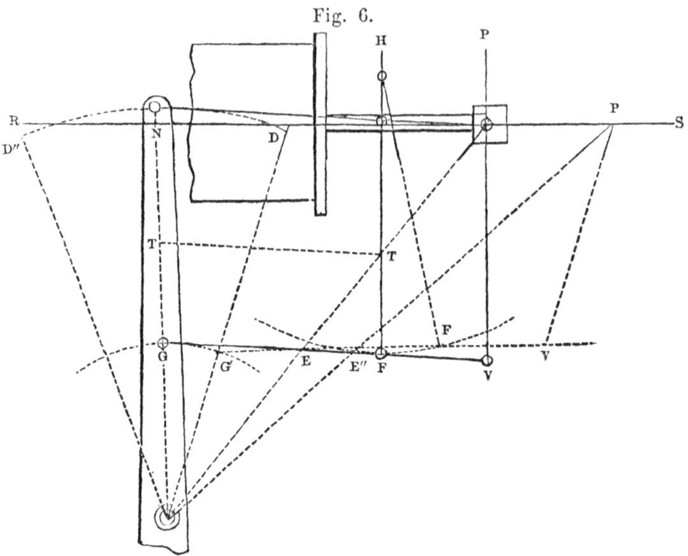


2. RULE 5.—The following dimensions are those of the Red Rov steamer: C G = 32 D P = 94 Q D = 74 C D = 65 P Q = 20.

By the rule we have, $A = (32)^2 \times 94 = 96256$ and

$$\frac{96256}{74 \times 65 - 94 \times 32} = \frac{96256}{1802} = 53 \cdot 4,$$

which is the required length of the radius bar.

3. RULE 6.—Take the same data as in the last example, on supposing that C G is not given, and that the centre H is fixed a horizontal distance from the main centre, equal to 83·5 inche Then the half versed sine of the arc D′ D D″ will be about inches, and we will have by the rule

$$A = (83 \cdot 5 + 2)^2 \times 94 = 705963 \cdot 5 \text{ and}$$

$$\frac{A}{85 \cdot 5 \times 94 + 65 \times 74} = \frac{705963 \cdot 5}{1284 \cdot 7} = 54 \cdot 8 \text{ inches,}$$

the required length of the radius bar in this case.

TABLE (A).

| This column gives $\frac{FH}{CG}$ when C G is the greater, and $\frac{CG}{FH}$ when F H is the greater. | Correction to be added to or subtracted from the calculated length of the radius bar, in decimal parts of its calculated length. |
|---|---|
| 1·0 | 0 |
| ·9 | ·0034 |
| ·8 | ·0075 |
| ·7 | ·0163 |
| ·6 | ·0270 |
| ·5 | ·0452 |
| ·4 | ·0817 |

In both of the last two examples $\frac{\text{C G}}{\text{H F}} = \cdot 6$ nearly. The correction found by Table (A), therefore, would be $54 \times \cdot 027 = 1\cdot 458$ inches, which must be subtracted from the lengths already found for the radius bar, because it is longer than C G. The corrected lengths will therefore be

In example 2......................F H = 51·94 inches.
In example 3......................F H = 53·34 inches.

Rule.—*To find the depth of the main beam at the centre.*—Divide the length in inches from the centre of motion to the point where the piston rod is attached, by the diameter of the cylinder in inches; multiply the quotient by the maximum pressure in pounds per square inch of the steam in the boiler; divide the product by 202 for cast iron, and 236 for malleable iron: in either case, the cube root of the quotient multiplied by the diameter of the cylinder in inches gives the depth in inches of the beam at the centre of motion. *To find the breadth at the centre.*—Divide the depth in inches by 16; the quotient is the breadth in inches.

An engine beam is three times the diameter of the cylinder, from the centre to the point where the piston rod acts on it; the force of the steam in the boiler when about to force open the safety valve is 10 lbs. per square inch. Required the depth and breadth when the beam is of cast iron.

In this case $n = 3$, and $P = 10$, and therefore

$$d = \text{D}\left\{\frac{30}{202}\right\}^{\frac{1}{3}} = \cdot 53\ \text{D}.$$

$$\text{The breadth} = \frac{\cdot 53}{16}\ \text{D} = \cdot 03\ \text{D}.$$

It will be observed that our rule gives the least value to the depth. In actual practice, however, it is necessary to make allowance for accidents, or for faultiness in the materials. This may be done by making the depth greater than that determined by the rule; or, perhaps more properly, by taking the pressure of the steam much greater than it can ever possibly be. As for the dimensions of the other parts of the beam, it is obvious that they ought to diminish towards the extremities; for the power of a beam to resist a cross strain varies inversely as its length. The dimensions may be determined from the formula $f\,b\,d^2 = 6\ \text{W}\ l$.

To apply the formula to cranks, we may assume the depth at the shaft to be equal to $n$ times the diameter of the shaft; hence, if $m \times \text{D}$ be the diameter of the shaft, the depth of the crank will be $n \times m \times \text{D}$. Substituting this in the formula $f\,b\,d^2 = 6\ \text{W}\ l$, and it becomes $f\,b \times n^2 \times m^2 \times \text{D}^2 = 6\ \text{W}\ l$. Now, as before, $\text{W} = \cdot 7854 \times \text{P} \times \text{D}^2$, so that the formula becomes $f \times b \times n^2 \times m^2 = 4\cdot 7124 \times \text{P} \times l$. The value of $n$ is arbitrary. In practice it may be made equal to $1\frac{1}{2}$ or 1·5. Taking this value, then, for cast iron, the formula becomes $15300 \times b \times \frac{9}{4} \times m^2 = 4{\cdot}7124 \times P \times l$, or $7305\, m^2\, b = P\, l$; but if L denote the length of the crank in feet, the formula becomes $609\, m^2\, b = P\, L$, and $\therefore b = P \times L \div 609\, m^2$. This formula may be put into the form of a rule, thus:—

RULE.—*To find the breadth at the shaft when the depth is equal to $1\frac{1}{2}$ times the diameter of the shaft.*—Divide the square of the diameter of the shaft in inches by the square of the diameter of the cylinder; multiply the quotient by 609, and reserve the product for a divisor; multiply the greatest elastic force of the steam in lbs. per square inch by the length of the crank in feet, and divide the product by the reserved divisor: the quotient is the breadth of the crank at the shaft.

A crank shaft is $\frac{1}{4}$ the diameter of the cylinder; the greatest possible force of the steam in the boiler is 20 lbs. per square inch; and the length of the shaft is 3 feet. Required the breadth of the crank at the shaft when its depth is equal to $1\frac{1}{2}$ times the diameter of the shaft.

In this case $m = \frac{1}{4}$, so that the reserved divisor $- \frac{609}{16} = 38$: again, elastic force of steam in lbs. per square inch $= 20$ lbs.; hence width of crank $= \frac{3 \times 20}{38} = 1{\cdot}6$ inches nearly.

RULE.—*To find the diameter of a revolving shaft.*—Form a reserved divisor thus: multiply the number of revolutions which the shaft makes for each double stroke of the piston by the number 1222 for cast iron, and the number 1376 for malleable iron. Then divide the radius of the crank, or the radius of the wheel, by the diameter of the cylinder; multiply the quotient by the greatest pressure of the steam in the boiler expressed in lbs. per square inch; divide the product by the reserved divisor; extract the cube root of the quotient, and multiply the result by the diameter of the cylinder in inches. The product is the diameter of the shaft in inches.

## STRENGTH OF RODS WHEN THE STRAIN IS WHOLLY TENSILE; SUCH AS THE PISTON ROD OF SINGLE ACTING ENGINES, PUMP RODS, ETC.

RULE.—*To find the diameter of a rod exposed to a tensile force only.*—Multiply the diameter of the piston in inches by the square root of the greatest elastic force of the steam in the boiler estimated in lbs. per square inch; the product, divided by 95, is the diameter of the rod in inches.

Required the diameter of the transverse section of a piston rod in a single acting engine, when the diameter of the cylinder is 50 inches, and the greatest possible force of the steam in the boiler is 16 lbs. per square inch. Here, according to the formula,

$$d = \frac{50}{95} \sqrt{16} = \frac{200}{95} = 2{\cdot}1 \text{ inches.}$$

RULE.—*To find the strength of rods alternately extended and compressed, such as the piston rods of double acting engines.*—Multiply the diameter of the piston in inches by the square root of the maximum pressure of the steam in lbs. per square inch; divide the product by

47 for cast iron,
50 for malleable iron.

This rule applies to the piston rods of double acting engines, parallel motion rods, air-pump and force-pump rods, and the like. The rule may also be applied to determine the strength of connecting rods, by taking, instead of P, a number P′, such that P′ × sine of the greatest angle which the connecting rod makes with the direction = P.

Supposing the greatest force of the steam in the boiler to be 16 lbs. per square inch, and the diameter of the cylinder 50 inches; required the diameter of the piston rod, supposing the engine to be double acting. In this case

$$\text{for cast iron } d = \frac{\text{D}}{47}\sqrt{\text{P}} = \frac{50 \times 4}{47} = 5 \text{ inches nearly;}$$

$$\text{for malleable iron } d = \frac{\text{D}}{50}\sqrt{\text{P}} = 4 \text{ inches.}$$

The pressure, however, is always taken in practice at more than 16 lbs. If the pressure be taken at 25 lbs., the diameter of a malleable iron piston rod will be 5 inches, which is the usual proportion. Piston rods are never made of cast iron, but air-pump rods are sometimes made of brass, and the connecting rods of land engines are cast iron in most cases.

FORMULAS FOR THE STRENGTH OF VARIOUS PARTS OF MARINE ENGINES.

The following general rules give the dimensions proper for the parts of marine engines, and we shall recapitulate, with all possible brevity, the data upon which the denominations rest.

Let pressure of the steam in boiler = $p$ lbs. per square inch,
Diameter of cylinder = D inches,
Length of stroke = 2 R inches.

The vacuum below the piston is never complete, so that there always remains a vapour of steam possessing a certain elasticity. We may suppose this vapour to be able to balance the weight of the piston. Hence the entire pressure on the square inch of piston in lbs. = $p$ + pressure of atmosphere = 15 + $p$. We shall substitute P for 15 + $p$. Hence

$$\text{Entire pressure on piston in lbs.} = \cdot 7854 \times (15 + p) \times \text{D}^2$$
$$= \cdot 7854 \times \text{P} \times \text{D}^2.$$

The dimensions of the paddle-shaft journal may be found from the following formulas, which are calculated so that the strain in ordinary working = $\frac{5}{6}$ elastic force.

$$\text{Diameter of paddle-shaft journal} = \cdot 08264\,\{\text{R} \times \text{P} \times \text{D}^2\}^{\frac{1}{3}}$$
$$\text{Length of ditto} = 1\tfrac{1}{4} \times \text{diameter.}$$

The dimensions of the several parts of the crank may be found from the following formulas, which are calculated so that the strain in ordinary working = one-half the elastic force; and when one paddle is suddenly brought up, the strain at shaft end of crank = $\frac{2}{3}$ elastic force, the strain at pin end of crank = elastic force.

Exterior diameter of large eye = diameter of paddle-shaft +

$$\left\{\frac{D\left[P \times 1{\cdot}561 \times R^2 + {\cdot}00494 \times D^2 \times P^2\right]^{\frac{1}{2}}}{75{\cdot}59 \times \sqrt{R}}\right\}^{\frac{2}{3}}$$

Length of ditto = diameter of paddle shaft.

Exterior diameter of small eye = diameter of crank pin + 02521 × $\sqrt{P}$ × D.

Length of ditto = ·0375 × $\sqrt{P}$ × D.

Thickness of web at paddle centre =

$$\left\{\frac{D^2 \times P \times \sqrt{\{1{\cdot}561 \times R^2 + {\cdot}00494 \times D^2 \times P\}}}{9000}\right\}^{\frac{1}{3}}$$

Breadth of ditto = 2 × thickness.

Thickness of web at pin centre — ·022 × $\sqrt{P}$ × D.

Breadth of ditto = $\frac{3}{2}$ × thickness.

As these formulas are rather complicated, we may show what they become when $p$ = 10 or P = 25.

Exterior diameter of large eye = diameter of paddle shaft +

$$\left\{\frac{D\sqrt{(1{\cdot}561 \times R^2 + {\cdot}1235 \times D^2)}}{15{\cdot}12 \times \sqrt{R}}\right\}^{\frac{2}{3}}$$

Length of ditto = diameter of paddle shaft.

Exterior diameter of small eye = equal diameter of crank pin + 126 × D.

Length of ditto = ·1875 × D.

Thickness of web at pin centre = ·11 × D.

Breadth of ditto = $\frac{3}{2}$ × thickness of web.

The dimensions of the crank pin journal may be found from the following formulas, which are calculated so that strain when bearing at outer end = elastic force, and in ordinary working strain = one-third of elastic force.

Diameter of crank-pin journal = ·02836 × $\sqrt{P}$ × D.
Length of ditto = $\frac{9}{8}$ × diameter.

The dimensions of the several parts of the cross head may be found from the following formulas, in which we have assumed, for the purpose of calculation, the length = 1·4 × D. The formulas have been calculated so as to give the strain of web = $\frac{1}{2{\cdot}225}$ × elastic force; strain of journal in ordinary working = $\frac{1}{2{\cdot}33}$ × elastic force, and when bearing at outer end = $\frac{1}{1{\cdot}165}$ × elastic force.

Exterior diameter of eye = diameter of hole + $\cdot 02827 \times P^{\frac{1}{3}} \times D$.

Depth of ditto = $\cdot 0979 \times P^{\frac{1}{3}} \times D$.

Diameter of journal = $\cdot 01716 \times \sqrt{P} \times D$.

Length of ditto = $\frac{9}{8}$ diameter of journal.

Thickness of web at middle = $\cdot 0245 \times P^{\frac{1}{3}} \times D$.

Breadth of ditto = $\cdot 09178 \times P^{\frac{1}{3}} \times D$.

Thickness of web at journal = $\cdot 0122 \times P^{\frac{1}{2}} \times D$.

Breadth of ditto = $\cdot 0203 \times P^{\frac{1}{2}} \times D$.

The dimensions of the several parts of the piston rod may be found from the following formulas, which are calculated so that the strain of piston rod = $\frac{1}{7}$ elastic force.

Diameter of the piston rod = $\dfrac{\sqrt{P} \times D.}{50}$

Length of part in piston = $\cdot 04 \times D \times P$.

Major diameter of part in crosshead = $\cdot 019 \times \sqrt{P} \times D$.

Minor diameter of ditto = $\cdot 018 \times \sqrt{P} \times D$.

Major diameter of part in piston = $\cdot 028 \times \sqrt{P} \times D$.

Minor diameter of ditto = $\cdot 023 \times \sqrt{P} \times D$.

Depth of gibs and cutter through crosshead = $\cdot 0358 \times P^{\frac{1}{3}} \times D$.

Thickness of ditto = $\cdot 007 \times P^{\frac{1}{3}} \times D$.

Depth of cutter through piston = $\cdot 017 \times \sqrt{P} \times D$.

Thickness of ditto = $\cdot 007 \times P^{\frac{1}{2}} \times D$.

The dimensions of the several parts of the connecting rod may be found from the following formulas, which are calculated so that the strain of the connecting rod and the strain of the strap are both equal to one-sixth of the elastic force.

Diameter of connecting rod at ends = $\cdot 019 \times P^{\frac{1}{2}} \times D$.

Diameter of ditto at middle = $\{1 + \cdot 0035 \times \text{length in inches}\} \times \cdot 019 \times \sqrt{P} \times D$.

Major diameter of part in crosstail = $\cdot 0196 \times P^{\frac{1}{2}} \times D$.

Minor ditto = $\cdot 018 \times P^{\frac{1}{2}} \times D$.

Breadth of butt = $\cdot 0313 \times P^{\frac{1}{2}} \times D$.

Thickness of ditto = $\cdot 025 \times P^{\frac{1}{2}} \times D$.

Mean thickness of strap at cutter = $\cdot 00854 \times \sqrt{P} \times D$.

Ditto above cutter = $\cdot 00634 \times \sqrt{P} \times D$.

Distance of cutter from end of strap = $\cdot 0097 \times \sqrt{P} \times D$.

Breadth of gibs and cutter through crosstail = $\cdot 0358 \times P^{\frac{1}{3}} \times D$.

Breadth of gibs and cutter through butt = $\cdot 022 \times P^{\frac{1}{2}} \times D$.

Thickness of ditto = $\cdot 00564 \times P^{\frac{1}{2}} \times D$.

The dimensions of the several parts of the side rods may be found from the following formulas, which are calculated so as to make the strain of the side rod = one-sixth of elastic force, and the strains of strap and cutter = one-fifth of elastic force.

Diameter of cylinder side rods at ends = $\cdot 0129 \times P^{\frac{1}{2}} \times D$.
Diameter of ditto at middle = $(1 + \cdot 0035 \times \text{length in inches})$.

$$\times \cdot 0129 \times P^{\frac{1}{2}} \times D.$$

Breadth of butt = $\cdot 0154 \times P^{\frac{1}{2}} \times D$.

Thickness of ditto = $\cdot 0122 \times P^{\frac{1}{2}} \times D$.

Diameter of journal at top end of side rod = $\cdot 01716 \times P^{\frac{1}{2}} \times D$.
Length of journal at top end = $\frac{9}{8}$ diameter.

Diameter of journal at bottom end = $\cdot 014 \times P^{\frac{1}{2}} \times D$.

Length of ditto = $\cdot 0152 \times P^{\frac{1}{2}} \times D$.

Mean thickness of strap at cutter = $\cdot 00643 \times P^{\frac{1}{2}} \times D$.

Ditto below cutter = $\cdot 0047 \times P^{\frac{1}{2}} \times D$.

Breadth of gibs and cutter = $\cdot 016 \times P^{\frac{1}{2}} \times D$.

Thickness of ditto = $\cdot 0033 \times P^{\frac{1}{2}} \times D$.

The dimensions of the main centre journal may be found from the following formulas, which are calculated so as to make the strain in ordinary working = one half elastic force.

Diameter of main centre journal = $\cdot 0367 \times P^{2} \times D$.
Length of ditto = $\frac{3}{2} \times$ diameter.

The dimensions of the several parts of the air-pump may be found from the corresponding formulas given above, by taking for D another number *d* the diameter of air-pump.

### DIMENSIONS OF THE SEVERAL PARTS OF FURNACES AND BOILERS.

Perhaps in none of the parts of a steam engine does the practice of engineers vary more than in those connected with furnaces and boilers. There are, no doubt, certain proportions for these, as well as for the others, which produce the maximum amount of useful effect for particular given purposes; but the determination of these proportions, from theoretical considerations, has hitherto been attended with insuperable difficulties, arising principally from our imperfect knowledge of the laws of combustion of fuel, and of the laws according to which caloric is imparted to the water in the boiler. In giving, therefore, the following proportions for the different parts, we desire to have it understood that we do not affirm them to be the best, absolutely considered; we give them only as the average practice of the best modern constructors. In most of the cases we have given the average value per nominal horse power. It is well known that the term horse power is a conventional unit for measuring the size of steam engines, just as a foot or a mile is a unit for the measurement of extension. There is this difference, however, in the two cases, that whereas the length of a foot is fixed definitively, and is known to every one, the dimensions proper to an engine horse power differ in the practice of every different maker: and the same kind of confusion is thereby introduced into engineering as if one person were to make his foot-rule eleven inches long, and another thirteen inches. It signifies very little what a horse power is defined to be; but when once defined, the measurement should be kept inviolable. The question now arises, what standard ought to be the accepted one. For our present purpose, it is necessary to connect by a formula the three quantities, nominal horses power, length of stroke, and diameter of cylinder. With this intention,

Let $S$ = length of stroke in feet,
$d$ = diameter of cylinder in inches;

Then the nominal horse power $= \dfrac{d^2 \times \sqrt[3]{S}}{47}$ nearly.

I. *Area of Fire Grate.*—The average practice is to give ·55 square feet for each nominal horse power. Hence the following rule:

RULE 1.—*To find the area of the fire grate.*—Multiply the number of horses power by ·55; the product is the area of the fire grate in square feet.

Required the total area of the fire grate for an engine of 400 horse power. Here total area of fire grate in square feet = $400 \times \cdot 55 = 220$.

A rule may also be found for expressing the area of the fire grate in terms of the length of stroke and the diameter of the cylinder. For this purpose we have,

$$\text{total area of fire grate} = \frac{\cdot 55 \times d^2 \times \sqrt[3]{S}}{47} \text{ feet} = \frac{d^2 \times \sqrt[3]{S}}{86} \text{ feet.}$$

This formula expressed in words gives the following rule.

RULE 2.—*To find the area of fire grate.*—Multiply the cube root of the length of stroke in feet by the square of the diameter in inches; divide the product by 86; the quotient is the area of fire grate in square feet.

Required the total area of the fire grate for an engine whose stroke = 8 feet, and diameter of cylinder = 50 inches.

Here, according to the rule,

$$\text{total area of fire grate in square feet} = \frac{50^2 \times \sqrt[3]{8}}{86} = \frac{2500 \times 2}{86} = \frac{5000}{86} = 59 \text{ nearly.}$$

In order to work this example by the first rule, we find the nominal horse power of the engine whose dimensions we have specified is 104·3; hence,

total area of fire grate in square feet = $106 \cdot 4 \times \cdot 55 = 58 \cdot 5$.

With regard to these rules we may remark, not only that they are founded on practice, and therefore empyrical, but they are only applicable to large engines. When an engine is very small, it requires a much larger area of fire grate in proportion to its size than a larger one. This depends upon the necessity of having a certain amount of fire grate for the proper combustion of the coal.

II. *Length of Furnace.*—The length of the furnace differs considerably, even in the practice of the same engineer. Indeed, all the dimensions of the furnace depend to a certain extent upon the peculiarity of its position. From the difficulty of firing long furnaces efficiently, it has been found more beneficial to restrict the length of the furnace to about six feet than to employ furnaces of greater length.

III. *Height of Furnace above Bars.*—This dimension is variable, but it is a common practice to make the height about two feet.

IV. *Capacity of Furnace Chamber above Bars.*—The average per horse power may be taken at 1·17 feet. Hence the following rule:

RULE.—*To find the capacity of furnace chamber above bars.*—Multiply the number of nominal horses power by 1·17; the product is the capacity of furnace chambers above bars in cubic feet.

V. *Areas of Flues or Tubes in smallest part.*—The average value of the area per horse power is 11·2 sq. in. Hence we have the following rule:

RULE.—*To find the total area of the flues or tubes in smallest part.*—Multiply the number of horse power by 11·2; the product is the total area in square inches of flues or tubes in smallest part.

Required total area of flues or tubes for the boiler of a steam engine when the horse power = 400.

For this example we have, according to the rule,

$$\text{Total area in square inches} = 400 \times 11{\cdot}2 = 4480.$$

We may also find a very convenient rule expressed in terms of the stroke and the diameter of cylinder. Thus,

$$\text{Total area of tubes or flues in square inches} = \frac{11{\cdot}2 \times d^2 \times \sqrt[3]{S}}{47}$$

$$= \frac{d^2 \times \sqrt[3]{S}}{4}$$

VI. *Effective Heating Surface.*—The effective heating surface of flue boilers is the whole of furnace surface above bars, the whole of tops of flues, half the sides of flues, and none of the bottoms; hence the effective flue surface is about half the total flue surface. In tubular boilers, however, the whole of the tube surface is reckoned effective surface.

### EFFECTIVE HEATING SURFACE OF FLUE BOILERS.

RULE 1.—*To find the effective heating surface of marine flue boilers of large size.*—Multiply the number of nominal horse power by 5; the product is the area of effective heating surface in square feet.

Required the effective heating surface of an engine of 400 nominal horse power.

In this case, according to the rule, effective heating surface in square feet $= 400 \times 5 = 2000$.

The effective heating surface may be expressed in terms of the length of stroke and the diameter of the cylinder.

Rule 2.—*To find the total effective heating surface of marine flue boilers.*—Multiply the square of the diameter of cylinder in inches by the cube root of the length of stroke in feet; divide the product by 10: the quotient expresses the number of square feet of effective heating surface.

Required the amount of effective heating surface for an engine whose stroke = 8 ft., and diameter of cylinder = 50 inches.

Here, according to Rule 2, effective heating surface in square feet

$$= \frac{50^2 \times \sqrt[3]{8}}{10} = \frac{2500 \times 2}{10} = \frac{5000}{10} = 500.$$

To solve this example according to the first rule, we have the nominal horse power of the engine equal to 106·4. Hence, according to Rule 2, total effective heating surface in square feet $= 106{\cdot}4 \times 4{\cdot}92 = 523\frac{1}{2}$.

EFFECTIVE HEATING SURFACE OF TUBULAR BOILERS.

The effective heating surface of tubular boilers is about equal to the total heating surface of flue boilers, or is double the effective surface; but then the total tube surface is reckoned effective surface.

It appears that the total heating surface of flue and tubular marine boilers is about the same, namely, about 10 square feet per horse power.

VII. *Area of Chimney.*—Rule 1.—*To find the area of chimney.*—Multiply the number of nominal horse power by 10·23; the product is the area of chimney in square inches.

Required the area of the chimney for an engine of 400 nominal horse power.

In this example we have, according to the rule,

area of chimney in square inches $= 400 \times 10{\cdot}23 = 4092$.

We may also find a rule for connecting together the area of the chimney, the length of the stroke, and the diameter of the cylinder.

Rule 2.—*To find the area of the chimney.*—Multiply the square of the diameter expressed in inches by the cube root of the stroke expressed in feet; divide the product by the number 5; the quotient expresses the number of square inches in the area of chimney.

Required the area of the chimney for an engine whose stroke = 8 feet, and diameter of cylinder = 50 inches.

We have in this example from the rule,

$$\text{area of chimney in square inches} = \frac{50^2 \times \sqrt[3]{8}}{5} = \frac{2500 \times 2}{5} = 1000.$$

To work this example according to the first rule, we find, that the nominal horse power of this engine is 104·6: hence,

area of chimney in square inches = 104·6 × 10·23 = 1070.

The latter value is greater than the former one by 70 inches. This difference arises from our taking too great a divisor in Rule 2. Either of the values, however, is near enough for all practical purposes.

VIII. *Water in Boiler.*—The quantity of water in the boiler differs not only for different boilers, but differs even for the same boiler at different times. It may be useful, however, to know the average quantity of water in the boiler for an engine of a given horse power.

RULE 1.—*To determine the average quantity of water in the boiler.*—Multiply the number of horse power by 5; the product expresses the cubic feet of water usually in the boiler.

This rule may be so modified as to make it depend upon the stroke and diameter of the cylinder of engine.

RULE 2.—*To determine the cubic feet of water usually in the boiler.*—Multiply together the cube root of the stroke in feet, the square of the diameter of the cylinder in inches, and the number 5; divide the continual product by 47; the quotient expresses the cubic feet of water usually in the boiler.

Required the usual quantity of water in the boilers of an engine whose stroke = 8 feet, and diameter of cylinder 50 inches.

Here we have from the rule,

$$\text{cubic feet of water in boiler} = \frac{5 \times 50^2 \times \sqrt[3]{8}}{47} = \frac{5 \times 2500 \times 2}{47}$$

$$= \frac{25000}{47} = 532 \text{ nearly.}$$

The engine, with the dimensions we have specified, is of 106·4 nominal horse power. Hence, according to Rule 1,

cubic feet of water in boiler = 106·4 × 5 = 532.

IX. *Area of Water Level.*—RULE 1.—*To find the area of water level.*—The area of water level contains the same number of square feet as there are units in the number expressing the nominal horse power of the engine.

Required the area of water level for an engine of 200 nominal horse power. According to the rule, the answer is 200 square feet.

We add a rule for finding the area of water level when the diameter of cylinder and the length of stroke is given.

RULE 2.—*To find the area of water level.*—Multiply the square of the diameter in inches by the cube root of the stroke in feet; divide the product by 47; the quotient expresses the number of square feet in the area of water level.

Required the area of the water level for an engine whose stroke is 8 feet, and diameter of cylinder 50 inches.

In this case, according to the rule,

$$\text{area of water level in square feet} = \frac{50^2 \times \sqrt[3]{8}}{47} = 106.$$

X. *Steam Room.*—It is obvious that the steam room, like the quantity of water, is an extremely variable quantity, differing, not only for different boilers, but even in the same boiler at different times. It is desirable, however, to know the content of that part of the boiler usually filled with steam.

RULE 1.—*To determine the average quantity of steam room.*—Multiply the number expressing the nominal horse power by 3; the product expresses the average number of cubic feet of steam room.

Required the average capacity of steam room for an engine of 460 nominal horse power.

According to the rule,

Average capacity of steam room = 460 × 3 cubic feet = 1380 cubic feet.

This rule may be so modified as to apply when the length of stroke and diameter of cylinder are given.

RULE 2.—Multiply the square of the diameter of the cylinder in inches by the cube root of the stroke in feet; divide the product by 15; the quotient expresses the number of cubic feet of steam room.

Required the average capacity of steam room for an engine whose stroke is 8 feet, and diameter of cylinder 5 inches.

In this case, according to the rule,

$$\text{Steam room in cubic feet} = \frac{50^2 \times \sqrt[3]{8}}{15} = \frac{2500 \times 2}{15} = \frac{5000}{15} = 333\tfrac{1}{3}.$$

We find that the nominal horse power of this engine is 106·4; hence, according to Rule 1,

average steam room in cubic feet = 106·4 × 3 = 320 nearly.

Before leaving these rules, we would again repeat that they ought not to be considered as rules founded upon considerations for giving the maximum effect from the combustion of a given amount of fuel; and consequently the engineer ought not to consider them as invariable, but merely to be followed as far as circumstances will permit. We give them, indeed, as the medium value of the very variable practice of several well-known constructors; consequently, although the proportions given by the rules may not be the best possible for producing the most useful effect, still the engineer who is guided by them is sure not to be very far from the common practice of most of our best engineers. It has often been lamented that the methods used by different engine makers for estimating the nominal powers of their engines have been so various that we can form no real estimate of the dimensions of the engine, from its reputed nominal horse power, unless we know its maker; but the same confusion exists, also, to some extent, in the construction of boilers. Indeed, many things may be mentioned, which have hitherto operated as a barrier to the practical application of any standard of engine power for proportioning the different parts of the boiler and furnace. The magnitude of furnace and the extent of heating surface necessary to produce any required rate of evaporation in the boiler are indeed known, yet each engine-maker has his own rule in these matters, and which he seems to think preferable to all others, and there are various circumstances influencing the result which render facts incomparable unless those circumstances are the same. Thus the circumstances that govern the rate of evaporation, as influenced by different degrees of draught, may be regarded as but imperfectly known. And, supposing the difficulty of ascertaining this rate of evaporation were surmounted, there would still remain some difficulty in ascertaining the amount of power absorbed by the condensation of the steam on its passage to the cylinder—the imperfect condensation of the same steam after it has worked the piston—the friction of the various moving parts of the machinery—and, especially, the difference of effect of these losses of power in engines constructed on different scales of magnitude. Practice must often vary, to a certain extent, in the construction of the different parts of the boiler and furnace of an engine; for, independently of the difficulty of solving the general problem in engineering, the determination of the maximum effect with the minimum of means, practice would still require to vary according as in any particular case the desired minimum of means was that of weight, or bulk, or expense of material. Again, in estimating the proper proportions for a boiler and its appendages, reference ought to be made to the distinction between the "power" or "effect" of the boiler, and its "duty." This is a distinction to be considered also in the engine itself. The power of an engine has reference to the time it takes to produce a certain mechanical effect without reference to the amount of fuel consumed; and, on the other hand, the duty of an engine has reference to the amount of mechanical effect produced by a certain consumption of fuel, and is independent of the time it takes to produce that effect. In expressing the duty of engines, it would have prevented much needless confusion if the duty of the boiler had been entirely separated from that of the engine, as, indeed, they are two very distinct things. The duty performed by ordinary land rotative steam engines is—

One horse power exerted by 10 lbs. of fuel an hour; or,

Quarter of a million of lbs. raised 1 foot high by 1 lb. of coal; or,

Twenty millions of lbs. raised one foot by each bushel of coals.

Though in the best class of rotative engines the consumption is not above half of this amount.

The constant aim of different engine makers is to increase the amount of the duty; that is, to make 10 lbs. of fuel exert a greater effect than one horse power; or, in other words, to make 1 lb. of coal raise more than a quarter of a million of lbs. one foot high. To a great extent they have been successful in this. They have caused 5 lbs. of coal to exert the force of one horse power, and even in some cases as little as 3½ lbs.; but in these latter cases the economy is due chiefly to expansive action. In some of the engines, however, working with a consumption of 10 lbs. of coal per nominal horse hower per hour, the power really exerted amounts to much more than that represented by 33,000 lbs. lifted one foot high in the minute for each horse power. Some engines lift 56,000 lbs. one foot high in the minute by each horse power, with a consumption of 10 lbs. of coal per horse power per hour; and even this performance has been somewhat exceeded without a recourse to expansive action. In all modern engines the actual performance much exceeds the nominal power; and reference must be had to this circumstance in contrasting the duty of different engines.

### MECHANICAL POWER OF STEAM.

We may here give a table of some of the properties of steam, and of its mechanical effects at different pressures. This table may help to solve many problems respecting the mechanical effect of steam, usually requiring much laborious calculation.

| Pressures. | | Temperature in degrees Fahren. | Weight of a Cubic Foot Steam. | Velocity of Exit. | Mechanical Effect in Horse Power of 1 Lb. of Steam. | | | | | | | |
|---|---|---|---|---|---|---|---|---|---|---|---|---|
| | | | | | Without Condensation. Expansion. | | | | Condensation. Expansion. | | | |
| Atmosphere. | Lbs. per Sq. Inch. | | | | 0 | ½ | ⅓ | ¼ | 0 | ½ | ⅓ | ¼ |
| 1·00 | 14·70 | 212·00 | 0·0364 | 0 | 0 | 32·4 | 95·2 | 170·5 | 91·3 | 150·1 | 178·6 | 194·6 |
| 1·25 | 18·38 | 223·88 | 0·0440 | 873 | 21·5 | 10·1 | 32·3 | 87·4 | 95·9 | 158·7 | 190·6 | 209·9 |
| 1·50 | 22·05 | 234·32 | 0·0529 | 1135 | 36·4 | 39·3 | 10·8 | 30·6 | 99·3 | 165·2 | 199·6 | 221·1 |
| 1·75 | 25·72 | 242·78 | 0·0609 | 1295 | 47·4 | 60·8 | 42·5 | 11·1 | 102·0 | 170·0 | 206·2 | 229·5 |
| 2·00 | 29·40 | 250·79 | 0·0688 | 1407 | 55·9 | 77·5 | 67·0 | 43·2 | 104·3 | 174·2 | 212·0 | 236·5 |
| 2·25 | 33·08 | 257·90 | 0·0766 | 1491 | 62·8 | 90·9 | 86·5 | 68·8 | 106·2 | 177·7 | 216·7 | 242·4 |
| 2·50 | 36·75 | 263·93 | 0·0344 | 1556 | 68·4 | 101·8 | 102·4 | 89·6 | 107·7 | 180·5 | 220·5 | 247·1 |
| 2·75 | 40·42 | 269·87 | 0·0921 | 1608 | 73·1 | 111·0 | 115·8 | 107·1 | 109·3 | 183·2 | 224·2 | 251·6 |
| 3·00 | 44·10 | 275·00 | 0·0998 | 1652 | 71·1 | 118·8 | 127·1 | 121·9 | 110·6 | 185·4 | 227·7 | 255·2 |
| 3·35 | 47·78 | 279·86 | 0·1073 | 1690 | 80·7 | 125·6 | 137·1 | 130·7 | 111·7 | 187·6 | 230·0 | 258·7 |
| 3·50 | 51·45 | 284·63 | 0·1148 | 1722 | 83·8 | 131·5 | 145·6 | 145·8 | 112·7 | 189·4 | 232·4 | 261·6 |
| 3·75 | 55·12 | 288·66 | 0·1225 | 1750 | 86·5 | 136·8 | 153·2 | 155·6 | 113·7 | 190·1 | 234·7 | 264·4 |
| 4·00 | 58·18 | 292·91 | 0·1298 | 1774 | 89·0 | 141·5 | 160 | 164·5 | 114·6 | 192·8 | 236·9 | 267·0 |
| 4·50 | 66·15 | 300·27 | 0·1445 | 1816 | 93·2 | 149·8 | 171·5 | 179·4 | 116·2 | 195·6 | 240·5 | 271·4 |
| 5·00 | 73·50 | 307·94 | 0·1590 | 1850 | 96·8 | 156·5 | 181·6 | 192·0 | 117·7 | 198·3 | 244·1 | 275·6 |
| 6·00 | 88·20 | 320·00 | 0·1878 | 1904 | 102·5 | 167·2 | 196·5 | 211·4 | 120·2 | 202·6 | 249·7 | 282·2 |
| 7·00 | 102·90 | 331·56 | 0·2159 | 1945 | 107·0 | 175·6 | 208·4 | 226·5 | 122·4 | 205·4 | 254·6 | 288·1 |
| 8·00 | 117·60 | 340·83 | 0·2436 | 1978 | 110·6 | 182·4 | 217·9 | 238·4 | 124·3 | 209 | 258·8 | 292·1 |
| 9·00 | 132·30 | 351·32 | 0·2708 | 2006 | 113·7 | 188·2 | 225·9 | 248·5 | 126·0 | 212 | 262·7 | 293·6 |
| 10·00 | 147·00 | 359·60 | 0·2977 | 2029 | 116·3 | 193·0 | 232·5 | 256·7 | 127·5 | 215 | 266·0 | 301·4 |
| 12·50 | 183·75 | 377·42 | 0·3642 | 2074 | 121·5 | 202·5 | 245·5 | 273·0 | 130·7 | 220 | 272·9 | 309·5 |
| 15·00 | 220·50 | 392·90 | 0·4283 | 2109 | 125·7 | 210·0 | 255·6 | 285·4 | 133·4 | 225 | 278·9 | 316·4 |
| 17·50 | 257·25 | 406·40 | 0·4924 | 2136 | 129·0 | 216·0 | 263·6 | 295·2 | 135·7 | 229 | 283·9 | 322·2 |
| 20 | 294·00 | 418·56 | 0·5549 | 2159 | 131·8 | 221·0 | 270·3 | 305·3 | 137·8 | 233 | 288·3 | 327·2 |
| 25 | 367·50 | 429·34 | 0·6775 | 2196 | 136·3 | 229·1 | 281·0 | 316·2 | 141·2 | 238 | 295·7 | 335·8 |
| 30 | 441·00 | 457·16 | 0·7970 | 2226 | 140·0 | 235·6 | 289·5 | 326·4 | 144·2 | 244 | 302·0 | 343·1 |

It is quite clear that although there is no theoretical limit to the benefit derivable from expansion, there must be a limit in practice, arising from the friction incidental to the use of very large cylinders, the magnitude of the deduction due to uncondensed vapour when the steam is of a very low pressure, and other circumstances which it is needless to relate. It is clear, too, that while the efficiency of the steam is increased by expansive action, the efficiency of the engine is diminished, unless the pressure of the steam or the speed of the piston be increased correspondingly; and that an engine of any given size will not exert the same power if made to operate expansively without any other alteration that would have been realized if the engine had been worked with the full pressure of the steam. In the Cornish engines, which work with steam of 40 lbs. on the inch, the steam is cut off at one-twelfth of the stroke; but if the steam were cut off at one-twelfth of the stroke in engines employing a very low pressure, it would probably be found that there would be a loss rather than a gain from carrying the expansion so far, as the benefit might be more than neutralized by the friction incidental to the use of so large a cylinder as would be necessary to accomplish this expansion; and unless the vacuum were a very good one, there would be but little difference between the pressure of the steam at the end of the stroke and the pressure of the vapour in the condenser, so that the urging force might not at that point be sufficient to overcome the friction. In practice, therefore, in particular cases, expansion may be carried too far, though theoretically the amount of the benefit increases with the amount of the expansion.

We must here introduce a simple practical rule to enable those who may not be familiar with mathematical symbols to determine the amount of benefit due to any particular measure of expansion. When expansion is performed by an expansion valve, it is an easy thing to ascertain at what point of the stroke the valve is shut by the cam, and where expansion is performed by the slide valve the amount of expansion is easily determinable when the lap and stroke of the valve are known.

RULE.—*To find the Increase of Efficiency arising from working Steam expansively.*—Divide the total length of the stroke by the distance (which call 1) through which the piston moves before the steam is cut off. The hyperbolic logarithm of the whole stroke expressed in terms of the part of the stroke performed with the full pressure of steam, represents the increase of efficiency due to expansion.

Suppose that the pressure of the steam working an engine is 45 lbs. on the square inch above the atmosphere, and that the steam is cut off at one-fourth of the stroke; what is the increase of efficiency due to this measure of expansion?

If one-fourth be reckoned as 1, then four-fourths must be taken as 4, and the hyperbolic logarithm of 4 will be found to be 1·386, which is the increase of efficiency. The total efficiency of the quantity of steam expended during a stroke, therefore, which without expansion would have been 1, becomes 2·386 when expanded into 4 times its bulk, or, in round numbers, 2·4.

Let the pressure of the steam be the same as in the last example, and let the steam be cut off at half-stroke: what, then, is the increase of efficiency?

Here half the stroke is to be reckoned as 1, and the whole stroke has therefore to be reckoned as 2. The hyperbolic logarithm of 2 is ·693, which is the increase of efficiency, and the total efficiency of the stroke is 1·693, or 1·7.

We may here give a table to illustrate the mechanical effect of steam under varying circumstances. The table shows the me-

| Total pressure in lbs. per Square Inch. | Corresponding Temperature. | Volume of Steam compared with Water. | Mechanical effect of Cubic Inch of Water. | Total pressure in lbs. per Square Inch. | Corresponding Temperature. | Volume of Steam compared with Water. | Mechanical effect of Cubic Inch of Water. |
|---|---|---|---|---|---|---|---|
| 1 | 103 | 20·868 | 1739 | 51 | 284 | 544 | 2312 |
| 2 | 126 | 10·874 | 1812 | 52 | 286 | 534 | 2316 |
| 3 | 141 | 7437 | 1859 | 53 | 287 | 525 | 2320 |
| 4 | 152 | 5685 | 1895 | 54 | 288 | 516 | 2324 |
| 5 | 161 | 4617 | 1924 | 55 | 289 | 508 | 2327 |
| 6 | 169 | 3897 | 1948 | 56 | 290½ | 500 | 2331 |
| 7 | 176 | 3376 | 1969 | 57 | 292 | 492 | 2335 |
| 8 | 182 | 2983 | 1989 | 58 | 293 | 484 | 2339 |
| 9 | 187 | 2674 | 2006 | 59 | 294 | 477 | 2343 |
| 10 | 192 | 2426 | 2022 | 60 | 296 | 470 | 2347 |
| 11 | 197 | 2221 | 2036 | 61 | 297 | 463 | 2351 |
| 12 | 201 | 2050 | 2050 | 62 | 298 | 456 | 2355 |
| 13 | 205 | 1904 | 2063 | 63 | 299 | 449 | 2359 |
| 14 | 209 | 1778 | 2074 | 64 | 300 | 443 | 2362 |
| 15 | 213 | 1669 | 2086 | 65 | 301 | 437 | 2365 |
| 16 | 216 | 1573 | 2097 | 66 | 302 | 431 | 2369 |
| 17 | 220 | 1488 | 2107 | 67 | 303 | 425 | 2372 |
| 18 | 223 | 1411 | 2117 | 68 | 304 | 419 | 2375 |
| 19 | 226 | 1343 | 2126 | 69 | 305 | 414 | 2378 |
| 20 | 228 | 1281 | 2135 | 70 | 306 | 408 | 2382 |
| 21 | 231 | 1225 | 2144 | 71 | 307 | 403 | 2385 |
| 22 | 234 | 1174 | 2152 | 72 | 308 | 398 | 2388 |
| 23 | 236 | 1127 | 2160 | 73 | 309 | 393 | 2391 |
| 24 | 239 | 1084 | 2168 | 74 | 310 | 388 | 2394 |
| 25 | 241 | 1044 | 2175 | 75 | 311 | 383 | 2397 |
| 26 | 243 | 1007 | 2182 | 76 | 312 | 379 | 2400 |
| 27 | 245 | 973 | 2189 | 77 | 313 | 374 | 2403 |
| 28 | 248 | 941 | 2196 | 78 | 314 | 370 | 2405 |
| 29 | 250 | 911 | 2202 | 79 | 315 | 366 | 2408 |
| 30 | 252 | 883 | 2209 | 80 | 316 | 362 | 2411 |
| 31 | 254 | 857 | 2215 | 81 | 317 | 358 | 2414 |
| 32 | 255 | 833 | 2221 | 82 | 318 | 354 | 2417 |
| 33 | 257 | 810 | 2226 | 83 | 318 | 350 | 2419 |
| 34 | 259 | 788 | 2232 | 84 | 319 | 346 | 2422 |
| 35 | 261 | 767 | 2238 | 85 | 320 | 342 | 2425 |
| 36 | 263 | 748 | 2243 | 86 | 321 | 339 | 2427 |
| 37 | 264 | 729 | 2248 | 87 | 322 | 335 | 2430 |
| 38 | 266 | 712 | 2253 | 88 | 323 | 332 | 2432 |
| 39 | 267 | 695 | 2259 | 89 | 323 | 328 | 2435 |
| 40 | 269 | 679 | 2264 | 90 | 324 | 325 | 2438 |
| 41 | 271 | 664 | 2268 | 91 | 325 | 322 | 2440 |
| 42 | 272 | 649 | 2273 | 92 | 326 | 319 | 2443 |
| 43 | 274 | 635 | 2278 | 93 | 327 | 316 | 2445 |
| 44 | 275 | 622 | 2282 | 94 | 327 | 313 | 2448 |
| 45 | 276 | 610 | 2287 | 95 | 328 | 310 | 2450 |
| 46 | 278 | 598 | 2291 | 96 | 329 | 307 | 2453 |
| 47 | 279 | 586 | 2296 | 97 | 330 | 304 | 2455 |
| 48 | 280 | 575 | 2300 | 98 | 330 | 301 | 2457 |
| 49 | 282 | 564 | 2304 | 99 | 331 | 298 | 2460 |
| 50 | 283 | 554 | 2308 | 100 | 332 | 295 | 2462 |

chanical effect of the steam generated from a cubic inch of water. Our formula gives the effect of a cubic foot of water; but it can be modified to give the effect of the steam of a cubic inch by dividing by 1728. In this manner we find, for the mechanical effect of the steam of a cubic inch of water, about 3 $(459 + t)$ lbs. raised one foot high. The table shows that the mechanical effect increases with the temperature. The increase is very rapid for temperatures below 212°; but for temperatures above this the increase is less; and for the temperatures used in practice we may consider, without any material error, the mechanical effect as constant.

## INDICATOR.

An instrument for ascertaining the amount of the pressure of steam and the state of the vacuum throughout the stroke of a steam engine. Fitzgerald and Neucumn long employed an instrument of this kind, the nature of which was for a long time not generally known. Boulton and Watt used an instrument acting upon the same principle and equally accurate; but much more portable. In peculiarity of construction it is simply a small cylinder truly bored, and into which a piston is inserted and loaded by a spring of suitable elasticity to the graduated scale thereon attached.

The action of an indicator is that of describing, on a piece of paper attached, a diagram or figure approximating more or less to that of a rectangle, varying of course with the merits or demerits of the engine's productive effect. The breadth or height of the diagram is the sum of the force of the steam and extent of the vacuum; the length being the amount of revolution given to the paper during the piston's performance of its stroke.

To render the indicator applicable, it is commonly screwed into the cylinder cover, and the motion to the paper obtained by means of a sufficient length of small twine attached to one of the radius bars; but such application cannot always be conveniently effected, more especially in engines on the marine principle; hence, other parts of such engines, and other means whereby to effect a proper degree of motion, must unavoidably be resorted to. In those of direct action the crosshead is the only convenient place of attachment; but because the length of the engine's stroke is considerably more than the movement required for the paper on the indicator, it is necessary to introduce a pulley and axle, by which means the various movements are qualified to suit each other.

When the indicator is fixed and the movement for the paper properly adjusted, allow the engine to make a few revolutions previous to opening the cock; by which means a horizontal line will be described upon the paper by the pencil attached, and denominated the atmospheric line, because it distinguishes between the effect of the steam and that of the vacuum. Open the cock, and if the engine be upon the descending stroke, the steam will instantly raise the piston of the indicator, and, by the motion of the paper with the pencil pressing thereon, the top side of the diagram will be formed.

At the termination of the stroke and immediately previous to its return, the piston of the indicator is pressed down by the surrounding atmosphere, consequently the bottom side of the diagram is described, and by the time the engine is about to make another descending stroke, the piston of the indicator is where it first started from, the diagram being completed; hence is delineated the mean elastic action of the steam above that of the atmospheric line, and also the mean extent of the vacuum underneath it.

But in order to elucidate more clearly by example, take the following diagram, taken from a marine engine, the steam being cut off after the piston had passed through two-thirds of its stroke, the graduated scale on the indicator, tenths of an inch, as shown at each end of the diagram annexed.

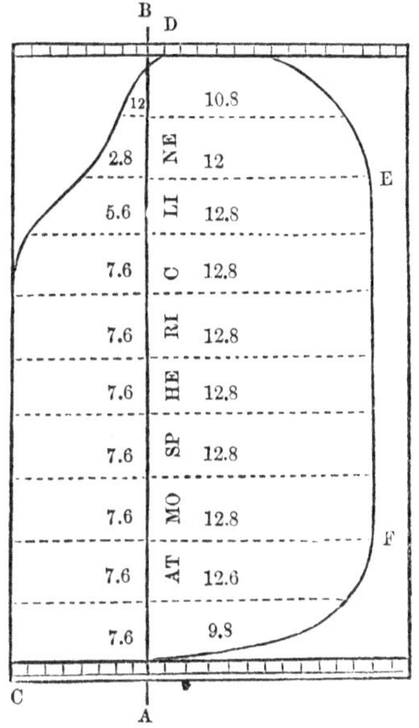


Previous to the cock being opened, the atmospheric line AB was formed, and, when opened, the pencil was instantly raised by the action of the steam on the piston to C, or what is generally termed the *starting corner;* by the movement of the paper and at the termination of the stroke the line CD was formed, showing the force of the steam and extent of expansion; from D to E show the moments of eduction; from E to F the quality of the vacuum; and from F to A the lead or advance of the valve; thus every change in the engine is exhibited, and every deviation from a rectangle, except that of *expansion* and *lead* of the valve show the extent of proportionate defect. Expansion produces apparently a defective diagram, but in reality such is not the case, because the diminished power of the engine is more than compensated by the saving in steam. Also the lead of the valve produces an apparent defect, but a certain amount must be given, as being found advantageous to the working of the engine, but the steam and eduction corners ought to be as square as possible; any rounding on the steam corner shows a defect from want of lead; and rounding on the eduction corner that of the passages or apertures being too small.

Rule.—*To compute the power of an Engine from the Indicator Diagram.*—Divide the diagram in the direction of its length into any convenient number of equal parts, through which draw lines at right angles to the atmospheric line, add together the lengths of all the spaces taken in measurements corresponding with the scale on the indicator, divide the sum by the number of spaces, and the quotient is the mean effective pressure on the piston in lbs. per square inch.

Let the result of the preceding diagram be taken as an example. Then, the whole sum of vacuum spaces = 1220 ÷ 10 = 12·2 lbs. mean effect obtained by the vacuum; and in a similar manner the mean effective pressure of steam is found to be 6·28 lbs., hence the total effective force = 18·48 lbs. per square inch. And supposing 2·5 lbs. per square inch be absorbed by friction, What is the actual power of the engine, the cylinder's diameter being 32 inches, and the velocity of the piston 226 feet per minute?

18·48 − 2·5 = 15·98 lbs. per square inch of net available force.

$$\text{Then } \frac{32^2 \times \cdot 7854 \times 15\cdot 98 \times 226}{33000} = 88 \text{ horses power.}$$

The line under the diagram and parallel to the atmospheric line is $\frac{15}{10}$ths distant, and represents the perfect vacuum line, the space between showing the amount of force with which the uncondensed steam or vapour resists the ascent or descent of the piston at every part of the stroke.

As the mean pressure of the atmosphere is 15 lbs. per square inch, and the mean specific gravity of mercury 13560, or 2·037 cubic inches equal 1 lb., it will of course rise in the barometer attached to the condenser about 2 inches for every lb. effect of vacuum, and as a pure vacuum would be indicated by 30 inches of mercury, the distance between the two lines shows whether there is or is not any amount of defect, as sometimes there is a considerable difference in extent of vacuum in the cylinder to that in the condenser.

*To estimate by means of an indicator the amount of effective power produced by a steam engine.*—Multiply the area of the piston in square inches by the average force of the steam in lbs. and by the velocity of the piston in feet per minute; divide the product by 33,000, and $\frac{7}{10}$ths of the quotient equal the effective power.

Suppose an engine with a cylinder of 37½ inches diameter, a stroke of 7 feet, and making 17 revolutions per minute, or 238 feet velocity, and the average indicated pressure of the steam 16·73 lbs. per square inch; required the effective power.

$$\frac{\text{Area} = 1104\cdot 4687 \text{ inches} \times 16\cdot 73 \text{ lbs.} \times 238 \text{ feet}}{33000}$$

$$= \frac{133\cdot 26 \times 7}{10} = 93\cdot 282 \text{ horse power.}$$

*To determine the proper velocity for the piston of a steam engine.*—Multiply the logarithm of the *n*th part of the stroke at which the steam is cut off by 2·3, and to the product of which add ·7. Multiply the sum by the distance in feet the piston has travelled when the steam is cut off, and 120 times the square root of the product equal the proper velocity for the piston in feet per minute.

## WEIGHT COMBINED WITH MASS, VELOCITY, FORCE, AND WORK DONE.

CALCULATIONS ON THE PRINCIPLE OF VIS VIVA.—MATERIALS EMPLOYED IN THE CONSTRUCTION OF MACHINES.—STRENGTH OF MATERIALS, THEIR PROPERTIES.—TORSION, DEFLEXION, ELASTICITY, TENACITIES, COMPRESSIONS, ETC.—FRICTION OF REST AND OF MOTION, COEFFICIENTS OF ALL SORTS OF MOTION.—BANDS.—ROPES.—WHEELS.—HYDRAULICS.—NEW TABLES FOR THE MOTION AND FRICTION OF WATER.—WATER-WHEELS.—WINDMILLS, ETC.

1. Suppose a body resting on a perfectly smooth table, and, when in motion, to present no impediment to the body in its course, but merely to counteract the force of gravity upon it; if this body weighing 800 lbs. be pressed by the force of 30 lbs. acting horizontally and continuously, the motion under such circumstances will be uniformly accelerated: what is the acceleration?

$$\frac{30}{800} \times 32{\cdot}2 = 1{\cdot}2075 \text{ feet the second.}$$

2. What force is necessary to move the above-mentioned heavy body, with a 23 feet acceleration, under the same circumstances?

$$\frac{23}{32{\cdot}2} \times 800 = 57{\cdot}14285 \text{ lbs.}$$

The second of these examples illustrates the principle that the force which impels a body with a certain acceleration is equal to the weight of the body multiplied by the ratio of its acceleration to that of gravity. The first illustrates the reverse, namely, the acceleration with which a body is moved forward with a given force, is equal to the acceleration of gravity multiplied by the ratio of the force to the weight.

3. A railway car, weighing 1120 lbs., moves with a 5 feet velocity upon horizontal rails, which, let us suppose, offer no impediment to the motion, and is constantly pushed by an invariable force of 50 lbs. during 20 seconds: with what velocity is it moving at the end of the 20th second, or at the beginning of the 21st second?

$$5 + 32{\cdot}2 \times \frac{50}{1120} \times 20 = 33{\cdot}75, \text{ the velocity.}$$

4. A carriage, circumstanced as in the last question, weighs 4000 lbs.; its initial velocity is 30 feet the second, and its terminal velocity is 70 feet: with which force is the body impelled, supposing it to be in motion 20 seconds?

$$\frac{(70 - 30) \times 4000}{32{\cdot}2 \times 20} = 242{\cdot}17 \text{ lbs.}$$

We have before noticed that the weight (W), divided by 32·2, or ($g$), gives the *mass;* that is,

$$\frac{\text{Weight}}{g} = \text{mass},$$

And, force = mass × acceleration.

5. Suppose a railway carriage, weighing 6440 lbs., moves on a horizontal plane offering no impediment, and is uniformly accelerated 4 feet the second, what continuous force is applied?

$$\frac{6440}{32 \cdot 2} = 200 \text{ lbs. mass.}$$

$$200 \times 4 = 800 \text{ lbs., the force applied.}$$

By the four succeeding formulas, all questions may be answered that may be proposed relative to the rectilinear motions of bodies by a constant force.

For uniformly accelerated motions:

$$v = a + 32 \cdot 2 \frac{F}{W} \times t;$$

$$s = at + 16 \cdot 1 \frac{F}{W} \times t^2.$$

For uniformly retarded motions:

$$v = a - 32 \cdot 2 \frac{F}{W} \times t;$$

$$s = at - 16 \cdot 1 \times \frac{F}{W} \times t^2;$$

$t$ = the time in seconds, W = the weight in lbs., F = the force in lbs., $a$ = the initial velocity, and $v$ = the terminal velocity.

6. A sleigh, weighing 2000 lbs., going at the rate of 20 feet a second, has to overcome by its motion a friction of 30 lbs.: what velocity has it after 10 seconds, and what distance has it described?

$$20 - 32 \cdot 2 \times \frac{30}{2000} \times 10 = 15 \cdot 17 \text{ feet velocity.}$$

$$20 \times 10 - 16 \cdot 1 \times \frac{30}{2000} \times (10)^2 = 175 \cdot 85 \text{ feet, distance de-}$$

scribed.

7. In order to find the mechanical work which a draught-horse performs in drawing a carriage, an instrument called a dynamometer, or measure of force, is thus used: it is put into communication on one side of the carriage, and on the other with the traces of the horse, and the force is observed from time to time. Let 126 lbs. be the initial force; after 40 feet is described, let 130 lbs. be the force given by the dynamometer; after 40 feet more is described, let 129 lbs. be the force; after 40 feet more is passed over, let 140 lbs. be the force; and let the next two spaces of 40 feet give forces of 130 and 120 lbs. respectively. What is the mechanical work done?

126 initial force.
120 terminal force.

2)246

123 mean.

$$\frac{123 + 130 + 129 + 140 + 130}{5} = 130{\cdot}4$$

$$130{\cdot}4 \times 40 \times 5 = 26080 \text{ units of work.}$$

The following rule, usually given to find the areas of irregular figures, may be applied where great accuracy is required.

RULE.—To the sum of the first and last, or extreme ordinates, add four times the sum of the 2d, 4th, 6th, or even ordinates, and twice the sum of the 3d, 5th, 7th, &c., or odd ordinates, not including the extreme ones; the result multiplied by $\frac{1}{3}$ the ordinates' equidistance will be the sum.

$$\begin{array}{r} 126 \\ 120 \\ \hline 246 \end{array} \text{ sum of first and last.}$$

$$246 + 4 \times 130 + 2 \times 129 + 4 \times 140 + 2 \times 130 = 1844.$$

$\frac{1844 \times 40}{3} = 24586{\cdot}66$ units of work or pounds raised one foot high. This rule of equidistant ordinates is of great use in the art of ship-building. This application we shall introduce in the proper place.

8. How many units of work are necessary to impart to a carriage of 3000 lbs. weight, resting on a perfectly smooth railroad, a velocity of 100 feet?

$$\frac{(100)^2}{2 \times 32{\cdot}2} \times 3000 = 465838{\cdot}2 \text{ units.}$$

A unit of work is that labour which is equal to the raising of a pound through the space of one foot. A unit of work is done when one pound pressure is exerted through a space of one foot, no matter in what direction that space may lie.

Kane Fitzgerald, the first that made steam turn a crank, and patented it, and the fly-wheel to regulate its motion, estimated that a horse could perform 33000 units of work in a minute, that is, raise 33000 lbs. one foot high in a minute. To perform 465838·2 units of work in 10 minutes would require the application 1·4116 horse power.

9. What work is done by a force, acting upon another carriage, under the same circumstances, weighing 5000 lbs., which transforms the velocity from 30 to 50 feet?

$$\frac{(30)^2}{64{\cdot}4} = 13{\cdot}9907, \text{ the height due to 30 feet velocity.}$$

$$\frac{(50)^2}{64{\cdot}4} = 38{\cdot}8043, \text{ the height due to 50 feet velocity.}$$

$$\begin{array}{lr} \text{From} & 38{\cdot}8043 \\ \text{Take} & 13{\cdot}9907 \\ \hline & 24{\cdot}8136 \\ & 5000 \\ \hline & 124068{\cdot}0000 \end{array}$$

x 2

∴ 124068 are the units of work, and just so much work will the carriage perform if a resistance be opposed to it, and it be gradually brought from a 50 feet velocity to a 30 feet velocity.

The following is without doubt a very simple formula, but the most useful one in mechanics; by it we have solved the last two questions:

$$Fs = (H - h)\,W.$$

This simple formula involves the principle technically termed the principle of VIS VIVA, or LIVING FORCES. H is the height due to one velocity, say $v$ or $H = \frac{v^2}{2g}$ and $h$, the height due to another $a$, or $h = \frac{a^2}{2g}$. The weight of the mass = W; the force F, and the space $s$.

To express this principle in words, we may say, that the working power ($Fs$) which a mass either acquires when it passes from a lesser velocity ($a$) to a greater velocity ($v$), or produces when it is compelled to pass from a greater velocity ($v$) into a less ($a$), is always equal to the product of the weight of the mass and the difference of the heights due to the velocities.

When we know the *units of work*, and the distance in which the change of velocity goes on, the force is easily found; and when the force is known, the distance is readily determined. Suppose, in the last example, that the change of velocity from 30 to 50 feet took place in a distance of 300 feet, then

$\frac{124068}{300} = 413{\cdot}56$ lbs. = F, the force constantly applied during 300 feet.

10. If a sleigh, weighing 2000 lbs., after describing a distance of 250 feet, has completely lost a velocity of 100 feet, what constant resistance does the friction offer?

Since the terminal velocity = 0, the height due to it = 0, hence

$$\frac{(100)^2}{64{\cdot}4} \times \frac{2000}{250} = 1242{\cdot}2352 \text{ lbs.}$$

We have been calculating upon the principle of *vis viva;* but the product of the *mass* and the square of the velocity, without attaching to it any definite idea, is termed the *vis viva,* or living force.

11. A body weighing 2300 lbs. moves with a velocity of 20 feet the second, required the *vis viva?*

$$\frac{2300}{32{\cdot}2} = 71{\cdot}42857 \text{ lbs., mass.}$$

$$71{\cdot}42857 \times (20)^2 = 28571{\cdot}428, \text{ the amount of } vis\ viva.$$

Hence, if a *mass* enters from a velocity $a$, into another $v$, the unit of work done is equal to half the difference of the *vis viva,* at the commencement and end of the change of velocity.

For if the *mass* be put = M, and W the weight,

Then $M = \frac{W}{g}$, and the *vis viva* to velocity $a = Ma^2 = \frac{Wa^2}{g}$;

and the *vis viva* to velocity $v = Mv^2 = \frac{Wv^2}{g}$.

Then $\frac{1}{2}\left\{\frac{Wv^2}{g} - \frac{Wa^2}{g}\right\} = \left(\frac{v^2}{2g} - \frac{a^2}{2g}\right) \times W = (H - h)\,W$, for $\frac{v^2}{2g}$ and $\frac{a^2}{2g}$, give the heights due to the velocities $v$ and $a$, respectively. The useful formula

$$Fs = (H - h)\,W,$$

before given, page 270, may be applied to variable as well as to constant forces, if, instead of the constant force F, the mean value of the force be applied.

---

## STRENGTH OF MATERIALS.

### ON MATERIAL EMPLOYED IN THE CONSTRUCTION OF MACHINES.

IN theoretical mechanics, we deal with imaginary quantities, which are perfect in all their properties; they are perfectly hard, and perfectly elastic; devoid of weight in statics and of friction in dynamics. In practical mechanics, we deal with real material objects, among which we find none which are perfectly hard, and none, except gaseous bodies, which are perfectly elastic; all have weight, and experience resistance in dynamical action. Practical mechanics is the science of automatic labour, and its objects are machines and their applications to the transmission, modification, and regulation of motive power. In this it takes as a basis the theoretical deductions of pure mechanics, but superadds to the formulæ of the mathematician a multitude of facts deduced from observation, and experimentally elaborates a new code of laws suited to the varied conditions to be fulfilled in the economy of the industrial arts.

In reference to the structure of machines, it is to be observed that however simple or complex the machine may be, it is of importance that its parts combine lightness with strength, and rigidity with uniformity of action; and that it communicates the power without shocks and sudden changes of motion, by which the passive resistances may be increased and the effect of the engine diminished.

To adjust properly the disposition and arrangement of the individual members of a machine, implies an exact knowledge and estimate of the amount of strain to which they are respectively subject in the working of the machine; and this skill, when exercised in conjunction with an intimate acquaintance with the nature of the materials of which the parts are themselves composed, must contribute to the production of a machine possessing the highest amount of capability attainable with the given conditions.

*Materials.*—The material most commonly employed in the construction of machinery is iron, in the two states of *cast* and *wrought* or *forged* iron; and of these, there are several varieties of quality. It becomes therefore a problem of much practical importance to determine, at least approximately, the capabilities of the particular material employed, to resist permanent alteration in the directions in which they are subjected to strain in the reception and transmission of the motive power.

To indicate briefly the fundamental conditions which determine the capability of a given weight and form of material to resist a given force, it must, in the first place, be observed, that rupture may take place either by tension or by compression in the direction of the length. To the former condition of strain is opposed the *tenacity* of the material; to the other is opposed the resistance to the *crushing of its substance.* Rupture, by *transverse strain*, is opposed both by the tenacity of the material and its capability to withstand compression together of its particles. Lastly, the bar may be ruptured by *torsion.* Mr. Oliver Byrne, the author of the present work, in his New Theory of the Strength of Materials has pointed out new elements of much importance.

The capabilities of a material to resist extension and compression are often different. Thus, the soft gray variety of cast iron offers a greater resistance to a force of extension than the white variety in a ratio of nearly eight to five; but the last offers the greatest resistance to a compressing force.

The resistance of cast iron to rupture by extension varies from 6 to 9 tons upon the square inch; and that to rupture by compression, from 36 to 65 tons. The resistance to extension of the best forged iron may be reckoned at 25 tons per inch; but the corresponding resistance to compression, although not satisfactorily ascertained, is generally considered to be greatly less than that of cast iron. Roudelet makes it 31½ tons on the square inch. Cast iron (and even wood) is therefore to be preferred for vertical supports.

The forces resisting rupture are as the areas of the sections of rupture, the material being the same; this principle holds not only in respect of iron, but also of wood. Many inquiries have been instituted to determine the commonly received principle, that the strength of rectangular beams of the same width to resist rupture by transverse strain is as the *squares of the depths* of the beams.

In these respects the experiments, although valuable on account of their extent and the care with which they were conducted, possess little novelty; but in directing attention to the elastic properties of the materials experimented upon, it was found that the received doctrine of relation between the limit of elasticity and weight requires modification. The common assumption is, that the destruction of the elastic properties of a material, that is, the displacement beyond the elastic limit, does not manifest itself until the load exceeds *one-third of the breaking weight.* It was found, however, on the contrary, that its effect was produced and manifested in a permanent *set* of the material when the load did not exceed *one-sixteenth* of that necessary to produce rupture. Thus a bar of one inch square, supported between props $4\frac{1}{2}$ feet apart, did not break till loaded with 496 lbs. but showed a permanent deflection or *set* when loaded with 16 lbs. In other cases, loads of 7 lbs. and 14 lbs. were found to produce permanent sets when the breaking weights were respectively 364 lbs. and 1120 lbs. These sets were therefore given by $\frac{1}{52}$d and $\frac{1}{80}$th of the breaking weights.

Since these results were obtained, it has been found that *time* and the *weight* of the material itself are sufficient to effect a permanent deflection in a beam supported between props, so that there would seem to be no such limits in respect to transverse strain as those known by the name of elastic limits, and consequently the principle of loading a beam within the elastic limit has no foundation in practice. The beam yields continually to the load, but with an exceedingly slow progression, until the load approximates to the breaking weight, when rupture speedily succeeds to a rapid deflection.

As respects the effect of tension and compression by transverse strain, it was ascertained by a very ingenious experiment that equal loads produced equal deflections in both cases.

Another most important principle developed by experiments, is that respecting the compression of supporting columns of different heights. When the height of the column exceeded a certain limit, it was found that the crushing force became constant, and did not increase as the height of the column increased, until it reached another limit at which it began to yield, not strictly by crushing, but by the bending of the material. The first limit was found to be a height of little less than three times the radius of the column; and the second double that height, or about six times the radius of the column. In columns of different heights between these limits, having equal diameters, the force producing rupture by compression was nearly constant. When the column was *less* than the lower limit, the crushing force became *greater*, and when it was greater than the higher limit, the crushing force became *less*. It was further found that in all cases, where the height of the column was exactly above the limits of three times the radius, the section of rupture was a plane inclined at nearly the same constant angle of 55 degrees to the axis of the column. These facts mutually explain each other; for in every height of column above the limit, the section of rupture being a plane at the same angle to the axis of the column, must of necessity be a plane of the same size, and therefore in each case the cohesion of the same number of particles must be overcome in producing rupture. And further, the same number of particles being to be overcome under the same circumstances for every different height, the same force will be required to overcome that amount of cohesion, until at double the height (three diameters) the column begins to bend under its load. This height being surpassed, it follows that a pressure which becomes continually less as the length of the column is increased, will be sufficient to break it.

This property, moreover, is not confined to cast iron; the experiments of M. Rondelet show that with columns of wrought iron, wood, and stone, similar results are obtained.

From these facts then, it appears that if supporting columns be taken of different diameters, and of heights so great as not to allow of their bending, yet sufficiently high to allow of a complete separation of the planes of fracture, that is, of heights intermediate to three times and six times their radius, then will their strengths be as the number of particles in their planes of fracture; and the planes of fracture being inclined at equal angles to the axes of the columns, their areas will be as the transverse sections of the columns, and consequently the *strengths* of the columns will be as their transverse sections respectively. Taking the mean of three experiments upon a column $\frac{1}{4}$ inch diameter, the crushing force was 6426 lbs.; whilst the mean of four experiments, conducted in exactly the same manner, upon a column of $\frac{3}{8}$ of an inch diameter, gave 14542 lbs. The diameters of the columns being 2 to 3, the areas of transverse section were therefore 4 to 9, which is very nearly the ratio of the crushing weights.

When the length of the column is so great that its fracture is produced wholly by bending of its material, the limit has been fixed for columns of cast iron, at 30 times the diameter when the ends are flat, and 15 times the diameter when the ends are rounded. In shorter columns, fracture takes place partly by crushing and partly by bending of the material. When the column is enlarged in the middle of its length from one and a half to two times the diameter of the ends, the strength was found by the same experimenter to be greater by one-seventh than in solid columns containing the same quantity of iron, in the same length, with their extremities rounded; and stronger by an eighth or a ninth when the extremities were flat and rendered immovable by disks.

The following formulas give the absolute strength of cylindrical columns to sustain pressure in the direction of their length. In these formulas

$D$ = the external diameter of the column in inches.
$d$ = the internal diameter of hollow columns in inches.
$L$ = the length of the column in feet.
$W$ = the breaking weight in tons.

| Character of the column. | Length of the column exceeding 15 times its diameter. | Length of the column exceeding 30 times its diameter. |
|---|---|---|
| | Both ends rounded. | Both ends flat. |
| Solid cylindrical column of cast iron, | $W = 14{\cdot}9 \frac{D^{3{\cdot}76}}{L^{1{\cdot}7}}$ | $W = 44{\cdot}16 \frac{D^{3{\cdot}55}}{L^{1{\cdot}7}}$ |
| Hollow cylindrical column of cast iron, | $W = 13 \frac{D^{3{\cdot}76} - d^{3{\cdot}76}}{L^{1{\cdot}7}}$ | $W = 44{\cdot}34 \frac{D^{3{\cdot}55} - d^{3{\cdot}55}}{L^{1{\cdot}7}}$ |
| Solid cylindrical column of wrought iron, | $W = 42{\cdot}8 \frac{D^{3{\cdot}76}}{L^{2}}$ | $W = 133{\cdot}75 \frac{D^{3{\cdot}55}}{L^{2}}$ |

For shorter columns, if $W'$ represent the weight in tons which would break the column by bending alone, as given by the preceding formulas, and W″ the weight in tons which would crush the column without bending it, as determined from the subjoined table, then the absolute breaking weight of the column W, is represented in tons by the formula,

$$W = \frac{W' \times W''}{W' + W''}$$

These rules require the use of logarithms in their applications.

When a beam is deflected by transverse strain, the material on that side of it on which it sustains the strain is *compressed*, and the material on the opposite side is *extended*. The imaginary surface at which the compression terminates and the extension begins—at which there is supposed to be neither extension nor compression—is termed the *neutral axis* of the beam. What constitutes the strength of a beam is its resistance to compression on the one side and to extension on the other side of that axis—the forces acting about the line of axis like antagonist force at the two extremities of a lever, so that if either of them yield, the beam will be broken. It becomes, however, a question of importance to determine the relation of these forces; in other words, to determine whether the beam of given form and material will yield first to compression or to extension. This point is settled by reference to the columns of the subsequent table, page 280, in which it will be observed that the *metals* require a much greater force to crush them than to tear them asunder, and that the *woods* require a much smaller force.

There is also another consideration which must not be overlooked. Bearing in mind the condition of antagonism of the forces, it is obvious, that the further these forces are placed from the neutral axis, that is, from the fulcrum of their leverage, the greater must be their effect. In other words, all the material resisting compression will produce its greatest effect when collected the farthest possible from the neutral axis at the top of the beam; and, in like manner, all the material resisting extension will produce its greatest effect when similarly disposed at the bottom of the beam. We are thus directed to the first general principle of the distribution of the material into two flanges—one forming the top and the other the bottom of the beam—joined by a comparatively slender rib. Associating with this principle the relation of the forces of extension and compression of the material employed, we arrive at a form of beam in which the material is so distributed, that at the instant it is about to break by extension on the one side, it is about to break by compression on the other, and consequently is of the strongest form. Thus, supposing that it is required to determine that form in a girder of cast iron: the ratio of the crushing force of that metal to the force of extension may be taken generally as $6\frac{1}{2}$ to 1, which is therefore also the ratio of the lower to the upper flange, as in the annexed sectional diagram.

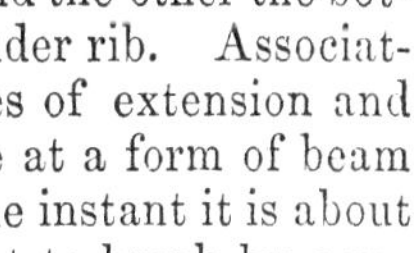

A series of nine castings were made, gradually increasing the lower flange at the expense of the upper one, and in the first eight experiments the beam broke by the tearing asunder of the lower flange; and in the last experiment the beam yielded by the crushing of the upper flange. In the eight experiments the upper flange was therefore the weakest, and in the ninth the strongest, so that the form of maximum strength was intermediate, and very closely allied to that form of beam employed in the last experiment, which was greatly the strongest. The circumstances of these experiments are contained in the following table.

| No. of experiments. | Ratio of surfaces of compression and extension. | Area of cross sections in sq. inches. | Strength per sq. inch of sections in lbs. |
|---|---|---|---|
| 1 | 1 to 1 | 2·82 | 2368 |
| 2 | 1 to 2 | 2·87 | 2567 |
| 3 | 1 to 4 | 3·02 | 2737 |
| 4 | 1 to $4\frac{1}{2}$ | 3·37 | 3183 |
| 5 | 1 to 4 | 4·50 | 3214 |
| 6 | 1 to $5\frac{1}{2}$ | 5·00 | 3346 |
| 7 | 1 to $3\frac{1}{5}$ | 4·628 | 3246 |
| 8 | 1 to 4·3 | 5·86 | 3317 |
| 9 | 1 to 6·1 | 6·4 | 4075 |

To determine the weight necessary to *break* beams cast according to the form described:

Multiply the area of the section of the lower flange by the depth of the beam, and divide the product by the distance between the two points on which the beam is supported: this quotient multiplied by 536 when the beams are cast erect, and by 514 when they are cast horizontally, will give the breaking weight in cwts.

From this it is not to be inferred that the beam ought to have the same transverse section throughout its length. On the contrary, it is clear that the section ought to have a definite relation to the leverage at which the load acts. From a mathematical consideration of the conditions, it indeed appears that the effect of a given load to break the beam varies when it is placed over different points of it, as the products of the distances of these points from the points of support of the beam. Thus the effect of a weight placed at the point $W_1$ is to the effect of the same weight acting upon the point $W_2$, as the product $AW_1 \times W_1B$ is to the product $AW_2 \times W_2B$; the points of support being at A and B. Since then the effect of a weight increases as it approaches the middle of the length of the beam, at which it is a maximum, it is plain that the beam does not require to have the same transverse section near to its extremities as in the middle; and, guided by the principle stated, it is easy to perceive that its strength at different points should in strictness vary as the products of the distances of these points from the points of support. By

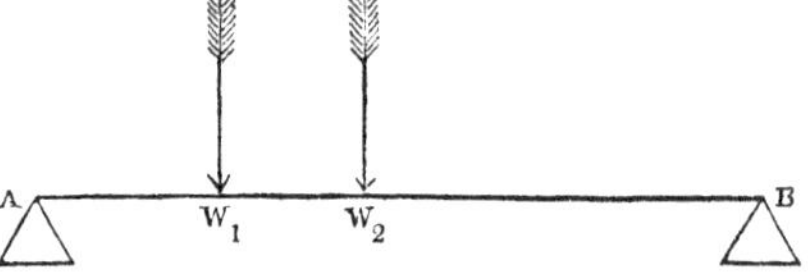

taking this law as a fundamental condition in the distribution of the strength of a beam, whose load we may conceive to be accumulated at the middle of its length, we arrive at the strongest form which can be attained under given circumstances, with a given amount of material; we arrive at that form which renders the beam equally liable to rupture at every point. Now this form of maximum strength may be attained in two ways; either by varying the depth of the beam according to the law stated, or by preserving the depth everywhere the same, and varying the dimensions of the upper and lower flanges according to the same law. The conditions are manifestly identical. We may therefore assume generally the condition that the section is *rectangular*, and that the thickness of the flanges is constant; then the outline determined by the law in question, in the one case of the *elevation* of the beam and in the other of the *plan* of the flanges, is the geometrical curve called a parabola—rather, two parabolas joined base to base at the middle between the points of support. The annexed diagram represents the plan of a cast-iron girder according to this form, the depth

being uniform throughout. Both flanges are of the same form, but the dimensions of the upper one are such as to give it only a sixth of the strength of the other.

This, it will be observed, is also the form, considered as an elevation, of the beam of a steam engine, which good taste and regard to economy of material have rendered common.

It must, however, be borne in mind, that in the actual practice of construction, materials cannot with safety be subjected to forces approaching to those which produce rupture. In machinery especially, they are liable to various and accidental pressures, besides those of a permanent kind, for which allowance must be made. The engineer must therefore in his practice depend much on experience and consideration of the species of work which the engine is designed to perform. If the engine be intended for spinning, pumping, blowing, or other *regular* work, the material may be subjected to pressures approaching two-thirds of that which would actually produce rupture; but in engines employed to drive bone-mills, stampers, breaking-down rollers, and the like, double that strength will often be found insufficient. In cases of that nature, experience is a better guide than theory.

It is also to be remarked that we are often obliged to depart from the form of strength which the calculation gives, on account of the partial strains which would be put upon some of the parts of a casting, in consequence of unequal cooling of the metal when the thicknesses are unequal. An expert founder can often reduce the irregular contractions which thus result; but, even under the best management, fracture is not unfrequently produced by irregularity of cooling, and it is at all times better to avoid the danger entirely, than to endeavour to obviate it by artifice. For this reason, the parts of a casting ought to be as nearly as possible of such thickness as to cool and contract regularly, and by that means all partial strain of the parts will be avoided.

With respect to design, it is also to be remarked, that mere theoretical properties of parts will not, under all the varieties of circumstances which arise in the working of a machine, insure that exact adjustment of material and propriety of form so much desired in constructive mechanics. Every design ought to take for its basis the mathematical conditions involved, and it would, perhaps, be impossible to arrive at the best forms and proportions by any more direct mode of calculation; but it is necessary to superadd to the mathematical demonstration, the exercise of a well-matured judgment, to secure that degree of adjustment and arrangement of parts in which the merits of a good design mainly consist. A purely theoretical engine would look strangely deficient to the practised eye of the engineer; and the merely theoretical contriver would speedily find himself lost, should he venture beyond his construction on paper. His nice calculations of the "work to be performed," of the *vis viva* of the mechanical organs of his machine, and of the *modulus of elasticity* of his material, would, in practice, alike deceive him.

The first consideration in the design of a machine is the quantity of work which each part has to perform—in other words, the forces, active and inactive, which it has to resist; the direction of the forces in relation to the cross-section and points of support; the velocity, and the changes of velocity to which the moving parts are subject. The calculations necessary to obtain these must not be confined to theory alone; neither should they be entirely deduced by "rule of thumb;" by the first mode the strength would, in all probability, be deficient from deficiency of material, and by the second *rule* the material would be injudiciously disposed; weight would be added often where least needed, merely from the determination to avoid fracture, and in consequence of a want of knowledge respecting the true forms best adapted to give strength.

To the following general principles, in practice, there are but few real exceptions:

I. *Direct Strain.*—To this a straight line must be opposed, and if the part be of considerable length, vibration ought to be counteracted by intersection of planes, (technically *feathers,*) as represented in the annexed diagrams, or some such form, consistent with the purpose for which the part is intended.

II. *Transverse Strain.*—To this a parabolic form of section must be opposed, or some simple figure including the parabolic form. For economy of material, the vertex of the curve ought to be at the point where the force is applied; and when the strain passes alternately from one side of the part to the other, the curve ought to be on both sides, as in the beam of a steam engine.

When a loaded piece is supported at one end only, if the breadth be everywhere the same, the form of equal strength is a triangle; but, if the section be a circle, then the solid will be that generated by the revolution of a semi-parabola about its longer axis. In practice, it will, however, be sufficient to employ the frustum of a cone, of which, in the case of cast iron, the diameter at the unsupported end is one-third of the diameter at the fixed end.

III. *Torsion.*—The section most commonly opposed to torsion is a circle; and, if the strain be applied to a cylinder, it is obvious the rupture must first take place at the surface, where the torsion is greatest, and that the further the material is placed from the neutral axis, the greater must be its power of resistance; and hence, the amount of materials being the same, a shaft is stronger when made hollow than if it were made solid.

It ought not, however, to be supposed that the circle is the only figure which gives an axis the property of offering, in every direction, the same resistance to flexure. On the contrary, a square section gives the same resistance in the direction of its sides, and of its diagonals; and, indeed, in every direction the resistance is equal. This is, moreover, the case with a great number of other figures, which may be formed by combining the circle and the square in a symmetrical manner; and hence, if the axis, strengthened by salient sides, as in feathered shafts, do not answer as well as cylindrical ones, it must arise from their not being so well disposed to resist torsion, and not from any irregularities of flexure about the axis inherent in the particular form of section.

This subject has been investigated with much care, and, according to M. Cauchy, the modulus of rupture by torsion, T, is connected with the modulus of rupture by transverse strain S, by the simple analogy $T = \frac{4}{5} S$.

The forms of all the parts of a machine, in whatever situation and under every variety of circumstances, may be deduced from these simple figures; and, if the calculations of their dimensions be correctly determined, the parts will not only possess the requisite degree of strength, but they will also accord with the general principles of good taste.

In arranging the details of a machine, two circumstances ought to be taken into consideration. The first is, that the parts subject to wear and influenced by strain, should be capable of adjustment; the second is, that every part should, in relation to the work it has to perform, be equally strong, and present to the eye a figure that is consistent with its degree of action. Theory, practice, and taste must all combine to produce such a combination. No formal law can be expressed, either by words or figures, by which a certain contour should be preferred to another; both may be equally strong and equally correct in reference to theory; custom, then, must be appealed to as the guide.

## TABLES OF THE MECHANICAL PROPERTIES OF THE MATERIALS MOST COMMONLY EMPLOYED IN THE CONSTRUCTION OF MACHINES AND FRAMINGS.

| NAMES. | Specific Gravity. | Weight of 1 cubic ft. in lbs. | Tenacity per square inch in lbs. | Crushing force per sq. in. in lbs. | Modulus of elasticity in lbs. | Mod. of rupture in lbs. | Crushing force to tenacity. |
|---|---|---|---|---|---|---|---|
| TABLE I.—*Mechanical Properties of the Common Metals.* | | | | | | | |
| Brass (cast) | 8·399 | 525·00 | 17968 | 10304 | 8930000 | — | 0·573 : 1 |
| Copper (cast) | 8·607 | 537·93 | 19072 | — | — | — | — |
| ditto (sheet) | 8·785 | 549·06 | — | — | — | — | — |
| ditto (wire-drawn) | 8·878 | 560·00 | 61228 | — | — | — | — |
| ditto (in bolts) | — | — | 48000 | — | — | — | — |
| Iron (English wrought) | 7·700 | 481·20 | 25½ tons | — | 24920000 | — | — |
| ditto (in bars) | 7·760<br>7·800 | 475·50<br>487·00 | —<br>25½ tons | — | — | — | — |
| ditto (hammered) | — | — | 30 tons | — | — | — | — |
| ditto (Russian) in bars | — | — | 27 tons | — | — | — | — |
| ditto (Swedish) in bars | — | — | 32 tons | — | — | — | — |
| ditto (English) in wire, 10th inch diam. | — | — | 36 to 43 tons | — | — | — | — |
| ditto (Russian) in wire, 1-20th to 1-30th inch diameter | — | — | 60 to 91 tons | — | — | — | — |
| ditto (rolled in sheets and cut lengthwise) | — | — | 14 tons | — | — | — | — |
| ditto cut crosswise | — | — | 18 tons | — | — | — | — |
| ditto in chains, oval links, 6 inches clear, iron ½ inch diameter | — | — | 21½ tons | — | — | — | — |
| ditto (Brunton's) with stay cross link | — | — | 25 tons | — | — | — | — |
| Cast-iron (Old Park) | — | — | — | — | 18014400 | 48240 | — |
| ditto (Adelphi) | — | — | — | — | 18353600 | 45360 | — |
| ditto (Alfreton) | — | — | — | — | 17686400 | 44046 | — |
| ditto (scrap) | — | — | — | — | 18032000 | 45828 | — |
| ditto (Carron, No. 2) hot blast | 7·046 | 440·37 | 13505 | 108540 | 16085000 | 37503 | 8·037 : 1 |
| ditto ( do. do. ) cold blast | 7·066 | 441·62 | 16683 | 106375 | 17270500 | 38556 | 6·376 : 1 |
| ditto ( do. No. 3) | 7·094 | 443·37 | 14200 | 115442 | 16246966 | 33980 | 8·129 : 1 |
| ditto ( do. do. ) hot blast | 7·056 | 441·00 | 17755 | 133440 | 17873100 | 42120 | 7·515 : 1 |
| ditto (Devon, No. 3) cold blast | 7·295 | 455·93 | — | — | 22907700 | 36288 | — |
| ditto ( do. do. ) hot blast | 7·229 | 451·81 | 21907 | 145435 | 22473650 | 43497 | — |
| ditto (Buffrey, No, 1) cold blast | 7·079 | 442·43 | 17466 | 93366 | 15381200 | 37503 | 5·346 : 1 |
| ditto ( do. do. ) hot blast | 6·998 | 437·37 | 13434 | 86397 | 13730500 | 35316 | 6·431 : 1 |
| ditto (Coed-Talon, No. 2) cold blast | 6·955 | 434·06 | 18855 | 81770 | 14313500 | 33104 | 4·337 : 1 |
| ditto ( do. do. ) hot blast | 6·968 | 435·50 | 16676 | 82739 | 14322500 | 33145 | 4·961 : 1 |
| ditto ( do. No. 3) cold blast | 7·104 | 449·62 | — | — | 17102000 | 43541 | — |
| ditto ( do. do. ) hot blast | 6·970 | 435·62 | — | — | 14707900 | 40159 | — |
| ditto (Milton, No. 1) hot blast | 6·976 | 436·00 | — | — | 11974500 | 28552 | — |
| ditto (Muirkirk, No. 1) cold blast | 7·113 | 444·56 | — | — | 14003550 | 35923 | — |
| ditto ( do. do. ) hot blast | 6·953 | 434·56 | — | — | 13294400 | 33850 | — |
| ditto (Elsicar, No. 1) cold blast | 7·030 | 439·37 | — | — | 13981000 | 34862 | — |
| Lead (English cast) | 11·446 | 717·45 | 1824 | — | 720000 | — | — |
| ditto (milled-sheet) | 11·407 | 712·93 | 3328 | — | — | — | — |
| ditto (wire-drawn | 11·317 | 705·12 | 2581 | — | — | — | — |
| Silver (standard) | 10·312 | 644·50 | 40902 | — | — | — | — |
| Mercury (at 32°) | 13·619 | 851·18 | — | — | — | — | — |
| ditto (at 60°) | 13·580 | 848·75 | — | — | — | — | — |
| Steel (soft) | 7·780 | 486·25 | 120000 | — | — | — | — |
| ditto (razor-tempered) | 7·840 | 490·00 | 150000 | — | 29000000 | — | — |
| Tin (cast) | 7·291 | 455·68 | 5322 | — | 4608000 | — | — |
| Zinc (cast) | 7·028 | 439·25 | — | — | 13680000 | — | — |
| ditto (rolled | 7·215 | 450·9 | — | — | — | — | — |
| TABLE II.—*Principal Woods.* | | | | | | | |
| Acacia (English) | 0·71 | 44·37 | 16000 | — | 1152000 | 11202 | — |
| Beech, New | 0·854 | 53·37 | 15784 | 7733 | | | |
| Beech, Dry | 0·690 | 43·12 | 17850 | 9363 | 13536000 | 93363 | 0·55 : 1 |
| Birch, Common | 0·792 | 49·50 | 15000 | 6402 | 1562400 | 10920 | 0·43 : 1 |
| Birch, American | 0·648 | 40·50 | — | 11633 | 1257600 | 9624 | — |
| Deal, Christiania middle | 0·698 | 43·62 | 12400 | — | 1672000 | 9864 | — |
| Deal, Memel middle | 0·590 | 36·87 | — | — | 1535200 | 10386 | — |
| Deal, Norway spruce | 0·340 | 21·25 | 17600 | — | — | — | — |
| Deal, English | 0·470 | 29·37 | 7000 | — | — | — | — |
| Elm (seasoned) | 0·588 | 36·75 | 13489 | 10331 | 699840 | 6078 | 0·79 : 1 |
| Fir, New England | 0·553 | 34·56 | — | — | 2191200 | 6612 | — |
| Fir, Riga | 0·753 | 47·06 | 12000 | 6000 | — | — | 0·50 : 1 |
| Larch (seasoned) | 0·522 | 32·62 | 10220 | 5568 | 1052800 | 6894 | 0·55 : 1 |
| Lignum-vitæ | 1·220 | 76·25 | 11800 | — | — | — | — |
| Mahogany (Spanish) | 0·800 | 50·00 | 16500 | 8198 | — | — | 0·50 : 1 |
| Oak, English | 0·934 | 58·37 | 17300 | 4684 wet<br>9504 dry | 1451200 | 10032 | 0·28 : 1<br>0·57 : 1 |
| Oak, Canadian | 0·872 | 54·50 | 10253 | 4231 wet<br>9509 dry | 2148800 | 10596 | 0·42 : 1<br>0·95 : 1 |
| Oak, Dantzic | 0·756 | 47·24 | 12780 | — | 1191200 | 8748 | — |
| Pine, Pitch | 0·660 | 41·25 | 7818 | — | 1225600 | 9792 | — |
| Pine, Red | 0·657 | 41·06 | — | 5375 | 1840000 | 8946 | — |
| Pine, Yellow | 0·461 | 28·81 | — | 5445 | 1600000 | — | — |
| Plane-tree | 0·64 | 40·00 | 11700 | — | — | — | — |
| Poplar | 0·383 | 23·93 | 7200 | 3107 wet<br>5124 dry | — | — | 0·43 : 1<br>0·74 : 1 |
| Teak (dry) | 0·657 | 41·06 | 15000 | 12101 | 2414400 | 14772 | 0·81 : 1 |
| Willow (dry) | 0·390 | 24·37 | 14000 | — | — | — | — |
| Yew (Spanish) | 0·807 | 50·43 | 8000 | — | — | — | — |

THE COHESIVE STRENGTH OF BODIES.

*The following* TABLE *contains the result of experiments on the cohesive strength of various bodies in avoirdupois pounds; also, one-third of the ultimate strength of each body, this being considered sufficient, in most cases, for a permanent load:*

| Names of Bodies. | Square Bar. | One-third. | Round Bar. | One-third. |
|---|---|---|---|---|
| WOODS. | *lbs.* | *lbs.* | *lbs.* | *lbs.* |
| Boxwood | 20000 | 6667 | 15708 | 5236 |
| Ash | 17000 | 5667 | 13357 | 4452 |
| Teak | 15000 | 5000 | 11781 | 3927 |
| Fir | 12000 | 4000 | 9424 | 3141 |
| Beach | 11500 | 3866 | 9032 | 3011 |
| Oak | 11000 | 3667 | 8639 | 2880 |
| METALS. | | | | |
| Cast iron | 18656 | 6219 | 14652 | 4884 |
| English wrought iron | 55872 | 18624 | 43881 | 14627 |
| Swedish do. do. | 72064 | 24021 | 56599 | 18866 |
| Blistered steel | 133152 | 44384 | 104577 | 34859 |
| Shear do. | 124400 | 41366 | 97703 | 32568 |
| Cast do. | 134256 | 44752 | 105454 | 35151 |
| Cast copper | 19072 | 6357 | 14979 | 4993 |
| Wrought do. | 33792 | 11264 | 26540 | 8827 |
| Yellow brass | 17968 | 5989 | 14112 | 4704 |
| Cast tin | 4736 | 1579 | 3719 | 1239 |
| Cast lead | 1824 | 608 | 1432 | 477 |

PROBLEM I.

RULE.—*To find the ultimate cohesive strength of square, round, and rectangular bars, of any of the various bodies, as specified in the table.*—Multiply the strength of an inch bar, (as in the table,) of the body required, by the cross sectional area of square and rectangular bars, or by the square of the diameter of round bars; and the product will be the ultimate cohesive strength.

A bar of cast iron being $1\frac{1}{2}$ inches square, required its cohesive power.

$$1{\cdot}5 \times 1{\cdot}5 \times 18656 = 41976 \text{ lbs.}$$

Required the cohesive force of a bar of English wrought iron, 2 inches broad, and $\frac{3}{8}$ of an inch in thickness.

$$2 \times {\cdot}375 \times 55872 = 41904 \text{ lbs.}$$

Required the ultimate cohesive strength of a round bar of wrought copper $\frac{3}{4}$ of an inch in diameter.

$${\cdot}75^2 \times 26540 = 14928{\cdot}75 \text{ lbs.}$$

PROBLEM II.

RULE.—*The weight of a body being given, to find the cross sectional dimensions of a bar or rod capable of sustaining that weight.*—For square and round bars, divide the weight given by one-third of the cohesive strength of an inch bar, (as specified in the table,) and the square root of the quotient will be the side of the square, or diameter of the bar in inches.

And if rectangular, divide the quotient by the breadth, and the result will be the thickness.

What must be the side of a square bar of Swedish iron to sustain a permanent weight of 18000 lbs?

$$\sqrt{\frac{18000}{24021}} = \cdot 86, \text{ or nearly } \tfrac{7}{8} \text{ of an inch square.}$$

Required the diameter of a round rod of cast copper to carry a weight of 6800 lbs.

$$\sqrt{\frac{6800}{4993}} = 1.16 \text{ inches diameter.}$$

A bar of English wrought iron is to be applied to carry a weight of 2760 lbs.; required the thickness, the breadth being two inches.

$$\frac{2760}{18624} = \cdot 142 \div 2 = \cdot 071 \text{ of an inch in thickness.}$$

A TABLE *showing the circumference of a rope equal to a chain made of iron of a given diameter, and the weight in tons that each is proved to carry; also, the weight of a foot of chain made from iron of that dimension.*

| Ropes. Cir. in Ins. | Chains. Diam. in Inches. | Proved to carry in tons. | Weight of a lineal foot in lbs. Avr. |
|---|---|---|---|
| 3 | $\frac{1}{4}$ and $\frac{1}{16}$ | 1 | 1·08 |
| 4 | $\frac{3}{8}$ | 2 | 1·5 |
| $4\frac{3}{4}$ | $\frac{3}{8}$ and $\frac{1}{16}$ | 3 | 2 |
| $5\frac{1}{4}$ | $\frac{1}{2}$ | 4 | 2·7 |
| 6 | $\frac{1}{2}$ and $\frac{1}{16}$ | 5 | 3·3 |
| $6\frac{1}{2}$ | $\frac{5}{8}$ | 6 | 4 |
| 7 | $\frac{5}{8}$ and $\frac{1}{16}$ | 8 | 4·6 |
| $7\frac{1}{2}$ | $\frac{3}{4}$ | $9\frac{3}{4}$ | 5·5 |
| 8 | $\frac{3}{4}$ and $\frac{1}{16}$ | $11\frac{1}{4}$ | 6·1 |
| 9 | $\frac{7}{8}$ | 13 | 7·2 |
| $9\frac{1}{2}$ | $\frac{7}{8}$ and $\frac{1}{16}$ | 15 | 8·4 |
| $10\frac{1}{2}$ | 1 inch. | 18 | 9·4 |

## ON THE TRANSVERSE STRENGTH OF BODIES.

The *tranverse strength* of a body is that power which it exerts in opposing any force acting in a perpendicular direction to its length, as in the case of beams, levers, &c., for the fundamental principles of which observe the following:—

That the transverse strength of beams, &c. is inversely as their lengths, and directly as their breadths, and square of their depths, and, if cylindrical, as the cubes of their diameters; that is, if a beam 6 feet long, 2 inches broad, and 4 inches deep, can carry 2000 lbs., another beam of the same material, 12 feet long, 2 inches broad, and 4 inches deep, will only carry 1000, being inversely as their lengths. Again, if a beam 6 feet long, 2 inches broad, and 4 inches deep, can support a weight of 2000 lbs., another beam of the same material, 6 feet long, 4 inches broad, and 4 inches deep, will support double that weight, being directly as their breadths; —but a beam of that material, 6 feet long, 2 inches broad, and 8 inches deep, will sustain a weight of 8000 lbs.; being as the square of their depths.

From a mean of experiments made, to ascertain the transverse strength of various bodies, it appears that the ultimate strength of an inch square, and an inch round bar of each, 1 foot long, loaded in the middle, and lying loose at both ends, is nearly as follows, in lbs. avoirdupois.

| Names of Bodies. | Square Bar. | One-third. | Round Bar. | One-third. |
|---|---|---|---|---|
| Oak | 800 | 267 | 628 | 209 |
| Ash | 1137 | 379 | 893 | 298 |
| Elm | 569 | 139 | 447 | 149 |
| Pitch pine | 916 | 305 | 719 | 239 |
| Deal | 566 | 188 | 444 | 148 |
| Cast iron | 2580 | 860 | 2026 | 675 |
| Wrought iron | 4013 | 1338 | 3152 | 1050 |

PROBLEM I.

RULE.—*To find the ultimate transverse strength of any rectangular beam, supported at both ends, and loaded in the middle; or supported in the middle, and loaded at both ends; also, when the weight is between the middle and the end; likewise when fixed at one end and loaded at the other.*—Multiply the strength of an inch square bar, 1 foot long, (as in the table,) by the breadth, and square of the depth in inches, and divide the product by the length in feet; the quotient will be the weight in lbs. avoirdupois.

What weight will break a beam of oak 4 inches broad, 8 inches deep, and 20 feet between the supports?

$$\frac{800 \times 4 \times 8^2}{20} = 10240 \text{ lbs.}$$

When a beam is supported in the middle, and loaded at each end, it will bear the same weight as when supported at both ends and loaded in the middle; that is, each end will bear half the weight.

When the weight is not situated in the middle of the beam, but placed somewhere between the middle and the end, multiply twice the length of the long end by twice the length of the short end, and divide the product by the whole length of the beam; the quotient will be the effectual length.

Required the ultimate transverse strength of a pitch pine plank 24 feet long, 3 inches broad, 7 inches deep, and the weight placed 8 feet from one end.

$$\frac{32 \times 16}{24} = 21{\cdot}3 \text{ effective length.}$$

$$\text{and } \frac{916 \times 3 \times 7^2}{21{\cdot}3} = 6321 \text{ lbs.}$$

Again, when a beam is fixed at one end and loaded at the other, it will only bear $\frac{1}{4}$ of the weight as when supported at both ends and loaded in the middle.

What is the weight requisite to break a deal beam 6 inches broad, 9 inches deep, and projecting 12 feet from the wall?

$$\frac{566 \times 6 \times 9^2}{12} = 22923 \div 4 = 5730 \cdot 7 \text{ lbs.}$$

The same rules apply as well to beams of a cylindrical form, with this exception, that the strength of a round bar (as in the table) is multiplied by the cube of the diameter, in place of the breadth, and square of the depth.

Required the ultimate transverse strength of a solid cylinder of cast iron 12 feet long and 5 inches diameter.

$$\frac{2026 \times 5^3}{12} = 21104 \text{ lbs.}$$

What is the ultimate transverse strength of a hollow shaft of cast iron 12 feet long, 8 inches diameter outside, and containing the same cross sectional area as a solid cylinder 5 inches diameter?

$$\sqrt{8^2 - 5^2} = 6 \cdot 24, \text{ and } 8^3 - 6 \cdot 24^3 = 269.$$

$$\text{Then, } \frac{2026 \times 269}{12} = 45416 \text{ lbs.}$$

When a beam is fixed at both ends, and loaded in the middle, it will bear one-half more than it will when loose at both ends.

And if a beam is loose at both ends, and the weight laid uniformly along its length, it will bear double; but if fixed at both ends, and the weight laid uniformly along its length, it will bear triple the weight.

PROBLEM II.

RULE.—*To find the breadth or depth of beams intended to support a permanent weight.*—Multiply the length between the supports, in feet, by the weight to be supported in lbs., and divide the product by one-third of the ultimate strength of an inch bar, (as in the table,) multiplied by the square of the depth; the quotient will be the breadth, or, multiplied by the breadth, the quotient will be the square of the depth, both in inches.

Required the breadth of a cast iron beam 16 feet long, 7 inches deep, and to support a weight of 4 tons in the middle.

$$4 \text{ tons} = 8960 \text{ lbs. and } \frac{8960 \times 16}{860 \times 7^2} = 3 \cdot 4 \text{ inches.}$$

What must be the depth of a cast iron beam 3·4 inches broad, 16 feet long, and to bear a permanent weight of four tons in the middle?

$$\sqrt{\frac{8960 \times 16}{860 \times 3 \cdot 4}} = 7 \text{ inches.}$$

When a beam is fixed at both ends, the divisor must be multiplied by 1·5, on account of it being capable of bearing one-half more.

When a beam is loaded uniformly throughout, and loose at both ends, the divisor must be multiplied by 2, because it will bear double the weight.

If a beam is fast at both ends, and loaded uniformly throughout, the divisor must be multipled by 3, on account that it will bear triple the weight.

Required the breadth of an oak beam 20 feet long, 12 inches deep, made fast at both ends, and to be capable of supporting a weight of 12 tons in the middle.

$$12 \text{ tons} = 26880 \text{ lbs., and } \frac{26880 \times 20}{266 \times 12^2 \times 1{\cdot}5} = 9{\cdot}7 \text{ inches.}$$

Again, when a beam is fixed at one end, and loaded at the other, the divisor must be multiplied by ·25; because it will only bear one-fourth of the weight.

Required the depth of a beam of ash 6 inches broad, 9 feet projecting from the wall, and to carry a weight of 47 cwt.

$$47 \text{ cwt.} = 5264 \text{ lbs., and } \sqrt{\frac{5264 \times 9}{379 \times 6 \times {\cdot}25}} = 9{\cdot}12 \text{ inches deep.}$$

And when the weight is not placed in the middle of a beam, the effective length must be found as in Problem I.

Required the depth of a deal beam 20 feet long, and to support a weight of 63 cwt. 6 feet from one end.

$$\frac{28 \times 12}{20} = 16{\cdot}8 \text{ effective length of beam, and}$$

$$63 \text{ cwt.} = 7056 \text{ lbs.; hence}$$

$$\sqrt{\frac{7056 \times 16{\cdot}8}{188 \times 6}} = 10{\cdot}24 \text{ inches deep.}$$

Beams or shafts exposed to lateral pressure are subject to all the foregoing rules, but in the case of water-wheel shafts, &c., some allowances must be made for wear; then the divisor may be changed from 675 to 600 for cast iron.

Required the diameter of bearings for a water-wheel shaft 12 feet long, to carry a weight of 10 tons in the middle.

$$10 \text{ tons} = 22400 \text{ lbs., and}$$

$$\frac{22400}{600} = \sqrt[3]{448} = 7{\cdot}65 \text{ inches diameter.}$$

And when the weight is equally distributed along its length, the cube root of half the quotient will be the diameter, thus:

$$\frac{448}{2} = \sqrt[3]{224} = 6{\cdot}07 \text{ inches diameter.}$$

Required the diameter of a solid cylinder of cast iron, for the shaft of a crane, to be capable of sustaining a weight of 10 tons;

one end of the shaft to be made fast in the ground, the other to projèct $6\frac{1}{2}$ feet; and the effective leverage of the jib as $1\frac{3}{4}$ to 1.

10 tons = 22400 lbs., and

$$\frac{22400 \times 6{\cdot}5 \times 1{\cdot}75}{675 \times {\cdot}25} = 1509$$

And $\sqrt[3]{1509} = 11{\cdot}47$ inches diameter.

The strength of cast iron to wrought iron, in this direction, is as 9 is to 14 nearly; hence, if wrought iron is taken in place of cast iron in the last example, what must be its diameter?

$$\sqrt[3]{\frac{1509 \times 9}{14}} = 9{\cdot}89 \text{ inches diameter.}$$

### ON TORSION OR TWISTING.

The strength of bodies to resist *torsion*, or wrenching asunder, is directly as the cubes of their diameters; or, if square, as the cube of one side; and inversely as the force applied multiplied into the length of the lever.

Hence the rule.—1. Multiply the strength of an inch bar, by experiment, (as in the following table,) by the cube of the diameter, or of one side in inches; and divide by the radius of the wheel, or length of the lever also in inches; and the quotient will be the ultimate strength of the shaft or bar, in lbs. avoirdupois.

2.—Multiply the force applied in pounds by the length of the lever in inches, and divide the product by one-third of the ultimate strength of an inch bar, (as in the table,) and the cube root of the quotient will be the diameter, or side of a square bar in inches; that is, capable of resisting that force permanently.

*The following* TABLE *contains the result of experiments on inch bars, of various metals, in lbs. avoirdupois.*

| Names of Bodies. | Round Bar. | One-third. | Square Bar. | One-third. |
|---|---|---|---|---|
| Cast iron............... | 11943 | 3981 | 15206 | 5069 |
| English wrought iron | 12063 | 4021 | 15360 | 5120 |
| Swedish do. do. | 11400 | 3800 | 14592 | 4864 |
| Blistered steel......... | 20025 | 6675 | 25497 | 8499 |
| Shear.......do......... | 20508 | 6836 | 26112 | 8704 |
| Cast.........do......... | 21111 | 7037 | 26880 | 8960 |
| Yellow brass.......... | 5549 | 1850 | 7065 | 2355 |
| Cast copper........... | 4825 | 1608 | 6144 | 2048 |
| Tin....................... | 1688 | 563 | 2150 | 717 |
| Lead..................... | 1206 | 402 | 1536 | 512 |

What weight, applied on the end of a 5 feet lever, will wrench asunder a 3 inch round bar of cast iron?

$$\frac{11943 \times 3^3}{60} = 5374 \text{ lbs. avoirdupois.}$$

Required the side of a square bar of wrought iron, capable of resisting the twist of 600 lbs. on the end of a lever 8 feet long.

$$\sqrt[3]{\frac{600 \times 96}{5120}} = 2\tfrac{1}{4} \text{ inches.}$$

In the case of revolving shafts for machinery, &c., the strength is directly as the cubes of their diameters, and revolutions, and inversely as the resistance they have to overcome; hence,

From *practice*, we find that a 40 horse power steam engine, making 25 revolutions per minute, requires a shaft (*if made of wrought-iron*) to be 8 inches diameter: now, the cube of 8, multiplied by 25, and divided by 40 = 320; which serves as a constant multiplier for all others in the same proportion.

What must be the diameter of a wrought iron shaft for an engine of 65 horse power, making 23 revolutions per minute?

$$\sqrt[3]{\frac{65 \times 320}{23}} = 9{\cdot}67 \text{ inches diameter.}$$

James Glenie, the mathematician, gives 400 as a constant multiplier for cast iron shafts that are intended for first movers in machinery;

200 for second movers; and

100 for shafts connecting smaller machinery, &c.

The velocity of a 30 horse power steam engine is intended to be 19 revolutions per minute. Required the diameter of bearings for the fly-wheel shaft.

$$\sqrt[3]{\frac{400 \times 30}{19}} = 8{\cdot}579 \text{ inches diameter.}$$

Required the diameter of the bearings of shafts, as second movers from a 30 horse engine; their velocity being 36 revolutions per minute.

$$\sqrt[3]{\frac{200 \times 30}{36}} = 5{\cdot}5 \text{ inches diameter.}$$

When shafting is intended to be of wrought iron, use 160 as the multiplier for second movers; and 80 for shafts connecting smaller machinery.

TABLE *of the Proportionate Length of Bearings, or Journals for Shafts of various diameters.*

| Dia. in Inches. | Len. in Inches. | Dia. in Inches. | Len. in Inches. |
|---|---|---|---|
| 1 | $1\frac{3}{4}$ | $6\frac{1}{2}$ | $8\frac{3}{4}$ |
| $1\frac{1}{2}$ | $2\frac{1}{4}$ | 7 | $9\frac{3}{8}$ |
| 2 | 3 | $7\frac{1}{2}$ | 10 |
| $2\frac{1}{4}$ | $3\frac{1}{4}$ | 8 | $10\frac{3}{4}$ |
| $2\frac{1}{2}$ | $3\frac{1}{2}$ | $8\frac{1}{2}$ | $11\frac{3}{8}$ |
| 3 | $4\frac{1}{4}$ | 9 | 12 |
| $3\frac{1}{2}$ | $4\frac{7}{8}$ | $9\frac{1}{2}$ | $12\frac{3}{4}$ |
| 4 | $5\frac{1}{2}$ | 10 | $13\frac{1}{4}$ |
| $4\frac{1}{2}$ | $6\frac{1}{8}$ | $10\frac{1}{2}$ | 14 |
| 5 | $6\frac{3}{4}$ | 11 | $14\frac{1}{2}$ |
| $5\frac{1}{2}$ | $7\frac{1}{2}$ | $11\frac{1}{2}$ | $15\frac{1}{4}$ |
| 6 | $8\frac{1}{4}$ | 12 | 16 |

*Tenacities, Resistances to Compression, and other Properties of the common Materials of Construction.*

| Names of Bodies. | Absolute. | | Compared with Cast Iron. | | |
|---|---|---|---|---|---|
| | Tenacity in lbs. per sq. inch. | Resistance to compression in lbs. per sq. in. | Its strength is | Its extensibility is | Its stiffness is |
| Ash | 14130 | — | 0·23 | 2·6 | 0·089 |
| Beech | 12225 | 8548 | 0·15 | 2·1 | 0·073 |
| Brass | 17968 | 10304 | 0·435 | 0·9 | 0·49 |
| Brick | 275 | 562 | — | — | — |
| Cast iron | 13434 | 86397 | 1·000 | 1·0 | 1·000 |
| Copper (wrought) | 33000 | — | — | — | — |
| Elm | 9720 | 1033 | 0·21 | 2·9 | 0·073 |
| Fir, or Pine, white | 12346 | 2028 | 0·23 | 2·4 | 0·1 |
| — — red | 11800 | 5375 | 0·3 | 2·4 | 0·1 |
| — — yellow | 11835 | 5445 | 0·25 | 2·9 | 0·087 |
| Granite, Aberdeen | — | 10910 | — | — | — |
| Gun-metal (copper 8, and tin 1) | 35838 | — | 0·65 | 1·25 | 0·535 |
| Malleable iron | 56000 | — | 1·12 | 0·86 | 1·3 |
| Larch | 12240 | 5568 | 0·136 | 2·3 | 0·058 |
| Lead | 1824 | — | 0·096 | 2·5 | 0·0385 |
| Mahogany, Honduras | 11475 | 8000 | 0·24 | 2·9 | 0·487 |
| Marble | 551 | 6060 | — | — | — |
| Oak | 11880 | 9504 | 0·25 | 2·8 | 0·093 |
| Rope (1 in. in circum.) | 200 | — | — | — | — |
| Steel | 128000 | — | — | — | — |
| Stone, Bath | 478 | — | — | — | — |
| — Craigleith | 772 | 5490 | — | — | — |
| — Dundee | 2661 | 6630 | — | — | — |
| — Portland | 857 | 3729 | — | — | — |
| Tin (cast) | 4736 | — | 0·182 | 0·75 | 0·25 |
| Zinc (sheet) | 9120 | — | 0·365 | 0·5 | 0·76 |

*Comparative Strength and Weight of Ropes and Chains.*

| Circum. of rope in inches. | Weight per fathom in lbs. | Diameter of chain in inches. | Weight per fathom in lbs. | Proof strength in tons & cwt. | Circum. of rope in inches. | Weight per fathom in lbs. | Diameter of chain in inches. | Weight per fathom in lbs. | Proof strength in tons & cwt. |
|---|---|---|---|---|---|---|---|---|---|
| $3\frac{1}{2}$ | $2\frac{3}{4}$ | $\frac{5}{16}$ | $5\frac{1}{2}$ | 1 $5\frac{1}{2}$ | 10 | 23 | $\frac{7}{8}$ | 43 | 10 0 |
| $4\frac{1}{4}$ | $4\frac{3}{4}$ | $\frac{3}{8}$ | 8 | 1 $16\frac{3}{4}$ | $10\frac{3}{4}$ | 28 | $\frac{15}{16}$ | 49 | 11 11 |
| 5 | $5\frac{3}{4}$ | $\frac{7}{16}$ | $10\frac{1}{2}$ | 2 10 | $11\frac{1}{2}$ | $30\frac{1}{2}$ | 1in. | 56 | 13 8 |
| $5\frac{3}{4}$ | 7 | $\frac{1}{2}$ | 14 | 3 $5\frac{1}{2}$ | $12\frac{1}{4}$ | 36 | $1\frac{1}{16}$ | 63 | 14 18 |
| $6\frac{1}{2}$ | $9\frac{3}{4}$ | $\frac{9}{16}$ | 18 | 4 $3\frac{1}{2}$ | 13 | 39 | $1\frac{1}{8}$ | 71 | 16 14 |
| 7 | $11\frac{1}{4}$ | $\frac{5}{8}$ | 22 | 5 2 | $13\frac{3}{4}$ | 45 | $1\frac{3}{16}$ | 79 | 18 11 |
| 8 | 15 | $\frac{11}{16}$ | 27 | 6 $4\frac{1}{2}$ | $14\frac{1}{2}$ | $48\frac{1}{2}$ | $1\frac{1}{4}$ | 87 | 20 8 |
| $8\frac{3}{4}$ | 19 | $\frac{3}{4}$ | 32 | 7 7 | $15\frac{1}{4}$ | 56 | $1\frac{5}{16}$ | 96 | 22 13 |
| $9\frac{1}{2}$ | 21 | $\frac{13}{16}$ | 37 | 8 $13\frac{1}{2}$ | 16 | 60 | $1\frac{3}{8}$ | 106 | 24 18 |

It must be understood and also borne in mind, that in estimating the amount of tensile strain to which a body is subjected, the weight of the body itself must also be taken into account; for according to its position so may it approximate to its whole weight, in tending to produce tension within itself; as in the almost constant application of ropes and chains to great depths, considerable heights, &c.

*Alloys that are of greater Tenacity than the sum of their Constituents, as determined by the Experiments of Muschenbroek.*

Swedish copper 6 parts, Malacca tin 1—tenacity per square inch 64,000 lbs.
Chili copper 6 parts, Malacca tin 1..........60,000
Japan copper 5 parts, Banca tin 1..........57,000
Anglesea copper 6 parts, Cornish tin 1..........41,000
Common block tin 4, lead 1, zinc 1..........13,000
Malacca tin 4, regulus of antimony 1..........12,000
Block tin 3, lead 1..........10,200
Block tin 8, zinc 1..........10,000
Lead 1, zinc 1.......... 4,500

TABLE *of Data, containing the Results of Experiments on the Elasticity and Strength of various Species of Timber.*

| Species of Timber. | Value of E. | Value of S. | Species of Timber. | Value of E. | Value of S. |
|---|---|---|---|---|---|
| Teak .......... | 174·7 | 2462 | Elm.......... | 50·64 | 1013 |
| Poona.......... | 122·26 | 2221 | Pitch pine.......... | 88·68 | 1632 |
| English oak.......... | 105 | 1672 | Red pine.......... | 133 | 1341 |
| Canadian do.......... | 155·5 | 1766 | New England fir | 158·5 | 1102 |
| Dantzic do.......... | 86·2 | 1457 | Riga fir.......... | 90 | 1100 |
| Adriatic do.......... | 70·5 | 1383 | Mar Forest do. | 63 | 1200 |
| Ash.......... | 119 | 2026 | Larch.......... | 76 | 900 |
| Beech.......... | 98 | 1556 | Norway spruce... | 105·47 | 1474 |

RULE.—*To find the dimensions of a beam capable of sustaining a given weight, with a given degree of deflection, when supported at both ends.*—Multiply the weight to be supported in lbs. by the cube of the length in feet; divide the product by 32 times the tabular value of E, multiplied into the given deflection in inches, and the quotient is the breadth multiplied by the cube of the depth in inches.

When the beam is intended to be square, then the fourth root of the quotient is the breadth and depth required.

If the beam is to be cylindrical, multiply the quotient by 1·7, and the fourth root of the product is the diameter.

The distance between the supports of a beam of Riga fir is 16 feet, and the weight it must be capable of sustaining in the middle of its length is 8000 lbs., with a deflection of not more than $\frac{3}{4}$ of an inch; what must be the depth of the beam, supposing the breadth 8 inches?

$$\frac{16 \times 8000}{90 \times 32 \times \cdot 75} = 15175 \div 8 = \sqrt[3]{1897} = 12\cdot 35 \text{ in. the depth.}$$

RULE.—*To determine the absolute strength of a rectangular beam of timber when supported at both ends, and loaded in the middle of its length, as beams in general ought to be calculated to, so that they may be rendered capable of withstanding all accidental cases of emergency.*—Multiply the tabular value of S by four times the depth of the beam in inches, and by the area of the cross section in inches; divide the product by the distance between the supports in inches, and the quotient will be the absolute strength of the beam in lbs.

If the beam be not laid horizontally, the distance between the supports, for calculation, must be the horizontal distance.

One-fourth of the weight obtained by the rule is the greatest weight that ought to be applied in practice as permanent load.

If the load is to be applied at any other point than the middle, then the strength will be, as the product of the two distances is to the square of half the length of the beam between the supports; or, twice the distance from one end, multiplied by twice from the other, and divided by the whole length, equal the effective length of the beam.

In a building 18 feet in width, an engine boiler of $5\frac{1}{2}$ tons is to be fixed, the centre of which to be 7 feet from the wall; and having two pieces of red pine 10 inches by 6, which I can lay across the two walls for the purpose of slinging it at each end,—may I with sufficient confidence apply them, so as to effect this object?

$$\frac{2240 \times 5 \cdot 5}{2} = 6160 \text{ lbs. to carry at each end.}$$

And 18 feet $- 7 = 11$, double each, or 14 and 22, then $\frac{14 \times 22}{18}$ $= 17$ feet, or 204 inches, effective length of beam.

Tabular value of S, red pine $= \frac{1341 \times 4 \times 10 \times 60}{204} = 15776$ lbs., the absolute strength of each piece of timber at that point.

RULE.—*To determine the dimensions of a rectangular beam capable of supporting a required weight, with a given degree of deflection, when fixed at one end.*—Divide the weight to be supported, in lbs., by the tabular value of E, multiplied by the breadth and deflection, both in inches; and the cube root of the quotient, multiplied by the length in feet, equal the depth required in inches.

A beam of ash is intended to bear a load of 700 lbs. at its extremity; its length being 5 feet, its breadth 4 inches, and the deflection not to exceed $\frac{1}{2}$ an inch.

Tabular value of E $= 119 \times 4 \times \cdot 5 = 238$, the divisor; then $700 \div 238 = \sqrt[3]{2 \cdot 94} \times 5 = 7 \cdot 25$ inches, depth of the beam.

RULE.—*To find the absolute strength of a rectangular beam, when fixed at one end, and loaded at the other.*—Multiply the value of S by the depth of the beam, and by the area of its section, both in inches; divide the product by the leverage in inches, and the quotient equal the absolute strength of the beam in lbs.

A beam of Riga fir, 12 inches by $4\frac{1}{2}$, and projecting $6\frac{1}{2}$ feet from the wall; what is the greatest weight it will support at the extremity of its length?

Tabular value of S $= 1100$

$12 \times 4 \cdot 5 = 54$ sectional area,

$$\text{Then, } \frac{1100 \times 12 \times 54}{78} = 9138 \cdot 4 \text{ lbs.}$$

When fracture of a beam is produced by vertical pressure, the fibres of the lower section of fracture are separated by extension, whilst at the same time those of the upper portion are destroyed by compression; hence exists a point in section where neither the one nor the other takes place, and which is distinguished as the point of neutral axis. Therefore, by the law of fracture thus established, and proper data of tenacity and compression given, as in the Table (p. 281), we are enabled to form metal beams of strongest section with the least possible material: thus, in cast iron the resistance to compression is nearly as $6\frac{1}{2}$ to 1 of tenacity; consequently a beam of cast iron, to be of strongest section, must be of the form TB, and a parabola in the direction of its length, the quantity of material in the bottom flange being about $6\frac{1}{2}$ times that of the upper: but such is not the case with beams of timber; for although the tenacity of timber be on an average twice that of its resistance to compression, its flexibility is so great, that any considerable length of beam, where columns cannot be situated to its support, requires to be strengthened or trussed by iron rods, as in the following manner:

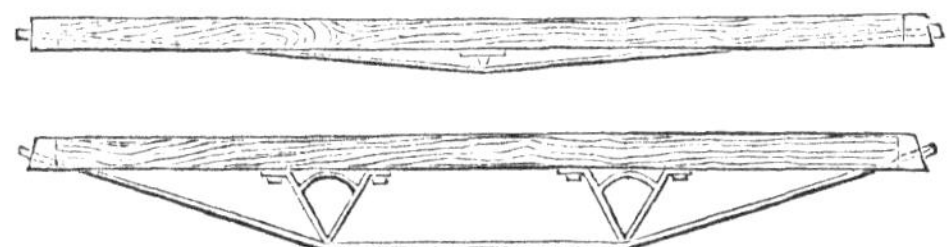

And these applications of principle not only tend to diminish deflection, but the required purpose is also more effectively attained, and that by lighter pieces of timber.

RULE.—*To ascertain the absolute strength of a cast iron beam of the preceding form, or that of strongest section.*—Multiply the sectional area of the bottom flange in inches by the depth of the beam in inches, and divide the product by the distance between the supports also in inches; and 514 times the quotient equal the absolute strength of the beam in cwts.

The strongest form in which any given quantity of matter can be disposed is that of a hollow cylinder; and it has been demonstrated that the maximum of strength is obtained in cast iron, when the thickness of the annulus or ring amounts to $\frac{1}{5}$th of the cylinder's external diameter; the relative strength of a solid to that of a hollow cylinder being as the diameters of their sections.

The following table shows the greatest weight that ever ought to be laid upon a beam for permanent load, and if there be any liability to jerks, &c., ample allowance must be made; also, the weight of the beam itself must be included.

RULE.—*To find the weight of a cast iron beam of given dimensions.*—Multiply the sectional area in inches by the length in feet, and by 3·2, the product equal the weight in lbs.

Required the weight of a uniform rectangular beam of cast iron, 16 feet in length, 11 inches in breadth, and $1\frac{1}{2}$ inch in thickness.

$$11 \times 1{\cdot}5 \times 16 \times 3{\cdot}2 = 844{\cdot}8 \text{ lbs.}$$

A Table *showing the Weight or Pressure a Beam of Cast Iron, 1 inch in breadth, will sustain without destroying its elastic force, when it is supported at each end, and loaded in the middle of its length, and also the deflection in the middle which that weight will produce.*

| Length. | 6 feet. | | 7 feet. | | 8 feet. | | 9 feet. | | 10 feet. | |
|---|---|---|---|---|---|---|---|---|---|---|
| Depth in in. | Wt. in lbs. | Defl. in in. | Wt. in lbs. | Defl. in in. | Wt. in lbs. | Defl. in in. | Wt. in lbs. | Defl. in in. | Wt. in lbs. | Defl. in in. |
| 3 | 1278 | ·24 | 1089 | ·33 | 954 | ·426 | 855 | ·54 | 765 | ·66 |
| 3½ | 1739 | ·205 | 1482 | ·28 | 1298 | ·365 | 1164 | ·46 | 1041 | ·57 |
| 4 | 2272 | ·18 | 1936 | ·245 | 1700 | ·32 | 1520 | ·405 | 1360 | ·5 |
| 4½ | 2875 | ·16 | 2450 | ·217 | 2146 | ·284 | 1924 | ·36 | 1721 | ·443 |
| 5 | 3560 | ·144 | 3050 | ·196 | 2650 | ·256 | 2375 | ·32 | 2125 | ·4 |
| 6 | 5112 | ·12 | 4356 | ·163 | 3816 | ·213 | 3420 | ·27 | 3060 | ·33 |
| 7 | 6958 | ·103 | 5929 | ·14 | 5194 | ·183 | 4655 | ·23 | 4165 | ·29 |
| 8 | 9088 | ·09 | 7744 | ·123 | 6784 | ·16 | 6080 | ·203 | 5440 | ·25 |
| 9 | — | — | 9801 | ·109 | 8586 | ·142 | 7695 | ·18 | 6885 | ·22 |
| 10 | — | — | 12100 | ·098 | 10600 | ·128 | 9500 | ·162 | 8500 | ·2 |
| 11 | — | — | — | — | 12826 | ·117 | 11495 | ·15 | 10285 | ·18 |
| 12 | — | — | — | — | 15264 | ·107 | 13680 | ·135 | 12240 | ·17 |
| 13 | — | — | — | — | — | — | 16100 | ·125 | 14400 | ·154 |
| 14 | — | — | — | — | — | — | 18600 | ·115 | 16700 | ·143 |
| | 12 feet. | | 14 feet. | | 16 feet. | | 18 feet. | | 20 feet. | |
| 6 | 2548 | ·48 | 2184 | ·65 | 1912 | ·85 | 1699 | 1·08 | 1530 | 1·34 |
| 7 | 3471 | ·41 | 2975 | ·58 | 2603 | ·73 | 2314 | ·93 | 2082 | 1·14 |
| 8 | 4532 | ·36 | 3884 | ·49 | 3396 | ·64 | 3020 | ·81 | 2720 | 1·00 |
| 9 | 5733 | ·32 | 4914 | ·44 | 4302 | ·57 | 3825 | ·72 | 3438 | ·89 |
| 10 | 7083 | ·28 | 6071 | ·39 | 5312 | ·51 | 4722 | ·64 | 4250 | ·8 |
| 11 | 8570 | ·26 | 7346 | ·36 | 6428 | ·47 | 5714 | ·59 | 5142 | ·73 |
| 12 | 10192 | ·24 | 8736 | ·33 | 7648 | ·43 | 6796 | ·54 | 6120 | ·67 |
| 13 | 11971 | ·22 | 10260 | ·31 | 8978 | ·39 | 7980 | ·49 | 7182 | ·61 |
| 14 | 13883 | ·21 | 11900 | ·28 | 10412 | ·36 | 9255 | ·46 | 8330 | ·57 |
| 15 | 15937 | ·19 | 13660 | ·26 | 11952 | ·34 | 10624 | ·43 | 9562 | ·53 |
| 16 | 18128 | ·18 | 15536 | ·24 | 13584 | ·32 | 12080 | ·40 | 10880 | ·5 |
| 17 | 20500 | ·17 | 17500 | ·23 | 15353 | ·3 | 13647 | ·38 | 12282 | ·47 |
| 18 | 22932 | ·16 | 19656 | ·21 | 17208 | ·28 | 15700 | ·36 | 13752 | ·44 |

*Resistance of Bodies to Flexure by Vertical Pressure.*—When a piece of timber is employed as a column or support, its tendency to yielding by compression is different according to the proportion between its length and area of its cross section; and supposing the form that of a cylinder whose length is less than seven or eight times its diameter, it is impossible to bend it by any force applied longitudinally, as it will be destroyed by splitting before that bending can take place; but when the length exceeds this, the column will bend under a certain load, and be ultimately destroyed by a similar kind of action to that which has place in the transverse strain.

Columns of cast iron and of other bodies are also similarly circumstanced.

When the length of a cast iron column with flat ends equals about thirty times its diameter, fracture will be produced wholly by bending of the material;—when of less length, fracture takes place partly by crushing and partly by bending: but, when the column is enlarged in the middle of its length from one and a half to twice its diameter at the ends, by being cast hollow, the strength is greater by $\frac{1}{7}$th than in a solid column containing the same quantity of material.

RULE.—*To determine the dimensions of a support or column to bear without sensible curvature a given pressure in the direction of its axis.*—Multiply the pressure to be supported in lbs. by the square of the column's length in feet, and divide the product by twenty times the tabular value of E; and the quotient will be equal to the breadth multiplied by the cube of the least thickness, both being expressed in inches.

When the pillar or support is a square, its side will be the fourth root of the quotient.

If the pillar or column be a cylinder, multiply the tabular value of E by 12, and the fourth root of the quotient equal the diameter.

What should be the least dimensions of an oak support, to bear a weight of 2240 lbs. without sensible flexure, its breadth being 3 inches, and its length 5 feet?

Tabular value of E = 105, and $\frac{2240 \times 5^2}{20 \times 105 \times 3} = \sqrt[3]{8{\cdot}888} =$ 2·05 inches.

Required the side of a square piece of Riga fir, 9 feet in length, to bear a permanent weight of 6000 lbs.

Tabular value of E = 96, and $\frac{6000 \times 9^2}{20 \times 96} = \sqrt[4]{253} =$ 4 inches nearly.

*Dimensions of Cylindrical Columns of Cast Iron to sustain a given load or pressure with safety.*

| Diameter in inches. | Length or height in feet. | | | | | | | | | | |
|---|---|---|---|---|---|---|---|---|---|---|---|
| | 4 | 6 | 8 | 10 | 12 | 14 | 16 | 18 | 20 | 22 | 21 |
| | Weight or load in cwts. | | | | | | | | | | |
| 2 | 72 | 60 | 49 | 40 | 32 | 26 | 22 | 18 | 15 | 13 | 11 |
| $2\frac{1}{2}$ | 119 | 105 | 91 | 77 | 65 | 55 | 47 | 40 | 34 | 29 | 25 |
| 3 | 178 | 163 | 145 | 128 | 111 | 97 | 84 | 73 | 64 | 56 | 49 |
| $3\frac{1}{2}$ | 247 | 232 | 214 | 191 | 172 | 156 | 135 | 119 | 106 | 94 | 83 |
| 4 | 326 | 310 | 288 | 266 | 242 | 220 | 198 | 178 | 160 | 144 | 130 |
| $4\frac{1}{2}$ | 418 | 400 | 379 | 354 | 327 | 301 | 275 | 251 | 229 | 208 | 189 |
| 5 | 522 | 501 | 479 | 452 | 427 | 394 | 365 | 337 | 310 | 285 | 262 |
| 6 | 607 | 592 | 573 | 550 | 525 | 497 | 469 | 440 | 413 | 386 | 360 |
| 7 | 1032 | 1013 | 989 | 959 | 924 | 887 | 848 | 808 | 765 | 725 | 686 |
| 8 | 1333 | 1315 | 1289 | 1259 | 1224 | 1185 | 1142 | 1097 | 1052 | 1005 | 959 |
| 9 | 1716 | 1697 | 1672 | 1640 | 1603 | 1561 | 1515 | 1467 | 1416 | 1364 | 1311 |
| 10 | 2119 | 2100 | 2077 | 2045 | 2007 | 1964 | 1916 | 1865 | 1811 | 1755 | 1697 |
| 11 | 2570 | 2550 | 2520 | 2490 | 2450 | 2410 | 2358 | 2305 | 2248 | 2189 | 2127 |
| 12 | 3050 | 3040 | 3020 | 2970 | 2930 | 2900 | 2830 | 2780 | 2730 | 2670 | 2600 |

*Practical utility of the preceding Table.*—Wanting to support the front of a building with cast iron columns 18 feet in length, 8 inches in diameter, and the metal 1 inch in thickness; what weight may

I confidently expect each column capable of supporting without tendency to deflection?

| | | |
|---|---|---|
| Opposite 8 inches diameter and under 18 feet | = | 1097 |
| Also opposite 6 in. diameter and under 18 feet | = | 440 |
| | = | 657 cwts. |

| | | | | |
|---|---|---|---|---|
| The strength of | cast iron as a column being | | = | 1·0000 |
| — | steel | — | = | 2·518 |
| — | wrought iron | — | = | 1·745 |
| — | oak (Dantzic) | — | = | ·1088 |
| — | red deal | — | = | ·0785 |

*Elasticity of torsion, or resistance of bodies to twisting.*—The angle of flexure by torsion is as the length and extensibility of the body directly, and inversely as the diameter; hence, the length of a bar or shaft being given, the power, and the leverage the power acts with, being known, and also the number of degrees of torsion that will not affect the action of the machine, to determine the diameter in cast iron with a given angle of flexure.

RULE.—Multiply the power in lbs. by the length of the shaft in feet, and by the leverage in feet; divide the product by fifty-five times the number of degrees in the angle of torsion, and the fourth root of the quotient equal the shaft's diameter in inches.

Required the diameters for a series of shafts 35 feet in length, and to transmit a power equal to 1245 lbs., acting at the circumference of a wheel $2\frac{1}{2}$ feet radius, so that the twist of the shafts on the application of the power may not exceed one degree.

$$\frac{1245 \times 35 \times 2·5}{55 \times 1} = \sqrt[4]{1981} = 6·67 \text{ inches in diameter.}$$

*Relative strength of metals to resist torsion.*

| | | | |
|---|---|---|---|
| Cast iron........... | = 1 | Swedish bar iron ... | = 1·05 |
| Copper.............. | = ·48 | English do. ...... | = 1·12 |
| Yellow brass........ | = ·511 | Shear steel.......... | = 1·96 |
| Gun-metal .......... | = ·55 | Cast do............ | = 2·1 |

### DEFLEXION OF RECTANGULAR BEAMS.

RULE.—*To ascertain the amount of deflexion of a uniform beam of cast iron, supported at both ends, and loaded in the middle to the extent of its elastic force.*—Multiply the square of the length in feet by ·02, and the product divided by the depth in inches equal the deflexion.

Required the deflection of a cast iron beam 18 feet long between the supports, 12·8 inches deep, 2·56 inches in breadth, and bearing a weight of 20,000 lbs. in the middle of its length.

$$\frac{18^2 \times ·02}{12·8} = ·506 \text{ inches from a straight line in the middle.}$$

For beams of a similar description, loaded uniformly, the rule is the same, only multiply by ·025 in place of ·02.

RULE.—*To find the deflection of a beam when fixed at one end and loaded at the other.*—Divide the length in feet of the fixed part of the beam by the length in feet of the part which yields to the force, and add 1 to the quotient; then multiply the square of the length in feet by the quotient so increased, and also by ·13; divide this product by the middle depth in inches, and the quotient will be the deflection, in inches also.

Multiply the deflection so obtained for cast iron by ·86, the product equal the deflection for wrought iron; for oak, multiply by 2·8; and for fir, 2·4.

A TABLE *of the Depths of Square Beams or Bars of Cast Iron, calculated to support from* 1 *Cwt. to* 14 *Tons in the Middle, the Deflection not to exceed* $\frac{1}{40}$*th of an Inch for each Foot in Length.*

| Lengths in Feet | | 4 | 6 | 8 | 10 | 12 | 14 | 16 | 18 | 20 | 22 | 24 | 26 | 28 | 30 |
|---|---|---|---|---|---|---|---|---|---|---|---|---|---|---|---|
| Weight in cwt. | Weight in lbs. | Depth. | Depth. | Depth. | Depth. | Depth. | Depth. | Depth. | Depth. | Depth. | Depth. | Depth. | Depth. | Depth. | Depth. |
| | | In. | In. | In. | In. | In. | In. | In. | In. | In. | In. | In. | In. | In. | In. |
| 1 cwt. | 112 | 1·2 | 1·4 | 1·7 | 1·9 | 2·0 | 2·2 | 2·4 | 2·5 | 2·6 | 2·7 | 2·9 | 3·0 | 3·1 | 3·2 |
| 2 | 124 | 1·4 | 1·7 | 2·0 | 2·2 | 2·4 | 2·6 | 2·8 | 3·0 | 3·1 | 3·3 | 3·4 | 3·6 | 3·7 | 3·8 |
| 3 | 336 | 1·6 | 1·9 | 2·2 | 2·4 | 2·7 | 2·9 | 3·1 | 3·3 | 3·4 | 3·6 | 3·8 | 3·9 | 4·1 | 4·2 |
| 4 | 448 | 1·7 | 2·0 | 2·4 | 2·6 | 2·9 | 3·1 | 3·3 | 3·5 | 3·7 | 3·9 | 4·0 | 4·2 | 4·3 | 4·5 |
| 5 | 560 | 1·8 | 2·2 | 2·5 | 2·8 | 3·0 | 3·3 | 3·5 | 3·7 | 3·9 | 4·1 | 4·3 | 4·4 | 4·6 | 4·8 |
| 6 | 672 | 1·8 | 2·2 | 2·6 | 2·9 | 3·2 | 3·4 | 3·7 | 3 9 | 4·1 | 4·3 | 4·5 | 4·6 | 4·8 | 5·0 |
| 7 | 784 | 1·9 | 2·3 | 2·7 | 3·0 | 3·3 | 3·6 | 3·8 | 4·1 | 4·2 | 4·4 | 4·6 | 4·8 | 5·0 | 5·2 |
| 8 | 896 | 2·0 | 2·4 | 2·8 | 3·1 | 3·4 | 3·7 | 3·9 | 4·2 | 4·4 | 4·6 | 4·8 | 5·0 | 5·2 | 5·4 |
| 9 | 1,008 | 2·0 | 2·5 | 2·9 | 3·2 | 3·5 | 3·8 | 4·0 | 4·3 | 4·5 | 4·7 | 4·9 | 5·1 | 5·3 | 5·5 |
| 10 | 1,120 | 2·1 | 2·6 | 3·0 | 3·3 | 3·6 | 3·9 | 4·2 | 4·4 | 4·7 | 4·9 | 5·2 | 5·3 | 5·4 | 5·7 |
| 11 | 1,232 | 2·1 | 2 6 | 3·0 | 3·4 | 3·7 | 4·0 | 4·3 | 4·5 | 4·8 | 5·0 | 5·3 | 5·4 | 5·6 | 5·8 |
| 12 | 1,344 | 2·2 | 2·7 | 3·1 | 3·5 | 3·8 | 4·1 | 4·4 | 4·7 | 4·9 | 5·1 | 5·3 | 5·5 | 5·7 | 5·9 |
| 13 | 1,456 | 2·2 | 2·7 | 3·1 | 3·5 | 3·8 | 4·2 | 4·4 | 4·7 | 4·9 | 5·2 | 5·4 | 5·6 | 5·9 | 6·0 |
| 14 | 1,568 | 2·3 | 2·8 | 3·2 | 3·6 | 3·9 | 4·2 | 4·5 | 4·8 | 5·0 | 5·3 | 5·5 | 5·7 | 6·0 | 6·1 |
| 15 | 1,680 | 2·3 | 2·8 | 3·2 | 3·6 | 4·0 | 4·3 | 4·6 | 4·9 | 5·2 | 5·4 | 5·6 | 5·8 | 6·1 | 6·2 |
| 16 | 1,792 | 2·4 | 2·9 | 3·3 | 3·7 | 4·0 | 4·4 | 4·7 | 5·0 | 5·2 | 5·5 | 5·7 | 5·9 | 6·2 | 6·4 |
| 17 | 1,904 | 2·4 | 2·9 | 3·4 | 3·8 | 4·1 | 4·4 | 4·7 | 5·0 | 5·3 | 5·5 | 5·8 | 6·0 | 6·2 | 6·5 |
| 18 | 2,016 | 2·4 | 3·0 | 3·4 | 3·8 | 4·2 | 4·5 | 4·8 | 5·1 | 5·4 | 5·6 | 5·9 | 6·1 | 6·4 | 6·6 |
| 19 | 2,128 | 2·5 | 3·0 | 3·5 | 3·9 | 4·2 | 4·6 | 4·9 | 5·2 | 5·4 | 5·7 | 6·0 | 6·2 | 6·5 | 6·7 |
| 1 ton. | 2,240 | 2·5 | 3·0 | 3·5 | 3·9 | 4·3 | 4·6 | 4·9 | 5·2 | 5·5 | 5·8 | 6·0 | 6·3 | 6·5 | 6·8 |
| 1¼ | 2,800 | 2·6 | 3·2 | 3·7 | 4·1 | 4·5 | 4·9 | 5·2 | 5·5 | 5·8 | 6·1 | 6·4 | 6·6 | 6·9 | 7·2 |
| 1½ | 3,360 | 2·8 | 3·4 | 3·9 | 4·3 | 4·7 | 5·1 | 5·5 | 5·8 | 6·1 | 6·4 | 6·7 | 7·0 | 7·2 | 7·5 |
| 1¾ | 3,920 | 2·9 | 3·5 | 4·0 | 4·5 | 4·9 | 5·3 | 5·7 | 6·0 | 6·3 | 6·7 | 6·9 | 7·2 | 7·5 | 7·7 |
| 2 | 4,480 | 2·9 | 3·5 | 4·1 | 4·7 | 5·1 | 5·5 | 5·9 | 6·2 | 6·5 | 6·8 | 7·2 | 7·6 | 7·7 | 8·0 |
| 2½ | 5,600 | 3·1 | 3·8 | 4·4 | 4·9 | 5·5 | 5·8 | 6·2 | 6·6 | 6·9 | 7·3 | 7·6 | 7·9 | 8·2 | 8·5 |
| 3 | 6,720 | 3·3 | 4·0 | 4·6 | 5·1 | 5·7 | 6·1 | 6·5 | 6·9 | 7·3 | 7·6 | 7·9 | 8·3 | 8·6 | 8·9 |
| 0½ | 7,840 | 3·4 | 4·1 | 4·8 | 5·3 | 5·8 | 6·3 | 6·7 | 7·1 | 7·5 | 7·9 | 8·2 | 8·6 | 8·9 | 9·2 |
| 4 | 8,960 | 3·5 | 4·3 | 4·9 | 5·5 | 6·0 | 6·5 | 7·0 | 7·4 | 7·8 | 8·2 | 8·5 | 8·9 | 9·2 | 9·5 |
| 4½ | 10,080 | — | 4·4 | 5·1 | 5·7 | 6·2 | 6·7 | 7·2 | 7·6 | 8·0 | 8·4 | 8·8 | 9·1 | 9·5 | 9·8 |
| 5 | 11,200 | — | 4·5 | 5·2 | 5·8 | 6·4 | 6·9 | 7·4 | 7·8 | 8·2 | 8·6 | 9·0 | 9·4 | 9·7 | 10·1 |
| 6 | 13,440 | — | — | 5·5 | 6·1 | 6·7 | 7·2 | 7·7 | 8·2 | 8·6 | 9·0 | 9·4 | 9·8 | 10·2 | 10·5 |
| 7 | 15,680 | — | — | 5·7 | 6·3 | 6·9 | 7·5 | 8·0 | 8·5 | 8·9 | 9·4 | 9·8 | 10·2 | 10·6 | 11·0 |
| 8 | 17,920 | — | — | 5·9 | 6·6 | 7·2 | 7·8 | 8·3 | 8·8 | 9·3 | 9·7 | 10·1 | 10·6 | 10·9 | 11·3 |
| 9 | 20,160 | — | — | 6·0 | 6·8 | 7·4 | 8·0 | 8·5 | 9·0 | 9·5 | 10·0 | 10·4 | 10·9 | 11·3 | 11·7 |
| 10 | 22,400 | — | — | — | 6·9 | 7·6 | 8·2 | 8·8 | 9·3 | 9·8 | 10·3 | 10·7 | 11·2 | 11·6 | 12·0 |
| 11 | 24,640 | — | — | — | 7·1 | 7·8 | 8·4 | 9·0 | 9·5 | 10·0 | 10·5 | 11·0 | 11·5 | 11·9 | 12·3 |
| 12 | 26,880 | — | — | — | 7·2 | 7·9 | 8·6 | 9·2 | 9·7 | 10·2 | 10·8 | 11·2 | 11·7 | 12·1 | 12·5 |
| 13 | 29,120 | — | — | — | 7·4 | 8·1 | 8·8 | 9·4 | 9·9 | 10·4 | 11·0 | 11·5 | 11·9 | 12·4 | 12·8 |
| 14 | 31,360 | — | — | — | 7·5 | 8·3 | 8·9 | 9·5 | 10·1 | 10·6 | 11·1 | 11·7 | 12·1 | 12·6 | 13·0 |
| Deflection in inches | | ·1 | ·15 | ·2 | ·25 | ·3 | ·35 | ·4 | ·45 | ·5 | ·55 | ·6 | ·65 | ·7 | ·75 |

| Lengths in Feet | | 10 | 12 | 14 | 16 | 18 | 20 | 22 | 24 | 26 | 28 | 30 | 32 | 34 | 36 |
|---|---|---|---|---|---|---|---|---|---|---|---|---|---|---|---|
| 15 | 33,600 | 7·7 | 8·4 | 9·1 | 9·7 | 10·3 | 10·8 | 11·4 | 11·9 | 12·3 | 12·8 | 13·2 | 13·7 | 14·1 | 14·5 |
| 16 | 35,840 | 7·8 | 8·5 | 9·2 | 9·8 | 10·4 | 11·0 | 11·5 | 12·0 | 12·5 | 13·0 | 13·5 | 13·9 | 14·3 | 14·7 |
| 17 | 38,080 | 7·9 | 8·7 | 9·4 | 10·0 | 10·6 | 11·2 | 11·7 | 12·2 | 12·7 | 13·2 | 13·7 | 14·1 | 14·5 | 14·9 |
| 18 | 40,320 | 8·0 | 8·8 | 9·5 | 10·1 | 10·8 | 11·3 | 11·9 | 12·4 | 12·9 | 13·4 | 13·9 | 14·3 | 14·7 | 15·1 |
| 19 | 42,560 | 8·1 | 8·9 | 9·6 | 10·3 | 10·9 | 11·5 | 12·2 | 12·6 | 13·1 | 13·6 | 14·1 | 14·5 | 15·0 | 15·4 |
| 20 | 44,800 | — | 9·0 | 9·7 | 10·4 | 11·0 | 11·6 | 12·5 | 12·7 | 13·2 | 13·8 | 14·2 | 14·7 | 15·1 | 15·6 |
| 22 | 49,280 | — | 9·2 | 10·0 | 10·7 | 11·3 | 11·9 | 12·8 | 13·0 | 13·6 | 14·1 | 14·6 | 15·1 | 15·5 | 15·9 |
| 24 | 53,760 | — | 9·4 | 10·2 | 10·9 | 11·5 | 12·2 | 13·0 | 13·4 | 13·9 | 14·4 | 14·9 | 15·4 | 15·9 | 16·3 |
| 26 | 58,240 | — | 9·6 | 10·4 | 11·1 | 11·8 | 12·4 | 13·3 | 13·6 | 14·2 | 14·7 | 15·2 | 15·7 | 16·2 | 16·7 |
| 28 | 62,720 | — | 9·8 | 10·6 | 11·4 | 12·0 | 12·7 | 13·5 | 13·9 | 14·4 | 15·0 | 15·5 | 16·0 | 16·5 | 17·0 |
| Deflection in inches | | ·25 | ·3 | ·35 | ·4 | ·45 | ·5 | ·55 | ·6 | ·66 | ·7 | ·75 | ·8 | ·85 | ·9 |

| Lengths in Feet | | 14 | 16 | 18 | 20 | 22 | 24 | 26 | 28 | 30 | 32 | 34 | 36 | 38 | 40 |
|---|---|---|---|---|---|---|---|---|---|---|---|---|---|---|---|
| Weight in tons. | Weight in lbs. | Depth. | Depth. | Depth. | Depth. | Depth. | Depth. | Depth. | Depth. | Depth. | Depth. | Depth. | Depth. | Depth. | Depth. |
| | | In. | In. | In. | In. | In. | In. | In. | In. | In. | In. | In. | In. | In. | In. |
| 30 | 67,200 | 10·8 | 11·5 | 12·2 | 12·9 | 13·5 | 14·1 | 14·7 | 15·2 | 15·7 | 16·3 | 16·8 | 17·3 | 17·7 | 18·2 |
| 32 | 71,680 | 11·0 | 11·7 | 12·4 | 13·1 | 13·7 | 14·3 | 14·9 | 15·5 | 16·0 | 16·5 | 17·0 | 17·5 | 18·0 | 18·5 |
| 34 | 76,160 | 11·1 | 11·9 | 12·6 | 13·3 | 13·9 | 14·5 | 15·1 | 15·7 | 16·2 | 16·8 | 17·3 | 17·8 | 18·3 | 18·8 |
| 36 | 80,640 | 11·3 | 12·0 | 12·8 | 13·4 | 14·1 | 14·7 | 15·3 | 15·9 | 16·5 | 17·0 | 17·5 | 18·0 | 18·5 | 19·0 |
| 38 | 85,120 | 11·4 | 12·2 | 13·0 | 13·6 | 14·3 | 14·9 | 15·5 | 16·1 | 16·7 | 17·2 | 17·8 | 18·3 | 18·8 | 19·3 |
| 40 | 89,600 | — | 12·4 | 13·1 | 13·8 | 14·5 | 15·1 | 15·7 | 16·4 | 16·9 | 17·5 | 18·0 | 18·5 | 19·1 | 19·5 |
| 42 | 94,080 | — | 12·5 | 13·3 | 14·0 | 14·7 | 15·3 | 15·9 | 16·5 | 17·1 | 17·7 | 18·2 | 18·7 | 19·3 | 19·8 |
| 44 | 98,560 | — | 12·7 | 13·5 | 14·2 | 14·9 | 15·5 | 16·1 | 16·8 | 17·4 | 17·9 | 18·5 | 19·0 | 19·5 | 20·0 |
| 46 | 103,040 | — | 12·8 | 13·6 | 14·3 | 15·0 | 15·7 | 16·3 | 17·0 | 17·6 | 18·1 | 18·7 | 19·2 | 19·8 | 20·3 |
| 48 | 107,520 | — | 13·0 | 13·7 | 14·5 | 15·2 | 15·9 | 16·5 | 17·1 | 17·7 | 18·3 | 18·8 | 19·4 | 20·0 | 20·5 |
| 50 | 112,000 | — | — | 13·8 | 14·6 | 15·3 | 16·0 | 16·6 | 17·3 | 17·9 | 18·5 | 19·0 | 19·6 | 20·1 | 20·7 |
| 52 | 116,480 | — | — | 14·0 | 14·7 | 15·5 | 16·2 | 16·8 | 17·5 | 18·1 | 18·7 | 19·2 | 19·8 | 20·3 | 21·0 |
| 54 | 120,960 | — | — | 14·1 | 14·9 | 15·7 | 16·3 | 17·0 | 17·6 | 18·2 | 18·8 | 19·4 | 19·9 | 20·5 | 21·1 |
| 56 | 125,440 | — | — | 14·3 | 15·0 | 15·8 | 16·5 | 17·1 | 17·8 | 18·4 | 19·0 | 19·6 | 20·1 | 20·7 | 21·3 |
| 58 | 129,920 | — | — | 14·4 | 15·1 | 15·9 | 16·6 | 17·3 | 17·9 | 18·5 | 19·2 | 19·7 | 20·3 | 20·9 | 21·4 |
| 60 | 134,400 | — | — | 14·5 | 15·3 | 16·0 | 16·7 | 17·4 | 18·1 | 18·7 | 19·3 | 19·9 | 20·5 | 21·1 | 21·6 |
| Deflection in inches | | ·35 | ·4 | ·45 | ·5 | ·55 | ·6 | ·65 | ·7 | ·75 | ·8 | ·85 | ·9 | ·95 | 1·0 |

*Examples illustrative of the Table.*—1. To find the depth of a rectangular bar of cast iron to support a weight of 10 tons in the middle of its length, the deflection not to exceed $\frac{1}{40}$ of an inch per foot in length, and its length 20 feet, also let the depth be 6 times the breadth.

Opposite 6 times the weight and under 20 feet in length is 15·3 inches, the depth, and $\frac{1}{6}$ of 15·3 = 2·6 inches, the breadth.

2. To find the diameter for a cast iron shaft or solid cylinder that will bear a given pressure, the flexure in the middle not to exceed $\frac{1}{40}$th of an inch for each foot of its length, the distance of the bearings being 20 feet, and the pressure on the middle equals 10 tons.

Constant multiplier 1·7 for round shafts, then $10 \times 1{\cdot}7 = 17$. And opposite 17 tons and under 20 feet is 11·2 inches for the diameter.

But half that flexure is quite enough for revolving shafts: hence $17 \times 2 = 34$ tons, and opposite 34 tons is 13·3 inches for the diameter.

3. A body 256 lbs. weight, presses against its horizontal support, so that it requires the force of 52 lbs. to overcome its friction; if the body be increased to 8750 lbs., what force will cause it to pass from a state of rest to one of motion?

$$\frac{52}{256} = {\cdot}203125 = \text{, in this case, the } \textit{coefficient of friction;}$$

$\therefore 8750 \times 203125 = 1777{\cdot}34375$ lbs., the force required.

This calculation is based upon the law, that friction is proportional to the normal pressure between the rubbing surfaces. Twice the pressure gives twice the friction; three times the pressure gives three times the friction; and so on. With light pressures, this law may not hold, but then it is to be attributed to the proportionately greater effect of adhesion.

4. If a sleigh, weighing 250 lbs., requires a force of 28 lbs. to draw it along; when 1120 lbs. are placed in it, required the units of work expended to move the whole 350 feet?

$$\frac{28}{250} = \cdot 112, \text{ the coefficient of friction.}$$

Then $(1120 + 250) \times \cdot 112 = 153\cdot 44$ lbs., the force required to move the whole.

$\therefore 153\cdot 44 \times 350 = 53704$, the units of work required.

A UNIT OF WORK is the labour which is equal to that of raising one pound a foot high. It is supposed that a horse can perform 33000 units of work in a minute.

It may also be remarked that friction is independent of the extent of the surfaces in contact, except with trifling pressures and large surfaces, which is on account of the effect of adhesion. The friction of motion is independent of velocity, and is generally less than that of quiescence.

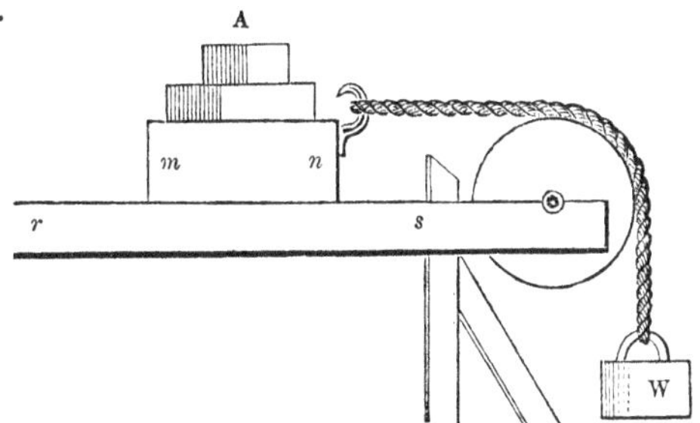


5. Required the coefficient of friction, for *a sliding motion*, of cast iron upon wrought, lubricated with Devlin's oil, and under the following circumstances: the load A, and sledge *nm*, weighs 8420 lbs., and requires a weight W, of 1200 lbs. to cause it to pass from a state of rest into one of motion: the sledge and load pass over 22 feet on the horizontal way *rs*, in 8 seconds.

In this case the coefficient of sliding motion will be

$$\frac{1200}{8420} - \frac{1200 + 8420}{8420} \times \frac{2 \times 22}{g \times 8^2},$$

in which $g = 32\cdot 2$ feet; the acceleration of the free descent of bodies brought about by gravity. The above expression becomes

$$142515 - 1\cdot 142515 \times \frac{44}{2060\cdot 8} = \cdot 118121.$$

Hence the coefficient of the friction of motion is ·118121, and the coefficient of the friction of quiescence is ·142515.

## OF FRICTION, OR RESISTANCE TO MOTION IN BODIES ROLLING OR RUBBING ON EACH OTHER.

In the years 1831, 1832, and 1833, a very extensive set of experiments were made at Metz, by M. Morin, under the sanction of the French government, to determine as nearly as possible the laws of friction; and by which the following were fully established:

1. When no unguent is interposed, the friction of any two surfaces (whether of quiescence or of motion) is directly proportional to the force with which they are pressed perpendicularly together; so that for any two given surfaces of contact there is a constant ratio of the friction to the perpendicular pressure of the one surface upon the other. Whilst this ratio is thus the same for the same

surfaces of contact, it is different for different surfaces of contact. The particular value of it in respect to any two given surfaces of contact is called the coefficient of friction in respect to those surfaces.

2. When no unguent is interposed, the amount of the friction is, in every case, wholly independent of the extent of the surfaces of contact; so that, the force with which two surfaces are pressed together being the same, their friction is the same, whatever may be the extent of their surfaces of contact.

3. That the friction of motion is wholly independent of the velocity of the motion.

4. That where unguents are interposed, the coefficient of friction depends upon the nature of the unguent, and upon the greater or less abundance of the supply. In respect to the supply of the unguent, there are two extreme cases, that in which the surfaces of contact are but slightly rubbed with the unctuous matter, as, for instance, with an oiled or greasy cloth, and that in which a continuous stratum of unguent remains continually interposed between the moving surfaces; and in this state the amount of friction is found to be dependent rather upon the nature of the unguent than upon that of the surfaces of contact. M. Morin found that with unguents (hog's lard and olive oil) interposed in a continuous stratum between surfaces of wood on metal, wood on wood, metal on wood, and metal on metal, when in motion, have all of them very near the same coefficient of friction, being in all cases included between ·07 and ·08.

The coefficient for the unguent tallow is the same, except in that of metals upon metals. This unguent appears to be less suited for metallic substances than the others, and gives for the mean value of its coefficient, under the same circumstances, ·10. Hence, it is evident, that where the extent of the surface sustaining a given pressure is so great as to make the pressure less than that which corresponds to a state of perfect separation, this greater extent of surface tends to increase the friction by reason of that adhesiveness of the unguent, dependent upon its greater or less viscosity, whose effect is proportional to the extent of the surfaces between which it is interposed.

It was found, from a mean of experiments with different unguents on axles, in motion and under different pressures, that, with the unguent tallow, under a pressure of from 1 to 5 cwt., the friction did not exceed $\frac{1}{39}$th of the whole pressure; when soft soap was applied, it became $\frac{1}{34}$th; and with the softer unguents applied, such as oil, hog's lard, &c., the ratio of the friction to the pressure increased; but with the harder unguents, as soft soap, tallow, and anti-attrition composition, the friction considerably diminished; consequently, to render an unguent of proper efficiency, the nature of the unguent must be measured by the pressure or weight tending to force the surfaces together.

TABLE *of the Results of Experiments on the Friction of Unctuous Surfaces.* By M. MORIN.

| Surfaces of Contact. | Coefficients of Friction. | |
|---|---|---|
| | Friction of Motion. | Friction of Quiescence. |
| Oak upon oak, the fibres being parallel to the motion | 0·018 | 0·390 |
| Ditto, the fibres of the moving body being perpendicular to the motion | 0·143 | 0·314 |
| Oak upon elm, fibres parallel | 0·136 | |
| Elm upon oak, do | 0·119 | 0·420 |
| Beech upon oak, do | 0·330 | |
| Elm upon elm, do | 0·140 | |
| Wrought iron upon elm, do | 0·138 | |
| Ditto upon wrought iron, do | 0·177 | |
| Ditto upon cast iron, do | ... | 0·118 |
| Cast iron upon wrought iron, do | 0·143 | |
| Wrought iron upon brass, do | 0·160 | |
| Brass upon wrought iron, do | 0·166 | |
| Cast iron upon oak, do | 0·107 | 0·100 |
| Ditto upon elm, do., the unguent being tallow | 0·125 | |
| Ditto, do., the unguent being hog's lard and black lead | 0·137 | |
| Elm upon cast iron | 0·135 | 0·098 |
| Cast iron upon cast iron | 0·144 | |
| Ditto upon brass | 0·132 | |
| Brass upon cast iron | 0·107 | |
| Ditto upon brass | 0·134 | 0·164 |
| Copper upon oak | 0·100 | |
| Yellow copper upon cast iron | 0·115 | |
| Leather (ox-hide), well tanned, upon cast iron, wetted | 0·229 | 0·267 |
| Ditto upon brass, wetted | 0·244 | |

In these experiments, the surfaces, after having been smeared with an unguent, were wiped, so that no interposing layer of the unguent prevented intimate contact.

TABLE *of the Results of Experiments on Friction, with Unguents interposed.* By M. MORIN.

| Surfaces of Contact. | Coefficients of Friction. | | Unguents. |
|---|---|---|---|
| | Friction of Motion. | Friction of Quiescence. | |
| Oak upon oak, fibres parallel | 0·164 | 0·440 | Dry soap. |
| Do. do | 0·075 | 0·164 | Tallow. |
| Do. do | 0·067 | ... | Hog's lard. |
| Do., fibres perpendicular | 0·083 | 0·254 | Tallow. |
| Do. do | 0·072 | ... | Hog's lard. |
| Do. do | 0·250 | ... | Water. |
| Do. upon elm, fibres parallel | 0·136 | ... | Dry soap. |
| Do. do | 0·073 | 0·178 | Tallow. |
| Do. do | 0·066 | ... | Hog's lard. |
| Do. upon cast iron | 0·080 | ... | Tallow. |
| Do. upon wrought iron | 0·098 | ... | Tallow. |
| Beech upon oak, fibres parallel | 0·055 | ... | Tallow. |
| Elm upon oak, do | 0·137 | 0·411 | Dry soap. |
| Do. do | 0·170 | 0·142 | Tallow. |
| Do. do | 0·060 | ... | Hog's lard. |
| Elm upon elm, do | 0·139 | 0·217 | Dry soap. |
| Do. upon cast iron | 0·066 | ... | Tallow. |
| Wrought iron upon oak, fibres parallel | 0·256 | 0·649 | Greased and saturated with water. |
| Do. do | 0·214 | ... | Dry soap. |

| Surfaces of Contact. | Coefficients of Friction. | | Unguents. |
|---|---|---|---|
| | Friction of Motion. | Friction of Quiescence. | |
| Wrought iron upon oak, fibres parallel | 0·085 | 0·108 | Tallow. |
| Do. upon elm, do | 0·078 | ... | Tallow. |
| Do. do | 0·076 | ... | Hog's lard. |
| Do. do | 0·055 | ... | Olive oil. |
| Do. upon cast iron, do | 0·103 | ... | Tallow. |
| Do. do | 0·076 | ... | Hog's lard. |
| Do. do | 0·066 | 0·100 | Olive oil. |
| Do. upon wrought iron, do | 0·082 | ... | Tallow. |
| Do. do | 0·081 | ... | Hog's lard. |
| Do. do | 0·070 | 0·115 | Olive oil. |
| Wrought iron upon brass, do | 0·103 | ... | Tallow. |
| Do. do | 0·075 | ... | Hog's lard. |
| Do. do | 0·078 | ... | Olive oil. |
| Cast iron upon oak, do | 0·189 | ... | Dry soap. |
| Do. do | 0·218 | 0·646 | Greased and saturated with water. |
| Do. do | 0·078 | 0·100 | Tallow. |
| Do. do | 0·075 | ... | Hog's lard. |
| Do. do | 0·075 | 0·100 | Olive oil. |
| Do. upon elm, do | 0·077 | ... | Tallow. |
| Do. do | 0·061 | ... | Olive oil. |
| Do. do | 0·091 | ... | Hog's lard and plumbago. |
| Do. upon wrought iron | ... | 0·100 | Tallow. |
| Do. upon cast iron | 0·314 | ... | Water. |
| Do. do. | 0·197 | ... | Soap. |
| Do. do. | 0·100 | 0·100 | Tallow. |
| Do. do. | 0·070 | 0·100 | Hog's lard. |
| Do. do | 0·064 | ... | Olive oil. |
| Do. do. | 0·055 | ... | Hog's lard and plumbago. |
| Do. upon brass | 0·103 | ... | Tallow. |
| Do. do | 0·075 | ... | Hog's lard. |
| Do. do | 0·078 | ... | Olive oil. |
| Copper upon oak, fibres parallel | 0·069 | 0·100 | Tallow. |
| Yellow copper upon cast iron | 0·072 | 0·103 | Tallow. |
| Do. do | 0·068 | ... | Hog's lard. |
| Do. do.. | 0·066 | ... | Olive oil. |
| Brass upon cast iron | 0·086 | 0·106 | Tallow. |
| Do. do | 0·077 | ... | Olive oil. |
| Do. upon wrought iron | 0·081 | ... | Tallow. |
| Do. do | 0·089 | ... | Lard and plumbago. |
| Do. do | 0·072 | ... | Olive oil. |
| Brass upon brass | 0·058 | ... | Olive oil. |
| Steel upon cast iron | 0·105 | 0·108 | Tallow. |
| Do. do | 0·081 | ... | Hog's lard. |
| Do. do | 0·079 | ... | Olive oil. |
| Do. upon wrought iron | 0·093 | ... | Tallow. |
| Do. do | 0·076 | ... | Hog's lard. |
| Do. upon brass | 0·056 | ... | Tallow. |
| Do. do | 0·053 | ... | Olive oil. |
| Do. do | 0·067 | ... | Lard and plumbago. |
| Tanned ox-hide upon cast iron | 0·365 | ... | Greased and saturated with water. |

The extent of the surfaces in these experiments bore such a relation to the pressure as to cause them to be separated from one another throughout by an interposed stratum of the unguent.

TABLE *of the Results of Experiments on the Friction of Gudgeons or Axle-ends, in motion upon their bearings.* By M. MORIN.

| Surfaces in Contact. | State of the Surfaces. | Coefficient of Friction. |
|---|---|---|
| Cast iron axles in cast iron bearings. | Coated with oil of olives, with hog's lard, tallow, and soft gome........... | 0·07 to 0·08 |
| | With the same and water... | 0·08 |
| | Coated with asphaltum..... | 0·054 |
| | Greasy.......................... | 0·14 |
| | Greasy and wetted.......... | 0·14 |
| Cast iron axles in cast iron bearings. | Coated with oil of olives, with hog's lard, tallow, and soft gome........... | 0·07 to 0·08 |
| | Greasy .......................... | 0·16 |
| | Greasy and damped......... | 0·16 |
| | Scarcely greasy............... | 0·19 |
| Wrought iron axles in cast iron bearings. | Coated with oil of olives, tallow, hog's lard, or soft gome................ | 0·07 to 0·08 |
| Wrought iron axles in brass bearings. | Coated with oil of olives, hog's lard, or tallow, | 0·07 to 0·08 |
| | Coated with hard gome..... | 0·09 |
| | Greasy and wetted.......... | 0·19 |
| | Scarcely greasy............... | 0·25 |
| Iron axles in lignum vitæ bearings. | Coated with oil or hog's lard ........................ | 0·11 |
| | Greasy .......................... | 0·19 |
| Brass axles in brass bearings. | Coated with oil .............. | 0·10 |
| | With hog's lard.............. | 0·09 |

TABLE *of Coefficients of Friction under Pressures increased continually up to limits of Abrasion.*

| Pressure per Square Inch. | Coefficients of Friction. | | | |
|---|---|---|---|---|
| | Wrought Iron upon Wrought Iron. | Wrought Iron upon Cast Iron. | Steel upon Cast Iron. | Brass upon Cast Iron. |
| 32·5 lbs. | ·140 | ·174 | ·166 | ·157 |
| 1·66 cwts. | ·250 | ·275 | ·300 | ·225 |
| 2·00 | ·271 | ·292 | ·333 | ·219 |
| 2·33 | ·285 | ·321 | ·340 | ·214 |
| 2·66 | ·297 | ·329 | ·344 | ·211 |
| 3·00 | ·312 | ·333 | ·347 | ·215 |
| 3·33 | ·350 | ·351 | ·351 | ·206 |
| 3·66 | ·376 | ·353 | ·353 | ·205 |
| 4·00 | ·395 | ·365 | ·354 | ·208 |
| 4·33 | ·403 | ·366 | ·356 | ·221 |
| 4·66 | ·409 | ·366 | ·357 | ·223 |
| 5·00 | ...... | ·367 | ·358 | ·233 |
| 5·33 | ...... | ·367 | ·359 | ·234 |
| 5·66 | ...... | ·367 | ·367 | ·235 |
| 6·00 | ...... | ·376 | ·403 | ·233 |
| 6·33 | ...... | ·434 | ...... | ·234 |
| 6·66 | ...... | ...... | ...... | ·235 |
| 7·00 | ...... | ...... | ...... | ·232 |
| 7·33 | ...... | ...... | ...... | ·273 |

Comparative friction of steam engines of different modifications, if the beam engine be taken as the standard of comparison :—

The vibrating engine..................has a gain of 1·1 per cent.
The direct-action engine, with slides — loss of 1·8 —
Ditto, with rollers...................... — gain of 0·8 —
Ditto, with a parallel motion......... — gain of 1·3 —

Excessive allowance for friction has hitherto been made in calculating the effective power of engines in general; as it is found practically, by experiments, that, where the pressure upon the piston is about 12 lbs. per square inch, the friction does not amount to more than $1\frac{1}{2}$ lbs.; and also that, by experiments with an indicator on an engine of 50 horse power, the whole amount of friction did not exceed 5 horse power, or one-tenth of the whole power of the engine.

RECENT EXPERIMENTS MADE BY M. MORIN ON THE STIFFNESS OF ROPES, OR THE RESISTANCE OF ROPES TO BENDING UPON A CIRCULAR ARC.

The experiments upon which the rules and table following are founded were made by Coulomb, with an apparatus the invention of Amonton, and Coulomb himself deduced from them the following results :—

1. That the resistance to bending could be represented by an expression consisting of two terms, the one constant for each rope and each roller, which we shall designate by the letter A, and which this philosopher named the natural stiffness, because it depends on the mode of fabrication of the rope, and the degree of tension of its yarns and strands; the other, proportional to the tension, T, of the end of the rope which is being bent, and which is expressed by the product, BT, in which B is also a number constant for each rope and each roller.

2. That the resistance to bending varied inversely as the diameter of the roller.

Thus the complete resistance is represented by the expression

$$\frac{A + BT}{D},$$

where D represents the diameter of the roller.

Coulomb supposed that for tarred ropes the stiffness was proportional to the number of yarns, and M. Navier inferred, from examination of Coulomb's experiments, that the coefficients A and B were proportional to a certain power of the diameter, which depended on the extent to which the cords were worn. M. Morin, however, deems this hypothesis inadmissible, and the following is an extract from his new work, "Leçons de Mécanique Pratique," December, 1846 :—

"To extend the results of the experiments of Coulomb to ropes of different diameters from those which had been experimented upon, M. Navier has allowed, very explicitly, what Coulomb had but surmised: that the coefficients, A, were proportional to a certain power of the diameter, which depended on the state of wear of the ropes; but this supposition appears to us neither borne out, nor even admissible, for it would lead to this consequence, that a worn rope of a metre diameter would have the same stiffness as a new rope, which is evidently wrong; and, besides, the comparison alone of the values of A and B shows that the power to which the diameter should be raised would not be the same for the two terms of the resistance."

Since, then, the form proposed by M. Navier for the expression of the resistance of ropes to bending cannot be admitted, it is necessary to search for another, and it appears natural to try if the factors A and B cannot be expressed for white ropes, simply according to the number of yarns in the ropes, as Coulomb has inferred for tarred ropes.

Now, dividing the values of A, obtained for each rope by M. Navier, by the number of yarns, we find for

$$n = 30 \quad d = 0^{m}\cdot200 \quad A = 0\cdot222460 \quad \frac{A}{n} = 0\cdot0074153.$$

$$n = 15 \quad d = 0^{m}\cdot144 \quad A = 0\cdot063514 \quad \frac{A}{n} = 0\cdot0042343.$$

$$n = 6 \quad d = 0^{m}\cdot0088 \quad A = 0\cdot010604 \quad \frac{A}{n} = 0\cdot0017673.$$

It is seen from this that the number A is not simply proportional to the number of yarns.

Comparing, then, the values of the ratio $\frac{A}{n}$ corresponding to the three ropes, we find the following results:—

| Number of yarns. | Values of $\frac{A}{n}$. | Differences of the numbers of yarns. | Differences of the values of $\frac{A}{n}$. | Differences of the values of $\frac{A}{n}$ for each yarn of difference. |
|---|---|---|---|---|
| 30 | 0·0074153 | From 30 to 15. 15 yarns | 0·0031810 | 0·000212 |
| 15 | 0·0042343 | — 15 to 6. 9 — | 0·0024770 | 0·000272 |
| 6 | 0·0017673 | — 30 to 6. 24 — | 0·0056400 | 0·000252 |

Mean difference per yarn, 0·000245

It follows, from the above, that the values of A, given by the experiments, will be represented with sufficient exactness for all practical purposes by the formula

$$A = n\,[0\cdot0017673 + 0\cdot000245\,(n - 6)].$$
$$= n\,[0\cdot0002973 + 0\cdot000245\,n].$$

An expression relating only to dry white ropes, such as were used by Coulomb in his experiments.

With regard to the number B, it appears to be proportional to the number of yarns, for we find for

$$n = 30 \; d = 0^{m}\cdot0200 \quad B = 0\cdot009738 \; \frac{B}{n} = 0\cdot0003246$$

$$n = 15 \; d = 0^{m}\cdot0144 \quad B = 0\cdot005518 \; \frac{B}{n} = 0\cdot0003678$$

$$n = 6 \; d = 0^{m}\cdot0088 \quad B = 0\cdot002380 \; \frac{B}{n} = 0\cdot0003967$$

Mean......................0·0003630

Whence

$$B = 0\cdot000363 \; n.$$

Consequently, the results of the experiments of Coulomb on dry white ropes will be represented with sufficient exactness for practical purposes by the formula

$$K = n \, [0\cdot000297 + 0\cdot000245 \; n + 0\cdot000363 \; T] \text{ kil.}$$

which will give the resistance to bending upon a drum of a metre in diameter, or by the formula

$$R = \frac{n}{D} \, [0\cdot000297 + 0\cdot000245 \; n + 0\cdot000363 \; T] \text{ kil.}$$

for a drum of diameter D metres.

These formulas, transformed into the American scale of weights and measures, become

$$R = n \, [0\cdot0021508 + 0\cdot0017724 \; n + 0\cdot00119096 \; T] \text{ lbs.}$$

for a drum of a foot in diameter, and

$$R = \frac{n}{D} \, [0\cdot0021508 + 0\cdot0017724 \; n + 0.00119096 \; T] \text{ lbs.}$$

for a drum of diameter D feet.

With respect to worn ropes, the rule given by M. Navier cannot be admitted, as we have shown above, because it would give for the stiffness of a rope of a diameter equal to unity the same stiffness as for a new rope.

The experiments of Coulomb on worn ropes not being sufficiently complete, and not furnishing any precise data, it is not possible, without new researches, to give a rule for calculating the stiffness of these ropes.

### TARRED ROPES.

In reducing the results of the experiments of Coulomb on tarred ropes, as we have done for white ropes, we find the following values :—

$n = 30$ yarns $A = 0\cdot34982$ $B = 0\cdot0125605$
$n = 15$ — $A = 0\cdot106003$ $B = 0\cdot006037$
$n = 6$ — $A = 0\cdot0212012$ $B = 0\cdot0025997$

which differ very slightly from those which M. Navier has given. But, if we look for the resistance corresponding to each yarn, we find

$$n = 30 \text{ yarns} \quad \frac{A}{n} = 0{\cdot}0116603 \quad \frac{B}{n} = 0{\cdot}000418683$$
$$n = 15 \quad — \quad \frac{A}{n} = 0{\cdot}0070662 \quad \frac{B}{n} = 0{\cdot}000402466$$
$$n = \ 6 \quad — \quad \frac{A}{n} = 0{\cdot}0035335 \quad \frac{B}{n} = 0{\cdot}000433283$$

Mean............0·000418144

We see by this that the value of B is for tarred ropes, as for white ropes, sensibly proportional to the number of yarns, but it is not so for that of A, as M. Navier has supposed.

Comparing, as we have done for white ropes, the values of $\frac{A}{n}$ corresponding to the three ropes of 30, 15, and 6 yarns, we obtain the following results:—

| Number of yarns. | Values of $\frac{A}{n}$. | Differences of the number of yarns. | Differences of the values of $\frac{A}{n}$. | Differences of the values of $\frac{A}{n}$ for each yarn of difference. |
|---|---|---|---|---|
| 30 | 0·0116603 | From 30 to 15. 15 yarns | 0·0045941 | 0·000306 |
| 15 | 0·0070662 | — 15 to 6. 9 — | 0·0035327 | 0·000392 |
| 6 | 0·0035335 | — 60 to 6. 25 — | 0·0081268 | 0·000339 |

Mean......................0·000346

It follows from this that the value of A can be represented by the formula

$$A = n\,[0{\cdot}0035335 + 0{\cdot}000346\,(n - 6)]$$
$$= n\,[0{\cdot}0014575 + 0{\cdot}000346\,n]$$

and the whole resistance on a roller of diameter D metres, by

$$R = \frac{n}{D}\,[0{\cdot}0014575 + 0{\cdot}000346\,n + 0{\cdot}000418144\,T] \text{ kil.}$$

Transforming this expression to the American scale of weights and measures, we have

$$R = \frac{n}{D}\,[0{\cdot}01054412 + 0{\cdot}00250309\,n + 0{\cdot}001371889\,T] \text{ lbs.}$$

for the resistance on a roller of diameter D feet.

This expression is exactly of the same form as that which relates to white ropes, and shows that the stiffness of tarred ropes is a little greater than that of new white ropes.

In the following table, the diameters corresponding to the different numbers of yarns are calculated from the data of Coulomb, by the formulas,

$$d \text{ cent.} = \sqrt{0{\cdot}1338\,n} \text{ for dry white ropes, and}$$
$$d \text{ cent.} = \sqrt{0.186\,n} \text{ for tarred ropes,}$$

which, reduced to the American scale, become

$$d \text{ inches} = \sqrt{0{\cdot}020739\,n} \text{ for dry white ropes, and}$$
$$d \text{ inches} = \sqrt{0{\cdot}02883} \text{ for tarred ropes.}$$

NOTE.—The diameter of the rope is to be included in D; thus, with an inch rope passing round a pulley, 8 inches in diameter in the groove, the diameter of the roller is to be considered as 9 inches.

| No. of yarns. | Dry White Ropes. | | | Tarred Ropes. | | |
|---|---|---|---|---|---|---|
| | Diameter. | Value of the natural stiffness, A. | Value of the stiffness proportional to the tension, B. | Diameter. | Value of the natural stiffness, A. | Value of the stiffness proportional to the tension, B. |
| | ft. | lbs. | | ft. | lbs. | |
| 6 | 0·0293 | 0·0767120 | 0·0071457 | 0·0347 | 0·153376 | 0·00823133 |
| 9 | 0.0360 | 0·1629234 | 0·0107186 | 0·0425 | 0·297647 | 0·01234700 |
| 12 | 0·0416 | 0·2810384 | 0·0142915 | 0·0490 | 0·486976 | 0·01646267 |
| 15 | 0·0465 | 0·4310571 | 0·0178644 | 0·0548 | 0·721357 | 0·02057834 |
| 18 | 0·0509 | 0·6129795 | 0·0214373 | 0·0600 | 0·000795 | 0·02469400 |
| 21 | 0·0550 | 0·8268054 | 0·0250102 | 0·0648 | 1·325289 | 0·02880967 |
| 24 | 0·0588 | 1.0725350 | 0·0285831 | 0·0693 | 1·694839 | 0·03292534 |
| 27 | 0·0622 | 1·3501682 | 0·0321559 | 0·0735 | 2·109444 | 0·03704100 |
| 30 | 0·0657 | 1·6597051 | 0·0357288 | 0·0775 | 2·569105 | 0·04115667 |
| 33 | 0·0689 | 2·0011455 | 0·0393017 | 0·0813 | 3·073821 | 0·04527234 |
| 36 | 0·0720 | 2·3744897 | 0·0428746 | 0·0849 | 3·623593 | 0·04938800 |
| 39 | 0·0749 | 2·7797375 | 0·0464475 | 6·0884 | 4·218416 | 0·05350367 |
| 42 | 0·0778 | 3·2168888 | 0·0500203 | 0·0917 | 4·858304 | 0·05761934 |
| 45 | 0·0805 | 3·6859438 | 0·0535932 | 0·0949 | 5·543242 | 0·06173501 |
| 48 | 0·0831 | 4·1869024 | 0·0571661 | 0·0980 | 6·273237 | 0·06585067 |
| 51 | 0·0857 | 4·7197647 | 0·0607390 | 0·1010 | 7·048287 | 0·06996634 |
| 54 | 0·0882 | 5·2845306 | 0·0643119 | 0·1040 | 7·868393 | 0·07408201 |
| 57 | 0·0908 | 5·8812001 | 0·0678847 | 0·1070 | 8·733554 | 0·07819767 |
| 60 | 0·0926 | 6·5097733 | 0·0714576 | 0·1099 | 9·643771 | 0·08231334 |
| $n$ | $\sqrt{0·000144n}$ | $\left\{\begin{array}{l}0·0021503n\\+0·0017724n\frac{2}{n}\end{array}\right.$ | $0·00119098n$ | $\sqrt{0·00020n}$ | $\left\{\begin{array}{l}0·01054412n\\+0·00250309n\frac{2}{n}\end{array}\right.$ | $0·001371889n$ |

### *Application of the preceding Tables or Formulas.*

To find the stiffness of a rope of a given diameter or number of yarns, we must first obtain from the table, or by the formulas, the values of the quantities A and B corresponding to these given quantities, and knowing the tension, T, of the end to be wound up, we shall have its resistance to bending on a drum of a foot in diameter, by the formula

$$R = A + BT.$$

Then, dividing this quantity by the diameter of the roller or pulley round which the rope is actually to be bent, we shall have the resistance to bending on this roller.

What is the stiffness of a dry white rope, in good condition, of 60 yarns, or ·0928 diameter, which passes over a pulley of 6 inches diameter in the groove, under a tension of 1000 lbs.? The table gives for a dry white rope of 60 yarns, in good condition, bent upon a drum of a foot in diameter,

$$A = 0·50977 \quad B = 0·0714576$$

and we have $D = 0·5 + 0·0928$; and consequently,

$$R = \frac{6·50977 + 0·0714576 \times 1000}{0·5928} = 128 \text{ lbs.}$$

The whole resistance to be overcome, not including the friction on the axis, is then

$$Q + R = 1000 + 128 = 1128 \text{ lbs.}$$

The stiffness in this case augments the resistance by more than one-eighth of its value.

FURTHER RECENT EXPERIMENTS MADE BY M. MORIN, ON THE TRACTION OF CARRIAGES, AND THE DESTRUCTIVE EFFECTS WHICH THEY PRODUCE UPON THE ROADS.

The study of the effects which are produced when a carriage is set in motion can be divided into two distinct parts: the traction of carriages, properly so called, and their action upon the roads.

The researches relative to the traction of carriages have for their object to determine the magnitude of the effort that the motive power ought to exercise according to the weight of the load, to the diameter and breadth of the wheels, to the velocity of the carriage, and to the state of repair and nature of the roads.

The first experiments on the resistance that cylindrical bodies offer to being rolled on a level surface are due to Coulomb, who determined the resistance offered by rollers of lignum vitæ and elm, on plane oak surfaces placed horizontally.

His experiments showed that the resistance was directly proportional to the pressure, and inversely proportional to the diameter of the rollers.

If, then, P represent the pressure, and $r$ the radius of the roller, the resistance to rolling, R, could, according to the laws of Coulomb, be expressed by the formula

$$R = A \frac{P}{r}$$

in which A would be a number, constant for each kind of ground, but varying with different kinds, and with the state of their surfaces.

The results of experiments made at Vincennes show that the law of Coulomb is approximately correct, but that the resistance increases as the width of the parts in contact diminishes.

Other experiments of the same nature have confirmed these conclusions; and we may allow, at least, as a law sufficiently exact for practical purposes, that for woods, plasters, leather, and generally for hard bodies, the resistance to rolling is nearly—

1st. Proportional to the pressure.

2d. Inversely proportional to the diameter of the wheels.

3d. Greater as the breadth of the zone in contact is smaller.

EXPERIMENTS UPON CARRIAGES TRAVELLING ON ORDINARY ROADS.

These experiments were not considered sufficient to authorize the extension of the foregoing conclusions to the motion of carriages on ordinary roads. It was necessary to operate directly on the carriages themselves, and in the usual circumstances in which they are placed. Experiments on this subject were therefore undertaken, first at Metz, in 1837 and 1838, and afterwards at Courbevoie, in 1839 and 1841, with carriages of every species; and attention was directed separately to the influence upon the magnitude of the traction, of the pressure, of the diameter of the wheels, of their breadth, of the speed, and of the state of the ground.

In heavily laden carriages, which it is most important to take into consideration, the weight of the wheels may be neglected in comparison with the total load; and the relation between the load and the traction, upon a level road, is approximately given by the equation—

$$\frac{F_1}{P_1} = \frac{2\,(A \times fr_1)}{r' \times r''} \text{ for carriages with four wheels,}$$

and
$$\frac{F_1}{P_1} = \frac{A \times fr_1}{r} \text{ for carriages with two wheels,}$$

in which $F_1$ represents the horizontal component of the traction;
$P_1$ the total pressure on the ground;
$r'$ and $r''$ the radii of the fore and hind wheels;
$r_1$ the mean radius of the boxes;
$f$ the coefficient of friction;
and A the constant multiplier in Coulomb's formula for the resistance to rolling.

These expressions will serve us hereafter to determine, by aid of experiment, the ratio of the traction to the load for the most usual cases.

### *Influence of the Pressure.*

To observe the influence of the pressure upon the resistance to rolling, the same carriages were made to pass with different loads over the same road in the same state.

The results of some of these experiments, made at a walking pace, are given in the following table:—

| Carriages employed. | Road traversed. | Pressure. | Traction. | Ratio of the traction to the load. |
|---|---|---|---|---|
| | | kil. | kil. | |
| Chariot porte corps d'artillerie. | Road from Courbevoie to Colomber, dry, in good repair, dusty. | 6992 | 180·71 | 1/38·6 |
| | | 6140 | 159·9 | 1/39·2 |
| | | 4580 | 113·7 | 1/40·2 |
| Chariot de roulage, without springs. | Road from Courbevoie to Bezous, solid, *hard gravel, very dry. | 7126 | 138·9 | 1/51·3 |
| | | 5458 | 115·5 | 1/48·9 |
| | | 4450 | 93·2 | 1/47·7 |
| | | 3430 | 68·4 | 1/50·2 |
| Chariot de roulage, with springs. | Road from Colomber to Courbevoie, pitched, in ordinary repair, †muddy | 1600 | 39·3 | 1/40·8 |
| | | 3292 | 89·2 | 1/36·9 |
| | | 4996 | 136·0 | 1/36·8 |
| Carriages with six equal wheels. | Road from Courbevoie to Colomber, deep ruts, with muddy detritus. | 3000 | 138·9 | 1/21·6 |
| | | 4692 | 224·0 | 1/21·0 |
| Two carriages with six equal wheels, hooked on, one behind the other. | | 6000 | 285·8 | 1/21·0 |
| | | 6000 | 286·7 | 1/21·0 |

From the examination of this table, it appears that on ‡solid gravel and on pitched roads the resistance of carriages to traction is sensibly proportional to the pressure.

* En gravier dur. † Pavé en état ordinaire. ‡ En empierrement solide.

We remark that the experiments made upon one and upon two six-wheeled carriages have given the same traction for a load of 6000 kilogrammes, including the vehicle, whether it was borne upon one carriage or upon two. It follows thence that the traction is, cæteris paribus and between certain limits, independent of the number of wheels.

### *Influence of the Diameter of the Wheels.*

To observe the influence of the diameter of the wheels on the traction, carriages loaded with the same weights, having wheels with tires of the same width, and of which the diameters only were varied between very extended limits, were made to traverse the same parts of roads in the same state. Some of the results obtained are given in the following table.

These examples show that on solid roads it may be admitted as a practical law that the traction is inversely proportional to the diameters of the wheels.

| Carriages employed. | Roads traversed. | Diameter of the wheels in metres. Fore wheels $2r'$ | Hind wheels $2r''$ | Diameter of the wheels in English feet. Fore wheels $2r'$ | Hind wheels $2r''$ | Total pressure, $P_1$. | Traction, $F_1$. | Ratio of the traction to the pressure. | Friction of the boxes on the axles. | Resistance to rolling, R. | Value of A for the French scale. | Value of A for the American scale. |
|---|---|---|---|---|---|---|---|---|---|---|---|---|
| | | m. | m. | | | kil. | kil. | | kil. | kil. | | |
| Chariot porte corps d'artillerie. | Road from Courbevoie to Colomber, *solid gravel, dusty. | 2·029 | 2·029 | 6·657 | 6·657 | 4928 | 81·6 | 1/60· | 9·6 | 72·0 | 0·0148 | 0·04856 |
| | | 1·453 | 1·453 | 4·767 | 4·767 | 4930 | 108·6 | 1/45·5 | 14·4 | 94·2 | 0·0139 | 0·04560 |
| | | 0·872 | 0·872 | 2·861 | 2·861 | 4924 | 179·0 | 1/27·4 | 25·3 | 153·7 | 0·0137 | 0·04494 |
| Porte corps d'artillerie. | †Pitched pavement of Fontainebleau. | 2·029 | 2·029 | 6·657 | 6·657 | 4692 | 51·45 | 1/90·45 | 9·0 | 42·45 | 0·0092 | 0·03018 |
| | | 1·453 | 1·453 | 4·767 | 4·767 | 4594 | 71·45 | 1/64·3 | 13·2 | 58·25 | 0·0092 | 0·03018 |
| Chariot comtois. | | 1·110 | 1·358 | 3·642 | 4·455 | 1871 | 32·10 | 1/58·4 | 4·7 | 27·40 | 0·0089 | 0·02920 |
| A six-wheeled carriage. | | 0·860 | 0·860 | 2·822 | 2·822 | 3270 | 81·05 | 1/40·4 | 9·7 | 71·35 | 0·0094 | 0·03084 |
| The same with four wheels. | | 0·860 | 0·860 | 2·822 | 2·822 | 3270 | 78·80 | 1/41·5 | 9·7 | 69·10 | 0·0091 | 0·02986 |
| Camion. | | 0·592 | 0·660 | 1·942 | 2·165 | 1500 | 52·30 | 1/28·8 | 8·8 | 43·50 | 0·0091 | 0·02986 |
| Camion. | | 0·420 | 0·597 | 1·378 | 1·959 | 1600 | 68·20 | 1/22·4 | 11·6 | 56·60 | 0·0089 | 0·02920 |

### *Influence of the Width of the Felloes.*

Experiments made upon wheels of different breadths, having the same diameter, show, 1st, that on soft ground the resistance to rolling *increases* as the width of the felloe; 2dly, on solid gravel and pitched roads, the resistance is very nearly *independent* of the width of the felloe.

### *Influence of the Velocity.*

To investigate the influence of the velocity on the traction of carriages, the same carriages were made to traverse different roads in various conditions; and in each series of experiments the velocities, while all other circumstances remained the same, underwent successive changes from a walk to a canter.

Some of the results of these experiments are given in the following table:—

* Empierrement solide. † Pavé en grès.

| Carriage employed. | Ground passed over. | Load. | Pace. | Rate of speed, in miles, per hour. | Traction. | Ratio of the traction to the load. |
|---|---|---|---|---|---|---|
| | | kil. | | miles. | kil. | |
| Apparatus upon a brass shaft. | Ground of the polygon at Metz, wet and soft. | 1042 | Walk........ | 3·13 | 165·0 | 1/6·32 |
| | | | Trot........ | 6·26 | 168·0 | 1/6·2 |
| | | 1335 | Walk........ | 2·860 | 215·0 | 1/6·21 |
| | | | Trot......... | 7·560 | 197·0 | 1/6·78 |
| A sixteen-pounder carriage and piece. | Road from Metz to Montigny, solid gravel, very even and very dry. | 3750 | Walk........ | 2·820 | 92· | 1/40·8 |
| | | | *Brisk walk | 3·400 | 92· | 1/40·8 |
| | | | Trot......... | 5·480 | 102· | 1/36·8 |
| | | | †Canter...... | 8·450 | 121· | 1/31· |
| Chariot des Messageries, suspended upon six springs. | Pitched road of Fontainebleau. | 3288 | Walk........ | 2·770 | 144· | 1/22·8 |
| | | 3353 | *Brisk walk | 3·82 | 153· | 1/21·9 |
| | | | Trot......... | 5·28 | 161· | 1/20·8 |
| | | | ‡Brisk trot. | 8·05 | 183·5 | 1/18·3 |

We see, by these examples, that the traction undergoes no sensible augmentation with the increase of velocity on soft grounds; but that on solid and uneven roads it increases with an increase of velocity, and in a greater degree as the ground is more uneven, and the carriage has less spring.

To find the relation between the resistance to rolling and the velocity, the velocities were set off as abscissas, and the values of A furnished by the experiments, as ordinates; and the points thus determined were, for each series of experiments, situated very nearly upon a straight line. The value of A, then, can be represented by the expression,

$$A = a + d\,(V - 2)$$

in which $a$ is a number constant for each particular state of each kind of ground, and which expresses the value of the number A for the velocity, V = 2 miles, (per hour,) which is that of a very slow walk.

$d$, a factor constant for each kind of ground and each sort of carriage.

The results of experiments made with a carriage of a siege train, with its piece, gave, on the Montigny road, §very good solid gravel,—

$$A = 0{\cdot}03215 \times 0{\cdot}00295\,(V - 2).$$

On the ||pitched road of Metz, $A = 0{\cdot}01936 \times 0{\cdot}08200\,(V - 2)$.

These examples are sufficient to show—

1st. That, at a walk, the resistance on a good pitched road is less than that on very good solid gravel, very dry.

2d. That, at high speeds, the resistance on the pitched road increases very rapidly with the velocity.

On rough roads the resistance increases with the velocity much more slowly, however, for carriages with springs.

* Pas allongé. † Grand trot. ‡ Trot allongé.
§ En très bon empierrement. || Pavé en grès de Sieack.

Thus, for a chariot des Messageries Générales, on a pitched road, the experiments gave $A = 0{\cdot}0117 \times 0{\cdot}00361\,(V - 2)$; while, with the springs wedged so as to prevent their action, the experiments gave, for the same carriage, on a similar road, $A = 0{\cdot}02723 \times 0{\cdot}01312\,(V - 2)$. At a speed of nine miles per hour, the springs diminish the resistance by one-half.

The experiments further showed that, while the pitched road was inferior to a *solid gravel road when dry and in good repair, the latter lost its superiority when muddy or out of repair.

### INFLUENCE OF THE INCLINATION OF THE TRACES.

The inclination of the traces, to produce the maximum effect, is given by the expression—

$$h f = \frac{A \times 0{\cdot}96 f r'}{r - 0{\cdot}4 f r'}$$

in which $h$ = the height of the fore extremity of the trace above the point where it is attached to the carriage; $b$ = the horizontal distance between these two points. $r'$ is the radius of interior of the boxes, and $r$ the radius of the wheel.

The inclination given by this expression for ordinary carriages is very small; and for trucks with wheels of small diameter it is much less than the construction generally permits.

It follows, from the preceding remarks, that it is advantageous to employ, for all carriages, wheels of as large a diameter as can be used, without interfering with the other essentials to the purposes to which they are to be adapted. Carts have, in this respect, the advantage over wagons; but, on the other hand, on rough roads, the thill horse, jerked about by the shafts, is soon fatigued. Now, by bringing the hind wheels as far forward as possible, and placing the load nearly over them, the wagon is, in effect, transformed into a cart; only care must be taken to place the centre of gravity of the load so far in front of the hind wheels that the wagon may not turn over in going up hill.

### ON THE DESTRUCTIVE EFFECTS PRODUCED BY CARRIAGES ON THE ROADS.

If we take stones of mean diameter from $2\frac{3}{4}$ to $3\frac{1}{4}$ inches, and, on a road slightly moist and soft, place them first under the small wheels of a diligence, and then under the large wheels, we find that, in the former case, the stones, pushed forward by the small wheels, penetrate the surface, ploughing and tearing it up; while in the latter, being merely pressed and leant upon by the large wheels, they undergo no displacement.

From this simple experiment we are enabled to conclude that the wear of the roads by the wheels of carriages is greater the smaller the diameter of the wheels.

Experiments having proved that on hard grounds the traction was but slightly increased when the breadths of the wheels was

---

* En empierrement.

diminished, we might also conclude that the wear of the road would be but slightly increased by diminishing the width of the felloes.

Lastly, the resistance to rolling increasing with the velocity, it was natural to think that carriages going at a trot would do more injury to the roads than those going at a walk. But springs, by diminishing the intensity of the impacts, are able to compensate, in certain proportions, for the effects of the velocity.

Experiments, made upon a grand scale, and having for their object to observe directly the destructive effects of carriages upon the roads, have confirmed these conclusions.

These experiments showed that with equal loads, on a solid gravel road, wheels of two inches breadth produced considerably more wear than those of 4½ inches, but that beyond the latter width there was scarcely any advantage, so far as the preservation of the road was concerned, in increasing the size of the tire of the wheel.

Experiments made with wheels of the same breadth, and of diameters of 2·86 ft., 4·77 ft., and 6·69 ft., showed that after the carriage of 10018·2 tons, over tracks 218·72 yards long, the track passed over by the carriage with the smallest wheels was by far the most worn; while, on that passed over by the carriage with the wheels of 6·69 ft. diameter, the wear was scarcely perceptible.

Experiments made upon two wagons exactly similar in all other respects, but one with and one without springs, showed that the wear of the roads, as well as the increase of traction, after the passage of 4577·36 tons over the same track, was sensibly the same for the carriage without springs, going at a walk of from 2·237 to 2·684 miles per hour, and for that with springs, going at a trot of from 7·158 to 8·053 miles per hour.

---

## HYDRAULICS.

### THE DISCHARGE OF WATER BY SIMPLE ORIFICES AND TUBES.

THE formulas for finding the quantities of water discharged in a given time are of an extensive and complicated nature. The more important and practical results are given in the following Deductions.

When an aperture is made in the bottom or side of a vessel containing water or other homogeneous fluid, the whole of the particles of fluid in the vessel will descend in lines nearly vertical, until they arrive within three or four inches of the place of discharge, when they will acquire a direction more or less oblique, and flow directly towards the orifice.

The particles, however, that are immediately over the orifice, descend vertically through the whole distance, while those nearer to the sides of the vessel, diverted into a direction more or less oblique as they approach the orifice, move with a less velocity than the former; and thus it is that there is produced a contraction in the size of the stream immediately beyond the opening, designated the *vena contracta*, and bearing a proportion to that of the orifice of about 5 to 8, if it pass through a thin plate, or of 6 to 8, if through a short cylindrical tube. But if the tube be conical to a length equal to half its larger diameter, having the issuing diameter less than the entering diameter in the proportion of 26 to 33, the stream does not become contracted.

If the vessel be kept constantly full, there will flow from the aperture twice the quantity that the vessel is capable of containing, in the same time in which it would have emptied itself if not kept supplied.

1. How many horse-power (H. P.) is required to raise 6000 cubic feet of water the hour from a depth of 300 feet?

A cubic foot of water weighs 62·5 lbs. avoirdupois.

$$\frac{6000 \times 62{\cdot}5}{60} = 6250,$$ the weight of water raised a minute.

$6250 \times 300 = 1875000$, the units of work each minute.

Then $\frac{1875000}{33000} = 56{\cdot}818$ = the horse-power required.

2. What quantity of water may be discharged through a cylindrical mouth-piece 2 inches in diameter, under a head of 25 feet?

$\frac{2}{12} = \frac{1}{6}$ of a foot; ∴ the area of the cross section of the mouth-piece, in feet, is $\frac{1}{6} \times \frac{1}{6} \times {\cdot}7854 = {\cdot}021816$.

Theory gives $\cdot021816 \sqrt{2\,g \times 25}$ the cubic feet discharged each second; but experiments show that the effective discharge is 97 per cent. of this theoretical quantity: $g = 32{\cdot}2$.

Hence, $\cdot97 \times \cdot021816 \sqrt{64{\cdot}4 \times 25} = \cdot84912$, the cubic feet discharged each second.

$\cdot84912 \times 62{\cdot}5 = 53{\cdot}0688$ lbs. of water discharged each second.

Effluent water produces, by its *vis viva*, about 6 per cent. less mechanical effect than does its weight by falling from the height of the head.

3. What quantity of water flows through a circular orifice in a thin horizontal plate, 3 inches in diameter, under a head of 49 feet?

Taking the contraction of the fluid vein into account, the velocity of the discharge is about 97 per cent. of that given by theory.

The theoretic velocity is $\sqrt{2g \times 49} = 7\sqrt{6{\cdot}44} = 56{\cdot}21$.

$\cdot97 \times 56{\cdot}21 = 54{\cdot}523$ = the velocity of the discharge.

The area of the transverse section of the contracted vein is ·64 of the transverse section of the orifice.

$\frac{3}{12} = \frac{1}{4} = \cdot25$, and $(\cdot25)^2 \times \cdot7854 = \cdot0490875$ = area of orifice.

∴ $\cdot64 \times \cdot0490875 = \cdot031416$, the area of the transverse section of the contracted vein.

Hence, $54{\cdot}523 \times {\cdot}031416 = 1{\cdot}7129$, the cubic feet of water discharged each second. The later experiments of Poncelet, Bidone, and Lesbros give ·563 for the coefficient of contraction. Water issuing through lesser orifices give greater coefficients of contraction, and become greater for elongated rectangles, than for those which approach the form of a square.

Observations show that the result above obtained is too great; $\frac{8}{13}$ of this result are found to be very near the truth.

$$\frac{8}{13} \text{ of } 1{\cdot}7129 = 1{\cdot}0541.$$

4. What quantity of water flows through a rectangular aperture 7·87 inches broad, and 3·94 inches deep, the surface of the water being 5 feet above the upper edge; the plate through which the water flows being ·125 of an inch thick.

$$\frac{7{\cdot}87}{12} = {\cdot}65583, \text{ decimal of a foot.}$$

$$\frac{3{\cdot}94}{12} = {\cdot}32833, \text{ decimal of a foot.}$$

5· and 5·32833 are the heads of water above the uppermost and lowest horizontal surfaces.

The theoretical discharge will be

$$\frac{2}{3} \times {\cdot}65583 \sqrt{2\,g}\left((5{\cdot}328)^{\frac{3}{2}} - (5)^{\frac{3}{2}}\right) = 3{\cdot}9268 \text{ cubic feet.}$$

Table I. gives the coefficient of efflux in this case, ·615, which is found opposite 5 feet and under 4 inches; for 3·94 is nearly equal 4.

$3{\cdot}9268 \times {\cdot}615 = 2{\cdot}415$ cubic feet, the effective discharge.

5. What water is discharged through a rectangular orifice in a thin plate 6 inches broad, 3 inches deep, under a head of 9 feet measured directly over the orifice?

$$\frac{6}{12} = {\cdot}5, \text{ decimal of a foot.}$$

$$\frac{3}{12} = {\cdot}25, \text{ decimal of a foot.}$$

The theoretical discharge will be

$$\frac{2}{3} \times {\cdot}5 \sqrt{2g}\left\{(9{\cdot}25)^{\frac{3}{2}} - (9)^{\frac{3}{2}}\right\} = 3{\cdot}033 \text{ cubic feet.}$$

Table II. gives the coefficient of efflux between ·604 and ·606; we shall take it at ·605, then

$3{\cdot}033 \times {\cdot}605 = 1{\cdot}833$ cubic feet, the effective discharge.

6. A weir ·82 feet broad, and 4·92 feet head of water, how many cubic feet are discharged each second?

The quantity will be

$$c \times {\cdot}82 \sqrt{2g\,(4{\cdot}92)^3};\ g = 32{\cdot}2;$$

TABLE I.—*The Coefficients for the Efflux through rectangular orifices in a thin vertical plate. The heads are measured where the water may be considered still.*

| Head of water, or distance of the surface of the water from the upper side of the orifice in feet. | Height of Orifice. | | | | | |
|---|---|---|---|---|---|---|
| | In. 8· | In. 4· | In. 2· | In. 1· | In. ·8 | In. ·4 |
| ·1 | ·579 | ·599 | ·619 | ·634 | ·656 | ·686 |
| ·2 | ·582 | ·601 | ·620 | ·638 | ·654 | ·681 |
| ·3 | ·585 | ·603 | ·621 | ·640 | ·653 | ·676 |
| ·4 | ·588 | ·605 | ·622 | ·639 | ·652 | ·671 |
| ·5 | ·591 | ·607 | ·623 | ·637 | ·650 | ·666 |
| ·6 | ·594 | ·609 | ·624 | ·635 | ·649 | ·662 |
| ·7 | ·596 | ·611 | ·625 | ·634 | ·648 | ·659 |
| ·8 | ·597 | ·613 | ·623 | ·632 | ·647 | ·656 |
| ·9 | ·598 | ·615 | ·627 | ·631 | ·645 | ·653 |
| 1·0 | ·599 | ·616 | ·628 | ·630 | ·644 | ·650 |
| 2·0 | ·600 | ·617 | ·628 | ·628 | ·641 | ·647 |
| 3·0 | ·601 | ·617 | ·626 | ·626 | ·638 | ·644 |
| 4·0 | ·602 | ·616 | ·624 | ·623 | ·634 | ·640 |
| 5·0 | ·604 | ·615 | ·621 | ·621 | ·630 | ·635 |
| 6·0 | ·603 | ·613 | ·618 | ·618 | ·625 | ·630 |
| 7·0 | ·602 | ·611 | ·615 | ·615 | ·621 | ·625 |
| 8·0 | ·601 | ·609 | ·612 | ·613 | ·617 | ·619 |
| 9·0 | ·600 | ·606 | ·609 | ·610 | ·614 | ·613 |
| 10·0 | ·600 | ·604 | ·606 | ·608 | ·611 | ·609 |

TABLE II.—*The Coefficients for the Efflux through rectangular orifices in a thin vertical plate, the heads of water being measured directly over the orifice.*

| Head of water, or distance of the surface of the water from the upper side of the orifice in feet. | EIGHT OF ORIFICE. | | | | | |
|---|---|---|---|---|---|---|
| | In. 8· | In. 4· | In. 2· | In. 1· | In. ·8 | In. 4 |
| ·1 | ·593 | ·613 | ·637 | ·659 | ·685 | ·708 |
| ·2 | ·593 | ·612 | ·636 | ·656 | ·680 | ·701 |
| ·3 | ·593 | ·613 | ·635 | ·653 | ·676 | ·694 |
| ·4 | ·594 | ·614 | ·634 | ·650 | ·672 | ·687 |
| ·5 | ·595 | ·614 | ·633 | ·647 | ·668 | ·681 |
| ·6 | ·597 | ·615 | ·632 | ·644 | ·664 | ·675 |
| ·7 | ·598 | ·615 | ·631 | ·641 | ·660 | ·669 |
| ·8 | ·599 | ·616 | ·630 | ·638 | ·655 | ·663 |
| ·9 | ·601 | ·616 | ·629 | ·635 | ·650 | ·657 |
| 1·0 | ·603 | ·617 | ·629 | ·632 | ·644 | ·651 |
| 2·0 | ·604 | ·617 | ·626 | ·628 | ·640 | ·646 |
| 3·0 | ·605 | ·616 | ·622 | ·627 | ·636 | ·641 |
| 4·0 | ·604 | ·614 | ·618 | ·624 | ·632 | ·636 |
| 5·0 | ·604 | ·613 | ·616 | ·621 | ·628 | ·631 |
| 6·0 | ·603 | ·612 | ·613 | ·618 | ·624 | ·626 |
| 7·0 | ·603 | ·610 | ·611 | ·616 | ·620 | ·621 |
| 8·0 | ·602 | ·608 | ·609 | ·614 | ·616 | ·617 |
| 9·0 | ·601 | ·607 | ·607 | ·612 | ·613 | ·613 |
| 10·0 | ·601 | ·603 | ·606 | ·610 | ·610 | ·609 |

$c$ is termed the coefficient of efflux, and on an average may be taken at ·4. It is found to vary from ·385 to ·444.

Then $\cdot 4 \times \cdot 82 \sqrt{(64 \cdot 4)\ (4 \cdot 92)^3} = 2 \cdot 670033$, the cubic feet discharged each second.

7. What breadth must be given to a notch, in a thin plate, with a head of water of 9 inches, to allow 10 cubic feet to flow each second?

The breadth will be represented by

$$\frac{10}{c \sqrt{2g} \times (\cdot 75)^3} = \frac{10}{\cdot 4 \times \sqrt{64 \cdot 4} \times (\cdot 75)^3} = 4 \cdot 7963 \text{ feet.}$$

Changes in the coefficients of efflux through convergent sides often present themselves in practice: they occur in dams which are inclined to the horizon.

Poncelet found the coefficient ·8, when the board was inclined 45°, and the coefficient ·74 for an inclination of 63° 34′, that is for a slope of 1 for a base, and 2 for a perpendicular.

8. If a sluice board, inclined at an angle of 50°, which goes across a channel 2·25 feet broad, is drawn out ·5 feet, what quantity of water will be discharged, the surface of the water standing 4· feet above the surface of the channel, and the coefficient of efflux taken at ·78?

The height of the aperture = ·5 sin. 50° = ·3830222; 4· and 4· − ·3830222 = 3·6169778, are the heads of water.

$$\therefore \frac{2}{3} \times 2 \cdot 25 \times \cdot 78 \times \sqrt{2g} \left\{ (4)^{\frac{3}{2}} - (3 \cdot 617)^{\frac{3}{2}} \right\} = 10 \cdot 5257 \text{ cu-}$$

bic feet, the quantity discharged.

The calculations just made appertain to those cases where the water flows from all sides towards the aperture, and forms a contracted vein on every side. We shall next calculate in cases where the water flows from one or more sides to the aperture, and hence produces a stream only partially contracted. $m$, $n$, $o$, $p$, are four orifices in the bottom ABCD of a vessel; the contraction by efflux through the orifice $o$, in the middle of the bottom, is general, as the water can flow to it from all sides; the contraction from the efflux through $m$, $n$, $p$, is partial, as the water can only flow to them from one, two, or three sides. Partial contraction gives an oblique direction to the stream, and increases the quantity discharged.

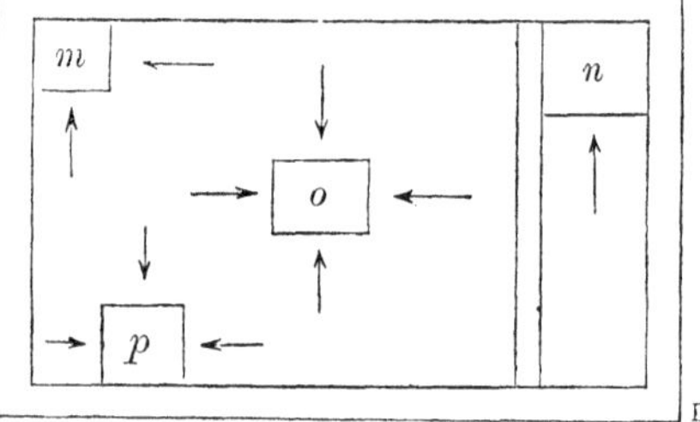


9. What quantity of water is delivered through a flow 4 feet broad, and 1 foot deep, vertical aperture, at a pressure of 2 feet above the upper edge, supposing the lower edge to coincide with

the lower side of the channel, so that there is no contraction at the bottom?

The theoretical discharge will be

$$\frac{2}{3} \times \frac{4}{1} \times \sqrt{2g} \left\{ (3)^{\frac{3}{2}} - (2)^{\frac{3}{2}} \right\} = 50{\cdot}668 \text{ cubic feet.}$$

The coefficient of contraction given in the table page 315, may be taken at ·603.

## I.—*Comparison of the Theoretical with the Real Discharges from an Orifice.*

| Constant height of the water in the reservoir above the centre of the orifice. | Theoretical discharge through a circular orifice one inch in diameter. | Real discharge in the same time through the same orifice. | Ratio of the theoretical to the real discharge. |
|---|---|---|---|
| Paris Feet. | Cubic Inches. | Cubic Inches. | |
| 1 | 4381 | 2722 | 1 to 0·62133 |
| 2 | 6196 | 3846 | 1 to 0·62073 |
| 3 | 7589 | 4710 | 1 to 0·62064 |
| 4 | 8763 | 5436 | 1 to 0·62034 |
| 5 | 9797 | 6075 | 1 to 0·62010 |
| 6 | 10732 | 6654 | 1 to 0·62000 |
| 7 | 11592 | 7183 | 1 to 0·61965 |
| 8 | 12392 | 7672 | 1 to 0·61911 |
| 9 | 13144 | 8135 | 1 to 0·61892 |
| 10 | 13855 | 8574 | 1 to 0·61883 |
| 11 | 14530 | 8990 | 1 to 0·61873 |
| 12 | 15180 | 9384 | 1 to 0·61819 |
| 13 | 15797 | 9764 | 1 to 0·61810 |
| 14 | 16393 | 10130 | 1 to 0·61795 |
| 15 | 16968 | 10472 | 1 to 0·61716 |

## II.—*Comparison of the Theoretical with the Real Discharges from a Tube.*

| Constant height of the water in the reservoir above the centre of the orifice. | Theoretical discharge through a circular orifice one inch in diameter. | Real discharge in the same time by a cylindrical tube one inch in diameter and two inches long. | Ratio of the theoretical to the real discharge. |
|---|---|---|---|
| Paris Feet. | Cubic Inches. | Cubic Inches. | |
| 1 | 4381 | 3539 | 1 to 0·81781 |
| 2 | 6196 | 5002 | 1 to 0·80729 |
| 3 | 7589 | 6126 | 1 to 0·80724 |
| 4 | 8763 | 7070 | 1 to 0·80681 |
| 5 | 9797 | 7900 | 1 to 0·80638 |
| 6 | 10732 | 8654 | 1 to 0·80638 |
| 7 | 11592 | 9340 | 1 to 0·80577 |
| 8 | 12392 | 9975 | 1 to 0·80496 |
| 9 | 13144 | 10579 | 1 to 0·80485 |
| 10 | 13855 | 11151 | 1 to 0·80483 |
| 11 | 14530 | 11693 | 1 to 0·80477 |
| 12 | 15180 | 12205 | 1 to 0·80403 |
| 13 | 15797 | 12699 | 1 to 0·80390 |
| 14 | 16393 | 13177 | 1 to 0·80382 |
| 15 | 16968 | 13620 | 1 to 0·80270 |

THE DISCHARGE BY DIFFERENT APERTURES AND TUBES, UNDER DIFFERENT HEADS OF WATER.

*The velocity of water flowing out of a horizontal aperture, is as the square root of the height of the head of the water.*—That is, the pressure, and consequently the height, is as the square of the velocity; for, the quantity flowing out in any short time is as the velocity; and the force required to produce a velocity in a certain quantity of matter in a given time is also as that velocity; therefore, the force must be as the square of the velocity.

Or, supposing a very small cylindrical plate of water, immediately over the orifice, to be put in motion at each instant, by the pressure of the whole cylinder upon it, employed only in generating its velocity; this plate would be urged by a force as much greater than its own weight as the column is higher than itself, through a space shorter in the same proportion than that height. But where the forces are inversely as the spaces described, the final velocities are equal. Therefore, the velocity of the water flowing out must be equal to that of a heavy body falling from the height of the head of water; which is found, very nearly, by multiplying the square root of that height in feet by 8, for the number of feet described in a second. Thus, a head of 1 foot gives 8; a head of 9 feet, 24. This is the theoretical velocity; but, in consequence of the contraction of the stream, we must, in order to obtain the actual velocity, multiply the square root of the height, in feet, by 5 instead of 8.

The velocity of a fluid issuing from an aperture is not affected by its density being greater or less. Mercury and water issue with equal velocities at equal altitudes.

The proportion of the theoretical to the actual velocity of a fluid issuing through an opening in a thin substance, according to M. Eytelwein, is as 1 to ·619; but more recent experiments make it as 1 to ·621 up to ·645.

APPLICATION OF THE TABLES IN THE PRECEDING PAGE.

TABLE I.—*To find the quantities of water discharged by orifices of different sizes under different altitudes of the fluid in the reservoir.*

To find the quantity of fluid discharged by a circular aperture 3 inches in diameter, the constant altitude being 30 feet.

As the real discharges are in the compound ratio of the area of the apertures and the square roots of the altitudes of the water, and as the theoretical quantity of water discharged by an orifice one inch in diameter from a height of 15 feet is, by the second column of the table, 16968 cubic inches in a minute, we have this proportion: $1 \sqrt{15} : 9 \sqrt{30} :: 16968 : 215961$ cubic inches; the theoretical quantity required. This quantity being diminished in the ratio of 1 to ·62, being the ratio of the theoretical to the actual discharge, according to the fourth column of the table, gives 133896 cubic inches for the actual quantity of water discharged by the given aperture. Hence, the quantity should be rather greater, because large orifices discharge more in proportion than small ones; while it should be rather less, because the altitude of the fluid being greater than that in the table with which it is compared, the flowing vein of water becomes rather more contracted. The quantity thus found, therefore, is nearly accurate as an average.

When the orifice and altitude are less than those in the table, a few cubic inches should be deducted from the result thus derived.

The altitude of the fluid being multiplied by the coefficient 8·016 will give its theoretical velocity; and as the velocities are as the quantities discharged, the real velocity may be deducted from the theoretical by means of the foregoing results.

TABLE II.—*To find the quantities of water discharged by tubes of different diameter, and under different heights of water.*

To find the quantity of water discharged by a cylindrical tube, 4 inches in diameter, and 8 inches long, the constant altitude of the water in the reservoir being 25 feet.

Find, in the same manner as by the example to Table I., the theoretical quantity discharged, which is furnished by this analogy. $1 \sqrt{15} : 16 \sqrt{25} :: 16968 : 350490$ cubic inches, the theoretical discharge. This, diminished in the ratio of 1 to ·81 by the 4th column, will give 28473 cubic inches for the actual quantity discharged. If the tube be *shorter* than twice its diameter, the quantity discharged will be diminished, and approximate to that from a simple orifice, as shown by the production of the *vena contracta* already described.

According to Eytelwein, the proportion of the theoretical to the real discharge through tubes, is as follows:

Through the shortest tube that will cause the stream to adhere everywhere to its sides, as 1 to 0·8125.

Through short tubes, having their lengths from two to four times their diameters, as 1 to 0·82.

Through a tube projecting within the reservoir, as 1 to 0·50.

It should, however, be stated, that in the contraction of the stream the ratio is not constant. It undergoes perceptible variations by altering the form and position of the orifice, the thickness of the plate, the form of the vessel, and the velocity of the issuing fluid.

*Deductions from experiments made by Bossut, Michelloti.*

1. That the quantities of fluid discharged in equal times from different-sized apertures, the altitude of the fluid in the reservoir being the same, are to each other nearly as the area of the apertures.

2. That the quantities of water discharged in equal times by the same orifice under different heads of water, are nearly as the square roots of the corresponding heights of water in the reservoir above the centre of the apertures.

3. That, in general, the quantities of water discharged, in the same time, by different apertures under different heights of water in the reservoir, are to one another in the compound ratio of the areas of the apertures, and the square roots of the altitudes of the water in the reservoirs.

4. That on account of the friction, the smallest orifice discharges proportionally less water than those which are larger and of a similar figure, under the same heads of water.

5. That, from the same cause, of several orifices whose areas are equal, that which has the smallest perimeter will discharge more water than the other, under the same altitudes of water in the reservoir. Hence, circular apertures are most advantageous, as they have less rubbing surface under the same area.

6. That, in consequence of a slight augmentation which the contraction of the fluid vein undergoes, in proportion as the height of the fluid in the reservoir increases, the expenditure ought to be a little diminished.

7. That the discharge of a fluid through a cylindrical horizontal tube, the diameter and length of which are equal to one another, is the same as through a simple orifice.

8. That if the cylindrical horizontal tube be of greater length than the extent of the diameter, the discharge of water is much increased.

9. That the length of the cylindrical horizontal tube may be increased with advantage to four times the diameter of the orifice.

10. That the diameters of the apertures and altitudes of water in the reservoir being the same, the theoretic discharge through a thin aperture, which is supposed to have no contraction in the vein, the discharge through an additional cylindrical tube of greater length than the extent of its diameter, and the actual discharge through an aperture pierced in a thin substance, are to each other as the numbers 16, 13, 10.

11. That the discharges by different additional cylindrical tubes, under the same head of water, are nearly proportional to the areas of the orifices, or to the squares of the diameters of the orifices.

12. That the discharges by additional cylindrical tubes of the same diameter, under different heads of water, are nearly proportional to the square roots of the head of water.

13. That from the two preceding corollaries it follows, in general, that the discharge during the same time, by different additional tubes, and under different heads of water in the reservoir, are to one another nearly in the compound ratio of the squares of the diameters of the tubes, and the square roots of the heads of water.

The discharge of fluids by additional tubes of a conical figure, when the inner to the outer diameter of the orifice is as 33 to 26, is augmented very nearly one-seventeenth and seven-tenths more than by cylindrical tubes, if the enlargement be not carried too far.

DISCHARGE BY COMPOUND TUBES.

*Deductions from the experiments of M. Venturi.*

In the discharge by compound tubes, if the part of the additional tube nearest the reservoir have the form of the contracted vein, the expenditure will be the same as if the fluid were not contracted at all; and if to the smallest diameter of this cone a cylindrical pipe be attached, of the same diameter as the least section of the contracted vein, the discharge of the fluid will, in a horizontal direction, be lessened by the friction of the water against the side of the pipe; but if the same tube be applied in a vertical direction, the expenditure will be augmented, on the principle of the gravitation of falling bodies; consequently, the greater the length of pipe, the more abundant is the discharge of fluid.

If the additional compound tube have a cone applied to the opposite extremity of the pipe, the expenditure will, under the same head of water, be increased, in comparison with that through a simple orifice, in the ratio of 24 to 10.

In order to produce this singular effect, the cone nearest to the reservoir must be of the form of the contracted vein, which will increase the expenditure in the ratio of 12·1 to 10. At the other extremity of the pipe, a truncated conical tube must be applied, of which the length must be nearly nine times the smaller diameter, and its outward diameter must be 1·8 times the smaller one. This additional cone will increase the discharge in the proportion of 24 to 10. But if a great length of pipe intervene, this additional tube has little or no effect on the quantity discharged.

According to M. Venturi's experiments on the discharge of water by bent tubes, it appears that while, with a height of water in the reservoir of 32·5 inches, 4 Paris cubic feet were discharged through a cylindrical horizontal tube in the space of 45 seconds, the discharge of the same quantity through a tube of the same diameter, with a curved end, occupied 50 seconds, and through a like tube bent at right angles, 70 seconds. Therefore, in making cocks or pipes for the discharge or conveyance of water, great attention should be paid to the nature and angle of the bendings; right angles should be studiously avoided.

The interruption of the discharge by various enlargements of the diameter of the tubes having been investigated by M. Venturi, by means of a tube with a diameter of 9 lines, enlarged in several parts to a diameter of 24 lines, the retardation was found to increase nearly in proportion to the number of enlargements; the motion of the fluid, in passing into the enlarged parts, being diverted from its direct course into eddies against the sides of the enlargements. From which it may be deduced, that if the internal roughness of a pipe diminish the expenditure, the friction of the water against these asperities does not form any considerable part of the cause. A right-lined tube may have its internal surface highly polished throughout its whole length, and it may everywhere possess a diameter greater than the orifice to which it is applied; but, nevertheless, the expenditure will be greatly retarded if the pipe should have enlarged parts or swellings. It is not enough that elbows and contractions be avoided; for it may happen, by an intermediate enlargement, that the whole of the other advantage may be lost. This will be obvious from the results in the following table, deduced from experiments with tubes having various enlargements of diameter.

| Head of water in inches. | Number of enlarged parts. | Seconds in which 4 cubic feet were discharged. |
|---|---|---|
| 32·5 | 0 | 109 |
| 32·5 | 1 | 147 |
| 32·5 | 3 | 192 |
| 32·5 | 5 | 240 |

### DISCHARGE BY CONDUIT PIPES.

On account of the friction against the sides, the less the diameter of the pipe, the less proportionally is the discharge of fluid. And, from the same cause, the greater the length of conduit pipe, the greater the diminution of the discharge. Hence, the discharges made in equal times by horizontal pipes of different lengths, but of the same diameter, and under the same altitude of water, are to one another in the inverse ratio of the square roots of the lengths. In order to have a perceptible and continuous discharge of fluid, the altitude of the water in the reservoir, above the axis of the conduit pipe, must not be less than $1\frac{2}{3}$ inch for every 180 feet of the length of the pipe.

The ratio of the difference of discharge in pipes, 16 and 24 lines diameter respectively, may be known by comparing the ratios of Table I. with the ratios of Table II., in the following page.

The greater the angle of inclination of a conduit pipe, the greater will be the discharge in a given time; but when the angle of the conduit pipe is 6° 31′, or the depression of the lower extremity of the pipe is one-eighth or one-ninth of its length, the relative gravity of the fluid will be counterbalanced by the resistance or friction against the sides; and the discharge is then the same as by an additional horizontal tube of the same diameter.

A curvilinear pipe, the altitude of the water in the reservoir being the same, discharges less water when the flexures lie horizontally, than a rectilinear pipe of the same diameter and length.

The discharge by a curvilinear pipe of the same diameter and length, and under the same head of water, is still further diminished when the flexures lie in a vertical instead of a horizontal plane.

When there is a number of contrary flexures in a large pipe, the air sometimes lodges in the highest parts of the flexures, and greatly retards the motion of the water, unless prevented by air-holes, or stopcocks.

TABLE I.—*Comparison of the discharge by conduit pipes of different lengths,* 16 *lines in diameter, with the discharge by additional tubes inserted in the same reservoir.*—By M. BOSSUT.

| Constant altitude of the Water above the centre of the aperture. | Length of the conduit pipe. | Quantity of Water discharged in a minute. | | Ratio between the quantities furnished by tube and pipe. |
|---|---|---|---|---|
| | | by additional tube, 16 lines in diameter. | by conduit pipe, 16 lines in diameter. | |
| Feet. | Feet. | Cubic Inches. | Cubic Inches. | |
| 1 | 30 | 6330 | 2778 | 100 to 43·39 |
| 1 | 60 | 6330 | 1957 | 100 to 30·91 |
| 1 | 90 | 6330 | 1587 | 100 to 25·07 |
| 1 | 120 | 6330 | 1351 | 100 to 21·34 |
| 1 | 150 | 6330 | 1178 | 100 to 18·61 |
| 1 | 180 | 6330 | 1052 | 100 to 16·62 |
| 2 | 30 | 8939 | 4066 | 100 to 45·48 |
| 2 | 60 | 8939 | 2888 | 100 to 32·31 |
| 2 | 90 | 8939 | 2352 | 100 to 26·31 |
| 2 | 120 | 8939 | 2011 | 100 to 22·50 |
| 2 | 150 | 8939 | 1762 | 100 to 19·71 |
| 2 | 180 | 8939 | 1583 | 100 to 17·70 |

TABLE II.—*Comparison of the discharge by conduit pipes of different lengths,* 24 *lines in diameter, with the discharge by additional tubes inserted in the same reservoir.*—By M. BOSSUT.

| Constant altitude of the Water above the centre of the aperture. | Length of the conduit pipe. | Quantity of Water discharged in a minute. | | Ratio between the quantities furnished by tube and pipe. |
|---|---|---|---|---|
| | | by additional tube, 24 lines in diameter. | by conduit pipe, 24 lines in diameter. | |
| Feet. | Feet. | Cubic Inches. | Cubic Inches. | |
| 1 | 30 | 14243 | 7680 | 100 to 53·92 |
| 1 | 60 | 14243 | 5564 | 100 to 39·06 |
| 1 | 90 | 14243 | 4534 | 100 to 31·83 |
| 1 | 120 | 14243 | 3944 | 100 to 27·69 |
| 1 | 150 | 14243 | 3486 | 100 to 24·48 |
| 1 | 180 | 14243 | 3119 | 100 to 21·90 |
| 2 | 30 | 20112 | 11219 | 100 to 55·78 |
| 2 | 60 | 20112 | 8190 | 100 to 40·72 |
| 2 | 90 | 20112 | 6812 | 100 to 33·87 |
| 2 | 120 | 20112 | 5885 | 100 to 29·26 |
| 2 | 150 | 20112 | 5232 | 100 to 26·01 |
| 2 | 180 | 20112 | 4710 | 100 to 23·41 |

### DISCHARGE BY WEIRS AND RECTANGULAR APERTURES.

*Rectangular orifices in the side of a reservoir, extending to the surface.*

The velocity varying nearly as the square root of the height, may here be represented by the ordinates of a parabola, and the quantity of water discharged by the area of the parabola, or two-thirds of that of the circumscribing rectangle. So that the quantity discharged may be found by taking two-thirds of the velocity due to the mean height, and allowing for the contraction of the stream, according to the form of the opening.

In a lake, for example, in the side of which a rectangular opening is made without any oblique lateral walls, three feet wide, and extending two feet below the surface of the water, the coefficient of the velocity, corrected for contraction, is 5·1, and the corrected mean velocity $\frac{2}{3}\sqrt{2} \times 5{\cdot}1 = 4{\cdot}8$; therefore the area being 6, the discharge of water in a second is 28·8 cubic feet, or nearly four hogsheads.

The same coefficient serves for determining the discharge over a weir of considerable breadth; and, hence, to deduce the depth or breadth requisite for the discharge of a given quantity of water. For example, a lake has a weir three feet in breadth, and the surface of the water stands at the height of five feet above it: it is required how much the weir must be widened, in order that the water may be a foot lower. Here the velocity is $\frac{2}{3}\sqrt{5} \times 5{\cdot}1$, and the quantity of water $\frac{2}{3}\sqrt{5} \times 5{\cdot}1 \times 3 \times 5$; but the velocity must be reduced to $\frac{2}{3}\sqrt{4} \times 5{\cdot}1$, and then the section will be $\frac{\frac{2}{3}\sqrt{5} \times 5{\cdot}1 \times 3 \times 5}{\frac{2}{3}\sqrt{4} \times 5{\cdot}1}$ $= \frac{\sqrt{5} \times 3 \times 5}{\sqrt{4}} = 7{\cdot}5 \times \sqrt{5}$; and the height being 4, the breadth must be $\frac{7{\cdot}5}{4}\sqrt{5} = 4{\cdot}19$ feet.

The discharge from reservoirs, with lateral orifices of considerable magnitude, and a constant head of water, may be found by determining the difference in the discharge by two open orifices of different heights; or, in most cases, with nearly equal accuracy, by considering the velocity due to the distance, below the surface, of the centre of gravity of the orifice.

Under the same height of water in the reservoir, the same quantity always flows in a canal, of whatever length and declivity; but in a tube, a difference in length and declivity has a great effect on the quantity of water discharged.

The velocity of water flowing in a river or stream varies at different parts of the same transverse section. It is found to be greatest where the water is deepest, at somewhat less than one-half the depth from the surface; diminishing towards the sides and shallow parts.

*Resistance to bodies moving in fluids.*—The deductions from the experiments of C. Colles, (who first planned the Croton Aqueduct, New York,) and others, on this intricate subject, are, as stated, thus:

1. The confirmation of the theory, that the resistance of fluids to passing bodies is as the squares of the velocities.

2. That, contrary to the received opinion, a cone will move through the water with much less resistance with its apex foremost, than with its base forward.

3. That the increasing the length of a solid, of almost any form, by the addition of a cylinder in the middle, diminishes the resistance with which it moves, provided the weight in the water remains the same.

4. That the greatest breadth of the moving body should be placed at the distance of two-fifths of the whole length from the bow, when applied to the ordinary forms in naval architecture.

5. That the bottom of a floating solid should be made triangular; as in that case it will meet with the least resistance when moving in the direction of its longest axis, and with the greatest resistance when moving with its broadside foremost.

*Friction of fluids.*—Some experiments have been made on this subject, with reference to the motion of bodies in water, upon a cylindrical model, 30 inches in length, 26 inches in diameter, and weighing 255 lbs. avoirdupois. The cylinder was placed in a cistern of salt water, and made to vibrate on knife-edges passing through its axis, and was deflected over to various angles by means of a weight attached to the arm of a lever. The experiments were then repeated without the water, and the following are the angles of deflection and vibration in the two cases.

| In the salt water. | | In the atmosphere. | |
|---|---|---|---|
| Angle of Deflection. | Angle to which it vibrated. | Angle of Deflection. | Angle to which it vibrated. |
| 22° 30′ | 22° 24′ | 22° 30′ | 20° 0′ |
| 22 10 | 22 6 | 21 36 | 21 3 |
| 21 54 | 21 48 | 20 48 | 20 16 |
| 21 36 | 21 30 | &c. | &c. |
| &c. | &c. | | |

Showing that the amplitude of vibration when oscillating in water is considerably less than when oscillating without water. In the experiments there is a falling off in the angle of 24′, or nearly half a degree. The amount of force acting on the surface of the cylinder necessary to cause the above difference was calculated; and the author thinks that it is not equally distributed on the surface of the cylinder, but that the amount on any particular part might vary as the depth. On this supposition, a constant pressure at a unit of depth is assumed, and this, multiplied by the depth of any other point of the cylinder immersed in the water, will give the pressure at that point. These forces or moments being summed by integration and equated with the sum of the moments given by the experiments, we have the value of the constant pressure at a unit of depth = ·0000469. This constant, in another experiment, the weight of the model being 197 lbs. avoirdupois, and consequently the part immersed in the water being different from that in the other experiment, was ·0000452, which differs very little from the former,—indicating the probability of the correctness of the assumption.

*The drainage of water through pipes.*—The experiments made under the direction of the Metropolitan Commissioners of Sewers, on the capacities of pipes for the drainage of towns, have presented some useful results for the guidance of those who have to make calculations for a similar purpose. The pipes, of various diameters, from 3 to 12 inches, were laid on a platform of 100 feet in length, the declivity of which could be varied from a horizontal level to a fall of 1 in 10. The water was admitted at the head of the pipe, and at five junctions, or tributary pipes on each side, so regulated as to keep the main pipe full.

The results were as follow:—

It was found—to mention only one result—that a line of 6-inch pipes, 100 feet long, at an inclination of 1 in 60, discharged 75 cubic feet per minute. The same experiment, repeated with the line of pipes reduced to 50 feet in length, gave very nearly the same result. Without the addition of junctions, the transverse sectional area of the stream of water near the discharging end was reduced to one-fifth of the corresponding area of the pipe, and it required a simple head of water of about 22 inches to give the same result as that accruing under the circumstances of the junctions. With regard to varying sizes and inclinations, it appears, sufficiently for practical purposes, that the squares of the discharges are as the fifth powers of the diameters; and again, that in steeper declivities than 1 in 70, the discharges are as the square roots of the inclinations; but at less declivities than 1 in 70, the ratios of the discharges diminish very rapidly, and are governed by no constant law. At a certain small declivity, the relative discharge is as the fifth root of the inclination; at a smaller declivity, it is found as the seventh root of the inclination; and so on, as it approaches the horizontal plane. This may be exemplified by the following results found by actual experiment:

*Discharges of a 6-inch pipe at several inclinations.*

| Inclination. | Discharges in 100 feet per minute. | Inclination. | Discharges in 100 feet per minute. |
|---|---|---|---|
| 1 in 60 | 75 | 1 in 320 | 49 |
| 1 in 80 | 68 | 1 in 400 | 48·5 |
| 1 in 100 | 63 | 1 in 480 | 48 |
| 1 in 120 | 59 | 1 in 640 | 47·5 |
| 1 in 160 | 54 | 1 in 800 | 47·2 |
| 1 in 200 | 52 | 1 in 1200 | 46·7 |
| 1 in 240 | 50 | Level | 46 |

The conclusion arrived at is, that the requisite sizes of drains and sewers can be determined (near enough for practical purposes, as an important circumstance has to be considered in providing for the deposition of solid matter, which disadvantageously alters the *form* of the aqueduct, and contracts the water-way) by taking the result of the 6-inch pipe, under the circumstances before mentioned as a *datum*, and assuming that the squares of the discharges are as the fifth powers of the diameters.

That at greater declivities than 1 in 70, the discharges are as the square roots of the inclinations.

That at less declivities than 1 in 70, the usual law will not obtain; but near approximations to the truth may be obtained by observing the relative discharges of a pipe laid at various small inclinations.

That increasing the number of junctions, at intervals, accelerates the velocity of the main stream in a ratio which increases as the square root of the inclination, and which is greater than the ratio of resistance due to a proportionable increase in the length of the aqueduct. The velocity at which the lateral streams enter the main line, is a most important circumstance governing the flow of water. In practice, these velocities are constantly variable, considered individually, and always different considered collectively, so that their united effect it is difficult to estimate. Again, the same sewer at different periods may be quite filled, but discharges in a given time very different quantities of water. It should be mentioned that in the case of the 6-inch pipe, which discharged 75 cubic feet per minute, the lateral streams had a velocity of a few feet per second, and the junctions were placed at an angle of about 35° with the main line. It is needless to say that all junctions should be made as nearly parallel with the main line as possible, otherwise the forces of the lateral currents may impede rather than maintain or accelerate the main streams.

---

## WATER WHEELS.

### THE UNDERSHOT WHEEL.

THE ratio between the power and effect of an undershot wheel is as 10 to 3·18; consequently 31·43 lbs. of water must be expended per second to produce a mechanical effect equal to that of the estimated labour of an active man.

The velocity of the periphery of the undershot wheel should be equal to half the velocity of the stream; the float-boards should be so constructed as to rise perpendicularly from the water; not more than one-half should ever be below the surface; and from 3 to 5 should be immersed at once, according to the magnitude of the wheel.

The following maxims have been deduced from experiments:—

1. The virtual or effective head of water being the same, the effect will be nearly as the quantity expended; that is, if a mill, driven by a fall of water, whose virtual head is 10 feet, and which discharges 30 cubic feet of water in a second, grind four bolls of corn in an hour; another mill having the same virtual head, but which discharges 60 cubic feet of water, will grind eight bolls of corn in an hour.

2. The expense of water being the same, the effect will be nearly as the height of the virtual or effective head.

3. The quantity of water expended being the same, the effect is nearly as the square of its velocity; that is, if a mill, driven by a

certain quantity of water, moving with the velocity of four feet per second, grind three bolls of corn in an hour; another mill, driven by the same quantity of water, moving with the velocity of five feet per second, will grind nearly $4\frac{7}{10}$ bolls in the hour, because $3 : 4\frac{7}{10} :: 4^2 : 5^2$ nearly.

4. The aperture being the same, the effect will be nearly as the cube of the velocity of the water; that is, if a mill driven by water, moving through a certain aperture, with the velocity of four feet per second, grind three bolls of corn in an hour; another mill, driven by water, moving through the same aperture with the velocity of five feet per second, will grind $5\frac{43}{50}$ bolls nearly in an hour; for as $3 : 5\frac{43}{50} :: 4^3 : 5^3$ nearly.

The height of the virtual head of water may be easily determined from the velocity of the water, for the heights are as the squares of the velocities, and, consequently, the velocities are as the square roots of the height.

*To calculate the proportions of undershot wheels.*—Find the perpendicular height of the fall of water above the bottom of the mill-course, and having diminished this number by one-half the depth of the water where it meets the wheel, call that *the height of the fall.*

Multiply the height of the fall, so found, by 64·348, and take the square root of the product, which will be *the velocity of the water.*

Take one-half of the velocity of the water, and it will be the velocity to be given to the float-boards, or the number of feet they must move through in a second, to produce a *maximum* effect. Divide the circumference of the wheel by the velocity of its float-boards per second, and the quotient will be the number of seconds in which the wheel revolves. Divide 60 by the quotient thus found, and the new quotient will be the number of revolutions made by the wheel in a minute.

Divide 90, the number of revolutions which a millstone, 5 feet in diameter, should make in a minute, by the number of revolutions made by the wheel in a minute, the quotient will be the number of turns the millstone ought to make for one turn of the wheel. Then, as the number of revolutions of the wheel in a minute is to the number of revolutions of the millstone in a minute, so must the number of staves in the trundle be to the number of teeth in the wheel, (the nearest in whole numbers.) Multiply the number of revolutions made by the wheel in a minute, by the number of revolutions made by the millstone for one turn of the wheel, and the product will be the number of revolutions made by the millstone in a minute.

The effect of the water wheel is a *maximum*, when its circumference moves with one-half, or, more accurately, with three-sevenths of the velocity of the stream.

### THE BREAST WHEEL.

The effect of a breast wheel is equal to the effect of an under shot wheel, whose head of water is equal to the difference of level between the surface of water in the reservoir, and the part where it strikes the wheel, added to that of an overshot, whose height is equal to the difference of level between the part where it strikes the wheel and the level of the tail water.

When the fall of water is between 4 and 10 feet, a breast wheel should be erected, provided there be enough of water; an undershot should be used when the fall is below 4 feet, and an overshot wheel when the fall exceeds 10 feet. Also, when the fall exceeds 10 feet, it should be divided into two, and two breast wheels be erected upon it.

TABLE *for breast wheels.*

| | Breadth of the float-boards. | Depth of the float-boards. | Radius of water wheel, reckoned from the extremity of float-boards. | Velocity of the wheel in a second. | Time in which the wheel performs one revolution. | Turns of the mill-stone for one of the wheels. | Force of the water upon the float-boards. | Water required in a second to turn the wheel. |
|---|---|---|---|---|---|---|---|---|
| | Feet. | Feet. | Feet. | Feet. | Sec. | ........ | lbs. avr. | Cubic ft. |
| 1 | 0·17 | 198·6 | 0·75 | 2·18 | 1·92 | 4·80 | 1536 | 74·30 |
| 2 | 0·34 | 35·1 | 1·50 | 3·09 | 2·72 | 6·80 | 1084 | 37·15 |
| 3 | 0·51 | 12·7 | 2·26 | 3·78 | 3·33 | 8·32 | 886 | 24·77 |
| 4 | 0·69 | 6·2 | 3·01 | 4·36 | 3·84 | 9·60 | 762 | 18·57 |
| 5 | 0·86 | 3·57 | 3·76 | 4·88 | 4·28 | 10·70 | 680 | 14·86 |
| 6 | 1·03 | 2·25 | 4·51 | 5·35 | 4·70 | 11·76 | 626 | 12·38 |
| 7 | 1·20 | 1·53 | 5·26 | 5·77 | 5·08 | 12·70 | 581 | 10·61 |
| 8 | 1·37 | 1·10 | 6·02 | 6·17 | 5·43 | 13·58 | 543 | 9·29 |
| 9 | 1·54 | 0·81 | 6·77 | 6·55 | 5·76 | 14·40 | 512 | 8·26 |
| 10 | 1·71 | 0·77 | 7·52 | 6·90 | 6·07 | 15·18 | 486 | 7·43 |

It is evident, from the preceding table, that when the height of the fall is less than 3 feet, the depth of the float-boards is so great, and their breadth so small, that the breast wheel cannot well be employed; and, on the contrary, when the height of the fall approaches to 10 feet, the depth of the float-boards is too small in proportion to their breadth; these two extremes, therefore, must be avoided in practice. The ninth column contains the quantity of water necessary for impelling the wheel; but the total expense of water should always exceed this by the quantity, at least, which escapes between the mill-course and the sides and extremities of the float-boards.

### THE OVERSHOT WHEEL.

The ratio between the power and effect of an overshot wheel, is as 10 to 6·6, when the water is delivered above the apex of the wheel, and is computed from the whole height of the fall; and as 10 to 8 when computed from the height of the wheel only; consequently, the quantity of water expended per second, to produce a mechanical effect equal to that of the aforesaid estimated labour of an active man, is, in the first instance, 15·15 lbs., and in the second instance, 12·5 lbs.

Hence, the effect of the overshot wheel, under the same circumstances of quantity and fall, is, at a medium, double that of the undershot.

The velocity of the periphery of an overshot wheel should be from 6½ to 8½ feet per second.

The higher the wheel is, in proportion to the whole descent, the greater will be the effect.

And from the equality of the ratio between the power and effect, subsisting where the constructions are similar, we must infer that the effects, as well as the powers, are as the quantities of water and perpendicular heights multiplied together respectively.

*Working machinery by hydraulic pressure.*—The vertical pressure of water, acting on a piston, for raising weights and driving machinery, is coming into use in many places where it can be advantageously applied. At Liverpool, Newcastle, Glasgow, and other places, it is applied to the working of cranes, drawing coal-wagons, and other purposes requiring continuous power. The presence of a natural fall, like that of Golway, Ireland, which can be conducted to the engine through pipes, is, of course, the most economical situation for the application of such power; in other situations, artificial power must be used to raise the water, which, even under this disadvantage, may, from its readiness and simplicity of action, be often serviceably employed. Wherever the contiguity of a steam engine would be dangerous, or otherwise objectionable, a water engine would afford the means of receiving and applying the power from any required distance, precautions being taken against the action of frost on the fluid.

Required the horse power of a centre discharging *Turbine* water wheel, the head of water being 25 feet, and the area of the opening 400 inches.

The following table shows the working horse power of both the inward and outward discharging Turbine water wheels; they are calculated to the square inch of opening.

| Centre Discharging Turbine. | | Outward Discharging Turbine. | Centre Discharging Turbine. | | Outward Discharging Turbine. |
|---|---|---|---|---|---|
| Head. | Horse Power. | Horse Power. | Head. | Horse Power. | Horse Power. |
| 3 | ·00821 | ·012611 | 22 | ·19523 | ·339972 |
| 4 | ·01483 | ·025145 | 23 | ·20787 | ·364182 |
| 5 | ·02137 | ·038124 | 24 | ·22315 | ·384615 |
| 6 | ·02685 | ·045618 | 25 | ·23667 | ·412013 |
| 7 | ·03414 | ·058314 | 26 | ·25125 | ·437519 |
| 8 | ·04198 | ·074413 | 27 | ·26482 | ·455698 |
| 9 | ·05206 | ·089025 | 28 | ·28135 | ·484427 |
| 10 | ·05883 | ·106215 | 29 | ·29563 | ·510833 |
| 11 | ·06921 | ·118127 | 30 | ·30817 | ·537721 |
| 12 | ·07851 | ·135610 | 31 | ·32316 | ·561425 |
| 13 | ·08882 | ·150638 | 32 | ·33617 | ·587148 |
| 14 | ·10054 | ·173158 | 33 | ·34823 | ·611013 |
| 15 | ·11002 | ·192234 | 34 | ·36154 | ·638174 |
| 16 | ·12093 | ·211592 | 35 | ·37123 | ·665164 |
| 17 | ·13196 | ·231161 | 36 | ·39874 | ·692156 |
| 18 | ·14275 | ·257145 | 37 | ·40118 | ·726148 |
| 19 | ·15613 | ·273325 | 38 | ·41762 | ·764115 |
| 20 | ·16927 | ·296618 | 39 | ·42156 | ·804479 |
| 21 | ·18109 | ·317167 | 40 | ·43718 | ·849814 |

Opposite 25 in the column marked "Head," the working horse power to the square inch is found to be ·25667, which, multiplied by 400, gives 94·668, the horse power required.

What is the working horse power of an outward discharging Turbine, under the effective head of 20 feet; the area of all the openings being 325 square inches. In the table, opposite 20, we find ·296618, then $\cdot 296618 \times 325 = 96 \cdot 4$, the required horse power.

What is the number of revolutions a minute of an outward discharging Turbine wheel, the head being 19 feet and the diameter of the wheel 60 inches?

In the table for the outward discharging wheel, opposite 19, and under 60 inches, we find 97, the number of revolutions required.

What is the number of revolutions a minute of an inward discharging Turbine, under a head of 21 feet, the diameter being 72 inches?

In the table for the inward discharging wheel, opposite 21 feet, and under 72 inches, we find 95, the number of revolutions a minute.

These Turbine tables were calculated by the author's brother, the late John O'Byrne, C. E., who died in New York, on the 6th of April, 1851.

*Outward discharging Turbine.*

| Head in feet. | Diameter in Inches. | | | | | | | | | | | | |
|---|---|---|---|---|---|---|---|---|---|---|---|---|---|
| | 24 | 30 | 36 | 42 | 48 | 54 | 60 | 66 | 72 | 78 | 84 | 90 | 96 |
| 3 | 100 | 80 | 70 | 60 | 52 | 42 | 37 | 35 | 32 | 30 | 28 | 27 | 21 |
| 4 | 111 | 89 | 73 | 63 | 57 | 49 | 44 | 41 | 37 | 34 | 32 | 30 | 28 |
| 5 | 123 | 100 | 82 | 71 | 62 | 55 | 51 | 45 | 42 | 38 | 37 | 33 | 31 |
| 6 | 135 | 109 | 91 | 78 | 68 | 62 | 55 | 50 | 45 | 42 | 38 | 37 | 36 |
| 7 | 146 | 118 | 96 | 84 | 73 | 65 | 59 | 53 | 49 | 47 | 42 | 40 | 38 |
| 8 | 156 | 125 | 105 | 90 | 79 | 71 | 63 | 57 | 52 | 49 | 43 | 42 | 39 |
| 9 | 166 | 133 | 111 | 95 | 83 | 75 | 67 | 61 | 57 | 50 | 49 | 45 | 41 |
| 10 | 175 | 140 | 117 | 100 | 87 | 79 | 70 | 64 | 59 | 55 | 51 | 47 | 46 |
| 11 | 183 | 147 | 122 | 105 | 92 | 81 | 74 | 67 | 62 | 57 | 54 | 49 | 48 |
| 12 | 191 | 156 | 127 | 110 | 96 | 85 | 79 | 70 | 64 | 59 | 55 | 53 | 51 |
| 13 | 200 | 159 | 133 | 115 | 100 | 89 | 81 | 73 | 67 | 62 | 57 | 55 | 53 |
| 14 | 206 | 166 | 138 | 118 | 104 | 92 | 83 | 75 | 69 | 64 | 59 | 57 | 55 |
| 15 | 213 | 171 | 142 | 122 | 107 | 95 | 86 | 78 | 72 | 66 | 61 | 58 | 56 |
| 16 | 222 | 177 | 148 | 126 | 111 | 98 | 89 | 82 | 74 | 69 | 64 | 59 | 57 |
| 17 | 227 | 182 | 152 | 131 | 115 | 101 | 91 | 83 | 77 | 71 | 66 | 62 | 59 |
| 18 | 234 | 187 | 156 | 134 | 117 | 105 | 94 | 85 | 78 | 73 | 67 | 63 | 61 |
| 19 | 238 | 193 | 161 | 138 | 120 | 107 | 97 | 88 | 81 | 74 | 69 | 64 | 63 |
| 20 | 247 | 197 | 164 | 141 | 124 | 110 | 99 | 90 | 84 | 76 | 71 | 66 | 64 |
| 21 | 252 | 202 | 168 | 145 | 126 | 114 | 101 | 92 | 85 | 78 | 73 | 68 | 65 |
| 22 | 259 | 208 | 172 | 149 | 129 | 115 | 105 | 94 | 87 | 80 | 74 | 69 | 67 |
| 23 | 263 | 212 | 176 | 151 | 133 | 119 | 106 | 96 | 89 | 84 | 77 | 72 | 70 |
| 24 | 270 | 216 | 180 | 155 | 135 | 120 | 109 | 98 | 92 | 85 | 78 | 74 | 72 |
| 25 | 277 | 222 | 184 | 158 | 138 | 123 | 111 | 101 | 93 | 86 | 80 | 76 | 74 |
| 26 | 282 | 226 | 189 | 161 | 141 | 125 | 113 | 103 | 95 | 87 | 81 | 78 | 76 |
| 27 | 286 | 229 | 191 | 165 | 143 | 129 | 116 | 105 | 97 | 88 | 83 | 79 | 77 |
| 28 | 291 | 233 | 195 | 167 | 146 | 130 | 118 | 107 | 99 | 91 | 85 | 80 | 78 |
| 29 | 297 | 237 | 199 | 170 | 149 | 132 | 119 | 109 | 100 | 92 | 86 | 81 | 80 |
| 30 | 303 | 241 | 202 | 174 | 152 | 135 | 122 | 111 | 102 | 94 | 88 | 82 | 81 |

*Inward discharging Turbine.*

| Head in feet. | Diameter in Inches. | | | | | | | | | | | | |
|---|---|---|---|---|---|---|---|---|---|---|---|---|---|
| | 24 | 30 | 36 | 42 | 48 | 54 | 60 | 66 | 72 | 78 | 84 | 90 | 96 |
| 3 | 111 | 86 | 74 | 62 | 54 | 48 | 47 | 40 | 36 | 32 | 31 | 30 | 27 |
| 4 | 125 | 96 | 83 | 70 | 62 | 55 | 51 | 45 | 41 | 37 | 36 | 34 | 31 |
| 5 | 141 | 112 | 94 | 78 | 69 | 61 | 55 | 50 | 46 | 43 | 40 | 37 | 36 |
| 6 | 152 | 122 | 101 | 86 | 76 | 67 | 62 | 55 | 51 | 47 | 43 | 42 | 38 |
| 7 | 166 | 131 | 108 | 93 | 82 | 72 | 65 | 60 | 54 | 51 | 47 | 44 | 42 |
| 8 | 175 | 139 | 116 | 99 | 87 | 76 | 71 | 63 | 57 | 54 | 49 | 47 | 45 |
| 9 | 186 | 149 | 123 | 106 | 93 | 81 | 74 | 68 | 63 | 57 | 53 | 51 | 47 |
| 10 | 195 | 156 | 129 | 111 | 99 | 86 | 78 | 71 | 66 | 61 | 56 | 52 | 49 |
| 11 | 208 | 167 | 136 | 117 | 102 | 91 | 82 | 74 | 68 | 63 | 58 | 56 | 52 |
| 12 | 217 | 169 | 142 | 122 | 107 | 97 | 85 | 78 | 71 | 66 | 61 | 57 | 54 |
| 13 | 221 | 178 | 148 | 127 | 112 | 99 | 89 | 82 | 74 | 69 | 64 | 61 | 56 |
| 14 | 231 | 184 | 153 | 133 | 116 | 104 | 92 | 85 | 76 | 71 | 66 | 62 | 58 |
| 15 | 238 | 191 | 159 | 136 | 119 | 107 | 95 | 87 | 80 | 73 | 68 | 64 | 61 |
| 16 | 245 | 198 | 165 | 144 | 123 | 111 | 99 | 90 | 83 | 76 | 71 | 66 | 63 |
| 17 | 252 | 203 | 168 | 148 | 127 | 114 | 102 | 92 | 85 | 78 | 73 | 68 | 64 |
| 18 | 269 | 209 | 173 | 150 | 132 | 116 | 104 | 95 | 87 | 82 | 75 | 69 | 66 |
| 19 | 267 | 215 | 176 | 153 | 134 | 120 | 108 | 98 | 89 | 83 | 77 | 72 | 67 |
| 20 | 276 | 222 | 183 | 157 | 138 | 122 | 111 | 101 | 93 | 85 | 79 | 74 | 69 |
| 21 | 288 | 226 | 186 | 162 | 141 | 125 | 113 | 103 | 95 | 86 | 80 | 75 | 71 |
| 22 | 290 | 230 | 192 | 164 | 145 | 129 | 116 | 107 | 96 | 89 | 83 | 77 | 73 |
| 23 | 299 | 235 | 196 | 167 | 146 | 133 | 118 | 109 | 97 | 91 | 84 | 79 | 74 |
| 24 | 303 | 240 | 201 | 171 | 151 | 135 | 122 | 111 | 101 | 93 | 86 | 80 | 75 |
| 25 | 310 | 247 | 206 | 176 | 155 | 138 | 123 | 112 | 104 | 96 | 88 | 82 | 76 |
| 26 | 314 | 248 | 210 | 180 | 157 | 139 | 126 | 115 | 106 | 97 | 90 | 84 | 79 |
| 27 | 319 | 254 | 213 | 183 | 162 | 142 | 128 | 117 | 108 | 99 | 92 | 85 | 80 |
| 28 | 327 | 261 | 218 | 186 | 164 | 146 | 129 | 119 | 109 | 102 | 93 | 87 | 82 |
| 29 | 333 | 265 | 221 | 189 | 166 | 148 | 133 | 121 | 111 | 103 | 95 | 89 | 83 |
| 30 | 336 | 271 | 224 | 193 | 168 | 151 | 136 | 124 | 114 | 105 | 97 | 90 | 85 |

## WINDMILLS.

1. The velocity of windmill sails, whether unloaded or loaded, so as to produce a maximum effect, is nearly as the velocity of the wind, their shape and position being the same.

2. The load at the maximum is nearly, but somewhat less than, as the square of the velocity of the wind, the shape and position of the sails being the same.

3. The effects of the same sails, at a maximum, are nearly, but somewhat less than, as the cubes of the velocity of the wind.

4. The load of the same sails, at the maximum, is nearly as the squares, and their effect as the cubes of their number of turns in a given time.

5. When sails are loaded so as to produce a maximum at a given velocity, and the velocity of the wind increases, the load continuing the same,—1st, the increase of effect, when the increase of the velocity of the wind is small, will be nearly as the squares of those velocities; 2dly, when the velocity of the wind is double, the effects will be nearly as 10 to $27\frac{1}{2}$; but, 3dly, when the velocities compared are more than double of that when the given load produces a maximum, the effects increase nearly in the simple ratio of the velocity of the wind.

6. In sails where the figure and position are similar, and the velocity of the wind the same, the number of turns, in a given time, will be reciprocally as the radius or length of the sail.

7. The load, at a maximum, which sails of a similar figure and position will overcome, at a given distance from the centre of motion, will be as the cube of the radius.

8. The effects of sails of similar figure and position are as the square of the radius.

9. The velocity of the extremities of Dutch sails, as well as of the enlarged sails, in all their usual positions when unloaded, or even loaded to a maximum, is considerably greater than that of the wind.

The results in Table 1 are for Dutch sails, in their common position, when the radius was 30 feet. Table 2 contains the most efficient angles.

1.

| Number of revolutions of wind-shaft in a minute. | Velocity of the wind in an hour. | Ratio between velocity of wind and revolutions of wind-shaft. |
|---|---|---|
| 3 | 2 miles | 0·666 |
| 5 | 4 miles | 0·800 |
| 6 | 5 miles | 0·833 |

2.

| Parts of the radius, which is divided into six parts. | Angle with the axis. | Angle of weather. |
|---|---|---|
| 1 | 72° | 18° |
| 2 | 71 | 19 |
| 3 | 72 | 18 middle |
| 4 | 74 | 16 |
| 5 | 77½ | 12½ |
| 6 | 83 | 7 |

Supposing the radius of the sail to be 30 feet, then the sail will commence at $\frac{1}{6}$, or 5 feet from the axis, where the angle of inclination will be 72 degrees; at $\frac{2}{6}$, or 10 feet from the axis, the angle will be 71 degrees, and so on.

*Results of Experiments on the effect of Windmill Sails in grinding corn.*—By M. Coulomb.

A windmill, with four sails, measuring 72 feet from the extremity of one sail to that of the opposite one, and 6 feet 7 inches wide, or a little more, was found capable of raising 1100 lbs. avoirdupois 238 feet in a minute, and of working, on an average, eight hours in a day. This is equivalent to the work of 34 men, 30 square feet of canvas performing about the daily work of a man.

When a vertical windmill is employed to grind corn, the millstone makes 5 revolutions in the same time that the sails and the arbor make 1.

The mill does not begin to turn till the velocity of the wind is about 13 feet per second.

When the velocity of the wind is 19 feet per second, the sails make from 11 to 12 turns in a minute, and the mill will grind from 880 to 990 lbs. avoirdupois in an hour, or about 22,000 lbs. in 24 hours.

# THE APPLICATION OF LOGARITHMS.

The practice of performing calculations by Logarithms is an exercise so useful to computers, that it requires a more particular explanation than could have been properly given in that part of the work allotted to Arithmetic.

A few of the various applications of logarithms, best suited to the calculations of the engineer and mechanic, have therefore been collected, and are, with other matter, given, in hopes that they will come into general use, as the certainty and accuracy of their results can be more safely relied upon and more easily obtained than with common arithmetic.

By a slight examination, the student will perceive, in some degree, the nature and effect of these calculations; and, by frequent exercise, will obtain a dexterity of operation in every case admitting of their use. He will also more readily penetrate the plans of the different devices employed in instrumental calculations, which are rendered obscure and perplexing to most practical men by their ignorance of the proper application of logarithms.

Logarithms are artificial numbers which stand for natural numbers, and are so contrived, that if the logarithm of one number be added to the logarithm of another, the sum will be the logarithm of the product of these numbers; and if the logarithm of one number be taken from the logarithm of another, the remainder is the logarithm of the latter divided by the former; and also, if the logarithm of a number be multiplied by 2, 3, 4, or 5, &c., we shall have the logarithm of the square, cube, &c., of that number; and, on the other hand, if divided by 2, 3, 4, or 5, &c., we have the logarithm of the square root, cube root, fourth root, &c., of the proposed number; so that with the aid of logarithms, multiplication and division are performed by addition and subtraction; and the raising of powers and extracting of roots are effected by multiplying or dividing by the indices of the powers and roots.

In the table at the end of this work, are given the logarithms of the natural numbers, from 1· to 1000000 by the help of differences; in large tables, only the decimal part of the logarithm is given, as the index is readily determined; for the index of the logarithm of any number greater than unity, is equal to one less than the number of figures on the left hand of the decimal point; thus,

The index of 12345· is 4·,
———— 1234·5 — 3·,
———— 123·45 — 2·,
———— 12·345 — 1·,
———— 1·2345 — 0·

The index of any decimal fraction is a negative number equal to one and the number of zeros immediately following the decimal point; thus,

The index of ·00012345 is −4· or $\bar{4}$·
———— ·0012345 is −3· or $\bar{3}$·
———— ·012345 is −2· or $\bar{2}$·
———— ·12345 is −1· or $\bar{1}$·

Because the decimal part of the logarithm is always positive, it is better to place the negative sign of the index above, instead of before it; thus, $\bar{3}$· instead of −3. For the log. of ·00012345 is better expressed by $\bar{4}$·0914911, than by −4·0914911, because only the index is negative—*i. e.*, 4 is negative and ·0914911 is positive, and may stand thus, −4· + ·0914911.

Sometimes, instead of employing negative indices, their complements to 10 are used:

for $\bar{4}$·0914911 is substituted 6·0914911
— $\bar{3}$·0914911 ———— 7·0914911
— $\bar{2}$·0914911 ———— 8·0914911
&c. &c.

When this is done, it is necessary to allow, at some subsequent stage, for the tens by which the indices have thus been increased.

It is so easy to take logarithms and their corresponding numbers out of tables of logarithms, that we need not dwell on the method of doing so, but proceed to their application.

MULTIPLICATION BY LOGARITHMS.

Take the logarithms of the factors from the table, and add them together; then the natural number answering to the sum is the product required: observing, in the addition, that what is to be carried from the decimal parts of the logarithms is always positive, and must therefore be added to the positive indices; the difference between this sum and the sum of the negative indices is the index of the logarithm of the product, to which prefix the sign of the greater.

This method will be found more convenient to those who have only a slight knowledge of logarithms, than that of using the arithmetical complements of the negative indices.

1. Multiply 37·153 by 4·086, by logarithms.

| *Nos.* | *Logs.* |
|---|---|
| 37·153 | 1·5699939 |
| 4·086 | 0·6112984 |
| Prod. 151·8071 | 2·1812923 |

2. Multiply 112·246 by 13·958, by logarithms.

| *Nos.* | *Logs.* |
|---|---|
| 112·246 | 2·0501709 |
| 13·958 | 1·1448232 |
| Prod. 1566·729 | 3·1949941 |

3. Multiply 46·7512 by ·3275, by logarithms.

| *Nos.* | *Logs.* |
|---|---|
| 46·7512 | 1·6697928 |
| ·3275 | $\bar{1}$·5152113 |
| Prod. 15·31102 | 1·1850041 |

Here the +1 that is to be carried from the decimals, cancels the −1, and consequently there remains 1 in the upper line to be set down.

4. Multiply ·37816 by ·04782, by logarithms.

| *Nos.* | *Logs.* |
|---|---|
| ·37816 | $\bar{1}$·5776756 |
| ·04782 | $\bar{2}$·6796096 |
| Prod. 0·0180836 | $\bar{2}$·2572852 |

Here the +1 that is to be carried from the decimals, destroys the −1 in the upper line, as before, and there remains the −2 to be set down.

5. Multiply 3·768, 2·053, and ·007693, together.

| *Nos.* | *Logs.* |
|---|---|
| 3·768 | 0·5761109 |
| 2·053 | 0·3123889 |
| ·007693 | $\bar{3}$·8860957 |
| Prod. ·0595108 | $\bar{2}$·7745955 |

Here the +1 that is to be carried from the decimals, when added to −3, makes −2 to be set down.

6. Multiply 3·586, 2·1046, ·8372, and ·0294, together.

| *Nos.* | *Logs.* |
|---|---|
| 3·586 | 0·5546103 |
| 2·1046 | 0·3231696 |
| ·8372 | $\bar{1}$·9228292 |
| ·0294 | $\bar{2}$·4683473 |
| Prod. ·1857618 | $\bar{1}$·2689564 |

Here the +2 that is to be carried, cancels the −2, and there remains the −1 to be set down.

## DIVISION BY LOGARITHMS.

From the logarithm of the dividend, subtract the logarithm of the divisor; the natural number answering to the remainder will be the quotient required.

Observing, that if the index of the logarithm to be subtracted is positive, it is to be counted as negative, and if negative, to be considered as positive; and if one has to be carried from the decimals, it is always negative: so that the index of the logarithm of the quotient is equal to the sum of the index of the dividend, the index of the divisor with its sign changed, and −1 when 1 is to be carried from the decimal part of the logarithms.

1. Divide 4768·2 by 36·954, by logarithms.

| *Nos.* | *Logs.* |
|---|---|
| 4768·2 | 3·6783545 |
| 36·954 | 1·5676615 |
| Quot. 129·032 | 2·1106930 |

2. Divide 21·754 by 2·4678, by logarithms.

| *Nos.* | *Logs.* |
|---|---|
| 21·754 | 1·3375391 |
| 2·4678 | 0·3923100 |
| Quot. 8·81514 | 0·9452291 |

3. Divide 4·6257 by ·17608, by logarithms.

| *Nos.* | *Logs.* |
|---|---|
| 4·6257 | 0·6651775 |
| ·17608 | $\bar{1}$·2457100 |
| Quot. 26·27045 | 1·4194675 |

Here the −1 in the lower index, is changed into +1, which is then taken for the index of the result.

4. Divide ·27684 by 5·1576, by logarithms.

| *Nos.* | *Logs.* |
|---|---|
| ·27684 | $\bar{1}$·4422288 |
| 5·1576 | 0·7124477 |
| Quot. ·0536761 | $\bar{2}$·7297811 |

Here the 1 that is to be carried from the decimals, is taken as −1, and then added to −1 in the upper index, which gives −2 for the index of the result.

5. Divide 6·9875 by ·075789, by logarithms.

| *Nos.* | *Logs.* |
|---|---|
| 6·9875 | 0·8443218 |
| ·075789 | $\bar{2}$·8796062 |
| Quot. 92·1967 | 1·9647156 |

Here the 1 that is to be carried from the decimals, is added to −2, which makes −1, and this put down, with its sign changed, is +1.

6. Divide ·19876 by ·0012345, by logarithms.

| *Nos.* | *Logs.* |
|---|---|
| ·19876 | $\bar{1}$·2983290 |
| ·0012345 | $\bar{3}$·0914911 |
| Quot. 161·0043 | 2·2068379 |

Here −3 in the lower index, is changed into +3, and this added to 1, the other index, gives +3 − 1, or 2.

### PROPORTION; OR, THE RULE OF THREE, BY LOGARITHMS.

From the sum of the logarithms of the numbers to be multiplied together, take the sum of the logarithms of the divisors: the remainder is the logarithm of the term sought.

Or the same may be performed more conveniently, for any single proportion, thus:—Find the complement of the logarithm of the first term, or what it wants of 10, by beginning at the left hand and taking each of the figures from 9, except the last figure on the right, which must be taken from 10; then add this result and the logarithms of the other two figures together: the sum, abating 10 in the index, will be the logarithm of the fourth term.

1. Find a fourth proportional to 37·125, 14·768, and 135·279, by logarithms.

| | |
|---|---|
| Log. of 37·125 | 1·5696665 |
| Complement | 8·4303335 |
| Log. of 14·768 | 1·1693217 |
| Log. of 135·279 | 2·1312304 |
| Ans. 53·8128 | 1·7308856 |

2. Find a fourth proportional to ·05764, ·7186, and ·34721, by logarithms.

| | |
|---|---|
| Log. of ·05764 | $\bar{2}$·7607240 |
| Complement | 11·2392760 |
| Log. of ·7186 | $\bar{1}$·8564872 |
| Log. of ·34721 | $\bar{1}$·5405922 |
| Ans. 4·32868 | 0·6363554 |

3. Find a third proportional to 12·796 and 3·24718, by logarithms.

| | |
|---|---|
| Log. of 12·796 | 1·1070742 |
| Complement | 8·8929258 |
| Log. of 3·24718 | 0·5115064 |
| Log. of 3·24718 | 0·5115064 |
| Ans. ·8240216 | $\bar{1}$·9159386 |

### INVOLUTION; OR, THE RAISING OF POWERS, BY LOGARITHMS.

Multiply the logarithm of the given number by the index of the proposed power; then the natural number answering to the result will be the power required. Observing, if the index be negative, the index of the product will be negative; but as what is to be carried from the decimal part will be affirmative, therefore the difference is the index of the result.

1. Find the square of 2·7568, by logarithms.

| | |
|---|---|
| Log. of 2·7568 | 0·4404053 |
| | 2 |
| Square 7·599947 | 0·8808106 |

2. Find the cube of 7·0851, by logarithms.

Log. of 7·0851..........................0·8503460
3

Cube 355·6625..........................2·5510380

Therefore 355·6625 is the answer.

3. Find the fifth power of ·87451, by logarithms.

Log. of ·87451..........................$\bar{1}$·9417648
5

Fifth power ·5114695 ..................$\bar{1}$·7088240

Where 5 times the negative index $\bar{1}$, being −5, and +4 to carry, the index of the power is $\bar{1}$.

4. Find the 365th power of 1·0045, by logarithms.

Log. of 1·0045..........................0·0019499
365

97495
116994
58497

Power 5·148888..................Log. 0·7117135

EVOLUTION; OR, THE EXTRACTION OF ROOTS, BY LOGARITHMS.

Divide the logarithm of the given number by 2 for the square root, 3 for the cube root, &c., and the natural number answering to the result will be the root required.

But if it be a compound root, or one that consists both of a root and a power, multiply the logarithm of the given number by the numerator of the index, and divide the product by the denominator, for the logarithm of the root sought.

Observing, in either case, when the index of the logarithm is negative, and cannot be divided without a remainder, to increase it by such a number as will render it exactly divisible; and then carry the units borrowed, as so many tens, to the first figure of the decimal part, and divide the whole accordingly.

1. Find the square root of 27·465, by logarithms.

Log. of 27·465......................2 ) 1·4387796

Root 5·2407..............................·7193898

2. Find the cube root of 35·6415, by logarithms.

Log. of 35·6415.....................3 ) 1·5519560

Root 3·29093 ............................·5173186

3. Find the fifth root of 7·0825, by logarithms.

Log. of 7·0825......................5 ) 0·8501866

Root 1·479235............................·1700373

4. Find the 365th root of 1·045, by logarithms.

Log. of 1·045......................365 ) 0·0191163

Root 1·000121............................0·0000524

5. Find the value of $(\cdot001234)^{\frac{2}{3}}$, by logarithms.

Log. of ·001234..........................$\overline{3}$·0913152
2

3 ) $\overline{6}$·1826304

Ans. ·00115047..........................$\overline{2}$·0608768

Here the divisor 3 being contained exactly twice in the negative index −6, the index of the quotient, to be put down, will be −2.

6. Find the value of $(\cdot024554)^{\frac{3}{2}}$, by logarithms.

Log. of ·024554........................$\overline{2}$·3901223
3

2 ) $\overline{6}$·1703669

Ans. ·00384754..........................$\overline{3}$·5851834

Here, 2 not being contained exactly in −5, 1 is added to it, which gives −3 for the quotient; and the 1 that is borrowed being carried to the next figure makes 11, which, divided by 2, gives ·5851834 for the decimal part of the logarithm.

METHOD OF CALCULATING THE LOGARITHM OF ANY GIVEN NUMBER, AND THE NUMBER CORRESPONDING TO ANY GIVEN LOGARITHM. DISCOVERED BY OLIVER BYRNE, THE AUTHOR OF THE PRESENT WORK.

The succeeding numbers possess a particular property, which is worth being remembered.

log. 1·371288574238542 = 0·1371288574238542
log. 10·00000000000000 = 1·000000000000000
log. 237·5812087593221 = 2·375812087593221
log. 3550·260181586591 = 3·550260181586591
log. 46692·46832877758 = 4·669246832877758
log. 576045·6934135527 = 5·760456934135527
log. 6834720·776754357 = 6·834720776754357
log. 78974890·31398144 = 7·897489031398144
log. 895191599·8267852 = 8·951915998267839
log. 9999999999·999999 = 9·999999999999999

In these numbers, if the decimal points be changed, it is evident the logarithms corresponding can also be set down without any calculation whatever.

Thus, the log. of 137·1288574238542 = 2·1371288574238542;
the log. of 35·50260181586591 = 1·550260181586591;
log. ·002375812087593221 = $\overline{3}$·375812087593221;
log. ·0008951915998267852 = $\overline{4}$·951915998267852;

and so on in similar cases, since the change of the decimal point in a number can only affect the whole number of its logarithm.

These numbers whose logarithms are made up of the same digits will be found extremely useful hereafter. We shall next give a simple method of multiplying any number by any power of 11, 101, 1001, 10001, 100001, &c.

This multiplication is performed by the aid of coefficients of a binomial raised to the proposed power.

$(x + y)^1 = x + y$, the coefficients are 1, 1.
$(x + y)^2 = x^2 + 2xy + y^2$, the coefficients are 1, 2, 1.
$(x + y)^3 = x^3 + 3x^2y + 3xy^2 + y^3$, the coefficients are 1, 3, 3 1.

The coefficients of $(x + y)^4$ are 1, 4, 6, 4, 1.
— — $(x + y)^5$ — 1, 5, 10, 10, 5, 1.
— — $(x + y)^6$ — 1, 6, 15, 20, 15, 6, 1.
— — $(x + y)^7$ — 1, 7, 21, 35, 35, 21, 7, 1.
— — $(x + y)^8$ — 1, 8, 28, 56, 70, 56, 28, 8, 1.
— — $(x + y)^9$ — 1, 9, 36, 84, 126, 126, 84, 36, 9, 1.

Let it be required to multiply 54247 by $(101)^6$.

The number must be divided into periods of two figures when the multiplier is 101; into periods of three figures when the multiplier is 1001; into periods of four figures when the multiplier is 10001; and so on.

| *e* | *d* | *c* | *b* | *a* | | | |
|---|---|---|---|---|---|---|---|
| | 54 | 24 | 70 | 00 | 00 | | 1 |
| | 3 | 25 | 48 | 20 | 00 | *a* | 6 |
| | | 8 | 13 | 70 | 50 | *b* | 15 |
| | | | 10 | 84 | 94 | *c* | 20 |
| | | | | 8 | 14 | *d* | 15 |
| | | | | | 3 | *e* | 6 |

$(54247) \times (101)^6 = 57\ 58\ 42\ 83\ 61$, true to 10 places of figures.

This operation is readily understood, since the multipliers for the 6th power are 1, 6, 15, 20, 15, 6, 1; we begin at *a*, a period in advance, and multiply by 6; then we commence at *b*, two periods in advance, and multiply by 15; at *c*, three periods in advance, and multiply by 20; at *d*, four periods in advance (counting from the right to the left), and multiply by 15; the period, *e*, should be multiplied by 6, but, as it is blank, we only set down the 3 carried from multiplying *d*, or its first figure by 6.

As it is extremely easy to operate with 1, 5, 10, 10, 5, 1, the multipliers for the 5th power, it may be more convenient first to multiply the given number by $(101)^5$, and then by $(101)^1$; because, to multiply any number by 5, we have only to affix a cipher (or suppose it affixed) and to take the half of the result.

The above example, if worked in the manner just described, will stand as follows:

| | *d* | *c* | *b* | *a* | | |
|---|---|---|---|---|---|---|
| | 54 | 24 | 70 | 00 | 00 | .....1 |
| | 2 | 71 | 23 | 50 | 00 | .....5..*a* |
| | | 5 | 42 | 47 | 00 | ...10..*b* |
| | | | 5 | 42 | 47 | ...10..*c* |
| | | | | 2 | 71 | .....5..*d* |
| | | | | | 1 | .....1 |
| $(54247) \times (101)^5 =$ | 57 | 01 | 41 | 42 | 19 | |
| | | 57 | 01 | 41 | 42 | |

57 58 42 83 61 $= (54247)^6 \times (101)^6$.

The truth of this is readily shown by common multiplication, but the process is cumbersome. However, for the sake of comparison, we shall in this instance multiply 54247 by (101) raised to the 6th power.

```
          101
          101
         ----
          101
         1010
        -----
        10201 = (101)².
          101
       ------
        10201
       102010
      -------
      1030301 = (101)³.
          101
     --------
      1030301
     10303010
    ---------
    104060301 = (101)⁴.
          101
   ----------
    104060401
   1040604010
  -----------
  10510100501 = (101)⁵.
          101
 ------------
  10510100501
 105101005010
 ------------
 1061520150601 = (101)⁶.
         54247
 -------------
 7430641054207
4246080602404
2123040301202
4246080602404
5307600753005
---------------------
5758428360|9652447 the required product,
```

which shows that the former process gives the result true to 10 places of figures, of which we shall add another example.

Multiply 34567812 by $(1001)^8$, so that the result may be true to 12 places of figures.

| c | b | a | | |
|---|---|---|---|---|
| | 3456 | 7812 | 0000 | .....1 |
| | 2 | 7654 | 2496 | .....8..*a* |
| | | 9 | 6790 | ...28..*b* |
| | | | 19 | ...56..*c* |

3459 5475 9305 the required product.

The remaining multipliers, 70, 56, 28, 8, 1, are not necessary in obtaining the first 12 figures of the product of 34567812 by 10001 in the 8th power.

As 28 and 56 are large multipliers, the work may stand thus

| c | b | a | | | |
|---|---|---|---|---|---|
| | 3456 | 7812 | 0000 | ...... 1 | |
| | 2 | 7654 | 2496 | ...*a*.. 8 | |
| | | 6 | 9136 | ...*b*..20 | 28 |
| | | 2 | 7654 | ...*b*.. 8 | |
| | | | 17 | ...*c*..50 | 56 |
| | | | 2 | ...*c*.. 6 | |

Result, = 345954759305 the same as before.

Perhaps this product might be obtained with greater ease by first multiplying 34567812 by $(10001)^5$, and the product by $(10001)^3$; the operation will stand thus:

```
345678120000...... 1
   172839060...... 5
       34568......10
           3......10
------------------
345850093631 = 34567812 × (10001)^5.
   103755298...... 3
       10376...... 3
------------------
```

345954759305 = twelve places of the product of 34567812 by $(10001)^5 \times (10001)^3 = (34567812) \times (10001)^8$.

Although these methods are extremely simple, yet cases will occur, when one of them will have the preference.

Our next object is to determine the logarithms 1·1; 1·01; 1·001; 1·0001; 1·00001; &c.

It is well known that

$$\log. (1 + n) = M \left(n - \tfrac{1}{2}n^2 + \tfrac{1}{3}n^3 - \tfrac{1}{4}n^4 + \tfrac{1}{5}n^5 - \tfrac{1}{6}n^6 + \&c.\right)$$

M being the modulus, = ·43294481903261827651l289, &c.

It is evident that when $n$ is $\frac{1}{10}$, $\frac{1}{100}$, $\frac{1}{1000}$, $\frac{1}{10000}$, &c., the calculation becomes very simple.

$$
\begin{aligned}
M &= \cdot 4342944819032518 \\
\tfrac{1}{2}M &= \cdot 2171472409516259 \\
\tfrac{1}{3}M &= \cdot 1447648273010839 \\
\tfrac{1}{4}M &= \cdot 1085736204758130 \\
\tfrac{1}{5}M &= \cdot 0868588963806504 \\
\tfrac{1}{6}M &= \cdot 0723824136505420 \\
\tfrac{1}{7}M &= \cdot 0720420788433217 \\
\tfrac{1}{8}M &= \cdot 0542868102379065 \\
\tfrac{1}{9}M &= \cdot 0482549424336946 \\
\tfrac{1}{10}M &= \cdot 0434294481903252
\end{aligned}
$$

&c. &c., are constants employed to determine the logarithms of 11, 101, 1001, 100001, &c.

To compute the log. of 1·001. In this case $n = \frac{1}{1000}$.

$$+ \quad \frac{M}{1000} = \cdot 0004342944819033 \text{ positive}$$

$$- \frac{\frac{1}{2}M}{(1000)^2} = \cdot 0000002171472410 \text{ negative}$$

$$\cdot 0004340773346623$$

$$+ \frac{\frac{1}{3}M}{(1000)^3} = \cdot 0000000001447648 \text{ positive}$$

$$\cdot 0004340774794271$$

$$- \frac{\frac{1}{4}M}{(1000)^4} = \cdot 0000000000001086 \text{ negative}$$

$$\cdot 0004340774793185$$

$$+ \frac{\frac{1}{5}M}{(1000)^5} = \cdot 0000000000000001 \text{ positive}$$

$$\cdot 0004340774793186 = \text{the log. of } 1\cdot 001;$$

true to sixteen places.

It is almost unnecessary to remark, that, instead of adding and subtracting alternately, as above, the positive and negative terms may be summed separately, which will render the operation more concise.

| *Positive Terms.* | *Negative Terms.* |
|---|---|
| ·0004342944819033 | ·0000002171472410 |
| 1447648 | 1086 |
| 1 | ·0000002171473496 |
| + ·0004342945266682 | |
| − 000000217473496 | |
| ·0004340774793186 = log. 1·001. | |

In a similar manner the succeeding logarithms may be obtained to almost any degree of accuracy.

| | | | | |
|---|---|---|---|---|
| Log. 1·1 | = ·041392685158225 | &c. which | we call | A |
| 1·01 | = ·004321373782643 | — | — | B |
| 1·001 | = ·000434077479319 | — | — | C |
| 1·0001 | = ·000043427276863 | — | — | D |
| 1·00001 | = ·000004342923104 | — | — | E |
| 1·000001 | = ·000000434294265 | — | — | F |
| 1·0000001 | = ·000000043429447 | — | — | G |
| 1·00000001 | = ·000000004342945 | — | — | H |
| 1·000000001 | = ·000000000434295 | — | — | I |
| 1·0000000001 | = ·000000000043430 | — | — | J |
| 1·00000000001 | = ·000000000004343 | — | — | K |
| 1·000000000001 | = ·000000000000434 | — | — | L |
| 1·0000000000001 | = ·000000000000043 | — | — | M |
| 1·00000000000001 | = ·000000000000004 | — | — | N |
| &c. | &c. | | | &c. |

Without further formality or paraphernalia, for it is presumed that such is not necessary, we shall commence operating, as the method can be acquired with ease, and put in a clearer point of view by proper examples.

Required the logarithm of 542470, to seven places of decimals.

```
          5 4|2 4|7 0|. .|
            3|2 5|4 8|2 0|
                8|1 3|7 1|
                 |1 0|8 5|
                       8|
          ---------------
          5 7 5 8|4 2 8 4 = 6 B = ·02592824
                1|7 2 7 5
                        3
          ---------------
     Take 5 7 6 0 1 5 6 2 = 3 D = ·00013028.
     From 5 7 6 0 4 5 6 9
          ---------------
     576) · · · · 3|0 0 7
                  2|8 8 0 = 5 E = ·00002171
                  -------
                    1|2 7
                    1|1 5 = 2 F = ·00000087
                    -----
                      1|2
                      1|2 = 2 G = ·00000009
                                -----------
                                ·02608119 Take
                               5·76045693 From
                               ----------
```

Hence we have log. 542470 = 5·73437574, which is correct to seven decimal places.

6 B is written to represent 6 times the log. of 1·01.

The nearest number to 542470, whose log. is composed of the same digits as itself, being 576045·6934, &c., our object was to raise 542470· to 576045·69 by multiplying 542470· by some power or powers of 1·1, 1·01, 1·001, 1·0001, &c.

It is here necessary to remark, that A is not employed, because the given number multiplied by 1·1, would exceed 576045·69; for a like reason C is omitted.

Again, when half the figures coincide, the process may be performed (as above) by common division; the part which coincides becoming the divisor; thus, in finding 5 E, 576 is divided into 3007, it goes 5 times, the E showing that there are five figures in each period at this step. For A, there is but one figure in each period; for B, there are two figures; for C, there are three figures in each period, and so on.

Let it be required to calculate the logarithm of 2785·9, true to seven places of decimals.

It will be found more convenient, in this instance, to bring the given number to 3550·26018, the log. of which is 3·55026908.

```
      2|7|8|5|9|0|0|0
       |5|5|7|1|8|0|0
       |  |2|7|8|5|9|0
      ---------------
      3 3|7 0|9 3|9 0 = 2 A = ·08278537
        1|6 8|5 4|7 0
             3|3 7|0 9
                 3|3 7
                   | 2
      ---------------
      3 5 4|2 8 9|0 8 = 5 B = ·02160687
           |7 0 8|5 8
                 |3 5
      ---------------
Take  3 5 4 9|9 8 0 1 = 2 C = ·00086815
From  3 5 5 0|2 6 0 2
      ---------------
355) · · · · 2|8 0 1 = 7 E = ·00003040
             2|4 8 5
             -------
              3|1 6 = 8 F = ·00000347
              2|8 4
              -----
               3|2 = 9 G = ·00000039
               3|2
                       Take  ·10529465
                       From 3·55026018
                       ---------------
           log. 2785·9 = 3·44496553
```

At the Observatory at Paris, $g = 9·80896$ metres, the second being the unit of time, what is the logarithm of 9·80896?

In this example, we shall bring 9·80896 to 9·99999, &c.

```
     98|08|96|00|00
       |98|08|96|00
     ---------------
     990|704|960|0 = 1 B = ·0043213738
        8|9163|44|6
            35|665|4
                83|2
     ---------------
     9996|5705|32 = 9 C = ·0039066973
          2|9989|72
                3|00
     ---------------
     99995|69804 = 3 D = ·0001302818
            3|99983
                  6
     ---------------
Take  9999969793 = 4 E = ·0000173717
From 10000000000
     ---------------
     · · · · · 30207
From which we have.........3 F = ·0000013029
                           2 H = ·0000000087
                           7 J = ·0000000003
                                 -----------
                          Take  ·0083770365
                          From 1·0000000000
                                -----------
               Log. 9·80896 = ·9916229635
```

As before observed, 9 C might have been obtained in the following manner:

```
         890|704|960|0 = 1 B, as above.
           4|953|524|8
               9|907|0
                    9|9
         ---------------
5 times  995|668|4017
           3|982|673|6
             5|973|9
                  4|0
         ---------------
4 times  9996570532 = 9 C.
```

A French metre is equal to 3·2808992 English feet, required the log. of 3·2808992.

```
e| d| c| b| a|
 |32|80|89|92|00...once
 | 2|29|66|29|44... 7 times from a
    | 6|88|98|88...21    —    b
       |11|48|31...35    —    c
          |11|48...35    —    d
             |  7...21   —    e
 ----------------
 35 17 56 80 18 = B 7.
```

The manner in which B 7 is obtained is worthy of remark: the multipliers being 1, 7, 21, 35, 35, 21, 7, 1, when 7 times the first line (commencing with the period marked *a*) is obtained, 21 times the same line (commencing with the period marked *b*) is determined by multiplying the 2d line by 3. If the 2d line be again multiplied by 5, we have the 4th line of the multiplier 35; but to multiply by 5, we have only to take the half the product produced by multiplying by 7, advancing the result one figure to the right. Hence, to find the result for 35 is almost as easy as to find the result for 5.

But the object in this case being to bring the proposed number to 35502601815, the process must be continued.

| | *c* | *b* | *a* | | |
|---|---|---|---|---|---|
| 1 | 351 | 756 | 801 | 8 | = B 7, as above. |
| 9 | 3 | 165 | 811 | 2 | |
| 36 | | 12 | 663 | 2 | |
| 84 | | | 29 | 6 | |
| | 354 | 935 | 305 | 8 | = C 9 |

The 2d (or 9) line is produced by beginning at *a*, but the multiplication may be performed by subtracting 3517568 from 35175680; the 36 line is produced by beginning at *b*, observing to carry from the preceding figure, making the usual allowance when the number is followed by 5, 6, 7, 8, or 9. The 36 line may be produced by multiplying the 9 line by 4, beginning one period more to the left. To multiply by 84 is not apparently so convenient, for 84 × 352 = 29|568; and as only one figure of the period 568 is required, when the proper allowance is made, the result becomes 29|6.

But, since 84 is equal to 36 × $2\frac{1}{3}$, we have only to multiply the 36 line by 2, and add $\frac{1}{3}$ of it; with such management, the work will stand thus:—

| | | | | | |
|---|---|---|---|---|---|
| 351 | 756 | 801 | 8 | = B 7, as before | |
| 3 | 165 | 811 | 2 | = 9 times | |
| | 12 | 663 | 2 | = 36 times | |
| | | 24 | 3 | = 72 times | = 84 times |
| | | 4 | 2 | = 12 times | |
| 354 | 935 | 305 | 8 | = C 9 | |

This amounts to very little more than adding the above numbers together.

Many other contractions will suggest themselves, when the mulpliers are large: thus, to multiply any number 57837 by 9, as alluded to above, is easily effected, by the following well-known process:—Subtract the first figure to the right from 10, the second from the first, the third from the second, and so on.

Thus, 57837 × 9 = {
578370...ten times
57837...once
520533...nine times

Such simple observations are to be found in every book on mental arithmetic, and therefore require but little attention here.

The whole work of the previous example will stand thus:—

```
               3 2|8 0|8 9|9 2|0 0
                 2|2 9|6 6|2 9|4 4
                    6|8 8|9 8|8 8
                      1 1|4 8|3 1
                         1 1|4 8 + 7
               ---------------------
B 7     = 3 5 1|7 5 6|8 0 1|8 = ·0302496165 = 7 B
              3|1 6 5|8 1 1|2
                  1 2|6 6 3|2
                        2 9|6
               ---------------------
C 9     = 3 5 4 9|3 5 3 0|5 8 = ·0039066973 = 9 C
                 |7 0 9 8|7 1
                         |3 5
               ---------------------
D 2     = 3 5 5 0 0|6 2 9 6 4 = ·0000868546 = 2 D
                  1|7 7 5 0 3
                            4
               ---------------------
Take E 5 = 3 5 5 0 2 4 0 4 7 1 = ·0000217146 = 5 E
From       3 5 5 0 2 6 0 1 8 2
               ---------------------
 3550). . . . . 1|9 7 1 1
F 5             1|7 7 5 0 = ·0000021715 = 5 F
                 ---------
                   1|9 6 1
G 5                1|7 7 5 = ·0000002172 = 5 G
                   -------
                     1|8 6
H 5                  1|7 8 = ·0000000217 = 5 H
                       ---
                        |8
I 2                     |7 = ·0000000009 = I 2
                        -
                       1|
J 3                    1| = ·0000000001 = J 3
                            -------------
                      Take  ·0342672944
                      From 3·5502601816
                           -------------
          Log. 3280·8992 = 3·5159928972
        ∴ log. 3·2808992 = 0·5159928972.
```

The constant sidereal year consists of 365·25636516 days; what is the log. of this number?

In this case it is better to bring *the constant* 35502601816 to 36525636516, instead of bringing the given number to the constant, as in the former examples.

```
              3 5 5 0 2 6 0 1 8 1 6
                7 1 0 0 5 2 0 3 6
                    3 5 5 0 2 6 0
B 2 =         3 6 2 1 6 2 0 4 1 1 2 = ·0086427476 = 2 B
                2 8 9 7 2 9 6 3 3
                    1 0 1 4 0 5 4
                          2 0 2 8
C 8 =         3 6 5 0 6 9 4 9 8 2 7 = ·0034726298 = 8 C
                  1 8 2 5 3 4 7 5
                          3 6 5 1
Take D 5 =    3 6 5 2 5 2 0 6 9 5 3 = ·0002171364 = 5 D
From          3 6 5 2 5 6 3 6 5 1 6
36525·2)              4 2 9 5 6 3
E 1 =                 3 6 5 2 5 2 = ·0000043429 = 1 E
                        6 4 3 1 1
F 1 =                   3 6 5 2 5 = ·0000004343 = 1 F
                        2 7 7 8 6
G 7 =                   2 5 5 6 8 = ·0000003040 = 7 G
                          2 2 1 8
H 6 =                     2 1 9 1 = ·0000000261 = 6 H
I 0                           2 7
J 7 =                         2 5 = ·0000000003 = 7 J
                                    ·0123376214
                                Add 3·5502601816
Hence, log. 3652·5636516 = 3·5625978030
  ∴    log. 365·25636516 = 2·562597803.
```

M. Regnault determined with the greatest care the density of mercury to be 13·59593 at the temperature 0°, centigrade. It is required to calculate the log. of 13·59593, to eight places of decimals.

In this case it is better to bring the given number to *the constant* 1371288574.

```
              1 3 5 9 5 9 3 0 0
                  1 0 8 7 6 7 4
                        3 8 0 7
                              8
C 8 =         1 3 7 0 5 0 7 8 8 = ·003472630 = 8 C
                      6 8 5 2 5
                            1 4
Subtract D 5 = 1 3 7 1 1 9 3 2 8 = ·000217136 = 5 D
From          1 3 7 1 2 8 8 5 7
                        9 5 2 9 = ·000026058 = E 6
E 6 =                   8 2 2 7
                        1 3 0 2
F 9 =                   1 2 3 4 = ·000003909 = F 9
                            6 8
H 5 =                       6 9 = ·000000022 = H 5
                                  ·003719755
```

Take ·003719755
From ·137128857

log. 1·359593 = ·133409102
∴ log. 13·59593 = 1·133409102.

TO DETERMINE THE NUMBER CORRESPONDING TO A GIVEN LOGARITHM.

This problem has been very much neglected—so much so, that none of our elementary books ever allude to a method of computing the number answering to a given logarithm. When an operation is performed by the use of logarithms, it is very seldom that the resulting logarithm can be found in the table; we have, therefore, to find the nearest less logarithm, and the next greater, and correct them by proportion, so that there may be found an intermediate number that will agree with the given logarithm, or nearly so. But although the *proportional parts of the difference* abridge this process, we can only find a number appertaining to any logarithm to seven places of figures when using our best modern tables. As, however, the tabular logarithms extend only to a degree of approximation, fixed generally at seven decimal places, all of which, except those answering to the number 10 and its powers, err, either in excess or defect, the maximum limit of which is $\frac{1}{2}$ in the last decimal, and since both errors may conspire, the 7th figure cannot be depended on as strictly true, unless the proposed logarithm falls between the limits of log. 10000 and log. 22200.

Indubitably we are now speaking of extreme cases, but since it is not an unfrequent occurrence that some calculations require the most rigid accuracy, and many resulting logarithms may be extended beyond the limits of the table, this subject ought to have a place in a work like the present. It is not part of the present design to enter into a strict or formal demonstration of the following mode of finding the number corresponding to a given logarithm, as the operation will be fully explained by suitable examples.

What number corresponds to the logarithm 3·44496555?

The next less constant log. to the one proposed is 2·37581209, or rather, 3·37581209, when the characteristic or index is increased by a unit.

First from 3·44496555
take 3·37581209

·06915346
·04139269 = 1 A

·02776077
·02592824 = 6 B

. . 183253
173631 = 4 C

. . . . 9622
8685 = 2 D

. . . . . 937

*Secondly.*

2|3 7 5 8 1 2 0 9 constant
|2 3 7 5 8 1 2 1 = A 1

2 6 1 3 3 9 3 3|0
1|5 6|8 0 3 6|0
3|9 2 0 0|9
5|2 2|7
3|9

2 7 7|4 1 6|9 6 5 = B 6
1|1 0 9|6 6 8
1|6 6 4
1

2 7 8 5|2 8 2 9|8 = C 4

```
.....937              2 7 8 5 2 8 2 9|8 = C 4
     869 = 2 E                 |5 5 7 0|6
     ---                               |3
......68               ------------------
      43 = 1 F        2 7 8 5 8|4|0|0|7 = D 2
      --                       |5|5|7|2 = E 2
......25                         |2|7|9 = F 1
      22 = 5 G                   1|3|9 = G 5
      --                            1|9 = H 7
.......3              -----------------
       3 = 7 H        2 7 8 5 9 0 0 1 6
```

∴ 2785·90016 is the number sought.

What number corresponds to the logarithm 5·73437574?

When the index of this log. is reduced by a unit, the nearest next less constant is 4·66924683.

```
From 4·73437574
Take 4·66924683
     ----------
      ·6512891
      4139269.........1 A
      -------
      ·2373622
      2160687.........5 B
      -------
      ··212035
       173631.........4 C
       ------
      ...39304
        39085.........9 D
        -----
      .....219  There is neither the equal of
           217.........5 F  this number, nor a
           ---
      .......2.........0 G  less, obtainable from
             2.........4 H  E, ∴ E 0, or E, is
             -              omitted.
```

```
Then, 4|6 6 9 2 4 6 8 3
       |4 6 6 9 2 4 6 8.........A 1
      ------------------
      5 1|3 6|1 7|1 5|1
        2|5 6|8 0|8 5|8
           5|1 3|6 1|7
              5|1 3|6
                   2|6
      ------------------
      5 3 9|8 1 6|7 8 8.........B 5
          2|1 5 9|2 6 7
              3|2 3 9
                    2
      ------------------
      5 4 1 9|7 9 2 9|6.........C 4
            4|8 7 7 8|1
                  1 9|5
      ------------------
      5 4 2 4 6 7 2|7|2.........D 9
                  |2 7|1|2.........F 5
                      |2|2.........H 4
      ------------------
      5 4 2 4 7 0 0 0 6
```

∴ 542470·006 is the number whose logarithm is 5·73437574.

Had the given logarithm represented a decimal with a positive index, the required number would be 0·000054247, &c.; or if written with a negative index, as $\bar{5}$·73437574, the result would be the same, for the characteristic $\bar{5}$, shows how many places the first significant figure is below unity.

Required the number corresponding to log. 2·3727451.

The constant 100000000 is the one to be employed in this case.

```
1·3727451 the given log. minus 1 in the index.
1·0000000
---------
 ·3727451
 3725342.........9 A
 -------
 ...2109
    1737.........4 D
    ----
 ....372
     347.........8 E
     ---
 .....25
      22.........5 F
      --
       3
       3.........7 G
       -
1 0 0 0 0 0 0 0 Constant.
  9 0 0 0 0 0 0
  3 6 0 0 0 0 0
    8 4 0 0 0 0
    1 2 6 0 0 0
      1 2 6 0 0
          8 4 0
            3 6
              9
---------------
  2 3 5 7 9 4 8 5  A 9
          9 4 3 2
                1
---------------
  2 3 5 8 8 9 1 8  D 4
          1 8 9 7  E 8
            1 1 8  F 5
              1 6  G 7
---------------
  2 3 5 9 0 9 4 9
```

∴ 235·90949 is the required number, and the seconds in the diurnal apparent motion of the stars.

$$235{\cdot}90949'' = 3'\ 55{\cdot}90949''.$$

Let it be required to find the *hyperbolic* logarithm of any number, as 3·1415926536. The common log. of this number is ·49714987269 (33), and the common log. of this log. is $\bar{1}$·6964873.

The modulus of the common system of logarithms is ·4342944819, &c.

∴ 1 : 4342944819 : : hyperbolic log. N : common log. N.

To distinguish the hyperbolic logarithm of the number N from its common logarithm, it is necessary to write the hyp. log. Log. N, and the common logarithm log. N.

Hence, 4342944819 × Log. N = log. N;
or log. (·4342944819) + log. (log. N) = log. (log. N).

∴ log. (Log. N) = log. (log. N) − $\bar{1}$·6377843; for $\bar{1}$·6377843 = log. ·4342944819.

Now, to work the above example, from $\bar{1}$·6964873
take $\bar{1}$·6377843
·0587030, the number corresponding to this *com. log.* will be the *hyp. log.* of 3·1415927. ·0587030 must be reduced to ·0000000 which is known to be the log. of 1.

| | |
|---|---|
| ·0587030 | |
| 0413927 | 1 A |
| .173103 | |
| 172855 | 4 B |
| ....248 | |
| 217 | 5 E |
| .....31 | |
| 30 | 7 F |
| ......$\bar{1}$ | 2 G |

| | |
|---|---|
| 1 A = 1 1 0 0 0 0 0 0 0 | |
| 4 4 0 0 0 0 0 | |
| 6 6 0 0 0 | |
| 4 4 0 | |
| 1 | |
| 1 1 4 4 6 6 4 4 1 | = B 4 |
| 5 7 2 3 | = E 5 |
| 8 0 1 | = F 7 |
| 2 3 | = G 2 |
| 1 1 4 4 7 2 9 8 8 | |

∴ 1·14472988 is the hyperbolic log. of 3·1415927, true to the last figure; for the hyp. log. 3·1415926535898 = 1·1447298858494.

The reason of this operation is very clear, because
$1 \times 1{\cdot}1 \times (1{\cdot}01)^4 \times (1{\cdot}00001)^5 \times (1{\cdot}000001)^7 \times (1{\cdot}0000001)^2 =$ 1·14472988.

This example answers the purpose of illustration, but the hyp. log. of 3·1415927 can be more readily found by dividing its com. log. ·49714987269 by the constant ·4342944819, which is termed the modulus of the common system of logarithms.

Suppose it is known that $\bar{1}$·3426139 is the log. of the decimal which a *French litre* is of an English gallon. Required the decimal.

The index, $\bar{1}$, may be changed to any other characteristic, so as to suit any of *the constants*, as the alteration is easily allowed for when the work is completed. In this instance, it is best to put +1 instead of $\bar{1}$.

| | |
|---|---|
| From 1·3426139 | |
| Take 1·0000000 | |
| ·3426139 | |
| 3311415 | = 8 A |
| ·0114724 | |
| ..86427 | = 2 B |
| 28297 | |
| 26045 | = 6 C |
| 2252 | |

| | |
|---|---|
| 1 0 0 0 0 0 0 0 0 | Constant |
| 8 0 0 0 0 0 0 0 | |
| 2 8 0 0 0 0 0 0 | |
| 5 6 0 0 0 0 0 | |
| 7 0 0 0 0 0 | |
| 5 6 0 0 0 | |
| 2 8 0 0 | |
| 8 0 | |
| 1 | |
| 2 1 4 3 5 8 8 8 1 | = A 8 |

```
2252
2171 = 5 D
  81
  43 = 1 E
  38
  35 = 8 F
   3
   3 = 7 G
```

```
2 1 4 3 5 8 8 8 1 = A 8
    4 2 8 7 1 7 8
        2 1 4 3 6
2 1 8 6 6 7 4 9 5 = B 2
      1 3 1 2 0 0 5
          3 2 8 0
                4
2 1 9 9 8 2 7 8 4 = C 6
      1 0 9 9 9 1
              2 2
2 2 0 0 9 2 7 9 7 = D 5
          2 2 0 1 = E 1
          1 7 6 1 = F 8
            7 5 4 = G 7
2 2 0 0 9 6 9 1 3
```

∴ The French litre = ·2200969 English gallons.

In measuring heights by the barometer, it is necessary to know the ratio of the density of the mercury to that of the air.

At Paris, a *litre* of air at 0° centigrade, under a pressure of 760 millimetres, weighs 1·293187 grammes. At the level of the sea, in latitude 45°, it weighs 1·292697 grammes. A *litre* of water, at its maximum density, weighs 1000 grammes, and a *litre* of mercury, at the temperature of 0° cent., weighs 13595·93 grammes:

$$\therefore \frac{13595{\cdot}93}{1{\cdot}292697} = \text{the ratio at } 45°$$

Now, log. 13595·93 = 4·133409102 (29)
and log. 1·292697 = 0·111496744 (30)
4·021912358 = the log. of the ratio at 45°.

To find the number corresponding to this log., it is necessary to reject the index for the present, and reduce the decimal part to zero. By this means the necessity of using any of the constants is superseded.

```
·021912358
·021606869 = 5 B
...305489
   303991 = 7 D
.....1498
     1303 = 3 F
......195
      174 = 4 G
.......21
       17 = 4 H
        4
        4 = 9 I
```

```
1 0 0 0 0 0 0 0 0 0
  5 0 0 0 0 0 0 0 0
      1 0 0 0 0 0 0
          1 0 0 0 0
                5 0
1 0 5 1 0 1 0 0 5   = B 5
      7 3 5 7 1
            2 2
1 0 5 1 7 4 5 9 8   = D 7
          3 1 6     = F 3
            4 2     = G 4
              4     = H 4
              1     = I 9
1 0 5 1 7 4 9 6 1
```

$\therefore$ by logarithms, $\frac{13595{\cdot}93}{1{\cdot}292697}$ = 10517·49, &c., which is easily verified by common division.

M. Regnault found that, at Paris, the litre of atmospheric air weighs 1·293187 grammes; the litre of nitrogen 1·256167 grammes; a litre of oxygen, 1·429802 grammes; of hydrogen, 0·089578 grammes; and of carbonic acid, 1·977414 grammes. But, strictly considered, these numbers are only correct for the locality in which the experiments were made; that is for the latitude of 48° 50′ 14″ and a height about 60 metres above the level of the sea; M. Regnault finds the weight of the litre of air under the parallel of 45° latitude, and at the same distance from the centre of the earth as that which the experiments were tried, to be 12·926697.

Assuming this as the standard, he deduces for any other latitude, any other distance from the centre of the earth, the formula,

$$w = \frac{1{\cdot}292697\ (1{\cdot}00001885)\ (1 - 0{\cdot}002837)\ \cos.\ 2\,\lambda}{1 + \frac{2\,h}{\mathrm{R}}}$$

Here, $w$ is the weight of the litre of air, R the mean radius of the earth = 6366198 metres, $h$ the height of the place of observation above the mean radius, and $\lambda$ the latitude of the place.

At Philadelphia, lat. 39° 56′ 51·5″, suppose the radius of the earth to be 6367653 metres, the weight of the litre of air will be 1·2914892 grammes. The ratio of the density of mercury to that of air at the level of the sea at Philadelphia is 10527·735 to 1; required the number of degrees in an arc whose length is equal to that of the radius.

$$\text{As } 3{\cdot}1415926535898 : 1 :: \frac{360}{2} : \text{the required degrees.}$$

| | |
|---:|---:|
| Log. 360 = | 2·556302500767 |
| log. 3·14159265359 = | 0·497149872694 |
| | 2·059452623073 |
| log. 2 = | 0·301029995664 |
| | 1·758122632409 = the log. of the |

number required.

When the index of this log. is changed into 4, the nearest next less constant is 4·669246832878.

| | | | |
|---|---:|---:|---|
| From | 4·758122632409 | 4 6 6 9 2 4 6 8 3 2 8 7 8 | = Constant |
| Take | 4·669246832878 | 9 3 3 8 4 9 3 6 6 5 7 6 | |
| | ·088875799531 | 4 6 6 9 2 4 6 8 3 2 9 | |
| 2 A = | ·82785370316 | 5 6 4 9 7 8 8 6 6 7 7 8 3 | = A 2 |
| | . . 6090429215 | 5 6 4 9 7 8 8 6 6 7 8 | |
| 1 B = | 4321373783 | 5 7 0 6 2 8 6 5 5 4 4 6 1 | = B 1 |
| | . . 1769055432 | 2 2 8 2 5 1 4 6 2 1 8 | |
| 4 C = | 1736309917 | 3 4 2 3 7 7 1 9 | |
| | . . . . 32745515 | 2 2 8 2 5 | |
| 7 E = | 30400462 | 6 | |
| | . . . . . 2345053 | 5 7 2 9 1 4 5 9 6 1 2 2 9 | = C 4 |

| | | | |
|---|---|---|---|
| | .....2345053 | 5 7 2 9 1 4 5 9 6 1 2 2 9 | = C 4 |
| 5 F = | 2171471 | 4 0 1 0 4 0 2 1 7 | |
| | ......173582 | 1 2 0 3 1 | |
| 3 G = | 130288 | 5 7 2 9 5 4 7 0 1 3 4 7 7 | = E 7 |
| | .......43294 | 2 8 6 4 7 7 3 5 | |
| 9 H = | 39087 | 5 7 | |
| | ........4207 | 5 7 2 9 5 7 5 6 6 1 2 6 9 | = F 5 |
| 9 I = | 3909 | 1 7 1 8 8 7 3 | = G 3 |
| | .........298 | 5 1 5 6 6 2 | = H 9 |
| 6 J = | 261 | 5 1 5 6 6 | = I 9 |
| | ..........37 | 3 4 3 8 | = J 6 |
| 8 K = | 35 | 4 5 8 | = K 8 |
| | ...........2 | 2 9 | = L 5 |
| 5 L = | 2 | 5 7 2 9 5 7 7 9 5 1 2 9 5 | = the number required. |

But the original index is 1; ∴ 57·29577951295° are the number of degrees in an arc the length of which is equal to that of the radius.

The above result may be easily verified by common division, a method, no doubt, which would be preferred by many, for logarithms are seldom used when the ordinary rules of arithmetic can be applied with any reasonable facility. However, this example, like many others, is introduced to show with what ease and correctness the number corresponding to a given log. can be obtained. The extent, also, by far exceeds that obtainable by any tables extant.

Other computations give,

$$r^{\circ} = 57{\cdot}2957795130^{\circ} = 57^{\circ}\ 17'\ 44''\ {\cdot}80624$$

the degrees in an arc = radius.

$$r' = 3437{\cdot}7467707849' = 3437'\ 44''\ {\cdot}80624$$

the minutes in an arc = radius.

$$r'' = 206264{\cdot}8062470963$$

the number of seconds in an arc = radius.

The relative mean motion of the moon from the sun in a Julian or fictitious year, of $365\frac{1}{4}$ days, is 12 cir. 4 signs, 12° 40′ 15·977315′ = 16029615·977315″.

∴ 16029615·977315″ : 1 circumference (= 129600″)
: : 365·25 days
: 29·5305889216 days = the mean synodic month.

This proportion may, for the sake of example, be found by logarithms.

| | |
|---|---|
| Log. 365·25......... | 2·56259022460634 |
| log. 1296000......... | 6·11260500153457 |
| | 8·67519522614091 |
| log. 16029615·977315 = | 7·20492311805406 |
| | 1·47027210808685 |

If the index of this log. be made 2 instead of 1, the nearest next less constant will be 2·375812087593221.

| | |
|---|---|
| From | 2·47027210808685 |
| Take | 2·37581208759322 |
| | ·09446002049363 |
| 2 A = | 08278537031645 |
| | .1167465017718 |
| 2 B = | 864274756529 |
| | ..303190261189 |
| 6 C = | 260446487591 |
| | ...42743773598 |
| 9 D = | 39084549177 |
| | ....3659224421 |
| 8 E = | 3474338483 |
| | .....184885938 |
| 4 F = | 173717706 |
| | ......11168232 |
| 2 G = | 8685889 |
| | .......2482343 |
| 5 H = | 2171473 |
| | ........310870 |
| 7 I = | 304006 |
| | ..........6863 |
| 1 J = | 4343 |
| | ..........2520 |
| 5 K = | 2172 |
| | ...........348 |
| 8 L = | 347 |
| 2 N = | 1 |

| | |
|---|---|
| 237581208759322 | Const. |
| 47516241751864 | |
| 2375812087593 | |
| 287473262598779 | = 2 A |
| 5749465251976 | |
| 28747326260 | |
| 293251475177015 | = 2 B |
| 1759508851062 | |
| 4398772128 | |
| 5865029 | |
| 4399 | |
| 2 | |
| 295015388669635 | = C 6 |
| 265513849803 | |
| 106205540 | |
| 24781 | |
| 4 | |
| 295281008749763 | = D 9 |
| 23622480700 | |
| 826787 | |
| 17 | |
| 295304632057267 | = E 8 |
| 1181218528 | |
| 1772 | |
| 295305813277567 | = F 4 |
| 59061163 | |
| 3 | |
| 295305872338733 | = G 2 |
| 14765294 | = H 5 |
| 2067141 | = I 7 |
| 29531 | = J 1 |
| 14765 | = K 5 |
| 2362 | = L 8 |
| 6 | = N 2 |
| 295305889217832 | |

∴ 29·5305889218 is the number required.

To perform, by logarithms, the ordinary operations of multiplication, division, proportion, or even the extraction of the square root, except in the way of illustration, is not the design of these pages; for such an application of logarithms, in a particular manner only, diminish the labour of the operator. It is not necessary, however, to examine minutely here the instances in which common arithmetic is preferable to artificial numbers; besides, much will depend on the skill and facility of the operator.

# TRIGONOMETRY.

ANGULAR MAGNITUDES.—TRIGONOMETRY.—HEIGHT AND DISTANCES.—SPHERICAL TRIGONOMETRY.—THE APPLICATION OF LOGARITHMS TO ANGULAR MAGNITUDES.

PLANE TRIGONOMETRY treats of the relations and calculations of the sides and angles of plane triangles.

The circumference of every circle is supposed to be divided into 360 equal parts, called degrees; also each degree into 60 minutes, each minute into 60 seconds, and so on.

Hence a semicircle contains 180 degrees, and a quadrant 90 degrees.

The measure of any angle is an arc of any circle contained between the two lines which form that angle, the angular point being the centre; and it is estimated by the number of degrees contained in that arc.

Hence, a right angle being measured by a quadrant, or quarter of the circle, is an angle of 90 degrees; and the sum of the three angles of every triangle, or two right angles, is equal to 180 degrees. Therefore, in a right-angled triangle, taking one of the acute angles from 90 degrees, leaves the other acute angle; and the sum of two angles, in any triangle, taken from 180 degrees, leaves the third angle; or one angle being taken from 180 degrees, leaves the sum of the other two angles.

Degrees are marked at the top of the figure with a small °, minutes with ′, seconds with ″, and so on. Thus, 57° 30′ 12″ denote 57 degrees 30 minutes and 12 seconds.

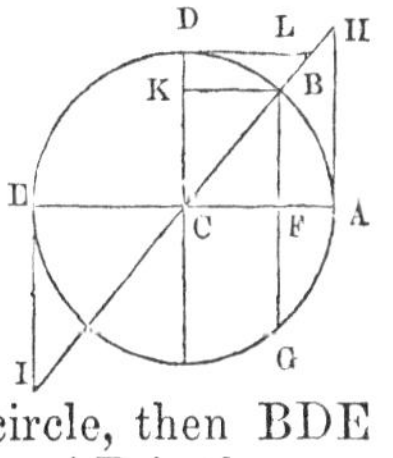

The complement of an arc, is what it wants of a quadrant or 90°. Thus, if AD be a quadrant, then BD is the complement of the arc AB; and, reciprocally, AB is the complement of BD. So that, if AB be an arc of 50°, then its complement BD will be 40°.

D L H K B E C F A I G

The supplement of an arc, is what it wants of a semicircle, or 180°. Thus, if ADE be a semicircle, then BDE is the supplement of the arc AB; and, reciprocally, AB is the supplement of the arc BDE. So that, if AB be an arc of 50°, then its supplement BDE will be 130°.

The sine, or right sine, of an arc, is the line drawn from one extremity of the arc, perpendicular to the diameter passing through the other extremity. Thus, BF is the sine of the arc AB, or of the arc BDE.

Hence the sine (BF) is half the chord (BG) of the double arc (BAG).

The versed sine of an arc, is the part of the diameter intercepted between the arc and its sine. So, AF is the versed sine of the arc AB, and EF the versed sine of the arc EDB.

The tangent of an arc is a line touching the circle in one extremity of that arc, continued from thence to meet a line drawn from the centre through the other extremity: which last line is called the secant of the same arc. Thus, AH is the tangent, and CH the secant, of the arc AB. Also, EI is the tangent, and CI the secant, of the supplemental arc BDE. And this latter tangent and secant are equal to the former, but are accounted negative, as being drawn in an opposite or contrary direction to the former.

The cosine, cotangent, and cosecant, of an arc, are the sine, tangent, and secant of the complement of that arc, the co being only a contraction of the word complement. Thus, the arcs AB, BD being the complements of each other, the sine, tangent or secant of the one of these, is the cosine, cotangent or cosecant of the other. So, BF, the sine of AB, is the cosine of BD; and BK, the sine of BD, is the cosine of AB: in like manner, AH, the tangent of AB, is the cotangent of BD; and DL, the tangent of DB, is the cotangent of AB: also, CH, the secant of AB, is the cosecant of BD; and CL, the secant of BD, is the cosecant of AB.

Hence several remarkable properties easily follow from these definitions; as,

That an arc and its supplement have the same sine, tangent, and secant; but the two latter, the tangent and secant, are accounted negative when the arc is greater than a quadrant or 90 degrees.

When the arc is 0, or nothing, the sine and tangent are nothing, but the secant is then the radius CA. But when the arc is a quadrant AD, then the sine is the greatest it can be, being the radius CD of the circle; and both the tangent and secant are infinite.

Of any arc AB, the versed sine AF, and cosine BK, or CF, together make up the radius CA of the circle. The radius CA, tangent AH, and secant CH, form a right-angled triangle CAH. So also do the radius, sine, and cosine, form another right-angled triangle CBF or CBK. As also the radius, cotangent, and cosecant, another right-angled triangle CDL. And all these right-angled triangles are similar to each other.

The sine, tangent, or secant of an angle, is the sine, tangent, or secant of the arc by which the angle is measured, or of the degrees, &c. in the same arc or angle.

The method of constructing the scales of chords, sines, tangents, and secants, usually engraven on instruments, for practice, is exhibited in the annexed figure.

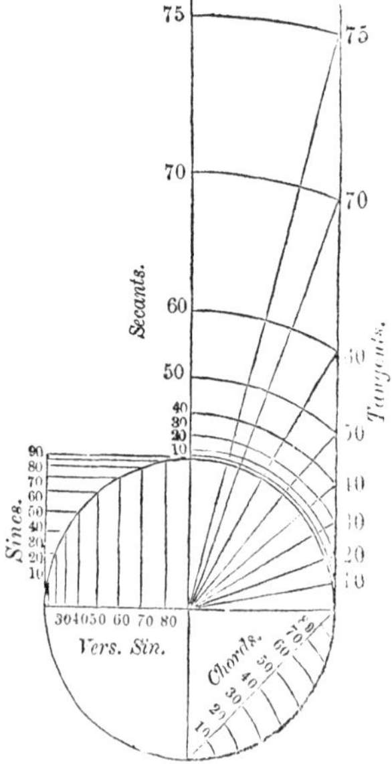

A trigonometrical canon, is a table exhibiting the length of the sine, tangent, and secant, to every degree and minute of the quadrant, with respect to the radius, which is expressed by unity, or 1, and conceived to be divided into 10000000 or more decimal parts. And further, the logarithms of these sines, tangents, and secants are also ranged in the tables; which are most commonly used, as they perform the calculations by only addition and subtraction, instead of the multiplication and division by the natural sines, &c., according to the nature of logarithms.

Upon this table depends the numeral solution of the several cases in trigonometry. It will therefore be proper to begin with the mode of constructing it, which may be done in the following manner:—

*To find the sine and cosine of a given arc.*

This problem is resolved after various ways. One of these is as follows, viz. by means of the ratio between the diameter and circumference of a circle, together with the known series for the sine and cosine, hereafter demonstrated. Thus, the semi-circumference of the circle, whose radius is 1, being 3·141592653589793, &c., the proportion will therefore be,

As the number of degrees or minutes in the semicircle,
Is to the degrees or minutes in the proposed arc,
So is 3·14159265, &c., to the length of the said arc.

This length of the arc being denoted by the letter $a$; also its sine and cosine by $s$ and $c$; then will these two be expressed by the two following series, viz.:—

$$s = a - \frac{a^3}{2.3} + \frac{a^5}{2.3.4.5} - \frac{a^7}{2.3.4.5.6.7} + \&c.$$

$$= a - \frac{a^3}{6} + \frac{a^5}{120} - \frac{a^7}{5040} + \&c.$$

$$c = 1 - \frac{a^2}{2} + \frac{a^4}{2.3.4} - \frac{a^6}{2.3.4.5.6} + \&c.$$

$$= 1 - \frac{a^2}{2} + \frac{a^4}{24} - \frac{a^6}{720} + \&c.$$

If it be required to find the sine and cosine of one minute. Then, the number of minutes in 180° being 10800, it will be first, as 10800 : 1 :: 3·14159265, &c. : ·000290888208665 = the length of an arc of one minute. Therefore, in this case,

$a =$ ·0002908882
and $\frac{1}{6}a^3 =$ ·000000000004, &c.
the difference is $s =$ ·0002908882 the sine of 1 minute.
Also, from 1·
take $\frac{1}{2}a^2 =$ 0·0000000423079, &c.
leaves $c =$ ·9999999577 the cosine of 1 minute.

For the sine and cosine of 5 degrees.

Here, as 180° : 5° : : 3·14159265, &c., : ·08726646 = $a$ the length of 5 degrees.

$$
\begin{aligned}
\text{Hence,} \quad a &= \cdot 08726646 \\
-\tfrac{1}{6}a^3 &= -\cdot 00011076 \\
+\tfrac{1}{120}a^5 &= \cdot 00000004 \\
\text{these collected give } s &= \cdot 08715574 \text{ the sine of } 5^\circ.
\end{aligned}
$$

$$
\begin{aligned}
\text{And, for the cosine,} \quad 1 &= 1\cdot \\
-\tfrac{1}{2}a^2 &= -\cdot 00380771 \\
+\tfrac{1}{24}a^4 &= \cdot 00000241 \\
\text{these collected, give } c &= \cdot 99619470 \text{ the consine of } 5^\circ.
\end{aligned}
$$

After the same manner, the sine and cosine of any other arc may be computed. But the greater the arc is, the slower the series will converge, in which case a greater number of terms must be taken to bring out the conclusion to the same degree of exactness.

Or, having found the sine, the cosine will be found from it, by the property of the right-angled triangle CBF, viz. the cosine $CF = \sqrt{CB^2 - BF^2}$, or $c = \sqrt{1 - s^2}$.

There are also other methods of constructing the canon of sines and cosines, which, for brevity's sake, are here omitted.

*To compute the tangents and secants.*

The sines and cosines being known, or found, by the foregoing problem; the tangents and secants will be easily found, from the principle of similar triangles, in the following manner:—

In the first figure, where, of the arc AB, BF is the sine, CF or BK the cosine, AH the tangent, CH the secant, DL the cotangent, and CL the cosecant, the radius being CA, or CB, or CD; the three similar triangles CFB, CAH, CDL, give the following proportions:

1. CF : FB : : CA : AH; whence the tangent is known, being a fourth proportional to the cosine, sine, and radius.

2. CF : CB : : CA : CH; whence the secant is known, being a third proportional to the cosine and radius.

3. BF : FC : : CD : DL; whence the cotangent is known, being a fourth proportional to the sine, cosine, and radius.

4. BF : BC : : CD : CL; whence the cosecant is known, being a third proportional to the sine and radius.

Having given an idea of the calculations of sines, tangents, and secants, we may now proceed to resolve the several cases of trigonometry; previous to which, however, it may be proper to add a few preparatory notes and observations, as below.

There are usually three methods of resolving triangles, or the cases of trigonometry—namely, geometrical construction, arithmetical computation, and instrumental operation.

*In the first method.*—The triangle is constructed by making the parts of the given magnitudes, namely, the sides from a scale of equal parts, and the angles from a scale of chords, or by some other instrument. Then, measuring the unknown parts by the same scales or instruments, the solution will be obtained near the truth.

*In the second method.*—Having stated the terms of the proportion according to the proper rule or theorem, resolve it like any other proportion, in which a fourth term is to be found from three given terms, by multiplying the second and third together, and dividing the product by the first, in working with the natural numbers; or, in working with the logarithms, add the logs. of the second and third terms together, and from the sum take the log. of the first term; then the natural number answering to the remainder is the fourth term sought.

*In the third method.*—Or instrumentally, as suppose by the log. lines on one side of the common two-foot scales; extend the compasses from the first term to the second or third, which happens to be of the same kind with it; then that extent will reach from the other term to the fourth term, as required, taking both extents towards the same end of the scale.

In every triangle, or case in trigonometry, there must be given three parts, to find the other three. And, of the three parts that are given, one of them at least must be a side; because the same angles are common to an infinite number of triangles.

All the cases in trigonometry may be comprised in three varieties only; viz.

1. When a side and its opposite angle are given.
2. When two sides and the contained angle are given.
3. When the three sides are given.

For there cannot possibly be more than these three varieties of cases; for each of which it will therefore be proper to give a separate theorem, as follows:

*When a side and its opposite angle are two of the given parts.*

Then the sides of the triangle have the same proportion to each other, as the sines of their opposite angles have.

That is,

As any one side,
Is to the sine of its opposite angle;
So is any other side,
To the sine of its opposite angle.

C D A E F B

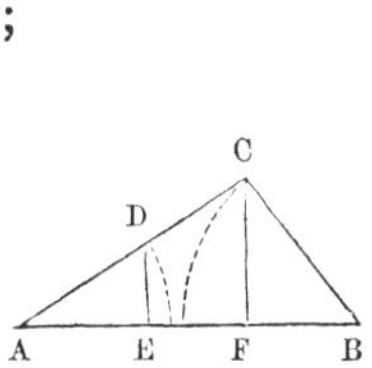

For, let ABC be the proposed triangle, having AB the greatest side, and BC the least. Take AD = BC, considering it as a radius; and let fall the perpendiculars DE, CF, which will evidently be the sines of the angles A and B, to the radius AD or BC. But the triangles ADE, ACF, are equiangular, and therefore AC : CF :: AD or BC : DE; that is, AC is to the sine of its opposite angle B, as BC to the sine of its opposite angle A.

In practice, to find an angle, begin the proportion with a side opposite a given angle. And to find a side, begin with an angle opposite a given side.

An angle found by this rule is ambiguous, or uncertain whether it be acute or obtuse, unless it be a right angle, or unless its magnitude be such as to prevent the ambiguity; because the sine answers to two angles, which are supplements to each other; and accordingly the geometrical construction forms two triangles with the same parts that are given, as in the example below; and when there is no restriction or limitation included in the question, either of them may be taken. The degrees in the table, answering to the sine, are the acute angle; but if the angle be obtuse, subtract those degrees from 180°, and the remainder will be the obtuse angle. When a given angle is obtuse, or a right one, there can be no ambiguity; for then neither of the other angles can be obtuse, and the geometrical construction will form only one triangle.

In the plane triangle ABC,

Given, { AB 345 yards; BC 232 yards; angle A 37° 20′ }

Required the other parts.

*Geometrically.*—Draw an indefinite line, upon which set off AB = 345, from some convenient scale of equal parts. Make the angle A = $37\frac{1}{3}$°. With a radius of 232, taken from the same scale of equal parts, and centre B, cross AC in the two points C, C. Lastly, join BC, BC, and the figure is constructed, which gives two triangles, showing that the case is ambiguous.

Then, the sides AC measured by the scale of equal parts, and the angles B and C measured by the line of chords, or other instrument, will be found to be nearly as below; viz.

| | | |
|---|---|---|
| AC 174 | angle B 27° | angle C $115\frac{1}{2}$° |
| or $374\frac{1}{2}$ | or $78\frac{1}{4}$ | or $64\frac{1}{2}$ |

*Arithmetically.*—First, to find the angles at C:

| | | |
|---|---|---|
| As side BC | 232 ........................log. | 2·3654880 |
| To sin. opp. angle A | 37° 20′ .................. | 9·7827958 |
| So side AB | 345 ........................ | 2·5378191 |
| To sin. opp. angle C | 115° 36′ or 64° 24...... | 9·9551269 |
| Add angle A | 37 20 — 37 20 | |
| The sum | 152 56 or 101 44 | |
| Taken from | 180 00 — 180 00 | |
| Leaves angle B | 27 04 or 78 16 | |

Then, to find the side AC:

| | | |
|---|---|---|
| As sine angle A | 37° 20′......................log. | 9·7827958 |
| To opposite side BC | 232 ...................... | 2·365488 |
| So sine angle B | 27° 04′..................... | 9·6580371 |
| | 78 16 ..................... | 9·9908291 |
| To opposite side AC | 174·07 ...................... | 2·2407293 |
| or, | 374·56 ...................... | 2·5735213 |

In the plane triangle ABC,

Given, { AB 365 poles; angle A 57° 12′; angle B 24 45 }

Required the other parts.

Ans. { angle C 98° 3′; AC 154·33; BC 309·86 }

In the plane triangle ABC,

Given, { AC 120 feet; BC 112 feet; angle A 57° 27′ }

Required the other parts.

Ans. { angle B 64° 34′ 21″ or, 115 25 39; angle C 57 58 39 or, 7 7 21; AB 112·65 feet or, 16·47 feet }

*When two sides and their contained angle are given.*

Then it will be,

As the sum of those two sides,
Is to the difference of the same sides;
So is the tang. of half the sum of their opposite angles,
To the tang. of half the difference of the same angles.

Hence, because it is known that the half sum of any two quantities increased by their half difference, gives the greater, and diminished by it gives the less, if the half difference of the angles, so found, be added to their half sum, it will give the greater angle, and subtracting it will leave the less angle.

Then, all the angles being now known, the unknown side will be found by the former theorem.

Let ABC be the proposed triangle, having the two given sides AC, BC, including the given angle C. With the centre C, and radius CA, the less of these two sides, describe a semicircle, meeting the other side BC produced in D and E. Join AE, AD, and draw DF parallel to AE.

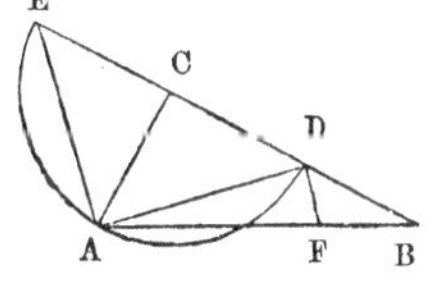


Then, BE is the sum, and BD the difference of the two given sides CB, CA. Also, the sum of the two angles CAB, CBA, is equal to the sum of the two CAD, CDA, these sums being each the supplement of the vertical angle C to two right angles: but the two latter CAD, CDA, are equal to each other, being opposite to the two equal sides CA, CD: hence, either of them, as CDA, is equal to half the sum of the two unknown angles CAB, CBA. Again, the exterior angle CDA is equal to the two interior angles B and DAB; therefore, the angle DAB is equal to the difference between CDA and B, or between CAD and B; consequently, the same angle DAB is equal to half the difference of the unknown angles B and CAB; of which it has been shown that CDA is the half sum.

Now the angle DAE, in a semicircle, is a right angle, or AE is perpendicular to AD; and DF, parallel to AE, is also perpendicular to AD: consequently, AE is the tangent of CDA the half sum and DF the tangent of DAB the half difference of the angles, to the same radius AD, by the definition of a tangent. But, the tangents AE, DF, being parallel, it will be as BE : BD :: AE : DF; that is, as the sum of the sides is to the difference of the sides, so is the tangent of half the sum of the opposite angles, to the tangent of half their difference.

The sum of the unknown angles is found, by taking the given angle from 180°.

In the plane triangle ABC,

Given, { AB 345 yards<br>AC 174·07 yards<br>angle A 37° 20′

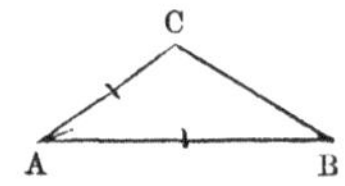


Required the other parts.

*Geometrically.*—Draw AB = 345 from a scale of equal parts. Make the angle A = 37° 20′. Set off AC = 174 by the scale of equal parts. Join BC, and it is done.

Then the other parts being measured, they are found to be nearly as follows, viz. the side BC 232 yards, the angle B 27°, and the angle C 115½°.

*Arithmetically.*

| | | |
|---|---|---|
| As sum of sides AB, AC.................. | 519·07 log. | 2·7152259 |
| To difference of sides AB, AC............ | 170·93 | 2·2328183 |
| So tangent half sum angles C and B..... | 71° 20′ | 10·4712979 |
| To tangent half difference angles C and B | 44 16 | 9·9888903 |
| Their sum gives angle C | 115 36 | |
| Their diff. gives angle B | 27 4 | |

Then, by the former theorem,

| | |
|---|---|
| As sine angle C 115° 36′, or 64° 24′......log. | 9·0551259 |
| To its opposite side AB 345.................. | 2·5378191 |
| So sine angle A 37° 20′...................... | 9·7827958 |
| To its opposite side BC 232................ | 2·3654890 |

In the plane triangle ABC,

Given, { AB 365 poles<br>AC 154·33<br>angle A 57° 12′

Required the other parts. { BC 309·86<br>angle B 24° 45′<br>angle C 98° 3′

In the plane triangle ABC,

Given, { AC 120 yards<br>BC 112 yards<br>angle C 57° 58′ 39″

Required the other parts. { AB 112·65<br>angle A 57° 27′ 0″<br>angle B 64 34 21

*When the three sides of the triangle are given.*

Then, having let fall a perpendicular from the greatest angle upon the opposite side, or base, dividing it into two segments, and the whole triangle into two right-angled triangles; it will be,

As the base, or sum of the segments,
Is to the sum of the other two sides;
So is the difference of those sides,
To the difference of the segments of the base.

Then, half the difference of the segments being added to the half sum, or the half base, gives the greater segment; and the same subtracted gives the less segment.

Hence, in each of the two right-angled triangles, there will be known two sides, and the angle opposite to one of them; consequently, the other angles will be found by the first problem.

The rectangle under the sum and difference of the two sides, is equal to the rectangle under the sum and difference of the two segments. Therefore, by forming the sides of these rectangles into a proportion, it will appear that the sums and differences are proportional, as in this theorem.

In the plane triangle ABC,

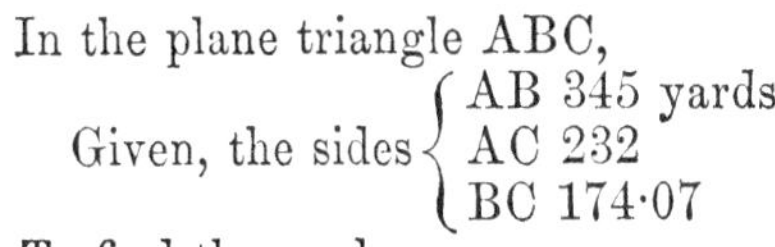

Given, the sides { AB 345 yards, AC 232, BC 174·07 }

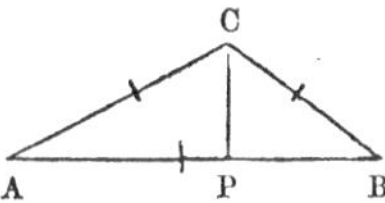


To find the angles.

*Geometrically.*—Draw the base AB = 345 by a scale of equal parts. With radius 232, and centre A, describe an arc; and with radius 174, and centre B, describe another arc, cutting the former in C. Join AC, BC, and it is done.

Then, by measuring the angles, they will be found to be nearly as follows, viz. angle A 27°, angle B $37\frac{1}{3}$°, and angle C $115\frac{1}{2}$°.

*Arithmetically.*—Having let fall the perpendicular CP, it will be,

As the base AB : AC + BC :: AC − BC : AP − BP
that is, as 345 : 406·07 :: 57·93 : 68·18 = AP − BP
its half is.......................... 34·09
the half base is..................172·50

the sum of these is...............206·59 = AP
and their difference..............138·41 = BP

Then, in the triangle APC, right-angled at P,

| | | |
|---|---|---|
| As the side AC..................... | 232 | ........log. 2·3654880 |
| To sine opposite angle............ | 90° | ........ 10·0000000 |
| So is side AP....................... | 206·59 | ........ 2·3151093 |
| To sine opposite angle ACP..... | 62° 56′ | ........ 9·9496213 |
| Which taken from............ | 90 00 | |
| Leaves the angle A......... | 27 04 | |

Again, in the triangle BPC, right-angled at P,

| | | | |
|---|---|---|---|
| As the side of BC.......... | 174·07 | .........log. | 2·2407239 |
| To sine opposite angle P... | 90° | ......... | 10·0000000 |
| So is side BP................ | 138·41 | ......... | 2·1411675 |
| To sin. opposite angle BCP | 52° 40′ | ......... | 9·9004436 |
| Which taken from..... | 90 00 | | |
| Leaves the angle B... | 37 20 | | |
| Also, the angle ACP... | 62° 56′ | | |
| Added to angle BCP... | 52 40 | | |
| Gives the whole angle ACB... | 115 36 | | |

So that all the three angles are as follow, viz. the angle A 27° 4′; the angle B 37° 20′; the angle C 115° 36′.

In the plane triangle ABC,

Given the sides, { AB 365 poles, AC 154·33, BC 309·86 }

To find the angles. { angle A 57° 12′, angle B 24 45, angle C 98 3 }

In the plane triangle ABC,

Given the sides, { AB 120, AC 112·65, BC 112 }

To find the angles. { angle A 57° 27′ 00″, angle B 57 58 39, angle C 64 34 21 }

The three foregoing theorems include all the cases of plane triangles, both right-angled and oblique; besides which, there are other theorems suited to some particular forms of triangles, which are sometimes more expeditious in their use than the general ones; one of which, as the case for which it serves so frequently occurs, may be here taken, as follows:—

*When, in a right-angled triangle, there are given one leg and the angles; to find the other leg or the hypothenuse; it will be,*

As radius, *i. e.* sine of 90° or tangent of 45°
Is to the given leg,
So is the tangent of its adjacent angle
To the other leg;
And so is the secant of the same angle
To the hypothenuse.

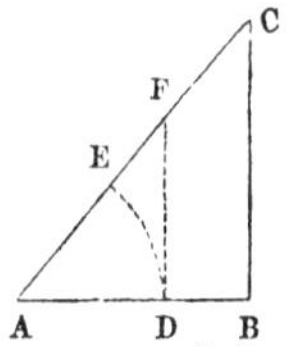

AB being the given leg, in the right-angled triangle ABC; with the centre A, and any assumed radius, AD, describe an arc DE, and draw DF perpendicular to AB, or parallel to BC. Now it is evident, from the definitions, that DF is the tangent, and AF the secant, of the arc DE, or of the angle A which is measured by that arc, to the radius AD. Then, because of the parallels BC, DF, it will be as AD : AB :: DF : BC :: AF : AC, which is the same as the theorem is in words.

In the right-angled triangle ABC,

Given $\left\{\begin{array}{l}\text{the leg AB } 162\\ \text{angle A } 53^\circ\ 7'\ 48''\end{array}\right\}$ to find AC and BC.

*Geometrically.*—Make AB = 162 equal parts, and the angle A = 53° 7′ 48″; then raise the perpendicular BC, meeting AC in C. So shall AC measure 270, and BC 216.

*Arithmetically.*

| | | | |
|---|---|---|---|
| As radius ..................... | tang. 45° | .........log. | 10·0000000 |
| To leg AB .................... | 162 | ......... | 2·2095150 |
| So tang. angle A............ | 53° 7′ 48″ | ......... | 10·1249371 |
| To leg BC .................... | 216 | ......... | 2·3344521 |
| So secant angle A........... | 53° 7′ 48″ | ......... | 10·2218477 |
| To hyp. AC ................. | 270 | ......... | 2·4313627 |

In the right-angled triangle ABC,

Given $\left\{\begin{array}{l}\text{the leg AB } 180\\ \text{the angle A } 62^\circ\ 40'\end{array}\right.$

To find the other two sides. $\left\{\begin{array}{l}\text{AC } 392{\cdot}0147\\ \text{BC } 348{\cdot}2464\end{array}\right.$

There is sometimes given another method for right-angled triangles, which is this:

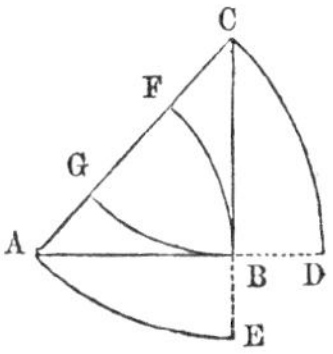

ABC being such a triangle, make one leg AB radius, that is, with centre A, and distance AB, describe an arc BF. Then it is evident that the other leg BC represents the tangent, and the hypothenuse AC the secant, of the arc BF, or of the angle A.

In like manner, if the leg BC be made radius; then the other leg AB will represent the tangent, and the hypothenuse AC the secant, of the arc BG or angle C.

But if the hypothenuse be made radius; then each leg will represent the sine of its opposite angle; namely, the leg AB the sine of the arc AE or angle C, and the leg BC the sine of the arc CD or angle A.

And then the general rule for all these cases is this, namely, that the sides of the triangle bear to each other the same proportion as the parts which they represent.

And this is called, Making every side radius.

---

## OF HEIGHTS AND DISTANCES.

By the mensuration and protraction of lines and angles, are determined the lengths, heights, depths, and distances of bodies or objects.

Accessible lines are measured by applying to them some certain measure a number of times, as an inch, or foot, or yard. But inaccessible lines must be measured by taking angles, or by some such method, drawn from the principles of geometry.

When instruments are used for taking the magnitude of the

angles in degrees, the lines are then calculated by trigonometry: in the other methods, the lines are calculated from the principle of similar triangles, without regard to the measure of the angles.

Angles of elevation, or of depression, are usually taken either with a theodolite, or with a quadrant, divided into degrees and minutes, and furnished with a plummet suspended from the centre, and two sides fixed on one of the radii, or else with telescopic sights.

*To take an angle of altitude and depression with the quadrant.*

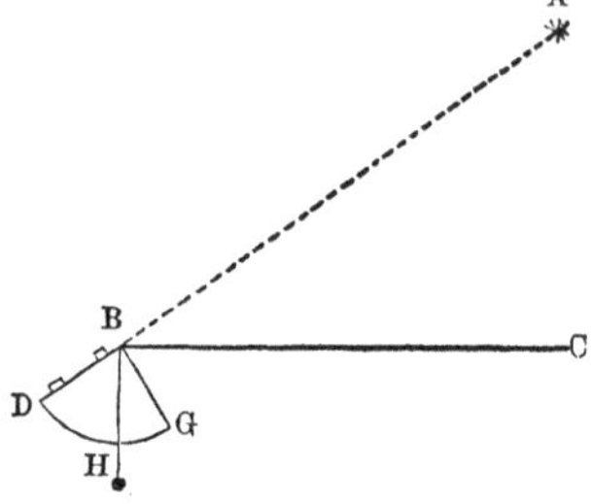


Let A be any object, as the sun, moon, or a star, or the top of a tower, or hill, or other eminence; and let it be required to find the measure of the angle ABC, which a line drawn from the object makes with the horizontal line BC.

Fix the centre of the quadrant in the angular point, and move it round there as a centre, till with one eye at D, the other being shut, you perceive the object A through the sights: then will the arc GH of the quadrant, cut off by the plumb line BH, be the measure of the angle ABC, as required.

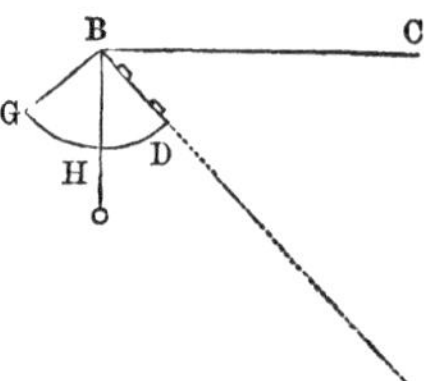


The angle ABC of depression of any object A, is taken in the same manner; except that here the eye is applied to the centre, and the measure of the angle is the arc GH, on the other side of the plumb line.

The following examples are to be constructed and calculated by the foregoing methods, treated of in trigonometry.

Having measured a distance of 200 feet, in a direct horizontal line, from the bottom of a steeple, the angle of elevation of its top, taken at that distance, was found to be 47° 30′: from hence it is required to find the height of the steeple.

*Construction.*—Draw an indefinite line, upon which set off AC = 200 equal parts, for the measured distance. Erect the indefinite perpendicular AB; and draw CB so as to make the angle C = 47° 30′, the angle of elevation; and it is done. Then AB, measured on the scale of equal parts, is nearly 218¼.

*Calculation.*

As radius.................................10·0000000
To AC 200................................ 2·3010300
So tang. angle C 47° 30′............10·0379475
To AB 218·26 required.............. 2·3389775

What was the perpendicular height of a cloud, or of a balloon, when its angles of elevation were 35° and 64°, as taken by two observers, at the same time, both on the same side of it, and in the same vertical plane; their distance, as under, being half a mile, or 880 yards. And what was its distance from the said two observers?

*Construction.*—Draw an indefinite ground line, upon which set off the given distance AB = 880; then A and B are the places of the observers. Make the angle A = 35°, and the angle B = 64°; and the intersection of the lines at C will be the place of the balloon; from whence the perpendicular CD, being let fall, will be its perpendicular height. Then, by measurement, are found the distances and height nearly, as follows, viz. AC 1631, BC 1041, DC 936.

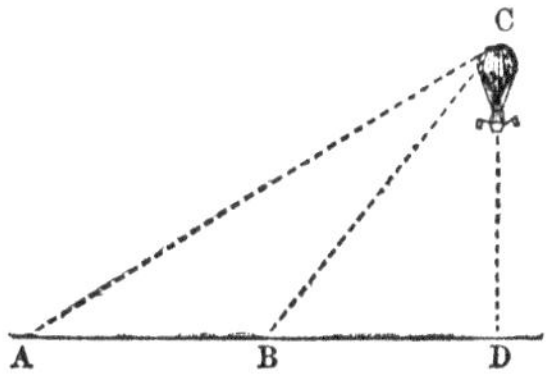

*Calculation.*

First, from angle B 64°
Take angle A 35
Leaves angle ACB 29

Then, in the triangle ABC,

| | | |
|---|---|---|
| As sine angle ACB | 29° | 9·6855712 |
| To opposite side AB | 880 | 2·9444827 |
| So sine angle A | 35° | 9·7585913 |
| To opposite side BC | 1041·125 | 3·0175028 |

| | | |
|---|---|---|
| As sine angle ACB | 29° | 9·6855712 |
| To opposite side AB | 880 | 2·9444827 |
| So sine angle B | 116° or 64° | 9·9536602 |
| To opposite side AC | 1631·442 | 3·2125717 |

And, in the triangle BCD,

| | | |
|---|---|---|
| As sine angle D | 90° | 10·0000000 |
| To opposite side BC | 1041·125 | 3·0175028 |
| So sine angle B | 64° | 9·9536602 |
| To opposite side CD | 935·757 | 2·9711630 |

Having to find the height of an obelisk standing on the top of a declivity, I first measured from its bottom, a distance of 40 feet, and there found the angle, formed by the oblique plane and a line imagined to go to top of the obelisk 41°; but, after measuring on in the same direction 60 feet farther, the like angle was only 23° 45′. What then was the height of the obelisk?

*Construction.*—Draw an indefinite line for the sloping plane or declivity, in which assume any point A for the bottom of the obelisk, from whence set off the distance AC = 40, and again CD = 60 equal parts. Then make the angle C = 41°, and the angle D = 23° 45′; and the point B, where the two lines meet, will be the top of the obelisk. Therefore AB, joined, will be its height.

*Calculation.*

| | |
|---|---|
| From the angle C | 41° 00′ |
| Take the angle D | 23 45 |
| Leaves the angle DBC | 17 15 |

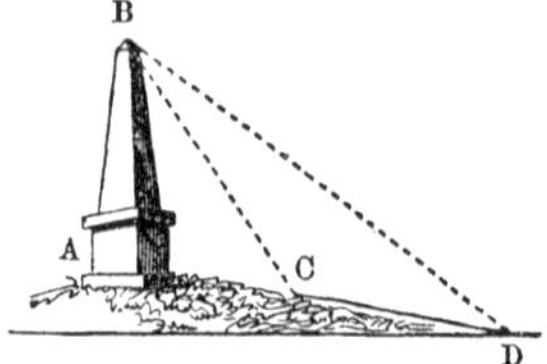


Then, in the triangle DBC,

| | | |
|---|---|---|
| As sine angle DBC | 17° 15′.................. | 9·4720856 |
| To opposite side DC | 60 .................. | 1·7781513 |
| So sine angle D | 24 45.................. | 9·6050320 |
| To opposite side CB | 81·488.................. | 1·9110977 |

And, in the triangle ABC,

| | | | |
|---|---|---|---|
| As sum of sides | CB, CA | 121·488 ..... | 2·0845333 |
| To difference of sides | CB, CA | 41·488 ..... | 1·6179225 |
| So tang. half sum angles | A, B | 69° 30′ ..... | 10·4272623 |
| To tang. half diff. angles | A, B | 42 24½..... | 9·9606516 |
| The diff. of these is angle | CBA | 27 5½ | |

| | | | |
|---|---|---|---|
| Lastly, as sine angle | CBA | 27° 5½′............... | 9·6582842 |
| To opposite side | CA | 40 ............... | 1·6020600 |
| So sine angle | C | 41° 0′ ............... | 9·8169429 |
| To opposite side | AB | 57·623............... | 1·7607187 |

Wanting to know the distance between two inaccessible trees, or other objects, from the top of a tower, 120 feet high, which lay in the same right line with the two objects, I took the angles formed by the perpendicular wall and lines conceived to be drawn from the top of the tower to the bottom of each tree, and found them to be 33° and 64½°. What then may be the distance between the two objects?

*Construction.*—Draw the indefinite ground line BD, and perpendicular to it BA = 120 equal parts. Then draw the two lines AC, AD, making the two angles BAC, BAD, equal to the given angles 33° and 64½°. So shall C and D be the places of the two objects.

*Calculation.*—First, In the right-angled triangle ABC,

| | | |
|---|---|---|
| As radius............................................ | | 10·0000000 |
| To AB.....................120 | ..................... | 2·0791812 |
| So tang. angle BAC..... 33° | ..................... | 9·8125174 |
| To BC................77·929 | ..................... | 1·8916986 |

And, in the right-angled triangle ABD,

| | | |
|---|---|---|
| As radius............................................ | | 10·0000000 |
| To AB.....................120 | .................... | 2·0791812 |
| So tang. angle BAD.... 64½° | .................... | 10·3215039 |
| To BD...............251·585 | .................... | 2·4006851 |

From which take BC 77·929

Leaves the dist. CD 173·656 as required.

Being on the side of a river, and wanting to know the distance to a house which was seen on the other side, I measured 200 yards in a straight line by the side of the river; and then at each end of this line of distance, took the horizontal angle formed between the house and the other end of the line; which angles were, the one of them 68° 2′, and the other 73° 15′. What then were the distances from each end to the house?

*Construction.*—Draw the line AB = 200 equal parts. Then draw AC so as to make the angle A = 68° 2′, and BC to make the angle B = 73° 15′. So shall the point C be the place of the house required.

*Calculation.*

| | | |
|---|---|---|
| To the given angle A | 68° | 2′ |
| Add the given angle B | 73 | 15 |
| Then their sum | 141 | 17 |
| Being taken from | 180 | 0 |
| Leaves the third angle C | 38 | 43 |

Hence, As sin. angle C 38° 43′..................9·7962062
To op. side AB 200 ..................2·3010300
So sin. angle A 68° 2′..................9·9672679
To op. side BC 296·54 ..................2·4720917

And, As sin. angle C 38° 43′..................9·7962062
To op. side AB 200 ..................2·3010300
So sin. angle B 73° 15′..................9·9811711
To op. side AC 306·19 ..................2·4859949

---

## SPHERICAL TRIGONOMETRY.

*This Article is taken from a short Practical Treatise on Spherical Trigonometry, by Oliver Byrne, the author of the present work. Published by J. A. Valpy. London,* 1835.

As the sides and angles of spherical triangles are measured by circular arcs, and as these arcs are often greater than 90°, it may be necessary to mention one or two particulars respecting them.

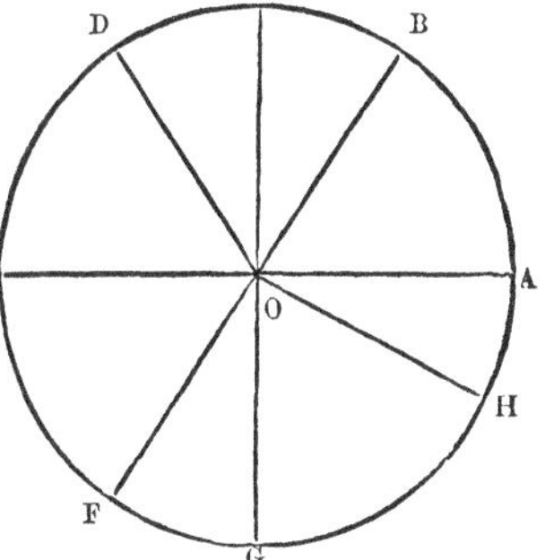


The arc CB, which when added to AB makes up a quadrant or 90°, is called the complement of the arc AB; every arc will have a complement, even those which are themselves greater than 90°, provided we consider the arcs measured in the direction ABCD, &c., as positive, and consequently those measured in the opposite direction as negative. The complement BC of the arc AB commences at B, where AB terminates, and may be considered as generated by the motion of B, the extremity of the radius OB, in the direction BC. But the complement of the arc AD or DC, commencing in like manner at the extremity D, must be generated by the motion of D in the opposite direction, and the angular magnitude AOD will here be diminished by the motion of OD, in generating the complement; therefore the complement of AOD or of AD may with propriety be considered negative.

Calling the arc AB or AD, $\theta$, the complement will be $90° - \theta$; the complement of 36° 44′ 33″ is 53° 15′ 27″; and the complement of 136° 27′ 39″ is negative 46° 27′ 39″.

The arc BE, which must be added to AB to make up a semicircle or 180°, is called the *supplement* of the arc AB. If the arc is greater than 180°, as the arc ADF its supplement, FE measured in the reverse direction is negative. The expression for the supplement of any arc $\theta$ is therefore $180° - \theta$; thus the supplement of 112° 29′ 35″ is 67° 30′ 25″, and the supplement of 205° 42′ is negative 25° 42′.

In the same manner as the complementary and supplementary arcs are considered as positive or negative, according to the direction in which they are measured, so are the arcs themselves positive or negative; thus, still taking A for the commencement, or *origin*, of the arcs, as AB is positive, AH will be negative. *In the doctrine of triangles*, we consider only positive angles or arcs, and the magnitudes of these are comprised between $\theta = 0$ and $\theta = 180°$; but in the general theory of angular quantity, we consider both positive and negative angles, according as they are situated above or below the fixed line AO, from which they are measured, that is, according as the arcs by which they are estimated are positive or negative. Thus the angle BOA is positive, and the angle AOH negative. Moreover, in this more extended theory of angular magnitude, an angle may consist of any number of degrees whatever; thus, if the revolving line OB set out from the fixed line OA, and make $n$ revolutions and a part, the angular magnitude generated is measured by $n$ times 360°, plus the degrees in the additional part.

In a right-angled spherical triangle we are to recognise but five

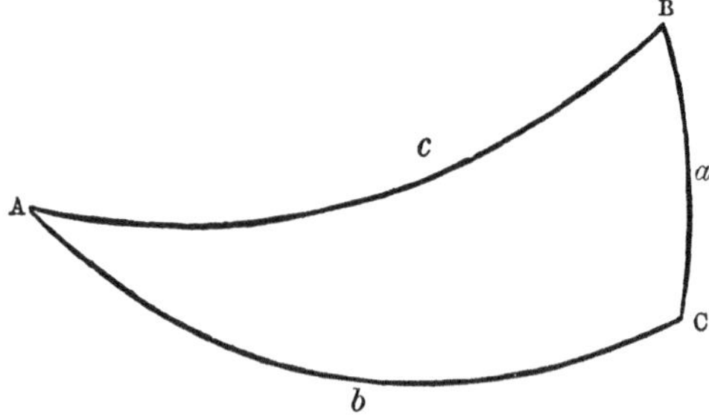


parts, namely, the three sides $a$, $b$, $c$, and the two angles A, B; so that the right angle C is omitted.

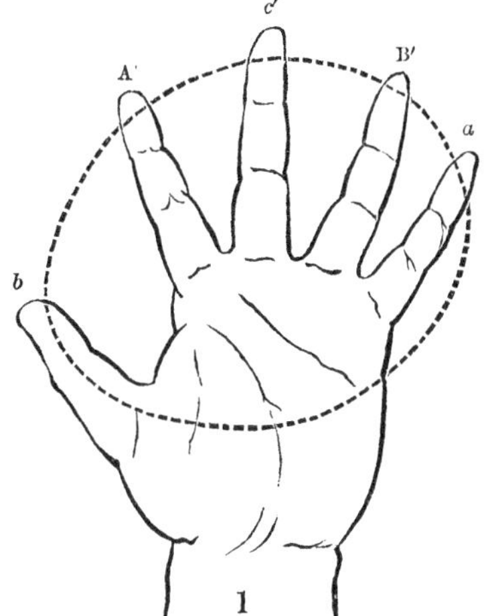


Let A', $c'$, B,' be the complements of A, $c$, B, respectively, and suppose $b$, $a$, B', $c'$, A', to be placed on the hand, as in the annexed figure, and that the fingers stand in a circular order, the parts represented by the fingers thus placed are called circular parts.

If we take any one of these as a middle part, the two which lie next to it, one on each side, will be *adjacent* parts. The two parts immediately beyond the adjacent parts, one on each side, are called the opposite parts.

Thus, taking A' for a middle part, $b$ and $c'$ will be *adjacent* parts, and $a$ and B' are opposite parts.

If we take $c'$ as a middle part, A' and B' are adjacent parts, and $b$, $a$, opposite parts.

When B' is a middle part, $c'$, $a$, become adjacent parts, and A', $b$, opposite parts.

Again, if we take $a$ as a middle part, then B', $b$, will be adjacent parts, and $c'$, A', opposite parts.

Lastly, taking $b$ as a middle part, A', $a$, are adjacent parts, and $c'$, B', opposite parts.

This being understood, Napier's two rules may be expressed as follows:—

I. Rad. × sin. middle part = product of tan. adjacent parts.

II. Rad. × sin. middle part = product of cos. opposite parts.

Both these rules may be comprehended in a single expression, thus,

Rad. sin. mid. = prod. tan. adja. = prod. cos. opp.;

and to retain this in the memory we have only to remember, that the vowels in the contractions *sin.*, *tan.*, *cos.*, are the same as those in the contractions *mid.*, *adja.*, *opp.*, to which they are joined.

These rules comprehend all the succeeding equations, reading from the centre, R = radius.

In the solution of right-angled spherical triangles, two parts are given to find a third, therefore it is necessary, in the application of this formula, to choose for the middle part that which causes the other two to become either adjacent parts or opposite parts.

In a right-angled spherical triangle, the hypothenuse

$c = 61° \; 4' \; 56''$; and the angle

$A = 61° \; 50' \; 29''$. Required the adjacent leg?

| | | | | | | | |
|---|---|---|---|---|---|---|---|
| | 90° | 0' | 00'' | | 90° | 0' | 00'' |
| = | 61 | 4 | 56 | A = | 61 | 50 | 29 |
| | 28 | 55 | 04 = $c'$. | | 28 | 9 | 31 = A. |

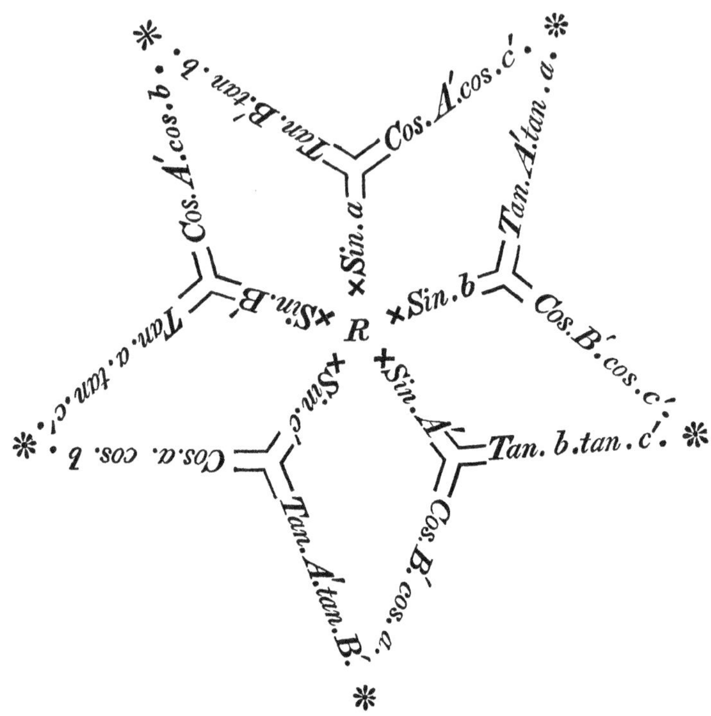


In this example, A′ is selected for the middle part, because then $b$ and $c'$ become adjacent parts, as in the annexed figure.

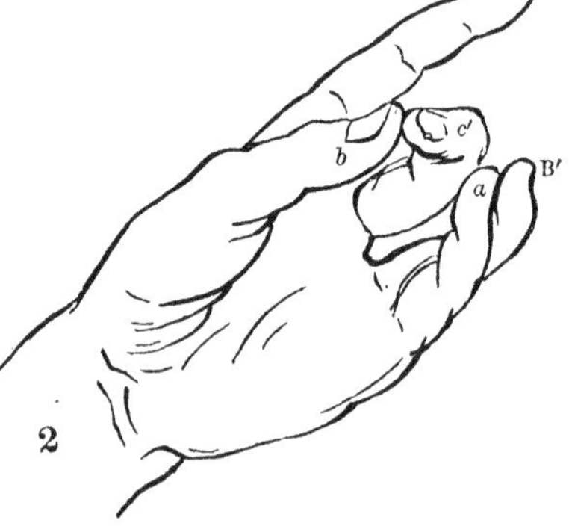


$$\text{Rad.} \times \sin. A' = \tan. b \times \tan. c'.$$

$$\therefore \tan. b = \frac{\text{rad.} \times \sin. A'}{\tan c'}.$$

*By Logarithms.*

Rad. — ........... — 10·0000000
Sin. A′—28°9′21″— 9·6738628

19·6738628
Tan. $c'$—28°55′4″— 9·7422808

Tan. $b'$—40°30′16″—9·9315820

The side adjacent to the *given* angle is acute or obtuse, according as the hypothenuse is of the same, or of different species with the *given* angle.

$$\therefore \text{the leg } b = 40^\circ\ 30'\ 16'', \text{ acute.}$$

Supposing the hypothenuse $c = 113^\circ\ 55'$, and the angle $A = 31^\circ\ 51'$, then the adjacent leg $b$ would be $117^\circ\ 34'$, obtuse.

In the right-angled spherical triangle ABC, are given the hypothenuse $c = 113° \ 55'$, and the angle $A = 104° \ 08'$; to find the opposite leg $a$.

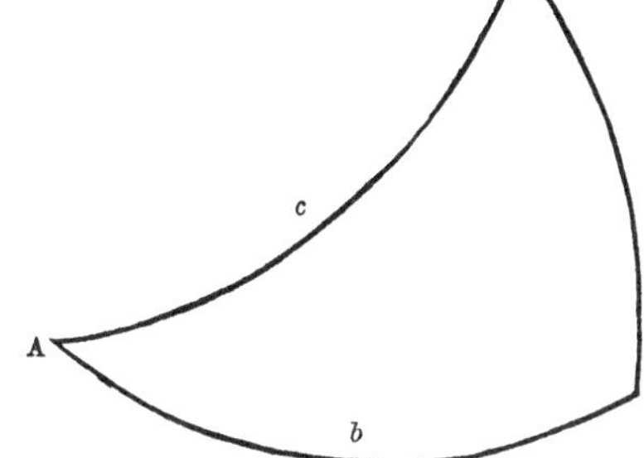


$$\begin{array}{rl} c = & 113° \ 55' \\ & \underline{\ 90 \quad 0\ } \\ & \ 23 \quad 55 = c'. \end{array}$$

$$\begin{array}{rl} A = & 104° \ 08' \\ & \underline{\ 90 \quad 0\ } \\ & \ 14 \quad 08 = A'. \end{array}$$

In this example, $a$ is taken for the middle part, then $A'$ and $c'$ are opposite parts. (See the subjoined figure.)

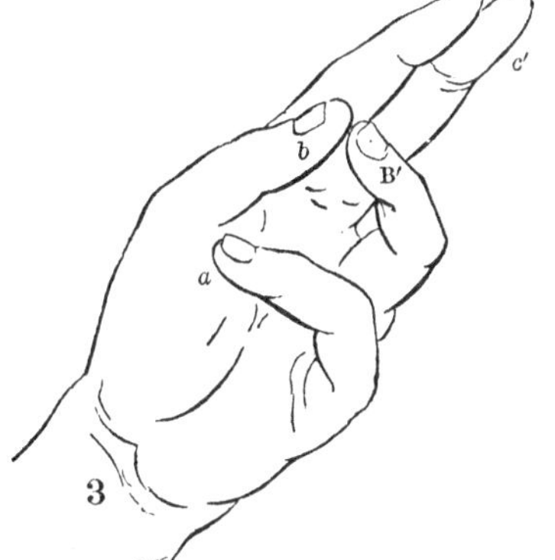


From the general formula, we have,

$$\text{Rad.} \times \sin. a = \cos. A' \times \cos. c'.$$

$$\therefore \sin. a = \frac{\cos. A' \times \cos. c'}{\text{Rad.}}.$$

*By Logarithms.*

$$\begin{array}{lr} \cos. A' - 14° \ 08' \ldots\ldots & 9{\cdot}9860509 \\ \cos. c' \ - 23 \quad 55 \ldots\ldots & \underline{9{\cdot}9610108} \\ & 19{\cdot}9476617 \\ \text{Radius}\ldots\ldots & \underline{10{\cdot}0000000} \\ \sin. a \begin{Bmatrix} 117° \ 34' \\ 62 \quad 26 \end{Bmatrix} \ldots & 9{\cdot}9476617 \end{array}$$

The obtuse side 117° 34′ is the leg required, for the side opposite to the given angle is always of the same species with the given angle.

If in a right-angled spherical triangle, the hypothenuse were 78° 20′, and the angle A = 37° 25′, then the opposite leg $a = 36° \ 31'$, and not 143° 29′, because the given angle is acute.

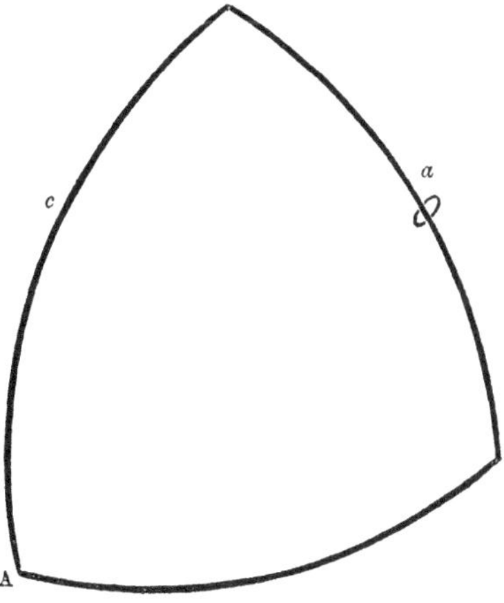


In a right-angled spherical triangle, are given $c = 78° \ 20'$, and $A = 37° \ 25'$, to find the angle B.

$$\begin{array}{rl} & 90° \quad 0' \\ c = & \underline{78 \quad 20} \\ & 11 \quad 40 = c'. \end{array}$$

$$\begin{array}{rl} & 90° \quad 0' \\ A = & \underline{37 \quad 25} \\ & 52 \quad 35 = A' \end{array}$$

Here the complement of the hypothenuse ($c'$) is the *middle part;* and the complement of the angle opposite the perpendicular ($A'$), and the complement of the angle opposite the base ($B'$) are the *adjacent* parts. This will readily be perceived by reference to the usual figure in the margin.

Rad. $\times$ sin. $c'$ = tan. $A'$ $\times$ tan. $B'$;

$$\therefore \text{tan. } B' = \frac{\text{Rad.} \times \text{sin. } c'}{\text{tan. } A'}.$$

*By Logarithms.*

| | |
|---|---|
| Rad. .................. | 10·0000000 |
| sin. $c'$ — 11° 40′. | 9·3058189 |
| | 19·3058189 |
| tan. $A'$ — 52° 35′ | 10·1163279 |
| ∴ tan. $B'$ — 8° 48′ | 9·1894910 |

But 90 — B = $B'$
hence 90 — $B'$ = B.

90° 0′
8 48
B = 81° 12′.

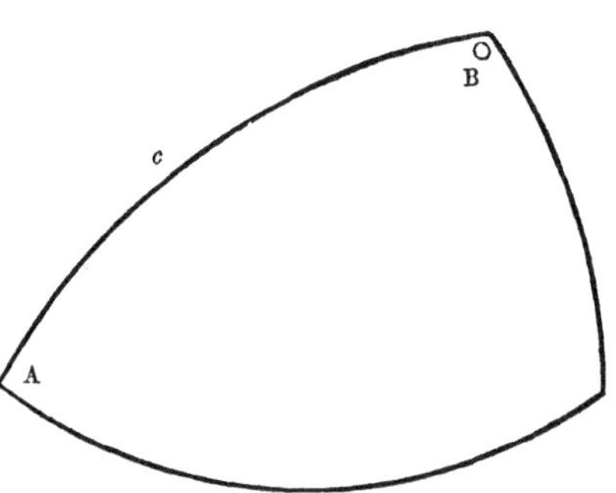


When the hypothenuse and an angle are given, the other angle is acute or obtuse, according as the given parts are of the same or of different species.

In the above example, both the given parts are acute, therefore the required angle is *acute;* but if one be acute and the other obtuse, then the angle found would be obtuse:—Thus, if the hypothenuse be 113° 55′, and the angle A = 31° 51′; then will $B'$ = 14° 08′, and the angle B = 104° 08′.

Given the hypothenuse $c$ = 61° 04′ 56″, and the side or leg, $a$ = 40° 30′ 20″, to find the angle adjacent to $a$.

90° 0′ 0″
$c$ = 61 04 56
28 55 04″ = $c''$.

The three parts are here connected; therefore the complement of B is the *middle part,* $a$ and the complement of $c$ are the adjacent parts.

Hence we have,

Rad. $\times$ sin. $B'$ = tan. $a$ $\times$ tan. $c'$.

$$\therefore \text{sin. } B' = \frac{\text{tan. } a \times \text{tan. } c'}{\text{Rad.}}$$

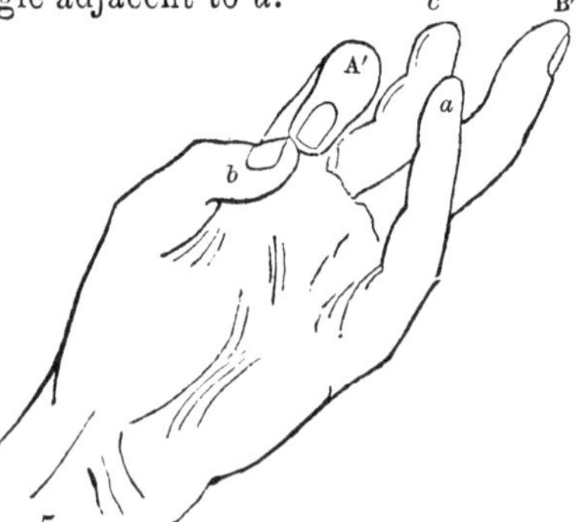

5

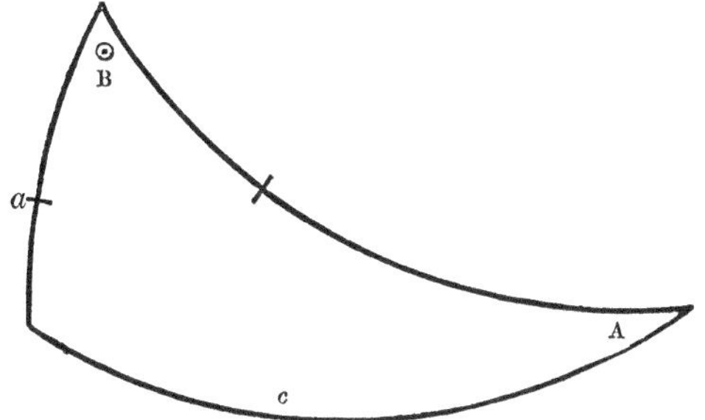


*By Logarithms.*

| | | |
|---|---|---|
| tan. $a$ — 40° 30′ 20″ | = | 9·9315841 |
| tan. $c'$ — 28 55 04 | = | 9·7422801 |
| | | 19·6738642 |
| Rad........................... | | 10·0000000 |
| sin. B′....28° 09′ 31″...... | | 9·6738642 |

$$\begin{array}{r} 90^\circ\ \ 0'\ \ 0'' \\ \text{B}' = \underline{28\ \ 09\ \ 31} \\ 61\ \ 50\ \ 29 = \text{B}. \end{array}$$

The angle adjacent to the given side is acute or obtuse according as the hypothenuse is of the same or of different species with the given side.

Before working the above example, it was easy to foresee that the angle B would be acute; but suppose the hypothenuse = 70° 20′, and the side $a$ = 117° 34′, then the angle B would be obtuse, because $a$ and $c$ are of different species.

RULE V.—In a spherical triangle, right-angled at $c$, are given $c$ = 78° 20′ and $b$ = 117° 34′, to find the angle B; opposite the given leg, (see the next diagram.)

In this example, $b$ becomes the middle part, and $c'$ and B′ opposite parts; and therefore, by the rule,

Rad. × sin. $b$ = cos. B′ × cos. $c'$; that is,

$$\cos. \text{B}' = \frac{\text{Rad.} \times \sin. b}{\cos. c'}.$$

$$90^\circ - 78^\circ\ 20' = 11^\circ\ 40' = c'.$$

*Hence, by Logarithms.*

| | |
|---|---|
| Rad......................... | 10·0000000 |
| sin. $b$ = sin. 117° 34′ or sin. 62 26 | 9·9476655 |
| | 19·9476655 |
| cos. $c'$ 11° 40′............. | 9·9909338 |
| cos. B′ 25° 09′............ | 9·9567317 |

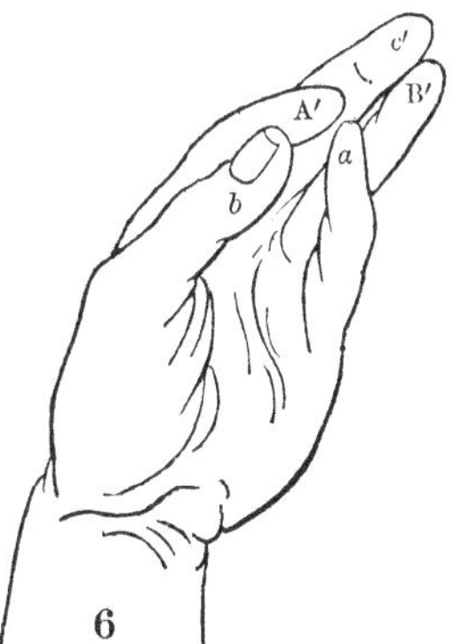

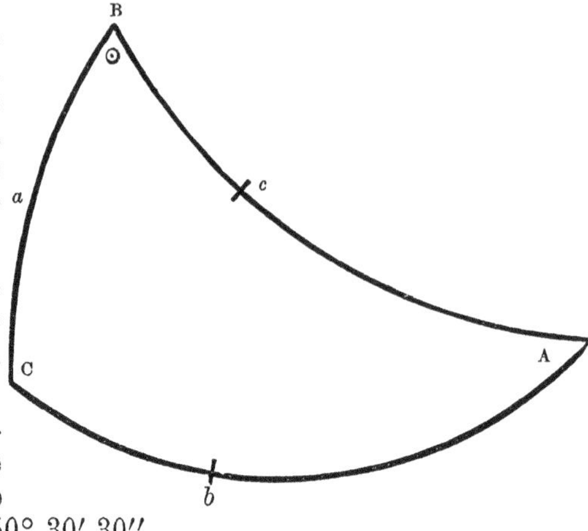


But since the angle opposite the given side is of the same species with the given side, 90° must be added to B′, to produce B:—viz. 90° + 25° 09′ = 115° 09′.

Given $c$ = 61° 04′ 56″, and $b$ = 40° 30′ 20″, to find the other side $a$.

Here $c'$ is the middle part, $a$ and $b$ the opposite parts; hence by position 4, $a$ = 50° 30′ 30″.

Given the side $b$ = 48° 24′ 16″, and the adjacent angle A = 66° 20′ 40″, to find the side $a$.

In this instance, $b$ is the middle part, the complement of A and $a$ are adjacent parts. Consequently, $a$ = 59° 38′ 27″.

In the right-angled spherical triangle ABC,

Given { The side $a$ = 59° 38′ 27″ / Its adjacent angle B = 52° 32′ 55″ } to find the angle A.

Answer, 66° 20′ 40″.

The required angle is of the same species as the *given* side, and *vice versa.*

Given the side $b$ = 49° 17′, and its adjacent angle A = 23° 28′, to find the hypothenuse.

Making A′ the middle part, the others will be adjacent parts, and, therefore, by the first rule we have $c$ = 51° 42′ 37″.

In a spherical triangle, right-angled at C, are given $b$ = 29° 12′ 50″, and B = 37° 26′ 21″, to find the side $a$.

Taking $a$ for the middle part, the other two will be adjacent parts; hence by the rule,

$$\text{Rad.} \times \sin. a = \tan. b \times \tan. B'$$
$$\text{that is, rad.} \times \sin. a = \tan. b \times \cot. B$$
$$\therefore \sin. a = \frac{\tan. b \times \cot. B}{\text{rad.}}$$

In this case, there are two solutions, i. e. $a$ and the supplement of $a$, because both of them have the same sine. As sin. $a$ is necessarily positive, $b$ and B must necessarily be always of the same species, so that, as observed before, the sides including the right angle are always of the same species as the opposite angles.

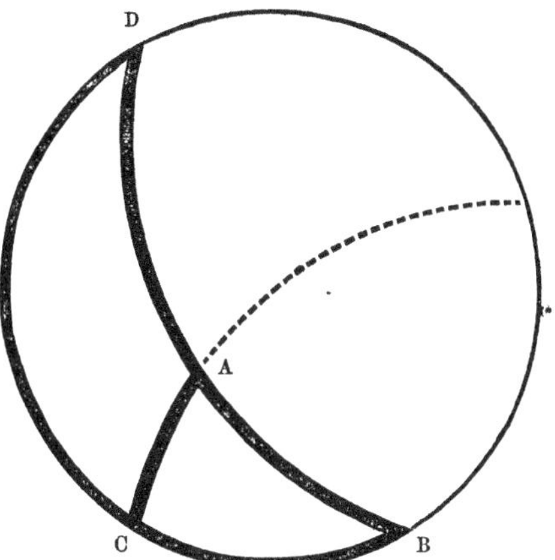


In working this example, we find the log. sin. $a$ = 9·8635411, which corresponds to 46° 55′ 02″,

or, 133° 04′ 58″.

It appears, therefore, that $a$ is *ambiguous*, for there exist two right-angled triangles, having an oblique angle, and the opposite side in the one equal to an oblique angle and an opposite side in the other, but the remaining oblique angle in the one the supplement of the remaining oblique angle in the other. These triangles are situated with respect to each other, on the sphere, as the triangles ABC, ADC, in the annexed diagram, in which, with the exception of the common side AC, and the equal angles B, D, the parts of the one triangle are supplements of the corresponding parts of the other.

In a right-angled spherical triangle are

Given $\left\{\begin{array}{l}\text{the side } a \ldots\ldots\ldots\ldots = 42° 12', \\ \text{its opposite angle A} = 48°\end{array}\right\}$ to find the adjacent angle B.

The complement of the given angle is the middle part; and neither $a$ nor B′ being joined to A′, they are consequently opposite parts; hence, the angle B = 64° 35′, or 115° 25′; this case, like the last, being ambiguous, or doubtful.

Given $a$ = 11° 30′, and A = 23° 30′, to find the hypothenuse $c$.

$c$ = 30°, or 150°, being ambiguous.

In a right-angled triangle, there are given the two perpendicular sides, viz. $a$ = 48° 24′ 16″, $b$ = 59° 38′ 27″, to find the angle A.

A = 66° 20′ 40″.

Given $a$ = 142° 31′, $b$ = 54° 22′, to find $c$.

$c$ = 117° 33′.

Given $\left\{\begin{array}{l}\text{A} = 37° 25' \\ \text{B} = 81\ 12\end{array}\right\}$ Required the side $a$.

$a$ = 36° 31′.

Given $\left\{\begin{array}{l}\text{A} = 66° 20' 40'' \\ \text{B} = 52\ 32\ 55\end{array}\right\}$ to find the hypothenuse $c$.

$c$ = 70° 23′ 42″.

## MEASUREMENT OF ANGLES.

*From the "Civil Engineer and Architect's Journal," for Oct. and Nov.* 1847.

A NEW METHOD OF MEASURING THE DEGREES, MINUTES, ETC., IN ANY RECTILINEAR ANGLE BY COMPASSES ONLY, WITHOUT USING SCALE OR PROTRACTOR.

APPLY AB = $x$, from B to 1; from 1 to 2; from 2 to 3; from 3 to 4; from 4 to 5. Then take B 5, in the compasses, and apply it from B to 6; from 6 to 7; from 7 to 8; from 8 to 9; and from 9 to 10, near the middle of the arc AB. With the same opening,

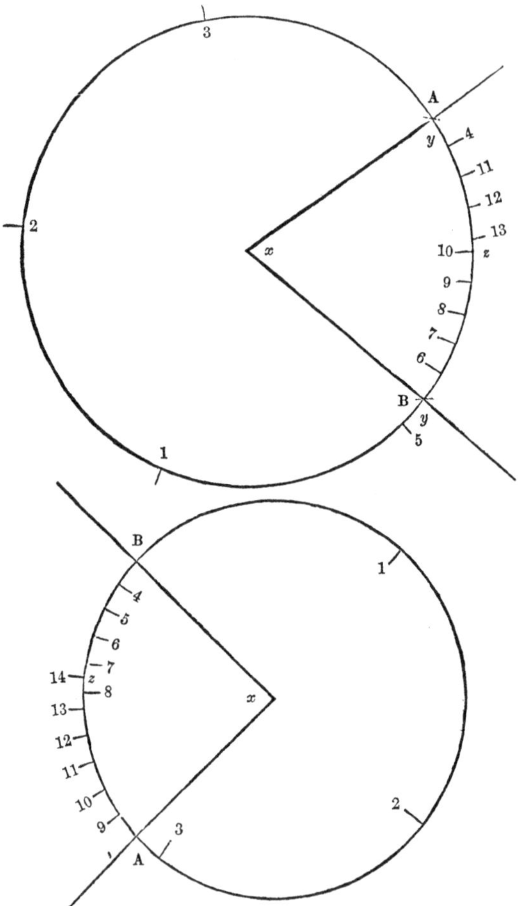


B 5 or A 4, or $y$, which we shall term it, lay off 4,11, 11,12, and 12,13. Then the arc between 13 and 10 is found to be contained 23 times in the arc AB.

Hence, we have,

$$5x - y = 360°;$$
$$9y + z = x;$$
$$23z = x; \text{ or, } z = \frac{x}{23}.$$

$$\therefore 9y + \frac{x}{23} = x, \qquad \therefore y = \frac{22x}{207}.$$

By substituting this value in the first equation, we obtain,

$$5x - \frac{22x}{207} = 360.$$

$$\frac{1013x}{207} = 360, \text{ and } x = \frac{360 \times 207}{1013} = 73° \ 33'{\cdot}82.$$

Apply AB $= x$, from B to 1; from 1 to 2; from 2 to 3; from 3 to 4. Then take B 4, in the compasses, and apply it on the arc, from B to 4; from 4 to 5; from 5 to 6; from 6 to 7; and from 7 to 8, near the middle of the arc AB. With the same opening, B 4 $= y$, lay off A 9, 9,10, 10,11, 11,12, 12,13, and 13,14. The arc between 14 and 8 is found to be contained nearly 24 times in the arc AB. Therefore, we have,

$$4x + y = 360;$$
$$11y - z = x;$$
$$24z = x; \text{ or, } z = \frac{x}{24}.$$

$$\therefore 11y - \frac{x}{24} = x; \qquad \therefore y = \frac{25x}{264}.$$

Substituting this value of $y$ in the first equation,

$$4x + \frac{25x}{264} = 360;$$

$$x = \frac{360 \times 264}{1071} = 88° \ 44'{\cdot}333.$$

*How to lay off an angle of any number of degrees, minutes, &c., with compasses only, without the use of scale or protractor.*

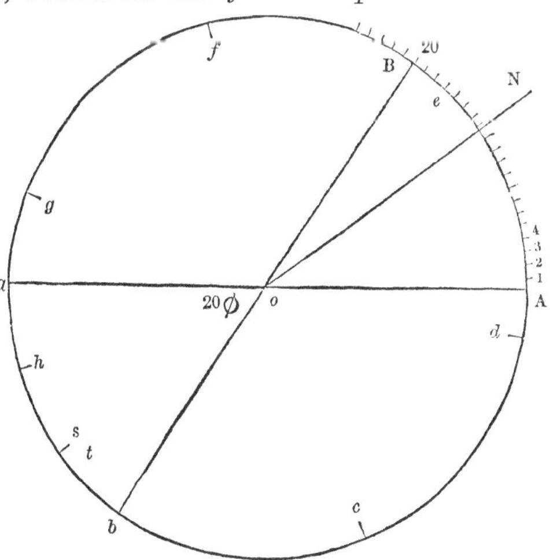


Let it be required to lay off an angle of $36° \ 40' = \beta$. Take any small opening of the compasses less than one-tenth of the radius, and lay off any number of equal small arcs, from A to 1; from 1 to 2; from 2 to 3, &c., until we have laid off an arc, AB, greater than the one required. Draw B $b$ through the centre $o$, then will the arc $a\,b$ = arc AB, which we shall put $= 20\,\phi$ in this example, and proceed to measure $a\,b$ as in the first example. Lay off $a\,b$ from $b$ to $c$; from $c$ to $d$; from $d$ to $e$; from $e$ to $f$; from $f$ to $g$. Putting $g\,a = \triangle_1$, then,

$$6 \times 20\,\phi + \triangle_1 = 360° = \frac{108}{11}\beta;\text{ because,}$$

$$\frac{360°}{36°\ 40'} = \frac{21600}{2200} = \frac{108}{11}.$$

Lay off, as before directed, $g\,a$, $= \triangle_1$, from $a$ to $h$, from $h$ to $s$, and $b$ to $t$; then calling $s\,t$, $\triangle_2$, we have

$$3\,\triangle_1 + \triangle_2 = 20\,\phi;$$

and we find that $s\,t$ is contained 28 times in the arc $a\,b$;

$$\therefore 120\,\phi + \triangle_1 = \frac{108}{11}\beta;\ 3\,\triangle_1 + \triangle_2 = 20\,\phi;\text{ and } 28\,\triangle_2 = 20\,\phi.$$

Eliminating $\triangle_1$ and $\triangle_2$, we find

$$\beta = \frac{29205}{2268}\phi = 12{\cdot}9\text{ times }\phi\text{ nearly};$$

$\therefore 36°\ 40' = \angle\,A\,o\,N$ is laid off with as much ease and certainty as by a protractor.

As a second example, let it be required to lay off an angle of 132° 27′. From 180° 0′ take 132° 27′ = 47° 33′, which put $= \beta$.

$$\frac{360°}{47°\ 33'} = \frac{2400}{317}\text{ when put } = \frac{\nu}{\delta},\text{ then }\frac{\nu}{\delta}\beta = 360° = \pi.$$

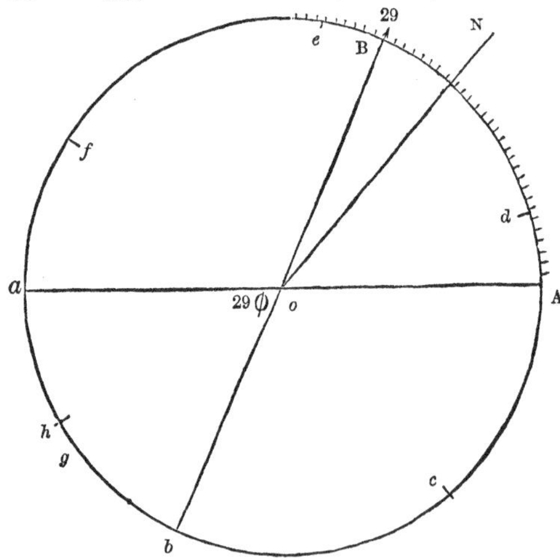


We have laid off 29 small arcs from A to B; $29 = \varepsilon$. $AB = a\,b = b\,c = c\,d = d\,e = e\,f$. And $a\,g = b\,h = a\,f = \triangle_1$; $h\,g = \triangle_2$.

$$\therefore 5 \times 29\,\phi + \triangle_1 = 360° = \frac{\nu}{\delta}\beta = m\,e\,\phi \pm \triangle_1 \qquad (1)$$

$$2\,\triangle_1 - \triangle_2 = 29\,\phi,\text{ or } n\,\triangle_1 \pm \triangle_2 = \varepsilon\,\phi \qquad (2)$$

$$13\,\triangle_2 = 29\,\phi,\quad\text{or}\quad q\,\triangle_2 = \varepsilon\,\phi \qquad (3)$$

Eliminating $\triangle_1$ and $\triangle_2$, we have

$$\beta = \frac{\{m\,n\,q \pm (q \mp 1)\}\,\varepsilon\,\delta}{\nu\,n\,q}\,\phi = \frac{\{5\cdot2\cdot13 + (13+1)\}\,29\cdot317}{2400\cdot2\cdot13}\,\phi = \frac{1323729}{62400}\,\phi = 21\tfrac{1}{4}$$ times $\phi$ very nearly. Hence the line $o$ N determines the angle $a\,o\,\mathrm{N} = 132^\circ\ 27'$.

In the expression

$$\beta = \frac{\{m\,n\,q \pm (q \mp 1)\}\,\varepsilon\,\delta}{\nu\,n\,q}\,\phi \quad (\mathrm{R})$$

substituting the numerals of the first example, then

$$\beta = \frac{\{6\cdot3\cdot28 + (28-1)\}\,20\cdot11}{108\cdot3\cdot28}\,\phi = \frac{29205}{2268}\,\phi = 12\cdot9$$ times $\phi$ nearly, the result before obtained.

The ambiguous signs of (R) cannot be mistaken or lead to error, if the manner in which it is deduced from (1), (2), (3), be attended to. From (3)

$\triangle_2 = \dfrac{\varepsilon\,\phi}{q}$; substituting this value of $\triangle_2$, in (2),

$n\,\triangle_1 = \varepsilon\,\phi \mp \triangle_2 = \varepsilon\,\phi \mp \dfrac{\varepsilon\,\phi}{q}$; which, when substituted for $\triangle_1$ in (1), gives

$\dfrac{\nu}{\delta}\,\beta = m\,\varepsilon\,\phi \pm \dfrac{1}{n}\left(\varepsilon\,\phi \mp \dfrac{\varepsilon\,\phi}{q}\right)$; from which (R) is found.

This method of measuring angles is more exact than it may appear; for if, in the first example, we take

$$5\,x - y = 360;\ 9\,y + z = x;\ \text{and}\ 20\,z = x,$$
$$\text{then}\ x = \frac{64800}{881} = 73^\circ\ 33'\ 85.$$

The first equations gave $73^\circ\ 33'\ 82$ when $23\,z = x$, so it does not matter much whether 20, 21, 22, 23, 24, or 25 times $z$ make $x$. This fact is particularly worth attention.

Given the three angles to find the three sides.

The following formulas give any side $a$ of any spherical triangle.

$$\sin.\ \tfrac{1}{2}\,a = \sqrt{\frac{-\cos.\ \tfrac{1}{2}\,\mathrm{S}\ \cos.\ (\tfrac{1}{2}\,\mathrm{S} - \mathrm{A})}{\sin.\ \mathrm{B}\ \sin.\ \mathrm{C}}},\ \text{and}$$

$$\cos.\ \tfrac{1}{2}\,a = \sqrt{\frac{\cos.\ (\tfrac{1}{2}\,\mathrm{S} - \mathrm{B})\ \cos.\ (\tfrac{1}{2}\,\mathrm{S} - \mathrm{C})}{\sin.\ \mathrm{B}\ \sin.\ \mathrm{C}.}}$$

Given the three sides to find the three angles.

$$\sin.\ \tfrac{1}{2}\,\mathrm{A} = \sqrt{\frac{\sin.\ (\tfrac{1}{2}\,\mathrm{S} - b)\ \sin.\ (\tfrac{1}{2}\,\mathrm{S} - c)}{\sin.\ b\ \sin.\ c.}}$$

$$\cos.\ \tfrac{1}{2}\,\mathrm{A} = \sqrt{\frac{\sin.\ \tfrac{1}{2}\,\mathrm{S}\ \sin.\ (\tfrac{1}{2}\,\mathrm{S} - a)}{\sin.\ b\ \sin.\ c.}}$$

## GRAVITY—WEIGHT—MASS.

SPECIFIC GRAVITY, CENTRE OF GRAVITY, AND OTHER CENTRES OF BODIES. —WEIGHTS OF ENGINEERING AND MECHANICAL MATERIALS.—BRASS, COPPER, STEEL, IRON, WATER, STONE, LEAD, TIN, ROUND, SQUARE, FLAT, ANGULAR, ETC.

1. IN a second, the acceleration of a body falling freely in vacuo is 32·2 feet; what velocity has it acquired at the end of 5 seconds?

$$32{\cdot}2 \times 5 = 161 \text{ feet, the velocity.}$$

2. A cylinder rolling down an inclined plane with an initial velocity of 24 feet a second, and suppose it to acquire each second 5 additional feet velocity; what is its velocity at the end of 3·7 seconds?

$$24 + 3{\cdot}7 \times 5 = 42{\cdot}5 \text{ feet.}$$

3. Suppose a locomotive, moving at the rate of 30 feet a second, (as it is usually termed, with a 30 feet velocity,) and suppose it to lose 5 feet velocity every second; what is its velocity at the end of 3·33 seconds?

The acceleration is — 3·33, negative.

$$\therefore 30 - 5 \times 3{\cdot}33 = 13{\cdot}35 \text{ feet.}$$

4. If a body has acquired a velocity of 36 feet in 11 seconds, by uniformly accelerated motion; what is the space described?

$$\frac{36 \times 11}{2} = 198 \text{ feet.}$$

5. A carriage at rest moves with an accelerated motion over a space of 200 feet in 45 seconds; at what velocity does it proceed at the beginning of the 46th second?

$$\frac{200 \times 2}{45} = 8{\cdot}8889 \text{ feet, the velocity at the end of the 45th second.}$$

The four fundamental formulas of uniformly accelerated motion are

$$v = p\,t; \quad s = \frac{v\,t}{2}; \quad s = \frac{p\,t^2}{2}; \quad s = \frac{v^2}{2p}.$$

$v$ the velocity, $p$ the acceleration, $t$ the time, and $s$ the space.

6. What space will a body describe that moves with an acceleration of 11·5 feet for 10 seconds.

$$\frac{11{\cdot}5 \times (10)^2}{2} = 575 \text{ feet.}$$

7. A body commences to move with an acceleration of 5·5 feet, and moves on until it is moving at the rate of 100 feet a second; what space has it described?

$$\frac{(100)^2}{2 \times 5{\cdot}5} = 909{\cdot}09 \text{ feet.}$$

8. A body is propelled with an initial velocity of 3 feet, and with an acceleration of 8 feet a second; what space is described in 13 seconds?

$$3 \times 13 + \frac{8 \times (13)^2}{2} = 715 \text{ feet.}$$

9. What distance will a body perform in 35 seconds, commencing with a velocity of 10 feet, and being accelerated to move with a velocity of 40 feet at the beginning of the 36th second?

$$\frac{10 + 40}{2} \times 35 = 875 \text{ feet, the distance.}$$

The formulas for a uniformly accelerated motion, commencing with a velocity $c$, are as follow:—

$$v = c + pt; \quad s = ct + \frac{pt^2}{2}; \quad s = \frac{c + v}{2} t; \quad s = \frac{v^2 - c^2}{2p}.$$

The succeeding formulas are applicable for a uniformly retarded motion with an initial velocity $c$.

$$v = c - pt; \quad s = ct - \frac{pt^2}{2}; \quad s = \frac{c + v}{2} t; \quad s = \frac{c^2 - v^2}{2p}.$$

10. A body rolls up an inclined plane, with an initial velocity of 50 feet, and suffers a retardation of 10 feet the second; to what height will it ascend?

$$\frac{50}{10} = 5 \text{ seconds, the time.}$$

$$\frac{(50)^2}{2 \times 10} = 125 \text{ feet, the height required.}$$

The free vertical descent of bodies in vacuo offers an important example of uniformly accelerated motion. The acceleration in the previous examples was designated by $p$, but in the particular motion, brought about by the force of gravity, the acceleration is designated by the letter $g$, and has the mean value of 32·2 feet.

If this value of $g$ be substituted for $p$, in the preceding formula, we have,

$$v = 32{\cdot}2 \times t; \; v = 8{\cdot}024964 \times \sqrt{s}; \; s = 16{\cdot}1 \times t^2; \; s = {\cdot}015528 \times v^2;$$
$$t = {\cdot}031056 \times v; \text{ and } t = {\cdot}2492224 \times \sqrt{s}.$$

11. What velocity will a body acquire at the end of 5 seconds, in its free descent?

$$32{\cdot}2 \times 5 = 161 \text{ feet.}$$

12. What velocity will a body acquire, after a free descent through a space of 400 feet?

$$8{\cdot}024964 \times \sqrt{400} = 160{\cdot}49928 \text{ feet.}$$

13. What space will a body pass over in its free descent during 10 seconds?

$$16{\cdot}1 \times (10)^2 = 1610 \text{ feet.}$$

14. A body falling freely in vacuo, has in its free descent acquired a velocity of 112 feet; what space is passed over?

$$\cdot 015528 \times (112)^2 = 194 \cdot 783232 \text{ feet.}$$

15. In what time will a body falling freely acquire the velocity of 30 feet?

$$\cdot 031056 \times 30 = \cdot 93168 \text{ seconds.}$$

16. In what time will a body pass over a space of 16 feet, falling freely in vacuo?

$$\cdot 2492224 \times \sqrt{16} = \cdot 9968896 \text{ seconds.}$$

If the free descent of bodies go on, with an initial velocity, which we may call $c$, the formulas are,

$$v = c + g\,t;\ v = c + 32 \cdot 2 \times t;\ v = \sqrt{c^2 + 2\,g\,s};\ v = \sqrt{c^2 + 64 \cdot 4 \times s};$$

$$s = c\,t + g\,\frac{t^2}{2} = c\,t + 16 \cdot 1 \times t^2;\ s = \frac{v^2 - c^2}{2\,g} = \cdot 015528\,(v^2 - c^2).$$

If a body be projected vertically to height, with a velocity which we shall term $c$, then the formulas become,

$$v = c - 32 \cdot 2 \times t;\ v = \sqrt{c^2 - 64 \cdot 4 \times s};\ s = c\,t - g\,\frac{t^2}{2} =$$

$$c\,t - 16 \cdot 1 \times t^2;\ s = \frac{c^2 - v^2}{2\,g} = \cdot 015528\,(c^2 - v^2).$$

17. What space is described by a body passing from 18 feet velocity to 30 feet velocity during its free descent in vacuo.

From the annexed table, we find that the height due to 30 feet velocity.............................................= 13·97516

The height due to 18...............................= 5·03106

Space described....................................... 8·94410

Since this problem and table are often required in practical mechanics, we shall enter into more particulars respecting it.

$$\text{As } s = \frac{v^2 - c^2}{2\,g} = \frac{v^2}{2\,g} - \frac{c^2}{2\,g},$$

if we put $h$ = height due to the initial velocity $c$; that is, $h = \frac{c^2}{2\,g}$; and $h_1$ = the height due to the terminal velocity $v$; that is, $h_1 = \frac{v^2}{2\,g}$; then,

$s = h_1 - h$, for falling bodies, as in the last example; and

$s = h - h_1$, for ascending bodies.

Although these formulas are only strictly true for a free descent in vacuo, they may be used in air, when the velocity is not great. The table will be found useful in hydraulics, and for other heights and velocities besides those set down, for by inspection it is seen that the height ·201242 answers to the velocity 3·6; and the height 20·12423 to 36; and the height 2012·423 to 360; and so on.

TABLE *of the Heights corresponding to different Velocities, in feet the second.*

| Velocity in Feet. | Corresponding Height in Feet. | | | | | | | | | |
|---|---|---|---|---|---|---|---|---|---|---|
| | 0 | 1 | 2 | 3 | 4 | 5 | 6 | 7 | 8 | 9 |
| 0 | ·000000 | ·000155 | ·000621 | ·001398 | ·002484 | ·003882 | ·005590 | ·007609 | ·009938 | ·012578 |
| 1 | ·015528 | ·018789 | ·020652 | ·026242 | ·0304348 | ·0349379 | ·039752 | ·044876 | ·050311 | ·056056 |
| 2 | ·062112 | ·068478 | ·075155 | ·082143 | ·089441 | ·097050 | ·104969 | ·113199 | ·121739 | ·130590 |
| 3 | ·139752 | ·149224 | ·159006 | ·169099 | ·187888 | ·190217 | ·201242 | ·212577 | ·224224 | ·236180 |
| 4 | ·248447 | ·261025 | ·273913 | ·285714 | ·300621 | ·314441 | ·328572 | ·343013 | ·357764 | ·372826 |
| 5 | ·388199 | ·403882 | ·419877 | ·436180 | ·452795 | ·469720 | ·486956 | ·504503 | ·522360 | ·550578 |
| 6 | ·559006 | ·577795 | ·596894 | ·616304 | ·636025 | ·656060 | ·676397 | ·697050 | ·718013 | ·739286 |
| 7 | ·760870 | ·782764 | ·804970 | ·827484 | ·850310 | ·873447 | ·896895 | ·920652 | ·944721 | ·969099 |
| 8 | ·993789 | 1·018790 | 1·044100 | 1·069720 | 1·095652 | 1·121895 | 1·148421 | 1·175311 | 1·201482 | 1·229971 |
| 9 | 1·257764 | 1·285869 | 1·314285 | 1·343012 | 1·372050 | 1·401400 | 1·431055 | 1·461025 | 1·491304 | 1·521894 |

The following extension is obtained from the foregoing table, by mere inspection, and moving the decimal point as before directed.

| Velocity in Feet. | Corresponding Height in Feet. | Velocity in Feet. | Corresponding Height in Feet. | Velocity in Feet. | Corresponding Height in Feet. | Velocity in Feet. | Corresponding Height in Feet. |
|---|---|---|---|---|---|---|---|
| 10 | 1·552795 | 19 | 5·60559 | 28 | 12·17392 | 37 | 21·25777 |
| 11 | 1·878882 | 20 | 6·21118 | 29 | 13·05901 | 38 | 22·42236 |
| 12 | 2·065218 | 21 | 6·84783 | 30 | 13·97516 | 39 | 23·61802 |
| 13 | 2·624224 | 22 | 7·51553 | 31 | 14·92237 | 40 | 24·84472 |
| 14 | 3·043478 | 23 | 8·21429 | 32 | 15·90062 | 41 | 26·10249 |
| 15 | 3·49379 | 24 | 8·94410 | 33 | 16·90994 | 42 | 27·39131 |
| 16 | 3·97516 | 25 | 9·70497 | 34 | 18·78883 | 43 | 28·57143 |
| 17 | 4·48758 | 26 | 10·49690 | 35 | 19·02174 | 44 | 30·06212 |
| 18 | 5·03106 | 27 | 11·31988 | 36 | 20·12423 | 45 | 31·4441 |

18. What *mass* does a body weighing 30268 lbs. contain?

$$\frac{30268}{32{\cdot}2} = \frac{302680}{322} = 940 \text{ lbs.}$$

For the mass is equal to the weight divided by $g$. And $g$ is taken equal to 32·2; but the acceleration of gravity is somewhat variable; it becomes greater the nearer we approach the poles of the earth. It is greatest at the poles and least at the equator, and also diminishes the more a body is above or below the level of the sea. The *mass*, so long as nothing is added to or taken from it, is invariable, whether at the centre of the earth or at any distance from it. If M be the *mass* and W the weight of a body,

$$\text{Then } M = \frac{W}{g} = \frac{W}{32{\cdot}2} = {\cdot}0310559 \text{ W.}$$

19. What is the *mass* of a body whose weight is 200 lbs?

$${\cdot}031055 \times 200 = 6{\cdot}21118 \text{ lbs.}$$

The weight of a body whose *mass* is 200 lbs. is 32·2 × 200 = 6440·0 lbs. It may be remarked, that one and the same steel spring is differently bent by one and the same weight at different places.

The force which accelerates the motion of a heavy body on an inclined plane, is to the force of gravity as the sine of the inclination of the plane to the radius, or as the height of the plane to its length.

The velocity acquired by a body in falling from rest through a given height, is the same, whether it fall freely, or descend on a plane at whatever inclination.

The space through which a body will descend on an inclined plane, is to the space through which it would fall freely in the same time, as the sine of the inclination of the plane to the radius.

The velocities which bodies acquire by descending along chords of the same circle, are as the lengths of those chords.

If the body descend in a curve, it suffers no loss of velocity.

*The centre of gravity of a body is a point about which all its parts are in equilibrio.*

Hence, if a body be suspended or supported by this point, the body will rest in any position into which it is put. We may, therefore, consider the whole weight of a body as centred in this point.

The common centre of gravity of two or more bodies, is the point about which they would equiponderate or rest in any position. If the centres of gravity of two bodies be connected by a right line, the distances from the common centre of gravity are reciprocally as the weights of the bodies.

If a line be drawn from the centre of gravity of a body, perpendicular to the horizon, it is called *the line of direction*, being the line that the centre of gravity would describe if the body fell freely.

*The centre of gyration is that part of a body revolving about an axis, into which if the whole quantity of matter were collected, the same moving force would generate the same angular velocity.*

*To find the centre of Gyration.*—Multiply the weight of the several particles by the squares of their distances from the centre of motion, and divide the sum of the products by the weight of the whole mass; the square root of the quotient will be the distance of the centre of gyration from the centre of motion.

The distances of the centre of gyration from the centre of motion, in different revolving bodies, are as follow:—

In a straight rod revolving about one end, the length × ·5773.

In a circular plate, revolving on its centre, the radius × ·7071.

In a circular plate, revolving about one diameter, the radius × ·5.

In a thin circular ring, revolving about one diameter, radius × ·7071.

In a solid sphere, revolving about one diameter, the radius × ·6325.

In a thin hollow sphere, revolving about one diameter, radius × ·8164.

In a cone, revolving about its axis, the radius of the base × ·5477.

In a right-angled cone, revolving about its vertex, the height × ·866.

In a paraboloid, revolving about its axis, the radius of the base $\times$ ·5773.

*The centre of percussion is that point in a body revolving about a fixed axis, into which the whole of the force or motion is collected.*

It is, therefore, that point of a revolving body which would strike any obstacle with the greatest effect; and, from this property, it has received the name of the centre of percussion.

The centres of oscillation and percussion are in the same point.

If a heavy straight bar, of uniform density, be suspended at one extremity, the distance of its centre of percussion is two-thirds of its length.

In a long slender rod of a cylindrical or prismatic shape, the centre of percussion is nearly two-thirds of the length from the axis of suspension.

In an isosceles triangle, suspended by its apex, the distance of the centre of percussion is three-fourths of its altitude. In a line or rod whose density varies as the distance from the point of suspension, also in a fly-wheel, and in wheels in general, the centre of percussion is distant from the centre of suspension three-fourths of the length.

In a very slender cone or pyramid, vibrating about its apex, the distance of its centre of percussion is nearly four-fifths of its length.

Pendulums of the same length vibrate slower, the nearer they are brought to the equator. A pendulum, therefore, to vibrate seconds at the equator, must be somewhat shorter than at the poles.

When we consider a simple pendulum as a ball, which is suspended by a rod or line, supposed to be inflexible, and without weight, we suppose the whole weight to be collected in the centre of gravity of the ball. But when a pendulum consists of a ball, or any other figure, suspended by a metallic or wooden rod, the length of the pendulum is the distance from the point of suspension to a point in the pendulum, called the *centre of oscillation*, which does not exactly coincide with the centre of gravity of the ball.

If a rod of iron were suspended, and made to vibrate, that point in which all its force would be collected is called its centre of oscillation, and is situated at two-thirds the length of the rod from the point of suspension.

---

## SPECIFIC GRAVITY.

THE comparative density of various substances, expressed by the term *specific gravity*, affords the means of readily determining the bulk from the known weight, or the weight from the known bulk; and this will be found more especially useful, in cases where the substance is too large to admit of being weighed, or too irregular in shape to allow of correct measurement. The standard with which all solids and liquids are thus compared, is that of distilled water, one cubic foot of which weighs 1000 ounces avoirdupois;

and the specific gravity of a *solid* body is determined by the difference between its weight in the air, and in water. Thus,

If the body be *heavier* than water, it will displace a quantity of fluid equal to it in *bulk*, and will lose as much weight on immersion as that of an equal bulk of the fluid. Let it be weighed first, therefore, in the air, and then in water, and its weight in the air be divided by the difference between the two weights, and the quotient will be its specific gravity, that of water being unity.

A piece of copper ore weighs $56\frac{1}{4}$ ounces in the air, and $43\frac{3}{4}$ ounces in water; required its specific gravity.

$56{\cdot}25 - 43{\cdot}75 = 12{\cdot}5$ and $56{\cdot}25 \div 12{\cdot}5 = 4{\cdot}5$, the specific gravity.

If the body be *lighter* than water, it will float, and displace a quantity of fluid equal to it in *weight*, the bulk of which will be equal to that only of the part immersed. A heavier substance must, therefore, be attached to it, so that the two may sink in the fluid. Then, the weight of the lighter substance in the air, must be added to that of the heavier substance in water, and the weight of both united, in water, be subtracted from the sum; the weight of the lighter body in the air must then be divided by the difference, and the quotient will be the specific gravity of the lighter substance required.

A piece of fir weighs 40 ounces in the air, and, being immersed in water attached to a piece of iron weighing 30 ounces, the *two* together are found to weigh 3·3 ounces in water, and the iron alone, 25·8 ounces in the water; required the specific gravity of the wood.

$40 + 25{\cdot}8 = 65{\cdot}8 - 3{\cdot}3 = 62{\cdot}5$; and $40 \div 62{\cdot}5 = 0{\cdot}64$, the specific gravity of the fir.

The specific gravity of a *fluid* may be determined by taking a solid body, heavy enough to sink in the fluid, and of known specific gravity, and weighing it both in the air and in the fluid. The difference between the two weights must be multiplied by the specific gravity of the solid body, and the product divided by the weight of the solid in the air: the quotient will be the specific gravity of the fluid, that of water being unity.

Required the specific gravity of a given mixture of muriatic acid and water; a piece of glass, the specific gravity of which is 3, weighing $3\frac{3}{4}$ ounces when immersed in it, and 6 ounces in the air.

$6 - 3{\cdot}75 = 2{\cdot}25 \times 3 = 6{\cdot}75 \div 6 = 1{\cdot}125$, the specific gravity.

Since the weight of a cubic foot of distilled water, at the temperature of 60 degrees, (Fahrenheit,) has been ascertained to be 1000 avoirdupois ounces, it follows that the specific gravities of all bodies compared with it, may be made to express the weight, in ounces, of a cubic foot of each, by multiplying these specific gravities (compared with that of water as unity) by 1000. Thus, that of water being 1, and that of silver, as compared with it, being 10·474, the multiplication of each by 1000 will give 1000 ounces for the cubic foot of water, and 10474 ounces for the cubic foot of silver.

In the following tables of *specific gravities*, the numbers in the *first* column, if taken as whole numbers, represent the weight of a cubic foot in ounces; but if the last *three* figures are taken as decimals, they indicate the specific gravity of the body, water being considered as unity, or 1.

To ascertain the number of cubic feet in a substance, from its weight, the whole weight in *pounds* avoirdupois must be divided by the figures against the name, in the *second* column of the table, taken as whole numbers and decimals, and the quotient will be the contents in cubic feet.

Required the cubic content of a mass of cast-iron, weighing 7 cwt. 1 qr. = 812 lbs.

$$812 \text{ lbs.} \div 450{\cdot}5 \text{ (the tabular weight)} = 1{\cdot}803 \text{ cubic feet.}$$

To find the weight from the measurement or cubic content of a substance, this operation must be reversed, and the number of cubic feet, found by the rules given under "*Mensuration of Solids*," *multiplied* by the figures in the *second* column, to obtain the weight in pounds avoirdupois.

Required the weight of a log of oak, 3 feet by 2 feet 6 inches, and 9 feet long.

$$9 \times 3 \times 2{\cdot}5 = 67{\cdot}5 \text{ cubic feet.}$$

And $67{\cdot}5 \times 58{\cdot}2$ (the tabular weight) $= 3928{\cdot}5$ lbs., or 35 cwt. 0 qr. $8\frac{1}{2}$ lbs.

The velocity $g$, which is the measure of the force of gravity, varies with the latitude of the place, and with its altitude above the level of the sea.

The force of gravity at the latitude of 45° = 32·1803 feet; at any other latitude $L$, $g = 32{\cdot}1803$ feet $- 0{\cdot}0821 \cos. 2L$. If $g'$ represents the force of gravity at the height $h$ above the sea, and $r$ the radius of the earth, the force of gravity at the level of the sea will be $g = g'\left(1 + \frac{5h}{4r}\right)$.

In the latitude of London, at the level of the sea, $g = 32{\cdot}191$ feet.
Do. Washington, do. do., $g = 32{\cdot}155$ feet.

The length of a pendulum vibrating seconds is in a constant ratio to the force of gravity.

$$\frac{g}{l} = 9{\cdot}8696044.$$

*Length of a pendulum vibrating seconds at the level of the sea, in various latitudes.*

| | | |
|---|---|---|
| At the Equator | | 39·0152 inches. |
| Washington, | lat. 38° 53′ 23″ | 39·0958 — |
| New York, | lat. 40° 42′ 40″ | 39·1017 — |
| London, | lat. 51° 31′ | 39·1393 — |
| | lat. 45° | 39·1270 — |
| | lat. $L$ | 39·1270 in.—0·09982 cos. 2 $L$. |

## *Specific Gravity of various Substances.*

| METALS. | Weight of a cubic foot in ounces. | Weight of a cubic foot in pounds. |
|---|---|---|
| Antimony, fused | 6,624 | 414·0 |
| Bismuth, cast | 9,823 | 614·0 |
| Brass, common, cast | 7,824 | 489·0 |
| cast | 8,396 | 524·8 |
| wire-drawn | 8,544 | 534·0 |
| Copper, cast | 8,788 | 549·2 |
| wire-drawn | 8,878 | 554·9 |
| Gold, pure, cast | 19,258 | 1203·6 |
| 22 carats, stand | 17,486 | 1093·0 |
| 20 carats, trinket | 15,709 | 982·0 |
| Iron, cast | 7,207 | 450·5 |
| bars | 7,788 | 486·8 |
| Lead, cast | 11,352 | 709·5 |
| litharge | 6,300 | 393·8 |
| Manganese | 7,000 | 437·5 |
| Mercury, solid, 40° below 0° | 15,632 | 977·0 |
| at 32 deg. Fahr. | 13,619 | 851·2 |
| at 60 deg. | 13,580 | 848·8 |
| at 212 deg. | 13,375 | 836·8 |
| Nickel, cast | 7,807 | 488·0 |
| Platina, crude, grains | 15,602 | 975·1 |
| purified | 19,500 | 1218·8 |
| hammered | 20,337 | 1271·1 |
| rolled | 22,069 | 1379·4 |
| wire-drawn | 21,042 | 1315·1 |
| Silver, cast, pure | 10,474 | 654·6 |
| Parisian standard | 10,175 | 636·0 |
| French coin | 10,048 | 628·0 |
| shilling, Geo. III. | 10,534 | 658·4 |
| Steel, soft | 7,833 | 489·6 |
| hardened | 7,840 | 490·0 |
| tempered | 7,816 | 488·5 |
| tempered and hard | 7,818 | 488·6 |
| Tin, pure Cornish | 7,291 | 455·6 |
| Tungsten | 6,066 | 379·1 |
| Uranium | 6,440 | 402·5 |
| Wolfram | 7,119 | 445·0 |
| Zinc, usual state | 6,862 | 429·0 |
| pure | 7,191 | 449·5 |
| **WOODS.** | | |
| Ash | 845 | 52·9 |
| Beech | 852 | 53·2 |
| Box, Dutch | 912 | 57·0 |
| French | 1,328 | 83·0 |
| Brazilian | 1,031 | 64·5 |
| Cedar, American | 561 | 35·1 |
| Indian | 1,315 | 82·2 |
| Cherry-tree | 715 | 44·8 |
| Cocoa | 1,040 | 65·0 |
| Cork | 240 | 15·0 |
| Ebony, Indian | 1,209 | 75·6 |
| American | 1,331 | 83·2 |
| Elm | 671 | 42·0 |
| Fir, yellow | 657 | 41·1 |
| white | 569 | 35·6 |
| Lignum-vitæ | 1,333 | 83·4 |
| Lime-tree | 604 | 37·8 |
| Logwood | 913 | 57·1 |
| Mahogany | 1,063 | 66·5 |
| Maple | 750 | 47·0 |
| Oak, heart of, old | 1,170 | 73·1 |
| dry | 932 | 58·2 |
| Vine | 1,327 | 83·0 |
| Walnut | 671 | 42·0 |
| Willow | 585 | 36·6 |
| Yew | 807 | 50·5 |
| **STONES, EARTHS, ETC.** | | |
| Alabaster, yellow | 2,699 | 168·8 |
| white | 2,730 | 170·6 |
| Borax | 1,714 | 107·1 |
| Brick earth | 2,000 | 125·0 |
| Chalk | 2,784 | 174·0 |
| Coal, Cannel | 1,270 | 79·4 |
| Newcastle | 1,270 | 79·4 |
| Staffordshire | 1,240 | 77·5 |
| Scotch | 1,300 | 81·2 |
| Emery | 4,000 | 250·0 |
| Flint, black | 2,582 | 162·0 |
| Glass, flint | 2,933 | 170·9 |
| white | 2,892 | 168·2 |
| Granite, Aberd. blue | 2,625 | 164·1 |
| Cornish | 2,662 | 166·4 |
| Egyptian, red | 2,654 | 165·9 |
| " gray | 2,728 | 170·5 |
| **STONES.—*Continued.*** | | |
| Grindstone | 2,143 | 134·0 |
| Gypsum, opaque | 2,168 | 135·5 |
| semi-transparent | 2,306 | 144·1 |
| Jet, bituminous | 1,259 | 78·8 |
| Lime-stone | 3,182 | 199·0 |
| Marble | 2,700 | 168·8 |
| Mill-stone | 2,484 | 155·2 |
| Porcelain, China | 2,385 | 149·1 |
| Portland-stone | 2,570 | 160·6 |
| Pumice stone | 915 | 57·2 |
| Paving-stone | 2,416 | 151·0 |
| Purbeck-stone | 2,601 | 162·6 |
| Rotten-stone | 1,981 | 124·0 |
| Slate, common | 2,672 | 167·0 |
| new | 2,854 | 178·4 |
| Stone, common | 2,520 | 157·5 |
| rag | 2,470 | 154·4 |
| Sulphur, native | 2,033 | 127·1 |
| melted | 1,991 | 124·5 |
| **LIQUIDS.** | | |
| Acetic acid | 1,007 | 63·0 |
| Acetous acid | 1,025 | 64·1 |
| Alcohol, commercial | 837 | 52·3 |
| highly rectified | 829 | 51·8 |
| Ammonia, liquid | 897 | 56·1 |
| Beer | 1,023 | 68·0 |
| Ether, sulphuric | 739 | 46·2 |
| Milk of cows | 1,032 | 64·5 |
| Muriatic acid | 1,194 | 74·6 |
| Nitric acid | 1,271 | 79·5 |
| highly concentrated | 1,583 | 99·0 |
| Oil of almonds, sweet | 917 | 57·4 |
| hemp-seed | 926 | 58·0 |
| linseed | 940 | 58·8 |
| olives | 915 | 57·3 |
| poppies | 924 | 57·8 |
| rape-seed | 919 | 57·5 |
| turpentine, essence | 870 | 54·4 |
| whales | 923 | 57·8 |
| Spirits of wine, commercial | 837 | 52·4 |
| highly rectified | 829 | 51·9 |
| Sulphuric acid | 1,841 | 115·1 |
| highly concentrated | 2,125 | 133·0 |
| Turpentine, liquid | 991 | 62·0 |
| Vinegar, distilled | 1,010 | 63·1 |
| Water, rain, or distilled | 1,000 | 62·5 |
| sea | 1,026 | 64·1 |
| **MISCELLANEOUS SUBSTANCES.** | | |
| Beeswax | 965 | 60·4 |
| Butter | 942 | 59·0 |
| Camphor | 989 | 62·0 |
| Fat, beef or mutton | 923 | 57·8 |
| hogs' | 937 | 58·6 |
| Honey | 1,450 | 90·6 |
| Indigo | 769 | 48·1 |
| Ivory | 1,826 | 114·1 |
| Lard | 948 | 59·2 |
| Opium | 1,336 | 83·5 |
| Spermaceti | 943 | 59·0 |
| Sugar, white | 1,606 | 100·4 |
| Tallow | 942 | 59·0 |
| **GASES.** *Atmospheric air being estimated as 1.* | | |
| Atmospheric, or common air | | 1·000 |
| Ammoniacal gas | | ·590 |
| Azote | | ·969 |
| Carbonic acid | | 1·520 |
| Carbonic oxide | | ·960 |
| Carburetted hydrogen | | ·491 |
| Chlorine | | ·470 |
| Hydrogen | | ·074 |
| Muriatic acid gas | | 1·278 |
| Nitrous gas | | 1·094 |
| Nitrous acid gas | | 2·427 |
| Oxygen | | 1·104 |
| Steam | | ·690 |
| Sulphuretted hydrogen | | 1·777 |
| Sulphurous acid | | 2·193 |

## TABLE *of the Weight of a Foot in length of Flat and Rolled Iron.*

| Thickness in inches and parts. | BREADTH IN INCHES AND PARTS OF AN INCH. | | | | | | | | | | | | | | | |
|---|---|---|---|---|---|---|---|---|---|---|---|---|---|---|---|---|
| | 4 | 3¾ | 3½ | 3¼ | 3 | 2¾ | 2½ | 2¼ | 2 | 1¾ | 1½ | 1⅜ | 1¼ | 1 | ¾ | ½ |
| ⅛ | 1·68 | 1·57 | 1·47 | 1·36 | 1·26 | 1·15 | 1·05 | 0·94 | 0·84 | 0·73 | 0·63 | 0·57 | 0·52 | 0·42 | 0·31 | 0·21 |
| 3/16 | 2·52 | 2·36 | 2·20 | 2·04 | 1·89 | 1·73 | 1·57 | 1·41 | 1·26 | 1·10 | 0·94 | 0·86 | 0·78 | 0·63 | 0·47 | 0·31 |
| ¼ | 3·36 | 3·15 | 2·94 | 2·73 | 2·52 | 2·31 | 2·10 | 1·89 | 1·68 | 1·47 | 1·26 | 1·18 | 1·05 | 0·84 | 0·63 | 0·42 |
| ⅜ | 5·04 | 4·72 | 4·41 | 4·09 | 3·78 | 3·46 | 3·15 | 2·83 | 2·52 | 2·20 | 1·89 | 1·73 | 1·57 | 1·26 | 0·94 | 0·63 |
| ½ | 6·72 | 6·30 | 5·88 | 5·46 | 5·04 | 4·62 | 4·20 | 3·78 | 3·36 | 2·94 | 2·52 | 2·31 | 2·10 | 1·68 | 1·26 | |
| ⅝ | 8·40 | 7·87 | 7·35 | 6·82 | 6·30 | 5·77 | 5·25 | 4·72 | 4·20 | 3·67 | 3·15 | 2·88 | 2·62 | 2·10 | 1·57 | |
| ¾ | 10·08 | 9·45 | 8·82 | 8·19 | 7·56 | 6·93 | 6·30 | 5·66 | 5·04 | 4·41 | 3·78 | 3·46 | 3·15 | 2·52 | | |
| ⅞ | 11·76 | 11·02 | 10·29 | 9·45 | 8·82 | 8·08 | 7·35 | 6·61 | 5·88 | 5·14 | 4·41 | 4·04 | 3·67 | 2·94 | | |
| 1 | 13·44 | 12·60 | 11·76 | 10·92 | 10·08 | 9·24 | 8·40 | 7·56 | 6·72 | 5·87 | 5·04 | 4·62 | 4·20 | | | |
| 1⅛ | 15·12 | 14·16 | 13·20 | 12·28 | 11·34 | 10·39 | 9·45 | 8·50 | 7·56 | 6·60 | 5·67 | 5·19 | 4·72 | | | |
| 1¼ | 16·80 | 15·75 | 14·70 | 13·65 | 12·60 | 11·55 | 10·50 | 9·45 | 8·40 | 7·35 | 6·30 | 5·77 | | | | |
| 1⅜ | 18·48 | 17·32 | 16·16 | 15·01 | 13·86 | 12·70 | 11·55 | 10·39 | 9·24 | 8·07 | | | | | | |
| 1½ | 20·18 | 18·90 | 17·64 | 16·38 | 15·12 | 13·86 | 12·60 | 11·34 | 10·08 | 8·80 | | | | | | |
| 1¾ | 23·54 | 22·05 | 20·58 | 19·11 | 17·64 | 16·17 | 14·70 | 13·22 | | | | | | | | |
| 2 | 26·88 | 25·20 | 23·52 | 21·84 | 20·16 | 18·48 | 16·80 | 15·12 | | | | | | | | |
| 2½ | 33·65 | 31·50 | 29·40 | 27·39 | 25·20 | 23·10 | | | | | | | | | | |
| 3 | 40·32 | 37·80 | 35·28 | 32·76 | | | | | | | | | | | | |
| 3½ | 47·04 | | | | | | | | | | | | | | | |

## TABLE *of the Weight of Cast-iron Pipes, in lengths.*

| Bore. | Thick. | Long. | Weight. | Bore. | Thick. | Long. | Weight. | Bore. | Thick. | Long. | Weight. |
|---|---|---|---|---|---|---|---|---|---|---|---|
| Inch. | Inch. | Feet. | C. qr. lb. | Inch. | Inch. | Feet. | C. qr. lb. | Inch. | Inch. | Feet. | C. qr. lb. |
| 1 | ¼ | 3½ | 12 | 6½ | ⅜ | 9 | 2 0 16 | 11½ | ½ | 9 | 5 0 7 |
| | ⅜ | 3½ | 21 | | ½ | 9 | 2 3 20 | | ⅝ | 9 | 6 1 12 |
| 1½ | ¼ | 4½ | 21 | | ⅝ | 9 | 3 2 21 | | ¾ | 9 | 7 2 8 |
| | ⅜ | 4½ | 1 4 | | ¾ | 9 | 4 1 21 | | 1 | 9 | 10 1 2 |
| 2 | ¼ | 6 | 1 8 | | 1 | 9 | 6 0 14 | 12 | ½ | 9 | 5 0 24 |
| | ⅜ | 6 | 2 0 | 7 | ½ | 9 | 3 0 7 | | ⅝ | 9 | 6 2 8 |
| 2½ | ¼ | 6 | 1 16 | | ⅝ | 9 | 3 3 20 | | ¾ | 9 | 7 3 20 |
| | ⅜ | 6 | 2 10 | | ¾ | 9 | 4 3 5 | | 1 | 9 | 10 3 0 |
| | ½ | 6 | 3 10 | | 1 | 9 | 6 2 4 | 12½ | ½ | 9 | 5 1 16 |
| 3 | ¼ | 9 | 2 20 | 7½ | ½ | 9 | 3 1 6 | | ⅝ | 9 | 6 3 9 |
| | ⅜ | 9 | 1 0 6 | | ⅝ | 9 | 4 0 22 | | ¾ | 9 | 8 1 0 |
| | ½ | 9 | 1 1 12 | | ¾ | 9 | 5 0 10 | | 1 | 9 | 11 0 21 |
| | ⅝ | 9 | 1 3 6 | | 1 | 9 | 7 0 0 | 13 | ½ | 9 | 5 2 20 |
| | ¾ | 9 | 2 1 0 | 8 | ½ | 9 | 3 2 4 | | ⅝ | 9 | 7 0 14 |
| 3½ | ¼ | 9 | 3 0 | | ⅝ | 9 | 4 1 25 | | ¾ | 9 | 8 2 7 |
| | ⅜ | 9 | 1 0 21 | | ¾ | 9 | 5 1 18 | | 1 | 9 | 11 2 12 |
| | ½ | 9 | 1 2 14 | | 1 | 0 | 7 1 16 | 13½ | ½ | 9 | 5 3 7 |
| | ⅝ | 9 | 2 0 8 | 8½ | ½ | 9 | 3 3 2 | | ⅝ | 9 | 7 1 12 |
| | ¾ | 9 | 2 2 0 | | ⅝ | 9 | 4 2 26 | | ¾ | 9 | 8 3 16 |
| 4 | ⅜ | 9 | 1 1 10 | | ¾ | 9 | 5 2 22 | | 1 | 9 | 11 3 24 |
| | ½ | 9 | 1 3 12 | | 1 | 9 | 7 3 8 | 14 | ½ | 9 | 6 0 4 |
| | ⅝ | 9 | 2 1 12 | 9 | ½ | 9 | 4 0 0 | | ⅝ | 9 | 7 2 16 |
| | ¾ | 9 | 2 2 21 | | ⅝ | 9 | 5 0 4 | | ¾ | 9 | 9 1 0 |
| 4½ | ⅜ | 9 | 1 2 2 | | ¾ | 9 | 6 0 2 | | 1 | 9 | 12 1 14 |
| | ½ | 9 | 2 0 4 | | 1 | 9 | 8 0 26 | 14½ | ½ | 9 | 6 0 24 |
| | ⅝ | 9 | 2 2 14 | 9½ | ½ | 9 | 4 0 18 | | ⅝ | 9 | 7 3 14 |
| | ¾ | 9 | 3 0 21 | | ⅝ | 9 | 5 1 0 | | ¾ | 9 | 9 2 2 |
| 5 | ⅜ | 9 | 1 2 22 | | ¾ | 9 | 6 1 6 | | 1 | 9 | 12 3 6 |
| | ½ | 9 | 2 1 10 | | 1 | 9 | 8 2 20 | 15 | ½ | 9 | 6 1 21 |
| | ⅝ | 9 | 2 3 17 | 10 | ½ | 9 | 4 1 10 | | ¾ | 9 | 9 3 7 |
| | ¾ | 9 | 3 1 24 | | ⅝ | 9 | 5 1 26 | | 1 | 9 | 13 0 26 |
| 5½ | ⅜ | 9 | 1 3 10 | | ¾ | 9 | 4 2 14 | | 1¼ | 9 | 16 3 5 |
| | ½ | 9 | 2 2 0 | | 1 | 9 | 9 0 8 | 15½ | ½ | 9 | 6 2 14 |
| | ⅝ | 9 | 3 0 18 | 10½ | ½ | 9 | 4 2 14 | | ¾ | 9 | 10 9 10 |
| | ¾ | 9 | 3 3 7 | | ⅝ | 9 | 5 3 7 | | 1 | 9 | 13 2 17 |
| | 1 | 9 | 5 0 12 | | ¾ | 9 | 7 0 0 | | 1¼ | 9 | 17 1 6 |
| 6 | ⅜ | 9 | 2 0 0 | | 1 | 9 | 9 2 0 | 16 | ½ | 9 | 7 0 22 |
| | ½ | 9 | 2 2 21 | 11 | ½ | 9 | 4 3 14 | | ¾ | 9 | 10 1 20 |
| | ⅝ | 9 | 3 1 17 | | ⅝ | 9 | 6 0 11 | | 1 | 9 | 14 0 8 |
| | ¾ | 9 | 4 0 16 | | ¾ | 9 | 7 1 7 | | 1¼ | 9 | 17 3 14 |
| | 1 | 9 | 5 2 20 | | 1 | 9 | 9 3 20 | | 1½ | 9 | 21 3 4 |

TABLE *of the Weight of one Foot Length of Malleable Iron.*

| SQUARE IRON. | | ROUND IRON. | | | |
|---|---|---|---|---|---|
| Scantling. | Weight. | Diameter. | Weight. | Circumference. | Weight. |
| Inches. | Pounds. | Inches. | Pounds. | Inches. | Pounds. |
| $\frac{1}{4}$ | 0·21 | $\frac{1}{4}$ | 0·16 | 1 | 0·26 |
| $\frac{3}{8}$ | 0·47 | $\frac{3}{8}$ | 0·37 | $1\frac{1}{4}$ | 0·41 |
| $\frac{1}{2}$ | 0·84 | $\frac{1}{2}$ | 0·66 | $1\frac{1}{2}$ | 0·59 |
| $\frac{5}{8}$ | 1·34 | $\frac{5}{8}$ | 1·03 | $1\frac{3}{4}$ | 0·82 |
| $\frac{3}{4}$ | 1·89 | $\frac{3}{4}$ | 1·48 | 2 | 1·05 |
| $\frac{7}{8}$ | 2·57 | $\frac{7}{8}$ | 2·02 | $2\frac{1}{4}$ | 1·34 |
| 1 | 3·36 | 1 | 2·63 | $2\frac{1}{2}$ | 1·65 |
| $1\frac{1}{8}$ | 4·25 | $1\frac{1}{8}$ | 3·33 | $2\frac{3}{4}$ | 2·01 |
| $1\frac{1}{4}$ | 5·25 | $1\frac{1}{4}$ | 4·12 | 3 | 2·37 |
| $1\frac{3}{8}$ | 6·35 | $1\frac{3}{8}$ | 4·98 | $3\frac{1}{4}$ | 2·79 |
| $1\frac{1}{2}$ | 7·56 | $1\frac{1}{2}$ | 5·93 | $3\frac{1}{2}$ | 3·24 |
| $1\frac{5}{8}$ | 8·87 | $1\frac{5}{8}$ | 6·96 | $3\frac{3}{4}$ | 3·69 |
| $1\frac{3}{4}$ | 10·29 | $1\frac{3}{4}$ | 8·08 | 4 | 4·23 |
| $1\frac{7}{8}$ | 11·81 | $1\frac{7}{8}$ | 9·27 | $4\frac{1}{2}$ | 5·35 |
| 2 | 13·44 | 2 | 10·55 | 5 | 6·61 |
| $2\frac{1}{4}$ | 17·01 | $2\frac{1}{4}$ | 13·35 | $5\frac{1}{2}$ | 7·99 |
| $2\frac{1}{2}$ | 21·00 | $2\frac{1}{2}$ | 16·48 | 6 | 9·51 |
| $2\frac{3}{4}$ | 25·41 | $2\frac{3}{4}$ | 19·95 | $6\frac{1}{2}$ | 11·18 |
| 3 | 30·24 | 3 | 23·73 | 7 | 12·96 |
| $3\frac{1}{2}$ | 41·16 | $3\frac{1}{4}$ | 27·85 | $7\frac{1}{2}$ | 14·78 |
| 4 | 53·76 | $3\frac{1}{2}$ | 32·32 | 8 | 16·92 |
| $4\frac{1}{2}$ | 68·04 | $3\frac{3}{4}$ | 37·09 | $8\frac{1}{2}$ | 19·21 |
| 5 | 84·00 | 4 | 42·21 | 9 | 21·53 |
| 6 | 120·96 | $4\frac{1}{2}$ | 53·41 | 10 | 26·43 |
| 7 | 164·64 | 5 | 65·93 | 12 | 31·99 |

The following tables are rendered of great utility by means of this table:—

| | | | |
|---|---|---|---|
| The weight of | Water | being | 1· |
| ———— | Copper | = | 8·8 |
| ———— | Brass | = | 8·4 |
| ———— | Iron, cast | = | 7·2 |
| ———— | Lead | = | 11·3 |
| ———— | Zinc | = | 7·2 |
| ———— | Gun-metal | = | 8·7 |
| ———— | Sand | = | 1·5 |
| ———— | Coal | = | 1·25 |
| ———— | Brick | = | 2·0 |
| ———— | Stone | = | 2·5 |
| ———— | Timber, average | = | 0·85 |

Suppose it be required to ascertain the weight of a cast iron pipe $26\frac{1}{4}$ inches outside and $23\frac{3}{4}$ inside, the length being $6\frac{1}{2}$ feet.

Opposite $26\frac{1}{4}$ in the table is

$$234·8576 \times 7·2 \times 6·5 = 10991·135.$$

And opposite $23\frac{3}{4}$ in the table is

$$192·2856 \times 7·2 \times 6·5 = 8998·966 \text{ subtract}$$

$$1992·169 \text{ lbs. avr.}$$

The succeeding table contains the surface and solidity of spheres, together with the edge or dimensions of equal cubes, the length of equal cylinders, and the weight of water in avoirdupois pounds:—

## *Surface and Solidity of Spheres.*

| Diameter. | Surface. | Solidity. | Cube. | Cylinder. | Water in lbs. |
|---|---|---|---|---|---|
| 1 in. | 3·1416 | ·5236 | ·8060 | ·6666 | ·0190 |
| $\frac{1}{16}$ | 3·5465 | ·6280 | ·8563 | ·7082 | ·0227 |
| $\frac{1}{8}$ | 3·9760 | ·7455 | ·9067 | ·7500 | ·0270 |
| $\frac{3}{16}$ | 4·4301 | ·8767 | ·9571 | ·7917 | ·0317 |
| $\frac{1}{4}$ | 4·9087 | 1·0226 | 1·0075 | ·8333 | ·0370 |
| $\frac{5}{16}$ | 5·4117 | 1·1838 | 1·0578 | ·8750 | ·0428 |
| $\frac{3}{8}$ | 5·9395 | 1·3611 | 1·1082 | ·9166 | ·0500 |
| $\frac{7}{16}$ | 6·4918 | 1·5553 | 1·1586 | ·9583 | ·0563 |
| $\frac{1}{2}$ | 7·0686 | 1·7671 | 1·2090 | 1·0000 | ·0640 |
| $\frac{9}{16}$ | 7·6699 | 2·0000 | 1·2593 | 1·0416 | ·0723 |
| $\frac{5}{8}$ | 8·2957 | 2·2467 | 1·3097 | 1·0833 | ·0813 |
| $\frac{11}{16}$ | 8·9461 | 2·5161 | 1·3601 | 1·1349 | ·0910 |
| $\frac{3}{4}$ | 9·6211 | 2·8061 | 1·4105 | 1·1666 | ·1015 |
| $\frac{13}{16}$ | 10·3206 | 3·1176 | 1·4608 | 1·2083 | ·1128 |
| $\frac{7}{8}$ | 11·0446 | 3·4514 | 1·5112 | 1·2500 | ·1250 |
| $\frac{15}{16}$ | 11·7932 | 3·8081 | 1·5616 | 1·2916 | ·1377 |
| 2 in. | 12·5664 | 4·1888 | 1·6020 | 1·3333 | ·1516 |
| $\frac{1}{16}$ | 13·3640 | 4·5938 | 1·6633 | 1·3750 | ·1662 |
| $\frac{1}{8}$ | 14·1862 | 5·0243 | 1·7127 | 1·4166 | ·1818 |
| $\frac{3}{16}$ | 15·0330 | 5·4807 | 1·7631 | 1·4582 | ·1982 |
| $\frac{1}{4}$ | 15·9043 | 6·9640 | 1·8135 | 1·5000 | ·2160 |
| $\frac{5}{16}$ | 16·8000 | 6·4749 | 1·8638 | 1·5516 | ·2342 |
| $\frac{3}{8}$ | 17·7205 | 7·0143 | 1·9142 | 1·5832 | ·2540 |
| $\frac{7}{16}$ | 18·6655 | 7·5828 | 1·9646 | 1·6250 | ·2743 |
| $\frac{1}{2}$ | 19·6350 | 8·1812 | 2·0150 | 1·6666 | ·2960 |
| $\frac{9}{16}$ | 20·6290 | 8·8103 | 2·0653 | 1·7082 | ·3187 |
| $\frac{5}{8}$ | 21·6475 | 9·4708 | 2·1157 | 1·7500 | ·3426 |
| $\frac{11}{16}$ | 22·6907 | 10·1634 | 2·1661 | 1·7915 | ·3676 |
| $\frac{3}{4}$ | 23·7583 | 10·8892 | 2·2165 | 1·8332 | ·3939 |
| $\frac{13}{16}$ | 24·8505 | 11·6485 | 2·2668 | 1·8750 | ·4213 |
| $\frac{7}{8}$ | 25·9672 | 12·4426 | 2·3172 | 1·9165 | ·4501 |
| $\frac{15}{16}$ | 27·1084 | 13·2718 | 2·3676 | 1·9582 | ·4800 |
| 3 in. | 28·2744 | 14·1372 | 2·4180 | 2·0000 | ·5114 |
| $\frac{1}{16}$ | 29·4647 | 15·0392 | 2·4683 | 2·0415 | ·5440 |
| $\frac{1}{8}$ | 30·6796 | 15·9790 | 2·5187 | 2·0832 | ·5780 |
| $\frac{3}{16}$ | 31·9191 | 16·9570 | 2·5691 | 2·1250 | ·6133 |
| $\frac{1}{4}$ | 33·1831 | 17·9742 | 2·6195 | 2·1665 | ·6401 |
| $\frac{5}{16}$ | 35·3715 | 19·0311 | 2·6698 | 2·2082 | ·6884 |
| $\frac{3}{8}$ | 35·7847 | 20·1289 | 2·7202 | 2·2500 | ·7281 |
| $\frac{7}{16}$ | 37·1224 | 21·2680 | 2·7706 | 2·2915 | ·7693 |
| $\frac{1}{2}$ | 38·4846 | 22·4493 | 2·8210 | 2·3332 | ·8120 |
| $\frac{9}{16}$ | 39·8713 | 23·6735 | 2·8713 | 2·3750 | ·8561 |
| $\frac{5}{8}$ | 41·2825 | 24·9415 | 2·9217 | 2·4166 | ·9021 |
| $\frac{11}{16}$ | 42·7183 | 26·2539 | 2·9712 | 2·4582 | ·9496 |
| $\frac{3}{4}$ | 44·1787 | 27·6117 | 3·0225 | 2·5000 | ·9987 |
| $\frac{13}{16}$ | 45·6636 | 29·0102 | 3·0728 | 2·5415 | 1·0493 |
| $\frac{7}{8}$ | 47·1730 | 30·4659 | 3·1232 | 2·5832 | 1·1020 |
| $\frac{15}{16}$ | 48·7070 | 31·9640 | 3·1730 | 2·6250 | 1·1561 |
| 4 in. | 50·2656 | 33·5104 | 3·2240 | 2·6665 | 1·1974 |
| $\frac{1}{16}$ | 51·8486 | 35·1058 | 3·2743 | 2·7082 | 1·2698 |
| $\frac{1}{8}$ | 53·4562 | 36·7511 | 3·3247 | 2·7500 | 1·3293 |
| $\frac{3}{16}$ | 55·0884 | 38·4471 | 3·3751 | 2·7915 | 1·3906 |
| $\frac{1}{4}$ | 56·7451 | 40·1944 | 3·4255 | 2·8332 | 1·4538 |
| $\frac{5}{16}$ | 58·4262 | 42·0461 | 3·4758 | 2·8750 | 1·5208 |
| $\frac{3}{8}$ | 60·1321 | 43·8463 | 3·5262 | 2·9165 | 1·5860 |
| $\frac{7}{16}$ | 61·8625 | 45·7524 | 3·5766 | 2·9582 | 1·6550 |

| Diameter. | Surface. | Solidity. | Cube. | Cylinder. | Water in lbs. |
|---|---|---|---|---|---|
| 1/2 | 63·6174 | 47·7127 | 3·6270 | 3·0000 | 1·7258 |
| 9/16 | 65·3968 | 49·7290 | 3·6773 | 3·0415 | 1·7987 |
| 5/8 | 67·2007 | 51·8006 | 3·7277 | 3·0832 | 1·8736 |
| 11/16 | 69·0352 | 53·9290 | 3·7781 | 3·1250 | 1·9506 |
| 3/4 | 70·8823 | 56·1151 | 3·8285 | 3·1665 | 2·0297 |
| 13/16 | 72·7599 | 58·3595 | 3·8788 | 3·2080 | 2·1109 |
| 7/8 | 74·6620 | 60·6629 | 3·9292 | 3·2500 | 2·1942 |
| 15/16 | 76·5887 | 62·9261 | 3·9796 | 3·2913 | 2·2760 |
| 5 in. | 78·5400 | 65·4500 | 4·0300 | 3·3332 | 2·3673 |
| 1/16 | 80·5157 | 67·9351 | 4·0803 | 3·3750 | 2·4572 |
| 1/8 | 82·5160 | 70·4824 | 4·1307 | 3·4155 | 2·5453 |
| 3/16 | 84·5409 | 73·0926 | 4·1811 | 3·4582 | 2·6438 |
| 1/4 | 86·5903 | 75·7664 | 4·2315 | 3·5000 | 2·7605 |
| 5/16 | 88·6641 | 78·5077 | 4·2818 | 3·5414 | 2·8396 |
| 3/8 | 90·7627 | 81·3083 | 4·3322 | 3·5832 | 2·9407 |
| 7/16 | 92·8858 | 84·1777 | 4·3820 | 3·6250 | 3·0447 |
| 1/2 | 95·0334 | 87·1139 | 4·4330 | 3·6665 | 3·1509 |
| 9/16 | 97·2053 | 90·1175 | 4·4633 | 3·7080 | 3·2595 |
| 5/8 | 99·4021 | 93·1875 | 4·5337 | 3·7500 | 3·3706 |
| 11/16 | 101·6233 | 96·3304 | 4·5841 | 3·7913 | 3·4843 |
| 3/4 | 103·8691 | 99·5412 | 4·6345 | 3·8330 | 3·6004 |
| 13/16 | 106·1394 | 102·8225 | 4·6848 | 3·8750 | 3·7191 |
| 7/8 | 108·4342 | 106·1754 | 4·7352 | 3·9163 | 3·8404 |
| 15/16 | 110·7536 | 109·5973 | 4·7856 | 3·9580 | 3·9641 |
| 6 in. | 113·0976 | 113·0976 | 4·8360 | 4·0000 | 4·0907 |
| 1/16 | 115·4660 | 116·6688 | 4·8863 | 4·0417 | 4·2200 |
| 1/8 | 117·8590 | 120·3139 | 4·9367 | 4·0833 | 4·3517 |
| 3/16 | 120·2771 | 124·0374 | 4·9871 | 4·1250 | 4·4874 |
| 1/4 | 122·7187 | 127·8320 | 5·0375 | 4·1666 | 4·6236 |
| 5/16 | 125·1852 | 131·7053 | 5·0878 | 4·2083 | 4·7638 |
| 3/8 | 127·6765 | 135·6563 | 5·1382 | 4·2500 | 4·9067 |
| 7/16 | 130·1923 | 139·6854 | 5·1886 | 4·2917 | 5·0524 |
| 1/2 | 132·7326 | 143·7936 | 5·2390 | 4·3332 | 5·2010 |
| 9/16 | 135·2974 | 147·9815 | 5·2893 | 4·3750 | 5·3525 |
| 5/8 | 137·8867 | 152·2499 | 5·3377 | 4·4165 | 5·5069 |
| 11/16 | 140·5006 | 156·5997 | 5·3901 | 4·4583 | 5·6786 |
| 3/4 | 143·1391 | 161·0315 | 5·4405 | 4·5000 | 5·8245 |
| 13/16 | 145·8021 | 167·5461 | 5·4908 | 4·5416 | 6·0601 |
| 7/8 | 148·4896 | 170·1682 | 5·5412 | 4·5832 | 6·1550 |
| 15/16 | 151·2017 | 174·8270 | 5·5916 | 4·6250 | 6·3235 |
| 7 in. | 153·9384 | 179·5948 | 5·6420 | 4·6665 | 6·4960 |
| 1/16 | 156·6995 | 184·4484 | 5·6923 | 4·7082 | 6·6725 |
| 1/8 | 159·4852 | 189·3882 | 5·7427 | 4·7500 | 6·8502 |
| 3/16 | 162·2955 | 194·1165 | 5·7931 | 4·7915 | 7·0212 |
| 1/4 | 165·1303 | 199·5325 | 5·8435 | 4·8332 | 7·2171 |
| 5/16 | 167·9895 | 204·7371 | 5·8938 | 4·8750 | 7·4053 |
| 3/8 | 170·8735 | 210·0331 | 5·9442 | 4·9166 | 7·5970 |
| 7/16 | 173·7520 | 215·4172 | 5·9946 | 4·9582 | 7·7916 |
| 1/2 | 176·7150 | 220·8937 | 6·0450 | 5·0000 | 7·9897 |
| 9/16 | 179·6725 | 226·7240 | 6·0953 | 5·0415 | 8·2006 |
| 5/8 | 182·6545 | 232·1235 | 6·1457 | 5·0832 | 8·3960 |
| 11/16 | 185·6611 | 237·8883 | 6·1961 | 5·1250 | 8·6044 |
| 3/4 | 188·6923 | 243·7276 | 6·2465 | 5·1665 | 8·8157 |
| 13/16 | 191·7480 | 249·4720 | 6·2968 | 5·2082 | 9·0234 |
| 7/8 | 194·8282 | 255·7121 | 6·3472 | 5·2500 | 9·2491 |
| 15/16 | 197·9330 | 261·9673 | 6·3976 | 5·2913 | 9·4753 |
| 8 in. | 201·0624 | 268·0832 | 6·4480 | 5·3330 | 9·6965 |
| 1/16 | 204·2162 | 274·4156 | 6·4983 | 5·3750 | 9·9260 |

| Diameter. | Surface. | Solidity. | Cube. | Cylinder. | Water in lbs. |
|---|---|---|---|---|---|
| 1/8 | 207·3946 | 280·8469 | 6·5487 | 5·4164 | 10·1583 |
| 3/16 | 210·5976 | 287·3780 | 6·5991 | 5·4581 | 10·3944 |
| 1/4 | 213·8251 | 294·0095 | 6·6495 | 5·5000 | 10·6343 |
| 5/16 | 217·0770 | 300·7422 | 6·6998 | 5·5414 | 10·8778 |
| 3/8 | 220·3537 | 307·5771 | 6·7502 | 5·5831 | 11·1250 |
| 7/16 | 223·6549 | 314·5147 | 6·8006 | 5·6250 | 11·3760 |
| 1/2 | 226·9806 | 321·5553 | 6·8510 | 5·6664 | 11·6306 |
| 9/16 | 230·3308 | 328·7012 | 6·9013 | 5·7080 | 11·8891 |
| 5/8 | 233·7055 | 335·9517 | 6·9517 | 5·7500 | 12·1514 |
| 11/16 | 237·1048 | 343·3079 | 7·0021 | 5·7913 | 12·4170 |
| 3/4 | 240·5287 | 350·7710 | 7·0525 | 5·8330 | 12·6874 |
| 13/16 | 243·9771 | 358·3412 | 7·1028 | 5·8750 | 12·9612 |
| 7/8 | 247·4500 | 366·0199 | 7·1532 | 5·9163 | 13·2390 |
| 15/16 | 250·9475 | 373·8073 | 7·2036 | 5·9580 | 13·5206 |
| 9 in. | 254·4696 | 381·7017 | 7·2540 | 6·0000 | 13·8062 |
| 1/16 | 258·0261 | 389·7118 | 7·3043 | 6·0417 | 14·0959 |
| 1/8 | 261·5872 | 397·8306 | 7·3547 | 6·0833 | 14·3895 |
| 3/16 | 265·1829 | 406·0613 | 7·4051 | 6·1250 | 14·6872 |
| 1/4 | 268·8031 | 414·4048 | 7·4555 | 6·1667 | 14·9890 |
| 5/16 | 272·4477 | 421·2907 | 7·5058 | 6·2083 | 15·2381 |
| 3/8 | 276·1171 | 431·4361 | 7·5562 | 6·2500 | 15·6050 |
| 7/16 | 279·8110 | 440·1294 | 7·6066 | 6·2916 | 15·9195 |
| 1/2 | 283·5294 | 448·9215 | 7·6570 | 6·3333 | 16·2375 |
| 9/16 | 287·2723 | 457·8500 | 7·7073 | 6·3750 | 16·5604 |
| 5/8 | 291·0397 | 466·8763 | 7·7557 | 6·4166 | 16·6869 |
| 11/16 | 294·8310 | 476·0304 | 7·8081 | 6·4582 | 17·2180 |
| 3/4 | 298·4483 | 485·3035 | 7·8585 | 6·5000 | 17·5534 |
| 13/16 | 302·4894 | 494·6952 | 7·9088 | 6·5415 | 17·8931 |
| 7/8 | 306·3550 | 504·2094 | 7·9592 | 6·5832 | 18·2373 |
| 15/16 | 310·9452 | 513·8436 | 8·0096 | 6·6250 | 18·5857 |
| 10 in. | 314·1600 | 523·6000 | 8·0600 | 6·6666 | 18·6786 |
| 1/16 | 318·0992 | 533·4789 | 8·1103 | 6·7083 | 19·2960 |
| 1/8 | 322·0630 | 543·4814 | 8·1607 | 6·7500 | 19·6577 |
| 3/16 | 326·0514 | 553·6081 | 8·2111 | 6·7916 | 20·0240 |
| 1/4 | 330·0643 | 563·8603 | 8·2615 | 6·8333 | 20·3948 |
| 5/16 | 334·1016 | 574·2371 | 8·3118 | 6·8750 | 20·6682 |
| 3/8 | 338·1637 | 584·7415 | 8·3622 | 6·9166 | 21·1501 |
| 7/16 | 342·2503 | 595·3677 | 8·4126 | 6·9582 | 21·5344 |
| 1/2 | 346·3614 | 606·1318 | 8·4630 | 7·0000 | 21·9238 |
| 9/16 | 350·4970 | 617·0207 | 8·5133 | 7·0416 | 22·3176 |
| 5/8 | 354·6571 | 628·0387 | 8·5637 | 7·0833 | 22·7162 |
| 11/16 | 358·8418 | 639·1871 | 8·6141 | 7·1250 | 23·1194 |
| 3/4 | 363·0511 | 650·4666 | 8·6645 | 7·1666 | 23·5274 |
| 13/16 | 367·2849 | 661·8580 | 8·7148 | 7·2082 | 23·9394 |
| 7/8 | 371·5432 | 673·4222 | 8·7652 | 7·2500 | 24·3577 |
| 15/16 | 375·8261 | 685·0997 | 8·8156 | 7·2915 | 24·7801 |
| 11 in. | 380·1336 | 696·9116 | 8·8660 | 7·3330 | 25·2073 |
| 1/16 | 384·4655 | 708·9106 | 8·9163 | 7·3750 | 25·6414 |
| 1/8 | 388·8220 | 720·9409 | 8·9667 | 7·4165 | 26·0764 |
| 3/16 | 393·2031 | 733·1599 | 9·0171 | 7·4582 | 26·5184 |
| 1/4 | 397·6087 | 745·5004 | 9·0675 | 7·5000 | 26·5657 |
| 5/16 | 402·0387 | 758·0104 | 9·1178 | 7·5414 | 27·4162 |
| 3/8 | 406·4935 | 770·6440 | 9·1682 | 7·5832 | 27·8742 |
| 7/16 | 410·7728 | 783·5787 | 9·2186 | 7·6250 | 28·3420 |
| 1/2 | 415·4766 | 796·3301 | 9·2690 | 7·6664 | 28·8033 |
| 9/16 | 420·0049 | 809·3844 | 9·3193 | 7·7080 | 29·2754 |
| 5/8 | 424·5576 | 822·5807 | 9·3697 | 7·7500 | 29·7527 |
| 11/16 | 429·1351 | 835·9695 | 9·4201 | 7·7913 | 30·2370 |

| Diameter. | Surface. | Solidity. | Cube. | Cylinder. | Water in lbs. |
|---|---|---|---|---|---|
| ¾ | 433·7371 | 849·4035 | 9·4705 | 7·8330 | 30·7229 |
| 13/16 | 438·3636 | 863·0283 | 9·5208 | 7·8750 | 31·2157 |
| ⅞ | 443·0146 | 876·7999 | 9·5772 | 7·9163 | 31·3883 |
| 15/16 | 447·6902 | 890·7070 | 9·6216 | 7·9580 | 32·2169 |
| 12 in. | 452·3904 | 904·7808 | 9·6720 | 8·0000 | 32·7259 |
| ¼ | 471·4363 | 962·5158 | 9·8735 | 8·1666 | 34·8142 |
| ½ | 490·8750 | 1022·656 | 10·0750 | 8·3332 | 36·9886 |
| ¾ | 506·7064 | 1085·251 | 10·2765 | 8·5000 | 39·2535 |
| 13 in. | 530·9304 | 1150·337 | 10·4780 | 8·6666 | 41·6077 |
| ¼ | 551·5471 | 1218·000 | 10·6790 | 8·8332 | 44·0551 |
| ½ | 572·5566 | 1288·252 | 10·8810 | 9·0000 | 46·5961 |
| ¾ | 593·9587 | 1361·346 | 11·0825 | 9·1665 | 49·2399 |
| 14 in. | 615·7536 | 1436·758 | 11·2840 | 9·3332 | 51·9675 |
| ¼ | 637·9411 | 1515·106 | 11·4855 | 9·5000 | 54·8014 |
| ½ | 660·5214 | 1596·260 | 11·6870 | 9·6665 | 57·7367 |
| ¾ | 683·4943 | 1680·265 | 11·8885 | 9·8332 | 60·7751 |
| 15 in. | 706·8600 | 1767·150 | 12·0900 | 10·0000 | 64·0178 |
| ¼ | 730·6183 | 1856·988 | 12·2915 | 10·1666 | 67·1672 |
| ½ | 754·7694 | 1949·821 | 12·4930 | 10·3332 | 70·5250 |
| ¾ | 779·3131 | 2045·697 | 12·6940 | 10·5000 | 73·9929 |
| 16 in. | 804·2496 | 2144·665 | 12·8960 | 10·6666 | 77·5725 |

TABLE *containing the Weight of Flat Bar Iron,* 1 *foot in length, of various breadths and thicknesses.*

| Breadth in Inches. | THICKNESS IN PARTS OF AN INCH. | | | | | | | | | |
|---|---|---|---|---|---|---|---|---|---|---|
| | ¼ | 5/16 | ⅜ | 7/16 | ½ | 9/16 | ⅝ | ¾ | ⅞ | 1 inch. |
| | Lbs. | Lbs. | Lbs. | Lbs. | Lbs. | Lbs. | Lbs. | Lbs. | Lbs. | Lbs. |
| 1 in. | 0·83 | 1·04 | 1·25 | 1·45 | 1·66 | 1·87 | 2·08 | 2·50 | 2·91 | 3·33 |
| 1⅛ | 0·93 | 1·17 | 1·40 | 1·64 | 1·87 | 2·00 | 2·34 | 2·81 | 3·28 | 3·75 |
| 1¼ | 1·04 | 1·30 | 1·56 | 1·82 | 2·08 | 2·34 | 2·60 | 3·12 | 3·74 | 4·16 |
| 1⅜ | 1·14 | 1·43 | 1·71 | 2·00 | 2·29 | 2·57 | 2·86 | 3·43 | 4·01 | 4·58 |
| 1½ | 1·25 | 1·56 | 1·87 | 2·18 | 2·50 | 2·81 | 3·12 | 3·75 | 4·37 | 5·00 |
| 1⅝ | 1·35 | 1·69 | 2·03 | 2·36 | 2·70 | 3·04 | 3·38 | 4·06 | 4·73 | 5·41 |
| 1¾ | 1·45 | 1·82 | 2·18 | 2·55 | 2·91 | 3·28 | 3·64 | 4·37 | 5·10 | 5·83 |
| 1⅞ | 1·56 | 1·95 | 2·34 | 2·73 | 3·12 | 3·51 | 3·90 | 4·68 | 5·46 | 6·25 |
| 2 in. | 1·66 | 2·08 | 2·50 | 2·91 | 3·33 | 3·75 | 4·16 | 5·00 | 5·83 | 6·66 |
| 2⅛ | 1·77 | 2·21 | 2·65 | 3·09 | 3·54 | 3·98 | 4·42 | 5·31 | 6·19 | 7·08 |
| 2¼ | 1·87 | 2·34 | 2·81 | 3·28 | 3·75 | 4·21 | 4·68 | 5·62 | 6·56 | 7·50 |
| 2⅜ | 1·97 | 2·47 | 2·96 | 3·46 | 3·95 | 4·45 | 4·94 | 5·93 | 6·92 | 7·91 |
| 2½ | 2·08 | 2·60 | 3·12 | 3·64 | 4·16 | 4·68 | 5·20 | 6·25 | 7·29 | 8·33 |
| 2⅝ | 2·18 | 2·73 | 3·28 | 3·82 | 4·37 | 4·92 | 5·46 | 6·56 | 7·65 | 8·75 |
| 2¾ | 2·29 | 2·86 | 3·43 | 4·01 | 4·58 | 5·15 | 5·72 | 6·87 | 8·02 | 9·16 |
| 2⅞ | 2·39 | 2·99 | 3·59 | 4·19 | 4·79 | 5·39 | 5·98 | 7·18 | 8·38 | 9·58 |
| 3 in. | 2·50 | 3·12 | 3·75 | 4·37 | 5·00 | 5·62 | 6·25 | 7·50 | 8·75 | 10·00 |
| 3¼ | 2·70 | 3·38 | 4·06 | 4·73 | 5·41 | 6·09 | 6·77 | 8·12 | 9·47 | 10·83 |
| 3½ | 2·91 | 3·64 | 4·37 | 5·10 | 5·83 | 6·56 | 7·29 | 8·75 | 10·20 | 11·66 |
| 3¾ | 3·12 | 3·90 | 4·68 | 5·46 | 6·25 | 7·03 | 7·81 | 9·37 | 10·93 | 12·50 |
| 4 in. | 3·33 | 4·16 | 5·00 | 5·83 | 6·66 | 7·50 | 8·33 | 10·00 | 11·66 | 13·33 |
| 4¼ | 3·54 | 4·42 | 5·31 | 6·19 | 7·08 | 7·96 | 8·85 | 10·62 | 12·39 | 14·16 |
| 4½ | 3·75 | 4·68 | 5·62 | 6·56 | 7·50 | 8·43 | 9·37 | 11·25 | 13·12 | 15·00 |
| 4¾ | 3·95 | 4·94 | 5·93 | 6·92 | 7·91 | 8·90 | 9·89 | 11·87 | 13·85 | 15·33 |
| 5 in. | 4·17 | 5·20 | 6·25 | 7·29 | 8·33 | 9·37 | 10·41 | 12·50 | 14·58 | 16·66 |
| 5¼ | 4·37 | 5·46 | 6·56 | 7·65 | 8·75 | 9·84 | 10·93 | 13·12 | 15·31 | 17·50 |
| 5½ | 4·58 | 5·72 | 6·87 | 8·02 | 9·16 | 10·31 | 11·45 | 13·75 | 16·04 | 18·33 |
| 5¾ | 4·79 | 5·98 | 7·18 | 8·38 | 9·58 | 10·78 | 11·97 | 14·37 | 16·77 | 19·16 |
| 6 in. | 5·00 | 6·26 | 7·50 | 8·75 | 10·00 | 11·25 | 12·50 | 15·00 | 17·50 | 20·00 |

TABLE *combining the Specific Gravities and other Properties of Bodies. Water the standard of comparison, or* 1000.

METALS.

| Names. | Specific gravity. | Melting points in degrees of Fahrenheit. | Contraction in parts of an inch per lineal foot from the fluid to the average temperature in solid state. | Ultimate cohesive strength of an inch sq. prism in tons. | Scale of wire-drawing ductility. | Scale of laminable ductility. | Ratio of hardness. | Scale as conductors of electricity. | Ratio of power in the conduction of heat. |
|---|---|---|---|---|---|---|---|---|---|
| Platinum . . | 19500 | 3280 | . . | . . | 3 | 5 | . . | . . | 3·8 |
| Pure Gold . | 19258 | 2016 | . . | . . | 1 | 1 | 1·8 | 3 | 10·0 |
| Mercury . . | 13500 | . . | . . | . . | . . | . . | . . | . . | . . |
| Lead . . . . | 11352 | 612 | ·319 | ·81 | 8 | 7 | 1·0 | 6 | 1·8 |
| Pure Silver . | 10474 | 1873 | . . | . . | 2 | 2 | 2·4 | 2 | 9·7 |
| Bismuth . . | 8923 | 476 | ·156 | 1·45 | . . | . . | 2·0 | . . | . . |
| Copper, cast . | 8788 | 1996 | ·193 | 8·51 | . . | . . | . . | . . | . . |
| " wrought | 8910 | . . | . . | 15·08 | 5 | 3 | 2·8 | 1 | 8·9 |
| Brass, cast . | 7824 | 1900 | ·210 | 8·01 | . . | . . | to any degree | . . | . . |
| " sheet . | 8396 | . . | . . | 12·23 | 6 | 6 | . . | . . | 8·6 |
| Iron, cast . | 7264 | 2786 | ·125 | 7·87 | . . | . . | to any degree | . . | . . |
| " bar . . | 7700 | . . | ·137 | 25·00 | 4 | 8 | 4·7 | 4 | 3·7 |
| Steel, soft . | 7833 | . . | ·133 | 58·91 | . . | . . | . . | . . | . . |
| " hard . | 7816 | . . | . . | . . | . . | . . | to any degree | . . | . . |
| Tin, cast . . | 7291 | 442 | ·278 | 2·11 | 8 | 4 | 1·2 | 5 | 3·0 |
| Zinc, cast . . | 7190 | 773 | ·329 | 5·06 | 7 | 8 | 1·6 | 7 | 3·6 |

STONES, EARTHS, ETC.

| Names. | Specific gravity. | Weight of a cubic foot in lbs. | Cubic feet in a ton. | Tons required to crush 1½-in. cubes. |
|---|---|---|---|---|
| Marble, average | 2730 | 170·00 | 13 | 9·25 |
| Granite, ditto . | 2651 | 165·68 | 13½ | 6·2 |
| Purbeck stone . | 2601 | 162·56 | 13¾ | 9·0 |
| Portland ditto . | 2570 | 160·62 | 14 | 4·5 |
| Bristol ditto . . | 2554 | 159·62 | 14 | . . |
| Millstone . . . . | 2484 | 155·25 | 14½ | . . |
| Paving stone . . | 2415 | 150·93 | 14¾ | 5·7 |
| Craigleith ditto | 2362 | 147·62 | 15 | 5·0 |
| Grindstone . . . | 2143 | 133·93 | 16¾ | 6·6 |
| Chalk, Brit. . . | 2781 | 173·81 | 12¾ | 0·5 |
| Brick . . . . . . | 2000 | 125·00 | 17 | 0·8 |
| Coal, Scotch . . | 1300 | 81·15 | 27½ | . . |
| " Newcastle | 1270 | 79·37 | 27¼ | . . |
| " Staffordsh. | 1240 | 77·50 | 29 | . . |
| " Cannel . . | 1238 | 77·37 | 29 | . . |

TABLE *containing the Weight of Columns of Water, each one foot in length, and of Various Diameters, in lbs. avoirdupois.*

| Diam | Weight. | Diam. | Weight. | Diam. | Weight. | Diam. | Weight. | Diam. | Weight. | Diam. | Weight. |
|---|---|---|---|---|---|---|---|---|---|---|---|
| 3 in. | 3·0672 | 9 in. | 27·6120 | 15 in. | 76·7004 | 21 in. | 150·2376 | 27 in. | 248·5116 | 33 in. | 371·2344 |
| ⅛ | 3·3288 | ⅛ | 28·3848 | ⅛ | 77·9844 | ⅛ | 152·1288 | ⅛ | 250·8180 | ⅛ | 374·0520 |
| ¼ | 3.6000 | ¼ | 29·1672 | ¼ | 79·2792 | ¼ | 153·9348 | ¼ | 253·1352 | ¼ | 376·8004 |
| ⅜ | 3·8820 | ⅜ | 29·9604 | ⅜ | 80·5836 | ⅜ | 155·7396 | ⅜ | 255·4632 | ⅜ | 379·4592 |
| ½ | 4·1748 | ½ | 30·7657 | ½ | 81·9000 | ½ | 157·5780 | ½ | 257·8008 | ½ | 382·5684 |
| ⅝ | 4·4784 | ⅝ | 31·6524 | ⅝ | 83·2260 | ⅝ | 159·4152 | ⅝ | 260·1504 | ⅝ | 385·4292 |
| ¾ | 4·7928 | ¾ | 32·4060 | ¾ | 84·5628 | ¾ | 161·2644 | ¾ | 262·5096 | ¾ | 388·2996 |
| ⅞ | 5·1180 | ⅞ | 33·2424 | ⅞ | 85·9104 | ⅞ | 163·1220 | ⅞ | 264·8796 | ⅞ | 391·1820 |
| 4 in. | 5·4540 | 10 in. | 34·0884 | 16 in. | 87·2688 | 22 in. | 164 9928 | 28 in. | 267·2616 | 34 in. | 394·0740 |
| ⅛ | 5·7996 | ⅛ | 34·9464 | ⅛ | 88·6368 | ⅛ | 166·8732 | ⅛ | 269·6532 | ⅛ | 396·9768 |
| ¼ | 6·1572 | ¼ | 35·8152 | ¼ | 90·0168 | ¼ | 168·7632 | ¼ | 272·0544 | ¼ | 399·8928 |
| ⅜ | 6·5244 | ⅜ | 36·6936 | ⅜ | 91·4176 | ⅜ | 170·6652 | ⅜ | 275·6672 | ⅜ | 402·8088 |
| ½ | 6·9024 | ½ | 37·5828 | ½ | 92·8080 | ½ | 172·5780 | ½ | 276·8916 | ½ | 405·7500 |
| ⅝ | 7·2912 | ⅝ | 38·4828 | ⅝ | 94·2192 | ⅝ | 174·5004 | ⅝ | 279·3252 | ⅝ | 408·6948 |
| ¾ | 7·6908 | ¾ | 39·3936 | ¾ | 95·6412 | ¾ | 176·4336 | ¾ | 281·7708 | ¾ | 411·4116 |
| ⅞ | 8·1012 | ⅞ | 40·3152 | ⅞ | 97·0740 | ⅞ | 178·3776 | ⅞ | 284·2260 | ⅞ | 414·6180 |
| 5 in. | 8·5212 | 11 in. | 41·2476 | 17 in. | 98·5176 | 23 in. | 180·3324 | 29 in. | 286·6920 | 35 in. | 417·5952 |
| ⅛ | 8·9532 | ⅛ | 42·1908 | ⅛ | 99·9720 | ⅛ | 182·2980 | ⅛ | 289·1688 | ⅛ | 420·5844 |
| ¼ | 9·3948 | ¼ | 43·1436 | ¼ | 101·4372 | ¼ | 184·2744 | ¼ | 291·6564 | ¼ | 423·5832 |
| ⅜ | 9·8484 | ⅜ | 44·1084 | ⅜ | 102·9120 | ⅜ | 186·2616 | ⅜ | 294·1548 | ⅜ | 426·5928 |
| ½ | 10·3126 | ½ | 45·0828 | ½ | 104·3988 | ½ | 188·2584 | ½ | 296·5548 | ½ | 429·6120 |
| ⅝ | 10·7856 | ⅝ | 46·0680 | ⅝ | 105·8952 | ⅝ | 190·2672 | ⅝ | 299·1828 | ⅝ | 432·6432 |
| ¾ | 11·2704 | ¾ | 47·0640 | ¾ | 107·4024 | ¾ | 192·2856 | ¾ | 301·7124 | ¾ | 435·6840 |
| ⅞ | 11·7660 | ⅞ | 48·0708 | ⅞ | 108·9204 | ⅞ | 194·3184 | ⅞ | 304·2540 | ⅞ | 438·7368 |
| 6 in. | 12·2712 | 12 in. | 49·0384 | 18 in. | 110·4492 | 24 in. | 196·3548 | 30 in. | 306·8052 | 36 in. | 441·7992 |
| ⅛ | 12·7884 | ⅛ | 50·1168 | ⅛ | 111·9888 | ⅛ | 198·4056 | ⅛ | 309·3672 | ¼ | 447·9573 |
| ¼ | 13·3152 | ¼ | 51·1548 | ¼ | 113·5392 | ¼ | 200·4672 | ¼ | 311·9400 | ½ | 454·1678 |
| ⅜ | 13·8540 | ⅜ | 52·2048 | ⅜ | 115·0992 | ⅜ | 203·5384 | ⅜ | 314·5224 | ¾ | 460·4105 |
| ½ | 14·4024 | ½ | 53·2644 | ½ | 116·6712 | ½ | 204·6216 | ½ | 317·1168 | 37 in. | 466·6960 |
| ⅝ | 14·9616 | ⅝ | 54·3348 | ⅝ | 118·2528 | ⅝ | 206·7144 | ⅝ | 319·7220 | ¼ | 473·0240 |
| ¾ | 15·5316 | ¾ | 55·4760 | ¾ | 119·8452 | ¾ | 208·8192 | ¾ | 322·3368 | ½ | 479·3946 |
| ⅞ | 16·1124 | ⅞ | 56·4804 | ⅞ | 121·4484 | ⅞ | 210·9336 | ⅞ | 324·9624 | ¾ | 485·8078 |
| 7 in. | 16·7028 | 13 in. | 57·6108 | 19 in. | 123·0624 | 25 in. | 213·0588 | 31 in. | 327·6000 | 38 in. | 492·2637 |
| ⅛ | 17·3052 | ⅛ | 58·7244 | ⅛ | 124·6872 | ⅛ | 215·1948 | ⅛ | 330·2472 | ¼ | 498·7621 |
| ¼ | 17·9172 | ¼ | 59·8476 | ¼ | 126·3228 | ¼ | 217·3416 | ¼ | 332·9052 | ½ | 505·3032 |
| ⅜ | 18·5412 | ⅜ | 60·9828 | ⅜ | 127·9680 | ⅜ | 219·4980 | ⅜ | 335·5728 | ¾ | 511·9979 |
| ½ | 19·1748 | ½ | 62·1276 | ½ | 129·6252 | ½ | 221·6664 | ½ | 338·2524 | 39 in. | 518·4132 |
| ⅝ | 19·8192 | ⅝ | 63·2832 | ⅝ | 131·5320 | ⅝ | 223·8444 | ⅝ | 340·9428 | ¼ | 525·1821 |
| ¾ | 20·4744 | ¾ | 64·4496 | ¾ | 132·9696 | ¾ | 226·0344 | ¾ | 343·6428 | ½ | 531·8936 |
| ⅞ | 21·1404 | ⅞ | 65·6268 | ⅞ | 134·6580 | ⅞ | 228·2340 | ⅞ | 346·3536 | ¾ | 538·6478 |
| 8 in. | 21·8172 | 14 in. | 66·8148 | 20 in. | 136·3562 | 26 in. | 230·4444 | 32 in. | 349·0764 | 40 in. | 545·4445 |
| ⅛ | 22·5036 | ⅛ | 68·0136 | ⅛ | 138·0672 | ⅛ | 232·6644 | ⅛ | 351·8088 | ¼ | 552·2839 |
| ¼ | 23·2020 | ¼ | 69·2220 | ¼ | 139·7880 | ¼ | 234·8576 | ¼ | 354·5520 | ½ | 559·1659 |
| ⅜ | 23·9100 | ⅜ | 70·4424 | ⅜ | 141·5184 | ⅜ | 237·1404 | ⅜ | 357·3048 | ⅓ | 566·0904 |
| ½ | 24·5288 | ½ | 71·6724 | ½ | 143·2608 | ½ | 239·3928 | ½ | 360·0696 | 41 in. | 573·0577 |
| ⅝ | 25·3524 | ⅝ | 72·9120 | ⅝ | 145·0128 | ⅝ | 241·6572 | ⅝ | 362·8452 | ½ | 587·1199 |
| ¾ | 26·0988 | ¾ | 74·1648 | ¾ | 146·7756 | ¾ | 243·9312 | ¾ | 365·6304 | 42 in. | 601·3526 |
| ⅞ | 26 8500 | ⅞ | 75·4272 | ⅞ | 148·5492 | ⅞ | 246·2160 | ⅞ | 368·4276 | 50 in. | 799·2426 |

TABLE *containing the Weight of Square Bar Iron, from* 1 *to* 10 *feet in length, and from* ¼ *of an inch to* 6 *inches square.*

| Inches square. | LENGTH OF THE BARS IN FEET. | | | | | | | | | |
|---|---|---|---|---|---|---|---|---|---|---|
| | 1 foot. | 2 feet. | 3 feet. | 4 feet. | 5 feet. | 6 feet. | 7 feet. | 8 feet. | 9 feet. | 10 feet. |
| | Lbs. | Lbs. | Lbs. | Lbs. | Lbs. | Lbs. | Lbs. | Lbs. | Lbs. | Lbs. |
| ¼ | 0·2 | 0·4 | 0·6 | 0·8 | 1·1 | 1·3 | 1·5 | 1·7 | 1·9 | 2·1 |
| ⅜ | 0·5 | 1·0 | 1·4 | 1·9 | 2·4 | 2·9 | 3·3 | 3·8 | 4·3 | 4·8 |
| ½ | 0·8 | 1·7 | 2·5 | 3·4 | 4·2 | 5·1 | 5·9 | 6·8 | 7·6 | 8·5 |
| ⅝ | 1·3 | 2·6 | 4·0 | 5·3 | 6·6 | 7·9 | 9·2 | 10·6 | 11·0 | 13·2 |
| ¾ | 1·9 | 3·8 | 5·7 | 7·6 | 9·5 | 11·4 | 13·3 | 15·2 | 17·1 | 19·0 |
| ⅞ | 2·6 | 5·2 | 7·8 | 10·4 | 12·9 | 15·5 | 18·1 | 20·7 | 23·3 | 25·9 |
| 1 in. | 3·4 | 6·8 | 10·1 | 13·5 | 16·9 | 20·3 | 23·7 | 27·0 | 30·4 | 33·8 |
| 1⅛ | 4·3 | 8·6 | 12·8 | 17·1 | 21·4 | 25·7 | 29·9 | 34·2 | 38·5 | 42·8 |
| 1¼ | 5·3 | 10·6 | 15·8 | 21·1 | 26·4 | 31·7 | 37·0 | 42·2 | 47·5 | 52·8 |
| 1⅜ | 6·4 | 12·8 | 19·2 | 25·6 | 32·0 | 38·3 | 44·7 | 51·1 | 57·5 | 63·9 |
| 1½ | 7·6 | 15·2 | 22·8 | 30·4 | 38·0 | 45·6 | 53·2 | 60·8 | 68·4 | 76·0 |
| 1⅝ | 8·9 | 17·9 | 26·8 | 35·7 | 44·6 | 53·6 | 62·5 | 71·4 | 80·3 | 89·3 |
| 1¾ | 10·4 | 20·7 | 31·1 | 41·4 | 51·8 | 62·1 | 72·5 | 82·8 | 93·2 | 103·5 |
| 1⅞ | 11·9 | 23·8 | 35·6 | 47·5 | 59·4 | 71·3 | 83·2 | 95·1 | 106·9 | 118·8 |
| 2 in. | 13·5 | 27·0 | 40·6 | 54·1 | 67·6 | 81·1 | 94·6 | 108·2 | 121·7 | 135·2 |
| 2⅛ | 15·3 | 30·5 | 45·8 | 61·1 | 76·3 | 91·6 | 106·8 | 122·1 | 137·4 | 152·6 |
| 2¼ | 17·1 | 34·2 | 51·3 | 68·4 | 85·6 | 102·7 | 119·8 | 136·9 | 154·0 | 171·1 |
| 2⅜ | 19·1 | 38·1 | 57·2 | 76·3 | 95·3 | 114·4 | 133·5 | 152·5 | 171·6 | 190·7 |
| 2½ | 21·1 | 42·8 | 63·4 | 84·5 | 105·6 | 126·7 | 147·8 | 169·0 | 190·1 | 211·2 |
| 2⅝ | 23·3 | 46·6 | 69·9 | 93·2 | 116·5 | 139·8 | 163·0 | 186·3 | 209·6 | 232·9 |
| 2¾ | 25·6 | 51·1 | 76·7 | 102·2 | 127·8 | 153·4 | 178·9 | 204·5 | 230·0 | 255·6 |
| 2⅞ | 27·9 | 55·9 | 83·8 | 111·8 | 139·7 | 167·6 | 195·7 | 223·5 | 251·5 | 279·4 |
| 3 in. | 30·4 | 60·8 | 91·2 | 121·7 | 152·1 | 182·5 | 212·9 | 243·3 | 273·7 | 304·2 |
| 3⅛ | 33·0 | 66·0 | 99·0 | 132·0 | 165·1 | 198·1 | 231·1 | 264·1 | 297·1 | 330·1 |
| 3¼ | 35·7 | 71·4 | 107·1 | 142·8 | 178·5 | 214·2 | 249·9 | 285·6 | 321·3 | 357·0 |
| 3⅜ | 38·5 | 77·0 | 115·5 | 154·0 | 192·5 | 231·0 | 269·5 | 308·0 | 346·5 | 385·0 |
| 3½ | 41·4 | 82·8 | 124·2 | 165·6 | 207·0 | 248·4 | 289·8 | 331·3 | 372·7 | 414·1 |
| 3⅝ | 44·4 | 88·8 | 133·3 | 177·7 | 222·1 | 266·5 | 310·9 | 355·3 | 399·8 | 444·2 |
| 3¾ | 47·5 | 95·1 | 142·6 | 190·1 | 237·7 | 285·2 | 332·7 | 380·3 | 427·8 | 475·3 |
| 3⅞ | 50·8 | 101·5 | 152·3 | 203·0 | 253·8 | 304·5 | 355·3 | 406·0 | 456·8 | 507·6 |
| 4 in. | 54·1 | 108·2 | 162·3 | 216·3 | 270·4 | 324·5 | 378·6 | 432·7 | 486·8 | 540·8 |
| 4⅛ | 57·5 | 115·0 | 172·6 | 230·1 | 287·6 | 345·1 | 402·6 | 460·1 | 517·7 | 575·2 |
| 4¼ | 61·1 | 122·1 | 183·2 | 244·2 | 305·3 | 366·3 | 427·4 | 488·4 | 549·5 | 610·6 |
| 4⅜ | 64·7 | 129·4 | 194·1 | 258·8 | 323·5 | 388·2 | 452·9 | 517·6 | 582·3 | 647·0 |
| 4½ | 68·4 | 136·9 | 205·3 | 273·8 | 342·2 | 410·7 | 479·1 | 547·6 | 616·0 | 684·5 |
| 4⅝ | 72·3 | 144·6 | 216·9 | 289·2 | 361·5 | 433·8 | 506·1 | 578·4 | 650·7 | 723·1 |
| 4¾ | 76·3 | 152·5 | 228·8 | 305·1 | 381·3 | 457·6 | 533·8 | 610·1 | 686·4 | 762·6 |
| 4⅞ | 80·3 | 160·7 | 241·0 | 321·3 | 401·7 | 482·0 | 562·3 | 642·7 | 723·0 | 803·3 |
| 5 in. | 84·5 | 169·0 | 253·4 | 337·9 | 422·4 | 506·9 | 591·4 | 675·8 | 760·3 | 844·8 |
| 5¼ | 93·2 | 186·3 | 279·5 | 372·7 | 465·8 | 559·0 | 652·2 | 745·3 | 838·5 | 931·7 |
| 5½ | 102·2 | 204·5 | 306·7 | 409·0 | 511·2 | 613·4 | 715·7 | 817·9 | 920·2 | 1022·4 |
| 5¾ | 111·8 | 223·5 | 335·3 | 447·0 | 558·8 | 670·5 | 782·3 | 894·0 | 1005·8 | 1117·6 |
| 6 in. | 121·7 | 243·3 | 365·0 | 486·7 | 608·3 | 730·0 | 841·6 | 973·3 | 1009·5 | 1216·6 |

TABLE *of the Weight of a Square Foot of Sheet Iron in lbs. avoirdupois, the thickness being the number on the wire-gauge. No.* 1 *is* $\frac{5}{16}$ *of an inch; No.* 4, ¼; *No.* 11, ⅛, *&c.*

| No. on wire-gauge | 1 | 2 | 3 | 4 | 5 | 6 | 7 | 8 | 9 | 10 | 11 |
|---|---|---|---|---|---|---|---|---|---|---|---|
| Pounds avoir....... | 12·5 | 12 | 11 | 10 | 9 | 8 | 7·5 | 7 | 6 | 5·68 | 5 |
| No. on wire-gauge | 12 | 13 | 14 | 15 | 16 | 17 | 18 | 19 | 20 | 21 | 22 |
| Pounds avoir....... | 4·62 | 4·31 | 4 | 3·95 | 3 | 2·5 | 2·18 | 1·93 | 1·62 | 1·5 | 1·37 |

TABLE *of the Weight of a Square Foot of Boiler Plate Iron, from* $\frac{1}{8}$ *to* 1 *inch thick, in lbs. avoirdupois.*

| $\frac{1}{8}$ | $\frac{3}{16}$ | $\frac{1}{4}$ | $\frac{5}{16}$ | $\frac{3}{8}$ | $\frac{7}{16}$ | $\frac{1}{2}$ | $\frac{9}{16}$ | $\frac{5}{8}$ | $\frac{11}{16}$ | $\frac{3}{4}$ | $\frac{13}{16}$ | $\frac{7}{8}$ | $\frac{15}{16}$ | 1 in. |
|---|---|---|---|---|---|---|---|---|---|---|---|---|---|---|
| 5 | 7·5 | 10 | 12·5 | 15 | 17·5 | 20 | 22·5 | 25 | 27·5 | 30 | 32·5 | 35 | 37·5 | 40 |

TABLE *containing the Weight of Round Bar Iron, from* 1 *to* 10 *feet in length, and from* $\frac{1}{4}$ *of an inch to* 6 *inches diameter.*

| Inches diameter. | LENGTH OF THE BARS IN FEET. | | | | | | | | | |
|---|---|---|---|---|---|---|---|---|---|---|
| | 1 foot. | 2 feet. | 3 feet. | 4 feet. | 5 feet. | 6 feet. | 7 feet. | 8 feet. | 9 feet. | 10 feet. |
| | Lbs. | Lbs. | Lbs. | Lbs. | Lbs. | Lbs. | Lbs. | Lbs. | Lbs. | Lbs. |
| $\frac{1}{4}$ | 0·2 | 0·3 | 0·5 | 0·7 | 0·8 | 1·0 | 1·2 | 1·3 | 1·5 | 1·7 |
| $\frac{3}{8}$ | 0·4 | 0·7 | 1·1 | 1·5 | 1·9 | 2·2 | 2·6 | 3·0 | 3·4 | 3·7 |
| $\frac{1}{2}$ | 0·7 | 1·3 | 2·0 | 2·7 | 3·3 | 4·0 | 4·6 | 5·3 | 6·0 | 6·6 |
| $\frac{5}{8}$ | 1·0 | 2·1 | 3·1 | 4·2 | 5·2 | 6·3 | 7·3 | 8·3 | 9·4 | 10·4 |
| $\frac{3}{4}$ | 1·5 | 3·0 | 4·5 | 6·0 | 7·5 | 9·0 | 10·5 | 11·9 | 13·4 | 14·9 |
| $\frac{7}{8}$ | 2·0 | 4·1 | 6·1 | 8·1 | 10·2 | 12·2 | 14·2 | 16·3 | 18·3 | 20·3 |
| 1 in. | 2·7 | 5·3 | 8·0 | 10·6 | 13·3 | 15·9 | 18·6 | 21·2 | 23·9 | 26·5 |
| $1\frac{1}{8}$ | 3·4 | 6·7 | 10·1 | 13·4 | 16·8 | 20·2 | 23·5 | 26·9 | 30·2 | 33·6 |
| $1\frac{1}{4}$ | 4·2 | 8·3 | 12·5 | 16·7 | 20·9 | 25·0 | 29·2 | 33·4 | 37·5 | 41·7 |
| $1\frac{3}{8}$ | 5·0 | 10·0 | 15·1 | 20·1 | 25·1 | 30·1 | 35·1 | 40·2 | 45·2 | 50·2 |
| $1\frac{1}{2}$ | 6·0 | 11·9 | 17·9 | 23·9 | 29·9 | 35·8 | 41·8 | 47·8 | 53·7 | 59·7 |
| $1\frac{5}{8}$ | 7·0 | 14·0 | 21·0 | 28·0 | 35·1 | 42·1 | 49·1 | 56·1 | 63·1 | 70·1 |
| $1\frac{3}{4}$ | 8·1 | 16·3 | 24·4 | 32·5 | 40·6 | 48·8 | 56·9 | 65·0 | 73·2 | 81·3 |
| $1\frac{7}{8}$ | 9·3 | 18·7 | 28·0 | 37·3 | 46·7 | 56·0 | 65·3 | 74·7 | 84·0 | 93·3 |
| 2 in. | 10·6 | 21·2 | 31·8 | 42·5 | 53·1 | 63·7 | 74·3 | 84·9 | 95·5 | 106·2 |
| $2\frac{1}{8}$ | 12·0 | 24·0 | 36·0 | 48·0 | 59·9 | 71·9 | 83·9 | 95·9 | 107·9 | 119·9 |
| $2\frac{1}{4}$ | 13·4 | 26·9 | 40·3 | 53·8 | 67·2 | 80·6 | 94·1 | 107·5 | 121·0 | 134·4 |
| $2\frac{3}{8}$ | 15·0 | 30·0 | 44·9 | 60·0 | 74·9 | 89·9 | 104·8 | 119·8 | 134·8 | 149·8 |
| $2\frac{1}{2}$ | 16·7 | 33·4 | 50·1 | 66·8 | 83·5 | 100·1 | 116·8 | 133·6 | 150·2 | 166·9 |
| $2\frac{5}{8}$ | 18·3 | 36·6 | 54·9 | 73·2 | 91·5 | 109·8 | 128·1 | 146·3 | 164·6 | 182·9 |
| $2\frac{3}{4}$ | 20·1 | 40·2 | 60·2 | 80·3 | 100·4 | 120·5 | 140·5 | 160·6 | 180·7 | 200·8 |
| $2\frac{7}{8}$ | 21·9 | 43·9 | 65·8 | 87·8 | 109·7 | 131·7 | 153·6 | 175·6 | 197·5 | 219·4 |
| 3 in. | 23·9 | 47·8 | 71·7 | 95·6 | 119·4 | 143·3 | 167·2 | 191·1 | 215·0 | 238·9 |
| $3\frac{1}{8}$ | 25·9 | 51·9 | 77·8 | 103·7 | 129·6 | 155·6 | 181·5 | 207·4 | 233·3 | 259·3 |
| $3\frac{1}{4}$ | 28·0 | 56·1 | 84·1 | 112·2 | 140·2 | 168·2 | 196·3 | 224·3 | 253·4 | 280·4 |
| $3\frac{3}{8}$ | 30·2 | 60·5 | 90·7 | 121·0 | 151·2 | 181·4 | 211·7 | 241·9 | 272·2 | 302·4 |
| $3\frac{1}{2}$ | 32·5 | 65·0 | 97·5 | 130·0 | 162·6 | 195·1 | 227·6 | 260·1 | 292·6 | 325·1 |
| $3\frac{5}{8}$ | 34·9 | 69·8 | 104·7 | 139·5 | 174·4 | 209·3 | 244·2 | 279·1 | 314·0 | 348·9 |
| $3\frac{3}{4}$ | 37·3 | 74·7 | 112·0 | 149·3 | 186·7 | 224·0 | 261·3 | 298·7 | 336·0 | 373·3 |
| $3\frac{7}{8}$ | 39·9 | 79·7 | 119·6 | 159·5 | 199·3 | 239·2 | 279·0 | 318·9 | 358·8 | 398·6 |
| 4 in. | 42·5 | 84·9 | 127·4 | 169·9 | 212·3 | 254·8 | 297·2 | 339·7 | 382·2 | 424·6 |
| $4\frac{1}{8}$ | 45·2 | 90·3 | 135·5 | 180·7 | 225·9 | 271·0 | 316·2 | 361·4 | 406·6 | 451·7 |
| $4\frac{1}{4}$ | 48·0 | 95·9 | 143·9 | 191·8 | 239·8 | 287·7 | 335·7 | 383·6 | 431·6 | 479·5 |
| $4\frac{3}{8}$ | 50·8 | 101·6 | 152·4 | 203·3 | 254·1 | 304·9 | 355·7 | 406·5 | 457·3 | 508·2 |
| $4\frac{1}{2}$ | 53·8 | 107·5 | 161·3 | 215·0 | 268·8 | 322·6 | 376·3 | 430·1 | 483·8 | 537·6 |
| $4\frac{5}{8}$ | 56·8 | 113·6 | 170·4 | 227·2 | 283·9 | 340·7 | 397·5 | 454·3 | 511·1 | 567·9 |
| $4\frac{3}{4}$ | 60·0 | 119·8 | 179·7 | 239·6 | 299·5 | 359·4 | 419·3 | 479·2 | 539·1 | 599·0 |
| $4\frac{7}{8}$ | 63·1 | 126·2 | 189·3 | 252·4 | 315·5 | 378·6 | 441·7 | 504·8 | 567·8 | 630·9 |
| 5 in. | 66·8 | 133·5 | 200·3 | 267·0 | 333·8 | 400·5 | 467·3 | 534·0 | 600·8 | 667·5 |
| $5\frac{1}{4}$ | 73·2 | 146·3 | 219·5 | 292·7 | 365·9 | 439·0 | 512·2 | 585·4 | 658·5 | 731·7 |
| $5\frac{1}{2}$ | 80·3 | 160·6 | 240·9 | 321·2 | 401·5 | 481·8 | 562·1 | 642·4 | 722·7 | 803·0 |
| $5\frac{3}{4}$ | 87·8 | 175·6 | 263·3 | 351·1 | 438·9 | 526·7 | 614·4 | 702·2 | 790·0 | 877·8 |
| 6 in. | 95·6 | 191·1 | 286·7 | 382·2 | 477·8 | 573·3 | 668·9 | 764·4 | 860·0 | 955·5 |

TABLE *of the Weight of Cast Iron Plates, per Superficial Foot, from one-eighth of an inch to one inch thick.*

| ⅛ inch. | ¼ inch. | ⅜ inch. | ½ inch. | ⅝ inch. | ¾ inch. | ⅞ inch. | 1 inch. |
|---|---|---|---|---|---|---|---|
| lbs. oz. | lbs. oz. | lbs. oz. | lbs. oz. | lbs. oz. | lbs. oz. | lbs. oz. | lbs. oz. |
| 4 $13\frac{3}{8}$ | 9 $10\frac{5}{8}$ | 14 8 | 19 $5\frac{3}{8}$ | 24 $2\frac{3}{4}$ | 29 0 | 33 $13\frac{3}{8}$ | 38 $10\frac{3}{4}$ |

TABLE *containing the Weight of Cast Iron Pipes,* 1 *foot in length.*

| Diameter of Bore in Inches. | THICKNESS IN INCHES. | | | | | | | |
|---|---|---|---|---|---|---|---|---|
| | 3/8 | 1/2 | 5/8 | 3/4 | 7/8 | 1 inch. | 1 1/8 | 1 1/4 |
| | Lbs. | Lbs. | Lbs. | Lbs. | Lbs. | Lbs. | Lbs. | Lbs. |
| 1 1/2 | 6·9 | 9·9 | ...... | ...... | ...... | ...... | ...... | ...... |
| 2 | 8·8 | 12·3 | 16·1 | 20·3 | ...... | ...... | ...... | ...... |
| 2 1/2 | 10·6 | 14·7 | 19·2 | 23·9 | ...... | ...... | ...... | ...... |
| 3 | 12·4 | 17·2 | 22·2 | 27·6 | 33·3 | 39·3 | 45·6 | ...... |
| 3 1/2 | 14·2 | 19·6 | 25·3 | 31·3 | 37·6 | 44·2 | 51·1 | ...... |
| 4 | 16·8 | 22·1 | 28·4 | 35·0 | 41·9 | 49·1 | 56·6 | 64·4 |
| 4 1/2 | 18·0 | 24·5 | 31·4 | 38·7 | 46·2 | 54·0 | 62·1 | 70·6 |
| 5 | 19·8 | 27·0 | 34·5 | 42·3 | 50·5 | 58·9 | 67·6 | 76·7 |
| 5 1/2 | 21·6 | 29·5 | 37·6 | 46·0 | 54·8 | 63·8 | 73·2 | 82·8 |
| 6 | 23·5 | 31·9 | 40·7 | 49·7 | 59·1 | 68·7 | 78·7 | 88·8 |
| 6 1/2 | 25·3 | 34·4 | 43·7 | 53·4 | 63·4 | 73·4 | 84·2 | 95·1 |
| 7 | 27·2 | 36·8 | 46·8 | 56·8 | 67·7 | 78·5 | 89·7 | 101·2 |
| 7 1/2 | 29·0 | 39·1 | 49·9 | 60·7 | 72·0 | 83·5 | 95·3 | 107·4 |
| 8 | 30·8 | 41·7 | 52·9 | 64·4 | 76·2 | 88·4 | 100·8 | 113·5 |
| 8 1/2 | 32·9 | 44·4 | 56·2 | 68·3 | 80·8 | 93·5 | 106·5 | 119·9 |
| 9 | 34·5 | 46·6 | 59·1 | 71·8 | 84·8 | 98·2 | 111·8 | 125·8 |
| 9 1/2 | 36·3 | 49·1 | 62·1 | 75·5 | 89·1 | 103·1 | 117·4 | 131·9 |
| 10 | 38·2 | 51·5 | 65·2 | 79·2 | 93·4 | 108·0 | 122·8 | 138·1 |
| 10 1/2 | ...... | 54·0 | 68·2 | 82·8 | 97·7 | 112·9 | 128·4 | 144·2 |
| 11 | ...... | 56·4 | 71·3 | 86·5 | 102·0 | 117·8 | 133·9 | 150·3 |
| 11 1/2 | ...... | 58·9 | 74·3 | 90·1 | 106·3 | 122·7 | 139·4 | 156·4 |
| 12 | ...... | 61·3 | 77·4 | 93·6 | 110·6 | 127·6 | 145·0 | 162·6 |
| 13 | ...... | ...... | 82·7 | 101·2 | 118·2 | 137·4 | 154·1 | 173·5 |
| 14 | ...... | ...... | 89·5 | 108·2 | 126·5 | 146·2 | 165·3 | 185·2 |
| 15 | ...... | ...... | 95·2 | 115·7 | 135·3 | 156·2 | 176·2 | 198·1 |
| 16 | ...... | ...... | ...... | 123·3 | 143·1 | 166·1 | 187·5 | 211·3 |
| 17 | ...... | ...... | ...... | 130·2 | 152·5 | 178·5 | 198·2 | 223·4 |
| 18 | ...... | ...... | ...... | 137·0 | 161·2 | 185·3 | 209·1 | 235·6 |
| 19 | ...... | ...... | ...... | ...... | 169·2 | 195·7 | 222·3 | 247·1 |
| 20 | ...... | ...... | ...... | ...... | 178·1 | 205·2 | 233·2 | 259·0 |
| 21 | ...... | ...... | ...... | ...... | ...... | 214·1 | 243·5 | 273·2 |
| 22 | ...... | ...... | ...... | ...... | ...... | 223·0 | 254·8 | 285·4 |
| 23 | ...... | ...... | ...... | ...... | ...... | 233·4 | 265·5 | 298·3 |
| 24 | ...... | ...... | ...... | ...... | ...... | 245·2 | 277·5 | 310·6 |

TABLE *containing the Weight of Solid Cylinders of Cast Iron, one foot in length, and from* 3/4 *of an inch to* 14 *inches diameter.*

| Diameter in Inches. | Weight in Lbs. | Diameter in Inches. | Weight in Lbs. | Diameter in Inches. | Weight in Lbs. | Diameter in Inches. | Weight in Lbs. |
|---|---|---|---|---|---|---|---|
| 3/4 | 1·39 | 2 7/8 | 20·48 | 4 7/8 | 58·72 | 7 3/4 | 148·87 |
| 7/8 | 1·88 | 3 in. | 22·35 | 5 in. | 61·96 | 8 in. | 158·63 |
| 1 in. | 2·47 | 3 1/8 | 24·20 | 5 1/8 | 64·66 | 8 1/4 | 168·15 |
| 1 1/8 | 3·13 | 3 1/4 | 26·18 | 5 1/4 | 68·31 | 8 1/2 | 179·08 |
| 1 1/4 | 3·87 | 3 3/8 | 28·23 | 5 3/8 | 71·00 | 8 3/4 | 189·00 |
| 1 3/8 | 4·68 | 3 1/2 | 30·36 | 5 1/2 | 74·98 | 9 in. | 200·77 |
| 1 1/2 | 5·57 | 3 5/8 | 32·57 | 5 5/8 | 78·65 | 9 1/4 | 211·12 |
| 1 5/8 | 6·54 | 3 3/4 | 34·85 | 5 3/4 | 81·95 | 9 1/2 | 223·70 |
| 1 3/4 | 7·59 | 3 7/8 | 37·21 | 5 7/8 | 85·81 | 9 3/4 | 235·31 |
| 1 7/8 | 8·71 | 4 in. | 39·66 | 6 in. | 89·23 | 10 in. | 247·87 |
| 2 in. | 9·91 | 4 1/8 | 41·80 | 6 1/4 | 96·82 | 10 1/2 | 273·27 |
| 2 1/8 | 11·19 | 4 1/4 | 44·77 | 6 1/2 | 104·72 | 11 in. | 299·92 |
| 2 1/4 | 12·54 | 4 3/8 | 47·00 | 6 3/4 | 112·93 | 11 1/2 | 327·81 |
| 2 3/8 | 13·98 | 4 1/2 | 50·19 | 7 in. | 121·45 | 12 in. | 356·93 |
| 2 1/2 | 15·49 | 4 5/8 | 52·71 | 7 1/4 | 130·28 | 13 | 418·90 |
| 2 5/8 | 17·08 | 4 3/4 | 55·92 | 7 1/2 | 139·42 | 14 | 485·83 |
| 2 3/4 | 18·74 | | | | | | |

TABLE *containing the Weight of a Square Foot of Copper and Lead, in lbs. avoirdupois, from $\frac{1}{32}$ to $\frac{1}{2}$ an inch in thickness, advancing by $\frac{1}{32}$.*

| Thickness. | Copper. | Lead. |
|---|---|---|
| $\frac{1}{32}$ | 1·45 | 1·85 |
| $\frac{1}{16}$ | 2·90 | 3·70 |
| $\frac{3}{32}$ | 4·35 | 5·54 |
| $\frac{1}{8}$ | 5·80 | 7·39 |
| $\frac{1}{8} + \frac{1}{32}$ | 7·26 | 9·24 |
| $\frac{1}{8} + \frac{1}{16}$ | 8·71 | 11·08 |
| $\frac{1}{8} + \frac{3}{32}$ | 10·16 | 12·93 |
| $\frac{1}{4}$ | 11·61 | 14·77 |
| $\frac{1}{4} + \frac{1}{32}$ | 13·07 | 16·62 |
| $\frac{1}{4} + \frac{1}{16}$ | 14·52 | 18·47 |
| $\frac{1}{4} + \frac{3}{32}$ | 15·97 | 20·31 |
| $\frac{3}{8}$ | 17·41 | 22·16 |
| $\frac{3}{8} + \frac{1}{32}$ | 18·87 | 24·00 |
| $\frac{3}{8} + \frac{1}{16}$ | 20·32 | 25·85 |
| $\frac{3}{8} + \frac{3}{32}$ | 21·77 | 27·70 |
| $\frac{1}{2}$ | 23·22 | 29·55 |

TABLE *for finding the Weight of Malleable Iron, Copper, and Lead Pipes, 12 inches long, of various thicknesses, and any diameter required.*

| Thickness. | Malleable Iron. | Copper. | Lead. |
|---|---|---|---|
| $\frac{1}{32}$ of an inch. | ·104 | ·121 | ·1539 |
| $\frac{1}{16}$ | ·208 | ·2419 | ·3078 |
| $\frac{3}{32}$ | ·3108 | ·3628 | ·4616 |
| $\frac{1}{8}$ | ·414 | ·4838 | ·6155 |
| $\frac{1}{8} + \frac{1}{32}$ | ·518 | ·6047 | ·7691 |
| $\frac{1}{8} + \frac{1}{16}$ | ·621 | ·7258 | ·9232 |
| $\frac{1}{8} + \frac{3}{32}$ | ·725 | ·8466 | 1·0771 |
| $\frac{1}{4}$ | ·828 | ·9678 | 1·231 |

RULE.—Multiply the circumference of the pipe in inches by the numbers opposite the thickness required, and by the length in feet; the product will be the weight in avoirdupois lbs. nearly.

Required the weight of a copper pipe 12 feet long, 15 inches in circumference, $\frac{1}{8} + \frac{1}{16}$ of an inch in thickness.

$$·7258 \times 15 = 10·817 \times 12 = 130·644 \text{ lbs. nearly.}$$

TABLE *of the Weight of a Square Foot of Millboard in lbs. avoirdupois.*

| Thickness in inches...... | $\frac{1}{8}$ | $\frac{3}{16}$ | $\frac{1}{4}$ | $\frac{5}{16}$ | $\frac{3}{8}$ |
|---|---|---|---|---|---|
| Weight in lbs ............ | ·688 | 1·032 | 1·376 | 1·72 | 2·064 |

TABLE *containing the Weight of Wrought Iron Bars* 12 *inches long in lbs. avoirdupois.*

| Inch. | Round. | Square. | Inch. | Round. | Square. |
|---|---|---|---|---|---|
| 1/4 | ·163 | ·208 | 2 1/2 | 16·32 | 20·80 |
| 3/8 | ·367 | ·467 | 2 5/8 | 18·00 | 22·89 |
| 1/2 | ·653 | ·830 | 2 3/4 | 19·76 | 25·12 |
| 5/8 | 1·02 | 1·30 | 2 7/8 | 21·59 | 27·46 |
| 3/4 | 1·47 | 1·87 | 3 | 23·52 | 29·92 |
| 7/8 | 2·00 | 2·55 | 3 1/4 | 27·60 | 35·12 |
| 1 | 2·61 | 3·32 | 3 1/2 | 32·00 | 40·80 |
| 1 1/8 | 3·31 | 4·21 | 3 3/4 | 36·72 | 46·72 |
| 1 1/4 | 4·08 | 5·20 | 4 | 41·76 | 53·12 |
| 1 3/8 | 4·94 | 6·28 | 4 1/4 | 47·25 | 60·00 |
| 1 1/2 | 5·88 | 7·48 | 4 1/2 | 52·93 | 67·24 |
| 1 5/8 | 6·90 | 8·78 | 4 3/4 | 58·92 | 74·95 |
| 1 3/4 | 8·00 | 10·20 | 5 | 65·28 | 83·20 |
| 1 7/8 | 9·18 | 11·68 | 5 1/4 | 72·00 | 91·56 |
| 2 | 10·44 | 13·28 | 5 1/2 | 79·04 | 100·48 |
| 2 1/8 | 11·80 | 15·00 | 5 3/4 | 86·36 | 109·82 |
| 2 1/4 | 13·23 | 16·81 | 6 | 94·08 | 119·68 |
| 2 3/8 | 14·73 | 18·74 | 7 | 128·00 | 163·20 |

TABLE *of the Proportional Dimensions of* 6-*sided Nuts for Bolts from* 1/4 *to* 2 1/2 *inches diameter.*

| | | | | | | | | | |
|---|---|---|---|---|---|---|---|---|---|
| Diameter of bolts........ | 1/4 | 3/8 | 1/2 | 5/8 | 3/4 | 7/8 | 1 | 1 1/8 | 1 1/4 |
| Breadth of nuts.......... | 11/16 | 13/16 | 1 | 1 3/16 | 1 3/8 | 1 9/16 | 1 3/4 | 1 15/16 | 2 1/8 |
| Breadth over the angles | 3/4 | 1 5/6 | 1 1/8 | 1 3/8 | 1 8/16 | 1 13/16 | 2 | 2 1/4 | 2 7/16 |
| Thickness.................. | 5/16 | 7/16 | 9/16 | 3/4 | 7/8 | 1 | 1 1/8 | 1 1/4 | 1 7/16 |
| Diameter of bolts........ | 1 3/8 | 1 1/2 | 1 5/8 | 1 3/4 | 1 7/8 | 2 | 2 1/4 | 2 1/2 | |
| Breadth of nuts.......... | 2 5/16 | 2 1/2 | 2 11/16 | 2 7/8 | 3 1/16 | 3 1/4 | 3 5/8 | 4 | |
| Breadth over the angles | 2 11/16 | 2 7/8 | 3 1/8 | 3 5/16 | 3 1/2 | 3 3/4 | 4 3/16 | 4 5/8 | |
| Thickness.................. | 1 9/16 | 1 11/16 | 1 13/16 | 2 | 2 1/8 | 2 1/4 | 2 1/2 | 2 3/4 | |

TABLE *of the Specific Gravity of Water at different temperatures, that at* 62° *being taken as unity.*

| | | | |
|---|---|---|---|
| 70° F. | ·99913 | 52° F. | 1·00076 |
| 68 | ·99936 | 50 | 1·00087 |
| 66 | ·99958 | 48 | 1·00095 |
| 64 | ·99980 | 46 | 1·00102 |
| 62 | 1· | 44 | 1·00107 |
| 58 | 1·00035 | 42 | 1·00111 |
| 56 | 1·00050 | 40 | 1·00113 |
| 54 | 1·00064 | 38 | 1·00115 |

The difference of temperatures between 62° and 39°·2, where water attains its greatest density, will vary the bulk of a gallon rather less than the third of a cubic inch.

TABLE *of the Weight of Cast Iron Balls in pounds avoirdupois, from* 1 *to* 12 *inches diameter, advancing by an eighth.*

| Inches. | Lbs. | Inches. | Lbs. | Inches. | Lbs. |
|---|---|---|---|---|---|
| 1 | ·14 | 4¾ | 14·76 | 8½ | 84·56 |
| 1⅛ | ·20 | 4⅞ | 15·95 | 8⅝ | 88·34 |
| 1¼ | ·27 | 5 | 17·12 | 8¾ | 92·24 |
| 1⅜ | ·37 | 5⅛ | 18·54 | 8⅞ | 96·26 |
| 1½ | ·47 | 5¼ | 19·93 | 9 | 100·39 |
| 1⅝ | ·59 | 5⅜ | 21·39 | 9⅛ | 104·62 |
| 1¾ | ·74 | 5½ | 22·91 | 9¼ | 108·98 |
| 1⅞ | ·91 | 5⅝ | 24·51 | 9⅜ | 113·46 |
| 2 | 1·10 | 5¾ | 26·18 | 9½ | 118·06 |
| 2⅛ | 1·32 | 5⅞ | 27·91 | 9⅝ | 122·77 |
| 2¼ | 1·57 | 6 | 29·72 | 9¾ | 127·63 |
| 2⅜ | 1·84 | 6⅛ | 31·64 | 9⅞ | 132·60 |
| 2½ | 2·15 | 6¼ | 33·62 | 10 | 137·71 |
| 2⅝ | 2·49 | 6⅜ | 35·67 | 10⅛ | 142·91 |
| 2¾ | 2·86 | 6½ | 37·80 | 10¼ | 148·28 |
| 2⅞ | 3·27 | 6⅝ | 40·10 | 10⅜ | 153·78 |
| 3 | 3·72 | 6¾ | 42·35 | 10½ | 159·40 |
| 3⅛ | 4·20 | 6⅞ | 44·74 | 10⅝ | 165·16 |
| 3¼ | 4·72 | 7 | 47·21 | 10¾ | 171·05 |
| 3⅜ | 5·29 | 7⅛ | 49·79 | 10⅞ | 177·10 |
| 3½ | 5·80 | 7¼ | 52·47 | 11 | 183·29 |
| 3⅝ | 6·56 | 7⅜ | 55·23 | 11⅛ | 189·60 |
| 3¾ | 7·26 | 7½ | 58·06 | 11¼ | 196·10 |
| 3⅞ | 8·01 | 7⅝ | 60·04 | 11⅜ | 202·67 |
| 4 | 8·81 | 7¾ | 64·09 | 11½ | 209·43 |
| 4⅛ | 9·67 | 7⅞ | 67·25 | 11⅝ | 216·32 |
| 4¼ | 10·57 | 8 | 70·49 | 11¾ | 223·40 |
| 4⅜ | 11·53 | 8⅛ | 73·85 | 11⅞ | 230·57 |
| 4½ | 12·55 | 8¼ | 77·32 | 12 | 237·94 |
| 4⅝ | 13·62 | 8⅜ | 80·88 | | |

TABLE *of the Weight of Flat Bar Iron,* 12 *inches long, in lbs. avoirdupois.*

| Breadth in inches. | Thickness. ⅛ | 3/16 | ¼ | ⅜ | ½ | ⅝ | ¾ | ⅞ | 1 inch. |
|---|---|---|---|---|---|---|---|---|---|
| ½ | ·21 | ·31 | ·42 | ·63 | | | | | |
| ¾ | ·31 | ·47 | ·63 | ·94 | 1·26 | 1·57 | | | |
| 1 | ·42 | ·63 | ·84 | 1·26 | 1·68 | 2·10 | 2·52 | 2·94 | |
| 1¼ | ·52 | ·78 | 1·05 | 1·57 | 2·10 | 2·62 | 3·15 | 3·67 | 4·20 |
| 1⅜ | ·57 | ·86 | 1·18 | 1·73 | 2·31 | 2·88 | 3·46 | 4·04 | 4·62 |
| 1½ | ·63 | ·94 | 1·26 | 1·89 | 2·52 | 3·15 | 3·78 | 4·41 | 5·04 |
| 1¾ | ·73 | 1·10 | 1·47 | 2·20 | 2·94 | 3·67 | 4·41 | 5·14 | 5·87 |
| 2 | ·84 | 1·26 | 1·68 | 2·52 | 3·36 | 4·20 | 5·06 | 5·88 | 6·72 |
| 2¼ | ·96 | 1·41 | 1·89 | 2·83 | 3·78 | 4·72 | 5·66 | 6·61 | 7·56 |
| 2½ | 1·05 | 1·57 | 2·10 | 3·15 | 4·20 | 5·25 | 6·30 | 7·35 | 8·40 |
| 2¾ | 1·15 | 1·73 | 2·31 | 3·46 | 4·62 | 5·77 | 6·93 | 8·08 | 9·24 |
| 3 | 1·26 | 1·89 | 2·52 | 3·78 | 5·04 | 6·30 | 7·56 | 8·82 | 10·08 |
| 3¼ | 1·36 | 2·04 | 2·73 | 4·09 | 5·46 | 6·82 | 8·19 | 9·55 | 10·92 |
| 3½ | 1·47 | 2·20 | 2·94 | 4·41 | 5·88 | 7·35 | 8·82 | 10·29 | 11·76 |
| 3¾ | 1·57 | 2·36 | 3·15 | 4·72 | 6·30 | 7·87 | 9·45 | 11·02 | 12·60 |
| 4 | 1·68 | 2·52 | 3·36 | 5·04 | 6·72 | 8·40 | 10·08 | 11·76 | 13·44 |
| 4½ | 1·89 | 2·83 | 3·73 | 5·67 | 7·56 | 9·45 | 11·34 | 13·23 | 15·12 |
| 5 | 2·10 | 3·15 | 4·12 | 6·30 | 8·40 | 10·50 | 12·60 | 16·70 | 17·80 |
| 6 | 2·52 | 3·78 | 5·04 | 7·56 | 10·08 | 12·60 | 15·12 | 17·64 | 20·16 |

Weight of a copper rod 12 inches long and 1 inch diameter = 3·039 lbs.
Weight of a brass rod 12 inches long and 1 inch diameter = 2·86 lbs.

## BRASS.—*Weight of a Lineal Foot of Round and Square.*

| Diameter. | Weight of round. | Weight of square. | Diameter. | Weight of round. | Weight of square. |
|---|---|---|---|---|---|
| Inches. | Lbs. | Lbs. | Inches. | Lbs. | Lbs. |
| ¼ | ·17 | ·22 | 1¾ | 8·66 | 11·03 |
| ⅜ | ·39 | ·50 | 1⅞ | 9·95 | 12·66 |
| ½ | ·70 | ·90 | 2 | 11·32 | 14·41 |
| ⅝ | 1·10 | 1·40 | 2⅛ | 12·78 | 16·27 |
| ¾ | 1·59 | 2·02 | 2¼ | 14·32 | 18·24 |
| ⅞ | 2·16 | 2·75 | 2⅜ | 15·96 | 20·32 |
| 1 | 2·83 | 3·60 | 2½ | 17·68 | 22·53 |
| 1⅛ | 3·58 | 4·56 | 2⅝ | 19·50 | 24·83 |
| 1¼ | 4·42 | 5·63 | 2¾ | 21·40 | 27·25 |
| 1⅜ | 5·35 | 6·81 | 2⅞ | 23·39 | 29·78 |
| 1½ | 6·36 | 8·00 | 3 | 25·47 | 32·43 |
| 1⅝ | 7·47 | 9·51 | | | |

## STEEL.—*Weight of One Foot of Round Steel.*

| Diameter in inches and parts. | ¼ | ⅜ | ½ | ⅝ | ¾ | ⅞ | 1 | 1⅛ | 1¼ | 1⅜ | 1½ | 1⅝ | 1¾ | 1⅞ | 2 |
|---|---|---|---|---|---|---|---|---|---|---|---|---|---|---|---|
| Weight in lbs. and decimal parts. | ·167 | ·376 | ·669 | 1·04 | 1·5 | 2·05 | 2·67 | 3·38 | 4·18 | 5·06 | 6·02 | 7·07 | 8·2 | 9·41 | 11·71 |

## TABLES OF THE WEIGHTS OF ROLLED IRON,

*Per lineal foot, of various sections, illustrated in the accompanying cuts, viz. Parallel Angle Iron, equal and unequal sides; Taper Angle Iron; Parallel* T *Iron, equal and unequal depth and width; Taper* T *Iron; Sash Iron; and Permanent and Temporary Rails.*

## TABLE I.—*Parallel Angle Iron, of equal sides.* (Fig. 1.)

| Length of sides A B, in inches. | Uniform thickness throughout. | Weight of one lineal foot in lbs. |
|---|---|---|
| Inches. | Inches. | |
| 3 | ⅜ | 8·0 |
| 2¾ | ⅜ | 7·0 |
| 2½ | ⅜ | 5·75 |
| 2¼ | 5-16ths | 4·5 |
| 2 | ¼ full | 3·75 |
| 1¾ | ¼ | 3·0 |
| 1½ | ¼ | 2·5 |
| 1⅜ | No. 6 wire-gauge | 1·75 |
| 1¼ | 8 | 1·5 |
| 1⅛ | 9 | 1·25 |
| 1 | 10 | 1·0 |
| ⅞ | 10 | ·875 |
| ¾ | 11 | ·625 |
| ⅝ | 11 | ·568 |
| ½ | 12 | ·5 |

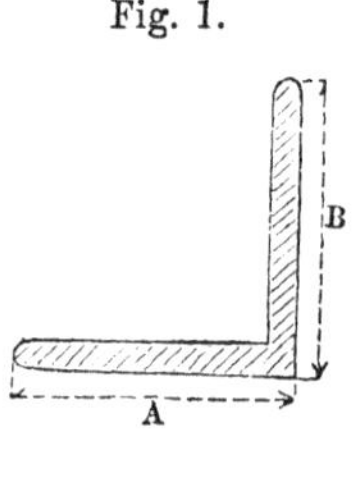

Fig. 1.

### TABLE II.—*Parallel Angle Iron, of unequal sides.* (Fig. 2.)

| Length of side A, in inches. | Length of side B, in inches. | Uniform thickness throughout. | Weight of one lineal foot in lbs. |
|---|---|---|---|
| Inches. | Inches. | Inches. | |
| $3\frac{1}{2}$ | 5 | $\frac{3}{8}$ | 9·75 |
| 3 | 5 | $\frac{3}{8}$ | 8·75 |
| 3 | 4 | 5-16ths | 7·5 |
| $2\frac{1}{4}$ | 4 | 5-16ths | 6·75 |
| $2\frac{1}{4}$ | 4 | $\frac{1}{4}$ | 5·75 |
| 2 | 4 | $\frac{1}{4}$ | 5·5 |
| $2\frac{1}{2}$ | 3 | $\frac{1}{4}$ | 4·75 |
| 2 | $2\frac{1}{2}$ | $\frac{1}{4}$ | 3·375 |
| $1\frac{1}{2}$ | 2 | $\frac{1}{4}$ | 2·875 |
| $1\frac{1}{2}$ | 2 | 3-16ths | 2·25 |

Fig. 2.

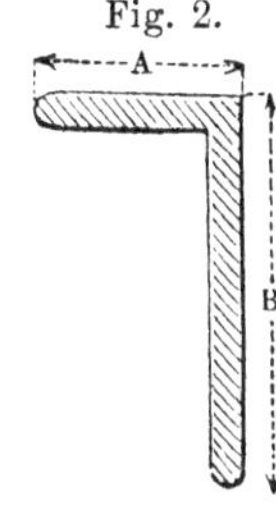


### TABLE III.—*Taper Angle Iron, of equal sides.* (Fig. 3.)

| Length of sides, A A, in inches. | Thickness of edges at B. | Thickness of root at C. | Weight of one lineal foot in lbs. |
|---|---|---|---|
| Inches. | Inches. | Inches. | |
| 4 | $\frac{1}{2}$ | $\frac{5}{8}$ | 14·0 |
| 3 | $\frac{1}{2}$ | $\frac{5}{8}$ | 10·375 |
| $2\frac{3}{4}$ | 7-16ths | 9-16ths | 8·25 |
| $2\frac{1}{2}$ | $\frac{3}{8}$ | $\frac{1}{2}$ | 6·5 |
| $2\frac{1}{4}$ | 5-16ths, full | 7-16ths | 5·0 |
| 2 | $\frac{1}{4}$ full | 5-16ths, full | 3·875 |
| $1\frac{3}{4}$ | $\frac{1}{4}$ | 5-16ths | 3·25 |
| $1\frac{1}{2}$ | $\frac{1}{4}$ bare | 5-16ths, bare | 2·625 |

Fig. 3.

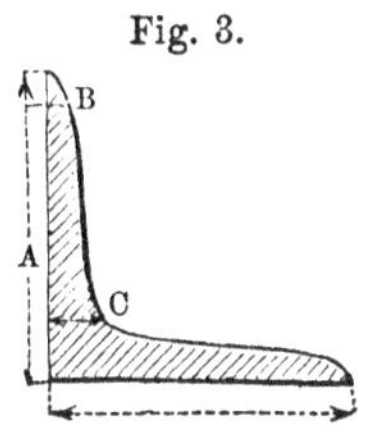


### TABLE IV.—*Parallel* T *Iron, of unequal width and depth.* (Fig. 4.)

| Width of top table A, in inches. | Total depth B, in inches. | Uniform thickness of top table C. | Uniform thickness of rib D. | Weight of one lineal foot in lbs. |
|---|---|---|---|---|
| Inches. | Inches. | Inches. | Inches. | |
| 5 | 6 | $\frac{1}{2}$ | $\frac{1}{2}$ | 15·75 |
| $4\frac{1}{2}$ | $3\frac{1}{4}$ | $\frac{1}{2}$ | 9.16ths | 13·25 |
| 4 | 3 | $\frac{3}{8}$ | $\frac{3}{8}$ | 8·875 |
| $3\frac{1}{2}$ | 3 | $\frac{3}{8}$ | $\frac{3}{8}$ | 8·25 |
| $3\frac{1}{2}$ | 4 | $\frac{1}{2}$ | $\frac{1}{2}$ | 12·5 |
| $2\frac{1}{2}$ | 3 | $\frac{3}{8}$ | $\frac{3}{8}$ | 7·0 |
| $2\frac{1}{4}$ | 2 | 5-16ths | $\frac{3}{8}$ full | 4·5 |
| 2 | $1\frac{1}{2}$ | 5-16ths | 5-16ths | 4·0 |
| $1\frac{3}{4}$ | 2 | $\frac{1}{4}$ | $\frac{1}{4}$ | 3·125 |
| $1\frac{1}{2}$ | 2 | $\frac{1}{4}$ | $\frac{1}{4}$ | 2·875 |
| $1\frac{1}{4}$ | $1\frac{1}{2}$ | $\frac{1}{4}$ | $\frac{1}{4}$ | 2·375 |
| 1 | $1\frac{1}{4}$ | 3-16ths | 3-16ths | 1·5 |
| $\frac{3}{4}$ | 1 | 3-16ths | 3-16ths | 1·125 |

Fig. 4.

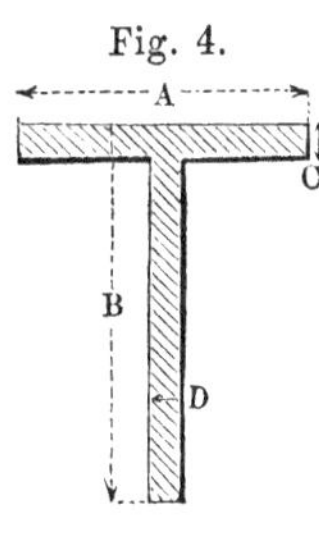

## Table V.—*Parallel* T *Iron, of equal depth and width.* (Fig. 5.)

| Width of top table, and total depth A A. | Uniform thickness throughout. | Weight of one lineal foot in lbs. |
|---|---|---|
| Inches. | Inches. | |
| 6 | ½ | |
| 5 | 7-16ths | 13·75 |
| 4 | ⅜ | 9·75 |
| 3½ | ⅜ | 8·5 |
| 3 | ⅜ | 7·5 |
| 2½ | 5-16ths | 4·625 |
| 2¼ | 5-16ths | 4·5 |
| 2 | 5-16ths | 3·75 |
| 1¾ | ¼ | 3·0 |
| 1½ | ¼ | 2·25 |
| 1¼ | ¼ | 1·75 |
| 1 | 3-16ths | 1·0 |
| ⅞ | ⅛ | ·725 |
| ¾ | ⅛ | ·625 |

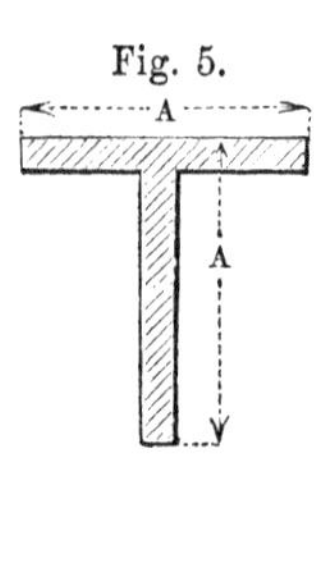

Fig. 5.

## Table VI.—*Taper* T *Iron.* (Fig. 6.)

| Width of top table A, in inches. | Total depth B, in inches. | Thickness of top table at root C. | Thickness of top table at edges D. | Uniform thickness of rib E. | Weight of one lineal foot in lbs. |
|---|---|---|---|---|---|
| Inches. | Inches. | Inches. | Inches. | Inches. | |
| 3 | 3¼ | ½ | ⅜ | 7-16ths | 8·0 |
| 3 | 2⅝ | 7-16ths | ⅜ | ½ | 8·0 |
| 2½ | 3 | 7-16ths | 5-16ths | 5-16ths | 5·25 |
| 2 | 2½ | ⅝ | ½ | ½ | 6·5 |
| 2 | 1½ | ⅜ full | 5-16ths | ⅜ | 3·5 |
| 2 | 1½ | 5-16ths | ¼ | ¼ | 2·875 |

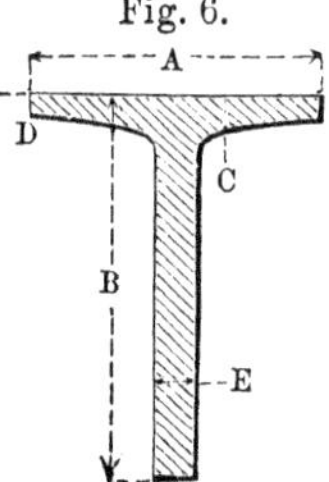

Fig. 6.

## Table VII.—*Sash Iron.* (Fig. 7.)

| Total depth A. | Depth of rebate B. | Width at edge C. | Greatest width D. | Weight of one lineal foot in lbs. |
|---|---|---|---|---|
| Inches. | Inches. | | Inches. | |
| 2 | 1 | No. 9 wire-gauge | 5-8ths | 1·75 |
| 1¾ | ¾ | 7 | 9-16ths | 1·625 |
| 1½ | ¾ | 6 | 9-16ths | 1·25 |
| 1⅜ | ⅝ | 10 | 9-16ths | 1·125 |
| 1¼ | ⅝ | 10 | 9-16ths | 1·0 |
| 1 | ½ | ⅛ | ½ | ·75 |

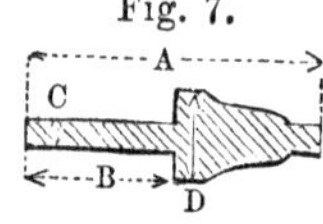

Fig. 7.

## Table VIII.—*Rails equal top and bottom Tables.* (Fig. 8.)

| Depth A, in inches. | Width across top and bottom B B, in inches. | Thickness of rib C. | Weight of one lineal foot in lbs. |
|---|---|---|---|
| Inches. | Inches. | Inches. | |
| 5 | 2⅝ | ¾ | 25·0 |
| 4½ | 2½ | ¾ | 23·33 |
| 4½ | 2½ | ⅝ | 21·66 |

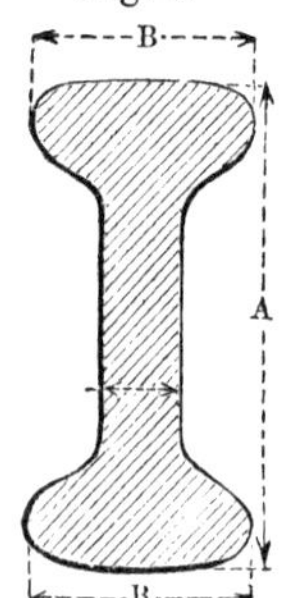

Fig. 8.

TABLE IX.—*Temporary Rails.* (Fig. 9.)

| Top width A, in inches. | Rib width B, in inches. | Bed width C, in inches. | Total depth D, in inches. | Thickness of bed E. | Weight of one lineal foot in lbs. |
|---|---|---|---|---|---|
| Inches. | Inches. | Inches. | Inches. | Inches. | |
| $1\frac{1}{2}$ | $\frac{5}{8}$ | 3 | 2 | 7-16ths | 9·0 |
| $1\frac{3}{4}$ | $\frac{5}{8}$ | 3 | $2\frac{1}{2}$ | $\frac{1}{2}$ | 12·0 |
| $1\frac{7}{8}$ | $\frac{5}{8}$ | 4 | 3 | $\frac{1}{2}$ | 16·0 |
| 2 | $\frac{5}{8}$ | 4 | 3 | $\frac{1}{2}$ | 17·33 |

Fig. 9.

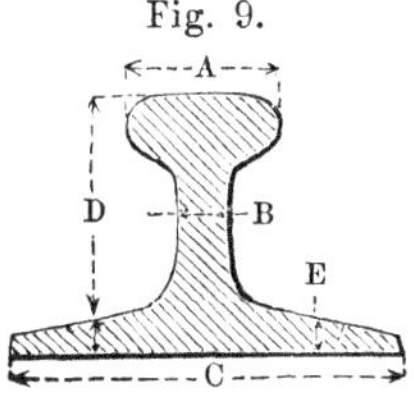


TABLE *of Natural Sines, Co-sines, Tangents, Co-tangents, Secants, and Co-secants, to every degree of the Quadrant.*

| Deg. | Sines. | Co-sines. | Tangents. | Co-tangents. | Secants. | Co-secants. | Degree. |
|---|---|---|---|---|---|---|---|
| 0 | ·00000 | 1·00000 | ·00000 | Infinite. | 1·00000 | Infinite. | 90 |
| 1 | ·01745 | ·99985 | ·01746 | 57·2900 | 1·00015 | 57·2987 | 89 |
| 2 | ·03490 | ·99939 | ·03492 | 28·6363 | 1·00061 | 28·6537 | 88 |
| 3 | ·05234 | ·99863 | ·05241 | 19·0811 | 1·00137 | 19·1073 | 87 |
| 4 | ·06976 | ·99756 | ·06993 | 14·3007 | 1·00244 | 14·3356 | 86 |
| 5 | ·08716 | ·99619 | ·08749 | 11·4301 | 1·00382 | 11·4737 | 85 |
| 6 | ·10453 | ·99452 | ·10510 | 9·51236 | 1·00551 | 9·56677 | 84 |
| 7 | ·12187 | ·99255 | ·12278 | 8·14435 | 1·00751 | 8·20551 | 83 |
| 8 | ·13917 | ·99027 | ·14054 | 7·11537 | 1·00983 | 7·18530 | 82 |
| 9 | ·15643 | ·98769 | ·15838 | 6·31375 | 1·01246 | 6·39245 | 81 |
| 10 | ·17365 | ·98481 | ·17633 | 5·67128 | 1·01543 | 5·75877 | 80 |
| 11 | ·19081 | ·98163 | ·19438 | 5·14455 | 1·01872 | 5·24084 | 79 |
| 12 | ·20791 | ·97815 | ·21256 | 4·70463 | 1·02234 | 4·80973 | 78 |
| 13 | ·22495 | ·97437 | ·23087 | 4·33148 | 1·02630 | 4·44541 | 77 |
| 14 | ·24192 | ·97030 | ·24933 | 4·01078 | 1·03061 | 4·13356 | 76 |
| 15 | ·25882 | ·96593 | ·26795 | 3·73205 | 1·03528 | 3·86370 | 75 |
| 16 | ·27564 | ·96126 | ·28675 | 3·48741 | 1·04030 | 3·62796 | 74 |
| 17 | ·29237 | ·95630 | ·30573 | 3·27085 | 1·04569 | 3·42030 | 73 |
| 18 | ·30902 | ·95106 | ·32492 | 3·07768 | 1·05146 | 3·23607 | 72 |
| 19 | ·32557 | ·94552 | ·34433 | 2·90421 | 1·05762 | 3·07155 | 71 |
| 20 | ·34202 | ·93969 | ·36397 | 2·74748 | 1·06418 | 2·92380 | 70 |
| 21 | ·35837 | ·93358 | ·38386 | 2·60509 | 1·07114 | 2·79043 | 69 |
| 22 | ·37461 | ·92718 | ·40403 | 2·47509 | 1·07853 | 2·66947 | 68 |
| 23 | ·39073 | ·92050 | ·42447 | 2·35585 | 1·08636 | 2·55930 | 67 |
| 24 | ·40674 | ·91355 | ·44523 | 2·24004 | 1·09464 | 2·45859 | 66 |
| 25 | ·42262 | ·90631 | ·46631 | 2·14451 | 1·10338 | 2·36620 | 65 |
| 26 | ·43837 | ·89879 | ·48773 | 2·05030 | 1·11260 | 2·28117 | 64 |
| 27 | ·45399 | ·89101 | ·50952 | 1·96261 | 1·12233 | 2·20869 | 63 |
| 28 | ·46947 | ·88295 | ·53171 | 1·88073 | 1·13257 | 2·13005 | 62 |
| 29 | ·48481 | ·87462 | ·55431 | 1·80405 | 1·14335 | 2·06266 | 61 |
| 30 | ·50000 | ·86603 | ·57735 | 1·73205 | 1·15470 | 2·00000 | 60 |
| 31 | ·51504 | ·85717 | ·60086 | 1·66428 | 1·16663 | 1·94160 | 59 |
| 32 | ·52992 | ·84805 | ·62487 | 1·60033 | 1·17918 | 1·88708 | 58 |
| 33 | ·54464 | ·83867 | ·64941 | 1·53986 | 1·19236 | 1·83608 | 57 |
| 34 | ·55919 | ·82904 | ·67451 | 1·48256 | 1·20622 | 1·78829 | 56 |
| 35 | ·57358 | ·81915 | ·70021 | 1·42815 | 1·22077 | 1·74345 | 55 |
| 36 | ·58778 | ·80902 | ·72654 | 1·37638 | 1·23607 | 1·70130 | 54 |
| 37 | ·60181 | ·79863 | ·75355 | 1·32704 | 1·25214 | 1·66164 | 53 |
| 38 | ·61566 | ·78801 | ·78129 | 1·27994 | 1·26902 | 1·62427 | 52 |
| 39 | ·62932 | ·77715 | ·80978 | 1·23490 | 1·28676 | 1·58902 | 51 |
| 40 | ·64279 | ·76604 | ·83910 | 1·19175 | 1·30541 | 1·55572 | 50 |
| 41 | ·65606 | ·75471 | ·86929 | 1·15037 | 1·32511 | 1·52425 | 49 |
| 42 | ·66913 | ·74314 | ·90040 | 1·11061 | 1·34561 | 1·49448 | 48 |
| 43 | ·68200 | ·73135 | ·93251 | 1·07237 | 1·36706 | 1·46628 | 47 |
| 44 | ·69466 | ·71934 | ·96569 | 1·03553 | 1·39012 | 1·43956 | 46 |
| 45 | ·70711 | ·70711 | 1·00000 | 1·00000 | 1·41421 | 1·41421 | 45 |
| Deg. | Co-sines. | Sines. | Co-tangents. | Tangents. | Co-secants. | Secants. | Degree. |

## MOMENT OF INERTIA.

CORDS, KNOTS, NODES, CHAIN-BRIDGE.—ANGULAR VELOCITY.—RADIUS OF GYRATION.

1. If the cord $q$ NB, be fixed at the extremity B, and stretched by a weight of 500 lbs. at the extremity $q$, and the middle knot or node N, by a force of 255 lbs. pulling upwards, under an angle $a$ N $b$ of 54°; what is the tension and position of NB.

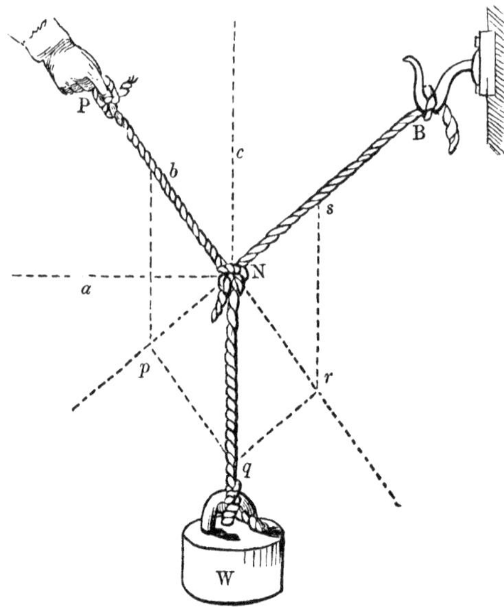


Angle $q$ N $r$ = 180° − angle $q$ NP; and 90° − $a$ N $b$ = $b$ N $c$ = $q$ N $r$ = 36°; cos. 36° = ·80902.

$\sqrt{500^2 + 255^2 - 2 \times 255 \times 500 \times \cos. 36°} = 329{\cdot}7$ lbs., the magnitude of the tension.

$\frac{500 \sin. 36°}{329{\cdot}7} = {\cdot}891386$ = sine of angles $b$ N $s$, or angle BN $r$ = 63° 2′.

2. Between the points A and B, a cord 10 feet in length is stretched by a weight W of 500 lbs. suspended to it by a ring; the horizontal distance AE = 6·6 feet, and the vertical distance BE = 3·2 feet; required the position of the ring C, the tensions, and directions of the rope.

The tensions of the cords AC, CB are equal, and angle AC $b$ = angle $b$ CB.

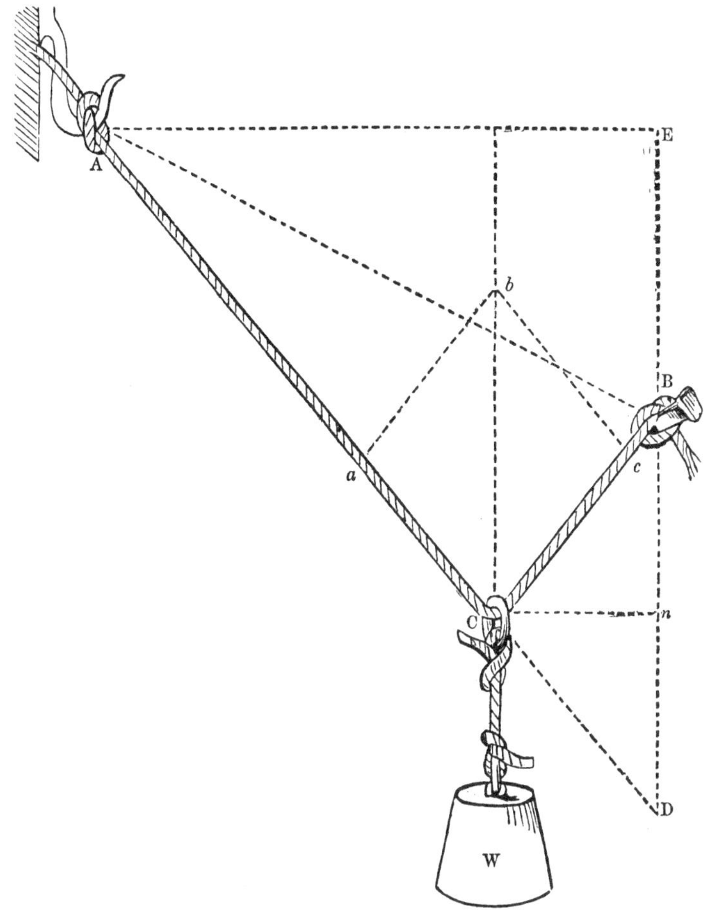


$$AD = AC + CB = 10 \text{ feet.}$$

$\sqrt{(10^2 - 6{\cdot}6^2)} = 7{\cdot}5126 = ED$; $BD = 7{\cdot}5126 - 3{\cdot}2 = 4{\cdot}3126$

$Dn = \frac{4{\cdot}3126}{2} = 2{\cdot}1563$; $7{\cdot}5126 : 2{\cdot}1563 :: 10 : \frac{21{\cdot}563}{7{\cdot}5126} = 2{\cdot}87 = CD = CB$; and $CA = 10 - 2{\cdot}87 = 7{\cdot}13$.

$$\frac{Bn}{Bc} = \text{cosine } b\,CB = \frac{2{\cdot}1563}{2{\cdot}87} = {\cdot}75132.$$

$\therefore \angle b\,CB = 41^\circ\ 18'$; $\frac{W}{2 \cos. 41^\circ\ 18'} = \frac{500}{1{\cdot}50264} = 332{\cdot}7$ lbs., the tension on the cord CB, which is equal to the tension on AC.

3. Let 500,000 lbs. be the whole weight on a chain-bridge whose span AB = 400 feet, and height of the arc CD = 40 feet; required the tensions and other circumstances respecting the chains.

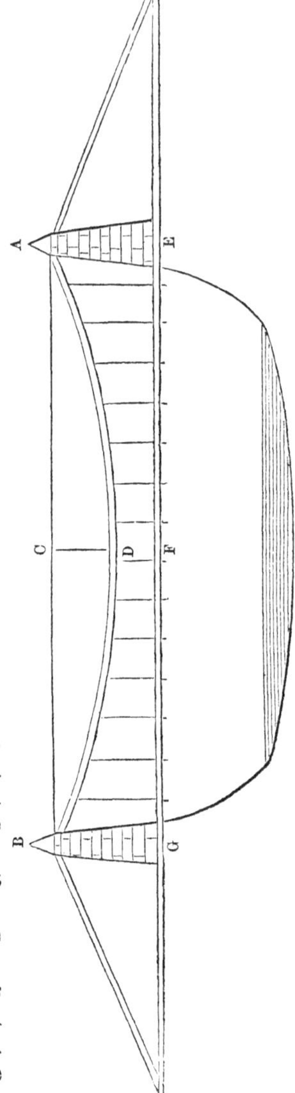


The tangent of the angles of inclination of the ends of the chain is equal

$\frac{40 \times 2}{200} = \cdot 40000$, the angle answering to this natural tangent is $21° \ 48'$.

The vertical tension at each point of suspension is = half the weight = 250000; the horizontal tension at the points of suspension = 250000 × cot. $21° \ 48' = \frac{250000}{\cdot 4} = 625000$ lbs.

The whole tension at one end will be

$\sqrt{625000^2 + 250000^2} = 673146$ lbs.

4. Suppose the piston of a steam engine, with its rod, weighs 1000 lbs.; it has no velocity at its highest and lowest positions, but in the middle the velocity is a maximum and equal 10 ft.; what effect will it accumulate by virtue of its inertia in the first half of its path, and give out again in the second half; and what is the mean force which would be requisite to accelerate the motion of the piston in the first half of its path, which is the same as that which it would exert in the second half by its retardation, the length of stroke being 8 feet.

According to the principle of *vis viva*, the effect which the piston will accumulate by virtue of its inertia in the first half of its path, and give out again in the second half = $\frac{10^2}{2 \times 32 \cdot 2} \times 1000 = 1552 \cdot 794$ units of work. Half the path of the piston = 4 feet; hence,

$$\frac{1552 \cdot 794}{4} = 388 \cdot 1985 \text{ lbs., the mean force.}$$

MOMENT OF INERTIA, or the MOMENT OF ROTATION, or the MOMENT OF THE MASS, is the sum of the products of the particles of the mass and the squares of their distances from the axis of rotation.

5. If a body at rest, but capable of turning round a fixed axis A, possesses a moment of inertia of 121 units of work, the measures taken in feet and pounds, made to turn by means of a cord and weight of 36 lbs., lying over a pulley in a path of 10 feet; what are the circumstances of the motion.

$\sqrt{\frac{2 \times 36 \times 10}{121}} = 2{\cdot}439347$ feet, the angular velocity of the body, which call $v$; so that each point at the distance of one foot from the axis of revolution will describe, after the accumulation of 121 units of work, 2·44 feet in a second.

$6{\cdot}2832 =$ circumference of a circle 2 feet in diameter,

$\frac{6{\cdot}2832}{2{\cdot}44} = 2{\cdot}6$ seconds, the time of one revolution.

6. If an angular velocity of 3 feet passes into a velocity of 7 feet; what mechanical effect will a *mass* produce so moving, supposing the *moment of inertia* to be 200, the measures taken in feet and pounds.

According to the principles of *vis viva*,

$(7^2 - 3^2)\frac{200}{2} = 4000$ units of work, which may be 40 lbs. raised 100 feet, 80 lbs. raised 50 feet, 400 lbs. raised 10 feet; and so on.

7. The weight of a rotating mass B is 500 lbs., its distance OB from the axis of rotation 3 feet, the weight W, constituting the moving force, 90 lbs., its arm AO = OC = 4 feet; required the circumstances of the motion that ensues.

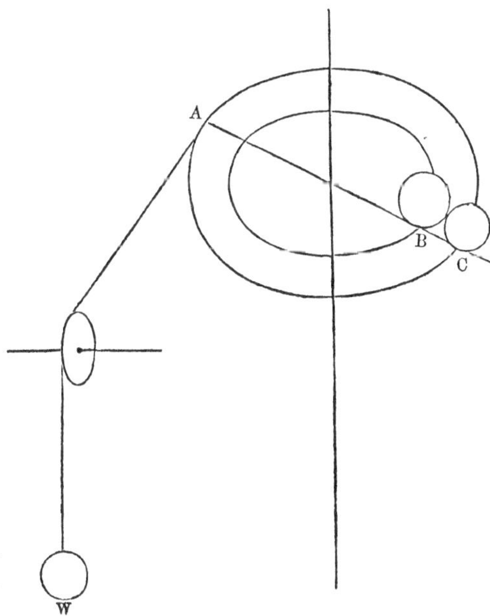


$\left\{90 + \frac{3^2}{4^2}500\right\} \div 32{\cdot}2 = 11{\cdot}53$ lbs., the inert *mass* accelerated by the force of W. And it is well known that the force divided by the *mass* gives the acceleration.

$\therefore \frac{90}{11 \cdot 53} = 7 \cdot 806$, the acceleration of the motion of W. The angular acceleration in a circle 1 foot from the axis $= \frac{7 \cdot 806}{4} = 1 \cdot 9515$.

After 10 seconds the acquired angular velocity will be

$$1 \cdot 9515 \times 10 = 19 \cdot 515.$$

And the corresponding distance $= \frac{1 \cdot 9515 \times 10^2}{2} = 97 \cdot 575$ feet, measured on a circle one foot from O.

The space described by the weight W is $\frac{7 \cdot 806 \times 10^2}{2} = 390 \cdot 3$ feet, which is the same as the space described by C. The circumference of a circle one foot from C $= 3 \cdot 1416$.

$$\therefore \frac{97 \cdot 575}{3 \cdot 1416} = 31 \cdot 059 \text{ revolutions.}$$

In the rotation of a body AB about a fixed axis O, all its points describe equal angles in equal times. If the body rotate in a certain time through the angle $\theta^\circ$, or arc $\phi = \frac{\theta^\circ}{180^\circ} \pi$, radius $= 1$; and hence, $\pi = 3 \cdot 141592$, &c.; the elements of the body, $a$, $b$, $c$, &c., at the distances $oa = x_1$, $ob = x_2$, &c. from the axis, will describe the arcs or spaces $aa_1 = \phi x_1$, $bb_1 = \phi x_2$, $cc_1 = \phi x_3$, &c. If the angular velocity, that is, the velocity of those points of the body which are distant a unit of length, a foot, from the axis of revolution, be put $= z$, then the simultaneous velocities of the elements of the mass at the distances $x_1$, $x_2$, $x_3$, &c., will be,

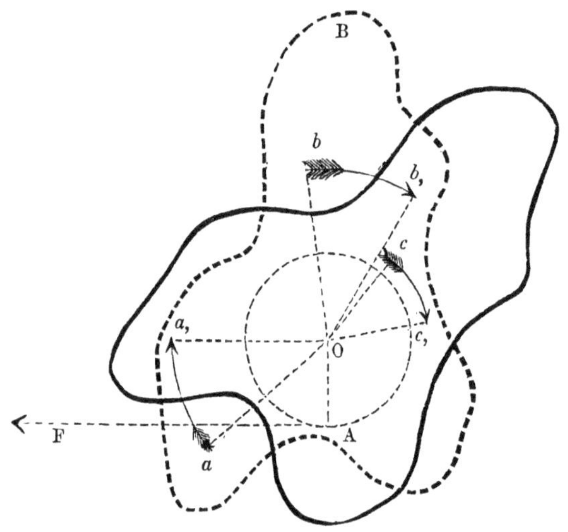


$$z x_1, \; z x_2, \; z x_3, \; \&c.$$

And if $a$ be the *mass* of the element at $a$; $b$ the *mass* of the element at $b$; $c$ the *mass* of the element at $c$, &c., their *vis viva* will be,

$$(z x_1)^2 a, \; (z x_2)^2 b, \; (z x_3)^2 c, \; \&c.$$

And the sum of the *vis viva* of the whole body $=$

$$z^2 (x_1^2 a + x_2^2 b + x_3^2 c, \; \&c.)$$

According to our definition, $x_1^2 a + x_2^2 b + x_3^2 c$, &c. is the *moment of inertia*, which may be represented by R; then $z^2$ R is the *vis viva* of a body revolving with the angular velocity $z$. Therefore, to communicate to a body in a state of rest an angular velocity $z$, a *mechanical effect* F $s$, or force × space = $\frac{1}{2}$ the *vis viva*, must be expended; that is, F $s = \frac{1}{2} z^2$ R, or, which is the same thing, a body performing the units of work F $s$, passes from the angular velocity $z$ to a state of rest. In general, if the initial angular velocity = $v$, and the terminal angular velocity = $z$, the units of work will be,

$$F s = \frac{z^2 - v^2}{2} \times R.$$

*The moment of inertia of a body about an axis not passing through the centre of gravity is equivalent to its moment of inertia about an axis running parallel to it through the centre of gravity, increased by the product of the mass of the body and the square of the distance of the two centres.*

It is necessary to know the moments of inertia of the principal geometrical bodies, because they very often come into application in mechanical investigations. If these bodies be homogeneous, as in the following we will always suppose to be the case, the particles of the mass $M_1$, $M_2$, &c. are proportional to the corresponding particles of the volume $V_1$, $V_2$, &c.; and hence the measure of the moment of inertia may be replaced by the sum of the particles of the volume, and the squares of their distances from the axis of revolution. In this sense, the moments of inertia of lines and surfaces may also be found.

If the whole mass of a body be supposed to be collected into one point, its distance from the axis may be determined on the supposition that the mass so concentrated possesses the same moment of inertia as if distributed over its space. This distance is called the *radius of gyration*, or *of inertia*. If R be the moment of inertia, M the mass, and $r$ the radius of gyration, we then have $M r^2 = R$, and hence $r = \sqrt{\frac{R}{M}}$. We must bear in mind that this radius by no means gives a determinate point, but a circle only, within whose circumference the mass may be considered as arbitrarily distributed.

If into the formula $R_1 = R + M e^2$, expressed in the words above printed in italics, we introduce $R = M r^2$ and $R_1 = M r_1^2$, we obtain $r_1^2 = r^2 + e^2$; that is, the square of the radius of gyration referred to a given axis = the square of the radius of gyration referred to a parallel line of gravity, plus the square of the distance between the two axes.

*Wheel and axle.*—The theory of the moment of inertia finds its most frequent application in machines and instruments, because in these rotary motions about a fixed axis are those which generally present themselves.

If two weights, P and Q, act on a wheel and axle ACDB, with the arms $CA = a$ and $DB = b$ through the medium of perfectly flexible strings, and if the radius of the gudgeons be so small that their friction may be neglected, it will remain in equilibrium if the statical moments P . CA and Q . DB are equal, and therefore $Pa = Qb$. But if the moment of the weight P is greater than that of Q, therefore $Pa > Qb$, P will descend and Q ascend; if $Pa < Qb$, P will ascend and Q descend. Let us now examine the

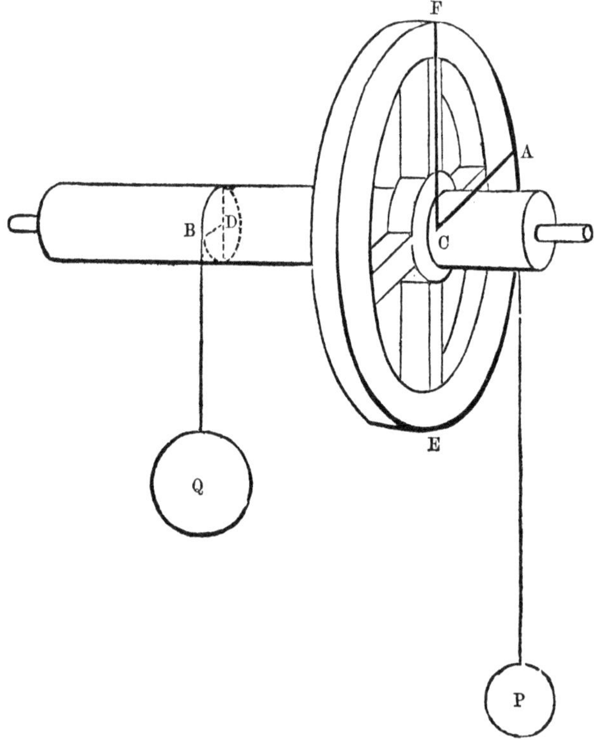


conditions of motion in the case that $Pa > Qb$. The force corresponding to the weight Q and acting at the arm $b$ generates at the arm $a$ a force $\frac{Qb}{a}$, which acts opposite to the force corresponding to the weight P, and hence there is a residuary moving force $P - \frac{Qb}{a}$ acting at A. The mass $\frac{Q}{g}$ is reduced by its transference from the distance $b$ to that of $a$ to $\frac{Qb^2}{ga^2}$; hence the mass moved by $P - \frac{Qb}{a}$ is $M = \left(P + \frac{Qb^2}{a^2}\right) \div g$, or, if the moment of inertia of the wheel and axle without the weights P and Q $= \frac{G\,y^2}{g}$, and, therefore, its inert mass reduced to A $= \frac{G\,y^2}{g\,a^2}$, we have, more exactly,

$$M = \left(P + \frac{Q\,b^2}{a^2} + \frac{G\,y^2}{a^2}\right) \div g = (P\,a^2 + Q\,b^2 + G\,y^2) \div g\,a^2.$$

From thence it follows that the accelerated motion of the weight P, together with that of the circumference of the wheel, namely,

$$p = \frac{\text{moving force}}{\text{mass}} = \frac{P - \frac{Q\,b}{a}}{P\,a^2 + Q\,b^2 + G\,y^2}\,g\,a^2 = \frac{P\,a - Q\,b}{P\,a^2 + Q\,b^2 + G\,y^2}\,g\,a;$$

on the other hand, the accelerated motion of the ascending weight Q, or of the circumference of the axle, is,

$$q = \frac{b}{a}\,p = \frac{P\,a - Q\,b}{P\,a^2 + Q\,b^2 + G\,y^2}\,g\,b.$$

The tension of the string by P is $S = P - \frac{P\,p}{g} = P\left(1 - \frac{p}{g}\right)$, that of the string by Q is $T = Q + \frac{Q\,q}{g} = Q\left(1 + \frac{q}{g}\right)$; hence the pressure on the gudgeon is,

$$S + T = P + Q - \frac{P\,p}{g} + \frac{Q\,q}{g} = P + Q - \frac{(P\,a - Q\,b)^2}{P\,a^2 + Q\,b^2 + G\,y^2};$$

the pressure, therefore, on the gudgeons for a revolving wheel and axle is less than for one in a state of equilibrium. Lastly, from the accelerating forces $p$ and $q$, the rest of the relations of motion may be found; after $t$ seconds, the velocity of P is $v = p\,t$, of Q is $v_1 = q\,t$, and the space described by P is $s = \frac{1}{2}\,p\,t^2$, by Q is $s_1 = \frac{1}{2}\,q\,t^2$.

Let the weight P at the wheel be = 60 lbs., that at the axle Q = 160 lbs., the arm of the first CA = $a$ = 20 inches, that of the second DB = $b$ = 6 inches; further, let the axle consist of a solid cylinder of 10 lbs. weight, and the wheel of two iron rings and four arms, the rings of 40 and 12 lbs., the arms together of 15 lbs. weight; lastly, let the radii of the greater ring AE = 20 and 19 inches, that of the less FG = 8 and 6 inches; required the conditions of motion of this machine. The moving force at the circumference of the wheel is,

$$P - \frac{b}{a}\,Q = 60 - \frac{6}{20}\,160 = 60 - 48 = 12 \text{ lbs.},$$

the moment of inertia of the machine, neglecting the masses of the gudgeons and the strings, is equivalent to the moment of inertia of the axle $= \frac{W\,b^2}{2} = \frac{10\,.\,6^2}{2} = 180$, plus the moment of the smaller ring $= \frac{R_1\,(r_1^{\,2} + r_2^{\,2})}{2} = \frac{12\,(8^2\,+\,6^2)}{2} = 600$, plus the moment of the larger ring $= \frac{40\,(20^2 + 19^2)}{2} = 15220$, plus the moment of the arms, approximately $= \frac{A\,(\rho_1^3 - \rho_2^3)}{3\,(\rho_1 - \rho_2)} = \frac{A\,\rho_1^2 + \rho_1\,\rho_2 + \rho_2^2)}{3} = \frac{15\,(19^2 + 19 \times 8 + 8^2)}{3} = 2885$; hence, collectively, $G\,y^2 = 180 + 600 + 15220 + 2885 = 18885$, or for foot measure $= \frac{18885}{144} = 131{\cdot}14$. The collective mass, reduced to the circumference of the wheel is,

$$= \left(P + \frac{Q\,b^2 + G\,y^2}{a^2}\right) \div g = \left[60 + 160\left(\frac{6}{20}\right)^2 + \frac{18885}{20^2}\right] \div g = \left(60 + 160 \times 0{\cdot}09 + \frac{18885}{400}\right) 0{\cdot}031 = 121{\cdot}61 \times 0{\cdot}031 = 337 \text{ lbs.}$$

Accordingly, the accelerated motion of the weight P, together with that of the circumference of the wheel, is,

$$p = \frac{P - \frac{b}{a}\,Q}{\frac{P + Q\,b^2 + G\,y^2}{a^2}}\,g = \frac{12}{3{\cdot}77} = 3{\cdot}183 \text{ feet};$$ on the other hand, that of Q is $q = \frac{b}{a}\,p = \frac{6}{20}\,3{\cdot}183 = 0{\cdot}954$ feet; further, the tension of the string by P is $= \left(1 - \frac{p}{g}\right) P = \left(1 - \frac{3{\cdot}133}{32{\cdot}2}\right) 60 = 54{\cdot}07$ lbs.; that by Q, on the other hand, $Q = \left(1 + \frac{q}{g}\right) Q = (1 + 0{\cdot}925 \times 0{\cdot}032)\,160 = 1{\cdot}030\;160 = 164{\cdot}8$ lbs.; and consequently the pressure on the gudgeons $S + T = 54{\cdot}06 + 164{\cdot}80 = 218{\cdot}86$ lbs., or inclusive of the weight of the machine $= 218{\cdot}86 + 77 = 295{\cdot}86$ lbs. After 10 seconds, P has acquired the velocity $p\,t = 3{\cdot}084 \times 10 = 30{\cdot}84$ feet, and described the space $s = \frac{v\,t}{2} = 30{\cdot}84 \times 5 = 154{\cdot}2$ feet, and Q has ascended a height $\frac{b}{a}\,s = 0{\cdot}3 \times 154{\cdot}2 = 46{\cdot}26$ feet.

The weight P which communicates to the weight Q the accelerated motion $q = \frac{P\,a\,b - Q\,b^2}{P\,a^2 + Q\,b^2 + G\,y^2}\,g$, may also be replaced by another weight $P_1$, without changing the acceleration of the motion Q, if it act at the arm $a_1$, for which,

$$\frac{P_1\,a_1 - Q\,b}{P_1\,a_1^2 + Q\,b^2 + G\,y^2} = \frac{P\,a - Q\,b}{P\,a^2 + Q\,b + G\,y^2}.$$

The magnitude $\frac{P\,a^2 + Q\,b^2 + G\,y^2}{P\,a - Q\,b}$, represented by $k$, and we obtain $a_1^2, - k\,a_1 = - \frac{Q\,b\,(b + k) + G\,y^2}{P_1}$, and the arm in question,

$$a_1 = \tfrac{1}{2} k \pm \sqrt{\left(\frac{k}{2}\right)^2 - \frac{Q b (b + k) + G y^2}{P_1}}.$$

We may also find by help of the differential calculus, that the motion of Q is most accelerated by the weight P, when the arm of the latter corresponds to the equation $P a^2 - 2 Q a b = Q b^2 + G y^2$, therefore,

$$a = \frac{b Q}{P} + \sqrt{\left(\frac{b Q}{P}\right)^2 + \frac{Q b^2 + G y^2}{P}}.$$

The formula found above assumes a complicated form if the friction of the gudgeons and the rigidity of the cord are taken into account. If we represent the statical moments of both resistances by $F r$, we must then substitute for the moving force $P - \frac{b}{a} Q$, the value $P - \frac{Q b + F r}{a}$, whence the acceleration of Q comes out,

$$q = \frac{(P a - F r) b - Q b^2}{P a^2 + Q b^2 + G y^2} g \text{ and } a = \frac{Q b + F r}{P} + \sqrt{\left(\frac{Q b + F r}{P}\right)^2 + \frac{Q b^2 + G y^2}{P}}.$$

The weights P = 30 lbs. Q = 80 lbs. act at the arms $a = 2$ feet. and $b = \frac{1}{2}$ foot of a wheel and axle, and their moments of inertia $G y^2$ amount to 60 lbs.; then the accelerated motion of the ascending weight Q is,

$$q = \frac{30 \times 2 \times \frac{1}{2} - 80 \times (\frac{1}{2})^2}{30 \times 2^2 + 80 \times (\frac{1}{2})^2 + 60} g = \frac{30 - 20}{120 + 20 + 60} 32 \cdot 2 = \frac{322}{200} =$$

1·61 feet. But if a weight $P_1 = 45$ lbs. generates the same acceleration in the motion of Q, the arm of $P_1$ is then,

$$a_1 = \frac{k}{2} \pm \sqrt{\left(\frac{k}{2}\right)^2 - \frac{80 \times \frac{1}{2} (\frac{1}{2} + k) + 60}{45}}, \text{ or as } k = \frac{200}{60 - 40} =$$

$$10,\ a_1 \text{ is } = 5 \pm \sqrt{25 - \frac{32}{3}} = 5 \pm \tfrac{1}{3} 11 \cdot 358 = 5 \pm 3 \cdot 786 = 8 \cdot 786$$

feet, or 1·214 feet.

The accelerated motion of Q comes out greatest if the arm of the force or radius of the wheel amount to,

$$a = \frac{\frac{1}{2} \times 80}{30} + \sqrt{\left(\frac{40}{30}\right)^2 + \frac{20 + 60}{30}} = \frac{4}{3} + \sqrt{\frac{16}{9} + \frac{24}{9}} = \frac{4 + \sqrt{40}}{3} =$$

3·4415 feet, and $q$ is $= \left(\frac{30 \times 1 \cdot 7207 - 20}{30 \times (3 \cdot 4415)^2 + 80}\right) g = \frac{31 \cdot 621}{435 \cdot 32} g =$ 2·339 feet.

The statical moment of the friction, together with the rigidity of the string, is $F r = 8$; then, instead of $Q b$, we must put $Q b + F r = 40 + 8 = 48$; whence it follows that,

$$a = \frac{48}{30} + \sqrt{\left(\frac{40}{30}\right)^2 + \frac{8}{3}} = 1 \cdot 6 + \sqrt{5 \cdot 227} = 3 \cdot 886,$$ and the correspondent maximum accelerating force

$$q = \frac{30 \times 1 \cdot 943 - 8 \times \frac{1}{2} - 20}{30 \times (3 \cdot 886)^2 + 80} g = \frac{34 \cdot 29}{533} \times 32 \cdot 2 = 2 \cdot 071 \text{ feet.}$$

## WEIGHT, ACCELERATION, AND MASS.

PARALLELOGRAM OF FORCES.—THE PRINCIPLE OF VIRTUAL VELOCITIES. —MECHANICAL POWERS: CONTINUOUS CIRCULAR MOTION, GEARING, TEETH OF WHEELS, DRUMS, PULLEYS, PUMPING ENGINES, ETC.

1. If a weight of 10 lbs., moved by the hand, ascends with a 3 feet acceleration, what is the pressure on the hand?

$$10\,(1 + \frac{3}{32 \cdot 2}) = 10 \cdot 93168 \text{ lbs.}$$

If a weight of 10 lbs., moved by the hand, descends with a 3 feet acceleration, the pressure on the hand will be 9·06832 lbs., for then

$$10\,(1 - \frac{3}{32 \cdot 2}) = 9 \cdot 06832.$$

If $w$ be the weight of the *mass* acted upon by the force of the hand, and also by the force of gravity, as $g = 32 \cdot 2$, the *mass* moved by the sum or difference of these forces will be $= \frac{w}{g}$. If P be the pressure on the hand, and $p$ its acceleration, the body falls with the force $\frac{w}{g}\,p$; it also falls with the force $w - \text{P}$; hence,

$$w - \text{P} = \frac{w}{g}\,p \quad \therefore \text{P} = (1 - \frac{p}{g})\,w.$$

When the body is ascending, then $p$ is negative,

$$\text{and } w + \text{P} = \frac{w}{g}\,(-\,p) \quad \therefore \text{P} = (1 + \frac{p}{g})\,w.$$

2. If a body of 200 lbs. be moved on a smooth horizontal track, by the joint action of two forces, and describes a space of 10 feet in the first second, what is the amount of each of these forces; the first makes an angle of 35° with the track upon which the body moves, and the other an angle of 50°?

In solving this question, the natural sines of the angles 35°, 50°, and of their sum 85°, will be required. We shall first take these from the table:

$$\begin{aligned} \sin. 35° &= \cdot 57358 \\ \sin. 50° &= \cdot 76604 \\ \sin. 85° &= \cdot 99619. \end{aligned}$$

The acceleration is = 20 feet, that is, twice the space passed over in the first second,

$\frac{200}{32 \cdot 2}$ = the *mass*, and $\frac{200}{32 \cdot 2} \times 20 = 124 \cdot 224$ lbs., the force of the resultant, in the direction of the track upon which the body moves.

$$\text{One of the components} = \frac{124{\cdot}224 \text{ sin. } 35^\circ}{\text{sin. } (35^\circ + 50^\circ)} = 71{\cdot}52 \text{ lbs.}$$

$$\text{The other component} = \frac{124{\cdot}224 \text{ sin. } 50^\circ}{\text{sin. } (35^\circ + 50^\circ)} = 95{\cdot}52 \text{ lbs.}$$

These, and the like results, may be obtained with greater ease by logarithms.

| | | |
|---|---|---|
| Log. 124·224 | = | 2·0942055 |
| Log. sin. 35° | = | 9·7585913 |
| | | 11·8527968 |
| Log. sin. 85° | = | 9·9983442 |
| Log. of 71·52413 | = | 1·8544526 |
| Log. 124·224 | = | 2·0942055 |
| Log. sin. 50° | = | 9·8842540 |
| | | 11·9784595 |
| Log. sin. (85°) | = | 9·9983442 |
| ·Log. of 95·5247 | = | 1·9801153 |

3. A carriage weighing 8000 lbs. is moved forward by a force $f_1$ of 500 lbs. upon a horizontal surface AB; during the motion, two resistances have to be overcome, one horizontal of 100 lbs., the amount of friction, represented in the figure by $f_3$, the other $f_2$ of

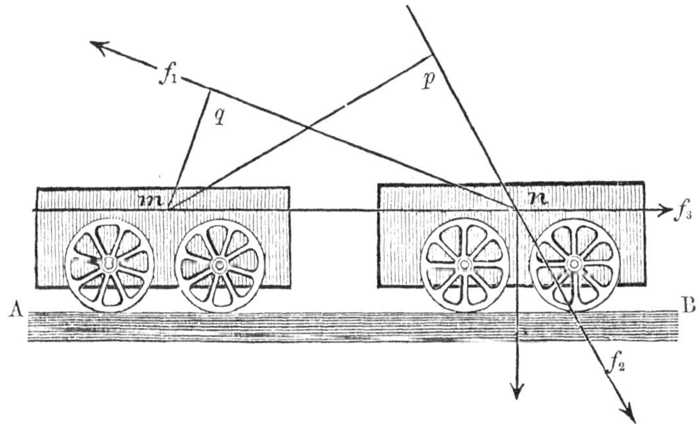


200 lbs. acting downwards; the angles $f_3\,n f_2$ and $f_1\,n\,m$, which the directions of these forces make with the horizon, are 61° and 21° respectively: it is required to know what work the force $f_1$ will perform by converting a 5 feet initial velocity of the carriage into a 20 feet velocity.

If we put $x = n\,m$, the distance the carriage moves in passing from a 5 to a 20 feet velocity,

The work of the force $f_1 = f_1 \times n\,q = 500 \times \text{cos. } 21^\circ \times x$.
The work of the force $f_3 = (-f_3) \times n\,m = -100 \times x$.
The work of the force $f_2 = (-f_2) \times n\,p = -200 \times \text{cos. } 61^\circ \times x$.

Consequently, the work of the effective force will be $269 \cdot 828 \times x = \{500 \times \cdot 94358 - 100 - 200 \times \cdot 48481\} x$, since the natural cosine of $21° = \cdot 93358$, and the natural cosine of $61° = \cdot 48481$.

But according to the principle of *vis viva*, the work done is equal to

$$\frac{20^2 - 5^2}{64 \cdot 4} \times 8000 = 46589 \cdot 82.$$

$$\therefore 269 \cdot 828 \times x = 46589 \cdot 82 \text{ and } x = \frac{46589 \cdot 82}{269 \cdot 828} =$$

772·665 feet, the space passed over by the carriage.

This question is solved on the PRINCIPLE OF VIRTUAL VELOCITIES, which we shall explain, as it is of essential service in practical mechanics.

This explanation depends on what is technically termed the "*Parallelogram of Forces.*"

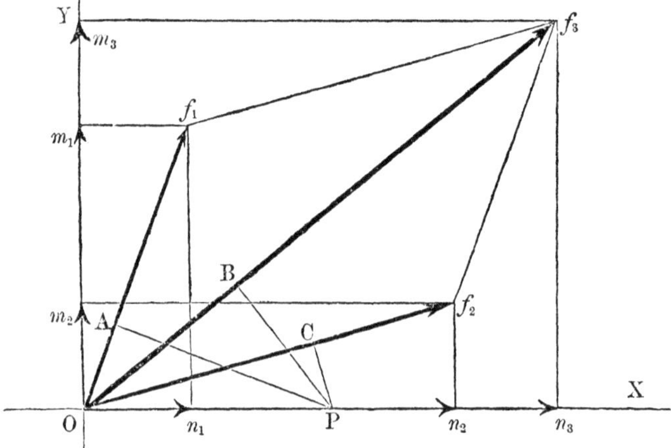


When a material point O, is acted upon by two forces $f_1$, $f_2$, whose directions $Of_1$, $Of_2$, make with each other an angle, if $Of_1$, $Of_2$ represent the magnitudes and directions of the forces, the diagonal of the parallelogram $O f_1 f_3 f_2$ represents the resultant in magnitude and direction; that is, the diagonal represents a single force equal to the combined actions of the forces represented by the sides. And if the sides of the parallelogram represent the accelerations of the forces, the diagonal represents the resultant acceleration. Draw through O, two axes OX and OY, at right angles to each other, and resolve the forces $f_1$ and $f_2$, as well as their resultant $f_3$, into components in the directions of these axes; namely, $f_1$ into $n_1$ and $m_1$; $f_2$ into $n_2$ and $m_2$; and $f_3$ into $n_3$ and $m_3$. The forces in one axis are $n_1$, $n_2$, and $n_3$; and those in the other $m_1$, $m_2$, and $m_3$. And by the parallelogram of forces it is well known that

$$n_3 = n_1 + n_2 \text{ and } m_3 = m_1 + m_2. \quad \text{(E).}$$

Now if we take in the axis OX any point P, and let fall from it the perpendiculars PA, PB, PC, on the directions of the forces $f_1$, $f_3$, $f_2$, we obtain the following similar right-angled triangles, namely,

OAP and $O\,n_1 f_1$ are similar;
OBP and $O\,n_3 f_3$ ———— ;
OCP and $O\,n_2 f_2$ ———— ;

$$\therefore \frac{O\,n_1}{O\,f_1} = \frac{OA}{OP} = \frac{n_1}{f_1} \text{ and } n_1 = \frac{AO}{OP} f_1.$$ It is easily seen also that $n_2 = \frac{CO}{OP} f_2$; and $n_3 = \frac{BO}{OP} f_3$.

If the values be substituted in (E), we obtain

$$BO \times f_3 = CO \times f_2 + AO \times f_1.$$

From the similarity of these triangles, and the remaining equation of (E), we can readily find that

$$PB \times f_3 = PA \times f_1 + PC \times f_2.$$

The equation becomes more compact by putting

OA, OC, OB, respectively equal $s_1$, $s_2$ $s_3$; and
PA, PC, PB, ———————— $q_1$, $q_2$, $q_3$.
Then $f_3\,s_3 = f_2\,s_2 + f_1\,s_1$ and $f_3\,q_2 = f_2\,q_2 + f_1\,q_1$.

The same holds good with any number of forces $f_1$, $f_2$, $f_3$, &c., and their resultant $f_n$, that is

$$f_n\,s_n = f_1\,s_1 + f_2\,s_2 + f_3\,s_3 + \text{\&c.}$$
$$\text{and } f_n\,q_n = f_1\,q_1 + f_2\,q_2 + f_3\,q_3 + \text{\&c.}$$

If the point of application O, move in a straight line to P, then OA $= s_1$ is called the space of the force $f_1$, and $f_1\,s_1$ the work done by the force $f_1$, in moving the body from O to P. OB is the space of the resultant, and the product $f_3\,s_3$, the work done by it. $f_2\,s_2$ is the work done by $f_2$ in moving the material point O from O to P. Hence the work done by the resultant is equal to all the work done by the component forces, as we have shown,

$$f_n\,s_n = f_1\,s_1 + f_2\,s_2 + f_3\,s_3 + \text{\&c.}$$

---

## PRINCIPLES AND PRACTICAL APPLICATIONS OF MECHANICAL POWERS.

MECHANICAL Powers, or the Elements of Machinery, are certain simple mechanical arrangements whereby weights may be raised or resistances overcome with the exertion of less power or strength than is necessary without them.

They are usually accounted six in number, viz. the *lever*, the *wheel* and *axle*, the *pulley*, the *inclined plane*, the *wedge*, and the *screw;* but properly two of these comprise the whole, namely, the *lever* and *inclined plane*,—the wheel and axle being only a lever of the first kind, and the pulley a lever of the second,—the wedge and the screw being also similarly allied to that of the inclined plane: however, although such seems to be the case in these respects, yet they each require, on account of their various modifications, a peculiar rule of estimation adapted expressly to the different circumstances in which they are individually required to act.

### THE LEVER.

Levers, according to mode of application, as the following, are distinguished as being of the first, second, or third kind; and although levers of equal lengths produce different effects, the general principles of estimation in all are the same; namely, the power is to the weight or resistance, as the distance of the one end to the fulcrum is to the distance of the other end to the same point.

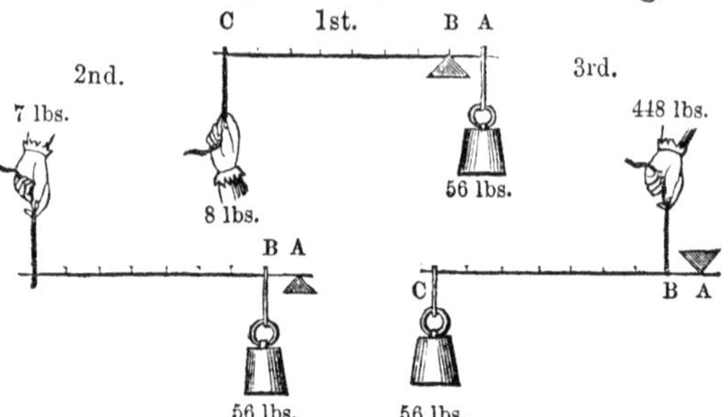


In the *first kind*, the power is to the resistance, as the distance AB is to the distance BC.

In the *second*, the power is to the resistance, as the distance AB is to that of AC; and,

In the *third*, the resistance is to the power, as the distance AB is to that of AC.

RULE, *first kind*.—Divide the longer by the shorter end of the lever from the fulcrum, and the quotient is the effective force that the power applied is equal to.

Let the handle of a pump equal 65 inches in length, and 10 inches from the shortest end to centre of motion; what is the amount of effective leverage thereby obtained?

$$65 - 10 = 55, \text{ and } \frac{55}{10} = 5\tfrac{1}{2} \text{ to } 1.$$

Required the situation of the fulcrum on which to rest a lever of 15 feet, so that $2\frac{1}{2}$ cwt. placed at one end may equipoise 30 cwt. at the other, the weight of the lever not being taken into account.

$\frac{15 \times 2{\cdot}5}{2{\cdot}5 + 30} = 1{\cdot}154$ feet from the end on which the 30 cwt. is to be placed.

It is by the second kind of lever that the greatest effect is obtained from any given amount of power; hence the propriety of the application of this principle to the working of force pumps, and shearing of iron, as by the lever of a punching-press, &c.

RULE, *second kind*.—Divide the whole length of lever, or distance from power to fulcrum, by the distance from fulcrum to weight, and the quotient is the proportion of effect that the power is to the weight or resistance to be overcome.

Required the amount of effect or force produced by a power of

50 lbs. on the ram of a Bramah's pump, the length of the lever being 3 feet, and distance from ram to fulcrum $4\frac{1}{2}$ inches.

3 feet = 36 inches, and $\frac{36}{4 \cdot 5} = 8$, or the power and resistance are to each other as 8 to 1; hence $50 \times 8 = 400$ lbs. force upon the ram.

The lever on the safety valve of a steam boiler is of the *third kind*, the action of the steam being the power, and the weight or spring-balance attached the resistance; but in such application the action of the lever's weight must also be taken into account.

### THE WHEEL AND PINION, OR CRANE.

The mechanical advantage of the wheel and axle, or crane, is as the velocity of the weight to the velocity of the power; and being only a modification of the first kind of lever, it of course partakes of the same principles.

RULE.—*To determine the amount of effective power produced from a given power by means of a crane with known peculiarities.*—Multiply together the diameter of the circle described by the winch, or handle, and the number of revolutions of the pinion to 1 of the wheel; divide the product by the barrel's diameter in equal terms of dimensions, and the quotient is the effective power to 1 of exertive force.

Let there be a crane the winch of which describes a circle of 30 inches in diameter; the pinion makes 8 revolutions for 1 of the wheel, and the barrel is 11 inches in diameter; required the effective power in principle, also the weight that 36 lbs. would raise, friction not being taken into account.

$\frac{30 \times 8}{11} = 21 \cdot 8$ to 1 of exertive force; and $21 \cdot 8 \times 36 = 784 \cdot 8$ lbs.

RULE.—*Given any two parts of a crane, to find the third, that shall produce any required proportion of mechanical effect.*—Multiply the two given parts together, and divide the product by the required proportion of effect; the quotient is the dimensions of the other parts in equal terms of unity.

Suppose that a crane is required, the ratio of power to effect being as 40 to 1, and that a wheel and pinion 11 to 1 is unavoidably compelled to be employed, also the throw of each handle to be 16 inches; what must be the barrel's diameter on which the rope or chain must coil?

$16 \times 2 = 32$ inches diameter described by the handle.

And $\frac{32 \times 11}{40} = 8 \cdot 8$ inches, the barrel's diameter.

### THE PULLEY.

The principle of the pulley, or, more practically, the block and tackle, is the distribution of weight on various points of support; the mechanical advantage derived depending entirely upon the flexibility and tension of the rope, and the number of pulleys or sheives in the lower or rising block: hence, by blocks and tackle of the usual kind, the power is to the weight as the number of cords attached to the lower block; whence the following rules.

Divide the weight to be raised by the number of cords leading *to*, *from*, or *attached* to the lower block; and the quotient is the power required to produce an equilibrium, provided friction did not exist.

Divide the weight to be raised by the power to be applied; the quotient is the number of sheives in, or cords attached to the rising block.

Required the power necessary to raise a weight of 3000 lbs. by a four and five-sheived block and tackle, the four being the movable or rising block.

Necessarily there are nine cords leading to and from the rising block.

$$\text{Consequently } \frac{3000}{9} = 333 \text{ lbs., the power required.}$$

I require to raise a weight of 1 ton 18 cwt., or 4256 lbs.; the amount of my power to effect this object being 500 lbs., what kind of block and tackle must I of necessity employ?

$\frac{4256}{500} = 8{\cdot}51$ cords; of necessity there must be 4 sheives or 9 cords in the rising block.

As the effective power of the crane may, by additional wheels and pinions, be increased to any required extent, so may the pulley and tackle be similarly augmented by purchase upon purchase.

### THE INCLINED PLANE.

The *inclined plane* is properly the second elementary power, and may be defined the lifting of a load by regular instalments. In principle it consists of any right line not coinciding with, but lying in a sloping direction to, that of the horizon; the standard of comparison of which commonly consists in referring the rise to so many parts in a certain length or distance, as 1 in 100, 1 in 200, &c.,—the first number representing the perpendicular height, and the latter the horizontal length in attaining such height, both numbers being of the same denomination, unless otherwise expressed; but it may be necessary to remark, that the inclination of a plane, the sine of inclination, the height per mile, or the height for any length, the ratio, &c., are all synonymous terms.

The advantage gained by the inclined plane, when the power acts in a parallel direction to the plane, is as the length to the height or angle of inclination: hence the rule. Divide the weight by the ratio of inclination, and the quotient equal the power that will just support that weight upon the plane. Or, multiply the weight by the height of the plane, and divide by the length,—the quotient is the power.

Required the power or equivalent weight capable of supporting a load of 350 lbs. upon a plane of 1 in 12, or 3 feet in height and 36 feet in length.

$$\frac{350}{12} = 29{\cdot}16 \text{ lbs., or } \frac{350 \times 3}{36} = 29{\cdot}16 \text{ lbs. power, as before.}$$

The weight multiplied by the length of the base, and the product divided by the length of the incline, the quotient equal the pressure or downward weight upon the incline.

TABLE *showing the Resistance opposed to the Motion of Carriages on different Inclinations of Ascending or Descending Planes, whatever part of the insistent weight they are drawn by.*

| Tens. | HUNDREDS. | | | | | | | | | |
|---|---|---|---|---|---|---|---|---|---|---|
| | | 100 | 200 | 300 | 400 | 500 | 600 | 700 | 800 | 900 |
| | | ·01 | ·005 | ·00333 | ·0025 | ·002 | ·00167 | ·00143 | ·00125 | ·00111 |
| 10 | ·1 | ·00909 | ·00476 | ·00322 | ·00244 | ·00196 | ·00164 | ·00141 | ·00123 | ·0011 |
| 20 | ·05 | ·00833 | ·00454 | ·00312 | ·00238 | ·00192 | ·00161 | ·00139 | ·00122 | ·00109 |
| 30 | ·0333 | ·00769 | ·00435 | ·00303 | ·00232 | ·00189 | ·00159 | ·00137 | ·0012 | ·00107 |
| 40 | ·025 | ·00714 | ·00417 | ·00294 | ·00227 | ·00185 | ·00156 | ·00135 | ·00119 | ·00106 |
| 50 | ·02 | ·00667 | ·004 | ·00286 | ·00222 | ·00182 | ·00154 | ·00133 | ·00118 | ·00105 |
| 60 | ·0166 | ·00625 | ·00385 | ·00278 | ·00217 | ·00178 | ·00151 | ·00131 | ·00116 | ·00104 |
| 70 | ·0143 | ·00588 | ·0037 | ·0027 | ·00213 | ·00175 | ·00149 | ·0013 | ·00115 | ·00103 |
| 80 | ·0125 | ·00555 | ·00357 | ·00263 | ·00208 | ·00172 | ·00147 | ·00128 | ·00114 | ·00102 |
| 90 | ·0111 | ·00526 | ·00345 | ·00256 | ·00204 | ·00169 | ·00145 | ·00126 | ·00112 | ·00101 |

Although this table has been calculated particularly for carriages on railway inclines, it may with equal propriety be applied to any other incline, the amount of traction on a level being known.

*Application of the preceding Table.*

What weight will a tractive power of 150 lbs. draw up an incline of 1 in 340, the resistance on the level being estimated at $\frac{1}{240}$th part of the insistent weight?

In a line with 40 in the left-hand column and under 200 is ·00417
Also in the same line and under 390 is........................ ·00294

Added together = ·00711

$$\text{Then } \frac{150}{\cdot 00711} = 21097 \text{ lbs. weight drawn up the plane.}$$

What weight would a force of 150 lbs. draw down the same plane, the fraction on the level being the same as before?

Friction on the level = ·00417
Gravity of the plane = ·00294 subtract
= ·00123

$$\text{And } \frac{150}{\cdot 00123} = 121915 \text{ lbs. weight drawn down the plane.}$$

*Example of incline when velocity is taken into account.*—A power of 230 lbs., at a velocity of 75 feet per minute, is to be employed for moving weights up an inclined plane 12 feet in height and 163 feet in length, the least velocity of the weight to be 8 feet per minute; required the greatest weight that the power is equal to.

$$\frac{230 \times 75 \times 163}{12 \times 8} = \frac{2811750}{96} = 29288 \text{ lbs., or } 13{\cdot}25 \text{ tons.}$$

TABLE *of Inclined Planes, showing the ascent or descent per yard, and the corresponding ascent or descent per chain, per mile; and also the ratio.*

| Per yard. | | Per chain. | Per mile. | Ratio. | Per yard. | | Per chain. | Per mile. | Ratio. |
|---|---|---|---|---|---|---|---|---|---|
| In parts of an in. | In dec'ls. of an inch. | Inches. | Feet. | 1 inch. | In parts of an in. | In decimals of an inch. | Inches. | Feet. | 1 inch. |
| $\frac{1}{64}$ | ·0156 | ·344 | 2·29 | 2304 | $\frac{7}{16}$ | ·4375 | 9·625 | 64·17 | 82 |
| $\frac{1}{48}$ | ·0208 | ·458 | 3·06 | 1728 | $\frac{1}{2}$ | ·5 | 11 | 73·33 | 72 |
| $\frac{1}{32}$ | ·0312 | ·687 | 4·58 | 1152 | $\frac{9}{16}$ | ·5625 | 12·375 | 82·5 | 64 |
| $\frac{1}{24}$ | ·0417 | ·917 | 6·11 | 864 | $\frac{7}{12}$ | ·5833 | 12·833 | 85·56 | 62 |
| $\frac{1}{16}$ | ·0625 | 1·375 | 9·17 | 576 | $\frac{3}{5}$ | ·6 | 13·2 | 88 | 60 |
| $\frac{1}{12}$ | ·0833 | 1·833 | 12·22 | 432 | $\frac{5}{8}$ | ·625 | 13·75 | 91·67 | 58 |
| $\frac{1}{10}$ | ·1 | 2·2 | 14·67 | 360 | $\frac{2}{3}$ | ·6667 | 14·667 | 97·78 | 54 |
| $\frac{1}{8}$ | ·125 | 2·75 | 18·33 | 288 | $\frac{11}{16}$ | ·6875 | 15·125 | 100·83 | 52 |
| $\frac{1}{6}$ | ·1667 | 3·667 | 24·44 | 216 | $\frac{7}{10}$ | ·7 | 15·4 | 102·67 | 51 |
| $\frac{3}{16}$ | ·1875 | 4·125 | 27·50 | 192 | $\frac{3}{4}$ | ·75 | 16·5 | 110 | 48 |
| $\frac{1}{5}$ | ·2 | 4·4 | 29·33 | 180 | $\frac{4}{5}$ | ·8 | 17·6 | 117·33 | 45 |
| $\frac{1}{4}$ | ·25 | 5·5 | 36·67 | 144 | $\frac{13}{16}$ | ·8125 | 17·875 | 119·17 | 44 |
| $\frac{3}{10}$ | ·3 | 6·6 | 44 | 120 | $\frac{5}{6}$ | ·8333 | 18·333 | 122·22 | 43 |
| $\frac{5}{16}$ | ·3125 | 6·875 | 45·83 | 115 | $\frac{7}{8}$ | ·875 | 19·25 | 128·33 | 41 |
| $\frac{1}{3}$ | ·3333 | 7·333 | 48·89 | 108 | $\frac{9}{10}$ | ·9 | 19·8 | 132 | 40 |
| $\frac{3}{8}$ | ·375 | 8·25 | 55 | 96 | $\frac{11}{12}$ | ·9167 | 20·167 | 134·44 | 39 |
| $\frac{2}{5}$ | ·4 | 8·8 | 58·67 | 20 | $\frac{15}{16}$ | ·9375 | 20·625 | 137·5 | 38 |
| $\frac{5}{12}$ | ·4167 | 9·167 | 61·11 | 86 | 1 | 1 | 22 | 146·67 | 36 |

THE WEDGE.

The wedge is a double inclined plane; consequently its principles are the same: hence, when two bodies are forced asunder by means of the wedge in a direction parallel to its head,—Multiply the resisting power by half the thickness of the head or back of the wedge, and divide the product by the length of one of its inclined sides; the quotient is the force equal to the resistance.

The breadth of the back or head of a wedge being 3 inches, and its inclined sides each 10 inches, required the power necessary to act upon the wedge so as to separate two substances whose resisting force is equal to 150 lbs.

$$\frac{150 \times 1{\cdot}5}{10} = 22{\cdot}5 \text{ lbs.}$$

When only one of the bodies is movable, the whole breadth of the wedge is taken for the multiplier.

THE SCREW.

The screw, in principle, is that of an inclined plane wound around a cylinder, which generates a spiral of uniform inclination, each revolution producing a rise or traverse motion equal to the pitch of the screw, or distance between two consecutive threads,—the pitch being the height or angle of inclination, and the circumference the length of the plane when a lever is not applied; but the lever being a necessary qualification of the screw, the circle which it describes is taken, instead of the screw's circumference, as the length of the plane: hence the mechanical advantage is, as the circumference of the circle described by the lever where the power acts, is to the pitch of the screw, so is the force to the resistance in principle.

Required the effective power obtained by a screw of $\frac{7}{8}$ inch pitch, and moved by a force equal to 50 lbs. at the extremity of a lever 30 inches in length.

$$\frac{30 \times 2 \times 3{\cdot}1416 \times 50}{{\cdot}875} = 10760 \text{ lbs.}$$

Required the power necessary to overcome a resistance equal to 7000 lbs. by a screw of $1\frac{1}{4}$ inch pitch, and moved by a lever 25 inches in length.

$$\frac{7000 \times 1{\cdot}25}{25 \times 2 \times 3{\cdot}1416} = 55{\cdot}73 \text{ lbs. power.}$$

In the case of a screw acting on the periphery of a toothed wheel, the power is to the resistance, as the product of the circle's circumference described by the winch or lever, and radius of the wheel, to the product of the screw's pitch, and radius of the axle, or point whence the power is transmitted; but observe, that if the screw consist of more than one helix or thread, the apparent pitch must be increased so many times as there are threads in the screw. Hence, *to find what weight a given power will equipoise:*

RULE.—Multiply together the radius of the wheel, the length of the lever at which the power acts, the magnitude of the power, and the constant number 6·2832; divide the product by the radius of the axle into the pitch of the screw, and the quotient is the weight that the power is equal to.

What weight will be sustained in equilibrio by a power of 100 lbs. acting at the end of a lever 24 inches in length, the radius of the axle, or point whence the power is transmitted, being 8 inches, the radius of the wheel 14 inches, the screw consisting of a double thread, and the apparent pitch equal $\frac{5}{8}$ of an inch?

$$\frac{14 \times 24 \times 100 \times 6{\cdot}2832}{{\cdot}625 \times 2 \times 8} = 21111{\cdot}55 \text{ lbs., or } 9{\cdot}4 \text{ tons, the}$$

power sustained.

If an endless screw be turned by a handle of 20 inches, the threads of the screw being distant half an inch; the screw turns a toothed wheel, the pinion of which turns another wheel, and the pinion of this another wheel, to the barrel of which a weight W is attached; it is required to find the weight a man will be able to sustain, who acts at the handle with a force of 150 lbs., the diameters of the wheels being 18 inches, and those of the pinions and barrel 2 inches.

$$150 \times 20 \times 3{\cdot}1416 \times 2 \times 18^3 = W \times 2^3 \times \tfrac{1}{2};$$
$$\therefore W = 12269 \text{ tons.}$$

## CONTINUOUS CIRCULAR MOTION.

In mechanics, circular motion is transmitted by means of *wheels*, *drums*, or *pulleys;* and accordingly as the driving and driven are of equal or unequal diameters, so are equal or unequal velocities produced: hence the principle on which the following rules are founded.

RULE.—*When time is not taken into account.*—Divide the greater diameter, or number of teeth, by the lesser diameter, or number of teeth, and the quotient is the number of revolutions the lesser will make for 1 of the greater.

How many revolutions will a pinion of 20 teeth make for 1 of a wheel with 125?

$$125 \div 20 = 6{\cdot}25, \text{ or } 6\tfrac{1}{4} \text{ revolutions.}$$

Intermediate wheels, of whatever diameters, so as to connect communication at any required distance apart, cause no variation of velocity more than otherwise would result were the first and last in immediate contact.

RULE.—*To find the number of revolutions of the last, to* 1 *of the first, in a train of wheels and pinions.*—Divide the product of all the teeth in the driving, by the product of all the teeth in the driven, and the quotient equal the ratio of velocity required.

Required the ratio of velocity of the last, to 1 of the first, in the following train of wheels and pinions; viz., *pinions driving*,—the first of which contains 10 teeth, the second 15, and third 18;—*wheels driven*,—first 15 teeth, second 25, and third 32.

$\frac{10 \times 15 \times 18}{15 \times 25 \times 32} = {\cdot}225$ of a revolution the wheel will make to 1 of the pinion.

A wheel of 42 teeth giving motion to one of 12, on which shaft is a pulley of 21 inches diameter, driving one of 6; required the number of revolutions of the last pulley to 1 of the first wheel.

$$\frac{42 \times 21}{12 \times 6} = 12{\cdot}25, \text{ or } 12\tfrac{1}{4} \text{ revolutions.}$$

Where increase or decrease of velocity is required to be communicated by wheel-work, it has been demonstrated that the number of teeth on each pinion should not be less than 1 to 6 of its wheel, unless there be some other important reason for a higher ratio.

RULE.—*When time must be regarded.*—Multiply the diameter, or number of teeth in the driver, by its velocity in any given time, and divide the product by the required velocity of the driven; the quotient equal the number of teeth, or diameter of the driven, to produce the velocity required.

If a wheel containing 84 teeth makes 20 revolutions per minute, how many must another contain to work in contact, and make 60 revolutions in the same time?

$$\frac{84 \times 20}{60} = 28 \text{ teeth.}$$

From a shaft making 45 revolutions per minute, and with a pinion 9 inches diameter at the pitch line, I wish to transmit motion at 15 revolutions per minute; what at the pitch line must be the diameter of the wheel?

$$\frac{45 \times 9}{15} = 27 \text{ inches.}$$

Required the diameter of a pulley to make 16 revolutions in the same time as one of 24 inches making 36.

$$\frac{24 \times 36}{16} = 54 \text{ inches.}$$

RULE.—*The distance between the centres and velocities of two wheels being given, to find their proper diameters.*—Divide the greatest velocity by the least; the quotient is the ratio of diameter the wheels must bear to each other. Hence, divide the distance between the centres by the ratio plus 1; the quotient equal the radius of the smaller wheel; and subtract the radius thus obtained from the distance between the centres; the remainder equal the radius of the other.

The distance of two shafts from centre to centre is 50 inches, and the velocity of the one 25 revolutions per minute, the other is to make 80 in the same time; the proper diameters of the wheels at the pitch lines are required.

$80 \div 25 = 3{\cdot}2$, ratio of velocity, and $\frac{50}{3{\cdot}2 + 1} = 11{\cdot}9$, the radius of the smaller wheel; then $50 - 11{\cdot}9 = 38{\cdot}1$, radius of larger; their diameters are $11{\cdot}9 \times 2 = 23{\cdot}8$, and $38{\cdot}1 \times 2 = 76{\cdot}2$ inches.

To obtain or diminish an accumulated velocity by means of wheels and pinions, or wheels, pinions, and pulleys, it is necessary that a proportional ratio of velocity should exist, and which is simply thus attained:—Multiply the given and required velocities together, and the square root of the product is the mean or proportionate velocity.

Let the given velocity of a wheel containing 54 teeth equal 16 revolutions per minute, and the given diameter of an intermediate pulley equal 25 inches, to obtain a velocity of 81 revolutions in a machine; required the number of teeth in the intermediate wheel, and diameter of the last pulley.

$\sqrt{81 \times 16} = 36$ mean velocity.

$\frac{54 \times 16}{36} = 24$ teeth, and $\frac{25 \times 36}{81} = 11{\cdot}1$ inches, diameter of pulley.

*To determine the proportion of wheels for screw cutting by a lathe.*—In a lathe properly adapted, screws to any degree of pitch, or number of threads in a given length, may be cut by means of a leading screw of any given pitch, accompanied with change wheels and pinions; course pitches being effected generally by means of one wheel and one pinion with a *carrier*, or *intermediate wheel*, which cause no variation or change of motion to take place: hence the following

RULE.—Divide the number of threads in a given length of the screw which is to be cut, by the number of threads in the same length of the leading screw attached to the lathe; and the quotient is the ratio that the wheel on the end of the screw must bear to that on the end of the lathe spindle.

Let it be required to cut a screw with 5 threads in an inch, the leading screw being of $\frac{1}{2}$ inch pitch, or containing 2 threads in an inch; what must be the ratio of wheels applied?

$$5 \div 2 = 2{\cdot}5, \text{ the ratio they must bear to each other.}$$

Then suppose a pinion of 40 teeth be fixed upon for the spindle,—

$$40 \times 2{\cdot}5 = 100 \text{ teeth for the wheel on the end of the screw.}$$

But screws of a greater degree of fineness than about 8 threads in an inch are more conveniently cut by an additional wheel and pinion, because of the proper degree of velocity being more effectively attained; and these, on account of revolving upon a stud, are commonly designated the *stud-wheels*, or *stud-wheel* and *pinion;* but the mode of calculation and ratio of screw are the same as in the preceding rule;—hence, all that is further necessary is to fix upon any 3 wheels at pleasure, as those for the spindle and stud-wheels,—then multiply the number of teeth in the spindle-wheel by the ratio of the screw, and by the number of teeth in that wheel or pinion which is in contact with the wheel on the end of the screw; divide the product by the stud-wheel in contact with the spindle-wheel, and the quotient is the number of teeth required in the wheel on the end of the leading screw.

Suppose a screw is required to be cut containing 25 threads in an inch, the leading screw as before having 2 threads in an inch, and that a wheel of 60 teeth is fixed upon for the end of the spindle, 20 for the pinion in contact with the screw-wheel, and 100 for that in contact with the wheel on the end of the spindle;—required the number of teeth in the wheel for the end of the leading screw.

$$25 \div 2 = 12{\cdot}5, \text{ and } \frac{60 \times 12{\cdot}5 \times 20}{100} = 150 \text{ teeth.}$$

Or, suppose the spindle and screw-wheels to be those fixed upon, also any one of the stud-wheels, to find the number of teeth in the other.

$$\frac{60 \times 12{\cdot}5}{150 \times 100} = 20 \text{ teeth, or } \frac{60 \times 12{\cdot}5 \times 20}{150} = 100 \text{ teeth.}$$

Table *of Change Wheels for Screw Cutting, the leading screw being of ½ inch pitch, or containing two threads in an inch.*

| Number of threads in inch of screw. | Number of teeth in | | Number of threads in inch of screw. | Number of teeth in | | | | Number of threads in inch of screw. | Number of teeth in | | | |
|---|---|---|---|---|---|---|---|---|---|---|---|---|
| | Lathe spindle-wheel. | Leading screw-wheel. | | Lathe spindle-wheel. | Wheel in contact with spindle-wheel. | Pinion in contact with screw-wheel. | Leading screw-wheel. | | Lathe spindle-wheel. | Wheel in contact with spindle-wheel. | Pinion in contact with screw-wheel. | Leading screw-wheel. |
| 1 | 80 | 40 | 8¼ | 40 | 55 | 20 | 60 | 19 | 50 | 95 | 20 | 100 |
| 1¼ | 80 | 50 | 8½ | 90 | 85 | 20 | 90 | 19½ | 80 | 120 | 20 | 130 |
| 1½ | 80 | 60 | 8¾ | 60 | 70 | 20 | 75 | 20 | 60 | 100 | 20 | 120 |
| 1¾ | 80 | 70 | 9½ | 90 | 90 | 20 | 95 | 20¼ | 40 | 90 | 20 | 90 |
| 2 | 80 | 90 | 9¾ | 40 | 60 | 20 | 65 | 21 | 80 | 120 | 20 | 140 |
| 2¼ | 80 | 90 | 10 | 60 | 75 | 20 | 80 | 22 | 60 | 110 | 20 | 120 |
| 2½ | 80 | 100 | 10½ | 50 | 70 | 20 | 75 | 22½ | 80 | 120 | 20 | 150 |
| 2¾ | 80 | 110 | 11 | 60 | 55 | 20 | 120 | 22¾ | 80 | 130 | 20 | 140 |
| 3 | 80 | 120 | 12 | 90 | 90 | 20 | 120 | 23¾ | 40 | 95 | 20 | 100 |
| 3¼ | 80 | 130 | 12¾ | 60 | 85 | 20 | 90 | 24 | 65 | 120 | 20 | 130 |
| 3½ | 80 | 140 | 13 | 90 | 90 | 20 | 130 | 25 | 60 | 100 | 20 | 150 |
| 3¾ | 80 | 150 | 13½ | 60 | 90 | 20 | 90 | 25½ | 30 | 85 | 20 | 90 |
| 4 | 40 | 80 | 13¾ | 80 | 100 | 20 | 110 | 26 | 70 | 130 | 20 | 140 |
| 4¼ | 40 | 85 | 14 | 90 | 90 | 20 | 140 | 27 | 40 | 90 | 20 | 120 |
| 4½ | 40 | 90 | 14¼ | 60 | 90 | 20 | 95 | 27½ | 40 | 100 | 20 | 110 |
| 4¾ | 40 | 95 | 15 | 90 | 90 | 20 | 150 | 28 | 75 | 140 | 20 | 150 |
| 5 | 40 | 100 | 16 | 60 | 80 | 20 | 120 | 28½ | 30 | 90 | 20 | 95 |
| 5½ | 40 | 110 | 16¼ | 80 | 100 | 20 | 130 | 30 | 70 | 140 | 20 | 150 |
| 6 | 40 | 120 | 16½ | 80 | 110 | 20 | 120 | 32 | 30 | 80 | 20 | 120 |
| 6½ | 40 | 130 | 17 | 45 | 85 | 20 | 90 | 33 | 40 | 110 | 20 | 120 |
| 7 | 40 | 140 | 17½ | 80 | 100 | 20 | 140 | 34 | 30 | 85 | 20 | 120 |
| 7½ | 40 | 150 | 18 | 40 | 60 | 20 | 120 | 35 | 60 | 140 | 20 | 150 |
| 8 | 30 | 120 | 18¾ | 80 | 100 | 20 | 150 | 36 | 30 | 90 | 20 | 120 |

Table *by which to determine the Number of Teeth, or Pitch of Small Wheels.*

| Diametral pitch. | Circular pitch. | Diametral pitch. | Circular pitch. |
|---|---|---|---|
| 3 | 1·047 | 9 | ·349 |
| 4 | ·785 | 10 | ·314 |
| 5 | ·628 | 12 | ·262 |
| 6 | ·524 | 14 | ·224 |
| 7 | ·449 | 16 | ·196 |
| 8 | ·393 | 20 | ·157 |

Required the number of teeth that a wheel of 16 inches diameter will contain of a 10 pitch.

$16 \times 10 = 160$ teeth, and the circular pitch $=$ ·314 inch.

What must be the diameter of a wheel for a 9 pitch of 126 teeth?

$\frac{126}{9} = 14$ inches diameter, circular pitch ·349 inch.

The pitch is reckoned on the diameter of the wheel instead of the circumference, and designated wheels of 8 pitch, 12 pitch, &c.

TABLE *of the Diameters of Wheels at their pitch circle, to contain a required number of teeth at a given pitch.*

| Number of teeth. | PITCH OF THE TEETH IN INCHES. | | | | | | | | | | | | | |
|---|---|---|---|---|---|---|---|---|---|---|---|---|---|---|
| | 1 in. | 1⅛ | 1¼ | 1⅜ | 1½ | 1⅝ | 1¾ | 1⅞ | 2 in. | 2⅛ | 2¼ | 2½ | 2¾ | 3 in. |
| | DIAMETER AT THE PITCH CIRCLE IN FEET AND INCHES. | | | | | | | | | | | | | |
| 10 | 0 3¼ | 0 3⅝ | 0 4 | 0 4½ | 0 4⅞ | 0 5¼ | 0 5⅝ | 0 6 | 0 6½ | 0 6⅞ | 0 7¼ | 0 8 | 0 8¾ | 0 9⅝ |
| 11 | 0 3½ | 0 4 | 0 4⅜ | 0 5 | 0 5⅜ | 0 5¾ | 0 6¼ | 0 6⅝ | 0 7 | 0 7½ | 0 7⅞ | 0 8¾ | 0 9¾ | 0 10⅝ |
| 12 | 0 3⅞ | 0 4⅜ | 0 4⅞ | 0 5⅜ | 0 5⅞ | 0 6⅜ | 0 6¾ | 0 7⅛ | 0 7⅝ | 0 8⅛ | 0 8⅝ | 0 9⅝ | 0 10⅝ | 0 11½ |
| 13 | 0 4⅛ | 0 4¾ | 0 5¼ | 0 5¾ | 0 6¼ | 0 6⅞ | 0 7⅜ | 0 7⅞ | 0 8⅜ | 0 8⅞ | 0 9⅜ | 0 10⅜ | 0 11½ | 1 0½ |
| 14 | 0 4½ | 0 5 | 0 5⅝ | 0 6¼ | 0 6¾ | 0 7⅜ | 0 7⅞ | 0 8½ | 0 9 | 0 9½ | 0 10 | 0 11¼ | 1 0⅜ | 1 1½ |
| 15 | 0 4⅞ | 0 5⅜ | 0 6 | 0 6⅝ | 0 7¼ | 0 7⅞ | 0 8½ | 0 9 | 0 9⅝ | 0 10¼ | 0 10¾ | 1 0 | 1 1¼ | 1 2⅜ |
| 16 | 0 5⅛ | 0 5¾ | 0 6⅜ | 0 7 | 0 7⅝ | 0 8⅜ | 0 9 | 0 9⅝ | 0 10¼ | 0 10⅞ | 0 11½ | 1 0¾ | 1 2 | 1 3⅜ |
| 17 | 0 5½ | 0 6⅛ | 0 6¾ | 0 7½ | 0 8¼ | 0 8⅞ | 0 9⅝ | 0 10¼ | 0 10⅞ | 0 11½ | 1 0¼ | 1 1½ | 1 2⅞ | 1 4¼ |
| 18 | 0 5¾ | 0 6½ | 0 7⅛ | 0 8 | 0 8⅝ | 0 9⅜ | 0 10 | 0 10¾ | 0 11½ | 1 0¼ | 1 0⅞ | 1 2⅜ | 1 3¾ | 1 5¼ |
| 19 | 0 6 | 0 6⅞ | 0 7½ | 0 8⅜ | 0 9⅛ | 0 9⅞ | 0 10⅝ | 0 11⅜ | 1 0⅛ | 1 0⅞ | 1 1⅝ | 1 3⅛ | 1 4⅝ | 1 6¼ |
| 20 | 0 6⅜ | 0 7⅛ | 0 8 | 0 8⅞ | 0 9⅝ | 0 10⅜ | 0 11¼ | 1 0 | 1 0¾ | 1 1½ | 1 2⅜ | 1 4 | 1 5½ | 1 7⅛ |
| 21 | 0 6¾ | 0 7½ | 0 8⅜ | 0 9¼ | 0 10 | 0 11 | 0 11¾ | 1 0½ | 1 1½ | 1 2¼ | 1 3 | 1 4¾ | 1 6⅜ | 1 8⅛ |
| 22 | 0 7 | 0 7⅞ | 0 8¾ | 0 9⅝ | 0 10⅝ | 0 11½ | 1 0⅜ | 1 1⅛ | 1 2 | 1 2⅞ | 1 3¾ | 1 5½ | 1 7¼ | 1 9 |
| 23 | 0 7⅜ | 0 8¼ | 0 9⅛ | 0 10 | 0 11 | 1 0 | 1 0⅞ | 1 1¾ | 1 2⅝ | 1 3½ | 1 4½ | 1 6⅜ | 1 8 | 1 10 |
| 24 | 0 7⅝ | 0 8⅝ | 0 9½ | 0 10½ | 0 11½ | 1 0½ | 1 1½ | 1 2⅜ | 1 3⅜ | 1 4¼ | 1 5¼ | 1 7⅛ | 1 9 | 1 10⅞ |
| 25 | 0 8 | 0 9 | 0 10 | 0 11 | 1 0 | 1 1 | 1 2 | 1 2⅞ | 1 3⅞ | 1 4⅞ | 1 6 | 1 8 | 1 9⅞ | 1 11⅞ |
| 26 | 0 8¼ | 0 9¼ | 0 10⅜ | 0 11½ | 1 0½ | 1 1½ | 1 2½ | 1 3½ | 1 4½ | 1 5½ | 1 6⅝ | 1 8¾ | 1 10¾ | 2 0⅞ |
| 27 | 0 8⅝ | 0 9⅝ | 0 10¾ | 0 11⅞ | 1 1 | 1 2 | 1 3 | 1 4⅛ | 1 5¼ | 1 6¼ | 1 7⅜ | 1 9½ | 1 11⅝ | 2 1¾ |
| 28 | 0 9 | 0 10 | 0 11¼ | 1 0¼ | 1 1¼ | 1 2½ | 1 3⅝ | 1 4⅝ | 1 5¾ | 1 6⅞ | 1 8 | 1 10¼ | 2 0½ | 2 2¾ |
| 29 | 0 9¼ | 0 10⅜ | 0 11⅜ | 1 0¾ | 1 1⅞ | 1 3 | 1 4⅛ | 1 5⅜ | 1 6½ | 1 7⅝ | 1 8¾ | 1 11⅛ | 2 1⅜ | 2 3¾ |
| 30 | 0 9½ | 0 10¾ | 1 0 | 1 1⅛ | 1 2⅜ | 1 3½ | 1 4½ | 1 6 | 1 7⅛ | 1 8¼ | 1 9½ | 2 0 | 2 2¼ | 2 4⅝ |
| 31 | 0 9⅞ | 0 11⅛ | 1 0⅜ | 1 1⅝ | 1 2⅞ | 1 4 | 1 5⅜ | 1 6½ | 1 7¾ | 1 9 | 1 10¼ | 2 0¾ | 2 3⅛ | 2 5⅝ |
| 32 | 0 10¼ | 0 11½ | 1 0¾ | 1 2 | 1 3⅜ | 1 4⅝ | 1 5⅞ | 1 7⅛ | 1 8⅜ | 1 9⅝ | 1 11 | 2 1½ | 2 4 | 2 6½ |
| 33 | 0 10½ | 0 11⅞ | 1 1⅛ | 1 2½ | 1 3¾ | 1 5⅛ | 1 6½ | 1 7¾ | 1 9 | 1 10⅜ | 1 11⅝ | 2 2¼ | 2 4⅞ | 2 7½ |
| 34 | 0 10⅞ | 1 0⅛ | 1 1⅝ | 1 3 | 1 4¼ | 1 5⅝ | 1 7 | 1 8⅜ | 1 9⅝ | 1 11 | 2 0⅜ | 2 3 | 2 5¾ | 2 8½ |
| 35 | 0 11⅛ | 1 0⅝ | 1 2 | 1 3⅜ | 1 4¾ | 1 6⅛ | 1 7½ | 1 9 | 1 10⅛ | 1 11¾ | 2 1 | 2 3⅞ | 2 6⅝ | 2 9½ |
| 36 | 0 11½ | 1 1 | 1 2⅜ | 1 3¾ | 1 5¼ | 1 6⅝ | 1 8 | 1 9½ | 1 10⅞ | 2 0⅜ | 2 2 | 2 4⅝ | 2 7½ | 2 10⅜ |
| 37 | 0 11¾ | 1 1¼ | 1 2¾ | 1 4¼ | 1 5⅝ | 1 7¼ | 1 8⅝ | 1 10 | 1 11½ | 2 1 | 2 2½ | 2 5½ | 2 8⅜ | 2 11⅜ |
| 38 | 1 0⅛ | 1 1⅝ | 1 3¼ | 1 4⅝ | 1 6¼ | 1 7⅝ | 1 9¼ | 1 10¾ | 2 0¼ | 2 1¾ | 2 3¼ | 2 6¼ | 2 9¼ | 3 0¼ |
| 39 | 1 0⅜ | 1 2 | 1 3½ | 1 5 | 1 6⅝ | 1 8¼ | 1 9¾ | 1 11⅜ | 2 0⅞ | 2 2⅜ | 2 4 | 2 7 | 2 10⅛ | 3 1¼ |
| 40 | 1 0¾ | 1 2⅜ | 1 4 | 1 5½ | 1 7⅜ | 1 8¾ | 1 10⅜ | 1 11⅞ | 2 1½ | 2 3 | 2 4⅝ | 2 7⅞ | 2 10¾ | 3 2¼ |
| 41 | 1 1 | 1 2¾ | 1 4⅜ | 1 6 | 1 7⅝ | 1 9¼ | 1 10⅞ | 2 0½ | 2 2⅛ | 2 3¾ | 2 5⅜ | 2 8⅝ | 2 11⅞ | 3 3⅛ |
| 42 | 1 1⅜ | 1 3 | 1 4¾ | 1 6⅜ | 1 8 | 1 9¾ | 1 11½ | 2 1 | 2 2¾ | 2 4½ | 2 6 | 2 9⅜ | 3 0¾ | 3 4⅛ |
| 43 | 1 1⅝ | 1 3½ | 1 5⅛ | 1 6⅞ | 1 8⅝ | 1 10¼ | 2 0 | 2 1⅝ | 2 3⅜ | 2 5 | 2 6¾ | 2 10¼ | 3 1⅝ | 3 5 |
| 44 | 1 2 | 1 3¾ | 1 5⅝ | 1 7¼ | 1 9 | 1 10¾ | 2 0½ | 2 2¼ | 2 4 | 2 5¾ | 2 7½ | 2 11 | 3 2½ | 3 6 |
| 45 | 1 2⅜ | 1 4⅛ | 1 6 | 1 7¾ | 1 9½ | 1 11⅜ | 2 1 | 2 2⅞ | 2 4⅝ | 2 6½ | 2 8¼ | 2 11¾ | 3 3⅜ | 3 7 |
| 46 | 1 2⅝ | 1 4½ | 1 6⅜ | 1 8⅛ | 1 10 | 1 11⅞ | 2 1⅝ | 2 3½ | 2 5¼ | 2 7⅛ | 2 9 | 3 0⅝ | 3 4¼ | 3 7⅞ |
| 47 | 1 2⅞ | 1 4⅞ | 1 6¾ | 1 8⅝ | 1 10½ | 2 0⅜ | 2 1⅞ | 2 4 | 2 6 | 2 7⅞ | 2 9⅝ | 3 1½ | 3 5¼ | 3 8⅞ |
| 48 | 1 3¼ | 1 5¼ | 1 7⅛ | 1 9 | 1 11 | 2 0⅞ | 2 2¾ | 2 4⅝ | 2 6½ | 2 8½ | 2 10⅜ | 3 2¼ | 3 6 | 3 9⅞ |
| 49 | 1 3½ | 1 5⅝ | 1 7½ | 1 9½ | 1 11½ | 2 1⅜ | 2 3⅜ | 2 5¼ | 2 7¼ | 2 9⅛ | 2 11 | 3 3 | 3 6⅞ | 3 10⅞ |
| 50 | 1 3⅞ | 1 6 | 1 8 | 1 9⅞ | 2 0 | 2 1⅞ | 2 3⅞ | 2 5⅞ | 2 7⅞ | 2 9¾ | 2 11¾ | 3 3¾ | 3 7¾ | 3 11¾ |
| 51 | 1 4¼ | 1 6¼ | 1 8⅜ | 1 10⅜ | 2 0⅜ | 2 2⅜ | 2 4½ | 2 6½ | 2 8½ | 2 10½ | 3 0½ | 3 4½ | 3 8⅝ | 4 0¾ |
| 52 | 1 4½ | 1 6⅝ | 1 8¾ | 1 10¾ | 2 0⅞ | 2 2⅞ | 2 4⅞ | 2 7⅛ | 2 9⅛ | 2 11⅛ | 3 1¼ | 3 5⅜ | 3 9½ | 4 1⅝ |
| 53 | 1 4⅞ | 1 6⅞ | 1 9⅛ | 1 11¼ | 2 1¼ | 2 3⅜ | 2 5⅜ | 2 7⅝ | 2 9¾ | 2 11⅞ | 3 2 | 3 6¼ | 3 10½ | 4 2⅝ |
| 54 | 1 5⅛ | 1 7⅜ | 1 9½ | 1 11⅝ | 2 1¾ | 2 3⅞ | 2 6 | 2 8¼ | 2 10⅜ | 3 0½ | 3 2⅝ | 3 7 | 3 11 | 4 3½ |
| 55 | 1 5½ | 1 7¾ | 1 9⅞ | 2 0 | 2 2¼ | 2 4½ | 2 6⅝ | 2 8⅞ | 2 11 | 3 1¼ | 3 3½ | 3 7¾ | 4 0⅛ | 4 4½ |
| 56 | 1 5⅞ | 1 8⅛ | 1 10¼ | 2 0½ | 2 2¾ | 2 4⅞ | 2 7⅛ | 2 9⅜ | 2 11⅝ | 3 1⅞ | 3 4⅛ | 3 8½ | 4 1 | 4 5½ |
| 57 | 1 6⅛ | 1 8½ | 1 10⅝ | 2 0⅞ | 2 3¼ | 2 5¼ | 2 7¾ | 2 10 | 3 0¼ | 3 2½ | 3 4⅞ | 3 9⅜ | 4 1⅞ | 4 6⅜ |
| 58 | 1 6½ | 1 8¾ | 1 11 | 2 1⅜ | 2 3⅝ | 2 6 | 2 8¼ | 2 10⅝ | 3 0⅞ | 3 3¼ | 3 5½ | 3 10⅛ | 4 2¾ | 4 7⅜ |
| 59 | 1 6¾ | 1 9⅛ | 1 11½ | 2 1⅞ | 2 4⅛ | 2 6½ | 2 8⅞ | 2 11¼ | 3 1½ | 3 4 | 3 6¼ | 3 11⅛ | 4 3⅜ | 4 8⅜ |
| 60 | 1 7⅛ | 1 9⅜ | 1 11⅞ | 2 2¼ | 2 4⅝ | 2 7 | 2 9⅜ | 2 11¾ | 3 2¼ | 3 4⅝ | 3 7 | 3 11¾ | 4 4½ | 4 9¼ |
| 61 | 1 7½ | 1 9⅞ | 2 0¼ | 2 2¾ | 2 5¼ | 2 7½ | 2 10 | 3 0⅜ | 3 2⅞ | 3 5¼ | 3 7¾ | 4 0½ | 4 5⅜ | 4 10¼ |
| 62 | 1 7¾ | 1 10¼ | 2 0⅝ | 2 3⅛ | 2 5⅝ | 2 8 | 2 10½ | 3 1 | 3 3½ | 3 6 | 3 8½ | 4 1⅜ | 4 6¼ | 4 11¼ |
| 63 | 1 8 | 1 10½ | 2 1 | 2 3½ | 2 6 | 2 8⅝ | 2 11 | 3 1⅜ | 3 4⅛ | 3 6⅝ | 3 9⅛ | 4 2⅛ | 4 7⅛ | 5 0⅛ |
| 64 | 1 8⅜ | 1 10⅞ | 2 1½ | 2 4 | 2 6½ | 2 9⅛ | 2 11⅝ | 3 2¼ | 3 4¾ | 3 7¼ | 3 9⅞ | 4 3 | 4 8 | 5 1⅛ |
| 65 | 1 8⅝ | 1 11¼ | 2 1⅞ | 2 4½ | 2 7 | 2 9⅝ | 3 0¼ | 3 2⅞ | 3 5⅜ | 3 8 | 3 10½ | 4 3¾ | 4 8⅞ | 5 2 |
| 66 | 1 9 | 1 11⅝ | 2 2¼ | 2 4⅞ | 2 7½ | 2 10⅛ | 3 0⅞ | 3 3⅜ | 3 6 | 3 8⅝ | 3 11¼ | 4 4½ | 4 9¾ | 5 3 |
| 67 | 1 9⅜ | 2 0 | 2 2⅝ | 2 5¼ | 2 8 | 2 10⅝ | 3 1⅜ | 3 4 | 3 6½ | 3 9⅜ | 4 0 | 4 5⅜ | 4 10⅝ | 5 4 |
| 68 | 1 9⅝ | 2 0⅜ | 2 3 | 2 5¾ | 2 8½ | 2 11⅛ | 3 1¾ | 3 4⅝ | 3 7¼ | 3 10 | 4 0¾ | 4 6⅛ | 4 11½ | 5 5 |
| 69 | 1 9⅞ | 2 0¾ | 2 3½ | 2 6¼ | 2 8⅞ | 2 11¾ | 3 2⅜ | 3 5⅛ | 3 7⅞ | 3 10⅝ | 4 1½ | 4 7 | 5 0⅝ | 5 6 |
| 70 | 1 10¼ | 2 1 | 2 3⅞ | 2 6⅝ | 2 9⅜ | 3 0¼ | 3 3 | 3 5⅞ | 3 8½ | 3 11⅜ | 4 2⅛ | 4 7⅞ | 5 1¼ | 5 6⅞ |

| Number of teeth. | PITCH OF THE TEETH IN INCHES. | | | | | | | | | | | | | |
|---|---|---|---|---|---|---|---|---|---|---|---|---|---|---|
| | 1 in. | 1⅛ | 1¼ | 1⅜ | 1½ | 1⅝ | 1¾ | 1⅞ | 2 in. | 2⅛ | 2¼ | 2½ | 2¾ | 3 in. |
| | DIAMETER AT THE PITCH CIRCLE IN FEET AND INCHES. | | | | | | | | | | | | | |
| 71 | 1 10⅝ | 2 1½ | 2 4¼ | 2 7 | 2 9⅞ | 3 0¾ | 3 3½ | 3 6⅜ | 3 9¼ | 4 0 | 4 2⅞ | 4 8½ | 5 2⅛ | 5 7¾ |
| 72 | 1 10⅞ | 2 1¾ | 2 4⅝ | 2 7½ | 2 10⅜ | 3 1¼ | 3 4⅛ | 3 6¾ | 3 9⅞ | 4 0¾ | 4 3½ | 4 9¼ | 5 3 | 5 8¾ |
| 73 | 1 11¼ | 2 2⅛ | 2 5 | 2 8 | 2 10⅞ | 3 1¾ | 3 4⅝ | 3 7½ | 3 10½ | 4 1⅜ | 4 4¼ | 4 10 | 5 3⅞ | 5 9¾ |
| 74 | 1 11½ | 2 2½ | 2 5½ | 2 8⅜ | 2 11⅜ | 3 2¼ | 3 5¼ | 3 7⅞ | 3 11⅛ | 4 2 | 4 5 | 4 10⅞ | 5 4¾ | 5 10⅝ |
| 75 | 1 11⅞ | 2 2⅞ | 2 5⅞ | 2 8⅞ | 2 11⅞ | 3 2¾ | 3 5¾ | 3 8¾ | 3 11¾ | 4 2¾ | 4 5¾ | 4 11¾ | 5 5⅜ | 5 11⅝ |
| 76 | 2 0⅛ | 2 3¼ | 2 6¼ | 2 9¼ | 3 0¼ | 3 3¼ | 3 6⅛ | 3 9⅜ | 4 0⅜ | 4 3½ | 4 6½ | 5 0½ | 5 6½ | 6 0½ |
| 77 | 2 0½ | 2 3½ | 2 6⅝ | 2 9¾ | 3 0¾ | 3 3⅞ | 3 6⅞ | 3 9⅞ | 4 1 | 4 4 | 4 7⅛ | 5 1¼ | 5 7⅜ | 6 1½ |
| 78 | 2 0⅞ | 2 3⅞ | 2 7 | 2 10⅛ | 3 1½ | 3 4⅜ | 3 7½ | 3 10½ | 4 1⅝ | 4 4¾ | 4 7⅞ | 5 2 | 5 8¼ | 6 2½ |
| 79 | 2 1⅛ | 2 4⅛ | 2 7⅜ | 2 10½ | 3 1¾ | 3 4⅞ | 3 8 | 3 11⅛ | 4 2¼ | 4 5½ | 4 8½ | 5 2⅞ | 5 9⅛ | 6 3½ |
| 80 | 2 1½ | 2 4⅝ | 2 7¾ | 2 11 | 3 2¼ | 3 5⅜ | 3 8½ | 3 11¾ | 4 3 | 4 6⅛ | 4 9¼ | 5 3⅝ | 5 10 | 6 4⅜ |
| 81 | 2 1¾ | 2 5 | 2 8¼ | 2 11½ | 3 2⅝ | 3 5⅞ | 3 9⅛ | 4 0⅜ | 4 3½ | 4 6⅝ | 4 10 | 5 4½ | 5 10⅞ | 6 5⅜ |
| 82 | 2 2⅛ | 2 5⅜ | 2 8⅝ | 2 11⅞ | 3 3⅛ | 3 6⅜ | 3 9⅝ | 4 0⅞ | 4 4¼ | 4 7½ | 4 10¾ | 5 5¼ | 5 11¾ | 6 6⅜ |
| 83 | 2 2½ | 2 5¾ | 2 9 | 3 0⅜ | 3 3⅝ | 3 6⅞ | 3 10¼ | 4 1½ | 4 4⅞ | 4 8⅛ | 4 11½ | 5 6 | 6 0⅝ | 6 7¼ |
| 84 | 2 2¾ | 2 6 | 2 9⅜ | 3 0¾ | 3 4 | 3 7½ | 3 10¾ | 4 2⅛ | 4 5½ | 4 8⅞ | 5 0⅛ | 5 6⅞ | 6 1½ | 6 8⅛ |
| 85 | 2 3 | 2 6⅜ | 2 9¾ | 3 1¼ | 3 4½ | 3 7⅞ | 3 11¼ | 4 2¾ | 4 6⅛ | 4 9½ | 5 0⅞ | 5 7⅜ | 6 2⅜ | 6 9⅛ |
| 86 | 2 3⅜ | 2 6¾ | 2 10¼ | 3 1⅝ | 3 5¼ | 3 8½ | 3 11⅞ | 4 3¼ | 4 6¾ | 4 10⅛ | 5 1⅝ | 5 8½ | 6 3¼ | 6 10½ |
| 87 | 2 3⅝ | 2 7⅛ | 2 10⅜ | 3 2 | 3 5½ | 3 9 | 4 0½ | 4 3⅞ | 4 7⅜ | 4 10⅞ | 5 2¼ | 5 9¼ | 6 4⅛ | 6 11 |
| 88 | 2 4 | 2 7½ | 2 11 | 3 2½ | 3 6 | 3 9½ | 4 1 | 4 4½ | 4 8 | 4 11½ | 5 3 | 5 10 | 6 5 | 7 0 |
| 89 | 2 4⅜ | 2 7⅞ | 2 11⅜ | 3 2⅞ | 3 6½ | 3 10 | 4 1½ | 4 5⅛ | 4 8⅝ | 5 0¼ | 5 3¼ | 5 10¾ | 6 5⅞ | 7 1 |
| 90 | 2 4⅝ | 2 8¼ | 2 11¾ | 3 3¼ | 3 7 | 3 10½ | 4 2⅛ | 4 5¾ | 4 9¼ | 5 0⅞ | 5 4¼ | 5 11⅝ | 6 6¾ | 7 2 |
| 91 | 2 4⅞ | 2 8½ | 3 0¼ | 3 3⅞ | 3 7½ | 3 11 | 4 2¾ | 4 6¼ | 4 9⅞ | 5 1½ | 5 5⅛ | 6 0⅜ | 6 7⅝ | 7 2⅞ |
| 92 | 2 5¼ | 2 8⅞ | 3 0⅝ | 3 4¼ | 3 7⅞ | 3 11⅝ | 4 3¼ | 4 7 | 4 10½ | 5 2⅛ | 5 5⅞ | 6 1 | 6 8½ | 7 3⅞ |
| 93 | 2 5⅝ | 2 9¼ | 3 1 | 3 4¾ | 3 8⅜ | 4 0⅛ | 4 3⅞ | 4 7½ | 4 11¼ | 5 2⅞ | 5 6⅝ | 6 2 | 6 9⅜ | 7 4⅞ |
| 94 | 2 5⅞ | 2 9⅝ | 3 1⅜ | 3 5⅛ | 3 8⅞ | 4 0⅝ | 4 4⅜ | 4 8⅛ | 4 11¾ | 5 3½ | 5 7⅛ | 6 2¾ | 6 10¼ | 7 5¾ |
| 95 | 2 6¼ | 2 10 | 3 1¾ | 3 5½ | 3 9⅜ | 4 1⅛ | 4 4⅞ | 4 8¾ | 5 0½ | 5 4¼ | 5 8 | 6 3½ | 6 11⅛ | 7 6¾ |
| 96 | 2 6½ | 2 10⅜ | 3 2⅛ | 3 6 | 3 9¾ | 4 1⅝ | 4 5½ | 4 9⅜ | 5 1⅛ | 5 5 | 5 8¾ | 6 4⅜ | 7 0 | 7 7⅝ |
| 97 | 2 6⅞ | 2 10¾ | 3 2⅝ | 3 6½ | 3 10¼ | 4 2⅛ | 4 6 | 4 10 | 5 1¾ | 5 5⅝ | 5 9½ | 6 5¼ | 7 0⅞ | 7 8⅝ |
| 98 | 2 7⅛ | 2 11 | 3 3 | 3 6⅞ | 3 10¾ | 4 2⅝ | 4 6½ | 4 10½ | 5 2⅜ | 5 6¼ | 5 10⅛ | 6 6 | 7 1¾ | 7 9½ |
| 99 | 2 7½ | 2 11⅜ | 3 3⅜ | 3 7⅜ | 3 11¼ | 4 3¼ | 4 7⅛ | 4 11 | 5 3 | 5 7 | 5 11 | 6 6¾ | 7 2⅝ | 7 10½ |
| 100 | 2 7⅞ | 2 11⅞ | 3 3¾ | 3 7¾ | 3 11¾ | 4 3¾ | 4 7¾ | 4 11⅝ | 5 3⅝ | 5 7⅝ | 5 11⅝ | 6 7½ | 7 3⅝ | 7 11½ |
| 101 | 2 8¼ | 3 0⅛ | 3 4⅛ | 3 8¼ | 4 0¼ | 4 4¼ | 4 8¼ | 5 0¼ | 5 4¼ | 5 8¼ | 6 0¼ | 6 8⅜ | 7 4⅜ | 8 0½ |
| 102 | 2 8½ | 3 0½ | 3 4½ | 3 8⅝ | 4 0⅝ | 4 4¾ | 4 8⅞ | 5 1 | 5 5 | 5 9 | 6 1 | 6 9⅛ | 7 5¼ | 8 1⅜ |

## TABLE *of the Strength of the Teeth of Cast Iron Wheels at a given velocity.*

| Pitch of teeth in inches. | Thickness of teeth in inches. | Breadth of teeth in inches. | Strength of teeth in horse power, at | | | |
|---|---|---|---|---|---|---|
| | | | 3 feet per second. | 4 feet per second. | 6 feet per second. | 8 feet per second. |
| 3·99 | 1·9 | 7·6 | 20·57 | 27·43 | 41·14 | 54·85 |
| 3·78 | 1·8 | 7·2 | 17·49 | 23·32 | 34·98 | 46·64 |
| 3·57 | 1·7 | 6·8 | 14·73 | 19·65 | 29·46 | 39·28 |
| 3·36 | 1·6 | 6·4 | 12·28 | 16·38 | 24·56 | 32·74 |
| 3·15 | 1·5 | 6 | 10·12 | 13·50 | 20·24 | 26·98 |
| 2·94 | 1·4 | 5·6 | 8·22 | 10·97 | 16·44 | 21·92 |
| 2·73 | 1·3 | 5·2 | 6·58 | 8·78 | 13·16 | 17·54 |
| 2·52 | 1·2 | 4·8 | 5·18 | 6·91 | 10·36 | 13·81 |
| 2·31 | 1·1 | 4·4 | 3·99 | 5·32 | 7·98 | 10·64 |
| 2·1 | 1·0 | 4 | 3·00 | 4·00 | 6·00 | 8·00 |
| 1·89 | ·9 | 3·6 | 2·18 | 2·91 | 4·36 | 5·81 |
| 1·68 | ·8 | 3·2 | 1·53 | 2·04 | 3·06 | 3·08 |
| 1·47 | ·7 | 2·8 | 1·027 | 1·37 | 2·04 | 2·72 |
| 1·26 | ·6 | 2·4 | ·64 | ·86 | 1·38 | 1·84 |
| 1·05 | ·5 | 2 | ·375 | ·50 | ·75 | 1·00 |

## ADDITIONAL EXAMPLES ON THE VELOCITY OF WHEELS, DRUMS, PULLEYS, ETC.

If a wheel that contains 75 teeth makes 16 revolutions per minute, required the number of teeth in another to work in it, and make 24 revolutions in the same time.

$$\frac{75 \times 16}{24} = 50 \text{ teeth.}$$

A wheel, 64 inches diameter, and making 42 revolutions per minute, is to give motion to a shaft at the rate of 77 revolutions in the same time: required the diameter of a wheel suitable for that purpose.

$$\frac{64 \times 42}{77} = 34{\cdot}9 \text{ inches.}$$

Required the number of revolutions per minute made by a wheel or pulley 20 inches diameter, when driven by another of 4 feet diameter, and making 46 revolutions per minute.

$$\frac{48 \times 46}{20} = 110{\cdot}4 \text{ revolutions.}$$

A shaft, at the rate of 22 revolutions per minute, is to give motion, by a pair of wheels, to another shaft at the rate of $15\frac{1}{2}$; the distance of the shafts from centre to centre is $45\frac{1}{2}$ inches; the diameters of the wheels at the pitch lines are required.

$$\frac{45{\cdot}5 \times 15{\cdot}5}{22 + 15{\cdot}5} = 18{\cdot}81 \text{ radius of the driving wheel.}$$

$$\text{And } \frac{45{\cdot}5 \times 22}{22 + 15{\cdot}5} = 26{\cdot}69 \text{ radius of the driven wheel.}$$

Suppose a drum to make 20 revolutions per minute, required the diameter of another to make 58 revolutions in the same time.

$58 \div 20 = 2{\cdot}9$, that is, their diameters must be as $2{\cdot}9$ to 1; thus, if the one making 20 revolutions be called 30 inches, the other will be $30 \div 2{\cdot}9 = 10{\cdot}345$ inches diameter.

Required the diameter of a pulley, to make $12\frac{1}{2}$ revolutions in the same time as one of 32 inches making 26.

$$\frac{32 \times 26}{12{\cdot}5} = 66{\cdot}56 \text{ inches diameter.}$$

A shaft, at the rate of 16 revolutions per minute, is to give motion to a piece of machinery at the rate of 81 revolutions in the same time; the motion is to be communicated by means of two wheels and two pulleys with an intermediate shaft; the driving wheel contains 54 feet, and the driving pulley is 25 inches diameter; required the number of teeth in the other wheel, and the diameter of the other pulley.

$\sqrt{81 \times 16} = 36$, the mean velocity between 16 and 81; then, $\frac{16 \times 54}{36} = 24$ teeth; and $\frac{36 \times 25}{81} = 11{\cdot}11$ inches, diameter of pulley.

Suppose in the last example the revolutions of one of the wheels to be given, the number of teeth in both, and likewise the diameter of each pulley, to find the revolutions of the last pulley.

$$\frac{16 \times 54}{24} = 36, \text{ velocity of the intermediate shaft;}$$

$$\text{and } \frac{36 \times 25}{11{\cdot}11} = 81, \text{ the velocity of the machine.}$$

TABLE *for finding the radius of a wheel when the pitch is given, or the pitch of a wheel when the radius is given, that shall contain from* 10 *to* 150 *teeth, and any pitch required.*

| Number of Teeth. | Radius. | Number of Teeth. | Radius. | Number of Teeth. | Radius. | Number of Teeth. | Radius. |
|---|---|---|---|---|---|---|---|
| 10 | 1·618 | 46 | 7·327 | 81 | 12·895 | 116 | 18·464 |
| 11 | 1·774 | 47 | 7·486 | 82 | 13·054 | 117 | 18·623 |
| 12 | 1·932 | 48 | 7·645 | 83 | 13·213 | 118 | 18·782 |
| 13 | 2·089 | 49 | 7·804 | 84 | 13·370 | 119 | 18·941 |
| 14 | 2·247 | 50 | 7·963 | 85 | 13·531 | 120 | 19·101 |
| 15 | 2·405 | 51 | 8·122 | 86 | 13·690 | 121 | 19·260 |
| 16 | 2·563 | 52 | 8·281 | 87 | 13·849 | 122 | 19·419 |
| 17 | 2·721 | 53 | 8·440 | 88 | 14·008 | 123 | 19·578 |
| 18 | 2·879 | 54 | 8·599 | 89 | 14·168 | 124 | 19·737 |
| 19 | 3·038 | 55 | 8·758 | 90 | 14·327 | 125 | 19·896 |
| 20 | 3·196 | 56 | 8·917 | 91 | 14·486 | 126 | 20·055 |
| 21 | 3·355 | 57 | 9·076 | 92 | 14·645 | 127 | 20·214 |
| 22 | 3·513 | 58 | 9·235 | 93 | 14·804 | 128 | 20·374 |
| 23 | 3·672 | 59 | 9·394 | 94 | 14·963 | 129 | 20·533 |
| 24 | 3·830 | 60 | 9·553 | 95 | 15·122 | 130 | 20·692 |
| 25 | 3·989 | 61 | 9·712 | 96 | 15·281 | 131 | 20·851 |
| 26 | 4·148 | 62 | 9·872 | 97 | 15·440 | 132 | 21·010 |
| 27 | 4·307 | 63 | 10·031 | 98 | 15·600 | 133 | 21·169 |
| 28 | 4·165 | 64 | 10·190 | 99 | 15·759 | 134 | 21·328 |
| 29 | 4·624 | 65 | 10·349 | 100 | 15·918 | 135 | 21·488 |
| 30 | 4·788 | 66 | 10·508 | 101 | 16·077 | 136 | 21·647 |
| 31 | 4·942 | 67 | 10·667 | 102 | 16·236 | 137 | 21·806 |
| 32 | 5·101 | 68 | 10·826 | 103 | 16·395 | 138 | 21·965 |
| 33 | 5·260 | 69 | 10·985 | 104 | 16·554 | 139 | 22·124 |
| 34 | 5·419 | 70 | 11·144 | 105 | 16·713 | 140 | 22·283 |
| 35 | 5·578 | 71 | 11·303 | 106 | 16·873 | 141 | 22·442 |
| 36 | 5·737 | 72 | 11·463 | 107 | 17·032 | 142 | 22·602 |
| 37 | 5·896 | 73 | 11·622 | 108 | 17·191 | 143 | 22·761 |
| 38 | 6·055 | 74 | 11·781 | 109 | 17·350 | 144 | 22·920 |
| 39 | 6·214 | 75 | 11·940 | 110 | 17·509 | 145 | 23·079 |
| 40 | 6·373 | 76 | 12·099 | 111 | 17·668 | 146 | 23·238 |
| 41 | 6·532 | 77 | 12·258 | 112 | 17·827 | 147 | 23·397 |
| 42 | 6·691 | 78 | 12·417 | 113 | 17·987 | 148 | 23·556 |
| 43 | 6·850 | 79 | 12·576 | 114 | 18·146 | 149 | 23·716 |
| 44 | 7·009 | 80 | 12·735 | 115 | 18·305 | 150 | 23·875 |
| 45 | 7·168 | | | | | | |

RULE.—Multiply the radius in the table by the pitch given, and the product will be the radius of the wheel required.

Or, divide the radius of the wheel by the radius in the table, and the quotient will be the pitch of the wheel required.

Required the radius of a wheel to contain 64 teeth, of 3 inch pitch.

$$10{\cdot}19 \times 3 = 30{\cdot}57 \text{ inches.}$$

What is the pitch of a wheel to contain 80 teeth, when the radius is 25·47 inches?

$$25{\cdot}47 \div 12{\cdot}735 = 2 \text{ inch pitch.}$$

Or, set off upon a straight line AB seven times the pitch AC given; divide that, or another exactly the same length, into eleven equal parts; call each of those divisions four, or each of those divisions will be equal to four teeth upon the radius. If a circle be made with any number (20) of these equal parts as radius, AC the pitch will go that number (20) of times round the circle.

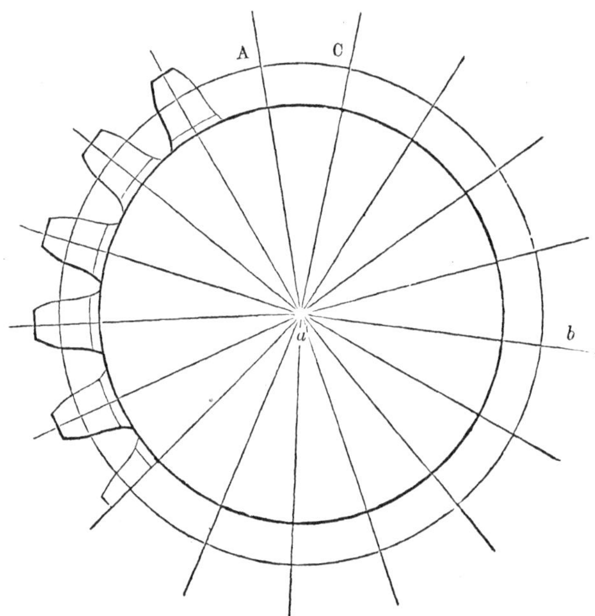


Were it required to find the diameter of a wheel to contain 17 teeth, the construction would be as follows:—

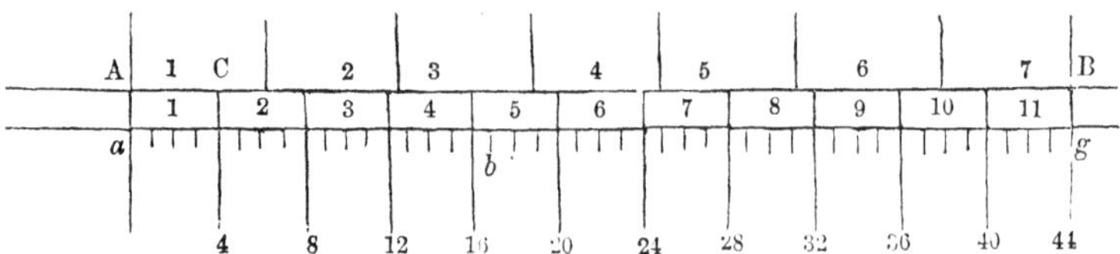


Thus, 4 divisions and $\frac{1}{4}$ of another equal the radius of the wheel, that is $a_1\, b_1 = a\, b$, and $A_1\, C_1 = AC$.

*Regular approved proportions for wheels with flat arms in the middle of the ring, and ribs or feathers on each side.*—The length of the teeth = $\frac{6}{9}$ the pitch, besides clearance, or $\frac{5}{7}$ the pitch, clearance included.

| | | |
|---|---|---|
| Thickness of the teeth ........................... | $\frac{4}{9}$ | the pitch. |
| Breadth on the face.............................. | $2\frac{1}{2}$ | — |
| Edge of the rim.................................. | $\frac{4}{9}$ | — |
| Rib projecting inside the rim.................... | $\frac{4}{9}$ | — |
| Thickness of the flat arms....................... | $\frac{4}{9}$ | — |

Breadth of the arms at the points = 2 teeth and $\frac{1}{4}$ the pitch, getting broader towards the centre of the wheel in the proportion of $\frac{1}{2}$ inch to every foot in length.

Thickness of the ribs, or feathers, $\frac{1}{4}$ the pitch.

Thickness of metal round the eye, or centre, $\frac{7}{9}$ the pitch.

Wheels made with plain arms, the teeth are in the same proportion as above; the ring and the arms are each equal to one cog or tooth in thickness, and the metal round the eye same as above, in feathered wheels.

These proportions differ, though slightly, in different works and in different localities; but they are the most commonly employed, and are besides the most consistent with good and accurate workmanship. For the sake of more easy reference, we collect them into a table, which the annexed diagram will serve fully to explain. They stand thus:—

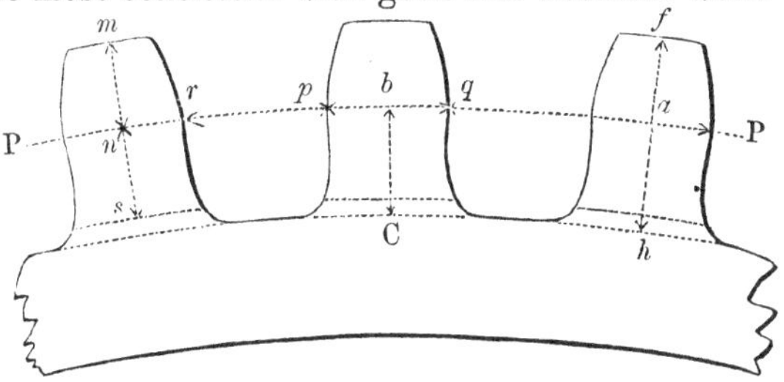

$$
\begin{aligned}
ab &= \text{Pitch of teeth} &&= 1 \text{ pitch.}\\
mn &- \text{Depth to pitch line, PP,} &&= \tfrac{3}{10} \text{ —.}\\
ns + nm &= \text{Working depth of tooth,} &&= \tfrac{6}{10} \text{ —.}\\
Cb - ns &= \text{Bottom clearance,} &&= \tfrac{1}{10} \text{ —.}\\
fh &= \text{Whole depth to root,} &&= \tfrac{7}{10} \text{ —.}\\
pq &= \text{Thickness of tooth,} &&= \tfrac{5}{11} \text{ —.}\\
rp &= \text{Width of space,} &&= \tfrac{6}{11} \text{ —.}
\end{aligned}
$$

The use of the following table is very evident, and the manner of applying it may be rendered still more obvious by the following examples:—

$$\pi = 3{\cdot}1416.$$

1. Given a wheel of 88 teeth, $2\frac{1}{2}$ inch pitch, to find the diameter of the pitch circle. Here the tabular number in the second column answering to the given pitch is ·7958, which multiplied by 88 gives 70·03 for the diameter required.

2. Given a wheel of 5 feet (60 inches) diameter, $2\frac{3}{4}$ inch pitch, to find the number of teeth. Here the factor in the third column

corresponding to the given pitch is 1·1333, which multiplied by 60 gives 68 for the number of teeth.

It may, however, so happen that the answer found in this manner contains a fraction—which being inadmissible by the nature of the question, it becomes necessary to alter slightly the diameter of the pitch circle. This is readily accomplished by taking the nearest whole number to the answer found, and finding the modified diameter by means of the second column. The following case will fully explain what is meant:

3. Given a wheel 33 inches diameter, $1\frac{3}{4}$ inch pitch, to find the number of teeth. The corresponding factor is 1·7952, which multiplied by 33 gives 59·242 for the number of teeth, that is, $59\frac{1}{4}$ teeth nearly. Now, 59 would here be the nearest whole number; but as a wheel of 60 teeth may be preferred for convenience of calculation of speeds, we may adopt that number and find the diameter corresponding. The factor in the second column answering to $1\frac{3}{4}$ pitch is ·557, and this multiplied by 60 gives 33·4 inches as the diameter which the wheel ought to have.

| Pitch in inches and parts of an inch. | $D = \frac{P}{\pi} \times N$ <br> RULE.—To find the diameter in inches, multiply the number of teeth by the tabular number answering to the given pitch. | $N = \frac{\pi}{P} \times D$ <br> RULE.—To find the number of teeth, multiply the given diameter in inches by the tabular number answering to the given pitch. |
|---|---|---|
| Values of P | Values of $\frac{P}{\pi}$ | Values of $\frac{\pi}{P}$ |
| 6 | 1·9095 | ·5236 |
| 5 | 1·5915 | ·6283 |
| $4\frac{1}{2}$ | 1·4270 | ·6981 |
| 4 | 1·2732 | ·7854 |
| $3\frac{1}{2}$ | 1·1141 | ·8976 |
| 3 | ·9547 | 1·0472 |
| $2\frac{3}{4}$ | ·8754 | 1·1333 |
| $2\frac{1}{2}$ | ·7958 | 1·2566 |
| $2\frac{1}{4}$ | ·7135 | 1·3963 |
| 2 | ·6366 | 1·5708 |
| $1\frac{7}{8}$ | ·5937 | 1·6755 |
| $1\frac{3}{4}$ | ·5570 | 1·7952 |
| $1\frac{5}{8}$ | ·5141 | 1·9264 |
| $1\frac{1}{2}$ | ·4774 | 2·0944 |
| $1\frac{3}{8}$ | ·4377 | 2·2848 |
| $1\frac{1}{4}$ | ·3979 | 2·5132 |
| $1\frac{1}{8}$ | ·3568 | 2·7926 |
| 1 | ·3183 | 3·1416 |
| $\frac{7}{8}$ | ·2785 | 3·5904 |
| $\frac{3}{4}$ | ·2387 | 4·1888 |
| $\frac{5}{8}$ | ·1989 | 5·0266 |
| $\frac{1}{2}$ | ·1592 | 6·2832 |
| $\frac{3}{8}$ | ·1194 | 8·3776 |
| $\frac{1}{4}$ | ·0796 | 12·5664 |

RULE.—*To find the power that a cast iron wheel is capable of transmitting at any given velocity.*—Multiply the breadth of the teeth, or face of the wheel, in inches, by the square of the thickness of one tooth, and divide the product by the length of the teeth, the quotient is the strength in horse power at a velocity of 136 feet per minute.

Required the power that a wheel of the following dimensions ought to transmit with safety, namely,

Breadth of teeth...............$7\frac{1}{2}$ inches,
Thickness.......................1·4
And length......................2

$$1{\cdot}4^2 = 1{\cdot}96, \text{ and } \frac{7{\cdot}5 \times 1{\cdot}96}{2} = 7{\cdot}35 \text{ horse power.}$$

The strength at any other velocity is found by multiplying the power so obtained by any other required velocity, and by ·0044, the quotient is the power at that velocity.

Suppose the wheel as above, at a velocity of 320 feet per minute.

$$7{\cdot}35 \times 320 \times {\cdot}0044 = 10{\cdot}3488 \text{ horse power.}$$

## ON THE MAXIMUM VELOCITY AND POWER OF WATER WHEELS.

### OF UNDERSHOT WHEELS.

THE term "undershot" is applied to a wheel when the water strikes at, or below, the centre; and the greatest effect is produced when the periphery of the wheels moves with a velocity of ·57 that of the water; hence, to find the velocity of the water, multiply the square root or the perpendicular height of the fall in feet by 8, and the product is the velocity in feet per second.

Required the maximum velocity of an undershot wheel, when propelled by a fall of water 6 feet in height.

$\sqrt{6} = 2{\cdot}45 \times 8 = 19{\cdot}6$ feet, velocity of water.

And $19{\cdot}6 \times {\cdot}57 = 11{\cdot}17$ feet per second for the wheel.

### OF BREAST AND OVERSHOT WHEELS.

Wheels that have the water applied between the centre and the vertex are styled breast wheels, and overshot when the water is brought over the wheel and laid on the opposite side; however, in either case the maximum velocity is $\frac{2}{3}$ that of the water; hence, to find the head of water proper for a wheel at any velocity, say:

As the square of 16·083, or 258·67, is to 4, so is the square of the velocity of the wheel in feet per second to the head of water required. By *head* is understood the distance between the aperture of the sluice and where the water strikes upon the wheel.

Required the head of water necessary for a wheel of 24 feet diameter, moving with a velocity of 5 feet per second.

$$\frac{5 \times 3}{2} = 7{\cdot}5 \text{ feet, velocity of the water.}$$

And $258{\cdot}67 : 4 :: 7{\cdot}5^2 : {\cdot}87$ feet, head of water required.

But one-tenth of a foot of head must be added for every foot of increase in the diameter of the wheel, from 15 to 20 feet, and ·05 more for every foot of increase from 20 to 30 feet, commencing with five-tenths for a 15 feet wheel.

This additional head is intended to compensate for the friction of water in the aperture of the sluice to keep the velocity as 3 to 2 of the wheel; thus, in place of ·87 feet head for a 24 feet wheel, it will be $\cdot 87 + 1{\cdot}2 = 2{\cdot}07$ feet head of water.

If the water flow from under the sluice, multiply the square root of the depth in feet by 5·4, and by the area of the orifice also in feet, and the product is the quantity discharged in cubic feet per second.

Again, if the water flow over the sluice, multiply the square root of the depth in feet by 5·4, and $\frac{2}{3}$ of the product multiplied by the length and depth, also in feet, gives the number of cubic feet discharged per second nearly.

Required the number of cubic feet per second that will issue from the orifice of a sluice 5 feet long, 9 inches wide, and 4 feet from the surface of the water.

$$\sqrt{4} = 2 \times 5{\cdot}4 = 10{\cdot}8 \text{ feet velocity.}$$

$$\text{And } 5 \times {\cdot}75 \times 10{\cdot}8 = 40{\cdot}5 \text{ cubic feet per second.}$$

What quantity of water per second will be expended over a wear, dam, or sluice, whose length is 10 feet, and depth 6 inches?

$$\sqrt{{\cdot}5} = {\cdot}2236 \times 5{\cdot}4 = \frac{1{\cdot}20744 \times 2}{3} = {\cdot}80496 \text{ feet velocity.}$$

Then $10 \times {\cdot}5 = 5$ feet, and ${\cdot}80496 \times 5 = 4{\cdot}0248$ cubic feet per second nearly.

In estimating the power of water wheels, half the head must be added to the whole fall, because 1 foot of fall is equal to 2 feet of head; call this the effective perpendicular descent; multiply the weight of the water per second by the effective perpendicular descent and by 60; divide the product by 33,000, and the quotient is the effect expressed in horse power.

Given 16 cubic feet of water per second, to be applied to an undershot wheel, the head being 12 feet; required the power produced.

$$12 \div 2 = 6 \text{ and } \frac{6 \times 16 \times 62{\cdot}5 \times 60}{33000} = 10{\cdot}9 \text{ horse power nearly.}$$

Given 16 cubic feet of water per second, to be applied to a high breast or an overshot wheel, with 2 feet head and 10 feet fall; required the power.

$$2 \div 2 = 1 \text{ and } \frac{\overline{1 + 10} \times 16 \times 62{\cdot}5 \times 60}{33000} = 20 \text{ horse power.}$$

Only about two-thirds of the above results can be taken as real communicative power to machinery.

### OF THE CIRCLE OF GYRATION IN WATER WHEELS.

The centre or circle of gyration is that point in a revolving body into which, if the whole quantity of matter were collected, the same moving force would generate the same angular velocity, which renders it of the utmost importance in the erection of water wheels, and the motion ought always to be communicated from that point when it is possible.

Rule.—*To find the circle of gyration.*—Add into one sum twice the weight of the shrouding, buckets, &c., multiplied by the square of the radius, $\frac{2}{3}$ of the weight of the arms, multiplied by the square of the radius, and the weight of the water multiplied by the square of the radius also; divide the sum by twice the weight of the shrouding, arms, &c., added to the weight of the water, and the square root of the quotient is the distance of the circle of gyration from the centre of suspension nearly.

Required the distance of the centre of gyration from the centre of suspension in a water wheel 22 feet diameter, shrouding, buckets, &c. = 18 tons, arms = 12 tons, and water = 10 tons.

$$22 \div 2 = 11 \text{ and } 11^2 = 121$$
$$\text{Then, } 18 \times 2 = 36 \times 121 = 4356$$
$$\tfrac{2}{3} \text{ of } 12 = 8 \times 121 = 968$$
$$\text{water} = 10 \times 121 = 1210$$
$$\overline{6534}$$

And $\overline{18 + 12} \times 2 = 60 + 10 = 70$; hence,

$$\sqrt{\frac{6534}{70}} = 9{\cdot}6 \text{ feet from the centre of suspension nearly.}$$

TABLE *of Angles for Windmill Sails.*

| Number. | Angle with the Plane of Motion. | |
|---|---|---|
| 1 | 18° | 24° |
| 2 | 19 | 21 |
| 3 | 18 | 18 |
| 4 | 16 | 14 |
| 5 | 12½ | 9 |
| 6 | 7 | 3 extremity. |

The radius is supposed to be divided into six equal parts, and $\frac{1}{6}$ from the centre is called 1, the extremity being denoted by 6.

The first column contains the angles according to an old custom; but experience has taught us that the angles in the second column are preferable.

THE VELOCITY OF THRESHING MACHINES, MILLSTONES, BORING IRON, ETC.

The drum or beaters of a threshing machine ought to move with a velocity of about 3000 feet per minute; hence, divide 11460 by the diameter of the drum in inches; or 955 by the diameter of the drum in feet; and the quotient is the number of revolutions required per minute. And the feeding rollers must make half the revolutions of the drum, when their diameters are about 3½ inches.

If the machine is driven by horses, their velocity ought to be from 2½ to 3 times round a 24 feet ring per minute.

Divide 500 by the diameter of a millstone, in feet, or 6000 by the diameter in inches, and the quotient is the number of revolutions required per minute.

In boring cast iron the cutters ought to have a velocity of about 108 inches per minute, or divide 36 by the diameter in inches, the quotient is the number of revolutions of the boring head per minute. And divide 100 by the diameter in inches, the quotient is the number of revolutions per minute, for turning wrought iron in general, and about half that velocity for cast iron.

## OF PUMPS AND PUMPING ENGINES.

PUMPS are chiefly designated by the names of lifting and force pumps; lifting pumps are applied to wells, &c., where the height of the bucket, from the surface of the water, must not exceed 33 feet; this being nearly equal to the pressure of the atmosphere, or the height to which water would be forced up into a vacuum by the pressure of the atmosphere. Force pumps are applicable on all other occasions, as raising water to any required height, supplying boilers against the force of the steam, hydrostatic presses, &c.

The power required to raise water to any height is as the weight and velocity of the water with an addition of about $\frac{1}{5}$ of the whole power for friction; hence the

RULE.—Multiply the perpendicular height of the water, in feet, by the velocity, also in feet, and by the square of the pump's diameter in inches, and again by ·341; (this being the weight of a column of water 1 inch diameter, and 12 inches high, in lbs. avoirdupois;) divide the product by 33,000, and $\frac{1}{5}$ of the quotient added to the whole quotient will be the number of horse power required.

Required the power necessary to overcome the resistance and friction of a column of water 4 inches diameter, 60 feet high, and flowing with a velocity of 130 feet per minute.

$$\frac{60 \times 130 \times 4^2 \times \cdot341}{33000} = \frac{1\cdot3}{5} = \cdot26 + 1\cdot3 = 1\cdot56 \text{ horse power nearly.}$$

Hot liquor pumps, or pumps to be employed in raising any fluid where steam is generated, require to be placed in the fluid, or as low as the bottom of it, on account of the steam filling the pipes, and acting as a counterpoise to the atmosphere; and the diameter of the pipes to and from a pump ought not to be less than $\frac{2}{3}$ of the pump's diameter.

RULE.—*The diameter of a pump and velocity of the water given, to find the quantity discharged in gallons, or cubic feet, in any given time.*—Multiply the velocity of the water, in feet per minute, by the square of the pump's diameter in inches, and by ·041 for gallons, or ·005454 for cubic feet, and the product will be the number of gallons, or cubic feet, discharged in the given time nearly.

What is the number of gallons of water discharged per hour by a pump 4 inches diameter, the water flowing at the rate of 130 feet per minute?

$$130 \times 60 = 7800 \text{ feet per hour.}$$
$$\text{And, } 7800 \times 4^2 \times \cdot041 = 5116\cdot8 \text{ gallons.}$$

RULE 1.—*The length of stroke and number of strokes given, to find the diameter of a pump, and number of horse power that will discharge a given quantity of water in a given time.*—Multiply the number of cubic feet by 2201, and divide the product by the velocity of the water, in inches, and the square root of the quotient will be the pump's diameter, in inches.

2. Multiply the number of cubic feet by 62·5, and by the perpendicular height of the water in feet, divide the product by 33,000, then will $\frac{1}{5}$ of the quotient, added to the whole quotient, be the number of horse power required.

Required the diameter of a pump, and number of horse power, capable of filling a cistern 20 feet long, 12 feet wide, and $6\frac{1}{2}$ feet deep, in 45 minutes, whose perpendicular height is 53 feet; the pump to have an effective stroke of 26 inches, and make 30 strokes per minute.

$$20 \times 12 \times 6{\cdot}5 = 1560 \text{ cubic feet, and}$$

$$\frac{1560}{45} = 34{\cdot}66 \text{ cubic feet per minute.}$$

$$\text{Then, } \frac{34{\cdot}66 \times 2201}{\surd 26 \times 30} = 9{\cdot}89 \text{ inches diameter of pump.}$$

$$\text{And } \frac{34{\cdot}66 \times 62{\cdot}5 \times 53}{33000} = \frac{3{\cdot}48}{5} = {\cdot}69 + 3{\cdot}48 = 4{\cdot}17 \text{ horse}$$

power.

Rule.—*To find the time a cistern will take in filling, when a known quantity of water is going in, and a known portion of that water is going out, in a given time.*—Divide the content of the cistern, in gallons, by the difference of the quantity going in, and the quantity going out, and the quotient is the time in hours and parts that the cistern will take in filling.

If 30 gallons per hour run in and $22\frac{1}{2}$ gallons per hour run out of a cistern capable of containing 200 gallons, in what time will the cistern be filled?

$30 - 22{\cdot}5 = 7{\cdot}5$, and $200 \div 7{\cdot}5 = 26{\cdot}666$, or 26 hours and 40 minutes.

*To find the time a vessel will take in emptying itself of water.*—Mr. O'Neill ascertained, from very accurate experiments, that a vessel, 3·166 feet long and 2·705 inches diameter, would empty itself in 3 minutes and 16 seconds, through an orifice in the bottom, whose area is ·0141 inches; and another 6·458 feet long, the diameter and orifice, as before, would do the same in 4 minutes and 40 seconds; hence, from these experiments, a rule is obtained, namely,

Multiply the square root of the depth in feet by the area of the falling surface in inches, divide the product by the area of the orifice, multiplied by 3·7, and the quotient is the time required in seconds, nearly.

How long will it require to empty a vessel of water, 9 feet high, and 20 inches diameter, through a hole $\frac{3}{4}$ inch in diameter?

$\surd 9 = 3$, the square root of the depth,
314·16 inches, area of the falling surface,
·4417 inches, area of the orifice;

Then, $\frac{314 \cdot 16 \times 3}{\cdot 4417 \times 3 \cdot 7} = 576 \cdot 7$ seconds, or 9 minutes and 36 seconds.

*On the pressure of fluids.*—The side of any vessel containing a fluid sustains a pressure equal to the area of the side, multiplied by half the depth; thus,

Suppose each side of a vessel to be 12 feet long and 5 feet deep, when filled with water, what pressure is upon each side?

$12 \times 5 = 60$ feet, the area of the side,
$2 \cdot 5$ feet = half the depth, and
$62 \cdot 5$ lbs. = the weight of a cubic foot of water.
Then, $60 \times 2 \cdot 5 \times 62 \cdot 5 = 9375$ lbs.

RULE.—*To find the weight that a given power can raise by a hydrostatic press.*—Multiply the square of the diameter of the ram in inches by the power applied in lbs., and by the effective leverage of the pump-handle; divide the product by the square of the pump's diameter, also in inches, and the quotient is the weight that the power is equal to.

What weight will a power of 50 lbs. raise by means of a hydrostatic press, whose ram is 7 inches diameter, pump $\frac{7}{8}$, and the effective leverage of the pump-handle being as 6 to 1?

$$\frac{7^2 \times 50 \times 6}{\cdot 875^2} = 19200 \text{ lbs., or 8 tons 11 cwt.}$$

In the following rules for pumping engines the boiler is supposed to be loaded with about $2\frac{1}{2}$ lbs. per square inch, and the barometer attached to the condenser indicating 26 inches on an average, or 13 lbs., = $15\frac{1}{2}$ lbs., from which deduct $\frac{1}{3}$ for friction, leaves a pressure of 10 lbs. nearly upon each square inch of the piston.

RULE.—*To find the diameter of a cylinder to work a pump of a given diameter for a given depth.*—Multiply the square of the pump's diameter in inches by $\frac{1}{3}$ of the depth of the pit in fathoms, and the square root of the product will be the cylinder's diameter in inches.

Required the diameter of a cylinder to work a pump 12 inches diameter and 27 fathoms deep.

$$\sqrt{(12^2 \times 9)} = 36 \text{ inches diameter.}$$

RULE.—*To find the diameter of a pump, that a cylinder of a given diameter can work at a given depth.*—Divide three times the square of the cylinder's diameter in inches by the depth of the pit in fathoms, and the square root of the quotient will be the pump's diameter in inches.

What diameter of a pump will a 36-inch cylinder be capable of working 27 fathoms deep?

$$\sqrt{\frac{36^2 \times 3}{27}} = 12 \text{ inches diameter.}$$

RULE.—*To find the depth from which a pump of a given diameter will work by means of a cylinder of a given diameter.*—Divide three times the square of the cylinder's diameter in inches by the square of the pump's diameter also in inches, and the quotient will be the depth of the pit in fathoms.

Required the depth that a cylinder of 36 inches diameter will work a pump of 12 inches diameter.

$$\sqrt{\frac{36^2 \times 3}{144}} = 27 \text{ fathoms.}$$

An inelastic body of 30 lbs. weight, moves with a 3 feet velocity, and is struck by another inelastic body having a 7 feet velocity, the two will then proceed, after the blow, with the velocity

$$v = \frac{50 \times 7 + 30 \times 3}{50 + 30} = \frac{350 + 90}{80} = \frac{44}{8} = \frac{11}{2} = 5\tfrac{1}{2} \text{ feet.}$$

To cause a body of 120 lbs. weight to pass from a velocity $c_2 = 1\frac{1}{2}$ feet into a 2 feet velocity $v$, it is struck by a heavy body of 50 lbs., what velocity will the body acquire? Here

$$c_1 = v + \frac{(v - c_2)\,M_2}{M_1} = 2 + \frac{(2 - 1{\cdot}5) \times 120}{50} = 2 + \frac{6}{5} = 3{\cdot}2$$

feet.

Two perfectly elastic spheres, the one of 10 lbs. the other of 16 lbs. weight, impinge with the velocities 12 and 6 feet against each other, what will be their velocities after impact? Here $M_1 = 10$ and $c_1 = 12$ feet, but $M_2 = 16$ and $c_2 = -6$ feet, hence the loss of velocity of the first body will be

$$c_1 - v_1 = \frac{2 \times 16\,(12 + 6)}{10 + 16} = \frac{2 \times 16 \times 18}{26} = 22{\cdot}154 \text{ feet; and}$$

the gain in velocity of the other, $v_2 - c_2 = \dfrac{2 \times 10 \times 18}{26} = 13{\cdot}846$ feet. From this the first body after impact will recoil with the velocity $v_1 - 12 \quad 22{\cdot}154 = -10{\cdot}154$ feet; and the other with that of $-6 + 13{\cdot}846 = 7{,}846$ feet. Moreover, the measure of *vis viva* of the two bodies after impact $= M_1 v_1^2 + M_2 v_2^2 = 10 \times 10{\cdot}154^2 + 16 \times 7{\cdot}846^2 = 1031 + 985 = 2016$, as likewise of that before impact, namely: $M_1 c_1^2 + M_2 c_2^2 = 10 \times 12^2 + 16 \times 6^2 = 1440 + 576 = 2016$. Were these bodies inelastic, the first would only lose in velocity $\dfrac{c_1 - v_1}{2} = 11{\cdot}077$ feet, and the other gain $\dfrac{v_2 - c_2}{2} = 6{\cdot}923$ feet; the first would still retain, after impact, the velocity $12 - 11{\cdot}077 = 0{\cdot}923$ feet, and the second take up the velocity $-6 + 6{\cdot}923 = 0{\cdot}923$, and the loss of mechanical effect would be $(2016 - (10 + 16)\,0{\cdot}923^2) \div 2g = (2016 - 2{\cdot}22) \times 0{\cdot}0155 = 29{\cdot}35$ ft. lbs.

## CENTRIPETAL AND CENTRIFUGAL FORCE.

1. WHAT is the centrifugal force of a body weighing 20 lbs. that describes a circle of 10 feet radius 200 times in a minute?

$\cdot000331 \times 200^2 \times 20 \times 10 = 2648$ lbs., the centrifugal force. $\cdot00331$ is a constant number.

It is a well established fact that the centrifugal force is to the weight of the body as double the height due to the velocity is to the radius of revolution. Hence, this question may be thus solved:

$20 \times 3\cdot1416 = 62\cdot832$, the circumference of the circle of 10 feet radius.

$62\cdot832 \times 200 = 12566\cdot4$ feet, the space passed over by the weight in one minute.

$\frac{12566\cdot4}{60} = 209\cdot44$ feet, the space described in a second, which is called the velocity.

$\frac{(209\cdot44)^2}{64\cdot4} = 681\cdot136$ feet, the height due to the velocity.

If F be the centrifugal force—

$$F : 20 :: 1362\cdot272 : 10.$$

$\therefore F = \frac{1362\cdot272 \times 20}{10} = 2724\cdot544$ lbs. The former rule gives 2648 lbs.

2. What is the centrifugal force at the equator on a body weighing 300 lbs., supposing the radius of the earth $= 21000000$ feet, and the time of rotation $= 86400'' = 24$ hours?

$F = 1\cdot224 \times \frac{21000000 \times 300}{86400^2} = 1\cdot03298$ lbs., or one pound very nearly. $1\cdot224$ is a constant multiplier.

$3\cdot1416 \times 21000000 = 65973600$ feet, $\frac{1}{2}$ the circumference of the earth at the equator.

$\frac{2 \times 65973600}{86400} = 1527\cdot16$ feet, the velocity of the weight each second.

$\frac{(1527\cdot16)^2}{64\cdot4} = 36214\cdot56$, the height due to the velocity.

$$F : 300 :: 72429\cdot12 : 21000000.$$

$F = \frac{72429\cdot12 \times 300}{21000000} = 1\cdot0347$ nearly, as by the former approximate method.

3. If a body weighing 100 lbs. describe a circle of 10 feet radius 300 times a minute, what is the diameter of a cast iron cylindrical

rod, connecting the body with the axis, that will safely support this weight? The centrifugal force will be,

$$\cdot 000331 \times 300^2 \times 100 \times 10 = 29790 \text{ lbs.}$$

From the strength of materials, page 281, we find that the ultimate cohesive strength for each circular inch of cross sectional area is 14652 lbs.; but one-third of this weight, or 4884 lbs., can only be applied with safety.

$$\therefore \sqrt{\frac{29790}{4884}} = 2\cdot 46982 \text{ inches, the diameter of the cylindrical rod.}$$

4. The dimensions, the density, and strength of a millstone ABDE are given; it is required to find the angular velocity $v$, in consequence of which rupture will take place on account of the centrifugal force.

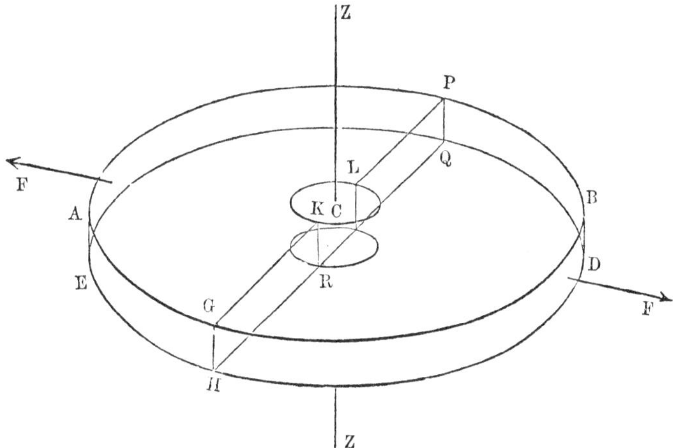


If we put the radius of the millstone $= r_1 = 24$ inches $=$ CG; the radius $=$ CK of its eye $= r_2 = 4$ inches; the height PQ $=$ GH $= l = 12$ inches; the density $= t = 2500 =$ specific gravity of the millstone; and the modulus of strength $=$ K $=$ 750 lbs. $=$ the ultimate cohesive strength of each square inch of cross sectional area in the section PH, supposing the centrifugal forces $-$ F and $+$ F to cause the separation in this section.

$$(r_1 - r_2)\, l = \text{area of parallelogram GR.}$$

Hence, the force in lbs. required to cause rupture will be,

$2\,(r_1 - r_2)\, l \times \text{K}$; the weight of the stone $\text{G} = \pi\,(r_1^2 - r_2^2)\, l\gamma$, and the radius of gyration of each half of the stone, i. e. the distance of its centre of gravity from the axis of rotation $r = \frac{4}{3\pi} \times \frac{r_1^3 - r_2^3}{r_1^2 - r_1^2}$.
At the moment of rupture, the centrifugal force of half the stone is equivalent to the strength; we hence obtain the equation of con-

dition $\omega \times \frac{1}{2} \frac{G r}{g} = 2 (r_1 - r_2) l K$, i. e. $\omega^2 \times \frac{2}{3} (r_1^3 - r_2^3) \frac{l \gamma}{g} = 2 (r_1 - r_2) l K$; or leaving out $2 l$ on both sides, it follows that

$$\omega = \sqrt{\frac{3 g (r_1 - r_2) K}{(r_1^3 - r_2^3) \gamma}} = \sqrt{\frac{3 g K}{(r_1^2 + r_1 r_2 + r_2^2) \gamma}}.$$

If $r_1 = 2$ feet $= 24$ inches, $r_2 = 4$ inches, $K = 750$ lbs., and the specific gravity of the millstone $= 2{\cdot}5$; therefore the weight of a cubic inch of its mass $= \frac{62{\cdot}5 \times 2{\cdot}5}{1728} = 0{\cdot}0903$ lbs.; it follows that the angular velocity at the moment of rupture is,

$$\omega = \sqrt{\frac{3 \times 12 \times 32{\cdot}2 \times 750}{688 \times 0{\cdot}9903}} = \sqrt{\frac{869400}{62{\cdot}1264}} = 112{\cdot}1 \text{ inches.}$$

If the number of rotations per minute $= n$, we have then $\omega = \frac{2 \pi n}{60}$; hence, inversely, $n = \frac{30 \omega}{\pi}$, but here $= \frac{30 \times 112{\cdot}1}{\pi} = 1070$. The average number of rotations of such a millstone is only 120, therefore 9 times less.

With what velocity must a body of 8 lbs. impinge against another at rest of 25 lbs., in order that the last may have a velocity of 2 feet? Were the bodies inelastic, we should then have to put: $v = \frac{M_1 c_1}{M_1 + M_2}$, i. e. $2 = \frac{8 \times c_1}{8 + 25}$, hence $c_1 = \frac{33}{4} = 8\frac{1}{4}$ feet, the required velocity; but were they elastic, we should have $v_2 = \frac{2 M_1 c_1}{M_1 + M_2}$; hence, $c_1 = \frac{33}{8} = 4\frac{1}{8}$ feet.

If in a machine, 16 blows per minute take place between two inelastic bodies $M_1 = \frac{1000}{g}$ lbs. and $M_2 = \frac{1200}{g}$ lbs., with the velocities $c_1 = 5$ feet, and $c_2 = 2$ feet, then the loss in mechanical effect from these blows will be: $L = \frac{16}{60} \times \frac{(5 - 2)^2}{2 g} \times \frac{1000{\cdot}1200}{2200} = \frac{4}{15} \times 9 \times \frac{1}{64{\cdot}4} \times \frac{6000}{11} = 0{\cdot}576 \times \frac{400}{11} = 20{\cdot}94$ units of work per second.

If two trains upon a railroad of 120000 lbs. and 160000 lbs. weight, come into collision with the velocities $c_1 = 20$, and $c_2 = 15$ feet, there will ensue a loss of mechanical effect expended upon the destruction of the locomotives and carriages, which in the case of perfect inelasticity of the impinging parts, will amount to

$$= \frac{(20 + 15)^2}{2 g} \times \frac{120000 \times 160000}{280000} = 35^2 \times \frac{1}{64{\cdot}4} \times \frac{1920000}{28} =$$

1344000 ft. lbs., or units of work.

## SHIP-BUILDING AND NAVAL ARCHITECTURE.

Two rules, by which the principal calculations in the art of ship-building are made, may be employed to measure the area or superficial space enclosed by a curve, and a straight line taken as a base.

Rule I.—If the area bounded by the curve line ABC and the straight line AC is required to be estimated, by the rule, the base AC is divided into an even number of equal parts, to give an odd number of points of division.

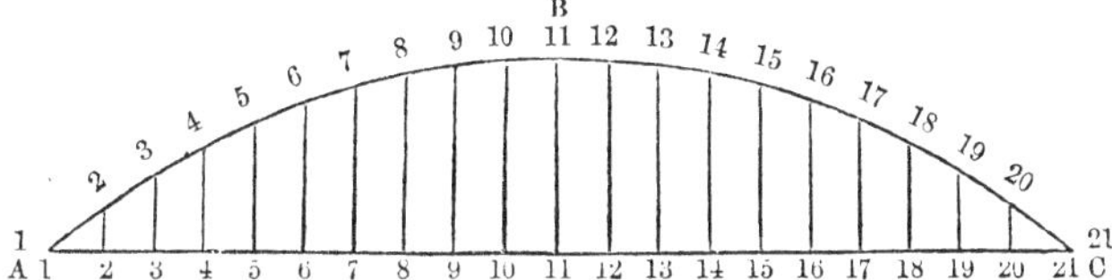


Where the base AC is divided into twenty equal parts, giving twenty-one points of division, and the lines 1·1, 2·2, 3·3, &c., are drawn from these points at right angles or square to AC, to meet the curve ABC, these lines, 1·1, 2·2, 3·3, &c., are denominated ordinates, and the linear measurement of them, on a scale of parts, is taken and used in the following general expression of the rule.

$$\text{Area} = \{A + 4P + 2Q\} \frac{r}{3}.$$

Where A = sum of the first and last ordinates, or 1·1 and 21·21.

4 P = sum of the even ordinates multiplied by 4.

Or, {2d + 4th + 6th + 8th + 10th + 12th + 14th + 16th + 18th + 20th} × 4.

2 Q = sum of the remaining ordinates; or,

{3d + 5th + 7th + 9th + 11th + 13th + 15th + 17th + 19th} × 2.

And $r$ is equal to the linear measurement of the common interval between the ordinates, or one of the equal divisions of the base AC. This rule, for determining the area contained under the curve and the base, may be put under another form; for as the

$\text{Area} = \{A + 4P + 2Q\} \times \frac{r}{3}$; it may be transferred into

$$\text{Area} = \left\{\frac{A}{2} + 2P + Q\right\} \times \frac{2r}{3}.$$

The practical advantages to be derived from this modification of the general rule will appear when the methods of calculation are further developed.

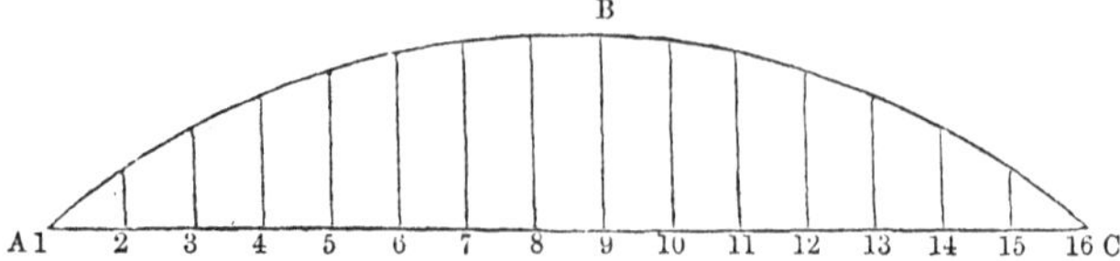


RULE II.—If the base AC be so divided that the equal intervals are in number a multiple of the numeral 3, then the total number of the points of division, and consequently the ordinates to the curve, will be a multiple of the numeral 3 with one added, and the area under the curve ABC, and the base AC, can be determined by the following general expression:

$$\text{Area} = \{A + 2\,P + 3\,Q\} \times \frac{3r}{8}.$$

Where A = sum of the first and last ordinates, or 1 and 16.

2 P = sum of the 4th, 7th, 10th, 13th, multiplied by 2, or ordinates bearing the distinction of being in position as multiples of the numeral 3, with one added.

3 Q, the sum of the remaining ordinates, multiplied by 3, or of the 2d, 3d, 5th, 6th, 7th, 8th, 9th, 11th, 12th, 14th, and 15th, multiplied by 3.

The number of equal divisions for this rule must be either 3, 6, 9, 12, or 15, &c., being multiples of the numeral 3, whence the ordinates will be in number under such divisions, multiples of the numeral 3, with one added.

This rule admits also of a modification in form, to make it more convenient of application.

$$\text{For area} = \{A + 2\,P + 3\,Q\} \times \frac{3}{8}\,r.$$

As before advanced for the change adopted in the general expression for the first rule, the utility of this modification of the second rule will be observable when the calculations on the immersed body are proceeded with.

The rules are formed under the supposition that in the first rule the curve ABC, which passes through the extremities of the ordinates, is a portion of a common parabola, while in the second rule the curve is assumed to be a cubic parabola; the results to be obtained from an indiscriminate use of either of these rules, differ from each other in so trifling a degree, (considered practically and not mathematically,) as not to sensibly affect the deductions derived by them.

William O'Neill, or, as English writers term him, William Neal, was the first to rectify a curve of any sort; this curve was the semi-cubical parabola; these rules, of such use in the art of shipbuilding, were first given by him, but as is usual, claimed by English pretenders.

The foregoing rules, when applied to the measurement of the

immersed portion of a floating body, as the displacement of a ship, are used as follows.

The ship is considered as being divided longitudinally by equidistant athwartship or transverse vertical planes, the boundaries of which planes give the external form of the vessel at the respective stations, and therefore the comparative forms of any intermediate portion of it.

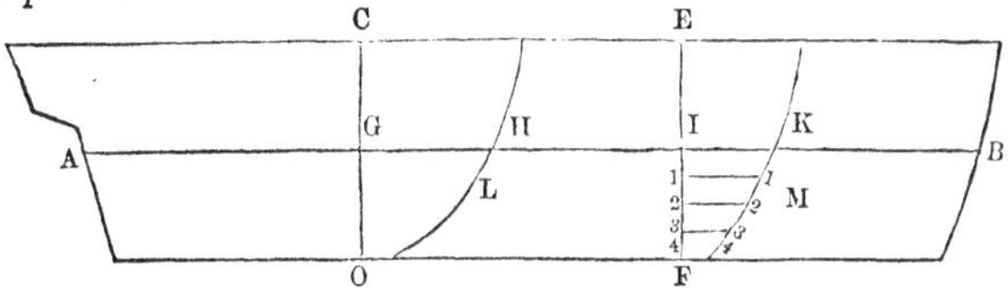

If the ship be immersed to the line AB, considered as the line of the proposed deepest immersion or lading, the curves HLO and KMF would give the external form of the ship at the positions G and I in that line; and the areas GHLO, IKMF contained under the curves HLO, KMF, the right lines GH, IK, (the half-breadths of the plane of proposed flotation AB at the points G and I,) and the right lines GO, IF, the immersed depths of the body at those points are the areas to be measured; and if the areas obtained be represented by linear measurements, and are set off on lines drawn at right angles to the line AB at their respective stations, a curve bounding the representative areas would be formed, and the measurement by the rules of the area contained under this curve, and the right line, AB, or length of the ship on the load-water line, would give the sum of the areas thus represented, and thence the solid contents of the immersed portion of the ship in cubic feet of space. In accordance with this application of those rules to measure the displacement of the ship, the usual practice is to divide the ship into equidistant vertical and longitudinal planes, the longitudinal planes being parallel to the load-water section or horizontal section formed by the proposed deepest immersion.

To measure the areas of these planes after they have been delineated by the draughtsman, the constructor divides the depth of each of the vertical sections, or the length of each horizontal section, into such a number of equal divisions as will make either one or the other of the rules 1 or 2 applicable. If the first rule be preferred, the equal divisions must be of an even number, so that there may be an odd number of ordinates; while the use of the second rule, to measure the area, will require the equal divisions of the base to be in number a multiple of the numeral 3, which will make the ordinates to be in number a multiple of the numeral 3, with one added. From the points of equal divisions in the respective sections thus determined, perpendicular ordinates are drawn to meet the curve, or the external form of the transverse planes of the body; and a table for the ordinates thus obtained, having been made, as shown page 467, the measures by scale of the respective ordinates are therein inserted.

For the area IKMF, the linear measurements of IK, 1·1, 2·2, 3·3, 4·4, are taken by a scale of parts, and inserted in the column marked 5, page 467, the whole length AB of the load-water line being divided into 10 equal divisions, and the area IKMF being supposed as the fifth from B, the fore extreme of the load-water line. To apply the first rule to the measurement of the area of No. 5 section, the ordinates are extracted from the table, page 467, and operated upon as directed by the rule; viz.

$$\text{Area} = \{A + 4\,P + 2\,Q\} \times \frac{r}{3}.$$

| IK, or first, | 1·1, or 2d, | 2·2 or 3d, |
|---|---|---|
| 4·4, or last, | 3·3, or 4th, | × 2. |
| ——— | added together | or 2 Q. |
| added together = A. | and × 4 = 4 P. | |

$$\text{By rule, area} = \{A + 4\,P + 2\,Q\} \times \frac{r}{3}.$$

Whence area $= \{(\text{IK} + 4{\cdot}4) + (1{\cdot}1 + 3{\cdot}3)\,4 + 2{\cdot}2 \times 2\} \times \frac{r}{3} =$ area IKMF; and, in a similar manner, may the several areas of the other transverse sections be determined.

When these areas have all been thus measured, they are to be summed by the same rules; the areas themselves being considered as lines, and the result will give the solid for displacement in cubic feet. To shorten this tedious application of the formula, the arrangement of having double-columned tables of ordinates was introduced, as shown on page 484, and for the more ready use of this enlarged table, the modifications in the formula 467, before alluded to, were adopted, that of

$$\text{Area} = \left\{ A + 4\,P + 2\,Q \right\} \times \frac{r}{3} = \left\{ \frac{A}{2} + 2\,P + Q \right\} \times \frac{2r}{3},$$

and that of

$$\text{Area} = \left\{ A + 2\,P + 3\,Q \right\} \times \frac{3r}{8} = \left\{ \frac{A}{2} + P + 1{\cdot}5\,Q \right\} \times \tfrac{3}{4}\,r,$$

as rendering the required number of figures much less, whereby accuracy of calculation is insured and time is saved.

In using a table of ordinates constructed for this method of calculation, the linear measurement of the several ordinates of vertical section 5 and the corresponding ones of all the others would be inserted in the double columns prepared for them, in the following order:—

In the first and last lines of the enlarged table for the ordinates, distinguishable by $\frac{A}{2}$, in the left-hand column of each pair, the measurements of the first and last ordinates of the respective areas are placed, and in the right-hand column of each pair one-half of such measurements, as being one-half of the first and last ordinates of each vertical section or area. In the lines distinguished by 2 P, in the left-hand column, the measurements of the even ordinates of each respective area are placed, which having been multiplied by two, the result is placed in the respective right-hand columns prepared for each vertical section; while in those lines of the table distinguished by Q, the measurements of the ordinates themselves are placed in the right-hand columns, as not requiring by the modification of the rules any operation to be used on them, before being taken into the sum forming the sub-multiple of the respective areas.

It may here with propriety be suggested, that in practice the insertion of the linear measurements of the ordinates in the table in red ink will be found useful, and that after such has been done, by the upper line of figures in the table of ordinates thus arranged, being divided by two, the second line of figures being multipled by two, and so on with the others as shown by the table, and the results thus obtained being inserted in their respective right-hand columns as before described, great facility and despatch of calculation are afforded to the constructor.

That this method will yield a correct measurement of the areas will be evident by an inspection of the terms of the general expression of area $= \left\{ \frac{A}{2} + 2P + Q \right\} \times \frac{2r}{3}$, which are placed against the several lines of the table of ordinates. And it will be equally apparent, that the sum total of the figures inserted in the right-hand columns appropriated to each section is a sub-multiple of the area of each section, and that these results arising from the use of the form for area of $\left\{ \frac{A}{2} + 2P + Q \right\}$ will be one-half of those that would be obtained by abstracting the ordinates from the table, page 467, and using them in the expression $A + 4P + 2Q$; and therefore to complete the calculation for the areas by the rule, the first results for the areas must be multiplied by $\frac{2r}{3}$, and the last by $\frac{r}{3}$, where $r$ is equal to the common interval or equal division of the base in linear feet; or the part of the expression for areas of $\left\{ \frac{A}{2} + 2P + Q \right\}$ must be multiplied by $\frac{2r}{3}$, to make it equivalent to $\{A + 4P + 2Q\} \times \frac{r}{3}$.

The sub-multiples of the areas of the vertical sections thus determined, require to be summed together for the solid of displacement, and by considering the sub-multiples of the areas to be, as before stated, represented by lines or proportionate ordinates, O'Neill's rules, by the same table of ordinates with an additional column, may be made available to the development of the solid of displacement. For the sectional areas being represented by lines, by the first rule, one-half the first and last areas, added to the sum of the products arising from multiplying the even ordinates or representative areas by two, together with the odd ordinates or the areas as given by the tables, and these being placed in the additional column of the table prepared for them, the sub-multiple of the solid of displacement will be given.

The operation will stand thus: Sub-multiple of each of the areas $= \left\{ \frac{A}{2} + 2P + Q \right\}$, or each area will be $\frac{2r}{3}$ less than the full result, and the representative lines for the areas will be diminished in that proportion; and having used these sub-multiples of the areas thus diminished in the second operation for obtaining the sub-multiple of the solid of displacement under the same rule, the results will again be $\frac{2r'}{3}$ less than the true result; therefore the sum thus determined will have to be multiplied by the quantity $\frac{2r}{3} \times \frac{2r'}{3}$, to give the solid required. In this expression, of $\frac{2r}{3} \times \frac{2r'}{3}$, $r =$ the equal distances taken in the vertical planes to obtain the respective vertical areas; $r' =$ the equal distances at which the vertical areas are apart on the longitudinal plane of the ship.

The displacement being thus determined, by an arrangement of an enlarged table of ordinates, the functions arising from the sub-multiples of the areas of the vertical sections being placed in O'Neill's rules to ascertain the displacement, may be used in the table of ordinates to find the distance of the centre of gravity of the immersed body from any assumed vertical plane; and also the distance that the same point—"the centre of gravity of displacement" —is in depth from the load-water or line of deepest immersion, and that from the considerations which follow:—

In a system of bodies, the centre of gravity of it is found by multiplying the magnitude or density of each body by its respective distance from the beginning of the system, and dividing the sum of such products by the sum of the magnitudes or densities. The displacement of a ship may be considered as made up of a succession of vertical immersed areas; and if it be assumed that the moments arising from multiplying the area of each section by its relative distance from an initial plane may be represented by successive lineal measurements, the general rules will furnish the summation of such moments; and the displacement or sum of the areas has been obtained by a similar process, from whence, by the rule for finding the centre of gravity of a system as before given, the distance of the common centre of gravity from the assumed initial plane would be ascertained, by dividing the sum of the moments of the areas by the sum of the areas or the solid of displacement.

To extend this reasoning to the enlarged table of ordinates used for the second method of calculation: The sub-multiples of the respective areas, when put into the formulas to obtain the proportionate solid of displacement, are relatively changed in value to give that solid, and consequently the moments of such functions of the vertical areas will be to each other in the same ratio; and the sum of these proportionate moments, if considered as lines, can be ascertained by multiplying the functions of the areas by their relative distances from the assumed initial plane, or by the number of the equal intervals of division they are respectively from it, and afterwards, by the rules, summing these results, forming the sum of the moments of the sub-multiples of the functions of the vertical areas: and the proportionate sub-multiple for the displacement is shown on the table; the division therefore of the former, or the sum of the proportional moments of the functions of the areas, by the proportionate sub-multiple for the displacement, will give the distance (in intervals of equal division) that the centre of gravity of the displacement is from the initial plane, which being multiplied by the value in feet of the equal intervals between the areas, will give the distance in feet from the assumed initial plane, or from the extremity of the base line of the proportional sectional areas for displacement. This reasoning will apply equally to finding the position of the centre of gravity of the body immersed, both as respects length and depth, and on the enlarged tables for construction given, (pages 484 and 485,) the constructor, by adopting this arrangement, will at once have under his observation the calculations *on*, and the results *of*, the most important elements of a naval construction.

The foregoing tabular system, for the application of O'Neill's rules to the calculations required on the immersed volume of a ship's bottom, led to a lineal delineation of the numerical results of the tables, and thence the development of a curve of sectional areas, on a base equivalent to the length of the immersed portion of the body, or of the length at the load-water line. To effect this, the sub-multiples of the sectional areas, taken from the tables for calculation, are severally divided by such a constant number as to make their delineation convenient; then these thus further reduced sub-multiples of the areas, being set off at their respective positions on the base, formed by the length of the load-water line, a curve passed through the extreme points of these measurements, will bound an area, that to the depth used for the common divisor would form a zone, representative of the solid of displacement. The accuracy of such a representation will be easily admitted, if the former reasoning is referred to.

To obtain the solid of displacement from this representative area, the load-water line or plane of deepest immersion is considered as being divided lengthwise into two equal parts, which assumption divides the base of the curve of sectional areas also into two equal portions: the line of representative area to that medial point is then drawn to the curve, and triangles are formed on each side of it by joining the point where it meets the curve with the extremities of the base line; this arrangement divides the representative area into four parts, two triangles which are equal, viz. 1 and 2, and two other areas which are contained under the hypothenuse of these triangles and the curves of sections, or 3 and 4 of the annexed diagram.

*Diagram of a Curve of Sectional Areas.*

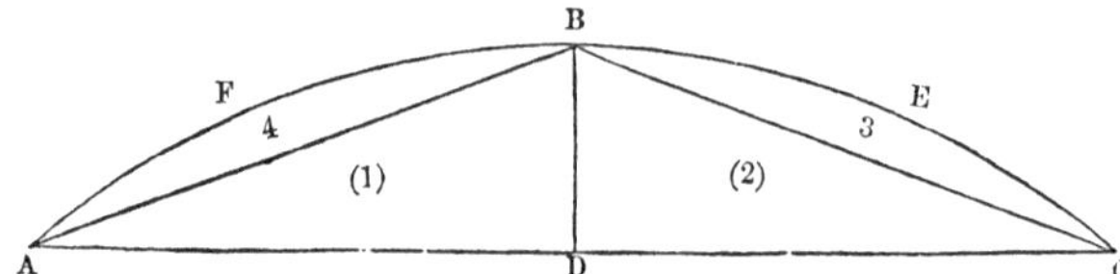


ABCDA equal sectional area, representative of the half displacement as a zone of a given common depth.

AC equal the length of the load-water section from the fore-side of the rabbet of the stem to the aft-side of the rabbet of the post, and D the point of equal division.

BD, the representative area of half the immersed vertical section at the medial point D, joining B with the points A and C, will complete the division of the representative area ABCDA.

ABD and CBD, under such considerations, are equal triangles.

BECB, BFAB, areas, bounded respectively by the hypothenuse AB or BC of the triangles and the curve of sectional areas; and, supposing the curves AFB and BEC to be portions of common parabolas, the solid of displacement will be in the following terms:

The area of each of the triangles is equal to $\frac{1}{4}$ of AC $\times$ BD; hence the sum of the two $= \frac{1}{2}$ of AC $\times$ BD: the hypothenuse AB or BC $= \sqrt{\left[\left(\frac{AC}{2}\right)^2 + BD^2\right]}$, and the area of BECB if considered as approximating to a common parabola $= \sqrt{\left[\left(\frac{AC}{2}\right)^2 + BD^2\right]}$ $\times$ $\frac{2}{3}$ of the greatest perpendicular on the hypothenuse BC.

Area of BFAB under the same assumption $= \sqrt{\left[\left(\frac{AC}{2}\right)^2 + BD^2\right]}$ $\times$ $\frac{2}{3}$ of the greatest perpendicular on the hypothenuse AB; whence the whole displacement will be expressed by $\frac{1}{2}$ AC $\times$ BD $\times$ $\sqrt{\left[\left(\frac{AC}{2}\right)^2 + BD^2\right]}$ $\times$ $\frac{2}{3}$ of the greatest perpendicular on the hypothenuse B C $+$ $\sqrt{\left[\left(\frac{AC}{2}\right)^2 + BD^2\right]}$ $\times$ $\frac{2}{3}$ of the greatest perpendicular on the hypothenuse AB.

By a similar method, from the light draught of water, or the depth of immersion on launching the ship, the light displacement, or the weight of the hull or fabric, may be delineated and estimated; and the representative curve for it being placed relatively on the same base as that used for the representative curve for the load displacement, the area contained between the curve bounding the representative area for the load displacement, and the curve bounding the representative area for the light displacement, will be a representative area of the sum of the weights to be received on board, and point out their position to bring the ship from the light line of flotation, or the line of immersion due to the weight of the hull when completed in every respect, to that of the deepest immersion, or the proposed load-water line of the constructor—a representation that would enable the constructor to apportion the weights to be placed on board to the upward pressure of the water, and thence approximate to the stowage that would insure the easiest movements of a ship in a sea.

By an inspection of the diagram of the curve of sectional areas, it will clearly be seen that the representative area for displacement under the division of it, into the triangles 1 and 2, and parabolic portions of the area 3 and 4, will point out the relative capacities of the displacement, under the fore and after half-lengths of the base or load-water line; for, by construction, the triangles ABD and CBD are equal, and therefore the comparative values of the areas BECB and BFAB, or of $\sqrt{\left[\left(\frac{AC}{2}\right)^2 + BD^2\right]} \times \frac{2}{3}$ of the greatest perpendicular on the hypothenuse BC, compared with $\sqrt{\left[\left(\frac{AC}{2}\right)^2 + BD^2\right]} \times \frac{2}{3}$ of the greatest perpendicular on the hypothenuse AB, or of the relative values of the greatest perpendiculars on the hypothenuses BC and AB, will give the relative capacities of the fore and after portions of the immersed body or the displacement.

The representative area ABCDA admits also of a measurement by the second rule.

Let BD, as before, be the representative area at the middle point.

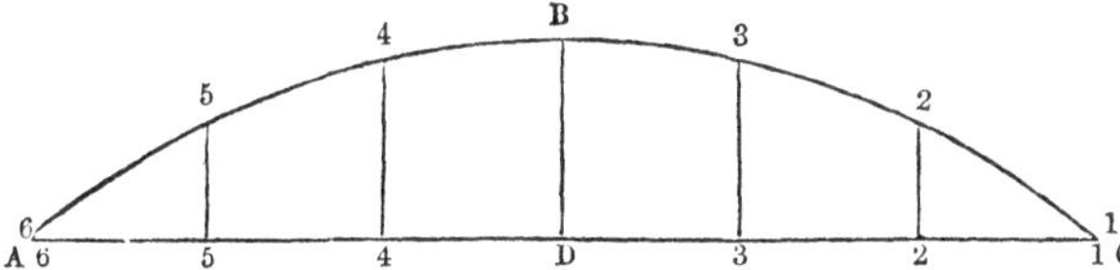


Divide AD or DC into three equal portions, then the equal divisions being a multiple of 3, the second rule is applicable to measure the areas ABDA or BCDB; for the area ABDA =

$$\left\{ 6{,}6 + BD + 2 \times 0 + 3\ \{\overline{4{,}4 + 5{,}5}\} \right\} \frac{3r}{8};\ 6{,}6 = 0;$$

$$= \left\{ BD + 3\ \{4{,}4 + 5{,}5\} \right\} \frac{DC}{8};\ \text{and area BCDB} =$$

$$\left\{ \overline{1{,}1 + BD} + 2 \times 0 + 3 \times \{2{,}2 + 3{,}3\} \right\} \frac{3r}{8},\ \text{where } 1{,}1 = 0$$

$$\left\{ BD + 3 \times \{\overline{2{,}2 + 3{,}3}\} \right\} \frac{AD}{8} = \text{BCDB, and the displace-}$$

$$\text{ment} = \left\{ BD + 3 \times \{4{,}4 + 5{,}5\} \right\} \frac{DC}{8} + \left\{ BD + 3 \times \{2{,}2 + 3{,}3\} \right\}$$

$\times \frac{AD}{8} \times$ by the constant divisor of the areas, or the depth of the zone in feet.

The rules given by O'Neill for the measurement of the immersed portion of the body of a ship, having been theoretically stated, the practical application of them will be given on the construction.

The immersed part of a ship, being a portion of the parallelopipedon formed by the three dimensions;—length on the load-water line, from the foreside of the rabbet of the stem to the aftside of the rabbet of the stern-post; extreme breadth in midships of the load-water section; and depth of immersion in midships from the lower edge of the rabbet of the keel;—it would seem that the first step towards the reduction of the parallelopipedon, or oblong, into the required form, would be to find what portion of it would be of the same contents as the proposed displacement of the ship—a knowledge of which would enable the constructor, by a comparison of the result with a similar element of an approved ship, to determine whether the principal dimensions assumed would (under the form intended) give an immersed body equal to carrying the proposed weights or lading.

The relative capacities of the immersed bodies contained under the fore and after lengths of the load-water line must next be fixed, and the constructor in this very important element of a construction will find little to guide him from the results of past experience and practice. From deductions on approved ships of rival constructors it will be developed, that in this essential element, "the relative difference between the two bodies," they vary from 1 to 13 per cent. on the whole displacement.

The relative capacities of the fore and after bodies of immersion under the proposed load-water line would seem at the first glance of the subject to be a fixed and determinate quantity, as being a conclusion easily arrived at from a knowledge of the proportions due to the superincumbent weights—under such a consideration, the weight of the anchors, bowsprit, and foremast would necessarily be supposed to require an excess in the body immersed under the fore half-length of the load-water line over that immersed under the after half-length of the same element.

In a ship, the necessary arrangement of the weights, to preserve the proposed relative immersion of the extremes or the intended draught of water, would be pointed out by a delineated curve of sectional areas, described as before directed; but a want of that system, or of some other, has often caused an error in the actual draught of water, and that under a great relative excess of the volumes of displacement in the fore and after portions of the immersed body.

The men-of-war brigs built to a construction-draught of water 12 ft. 9 in. forward, 14 ft. 4 in. abaft, giving 1 ft. 7 in. difference, had under such a construction a difference of displacement between the immersed bodies under the fore and after half-lengths of the load-water line that was equivalent to 10·4 tons for every 100 tons of the vessel's total displacement or weight; but these ships, when stowed and equipped for sea, came to the load-draught of water of 14 ft. 2 in. forward, 14 ft. 3 in. aft,—difference 1 inch, or an immersion of the fore extreme of 18 inches more than was intended by the constructor. The reason of this practical departure from the proposed line of flotation of the constructor was, that the internal space or hold of the ship necessarily followed the external form, giving a hold proportionate to the displacement contained under the several portions of the body; but an injudicious disposal of the stores (in placing the weights too far forward) made them more than equivalent to the upward pressure of the water at the respective portions of the proposed immersion of the body, and thence arose the error or excess in the fore immersion by giving a greater draught of water than was designed. The stowage of a ship's hold, under a reference to the representative area for the displacement, contained between the curves of sectional areas developed for the light and load displacements, would prevent similar errors under any extent to which the relative capacity of the two bodies might be carried. This relative capacity of the two bodies will affect the form of the vessel's extremes, giving a short or long bow, a clear or full run to the rudder; for the whole displacement *being a fixed quantity, if the portion of it under the fore half-length of the load-water line be increased*, it must be followed *by a proportionate diminution of the portion of the displacement under the after half-length of the load-water line, so that the total volume of the displacement may remain the same*, which arrangement will give a proportionately full bow and clean run, and *vice versâ.*

The curve of sectional areas under the foregoing considerations is also applicable to a comparison of the relative qualities of ships of the same rate, by showing at one view the distribution of the volume of displacement in each ship, under the draught of water which has been found on trial to give the greatest velocity; based on which, deductions may be made from the relative capacities of the bodies pointed out by the sectional curves, that will serve to guide the naval constructor in future constructions.

The curve of sectional areas is also available for forming a scale to measure the amount of displacement of a ship to any assumed or given draught of water. To effect this, on the sheer draught or longitudinal plan of the ship between the load-water line, or that of deepest immersion, and the line denoting the upper edge of the rabbet of the keel, draw intermediate lines parallel to the load-water line as denoting lines of intermediate immersion between the keel and load-water line; these lines may be placed equidistant from each other, but they are not necessarily required to be so. Find the curve of sectional areas, due to each immersion of the ship denoted by these lines, and measure the areas bounded respectively by these curves, in the manner as before directed for the load displacement: these results will give the magnitudes of the immersed portions of the body in cubic feet, which being divided by 35, the mean of the number of cubic feet of salt or fresh water that are equivalent to a ton in weight, will give their respective weights in tons.

Assume a line of scale for depth, or mean draught of water, the lower part of which is to be considered the underside of the false keel of the ship, and set off on this line, by means of a scale of parts, the depths of the immersions at the middle section of the longitudinal plan; draw lines (at the points thus obtained) perpendicular to this assumed line for depth or draught of water, and having determined a scale to denote the tons, set off on each line by this scale the tons ascertained by the curves of sectional areas to be due to the respective immersions of the body; then a curve passed through these points will be one on which the weights in tons due to the intermediate immersions of the body may be ascertained; or, the displacement of a ship to the mean of a given draught may be found by setting up the mean depth on the scale, showing the draught of water—transferring that depth to the curve for tonnage, and then carrying the point thus obtained on the curve for tonnage to the scale of tons, which will give the number of tons of displacement to that depth of immersion or draught of water.

*Description of the several plans to be delineated by the draughtsman, previous to the commencement of the calculations.*

*Sheer Plan.*—A projection of the form of the vessel on a longitudinal and vertical plane, assumed to pass through the middle of the ship, and on which the position of any point in her may be fixed with respect to height and length.

*Half-breadth Plan.*—The form of the vessel projected on to a longitudinal and horizontal plane, assumed to pass through the extreme length of ship, and on which the position of any point in the ship may be fixed for length and breadth.

*Body Plan.*—The forms of the vertical and athwartship sections of the ship, projected on to a vertical and athwartship plane, assumed to pass through the largest athwartship and vertical section of her, and on which plan the position of any point in the ship may be fixed for height and breadth.

These plans conjointly will determine every possible point required; for, by inspection, it will be found—

That the sheer and half-breadth plans have
one dimension common to both, viz.:..........Length.
Half-breadth and body plane.......................Breadth.
Sheer and body plane...............................Height.

For sheer plan gives length and height...... }
Half-breadth plan gives length and breadth } of the same point.
Body plan gives breadth and height......... }

Which dimensions form the co-ordinates for any point in the solid, and must determine the position of it.

The point C in the load-water section AB, has for its co-ordinates to fix its position,

The length, 1·5 of the half-breadth plan.
Height, 5·C of the sheer plan,
And the breadth, 1·C of the body plan of section.

And the same for any other point of the solid or of the ship.

In the sheer plan, AB represents the line of deepest immersion, *a a*, *b b*, *c c*, *d d*, lines drawn parallel to that line at a distance of ·9 feet, making with AB an odd number of ordinates for the use of the first general rule for the area, where area $= \{A + 4P + 2Q\} \times \frac{r}{3}$,
and A = the sum of the first and last ordinates.
P = the sum of the even ordinates, as 2, 4.
Q = the sum of the odd ordinates, as 3, &c.

The line AB, or length of the load-water line, is bisected at C, and AC, CB are thence equal; C being the middle point of the load-water line, the spaces BC, AC are again divided into four equal divisions, giving five ordinates for each space, at a distance apart of 5·5 feet.

This arrangement will give the immersed body of the vessel, as being divided into two parts under an equal division of the load-water line, and an odd number of ordinates in each section of the body for the application of the first general rule given for finding the areas of the vertical sections and thence the displacement.

The half-breadth plan delineates the form of the body immersed for length and breadth, the line AB of the sheer plan being represented in the half-breadth plan by the line marked AB, and *a a*, *b b*, *c c*, *d d*, of the sheer plan by the lines similarly distinguished in the half-breadth plan.

The body plan gives the form of the body in the depth, the lines distinguished 5·5 in the sheer and half-breadth plans being in the body plan developed by the curve 5·5·5, giving the external form of the ship at the section 5·5; the same reasoning applies to the other divisions of the load-water line AB.

---

A pile of 400 lbs. weight is driven by the last round of 20 blows of a 500 lbs. heavy ram, falling from a height of 5 feet; 6 inches deeper, what resistance will the ground offer, or what load will the pile sustain without penetrating deeper?

Here $G = 400$, $G_1 = 700$ lbs., $H = 5$, and $s = \frac{0{\cdot}5}{20} = 0{\cdot}025$ feet, whereby it is supposed that the pile penetrates equally far for each blow.

$$P = \left(\frac{700}{700 + 400}\right)^2 \frac{400 \times 5}{0{\cdot}025} = \left(\frac{7}{11}\right)^2 \times 80000 = 32400 \text{ lbs.},$$

the ram, not during penetration, remaining upon the pile.

$$P = \frac{700^2 \times 5}{1100 \times 0{\cdot}025} = \frac{4900}{11} \times 200 = 89100 \text{ lbs.},$$ the ram remaining upon the pile during penetration.

For duration, with security, such piles are only loaded from $\frac{1}{100}$ to $\frac{1}{10}$ of their strength.

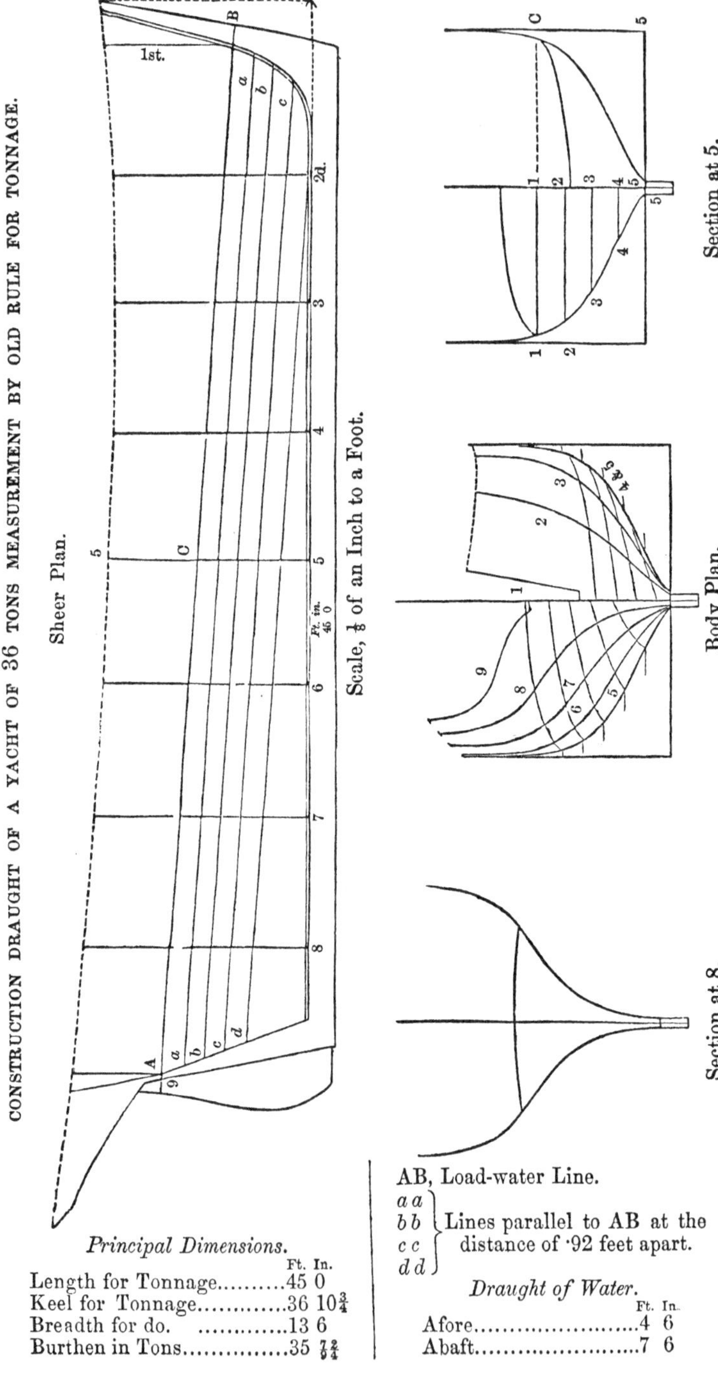


CONSTRUCTION DRAUGHT OF A YACHT OF 36 TONS MEASUREMENT BY OLD RULE FOR TONNAGE.

*Principal Dimensions.*

| | Ft. | In. |
|---|---|---|
| Length for Tonnage.......... | 45 | 0 |
| Keel for Tonnage............. | 36 | 10¾ |
| Breadth for do. ............. | 13 | 6 |
| Burthen in Tons............... | 35 | $\frac{72}{94}$ |

AB, Load-water Line.

*a a*, *b b*, *c c*, *d d* } Lines parallel to AB at the distance of ·92 feet apart.

*Draught of Water.*

| | Ft. | In. |
|---|---|---|
| Afore........................ | 4 | 6 |
| Abaft........................ | 7 | 6 |

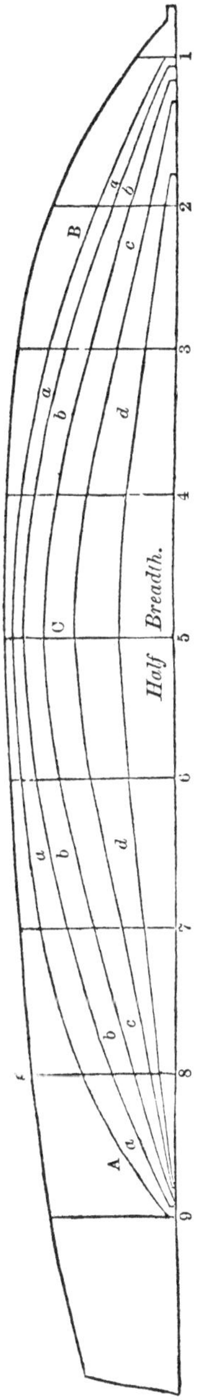

*Calculations required for the construction drawing of a yacht of 36 tons.—1st. Usual mode of calculating the displacement by vertical and horizontal sections.*

TABLE *of Ordinates for Yacht of* 36 *Tons.*

| Distinguishing No. of the sections. | 1 | 2 | 3 | 4 | (5) | 6 | 7 | 8 | 9 | |
|---|---|---|---|---|---|---|---|---|---|---|
| 1′ A | ·4 | 3·0 | 5·0 | 6·0 | 6·3 | 6·1 | 5·4 | 3·7 | ·4 | $r$ = the distance between the ordinates used for the vertical section = ·92 feet. |
| 2′ P | ·35 | 2·4 | 4·2 | 5·6 | 5·6 | 5·5 | 4·4 | 2·6 | ·35 | |
| 3′ Q | ·3 | 1·7 | 3·2 | 4·4 | 5·0 | 4·6 | 3·4 | 1·7 | ·3 | $r'$ = the distance between the ordinates used for the horizontal sections = 5·5 feet. |
| 4′ P | ·25 | 1·0 | 2·2 | 3·2 | 3·8 | 3·4 | 2·4 | 1·1 | ·25 | |
| 5′ A | ·2 | ·4 | 1·3 | 2·0 | 2·4 | 2·0 | 1·4 | ·6 | ·2 | |

From this Table the following application of O'Neill's rule, No. 1, is usually made to obtain the volume of displacement to the draught of water shown on the drawing as the load-water line, or line of proposed deepest immersion, designated by AB.

General terms of the rule :—

$$\text{Area} = \left\{ A + 4P + 2Q \right\} \times \frac{r}{3}.$$

To find $\frac{1}{2}$ the area of vertical section 1, fore body :—

A = sum of the first and last { ·4, ·2 } ·6 = A

4 P = four times the sum of the even ordinates, or of (2) and (4)...... { ·35, ·25 } ·60 = P; × 4· = 2·4 = 4 P

2 Q = twice the sum of the odd ordinates, or of (3) } ·3 = Q; × 2 = ·60 = 2 Q

Whence the area, which is equal to

$$\left\{ A + 4P + 2Q \right\} \times \frac{r}{3} = \left\{ \cdot 6 + 2{\cdot}4 + \cdot 6 \right\} \times \frac{\cdot 92}{3}.$$

$3{\cdot}6 \times \frac{\cdot 92}{3} = 1{\cdot}2 \times \cdot 92 = 1{\cdot}104 = \frac{1}{2}$ area of section 1.

Which sum is half the area of the section 1, and is kept in that form of the half-measurement for the convenience of calculation.

FORE BODY.

*Vertical Section 2.*

| | | |
|---|---|---|
| 3·0 | 2·4 | 1·7 |
| ·4 | 1·0 | 2 |
| 3·4 = A | 3·4 = P | 3·4 = 2 Q |
| | 4 | |
| | 13·6 = 4 P | |
| | 3·4 = A | |
| | 3·4 = 2 Q | |
| | 20·4 = A + 4 P + 2 Q | |
| | ·92 = $r$ | |
| | 408 | |
| | 1836 | |
| | 3) 18·768 | |
| | 6·256 = ½ area of Section 2. | |

*Vertical Section 3.*

| | | |
|---|---|---|
| 5·0 | 4·2 | 3·2 |
| 1·3 | 2·2 | 2 |
| 6·3 = A | 6·4 = P | 6·4 = 2 Q |
| | 4 | |
| | 25·6 = 4 P | |
| | 6·3 = A | |
| | 6·4 = 2 Q | |
| | 38·3 = A + 4 P + 2 Q | |
| | ·92 = $r$ | |
| | 766 | |
| | 3447 | |
| | 3) 35·236 | |
| | 11·745 = $\overline{A + 4P + 2Q} \times \frac{r}{3}$ = ½ area of Section 3. | |

*Vertical Section 4.*

| | | |
|---|---|---|
| 6·0 | 5·6 | 4·4 |
| 2·0 | 3·2 | 2 |
| 8·0 = A | 8·8 = P | 8·8 = 2 Q |
| | 4 | |
| | 35·2 = 4 P | |
| | 8·0 = A | |
| | 8·8 = 2 Q | |
| | 52·0 = A + 4 P + 2 Q | |
| | ·92 = $r$ | |
| | 1040 | |
| | 4680 | |
| | 3) 47·840 | |
| | 15·946 = $\overline{A + 4P + 2Q} \times \frac{r}{3}$ = ½ area of Section 4. | |

*Vertical Section 5.*

$$\begin{array}{l} 6{\cdot}3 \\ 2{\cdot}4 \\ \hline 8{\cdot}7 = A \end{array} \qquad \begin{array}{r} 5{\cdot}6 \\ 3{\cdot}8 \\ \hline 9{\cdot}4 = P \end{array} \qquad \begin{array}{r} 5{\cdot}0 \\ 2 \\ \hline 10{\cdot}0 = 2\,Q \end{array}$$

$$\begin{array}{rl} 9{\cdot}4 & = P \\ 4 & \\ \hline 37{\cdot}6 & = 4\,P \\ 8{\cdot}7 & = A \\ 10{\cdot}0 & = 2\,Q \\ \hline 56{\cdot}3 & = A + 4\,P + 2\,Q \\ {\cdot}92 & = r \\ \hline 1126 & \\ 5067\phantom{0} & \\ \hline 3\,)\,51{\cdot}796 & \\ \hline 17{\cdot}265 & = \overline{A + 4\,P + 2\,Q} \times \frac{r}{3} = \tfrac{1}{2} \text{ area of Section 5.} \end{array}$$

Half areas of Vertical Sections 1, 2, 3, 4, and 5.

| | |
|---|---|
| No. 1 | 1·104 feet. |
| 2 | 6·256 |
| 3 | 11·745 |
| 4 | 15·946 |
| 5 | 17·265 |

Displacement of the body under the fore half-length of the load-water line by the vertical sections, or the summation of the vertical areas 1, 2, 3, 4, and 5, by the formula for the solid, as being equal to

$$\left\{ A' + 4\,P' + 2\,Q' \right\} \times \frac{r'}{3} \text{ where } A' = \text{sum of 1st and 5th areas.}$$

$P' =$ " 2d and 4th areas.

$Q' =$ " 3d area.

And $r'$ = distance between the vertical sections, or 5·5 feet.

$$\begin{array}{l} 1\ldots\ 1{\cdot}104 \\ 5\ldots 17{\cdot}265 \\ \hline 18{\cdot}369 = A' \end{array} \qquad \begin{array}{l} 2\ldots\ 6{\cdot}256 \\ 4\ldots 15{\cdot}946 \\ \hline 22{\cdot}202 = P' \end{array} \qquad \begin{array}{r} 3\ldots 11{\cdot}745 - Q' \\ 2\phantom{- Q'} \\ \hline 23{\cdot}490 = 2\,Q' \end{array}$$

$$\begin{array}{rl} 22{\cdot}202 & = P' \\ 4 & \\ \hline 88{\cdot}808 & = 4\,P' \\ 18{\cdot}369 & = A' \\ 23{\cdot}490 & = 2\,Q' \\ \hline 130{\cdot}667 & = A' + 4\,P' + 2\,Q' \\ 5{\cdot}5 & = r' \\ \hline 653335 & \\ 653335\phantom{0} & \\ \hline 3\,)\,718{\cdot}6685 & \\ \hline 239{\cdot}556 & = \overline{A' + 4\,P' + 2\,Q'} \times \frac{r'}{3} = \text{cubic ft. of space in } \tfrac{1}{2} \text{ fore-body.} \\ 2 & \\ \hline 479{\cdot}112 & = \text{cubic feet of space in fore-body.} \end{array}$$

Displacement of the body immersed under the after half-length of the load-water line by the vertical areas 5, 6, 7, 8, and 9 of the Table of ordinates.

*Vertical Section* 6.

5, as fore body. 17·265

| | | |
|---|---|---|
| 6·1 | 5·5 | 4·6 = Q |
| 2·0 | 3·4 | 2 |
| 8·1 = A | 8·9 = P | 9·2 = 2 Q |

4
35·6 = 4 P
8·1 = A
9·2 = 2 Q
52·9 = A + 4 P + 2 Q
·92 = $r$
1058
4761
3 ) 48·668
16·222 = $\overline{A+4P+2Q}\times\frac{r}{3}$ = { ½ area of Section 6.

*Vertical Section* 7.

| | | |
|---|---|---|
| 5·4 | 4·4 | 3·4 = Q |
| 1·4 | 2·4 | 2 |
| 6·8 = A | 6·8 = P | 6·8 = 2 Q |

4
27·2 = 4 P
6·8 = 2 Q
6·8 = A
40·8 = A + 4 P + 2 Q
·92 = $r$
816
3672
3 ) 37·536
12·512 = $\overline{A+4P+2Q}\times\frac{r}{3}$ = { ½ area of Section 7.

*Vertical Section* 8.

| | | |
|---|---|---|
| 3·7 | 2·6 | 1·7 = Q |
| ·6 | 1·1 | ·2 |
| 4·3 = A | 3·7 = P | 3·4 = 2 Q |

4
14·8 = 4 P
4·3 = A
3·4 = 2 Q
22·5 = A + 4 P + 2 Q
·92 = $r$
450
2025
3 ) 20·700
6·9 = $\overline{A+4P+2Q}\times\frac{r}{3}$ = { ½ area of Section 8.

*Vertical Section* 9.

$$\begin{array}{l} \cdot 4 \\ \cdot 2 \\ \hline \cdot 6 = A \end{array} \qquad \begin{array}{r} \cdot 35 \\ \cdot 25 \\ \hline \cdot 60 = P \\ 4 \\ \hline 2\cdot 4 = 4\,P \\ \cdot 6 = A \\ \cdot 6 = 2\,Q \\ \hline 3\cdot 6 = A + 4\,P + 2\,Q \\ \cdot 92 = r \\ \hline 72 \\ 324 \\ \hline 3\,)\,3\cdot 312 \\ \hline 1\cdot 104 = \overline{A + 4\,P + 2\,Q} \times \frac{r}{3} = \tfrac{1}{2} \text{ area of Section 9.} \end{array} \qquad \begin{array}{r} \cdot 3 = Q \\ 2 \\ \hline \cdot 6 = 2\,Q \end{array}$$

Half areas of the vertical sections 5, 6, 7, 8, and 9.

| Sections. | Areas. |
|---|---|
| 5 | 17·265 |
| 6 | 16·22 |
| 7 | 12·512 |
| 8 | 6·9 |
| 9 | 1·104 |

Displacement of the after-body under the after half-length of the load-water line by the vertical sections, or the summation of the immersed areas of the vertical sections 5, 6, 7, 8, and 9 by the formula for the solid as being equal to

$$\overline{A' + 4\,P' + 2\,Q'} \times \frac{r'}{3}$$

where $A'$ = sum of the 5th and 9th areas.
$P'$ = " 6th and 8th areas.
$Q'$ = " 7th area.

and $r'$ = the distance between the vertical sections, or 5·5 ft.

$$\begin{array}{l} 5 \ldots 17\cdot 265 \\ 9 \ldots\ \ 1\cdot 104 \\ \hline 18\cdot 369 = A' \end{array} \qquad \begin{array}{r} 6 \ldots 16\cdot 22 \\ 8 \ldots\ 6\cdot 900 \\ \hline 23\cdot 120 = P' \\ 4 \\ \hline 92\cdot 480 = 4\,P' \\ 25\cdot 024 = 2\,Q' \\ 18\cdot 369 = A' \\ \hline 135\cdot 873 = A' + 4\,P' + 2\,Q' \\ 5\cdot 5 = r \\ \hline 679\cdot 365 \\ 67\cdot 936 \\ \hline 3\,)\,747\cdot 3015 \\ \hline 249\cdot 1005 = \overline{A' + 4\,P' + 2\,Q'} \times \frac{r'}{3} = \text{cubic ft. in } \tfrac{1}{2} \text{ after-body.} \\ 2 \\ \hline 498\cdot 2010 \text{ After-body in cubic ft. of space.} \end{array} \qquad \begin{array}{r} 7 \ldots 12\cdot 512 = Q' \\ 2 \\ \hline 25\cdot 024 = 2\,Q' \end{array}$$

Displacement of Fore-body by Horizontal Sections.

*Horizontal Section* 1′.

$$\begin{array}{r} 0·4 \\ 6·3 \\ \hline 6·7 = A' \end{array} \qquad \begin{array}{r} 6·0 \\ 3·0 \\ \hline 9·0 = P \\ 4 \\ \hline 36·00 = 4\,P \\ 10·00 = 2\,Q \\ 6·70 = A \\ \hline 52·70 = A + 4\,P + 2\,Q \\ 5·5 = r \\ \hline 2635 \\ 2635 \\ \hline 3\,)\,289·85 \\ \hline \end{array} \qquad \begin{array}{r} 5·0 = Q \\ 2 \\ \hline 10·0 = Q \end{array}$$

$$96·61 = \overline{A + 4\,P + 2\,Q} \times \frac{r}{3} = \tfrac{1}{2} \text{ area of Section } 1'.$$

*Horizontal Section* 2′.

$$\begin{array}{r} ·35 \\ 5·60 \\ \hline 5·95 = A \end{array} \qquad \begin{array}{r} 5·7 \\ 2·4 \\ \hline 8·1 = P \\ 4 \\ \hline 32·4 = 4\,P \\ 8·4 = 2\,Q \\ 5·95 = A \\ \hline 46·75 = A + 4\,P + 2\,Q \\ 5·5 = r \\ \hline 23375 \\ 23375 \\ \hline 3\,)\,257·125 \\ \hline \end{array} \qquad \begin{array}{r} 4·2 = Q \\ 2 \\ \hline 8·4 = 2\,Q \end{array}$$

$$85·708 = \overline{A + 4\,P + 2\,Q} \times \frac{r}{3} = \tfrac{1}{2} \text{ area of Section } 2'.$$

*Horizontal Section* 3′.

$$\begin{array}{r} ·3 \\ 5·0 \\ \hline 5·3 = A \end{array} \qquad \begin{array}{r} 4·4 \\ 1·7 \\ \hline 6·1 = P \\ 4 \\ \hline 24·4 = 4\,P \\ 5·3 = A \\ 6·4 = 2\,Q \\ \hline 36·1 = A + 4\,P + 2\,Q \\ 5·5 = r \\ \hline 1805 \\ 1805 \\ \hline 3\,)\,198·55 \\ \hline \end{array} \qquad \begin{array}{r} 3·2 = Q \\ 2 \\ \hline 6·4 = 2\,Q \end{array}$$

$$66·18 = \overline{A + 4\,P + 2\,Q} \times \frac{r}{3} = \tfrac{1}{2} \text{ area of Section } 3'.$$

*Horizontal Section 4′.*

| | | |
|---|---|---|
| ·25 | 3·2 | 2·2 = Q |
| 3·8 | 1·0 | 2 |
| 4·05 = A | 4·2 = P | 4·4 = 2 Q |

4
16·8 = 4 P
4·05 = A
4·40 = 2 Q
25·25 = A + 4 P + 2 Q
5·5 = $r$
12625
12625
3 ) 138·875

$$46{\cdot}291 = \overline{A + 4P + 2Q} \times \frac{r}{3} = \left\{ \begin{array}{l} \tfrac{1}{2} \text{ area of} \\ \text{Section } 4'. \end{array} \right.$$

*Horizontal Section 5′.*

| | | |
|---|---|---|
| ·2 | 2·0 | 1·3 = Q |
| 2·4 | ·4 | 2 |
| 2·6 = A | 2·4 = P | 2·6 = 2 Q |

4
9·6 = 4 P
2·6 = A
2·6 = 2 Q
14·8 = A + 4 P + 2 Q
5·5 = $r$
740
740
3 ) 81·40

$$27{\cdot}13 = \overline{A + 4P + 2Q} \times \frac{r}{3} = \left\{ \begin{array}{l} \tfrac{1}{2} \text{ area of} \\ \text{Section } 5'. \end{array} \right.$$

Displacement of the fore-body under the fore half-length of the load-water line by horizontal sections, or the summation of the horizontal sections of the fore-body 1′, 2′, 3′, 4′, and 5′, by the formula for the solid, as being equal to

$$\overline{A' + 4P' + 2Q'} \times \frac{r}{3};$$

where A′ = sum of the 1′st and 5′th areas;
P′ = " 2′d and 4′th areas;
Q′ = " 3′d area;
and $r$ = the distance between the horizontal sections, or ·92 feet.

Half areas of the Horizontal Sections 1′, 2′, 3′, 4′, and 5′.

| | |
|---|---|
| 1′ = 96·61. | 4′ = 46·29. |
| 2′ = 85·708. | 5′ = 27·13. |
| 3′ = 66·18. | |

| Areas. | Areas. | Areas. |
|---|---|---|
| 1′...96·61 | 2′...85·708 | 3′...66·18 = Q′ |
| 5′...27·13 | 4′...46·290 | 2 |
| 123·74 = A′ | 131·998 = P′ | 132·36 = 2 Q′ |

131·998 = P′
4

527·992 = 4 P′
123·740 = A′
132·360 = 2 Q′

784·092 = A′ + 4 P′ + 2 Q′
·92 = $r$

1568184
7056828

3) 721·36464

240·45 $= \overline{A'+4P'+2Q'} \times \frac{r}{3} =$ { cubic ft. in ½ fore-body.
2

480·90 = fore-body by horizontal sections in cubic feet of space.

Displacement, by horizontal sections of the body immersed under the after half-length of the load-water line, or by the horizontal areas 1′, 2′, 3′, 4′, and 5′, of the table of ordinates.

Calculated areas of 1′, 2,′ 3′, 4′, and 5′.

*Section* 1′ *After-body.*

| | | |
|---|---|---|
| 6·3 | 6·1 | 5·4 = Q |
| ·4 | 3·7 | 2 |
| 6·7 = A | 9·8 = P | 10·8 = 2 Q |

9·8 = P
4

39·2 = 4 P
10·8 = 2 Q
6·7 = A

56·7 = A + 4 P + 2 Q
5·5 = $r$

2835
2835

3) 311·85

103·95 $= \overline{A + 4P + 2Q} \times \frac{r'}{3} =$ { ½ area of Section 1′.

*Section 2′ After-body.*

$$\begin{array}{lll}
\begin{array}{r} 5{\cdot}6 \\ {\cdot}35 \\ \hline 5{\cdot}95 = \mathrm{A} \end{array} &
\begin{array}{r} 5{\cdot}5 \\ 2{\cdot}6 \\ \hline 8{\cdot}1 = \mathrm{P} \end{array} &
\begin{array}{r} 4{\cdot}4 \\ 2 \\ \hline 8{\cdot}8 = 2\,\mathrm{Q} \end{array}
\end{array}$$

$$\begin{array}{rl}
8{\cdot}1 & = \mathrm{P} \\
4 & \\ \hline
32{\cdot}40 & = 4\,\mathrm{P} \\
5{\cdot}95 & = \mathrm{A} \\
8{\cdot}80 & = 2\,\mathrm{Q} \\ \hline
47{\cdot}15 & = \mathrm{A} + 4\,\mathrm{P} + 2\,\mathrm{Q} \\
5{\cdot}5 & = r \\ \hline
23575 & \\
23575\phantom{0} & \\ \hline
3\,)\,259{\cdot}325 & \\ \hline
86{\cdot}441 & = \overline{\mathrm{A} + 4\,\mathrm{P} + 2\,\mathrm{Q}} \times \dfrac{r'}{3} = \tfrac{1}{2} \text{ area of Section } 2'.
\end{array}$$

*Section 3′ After-body.*

$$\begin{array}{lll}
\begin{array}{r} 5{\cdot}0 \\ {\cdot}3 \\ \hline 5{\cdot}3 = \mathrm{A} \end{array} &
\begin{array}{r} 4{\cdot}6 \\ 1{\cdot}7 \\ \hline 6{\cdot}3 = \mathrm{P} \end{array} &
\begin{array}{r} 3{\cdot}4 = \mathrm{Q} \\ {\cdot}2 \\ \hline 6{\cdot}8 = 2\,\mathrm{Q} \end{array}
\end{array}$$

$$\begin{array}{rl}
6{\cdot}3 & = \mathrm{P} \\
4 & \\ \hline
25{\cdot}2 & = 4\,\mathrm{P} \\
5{\cdot}3 & = \mathrm{A} \\
6{\cdot}8 & = 2\,\mathrm{Q} \\ \hline
37{\cdot}3 & = \mathrm{A} + 4\,\mathrm{P} + 2\,\mathrm{Q} \\
5{\cdot}5 & = r \\ \hline
1865 & \\
1865\phantom{0} & \\ \hline
3\,)\,205{\cdot}15 & \\ \hline
68{\cdot}38 & = \overline{\mathrm{A} + 4\mathrm{P} + 2\mathrm{Q}} \times \dfrac{r'}{3} = \tfrac{1}{2} \text{ area of Section } 3'.
\end{array}$$

*Section 4′ After-body.*

$$\begin{array}{lll}
\begin{array}{r} 3{\cdot}8 \\ {\cdot}25 \\ \hline 4{\cdot}05 = \mathrm{A} \end{array} &
\begin{array}{r} 3{\cdot}4 \\ 1{\cdot}1 \\ \hline 4{\cdot}5 = \mathrm{P} \end{array} &
\begin{array}{r} 2{\cdot}4 = \mathrm{Q} \\ 2 \\ \hline 4{\cdot}8 = 2\,\mathrm{Q} \end{array}
\end{array}$$

$$\begin{array}{rl}
4{\cdot}5 & = \mathrm{P} \\
4 & \\ \hline
18{\cdot}00 & = 4\,\mathrm{P} \\
4{\cdot}05 & = \mathrm{A} \\
4{\cdot}80 & = 2\,\mathrm{Q} \\ \hline
26{\cdot}85 & = \mathrm{A} + 4\,\mathrm{P} + 2\,\mathrm{Q} \\
5{\cdot}5 & = r' \\ \hline
13425 & \\
13425\phantom{0} & \\ \hline
3\,)\,147{\cdot}675 & \\ \hline
49{\cdot}225 & = \overline{\mathrm{A} + 4\,\mathrm{P} + 2\,\mathrm{Q}} \times \dfrac{r'}{3} = \tfrac{1}{2} \text{ area of Section } 4'.
\end{array}$$

*Section 5′ After-body.*

$$\begin{array}{r} 2{\cdot}4 \\ {\cdot}2 \\ \hline 2{\cdot}6 = A \end{array} \qquad \begin{array}{rl} 2{\cdot}0 & \\ {\cdot}6 & \\ \hline 2{\cdot}6 & = P \\ 4 & \\ \hline 10{\cdot}4 & = 4\,P \\ 2{\cdot}8 & = 2\,Q \\ 2{\cdot}6 & = A \\ \hline 15{\cdot}8 & = A + 4\,P + 2\,Q \\ 5{\cdot}5 & = r' \\ \hline 790 & \\ 790 & \\ 3\,)\,\overline{86{\cdot}90} & \\ \hline 28{\cdot}96 & = \overline{A + 4P + 2\,Q} \times \frac{r'}{3} = \tfrac{1}{2} \text{ area of Section } 5'. \end{array} \qquad \begin{array}{rl} 1{\cdot}4 & = Q \\ 2 & \\ \hline 2{\cdot}8 & = 2\,Q \end{array}$$

Displacement by horizontal sections of the after-body under the after half-length of the load-water line, or the summation of the horizontal sections of the after-body, 1′, 2′, 3′, 4′, and 5′, by the formula of the solid, as being equal to

$$\overline{A' + 4\,P' + 2\,Q'} \times \frac{r'}{3}.$$

Half areas of the After Horizontal Sections, 1′, 2′, 3′, 4′, and 5′.

| Sections. | Areas. |
|---|---|
| 1′ | 103·95. |
| 2′ | 86·44. |
| 3′ | 68·38. |
| 4′ | 49·22. |
| 5′ | 28·96. |

$$\begin{array}{r} \text{Areas.} \\ 1' \ldots 103{\cdot}95 \\ 5' \ldots\ 28{\cdot}96 \\ \hline 132{\cdot}91 = A' \end{array} \qquad \begin{array}{rl} \text{Areas.} & \\ 2' \ldots 86{\cdot}44 & \\ 4' \ldots 49{\cdot}22 & \\ \hline 135{\cdot}66 & = P' \\ 4 & \\ \hline 542{\cdot}64 & = 4\,P' \\ 132{\cdot}91 & = A' \\ 136{\cdot}76 & = 2\,Q' \\ \hline 812{\cdot}31 & = A' + 4\,P' + 2\,Q' \\ {\cdot}92 & = r \\ \hline 162462 & \\ 731079 & \\ 3\,)\,\overline{747{\cdot}3252} & \\ \hline 249{\cdot}1084 & = \overline{A' + 4\,P' + 2\,Q'} \times \frac{r}{3} = \text{cubic ft. of} \\ 2 & \tfrac{1}{2} \text{ after-body by horizontal sections.} \\ \hline 498{\cdot}2168 & = \text{After-body by horizontal sections in cubic feet of space.} \end{array} \qquad \begin{array}{rl} \text{Areas.} & \\ 3' \ldots 68{\cdot}38 & = Q' \\ 2 & \\ \hline 136{\cdot}76 & = 2\,Q' \end{array}$$

DISPLACEMENT.

| *By Vertical Sections.* | Cubic Feet. | *By Horizontal Sections.* | Cubic Feet. |
|---|---|---|---|
| Fore-body (p. 469) | 479·11 | Fore-body (p. 474) | 480·900 |
| After-body (p. 471) | 498·20 | After-body (p. 476) | 498·216 |
| Sum | 977·30 | Sum | 979·116 |
| Half | 488·65 | Half | 489·558 |

| | Cubic Feet. |
|---|---|
| By Horizontal Sections | 979·116 |
| By Vertical Sections | 977·300 |
| Difference | 1·816 cubic feet. |

Cubic Feet.
979·49 = capacity or displacement in cubic feet of space.

The mean weight of salt and fresh water gives 35 cubic feet of space, when filled with water, to be equivalent to a ton avoirdupois; thence the displacement in cubic feet of space being divided by 35 will give the weight of the volume displaced in tons avoirdupois; or 979·49 being divided by 35 gives

5) 979·49
7) 195·898
27·985 tons, the weight of the calculated immersed body in tons.

AREA OF THE MIDSHIP SECTION, OR OF THE GREATEST TRANSVERSE SECTION.

*Section at 5.*

1·1...6·3   2·2...6·0   3·3...4·8 = Q
5·5... ·2   4·4...2·3   2
6·5 = A   8·3 = P   9·6 = 2 Q
4
33·2 = 4 P
6·5 = A
9·6 = 2 Q
49·3 = A + 4 P + 2 Q
1·25 = $\frac{r}{3}$ where $r$ = the depth, from 1 to 5, divided by 4 = 5 ft. by 4 = 1·25 ft.
2465
986
493
3) 61·625
20·541 = $\overline{A + 4P + 2Q} \times \frac{r}{3}$ = ½ area of midship section.
2
41·082 = Area of midship section without keel.

### LOAD-WATER LINE.

Area of the load-water line, or area of the assumed deepest plane of immersion, delineated on the half-breadth plan, and marked by the curve AB. From the table of ordinates, p. 467, we have—

$$\begin{array}{l} \cdot 4 \\ \underline{\cdot 4} \\ \cdot 8 = A \end{array} \qquad \begin{array}{r} 3\cdot 0 \\ 6\cdot 0 \\ 6\cdot 1 \\ \underline{3\cdot 7} \\ 18\cdot 8 = P \\ \underline{4} \\ 75\cdot 2 = 4\,P \\ \cdot 8 = A \\ \underline{33\cdot 4 = 2\,Q} \\ 109\cdot 4 = A + 4\,P + 2\,Q \\ \underline{5\cdot 5 = r'} \\ 5470 \\ \underline{5470} \\ 3\,)\,601\cdot 70 \\ 200\cdot 56 = \overline{A + 4\,P + 2\,Q} \times \frac{r'}{3} = \left\{ \begin{array}{l} \frac{1}{2} \text{ area of load-} \\ \text{water line.} \end{array} \right. \end{array} \qquad \begin{array}{r} 5\cdot 0 \\ 6\cdot 3 \\ \underline{5\cdot 4} \\ 16\cdot 7 = Q \\ \underline{2} \\ 33\cdot 4 = 2\,Q \end{array}$$

$200\cdot 56 = \frac{1}{2}$ area of load-water section in superficial feet.
$\underline{\quad 2 \quad}$

$401\cdot 12 =$ area of load-water section, which amount of area being divided by 12, will give the number of cubic feet of space that would be contained in a zone of that area of an inch in depth, and that result being again divided by 35, as the number of cubic feet of water equivalent to a ton in weight, will give the number of tons that will immerse the vessel one inch at that line of immersion.

$12\,)\,401\cdot 12 =$ area of load-water section in superficial feet.
$5\,)\,33\cdot 42 =$ cubic feet in zone of one inch in depth.
$7\,)\,6\cdot 684$
$\cdot 955 =$ tons to the inch of immersion at load-water line.

### CENTRE OF GRAVITY OF THE DISPLACEMENT.

Estimated from Section 1, considered as the Initial Plane.

| | Distinguishing No. of the Areas. | ½ Vertical Areas. | | Moments. |
|---|---|---|---|---|
| From p. 469. | 1 | 1·104 | × 0 | 000·000 |
| | 2 | 6·256 | × 1 | 6·256 |
| | 3 | 11·745 | × 2 | 23·490 |
| | 4 | 16·069 | × 3 | 48·207 |
| | 5 | 17·265 | × 4 | 69·060 |
| From p. 471. | 6 | 16·222 | × 5 | 81·110 |
| | 7 | 12·512 | × 6 | 75·072 |
| | 8 | 6·900 | × 7 | 48·300 |
| | 9 | 1·104 | × 8 | 8·832 |

Moments placed in the Rule.

$$\text{Sum} = \overline{A + 4P + 2Q} \times \frac{r'}{3}$$

| | | |
|---|---|---|
| 000·000 | 6·256 | 23·490 |
| 8·832 | 48·207 | 69·060 |
| 8·832 = A | 81·110 | 75·072 |
| | 48·300 | 167·622 = Q |
| | 183·873 = P | 2 |
| | 4 | 335·244 = 2 Q |
| | 735·492 = 4 P | |

735·492 = 4 P
8·832 = A
335·244 = 2 Q
1079·568 = A + 4 P + 2 Q
5·5 = $r'$
5397840
5397840
3 ) 5937·6240
1979·208 $= \overline{A + 4P + 2Q} \times \frac{r'}{3} =$

sum of the moments of half the displacement from section 1, in intervals of space of 5·5 ft.; and the half displacement in cubic feet by vertical sections is 488·650 (p. 477) cubic ft.; whence it is found, by dividing the moment 1979·208 by 488·650, that the distance of the centre of gravity of displacement from the section 1 is as follows:—

488·65 ) 1979·208 ( 4·05 intervals from 1.
195460 interval = 5·5 ft.
246080
244325
1755 therefore 4·05 × 5·5 = 22·27 ft. = distance of the centre of gravity of the calculated immersed body from 1.

### DEPTH OF THE CENTRE OF GRAVITY OF THE DISPLACEMENT BELOW THE LOAD-WATER SECTION.

| Section. | Fore-body. Areas. (From p. 473.) | After-body. Areas. (From p. 475.) | Sum of the Areas. | | Moments. |
|---|---|---|---|---|---|
| 1′ | 96·61 | 103·95 | 200·56 | × 0 | = 000·000 |
| 2′ | 85·708 | 86·44 | 172·148 | × 1 | = 172·148 |
| 3′ | 66·18 | 68·38 | 134·56 | × 2 | = 269·12 |
| 4′ | 46·29 | 49·22 | 95·51 | × 3 | = 286·53 |
| 5′ | 27·13 | 28·96 | 56·09 | × 4 | = 224·36 |

$$
\begin{array}{rl}
000 \cdot 00 & \\
224 \cdot 36 & \\
\hline
224 \cdot 36 & = A
\end{array}
\qquad
\begin{array}{rl}
172 \cdot 148 & \\
286 \cdot 530 & \\
\hline
458 \cdot 678 & = P \\
4 & \\
\hline
1834 \cdot 712 & = 4\,P \\
224 \cdot 360 & = A \\
538 \cdot 240 & = 2\,Q \\
\hline
2597 \cdot 312 & = A + 4\,P + 2\,Q \\
\cdot 92 & = r \\
\hline
5194624 & \\
23375808 & \\
\hline
3\,)\,2389 \cdot 52704 & \\
\hline
796 \cdot 509 & = \overline{A + 4\,P + 2\,Q} \times \frac{r}{3} =
\end{array}
\qquad
\begin{array}{rl}
269 \cdot 12 & = Q \\
2 & \\
\hline
538 \cdot 24 & = 2\,Q
\end{array}
$$

sum of the moments of the half displacement calculated from the load-water line: the half displacement by horizontal sections is 489·588 (p. 477) cubic feet; the sum of the moments of the half displacement 796·509 ft., being divided by that quantity, will give the distance in intervals of ·92 ft.; the centre of gravity of displacement is below the load-water line.

Half solid of displacement. Moments.

489·558 ) 796·509 ) 1·62 intervals of ·92 feet; therefore

$$
\begin{array}{r}
489558 \\
\hline
3069510 \\
2937348 \\
\hline
1321620 \\
979116 \\
\hline
342504
\end{array}
\qquad
\begin{array}{r}
1 \cdot 62 \\
\times \cdot 92 \\
\hline
324 \\
1458 \\
\hline
1 \cdot 4904
\end{array}
$$

1·4904 ft. = the distance the centre of gravity of the calculated immersed body is below the load-water section.

DISTANCE OF THE CENTRE OF GRAVITY OF THE AREA OF THE LOAD-WATER SECTION FROM SECTION 1.

| No. of Section. | Ordinates of Section 1 from the Table, p. 467. | Distances of them in intervals of 5·5 ft. from Section 1. | Moments: being the Product of the Areas by the respective Distances. |
|---|---|---|---|
| 1 | ·4 | 0 | 000·00 |
| 2 | 3·0 | 1 | 3·0 |
| 3 | 5·0 | 2 | 10·0 |
| 4 | 6·0 | 3 | 18·0 |
| 5 | 6·3 | 4 | 25·2 |
| 6 | 6·1 | 5 | 30·5 |
| 7 | 5·4 | 6 | 32·4 |
| 8 | 3·7 | 7 | 25·9 |
| 9 | ·4 | 8 | 3·2 |

The moments, for summation, put into the rule.

| | | |
|---|---|---|
| 00·0 | 3·0 | 10·0 |
| 3·2 | 18·0 | 25·2 |
| 3·2 = A | 30·5 | 32·4 |
| | 25·9 | 67·6 = Q |
| | 77·4 = P | 2 |
| | 4 | 135·2 = 2 Q |
| | 309·6 = 4 P | |
| | 3·2 = A | |
| | 135·2 = 2 Q | |
| | 448·0 = A + 4 P + 2 Q | |
| | 5·5 = $r'$ | |
| | 2240 | |
| | 2240 | |
| | 3 ) 2464·0 | |

$$821·3 = \overline{A + 4P + 2Q} \times \frac{r'}{3} =$$

sum of the moments of the half area of the load-water section reckoned from 1; the half area of the load-water section is 200·56 feet (p. 478); the distance, therefore, of the centre of gravity of the load-water section from 1 will be found in intervals of space of 5·5 feet, by dividing the sum of these moments by the half area, thus:—

| Half Area. | Moments. | No. |
|---|---|---|
| 200·56 ) | 821·3333 | ( 4·09 intervals, each 5·5 ft. in length. |
| | 80224 | |
| | 190933 | |
| | 180504 | |
| | 10429 | |

and 4·09 × 5·5 = 22·5 ft. gives the distance of the centre of gravity of the load-water section from section 1 of the drawing.

Relative capacities of the bodies immersed under the fore and after lengths of equal division of the load-water line—

By former calculations.

After-body immersed contains........497·79 cubic ft. of space.
Fore-body " " ........481·70 cubic ft. of space.
Difference........ 16·09 =

the excess in cubic feet of space of the body displaced under the after half-length of the load-water line over that under the fore-half of the same line—

Sum of the bodies (by former calculation) or whole displacement in cubic feet of space (p. 477)...... } 979·49

equal to 9·7949 hundreds of cubic feet of space, whence 16·09, or the difference between the two bodies in cubic feet, being divided by 9·7949, or the displacement expressed in terms of the hundreds of cubic feet of space, will give the excess for every hundred cubic feet of the whole displacement.

| Displacement in Hundreds of Cubic Feet of Space. | Excess in Cubic Feet of Space. | |
|---|---|---|
| 9·7949 ) | 16·09000 ( 1·6 | = Ratio of the excess of the after-body of displacement over the fore-body of the same, denoted by a per-centage of the whole displacement. |
| | 97949 | |
| | 629510 | |
| | 587694 | |
| | ·41816 | |

METACENTRE.

A measure of the comparative stability of a ship, or the height of the metacentre above the centre of gravity of displacement estimated, from the expression $\frac{2}{3} \int \frac{y^3 \, dx}{D}$, in which $\int$ is the sign of integration and signifies sum:—

$y$ = the ordinates of the half-breadth load-water section.
dx = the differential increment of the length of load-water section.
D = displacement of the immersed portion of the body in cubic feet of space.

| Ordinates from the table. (Page 467.) | Cubes of the Ordinates. |
|---|---|
| ·4 | 00·064 |
| 3·0 | 27·000 |
| 5·0 | 125·000 |
| 6·0 | 216·000 |
| 6·3 | 250·047 |
| 6·1 | 226·981 |
| 5·4 | 157·464 |
| 3·7 | 50·653 |
| ·4 | 0·064 |

Cubes placed in O'Neill's rule for summation of

$$\text{Area} = (A + 4P + 2Q) \times \frac{r}{3}$$

| | | |
|---|---|---|
| 00·064 | 27·000 | 125·000 |
| 00·064 | 216·000 | 250·047 |
| ·128 = A | 226·981 | 157·464 |
| | 50·653 | 532·511 = Q |
| | 520·634 = P | 2 |
| | 4 | 1065·022 = 2 Q |
| | 2082·536 = 4 P | |
| | 1065·022 = 2 Q | |
| | 000·128 = A | |
| | 3147·686 = A + 4 P + 2 Q | |
| | 5·5 = $r'$ | |
| | 15738430 | |
| | 15738430 | |
| | 3 ) 17312·2730 | |

$$\int y^3 \, dx = 5770·7576 = \overline{A + 4P + 2Q} \times \frac{r'}{3} =$$

summation of the cubes of the ordinates of the load-water section: and the height of the metacentre above the centre of gravity of displacement is expressed by $\frac{2}{3} \int \frac{y^3\,dx}{D}$, in which expression $y^3\,dx =$ 5770·75 and D = 979·1 (p. 477) whence $\frac{2}{3} \times \frac{5770{\cdot}75}{979{\cdot}1} = 3{\cdot}98$ feet is the height of the metacentre above the centre of gravity of the displacement.

RESULTS OF THE CALCULATIONS.

*1st Method.*

| | |
|---|---|
| Displacement in cubic feet of space | = 979·149. |
| Displacement in tons of 35 cubic feet of water to a ton.............. | = 27·974. |
| Area of midship section............... | = 41·08 superficial feet. |
| Area of load-water line or plane at the proposed deepest immersion.. | = 401·12 superficial feet. |
| Tons to one inch of immersion at that flotation......................... | = ·955 tons. |
| Longitudinal distance of the centre of gravity of displacement from section 1. | = 22·22 feet. |
| Depth of the centre of gravity of displacement below the load-water section................................. | = 1·4904 feet. |
| Distance of the centre of gravity of the load-water section from vertical section 1......................... | = 22·5 feet. |
| Relative capacity of the after-body in excess of the fore-body in cubic feet of space......................... | = 16·09 |
| Per-centage on the whole displacement.............................. | = 1·06. |
| Height of the metacentre above the centre of gravity of displacement, estimated from the expression $\frac{2}{3} \int \frac{y^3\,dx}{D}$. | = 3·98 feet. |

The young naval architect has thus been led through the essential calculations on the immersed portion of a ship considered as a floating body. The term essential has here been used under a knowledge that the table of results might have been swollen to a small volume by a lengthened comparison of the elements of the naval construction, such as the ratio of the area of the midship section to the area of the load-water section, and that of the area of the midship section to the circumscribing parallelogram; data that will always suggest themselves to the mind, and furnish salutary exercise for his judgment, while the introduction of such comparisons into these rudiments might deter the novice from entering

FORE-BODY.

| Moments for Centre of Gravity of Displacement. | | C Functions of the Areas for the Solid. | Multipliers for Solid. | B Functions of Vertical Areas. | $\frac{A}{2}$ | | | 2 P | | | Q | | | 2 P | | | $\frac{A}{2}$ | | | | Moment for Centre of Gravity of Load-water Line. | Cubes of the Ordinates of Load-water Section. | | Summation of the Cubes for the Value of $\int y^3 dx$. |
|---|---|---|---|---|---|---|---|---|---|---|---|---|---|---|---|---|---|---|---|---|---|---|---|---|
| 000·00 | 0 | 0·90 | ½ | 1·80 | ·2 | 0·10 | 0·10 | ·25 | 0·50 | 0·12 | ·3 | 0·30 | 0·15 | ·35 | 0·70 | 0·17 | ·4 | 0·20 | 0·20 | 0 | 000·00 | 00·64 | ½ | 00·032 |
| 20·40 | 1 | 20·40 | 2 | 10·20 | ·4 | ·20 | 0·80 | 1·0 | 2·00 | 2·00 | 1·7 | 1·70 | 3·40 | 2·4 | 4·80 | 4·80 | 3·0 | 1·50 | 6·00 | 1 | 6·00 | 27·000 | 2 | 54·000 |
| 38·30 | 2 | 19·15 | 1 | 19·15 | 1·3 | ·65 | 1·30 | 2·2 | 4·40 | 2·20 | 3·2 | 3·20 | 3·20 | 4·2 | 8·40 | 4·20 | 5·0 | 2·50 | 5·00 | 2 | 10·00 | 125·000 | 1 | 125·000 |
| 156·00 | 3 | 52·00 | 2 | 26·00 | 2·0 | 1·00 | 4·00 | 3·2 | 6·40 | 6·40 | 4·4 | 4·40 | 8·80 | 5·6 | 11·20 | 11·20 | 6·0 | 3·00 | 12·00 | 3 | 36·00 | 216·000 | 2 | 432·000 |
| 56·28 | 4 | 14·07 | ½ | 28·15 | 2·4 | 1·20 | 1·20 | 3·8 | 7·60 | 1·90 | 5·0 | 5·00 | 2·50 | 5·6 | 11·20 | 2·80 | 6·3 | 3·15 | 3·15 | 4 | 12·60 | 250·047 | 1 | 250·047 |
| Function of the Solid by Vertical Areas | | 106·52 | | Function of the Solid by Longitudinal Areas 106·50. | | | 7·40 | | | 12·62 | | | 18·05 | | | 23·17 | | | 26·35 | Functions of Longitudinal Areas. | | | | |
| | | | | | | | 3·70 | | | 25·24 | | | 18.05 | | | 46·34 | | | 13·17 | | | | | |

## AFTER-BODY.

| Moments for Centre of Gravity of Displacement. | | C Functions of the Areas for the Solid. | Multipliers for Solid. | B Functions of Vertical Areas. | $\frac{A}{2}$ | | | 2 P | | | Q | | | 2 P | | | $\frac{A}{2}$ | | | | Moment for Centre of Gravity of Load-water Line. | Cubes of the Ordinates of Load-water Section. | | Summation of the Cubes for the Value of $\int y^3 dx$. |
|---|---|---|---|---|---|---|---|---|---|---|---|---|---|---|---|---|---|---|---|---|---|---|---|---|
| 56·28 | 4 | 14·07 | ½ | 23·15 | 2·4 | 1·20 | 1·50 | 3·8 | 7·60 | 1·90 | 5·0 | 5·00 | 2·50 | 5·6 | 11·20 | 2·80 | 6·3 | 3·15 | 3·15 | 4 | 12·60 | 250·047 | 1 | |
| 264·50 | 5 | 52·90 | 2 | 26·45 | 2·0 | 1·00 | 4·[illegible]0 | 3·4 | 6·80 | 6·80 | 4·6 | 4·60 | 9·20 | 5·5 | 11·00 | 11·00 | 6·1 | 3·05 | 12·20 | 5 | 61·00 | 226·981 | 2 | 453·962 |
| 122·40 | 6 | 20·40 | 1 | 20·40 | 1·4 | ·70 | 1·40 | 2·4 | 4·80 | 2·40 | 3·4 | 3·40 | 3·40 | 4·4 | 8·80 | 4·40 | 5·4 | 2·70 | 5·40 | 6 | 32·40 | 157·464 | 1 | 157·464 |
| 157·50 | 7 | 22·50 | 2 | 11·25 | ·6 | ·30 | 1·20 | 1·1 | 2·20 | 2·20 | 1·7 | 1·70 | 3·40 | 2·6 | 5·20 | 5·20 | 3·7 | 1·85 | 7·40 | 7 | 51·80 | 50·653 | 2 | 191·306 |
| 7·20 | 8 | ·90 | ½ | 1·80 | ·2 | ·10 | 0·10 | ·25 | 0·50 | 0·12 | ·3 | 0·30 | 0·15 | ·35 | ·70 | 0·17 | ·4 | 0·20 | 0·20 | 8 | 1·60 | 00·064 | ½ | 00·032 |
| 878·86 | | 110·77 Function of the Solid. | | | | | 7·90 | | | 13·42 | | | 18·65 | | | 23·57 | | | 28·35 | | 224·00 | | | 1573·843 |

| | Multipliers | Products | Sum of the Functions of Fore and After-bodies. | | | Moments for the Centre of Gravity of Displacement. |
|---|---|---|---|---|---|---|
| $\frac{A}{2}$ | ½ | 3·95 | 3·95, 3·70 | 7·65 | 4 | 30·60 |
| 2 P | 2 | 26·84 | 26·84, 25·24 | 52·08 | 3 | 156·24 |
| Q | 1 | 18·65 | 18·65, 18·05 | 36·70 | 2 | 73·40 |
| 2 P | 2 | 47·14 | 47·14, 46·34 | 93·48 | 1 | 93·48 |
| $\frac{A}{2}$ | ½ | 14·7 | 14·17, 13·17 | 27·34 | 0 | 000·00 |
| | | 110·75 | | 217·25 | | 353·72 |

$r = \cdot 92$ feet.

$r' = 5\cdot 5$ feet.

on a task that would thence seem to be involved in such voluminous results. For the second method of calculation, the table of ordinates is in two portions, viz. the fore and after-bodies under the division of the load-water section into two equal parts, the length of such section being restricted to the distance from the fore-edge of the rabbet of the stem to the after-edge of the rabbet of the post. The enlarged tables are shown at pages 484 and 485, and the directions for the working of these tables have been given at page 459, observing only that the ordinates have not been herein inserted in red, as it was there suggested, to insure perspicuity and accuracy.

RESULTS FROM THE TABLES.

By modified rule. Area $= \left\{\frac{A}{2} + 2\,P + Q\right\}\frac{2\,r}{3}$

And solid = areas for ordinates summed by rule $\Big\} = \left\{\frac{A'}{2} + 2\,P' + Q'\right\} + \frac{2\,r'}{3}$

Functions of the areas marked B $= \left\{\frac{A}{2} + 2\,P + Q\right\}$

Function of the solid equal to B, placed in O'Neill's rules = $A' + 2\,P' + Q' = E$

Whence displacement $= E \times \frac{2\,r}{3} \times \frac{2\,r'}{3}$, in the example $r = \cdot 92$ $r' = 5\cdot 5$.

Therefore ½ displacement $= E \times \frac{2\,r}{3} \times \frac{2\,r'}{3} = E \times \frac{1\cdot 84}{3} \times \frac{11}{3} = E \times \frac{20\cdot 24}{9}$.

VALUE OF E FROM THE TABLES BY VERTICAL SECTIONS.

Table 1...106·50 = submultiple of the fore-body by vertical sections.
Table 2...110·77 = " after-body " "
217·27 = sum of the submultiples = E.

½ displacement $= E \times \frac{20\cdot 24}{9} = \frac{217\cdot 27 \times 20\cdot 24}{9} = 24\cdot 14 \times 20\cdot 24 =$

488·5936 = ½ solid of displacement by the summation of the vertical areas given in cubic feet of space.
2
5 ) 977·1872
7 ) 195·4374
27·92 = Displacement by vertical sections in tons of 35 cubic feet of space.

VALUE OF E FROM THE TABLES BY HORIZONTAL SECTIONS.

Table 1...106·50 = submultiple of the fore-body by horizontal sections.

Table 2...110·75 = submultiple of the after-body by horizontal sections.

From whence the same results will be obtained.

AREA OF MIDSHIP SECTION.

From table 1...28·15 = submultiple of the area of Section 5.
1·84 = 2 $r$

11260
22520
2015

3) 51·7960

17·265 = ½ area of upper space of midship section.
3·275 = ½ area of the lower " " below $d$ $d$,

20·540 = ½ area of midship section.
2

41·08 = area of midship section.

AREA OF THE LOAD-WATER LINE.

From table 1...26·35 = submultiple of the area of the fore-body.
From table 2...28·35 = " " after-body.

54·70 = submultiple for ½ area of load-water line.
11 = 2 $r'$

3) 601·7

200·56 = ½ area = $\overline{\frac{A}{2} + 2P + Q} \times \frac{2r'}{3}$
2

12) 401·12 = area of load-water line.

5) 33·42

7) 6·684

·955 = tons per inch of immersion at the load-water line.

POSITION OF THE CENTRE OF GRAVITY OF DISPLACEMENT.

By table 2...878·86 = moments from Section 1.
and E........217·27 = corresponding function of the displacement.

217·27) 878·86 ( ·404 intervals of 5·5 feet, giving 4·04 × 5·5 = 22·22 feet as the distance of the centre of gravity of the displacement from Section 1.
869·08

97800
86908

10892

DEPTH OF THE CENTRE OF GRAVITY OF THE DISPLACEMENT BELOW THE LOAD-WATER SECTION.

By table 2...353·72 = moments from load-water line.
and E........217·25 = corresponding function of the displacement.

217·25 ) 353·72 ( 1·62 intervals of ·92 feet, giving 1·62 × ·92 = 1·4904 as the distance that the centre of gravity of displacement is below the load-water line.

217·25
136·470
130·350
61200
43450
17750

POSITION OF THE CENTRE OF GRAVITY OF THE LOAD-WATER LINE OF DEEPEST IMMERSION.

From table 1.......26·35 ft. From table 2...224·000 = moments
" 2.......28·35 from 1st section.

Function for area..54·7 ) 224·0 ( 4·09 intervals of 5·5 feet, giving 4·09 × 5·5 = 22·495 feet as the distance that the centre of gravity of the load-water section is from vertical section 1.

218·8
5200
4923
·277

RELATIVE CAPACITIES OF THE CALCULATED IMMERSED BODIES CONTAINED UNDER THE FORE AND AFTER-LENGTHS OF EQUAL DIVISION OF THE LOAD-WATER LINE.

| | Feet. |
|---|---|
| From table 1...Function for the fore-solid...... | 106·50 |
| From table 2...Function for the after-solid...... | 110·75 |
| | 4·25 |
| Sum of the functions...... | 217·25 |

The difference, 4·25 feet, expresses the excess in cubic feet of space of the body, displaced under the after half-length of the load-water line, over that under the fore half-length of the same line, and the sum of the functions, 217·25, is equal to 2·1725 hundreds of cubic feet of space; whence, 4·25 feet, or the difference between the functions for the two bodies, being divided by the function 2·1725, or the function for the displacement of the calculated body expressed in terms of hundreds of cubic feet of space, will give the excess for every hundred cubic feet of that displacement:

Function of Displacement. Excess in Cubic Feet of Space.

2·1725 ) 4·25000 ( 1·9 ratio of the excess of the after-body of calculation over the fore-body of the same, denoted by a per-centage of the displacement calculated by the table of ordinates.

2·1725
207750
195525
·12225

### HEIGHT OF THE METACENTRE ABOVE THE CENTRE OF GRAVITY OF DISPLACEMENT.

From table 2...The summation of the functions of the cubes of the ordinates for the value of the $\int y^3 dx$.......... } = 1573·843.

The corresponding function for the solid.......... = 217·25.

from whence the height of the metacentre above the centre of gravity of displacement, expressed by $\frac{2}{3}\int\frac{y^3 dx}{D}$ is as follows:

$$\int y^3 dx = 1573{\cdot}843 \times \frac{2r'}{3} \text{ where } r' = 5{\cdot}5 \text{ feet} =$$

$$\frac{1573{\cdot}843 \times 11}{3} = \frac{17312{\cdot}273}{3} = 5770{\cdot}75 \text{ feet.}$$

(Page 485) $217{\cdot}27 \times \frac{2r}{3} \times \frac{2r'}{3} = \frac{1}{2}$ displacement $= 488{\cdot}5936$ feet,

whence displacement or D = 977·1872; and thence

$$\frac{2}{3}\int\frac{y^3 dx}{D} = \frac{2}{3} \times \frac{5770{\cdot}75}{977{\cdot}1872} = \frac{11541{\cdot}53}{2931{\cdot}5616} = 3{\cdot}98 \text{ feet.}$$

### RESULTS OBTAINED UNDER THE TWO METHODS OF CALCULATION CONTRASTED.

| | Old Method. | Second Method. |
|---|---|---|
| Displacement in cubic feet of space... | 979·139 | 977·187 |
| Displacement in tons of 35 cubic feet of water to a ton...................... | 27·985 | 27·92 |
| | Superficial ft. | Superficial ft. |
| Area of midship section................. | 41·08 | 41·08 |
| Area of load-water line or plane at the proposed deepest immersion..... | 401·12 | 401·12 |
| Tons to one inch of immersion at line of flotation............................. | ·9526 tons. | ·955 tons. |
| Longitudinal distance of the centre of gravity of the displacement from section 1................................ | 22·22 ft. | 22·22 ft. |
| Depth of the centre of gravity of displacement below the load-water section...................................... | 1·4812 ft. | 1·4904 ft. |
| Relative capacities of the bodies....... | 1·6 per cent. | 1·9 per ct. |
| Height of the metacentre above the centre of gravity of displacement... | 3·98 ft. | 3·98 ft. |

## THIRD METHOD OF CALCULATION.

### CALCULATIONS ON THE DRAUGHT OF THE YACHT OF 36 TONS USING THE CURVE OF SECTIONAL AREAS.

The load-water line AB, in the sheer plan, is divided into two equal parts at the point C, and those equal parts are again subdivided at the points D and E; at the points C, D, and E,

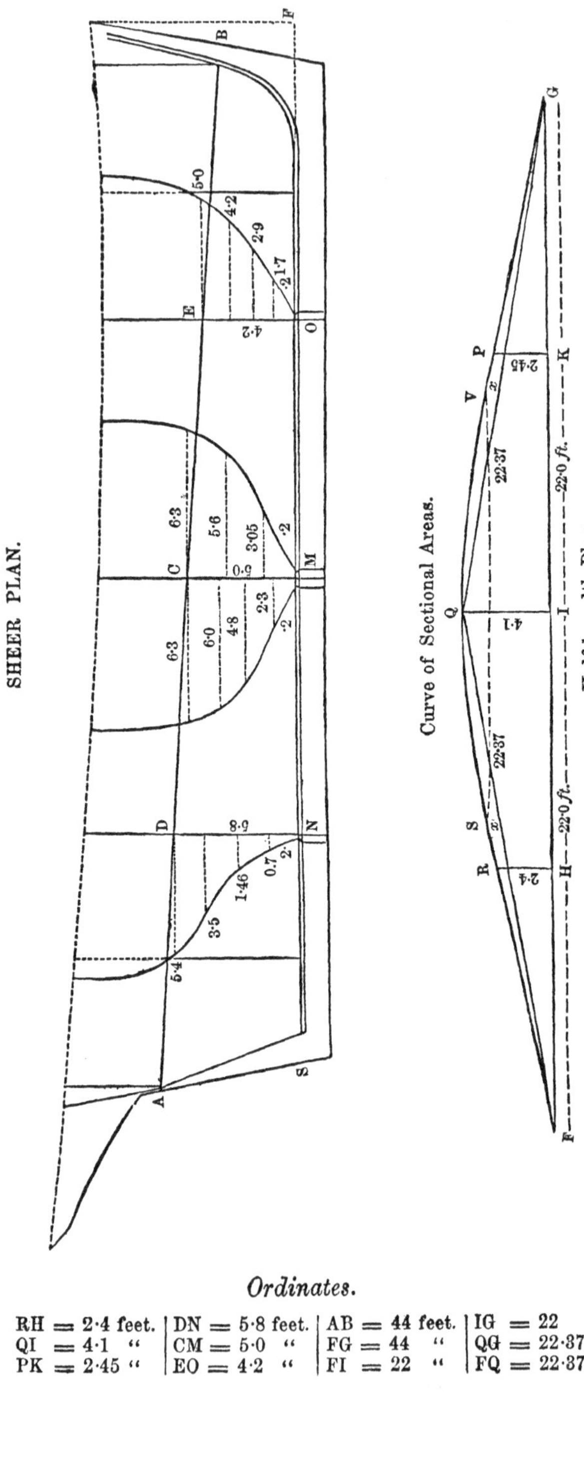

SHEER PLAN.

Curve of Sectional Areas.

Half-breadth Plan.

*Ordinates.*

| | | | |
|---|---|---|---|
| RH = 2·4 feet. | DN = 5·8 feet. | AB = 44 feet. | IG = 22 feet. |
| QI = 4·1 " | CM = 5·0 " | FG = 44 " | QG = 22·37 " |
| PK = 2·45 " | EO = 4·2 " | FI = 22 " | FQ = 22·37 " |

thus obtained, the transverse vertical sections of the vessel are delineated.

The length of the load-water line from the fore edge of the rabbet of the stem B, to the after edge of the rabbet of the post A, is next drawn below and parallel to the base line SF of the sheer plan; this line, FG, becomes the base line of the curve of the sectional areas. The common sections of the transverse vertical sections of C, D, and E, (which will be straight lines,) with this horizontal and longitudinal plan, are drawn from their respective points of division, H, I, and K, in half-breadth plan. The areas of these transverse vertical sections at D, C, and E, are then calculated, as before, thus:—

$$\text{Area} = \left\{ A + 4P + 2Q \right\} \times \frac{r}{3} = \left\{ \frac{A}{2} + 2P + Q \right\} \times \frac{2r}{3}; \text{ or,}$$

$$\text{Area} = \left\{ A + 2P + 3Q \right\} \times \frac{3}{8} r = \left\{ \frac{A}{2} + P + 1{\cdot}5\,Q \right\} \times \frac{3}{4} r.$$

*Half Area of Transverse Vertical Section, at C, by Rule* 1,

$$\text{or, } \tfrac{1}{2} \text{ Area} = \left\{ \frac{A}{2} + 2P + Q \right\} \times \frac{2r}{3}.$$

1st. ...6·3 2d ...6·0 3d...4·8 = Q
Last... ·2 4th...2·3

$2)\,6{\cdot}5$

$3{\cdot}25 = \frac{A}{2}$

$8{\cdot}3 = P$

2

$16{\cdot}60 = 2P$

$3{\cdot}25 = \frac{A}{2}$

$4{\cdot}80 = Q$

$24{\cdot}65 = \frac{A}{2} + 2P + Q$

$\cdot 83 = \frac{2r}{3}$

7395
19720

$20{\cdot}4595 = \overline{\frac{A}{2} + 2P + Q} \times \frac{2r}{3} = \tfrac{1}{2}$ area of section C in feet.

CM, or depth $= 5{\cdot}0$ feet, whence $\frac{CM}{4}$, or $\frac{5{\cdot}0}{4} = 1{\cdot}25 = r =$ distance between the ordinates, and $\frac{2r}{3} = \frac{2 \times 1{\cdot}25}{3} = \frac{2{\cdot}5}{3} = {\cdot}83$ feet.

*Half Area of Section C, by Rule 2,*

$$\text{or, } \tfrac{1}{2} \text{ area} = \left\{ \frac{A}{2} + P + 1{\cdot}5\,Q \right\} \times \frac{3}{4} r.$$

1st. ...6·3 P = 0 5·6 2d.
Last... ·2 3·05 3d.
2 ) 6·5 8·65 = Q.
$3{\cdot}25 = \frac{A}{2}$ 4·32 = ½ Q.
12·97 = 1·5 Q.

$$\left.\begin{array}{r} 3{\cdot}25 \\ 12{\cdot}97 \end{array}\right\} = \frac{A}{2} + P + 1{\cdot}5\,Q$$

16·22
5 = 3 r = CM = 5·0 feet.
4 ) 81·10

$$20{\cdot}275 = \tfrac{1}{2} \text{ area} = \overline{\frac{A}{2} + P + 1{\cdot}5\,Q} \times \frac{3}{4} r.$$

*Half Area of the Transverse Vertical Section at E.*

1st. ...5·0 2d. ...4·2 3d. ...2·9 = Q
Last... ·2 4th. ...1·7
2 ) 5·2 5·9 = P
$2{\cdot}6 = \frac{A}{2}$ 2
11·8 = 2 P
$2{\cdot}6 = \frac{A}{2}$
2·9 = Q
$17{\cdot}3 = \frac{A}{2} + 2P + Q$

EO, or depth = 4·2 feet, whence $\frac{EO}{4} = \frac{4{\cdot}2}{4} = 1{\cdot}05 = r =$ distance between the ordinates, and $\frac{2r}{3} = \frac{1{\cdot}05 \times 2}{3} = \frac{2{\cdot}1}{3} = {\cdot}7$ feet; therefore,

Area $= \left\{ \frac{A}{2} + 2P + Q \right\} \times \frac{2r}{3} = 17{\cdot}3 \times {\cdot}7 = 12{\cdot}11 =$ half area of transverse vertical section at E.

*Half Area of the Transverse Vertical Section at D.*

1st. ...5·40 2d. ...3·5 3d. ...1·46 = Q
Last...9·2 4th....0·7
2 ) 5·6 4·2 = P
$2{\cdot}8 = \frac{A}{2}$ 2
8·4 = 2 P
$2{\cdot}8 = \frac{A}{2}$
1·46 = Q
$12{\cdot}66 = \frac{A}{2} + 2P + Q$

DN, or depth = 5·8 feet, whence $\frac{DN}{4} = \frac{5 \cdot 8}{4} = 1 \cdot 45$ feet = $r$ = distance between the ordinates, and $\frac{2\,r}{3} = \frac{2 \times 1 \cdot 45}{3} = \frac{2 \cdot 9}{3} =$ ·97 feet; therefore,

$$\text{Area} = \left\{ \frac{A}{2} 2 + P + Q \right\} \times \frac{2\,r}{3} = 12 \cdot 66 \times \cdot 97 = 12 \cdot 28 \text{ feet} =$$

half area of transverse vertical section at D.

*Half Areas of the Transverse Vertical Sections.*

| | Feet. | | Feet. |
|---|---|---|---|
| At | E = 12·11 | Divided by 5 as the depth assumed for | 2·42 |
| | C = 20·20 | the zone, give the ordinates for the curve | 4·04 |
| | D = 12·28 | of sectional areas, as........................ | 2·45 |

of which 2·42 is set off from H as HR, 4·04 feet from I as IQ, and 2·45 feet from K as KP; the curve IRQPG, passing through the extremities P, Q, and R of the ordinates PK, QI, and RH, is the curve bounding the area of a zone, which, to the depth of 5 feet for a solid, will give in cubic feet of space the half displacement of the immersed body, or the displacement of the yacht to the line AB of proposed deepest immersion.

To measure this representative area, and from thence the solid, join the points Q, G, and I by the straight lines QG, QF, dividing the curvilinear area FRQPGF into the two triangles QGI, QFI, and the two areas GPQG, FRQF. The triangles by construction are equal, and the area of each one of them is equivalent to $\frac{GI \times QI}{2}$, or the whole area GQFIG = $\frac{GI \times QI}{2} \times 2 = GI \times QI$ or FI × IQ, FI being equal to IG, each being the half-length of the same element, the load-water line or line of deepest immersion. The areas QPGQ, QRFQ, are bounded by the curve lines QPG, QRF, which are assumed as portions of common parabolas, and under such an assumption their respective areas are equal to $\frac{2}{3}$ of the circumscribing parallelograms, or the area QPGQ = $\frac{2}{3}$ of GQ × $x$, and the area FRQF = $\frac{2}{3}$ of FQ × $x'$, where $x$ and $x'$ are the greatest perpendiculars that can be drawn from the bases QG and QF to meet the curves QPG, QRF.

DISPLACEMENT.

AB by a scale of parts = 44 feet, whence FI or IG equal $\frac{AB}{2} = \frac{44}{2}$ feet = 22 feet; ordinate QI of the medial section = 4·04 feet; and QG = FQ, being the respective hypothenuses of the equal triangles QGI, QFI, are each equal to $\sqrt{IG^2 + QI^2} = \sqrt{22^2 + 4 \cdot 04^2} = \sqrt{484 + 16 \cdot 32} = \sqrt{500 \cdot 32} = 22 \cdot 37$ feet; and $x$, by measurement with a scale of parts, = ·6 foot, and $x'$ also ·6 foot, from which the half displacement in cubic feet of space will be obtained as follows:—

Area FQGIF = GI × IQ. Cubic feet.

Solid under the area FQGIF } $= GI \times IQ \times 5 = 22 \times 4{\cdot}1 \times 5 =$ 451·00

Area QPGQ = $\frac{2}{3}$ of GQ × $x$

Solid under the area QPGQ } $= \frac{2}{3}$ of $GQ \times x \times 5 = \frac{2}{3} \times 22{\cdot}37 \times {\cdot}6 \times 5 =$ 44·74

Area FRQF = $\frac{2}{3}$ of FQ × $x'$

Solid under the area FRQF } $= \frac{2}{3}$ of $FQ \times x' \times 5 = \frac{2}{3} \times = 22{\cdot}37 \times {\cdot}6 \times 5 =$ 44·74

540·48

or area of the triangle QGI + area of the triangle QFI + area of the space QPGQ + area of the space FRQF = to the representative area FRQPG, which being multiplied by the assumed depth of 5 feet for the zone of half displacement gives 540·48 cubic feet of space, which divided by 35, as the number of such cubic feet that are equivalent to one ton of medium water, gives

3 ) 540·48

7 ) 108·09

15·44 tons for half displacement,

and that the whole weight of the body is equal to 15·54 × 2 = 30·88 tons = displacement to the line of proposed deepest immersion AB.

RELATIVE CAPACITIES OF THE BODIES IMMERSED UNDER THE FORE AND AFTER HALF-LENGTHS OF THE LOAD-WATER LINE, AS GIVEN BY THE DELINEATED CURVE OF SECTIONAL AREAS.

The triangles QGI and QFI being equal, the relative capacities of the fore and after-bodies will be determined by the proportion that the area QPGI bears to the area QRFI; and as these areas involve two equal terms, or that the base FQ = the base QG, it follows, that the relation of these areas to each other will be expressed by the proportion that the perpendiculars $x$ and $x'$ bear to each other. In the example given, the fore and after-bodies, or the displacements under the fore and after half-lengths of the load-water AB, are equal; as the perpendiculars $x$ and $x'$ taken from the diagram, on a scale of equal parts, are each ·6 of a foot.

The area of the midship section is denoted relatively by the medial ordinate of the curve of sections QI, and the full amount of it is obtained by multiplying the function QI by the depth of the zone M. In the example:

M = 5; QI = 4·04; then half area of medial section = 4·04 × 5
5

Area of midship section......20·20

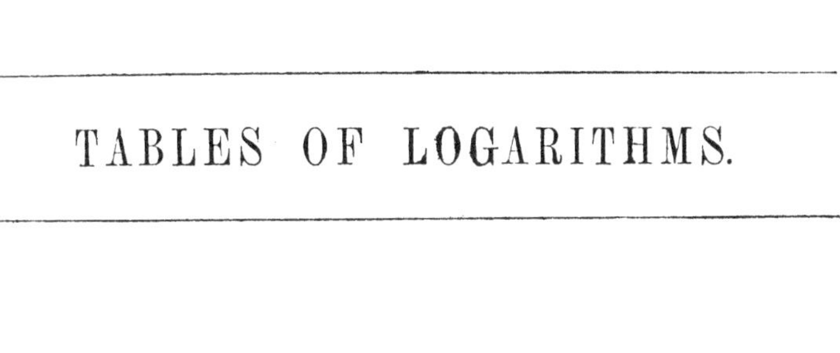

# TABLES OF LOGARITHMS.

| No. | Log. | Prop. Part. | No. | Log. | Prop. Part. | No. | Log. | Prop. Part. | No. | Log. | Prop. Part. |
|---|---|---|---|---|---|---|---|---|---|---|---|
| 1000 | 000000 | | 1060 | 025306 | | 1120 | 049218 | | 1180 | 071882 | |
| 1 | 000434 | 43 | 1 | 025715 | 41 | 1 | 049606 | 39 | 1 | 072250 | 37 |
| 2 | 000868 | 86 | 2 | 026124 | 82 | 2 | 049993 | 77 | 2 | 072617 | 73 |
| 3 | 001301 | 130 | 3 | 026533 | 122 | 3 | 050380 | 116 | 3 | 072985 | 110 |
| 4 | 001734 | 173 | 4 | 026942 | 163 | 4 | 050766 | 154 | 4 | 073352 | 147 |
| 5 | 002166 | 216 | 5 | 027350 | 204 | 5 | 051152 | 193 | 5 | 073718 | 183 |
| 6 | 002598 | 259 | 6 | 027757 | 245 | 6 | 051538 | 232 | 6 | 074085 | 220 |
| 7 | 003029 | 303 | 7 | 028164 | 286 | 7 | 051924 | 270 | 7 | 074451 | 256 |
| 8 | 003460 | 346 | 8 | 028571 | 326 | 8 | 052309 | 309 | 8 | 074816 | 293 |
| 9 | 003891 | 389 | 9 | 028978 | 367 | 9 | 052694 | 347 | 9 | 075182 | 330 |
| 1010 | 004321 | · | 1070 | 029384 | | 1130 | 053078 | | 1190 | 075547 | |
| 1 | 004751 | 43 | 1 | 029789 | 40 | 1 | 053463 | 38 | 1 | 675912 | 36 |
| 2 | 005180 | 86 | 2 | 030195 | 81 | 2 | 053846 | 77 | 2 | 076276 | 73 |
| 3 | 005609 | 128 | 3 | 030600 | 121 | 3 | 054230 | 115 | 3 | 076640 | 109 |
| 4 | 006038 | 171 | 4 | 031004 | 162 | 4 | 054613 | 153 | 4 | 077004 | 145 |
| 5 | 006466 | 214 | 5 | 031408 | 202 | 5 | 054996 | 191 | 5 | 077368 | 181 |
| 6 | 006894 | 257 | 6 | 031812 | 242 | 6 | 055378 | 230 | 6 | 077731 | 218 |
| 7 | 007321 | 300 | 7 | 032216 | 283 | 7 | 055760 | 268 | 7 | 078094 | 254 |
| 8 | 007748 | 343 | 8 | 032619 | 323 | 8 | 056142 | 306 | 8 | 078457 | 290 |
| 9 | 008174 | 385 | 9 | 033021 | 364 | 9 | 056524 | 345 | 9 | 078819 | 327 |
| 1020 | 008600 | | 1080 | 033424 | | 1140 | 056905 | | 1200 | 079181 | |
| 1 | 009026 | 42 | 1 | 033826 | 40 | 1 | 057286 | 38 | 1 | 079543 | 36 |
| 2 | 009451 | 85 | 2 | 034227 | 80 | 2 | 057666 | 76 | 2 | 079904 | 72 |
| 3 | 009876 | 127 | 3 | 034628 | 120 | 3 | 058046 | 114 | 3 | 080266 | 108 |
| 4 | 010300 | 170 | 4 | 035029 | 160 | 4 | 058426 | 152 | 4 | 080626 | 144 |
| 5 | 010724 | 212 | 5 | 035430 | 200 | 5 | 058805 | 190 | 5 | 080987 | 180 |
| 6 | 011147 | 254 | 6 | 035830 | 240 | 6 | 059185 | 228 | 6 | 081347 | 216 |
| 7 | 011570 | 297 | 7 | 036229 | 280 | 7 | 059563 | 266 | 7 | 081707 | 252 |
| 8 | 011993 | 339 | 8 | 036629 | 321 | 8 | 059942 | 304 | 8 | 082067 | 288 |
| 9 | 012415 | 382 | 9 | 037028 | 361 | 9 | 060320 | 342 | 9 | 082426 | 324 |
| 1030 | 012837 | | 1090 | 037426 | | 1150 | 060698 | | 1210 | 082785 | |
| 1 | 013259 | 42 | 1 | 037825 | 40 | 1 | 061075 | 38 | 1 | 083144 | 36 |
| 2 | 013680 | 84 | 2 | 038223 | 79 | 2 | 061452 | 75 | 2 | 083503 | 71 |
| 3 | 014100 | 126 | 3 | 038620 | 119 | 3 | 061829 | 113 | 3 | 083861 | 107 |
| 4 | 014520 | 168 | 4 | 039017 | 159 | 4 | 062206 | 160 | 4 | 084219 | 143 |
| 5 | 014940 | 210 | 5 | 039414 | 198 | 5 | 062582 | 188 | 5 | 084576 | 179 |
| 6 | 015360 | 252 | 6 | 039811 | 238 | 6 | 062958 | 226 | 6 | 084934 | 214 |
| 7 | 015779 | 294 | 7 | 040207 | 278 | 7 | 063333 | 263 | 7 | 085291 | 250 |
| 8 | 016197 | 336 | 8 | 040602 | 318 | 8 | 063709 | 301 | 8 | 085647 | 286 |
| 9 | 016615 | 378 | 9 | 040998 | 357 | 9 | 064083 | 338 | 9 | 086004 | 322 |
| 1040 | 017033 | | 1100 | 041393 | | 1160 | 064458 | | 1220 | 086360 | |
| 1 | 017451 | 42 | 1 | 041787 | 39 | 1 | 064832 | 37 | 1 | 086716 | 35 |
| 2 | 017868 | 83 | 2 | 042182 | 79 | 2 | 065206 | 75 | 2 | 087071 | 71 |
| 3 | 018284 | 125 | 3 | 042575 | 118 | 3 | 065580 | 112 | 3 | 087426 | 106 |
| 4 | 018700 | 166 | 4 | 042969 | 157 | 4 | 065953 | 149 | 4 | 087781 | 142 |
| 5 | 019116 | 208 | 5 | 043362 | 196 | 5 | 066326 | 186 | 5 | 088136 | 177 |
| 6 | 019532 | 250 | 6 | 043755 | 236 | 6 | 066699 | 224 | 6 | 088490 | 213 |
| 7 | 019947 | 291 | 7 | 044148 | 275 | 7 | 067071 | 261 | 7 | 088845 | 248 |
| 8 | 020361 | 333 | 8 | 044540 | 314 | 8 | 067443 | 298 | 8 | 089198 | 284 |
| 9 | 020775 | 374 | 9 | 044931 | 354 | 9 | 067814 | 336 | 9 | 089552 | 319 |
| 1050 | 021189 | | 1110 | 045323 | | 1170 | 068186 | | 1230 | 089905 | |
| 1 | 021603 | 41 | 1 | 045714 | 39 | 1 | 068557 | 37 | 1 | 090258 | 35 |
| 2 | 022016 | 82 | 2 | 046105 | 78 | 2 | 068928 | 74 | 2 | 090611 | 70 |
| 3 | 022428 | 124 | 3 | 046495 | 117 | 3 | 069298 | 111 | 3 | 090963 | 106 |
| 4 | 022841 | 165 | 4 | 046885 | 156 | 4 | 069668 | 148 | 4 | 091315 | 141 |
| 5 | 023252 | 206 | 5 | 047275 | 195 | 5 | 070038 | 185 | 5 | 091667 | 176 |
| 6 | 023664 | 247 | 6 | 047664 | 234 | 6 | 070407 | 222 | 6 | 092018 | 211 |
| 7 | 024075 | 288 | 7 | 048053 | 273 | 7 | 070776 | 259 | 7 | 092370 | 246 |
| 8 | 024486 | 330 | 8 | 048442 | 312 | 8 | 071145 | 296 | 8 | 092721 | 282 |
| 9 | 024896 | 371 | 9 | 048830 | 351 | 9 | 071514 | 333 | 9 | 093071 | 317 |

| No. | Log. | Prop. Part. | No. | Log. | Prop. Part. | No. | Log. | Prop. Part. | No. | Log. | Prop. Part. |
|---|---|---|---|---|---|---|---|---|---|---|---|
| 1240 | 093422 | | 1300 | 113943 | | 1360 | 133539 | | 1420 | 152288 | |
| 1 | 093772 | 35 | 1 | 114277 | 33 | 1 | 133858 | 32 | 1 | 152594 | 30 |
| 2 | 094122 | 70 | 2 | 114611 | 67 | 2 | 134177 | 64 | 2 | 152900 | 61 |
| 3 | 094471 | 105 | 3 | 114944 | 100 | 3 | 134496 | 96 | 3 | 153205 | 91 |
| 4 | 094820 | 140 | 4 | 115278 | 133 | 4 | 134814 | 127 | 4 | 153510 | 122 |
| 5 | 095169 | 175 | 5 | 115610 | 167 | 5 | 135133 | 159 | 5 | 153815 | 152 |
| 6 | 095518 | 210 | 6 | 115943 | 200 | 6 | 135451 | 191 | 6 | 154119 | 183 |
| 7 | 095866 | 245 | 7 | 116276 | 233 | 7 | 135768 | 223 | 7 | 154424 | 213 |
| 8 | 096215 | 280 | 8 | 116608 | 267 | 8 | 136086 | 255 | 8 | 154728 | 244 |
| 9 | 096562 | 315 | 9 | 116940 | 300 | 9 | 136403 | 287 | 9 | 155032 | 274 |
| 1250 | 096910 | | 1310 | 117271 | | 1370 | 136721 | | 1430 | 155336 | |
| 1 | 097257 | 35 | 1 | 117603 | 33 | 1 | 137037 | 32 | 1 | 155640 | 30 |
| 2 | 097604 | 69 | 2 | 117934 | 66 | 2 | 137354 | 63 | 2 | 155943 | 60 |
| 3 | 097951 | 104 | 3 | 118265 | 99 | 3 | 137670 | 94 | 3 | 156246 | 91 |
| 4 | 098297 | 138 | 4 | 118595 | 132 | 4 | 137987 | 126 | 4 | 156549 | 121 |
| 5 | 098644 | 173 | 5 | 118926 | 165 | 5 | 138303 | 158 | 5 | 156852 | 151 |
| 6 | 098990 | 208 | 6 | 119256 | 198 | 6 | 138618 | 189 | 6 | 157154 | 181 |
| 7 | 099335 | 242 | 7 | 119586 | 231 | 7 | 138934 | 221 | 7 | 157457 | 211 |
| 8 | 099681 | 277 | 8 | 119915 | 264 | 8 | 139249 | 252 | 8 | 157759 | 242 |
| 9 | 100026 | 311 | 9 | 120245 | 297 | 9 | 139564 | 284 | 9 | 158061 | 272 |
| 1260 | 100370 | | 1320 | 120574 | | 1380 | 139879 | | 1440 | 158362 | |
| 1 | 100715 | 34 | 1 | 120903 | 33 | 1 | 140194 | 31 | 1 | 158664 | 30 |
| 2 | 101059 | 69 | 2 | 121231 | 66 | 2 | 140508 | 63 | 2 | 158965 | 60 |
| 3 | 101403 | 103 | 3 | 121560 | 98 | 3 | 140822 | 94 | 3 | 159266 | 90 |
| 4 | 101747 | 137 | 4 | 121888 | 131 | 4 | 141136 | 125 | 4 | 159567 | 120 |
| 5 | 102090 | 172 | 5 | 122216 | 164 | 5 | 141450 | 157 | 5 | 159868 | 150 |
| 6 | 102434 | 206 | 6 | 122543 | 197 | 6 | 141763 | 188 | 6 | 160168 | 180 |
| 7 | 102777 | 240 | 7 | 122871 | 230 | 7 | 142076 | 219 | 7 | 160468 | 210 |
| 8 | 103119 | 275 | 8 | 123198 | 262 | 8 | 142389 | 251 | 8 | 160769 | 240 |
| 9 | 103462 | 309 | 9 | 123525 | 295 | 9 | 142702 | 282 | 9 | 161068 | 270 |
| 1270 | 103804 | | 1330 | 123852 | | 1390 | 143015 | | 1450 | 161368 | |
| 1 | 104146 | 34 | 1 | 124178 | 33 | 1 | 143327 | 31 | 1 | 161667 | 30 |
| 2 | 104487 | 68 | 2 | 124504 | 65 | 2 | 143639 | 62 | 2 | 161967 | 60 |
| 3 | 104828 | 102 | 3 | 124830 | 98 | 3 | 143951 | 93 | 3 | 162266 | 89 |
| 4 | 105169 | 136 | 4 | 125156 | 130 | 4 | 144263 | 125 | 4 | 162564 | 119 |
| 5 | 105510 | 170 | 5 | 125481 | 163 | 5 | 144574 | 156 | 5 | 162863 | 149 |
| 6 | 105851 | 204 | 6 | 125806 | 195 | 6 | 144885 | 187 | 6 | 163161 | 179 |
| 7 | 106191 | 238 | 7 | 126131 | 228 | 7 | 145196 | 218 | 7 | 163460 | 209 |
| 8 | 106531 | 272 | 8 | 126456 | 260 | 8 | 145507 | 249 | 8 | 163757 | 239 |
| 9 | 106870 | 306 | 9 | 126781 | 293 | 9 | 145818 | 280 | 9 | 164055 | 269 |
| 1280 | 107210 | | 1340 | 127105 | | 1400 | 146128 | | 1460 | 164353 | |
| 1 | 107549 | 34 | 1 | 127429 | 32 | 1 | 146438 | 31 | 1 | 164650 | 30 |
| 2 | 107888 | 67 | 2 | 127752 | 65 | 2 | 146748 | 62 | 2 | 164947 | 59 |
| 3 | 108227 | 101 | 3 | 128076 | 97 | 3 | 147058 | 93 | 3 | 165244 | 89 |
| 4 | 108565 | 135 | 4 | 128399 | 129 | 4 | 147367 | 124 | 4 | 165541 | 119 |
| 5 | 108903 | 169 | 5 | 128722 | 161 | 5 | 147676 | 155 | 5 | 165838 | 148 |
| 6 | 109241 | 203 | 6 | 129045 | 194 | 6 | 147985 | 186 | 6 | 166134 | 178 |
| 7 | 109578 | 237 | 7 | 129368 | 226 | 7 | 148294 | 217 | 7 | 166430 | 207 |
| 8 | 109916 | 270 | 8 | 129690 | 258 | 8 | 148603 | 248 | 8 | 166726 | 237 |
| 9 | 110253 | 304 | 9 | 130012 | 291 | 9 | 148911 | 279 | 9 | 167022 | 267 |
| 1290 | 110590 | | 1350 | 130334 | | 1410 | 149219 | | 1470 | 167317 | |
| 1 | 110926 | 34 | 1 | 130655 | 32 | 1 | 149527 | 31 | 1 | 167613 | 29 |
| 2 | 111262 | 67 | 2 | 130977 | 64 | 2 | 149835 | 61 | 2 | 167908 | 59 |
| 3 | 111598 | 101 | 3 | 131298 | 96 | 3 | 150142 | 92 | 3 | 168203 | 88 |
| 4 | 111934 | 134 | 4 | 131619 | 128 | 4 | 150449 | 123 | 4 | 168497 | 118 |
| 5 | 112270 | 168 | 5 | 131939 | 160 | 5 | 150756 | 154 | 5 | 168792 | 147 |
| 6 | 112605 | 201 | 6 | 132260 | 192 | 6 | 151063 | 184 | 6 | 169086 | 177 |
| 7 | 112940 | 235 | 7 | 132580 | 224 | 7 | 151370 | 215 | 7 | 169380 | 206 |
| 8 | 113275 | 268 | 8 | 132900 | 256 | 8 | 151676 | 246 | 8 | 169674 | 236 |
| 9 | 113609 | 302 | 9 | 133219 | 288 | 9 | 151982 | 277 | 9 | 169968 | 265 |

| No. | Log. | Prop. Part. | No. | Log. | Prop. Part. | No. | Log. | Prop. Part. | No. | Log. | Prop. Part. |
|---|---|---|---|---|---|---|---|---|---|---|---|
| 1480 | 170262 | | 1540 | 187521 | | 1600 | 204120 | | 1660 | 220108 | |
| 1 | 170555 | 29 | 1 | 187803 | 28 | 1 | 204391 | 27 | 1 | 220370 | 26 |
| 2 | 170848 | 58 | 2 | 188084 | 56 | 2 | 204662 | 54 | 2 | 220631 | 52 |
| 3 | 171141 | 88 | 3 | 188366 | 84 | 3 | 204933 | 81 | 3 | 220892 | 78 |
| 4 | 171434 | 117 | 4 | 188647 | 113 | 4 | 205204 | 108 | 4 | 221153 | 104 |
| 5 | 171726 | 146 | 5 | 188928 | 141 | 5 | 205475 | 135 | 5 | 221414 | 130 |
| 6 | 172019 | 175 | 6 | 189209 | 169 | 6 | 205745 | 162 | 6 | 221675 | 157 |
| 7 | 172311 | 204 | 7 | 189490 | 197 | 7 | 206016 | 189 | 7 | 221936 | 183 |
| 8 | 172603 | 234 | 8 | 189771 | 225 | 8 | 206286 | 216 | 8 | 222196 | 209 |
| 9 | 172895 | 263 | 9 | 190051 | 253 | 9 | 206556 | 243 | 9 | 222456 | 235 |
| 1490 | 173186 | | 1550 | 190332 | | 1610 | 206826 | | 1670 | 222716 | |
| 1 | 173478 | 29 | 1 | 190612 | 28 | 1 | 207095 | 27 | 1 | 222976 | 26 |
| 2 | 173769 | 58 | 2 | 190892 | 56 | 2 | 207365 | 54 | 2 | 223236 | 52 |
| 3 | 174060 | 87 | 3 | 191171 | 84 | 3 | 207634 | 81 | 3 | 223496 | 78 |
| 4 | 174351 | 116 | 4 | 191451 | 112 | 4 | 207903 | 108 | 4 | 223755 | 104 |
| 5 | 174641 | 145 | 5 | 191730 | 140 | 5 | 208172 | 135 | 5 | 224015 | 130 |
| 6 | 174932 | 175 | 6 | 192010 | 168 | 6 | 208441 | 162 | 6 | 224274 | 156 |
| 7 | 175222 | 204 | 7 | 192289 | 196 | 7 | 208710 | 188 | 7 | 224533 | 182 |
| 8 | 175512 | 233 | 8 | 192567 | 224 | 8 | 208978 | 215 | 8 | 224792 | 208 |
| 9 | 175802 | 261 | 9 | 192846 | 252 | 9 | 209247 | 241 | 9 | 225051 | 234 |
| 1500 | 176091 | | 1560 | 193125 | | 1620 | 209515 | | 1680 | 225309 | |
| 1 | 176381 | 29 | 1 | 193403 | 28 | 1 | 209783 | 27 | 1 | 225568 | 26 |
| 2 | 176670 | 58 | 2 | 193681 | 56 | 2 | 210051 | 54 | 2 | 225826 | 52 |
| 3 | 176959 | 86 | 3 | 193959 | 83 | 3 | 210318 | 80 | 3 | 226084 | 77 |
| 4 | 177248 | 115 | 4 | 194237 | 111 | 4 | 210586 | 107 | 4 | 226342 | 103 |
| 5 | 177536 | 144 | 5 | 194514 | 139 | 5 | 210853 | 134 | 5 | 226600 | 129 |
| 6 | 177825 | 173 | 6 | 194792 | 166 | 6 | 211120 | 161 | 6 | 226858 | 155 |
| 7 | 178113 | 202 | 7 | 195069 | 194 | 7 | 211388 | 187 | 7 | 227115 | 181 |
| 8 | 178401 | 231 | 8 | 195346 | 222 | 8 | 211654 | 214 | 8 | 227372 | 206 |
| 9 | 178689 | 259 | 9 | 195623 | 250 | 9 | 211921 | 240 | 9 | 227630 | 232 |
| 1510 | 178977 | | 1570 | 195900 | | 1630 | 212188 | | 1690 | 227887 | |
| 1 | 179264 | 29 | 1 | 196176 | 27 | 1 | 212454 | 27 | 1 | 228144 | 26 |
| 2 | 179552 | 57 | 2 | 196452 | 55 | 2 | 212720 | 53 | 2 | 228400 | 51 |
| 3 | 179839 | 86 | 3 | 196729 | 83 | 3 | 212986 | 80 | 3 | 228657 | 77 |
| 4 | 180126 | 115 | 4 | 197005 | 110 | 4 | 213252 | 106 | 4 | 228913 | 102 |
| 5 | 180413 | 144 | 5 | 197281 | 138 | 5 | 213518 | 133 | 5 | 229170 | 128 |
| 6 | 180699 | 172 | 6 | 197556 | 166 | 6 | 213783 | 159 | 6 | 229426 | 154 |
| 7 | 180986 | 201 | 7 | 197832 | 193 | 7 | 214049 | 186 | 7 | 229682 | 179 |
| 8 | 181272 | 230 | 8 | 198107 | 221 | 8 | 214314 | 212 | 8 | 229938 | 205 |
| 9 | 181558 | 258 | 9 | 108382 | 248 | 9 | 214579 | 239 | 9 | 230193 | 231 |
| 1520 | 181844 | | 1580 | 198657 | | 1640 | 214844 | | 1700 | 230449 | |
| 1 | 182129 | 28 | 1 | 198932 | 27 | 1 | 215109 | 26 | 1 | 230704 | 25 |
| 2 | 182415 | 57 | 2 | 199206 | 55 | 2 | 215373 | 53 | 2 | 230960 | 51 |
| 3 | 182700 | 86 | 3 | 199481 | 82 | 3 | 215638 | 79 | 3 | 231215 | 76 |
| 4 | 182985 | 114 | 4 | 199755 | 110 | 4 | 215902 | 106 | 4 | 231470 | 102 |
| 5 | 183270 | 143 | 5 | 200029 | 137 | 5 | 216166 | 132 | 5 | 231724 | 127 |
| 6 | 183554 | 171 | 6 | 200303 | 164 | 6 | 216430 | 158 | 6 | 231979 | 153 |
| 7 | 183839 | 200 | 7 | 200577 | 192 | 7 | 216694 | 185 | 7 | 232233 | 178 |
| 8 | 184123 | 228 | 8 | 200850 | 219 | 8 | 216957 | 211 | 8 | 232488 | 204 |
| 9 | 184407 | 256 | 9 | 201124 | 247 | 9 | 217221 | 238 | 9 | 232742 | 229 |
| 1530 | 184691 | | 1590 | 201397 | | 1650 | 217484 | | 1710 | 232996 | |
| 1 | 184975 | 28 | 1 | 201670 | 27 | 1 | 217747 | 26 | 1 | 233250 | 25 |
| 2 | 185259 | 57 | 2 | 201943 | 54 | 2 | 218010 | 52 | 2 | 233504 | 51 |
| 3 | 185542 | 85 | 3 | 202216 | 82 | 3 | 218273 | 79 | 3 | 233757 | 76 |
| 4 | 185825 | 113 | 4 | 202488 | 109 | 4 | 218535 | 105 | 4 | 234011 | 101 |
| 5 | 186108 | 142 | 5 | 202761 | 136 | 5 | 218798 | 131 | 5 | 234264 | 127 |
| 6 | 186391 | 170 | 6 | 203033 | 163 | 6 | 219060 | 157 | 6 | 234517 | 152 |
| 7 | 186674 | 198 | 7 | 203305 | 191 | 7 | 219322 | 183 | 7 | 234770 | 177 |
| 8 | 186956 | 227 | 8 | 203577 | 218 | 8 | 219584 | 210 | 8 | 235023 | 202 |
| 9 | 187239 | 255 | 9 | 203848 | 245 | 9 | 219846 | 236 | 9 | 235276 | 228 |

| No. | Log. | Prop. Part. | No. | Log. | Prop. Part. | No. | Log. | Prop. Part. | No. | Log. | Prop. Part. |
|---|---|---|---|---|---|---|---|---|---|---|---|
| 1720 | 235528 | | 1780 | 250420 | | 1840 | 264818 | | 1900 | 278754 | |
| 1 | 235781 | 25 | 1 | 250664 | 24 | 1 | 265054 | 23 | 1 | 278982 | 23 |
| 2 | 236033 | 50 | 2 | 250908 | 49 | 2 | 265290 | 47 | 2 | 279210 | 45 |
| 3 | 236285 | 76 | 3 | 251151 | 73 | 3 | 265525 | 70 | 3 | 279439 | 68 |
| 4 | 236537 | 101 | 4 | 251395 | 97 | 4 | 265761 | 94 | 4 | 279667 | 91 |
| 5 | 236789 | 126 | 5 | 251638 | 121 | 5 | 265996 | 117 | 5 | 279895 | 114 |
| 6 | 237041 | 151 | 6 | 251881 | 146 | 6 | 266232 | 141 | 6 | 280123 | 137 |
| 7 | 237292 | 176 | 7 | 252125 | 171 | 7 | 266467 | 164 | 7 | 280351 | 160 |
| 8 | 237544 | 202 | 8 | 252367 | 195 | 8 | 266702 | 188 | 8 | 280578 | 182 |
| 9 | 237795 | 227 | 9 | 252610 | 219 | 9 | 266937 | 211 | 9 | 280806 | 205 |
| 1730 | 238046 | | 1790 | 252853 | | 1850 | 267172 | | 1910 | 281033 | |
| 1 | 238297 | 25 | 1 | 253096 | 24 | 1 | 267406 | 23 | 1 | 281261 | 23 |
| 2 | 238548 | 50 | 2 | 253338 | 48 | 2 | 267641 | 47 | 2 | 281488 | 45 |
| 3 | 238799 | 75 | 3 | 253580 | 73 | 3 | 267875 | 70 | 3 | 281715 | 68 |
| 4 | 239049 | 100 | 4 | 253822 | 97 | 4 | 268110 | 94 | 4 | 281942 | 91 |
| 5 | 239299 | 125 | 5 | 254064 | 121 | 5 | 268344 | 117 | 5 | 282169 | 113 |
| 6 | 239550 | 150 | 6 | 254306 | 145 | 6 | 268578 | 141 | 6 | 282395 | 136 |
| 7 | 239800 | 175 | 7 | 254548 | 170 | 7 | 268812 | 164 | 7 | 282622 | 159 |
| 8 | 240050 | 200 | 8 | 254790 | 194 | 8 | 269046 | 188 | 8 | 282849 | 181 |
| 9 | 240300 | 225 | 9 | 255031 | 218 | 9 | 269279 | 211 | 9 | 283075 | 204 |
| 1740 | 240549 | | 1800 | 255273 | | 1860 | 269513 | | 1920 | 283301 | |
| 1 | 240799 | 25 | 1 | 255514 | 24 | 1 | 269746 | 23 | 1 | 283527 | 23 |
| 2 | 241048 | 50 | 2 | 255755 | 48 | 2 | 269980 | 47 | 2 | 283753 | 45 |
| 3 | 241297 | 75 | 3 | 255996 | 72 | 3 | 270213 | 70 | 3 | 283979 | 68 |
| 4 | 241546 | 100 | 4 | 256236 | 96 | 4 | 270446 | 93 | 4 | 284205 | 90 |
| 5 | 241795 | 124 | 5 | 256477 | 120 | 5 | 270679 | 116 | 5 | 284431 | 113 |
| 6 | 242044 | 149 | 6 | 256718 | 144 | 6 | 270912 | 140 | 6 | 284656 | 135 |
| 7 | 242293 | 174 | 7 | 256958 | 168 | 7 | 271144 | 163 | 7 | 284882 | 158 |
| 8 | 242541 | 199 | 8 | 257198 | 192 | 8 | 271377 | 186 | 8 | 285107 | 180 |
| 9 | 242790 | 223 | 9 | 257439 | 216 | 9 | 271609 | 210 | 9 | 285332 | 203 |
| 1750 | 243038 | | 1810 | 257679 | | 1870 | 271842 | | 1930 | 285557 | |
| 1 | 243286 | 25 | 1 | 257918 | 24 | 1 | 272074 | 23 | 1 | 285782 | 22 |
| 2 | 243534 | 50 | 2 | 258158 | 48 | 2 | 272306 | 46 | 2 | 286007 | 45 |
| 3 | 243782 | 74 | 3 | 258398 | 72 | 3 | 272538 | 70 | 3 | 286232 | 67 |
| 4 | 244030 | 99 | 4 | 258637 | 96 | 4 | 272776 | 93 | 4 | 286456 | 89 |
| 5 | 244277 | 124 | 5 | 258877 | 120 | 5 | 273001 | 116 | 5 | 286681 | 112 |
| 6 | 244524 | 149 | 6 | 259116 | 144 | 6 | 273233 | 139 | 6 | 286905 | 134 |
| 7 | 244772 | 174 | 7 | 259355 | 167 | 7 | 273464 | 162 | 7 | 287130 | 157 |
| 8 | 245019 | 198 | 8 | 259594 | 192 | 8 | 273696 | 186 | 8 | 287354 | 179 |
| 9 | 245266 | 222 | 9 | 259833 | 215 | 9 | 273927 | 209 | 9 | 287578 | 202 |
| 1760 | 245513 | | 1820 | 260071 | | 1880 | 274158 | | 1940 | 287802 | |
| 1 | 245759 | 25 | 1 | 260310 | 24 | 1 | 274389 | 23 | 1 | 288025 | 22 |
| 2 | 246006 | 49 | 2 | 260548 | 48 | 2 | 274620 | 46 | 2 | 288249 | 45 |
| 3 | 246252 | 74 | 3 | 260787 | 71 | 3 | 274850 | 69 | 3 | 288473 | 67 |
| 4 | 246499 | 98 | 4 | 261025 | 95 | 4 | 275081 | 92 | 4 | 288696 | 89 |
| 5 | 246745 | 123 | 5 | 261263 | 119 | 5 | 275311 | 115 | 5 | 288920 | 112 |
| 6 | 246991 | 148 | 6 | 261501 | 143 | 6 | 275542 | 138 | 6 | 289143 | 134 |
| 7 | 247236 | 173 | 7 | 261738 | 167 | 7 | 275772 | 161 | 7 | 289366 | 156 |
| 8 | 247482 | 197 | 8 | 261976 | 191 | 8 | 276002 | 184 | 8 | 289589 | 178 |
| 9 | 247728 | 221 | 9 | 262214 | 214 | 9 | 276232 | 207 | 9 | 289812 | 201 |
| 1770 | 247973 | | 1830 | 262451 | | 1890 | 276462 | | 1950 | 290035 | |
| 1 | 248219 | 25 | 1 | 262688 | 24 | 1 | 276691 | 23 | 1 | 290257 | 22 |
| 2 | 248464 | 49 | 2 | 262925 | 47 | 2 | 276921 | 46 | 2 | 290480 | 44 |
| 3 | 248709 | 74 | 3 | 263162 | 71 | 3 | 277151 | 69 | 3 | 290702 | 67 |
| 4 | 248954 | 98 | 4 | 263399 | 95 | 4 | 277380 | 92 | 4 | 290925 | 89 |
| 5 | 249198 | 123 | 5 | 263636 | 118 | 5 | 277609 | 115 | 5 | 291147 | 111 |
| 6 | 249443 | 147 | 6 | 263873 | 142 | 6 | 277838 | 138 | 6 | 291369 | 133 |
| 7 | 249687 | 172 | 7 | 264109 | 166 | 7 | 278067 | 161 | 7 | 291591 | 156 |
| 8 | 249932 | 196 | 8 | 264345 | 190 | 8 | 278296 | 183 | 8 | 291813 | 178 |
| 9 | 250176 | 220 | 9 | 264582 | 213 | 9 | 278525 | 206 | 9 | 292034 | 200 |

| No. | Log. | Prop. Part. | No. | Log. | Prop. Part. | No. | Log. | Prop. Part. | No. | Log. | Prop. Part. |
|---|---|---|---|---|---|---|---|---|---|---|---|
| 1960 | 292256 | | 2020 | 305351 | | 2080 | 318063 | | 2140 | 330414 | |
| 1 | 292478 | 22 | 1 | 305566 | 21 | 1 | 318272 | 21 | 1 | 330617 | 20 |
| 2 | 292699 | 44 | 2 | 305781 | 43 | 2 | 318481 | 42 | 2 | 330819 | 40 |
| 3 | 292920 | 66 | 3 | 305996 | 64 | 3 | 318689 | 63 | 3 | 331022 | 61 |
| 4 | 293141 | 88 | 4 | 306211 | 86 | 4 | 318898 | 83 | 4 | 331225 | 81 |
| 5 | 293363 | 110 | 5 | 306425 | 107 | 5 | 319106 | 104 | 5 | 331427 | 101 |
| 6 | 293583 | 133 | 6 | 306639 | 129 | 6 | 319314 | 125 | 6 | 331630 | 121 |
| 7 | 293804 | 155 | 7 | 306854 | 150 | 7 | 319522 | 146 | 7 | 331832 | 141 |
| 8 | 294025 | 177 | 8 | 307068 | 172 | 8 | 319730 | 167 | 8 | 332034 | 162 |
| 9 | 294246 | 199 | 9 | 307282 | 193 | 9 | 319938 | 188 | 9 | 332236 | 182 |
| 1970 | 294466 | | 2030 | 307496 | | 2090 | 320146 | | 2150 | 332438 | |
| 1 | 294687 | 22 | 1 | 307710 | 21 | 1 | 320354 | 21 | 1 | 332640 | 20 |
| 2 | 294907 | 44 | 2 | 307924 | 43 | 2 | 320562 | 41 | 2 | 332842 | 40 |
| 3 | 295127 | 66 | 3 | 308137 | 64 | 3 | 320769 | 62 | 3 | 333044 | 60 |
| 4 | 295347 | 88 | 4 | 308351 | 85 | 4 | 320977 | 83 | 4 | 333246 | 81 |
| 5 | 295567 | 110 | 5 | 308564 | 107 | 5 | 321184 | 104 | 5 | 333447 | 101 |
| 6 | 295787 | 132 | 6 | 308778 | 128 | 6 | 321391 | 125 | 6 | 333649 | 121 |
| 7 | 296007 | 154 | 7 | 308991 | 149 | 7 | 321598 | 145 | 7 | 333850 | 141 |
| 8 | 296226 | 176 | 8 | 309204 | 171 | 8 | 321805 | 166 | 8 | 334051 | 161 |
| 9 | 296446 | 198 | 9 | 309417 | 192 | 9 | 322012 | 187 | 9 | 334253 | 181 |
| 1980 | 296665 | | 2040 | 309630 | | 2100 | 322219 | | 2160 | 334454 | |
| 1 | 296884 | 22 | 1 | 309843 | 21 | 1 | 322426 | 21 | 1 | 334655 | 20 |
| 2 | 297104 | 44 | 2 | 310056 | 43 | 2 | 322633 | 41 | 2 | 334856 | 40 |
| 3 | 297323 | 66 | 3 | 310268 | 64 | 3 | 322839 | 62 | 3 | 335056 | 60 |
| 4 | 297542 | 88 | 4 | 310481 | 85 | 4 | 323046 | 82 | 4 | 335257 | 80 |
| 5 | 297761 | 109 | 5 | 310693 | 106 | 5 | 323252 | 103 | 5 | 335458 | 100 |
| 6 | 297979 | 131 | 6 | 310906 | 127 | 6 | 323458 | 124 | 6 | 335658 | 120 |
| 7 | 298198 | 153 | 7 | 311118 | 148 | 7 | 323665 | 144 | 7 | 335859 | 140 |
| 8 | 298416 | 175 | 8 | 311330 | 170 | 8 | 323871 | 165 | 8 | 336059 | 160 |
| 9 | 298635 | 197 | 9 | 311542 | 191 | 9 | 324077 | 186 | 9 | 336260 | 180 |
| 1990 | 298853 | | 2050 | 311754 | | 2110 | 324282 | | 2170 | 336460 | |
| 1 | 299071 | 22 | 1 | 311966 | 21 | 1 | 324488 | 21 | 1 | 336660 | 20 |
| 2 | 299289 | 44 | 2 | 312177 | 42 | 2 | 324694 | 41 | 2 | 336860 | 40 |
| 3 | 299507 | 65 | 3 | 312389 | 63 | 3 | 324899 | 62 | 3 | 337060 | 60 |
| 4 | 299725 | 87 | 4 | 312600 | 84 | 4 | 325105 | 82 | 4 | 337260 | 80 |
| 5 | 299943 | 109 | 5 | 312812 | 106 | 5 | 325310 | 103 | 5 | 337459 | 100 |
| 6 | 300160 | 131 | 6 | 313023 | 127 | 6 | 325516 | 123 | 6 | 337659 | 120 |
| 7 | 300378 | 153 | 7 | 313234 | 148 | 7 | 325721 | 144 | 7 | 337858 | 140 |
| 8 | 300595 | 174 | 8 | 313445 | 160 | 8 | 325926 | 164 | 8 | 338058 | 160 |
| 9 | 300813 | 196 | 9 | 313656 | 190 | 9 | 326131 | 185 | 9 | 338257 | 180 |
| 2000 | 301030 | | 2060 | 313867 | | 2120 | 326336 | | 2180 | 338456 | |
| 1 | 301247 | 22 | 1 | 314078 | 21 | 1 | 326541 | 20 | 1 | 338656 | 20 |
| 2 | 301464 | 43 | 2 | 314289 | 42 | 2 | 326745 | 41 | 2 | 338855 | 40 |
| 3 | 301681 | 65 | 3 | 314499 | 63 | 3 | 326950 | 61 | 3 | 339054 | 60 |
| 4 | 301898 | 87 | 4 | 314710 | 84 | 4 | 327155 | 82 | 4 | 339253 | 80 |
| 5 | 302114 | 108 | 5 | 314920 | 105 | 5 | 327359 | 102 | 5 | 339451 | 100 |
| 6 | 302331 | 130 | 6 | 315130 | 126 | 6 | 327563 | 123 | 6 | 339650 | 119 |
| 7 | 302547 | 152 | 7 | 315340 | 147 | 7 | 327767 | 143 | 7 | 339849 | 139 |
| 8 | 302764 | 173 | 8 | 315550 | 168 | 8 | 327972 | 164 | 8 | 340047 | 159 |
| 9 | 302980 | 195 | 9 | 315760 | 189 | 9 | 328176 | 184 | 9 | 340246 | 179 |
| 2010 | 303196 | | 2070 | 315970 | | 2130 | 328380 | | 2190 | 340444 | |
| 1 | 303412 | 22 | 1 | 316180 | 21 | 1 | 328583 | 20 | 1 | 340642 | 20 |
| 2 | 303628 | 43 | 2 | 316390 | 42 | 2 | 328787 | 41 | 2 | 340841 | 40 |
| 3 | 303844 | 65 | 3 | 316599 | 63 | 3 | 328991 | 61 | 3 | 341039 | 59 |
| 4 | 304059 | 86 | 4 | 316809 | 84 | 4 | 329194 | 81 | 4 | 341237 | 79 |
| 5 | 304275 | 108 | 5 | 317018 | 105 | 5 | 329398 | 102 | 5 | 341435 | 99 |
| 6 | 304490 | 129 | 6 | 317227 | 126 | 6 | 329601 | 122 | 6 | 341632 | 119 |
| 7 | 304706 | 151 | 7 | 317436 | 147 | 7 | 329805 | 142 | 7 | 341830 | 139 |
| 8 | 304921 | 172 | 8 | 317645 | 168 | 8 | 330008 | 163 | 8 | 342028 | 158 |
| 9 | 305136 | 194 | 9 | 317854 | 189 | 9 | 330211 | 183 | 9 | 342225 | 178 |

| No. | Log. | Prop. Part. | No. | Log. | Prop. Part. | No. | Log. | Prop. Part. | No. | Log. | Prop. Part. |
|---|---|---|---|---|---|---|---|---|---|---|---|
| 2200 | 342423 | | 2260 | 354108 | | 2320 | 365488 | | 2380 | 376577 | |
| 1 | 342620 | 20 | 1 | 354301 | 19 | 1 | 365675 | 19 | 1 | 376759 | 18 |
| 2 | 342817 | 39 | 2 | 354493 | 38 | 2 | 365862 | 37 | 2 | 376942 | 36 |
| 3 | 343014 | 59 | 3 | 354685 | 58 | 3 | 366049 | 56 | 3 | 377124 | 55 |
| 4 | 343212 | 79 | 4 | 354876 | 77 | 4 | 366236 | 75 | 4 | 377306 | 73 |
| 5 | 343409 | 99 | 5 | 355068 | 96 | 5 | 366423 | 93 | 5 | 377488 | 91 |
| 6 | 343606 | 118 | 6 | 355260 | 115 | 6 | 366610 | 112 | 6 | 377670 | 109 |
| 7 | 343802 | 138 | 7 | 355452 | 134 | 7 | 366796 | 131 | 7 | 377852 | 127 |
| 8 | 343999 | 158 | 8 | 355643 | 154 | 8 | 366983 | 150 | 8 | 378034 | 146 |
| 9 | 344196 | 178 | 9 | 355834 | 173 | 9 | 367169 | 168 | 9 | 378216 | 164 |
| 2210 | 344392 | | 2270 | 356026 | | 2330 | 367356 | | 2390 | 378398 | |
| 1 | 344589 | 20 | 1 | 356217 | 19 | 1 | 367542 | 19 | 1 | 378580 | 18 |
| 2 | 344785 | 39 | 2 | 356408 | 38 | 2 | 367729 | 37 | 2 | 378761 | 36 |
| 3 | 344981 | 59 | 3 | 356599 | 57 | 3 | 367915 | 56 | 3 | 378943 | 55 |
| 4 | 345178 | 78 | 4 | 356790 | 76 | 4 | 368101 | 75 | 4 | 379124 | 73 |
| 5 | 345374 | 98 | 5 | 356981 | 95 | 5 | 368287 | 93 | 5 | 379306 | 91 |
| 6 | 345570 | 118 | 6 | 357172 | 115 | 6 | 368473 | 112 | 6 | 379487 | 109 |
| 7 | 345766 | 137 | 7 | 357363 | 134 | 7 | 368659 | 130 | 7 | 379668 | 127 |
| 8 | 345962 | 157 | 8 | 357554 | 153 | 8 | 368844 | 149 | 8 | 379849 | 146 |
| 9 | 346157 | 176 | 9 | 357744 | 172 | 9 | 369030 | 167 | 9 | 380030 | 164 |
| 2220 | 346353 | | 2280 | 357935 | | 2340 | 369216 | | 2400 | 380211 | |
| 1 | 346549 | 19 | 1 | 358125 | 19 | 1 | 369401 | 19 | 1 | 380392 | 18 |
| 2 | 346744 | 39 | 2 | 358316 | 38 | 2 | 369587 | 37 | 2 | 380573 | 36 |
| 3 | 346939 | 58 | 3 | 358506 | 57 | 3 | 369772 | 56 | 3 | 380754 | 55 |
| 4 | 347135 | 78 | 4 | 358696 | 76 | 4 | 369958 | 74 | 4 | 380934 | 73 |
| 5 | 347330 | 97 | 5 | 358886 | 95 | 5 | 370143 | 93 | 5 | 381115 | 91 |
| 6 | 347525 | 117 | 6 | 359076 | 114 | 6 | 370328 | 111 | 6 | 381296 | 109 |
| 7 | 347720 | 137 | 7 | 359266 | 133 | 7 | 370513 | 130 | 7 | 381476 | 127 |
| 8 | 347915 | 156 | 8 | 359456 | 152 | 8 | 370698 | 148 | 8 | 381656 | 145 |
| 9 | 348110 | 175 | 9 | 359646 | 171 | 9 | 370883 | 167 | 9 | 381837 | 163 |
| 2230 | 848305 | | 2290 | 359835 | | 2350 | 371068 | | 2410 | 382017 | |
| 1 | 848500 | 19 | 1 | 360025 | 19 | 1 | 371253 | 18 | 1 | 382197 | 18 |
| 2 | 848694 | 39 | 2 | 360215 | 38 | 2 | 371437 | 37 | 2 | 382377 | 36 |
| 3 | 348889 | 58 | 3 | 360404 | 57 | 3 | 371622 | 55 | 3 | 382557 | 54 |
| 4 | 349083 | 78 | 4 | 360593 | 76 | 4 | 371806 | 74 | 4 | 382737 | 72 |
| 5 | 349278 | 97 | 5 | 360783 | 95 | 5 | 371991 | 92 | 5 | 382917 | 90 |
| 6 | 349472 | 117 | 6 | 360972 | 114 | 6 | 372175 | 111 | 6 | 383097 | 108 |
| 7 | 349666 | 137 | 7 | 361161 | 133 | 7 | 372360 | 129 | 7 | 383277 | 126 |
| 8 | 349860 | 156 | 8 | 361350 | 152 | 8 | 372544 | 148 | 8 | 383456 | 144 |
| 9 | 350054 | 175 | 9 | 361539 | 171 | 9 | 372728 | 166 | 9 | 383636 | 162 |
| 2240 | 350248 | | 2300 | 361728 | | 2360 | 372912 | | 2420 | 383815 | |
| 1 | 350442 | 19 | 1 | 361917 | 19 | 1 | 373096 | 18 | 1 | 383995 | 18 |
| 2 | 350636 | 39 | 2 | 362105 | 38 | 2 | 373280 | 37 | 2 | 384174 | 36 |
| 3 | 350829 | 58 | 3 | 362294 | 56 | 3 | 373464 | 55 | 3 | 384353 | 54 |
| 4 | 351023 | 77 | 4 | 362482 | 75 | 4 | 373647 | 74 | 4 | 384533 | 72 |
| 5 | 351216 | 97 | 5 | 362671 | 94 | 5 | 373831 | 92 | 5 | 384712 | 90 |
| 6 | 351410 | 116 | 6 | 362859 | 113 | 6 | 374015 | 110 | 6 | 384891 | 108 |
| 7 | 351603 | 135 | 7 | 363048 | 132 | 7 | 374198 | 129 | 7 | 385070 | 126 |
| 8 | 351796 | 155 | 8 | 363236 | 151 | 8 | 374382 | 147 | 8 | 385249 | 144 |
| 9 | 351989 | 174 | 9 | 363424 | 170 | 9 | 374565 | 166 | 9 | 385428 | 162 |
| 2250 | 352182 | | 2310 | 363612 | | 2370 | 374748 | | 2430 | 385606 | |
| 1 | 352375 | 19 | 1 | 363800 | 19 | 1 | 374932 | 18 | 1 | 385785 | 18 |
| 2 | 352568 | 38 | 2 | 363988 | 37 | 2 | 375115 | 37 | 2 | 385964 | 35 |
| 3 | 352761 | 58 | 3 | 364176 | 56 | 3 | 375298 | 55 | 3 | 386142 | 53 |
| 4 | 352954 | 77 | 4 | 364363 | 75 | 4 | 375481 | 73 | 4 | 386321 | 71 |
| 5 | 353147 | 96 | 5 | 364551 | 94 | 5 | 375664 | 92 | 5 | 386499 | 89 |
| 6 | 353339 | 115 | 6 | 364739 | 112 | 6 | 375846 | 110 | 6 | 386677 | 107 |
| 7 | 353532 | 134 | 7 | 364926 | 131 | 7 | 376029 | 128 | 7 | 386856 | 125 |
| 8 | 353724 | 154 | 8 | 365113 | 150 | 8 | 376212 | 147 | 8 | 387034 | 143 |
| 9 | 353916 | 173. | 9 | 365301 | 169 | 9 | 376394 | 165 | 9 | 387212 | 161 |

| No. | Log. | Prop. Part. | No. | Log. | Prop. Part. | No. | Log. | Prop. Part. | No. | Log. | Prop. Part. |
|---|---|---|---|---|---|---|---|---|---|---|---|
| 2440 | 387390 | | 2500 | 397940 | | 2560 | 408240 | | 2620 | 418301 | |
| 1 | 387568 | 18 | 1 | 398114 | 17 | 1 | 408410 | 17 | 1 | 418467 | 17 |
| 2 | 387746 | 36 | 2 | 398287 | 35 | 2 | 408579 | 34 | 2 | 418633 | 33 |
| 3 | 387923 | 53 | 3 | 398461 | 53 | 3 | 408749 | 51 | 3 | 418798 | 50 |
| 4 | 388101 | 71 | 4 | 398634 | 69 | 4 | 408918 | 68 | 4 | 418964 | 66 |
| 5 | 388279 | 89 | 5 | 398808 | 87 | 5 | 409087 | 85 | 5 | 419129 | 83 |
| 6 | 388456 | 107 | 6 | 398981 | 104 | 6 | 409257 | 102 | 6 | 419295 | 99 |
| 7 | 388634 | 125 | 7 | 399154 | 121 | 7 | 409426 | 119 | 7 | 419460 | 116 |
| 8 | 388811 | 142 | 8 | 399327 | 138 | 8 | 409595 | 136 | 8 | 419625 | 132 |
| 9 | 388989 | 160 | 9 | 399501 | 156 | 9 | 409764 | 153 | 9 | 419791 | 149 |
| 2450 | 389166 | | 2510 | 399674 | | 2570 | 409933 | | 2630 | 419956 | |
| 1 | 389343 | 18 | 1 | 399847 | 17 | 1 | 410102 | 17 | 1 | 420121 | 16 |
| 2 | 389520 | 36 | 2 | 400020 | 35 | 2 | 410271 | 34 | 2 | 420286 | 33 |
| 3 | 389697 | 53 | 3 | 400192 | 53 | 3 | 410440 | 50 | 3 | 420451 | 49 |
| 4 | 389875 | 71 | 4 | 400365 | 69 | 4 | 410608 | 67 | 4 | 420616 | 66 |
| 5 | 390051 | 89 | 5 | 400538 | 87 | 5 | 410777 | 84 | 5 | 420781 | 82 |
| 6 | 390228 | 107 | 6 | 400711 | 104 | 6 | 410946 | 101 | 6 | 420945 | 99 |
| 7 | 390405 | 125 | 7 | 400883 | 121 | 7 | 411114 | 118 | 7 | 421110 | 115 |
| 8 | 390582 | 142 | 8 | 401056 | 138 | 8 | 411283 | 135 | 8 | 421275 | 132 |
| 9 | 390759 | 160 | 9 | 401228 | 156 | 9 | 411451 | 152 | 9 | 421439 | 148 |
| 2460 | 390935 | | 2520 | 401400 | | 2580 | 411620 | | 2640 | 421604 | |
| 1 | 391112 | 18 | 1 | 401573 | 17 | 1 | 411788 | 17 | 1 | 421768 | 16 |
| 2 | 391288 | 35 | 2 | 401745 | 34 | 2 | 411956 | 34 | 2 | 421933 | 33 |
| 3 | 391464 | 53 | 3 | 401917 | 52 | 3 | 412124 | 50 | 3 | 422097 | 49 |
| 4 | 391641 | 70 | 4 | 402089 | 69 | 4 | 412292 | 67 | 4 | 422261 | 66 |
| 5 | 391817 | 88 | 5 | 402261 | 86 | 5 | 412460 | 84 | 5 | 422426 | 82 |
| 6 | 391993 | 106 | 6 | 402433 | 103 | 6 | 412628 | 101 | 6 | 422590 | 99 |
| 7 | 392169 | 123 | 7 | 402605 | 120 | 7 | 412796 | 118 | 7 | 422754 | 115 |
| 8 | 392345 | 141 | 8 | 402777 | 138 | 8 | 412964 | 135 | 8 | 422918 | 132 |
| 9 | 392521 | 158 | 9 | 402949 | 155 | 9 | 413132 | 152 | 9 | 423082 | 148 |
| 2470 | 392697 | | 2530 | 403120 | | 2590 | 413300 | | 2650 | 423246 | |
| 1 | 392873 | 18 | 1 | 403292 | 17 | 1 | 413467 | 17 | 1 | 423410 | 16 |
| 2 | 393048 | 35 | 2 | 403464 | 34 | 2 | 413635 | 33 | 2 | 423573 | 33 |
| 3 | 393224 | 53 | 3 | 403635 | 52 | 3 | 413802 | 50 | 3 | 423737 | 49 |
| 4 | 393400 | 70 | 4 | 403807 | 69 | 4 | 413970 | 67 | 4 | 423001 | 65 |
| 5 | 393575 | 88 | 5 | 403978 | 86 | 5 | 414137 | 84 | 5 | 424064 | 81 |
| 6 | 393751 | 106 | 6 | 404149 | 103 | 6 | 414305 | 101 | 6 | 424228 | 98 |
| 7 | 393926 | 123 | 7 | 404320 | 120 | 7 | 414472 | 117 | 7 | 424392 | 114 |
| 8 | 394101 | 141 | 8 | 404492 | 137 | 8 | 414639 | 134 | 8 | 424555 | 131 |
| 9 | 394276 | 158 | 9 | 404663 | 154 | 9 | 414806 | 151 | 9 | 424718 | 147 |
| 2480 | 394452 | | 2540 | 404834 | | 2600 | 414973 | | 2660 | 424882 | |
| 1 | 394627 | 17 | 1 | 405005 | 17 | 1 | 415140 | 17 | 1 | 425045 | 16 |
| 2 | 394802 | 35 | 2 | 405175 | 34 | 2 | 415307 | 33 | 2 | 425208 | 33 |
| 3 | 394977 | 53 | 3 | 405346 | 51 | 3 | 415474 | 50 | 3 | 425371 | 49 |
| 4 | 395152 | 70 | 4 | 405517 | 68 | 4 | 415641 | 67 | 4 | 425534 | 65 |
| 5 | 395326 | 87 | 5 | 405688 | 85 | 5 | 415808 | 84 | 5 | 425697 | 81 |
| 6 | 395501 | 104 | 6 | 405858 | 102 | 6 | 415974 | 101 | 6 | 425860 | 98 |
| 7 | 395676 | 122 | 7 | 406029 | 119 | 7 | 416141 | 117 | 7 | 426023 | 114 |
| 8 | 395850 | 139 | 8 | 406199 | 136 | 8 | 416308 | 134 | 8 | 426186 | 130 |
| 9 | 396025 | 157 | 9 | 406370 | 153 | 9 | 416474 | 150 | 9 | 426349 | 147 |
| 2490 | 396199 | | 2550 | 406540 | | 2610 | 416640 | | 2670 | 426511 | |
| 1 | 396374 | 17 | 1 | 406710 | 17 | 1 | 416807 | 17 | 1 | 426674 | 16 |
| 2 | 396548 | 35 | 2 | 406881 | 34 | 2 | 416973 | 33 | 2 | 426836 | 33 |
| 3 | 396722 | 53 | 3 | 407051 | 51 | 3 | 417139 | 50 | 3 | 426999 | 49 |
| 4 | 396896 | 70 | 4 | 407221 | 68 | 4 | 417306 | 66 | 4 | 427161 | 65 |
| 5 | 397070 | 87 | 5 | 407391 | 85 | 5 | 417472 | 83 | 5 | 427324 | 81 |
| 6 | 397245 | 104 | 6 | 407561 | 102 | 6 | 417638 | 100 | 6 | 427486 | 98 |
| 7 | 397418 | 122 | 7 | 407731 | 119 | 7 | 417804 | 116 | 7 | 427648 | 114 |
| 8 | 397592 | 139 | 8 | 407900 | 136 | 8 | 417970 | 133 | 8 | 427811 | 130 |
| 9 | 397766 | 157 | 9 | 408070 | 153 | 9 | 418135 | 149 | 9 | 427973 | 147 |

| No. | Log. | Prop. Part. | No. | Log. | Prop. Part. | No. | Log. | Prop. Part. | No. | Log. | Prop. Part. |
|---|---|---|---|---|---|---|---|---|---|---|---|
| 2680 | 428135 | | 2740 | 437751 | | 2800 | 447158 | | 2860 | 456366 | |
| 1 | 428297 | 16 | 1 | 437909 | 16 | 1 | 447313 | 15 | 1 | 456518 | 15 |
| 2 | 428459 | 32 | 2 | 438067 | 32 | 2 | 447468 | 31 | 2 | 456670 | 30 |
| 3 | 428621 | 48 | 3 | 438226 | 47 | 3 | 447623 | 46 | 3 | 456821 | 46 |
| 4 | 428782 | 65 | 4 | 438384 | 63 | 4 | 447778 | 62 | 4 | 456973 | 61 |
| 5 | 428944 | 81 | 5 | 438542 | 79 | 5 | 447933 | 77 | 5 | 457125 | 76 |
| 6 | 429106 | 97 | 6 | 438700 | 95 | 6 | 448088 | 93 | 6 | 457276 | 91 |
| 7 | 429268 | 113 | 7 | 438859 | 111 | 7 | 448242 | 108 | 7 | 457428 | 106 |
| 8 | 429429 | 129 | 8 | 439017 | 127 | 8 | 448397 | 124 | 8 | 457579 | 122 |
| 9 | 429591 | 145 | 9 | 439175 | 143 | 9 | 448552 | 139 | 9 | 457730 | 137 |
| 2690 | 429752 | | 2750 | 439333 | | 2810 | 448706 | | 2870 | 457882 | |
| 1 | 429914 | 16 | 1 | 439491 | 16 | 1 | 448861 | 15 | 1 | 458033 | 15 |
| 2 | 430075 | 32 | 2 | 439648 | 32 | 2 | 449015 | 31 | 2 | 458184 | 30 |
| 3 | 430236 | 48 | 3 | 439806 | 47 | 3 | 449170 | 46 | 3 | 458336 | 45 |
| 4 | 430398 | 65 | 4 | 439964 | 63 | 4 | 449324 | 62 | 4 | 458487 | 61 |
| 5 | 430559 | 81 | 5 | 440122 | 79 | 5 | 449478 | 77 | 5 | 458638 | 76 |
| 6 | 430720 | 97 | 6 | 440279 | 95 | 6 | 449633 | 92 | 6 | 458789 | 91 |
| 7 | 430881 | 113 | 7 | 440437 | 111 | 7 | 449787 | 108 | 7 | 458940 | 106 |
| 8 | 431042 | 129 | 8 | 440594 | 126 | 8 | 449941 | 123 | 8 | 459091 | 121 |
| 9 | 431203 | 145 | 9 | 440752 | 142 | 9 | 450095 | 139 | 9 | 459242 | 136 |
| 2700 | 431364 | | 2760 | 440909 | | 2820 | 450249 | | 2880 | 459392 | |
| 1 | 431525 | 16 | 1 | 441066 | 16 | 1 | 450403 | 15 | 1 | 459543 | 15 |
| 2 | 431685 | 32 | 2 | 441224 | 31 | 2 | 450557 | 31 | 2 | 459694 | 30 |
| 3 | 431846 | 48 | 3 | 441381 | 47 | 3 | 450711 | 46 | 3 | 459845 | 45 |
| 4 | 432007 | 64 | 4 | 441538 | 63 | 4 | 450865 | 62 | 4 | 459995 | 61 |
| 5 | 432167 | 80 | 5 | 441695 | 78 | 5 | 451018 | 77 | 5 | 460146 | 76 |
| 6 | 432328 | 96 | 6 | 441852 | 94 | 6 | 451172 | 92 | 6 | 460296 | 91 |
| 7 | 432488 | 112 | 7 | 442009 | 110 | 7 | 451326 | 108 | 7 | 460447 | 106 |
| 8 | 432649 | 128 | 8 | 442166 | 126 | 8 | 451479 | 123 | 8 | 460597 | 121 |
| 9 | 432809 | 144 | 9 | 442323 | 141 | 9 | 451633 | 139 | 9 | 460747 | 136 |
| 2710 | 432969 | | 2770 | 442480 | | 2830 | 451786 | | 2890 | 460898 | |
| 1 | 433129 | 16 | 1 | 442636 | 16 | 1 | 451940 | 15 | 1 | 461048 | 15 |
| 2 | 433290 | 32 | 2 | 442793 | 31 | 2 | 452093 | 31 | 2 | 461198 | 30 |
| 3 | 433450 | 48 | 3 | 442950 | 47 | 3 | 452247 | 46 | 3 | 461348 | 45 |
| 4 | 433610 | 64 | 4 | 443106 | 63 | 4 | 452400 | 61 | 4 | 461498 | 60 |
| 5 | 433770 | 80 | 5 | 443263 | 78 | 5 | 452553 | 77 | 5 | 461649 | 75 |
| 6 | 433930 | 96 | 6 | 443419 | 94 | 6 | 452706 | 92 | 6 | 461799 | 90 |
| 7 | 434090 | 112 | 7 | 443576 | 110 | 7 | 452859 | 107 | 7 | 461948 | 105 |
| 8 | 434249 | 128 | 8 | 443732 | 126 | 8 | 453012 | 123 | 8 | 462098 | 120 |
| 9 | 434409 | 144 | 9 | 443888 | 141 | 9 | 453165 | 138 | 9 | 462248 | 135 |
| 2720 | 434569 | | 2780 | 444045 | | 2840 | 453318 | | 2900 | 462398 | |
| 1 | 434728 | 16 | 1 | 444201 | 16 | 1 | 453471 | 15 | 1 | 462548 | 15 |
| 2 | 434888 | 32 | 2 | 444357 | 31 | 2 | 453624 | 31 | 2 | 462697 | 30 |
| 3 | 435048 | 48 | 3 | 444513 | 47 | 3 | 453777 | 46 | 3 | 462847 | 45 |
| 4 | 435207 | 64 | 4 | 444669 | 62 | 4 | 453930 | 61 | 4 | 462997 | 60 |
| 5 | 435366 | 80 | 5 | 444825 | 78 | 5 | 454082 | 77 | 5 | 463146 | 75 |
| 6 | 435526 | 96 | 6 | 444981 | 94 | 6 | 454235 | 92 | 6 | 463296 | 90 |
| 7 | 435685 | 112 | 7 | 445137 | 109 | 7 | 454387 | 107 | 7 | 463445 | 105 |
| 8 | 435844 | 128 | 8 | 445293 | 125 | 8 | 454540 | 123 | 8 | 463594 | 120 |
| 9 | 436003 | 144 | 9 | 445448 | 140 | 9 | 454692 | 138 | 9 | 463744 | 135 |
| 2730 | 436163 | | 2790 | 445604 | | 2850 | 454845 | | 2910 | 463893 | |
| 1 | 436322 | 16 | 1 | 445760 | 16 | 1 | 454997 | 15 | 1 | 464042 | 15 |
| 2 | 436481 | 32 | 2 | 445915 | 31 | 2 | 455149 | 30 | 2 | 464191 | 30 |
| 3 | 436640 | 47 | 3 | 446071 | 47 | 3 | 455302 | 46 | 3 | 464340 | 45 |
| 4 | 436798 | 63 | 4 | 446226 | 62 | 4 | 455454 | 61 | 4 | 464489 | 60 |
| 5 | 436957 | 79 | 5 | 446382 | 78 | 5 | 455606 | 76 | 5 | 464639 | 75 |
| 6 | 437116 | 95 | 6 | 446537 | 94 | 6 | 455758 | 91 | 6 | 464787 | 90 |
| 7 | 437275 | 111 | 7 | 446692 | 109 | 7 | 455910 | 106 | 7 | 464936 | 105 |
| 8 | 437433 | 127 | 8 | 446848 | 125 | 8 | 456062 | 122 | 8 | 465085 | 120 |
| 9 | 437592 | 143 | 9 | 447003 | 140 | 9 | 456214 | 137 | 9 | 465234 | 135 |

| No. | Log. | Prop. Part. | No. | Log. | Prop. Part. | No. | Log. | Prop. Part. | No. | Log. | Prop. Part. |
|---|---|---|---|---|---|---|---|---|---|---|---|
| 2920 | 465383 | | 2980 | 474216 | | 3040 | 482874 | | 3100 | 491362 | |
| 1 | 465532 | 15 | 1 | 474362 | 15 | 1 | 483016 | 14 | 1 | 491502 | 14 |
| 2 | 465680 | 30 | 2 | 474508 | 29 | 2 | 483159 | 28 | 2 | 491642 | 28 |
| 3 | 465829 | 44 | 3 | 474653 | 44 | 3 | 483302 | 43 | 3 | 491782 | 42 |
| 4 | 465977 | 59 | 4 | 474799 | 58 | 4 | 483445 | 57 | 4 | 491922 | 56 |
| 5 | 466126 | 74 | 5 | 474944 | 73 | 5 | 483587 | 71 | 5 | 492062 | 70 |
| 6 | 466274 | 89 | 6 | 475090 | 88 | 6 | 483730 | 85 | 6 | 492201 | 84 |
| 7 | 466423 | 104 | 7 | 475235 | 102 | 7 | 483872 | 99 | 7 | 492341 | 98 |
| 8 | 466571 | 118 | 8 | 475381 | 117 | 8 | 484015 | 114 | 8 | 492481 | 112 |
| 9 | 466719 | 133 | 9 | 475526 | 131 | 9 | 484157 | 128 | 9 | 492621 | 126 |
| 2930 | 466868 | | 2990 | 475671 | | 3050 | 484300 | | 3110 | 492760 | |
| 1 | 467016 | 15 | 1 | 475816 | 15 | 1 | 484442 | 14 | 1 | 492900 | 14 |
| 2 | 467164 | 30 | 2 | 475962 | 29 | 2 | 484584 | 28 | 2 | 493040 | 28 |
| 3 | 467312 | 44 | 3 | 476107 | 43 | 3 | 484727 | 43 | 3 | 493179 | 42 |
| 4 | 467460 | 59 | 4 | 476252 | 58 | 4 | 484869 | 57 | 4 | 493319 | 56 |
| 5 | 467608 | 74 | 5 | 476397 | 72 | 5 | 485011 | 71 | 5 | 493458 | 70 |
| 6 | 467756 | 89 | 6 | 476542 | 87 | 6 | 485153 | 85 | 6 | 493597 | 84 |
| 7 | 467904 | 104 | 7 | 476687 | 101 | 7 | 485295 | 99 | 7 | 493737 | 98 |
| 8 | 468052 | 118 | 8 | 476832 | 116 | 8 | 485437 | 114 | 8 | 493876 | 142 |
| 9 | 468200 | 133 | 9 | 476976 | 130 | 9 | 485579 | 128 | 9 | 494015 | 126 |
| 2940 | 468347 | | 3000 | 477121 | | 3060 | 485721 | | 3120 | 494155 | |
| 1 | 468495 | 15 | 1 | 477266 | 14 | 1 | 485863 | 14 | 1 | 494294 | 14 |
| 2 | 468643 | 30 | 2 | 477411 | 29 | 2 | 486005 | 28 | 2 | 494433 | 28 |
| 3 | 468790 | 44 | 3 | 477555 | 43 | 3 | 486147 | 43 | 3 | 494572 | 41 |
| 4 | 868938 | 59 | 4 | 477700 | 58 | 4 | 486289 | 57 | 4 | 494711 | 56 |
| 5 | 469085 | 74 | 5 | 477844 | 72 | 5 | 486430 | 71 | 5 | 494850 | 69 |
| 6 | 469233 | 89 | 6 | 477989 | 87 | 6 | 486572 | 85 | 6 | 494989 | 83 |
| 7 | 469380 | 104 | 7 | 478133 | 101 | 7 | 486714 | 99 | 7 | 495128 | 97 |
| 8 | 469527 | 118 | 8 | 478278 | 116 | 8 | 486855 | 114 | 8 | 495267 | 111 |
| 9 | 469675 | 133 | 9 | 478422 | 130 | 9 | 486997 | 128 | 9 | 495406 | 125 |
| 2950 | 469822 | | 3010 | 478566 | | 3070 | 487138 | | 3130 | 495544 | |
| 1 | 469969 | 15 | 1 | 478711 | 14 | 1 | 487280 | 14 | 1 | 495683 | 14 |
| 2 | 470116 | 29 | 2 | 478855 | 29 | 2 | 487421 | 28 | 2 | 495822 | 28 |
| 3 | 470263 | 44 | 3 | 478999 | 43 | 3 | 487563 | 42 | 3 | 495960 | 41 |
| 4 | 470410 | 59 | 4 | 479143 | 58 | 4 | 487704 | 57 | 4 | 496099 | 56 |
| 5 | 470557 | 74 | 5 | 479287 | 72 | 5 | 487845 | 71 | 5 | 496237 | 69 |
| 6 | 470704 | 88 | 6 | 479431 | 86 | 6 | 487986 | 85 | 6 | 496376 | 83 |
| 7 | 470851 | 103 | 7 | 479575 | 101 | 7 | 488127 | 99 | 7 | 496514 | 97 |
| 8 | 470998 | 118 | 8 | 479719 | 115 | 8 | 488269 | 113 | 8 | 496653 | 111 |
| 9 | 471145 | 132 | 9 | 479863 | 130 | 9 | 488410 | 127 | 9 | 496791 | 125 |
| 2960 | 471292 | | 3020 | 480007 | | 3080 | 488551 | | 3140 | 496930 | |
| 1 | 471438 | 15 | 1 | 480151 | 14 | 1 | 488692 | 14 | 1 | 497068 | 14 |
| 2 | 471585 | 29 | 2 | 480294 | 29 | 2 | 488833 | 28 | 2 | 497206 | 28 |
| 3 | 471732 | 44 | 3 | 480438 | 43 | 3 | 488973 | 42 | 3 | 497344 | 41 |
| 4 | 471878 | 59 | 4 | 480582 | 58 | 4 | 489114 | 56 | 4 | 497482 | 55 |
| 5 | 472025 | 73 | 5 | 480725 | 72 | 5 | 489255 | 70 | 5 | 497621 | 69 |
| 6 | 472171 | 88 | 6 | 480869 | 86 | 6 | 489396 | 84 | 6 | 497759 | 83 |
| 7 | 472317 | 102 | 7 | 481012 | 101 | 7 | 489537 | 98 | 7 | 497897 | 97 |
| 8 | 472464 | 117 | 8 | 481156 | 115 | 8 | 489677 | 112 | 8 | 498035 | 110 |
| 9 | 472610 | 132 | 9 | 481299 | 130 | 9 | 489818 | 126 | 9 | 498173 | 124 |
| 2970 | 472756 | | 3030 | 481443 | | 3090 | 489958 | | 3150 | 498311 | |
| 1 | 472903 | 15 | 1 | 481586 | 14 | 1 | 490099 | 14 | 1 | 498448 | 14 |
| 2 | 473049 | 29 | 2 | 481729 | 29 | 2 | 490239 | 28 | 2 | 498586 | 28 |
| 3 | 473195 | 44 | 3 | 481872 | 43 | 3 | 490380 | 42 | 3 | 498724 | 41 |
| 4 | 473341 | 59 | 4 | 482016 | 57 | 4 | 490520 | 56 | 4 | 498862 | 55 |
| 5 | 473487 | 73 | 5 | 482159 | 71 | 5 | 490661 | 70 | 5 | 498999 | 69 |
| 6 | 473633 | 88 | 6 | 482302 | 86 | 6 | 490801 | 84 | 6 | 499137 | 83 |
| 7 | 473779 | 102 | 7 | 482445 | 100 | 7 | 490941 | 98 | 7 | 499275 | 97 |
| 8 | 473925 | 117 | 8 | 482588 | 114 | 8 | 491081 | 112 | 8 | 499412 | 110 |
| 9 | 474070 | 132 | 9 | 482731 | 129 | 9 | 491222 | 126 | 9 | 499550 | 124 |

| No. | Log. | Prop. Part. | No. | Log. | Prop. Part. | No. | Log. | Prop. Part. | No. | Log. | Prop. Part. |
|---|---|---|---|---|---|---|---|---|---|---|---|
| 3160 | 499687 | | 3220 | 507856 | | 3280 | 515874 | | 3340 | 523746 | |
| 1 | 499824 | 14 | 1 | 507991 | 13 | 1 | 516006 | 13 | 1 | 523876 | 13 |
| 2 | 499962 | 27 | 2 | 508125 | 27 | 2 | 516139 | 26 | 2 | 524006 | 26 |
| 3 | 500099 | 41 | 3 | 508260 | 40 | 3 | 516271 | 40 | 3 | 524136 | 39 |
| 4 | 500236 | 55 | 4 | 508395 | 54 | 4 | 516403 | 53 | 4 | 524266 | 52 |
| 5 | 500374 | 68 | 5 | 508530 | 67 | 5 | 516535 | 66 | 5 | 524396 | 65 |
| 6 | 500511 | 82 | 6 | 508664 | 81 | 6 | 516668 | 79 | 6 | 524526 | 78 |
| 7 | 500648 | 96 | 7 | 508799 | 94 | 7 | 516800 | 92 | 7 | 524656 | 91 |
| 8 | 500785 | 110 | 8 | 508933 | 108 | 8 | 516932 | 106 | 8 | 524785 | 104 |
| 9 | 500922 | 123 | 9 | 509068 | 121 | 9 | 517064 | 119 | 9 | 524915 | 117 |
| 3170 | 501059 | | 3230 | 509202 | | 3290 | 517196 | | 3350 | 525045 | |
| 1 | 501196 | 14 | 1 | 509337 | 13 | 1 | 517328 | 13 | 1 | 525174 | 13 |
| 2 | 501333 | 27 | 2 | 509471 | 27 | 2 | 517460 | 26 | 2 | 525304 | 26 |
| 3 | 501470 | 41 | 3 | 509606 | 40 | 3 | 517592 | 40 | 3 | 525434 | 39 |
| 4 | 501607 | 55 | 4 | 509740 | 54 | 4 | 517724 | 53 | 4 | 525563 | 52 |
| 5 | 501744 | 68 | 5 | 509874 | 67 | 5 | 517855 | 66 | 5 | 525693 | 65 |
| 6 | 501880 | 82 | 6 | 510008 | 81 | 6 | 517987 | 79 | 6 | 525822 | 78 |
| 7 | 502017 | 96 | 7 | 510143 | 94 | 7 | 518119 | 92 | 7 | 525951 | 91 |
| 8 | 502154 | 110 | 8 | 510277 | 108 | 8 | 518251 | 106 | 8 | 526081 | 104 |
| 9 | 502290 | 123 | 9 | 510411 | 121 | 9 | 518382 | 119 | 9 | 526210 | 117 |
| 3180 | 502427 | | 3240 | 510545 | | 3300 | 518514 | | 3360 | 526339 | |
| 1 | 502564 | 14 | 1 | 510679 | 13 | 1 | 518645 | 13 | 1 | 526468 | 13 |
| 2 | 502700 | 27 | 2 | 510813 | 27 | 2 | 518777 | 26 | 2 | 526598 | 26 |
| 3 | 502837 | 41 | 3 | 510947 | 40 | 3 | 518909 | 39 | 3 | 526727 | 39 |
| 4 | 502973 | 54 | 4 | 511081 | 54 | 4 | 519040 | 52 | 4 | 526856 | 52 |
| 5 | 503109 | 68 | 5 | 511215 | 67 | 5 | 519171 | 66 | 5 | 526985 | 65 |
| 6 | 503246 | 82 | 6 | 511348 | 80 | 6 | 519303 | 79 | 6 | 527114 | 78 |
| 7 | 503382 | 95 | 7 | 511482 | 94 | 7 | 519434 | 92 | 7 | 527243 | 91 |
| 8 | 503518 | 109 | 8 | 511616 | 107 | 8 | 519565 | 105 | 8 | 527372 | 104 |
| 9 | 503654 | 123 | 9 | 511750 | 121 | 9 | 519697 | 118 | 9 | 527501 | 117 |
| 3190 | 503791 | | 3250 | 511883 | | 3310 | 519828 | | 3370 | 527630 | |
| 1 | 503927 | 14 | 1 | 512017 | 13 | 1 | 519959 | 13 | 1 | 527759 | 13 |
| 2 | 504063 | 27 | 2 | 512150 | 27 | 2 | 520090 | 26 | 2 | 527888 | 26 |
| 3 | 504199 | 41 | 3 | 512284 | 40 | 3 | 520221 | 39 | 3 | 528016 | 38 |
| 4 | 504335 | 54 | 4 | 512417 | 53 | 4 | 520352 | 52 | 4 | 528145 | 51 |
| 5 | 504471 | 68 | 5 | 512551 | 67 | 5 | 520483 | 66 | 5 | 528274 | 64 |
| 6 | 504607 | 82 | 6 | 512684 | 80 | 6 | 520614 | 79 | 6 | 528402 | 77 |
| 7 | 504743 | 95 | 7 | 512818 | 93 | 7 | 520745 | 92 | 7 | 528531 | 90 |
| 8 | 504878 | 109 | 8 | 512951 | 107 | 8 | 520876 | 105 | 8 | 528660 | 103 |
| 9 | 505014 | 122 | 9 | 513084 | 120 | 9 | 521007 | 118 | 9 | 528788 | 116 |
| 3200 | 505150 | | 3260 | 513218 | | 3320 | 521138 | | 3380 | 528917 | |
| 1 | 505286 | 14 | 1 | 513351 | 13 | 1 | 521269 | 13 | 1 | 529045 | 13 |
| 2 | 505421 | 27 | 2 | 513484 | 27 | 2 | 521400 | 26 | 2 | 529174 | 26 |
| 3 | 505557 | 41 | 3 | 513617 | 40 | 3 | 521530 | 39 | 3 | 529302 | 38 |
| 4 | 505692 | 54 | 4 | 513750 | 53 | 4 | 521661 | 52 | 4 | 529430 | 51 |
| 5 | 505828 | 68 | 5 | 513883 | 66 | 5 | 521792 | 65 | 5 | 529559 | 64 |
| 6 | 505963 | 82 | 6 | 514016 | 80 | 6 | 521922 | 78 | 6 | 529687 | 77 |
| 7 | 506099 | 95 | 7 | 514149 | 93 | 7 | 522053 | 97 | 7 | 529815 | 90 |
| 8 | 506234 | 109 | 8 | 514282 | 106 | 8 | 522183 | 104 | 8 | 529943 | 103 |
| 9 | 506370 | 122 | 9 | 514415 | 120 | 9 | 522314 | 117 | 9 | 530072 | 116 |
| 3210 | 506505 | | 3270 | 514548 | | 3330 | 522444 | | 3390 | 530200 | |
| 1 | 506640 | 13 | 1 | 514680 | 13 | 1 | 522575 | 13 | 1 | 530328 | 13 |
| 2 | 506775 | 27 | 2 | 514813 | 27 | 2 | 522705 | 26 | 2 | 530456 | 26 |
| 3 | 506911 | 40 | 3 | 514946 | 40 | 3 | 522835 | 39 | 3 | 530584 | 38 |
| 4 | 507046 | 54 | 4 | 515079 | 53 | 4 | 522966 | 52 | 4 | 530712 | 51 |
| 5 | 507181 | 67 | 5 | 515211 | 66 | 5 | 523096 | 65 | 5 | 530840 | 64 |
| 6 | 507316 | 81 | 6 | 515344 | 80 | 6 | 523226 | 78 | 6 | 530968 | 77 |
| 7 | 507451 | 94 | 7 | 515476 | 93 | 7 | 523356 | 97 | 7 | 531095 | 90 |
| 8 | 507586 | 108 | 8 | 515609 | 106 | 8 | 523486 | 104 | 8 | 531223 | 102 |
| 9 | 507721 | 121 | 9 | 515741 | 120 | 9 | 523616 | 117 | 9 | 531351 | 115 |

| No. | Log. | Prop. Part. | No. | Log. | Prop. Part. | No. | Log. | Prop. Part. | No. | Log. | Prop. Part. |
|---|---|---|---|---|---|---|---|---|---|---|---|
| 3400 | 531479 | | 3460 | 539076 | | 3520 | 546543 | | 3580 | 553883 | |
| 1 | 531607 | 13 | 1 | 539202 | 13 | 1 | 546666 | 12 | 1 | 554004 | 12 |
| 2 | 531734 | 25 | 2 | 539327 | 25 | 2 | 546789 | 25 | 2 | 554126 | 24 |
| 3 | 531862 | 38 | 3 | 539452 | 38 | 3 | 546913 | 37 | 3 | 554247 | 36 |
| 4 | 531990 | 51 | 4 | 539578 | 50 | 4 | 547036 | 49 | 4 | 554368 | 49 |
| 5 | 532117 | 63 | 5 | 539703 | 63 | 5 | 547159 | 62 | 5 | 554489 | 61 |
| 6 | 532245 | 76 | 6 | 539829 | 75 | 6 | 547282 | 74 | 6 | 554610 | 73 |
| 7 | 532372 | 89 | 7 | 539954 | 88 | 7 | 547405 | 86 | 7 | 554731 | 85 |
| 8 | 532500 | 102 | 8 | 540079 | 100 | 8 | 547529 | 99 | 8 | 554852 | 97 |
| 9 | 532627 | 114 | 9 | 540204 | 113 | 9 | 547652 | 111 | 9 | 554973 | 109 |
| 3410 | 532754 | | 3470 | 540329 | | 3530 | 547775 | | 3590 | 555094 | |
| 1 | 532882 | 13 | 1 | 540455 | 12 | 1 | 547898 | 12 | 1 | 555215 | 12 |
| 2 | 533009 | 25 | 2 | 540580 | 25 | 2 | 548021 | 25 | 2 | 555336 | 24 |
| 3 | 533136 | 38 | 3 | 540705 | 37 | 3 | 548144 | 37 | 3 | 555457 | 36 |
| 4 | 533263 | 51 | 4 | 540830 | 50 | 4 | 548266 | 49 | 4 | 555578 | 48 |
| 5 | 533391 | 63 | 5 | 540955 | 62 | 5 | 548389 | 61 | 5 | 555699 | 60 |
| 6 | 533518 | 76 | 6 | 541080 | 75 | 6 | 548512 | 74 | 6 | 555820 | 72 |
| 7 | 533645 | 89 | 7 | 541205 | 87 | 7 | 548635 | 86 | 7 | 555940 | 84 |
| 8 | 533772 | 102 | 8 | 541330 | 100 | 8 | 548758 | 98 | 8 | 556061 | 96 |
| 9 | 533899 | 114 | 9 | 541454 | 112 | 9 | 548881 | 111 | 9 | 556182 | 108 |
| 3420 | 534026 | | 3480 | 541579 | | 3540 | 549003 | | 3600 | 556302 | |
| 1 | 534153 | 13 | 1 | 541704 | 12 | 1 | 549126 | 12 | 1 | 556423 | 12 |
| 2 | 534280 | 25 | 2 | 541829 | 25 | 2 | 549249 | 25 | 2 | 556544 | 24 |
| 3 | 534407 | 38 | 3 | 541953 | 37 | 3 | 549371 | 37 | 3 | 556664 | 36 |
| 4 | 534534 | 51 | 4 | 542078 | 50 | 4 | 549494 | 49 | 4 | 556785 | 48 |
| 5 | 534661 | 63 | 5 | 542203 | 62 | 5 | 549616 | 61 | 5 | 556905 | 60 |
| 6 | 534787 | 76 | 6 | 542327 | 75 | 6 | 549739 | 74 | 6 | 557026 | 72 |
| 7 | 534914 | 89 | 7 | 542452 | 87 | 7 | 549861 | 86 | 7 | 557146 | 84 |
| 8 | 535041 | 102 | 8 | 542576 | 100 | 8 | 549984 | 98 | 8 | 557267 | 96 |
| 9 | 535167 | 114 | 9 | 542701 | 112 | 9 | 550106 | 111 | 9 | 557387 | 108 |
| 3430 | 535294 | | 3490 | 542825 | | 3550 | 550228 | | 3610 | 557507 | |
| 1 | 535421 | 13 | 1 | 542950 | 12 | 1 | 550351 | 12 | 1 | 557627 | 12 |
| 2 | 535547 | 25 | 2 | 543074 | 25 | 2 | 550473 | 24 | 2 | 557748 | 24 |
| 3 | 535674 | 38 | 3 | 543199 | 37 | 3 | 550595 | 37 | 3 | 557868 | 36 |
| 4 | 535800 | 50 | 4 | 543323 | 50 | 4 | 550717 | 49 | 4 | 557988 | 48 |
| 5 | 535927 | 63 | 5 | 543447 | 62 | 5 | 550840 | 61 | 5 | 558108 | 60 |
| 6 | 536053 | 76 | 6 | 543571 | 75 | 6 | 550962 | 73 | 6 | 558228 | 72 |
| 7 | 536179 | 88 | 7 | 543696 | 87 | 7 | 551084 | 86 | 7 | 558348 | 84 |
| 8 | 536306 | 101 | 8 | 543820 | 100 | 8 | 551206 | 98 | 8 | 558469 | 96 |
| 9 | 536432 | 114 | 9 | 543944 | 112 | 9 | 551328 | 110 | 9 | 558589 | 108 |
| 3440 | 536558 | | 3500 | 544068 | | 3560 | 551450 | | 3620 | 558709 | |
| 1 | 536685 | 13 | 1 | 544192 | 12 | 1 | 551572 | 12 | 1 | 558829 | 12 |
| 2 | 536811 | 25 | 2 | 544316 | 25 | 2 | 551694 | 24 | 2 | 558948 | 24 |
| 3 | 536937 | 38 | 3 | 544440 | 37 | 3 | 551816 | 37 | 3 | 559068 | 36 |
| 4 | 537063 | 50 | 4 | 544564 | 50 | 4 | 551938 | 49 | 4 | 559188 | 48 |
| 5 | 537189 | 63 | 5 | 544688 | 62 | 5 | 552059 | 61 | 5 | 559308 | 60 |
| 6 | 537315 | 76 | 6 | 544812 | 74 | 6 | 552181 | 73 | 6 | 559428 | 72 |
| 7 | 537441 | 88 | 7 | 544936 | 87 | 7 | 552303 | 86 | 7 | 559548 | 84 |
| 8 | 537567 | 101 | 8 | 545060 | 99 | 8 | 552425 | 98 | 8 | 559667 | 96 |
| 9 | 537693 | 114 | 9 | 545183 | 112 | 9 | 552546 | 110 | 9 | 559787 | 108 |
| 3450 | 537819 | | 3510 | 545307 | | 3570 | 552668 | | 3630 | 559907 | |
| 1 | 537945 | 13 | 1 | 545431 | 12 | 1 | 552790 | 12 | 1 | 560026 | 12 |
| 2 | 538071 | 25 | 2 | 545554 | 25 | 2 | 552911 | 24 | 2 | 560146 | 24 |
| 3 | 538197 | 38 | 3 | 545678 | 37 | 3 | 553033 | 36 | 3 | 560265 | 36 |
| 4 | 538322 | 50 | 4 | 545802 | 49 | 4 | 553154 | 49 | 4 | 560385 | 48 |
| 5 | 538448 | 63 | 5 | 545925 | 62 | 5 | 553276 | 61 | 5 | 560504 | 60 |
| 6 | 538574 | 76 | 6 | 546049 | 74 | 6 | 553398 | 73 | 6 | 560624 | 72 |
| 7 | 538699 | 88 | 7 | 546172 | 86 | 7 | 553519 | 85 | 7 | 560743 | 84 |
| 8 | 538825 | 101 | 8 | 546296 | 99 | 8 | 553640 | 97 | 8 | 560863 | 96 |
| 9 | 538951 | 114 | 9 | 546419 | 111 | 9 | 553762 | 109 | 9 | 560982 | 108 |

| No. | Log. | Prop. Part. | No. | Log. | Prop. Part. | No. | Log. | Prop. Part. | No. | Log. | Prop. Part. |
|---|---|---|---|---|---|---|---|---|---|---|---|
| 3640 | 561101 | | 3700 | 568202 | | 3760 | 575188 | | 3820 | 582063 | |
| 1 | 561221 | 12 | 1 | 568319 | 12 | 1 | 575303 | 12 | 1 | 582177 | 11 |
| 2 | 561340 | 24 | 2 | 568436 | 23 | 2 | 575419 | 23 | 2 | 582291 | 23 |
| 3 | 561459 | 36 | 3 | 568554 | 35 | 3 | 575534 | 35 | 3 | 582404 | 34 |
| 4 | 561578 | 48 | 4 | 568671 | 47 | 4 | 575650 | 46 | 4 | 582518 | 45 |
| 5 | 561698 | 60 | 5 | 568788 | 58 | 5 | 575765 | 58 | 5 | 582631 | 56 |
| 6 | 561817 | 72 | 6 | 568905 | 70 | 6 | 575880 | 69 | 6 | 582745 | 68 |
| 7 | 561936 | 84 | 7 | 569023 | 82 | 7 | 575996 | 80 | 7 | 582858 | 79 |
| 8 | 562055 | 96 | 8 | 569140 | 94 | 8 | 576111 | 92 | 8 | 582972 | 90 |
| 9 | 562174 | 108 | 9 | 569257 | 106 | 9 | 576226 | 104 | 9 | 583085 | 102 |
| 3650 | 562293 | | 3710 | 569374 | | 3770 | 576341 | | 3830 | 583199 | |
| 1 | 562412 | 12 | 1 | 569491 | 12 | 1 | 576457 | 12 | 1 | 583312 | 11 |
| 2 | 562531 | 24 | 2 | 569608 | 23 | 2 | 576572 | 23 | 2 | 583426 | 23 |
| 3 | 562650 | 36 | 3 | 569725 | 35 | 3 | 576687 | 35 | 3 | 583539 | 34 |
| 4 | 562768 | 48 | 4 | 569842 | 47 | 4 | 576802 | 46 | 4 | 583652 | 45 |
| 5 | 562887 | 60 | 5 | 569959 | 58 | 5 | 576917 | 58 | 5 | 583765 | 56 |
| 6 | 563006 | 71 | 6 | 570076 | 70 | 6 | 577032 | 69 | 6 | 583879 | 68 |
| 7 | 563125 | 83 | 7 | 570193 | 82 | 7 | 577147 | 80 | 7 | 583992 | 79 |
| 8 | 563244 | 95 | 8 | 570309 | 94 | 8 | 577262 | 92 | 8 | 584105 | 90 |
| 9 | 563362 | 107 | 9 | 570426 | 106 | 9 | 577377 | 104 | 9 | 584218 | 102 |
| 3660 | 563481 | | 3720 | 570543 | | 3780 | 577492 | | 3840 | 584331 | |
| 1 | 563600 | 12 | 1 | 570660 | 12 | 1 | 577607 | 11 | 1 | 584444 | 11 |
| 2 | 563718 | 24 | 2 | 570776 | 23 | 2 | 577721 | 23 | 2 | 584557 | 23 |
| 3 | 563837 | 36 | 3 | 570893 | 35 | 3 | 577836 | 34 | 3 | 584670 | 34 |
| 4 | 563955 | 48 | 4 | 571010 | 47 | 4 | 577951 | 46 | 4 | 584783 | 45 |
| 5 | 564074 | 60 | 5 | 571126 | 58 | 5 | 578066 | 57 | 5 | 584896 | 56 |
| 6 | 564192 | 71 | 6 | 571243 | 70 | 6 | 578181 | 68 | 6 | 585009 | 68 |
| 7 | 564311 | 83 | 7 | 571359 | 81 | 7 | 578295 | 80 | 7 | 585122 | 79 |
| 8 | 564429 | 95 | 8 | 571476 | 93 | 8 | 578410 | 91 | 8 | 585235 | 90 |
| 9 | 564548 | 107 | 9 | 571592 | 105 | 9 | 578525 | 103 | 9 | 585348 | 102 |
| 3670 | 564666 | | 3730 | 571709 | | 3790 | 578639 | | 3850 | 585461 | |
| 1 | 564784 | 12 | 1 | 571825 | 12 | 1 | 578754 | 11 | 1 | 585574 | 11 |
| 2 | 564903 | 24 | 2 | 571942 | 23 | 2 | 578868 | 23 | 2 | 585686 | 22 |
| 3 | 565021 | 36 | 3 | 572058 | 35 | 3 | 578983 | 34 | 3 | 585799 | 34 |
| 4 | 565139 | 47 | 4 | 572174 | 47 | 4 | 579097 | 46 | 4 | 585912 | 45 |
| 5 | 565257 | 59 | 5 | 572291 | 58 | 5 | 579212 | 57 | 5 | 586024 | 56 |
| 6 | 565376 | 71 | 6 | 572407 | 70 | 6 | 579326 | 68 | 6 | 586137 | 67 |
| 7 | 565494 | 83 | 7 | 572523 | 81 | 7 | 579441 | 80 | 7 | 586250 | 78 |
| 8 | 565612 | 95 | 8 | 572639 | 93 | 8 | 579555 | 91 | 8 | 586362 | 90 |
| 9 | 565730 | 107 | 9 | 572755 | 105 | 9 | 579669 | 103 | 9 | 586475 | 101 |
| 3680 | 565848 | | 3740 | 572872 | | 3800 | 579784 | | 3860 | 586587 | |
| 1 | 565966 | 12 | 1 | 572988 | 12 | 1 | 579898 | 11 | 1 | 586700 | 11 |
| 2 | 566084 | 24 | 2 | 573104 | 23 | 2 | 580012 | 23 | 2 | 586812 | 22 |
| 3 | 566202 | 35 | 3 | 573220 | 35 | 3 | 580126 | 34 | 3 | 586925 | 34 |
| 4 | 566320 | 47 | 4 | 573336 | 46 | 4 | 580240 | 46 | 4 | 587037 | 45 |
| 5 | 566437 | 59 | 5 | 573452 | 58 | 5 | 580355 | 57 | 5 | 587149 | 56 |
| 6 | 566555 | 71 | 6 | 573568 | 70 | 6 | 580469 | 68 | 6 | 587262 | 67 |
| 7 | 566673 | 83 | 7 | 573684 | 81 | 7 | 580583 | 80 | 7 | 587374 | 78 |
| 8 | 566791 | 94 | 8 | 573800 | 93 | 8 | 580697 | 91 | 8 | 587486 | 90 |
| 9 | 566909 | 106 | 9 | 573915 | 104 | 9 | 580811 | 103 | 9 | 587599 | 101 |
| 3690 | 567026 | | 3750 | 574031 | | 3810 | 580925 | | 3870 | 587711 | |
| 1 | 567144 | 12 | 1 | 574147 | 12 | 1 | 581039 | 11 | 1 | 587823 | 11 |
| 2 | 567262 | 24 | 2 | 574263 | 23 | 2 | 581153 | 23 | 2 | 587935 | 22 |
| 3 | 567379 | 35 | 3 | 574379 | 35 | 3 | 581267 | 34 | 3 | 588047 | 34 |
| 4 | 567497 | 47 | 4 | 574494 | 46 | 4 | 581381 | 46 | 4 | 588160 | 45 |
| 5 | 567614 | 59 | 5 | 574610 | 58 | 5 | 581495 | 57 | 5 | 588272 | 56 |
| 6 | 567732 | 71 | 6 | 574726 | 70 | 6 | 581608 | 68 | 6 | 588384 | 67 |
| 7 | 567849 | 83 | 7 | 574841 | 81 | 7 | 581722 | 80 | 7 | 588496 | 78 |
| 8 | 567967 | 94 | 8 | 574957 | 93 | 8 | 581836 | 91 | 8 | 588608 | 90 |
| 9 | 568084 | 106 | 9 | 575072 | 104 | 9 | 581950 | 103 | 9 | 588720 | 101 |

| No. | Log. | Prop. Part. | No. | Log. | Prop. Part. | No. | Log. | Prop. Part. | No. | Log. | Prop. Part. |
|---|---|---|---|---|---|---|---|---|---|---|---|
| 3880 | 588832 | | 3940 | 595496 | | 4000 | 602060 | | 4060 | 608526 | |
| 1 | 588944 | 11 | 1 | 595606 | 11 | 1 | 602169 | 11 | 1 | 608633 | 11 |
| 2 | 589056 | 22 | 2 | 595717 | 22 | 2 | 602277 | 22 | 2 | 608740 | 21 |
| 3 | 589167 | 33 | 3 | 595827 | 33 | 3 | 602386 | 33 | 3 | 608847 | 32 |
| 4 | 589279 | 44 | 4 | 595937 | 44 | 4 | 602494 | 43 | 4 | 608954 | 43 |
| 5 | 589391 | 56 | 5 | 596047 | 55 | 5 | 602603 | 54 | 5 | 609061 | 53 |
| 6 | 589503 | 67 | 6 | 596157 | 66 | 6 | 602711 | 65 | 6 | 609167 | 64 |
| 7 | 589615 | 78 | 7 | 596267 | 77 | 7 | 602819 | 76 | 7 | 609274 | 75 |
| 8 | 589726 | 89 | 8 | 596377 | 88 | 8 | 602928 | 87 | 8 | 609381 | 86 |
| 9 | 589838 | 100 | 9 | 596487 | 99 | 9 | 603036 | 98 | 9 | 609488 | 96 |
| 3890 | 589950 | | 3950 | 596597 | | 4010 | 603144 | | 4070 | 609594 | |
| 1 | 590061 | 11 | 1 | 596707 | 11 | 1 | 603253 | 11 | 1 | 609701 | 11 |
| 2 | 590173 | 22 | 2 | 596817 | 22 | 2 | 603361 | 22 | 2 | 609808 | 21 |
| 3 | 590284 | 33 | 3 | 596927 | 33 | 3 | 603469 | 33 | 3 | 609914 | 32 |
| 4 | 590396 | 44 | 4 | 597037 | 44 | 4 | 603577 | 43 | 4 | 610021 | 43 |
| 5 | 590507 | 56 | 5 | 597146 | 55 | 5 | 603686 | 54 | 5 | 610128 | 53 |
| 6 | 590619 | 67 | 6 | 597256 | 66 | 6 | 603794 | 65 | 6 | 610234 | 64 |
| 7 | 590730 | 78 | 7 | 597366 | 77 | 7 | 603902 | 76 | 7 | 610341 | 75 |
| 8 | 590842 | 89 | 8 | 597476 | 88 | 8 | 604010 | 87 | 8 | 610447 | 86 |
| 9 | 590953 | 100 | 9 | 597585 | 99 | 9 | 604118 | 98 | 9 | 610554 | 96 |
| 3900 | 591065 | | 3960 | 597695 | | 4020 | 604226 | | 4080 | 610660 | |
| 1 | 591176 | 11 | 1 | 597805 | 11 | 1 | 604334 | 11 | 1 | 610767 | 11 |
| 2 | 591287 | 22 | 2 | 597914 | 22 | 2 | 604442 | 22 | 2 | 610873 | 21 |
| 3 | 591399 | 33 | 3 | 598024 | 33 | 3 | 604550 | 32 | 3 | 610979 | 32 |
| 4 | 591510 | 44 | 4 | 598134 | 44 | 4 | 604658 | 43 | 4 | 611086 | 42 |
| 5 | 591621 | 56 | 5 | 598243 | 55 | 5 | 604766 | 54 | 5 | 611192 | 53 |
| 6 | 591732 | 67 | 6 | 598353 | 66 | 6 | 604874 | 65 | 6 | 611298 | 64 |
| 7 | 591843 | 78 | 7 | 598462 | 77 | 7 | 604982 | 76 | 7 | 611405 | 74 |
| 8 | 591955 | 89 | 8 | 598572 | 88 | 8 | 605089 | 86 | 8 | 611511 | 85 |
| 9 | 592066 | 100 | 9 | 598681 | 99 | 9 | 605197 | 97 | 9 | 611617 | 95 |
| 3910 | 592177 | | 3970 | 598790 | | 4030 | 605305 | | 4090 | 611723 | |
| 1 | 592288 | 11 | 1 | 598900 | 11 | 1 | 605413 | 11 | 1 | 611829 | 11 |
| 2 | 592399 | 22 | 2 | 599009 | 22 | 2 | 605521 | 22 | 2 | 611936 | 21 |
| 3 | 592510 | 33 | 3 | 599119 | 33 | 3 | 605628 | 32 | 3 | 612042 | 32 |
| 4 | 592621 | 44 | 4 | 599228 | 44 | 4 | 605736 | 43 | 4 | 612148 | 42 |
| 5 | 592732 | 55 | 5 | 599337 | 55 | 5 | 605844 | 54 | 5 | 612254 | 53 |
| 6 | 592843 | 67 | 6 | 599446 | 66 | 6 | 605951 | 65 | 6 | 612360 | 64 |
| 7 | 592954 | 78 | 7 | 599556 | 77 | 7 | 606059 | 76 | 7 | 612466 | 74 |
| 8 | 593064 | 89 | 8 | 599665 | 88 | 8 | 606166 | 86 | 8 | 612572 | 85 |
| 9 | 593175 | 100 | 9 | 599774 | 99 | 9 | 606274 | 97 | 9 | 612678 | 95 |
| 3920 | 593286 | | 3980 | 599883 | | 4040 | 606381 | | 4100 | 612784 | |
| 1 | 593397 | 11 | 1 | 599992 | 11 | 1 | 606489 | 11 | 1 | 612890 | 11 |
| 2 | 593508 | 22 | 2 | 600101 | 22 | 2 | 606596 | 21 | 2 | 612996 | 21 |
| 3 | 593618 | 33 | 3 | 600210 | 33 | 3 | 606704 | 32 | 3 | 613101 | 32 |
| 4 | 593729 | 44 | 4 | 600319 | 44 | 4 | 606811 | 43 | 4 | 613207 | 42 |
| 5 | 593840 | 55 | 5 | 600428 | 54 | 5 | 606919 | 54 | 5 | 613313 | 53 |
| 6 | 593950 | 66 | 6 | 600537 | 65 | 6 | 607026 | 64 | 6 | 613419 | 64 |
| 7 | 594061 | 77 | 7 | 600646 | 76 | 7 | 607133 | 75 | 7 | 613525 | 74 |
| 8 | 594171 | 88 | 8 | 600755 | 87 | 8 | 607241 | 86 | 8 | 613630 | 85 |
| 9 | 594282 | 99 | 9 | 600864 | 95 | 9 | 607348 | 96 | 9 | 613736 | 95 |
| 3930 | 594393 | | 3990 | 600973 | | 4050 | 607455 | | 4110 | 613842 | |
| 1 | 594503 | 11 | 1 | 601082 | 11 | 1 | 607562 | 11 | 1 | 613947 | 11 |
| 2 | 594613 | 22 | 2 | 601190 | 22 | 2 | 607669 | 21 | 2 | 614053 | 21 |
| 3 | 594724 | 33 | 3 | 601299 | 33 | 3 | 607777 | 32 | 3 | 614159 | 32 |
| 4 | 594834 | 44 | 4 | 601408 | 44 | 4 | 607884 | 43 | 4 | 614264 | 42 |
| 5 | 594945 | 55 | 5 | 601517 | 54 | 5 | 607991 | 54 | 5 | 614370 | 53 |
| 6 | 595055 | 66 | 6 | 601625 | 65 | 6 | 608098 | 64 | 6 | 614475 | 63 |
| 7 | 595165 | 77 | 7 | 601734 | 76 | 7 | 608205 | 75 | 7 | 614581 | 74 |
| 8 | 595276 | 88 | 8 | 601843 | 87 | 8 | 608312 | 86 | 8 | 614686 | 84 |
| 9 | 595386 | 99 | 9 | 601951 | 98 | 9 | 608419 | 96 | 9 | 614792 | 95 |

| No. | Log. | Prop. Part. | No. | Log. | Prop. Part. | No. | Log. | Prop. Part. | No. | Log. | Prop. Part. |
|---|---|---|---|---|---|---|---|---|---|---|---|
| 4120 | 614897 | | 4180 | 621176 | | 4240 | 627366 | | 4300 | 633468 | |
| 1 | 615003 | 11 | 1 | 621280 | 10 | 1 | 627468 | 10 | 1 | 633569 | 10 |
| 2 | 615108 | 21 | 2 | 621384 | 21 | 2 | 627571 | 20 | 2 | 633670 | 20 |
| 3 | 615213 | 31 | 3 | 621488 | 31 | 3 | 627673 | 31 | 3 | 633771 | 30 |
| 4 | 615319 | 42 | 4 | 621592 | 42 | 4 | 627775 | 41 | 4 | 633872 | 40 |
| 5 | 615424 | 52 | 5 | 621695 | 52 | 5 | 627878 | 51 | 5 | 633973 | 50 |
| 6 | 615529 | 63 | 6 | 621799 | 62 | 6 | 627980 | 61 | 6 | 634074 | 61 |
| 7 | 615634 | 73 | 7 | 621903 | 73 | 7 | 628082 | 72 | 7 | 634175 | 71 |
| 8 | 615740 | 84 | 8 | 622007 | 83 | 8 | 628184 | 82 | 8 | 634276 | 81 |
| 9 | 615845 | 95 | 9 | 622110 | 94 | 9 | 628287 | 92 | 9 | 634376 | 91 |
| 4130 | 615950 | | 4190 | 622214 | | 4250 | 628389 | | 4310 | 634477 | |
| 1 | 616055 | 11 | 1 | 622318 | 10 | 1 | 628491 | 10 | 1 | 634578 | 10 |
| 2 | 616160 | 21 | 2 | 622421 | 21 | 2 | 628593 | 20 | 2 | 634679 | 20 |
| 3 | 616265 | 31 | 3 | 622525 | 31 | 3 | 628695 | 31 | 3 | 634779 | 30 |
| 4 | 616370 | 42 | 4 | 622628 | 41 | 4 | 628797 | 41 | 4 | 634880 | 40 |
| 5 | 616475 | 52 | 5 | 622732 | 52 | 5 | 628900 | 51 | 5 | 634981 | 50 |
| 6 | 616580 | 63 | 6 | 622835 | 62 | 6 | 629002 | 61 | 6 | 635081 | 61 |
| 7 | 616685 | 73 | 7 | 622939 | 72 | 7 | 629104 | 72 | 7 | 635182 | 71 |
| 8 | 616790 | 84 | 8 | 623042 | 83 | 8 | 629206 | 82 | 8 | 635283 | 81 |
| 9 | 616895 | 95 | 9 | 623146 | 93 | 9 | 629308 | 92 | 9 | 635383 | 91 |
| 4140 | 617000 | | 4200 | 623249 | | 4260 | 629410 | | 4320 | 635484 | |
| 1 | 617105 | 10 | 1 | 623353 | 10 | 1 | 629511 | 10 | 1 | 635584 | 10 |
| 2 | 617210 | 21 | 2 | 623456 | 21 | 2 | 629613 | 20 | 2 | 635685 | 20 |
| 3 | 617315 | 31 | 3 | 623559 | 31 | 3 | 629715 | 30 | 3 | 635785 | 30 |
| 4 | 617420 | 42 | 4 | 623663 | 41 | 4 | 629817 | 41 | 4 | 635886 | 40 |
| 5 | 617524 | 52 | 5 | 623766 | 52 | 5 | 629919 | 51 | 5 | 635986 | 50 |
| 6 | 617629 | 63 | 6 | 623869 | 62 | 6 | 630021 | 61 | 6 | 636086 | 60 |
| 7 | 617734 | 73 | 7 | 623972 | 72 | 7 | 630123 | 71 | 7 | 636187 | 70 |
| 8 | 617839 | 84 | 8 | 624076 | 83 | 8 | 630224 | 81 | 8 | 636287 | 80 |
| 9 | 617943 | 94 | 9 | 624179 | 93 | 9 | 630326 | 91 | 9 | 636388 | 90 |
| 4150 | 618048 | | 4210 | 624282 | | 4270 | 630428 | | 4330 | 636488 | |
| 1 | 618153 | 10 | 1 | 624385 | 10 | 1 | 630530 | 10 | 1 | 636588 | 10 |
| 2 | 618257 | 21 | 2 | 624488 | 21 | 2 | 630631 | 20 | 2 | 636688 | 20 |
| 3 | 618362 | 31 | 3 | 624591 | 31 | 3 | 630733 | 30 | 3 | 636789 | 30 |
| 4 | 618466 | 42 | 4 | 624694 | 41 | 4 | 630834 | 41 | 4 | 636889 | 40 |
| 5 | 618571 | 52 | 5 | 624798 | 51 | 5 | 630936 | 51 | 5 | 636989 | 50 |
| 6 | 618675 | 62 | 6 | 624901 | 62 | 6 | 631038 | 61 | 6 | 637089 | 60 |
| 7 | 618780 | 73 | 7 | 625004 | 72 | 7 | 631139 | 71 | 7 | 637189 | 70 |
| 8 | 618884 | 83 | 8 | 625107 | 82 | 8 | 631241 | 81 | 8 | 637289 | 80 |
| 9 | 618989 | 94 | 9 | 625209 | 93 | 9 | 631342 | 91 | 9 | 637390 | 90 |
| 4160 | 619093 | | 4220 | 625312 | | 4280 | 631444 | | 4340 | 637490 | |
| 1 | 619198 | 10 | 1 | 625415 | 10 | 1 | 631545 | 10 | 1 | 637590 | 10 |
| 2 | 619302 | 21 | 2 | 625518 | 21 | 2 | 631647 | 20 | 2 | 637690 | 20 |
| 3 | 619406 | 31 | 3 | 625621 | 31 | 3 | 631748 | 30 | 3 | 637790 | 30 |
| 4 | 619511 | 42 | 4 | 625724 | 41 | 4 | 631849 | 41 | 4 | 637890 | 40 |
| 5 | 619615 | 52 | 5 | 625827 | 51 | 5 | 631951 | 51 | 5 | 637990 | 50 |
| 6 | 619719 | 62 | 6 | 625929 | 62 | 6 | 632052 | 61 | 6 | 638090 | 60 |
| 7 | 619823 | 73 | 7 | 626032 | 72 | 7 | 632153 | 71 | 7 | 638190 | 70 |
| 8 | 619928 | 83 | 8 | 626135 | 82 | 8 | 632255 | 81 | 8 | 638289 | 80 |
| 9 | 620032 | 94 | 9 | 626238 | 93 | 9 | 632356 | 91 | 9 | 638389 | 90 |
| 4170 | 620136 | | 4230 | 626340 | | 4290 | 632457 | | 4350 | 638489 | |
| 1 | 620240 | 10 | 1 | 626443 | 10 | 1 | 632558 | 10 | 1 | 638589 | 10 |
| 2 | 620344 | 21 | 2 | 626546 | 21 | 2 | 632660 | 20 | 2 | 638689 | 20 |
| 3 | 620448 | 31 | 3 | 626648 | 31 | 3 | 632761 | 30 | 3 | 638789 | 30 |
| 4 | 620552 | 42 | 4 | 626751 | 41 | 4 | 632862 | 41 | 4 | 638888 | 40 |
| 5 | 620656 | 52 | 5 | 626853 | 51 | 5 | 632963 | 51 | 5 | 638988 | 50 |
| 6 | 620760 | 62 | 6 | 626956 | 62 | 6 | 633064 | 61 | 6 | 639088 | 60 |
| 7 | 620864 | 73 | 7 | 627058 | 72 | 7 | 633165 | 71 | 7 | 639188 | 70 |
| 8 | 620968 | 83 | 8 | 627161 | 82 | 8 | 633266 | 81 | 8 | 639287 | 80 |
| 9 | 621072 | 94 | 9 | 627263 | 93 | 9 | 633367 | 91 | 9 | 639387 | 90 |

| No. | Log. | Prop. Part. | No. | Log. | Prop. Part. | No. | Log. | Prop. Part. | No. | Log. | Prop. Part. |
|---|---|---|---|---|---|---|---|---|---|---|---|
| 4360 | 639486 | | 4420 | 645422 | | 4480 | 651278 | | 4540 | 657056 | |
| 1 | 639586 | 10 | 1 | 645520 | 10 | 1 | 651375 | 10 | 1 | 657151 | 10 |
| 2 | 639686 | 20 | 2 | 645619 | 20 | 2 | 651472 | 19 | 2 | 657247 | 19 |
| 3 | 639785 | 30 | 3 | 645717 | 30 | 3 | 651569 | 29 | 3 | 657343 | 28 |
| 4 | 639885 | 40 | 4 | 645815 | 39 | 4 | 651666 | 38 | 4 | 657438 | 38 |
| 5 | 639984 | 50 | 5 | 645913 | 49 | 5 | 651762 | 48 | 5 | 657534 | 47 |
| 6 | 640084 | 60 | 6 | 646011 | 59 | 6 | 651859 | 58 | 6 | 657629 | 57 |
| 7 | 640183 | 70 | 7 | 646109 | 69 | 7 | 651956 | 67 | 7 | 657725 | 67 |
| 8 | 640283 | 80 | 8 | 646208 | 79 | 8 | 652053 | 77 | 8 | 657820 | 76 |
| 9 | 640382 | 90 | 9 | 646306 | 89 | 9 | 652150 | 87 | 9 | 657916 | 86 |
| 4370 | 640481 | | 4430 | 646404 | | 4490 | 652246 | | 4550 | 658011 | |
| 1 | 940581 | 10 | 1 | 646502 | 10 | 1 | 652343 | 10 | 1 | 658107 | 10 |
| 2 | 640680 | 20 | 2 | 646600 | 20 | 2 | 652440 | 19 | 2 | 658202 | 19 |
| 3 | 640779 | 30 | 3 | 646698 | 29 | 3 | 652536 | 29 | 3 | 658298 | 28 |
| 4 | 640879 | 40 | 4 | 646796 | 39 | 4 | 652633 | 38 | 4 | 658393 | 38 |
| 5 | 640978 | 50 | 5 | 646894 | 49 | 5 | 652730 | 48 | 5 | 658488 | 47 |
| 6 | 641077 | 60 | 6 | 646991 | 59 | 6 | 652826 | 58 | 6 | 658584 | 57 |
| 7 | 641176 | 70 | 7 | 647089 | 69 | 7 | 652923 | 67 | 7 | 658679 | 67 |
| 8 | 641276 | 80 | 8 | 647187 | 78 | 8 | 653019 | 77 | 8 | 658774 | 76 |
| 9 | 641375 | 90 | 9 | 647285 | 88 | 9 | 653116 | 87 | 9 | 658870 | 86 |
| 4380 | 641474 | | 4440 | 647383 | | 4500 | 653213 | | 4560 | 658965 | |
| 1 | 641573 | 10 | 1 | 647481 | 10 | 1 | 653309 | 10 | 1 | 659060 | 10 |
| 2 | 641672 | 20 | 2 | 647579 | 20 | 2 | 653405 | 19 | 2 | 659155 | 19 |
| 3 | 641771 | 30 | 3 | 647676 | 29 | 3 | 653502 | 29 | 3 | 659250 | 28 |
| 4 | 641870 | 40 | 4 | 647774 | 39 | 4 | 653598 | 38 | 4 | 659346 | 38 |
| 5 | 641970 | 50 | 5 | 647872 | 49 | 5 | 653695 | 48 | 5 | 659441 | 47 |
| 6 | 642069 | 59 | 6 | 647969 | 59 | 6 | 653791 | 58 | 6 | 659536 | 57 |
| 7 | 642168 | 69 | 7 | 648067 | 69 | 7 | 653888 | 67 | 7 | 659631 | 67 |
| 8 | 642267 | 79 | 8 | 648165 | 78 | 8 | 653984 | 77 | 8 | 659726 | 76 |
| 9 | 642366 | 89 | 9 | 648262 | 88 | 9 | 654080 | 87 | 9 | 659821 | 86 |
| 4390 | 642464 | | 4450 | 648360 | | 4510 | 654176 | | 4570 | 659916 | |
| 1 | 642563 | 10 | 1 | 648458 | 10 | 1 | 654273 | 10 | 1 | 660011 | 10 |
| 2 | 642662 | 20 | 2 | 648555 | 19 | 2 | 654369 | 19 | 2 | 660106 | 19 |
| 3 | 642761 | 30 | 3 | 648653 | 29 | 3 | 654465 | 29 | 3 | 660201 | 28 |
| 4 | 642860 | 40 | 4 | 648750 | 39 | 4 | 654562 | 38 | 4 | 660296 | 38 |
| 5 | 642959 | 49 | 5 | 648848 | 49 | 5 | 654558 | 48 | 5 | 660391 | 47 |
| 6 | 643058 | 59 | 6 | 648945 | 58 | 6 | 654754 | 58 | 6 | 660486 | 57 |
| 7 | 643156 | 69 | 7 | 649043 | 68 | 7 | 654850 | 67 | 7 | 660581 | 67 |
| 8 | 643255 | 79 | 8 | 649140 | 78 | 8 | 654946 | 77 | 8 | 660676 | 76 |
| 9 | 643354 | 89 | 9 | 649237 | 88 | 9 | 655042 | 86 | 9 | 660771 | 86 |
| 4400 | 643453 | | 4460 | 649335 | | 4520 | 655138 | | 4580 | 660865 | |
| 1 | 643551 | 10 | 1 | 649432 | 10 | 1 | 655234 | 10 | 1 | 660960 | 9 |
| 2 | 643650 | 20 | 2 | 649530 | 19 | 2 | 655331 | 19 | 2 | 661055 | 19 |
| 3 | 643749 | 30 | 3 | 649627 | 29 | 3 | 655427 | 29 | 3 | 661150 | 28 |
| 4 | 643847 | 39 | 4 | 649724 | 39 | 4 | 655523 | 38 | 4 | 661245 | 38 |
| 5 | 643946 | 49 | 5 | 649821 | 49 | 5 | 655619 | 48 | 5 | 661339 | 47 |
| 6 | 644044 | 59 | 6 | 649919 | 58 | 6 | 655714 | 58 | 6 | 661434 | 57 |
| 7 | 644143 | 69 | 7 | 650016 | 68 | 7 | 655810 | 67 | 7 | 661529 | 66 |
| 8 | 644242 | 79 | 8 | 650113 | 78 | 8 | 655906 | 77 | 8 | 661623 | 76 |
| 9 | 644340 | 89 | 9 | 650210 | 88 | 9 | 656002 | 86 | 9 | 661718 | 85 |
| 4410 | 644439 | | 4470 | 650307 | | 4530 | 656098 | | 4590 | 661813 | |
| 1 | 644537 | 10 | 1 | 650405 | 10 | 1 | 656194 | 10 | 1 | 661907 | 9 |
| 2 | 644635 | 20 | 2 | 650502 | 19 | 2 | 656290 | 19 | 2 | 662002 | 19 |
| 3 | 644734 | 30 | 3 | 650599 | 29 | 3 | 656386 | 29 | 3 | 662096 | 28 |
| 4 | 644832 | 39 | 4 | 650696 | 39 | 4 | 656481 | 38 | 4 | 662191 | 38 |
| 5 | 644931 | 49 | 5 | 650793 | 49 | 5 | 656577 | 48 | 5 | 662285 | 47 |
| 6 | 645029 | 59 | 6 | 650890 | 58 | 6 | 656673 | 58 | 6 | 662380 | 57 |
| 7 | 645127 | 69 | 7 | 650987 | 68 | 7 | 656769 | 67 | 7 | 662474 | 66 |
| 8 | 645226 | 79 | 8 | 651084 | 78 | 8 | 656864 | 77 | 8 | 662569 | 76 |
| 9 | 645324 | 89 | 9 | 651181 | 88 | 9 | 656960 | 86 | 9 | 662663 | 85 |

| No. | Log. | Prop. Part. | No. | Log. | Prop. Part. | No. | Log. | Prop. Part. | No. | Log. | Prop. Part. |
|---|---|---|---|---|---|---|---|---|---|---|---|
| 4600 | 662758 | | 4660 | 668386 | | 4720 | 673942 | | 4780 | 679428 | |
| 1 | 662852 | 9 | 1 | 668479 | 9 | 1 | 674034 | 9 | 1 | 679519 | 9 |
| 2 | 662947 | 19 | 2 | 668572 | 19 | 2 | 674126 | 18 | 2 | 679610 | 18 |
| 3 | 663041 | 28 | 3 | 668665 | 28 | 3 | 674218 | 28 | 3 | 679700 | 27 |
| 4 | 663135 | 38 | 4 | 668758 | 37 | 4 | 674310 | 37 | 4 | 679791 | 36 |
| 5 | 663230 | 47 | 5 | 668852 | 47 | 5 | 674402 | 46 | 5 | 679882 | 45 |
| 6 | 663324 | 57 | 6 | 668945 | 56 | 6 | 674494 | 55 | 6 | 679973 | 55 |
| 7 | 663418 | 66 | 7 | 669038 | 65 | 7 | 674586 | 64 | 7 | 680063 | 64 |
| 8 | 663512 | 76 | 8 | 669131 | 74 | 8 | 674677 | 74 | 8 | 680154 | 73 |
| 9 | 663607 | 85 | 9 | 669224 | 84 | 9 | 674769 | 83 | 9 | 680245 | 82 |
| 4610 | 663701 | | 4670 | 669317 | | 4730 | 674861 | | 4790 | 680335 | |
| 1 | 663795 | 9 | 1 | 669410 | 9 | 1 | 674953 | 9 | 1 | 680426 | 9 |
| 2 | 663889 | 19 | 2 | 669503 | 19 | 2 | 675045 | 18 | 2 | 680517 | 18 |
| 3 | 663983 | 28 | 3 | 669596 | 28 | 3 | 675136 | 28 | 3 | 680607 | 27 |
| 4 | 664078 | 38 | 4 | 669689 | 37 | 4 | 675228 | 37 | 4 | 680698 | 36 |
| 5 | 664172 | 47 | 5 | 669782 | 47 | 5 | 675320 | 46 | 5 | 680789 | 45 |
| 6 | 664266 | 56 | 6 | 669875 | 56 | 6 | 675412 | 55 | 6 | 680879 | 55 |
| 7 | 664360 | 66 | 7 | 669967 | 65 | 7 | 675503 | 64 | 7 | 680970 | 64 |
| 8 | 664454 | 75 | 8 | 670060 | 74 | 8 | 675595 | 74 | 8 | 681060 | 73 |
| 9 | 664548 | 85 | 9 | 670153 | 84 | 9 | 675687 | 83 | 9 | 681151 | 82 |
| 4620 | 664642 | | 4680 | 670246 | | 4740 | 675778 | | 4800 | 681241 | |
| 1 | 664736 | 9 | 1 | 670339 | 9 | 1 | 675870 | 9 | 1 | 681332 | 9 |
| 2 | 664830 | 19 | 2 | 670431 | 18 | 2 | 675962 | 18 | 2 | 681422 | 18 |
| 3 | 664924 | 28 | 3 | 670524 | 28 | 3 | 676053 | 27 | 3 | 681513 | 27 |
| 4 | 665018 | 38 | 4 | 670617 | 37 | 4 | 676145 | 36 | 4 | 681603 | 36 |
| 5 | 665112 | 47 | 5 | 670710 | 46 | 5 | 676236 | 46 | 5 | 681693 | 45 |
| 6 | 665206 | 56 | 6 | 670802 | 55 | 6 | 676328 | 55 | 6 | 681784 | 54 |
| 7 | 665299 | 66 | 7 | 670895 | 64 | 7 | 676419 | 64 | 7 | 681874 | 63 |
| 8 | 665393 | 75 | 8 | 670988 | 74 | 8 | 676511 | 73 | 8 | 681964 | 72 |
| 9 | 665487 | 85 | 9 | 671080 | 83 | 9 | 676602 | 82 | 9 | 682055 | 81 |
| 4630 | 665581 | | 4690 | 671173 | | 4750 | 676694 | | 4810 | 682145 | |
| 1 | 665675 | 9 | 1 | 671265 | 9 | 1 | 676785 | 9 | 1 | 682235 | 9 |
| 2 | 665769 | 19 | 2 | 671358 | 18 | 2 | 676876 | 18 | 2 | 682326 | 18 |
| 3 | 665862 | 28 | 3 | 671451 | 28 | 3 | 676968 | 27 | 3 | 682416 | 27 |
| 4 | 665956 | 38 | 4 | 671543 | 37 | 4 | 677059 | 36 | 4 | 682506 | 36 |
| 5 | 666050 | 47 | 5 | 671636 | 46 | 5 | 677151 | 46 | 5 | 682596 | 45 |
| 6 | 666143 | 56 | 6 | 671728 | 55 | 6 | 677242 | 55 | 6 | 682686 | 54 |
| 7 | 666237 | 66 | 7 | 671821 | 64 | 7 | 677333 | 64 | 7 | 682777 | 63 |
| 8 | 666331 | 75 | 8 | 671913 | 74 | 8 | 677424 | 73 | 8 | 682867 | 72 |
| 9 | 666424 | 85 | 9 | 672005 | 83 | 9 | 677516 | 82 | 9 | 682957 | 81 |
| 4640 | 666518 | | 4700 | 672098 | | 4760 | 677607 | | 4820 | 683047 | |
| 1 | 666612 | 9 | 1 | 672190 | 9 | 1 | 677698 | 9 | 1 | 683137 | 9 |
| 2 | 666705 | 19 | 2 | 672283 | 18 | 2 | 677789 | 18 | 2 | 683227 | 18 |
| 3 | 666799 | 28 | 3 | 672375 | 28 | 3 | 677881 | 27 | 3 | 683317 | 27 |
| 4 | 666892 | 37 | 4 | 672467 | 37 | 4 | 677972 | 36 | 4 | 683407 | 36 |
| 5 | 666986 | 47 | 5 | 672560 | 46 | 5 | 678063 | 45 | 5 | 683497 | 45 |
| 6 | 667079 | 56 | 6 | 672652 | 55 | 6 | 678154 | 55 | 6 | 683587 | 54 |
| 7 | 667173 | 65 | 7 | 672744 | 64 | 7 | 678245 | 64 | 7 | 683677 | 63 |
| 8 | 667266 | 74 | 8 | 672836 | 74 | 8 | 678336 | 73 | 8 | 683767 | 72 |
| 9 | 667359 | 84 | 9 | 672929 | 83 | 9 | 678427 | 82 | 9 | 683857 | 81 |
| 4650 | 667453 | | 4710 | 673021 | | 4770 | 678518 | | 4830 | 683947 | |
| 1 | 667546 | 9 | 1 | 673113 | 9 | 1 | 678609 | 9 | 1 | 684037 | 9 |
| 2 | 667640 | 19 | 2 | 673205 | 18 | 2 | 678700 | 18 | 2 | 684127 | 18 |
| 3 | 667733 | 28 | 3 | 673297 | 28 | 3 | 678791 | 27 | 3 | 684217 | 27 |
| 4 | 667826 | 37 | 4 | 673390 | 37 | 4 | 678882 | 36 | 4 | 684307 | 36 |
| 5 | 667920 | 47 | 5 | 673482 | 46 | 5 | 678973 | 45 | 5 | 684396 | 45 |
| 6 | 668013 | 56 | 6 | 673574 | 55 | 6 | 679064 | 55 | 6 | 684486 | 54 |
| 7 | 668106 | 65 | 7 | 673666 | 64 | 7 | 679155 | 64 | 7 | 684576 | 63 |
| 8 | 668199 | 74 | 8 | 673758 | 74 | 8 | 679246 | 73 | 8 | 684666 | 72 |
| 9 | 668293 | 84 | 9 | 673850 | 83 | 9 | 679337 | 82 | 9 | 684756 | 81 |

| No. | Log. | Prop. Part. | No. | Log. | Prop. Part. | No. | Log. | Prop. Part. | No. | Log. | Prop. Part. |
|---|---|---|---|---|---|---|---|---|---|---|---|
| 4840 | 684845 | | 4900 | 690196 | | 4960 | 695482 | | 5020 | 700704 | |
| 1 | 684935 | 9 | 1 | 690285 | 9 | 1 | 695569 | 9 | 1 | 700790 | 9 |
| 2 | 685025 | 18 | 2 | 690373 | 18 | 2 | 695657 | 17 | 2 | 700877 | 17 |
| 3 | 685114 | 27 | 3 | 690462 | 27 | 3 | 695744 | 26 | 3 | 700963 | 26 |
| 4 | 685204 | 36 | 4 | 690550 | 35 | 4 | 695832 | 35 | 4 | 701050 | 35 |
| 5 | 685294 | 45 | 5 | 690639 | 44 | 5 | 695919 | 44 | 5 | 701136 | 43 |
| 6 | 685383 | 54 | 6 | 690727 | 53 | 6 | 696007 | 52 | 6 | 701222 | 52 |
| 7 | 685473 | 63 | 7 | 690816 | 62 | 7 | 696094 | 61 | 7 | 701309 | 61 |
| 8 | 685563 | 72 | 8 | 690905 | 71 | 8 | 696182 | 70 | 8 | 701395 | 70 |
| 9 | 685652 | 81 | 9 | 690993 | 80 | 9 | 696269 | 79 | 9 | 701482 | 78 |
| 4850 | 685742 | | 4910 | 691081 | | 4970 | 696356 | | 5030 | 701568 | |
| 1 | 685831 | 9 | 1 | 691170 | 9 | 1 | 696444 | 9 | 1 | 701654 | 9 |
| 2 | 685921 | 18 | 2 | 691258 | 18 | 2 | 696531 | 17 | 2 | 701741 | 17 |
| 3 | 686010 | 27 | 3 | 691347 | 27 | 3 | 696618 | 26 | 3 | 701827 | 26 |
| 4 | 686100 | 36 | 4 | 691435 | 35 | 4 | 696706 | 35 | 4 | 701913 | 35 |
| 5 | 686189 | 45 | 5 | 691524 | 44 | 5 | 696793 | 44 | 5 | 701999 | 43 |
| 6 | 686279 | 54 | 6 | 691612 | 53 | 6 | 696880 | 52 | 6 | 702086 | 52 |
| 7 | 686368 | 63 | 7 | 691700 | 62 | 7 | 696968 | 61 | 7 | 702172 | 61 |
| 8 | 686457 | 72 | 8 | 691789 | 71 | 8 | 697055 | 70 | 8 | 702258 | 70 |
| 9 | 686547 | 81 | 9 | 691877 | 80 | 9 | 697142 | 79 | 9 | 702344 | 78 |
| 4860 | 686636 | | 4920 | 691965 | | 4980 | 697229 | | 5040 | 702430 | |
| 1 | 686726 | 9 | 1 | 692053 | 9 | 1 | 697317 | 9 | 1 | 702517 | 9 |
| 2 | 686815 | 18 | 2 | 692142 | 18 | 2 | 697404 | 17 | 2 | 702603 | 17 |
| 3 | 686904 | 27 | 3 | 692230 | 27 | 3 | 697491 | 26 | 3 | 702689 | 26 |
| 4 | 686994 | 36 | 4 | 692318 | 35 | 4 | 697578 | 35 | 4 | 702775 | 34 |
| 5 | 687083 | 45 | 5 | 692406 | 44 | 5 | 697665 | 44 | 5 | 702861 | 43 |
| 6 | 687172 | 54 | 6 | 692494 | 53 | 6 | 697752 | 52 | 6 | 702947 | 52 |
| 7 | 687261 | 63 | 7 | 692583 | 62 | 7 | 697839 | 61 | 7 | 703033 | 60 |
| 8 | 687351 | 72 | 8 | 692671 | 71 | 8 | 697926 | 70 | 8 | 703119 | 69 |
| 9 | 687440 | 81 | 9 | 692759 | 80 | 9 | 698013 | 79 | 9 | 703205 | 77 |
| 4870 | 687529 | | 4930 | 692847 | | 4990 | 698100 | | 5050 | 703291 | |
| 1 | 687618 | 9 | 1 | 692935 | 9 | 1 | 698188 | 9 | 1 | 703377 | 9 |
| 2 | 687707 | 18 | 2 | 693023 | 18 | 2 | 698275 | 17 | 2 | 703463 | 17 |
| 3 | 687796 | 27 | 3 | 693111 | 26 | 3 | 698362 | 26 | 3 | 703549 | 26 |
| 4 | 687886 | 36 | 4 | 693199 | 35 | 4 | 698448 | 35 | 4 | 703635 | 34 |
| 5 | 687975 | 45 | 5 | 693287 | 44 | 5 | 698535 | 44 | 5 | 703721 | 43 |
| 6 | 688064 | 54 | 6 | 693375 | 53 | 6 | 698622 | 52 | 6 | 703807 | 52 |
| 7 | 688153 | 62 | 7 | 693463 | 62 | 7 | 698709 | 61 | 7 | 703893 | 60 |
| 8 | 688242 | 72 | 8 | 693551 | 70 | 8 | 698796 | 70 | 8 | 703979 | 69 |
| 9 | 688331 | 80 | 9 | 693639 | 79 | 9 | 698883 | 79 | 9 | 704065 | 77 |
| 4880 | 688420 | | 4940 | 693727 | | 5000 | 698970 | | 5060 | 704150 | |
| 1 | 688509 | 9 | 1 | 693815 | 9 | 1 | 699057 | 9 | 1 | 704236 | 9 |
| 2 | 688598 | 18 | 2 | 693903 | 18 | 2 | 699144 | 17 | 2 | 704322 | 17 |
| 3 | 688687 | 27 | 3 | 693991 | 26 | 3 | 699231 | 26 | 3 | 704408 | 26 |
| 4 | 688776 | 36 | 4 | 694078 | 35 | 4 | 699317 | 35 | 4 | 704494 | 34 |
| 5 | 688865 | 45 | 5 | 694166 | 44 | 5 | 699404 | 43 | 5 | 704579 | 43 |
| 6 | 688953 | 54 | 6 | 694254 | 53 | 6 | 699491 | 52 | 6 | 704665 | 52 |
| 7 | 689042 | 62 | 7 | 694342 | 62 | 7 | 699578 | 61 | 7 | 704751 | 60 |
| 8 | 689131 | 72 | 8 | 694430 | 70 | 8 | 699664 | 70 | 8 | 704837 | 69 |
| 9 | 689220 | 80 | 9 | 694517 | 79 | 9 | 699751 | 78 | 9 | 704922 | 77 |
| 4890 | 689309 | | 4950 | 694605 | | 5010 | 699838 | | 5070 | 705008 | |
| 1 | 689398 | 9 | 1 | 694693 | 9 | 1 | 699924 | 9 | 1 | 705094 | 9 |
| 2 | 689486 | 18 | 2 | 694781 | 18 | 2 | 700011 | 17 | 2 | 705179 | 17 |
| 3 | 689575 | 27 | 3 | 694868 | 26 | 3 | 700098 | 26 | 3 | 705265 | 26 |
| 4 | 689664 | 36 | 4 | 694956 | 35 | 4 | 700184 | 35 | 4 | 705350 | 34 |
| 5 | 689753 | 45 | 5 | 695044 | 44 | 5 | 700271 | 43 | 5 | 705436 | 43 |
| 6 | 689841 | 54 | 6 | 695131 | 53 | 6 | 700358 | 52 | 6 | 705522 | 52 |
| 7 | 689930 | 62 | 7 | 695219 | 62 | 7 | 700444 | 61 | 7 | 705607 | 60 |
| 8 | 690019 | 72 | 8 | 695307 | 70 | 8 | 700531 | 70 | 8 | 705693 | 69 |
| 9 | 690107 | 80 | 9 | 695394 | 79 | 9 | 700617 | 78 | 9 | 705778 | 77 |

| No. | Log. | Prop. Part. | No. | Log. | Prop. Part. | No. | Log. | Prop. Part. | No. | Log. | Prop. Part. |
|---|---|---|---|---|---|---|---|---|---|---|---|
| 5080 | 705864 | | 5140 | 710963 | | 5200 | 716003 | | 5260 | 720986 | |
| 1 | 705949 | 9 | 1 | 711048 | 8 | 1 | 716087 | 8 | 1 | 721068 | 8 |
| 2 | 706035 | 17 | 2 | 711132 | 17 | 2 | 716170 | 17 | 2 | 721151 | 16 |
| 3 | 706120 | 26 | 3 | 711216 | 25 | 3 | 716254 | 25 | 3 | 721233 | 25 |
| 4 | 706206 | 34 | 4 | 711301 | 34 | 4 | 716337 | 34 | 4 | 721316 | 33 |
| 5 | 706291 | 43 | 5 | 711385 | 42 | 5 | 716421 | 42 | 5 | 721398 | 41 |
| 6 | 706376 | 51 | 6 | 711470 | 51 | 6 | 716504 | 50 | 6 | 721481 | 49 |
| 7 | 706462 | 60 | 7 | 711554 | 59 | 7 | 716588 | 59 | 7 | 721563 | 58 |
| 8 | 706547 | 68 | 8 | 711638 | 68 | 8 | 716671 | 67 | 8 | 721646 | 66 |
| 9 | 706632 | 77 | 9 | 711723 | 76 | 9 | 716754 | 76 | 9 | 721728 | 74 |
| 5090 | 706718 | | 5150 | 711807 | | 5210 | 716838 | | 5270 | 721811 | |
| 1 | 706803 | 9 | 1 | 711892 | 8 | 1 | 716921 | 8 | 1 | 721893 | 8 |
| 2 | 706888 | 17 | 2 | 711976 | 17 | 2 | 717004 | 17 | 2 | 721975 | 16 |
| 3 | 706974 | 26 | 3 | 712060 | 25 | 3 | 717088 | 25 | 3 | 722058 | 25 |
| 4 | 707059 | 34 | 4 | 712144 | 34 | 4 | 717171 | 33 | 4 | 722140 | 33 |
| 5 | 707144 | 43 | 5 | 712229 | 42 | 5 | 717254 | 42 | 5 | 722222 | 41 |
| 6 | 707229 | 51 | 6 | 712313 | 51 | 6 | 717338 | 50 | 6 | 722305 | 49 |
| 7 | 707315 | 60 | 7 | 712397 | 59 | 7 | 717421 | 58 | 7 | 722387 | 58 |
| 8 | 707400 | 68 | 8 | 712481 | 68 | 8 | 717504 | 66 | 8 | 722469 | 66 |
| 9 | 707485 | 77 | 9 | 712566 | 76 | 9 | 717587 | 75 | 9 | 722552 | 74 |
| 5100 | 707570 | | 5160 | 712650 | | 5220 | 717671 | | 5280 | 722634 | |
| 1 | 707655 | 9 | 1 | 712734 | 8 | 1 | 717754 | 8 | 1 | 722716 | 8 |
| 2 | 707740 | 17 | 2 | 712818 | 17 | 2 | 717837 | 17 | 2 | 722798 | 16 |
| 3 | 707826 | 26 | 3 | 712902 | 25 | 3 | 717920 | 25 | 3 | 722881 | 25 |
| 4 | 707911 | 34 | 4 | 712986 | 34 | 4 | 718003 | 33 | 4 | 722963 | 33 |
| 5 | 707996 | 43 | 5 | 713070 | 42 | 5 | 718086 | 42 | 5 | 723045 | 41 |
| 6 | 708081 | 51 | 6 | 713154 | 50 | 6 | 718169 | 50 | 6 | 723127 | 49 |
| 7 | 708166 | 60 | 7 | 713238 | 59 | 7 | 718253 | 58 | 7 | 723209 | 58 |
| 8 | 708251 | 68 | 8 | 713322 | 67 | 8 | 718336 | 66 | 8 | 723291 | 66 |
| 9 | 708336 | 77 | 9 | 713406 | 76 | 9 | 718419 | 75 | 9 | 723374 | 74 |
| 5110 | 708421 | | 5170 | 713490 | | 5230 | 718502 | | 5290 | 723456 | |
| 1 | 708506 | 9 | 1 | 713574 | 8 | 1 | 718585 | 8 | 1 | 723538 | 8 |
| 2 | 708591 | 17 | 2 | 713658 | 17 | 2 | 718668 | 17 | 2 | 723620 | 16 |
| 3 | 708676 | 26 | 3 | 713742 | 25 | 3 | 718751 | 25 | 3 | 723702 | 25 |
| 4 | 708761 | 34 | 4 | 713826 | 34 | 4 | 718834 | 33 | 4 | 723784 | 33 |
| 5 | 708846 | 43 | 5 | 713910 | 42 | 5 | 718917 | 42 | 5 | 723866 | 41 |
| 6 | 708931 | 51 | 6 | 713994 | 50 | 6 | 719000 | 50 | 6 | 723948 | 49 |
| 7 | 709015 | 60 | 7 | 714078 | 59 | 7 | 719083 | 58 | 7 | 724030 | 57 |
| 8 | 709100 | 68 | 8 | 714162 | 67 | 8 | 719165 | 66 | 8 | 724112 | 66 |
| 9 | 709185 | 77 | 9 | 714246 | 76 | 9 | 719248 | 75 | 9 | 724194 | 74 |
| 5120 | 709270 | | 5180 | 714330 | | 5240 | 719331 | | 5300 | 724276 | |
| 1 | 709355 | 8 | 1 | 714414 | 8 | 1 | 719414 | 8. | 1 | 724358 | 8 |
| 2 | 709440 | 17 | 2 | 714497 | 17 | 2 | 719497 | 17 | 2 | 724440 | 16 |
| 3 | 709524 | 25 | 3 | 714581 | 25 | 3 | 719580 | 25 | 3 | 724522 | 25 |
| 4 | 709609 | 34 | 4 | 714665 | 34 | 4 | 719663 | 33 | 4 | 724603 | 33 |
| 5 | 709694 | 42 | 5 | 714749 | 42 | 5 | 719745 | 41 | 5 | 724685 | 41 |
| 6 | 709779 | 51 | 6 | 714832 | 50 | 6 | 719828 | 50 | 6 | 724767 | 49 |
| 7 | 709863 | 59 | 7 | 714916 | 59 | 7 | 719911 | 58 | 7 | 724849 | 57 |
| 8 | 709948 | 68 | 8 | 715000 | 67 | 8 | 719994 | 66 | 8 | 724931 | 66 |
| 9 | 710033 | 76 | 9 | 715084 | 76 | 9 | 720077 | 75 | 9 | 725013 | 74 |
| 5130 | 710117 | | 5190 | 715167 | | 5250 | 720159 | | 5310 | 725095 | |
| 1 | 710202 | 8 | 1 | 715251 | 8 | 1 | 720242 | 8 | 1 | 725176 | 8 |
| 2 | 710287 | 17 | 2 | 715335 | 17 | 2 | 720325 | 17 | 2 | 725258 | 16 |
| 3 | 710371 | 25 | 3 | 715418 | 25 | 3 | 720407 | 25 | 3 | 725340 | 25 |
| 4 | 710456 | 34 | 4 | 715502 | 34 | 4 | 720490 | 33 | 4 | 725422 | 33 |
| 5 | 710540 | 42 | 5 | 715586 | 42 | 5 | 720573 | 41 | 5 | 725503 | 41 |
| 6 | 710625 | 51 | 6 | 715669 | 50 | 6 | 720655 | 50 | 6 | 725585 | 49 |
| 7 | 710710 | 59 | 7 | 715753 | 59 | 7 | 720738 | 58 | 7 | 725667 | 57 |
| 8 | 710794 | 67 | 8 | 715836 | 67 | 8 | 720821 | 66 | 8 | 725748 | 66 |
| 9 | 710879 | 76 | 9 | 715920 | 76 | 9 | 720903 | 75 | 9 | 725830 | 74 |

| No. | Log. | Prop. Part. | No. | Log. | Prop. Part. | No. | Log. | Prop. Part. | No. | Log. | Prop. Part. |
|---|---|---|---|---|---|---|---|---|---|---|---|
| 5320 | 725912 | | 5380 | 730782 | | 5440 | 735599 | | 5500 | 740363 | |
| 1 | 725993 | 8 | 1 | 730863 | 8 | 1 | 735679 | 8 | 1 | 740442 | 8 |
| 2 | 726075 | 16 | 2 | 730944 | 16 | 2 | 735759 | 16 | 2 | 740521 | 16 |
| 3 | 726156 | 24 | 3 | 731024 | 24 | 3 | 735838 | 24 | 3 | 740599 | 24 |
| 4 | 726238 | 33 | 4 | 731105 | 32 | 4 | 735918 | 32 | 4 | 740678 | 32 |
| 5 | 726320 | 41 | 5 | 731186 | 40 | 5 | 735998 | 40 | 5 | 740757 | 40 |
| 6 | 726401 | 49 | 6 | 731266 | 49 | 6 | 736078 | 48 | 6 | 740836 | 47 |
| 7 | 726483 | 57 | 7 | 731347 | 57 | 7 | 736157 | 56 | 7 | 740915 | 55 |
| 8 | 726564 | 65 | 8 | 731428 | 65 | 8 | 736237 | 64 | 8 | 740994 | 63 |
| 9 | 726646 | 73 | 9 | 731508 | 73 | 9 | 736317 | 72 | 9 | 741073 | 71 |
| 5330 | 726727 | | 5390 | 731589 | | 5450 | 736396 | | 5510 | 741152 | |
| 1 | 726809 | 8 | 1 | 731669 | 8 | 1 | 736476 | 8 | 1 | 741230 | 8 |
| 2 | 726890 | 16 | 2 | 731750 | 16 | 2 | 736556 | 16 | 2 | 741309 | 16 |
| 3 | 726972 | 24 | 3 | 731830 | 24 | 3 | 736635 | 24 | 3 | 741388 | 24 |
| 4 | 727053 | 33 | 4 | 731911 | 32 | 4 | 736715 | 32 | 4 | 741467 | 32 |
| 5 | 727134 | 41 | 5 | 731991 | 40 | 5 | 736795 | 40 | 5 | 741546 | 40 |
| 6 | 727216 | 49 | 6 | 732072 | 48 | 6 | 736874 | 48 | 6 | 741624 | 47 |
| 7 | 727297 | 57 | 7 | 732152 | 56 | 7 | 736954 | 56 | 7 | 741703 | 55 |
| 8 | 727379 | 65 | 8 | 732233 | 64 | 8 | 737034 | 64 | 8 | 741782 | 63 |
| 9 | 727460 | 73 | 9 | 732313 | 72 | 9 | 737113 | 72 | 9 | 741860 | 71 |
| 5340 | 727541 | | 5400 | 732394 | | 5460 | 737193 | | 5520 | 741939 | |
| 1 | 727623 | 8 | 1 | 732474 | 8 | 1 | 737272 | 8 | 1 | 742018 | 8 |
| 2 | 727704 | 16 | 2 | 732555 | 16 | 2 | 737352 | 16 | 2 | 742096 | 16 |
| 3 | 727785 | 24 | 3 | 732635 | 24 | 3 | 737431 | 24 | 3 | 742175 | 23 |
| 4 | 727866 | 33 | 4 | 732715 | 32 | 4 | 737511 | 32 | 4 | 742254 | 31 |
| 5 | 727948 | 41 | 5 | 732796 | 40 | 5 | 737590 | 40 | 5 | 742332 | 39 |
| 6 | 728029 | 49 | 6 | 732876 | 48 | 6 | 737670 | 48 | 6 | 742411 | 47 |
| 7 | 728110 | 57 | 7 | 732956 | 56 | 7 | 737749 | 56 | 7 | 742489 | 55 |
| 8 | 728191 | 65 | 8 | 733037 | 64 | 8 | 737829 | 64 | 8 | 742568 | 63 |
| 9 | 728273 | 73 | 9 | 733117 | 72 | 9 | 737908 | 72 | 9 | 742647 | 71 |
| 5350 | 728354 | | 5410 | 733197 | | 5470 | 737987 | | 5530 | 742725 | |
| 1 | 728435 | 8 | 1 | 733278 | 8 | 1 | 738067 | 8 | 1 | 742804 | 8 |
| 2 | 728516 | 16 | 2 | 733358 | 16 | 2 | 738146 | 16 | 2 | 742882 | 16 |
| 3 | 728597 | 24 | 3 | 733438 | 24 | 3 | 738225 | 24 | 3 | 742961 | 23 |
| 4 | 728678 | 33 | 4 | 733518 | 32 | 4 | 738305 | 32 | 4 | 743039 | 31 |
| 5 | 728759 | 41 | 5 | 733598 | 40 | 5 | 738384 | 40 | 5 | 743118 | 39 |
| 6 | 728841 | 49 | 6 | 733679 | 48 | 6 | 738463 | 48 | 6 | 743196 | 47 |
| 7 | 728922 | 57 | 7 | 733759 | 56 | 7 | 738543 | 56 | 7 | 743275 | 55 |
| 8 | 729003 | 65 | 8 | 733839 | 64 | 8 | 738622 | 64 | 8 | 743353 | 63 |
| 9 | 729084 | 73 | 9 | 733919 | 72 | 0 | 738701 | 72 | 9 | 743431 | 71 |
| 5360 | 729165 | | 5420 | 733999 | | 5480 | 738781 | | 5540 | 743510 | |
| 1 | 729246 | 8 | 1 | 734079 | 8 | 1 | 738860 | 8 | 1 | 743588 | 8 |
| 2 | 729327 | 16 | 2 | 734159 | 16 | 2 | 738939 | 16 | 2 | 743667 | 16 |
| 3 | 729408 | 24 | 3 | 734240 | 24 | 3 | 739018 | 24 | 3 | 743745 | 23 |
| 4 | 729489 | 32 | 4 | 734320 | 32 | 4 | 739097 | 32 | 4 | 743823 | 31 |
| 5 | 729570 | 41 | 5 | 734400 | 40 | 5 | 739177 | 40 | 5 | 743902 | 39 |
| 6 | 729651 | 49 | 6 | 734480 | 48 | 6 | 739256 | 47 | 6 | 743980 | 47 |
| 7 | 729732 | 57 | 7 | 734560 | 56 | 7 | 739335 | 55 | 7 | 744058 | 55 |
| 8 | 729813 | 65 | 8 | 734640 | 64 | 8 | 739414 | 63 | 8 | 744136 | 63 |
| 9 | 729893 | 73 | 9 | 734720 | 72 | 9 | 739493 | 71 | 9 | 744215 | 71 |
| 5370 | 729974 | | 5430 | 734800 | | 5490 | 739572 | | 5550 | 744293 | |
| 1 | 730055 | 8 | 1 | 734880 | 8 | 1 | 739651 | 8 | 1 | 744371 | 8 |
| 2 | 730136 | 16 | 2 | 734960 | 16 | 2 | 739730 | 16 | 2 | 744449 | 16 |
| 3 | 730217 | 24 | 3 | 735040 | 24 | 3 | 739810 | 24 | 3 | 744528 | 23 |
| 4 | 730298 | 32 | 4 | 735120 | 32 | 4 | 739889 | 32 | 4 | 744606 | 31 |
| 5 | 730378 | 40 | 5 | 735200 | 40 | 5 | 739968 | 40 | 5 | 744684 | 39 |
| 6 | 730459 | 49 | 6 | 735279 | 48 | 6 | 740047 | 47 | 6 | 744762 | 47 |
| 7 | 730540 | 57 | 7 | 735359 | 56 | 7 | 740126 | 55 | 7 | 744840 | 55 |
| 8 | 730621 | 65 | 8 | 735439 | 64 | 8 | 740205 | 63 | 8 | 744919 | 63 |
| 9 | 730702 | 73 | 9 | 735519 | 72 | 9 | 740284 | 71 | 9 | 744997 | 71 |

| No. | Log. | Prop. Part. | No. | Log. | Prop. Part. | No. | Log. | Prop. Part. | No. | Log. | Prop. Part. |
|---|---|---|---|---|---|---|---|---|---|---|---|
| 5560 | 745075 | | 5620 | 749736 | | 5680 | 754348 | | 5740 | 758912 | |
| 1 | 745153 | 8 | 1 | 749814 | 8 | 1 | 754425 | 8 | 1 | 758988 | 8 |
| 2 | 745231 | 16 | 2 | 749891 | 16 | 2 | 754501 | 15 | 2 | 759063 | 15 |
| 3 | 745309 | 23 | 3 | 749968 | 23 | 3 | 754578 | 23 | 3 | 759139 | 23 |
| 4 | 745387 | 31 | 4 | 750045 | 31 | 4 | 754654 | 30 | 4 | 759214 | 30 |
| 5 | 745465 | 39 | 5 | 750123 | 39 | 5 | 754730 | 38 | 5 | 759290 | 38 |
| 6 | 745543 | 47 | 6 | 750200 | 47 | 6 | 754807 | 46 | 6 | 759366 | 45 |
| 7 | 745621 | 55 | 7 | 750277 | 54 | 7 | 754883 | 53 | 7 | 759441 | 53 |
| 8 | 745699 | 62 | 8 | 750354 | 62 | 8 | 754960 | 61 | 8 | 759517 | 60 |
| 9 | 745777 | 70 | 9 | 750431 | 70 | 9 | 755036 | 69 | 9 | 759592 | 68 |
| 5570 | 745855 | | 5630 | 750508 | | 5690 | 755112 | | 5750 | 759668 | |
| 1 | 745933 | 8 | 1 | 750586 | 8 | 1 | 755189 | 8 | 1 | 759743 | 8 |
| 2 | 746011 | 16 | 2 | 750663 | 16 | 2 | 755265 | 15 | 2 | 759819 | 15 |
| 3 | 746089 | 23 | 3 | 750740 | 23 | 3 | 755341 | 23 | 3 | 759894 | 23 |
| 4 | 746167 | 31 | 4 | 750817 | 31 | 4 | 755417 | 30 | 4 | 759970 | 30 |
| 5 | 746245 | 39 | 5 | 750894 | 39 | 5 | 755494 | 38 | 5 | 760045 | 38 |
| 6 | 746323 | 47 | 6 | 750971 | 47 | 6 | 755570 | 46 | 6 | 760121 | 45 |
| 7 | 746401 | 55 | 7 | 751048 | 54 | 7 | 755646 | 53 | 7 | 760196 | 53 |
| 8 | 746479 | 62 | 8 | 751125 | 62 | 8 | 755722 | 61 | 8 | 760272 | 60 |
| 9 | 746556 | 70 | 9 | 751202 | 70 | 9 | 755799 | 69 | 9 | 760347 | 68 |
| 5580 | 746634 | | 5640 | 751279 | | 5700 | 755875 | | 5760 | 760422 | |
| 1 | 746712 | 8 | 1 | 751356 | 8 | 1 | 755951 | 8 | 1 | 760498 | 8 |
| 2 | 746790 | 16 | 2 | 751433 | 15 | 2 | 756027 | 15 | 2 | 760573 | 15 |
| 3 | 746868 | 23 | 3 | 751510 | 23 | 3 | 756103 | 23 | 3 | 760649 | 23 |
| 4 | 746945 | 31 | 4 | 751587 | 30 | 4 | 756180 | 30 | 4 | 760724 | 30 |
| 5 | 747023 | 39 | 5 | 751664 | 38 | 5 | 756256 | 38 | 5 | 760799 | 38 |
| 6 | 747101 | 47 | 6 | 751741 | 46 | 6 | 756332 | 46 | 6 | 760875 | 45 |
| 7 | 747179 | 55 | 7 | 751818 | 54 | 7 | 756408 | 53 | 7 | 760950 | 53 |
| 8 | 747256 | 62 | 8 | 751895 | 62 | 8 | 756484 | 61 | 8 | 761025 | 60 |
| 9 | 747334 | 70 | 9 | 751972 | 70 | 9 | 756560 | 69 | 9 | 761100 | 68 |
| 5590 | 747412 | | 5650 | 752048 | | 5710 | 756636 | | 5770 | 761176 | |
| 1 | 747489 | 8 | 1 | 752125 | 8 | 1 | 756712 | 8 | 1 | 761251 | 8 |
| 2 | 747567 | 16 | 2 | 752202 | 15 | 2 | 756788 | 15 | 2 | 761326 | 15 |
| 3 | 747645 | 23 | 3 | 752279 | 23 | 3 | 756864 | 23 | 3 | 761402 | 23 |
| 4 | 747722 | 31 | 4 | 752356 | 30 | 4 | 756940 | 30 | 4 | 761477 | 30 |
| 5 | 747800 | 39 | 5 | 752433 | 38 | 5 | 757016 | 38 | 5 | 761552 | 38 |
| 6 | 747878 | 47 | 6 | 752509 | 46 | 6 | 757092 | 46 | 6 | 761627 | 45 |
| 7 | 747955 | 54 | 7 | 752586 | 54 | 7 | 757168 | 53 | 7 | 761702 | 53 |
| 8 | 748033 | 62 | 8 | 752663 | 62 | 8 | 757244 | 61 | 8 | 761778 | 60 |
| 9 | 748110 | 70 | 9 | 752740 | 70 | 9 | 757320 | 69 | 9 | 761853 | 68 |
| 5600 | 748188 | | 5660 | 752816 | | 5720 | 757396 | | 5780 | 761928 | |
| 1 | 748266 | 8 | 1 | 752893 | 8 | 1 | 757472 | 8 | 1 | 762003 | 8 |
| 2 | 748343 | 16 | 2 | 752970 | 15 | 2 | 757548 | 15 | 2 | 763078 | 15 |
| 3 | 748421 | 23 | 3 | 753047 | 23 | 3 | 757624 | 23 | 3 | 762153 | 22 |
| 4 | 748498 | 31 | 4 | 753123 | 30 | 4 | 757700 | 30 | 4 | 762228 | 30 |
| 5 | 748576 | 39 | 5 | 753200 | 38 | 5 | 757775 | 38 | 5 | 762303 | 38 |
| 6 | 748653 | 47 | 6 | 753277 | 46 | 6 | 757851 | 46 | 6 | 762378 | 45 |
| 7 | 748731 | 54 | 7 | 753353 | 54 | 7 | 757927 | 53 | 7 | 762453 | 52 |
| 8 | 748808 | 62 | 8 | 753430 | 62 | 8 | 758003 | 61 | 8 | 762529 | 60 |
| 9 | 748885 | 70 | 9 | 753506 | 70 | 9 | 758079 | 68 | 9 | 762604 | 68 |
| 5610 | 748963 | | 5670 | 753583 | | 5730 | 758155 | | 5790 | 762679 | |
| 1 | 749040 | 8 | 1 | 753660 | 8 | 1 | 758230 | 8 | 1 | 762754 | 8 |
| 2 | 749118 | 16 | 2 | 753736 | 15 | 2 | 758306 | 15 | 2 | 762829 | 15 |
| 3 | 749195 | 23 | 3 | 753813 | 23 | 3 | 758382 | 23 | 3 | 762904 | 22 |
| 4 | 749272 | 31 | 4 | 753889 | 30 | 4 | 758458 | 30 | 4 | 762978 | 30 |
| 5 | 749350 | 39 | 5 | 753966 | 38 | 5 | 758533 | 38 | 5 | 763053 | 38 |
| 6 | 749427 | 47 | 6 | 754042 | 46 | 6 | 758609 | 46 | 6 | 763128 | 45 |
| 7 | 749504 | 54 | 7 | 754119 | 54 | 7 | 758685 | 53 | 7 | 763203 | 52 |
| 8 | 749582 | 62 | 8 | 754195 | 62 | 8 | 758760 | 61 | 8 | 763278 | 60 |
| 9 | 749659 | 70 | 9 | 754272 | 70 | 9 | 758836 | 68 | 9 | 763353 | 68 |

| No. | Log. | Prop. Part. | No. | Log. | Prop. Part. | No. | Log. | Prop. Part. | No. | Log. | Prop. Part. |
|---|---|---|---|---|---|---|---|---|---|---|---|
| 5800 | 763428 | | 5860 | 767898 | | 5920 | 772322 | | 5980 | 776701 | |
| 1 | 763503 | 7 | 1 | 767972 | 7 | 1 | 772395 | 7 | 1 | 776774 | 7 |
| 2 | 763578 | 15 | 2 | 768046 | 15 | 2 | 772468 | 15 | 2 | 776846 | 14 |
| 3 | 763653 | 22 | 3 | 768120 | 22 | 3 | 772542 | 22 | 3 | 776919 | 22 |
| 4 | 763727 | 30 | 4 | 768194 | 30 | 4 | 772615 | 29 | 4 | 776992 | 29 |
| 5 | 763802 | 37 | 5 | 768268 | 37 | 5 | 772688 | 37 | 5 | 777064 | 36 |
| 6 | 763877 | 45 | 6 | 768342 | 45 | 6 | 772762 | 44 | 6 | 777137 | 43 |
| 7 | 763952 | 52 | 7 | 768416 | 52 | 7 | 772835 | 51 | 7 | 777209 | 51 |
| 8 | 764027 | 60 | 8 | 768490 | 59 | 8 | 772908 | 59 | 8 | 777282 | 58 |
| 9 | 764101 | 67 | 9 | 768564 | 67 | 9 | 772981 | 66 | 9 | 777354 | 65 |
| | | | | | | | | | | | |
| 5810 | 764176 | | 5870 | 768638 | | 5930 | 773055 | | 5990 | 777427 | |
| 1 | 764251 | 7 | 1 | 768712 | 7 | 1 | 773128 | 7 | 1 | 777499 | 7 |
| 2 | 764326 | 15 | 2 | 768786 | 15 | 2 | 773201 | 15 | 2 | 777572 | 14 |
| 3 | 764400 | 22 | 3 | 768860 | 22 | 3 | 773274 | 22 | 3 | 777644 | 22 |
| 4 | 764475 | 30 | 4 | 768934 | 30 | 4 | 773348 | 29 | 4 | 777717 | 29 |
| 5 | 764550 | 37 | 5 | 769008 | 37 | 5 | 773421 | 37 | 5 | 777789 | 36 |
| 6 | 764624 | 45 | 6 | 769082 | 45 | 6 | 773494 | 44 | 6 | 777862 | 43 |
| 7 | 764699 | 52 | 7 | 769156 | 52 | 7 | 773567 | 51 | 7 | 777934 | 51 |
| 8 | 764774 | 60 | 8 | 769230 | 59 | 8 | 773640 | 59 | 8 | 778006 | 58 |
| 9 | 764848 | 67 | 9 | 769303 | 67 | 9 | 773713 | 66 | 9 | 778079 | 65 |
| | | | | | | | | | | | |
| 5820 | 764923 | | 5880 | 769377 | | 5940 | 773786 | | 6000 | 778151 | |
| 1 | 764998 | 7 | 1 | 769451 | 7 | 1 | 773860 | 7 | 1 | 778224 | 7 |
| 2 | 765072 | 15 | 2 | 769525 | 15 | 2 | 773933 | 15 | 2 | 778296 | 14 |
| 3 | 765147 | 22 | 3 | 769599 | 22 | 3 | 774006 | 22 | 3 | 778368 | 22 |
| 4 | 765221 | 30 | 4 | 769673 | 30 | 4 | 774079 | 29 | 4 | 778441 | 29 |
| 5 | 765296 | 37 | 5 | 769746 | 37 | 5 | 774152 | 37 | 5 | 778513 | 36 |
| 6 | 765370 | 45 | 6 | 769820 | 45 | 6 | 774225 | 44 | 6 | 778585 | 43 |
| 7 | 765445 | 52 | 7 | 769894 | 52 | 7 | 774298 | 51 | 7 | 778658 | 51 |
| 8 | 765520 | 60 | 8 | 769968 | 59 | 8 | 774371 | 59 | 8 | 778730 | 58 |
| 9 | 765594 | 67 | 9 | 770042 | 67 | 9 | 774444 | 66 | 9 | 778802 | 65 |
| | | | | | | | | | | | |
| 5830 | 765669 | | 5890 | 770115 | | 5950 | 774517 | | 6010 | 778874 | |
| 1 | 765743 | 7 | 1 | 770189 | 7 | 1 | 774590 | 7 | 1 | 778947 | 7 |
| 2 | 765818 | 15 | 2 | 770263 | 15 | 2 | 774663 | 15 | 2 | 779019 | 14 |
| 3 | 765892 | 22 | 3 | 770336 | 22 | 3 | 774736 | 22 | 3 | 779091 | 22 |
| 4 | 765966 | 30 | 4 | 770410 | 30 | 4 | 774809 | 29 | 4 | 779163 | 29 |
| 5 | 766041 | 37 | 5 | 770484 | 37 | 5 | 774882 | 37 | 5 | 779236 | 36 |
| 6 | 766115 | 45 | 6 | 770557 | 45 | 6 | 774955 | 44 | 6 | 779308 | 43 |
| 7 | 766190 | 52 | 7 | 770631 | 52 | 7 | 775028 | 51 | 7 | 779380 | 51 |
| 8 | 766264 | 60 | 8 | 770705 | 59 | 8 | 775100 | 59 | 8 | 779452 | 58 |
| 9 | 766338 | 67 | 9 | 770778 | 67 | 9 | 775173 | 66 | 9 | 779524 | 65 |
| | | | | | | | | | | | |
| 5840 | 766413 | | 5900 | 770852 | | 5960 | 775246 | | 6020 | 779596 | |
| 1 | 766487 | 7 | 1 | 770926 | 7 | 1 | 775319 | 7 | 1 | 779669 | 7 |
| 2 | 766562 | 15 | 2 | 770999 | 15 | 2 | 775392 | 15 | 2 | 779741 | 14 |
| 3 | 766636 | 22 | 3 | 771073 | 22 | 3 | 775465 | 22 | 3 | 779813 | 22 |
| 4 | 766710 | 30 | 4 | 771146 | 30 | 4 | 775538 | 29 | 4 | 779885 | 29 |
| 5 | 766785 | 37 | 5 | 771220 | 37 | 5 | 775610 | 37 | 5 | 779957 | 36 |
| 6 | 766859 | 45 | 6 | 771293 | 45 | 6 | 775683 | 44 | 6 | 780029 | 43 |
| 7 | 766933 | 52 | 7 | 771367 | 52 | 7 | 775756 | 51 | 7 | 780101 | 50 |
| 8 | 767007 | 60 | 8 | 771440 | 59 | 8 | 775829 | 59 | 8 | 780173 | 58 |
| 9 | 767082 | 67 | 9 | 771514 | 67 | 9 | 775902 | 66 | 9 | 780245 | 65 |
| | | | | | | | | | | | |
| 5850 | 767156 | | 5910 | 771587 | | 5970 | 775974 | | 6030 | 780317 | |
| 1 | 767230 | 7 | 1 | 771661 | 7 | 1 | 776047 | 7 | 1 | 780389 | 7 |
| 2 | 767304 | 15 | 2 | 771734 | 15 | 2 | 776120 | 15 | 2 | 780461 | 14 |
| 3 | 767379 | 22 | 3 | 771808 | 22 | 3 | 776193 | 22 | 3 | 780533 | 22 |
| 4 | 767453 | 30 | 4 | 771881 | 30 | 4 | 776265 | 29 | 4 | 780605 | 29 |
| 5 | 767527 | 37 | 5 | 771955 | 37 | 5 | 776338 | 37 | 5 | 780677 | 36 |
| 6 | 767601 | 45 | 6 | 772028 | 44 | 6 | 776411 | 44 | 6 | 780749 | 43 |
| 7 | 767675 | 52 | 7 | 772102 | 52 | 7 | 776483 | 51 | 7 | 780821 | 50 |
| 8 | 767749 | 59 | 8 | 772175 | 59 | 8 | 776556 | 59 | 8 | 780893 | 58 |
| 9 | 767823 | 67 | 9 | 772248 | 67 | 9 | 776629 | 66 | 9 | 780965 | 65 |

| No. | Log. | Prop. Part. | No. | Log. | Prop. Part. | No. | Log. | Prop. Part. | No. | Log. | Prop. Part. |
|---|---|---|---|---|---|---|---|---|---|---|---|
| 6040 | 781037 | | 6100 | 785330 | | 6160 | 789581 | | 6220 | 793790 | |
| 1 | 781109 | 7 | 1 | 785401 | 7 | 1 | 789651 | 7 | 1 | 793860 | 7 |
| 2 | 781181 | 14 | 2 | 785472 | 14 | 2 | 789722 | 14 | 2 | 793930 | 14 |
| 3 | 781253 | 22 | 3 | 785543 | 21 | 3 | 789792 | 21 | 3 | 794000 | 21 |
| 4 | 781324 | 29 | 4 | 785615 | 28 | 4 | 789863 | 28 | 4 | 794070 | 28 |
| 5 | 781396 | 36 | 5 | 785686 | 36 | 5 | 789933 | 35 | 5 | 794139 | 35 |
| 6 | 781468 | 43 | 6 | 785757 | 43 | 6 | 790004 | 42 | 6 | 794209 | 42 |
| 7 | 781540 | 50 | 7 | 785828 | 50 | 7 | 790074 | 49 | 7 | 794279 | 49 |
| 8 | 781612 | 58 | 8 | 785899 | 57 | 8 | 790144 | 56 | 8 | 794349 | 56 |
| 9 | 781684 | 65 | 9 | 785970 | 64 | 9 | 790215 | 63 | 9 | 794418 | 63 |
| 6050 | 781755 | | 6110 | 786041 | | 6170 | 790285 | | 6230 | 794488 | |
| 1 | 781827 | 7 | 1 | 786112 | 7 | 1 | 790356 | 7 | 1 | 794558 | 7 |
| 2 | 781899 | 14 | 2 | 786183 | 14 | 2 | 790426 | 14 | 2 | 794627 | 14 |
| 3 | 781971 | 22 | 3 | 786254 | 21 | 3 | 790496 | 21 | 3 | 794697 | 21 |
| 4 | 782042 | 29 | 4 | 786325 | 28 | 4 | 790567 | 28 | 4 | 794767 | 28 |
| 5 | 782114 | 36 | 5 | 786396 | 36 | 5 | 790637 | 35 | 5 | 794836 | 35 |
| 6 | 782186 | 43 | 6 | 786467 | 43 | 6 | 790707 | 42 | 6 | 794906 | 42 |
| 7 | 782258 | 50 | 7 | 786538 | 50 | 7 | 790778 | 49 | 7 | 794976 | 49 |
| 8 | 782329 | 58 | 8 | 786609 | 57 | 8 | 790848 | 56 | 8 | 795045 | 56 |
| 9 | 782401 | 65 | 9 | 786680 | 64 | 9 | 790918 | 63 | 9 | 795115 | 63 |
| 6060 | 782473 | | 6120 | 786751 | | 6180 | 790988 | | 6240 | 795185 | |
| 1 | 782544 | 7 | 1 | 786822 | 7 | 1 | 791059 | 7 | 1 | 795254 | 7 |
| 2 | 782616 | 14 | 2 | 786893 | 14 | 2 | 791129 | 14 | 2 | 795324 | 14 |
| 3 | 782688 | 21 | 3 | 786964 | 21 | 3 | 791199 | 21 | 3 | 795393 | 21 |
| 4 | 782759 | 29 | 4 | 787035 | 28 | 4 | 791269 | 28 | 4 | 795463 | 28 |
| 5 | 782831 | 36 | 5 | 787106 | 36 | 5 | 791340 | 35 | 5 | 795532 | 35 |
| 6 | 782902 | 43 | 6 | 787177 | 43 | 6 | 791410 | 42 | 6 | 795602 | 42 |
| 7 | 782974 | 50 | 7 | 787248 | 50 | 7 | 791480 | 49 | 7 | 795671 | 49 |
| 8 | 783046 | 57 | 8 | 787319 | 57 | 8 | 791550 | 56 | 8 | 795741 | 56 |
| 9 | 783117 | 64 | 9 | 787390 | 64 | 9 | 791620 | 63 | 9 | 795810 | 63 |
| 6070 | 783189 | | 6130 | 787460 | | 6190 | 791691 | | 6250 | 795880 | |
| 1 | 783260 | 7 | 1 | 787531 | 7 | 1 | 791761 | 7 | 1 | 795949 | 7 |
| 2 | 783332 | 14 | 2 | 787602 | 14 | 2 | 791831 | 14 | 2 | 796019 | 14 |
| 3 | 783403 | 21 | 3 | 787673 | 21 | 3 | 791901 | 21 | 3 | 796088 | 21 |
| 4 | 783475 | 29 | 4 | 787744 | 28 | 4 | 791971 | 28 | 4 | 796158 | 28 |
| 5 | 783546 | 36 | 5 | 787815 | 35 | 5 | 792041 | 35 | 5 | 796227 | 35 |
| 6 | 783618 | 43 | 6 | 787885 | 42 | 6 | 792111 | 42 | 6 | 796297 | 42 |
| 7 | 783689 | 50 | 7 | 787956 | 49 | 7 | 792181 | 49 | 7 | 796366 | 49 |
| 8 | 783761 | 57 | 8 | 788027 | 56 | 8 | 792252 | 56 | 8 | 796436 | 56 |
| 9 | 783832 | 64 | 9 | 788098 | 63 | 9 | 792322 | 63 | 9 | 796505 | 63 |
| 6080 | 783904 | | 6140 | 788168 | | 6200 | 792392 | | 6260 | 796574 | |
| 1 | 783975 | 7 | 1 | 788239 | 7 | 1 | 792462 | 7 | 1 | 796644 | 7 |
| 2 | 784046 | 14 | 2 | 788310 | 14 | 2 | 792532 | 14 | 2 | 796713 | 14 |
| 3 | 784118 | 21 | 3 | 788381 | 21 | 3 | 792602 | 21 | 3 | 796782 | 21 |
| 4 | 784189 | 29 | 4 | 788451 | 28 | 4 | 792672 | 28 | 4 | 796852 | 27 |
| 5 | 784261 | 36 | 5 | 788522 | 35 | 5 | 792742 | 35 | 5 | 796921 | 35 |
| 6 | 784332 | 43 | 6 | 788593 | 42 | 6 | 792812 | 42 | 6 | 796990 | 42 |
| 7 | 784403 | 50 | 7 | 788663 | 49 | 7 | 792882 | 49 | 7 | 797060 | 49 |
| 8 | 784475 | 57 | 8 | 788734 | 56 | 8 | 792952 | 56 | 8 | 797129 | 56 |
| 9 | 784546 | 64 | 9 | 788804 | 63 | 9 | 793022 | 63 | 9 | 797198 | 62 |
| 6090 | 784617 | | 6150 | 788875 | | 6210 | 793092 | | 6270 | 797268 | |
| 1 | 784689 | 7 | 1 | 788946 | 7 | 1 | 793162 | 7 | 1 | 797337 | 7 |
| 2 | 784760 | 14 | 2 | 789016 | 14 | 2 | 793231 | 14 | 2 | 797406 | 14 |
| 3 | 784831 | 21 | 3 | 789087 | 21 | 3 | 793301 | 21 | 3 | 797475 | 21 |
| 4 | 784902 | 29 | 4 | 789157 | 28 | 4 | 793371 | 28 | 4 | 797545 | 27 |
| 5 | 784974 | 36 | 5 | 789228 | 35 | 5 | 793441 | 35 | 5 | 797614 | 35 |
| 6 | 785045 | 43 | 6 | 789299 | 42 | 6 | 793511 | 42 | 6 | 797683 | 42 |
| 7 | 785116 | 50 | 7 | 789369 | 49 | 7 | 793581 | 49 | 7 | 797752 | 49 |
| 8 | 785187 | 57 | 8 | 789440 | 56 | 8 | 793651 | 56 | 8 | 797821 | 56 |
| 9 | 785259 | 64 | 9 | 789510 | 63 | 9 | 793721 | 63 | 9 | 797890 | 62 |

| No. | Log. | Prop. Part. | No. | Log. | Prop. Part. | No. | Log. | Prop. Part. | No. | Log. | Prop. Part. |
|---|---|---|---|---|---|---|---|---|---|---|---|
| 6280 | 797960 | | 6340 | 802089 | | 6400 | 806180 | | 6460 | 810233 | |
| 1 | 798029 | 7 | 1 | 802158 | 7 | 1 | 806248 | 7 | 1 | 810300 | 7 |
| 2 | 798098 | 14 | 2 | 802226 | 14 | 2 | 806316 | 14 | 2 | 810367 | 13 |
| 3 | 798167 | 21 | 3 | 802295 | 21 | 3 | 806384 | 20 | 3 | 810434 | 20 |
| 4 | 798236 | 28 | 4 | 802363 | 27 | 4 | 806451 | 27 | 4 | 810501 | 27 |
| 5 | 798305 | 34 | 5 | 802432 | 34 | 5 | 806519 | 34 | 5 | 810569 | 33 |
| 6 | 798374 | 41 | 6 | 802500 | 41 | 6 | 806587 | 41 | 6 | 810636 | 40 |
| 7 | 798443 | 48 | 7 | 802568 | 48 | 7 | 806655 | 48 | 7 | 810703 | 47 |
| 8 | 798512 | 55 | 8 | 802637 | 55 | 8 | 806723 | 54 | 8 | 810770 | 54 |
| 9 | 798582 | 62 | 9 | 802705 | 62 | 9 | 806790 | 61 | 9 | 810837 | 60 |
| 6290 | 798651 | | 6350 | 802774 | | 6410 | 806858 | | 6470 | 810904 | |
| 1 | 798720 | 7 | 1 | 802842 | 7 | 1 | 806926 | 7 | 1 | 810971 | 7 |
| 2 | 798789 | 14 | 2 | 802910 | 14 | 2 | 806994 | 14 | 2 | 811038 | 13 |
| 3 | 798858 | 21 | 3 | 802979 | 21 | 3 | 807061 | 20 | 3 | 811106 | 20 |
| 4 | 798927 | 28 | 4 | 803047 | 27 | 4 | 807129 | 27 | 4 | 811173 | 27 |
| 5 | 798996 | 34 | 5 | 803116 | 34 | 5 | 807197 | 34 | 5 | 811240 | 33 |
| 6 | 799065 | 41 | 6 | 803184 | 41 | 6 | 807264 | 41 | 6 | 811307 | 40 |
| 7 | 799134 | 48 | 7 | 803252 | 48 | 7 | 807332 | 48 | 7 | 811374 | 47 |
| 8 | 799203 | 55 | 8 | 803320 | 55 | 8 | 807400 | 54 | 8 | 811441 | 54 |
| 9 | 799272 | 62 | 9 | 803389 | 62 | 9 | 807467 | 61 | 9 | 811508 | 60 |
| 6300 | 799341 | | 6360 | 803457 | | 6420 | 807535 | | 6480 | 811575 | |
| 1 | 799409 | 7 | 1 | 803525 | 7 | 1 | 807603 | 7 | 1 | 811642 | 7 |
| 2 | 799478 | 14 | 2 | 803594 | 14 | 2 | 807670 | 14 | 2 | 811709 | 13 |
| 3 | 799547 | 21 | 3 | 803662 | 21 | 3 | 807738 | 20 | 3 | 811776 | 20 |
| 4 | 799616 | 28 | 4 | 803730 | 27 | 4 | 807806 | 27 | 4 | 811843 | 27 |
| 5 | 799685 | 34 | 5 | 803798 | 34 | 5 | 807873 | 34 | 5 | 811910 | 33 |
| 6 | 799754 | 41 | 6 | 803867 | 41 | 6 | 807941 | 41 | 6 | 811977 | 40 |
| 7 | 799823 | 48 | 7 | 803935 | 48 | 7 | 808008 | 48 | 7 | 812044 | 47 |
| 8 | 799892 | 55 | 8 | 804003 | 55 | 8 | 808076 | 54 | 8 | 812111 | 54 |
| 9 | 799961 | 62 | 9 | 804071 | 62 | 9 | 808143 | 61 | 9 | 812178 | 60 |
| 6310 | 800029 | | 6370 | 804139 | | 6430 | 808211 | | 6490 | 812245 | |
| 1 | 800098 | 7 | 1 | 804208 | 7 | 1 | 808279 | 7 | 1 | 812312 | 7 |
| 2 | 800167 | 14 | 2 | 804276 | 14 | 2 | 808346 | 14 | 2 | 812378 | 13 |
| 3 | 800236 | 21 | 3 | 804344 | 21 | 3 | 808414 | 20 | 3 | 812445 | 20 |
| 4 | 800305 | 28 | 4 | 804412 | 27 | 4 | 808481 | 27 | 4 | 812512 | 27 |
| 5 | 800373 | 34 | 5 | 804480 | 34 | 5 | 808549 | 34 | 5 | 812579 | 33 |
| 6 | 800442 | 41 | 6 | 804548 | 41 | 6 | 808616 | 41 | 6 | 812646 | 40 |
| 7 | 800511 | 48 | 7 | 804616 | 48 | 7 | 808684 | 48 | 7 | 812713 | 47 |
| 8 | 800580 | 55 | 8 | 804685 | 55 | 8 | 808751 | 54 | 8 | 812780 | 54 |
| 9 | 800648 | 62 | 9 | 804753 | 62 | 9 | 808818 | 61 | 9 | 812847 | 60 |
| 6320 | 800717 | | 6380 | 804821 | | 6440 | 808886 | | 6500 | 812913 | |
| 1 | 800786 | 7 | 1 | 804889 | 7 | 1 | 808953 | 7 | 1 | 812980 | 7 |
| 2 | 800854 | 14 | 2 | 804957 | 14 | 2 | 809021 | 13 | 2 | 813047 | 13 |
| 3 | 800923 | 21 | 3 | 805025 | 20 | 3 | 809088 | 20 | 3 | 813114 | 20 |
| 4 | 800992 | 28 | 4 | 805093 | 27 | 4 | 809156 | 27 | 4 | 813181 | 27 |
| 5 | 801060 | 34 | 5 | 805161 | 34 | 5 | 809223 | 34 | 5 | 813247 | 33 |
| 6 | 801129 | 41 | 6 | 805229 | 41 | 6 | 809290 | 40 | 6 | 813314 | 40 |
| 7 | 801198 | 48 | 7 | 805297 | 48 | 7 | 809358 | 47 | 7 | 813381 | 47 |
| 8 | 801266 | 55 | 8 | 805365 | 54 | 8 | 809425 | 54 | 8 | 813448 | 54 |
| 9 | 801335 | 62 | 9 | 805433 | 61 | 9 | 809492 | 61 | 9 | 813514 | 60 |
| 6330 | 801404 | | 6390 | 805501 | | 6450 | 809560 | | 6510 | 813581 | |
| 1 | 801472 | 7 | 1 | 805569 | 7 | 1 | 809627 | 7 | 1 | 813648 | 7 |
| 2 | 801541 | 14 | 2 | 805637 | 14 | 2 | 809694 | 13 | 2 | 813714 | 13 |
| 3 | 801609 | 21 | 3 | 805705 | 20 | 3 | 809762 | 20 | 3 | 813781 | 20 |
| 4 | 801678 | 27 | 4 | 805773 | 27 | 4 | 809829 | 27 | 4 | 813848 | 27 |
| 5 | 801747 | 34 | 5 | 805841 | 34 | 5 | 809896 | 34 | 5 | 813914 | 33 |
| 6 | 801815 | 41 | 6 | 805908 | 41 | 6 | 809964 | 40 | 6 | 813981 | 40 |
| 7 | 801884 | 48 | 7 | 805976 | 48 | 7 | 810031 | 47 | 7 | 814048 | 47 |
| 8 | 801952 | 55 | 8 | 806044 | 54 | 8 | 810098 | 54 | 8 | 814114 | 54 |
| 9 | 802021 | 62 | 9 | 806112 | 61 | 9 | 810165 | 61 | 9 | 814181 | 60 |

| No. | Log. | Prop. Part. | No. | Log. | Prop. Part. | No. | Log. | Prop. Part. | No. | Log. | Prop. Part. |
|---|---|---|---|---|---|---|---|---|---|---|---|
| 6520 | 814248 | | 6580 | 818226 | | 6640 | 822168 | | 6700 | 826075 | |
| 1 | 814314 | 7 | 1 | 818292 | 7 | 1 | 822233 | 7 | 1 | 826140 | 6 |
| 2 | 814381 | 13 | 2 | 818358 | 13 | 2 | 822299 | 13 | 2 | 826204 | 13 |
| 3 | 814447 | 20 | 3 | 818424 | 20 | 3 | 822364 | 20 | 3 | 826269 | 19 |
| 4 | 814514 | 26 | 4 | 818490 | 26 | 4 | 822430 | 26 | 4 | 826334 | 26 |
| 5 | 814581 | 33 | 5 | 818556 | 33 | 5 | 822495 | 33 | 5 | 826399 | 32 |
| 6 | 814647 | 40 | 6 | 818622 | 40 | 6 | 822560 | 39 | 6 | 826464 | 39 |
| 7 | 814714 | 46 | 7 | 818688 | 46 | 7 | 822626 | 46 | 7 | 826528 | 45 |
| 8 | 814780 | 53 | 8 | 818754 | 53 | 8 | 822691 | 52 | 8 | 826593 | 52 |
| 9 | 814847 | 60 | 9 | 818819 | 59 | 9 | 822756 | 59 | 9 | 826658 | 58 |
| 6530 | 814913 | | 6590 | 818885 | | 6650 | 822823 | | 6710 | 826722 | |
| 1 | 814980 | 7 | 1 | 818951 | 7 | 1 | 822887 | 7 | 1 | 826787 | 6 |
| 2 | 815046 | 13 | 2 | 819017 | 13 | 2 | 822952 | 13 | 2 | 826852 | 13 |
| 3 | 815113 | 20 | 3 | 819083 | 20 | 3 | 823018 | 20 | 3 | 826917 | 19 |
| 4 | 815179 | 26 | 4 | 819149 | 26 | 4 | 823083 | 26 | 4 | 826981 | 26 |
| 5 | 815246 | 33 | 5 | 819215 | 33 | 5 | 823148 | 33 | 5 | 827046 | 32 |
| 6 | 815312 | 40 | 6 | 819281 | 40 | 6 | 823213 | 39 | 6 | 827111 | 39 |
| 7 | 815378 | 46 | 7 | 819346 | 46 | 7 | 823279 | 46 | 7 | 827175 | 45 |
| 8 | 815445 | 53 | 8 | 819412 | 53 | 8 | 823344 | 52 | 8 | 827240 | 52 |
| 9 | 815511 | 60 | 9 | 819478 | 59 | 9 | 823409 | 59 | 9 | 827305 | 58 |
| 6540 | 815578 | | 6600 | 819544 | | 6660 | 823474 | | 6720 | 827369 | |
| 1 | 815644 | 7 | 1 | 819610 | 7 | 1 | 823539 | 7 | 1 | 827434 | 6 |
| 2 | 815711 | 13 | 2 | 819675 | 13 | 2 | 823605 | 13 | 2 | 827498 | 13 |
| 3 | 815777 | 20 | 3 | 819741 | 20 | 3 | 823670 | 20 | 3 | 827563 | 19 |
| 4 | 815843 | 26 | 4 | 819807 | 26 | 4 | 823735 | 26 | 4 | 827628 | 26 |
| 5 | 815910 | 33 | 5 | 819873 | 33 | 5 | 823800 | 33 | 5 | 827692 | 32 |
| 6 | 815976 | 40 | 6 | 819939 | 40 | 6 | 823865 | 39 | 6 | 827757 | 39 |
| 7 | 816042 | 46 | 7 | 820004 | 46 | 7 | 823930 | 46 | 7 | 827821 | 45 |
| 8 | 816109 | 53 | 8 | 820070 | 53 | 8 | 823996 | 52 | 8 | 827886 | 52 |
| 9 | 816175 | 60 | 9 | 820136 | 59 | 9 | 824061 | 59 | 9 | 827951 | 58 |
| 6550 | 816241 | | 6610 | 820201 | | 6670 | 824126 | | 6730 | 828015 | |
| 1 | 816308 | 7 | 1 | 820267 | 7 | 1 | 824191 | 6 | 1 | 828080 | 6 |
| 2 | 816374 | 13 | 2 | 820333 | 13 | 2 | 824256 | 13 | 2 | 828144 | 13 |
| 3 | 816440 | 20 | 3 | 820399 | 20 | 3 | 824321 | 19 | 3 | 828209 | 19 |
| 4 | 816506 | 26 | 4 | 820464 | 26 | 4 | 824386 | 26 | 4 | 828273 | 26 |
| 5 | 816573 | 33 | 5 | 820530 | 33 | 5 | 824451 | 32 | 5 | 828338 | 32 |
| 6 | 816639 | 40 | 6 | 820595 | 40 | 6 | 824516 | 39 | 6 | 828402 | 39 |
| 7 | 816705 | 46 | 7 | 820661 | 46 | 7 | 824581 | 45 | 7 | 828467 | 45 |
| 8 | 816771 | 53 | 8 | 820727 | 53 | 8 | 824646 | 52 | 8 | 828531 | 52 |
| 9 | 816838 | 60 | 9 | 820792 | 59 | 9 | 824711 | 58 | 9 | 828595 | 58 |
| 6560 | 816904 | | 6620 | 820858 | | 6680 | 824776 | | 6740 | 828660 | |
| 1 | 816970 | 7 | 1 | 820924 | 7 | 1 | 824841 | 6 | 1 | 828724 | 6 |
| 2 | 817036 | 13 | 2 | 820989 | 13 | 2 | 824906 | 13 | 2 | 828789 | 13 |
| 3 | 817102 | 20 | 3 | 821055 | 20 | 3 | 824971 | 19 | 3 | 828853 | 19 |
| 4 | 817169 | 26 | 4 | 821120 | 26 | 4 | 825036 | 26 | 4 | 828918 | 26 |
| 5 | 817235 | 33 | 5 | 821186 | 33 | 5 | 825101 | 32 | 5 | 828982 | 32 |
| 6 | 817301 | 40 | 6 | 821251 | 40 | 6 | 825166 | 39 | 6 | 829046 | 39 |
| 7 | 817367 | 46 | 7 | 821317 | 46 | 7 | 825231 | 45 | 7 | 829111 | 45 |
| 8 | 817433 | 53 | 8 | 821382 | 53 | 8 | 825296 | 52 | 8 | 829175 | 52 |
| 9 | 817499 | 59 | 9 | 821448 | 59 | 9 | 825361 | 58 | 9 | 829239 | 58 |
| 6570 | 817565 | | 6630 | 821514 | | 6690 | 825426 | | 6750 | 829304 | |
| 1 | 817631 | 7 | 1 | 821579 | 7 | 1 | 825491 | 6 | 1 | 829368 | 6 |
| 2 | 817698 | 13 | 2 | 821644 | 13 | 2 | 825556 | 13 | 2 | 829432 | 13 |
| 3 | 817764 | 20 | 3 | 821710 | 20 | 3 | 825621 | 19 | 3 | 829497 | 19 |
| 4 | 817830 | 26 | 4 | 821775 | 26 | 4 | 825686 | 26 | 4 | 829561 | 26 |
| 5 | 817896 | 33 | 5 | 821841 | 33 | 5 | 825751 | 32 | 5 | 829625 | 32 |
| 6 | 817962 | 40 | 6 | 821906 | 39 | 6 | 825815 | 39 | 6 | 829690 | 39 |
| 7 | 818028 | 46 | 7 | 821972 | 46 | 7 | 825880 | 45 | 7 | 829754 | 45 |
| 8 | 818094 | 53 | 8 | 822037 | 52 | 8 | 825945 | 52 | 8 | 829818 | 52 |
| 9 | 818160 | 59 | 9 | 822103 | 59 | 9 | 826010 | 58 | 9 | 829882 | 58 |

| No. | Log. | Prop. Part. | No. | Log. | Prop. Part. | No. | Log | Prop. Part. | No. | Log. | Prop. Part. |
|---|---|---|---|---|---|---|---|---|---|---|---|
| 6760 | 829947 | | 6820 | 833784 | | 6880 | 837588 | | 6940 | 841359 | |
| 1 | 830011 | 6 | 1 | 833848 | 6 | 1 | 837652 | 6 | 1 | 841422 | 6 |
| 2 | 830075 | 13 | 2 | 833912 | 13 | 2 | 837715 | 13 | 2 | 841485 | 13 |
| 3 | 830139 | 19 | 3 | 833975 | 19 | 3 | 837778 | 19 | 3 | 841547 | 19 |
| 4 | 830204 | 26 | 4 | 834039 | 26 | 4 | 837841 | 25 | 4 | 841610 | 25 |
| 5 | 830268 | 32 | 5 | 834103 | 32 | 5 | 837904 | 32 | 5 | 841672 | 31 |
| 6 | 830332 | 38 | 6 | 834166 | 38 | 6 | 837967 | 38 | 6 | 841735 | 38 |
| 7 | 830396 | 45 | 7 | 834230 | 45 | 7 | 838030 | 44 | 7 | 841797 | 44 |
| 8 | 830460 | 51 | 8 | 834293 | 51 | 8 | 838093 | 50 | 8 | 841860 | 50 |
| 9 | 830525 | 58 | 9 | 834357 | 58 | 9 | 838156 | 57 | 9 | 841922 | 56 |
| 6770 | 830589 | | 6830 | 834421 | | 6890 | 838219 | | 6950 | 841985 | |
| 1 | 830653 | 6 | 1 | 834484 | 6 | 1 | 838282 | 6 | 1 | 842047 | 6 |
| 2 | 830717 | 13 | 2 | 834548 | 13 | 2 | 838345 | 13 | 2 | 842110 | 12 |
| 3 | 830781 | 19 | 3 | 834611 | 19 | 3 | 838408 | 19 | 3 | 842172 | 19 |
| 4 | 830845 | 26 | 4 | 834675 | 26 | 4 | 838471 | 25 | 4 | 842235 | 25 |
| 5 | 830909 | 32 | 5 | 834739 | 32 | 5 | 838534 | 32 | 5 | 842297 | 31 |
| 6 | 830973 | 38 | 6 | 834802 | 38 | 6 | 838597 | 38 | 6 | 842360 | 37 |
| 7 | 831037 | 45 | 7 | 834866 | 45 | 7 | 838660 | 44 | 7 | 842422 | 44 |
| 8 | 831102 | 51 | 8 | 834929 | 51 | 8 | 838723 | 50 | 8 | 842484 | 50 |
| 9 | 831166 | 58 | 9 | 834993 | 58 | 9 | 838786 | 57 | 9 | 842547 | 56 |
| 6780 | 831230 | | 6840 | 835056 | | 6900 | 838849 | | 6960 | 842609 | |
| 1 | 831294 | 6 | 1 | 835120 | 6 | 1 | 838912 | 6 | 1 | 842672 | 6 |
| 2 | 831358 | 13 | 2 | 835183 | 13 | 2 | 838975 | 13 | 2 | 842734 | 12 |
| 3 | 831422 | 19 | 3 | 835247 | 19 | 3 | 839038 | 19 | 3 | 842796 | 19 |
| 4 | 831486 | 26 | 4 | 835310 | 26 | 4 | 839101 | 25 | 4 | 842859 | 25 |
| 5 | 831550 | 32 | 5 | 835373 | 32 | 5 | 839164 | 31 | 5 | 842921 | 31 |
| 6 | 831614 | 38 | 6 | 835437 | 38 | 6 | 839227 | 38 | 6 | 842983 | 37 |
| 7 | 831678 | 45 | 7 | 835500 | 45 | 7 | 839289 | 44 | 7 | 843046 | 44 |
| 8 | 831742 | 51 | 8 | 835564 | 51 | 8 | 839352 | 50 | 8 | 843108 | 50 |
| 9 | 831806 | 58 | 9 | 835627 | 58 | 9 | 839415 | 57 | 9 | 843170 | 56 |
| 6790 | 831870 | | 6850 | 835691 | | 6910 | 839478 | | 6970 | 843233 | |
| 1 | 831934 | 6 | 1 | 835754 | 6 | 1 | 839541 | 6 | 1 | 843295 | 6 |
| 2 | 831998 | 13 | 2 | 835817 | 13 | 2 | 839604 | 13 | 2 | 843357 | 12 |
| 3 | 832062 | 19 | 3 | 835881 | 19 | 3 | 839667 | 19 | 3 | 843420 | 19 |
| 4 | 832126 | 26 | 4 | 835944 | 26 | 4 | 839729 | 25 | 4 | 843482 | 25 |
| 5 | 832189 | 32 | 5 | 836007 | 32 | 5 | 839792 | 31 | 5 | 843544 | 31 |
| 6 | 832253 | 38 | 6 | 836071 | 38 | 6 | 839855 | 38 | 6 | 843606 | 37 |
| 7 | 832317 | 45 | 7 | 836134 | 45 | 7 | 839918 | 44 | 7 | 843669 | 43 |
| 8 | 832381 | 51 | 8 | 836197 | 51 | 8 | 839981 | 50 | 8 | 843731 | 50 |
| 9 | 832445 | 58 | 9 | 836261 | 58 | 9 | 840043 | 57 | 9 | 843793 | 56 |
| 6800 | 832509 | | 6860 | 836324 | | 6920 | 840106 | | 6980 | 843855 | |
| 1 | 832573 | 6 | 1 | 836387 | 6 | 1 | 840169 | 6 | 1 | 843918 | 6 |
| 2 | 832637 | 13 | 2 | 836451 | 13 | 2 | 840232 | 13 | 2 | 843980 | 12 |
| 3 | 832700 | 19 | 3 | 836514 | 19 | 3 | 840294 | 19 | 3 | 844042 | 19 |
| 4 | 832764 | 26 | 4 | 836577 | 26 | 4 | 840357 | 25 | 4 | 844104 | 25 |
| 5 | 832828 | 32 | 5 | 836641 | 32 | 5 | 840420 | 31 | 5 | 844166 | 31 |
| 6 | 832892 | 38 | 6 | 836704 | 38 | 6 | 840482 | 38 | 6 | 844229 | 37 |
| 7 | 832956 | 45 | 7 | 836767 | 45 | 7 | 840545 | 44 | 7 | 844291 | 43 |
| 8 | 833020 | 51 | 8 | 836830 | 51 | 8 | 840608 | 50 | 8 | 844353 | 50 |
| 9 | 833083 | 58 | 9 | 836894 | 58 | 9 | 840671 | 57 | 9 | 844415 | 56 |
| 6810 | 833147 | | 6870 | 836957 | | 6930 | 840733 | | 6990 | 844477 | |
| 1 | 833211 | 6 | 1 | 837020 | 6 | 1 | 840796 | 6 | 1 | 844539 | 6 |
| 2 | 833275 | 13 | 2 | 837083 | 13 | 2 | 840859 | 13 | 2 | 844601 | 12 |
| 3 | 833338 | 19 | 3 | 837146 | 19 | 3 | 840921 | 19 | 3 | 844664 | 19 |
| 4 | 833402 | 26 | 4 | 837210 | 25 | 4 | 840984 | 25 | 4 | 844726 | 25 |
| 5 | 833466 | 32 | 5 | 837273 | 32 | 5 | 841046 | 31 | 5 | 844788 | 31 |
| 6 | 833530 | 38 | 6 | 837336 | 38 | 6 | 841109 | 38 | 6 | 844850 | 37 |
| 7 | 833593 | 45 | 7 | 837399 | 44 | 7 | 841172 | 44 | 7 | 844912 | 43 |
| 8 | 833657 | 51 | 8 | 837462 | 51 | 8 | 841234 | 50 | 8 | 844974 | 50 |
| 9 | 833721 | 58 | 9 | 837525 | 57 | 9 | 841297 | 56 | 9 | 845036 | 56 |

| No. | Log. | Prop. Part. | No. | Log. | Prop. Part. | No. | Log. | Prop. Part. | No. | Log. | Prop. Part. |
|---|---|---|---|---|---|---|---|---|---|---|---|
| 7000 | 845098 | | 7060 | 848805 | | 7120 | 852480 | | 7180 | 856124 | |
| 1 | 845160 | 6 | 1 | 848866 | 6 | 1 | 852541 | 6 | 1 | 856185 | 6 |
| 2 | 845222 | 12 | 2 | 848928 | 12 | 2 | 852602 | 12 | 2 | 856245 | 12 |
| 3 | 845284 | 19 | 3 | 848989 | 18 | 3 | 852663 | 18 | 3 | 856306 | 18 |
| 4 | 845346 | 25 | 4 | 849051 | 25 | 4 | 852724 | 24 | 4 | 856366 | 24 |
| 5 | 845408 | 31 | 5 | 849112 | 31 | 5 | 852785 | 30 | 5 | 856427 | 30 |
| 6 | 845470 | 37 | 6 | 849174 | 37 | 6 | 852846 | 37 | 6 | 856487 | 36 |
| 7 | 845532 | 43 | 7 | 849235 | 43 | 7 | 852907 | 43 | 7 | 856548 | 42 |
| 8 | 845594 | 50 | 8 | 849296 | 49 | 8 | 852968 | 49 | 8 | 856608 | 48 |
| 9 | 845656 | 56 | 9 | 849358 | 55 | 9 | 853029 | 55 | 9 | 856668 | 54 |
| 7010 | 845718 | | 7070 | 849419 | | 7130 | 853090 | | 7190 | 856729 | |
| 1 | 845780 | 6 | 1 | 849481 | 6 | 1 | 853150 | 6 | 1 | 856789 | 6 |
| 2 | 845842 | 12 | 2 | 849542 | 12 | 2 | 853211 | 12 | 2 | 856850 | 12 |
| 3 | 845904 | 19 | 3 | 849604 | 18 | 3 | 853272 | 18 | 3 | 856910 | 18 |
| 4 | 845966 | 25 | 4 | 849665 | 25 | 4 | 853333 | 24 | 4 | 856970 | 24 |
| 5 | 846028 | 31 | 5 | 849726 | 31 | 5 | 853394 | 30 | 5 | 857031 | 30 |
| 6 | 846090 | 37 | 6 | 849788 | 37 | 6 | 853455 | 37 | 6 | 857091 | 36 |
| 7 | 846151 | 43 | 7 | 849849 | 43 | 7 | 853516 | 43 | 7 | 857151 | 42 |
| 8 | 846213 | 50 | 8 | 849911 | 49 | 8 | 853576 | 49 | 8 | 857212 | 48 |
| 9 | 846275 | 56 | 9 | 849972 | 55 | 9 | 853637 | 55 | 9 | 857272 | 54 |
| 7020 | 846337 | | 7080 | 850033 | | 7140 | 853698 | | 7200 | 857332 | |
| 1 | 846399 | 6 | 1 | 850095 | 6 | 1 | 853759 | 6 | 1 | 857393 | 6 |
| 2 | 846461 | 12 | 2 | 850156 | 12 | 2 | 853820 | 12 | 2 | 857453 | 12 |
| 3 | 846523 | 19 | 3 | 850217 | 18 | 3 | 853881 | 18 | 3 | 857513 | 18 |
| 4 | 846584 | 25 | 4 | 850279 | 25 | 4 | 853941 | 24 | 4 | 857574 | 24 |
| 5 | 846646 | 31 | 5 | 850340 | 31 | 5 | 854002 | 30 | 5 | 857634 | 30 |
| 6 | 846708 | 37 | 6 | 850401 | 37 | 6 | 854063 | 37 | 6 | 857694 | 36 |
| 7 | 846770 | 43 | 7 | 850462 | 43 | 7 | 854124 | 43 | 7 | 857754 | 42 |
| 8 | 846832 | 50 | 8 | 850524 | 49 | 8 | 854185 | 49 | 8 | 857815 | 48 |
| 9 | 846894 | 56 | 9 | 850585 | 55 | 9 | 854245 | 55 | 9 | 857875 | 54 |
| 7030 | 846955 | | 7090 | 850646 | | 7150 | 854306 | | 7210 | 857935 | |
| 1 | 847017 | 6 | 1 | 850707 | 6 | 1 | 854367 | 6 | 1 | 857995 | 6 |
| 2 | 847079 | 12 | 2 | 850769 | 12 | 2 | 854427 | 12 | 2 | 858056 | 12 |
| 3 | 847141 | 19 | 3 | 850830 | 18 | 3 | 854488 | 18 | 3 | 858116 | 18 |
| 4 | 847202 | 25 | 4 | 850891 | 25 | 4 | 854549 | 24 | 4 | 858176 | 24 |
| 5 | 847264 | 31 | 5 | 850952 | 31 | 5 | 854610 | 30 | 5 | 858236 | 30 |
| 6 | 847326 | 37 | 6 | 851014 | 37 | 6 | 854670 | 36 | 6 | 858297 | 36 |
| 7 | 847388 | 43 | 7 | 851075 | 43 | 7 | 854731 | 42 | 7 | 858357 | 42 |
| 8 | 847449 | 50 | 8 | 851136 | 49 | 8 | 854792 | 48 | 8 | 858417 | 48 |
| 9 | 847511 | 56 | 9 | 851197 | 55 | 9 | 854852 | 54 | 9 | 858477 | 54 |
| 7040 | 847573 | | 7100 | 851258 | | 7160 | 854913 | | 7220 | 858537 | |
| 1 | 847634 | 6 | 1 | 851320 | 6 | 1 | 854974 | 6 | 1 | 858597 | 6 |
| 2 | 847696 | 12 | 2 | 851381 | 12 | 2 | 855034 | 12 | 2 | 858657 | 12 |
| 3 | 847758 | 18 | 3 | 851442 | 18 | 3 | 855095 | 18 | 3 | 858718 | 18 |
| 4 | 847819 | 25 | 4 | 851503 | 25 | 4 | 855156 | 24 | 4 | 858778 | 24 |
| 5 | 847881 | 31 | 5 | 851564 | 31 | 5 | 855216 | 30 | 5 | 858838 | 30 |
| 6 | 847943 | 37 | 6 | 851625 | 37 | 6 | 855277 | 36 | 6 | 858898 | 36 |
| 7 | 848004 | 43 | 7 | 851686 | 43 | 7 | 855337 | 42 | 7 | 858958 | 42 |
| 8 | 848066 | 49 | 8 | 851747 | 49 | 8 | 855398 | 48 | 8 | 859018 | 48 |
| 9 | 848127 | 55 | 9 | 851808 | 55 | 9 | 855459 | 54 | 9 | 859078 | 54 |
| 7050 | 848189 | | 7110 | 851870 | | 7170 | 855519 | | 7230 | 859138 | |
| 1 | 848251 | 6 | 1 | 851931 | 6 | 1 | 855580 | 6 | 1 | 859198 | 6 |
| 2 | 848312 | 12 | 2 | 851992 | 12 | 2 | 855640 | 12 | 2 | 859258 | 12 |
| 3 | 848374 | 18 | 3 | 852053 | 18 | 3 | 855701 | 18 | 3 | 859318 | 18 |
| 4 | 848435 | 25 | 4 | 852114 | 25 | 4 | 855761 | 24 | 4 | 859378 | 24 |
| 5 | 848497 | 31 | 5 | 852175 | 31 | 5 | 855822 | 30 | 5 | 859438 | 30 |
| 6 | 848559 | 37 | 6 | 852236 | 37 | 6 | 855882 | 36 | 6 | 859499 | 36 |
| 7 | 848620 | 43 | 7 | 852297 | 43 | 7 | 855943 | 42 | 7 | 859559 | 42 |
| 8 | 848682 | 49 | 8 | 852358 | 49 | 8 | 856003 | 48 | 8 | 859619 | 48 |
| 9 | 848743 | 55 | 9 | 852419 | 55 | 9 | 856064 | 54 | 9 | 859679 | 54 |

| No. | Log. | Prop. Part. | No. | Log. | Prop. Part. | No. | Log. | Prop. Part. | No. | Log. | Prop. Part. |
|---|---|---|---|---|---|---|---|---|---|---|---|
| 7240 | 859739 | | 7300 | 863323 | | 7360 | 866878 | | 7420 | 870404 | |
| 1 | 859799 | 6 | 1 | 863382 | 6 | 1 | 866937 | 6 | 1 | 870462 | 6 |
| 2 | 859858 | 12 | 2 | 863442 | 12 | 2 | 866996 | 12 | 2 | 870521 | 12 |
| 3 | 859918 | 18 | 3 | 863501 | 18 | 3 | 867055 | 18 | 3 | 870579 | 18 |
| 4 | 859978 | 24 | 4 | 863561 | 24 | 4 | 867114 | 24 | 4 | 870638 | 24 |
| 5 | 860038 | 30 | 5 | 863620 | 30 | 5 | 867173 | 29 | 5 | 870696 | 29 |
| 6 | 860098 | 36 | 6 | 863680 | 36 | 6 | 867232 | 35 | 6 | 870755 | 35 |
| 7 | 860158 | 42 | 7 | 863739 | 42 | 7 | 867291 | 41 | 7 | 870813 | 41 |
| 8 | 860218 | 48 | 8 | 863798 | 48 | 8 | 867350 | 47 | 8 | 870872 | 47 |
| 9 | 860278 | 54 | 9 | 863858 | 54 | 9 | 867409 | 53 | 9 | 870930 | 53 |
| 7250 | 860338 | | 7310 | 863917 | | 7370 | 867467 | | 7430 | 870989 | |
| 1 | 860398 | 6 | 1 | 863977 | 6 | 1 | 867526 | 6 | 1 | 871047 | 6 |
| 2 | 860458 | 12 | 2 | 864036 | 12 | 2 | 867585 | 12 | 2 | 871106 | 12 |
| 3 | 860518 | 18 | 3 | 864096 | 18 | 3 | 867644 | 18 | 3 | 871164 | 18 |
| 4 | 860578 | 24 | 4 | 864155 | 24 | 4 | 867703 | 24 | 4 | 871223 | 24 |
| 5 | 860637 | 30 | 5 | 864214 | 30 | 5 | 867762 | 29 | 5 | 871281 | 29 |
| 6 | 860697 | 36 | 6 | 864274 | 36 | 6 | 867821 | 35 | 6 | 871339 | 35 |
| 7 | 860757 | 42 | 7 | 864333 | 42 | 7 | 867880 | 41 | 7 | 871398 | 41 |
| 8 | 860817 | 48 | 8 | 864392 | 48 | 8 | 867939 | 47 | 8 | 871456 | 47 |
| 9 | 860877 | 54 | 9 | 864452 | 54 | 9 | 867998 | 53 | 9 | 871515 | 53 |
| 7260 | 860937 | | 7320 | 864511 | | 7380 | 868056 | | 7440 | 871573 | |
| 1 | 860996 | 6 | 1 | 864570 | 6 | 1 | 868115 | 6 | 1 | 871631 | 6 |
| 2 | 861056 | 12 | 2 | 864630 | 12 | 2 | 868174 | 12 | 2 | 871690 | 12 |
| 3 | 861116 | 18 | 3 | 864689 | 18 | 3 | 868233 | 18 | 3 | 871748 | 18 |
| 4 | 861176 | 24 | 4 | 864748 | 24 | 4 | 868292 | 24 | 4 | 871806 | 23 |
| 5 | 861236 | 30 | 5 | 864808 | 30 | 5 | 868350 | 29 | 5 | 871865 | 29 |
| 6 | 861295 | 36 | 6 | 864867 | 36 | 6 | 868409 | 35 | 6 | 871923 | 35 |
| 7 | 861355 | 42 | 7 | 864926 | 42 | 7 | 868468 | 41 | 7 | 871981 | 41 |
| 8 | 861415 | 48 | 8 | 864985 | 48 | 8 | 868527 | 47 | 8 | 872040 | 47 |
| 9 | 861475 | 54 | 9 | 865045 | 54 | 9 | 868586 | 53 | 9 | 872098 | 53 |
| 7270 | 861534 | | 7330 | 865104 | | 7390 | 868644 | | 7450 | 872156 | |
| 1 | 861594 | 6 | 1 | 865163 | 6 | 1 | 868703 | 6 | 1 | 872215 | 6 |
| 2 | 861654 | 12 | 2 | 865222 | 12 | 2 | 868762 | 12 | 2 | 872273 | 12 |
| 3 | 861714 | 18 | 3 | 865282 | 18 | 3 | 868821 | 18 | 3 | 872331 | 18 |
| 4 | 861773 | 24 | 4 | 865341 | 24 | 4 | 868879 | 24 | 4 | 872389 | 23 |
| 5 | 861833 | 30 | 5 | 865400 | 30 | 5 | 868938 | 29 | 5 | 872448 | 29 |
| 6 | 861893 | 36 | 6 | 865459 | 36 | 6 | 868997 | 35 | 6 | 872506 | 35 |
| 7 | 861952 | 42 | 7 | 865518 | 42 | 7 | 869056 | 41 | 7 | 872564 | 41 |
| 8 | 862012 | 48 | 8 | 865578 | 48 | 8 | 869114 | 47 | 8 | 872622 | 47 |
| 9 | 862072 | 54 | 9 | 865637 | 54 | 9 | 869173 | 53 | 9 | 872681 | 53 |
| 7280 | 862131 | | 7340 | 865696 | | 7400 | 869232 | | 7460 | 872739 | |
| 1 | 862191 | 6 | 1 | 865755 | 6 | 1 | 869290 | 6 | 1 | 872797 | 6 |
| 2 | 862251 | 12 | 2 | 865814 | 12 | 2 | 869349 | 12 | 2 | 872855 | 12 |
| 3 | 862310 | 18 | 3 | 865874 | 18 | 3 | 869408 | 18 | 3 | 872913 | 18 |
| 4 | 862370 | 24 | 4 | 865933 | 24 | 4 | 869466 | 24 | 4 | 872972 | 23 |
| 5 | 862430 | 30 | 5 | 865992 | 30 | 5 | 869525 | 29 | 5 | 873030 | 29 |
| 6 | 862489 | 36 | 6 | 866051 | 36 | 6 | 869584 | 35 | 6 | 873088 | 35 |
| 7 | 862549 | 42 | 7 | 866110 | 42 | 7 | 869642 | 41 | 7 | 873146 | 41 |
| 8 | 862608 | 48 | 8 | 866169 | 48 | 8 | 869701 | 47 | 8 | 873204 | 47 |
| 9 | 862668 | 54 | 9 | 866228 | 54 | 9 | 869760 | 53 | 9 | 873262 | 53 |
| 7290 | 862728 | | 7350 | 866287 | | 7410 | 869818 | | 7470 | 873321 | |
| 1 | 862787 | 6 | 1 | 866346 | 6 | 1 | 869877 | 6 | 1 | 873379 | 6 |
| 2 | 862847 | 12 | 2 | 866405 | 12 | 2 | 869935 | 12 | 2 | 873437 | 12 |
| 3 | 862906 | 18 | 3 | 866465 | 18 | 3 | 869994 | 18 | 3 | 873495 | 18 |
| 4 | 862966 | 24 | 4 | 866524 | 24 | 4 | 870053 | 24 | 4 | 873553 | 23 |
| 5 | 863025 | 30 | 5 | 866583 | 30 | 5 | 870111 | 29 | 5 | 873611 | 29 |
| 6 | 863085 | 36 | 6 | 866642 | 35 | 6 | 870170 | 35 | 6 | 873669 | 35 |
| 7 | 863144 | 42 | 7 | 866701 | 41 | 7 | 870228 | 41 | 7 | 873727 | 41 |
| 8 | 863204 | 48 | 8 | 866760 | 47 | 8 | 870287 | 47 | 8 | 873785 | 47 |
| 9 | 863263 | 54 | 9 | 866819 | 53 | 9 | 870345 | 53 | 9 | 873844 | 53 |

| No. | Log. | Prop. Part. | No. | Log. | Prop. Part. | No. | Log. | Prop. Part. | No. | Log. | Prop. Part. |
|---|---|---|---|---|---|---|---|---|---|---|---|
| 7480 | 873902 | | 7540 | 877371 | | 7600 | 880814 | | 7660 | 884229 | |
| 1 | 873960 | 6 | 1 | 877429 | 6 | 1 | 880871 | 6 | 1 | 884285 | 6 |
| 2 | 874018 | 12 | 2 | 877486 | 12 | 2 | 880928 | 11 | 2 | 884342 | 11 |
| 3 | 874076 | 17 | 3 | 877544 | 17 | 3 | 880985 | 17 | 3 | 884399 | 17 |
| 4 | 874134 | 23 | 4 | 877602 | 23 | 4 | 881042 | 23 | 4 | 884455 | 23 |
| 5 | 874192 | 29 | 5 | 877659 | 29 | 5 | 881099 | 28 | 5 | 884512 | 28 |
| 6 | 874250 | 35 | 6 | 877717 | 34 | 6 | 881156 | 34 | 6 | 884569 | 34 |
| 7 | 874308 | 41 | 7 | 877774 | 40 | 7 | 881213 | 40 | 7 | 884625 | 40 |
| 8 | 874366 | 46 | 8 | 877832 | 46 | 8 | 881270 | 46 | 8 | 884682 | 46 |
| 9 | 874424 | 52 | 9 | 877889 | 52 | 9 | 881328 | 51 | 9 | 884739 | 51 |
| 7490 | 874482 | | 7550 | 877947 | | 7610 | 881385 | | 7670 | 884795 | |
| 1 | 874540 | 6 | 1 | 878004 | 6 | 1 | 881442 | 6 | 1 | 884852 | 6 |
| 2 | 874598 | 12 | 2 | 878062 | 12 | 2 | 881499 | 11 | 2 | 884909 | 11 |
| 3 | 874656 | 17 | 3 | 878119 | 17 | 3 | 881556 | 17 | 3 | 884965 | 17 |
| 4 | 874714 | 23 | 4 | 878177 | 23 | 4 | 881613 | 23 | 4 | 885022 | 23 |
| 5 | 874772 | 29 | 5 | 878234 | 29 | 5 | 881670 | 28 | 5 | 885078 | 28 |
| 6 | 874830 | 35 | 6 | 878292 | 34 | 6 | 881727 | 34 | 6 | 885135 | 34 |
| 7 | 874887 | 41 | 7 | 878349 | 40 | 7 | 881784 | 40 | 7 | 885192 | 40 |
| 8 | 874945 | 46 | 8 | 878407 | 46 | 8 | 881841 | 46 | 8 | 885248 | 46 |
| 9 | 875003 | 52 | 9 | 878464 | 52 | 9 | 881898 | 51 | 9 | 885305 | 51 |
| 7500 | 875061 | | 7560 | 878522 | | 7620 | 881955 | | 7680 | 885361 | |
| 1 | 875119 | 6 | 1 | 878579 | 6 | 1 | 882012 | 6 | 1 | 885418 | 6 |
| 2 | 875177 | 12 | 2 | 878637 | 12 | 2 | 882069 | 11 | 2 | 885474 | 11 |
| 3 | 875235 | 17 | 3 | 878694 | 17 | 3 | 882126 | 17 | 3 | 885531 | 17 |
| 4 | 875293 | 23 | 4 | 878751 | 23 | 4 | 882183 | 23 | 4 | 885587 | 23 |
| 5 | 875351 | 29 | 5 | 878809 | 29 | 5 | 882240 | 28 | 5 | 885644 | 28 |
| 6 | 875409 | 35 | 6 | 878866 | 34 | 6 | 882297 | 34 | 6 | 885700 | 34 |
| 7 | 875466 | 41 | 7 | 878924 | 40 | 7 | 882354 | 40 | 7 | 885757 | 39 |
| 8 | 875524 | 46 | 8 | 878981 | 46 | 8 | 882411 | 46 | 8 | 885813 | 45 |
| 9 | 875582 | 52 | 9 | 879038 | 52 | 9 | 882468 | 51 | 9 | 885870 | 51 |
| 7510 | 875640 | | 7570 | 879096 | | 7630 | 882524 | | 7690 | 885926 | |
| 1 | 875698 | 6 | 1 | 879153 | 6 | 1 | 882581 | 6 | 1 | 885983 | 6 |
| 2 | 875756 | 12 | 2 | 879211 | 12 | 2 | 882638 | 11 | 2 | 886039 | 11 |
| 3 | 875813 | 17 | 3 | 879268 | 17 | 3 | 882695 | 17 | 3 | 886096 | 17 |
| 4 | 875871 | 23 | 4 | 879325 | 23 | 4 | 882752 | 23 | 4 | 886152 | 23 |
| 5 | 875929 | 29 | 5 | 879383 | 29 | 5 | 882809 | 28 | 5 | 886209 | 28 |
| 6 | 875987 | 35 | 6 | 879440 | 34 | 6 | 882866 | 34 | 6 | 886265 | 34 |
| 7 | 876045 | 41 | 7 | 879497 | 40 | 7 | 882923 | 40 | 7 | 886321 | 39 |
| 8 | 876102 | 46 | 8 | 879555 | 46 | 8 | 882980 | 46 | 8 | 886378 | 45 |
| 9 | 876160 | 52 | 9 | 879612 | 52 | 9 | 883037 | 51 | 9 | 886434 | 51 |
| 7520 | 876218 | | 7580 | 879669 | | 7640 | 883093 | | 7700 | 886491 | |
| 1 | 876276 | 6 | 1 | 879726 | 6 | 1 | 883150 | 6 | 1 | 886547 | 6 |
| 2 | 876333 | 12 | 2 | 879784 | 11 | 2 | 883207 | 11 | 2 | 886604 | 11 |
| 3 | 876391 | 17 | 3 | 879841 | 17 | 3 | 883264 | 17 | 3 | 886660 | 17 |
| 4 | 876449 | 23 | 4 | 879898 | 23 | 4 | 883321 | 23 | 4 | 886716 | 23 |
| 5 | 876507 | 29 | 5 | 879956 | 28 | 5 | 883377 | 28 | 5 | 886773 | 28 |
| 6 | 876564 | 34 | 6 | 880013 | 34 | 6 | 883434 | 34 | 6 | 886829 | 34 |
| 7 | 876622 | 40 | 7 | 880070 | 40 | 7 | 883491 | 40 | 7 | 886885 | 39 |
| 8 | 876680 | 46 | 8 | 880127 | 46 | 8 | 883548 | 46 | 8 | 886942 | 45 |
| 9 | 876737 | 52 | 9 | 880185 | 51 | 9 | 883605 | 51 | 9 | 886998 | 51 |
| 7530 | 876795 | | 7590 | 880242 | | 7650 | 883661 | | 7710 | 887054 | |
| 1 | 876853 | 6 | 1 | 880299 | 6 | 1 | 883718 | 6 | 1 | 887111 | 6 |
| 2 | 876910 | 12 | 2 | 880356 | 11 | 2 | 883775 | 11 | 2 | 887167 | 11 |
| 3 | 876968 | 17 | 3 | 880413 | 17 | 3 | 883832 | 17 | 3 | 887223 | 17 |
| 4 | 877026 | 23 | 4 | 880471 | 23 | 4 | 883888 | 23 | 4 | 887280 | 23 |
| 5 | 877083 | 29 | 5 | 880528 | 28 | 5 | 883945 | 28 | 5 | 887336 | 28 |
| 6 | 877141 | 34 | 6 | 880585 | 34 | 6 | 884002 | 34 | 6 | 887392 | 34 |
| 7 | 877198 | 40 | 7 | 880642 | 40 | 7 | 884059 | 40 | 7 | 887449 | 39 |
| 8 | 877256 | 46 | 8 | 880699 | 46 | 8 | 884115 | 46 | 8 | 887505 | 45 |
| 9 | 877314 | 52 | 9 | 880756 | 51 | 9 | 884172 | 51 | 9 | 887561 | 51 |

| No. | Log. | Prop. Part. | No. | Log. | Prop. Part. | No. | Log. | Prop. Part. | No. | Log. | Prop. Part. |
|---|---|---|---|---|---|---|---|---|---|---|---|
| 7720 | 887617 | | 7780 | 890980 | | 7840 | 894316 | | 7900 | 897627 | |
| 1 | 887674 | 6 | 1 | 891035 | 6 | 1 | 894371 | 6 | 1 | 897682 | 6 |
| 2 | 887730 | 11 | 2 | 891091 | 11 | 2 | 894427 | 11 | 2 | 897737 | 11 |
| 3 | 887786 | 17 | 3 | 891147 | 17 | 3 | 894482 | 17 | 3 | 897792 | 17 |
| 4 | 887842 | 23 | 4 | 891203 | 22 | 4 | 894538 | 22 | 4 | 897847 | 22 |
| 5 | 887898 | 28 | 5 | 891259 | 28 | 5 | 894593 | 27 | 5 | 897902 | 27 |
| 6 | 887955 | 34 | 6 | 891314 | 34 | 6 | 894648 | 33 | 6 | 897957 | 33 |
| 7 | 888011 | 39 | 7 | 891370 | 39 | 7 | 894704 | 39 | 7 | 898012 | 39 |
| 8 | 888067 | 45 | 8 | 891426 | 45 | 8 | 894759 | 44 | 8 | 898067 | 44 |
| 9 | 888123 | 51 | 9 | 891482 | 50 | 9 | 894814 | 50 | 9 | 898122 | 50 |
| 7730 | 888179 | | 7790 | 891537 | | 7850 | 894870 | | 7910 | 898176 | |
| 1 | 888236 | 6 | 1 | 891593 | 6 | 1 | 894925 | 6 | 1 | 898231 | 6 |
| 2 | 888292 | 11 | 2 | 891649 | 11 | 2 | 894980 | 11 | 2 | 898286 | 11 |
| 3 | 888348 | 17 | 3 | 891705 | 17 | 3 | 895036 | 17 | 3 | 898341 | 17 |
| 4 | 888404 | 22 | 4 | 891760 | 22 | 4 | 895091 | 22 | 4 | 898396 | 22 |
| 5 | 888460 | 28 | 5 | 891816 | 28 | 5 | 895146 | 27 | 5 | 898451 | 27 |
| 6 | 888516 | 34 | 6 | 891872 | 33 | 6 | 895201 | 33 | 6 | 898506 | 33 |
| 7 | 888573 | 39 | 7 | 891928 | 39 | 7 | 895257 | 39 | 7 | 898561 | 39 |
| 8 | 888629 | 45 | 8 | 891983 | 44 | 8 | 895312 | 44 | 8 | 898615 | 44 |
| 9 | 888685 | 50 | 9 | 892039 | 50 | 9 | 895367 | 50 | 9 | 898670 | 50 |
| 7740 | 888741 | | 7800 | 892095 | | 7860 | 895423 | | 7920 | 898725 | |
| 1 | 888797 | 6 | 1 | 892150 | 6 | 1 | 895478 | 6 | 1 | 898780 | 5 |
| 2 | 888853 | 11 | 2 | 892206 | 11 | 2 | 895533 | 11 | 2 | 898835 | 11 |
| 3 | 888909 | 17 | 3 | 892262 | 17 | 3 | 895588 | 17 | 3 | 898890 | 17 |
| 4 | 888965 | 22 | 4 | 892317 | 22 | 4 | 895643 | 22 | 4 | 898944 | 22 |
| 5 | 889021 | 28 | 5 | 892373 | 28 | 5 | 895699 | 27 | 5 | 898999 | 27 |
| 6 | 889077 | 34 | 6 | 892429 | 33 | 6 | 895754 | 33 | 6 | 899054 | 33 |
| 7 | 889134 | 39 | 7 | 892484 | 39 | 7 | 895809 | 39 | 7 | 899109 | 38 |
| 8 | 889190 | 45 | 8 | 892540 | 44 | 8 | 895864 | 44 | 8 | 899164 | 44 |
| 9 | 889246 | 50 | 9 | 892595 | 50 | 9 | 895920 | 50 | 9 | 899218 | 50 |
| 7750 | 889302 | | 7810 | 892651 | | 7870 | 895975 | | 7930 | 899273 | |
| 1 | 889358 | 6 | 1 | 892707 | 6 | 1 | 896030 | 6 | 1 | 899328 | 5 |
| 2 | 889414 | 11 | 2 | 892762 | 11 | 2 | 896085 | 11 | 2 | 899383 | 11 |
| 3 | 889470 | 17 | 3 | 892818 | 17 | 3 | 896140 | 17 | 3 | 899437 | 17 |
| 4 | 889526 | 22 | 4 | 892873 | 22 | 4 | 896195 | 22 | 4 | 899492 | 22 |
| 5 | 889582 | 28 | 5 | 892929 | 28 | 5 | 896251 | 27 | 5 | 899547 | 27 |
| 6 | 889638 | 34 | 6 | 892985 | 33 | 6 | 896306 | 33 | 6 | 899602 | 33 |
| 7 | 889694 | 39 | 7 | 893040 | 39 | 7 | 896361 | 39 | 7 | 899656 | 38 |
| 8 | 889750 | 45 | 8 | 893096 | 44 | 8 | 896416 | 44 | 8 | 899711 | 44 |
| 9 | 889806 | 50 | 9 | 893151 | 50 | 9 | 896471 | 50 | 9 | 899766 | 50 |
| 7760 | 889862 | | 7820 | 893207 | | 7880 | 896526 | | 7940 | 899820 | |
| 1 | 889918 | 6 | 1 | 893262 | 6 | 1 | 896581 | 6 | 1 | 899875 | 5 |
| 2 | 889974 | 11 | 2 | 893318 | 11 | 2 | 896636 | 11 | 2 | 899930 | 11 |
| 3 | 890030 | 17 | 3 | 893373 | 17 | 3 | 896692 | 17 | 3 | 899985 | 17 |
| 4 | 890086 | 22 | 4 | 893429 | 22 | 4 | 896747 | 22 | 4 | 900039 | 22 |
| 5 | 890141 | 28 | 5 | 893484 | 28 | 5 | 896802 | 27 | 5 | 900094 | 27 |
| 6 | 890197 | 34 | 6 | 893540 | 33 | 6 | 896857 | 33 | 6 | 900149 | 33 |
| 7 | 890253 | 39 | 7 | 893595 | 39 | 7 | 896912 | 39 | 7 | 900203 | 38 |
| 8 | 890309 | 45 | 8 | 893651 | 44 | 8 | 896967 | 44 | 8 | 900258 | 44 |
| 9 | 890365 | 50 | 9 | 893706 | 50 | 9 | 897022 | 50 | 9 | 900312 | 50 |
| 7770 | 890421 | | 7830 | 893762 | | 7890 | 897077 | | 7950 | 900367 | |
| 1 | 890477 | 6 | 1 | 893817 | 6 | 1 | 897132 | 6 | 1 | 900422 | 5 |
| 2 | 890533 | 11 | 2 | 893873 | 11 | 2 | 897187 | 11 | 2 | 900476 | 11 |
| 3 | 890589 | 17 | 3 | 893928 | 17 | 3 | 897242 | 17 | 3 | 900531 | 17 |
| 4 | 890644 | 22 | 4 | 893984 | 22 | 4 | 897297 | 22 | 4 | 900586 | 22 |
| 5 | 890700 | 28 | 5 | 894039 | 28 | 5 | 897352 | 27 | 5 | 900640 | 27 |
| 6 | 890756 | 34 | 6 | 894094 | 33 | 6 | 897407 | 33 | 6 | 900695 | 33 |
| 7 | 890812 | 39 | 7 | 894150 | 39 | 7 | 897462 | 39 | 7 | 900749 | 38 |
| 8 | 890868 | 45 | 8 | 894205 | 44 | 8 | 897517 | 44 | 8 | 900804 | 44 |
| 9 | 890924 | 50 | 9 | 894261 | 50 | 9 | 897572 | 50 | 9 | 900858 | 50 |

| No. | Log. | Prop. Part. | No. | Log. | Prop. Part. | No. | Log. | Prop. Part. | No. | Log. | Prop. Part. |
|---|---|---|---|---|---|---|---|---|---|---|---|
| 7960 | 900913 | | 8020 | 904174 | | 8080 | 907411 | | 8140 | 910624 | |
| 1 | 900968 | 5 | 1 | 904228 | 5 | 1 | 907465 | 5 | 1 | 910678 | 5 |
| 2 | 901022 | 11 | 2 | 904283 | 11 | 2 | 907519 | 11 | 2 | 910731 | 11 |
| 3 | 901077 | 16 | 3 | 904337 | 16 | 3 | 907573 | 16 | 3 | 910784 | 16 |
| 4 | 901131 | 22 | 4 | 904391 | 22 | 4 | 907626 | 22 | 4 | 910838 | 21 |
| 5 | 901186 | 27 | 5 | 904445 | 27 | 5 | 907680 | 27 | 5 | 910891 | 27 |
| 6 | 901240 | 33 | 6 | 904499 | 32 | 6 | 907734 | 32 | 6 | 910944 | 32 |
| 7 | 901295 | 38 | 7 | 904553 | 38 | 7 | 907787 | 38 | 7 | 910998 | 37 |
| 8 | 901349 | 44 | 8 | 904607 | 43 | 8 | 907841 | 43 | 8 | 911051 | 43 |
| 9 | 901404 | 49 | 9 | 904661 | 49 | 9 | 907895 | 49 | 9 | 911104 | 48 |
| 7970 | 901458 | | 8030 | 904715 | | 8090 | 907948 | | 8150 | 911158 | |
| 1 | 901513 | 5 | 1 | 904770 | 5 | 1 | 908002 | 5 | 1 | 911211 | 5 |
| 2 | 901567 | 11 | 2 | 904824 | 11 | 2 | 908056 | 11 | 2 | 911264 | 11 |
| 3 | 901622 | 16 | 3 | 904878 | 16 | 3 | 908109 | 16 | 3 | 911317 | 16 |
| 4 | 901676 | 22 | 4 | 904932 | 22 | 4 | 908163 | 22 | 4 | 911371 | 21 |
| 5 | 901731 | 27 | 5 | 904986 | 27 | 5 | 908217 | 27 | 5 | 911424 | 27 |
| 6 | 901785 | 33 | 6 | 905040 | 32 | 6 | 908270 | 32 | 6 | 911477 | 32 |
| 7 | 901840 | 38 | 7 | 905094 | 38 | 7 | 908324 | 38 | 7 | 911530 | 37 |
| 8 | 901894 | 44 | 8 | 905148 | 43 | 8 | 908378 | 43 | 8 | 911584 | 42 |
| 9 | 901948 | 49 | 9 | 905202 | 49 | 9 | 908431 | 49 | 9 | 911637 | 48 |
| 7980 | 902003 | | 8040 | 905256 | | 8100 | 908485 | | 8160 | 911690 | |
| 1 | 902057 | 5 | 1 | 905310 | 5 | 1 | 908539 | 5 | 1 | 911743 | 5 |
| 2 | 902112 | 11 | 2 | 905364 | 11 | 2 | 908592 | 11 | 2 | 911797 | 11 |
| 3 | 902166 | 16 | 3 | 905418 | 16 | 3 | 908646 | 16 | 3 | 911850 | 16 |
| 4 | 902221 | 22 | 4 | 905472 | 22 | 4 | 908699 | 21 | 4 | 911903 | 21 |
| 5 | 902275 | 27 | 5 | 905526 | 27 | 5 | 908753 | 27 | 5 | 911956 | 27 |
| 6 | 902329 | 33 | 6 | 905580 | 32 | 6 | 908807 | 32 | 6 | 912009 | 32 |
| 7 | 902384 | 38 | 7 | 905634 | 38 | 7 | 908860 | 37 | 7 | 912063 | 37 |
| 8 | 902438 | 44 | 8 | 905688 | 43 | 8 | 908914 | 43 | 8 | 912116 | 42 |
| 9 | 902492 | 49 | 9 | 905742 | 49 | 9 | 908967 | 48 | 9 | 912169 | 48 |
| 7990 | 902547 | | 8050 | 905796 | | 8110 | 909021 | | 8170 | 912222 | |
| 1 | 902601 | 5 | 1 | 905850 | 5 | 1 | 909074 | 5 | 1 | 912275 | 5 |
| 2 | 902655 | 11 | 2 | 905904 | 11 | 2 | 909128 | 11 | 2 | 912328 | 11 |
| 3 | 902710 | 16 | 3 | 005958 | 16 | 3 | 909181 | 16 | 3 | 912381 | 16 |
| 4 | 902764 | 22 | 4 | 906012 | 22 | 4 | 909235 | 21 | 4 | 912435 | 21 |
| 5 | 902818 | 27 | 5 | 906065 | 27 | 5 | 909288 | 27 | 5 | 912488 | 27 |
| 6 | 902873 | 33 | 6 | 906119 | 32 | 6 | 909342 | 32 | 6 | 912541 | 32 |
| 7 | 902927 | 38 | 7 | 906173 | 38 | 7 | 909395 | 37 | 7 | 912594 | 37 |
| 8 | 902981 | 44 | 8 | 906227 | 43 | 8 | 909449 | 43 | 8 | 912647 | 42 |
| 9 | 903036 | 49 | 9 | 906281 | 49 | 9 | 909502 | 48 | 9 | 912700 | 48 |
| 8000 | 903090 | | 8060 | 906335 | | 8120 | 909556 | | 8180 | 912753 | |
| 1 | 903144 | 5 | 1 | 906389 | 5 | 1 | 909609 | 5 | 1 | 912806 | 5 |
| 2 | 903198 | 11 | 2 | 906443 | 11 | 2 | 909663 | 11 | 2 | 912859 | 11 |
| 3 | 903253 | 16 | 3 | 906497 | 16 | 3 | 909716 | 16 | 3 | 912913 | 16 |
| 4 | 903307 | 22 | 4 | 906550 | 22 | 4 | 909770 | 21 | 4 | 912966 | 21 |
| 5 | 903361 | 27 | 5 | 906604 | 27 | 5 | 909823 | 27 | 5 | 913019 | 27 |
| 6 | 903416 | 32 | 6 | 906658 | 32 | 6 | 909877 | 32 | 6 | 913072 | 32 |
| 7 | 903470 | 38 | 7 | 906712 | 38 | 7 | 909930 | 37 | 7 | 913125 | 37 |
| 8 | 903524 | 43 | 8 | 906766 | 43 | 8 | 909984 | 43 | 8 | 913178 | 42 |
| 9 | 903578 | 49 | 9 | 906820 | 49 | 9 | 910037 | 48 | 9 | 913231 | 48 |
| 8010 | 903632 | | 8070 | 906873 | | 8130 | 910090 | | 8190 | 913284 | |
| 1 | 903687 | 5 | 1 | 906927 | 5 | 1 | 910144 | 5 | 1 | 913337 | 5 |
| 2 | 903741 | 11 | 2 | 906981 | 11 | 2 | 910197 | 11 | 2 | 913390 | 11 |
| 3 | 903795 | 16 | 3 | 907035 | 16 | 3 | 910251 | 16 | 3 | 913443 | 16 |
| 4 | 903849 | 22 | 4 | 907089 | 22 | 4 | 910304 | 21 | 4 | 913496 | 21 |
| 5 | 903903 | 27 | 5 | 907142 | 27 | 5 | 910358 | 27 | 5 | 913549 | 27 |
| 6 | 903958 | 32 | 6 | 907196 | 32 | 6 | 910411 | 32 | 6 | 913602 | 32 |
| 7 | 904012 | 38 | 7 | 907250 | 38 | 7 | 910464 | 37 | 7 | 913655 | 37 |
| 8 | 904066 | 43 | 8 | 907304 | 43 | 8 | 910518 | 43 | 8 | 913708 | 42 |
| 9 | 904120 | 49 | 9 | 907358 | 49 | 9 | 910571 | 48 | 9 | 913761 | 48 |

| No. | Log. | Prop. Part. | No. | Log. | Prop. Part. | No. | Log. | Prop. Part. | No. | Log. | Prop. Part. |
|---|---|---|---|---|---|---|---|---|---|---|---|
| 8200 | 913814 | | 8260 | 916980 | | 8320 | 920123 | | 8380 | 923244 | |
| 1 | 913867 | 5 | 1 | 917033 | 5 | 1 | 920175 | 5 | 1 | 923296 | 5 |
| 2 | 913920 | 11 | 2 | 917085 | 11 | 2 | 920228 | 10 | 2 | 923348 | 10 |
| 3 | 913973 | 16 | 3 | 917138 | 16 | 3 | 920280 | 16 | 3 | 923399 | 16 |
| 4 | 914026 | 21 | 4 | 917190 | 21 | 4 | 920332 | 21 | 4 | 923451 | 21 |
| 5 | 914079 | 27 | 5 | 917243 | 26 | 5 | 920384 | 26 | 5 | 923503 | 26 |
| 6 | 914131 | 32 | 6 | 917295 | 31 | 6 | 920436 | 31 | 6 | 923555 | 31 |
| 7 | 914184 | 37 | 7 | 917348 | 37 | 7 | 920489 | 36 | 7 | 923607 | 36 |
| 8 | 914237 | 42 | 8 | 917400 | 42 | 8 | 920541 | 42 | 8 | 923658 | 42 |
| 9 | 914290 | 48 | 9 | 917453 | 47 | 9 | 920593 | 47 | 9 | 923710 | 47 |
| 8210 | 914343 | | 8270 | 917505 | | 8330 | 920645 | | 8390 | 923762 | |
| 1 | 914396 | 5 | 1 | 917558 | 5 | 1 | 920697 | 5 | 1 | 923814 | 5 |
| 2 | 914449 | 11 | 2 | 917610 | 11 | 2 | 920749 | 10 | 2 | 923865 | 10 |
| 3 | 914502 | 16 | 3 | 917663 | 16 | 3 | 920801 | 16 | 3 | 923917 | 16 |
| 4 | 914555 | 21 | 4 | 917715 | 21 | 4 | 920853 | 21 | 4 | 923969 | 21 |
| 5 | 914608 | 27 | 5 | 917768 | 26 | 5 | 920906 | 26 | 5 | 924021 | 26 |
| 6 | 914660 | 32 | 6 | 917820 | 31 | 6 | 920958 | 31 | 6 | 924072 | 31 |
| 7 | 914713 | 37 | 7 | 917873 | 37 | 7 | 921010 | 36 | 7 | 924124 | 36 |
| 8 | 914766 | 42 | 8 | 917925 | 42 | 8 | 921062 | 42 | 8 | 924176 | 42 |
| 9 | 914819 | 48 | 9 | 917978 | 47 | 9 | 921114 | 47 | 9 | 924228 | 47 |
| 8220 | 914872 | | 8280 | 918030 | | 8340 | 921166 | | 8400 | 924279 | |
| 1 | 914925 | 5 | 1 | 918083 | 5 | 1 | 921218 | 5 | 1 | 924331 | 5 |
| 2 | 914977 | 11 | 2 | 918135 | 11 | 2 | 921270 | 10 | 2 | 924383 | 10 |
| 3 | 915030 | 16 | 3 | 918188 | 16 | 3 | 921322 | 16 | 3 | 924434 | 15 |
| 4 | 915083 | 21 | 4 | 918240 | 21 | 4 | 921374 | 21 | 4 | 924486 | 21 |
| 5 | 915136 | 27 | 5 | 918292 | 26 | 5 | 921426 | 26 | 5 | 924538 | 26 |
| 6 | 915189 | 32 | 6 | 918345 | 31 | 6 | 921478 | 31 | 6 | 924589 | 31 |
| 7 | 915241 | 37 | 7 | 918397 | 37 | 7 | 921530 | 36 | 7 | 924641 | 36 |
| 8 | 915294 | 42 | 8 | 918450 | 42 | 8 | 921582 | 42 | 8 | 924693 | 41 |
| 9 | 915347 | 48 | 9 | 918502 | 47 | 9 | 921634 | 47 | 9 | 924744 | 46 |
| 8230 | 915400 | | 8290 | 918555 | | 8350 | 921686 | | 8410 | 924796 | |
| 1 | 915453 | 5 | 1 | 918607 | 5 | 1 | 921738 | 5 | 1 | 924848 | 5 |
| 2 | 915505 | 11 | 2 | 918659 | 11 | 2 | 921790 | 10 | 2 | 924899 | 10 |
| 3 | 915558 | 16 | 3 | 918712 | 16 | 3 | 921842 | 16 | 3 | 924951 | 15 |
| 4 | 915611 | 21 | 4 | 918764 | 21 | 4 | 921894 | 21 | 4 | 925002 | 21 |
| 5 | 915664 | 27 | 5 | 918816 | 26 | 5 | 921946 | 26 | 5 | 925054 | 26 |
| 6 | 915716 | 32 | 6 | 918869 | 31 | 6 | 921998 | 31 | 6 | 925106 | 31 |
| 7 | 915769 | 37 | 7 | 918921 | 37 | 7 | 922050 | 36 | 7 | 925157 | 36 |
| 8 | 915822 | 42 | 8 | 918973 | 42 | 8 | 922102 | 42 | 8 | 925209 | 41 |
| 9 | 915874 | 48 | 9 | 919026 | 47 | 9 | 922154 | 17 | 9 | 925260 | 46 |
| 8240 | 915927 | | 8300 | 919078 | | 8360 | 922206 | | 8420 | 925312 | |
| 1 | 915980 | 5 | 1 | 919130 | 5 | 1 | 922258 | 5 | 1 | 925364 | 5 |
| 2 | 916033 | 11 | 2 | 919183 | 11 | 2 | 922310 | 10 | 2 | 925415 | 10 |
| 3 | 916085 | 16 | 3 | 919235 | 16 | 3 | 922362 | 16 | 3 | 925467 | 15 |
| 4 | 916138 | 21 | 4 | 919287 | 21 | 4 | 922414 | 21 | 4 | 925518 | 21 |
| 5 | 916191 | 27 | 5 | 919340 | 26 | 5 | 922466 | 26 | 5 | 925570 | 26 |
| 6 | 916243 | 32 | 6 | 919392 | 31 | 6 | 922518 | 31 | 6 | 925621 | 31 |
| 7 | 916296 | 37 | 7 | 919444 | 37 | 7 | 922570 | 36 | 7 | 925673 | 36 |
| 8 | 916349 | 42 | 8 | 919496 | 42 | 8 | 922622 | 42 | 8 | 925724 | 41 |
| 9 | 916401 | 48 | 9 | 919549 | 47 | 9 | 922674 | 47 | 9 | 925776 | 46 |
| 8250 | 916454 | | 8310 | 919601 | | 8370 | 922725 | | 8430 | 925828 | |
| 1 | 916507 | 5 | 1 | 919653 | 5 | 1 | 922777 | 5 | 1 | 925879 | 5 |
| 2 | 916559 | 11 | 2 | 919705 | 11 | 2 | 922829 | 10 | 2 | 925931 | 10 |
| 3 | 916612 | 16 | 3 | 919758 | 16 | 3 | 922881 | 16 | 3 | 925982 | 15 |
| 4 | 916664 | 21 | 4 | 919810 | 21 | 4 | 922933 | 21 | 4 | 926034 | 21 |
| 5 | 916717 | 26 | 5 | 919862 | 26 | 5 | 922985 | 26 | 5 | 926085 | 26 |
| 6 | 916770 | 31 | 6 | 919914 | 31 | 6 | 923037 | 31 | 6 | 926137 | 31 |
| 7 | 916822 | 37 | 7 | 919967 | 37 | 7 | 923088 | 36 | 7 | 926188 | 36 |
| 8 | 916875 | 42 | 8 | 920019 | 42 | 8 | 923140 | 42 | 8 | 926239 | 41 |
| 9 | 916927 | 47 | 9 | 920071 | 47 | 9 | 923192 | 47 | 9 | 926291 | 46 |

| No. | Log. | Prop. Part. | No. | Log. | Prop. Part. | No. | Log. | Prop. Part. | No. | Log. | Prop. Part. |
|---|---|---|---|---|---|---|---|---|---|---|---|
| 8440 | 926342 | | 8500 | 929419 | | 8560 | 932474 | | 8620 | 935507 | |
| 1 | 926394 | 5 | 1 | 929470 | 5 | 1 | 932524 | 5 | 1 | 935558 | 5 |
| 2 | 926445 | 10 | 2 | 929521 | 10 | 2 | 932575 | 10 | 2 | 935608 | 10 |
| 3 | 926497 | 15 | 3 | 929572 | 15 | 3 | 932626 | 15 | 3 | 935658 | 15 |
| 4 | 926548 | 21 | 4 | 929623 | 20 | 4 | 932677 | 20 | 4 | 935709 | 20 |
| 5 | 926600 | 26 | 5 | 929674 | 26 | 5 | 932727 | 25 | 5 | 935759 | 25 |
| 6 | 926651 | 31 | 6 | 929725 | 31 | 6 | 932778 | 30 | 6 | 935809 | 30 |
| 7 | 926702 | 36 | 7 | 929776 | 36 | 7 | 932829 | 35 | 7 | 935860 | 35 |
| 8 | 926754 | 41 | 8 | 929827 | 41 | 8 | 932879 | 40 | 8 | 935910 | 40 |
| 9 | 926805 | 46 | 9 | 929878 | 46 | 9 | 932930 | 45 | 9 | 935960 | 45 |
| 8450 | 926857 | | 8510 | 929930 | | 8570 | 932981 | | 8630 | 936011 | |
| 1 | 926908 | 5 | 1 | 929981 | 5 | 1 | 933031 | 5 | 1 | 936061 | 5 |
| 2 | 926959 | 10 | 2 | 930032 | 10 | 2 | 933082 | 10 | 2 | 936111 | 10 |
| 3 | 927011 | 15 | 3 | 930083 | 15 | 3 | 933133 | 15 | 3 | 936162 | 15 |
| 4 | 927062 | 21 | 4 | 930134 | 20 | 4 | 933183 | 20 | 4 | 936212 | 20 |
| 5 | 927114 | 26 | 5 | 930185 | 26 | 5 | 933234 | 25 | 5 | 936262 | 25 |
| 6 | 927165 | 31 | 6 | 930236 | 31 | 6 | 933285 | 30 | 6 | 936313 | 30 |
| 7 | 927216 | 36 | 7 | 930287 | 36 | 7 | 933335 | 35 | 7 | 936363 | 35 |
| 8 | 927268 | 41 | 8 | 930338 | 41 | 8 | 933386 | 40 | 8 | 936413 | 40 |
| 9 | 927319 | 46 | 9 | 930389 | 46 | 9 | 933437 | 45 | 9 | 936463 | 45 |
| 8460 | 927370 | | 8520 | 930440 | | 8580 | 933487 | | 8640 | 936514 | |
| 1 | 927422 | 5 | 1 | 930491 | 5 | 1 | 933538 | 5 | 1 | 936564 | 5 |
| 2 | 927473 | 10 | 2 | 930541 | 10 | 2 | 933588 | 10 | 2 | 936614 | 10 |
| 3 | 927524 | 15 | 3 | 930592 | 15 | 3 | 933639 | 15 | 3 | 936664 | 15 |
| 4 | 927576 | 21 | 4 | 930643 | 20 | 4 | 933690 | 20 | 4 | 936715 | 20 |
| 5 | 927627 | 26 | 5 | 930694 | 25 | 5 | 933740 | 25 | 5 | 936765 | 25 |
| 6 | 927678 | 31 | 6 | 930745 | 31 | 6 | 933791 | 30 | 6 | 936815 | 30 |
| 7 | 927730 | 36 | 7 | 930796 | 36 | 7 | 933841 | 35 | 7 | 936865 | 35 |
| 8 | 927781 | 41 | 8 | 930847 | 41 | 8 | 933892 | 40 | 8 | 936916 | 40 |
| 9 | 927832 | 46 | 9 | 930898 | 46 | 9 | 933943 | 45 | 9 | 936966 | 45 |
| 8470 | 927883 | | 8530 | 930949 | | 8590 | 933993 | | 8650 | 937016 | |
| 1 | 927935 | 5 | 1 | 931000 | 5 | 1 | 934044 | 5 | 1 | 937066 | 5 |
| 2 | 927986 | 10 | 2 | 931051 | 10 | 2 | 934094 | 10 | 2 | 937116 | 10 |
| 3 | 928037 | 15 | 3 | 931102 | 15 | 3 | 934145 | 15 | 3 | 937167 | 15 |
| 4 | 928088 | 21 | 4 | 931153 | 20 | 4 | 934195 | 20 | 4 | 937217 | 20 |
| 5 | 928140 | 26 | 5 | 931203 | 25 | 5 | 934246 | 25 | 5 | 937267 | 25 |
| 6 | 928191 | 31 | 6 | 931254 | 31 | 6 | 934296 | 30 | 6 | 937317 | 30 |
| 7 | 928242 | 36 | 7 | 931305 | 36 | 7 | 934347 | 35 | 7 | 937367 | 35 |
| 8 | 928293 | 41 | 8 | 931356 | 41 | 8 | 934397 | 40 | 8 | 937418 | 40 |
| 9 | 928345 | 46 | 9 | 931407 | 46 | 9 | 934448 | 45 | 9 | 937468 | 45 |
| 8480 | 928396 | | 8540 | 931458 | | 8600 | 934498 | | 8660 | 937518 | |
| 1 | 928447 | 5 | 1 | 931509 | 5 | 1 | 934549 | 5 | 1 | 937568 | 5 |
| 2 | 928498 | 10 | 2 | 931560 | 10 | 2 | 934599 | 10 | 2 | 937618 | 10 |
| 3 | 928549 | 15 | 3 | 931610 | 15 | 3 | 934650 | 15 | 3 | 937668 | 15 |
| 4 | 928601 | 21 | 4 | 931661 | 20 | 4 | 934700 | 20 | 4 | 937718 | 20 |
| 5 | 928652 | 26 | 5 | 931712 | 25 | 5 | 934751 | 25 | 5 | 937769 | 25 |
| 6 | 928703 | 31 | 6 | 931763 | 31 | 6 | 934801 | 30 | 6 | 937819 | 30 |
| 7 | 928754 | 36 | 7 | 931814 | 36 | 7 | 934852 | 35 | 7 | 937869 | 35 |
| 8 | 928805 | 41 | 8 | 931864 | 41 | 8 | 934902 | 40 | 8 | 937919 | 40 |
| 9 | 928856 | 46 | 9 | 931915 | 46 | 9 | 934953 | 45 | 9 | 937969 | 45 |
| 8490 | 928908 | | 8550 | 931966 | | 8610 | 935003 | | 8670 | 938019 | |
| 1 | 928959 | 5 | 1 | 932017 | 5 | 1 | 935054 | 5 | 1 | 938069 | 5 |
| 2 | 929010 | 10 | 2 | 932068 | 10 | 2 | 935104 | 10 | 2 | 938119 | 10 |
| 3 | 929061 | 15 | 3 | 932118 | 15 | 3 | 935154 | 15 | 3 | 938169 | 15 |
| 4 | 929112 | 20 | 4 | 932169 | 20 | 4 | 935205 | 20 | 4 | 938219 | 20 |
| 5 | 929163 | 26 | 5 | 932220 | 25 | 5 | 935255 | 25 | 5 | 938269 | 25 |
| 6 | 929214 | 31 | 6 | 932271 | 30 | 6 | 935306 | 30 | 6 | 938319 | 30 |
| 7 | 929266 | 36 | 7 | 932321 | 35 | 7 | 935356 | 35 | 7 | 938370 | 35 |
| 8 | 929317 | 41 | 8 | 932372 | 40 | 8 | 935406 | 40 | 8 | 938420 | 40 |
| 9 | 929368 | 46 | 9 | 932423 | 45 | 9 | 935457 | 45 | 9 | 938470 | 45 |

| No. | Log. | Prop. Part. | No. | Log. | Prop. Part. | No. | Log. | Prop. Part. | No. | Log. | Prop. Part. |
|---|---|---|---|---|---|---|---|---|---|---|---|
| 8680 | 938520 | | 8740 | 941511 | | 8800 | 944483 | | 8860 | 947434 | |
| 1 | 938570 | 5 | 1 | 941561 | 5 | 1 | 944532 | 5 | 1 | 947483 | 5 |
| 2 | 938620 | 10 | 2 | 941611 | 10 | 2 | 944581 | 10 | 2 | 947532 | 10 |
| 3 | 938670 | 15 | 3 | 941660 | 15 | 3 | 944631 | 15 | 3 | 947581 | 15 |
| 4 | 938720 | 20 | 4 | 941710 | 20 | 4 | 944680 | 20 | 4 | 947630 | 20 |
| 5 | 938770 | 25 | 5 | 941760 | 25 | 5 | 944729 | 25 | 5 | 947679 | 25 |
| 6 | 938820 | 30 | 6 | 941809 | 30 | 6 | 944779 | 30 | 6 | 947728 | 29 |
| 7 | 938870 | 35 | 7 | 941859 | 35 | 7 | 944828 | 35 | 7 | 947777 | 34 |
| 8 | 938920 | 40 | 8 | 941909 | 40 | 8 | 944877 | 40 | 8 | 947826 | 39 |
| 9 | 938970 | 45 | 9 | 941958 | 45 | 9 | 944927 | 45 | 9 | 947875 | 44 |
| 8690 | 939020 | | 8750 | 942008 | | 8810 | 944976 | | 8870 | 947924 | |
| 1 | 939070 | 5 | 1 | 942058 | 5 | 1 | 945025 | 5 | 1 | 947973 | 5 |
| 2 | 939120 | 10 | 2 | 942107 | 10 | 2 | 945074 | 10 | 2 | 948021 | 10 |
| 3 | 939170 | 15 | 3 | 942157 | 15 | 3 | 945124 | 15 | 3 | 948070 | 15 |
| 4 | 939220 | 20 | 4 | 942206 | 20 | 4 | 945173 | 20 | 4 | 948119 | 20 |
| 5 | 939270 | 25 | 5 | 942256 | 25 | 5 | 945222 | 25 | 5 | 948168 | 25 |
| 6 | 939319 | 30 | 6 | 942306 | 30 | 6 | 945272 | 30 | 6 | 948217 | 29 |
| 7 | 939369 | 35 | 7 | 942355 | 35 | 7 | 945321 | 35 | 7 | 948266 | 34 |
| 8 | 939419 | 40 | 8 | 942405 | 40 | 8 | 945370 | 40 | 8 | 948315 | 39 |
| 9 | 939469 | 45 | 9 | 942454 | 45 | 9 | 945419 | 45 | 9 | 948364 | 44 |
| 8700 | 939519 | | 8760 | 942504 | | 8820 | 945469 | | 8880 | 948413 | |
| 1 | 939569 | 5 | 1 | 942554 | 5 | 1 | 945518 | 5 | 1 | 948462 | 5 |
| 2 | 939619 | 10 | 2 | 942603 | 10 | 2 | 945567 | 10 | 2 | 948511 | 10 |
| 3 | 939669 | 15 | 3 | 942653 | 15 | 3 | 945616 | 15 | 3 | 948560 | 15 |
| 4 | 939719 | 20 | 4 | 942702 | 20 | 4 | 945665 | 20 | 4 | 948608 | 20 |
| 5 | 939769 | 25 | 5 | 942752 | 25 | 5 | 945715 | 25 | 5 | 948657 | 25 |
| 6 | 939819 | 30 | 6 | 942801 | 30 | 6 | 945764 | 29 | 6 | 948706 | 29 |
| 7 | 939868 | 35 | 7 | 942851 | 35 | 7 | 945813 | 34 | 7 | 948755 | 34 |
| 8 | 939918 | 40 | 8 | 942900 | 40 | 8 | 945862 | 39 | 8 | 948804 | 39 |
| 9 | 939968 | 45 | 9 | 942950 | 45 | 9 | 945911 | 44 | 9 | 948853 | 44 |
| 8710 | 940018 | | 8770 | 943000 | | 8830 | 945961 | | 8890 | 948902 | |
| 1 | 940068 | 5 | 1 | 943049 | 5 | 1 | 946010 | 5 | 1 | 948951 | 5 |
| 2 | 940118 | 10 | 2 | 943099 | 10 | 2 | 946059 | 10 | 2 | 948999 | 10 |
| 3 | 940168 | 15 | 3 | 943148 | 15 | 3 | 946108 | 15 | 3 | 949048 | 15 |
| 4 | 940218 | 20 | 4 | 943198 | 20 | 4 | 946157 | 20 | 4 | 949097 | 20 |
| 5 | 940267 | 25 | 5 | 943247 | 25 | 5 | 946207 | 25 | 5 | 949146 | 25 |
| 6 | 940317 | 30 | 6 | 943297 | 30 | 6 | 946256 | 29 | 6 | 949195 | 29 |
| 7 | 940367 | 35 | 7 | 943346 | 35 | 7 | 946305 | 34 | 7 | 949244 | 34 |
| 8 | 940417 | 40 | 8 | 943396 | 40 | 8 | 946354 | 39 | 8 | 949292 | 39 |
| 9 | 940467 | 45 | 9 | 943445 | 45 | 9 | 946403 | 44 | 9 | 010341 | 44 |
| 8720 | 940516 | | 8780 | 943494 | | 8840 | 946452 | | 8900 | 949390 | |
| 1 | 940566 | 5 | 1 | 943544 | 5 | 1 | 946501 | 5 | 1 | 949439 | 5 |
| 2 | 940616 | 10 | 2 | 943593 | 10 | 2 | 946550 | 10 | 2 | 949488 | 10 |
| 3 | 940666 | 15 | 3 | 943643 | 15 | 3 | 946600 | 15 | 3 | 949536 | 15 |
| 4 | 940716 | 20 | 4 | 943692 | 20 | 4 | 946649 | 20 | 4 | 949585 | 20 |
| 5 | 940765 | 25 | 5 | 943742 | 25 | 5 | 946698 | 25 | 5 | 949634 | 25 |
| 6 | 940815 | 30 | 6 | 943791 | 30 | 6 | 946747 | 29 | 6 | 949683 | 29 |
| 7 | 940865 | 35 | 7 | 943841 | 35 | 7 | 946796 | 34 | 7 | 949731 | 34 |
| 8 | 940915 | 40 | 8 | 943890 | 40 | 8 | 946845 | 39 | 8 | 949780 | 39 |
| 9 | 940964 | 45 | 9 | 943939 | 45 | 9 | 946894 | 44 | 9 | 949829 | 44 |
| 8730 | 941014 | | 8790 | 943989 | | 8850 | 946943 | | 8910 | 949878 | |
| 1 | 941064 | 5 | 1 | 944038 | 5 | 1 | 946992 | 5 | 1 | 949926 | 5 |
| 2 | 941114 | 10 | 2 | 944088 | 10 | 2 | 947041 | 10 | 2 | 949975 | 10 |
| 3 | 941163 | 15 | 3 | 944137 | 15 | 3 | 947090 | 15 | 3 | 950024 | 15 |
| 4 | 941213 | 20 | 4 | 944186 | 20 | 4 | 947139 | 20 | 4 | 950073 | 20 |
| 5 | 941263 | 25 | 5 | 944236 | 25 | 5 | 947189 | 25 | 5 | 950121 | 25 |
| 6 | 941313 | 30 | 6 | 944285 | 30 | 6 | 947238 | 29 | 6 | 950170 | 29 |
| 7 | 941362 | 35 | 7 | 944335 | 35 | 7 | 947287 | 34 | 7 | 950219 | 34 |
| 8 | 941412 | 40 | 8 | 944384 | 40 | 8 | 947336 | 39 | 8 | 950267 | 39 |
| 9 | 941462 | 45 | 9 | 944433 | 45 | 9 | 947385 | 44 | 9 | 950316 | 44 |

| No. | Log. | Prop. Part. | No. | Log. | Prop. Part. | No. | Log. | Prop. Part. | No. | Log. | Prop. Part. |
|---|---|---|---|---|---|---|---|---|---|---|---|
| 8920 | 950365 | | 8980 | 953276 | | 9040 | 956168 | | 9100 | 959041 | |
| 1 | 950413 | 5 | 1 | 953325 | 5 | 1 | 956216 | 5 | 1 | 959089 | 5 |
| 2 | 950462 | 10 | 2 | 953373 | 10 | 2 | 956264 | 10 | 2 | 959137 | 10 |
| 3 | 950511 | 15 | 3 | 953421 | 15 | 3 | 956312 | 14 | 3 | 959184 | 14 |
| 4 | 950560 | 19 | 4 | 953470 | 19 | 4 | 956361 | 19 | 4 | 959232 | 19 |
| 5 | 950608 | 24 | 5 | 953518 | 24 | 5 | 956409 | 24 | 5 | 959280 | 24 |
| 6 | 950657 | 29 | 6 | 953566 | 29 | 6 | 956457 | 29 | 6 | 959328 | 29 |
| 7 | 950705 | 34 | 7 | 953615 | 34 | 7 | 956505 | 34 | 7 | 959375 | 34 |
| 8 | 950754 | 39 | 8 | 953663 | 39 | 8 | 956553 | 38 | 8 | 959423 | 38 |
| 9 | 950803 | 44 | 9 | 953711 | 44 | 9 | 956601 | 43 | 9 | 959471 | 43 |
| 8930 | 950851 | | 8990 | 953760 | | 9050 | 956649 | | 9110 | 959518 | |
| 1 | 950900 | 5 | 1 | 953808 | 5 | 1 | 956697 | 5 | 1 | 959566 | 5 |
| 2 | 950949 | 10 | 2 | 953856 | 10 | 2 | 956745 | 10 | 2 | 959614 | 10 |
| 3 | 950997 | 15 | 3 | 953905 | 15 | 3 | 956792 | 14 | 3 | 959661 | 14 |
| 4 | 951046 | 19 | 4 | 953953 | 19 | 4 | 956840 | 19 | 4 | 959709 | 19 |
| 5 | 951095 | 24 | 5 | 954001 | 24 | 5 | 956888 | 24 | 5 | 959757 | 24 |
| 6 | 951143 | 29 | 6 | 954049 | 29 | 6 | 956936 | 29 | 6 | 959804 | 29 |
| 7 | 951192 | 34 | 7 | 954098 | 34 | 7 | 956984 | 34 | 7 | 959852 | 34 |
| 8 | 951240 | 39 | 8 | 954146 | 39 | 8 | 957032 | 38 | 8 | 959900 | 38 |
| 9 | 951289 | 44 | 9 | 954194 | 44 | 9 | 957080 | 43 | 9 | 959947 | 43 |
| 8940 | 951337 | | 9000 | 954242 | | 9060 | 957128 | | 9120 | 959995 | |
| 1 | 951386 | 5 | 1 | 954291 | 5 | 1 | 957176 | 5 | 1 | 960042 | 5 |
| 2 | 951435 | 10 | 2 | 954339 | 10 | 2 | 957224 | 10 | 2 | 960090 | 10 |
| 3 | 951483 | 15 | 3 | 954387 | 14 | 3 | 957272 | 14 | 3 | 960138 | 14 |
| 4 | 951532 | 19 | 4 | 954435 | 19 | 4 | 957320 | 19 | 4 | 960185 | 19 |
| 5 | 951580 | 24 | 5 | 954484 | 24 | 5 | 957368 | 24 | 5 | 960233 | 24 |
| 6 | 951629 | 29 | 6 | 954532 | 29 | 6 | 957416 | 29 | 6 | 960280 | 28 |
| 7 | 951677 | 34 | 7 | 954580 | 34 | 7 | 957464 | 34 | 7 | 960328 | 33 |
| 8 | 951726 | 39 | 8 | 954628 | 38 | 8 | 957511 | 38 | 8 | 960376 | 38 |
| 9 | 951774 | 44 | 9 | 954677 | 43 | 9 | 957559 | 43 | 9 | 960423 | 43 |
| 8950 | 951823 | | 9010 | 954725 | | 9070 | 957607 | | 9130 | 960471 | |
| 1 | 951872 | 5 | 1 | 954773 | 5 | 1 | 957655 | 5 | 1 | 960518 | 5 |
| 2 | 951920 | 10 | 2 | 954821 | 10 | 2 | 957703 | 10 | 2 | 960566 | 10 |
| 3 | 951969 | 15 | 3 | 954869 | 14 | 3 | 957751 | 14 | 3 | 960613 | 14 |
| 4 | 952017 | 19 | 4 | 954918 | 19 | 4 | 957799 | 19 | 4 | 960661 | 19 |
| 5 | 952066 | 24 | 5 | 954966 | 24 | 5 | 957847 | 24 | 5 | 960709 | 24 |
| 6 | 952114 | 29 | 6 | 955014 | 29 | 6 | 957894 | 29 | 6 | 960756 | 28 |
| 7 | 952163 | 34 | 7 | 955062 | 34 | 7 | 957942 | 34 | 7 | 960804 | 33 |
| 8 | 952211 | 39 | 8 | 955110 | 38 | 8 | 957990 | 38 | 8 | 960851 | 38 |
| 9 | 952259 | 44 | 9 | 955158 | 43 | 9 | 958038 | 43 | 9 | 960899 | 43 |
| 8960 | 952308 | | 9020 | 955206 | | 9080 | 958086 | | 9140 | 960946 | |
| 1 | 952356 | 5 | 1 | 955255 | 5 | 1 | 958134 | 5 | 1 | 960994 | 5 |
| 2 | 952405 | 10 | 2 | 955303 | 10 | 2 | 958181 | 10 | 2 | 961041 | 10 |
| 3 | 952453 | 15 | 3 | 955351 | 14 | 3 | 958229 | 14 | 3 | 961089 | 14 |
| 4 | 952502 | 19 | 4 | 955399 | 19 | 4 | 958277 | 19 | 4 | 961136 | 19 |
| 5 | 952550 | 24 | 5 | 955447 | 24 | 5 | 958325 | 24 | 5 | 961184 | 24 |
| 6 | 952599 | 29 | 6 | 955495 | 29 | 6 | 958373 | 29 | 6 | 961231 | 28 |
| 7 | 952647 | 34 | 7 | 955543 | 34 | 7 | 958420 | 34 | 7 | 961279 | 33 |
| 8 | 952696 | 39 | 8 | 955592 | 38 | 8 | 958468 | 38 | 8 | 961326 | 38 |
| 9 | 952744 | 44 | 9 | 955640 | 43 | 9 | 958516 | 43 | 9 | 961374 | 43 |
| 8970 | 952792 | | 9030 | 955688 | | 9090 | 958564 | | 9150 | 961421 | |
| 1 | 952841 | 5 | 1 | 955736 | 5 | 1 | 958612 | 5 | 1 | 961469 | 5 |
| 2 | 952889 | 10 | 2 | 955784 | 10 | 2 | 958659 | 10 | 2 | 961516 | 10 |
| 3 | 952938 | 15 | 3 | 955832 | 14 | 3 | 958707 | 14 | 3 | 961563 | 14 |
| 4 | 952986 | 19 | 4 | 955880 | 19 | 4 | 958755 | 19 | 4 | 961611 | 19 |
| 5 | 953034 | 24 | 5 | 955928 | 24 | 5 | 958803 | 24 | 5 | 961658 | 24 |
| 6 | 953083 | 29 | 6 | 955976 | 29 | 6 | 958850 | 29 | 6 | 961706 | 28 |
| 7 | 953131 | 34 | 7 | 956024 | 34 | 7 | 958898 | 34 | 7 | 961753 | 33 |
| 8 | 953180 | 39 | 8 | 956072 | 38 | 8 | 958946 | 38 | 8 | 961801 | 38 |
| 9 | 953228 | 44 | 9 | 956120 | 43 | 9 | 958994 | 43 | 9 | 961848 | 43 |

| No. | Log. | Prop. Part. | No. | Log. | Prop. Part. | No. | Log. | Prop. Part. | No. | Log. | Prop. Part. |
|---|---|---|---|---|---|---|---|---|---|---|---|
| 9160 | 961895 | | 9220 | 964731 | | 9280 | 967548 | | 9340 | 970347 | |
| 1 | 961943 | 5 | 1 | 964778 | 5 | 1 | 967595 | 5 | 1 | 970393 | 5 |
| 2 | 961990 | 10 | 2 | 964825 | 9 | 2 | 967642 | 9 | 2 | 970440 | 9 |
| 3 | 962038 | 14 | 3 | 964872 | 14 | 3 | 967688 | 14 | 3 | 970486 | 14 |
| 4 | 962085 | 19 | 4 | 964919 | 19 | 4 | 967735 | 19 | 4 | 970533 | 19 |
| 5 | 962132 | 24 | 5 | 964966 | 24 | 5 | 967782 | 23 | 5 | 970579 | 23 |
| 6 | 962180 | 28 | 6 | 965013 | 28 | 6 | 967829 | 28 | 6 | 970626 | 28 |
| 7 | 962227 | 33 | 7 | 965060 | 33 | 7 | 967875 | 33 | 7 | 970672 | 33 |
| 8 | 962275 | 38 | 8 | 965108 | 38 | 8 | 967922 | 38 | 8 | 970719 | 37 |
| 9 | 962322 | 48 | 9 | 965155 | 42 | 9 | 967969 | 42 | 9 | 970765 | 42 |
| 9170 | 962369 | | 9230 | 965202 | | 9290 | 968016 | | 9350 | 970812 | |
| 1 | 962417 | 5 | 1 | 965249 | 5 | 1 | 968062 | 5 | 1 | 970858 | 5 |
| 2 | 962464 | 9 | 2 | 965296 | 9 | 2 | 968109 | 9 | 2 | 970904 | 9 |
| 3 | 962511 | 14 | 3 | 965343 | 14 | 3 | 968156 | 14 | 3 | 970951 | 14 |
| 4 | 962559 | 19 | 4 | 965390 | 19 | 4 | 968203 | 19 | 4 | 970997 | 19 |
| 5 | 962606 | 24 | 5 | 965437 | 24 | 5 | 968249 | 23 | 5 | 971044 | 23 |
| 6 | 962653 | 28 | 6 | 965484 | 28 | 6 | 968296 | 28 | 6 | 971090 | 28 |
| 7 | 962701 | 33 | 7 | 965531 | 33 | 7 | 968343 | 33 | 7 | 971137 | 33 |
| 8 | 962748 | 38 | 8 | 965578 | 38 | 8 | 968389 | 38 | 8 | 971183 | 37 |
| 9 | 962795 | 42 | 9 | 965625 | 42 | 9 | 968436 | 42 | 9 | 971229 | 42 |
| 9180 | 962843 | | 9240 | 965672 | | 9300 | 968483 | | 9360 | 971276 | |
| 1 | 962890 | 5 | 1 | 965719 | 5 | 1 | 968530 | 5 | 1 | 971322 | 5 |
| 2 | 962937 | 9 | 2 | 965766 | 9 | 2 | 968576 | 9 | 2 | 971369 | 9 |
| 3 | 962985 | 14 | 3 | 965813 | 14 | 3 | 968623 | 14 | 3 | 971415 | 14 |
| 4 | 963032 | 19 | 4 | 965860 | 19 | 4 | 968670 | 19 | 4 | 971461 | 19 |
| 5 | 963079 | 24 | 5 | 965907 | 24 | 5 | 968716 | 23 | 5 | 971508 | 23 |
| 6 | 963126 | 28 | 6 | 965954 | 28 | 6 | 968763 | 28 | 6 | 971554 | 28 |
| 7 | 963174 | 33 | 7 | 966001 | 33 | 7 | 968810 | 33 | 7 | 971600 | 33 |
| 8 | 963221 | 38 | 8 | 966048 | 38 | 8 | 968856 | 37 | 8 | 971647 | 37 |
| 9 | 963268 | 42 | 9 | 966095 | 42 | 9 | 968903 | 42 | 9 | 971693 | 42 |
| 9190 | 963315 | | 9250 | 966142 | | 9310 | 968950 | | 9370 | 971740 | |
| 1 | 963363 | 5 | 1 | 966189 | 5 | 1 | 968996 | 5 | 1 | 971786 | 5 |
| 2 | 963410 | 9 | 2 | 966236 | 9 | 2 | 969043 | 9 | 2 | 971832 | 9 |
| 3 | 963457 | 14 | 3 | 966283 | 14 | 3 | 969090 | 14 | 3 | 971879 | 14 |
| 4 | 963504 | 19 | 4 | 966329 | 19 | 4 | 969136 | 19 | 4 | 971925 | 19 |
| 5 | 963552 | 24 | 5 | 966376 | 24 | 5 | 969183 | 23 | 5 | 971971 | 23 |
| 6 | 963599 | 28 | 6 | 966423 | 28 | 6 | 969229 | 28 | 6 | 972018 | 28 |
| 7 | 963646 | 33 | 7 | 966470 | 33 | 7 | 969276 | 33 | 7 | 972064 | 33 |
| 8 | 963693 | 38 | 8 | 966517 | 38 | 8 | 969323 | 37 | 8 | 972110 | 37 |
| 9 | 963741 | 42 | 9 | 966564 | 42 | 9 | 969369 | 42 | 9 | 972156 | 12 |
| 9200 | 963788 | | 9260 | 966611 | | 9320 | 969416 | | 9380 | 972203 | |
| 1 | 963835 | 5 | 1 | 966658 | 5 | 1 | 969462 | 5 | 1 | 972249 | 5 |
| 2 | 963882 | 9 | 2 | 966705 | 9 | 2 | 969509 | 9 | 2 | 972295 | 9 |
| 3 | 963929 | 14 | 3 | 966752 | 14 | 3 | 969556 | 14 | 3 | 972342 | 14 |
| 4 | 963977 | 19 | 4 | 966798 | 19 | 4 | 969602 | 19 | 4 | 972388 | 18 |
| 5 | 964024 | 24 | 5 | 966845 | 24 | 5 | 969649 | 23 | 5 | 972434 | 23 |
| 6 | 964071 | 28 | 6 | 966892 | 28 | 6 | 969695 | 28 | 6 | 972480 | 28 |
| 7 | 964118 | 33 | 7 | 966939 | 33 | 7 | 969742 | 33 | 7 | 972527 | 32 |
| 8 | 964165 | 38 | 8 | 966986 | 38 | 8 | 969788 | 37 | 8 | 972573 | 37 |
| 9 | 964212 | 42 | 9 | 967033 | 42 | 9 | 969835 | 42 | 9 | 972619 | 41 |
| 9210 | 964260 | | 9270 | 967080 | | 9330 | 969882 | | 9390 | 972666 | |
| 1 | 964307 | 5 | 1 | 967127 | 5 | 1 | 969928 | 5 | 1 | 972712 | 5 |
| 2 | 964354 | 9 | 2 | 967173 | 9 | 2 | 969975 | 9 | 2 | 972758 | 9 |
| 3 | 964401 | 14 | 3 | 967220 | 14 | 3 | 970021 | 14 | 3 | 972804 | 14 |
| 4 | 964448 | 19 | 4 | 967267 | 19 | 4 | 970068 | 19 | 4 | 972851 | 18 |
| 5 | 964495 | 24 | 5 | 967314 | 24 | 5 | 970114 | 23 | 5 | 972897 | 23 |
| 6 | 964542 | 28 | 6 | 967361 | 28 | 6 | 970161 | 28 | 6 | 972943 | 28 |
| 7 | 964590 | 33 | 7 | 967408 | 33 | 7 | 970207 | 33 | 7 | 972989 | 32 |
| 8 | 964637 | 38 | 8 | 967454 | 38 | 8 | 970254 | 37 | 8 | 973035 | 37 |
| 9 | 964684 | 42 | 9 | 967501 | 42 | 9 | 970300 | 42 | 9 | 973082 | 41 |

| No. | Log. | Prop. Part. | No. | Log. | Prop. Part. | No. | Log. | Prop. Part. | No. | Log. | Prop. Part. |
|---|---|---|---|---|---|---|---|---|---|---|---|
| 9400 | 973128 | | 9460 | 975891 | | 9520 | 978637 | | 9580 | 981365 | |
| 1 | 973174 | 5 | 1 | 975937 | 5 | 1 | 978683 | 5 | 1 | 981411 | 5 |
| 2 | 973220 | 9 | 2 | 975983 | 9 | 2 | 978728 | 9 | 2 | 981456 | 9 |
| 3 | 973266 | 14 | 3 | 976029 | 14 | 3 | 978774 | 14 | 3 | 981501 | 14 |
| 4 | 973313 | 18 | 4 | 976075 | 18 | 4 | 978819 | 18 | 4 | 981547 | 18 |
| 5 | 973359 | 23 | 5 | 976121 | 23 | 5 | 978865 | 23 | 5 | 981592 | 23 |
| 6 | 973405 | 28 | 6 | 976166 | 28 | 6 | 978911 | 27 | 6 | 981637 | 27 |
| 7 | 973451 | 32 | 7 | 976212 | 32 | 7 | 978956 | 32 | 7 | 981683 | 32 |
| 8 | 973497 | 37 | 8 | 976258 | 37 | 8 | 979002 | 36 | 8 | 981728 | 36 |
| 9 | 973543 | 41 | 9 | 976304 | 41 | 9 | 979047 | 41 | 9 | 981773 | 41 |
| 9410 | 973590 | | 9470 | 976350 | | 9530 | 979093 | | 9590 | 981819 | |
| 1 | 973636 | 5 | 1 | 976396 | 5 | 1 | 979138 | 5 | 1 | 981864 | 5 |
| 2 | 973682 | 9 | 2 | 976442 | 9 | 2 | 979184 | 9 | 2 | 981909 | 9 |
| 3 | 973728 | 14 | 3 | 976487 | 14 | 3 | 979230 | 14 | 3 | 981954 | 14 |
| 4 | 973774 | 18 | 4 | 976533 | 18 | 4 | 979275 | 18 | 4 | 982000 | 18 |
| 5 | 973820 | 23 | 5 | 976579 | 23 | 5 | 979321 | 23 | 5 | 982045 | 23 |
| 6 | 973866 | 28 | 6 | 976625 | 28 | 6 | 979366 | 27 | 6 | 982090 | 27 |
| 7 | 973913 | 32 | 7 | 976671 | 32 | 7 | 979412 | 32 | 7 | 982135 | 32 |
| 8 | 973959 | 37 | 8 | 976717 | 37 | 8 | 979457 | 36 | 8 | 982181 | 36 |
| 9 | 974005 | 41 | 9 | 976762 | 41 | 9 | 979503 | 41 | 9 | 982226 | 41 |
| 9420 | 974051 | | 9480 | 976808 | | 9540 | 979548 | | 9600 | 982271 | |
| 1 | 974097 | 5 | 1 | 976854 | 5 | 1 | 979594 | 5 | 1 | 982316 | 5 |
| 2 | 974143 | 9 | 2 | 976900 | 9 | 2 | 979639 | 9 | 2 | 982362 | 9 |
| 3 | 974189 | 14 | 3 | 976946 | 14 | 3 | 979685 | 14 | 3 | 982407 | 14 |
| 4 | 974235 | 18 | 4 | 976991 | 18 | 4 | 979730 | 18 | 4 | 982452 | 18 |
| 5 | 974281 | 23 | 5 | 977037 | 23 | 5 | 979776 | 23 | 5 | 982497 | 23 |
| 6 | 974327 | 28 | 6 | 977083 | 27 | 6 | 979821 | 27 | 6 | 982543 | 27 |
| 7 | 974373 | 32 | 7 | 977129 | 32 | 7 | 979867 | 32 | 7 | 982588 | 32 |
| 8 | 974420 | 37 | 8 | 977175 | 37 | 8 | 979912 | 36 | 8 | 982633 | 36 |
| 9 | 974466 | 41 | 9 | 977220 | 41 | 9 | 979958 | 41 | 9 | 982678 | 41 |
| 9430 | 974512 | | 9490 | 977266 | | 9550 | 980003 | | 9610 | 982723 | |
| 1 | 974558 | 5 | 1 | 977312 | 5 | 1 | 980049 | 5 | 1 | 982769 | 5 |
| 2 | 974604 | 9 | 2 | 977358 | 9 | 2 | 980094 | 9 | 2 | 982814 | 9 |
| 3 | 974650 | 14 | 3 | 977403 | 14 | 3 | 980140 | 14 | 3 | 982859 | 14 |
| 4 | 974696 | 18 | 4 | 977449 | 18 | 4 | 980185 | 18 | 4 | 982904 | 18 |
| 5 | 974742 | 23 | 5 | 977495 | 23 | 5 | 980231 | 23 | 5 | 982949 | 23 |
| 6 | 974788 | 28 | 6 | 977541 | 27 | 6 | 980276 | 27 | 6 | 982994 | 27 |
| 7 | 974834 | 32 | 7 | 977586 | 32 | 7 | 980322 | 32 | 7 | 983040 | 32 |
| 8 | 974880 | 37 | 8 | 977632 | 37 | 8 | 980367 | 36 | 8 | 983085 | 36 |
| 9 | 974926 | 41 | 9 | 977678 | 41 | 9 | 980412 | 41 | 9 | 983130 | 41 |
| 9440 | 974972 | | 9500 | 977724 | | 9560 | 980458 | | 9620 | 983175 | |
| 1 | 975018 | 5 | 1 | 977769 | 5 | 1 | 980503 | 5 | 1 | 983220 | 5 |
| 2 | 975064 | 9 | 2 | 977815 | 9 | 2 | 980549 | 9 | 2 | 983265 | 9 |
| 3 | 975110 | 14 | 3 | 977861 | 14 | 3 | 980594 | 14 | 3 | 983310 | 14 |
| 4 | 975156 | 18 | 4 | 977906 | 18 | 4 | 980640 | 18 | 4 | 983356 | 18 |
| 5 | 975202 | 23 | 5 | 977952 | 23 | 5 | 980685 | 23 | 5 | 983401 | 23 |
| 6 | 975248 | 28 | 6 | 977998 | 27 | 6 | 980730 | 27 | 6 | 983446 | 27 |
| 7 | 975294 | 32 | 7 | 978043 | 32 | 7 | 980776 | 32 | 7 | 983491 | 32 |
| 8 | 975340 | 37 | 8 | 978089 | 37 | 8 | 980821 | 36 | 8 | 983536 | 36 |
| 9 | 975386 | 41 | 9 | 978135 | 41 | 9 | 980867 | 41 | 9 | 983581 | 41 |
| 9450 | 975432 | | 9510 | 978180 | | 9570 | 980912 | | 9630 | 983626 | |
| 1 | 975478 | 5 | 1 | 978226 | 5 | 1 | 980957 | 5 | 1 | 983671 | 5 |
| 2 | 975524 | 9 | 2 | 978272 | 9 | 2 | 981003 | 9 | 2 | 983716 | 9 |
| 3 | 975570 | 14 | 3 | 978317 | 14 | 3 | 981048 | 14 | 3 | 983762 | 14 |
| 4 | 975616 | 18 | 4 | 978363 | 18 | 4 | 981093 | 18 | 4 | 983807 | 18 |
| 5 | 975661 | 23 | 5 | 978409 | 23 | 5 | 981139 | 23 | 5 | 983852 | 23 |
| 6 | 975707 | 28 | 6 | 978454 | 27 | 6 | 981184 | 27 | 6 | 983897 | 27 |
| 7 | 975753 | 32 | 7 | 978500 | 32 | 7 | 981229 | 32 | 7 | 983942 | 32 |
| 8 | 975799 | 37 | 8 | 978546 | 37 | 8 | 981275 | 36 | 8 | 983987 | 36 |
| 9 | 975845 | 41 | 9 | 978591 | 41 | 9 | 981320 | 41 | 9 | 984032 | 41 |

| No. | Log. | Prop. Part. | No. | Log. | Prop. Part. | No. | Log. | Prop. Part. | No. | Log. | Prop. Part. |
|---|---|---|---|---|---|---|---|---|---|---|---|
| 9640 | 984077 | | 9700 | 986772 | | 9760 | 989450 | | 9820 | 992111 | |
| 1 | 984122 | 5 | 1 | 986816 | 4 | 1 | 989494 | 4 | 1 | 992156 | 4 |
| 2 | 984167 | 9 | 2 | 986861 | 9 | 2 | 989539 | 9 | 2 | 992200 | 9 |
| 3 | 984212 | 14 | 3 | 986906 | 13 | 3 | 989583 | 13 | 3 | 992244 | 13 |
| 4 | 984257 | 18 | 4 | 986951 | 18 | 4 | 989628 | 18 | 4 | 992288 | 18 |
| 5 | 984302 | 23 | 5 | 986995 | 22 | 5 | 989672 | 22 | 5 | 992333 | 22 |
| 6 | 984347 | 27 | 6 | 987040 | 27 | 6 | 989717 | 27 | 6 | 992377 | 26 |
| 7 | 984392 | 32 | 7 | 987085 | 31 | 7 | 989761 | 31 | 7 | 992421 | 31 |
| 8 | 984437 | 36 | 8 | 987130 | 36 | 8 | 989806 | 36 | 8 | 992465 | 35 |
| 9 | 984482 | 41 | 9 | 987174 | 40 | 9 | 989850 | 40 | 9 | 992509 | 40 |
| 9650 | 984527 | | 9710 | 987219 | | 9770 | 989895 | | 9830 | 992553 | |
| 1 | 984572 | 5 | 1 | 987264 | 4 | 1 | 989939 | 4 | 1 | 992598 | 4 |
| 2 | 984617 | 9 | 2 | 987309 | 9 | 2 | 989983 | 9 | 2 | 992642 | 9 |
| 3 | 984662 | 14 | 3 | 987353 | 13 | 3 | 990028 | 13 | 3 | 992686 | 13 |
| 4 | 984707 | 18 | 4 | 987398 | 18 | 4 | 990072 | 18 | 4 | 992730 | 18 |
| 5 | 984752 | 23 | 5 | 987443 | 22 | 5 | 990117 | 22 | 5 | 992774 | 22 |
| 6 | 984797 | 27 | 6 | 987487 | 27 | 6 | 990161 | 27 | 6 | 992818 | 26 |
| 7 | 984842 | 32 | 7 | 987532 | 31 | 7 | 990206 | 31 | 7 | 992863 | 31 |
| 8 | 984887 | 36 | 8 | 987577 | 36 | 8 | 990250 | 36 | 8 | 992907 | 35 |
| 9 | 984932 | 41 | 9 | 987622 | 40 | 9 | 990294 | 40 | 9 | 992951 | 40 |
| 9660 | 984977 | | 9720 | 987666 | | 9780 | 990339 | | 9840 | 992995 | |
| 1 | 985022 | 5 | 1 | 987711 | 4 | 1 | 990383 | 4 | 1 | 993039 | 4 |
| 2 | 985067 | 9 | 2 | 987756 | 9 | 2 | 990428 | 9 | 2 | 993083 | 9 |
| 3 | 985112 | 14 | 3 | 987800 | 13 | 3 | 990472 | 13 | 3 | 993127 | 13 |
| 4 | 985157 | 18 | 4 | 987845 | 18 | 4 | 990516 | 18 | 4 | 993172 | 18 |
| 5 | 985202 | 23 | 5 | 987890 | 22 | 5 | 990561 | 22 | 5 | 993216 | 22 |
| 6 | 985247 | 27 | 6 | 987934 | 27 | 6 | 990605 | 27 | 6 | 993260 | 26 |
| 7 | 985292 | 32 | 7 | 987979 | 31 | 7 | 990650 | 31 | 7 | 993304 | 31 |
| 8 | 985337 | 36 | 8 | 988024 | 36 | 8 | 990694 | 36 | 8 | 993348 | 35 |
| 9 | 985382 | 41 | 9 | 988068 | 40 | 9 | 990738 | 40 | 9 | 993392 | 40 |
| 9670 | 985426 | | 9730 | 988113 | | 9790 | 990783 | | 9850 | 993436 | |
| 1 | 985471 | 4 | 1 | 988157 | 4 | 1 | 990827 | 4 | 1 | 993480 | 4 |
| 2 | 985516 | 9 | 2 | 988202 | 9 | 2 | 990871 | 9 | 2 | 993524 | 9 |
| 3 | 985561 | 13 | 3 | 988247 | 13 | 3 | 990916 | 13 | 3 | 993568 | 13 |
| 4 | 985606 | 18 | 4 | 988291 | 18 | 4 | 990960 | 18 | 4 | 993613 | 18 |
| 5 | 985651 | 22 | 5 | 988336 | 22 | 5 | 991004 | 22 | 5 | 993657 | 22 |
| 6 | 985696 | 27 | 6 | 988381 | 27 | 6 | 991049 | 27 | 6 | 993701 | 26 |
| 7 | 985741 | 31 | 7 | 988425 | 31 | 7 | 991093 | 31 | 7 | 993745 | 31 |
| 8 | 985786 | 36 | 8 | 988470 | 36 | 8 | 991137 | 36 | 8 | 993789 | 35 |
| 9 | 985830 | 40 | 9 | 988514 | 40 | 9 | 991182 | 40 | 9 | 993833 | 40 |
| 9680 | 985875 | | 9740 | 988559 | | 9800 | 991226 | | 9860 | 993877 | |
| 1 | 985920 | 4 | 1 | 988603 | 4 | 1 | 991270 | 4 | 1 | 993921 | 4 |
| 2 | 985965 | 9 | 2 | 988648 | 9 | 2 | 991315 | 9 | 2 | 993965 | 9 |
| 3 | 986010 | 13 | 3 | 988693 | 13 | 3 | 991359 | 13 | 3 | 994009 | 13 |
| 4 | 986055 | 18 | 4 | 988737 | 18 | 4 | 991403 | 18 | 4 | 994053 | 18 |
| 5 | 986100 | 22 | 5 | 988782 | 22 | 5 | 991448 | 22 | 5 | 994097 | 22 |
| 6 | 986144 | 27 | 6 | 988826 | 27 | 6 | 991492 | 27 | 6 | 994141 | 26 |
| 7 | 986189 | 31 | 7 | 988871 | 31 | 7 | 991536 | 31 | 7 | 994185 | 31 |
| 8 | 986234 | 36 | 8 | 988915 | 36 | 8 | 991580 | 36 | 8 | 994229 | 35 |
| 9 | 986279 | 40 | 9 | 988960 | 40 | 9 | 991625 | 40 | 9 | 994273 | 40 |
| 9690 | 986324 | | 9750 | 989005 | | 9810 | 991669 | | 9870 | 994317 | |
| 1 | 986369 | 4 | 1 | 989049 | 4 | 1 | 991713 | 4 | 1 | 994361 | 4 |
| 2 | 986413 | 9 | 2 | 989094 | 9 | 2 | 991757 | 9 | 2 | 994405 | 9 |
| 3 | 986458 | 13 | 3 | 989138 | 13 | 3 | 991802 | 13 | 3 | 994449 | 13 |
| 4 | 986503 | 18 | 4 | 989183 | 18 | 4 | 991846 | 18 | 4 | 994493 | 18 |
| 5 | 986548 | 22 | 5 | 989227 | 22 | 5 | 991890 | 22 | 5 | 994537 | 22 |
| 6 | 986593 | 27 | 6 | 989272 | 27 | 6 | 991934 | 27 | 6 | 994581 | 26 |
| 7 | 986637 | 31 | 7 | 989316 | 31 | 7 | 991979 | 31 | 7 | 994625 | 31 |
| 8 | 986682 | 36 | 8 | 989361 | 36 | 8 | 992023 | 36 | 8 | 994669 | 35 |
| 9 | 986727 | 40 | 9 | 989405 | 40 | 9 | 992067 | 40 | 9 | 994713 | 40 |

| No. | Log. | Prop. Part. | No. | Log. | Prop. Part. | No. | Log. | Prop. Part. | No. | Log. | Prop. Part. |
|---|---|---|---|---|---|---|---|---|---|---|---|
| 9880 | 994757 | | 9910 | 996074 | | 9940 | 997386 | | 9970 | 998695 | |
| 1 | 994801 | 4 | 1 | 996117 | 4 | 1 | 997430 | 4 | 1 | 998739 | 4 |
| 2 | 994845 | 9 | 2 | 996161 | 9 | 2 | 997474 | 9 | 2 | 998782 | 9 |
| 3 | 994889 | 13 | 3 | 996205 | 13 | 3 | 997517 | 13 | 3 | 998826 | 13 |
| 4 | 994933 | 18 | 4 | 996249 | 18 | 4 | 997561 | 17 | 4 | 998869 | 17 |
| 5 | 994977 | 22 | 5 | 996293 | 22 | 5 | 997605 | 22 | 5 | 998913 | 22 |
| 6 | 995021 | 26 | 6 | 996336 | 26 | 6 | 997648 | 26 | 6 | 998956 | 26 |
| 7 | 995064 | 31 | 7 | 996380 | 31 | 7 | 997692 | 30 | 7 | 999000 | 30 |
| 8 | 995108 | 35 | 8 | 996424 | 35 | 8 | 997736 | 35 | 8 | 999043 | 35 |
| 9 | 995152 | 40 | 9 | 996468 | 40 | 9 | 997779 | 39 | 9 | 999087 | 39 |
| 9890 | 995196 | | 9920 | 996512 | | 9950 | 997823 | | 9980 | 999130 | |
| 1 | 995240 | 4 | 1 | 996555 | 4 | 1 | 997867 | 4 | 1 | 999174 | 4 |
| 2 | 995284 | 9 | 2 | 996599 | 9 | 2 | 997910 | 9 | 2 | 999218 | 9 |
| 3 | 995328 | 13 | 3 | 996643 | 13 | 3 | 997954 | 13 | 3 | 999261 | 13 |
| 4 | 995372 | 18 | 4 | 996687 | 18 | 4 | 997998 | 17 | 4 | 999305 | 17 |
| 5 | 995416 | 22 | 5 | 996730 | 22 | 5 | 998041 | 22 | 5 | 999348 | 22 |
| 6 | 995460 | 26 | 6 | 996774 | 26 | 6 | 998085 | 26 | 6 | 999392 | 26 |
| 7 | 995504 | 31 | 7 | 996818 | 31 | 7 | 998128 | 30 | 7 | 999435 | 30 |
| 8 | 995547 | 35 | 8 | 996862 | 35 | 8 | 998172 | 35 | 8 | 999478 | 35 |
| 9 | 995591 | 40 | 9 | 996905 | 40 | 9 | 998216 | 39 | 9 | 999522 | 39 |
| 9900 | 995635 | | 9930 | 996949 | | 9960 | 998259 | | 9990 | 999565 | |
| 1 | 995679 | 4 | 1 | 996993 | 4 | 1 | 998303 | 4 | 1 | 999609 | 4 |
| 2 | 995723 | 9 | 2 | 997037 | 9 | 2 | 998346 | 9 | 2 | 999652 | 9 |
| 3 | 995767 | 13 | 3 | 997080 | 13 | 3 | 998390 | 13 | 3 | 999696 | 13 |
| 4 | 995811 | 18 | 4 | 997124 | 18 | 4 | 998434 | 17 | 4 | 999739 | 17 |
| 5 | 995854 | 22 | 5 | 997168 | 22 | 5 | 998477 | 22 | 5 | 999783 | 22 |
| 6 | 995898 | 26 | 6 | 997212 | 26 | 6 | 998521 | 26 | 6 | 999826 | 26 |
| 7 | 995942 | 31 | 7 | 997255 | 31 | 7 | 998564 | 30 | 7 | 999870 | 30 |
| 8 | 995986 | 35 | 8 | 997299 | 35 | 8 | 998608 | 35 | 8 | 999913 | 35 |
| 9 | 996030 | 40 | 9 | 997343 | 39 | 9 | 998652 | 39 | 9 | 999957 | 39 |

| No. | Logarithms to 50 Decimal Places. |
|---|---|
| 1 | 0·00000000000000000000000000000000000000000000000000 |
| 2 | 0·30102999566398119521373889472449302676818988146211 |
| 3 | 0·47712125471966243729502790325511530920012886419069 |
| 4 | 0·60205999132796239042747778944898605353637976292422 |
| 5 | 0·69897000433601880478626110527550697323181011853789 |
| 6 | 0·77815125038364363250876679797960833596831874565280 |
| 7 | 0·84509804001425683071221625859263619348357239632397 |
| 8 | 0·90308998699194358564121668417347908030456964438633 |
| 9 | 0·95424250943932487459005580651023061840025772838139 |
| 10 | 1·00000000000000000000000000000000000000000000000000 |
| 11 | 1·04139268515822504075019997124302424170670219046645 |
| 12 | 1·07918124604762482772250569270410136273650862711491 |
| 13 | 1·11394335230683676920650515794232843082972918838707 |
| 14 | 1·14612803567823802592595551533171292202517622778607 |
| 15 | 1·17609125905568124208128900853062228243193898272859 |
| 16 | 1·20411998265592478085495557889797210707275952584843 |
| 17 | 1·23044892137827392854016989432833703000756737842505 |
| 18 | 1·25527250510330606980379470123472364516844760984350 |
| 19 | 1·27875360095282896153633347575692931795112933739450 |
| 20 | 1·30102999566398119521373889472449302676818988146211 |
| 21 | 1·32221929473391926800724416184775150268370126051866 |
| 22 | 1·34242268082220623596393886596751726847489207192856 |
| 23 | 1·36172783601759287886777711225118954969751103433610 |
| 24 | 1·38021124171160602293624458742859438950469850857702 |
| 25 | 1·39794000867203760957252221055101394646362023707578 |

0 DEG.

| ′ | Sine. | Diff. 100″ | Cosecant. | Tangent. | Diff. 100″ | Cotangent. | Secant. | Cosine. | ′ |
|---|---|---|---|---|---|---|---|---|---|
| 0 | | | Infinite. | | | Infinite. | ·000000 | 10·000000 | 60 |
| 1 | 6·463726 | | 3·536274 | 6·463726 | | 13·536274 | ·000000 | 10·000000 | 59 |
| 2 | 6·764756 | 501717 | 3·235244 | 6·764756 | 501717 | 13·235244 | ·000000 | 10·000000 | 58 |
| 3 | 6·940847 | 293485 | 3·059153 | 6·940847 | 293485 | 13·059153 | ·000000 | 10·000000 | 57 |
| 4 | 7·065786 | 208231 | 2·934214 | 7·065786 | 208231 | 12·934214 | ·000000 | 10·000000 | 56 |
| 5 | 7·162696 | 161517 | 2·837304 | 7·162696 | 161517 | 12·837304 | ·000000 | 10·000000 | 55 |
| 6 | 7·241877 | 131968 | 2·758123 | 7·241878 | 131969 | 12·758122 | ·000001 | 9·999999 | 54 |
| 7 | 7·308824 | 111578 | 2·691176 | 7·308825 | 111578 | 12·691175 | ·000001 | 9·999999 | 53 |
| 8 | 7·366816 | 96653 | 2·633184 | 7·366817 | 96653 | 12·633183 | ·000001 | 9·999999 | 52 |
| 9 | 7·417968 | 85254 | 2·582032 | 7·417970 | 85254 | 12·582030 | ·000001 | 9·999999 | 51 |
| 10 | 7·463726 | 76262 | 2·536274 | 7·463727 | 76263 | 12·536273 | ·000002 | 9·999998 | 50 |
| 11 | 7·505118 | 68988 | 2·494882 | 7·505120 | 68988 | 12·494880 | ·000002 | 9·999998 | 49 |
| 12 | 7·542906 | 62981 | 2·457094 | 7·542909 | 62981 | 12·457091 | ·000003 | 9·999997 | 48 |
| 13 | 7·577668 | 57936 | 2·422332 | 7·577672 | 57937 | 12·422328 | ·000003 | 9·999997 | 47 |
| 14 | 7·609853 | 53641 | 2·390147 | 7·609857 | 53642 | 12·390143 | ·000004 | 9·999996 | 46 |
| 15 | 7·639816 | 49938 | 2·360184 | 7·639820 | 49939 | 12·360180 | ·000004 | 9·999996 | 45 |
| 16 | 7·667845 | 46714 | 2·332155 | 7·667849 | 46715 | 12·332151 | ·000005 | 9·999995 | 44 |
| 17 | 7·694173 | 43881 | 2·305827 | 7·694179 | 43882 | 12·305821 | ·000005 | 9·999995 | 43 |
| 18 | 7·718997 | 41372 | 2·281003 | 7·719003 | 41373 | 12·280997 | ·000006 | 9·999994 | 42 |
| 19 | 7·742478 | 39135 | 2·257522 | 7·742484 | 39136 | 12·257516 | ·000007 | 9·999993 | 41 |
| 20 | 7·764754 | 37127 | 2·235246 | 7·764761 | 37128 | 12·235239 | ·000007 | 9·999993 | 40 |
| 21 | 7·785943 | 35315 | 2·214057 | 7·785951 | 35315 | 12·214049 | ·000008 | 9·999992 | 39 |
| 22 | 7·806146 | 33672 | 2·193854 | 7·806155 | 33673 | 12·193845 | ·000009 | 9·999991 | 38 |
| 23 | 7·825451 | 32175 | 2·174549 | 7·825460 | 32176 | 12·174540 | ·000010 | 9·999990 | 37 |
| 24 | 7·843934 | 30805 | 2·156066 | 7·843944 | 30807 | 12·156056 | ·000011 | 9·999989 | 36 |
| 25 | 7·861662 | 29547 | 2·138338 | 7·861674 | 29549 | 12·138326 | ·000011 | 9·999989 | 35 |
| 26 | 7·878695 | 28388 | 2·121305 | 7·878708 | 28390 | 12·121292 | ·000012 | 9·999988 | 34 |
| 27 | 7·895085 | 27317 | 2·104915 | 7·895099 | 27318 | 12·104901 | ·000013 | 9·999987 | 33 |
| 28 | 7·910879 | 26323 | 2·089121 | 7·910894 | 26325 | 12·089106 | ·000014 | 9·999986 | 32 |
| 29 | 7·926119 | 25399 | 2·073881 | 7·926134 | 25401 | 12·073866 | ·000015 | 9·999985 | 31 |
| 30 | 7·940842 | 24538 | 2·059158 | 7·940858 | 24510 | 12·059142 | ·000017 | 9·999983 | 30 |
| 31 | 7·955082 | 23733 | 2·044918 | 7·955100 | 23735 | 12·044900 | ·000018 | 9·999982 | 29 |
| 32 | 7·968870 | 22980 | 2·031130 | 7·968889 | 22982 | 12·031111 | ·000019 | 9·999981 | 28 |
| 33 | 7·982233 | 22273 | 2·017767 | 7·982253 | 22275 | 12·017747 | ·000020 | 9·999980 | 27 |
| 34 | 7·995198 | 21608 | 2·004802 | 7·995219 | 21610 | 12·004781 | ·000021 | 9·999979 | 26 |
| 35 | 8·007787 | 20981 | 1·992213 | 8·007809 | 20983 | 11·992191 | ·000023 | 9·999977 | 25 |
| 36 | 8·020021 | 20390 | 1·979979 | 8·020045 | 20392 | 11·979955 | ·000024 | 9·999976 | 24 |
| 37 | 8·031919 | 19831 | 1·968081 | 8·031945 | 19833 | 11·968055 | ·000025 | 9·999975 | 23 |
| 38 | 8·043501 | 19302 | 1·956499 | 8·043527 | 19305 | 11·956473 | ·000027 | 9·999973 | 22 |
| 39 | 8·054781 | 18801 | 1·945219 | 8·054809 | 18803 | 11·945191 | ·000028 | 9·999972 | 21 |
| 40 | 8·065776 | 18325 | 1·934224 | 8·065806 | 18327 | 11·934194 | ·000029 | 9·999971 | 20 |
| 41 | 8·076500 | 17872 | 1·923500 | 8·076531 | 17875 | 11·923469 | ·000031 | 9·999969 | 19 |
| 42 | 8·086965 | 17441 | 1·913035 | 8·086997 | 17444 | 11·913003 | ·000032 | 9·999968 | 18 |
| 43 | 8·097183 | 17031 | 1·902817 | 8·097217 | 17034 | 11·902783 | ·000034 | 9·999966 | 17 |
| 44 | 8·107167 | 16639 | 1·892833 | 8·107202 | 16642 | 11·892798 | ·000036 | 9·999964 | 16 |
| 45 | 8·116926 | 16265 | 1·883074 | 8·116963 | 16268 | 11·883037 | ·000037 | 9·999963 | 15 |
| 46 | 8·126471 | 15908 | 1·873529 | 8·126510 | 15911 | 11·873490 | ·000039 | 9·999961 | 14 |
| 47 | 8·135810 | 15566 | 1·864190 | 8·135851 | 15568 | 11·864149 | ·000041 | 9·999959 | 13 |
| 48 | 8·144953 | 15238 | 1·855047 | 8·144996 | 15241 | 11·855004 | ·000042 | 9·999958 | 12 |
| 49 | 8·153907 | 14924 | 1·846093 | 8·153952 | 14927 | 11·846048 | ·000044 | 9·999956 | 11 |
| 50 | 8·162681 | 14622 | 1·837319 | 8·162727 | 14625 | 11·837273 | ·000046 | 9·999954 | 10 |
| 51 | 8·171280 | 14333 | 1·828720 | 8·171328 | 14336 | 11·828672 | ·000048 | 9·999952 | 9 |
| 52 | 8·179713 | 14054 | 1·820287 | 8·179763 | 14057 | 11·820237 | ·000050 | 9·999950 | 8 |
| 53 | 8·187985 | 13786 | 1·812015 | 8·188036 | 13790 | 11·811964 | ·000052 | 9·999948 | 7 |
| 54 | 8·196102 | 13529 | 1·803898 | 8·196156 | 13532 | 11·803844 | ·000054 | 9·999946 | 6 |
| 55 | 8·204070 | 13280 | 1·795930 | 8·204126 | 13284 | 11·795874 | ·000056 | 9·999944 | 5 |
| 56 | 8·211895 | 13041 | 1·788105 | 8·211953 | 13044 | 11·788047 | ·000058 | 9·999942 | 4 |
| 57 | 8·219581 | 12810 | 1·780419 | 8·219641 | 12814 | 11·780359 | ·000060 | 9·999940 | 3 |
| 58 | 8·227134 | 12587 | 1·772866 | 8·227195 | 12591 | 11·772805 | ·000062 | 9·999938 | 2 |
| 59 | 8·234557 | 12372 | 1·765443 | 8·234621 | 12376 | 11·765379 | ·000064 | 9·999936 | 1 |
| 60 | 8·241855 | 12164 | 1·758145 | 8·241922 | 12168 | 11·758078 | ·000066 | 9·999934 | 0 |
| ′ | Cosine. | | Secant. | Cotangent. | | Tangent. | Cosecant. | Sine. | ′ |

89 DEG.

1 DEG.

| ′ | Sine. | Diff. 100″ | Cosecant. | Tangent. | Diff. 100″ | Cotangent. | Secant. | Cosine. | ′ |
|---|---|---|---|---|---|---|---|---|---|
| 0 | 8·241855 | | 1·758145 | 8·241921 | | 11·758079 | ·000066 | 9·999934 | 60 |
| 1 | 8·249033 | 11963 | 1·750967 | 8·249102 | 11967 | 11·750898 | ·000068 | 9·999932 | 59 |
| 2 | 8·256094 | 11768 | 1·743906 | 8·256165 | 11772 | 11·743835 | ·000071 | 9·999929 | 58 |
| 3 | 8·263042 | 11580 | 1·736958 | 8·263115 | 11584 | 11·736885 | ·000073 | 9·999927 | 57 |
| 4 | 8·269881 | 11397 | 1·730119 | 8·269956 | 11402 | 11·730044 | ·000075 | 9·999925 | 56 |
| 5 | 8·276614 | 11221 | 1·723386 | 8·276691 | 11225 | 11·723309 | ·000078 | 9·999922 | 55 |
| 6 | 8·283243 | 11050 | 1·716757 | 8·283323 | 11054 | 11·716677 | ·000080 | 9·999920 | 54 |
| 7 | 8·289773 | 10883 | 1·710227 | 8·289856 | 10887 | 11·710144 | ·000082 | 9·999918 | 53 |
| 8 | 8·296207 | 10722 | 1·703793 | 8·296292 | 10726 | 11·703708 | ·000085 | 9·999915 | 52 |
| 9 | 8·302546 | 10565 | 1·697454 | 8·302634 | 10570 | 11·697366 | ·000087 | 9·999913 | 51 |
| 10 | 8·308794 | 10413 | 1·691206 | 8·308884 | 10418 | 11·691116 | ·000090 | 9·999910 | 50 |
| 11 | 8·314954 | 10266 | 1·685046 | 8·315046 | 10270 | 11·684954 | ·000093 | 9·999907 | 49 |
| 12 | 8·321027 | 10122 | 1·678973 | 8·321122 | 10126 | 11·678878 | ·000095 | 9·999905 | 48 |
| 13 | 8·327016 | 9982 | 1·672984 | 8·327114 | 9987 | 11·672886 | ·000098 | 9·999902 | 47 |
| 14 | 8·332924 | 9847 | 1·667076 | 8·333025 | 9851 | 11·666975 | ·000101 | 9·999899 | 46 |
| 15 | 8·338753 | 9714 | 1·661247 | 8·338856 | 9719 | 11·661144 | ·000103 | 9·999897 | 45 |
| 16 | 8·344504 | 9586 | 1·655496 | 8·344610 | 9590 | 11·655390 | ·000106 | 9·999894 | 44 |
| 17 | 8·350181 | 9460 | 1·649819 | 8·350289 | 9465 | 11·649711 | ·000109 | 9·999891 | 43 |
| 18 | 8·355783 | 9338 | 1·644217 | 8·355895 | 9343 | 11·644105 | ·000112 | 9·999888 | 42 |
| 19 | 8·361315 | 9219 | 1·638685 | 8·361430 | 9224 | 11·638570 | ·000115 | 9·999885 | 41 |
| 20 | 8·366777 | 9103 | 1·633223 | 8·366895 | 9108 | 11·633105 | ·000118 | 9·999882 | 40 |
| 21 | 8·372171 | 8990 | 1·627829 | 8·372292 | 8995 | 11·627708 | ·000121 | 9·999879 | 39 |
| 22 | 8·377499 | 8880 | 1·622501 | 8·377622 | 8885 | 11·622378 | ·000124 | 9·999876 | 38 |
| 23 | 8·382762 | 8772 | 1·617238 | 8·382889 | 8777 | 11·617111 | ·000127 | 9·999873 | 37 |
| 24 | 8·387962 | 8667 | 1·612038 | 8·388092 | 8672 | 11·611908 | ·000130 | 9·999870 | 36 |
| 25 | 8·393101 | 8564 | 1·606899 | 8·393234 | 8570 | 11·606766 | ·000133 | 9·999867 | 35 |
| 26 | 8·398179 | 8464 | 1·601821 | 8·398315 | 8470 | 11·601685 | ·000136 | 9·999864 | 34 |
| 27 | 8·403199 | 8366 | 1·596801 | 8·403338 | 8371 | 11·596662 | ·000139 | 9·999861 | 33 |
| 28 | 8·408161 | 8271 | 1·591839 | 8·408304 | 8276 | 11·591696 | ·000142 | 9·999858 | 32 |
| 29 | 8·413068 | 8177 | 1·586932 | 8·413213 | 8182 | 11·586787 | ·000146 | 9·999854 | 31 |
| 30 | 8·417919 | 8086 | 1·582081 | 8·418068 | 8091 | 11·581932 | ·000149 | 9·999851 | 30 |
| 31 | 8·422717 | 7996 | 1·577283 | 8·422869 | 8002 | 11·577131 | ·000152 | 9·999848 | 29 |
| 32 | 8·427462 | 7909 | 1·572538 | 8·427618 | 7914 | 11·572382 | ·000156 | 9·999844 | 28 |
| 33 | 8·432156 | 7823 | 1·567844 | 8·432315 | 7829 | 11·567685 | ·000159 | 9·999841 | 27 |
| 34 | 8·436800 | 7740 | 1·563200 | 8·436962 | 7745 | 11·563038 | ·000162 | 9·999838 | 26 |
| 35 | 8·441394 | 7657 | 1·558606 | 8·441560 | 7663 | 11·558440 | ·000166 | 9·999834 | 25 |
| 36 | 8·445941 | 7577 | 1·554059 | 8·446110 | 7583 | 11·553890 | ·000169 | 9·999831 | 24 |
| 37 | 8·450440 | 7499 | 1·549560 | 8·450613 | 7505 | 11·549387 | ·000173 | 9·999827 | 23 |
| 38 | 8·454893 | 7422 | 1·545107 | 8·455070 | 7428 | 11·544930 | ·000176 | 9·999824 | 22 |
| 39 | 8·459301 | 7346 | 1·540699 | 8·459481 | 7352 | 11·540519 | ·000180 | 9·999820 | 21 |
| 40 | 8·463665 | 7273 | 1·536335 | 8·463849 | 7279 | 11·536151 | ·000184 | 9·999816 | 20 |
| 41 | 8·467985 | 7200 | 1·532015 | 8·468172 | 7206 | 11·531828 | ·000187 | 9·999813 | 19 |
| 42 | 8·472263 | 7129 | 1·527737 | 8·472454 | 7135 | 11·527546 | ·000191 | 9·999809 | 18 |
| 43 | 8·476498 | 7060 | 1·523502 | 8·476693 | 7066 | 11·523307 | ·000195 | 9·999805 | 17 |
| 44 | 8·480693 | 6991 | 1·519307 | 8·480892 | 6998 | 11·519108 | ·000199 | 9·999801 | 16 |
| 45 | 8·484848 | 6924 | 1·515152 | 8·485050 | 6931 | 11·514950 | ·000203 | 9·999797 | 15 |
| 46 | 8·488963 | 6859 | 1·511037 | 8·489170 | 6865 | 11·510830 | ·000206 | 9·999794 | 14 |
| 47 | 8·493040 | 6794 | 1·506960 | 8·493250 | 6801 | 11·506750 | ·000210 | 9·999790 | 13 |
| 48 | 8·497078 | 6731 | 1·502922 | 8·497293 | 6738 | 11·502707 | ·000214 | 9·999786 | 12 |
| 49 | 8·501080 | 6669 | 1·498920 | 8·501298 | 6676 | 11·498702 | ·000218 | 9·999782 | 11 |
| 50 | 8·505045 | 6608 | 1·494955 | 8·505267 | 6615 | 11·494733 | ·000222 | 9·999778 | 10 |
| 51 | 8·508974 | 6548 | 1·491026 | 8·509200 | 6555 | 11·490800 | ·000226 | 9·999774 | 9 |
| 52 | 8·512867 | 6489 | 1·487133 | 8·513098 | 6496 | 11·486902 | ·000231 | 9·999769 | 8 |
| 53 | 8·516726 | 6432 | 1·483274 | 8·516961 | 6439 | 11·483039 | ·000235 | 9·999765 | 7 |
| 54 | 8·520551 | 6375 | 1·479449 | 8·520790 | 6382 | 11·479210 | ·000239 | 9·999761 | 6 |
| 55 | 8·524343 | 6319 | 1·475657 | 8·524586 | 6326 | 11·475414 | ·000243 | 9·999757 | 5 |
| 56 | 8·528102 | 6264 | 1·471898 | 8·528349 | 6272 | 11·471651 | ·000247 | 9·999753 | 4 |
| 57 | 8·531828 | 6211 | 1·468172 | 8·532080 | 6218 | 11·467920 | ·000252 | 9·999748 | 3 |
| 58 | 8·535523 | 6158 | 1·464477 | 8·535779 | 6165 | 11·464221 | ·000256 | 9·999744 | 2 |
| 59 | 8·539186 | 6106 | 1·460814 | 8·539447 | 6113 | 11·460553 | ·000260 | 9·999740 | 1 |
| 60 | 8·542819 | 6055 | 1·457181 | 8·543084 | 6062 | 11·456916 | ·000265 | 9·999735 | 0 |
| ′ | Cosine. | | Secant. | Cotangent. | | Tangent. | Cosecant. | Sine. | ′ |

88 DEG.

2 DEG.

| ′ | Sine. | Diff. 100″ | Cosecant. | Tangent. | Diff. 100″ | Cotangent. | Secant. | Diff. 100″ | Cosine. | ′ |
|---|---|---|---|---|---|---|---|---|---|---|
| 0 | 8·542819 | | 1·457181 | 8·543084 | | 11·456916 | ·000265 | | 9·999735 | 60 |
| 1 | 8·546422 | 6004 | 1·453578 | 8·546691 | 6012 | 11·453309 | ·000269 | 07 | 9·999731 | 59 |
| 2 | 8·549995 | 5955 | 1·450005 | 8·550268 | 5962 | 11·449732 | ·000274 | 07 | 9·999726 | 58 |
| 3 | 8·553539 | 5906 | 1·446461 | 8·553817 | 5914 | 11·446183 | ·000278 | 08 | 9·999722 | 57 |
| 4 | 8·557054 | 5858 | 1·442946 | 8·557336 | 5866 | 11·442664 | ·000283 | 08 | 9·999717 | 56 |
| 5 | 8·560540 | 5811 | 1·439460 | 8·560828 | 5819 | 11·439172 | ·000287 | 07 | 9·999713 | 55 |
| 6 | 8·563999 | 5765 | 1·436001 | 8·564291 | 5773 | 11·435709 | ·000292 | 08 | 9·999708 | 54 |
| 7 | 8·567431 | 5719 | 1·432569 | 8·567727 | 5727 | 11·432273 | ·000296 | 07 | 9·999704 | 53 |
| 8 | 8·570836 | 5674 | 1·429164 | 8·571137 | 5682 | 11·428863 | ·000301 | 08 | 9·999699 | 52 |
| 9 | 8·574214 | 5630 | 1·425786 | 8·574520 | 5638 | 11·425480 | ·000306 | 08 | 9·999694 | 51 |
| 10 | 8·577566 | 5587 | 1·422434 | 8·577877 | 5595 | 11·422123 | ·000311 | 08 | 9·999689 | 50 |
| 11 | 8·580892 | 5544 | 1·419108 | 8·581208 | 5552 | 11·418792 | ·000315 | 07 | 9·999685 | 49 |
| 12 | 8·584193 | 5502 | 1·415807 | 8·584514 | 5510 | 11·415486 | ·000320 | 08 | 9·999680 | 48 |
| 13 | 8·587469 | 5460 | 1·412531 | 8·587795 | 5468 | 11·412205 | ·000325 | 08 | 9·999675 | 47 |
| 14 | 8·590721 | 5419 | 1·409279 | 8·591051 | 5427 | 11·408949 | ·000330 | 08 | 9·999670 | 46 |
| 15 | 8·593948 | 5379 | 1·406052 | 8·594283 | 5387 | 11·405717 | ·000335 | 08 | 9·999665 | 45 |
| 16 | 8·597152 | 5339 | 1·402848 | 8·597492 | 5347 | 11·402508 | ·000340 | 08 | 9·999660 | 44 |
| 17 | 8·600332 | 5300 | 1·399668 | 8·600677 | 5308 | 11·399323 | ·000345 | 08 | 9·999655 | 43 |
| 18 | 8·603489 | 5261 | 1·396511 | 8·603839 | 5270 | 11·396161 | ·000350 | 08 | 9·999650 | 42 |
| 19 | 8·606623 | 5223 | 1·393377 | 8·606978 | 5232 | 11·393022 | ·000355 | 08 | 9·999645 | 41 |
| 20 | 8·609734 | 5186 | 1·390266 | 8·610094 | 5194 | 11·389906 | ·000360 | 08 | 9·999640 | 40 |
| 21 | 8·612823 | 5149 | 1·387177 | 8·613189 | 5158 | 11·386811 | ·000365 | 08 | 9·999635 | 39 |
| 22 | 8·615891 | 5112 | 1·384109 | 8·616262 | 5121 | 11·383738 | ·000371 | 10 | 9·999629 | 38 |
| 23 | 8·618937 | 5076 | 1·381063 | 8·619313 | 5085 | 11·380687 | ·000376 | 08 | 9·999624 | 37 |
| 24 | 8·621962 | 5041 | 1·378038 | 8·622343 | 5050 | 11·377657 | ·000381 | 08 | 9·999619 | 36 |
| 25 | 8·624965 | 5006 | 1·375035 | 8·625352 | 5015 | 11·374648 | ·000386 | 08 | 9·999614 | 35 |
| 26 | 8·627948 | 4972 | 1·372052 | 8·628340 | 4981 | 11·371660 | ·000392 | 10 | 9·999608 | 34 |
| 27 | 8·630911 | 4938 | 1·369089 | 8·631308 | 4947 | 11·368692 | ·000397 | 08 | 9·999603 | 33 |
| 28 | 8·633854 | 4904 | 1·366146 | 8·634256 | 4913 | 11·365744 | ·000403 | 10 | 9·999597 | 32 |
| 29 | 8·636776 | 4871 | 1·363224 | 8·637184 | 4880 | 11·362816 | ·000408 | 08 | 9·999592 | 31 |
| 30 | 8·639680 | 4839 | 1·360320 | 8·640093 | 4848 | 11·359907 | ·000414 | 08 | 9·999586 | 30 |
| 31 | 8·642563 | 4806 | 1·357437 | 8·642982 | 4816 | 11·357018 | ·000419 | 10 | 9·999581 | 29 |
| 32 | 8·645428 | 4775 | 1·354572 | 8·645853 | 4784 | 11·354147 | ·000425 | 10 | 9·999575 | 28 |
| 33 | 8·648274 | 4743 | 1·351726 | 8·648704 | 4753 | 11·351296 | ·000430 | 08 | 9·999570 | 27 |
| 34 | 8·651102 | 4712 | 1·348898 | 8·651537 | 4722 | 11·348463 | ·000436 | 10 | 9·999564 | 26 |
| 35 | 8·653911 | 4682 | 1·346089 | 8·654352 | 4691 | 11·345648 | ·000442 | 10 | 9·999558 | 25 |
| 36 | 8·656702 | 4652 | 1·343298 | 8·657149 | 4661 | 11·342851 | ·000447 | 08 | 9·999553 | 24 |
| 37 | 8·659475 | 4622 | 1·340525 | 8·659928 | 4631 | 11·340072 | ·000453 | 10 | 9·999547 | 23 |
| 38 | 8·662230 | 4592 | 1·337770 | 8·662689 | 4602 | 11·337311 | ·000459 | 10 | 9·999541 | 22 |
| 39 | 8·664968 | 4563 | 1·335032 | 8·665433 | 4573 | 11·334567 | ·000465 | 10 | 9·999535 | 21 |
| 40 | 8·667689 | 4535 | 1·332311 | 8·668160 | 4544 | 11·331840 | ·000471 | 10 | 9·999529 | 20 |
| 41 | 8·670393 | 4506 | 1·329607 | 8·670870 | 4516 | 11·329130 | ·000476 | 08 | 9·999524 | 19 |
| 42 | 8·673080 | 4479 | 1·326920 | 8·673563 | 4488 | 11·326437 | ·000482 | 10 | 9·999518 | 18 |
| 43 | 8·675751 | 4451 | 1·324249 | 8·676239 | 4461 | 11·323761 | ·000488 | 10 | 9·999512 | 17 |
| 44 | 8·678405 | 4424 | 1·321595 | 8·678900 | 4434 | 11·321100 | ·000494 | 10 | 9·999506 | 16 |
| 45 | 8·681043 | 4397 | 1·318957 | 8·681544 | 4407 | 11·318456 | ·000500 | 10 | 9·999500 | 15 |
| 46 | 8·683665 | 4370 | 1·316335 | 8·684172 | 4380 | 11·315828 | ·000507 | 10 | 9·999493 | 14 |
| 47 | 8·686272 | 4344 | 1·313728 | 8·686784 | 4354 | 11·313216 | ·000513 | 12 | 9·999487 | 13 |
| 48 | 8·688863 | 4318 | 1·311137 | 8·689381 | 4328 | 11·310619 | ·000519 | 10 | 9·999481 | 12 |
| 49 | 8·691438 | 4292 | 1·308562 | 8·691963 | 4303 | 11·308037 | ·000525 | 10 | 9·999475 | 11 |
| 50 | 8·693998 | 4267 | 1·306002 | 8·694529 | 4277 | 11·305471 | ·000531 | 10 | 9·999469 | 10 |
| 51 | 8·696543 | 4242 | 1·303457 | 8·697081 | 4252 | 11·302919 | ·000537 | 12 | 9·999463 | 9 |
| 52 | 8·699073 | 4217 | 1·300927 | 8·699617 | 4228 | 11·300383 | ·000544 | 10 | 9·999456 | 8 |
| 53 | 8·701589 | 4192 | 1·298411 | 8·702139 | 4203 | 11·297861 | ·000550 | 10 | 9·999450 | 7 |
| 54 | 8·704090 | 4168 | 1·295910 | 8·704646 | 4179 | 11·295354 | ·000557 | 10 | 9·999443 | 6 |
| 55 | 8·706577 | 4144 | 1·293423 | 8·707140 | 4155 | 11·292860 | ·000563 | 12 | 9·999437 | 5 |
| 56 | 8·709049 | 4121 | 1·290951 | 8·709618 | 4132 | 11·290382 | ·000569 | 10 | 9·999431 | 4 |
| 57 | 8·711507 | 4097 | 1·288493 | 8·712083 | 4108 | 11·287917 | ·000576 | 12 | 9·999424 | 3 |
| 58 | 8·713952 | 4074 | 1·286048 | 8·714534 | 4085 | 11·285466 | ·000582 | 10 | 9·999418 | 2 |
| 59 | 8·716383 | 4052 | 1·283617 | 8·716972 | 4062 | 11·283028 | ·000589 | 12 | 9·999411 | 1 |
| 60 | 8·718800 | 4029 | 1·281200 | 8·719396 | 4040 | 11·280604 | ·000596 | 12 | 9·999404 | 0 |
| ′ | Cosine. | | Secant. | Cotangent. | | Tangent. | Cosecant. | | Sine. | ′ |

87 DEG.

3 DEG.

| ′ | Sine. | Diff. 100″ | Cosecant. | Tangent. | Diff. 100″ | Cotangent. | Secant. | Diff. 100″ | Cosine. | ′ |
|---|---|---|---|---|---|---|---|---|---|---|
| 0 | 8·718800 | | 1·281200 | 8·719396 | | 11·280604 | ·000596 | | 9·999404 | 60 |
| 1 | 8·721204 | 4006 | 1·278796 | 8·721806 | 4017 | 11·278194 | ·000602 | 10 | 9·999398 | 59 |
| 2 | 8·723595 | 3984 | 1·276405 | 8·724204 | 3995 | 11·275796 | ·000609 | 12 | 9·999391 | 58 |
| 3 | 8·725972 | 3962 | 1·274028 | 8·726588 | 3974 | 11·273412 | ·000616 | 12 | 9·999384 | 57 |
| 4 | 8·728337 | 3941 | 1·271663 | 8·728959 | 3952 | 11·271041 | ·000622 | 10 | 9·999378 | 56 |
| 5 | 8·730688 | 3919 | 1·269312 | 8·731317 | 3931 | 11·268683 | ·000629 | 12 | 9·999371 | 55 |
| 6 | 8·733027 | 3898 | 1·266973 | 8·733663 | 3909 | 11·266337 | ·000636 | 12 | 9·999364 | 54 |
| 7 | 8·735354 | 3877 | 1·264646 | 8·735996 | 3889 | 11·264004 | ·000643 | 12 | 9·999357 | 53 |
| 8 | 8·737667 | 3857 | 1·262333 | 8·738317 | 3868 | 11·261683 | ·000650 | 12 | 9·999350 | 52 |
| 9 | 8·739969 | 3836 | 1·260031 | 8·740626 | 3848 | 11·259374 | ·000657 | 12 | 9·999343 | 51 |
| 10 | 8·742259 | 3816 | 1·257741 | 8·742922 | 3827 | 11·257078 | ·000664 | 12 | 9·999336 | 50 |
| 11 | 8·744536 | 3796 | 1·255464 | 8·745207 | 3807 | 11·254793 | ·000671 | 12 | 9·999329 | 49 |
| 12 | 8·746802 | 3776 | 1·253198 | 8·747479 | 3787 | 11·252521 | ·000678 | 12 | 9·999322 | 48 |
| 13 | 8·749055 | 3756 | 1·250945 | 8·749740 | 3768 | 11·250260 | ·000685 | 12 | 9·999315 | 47 |
| 14 | 8·751297 | 3737 | 1·248703 | 8·751989 | 3749 | 11·248011 | ·000692 | 12 | 9·999308 | 46 |
| 15 | 8·753528 | 3717 | 1·246472 | 8·754227 | 3729 | 11·245773 | ·000699 | 12 | 9·999301 | 45 |
| 16 | 8·755747 | 3698 | 1·244253 | 8·756453 | 3710 | 11·243547 | ·000706 | 12 | 9·999294 | 44 |
| 17 | 8·757955 | 3679 | 1·242045 | 8·758668 | 3692 | 11·241332 | ·000713 | 13 | 9·999287 | 43 |
| 18 | 8·760151 | 3661 | 1·239849 | 8·760872 | 3673 | 11·239128 | ·000721 | 12 | 9·999279 | 42 |
| 19 | 8·762337 | 3642 | 1·237663 | 8·763065 | 3655 | 11·236935 | ·000728 | 12 | 9·999272 | 41 |
| 20 | 8·764511 | 3624 | 1·235489 | 8·765246 | 3636 | 11·234754 | ·000735 | 12 | 9·999265 | 40 |
| 21 | 8·766675 | 3606 | 1·233325 | 8·767417 | 3618 | 11·232583 | ·000743 | 13 | 9·999257 | 39 |
| 22 | 8·768828 | 3588 | 1·231172 | 8·769578 | 3600 | 11·230422 | ·000750 | 12 | 9·999250 | 38 |
| 23 | 8·770970 | 3570 | 1·229030 | 8·771727 | 3583 | 11·228273 | ·000758 | 13 | 9·999242 | 37 |
| 24 | 8·773101 | 3553 | 1·226899 | 8·773866 | 3565 | 11·226134 | ·000765 | 12 | 9·999235 | 36 |
| 25 | 8·775223 | 3535 | 1·224777 | 8·775995 | 3548 | 11·224005 | ·000773 | 13 | 9·999227 | 35 |
| 26 | 8·777333 | 3518 | 1·222667 | 8·778114 | 3531 | 11·221886 | ·000780 | 12 | 9·999220 | 34 |
| 27 | 8·779434 | 3501 | 1·220566 | 8·780222 | 3514 | 11·219778 | ·000788 | 13 | 9·999212 | 33 |
| 28 | 8·781524 | 3484 | 1·218476 | 8·782320 | 3497 | 11·217680 | ·000795 | 13 | 9·999205 | 32 |
| 29 | 8·783605 | 3467 | 1·216395 | 8·784408 | 3480 | 11·215592 | ·000803 | 13 | 9·999197 | 31 |
| 30 | 8·785675 | 3451 | 1·214325 | 8·786486 | 3464 | 11·213514 | ·000811 | 13 | 9·999189 | 30 |
| 31 | 8·787736 | 3434 | 1·212264 | 8·788554 | 3447 | 11·211446 | ·000819 | 13 | 9·999181 | 29 |
| 32 | 8·789787 | 3418 | 1·210213 | 8·790613 | 3431 | 11·209387 | ·000826 | 12 | 9·999174 | 28 |
| 33 | 8·791828 | 3402 | 1·208172 | 8·792662 | 3415 | 11·207338 | ·000834 | 13 | 9·999166 | 27 |
| 34 | 8·793859 | 3386 | 1·206141 | 8·794701 | 3399 | 11·205299 | ·000842 | 13 | 9·999158 | 26 |
| 35 | 8·795881 | 3370 | 1·204119 | 8·796731 | 3383 | 11·203269 | ·000850 | 13 | 9·999150 | 25 |
| 36 | 8·797894 | 3354 | 1·202106 | 8·798752 | 3368 | 11·201248 | ·000858 | 13 | 9·999142 | 24 |
| 37 | 8·799897 | 3339 | 1·200103 | 8·800763 | 3352 | 11·199237 | ·000866 | 13 | 9·999134 | 23 |
| 38 | 8·801892 | 3323 | 1·198108 | 8·802765 | 3337 | 11·197235 | ·000874 | 13 | 9·999126 | 22 |
| 39 | 8·803876 | 3308 | 1·196124 | 8·804758 | 3322 | 11·195242 | ·000882 | 13 | 9·999118 | 21 |
| 40 | 8·805852 | 3293 | 1·194148 | 8·806742 | 3306 | 11·193258 | ·000890 | 13 | 9·999110 | 20 |
| 41 | 8·807819 | 3278 | 1·192181 | 8·808717 | 3292 | 11·191283 | ·000898 | 13 | 9·999102 | 19 |
| 42 | 8·809777 | 3263 | 1·190223 | 8·810683 | 3277 | 11·189317 | ·000906 | 13 | 9·999094 | 18 |
| 43 | 8·811726 | 3249 | 1·188274 | 8·812641 | 3262 | 11·187359 | ·000914 | 13 | 9·999086 | 17 |
| 44 | 8·813667 | 3234 | 1·186333 | 8·814589 | 3248 | 11·185411 | ·000923 | 15 | 9·999077 | 16 |
| 45 | 8·815599 | 3219 | 1·184401 | 8·816529 | 3233 | 11·183471 | ·000931 | 13 | 9·999069 | 15 |
| 46 | 8·817522 | 3205 | 1·182478 | 8·818461 | 3219 | 11·181539 | ·000939 | 13 | 9·999061 | 14 |
| 47 | 8·819436 | 3191 | 1·180564 | 8·820384 | 3205 | 11·179616 | ·000947 | 15 | 9·999053 | 13 |
| 48 | 8·821343 | 3177 | 1·178657 | 8·822298 | 3191 | 11·177702 | ·000956 | 13 | 9·999044 | 12 |
| 49 | 8·823240 | 3163 | 1·176760 | 8·824205 | 3177 | 11·175795 | ·000964 | 13 | 9·999036 | 11 |
| 50 | 8·825130 | 3149 | 1·174870 | 8·826103 | 3163 | 11·173897 | ·000973 | 15 | 9·999027 | 10 |
| 51 | 8·827011 | 3135 | 1·172989 | 8·827992 | 3150 | 11·172008 | ·000981 | 13 | 9·999019 | 9 |
| 52 | 8·828884 | 3122 | 1·171116 | 8·829874 | 3136 | 11·170126 | ·000990 | 15 | 9·999010 | 8 |
| 53 | 8·830749 | 3108 | 1·169251 | 8·831748 | 3123 | 11·168252 | ·000998 | 13 | 9·999002 | 7 |
| 54 | 8·832607 | 3095 | 1·167393 | 8·833613 | 3109 | 11·166387 | ·001007 | 15 | 9·998993 | 6 |
| 55 | 8·834456 | 3082 | 1·165544 | 8·835471 | 3096 | 11·164529 | ·001016 | 15 | 9·998984 | 5 |
| 56 | 8·836297 | 3069 | 1·163703 | 8·837321 | 3083 | 11·162679 | ·001024 | 13 | 9·998976 | 4 |
| 57 | 8·838130 | 3056 | 1·161870 | 8·839163 | 3070 | 11·160837 | ·001033 | 15 | 9·998967 | 3 |
| 58 | 8·839956 | 3043 | 1·160044 | 8·840998 | 3057 | 11·159002 | ·001042 | 15 | 9·998958 | 2 |
| 59 | 8·841774 | 3030 | 1·158226 | 8·842825 | 3045 | 11·157175 | ·001050 | 13 | 9·998950 | 1 |
| 60 | 8·843585 | 3017 | 1·156415 | 8·844644 | 3032 | 11·155356 | ·001059 | 15 | 9·998941 | 0 |
| ′ | Cosine. | | Secant. | Cotangent. | | Tangent. | Cosecant. | | Sine. | ′ |

86 DEG.

4 DEG.

| ′ | Sine. | Diff. 100″ | Cosecant. | Tangent. | Diff. 100″ | Cotangent. | Secant. | Diff. 100″ | Cosine. | ′ |
|---|---|---|---|---|---|---|---|---|---|---|
| 0 | 8·843585 | | 1·156415 | 8·844644 | | 11·155356 | ·001059 | | 9·998941 | 60 |
| 1 | 8·845387 | 3005 | 1·154613 | 8·846455 | 3019 | 11·153545 | ·001068 | 15 | 9·998932 | 59 |
| 2 | 8·847183 | 2992 | 1·152817 | 8·848260 | 3007 | 11·151740 | ·001077 | 15 | 9·998923 | 58 |
| 3 | 8·848971 | 2980 | 1·151029 | 8·850057 | 2995 | 11·149943 | ·001086 | 15 | 9·998914 | 57 |
| 4 | 8·850751 | 2967 | 1·149249 | 8·851846 | 2982 | 11·148154 | ·001095 | 15 | 9·998905 | 56 |
| 5 | 8·852525 | 2955 | 1·147475 | 8·853628 | 2970 | 11·146372 | ·001104 | 15 | 9·998896 | 55 |
| 6 | 8·854291 | 2943 | 1·145709 | 8·855403 | 2958 | 11·144597 | ·001113 | 15 | 9·998887 | 54 |
| 7 | 8·856049 | 2931 | 1·143951 | 8·857171 | 2946 | 11·142829 | ·001122 | 15 | 9·998878 | 53 |
| 8 | 8·857801 | 2919 | 1·142199 | 8·858932 | 2935 | 11·141068 | ·001131 | 15 | 9·998869 | 52 |
| 9 | 8·859546 | 2908 | 1·140454 | 8·860686 | 2923 | 11·139314 | ·001140 | 15 | 9·998860 | 51 |
| 10 | 8·861283 | 2896 | 1·138717 | 8·862433 | 2911 | 11·137567 | ·001149 | 15 | 9·998851 | 50 |
| 11 | 8·863014 | 2884 | 1·136986 | 8·864173 | 2900 | 11·135827 | ·001159 | 17 | 9·998841 | 49 |
| 12 | 8·864738 | 2873 | 1·135262 | 8·865906 | 2888 | 11·134094 | ·001168 | 15 | 9·998832 | 48 |
| 13 | 8·866455 | 2861 | 1·133545 | 8·867632 | 2877 | 11·132368 | ·001177 | 15 | 9·998823 | 47 |
| 14 | 8·868165 | 2850 | 1·131835 | 8·869351 | 2866 | 11·130649 | ·001187 | 17 | 9·998813 | 46 |
| 15 | 8·869868 | 2839 | 1·130132 | 8·871064 | 2854 | 11·128936 | ·001196 | 15 | 9·998804 | 45 |
| 16 | 8·871565 | 2828 | 1·128435 | 8·872770 | 2843 | 11·127230 | ·001205 | 15 | 9·998795 | 44 |
| 17 | 8·873255 | 2817 | 1·126745 | 8·874469 | 2832 | 11·125531 | ·001215 | 17 | 9·998785 | 43 |
| 18 | 8·874938 | 2806 | 1·125062 | 8·876162 | 2821 | 11·123838 | ·001224 | 15 | 9·998776 | 42 |
| 19 | 8·876615 | 2795 | 1·123385 | 8·877849 | 2811 | 11·122151 | ·001234 | 17 | 9·998766 | 41 |
| 20 | 8·878285 | 2784 | 1·121715 | 8·879529 | 2800 | 11·120471 | ·001243 | 15 | 9·998757 | 40 |
| 21 | 8·879949 | 2773 | 1·120051 | 8·881202 | 2789 | 11·118798 | ·001253 | 17 | 9·998747 | 39 |
| 22 | 8·881607 | 2763 | 1·118393 | 8·882869 | 2779 | 11·117131 | ·001262 | 17 | 9·998738 | 38 |
| 23 | 8·883258 | 2752 | 1·116742 | 8·884530 | 2768 | 11·115470 | ·001272 | 15 | 9·998728 | 37 |
| 24 | 8·884903 | 2742 | 1·115097 | 8·886185 | 2758 | 11·113815 | ·001282 | 17 | 9·998718 | 36 |
| 25 | 8·886542 | 2731 | 1·113458 | 8·887833 | 2747 | 11·112167 | ·001292 | 17 | 9·998708 | 35 |
| 26 | 8·888174 | 2721 | 1·111826 | 8·889476 | 2737 | 11·110524 | ·001301 | 15 | 9·998699 | 34 |
| 27 | 8·889801 | 2711 | 1·110199 | 8·891112 | 2727 | 11·108888 | ·001311 | 17 | 9·998689 | 33 |
| 28 | 8·891421 | 2700 | 1·108579 | 8·892742 | 2717 | 11·107258 | ·001321 | 17 | 9·998679 | 32 |
| 29 | 8·893035 | 2690 | 1·106965 | 8·894366 | 2707 | 11·105634 | ·001332 | 17 | 9·998669 | 31 |
| 30 | 8·894643 | 2680 | 1·105357 | 8·895984 | 2697 | 11·104016 | ·001341 | 17 | 9·998659 | 30 |
| 31 | 8·896246 | 2670 | 1·103754 | 8·897596 | 2687 | 11·102404 | ·001351 | 17 | 9·998649 | 29 |
| 32 | 8·897842 | 2660 | 1·102158 | 8·899203 | 2677 | 11·100797 | ·001361 | 17 | 9·998639 | 28 |
| 33 | 8·899432 | 2651 | 1·100568 | 8·900803 | 2667 | 11·099197 | ·001371 | 17 | 9·998629 | 27 |
| 34 | 8·901017 | 2641 | 1·098983 | 8·902398 | 2658 | 11·097602 | ·001381 | 17 | 9·998619 | 26 |
| 35 | 8·902596 | 2631 | 1·097404 | 8·903987 | 2648 | 11·096013 | ·001391 | 17 | 9·998609 | 25 |
| 36 | 8·904169 | 2622 | 1·095831 | 8·905570 | 2638 | 11·094430 | ·001401 | 17 | 9·998599 | 24 |
| 37 | 8·905736 | 2612 | 1·094264 | 8·907147 | 2629 | 11·092853 | ·001411 | 17 | 9·998589 | 23 |
| 38 | 8·907297 | 2603 | 1·092703 | 8·908719 | 2620 | 11·091281 | ·001422 | 18 | 9·998578 | 22 |
| 39 | 8·908853 | 2593 | 1·091147 | 8·910285 | 2610 | 11·089715 | ·001432 | 17 | 9·998568 | 21 |
| 40 | 8·910404 | 2584 | 1·089596 | 8·911846 | 2601 | 11·088154 | ·001442 | 17 | 9·998558 | 20 |
| 41 | 8·911949 | 2575 | 1·088051 | 8·913401 | 2592 | 11·086599 | ·001452 | 17 | 9·998548 | 19 |
| 42 | 8·913488 | 2566 | 1·086512 | 8·914951 | 5283 | 11·085049 | ·001463 | 18 | 9·998537 | 18 |
| 43 | 8·915022 | 2556 | 1·084978 | 8·916495 | 2574 | 11·083505 | ·001473 | 17 | 9·998527 | 17 |
| 44 | 8·916550 | 2547 | 1·083450 | 8·918034 | 2565 | 11·081966 | ·001484 | 18 | 9·998516 | 16 |
| 45 | 8·918073 | 2538 | 1·081927 | 8·919568 | 2556 | 11·080432 | ·001494 | 17 | 9·998506 | 15 |
| 46 | 8·919591 | 2529 | 1·080409 | 8·921096 | 2547 | 11·078904 | ·001505 | 18 | 9·998495 | 14 |
| 47 | 8·921103 | 2520 | 1·078897 | 8·922619 | 2538 | 11·077381 | ·001515 | 17 | 9·998485 | 13 |
| 48 | 8·922610 | 2512 | 1·077390 | 8·924136 | 2530 | 11·075864 | ·001526 | 18 | 9·998474 | 12 |
| 49 | 8·924112 | 2503 | 1·075888 | 8·925649 | 2521 | 11·074351 | ·001536 | 17 | 9·998464 | 11 |
| 50 | 8·925609 | 2494 | 1·074391 | 8·927156 | 2512 | 11·072844 | ·001547 | 18 | 9·998453 | 10 |
| 51 | 8·927100 | 2486 | 1·072900 | 8·928658 | 2503 | 11·071342 | ·001558 | 18 | 9·998442 | 9 |
| 52 | 8·928587 | 2477 | 1·071413 | 8·930155 | 2495 | 11·069845 | ·001569 | 18 | 9·998431 | 8 |
| 53 | 8·930068 | 2469 | 1·069932 | 8·931647 | 2486 | 11·068353 | ·001579 | 17 | 9·998421 | 7 |
| 54 | 8·931544 | 2460 | 1·068456 | 8·933134 | 2478 | 11·066866 | ·001590 | 18 | 9·998410 | 6 |
| 55 | 8·933015 | 2452 | 1·066985 | 8·934616 | 2470 | 11·065384 | ·001601 | 18 | 9·998399 | 5 |
| 56 | 8·934481 | 2443 | 1·065519 | 8·936093 | 2461 | 11·063907 | ·001612 | 18 | 9·998388 | 4 |
| 57 | 8·935942 | 2435 | 1·064058 | 8·937565 | 2453 | 11·062435 | ·001623 | 18 | 9·998377 | 3 |
| 58 | 8·937398 | 2427 | 1·062602 | 8·939032 | 2445 | 11·060968 | ·001634 | 18 | 9·998366 | 2 |
| 59 | 8·938850 | 2419 | 1·061150 | 8·940494 | 2437 | 11·059506 | ·001645 | 18 | 9·998355 | 1 |
| 60 | 8·940296 | 2411 | 1·059704 | 8·941952 | 2429 | 11·058048 | ·001656 | 18 | 9·998344 | 0 |
| ′ | Cosine. | | Secant. | Cotangent. | | Tangent. | Cosecant. | | Sine. | ′ |

85 DEG.

5 DEG.

| ′ | Sine. | Diff. 100″ | Cosecant. | Tangent. | Diff. 100″ | Cotangent. | Secant. | Diff. 100″ | Cosine. | ′ |
|---|---|---|---|---|---|---|---|---|---|---|
| 0 | 8·940296 | | 1·059704 | 8·941952 | | 11·058048 | ·001656 | | 9·998344 | 60 |
| 1 | 8·941738 | 2403 | 1·058262 | 8·943404 | 2421 | 11·056596 | ·001667 | 18 | 9·998333 | 59 |
| 2 | 8·943174 | 2394 | 1·056826 | 8·944852 | 2413 | 11·055148 | ·001678 | 18 | 9·998322 | 58 |
| 3 | 8·944606 | 2387 | 1·055394 | 8·946295 | 2405 | 11·053705 | ·001689 | 18 | 9·998311 | 57 |
| 4 | 8·946034 | 2379 | 1·053966 | 8·947734 | 2397 | 11·052266 | ·001700 | 18 | 9·998300 | 56 |
| 5 | 8·947456 | 2371 | 1·052544 | 8·949168 | 2390 | 11·050832 | ·001711 | 18 | 9·998289 | 55 |
| 6 | 8·948874 | 2363 | 1·051126 | 8·950597 | 2382 | 11·049403 | ·001723 | 20 | 9·998277 | 54 |
| 7 | 8·950287 | 2355 | 1·049713 | 8·952021 | 2374 | 11·047979 | ·001734 | 18 | 9·998266 | 53 |
| 8 | 8·951696 | 2348 | 1·048304 | 8·953441 | 2366 | 11·046559 | ·001745 | 18 | 9·998255 | 52 |
| 9 | 8·953100 | 2340 | 1·046900 | 8·954856 | 2359 | 11·045144 | ·001757 | 20 | 9·998243 | 51 |
| 10 | 8·954499 | 2332 | 1·045501 | 8·956267 | 2351 | 11·043733 | ·001768 | 18 | 9·998232 | 50 |
| 11 | 8·955894 | 2325 | 1·044106 | 8·957674 | 2344 | 11·042326 | ·001780 | 20 | 9·998220 | 49 |
| 12 | 8·957284 | 2317 | 1·042716 | 8·959075 | 2336 | 11·040925 | ·001791 | 18 | 9·998209 | 48 |
| 13 | 8·958670 | 2310 | 1·041330 | 8·960473 | 2329 | 11·039527 | ·001803 | 20 | 9·998197 | 47 |
| 14 | 8·960052 | 2302 | 1·039948 | 8·961866 | 2322 | 11·038134 | ·001814 | 18 | 9·998186 | 46 |
| 15 | 8·961429 | 2295 | 1·038571 | 8·963255 | 2314 | 11·036745 | ·001826 | 20 | 9·998174 | 45 |
| 16 | 8·962801 | 2288 | 1·037199 | 8·964639 | 2307 | 11·035361 | ·001837 | 18 | 9·998163 | 44 |
| 17 | 8·964170 | 2280 | 1·035830 | 8·966019 | 2300 | 11·033981 | ·001849 | 20 | 9·998151 | 43 |
| 18 | 8·965534 | 2273 | 1·034466 | 8·967394 | 2293 | 11·032606 | ·001861 | 20 | 9·998139 | 42 |
| 19 | 8·966893 | 2266 | 1·033107 | 8·968766 | 2286 | 11·031234 | ·001872 | 18 | 9·998128 | 41 |
| 20 | 8·968249 | 2259 | 1·031751 | 8·970133 | 2279 | 11·029867 | ·001884 | 20 | 9·998116 | 40 |
| 21 | 8·969600 | 2252 | 1·030400 | 8·971496 | 2271 | 11·028504 | ·001896 | 20 | 9·998104 | 39 |
| 22 | 8·970947 | 2245 | 1·029053 | 8·972855 | 2265 | 11·027145 | ·001908 | 20 | 9·998092 | 38 |
| 23 | 8·972289 | 2238 | 1·027711 | 8·974209 | 2257 | 11·025791 | ·001920 | 20 | 9·998080 | 37 |
| 24 | 8·973628 | 2231 | 1·026372 | 8·975560 | 2251 | 11·024440 | ·001932 | 20 | 9·998068 | 36 |
| 25 | 8·974962 | 2224 | 1·025038 | 8·976906 | 2244 | 11·023094 | ·001944 | 20 | 9·998056 | 35 |
| 26 | 8·976293 | 2217 | 1·023707 | 8·978248 | 2237 | 11·021752 | ·001956 | 20 | 9·998044 | 34 |
| 27 | 8·977619 | 2210 | 1·022381 | 8·979586 | 2230 | 11·020414 | ·001968 | 20 | 9·998032 | 33 |
| 28 | 8·978941 | 2203 | 1·021059 | 8·980921 | 2223 | 11·019079 | ·001980 | 20 | 9·998020 | 32 |
| 29 | 8·980259 | 2197 | 1·019741 | 8·982251 | 2217 | 11·017749 | ·001992 | 20 | 9·998008 | 31 |
| 30 | 8·981573 | 2190 | 1·018427 | 8·983577 | 2210 | 11·016423 | ·002004 | 20 | 9·997996 | 30 |
| 31 | 8·982883 | 2183 | 1·017117 | 8·984899 | 2204 | 11·015101 | ·002016 | 20 | 9·997984 | 29 |
| 32 | 8·984189 | 2177 | 1·015811 | 8·986217 | 2197 | 11·013783 | ·002028 | 20 | 9·997972 | 28 |
| 33 | 8·985491 | 2170 | 1·014509 | 8·987532 | 2191 | 11·012468 | ·002041 | 22 | 9·997959 | 27 |
| 34 | 8·986789 | 2163 | 1·013211 | 8·988842 | 2184 | 11·011158 | ·002053 | 20 | 9·997947 | 26 |
| 35 | 8·988083 | 2157 | 1·011917 | 8·990149 | 2178 | 11·009851 | ·002065 | 20 | 9·997935 | 25 |
| 36 | 8·989374 | 2150 | 1·010626 | 8·991451 | 2171 | 11·008549 | ·002078 | 22 | 9·997922 | 24 |
| 37 | 8·990660 | 2144 | 1·009340 | 8·992750 | 2165 | 11·007250 | ·002090 | 20 | 9·997910 | 23 |
| 38 | 8·991943 | 2138 | 1·008057 | 8·994045 | 2158 | 11·005955 | ·002103 | 22 | 9·997897 | 22 |
| 39 | 8·993222 | 2131 | 1·006778 | 8·995337 | 2152 | 11·004663 | ·002115 | 20 | 9·997885 | 21 |
| 40 | 8·994497 | 2125 | 1·005503 | 8·996624 | 2146 | 11·003376 | ·002128 | 22 | 9·997872 | 20 |
| 41 | 8·995768 | 2119 | 1·004232 | 8·997908 | 2140 | 11·002092 | ·002140 | 20 | 9·997860 | 19 |
| 42 | 8·997036 | 2112 | 1·002964 | 8·999188 | 2134 | 11·000812 | ·002153 | 22 | 9·997847 | 18 |
| 43 | 8·998299 | 2106 | 1·001701 | 9·000465 | 2127 | 10·999535 | ·002165 | 20 | 9·997835 | 17 |
| 44 | 8·999560 | 2100 | 1·000440 | 9·001738 | 2121 | 10·998262 | ·002178 | 22 | 9·997822 | 16 |
| 45 | 9·000816 | 2094 | 0·999184 | 9·003007 | 2115 | 10·996993 | ·002191 | 22 | 9·997809 | 15 |
| 46 | 9·002069 | 2088 | 0·997931 | 9·004272 | 2109 | 10·995728 | ·002203 | 20 | 9·997797 | 14 |
| 47 | 9·003318 | 2082 | 0·996682 | 9·005534 | 2103 | 10·994466 | ·002216 | 22 | 9·997784 | 13 |
| 48 | 9·004563 | 2076 | 0·995437 | 9·006792 | 2097 | 10·993208 | ·002229 | 22 | 9·997771 | 12 |
| 49 | 9·005805 | 2070 | 0·994195 | 9·008047 | 2091 | 10·991953 | ·002242 | 22 | 9·997758 | 11 |
| 50 | 9·007044 | 2064 | 0·992956 | 9·009298 | 2085 | 10·990702 | ·002255 | 22 | 9·997745 | 10 |
| 51 | 9·008278 | 2058 | 0·991722 | 9·010546 | 2079 | 10·989454 | ·002268 | 22 | 9·997732 | 9 |
| 52 | 9·009510 | 2052 | 0·990490 | 9·011790 | 2074 | 10·988210 | ·002281 | 22 | 9·997719 | 8 |
| 53 | 9·010737 | 2046 | 0·989263 | 9·013031 | 2068 | 10·986969 | ·002294 | 22 | 9·997706 | 7 |
| 54 | 9·011962 | 2040 | 0·988038 | 9·014268 | 2062 | 10·985732 | ·002307 | 22 | 9·997693 | 6 |
| 55 | 9·013182 | 2034 | 0·986818 | 9·015502 | 2056 | 10·984498 | ·002320 | 22 | 9·997680 | 5 |
| 56 | 9·014400 | 2029 | 0·985600 | 9·016732 | 2051 | 10·983268 | ·002333 | 22 | 9·997667 | 4 |
| 57 | 9·015613 | 2023 | 0·984387 | 9·017959 | 2045 | 10·982041 | ·002346 | 22 | 9·997654 | 3 |
| 58 | 9·016824 | 2017 | 0·983176 | 9·019183 | 2039 | 10·980817 | ·002359 | 22 | 9·997641 | 2 |
| 59 | 9·018031 | 2012 | 0·981969 | 9·020403 | 2034 | 10·979597 | ·002372 | 22 | 9·997628 | 1 |
| 60 | 9·019235 | 2006 | 0·980765 | 9·021620 | 2028 | 10·978380 | ·002386 | 23 | 9·997614 | 0 |
| ′ | Cosine. | | Secant. | Cotangent. | | Tangent. | Cosecant. | | Sine. | ′ |

84 DEG.

6 DEG.

| ′ | Sine. | Diff. 100″ | Cosecant. | Tangent. | Diff. 100″ | Cotangent. | Secant. | Diff. 100″ | Cosine. | ′ |
|---|---|---|---|---|---|---|---|---|---|---|
| 0 | 9·019235 | | ·980765 | 9·021620 | | 10·978380 | ·002386 | | 9·997614 | 60 |
| 1 | 9·020435 | 2000 | ·979565 | 9·022834 | 2023 | 10·977166 | ·002399 | 22 | 9·997601 | 59 |
| 2 | 9·021632 | 1995 | ·978368 | 9·024044 | 2017 | 10·975956 | ·002412 | 22 | 9·997588 | 58 |
| 3 | 9·022825 | 1989 | ·977175 | 9·025251 | 2011 | 10·974749 | ·002426 | 23 | 9·997574 | 57 |
| 4 | 9·024016 | 1984 | ·975984 | 9·026455 | 2006 | 10·973545 | ·002439 | 22 | 9·997561 | 56 |
| 5 | 9·025203 | 1978 | ·974797 | 9·027655 | 2001 | 10·972345 | ·002453 | 23 | 9·997547 | 55 |
| 6 | 9·026386 | 1973 | ·973614 | 9·028852 | 1995 | 10·971148 | ·002466 | 22 | 9·997534 | 54 |
| 7 | 9·027567 | 1967 | ·972433 | 9·030046 | 1990 | 10·969954 | ·002480 | 23 | 9·997520 | 53 |
| 8 | 9·028744 | 1962 | ·971256 | 9·031237 | 1985 | 10·968763 | ·002493 | 22 | 9·997507 | 52 |
| 9 | 9·029918 | 1957 | ·970082 | 9·032425 | 1979 | 10·967575 | ·002507 | 23 | 9·997493 | 51 |
| 10 | 9·031089 | 1951 | ·968911 | 9·033609 | 1974 | 10·966391 | ·002520 | 22 | 9·997480 | 50 |
| 11 | 9·032257 | 1946 | ·967743 | 9·034791 | 1969 | 10·965209 | ·002534 | 23 | 9·997466 | 49 |
| 12 | 9·033421 | 1941 | ·966579 | 9·035969 | 1964 | 10·964031 | ·002548 | 23 | 9·997452 | 48 |
| 13 | 9·034582 | 1936 | ·965418 | 9·037144 | 1958 | 10·962856 | ·002561 | 22 | 9·997439 | 47 |
| 14 | 9·035741 | 1930 | ·964259 | 9·038316 | 1953 | 10·961684 | ·002575 | 23 | 9·997425 | 46 |
| 15 | 9·036896 | 1925 | ·963104 | 9·039485 | 1948 | 10·960515 | ·002589 | 23 | 9·997411 | 45 |
| 16 | 9·038048 | 1920 | ·961952 | 9·040651 | 1943 | 10·959349 | ·002603 | 23 | 9·997397 | 44 |
| 17 | 9·039197 | 1915 | ·960803 | 9·041813 | 1938 | 10·958187 | ·002617 | 23 | 9·997383 | 43 |
| 18 | 9·040342 | 1910 | ·959658 | 9·042973 | 1933 | 10·957027 | ·002631 | 23 | 9·997369 | 42 |
| 19 | 9·041485 | 1905 | ·958515 | 9·044130 | 1928 | 10·955870 | ·002645 | 23 | 9·997355 | 41 |
| 20 | 9·042625 | 1899 | ·957375 | 9·045284 | 1923 | 10·954716 | ·002659 | 23 | 9·997341 | 40 |
| 21 | 9·043762 | 1895 | ·956238 | 9·046434 | 1918 | 10·953566 | ·002673 | 23 | 9·997327 | 39 |
| 22 | 9·044895 | 1889 | ·955105 | 9·047582 | 1913 | 10·952418 | ·002687 | 23 | 9·997313 | 38 |
| 23 | 9·046026 | 1884 | ·953974 | 9·048727 | 1908 | 10·951273 | ·002701 | 23 | 9·997299 | 37 |
| 24 | 9·047154 | 1879 | ·952846 | 9·049869 | 1903 | 10·950131 | ·002715 | 23 | 9·997285 | 36 |
| 25 | 9·048279 | 1875 | ·951721 | 9·051008 | 1898 | 10·948992 | ·002729 | 23 | 9·997271 | 35 |
| 26 | 9·049400 | 1870 | ·950600 | 9·052144 | 1893 | 10·947856 | ·002743 | 23 | 9.997257 | 34 |
| 27 | 9·050519 | 1865 | ·949481 | 9·053277 | 1889 | 10·946723 | ·002758 | 25 | 9·997242 | 33 |
| 28 | 9·051635 | 1860 | ·948365 | 9·054407 | 1884 | 10·945593 | ·002772 | 23 | 9·997228 | 32 |
| 29 | 9·052749 | 1855 | ·947251 | 9·055535 | 1879 | 10·944465 | ·002786 | 23 | 9·997214 | 31 |
| 30 | 9·053859 | 1850 | ·946141 | 9·056659 | 1874 | 10·943341 | ·002801 | 25 | 9·997199 | 30 |
| 31 | 9·054966 | 1845 | ·945034 | 9·057781 | 1870 | 10·942219 | ·002815 | 23 | 9·997185 | 29 |
| 32 | 9·056071 | 1841 | ·943929 | 9·058900 | 1865 | 10·941100 | ·002830 | 25 | 9·997170 | 28 |
| 33 | 9·057172 | 1836 | ·942828 | 9·060016 | 1860 | 10·939984 | ·002844 | 23 | 9·997156 | 27 |
| 34 | 9·058271 | 1831 | ·941729 | 9·061130 | 1855 | 10·938870 | ·002859 | 25 | 9·997141 | 26 |
| 35 | 9·059367 | 1827 | ·940633 | 9·062240 | 1851 | 10·937760 | ·002873 | 23 | 9·997127 | 25 |
| 36 | 9·060460 | 1822 | ·939540 | 9·063348 | 1846 | 10·936652 | ·002888 | 25 | 9·997112 | 24 |
| 37 | 9·061551 | 1817 | ·938449 | 9·064453 | 1842 | 10·935547 | ·002902 | 23 | 9·997098 | 23 |
| 38 | 9·062639 | 1813 | ·937361 | 9·065556 | 1837 | 10·934444 | ·002917 | 25 | 9·997083 | 22 |
| 39 | 9·063724 | 1808 | ·936276 | 9·066655 | 1833 | 10·933345 | ·002932 | 25 | 9·997068 | 21 |
| 40 | 9·064806 | 1804 | ·935194 | 9·067752 | 1828 | 10·932248 | ·002947 | 25 | 9·997053 | 20 |
| 41 | 9·065885 | 1799 | ·934115 | 9·068846 | 1824 | 10·931154 | ·002961 | 23 | 9·997039 | 19 |
| 42 | 9·066962 | 1794 | ·933038 | 9·069938 | 1819 | 10·930062 | ·002976 | 25 | 9·997024 | 18 |
| 43 | 9·068036 | 1790 | ·931964 | 9·071027 | 1815 | 10·928973 | ·002991 | 25 | 9·997009 | 17 |
| 44 | 9·069107 | 1786 | ·930893 | 9·072113 | 1810 | 10·927887 | ·003006 | 25 | 9·996994 | 16 |
| 45 | 9·070176 | 1781 | ·929824 | 9·073197 | 1806 | 10·926803 | ·003021 | 25 | 9·996979 | 15 |
| 46 | 9·071242 | 1777 | ·928758 | 9·074278 | 1802 | 10·925722 | ·003036 | 25 | 9·996964 | 14 |
| 47 | 9·072306 | 1772 | ·927694 | 9·075356 | 1797 | 10·024644 | ·003051 | 25 | 9·996949 | 13 |
| 48 | 9·073366 | 1768 | ·926634 | 9·076432 | 1793 | 10·923568 | ·003066 | 25 | 9·996934 | 12 |
| 49 | 9·074424 | 1763 | ·925576 | 9·077505 | 1789 | 10·922495 | ·003081 | 25 | 9·996919 | 11 |
| 50 | 9·075480 | 1759 | ·924520 | 9.078576 | 1784 | 10·921424 | ·003096 | 25 | 9·996904 | 10 |
| 51 | 9·076533 | 1755 | ·923467 | 9·079644 | 1780 | 10·920356 | ·003111 | 25 | 9·996889 | 9 |
| 52 | 9·077583 | 1750 | ·922417 | 9·080710 | 1776 | 10·919290 | ·003126 | 25 | 9·996874 | 8 |
| 53 | 9·078631 | 1746 | ·921369 | 9·081773 | 1772 | 10·918227 | ·003142 | 27 | 9·996858 | 7 |
| 54 | 9·079676 | 1742 | ·920324 | 9·082833 | 1767 | 10·917167 | ·003157 | 25 | 9·996843 | 6 |
| 55 | 9·080719 | 1738 | ·919281 | 9·083891 | 1763 | 10·916109 | ·003172 | 25 | 9·996828 | 5 |
| 56 | 9·081759 | 1733 | ·918241 | 9·084947 | 1759 | 10 915053 | ·003188 | 27 | 9·996812 | 4 |
| 57 | 9·082797 | 1729 | ·917203 | 9·086000 | 1755 | 10·914000 | ·003203 | 25 | 9·996797 | 3 |
| 58 | 9·083232 | 1725 | ·916168 | 9·087050 | 1751 | 10·912950 | ·003218 | 25 | 9·996782 | 2 |
| 59 | 9·084864 | 1721 | ·915136 | 9·088098 | 1747 | 10·911902 | ·003234 | 27 | 9·996766 | 1 |
| 60 | 9·085894 | 1717 | ·914106 | 9·089144 | 1743 | 10·910856 | ·003249 | 25 | 9·996751 | 0 |
| ′ | Cosine. | | Secant. | Cotangent. | | Tangent. | Cosecant. | | Sine. | ′ |

83 DEG.

7 DEG.

| ' | Sine. | Diff. 100" | Cosecant. | Tangent. | Diff. 100" | Cotangent. | Secant. | Diff. 100" | Cosine. | ' |
|---|---|---|---|---|---|---|---|---|---|---|
| 0 | 9·085894 | | ·914106 | 9·089144 | | 10·910856 | ·003249 | | 9·996751 | 60 |
| 1 | 9·086922 | 1713 | ·913078 | 9·090187 | 1738 | 10·909813 | ·003265 | 27 | 9·996735 | 59 |
| 2 | 9·087947 | 1709 | ·912053 | 9·091228 | 1735 | 10·908772 | ·003280 | 25 | 9·996720 | 58 |
| 3 | 9·088970 | 1704 | ·911030 | 9·092266 | 1731 | 10·907734 | ·003296 | 27 | 9·996704 | 57 |
| 4 | 9·089990 | 1700 | ·910010 | 9·093302 | 1727 | 10·906698 | ·003312 | 27 | 9·996688 | 56 |
| 5 | 9·091008 | 1696 | ·908992 | 9·094336 | 1722 | 10·905664 | ·003327 | 25 | 9·996673 | 55 |
| 6 | 9·092024 | 1692 | ·907976 | 9·095367 | 1719 | 10·904633 | ·003343 | 27 | 9·996657 | 54 |
| 7 | 9·093037 | 1688 | ·906963 | 9·096395 | 1715 | 10·903605 | ·003359 | 27 | 9·996641 | 53 |
| 8 | 9·094047 | 1684 | ·905953 | 9·097422 | 1711 | 10·902578 | ·003375 | 27 | 9·996625 | 52 |
| 9 | 9·095056 | 1680 | ·904944 | 9·098446 | 1707 | 10·901554 | ·003390 | 25 | 9·996610 | 51 |
| 10 | 9·096062 | 1676 | ·903938 | 9·099468 | 1703 | 10·900532 | ·003406 | 27 | 9·996594 | 50 |
| 11 | 9·097065 | 1673 | ·902935 | 9·100487 | 1699 | 10·899513 | ·003422 | 27 | 9·996578 | 49 |
| 12 | 9·098066 | 1668 | ·901934 | 9·101504 | 1695 | 10·898496 | ·003438 | 27 | 9·996562 | 48 |
| 13 | 9·099065 | 1665 | ·900935 | 9·102519 | 1691 | 10·897481 | ·003454 | 27 | 9·996546 | 47 |
| 14 | 9·100062 | 1661 | ·899938 | 9·103532 | 1687 | 10·896468 | ·003470 | 27 | 9·996530 | 46 |
| 15 | 9·101056 | 1657 | ·898944 | 9·104542 | 1684 | 10·895458 | ·003486 | 27 | 9·996514 | 45 |
| 16 | 9·102048 | 1653 | ·897952 | 9·105550 | 1680 | 10·894450 | ·003502 | 27 | 9·996498 | 44 |
| 17 | 9·103037 | 1649 | ·896963 | 9·106556 | 1676 | 10·893444 | ·003518 | 27 | 9·996482 | 43 |
| 18 | 9·104025 | 1645 | ·895975 | 9·107559 | 1672 | 10·892441 | ·003535 | 28 | 9·996465 | 42 |
| 19 | 9·105010 | 1642 | ·894990 | 9·108560 | 1669 | 10·891440 | ·003551 | 27 | 9·996449 | 41 |
| 20 | 9·105992 | 1638 | ·894008 | 9·109559 | 1665 | 10·890441 | ·003567 | 27 | 9·996433 | 40 |
| 21 | 9·106973 | 1634 | ·893027 | 9·110556 | 1661 | 10·889444 | ·003583 | 27 | 9·996417 | 39 |
| 22 | 9·107951 | 1630 | ·892049 | 9·111551 | 1658 | 10·888449 | ·003600 | 28 | 9·996400 | 38 |
| 23 | 9·108927 | 1627 | ·891073 | 9·112543 | 1654 | 10·887457 | ·003616 | 27 | 9·996384 | 37 |
| 24 | 9·109901 | 1623 | ·890099 | 9·113533 | 1650 | 10·886467 | ·003632 | 27 | 9·996368 | 36 |
| 25 | 9·110873 | 1619 | ·889127 | 9·114521 | 1647 | 10·885479 | ·003649 | 28 | 9·996351 | 35 |
| 26 | 9·111842 | 1616 | ·888158 | 9·115507 | 1643 | 10·884493 | ·003665 | 27 | 9·996335 | 34 |
| 27 | 9·112809 | 1612 | ·887191 | 9·116491 | 1639 | 10·883509 | ·003682 | 28 | 9·996318 | 33 |
| 28 | 9·113774 | 1608 | ·886226 | 9·117472 | 1636 | 10·882528 | ·003698 | 27 | 9·996302 | 32 |
| 29 | 9·114737 | 1605 | ·885263 | 9·118452 | 1632 | 10·881548 | ·003715 | 28 | 9·996285 | 31 |
| 30 | 9·115698 | 1601 | ·884302 | 9·119429 | 1629 | 10·880571 | ·003731 | 27 | 9·996269 | 30 |
| 31 | 9·116656 | 1597 | ·883344 | 9·120404 | 1625 | 10·879596 | ·003748 | 28 | 9·996252 | 29 |
| 32 | 9·117613 | 1594 | ·882387 | 9·121377 | 1622 | 10·878623 | ·003765 | 28 | 9·996235 | 28 |
| 33 | 9·118567 | 1590 | ·881433 | 9·122348 | 1618 | 10·877652 | ·003781 | 27 | 9·996219 | 27 |
| 34 | 9·119519 | 1587 | ·880481 | 9·123317 | 1615 | 10·876683 | ·003798 | 28 | 9·996202 | 26 |
| 35 | 9·120469 | 1583 | ·879531 | 9·124284 | 1611 | 10·875716 | ·003815 | 28 | 9·996185 | 25 |
| 36 | 9·121417 | 1580 | ·878583 | 9·125249 | 1608 | 10·874751 | ·003832 | 28 | 9·996168 | 24 |
| 37 | 9·122362 | 1576 | ·877638 | 9·126211 | 1604 | 10·873789 | ·003849 | 28 | 9·996151 | 23 |
| 38 | 9·123306 | 1573 | ·876694 | 9·127172 | 1601 | 10·872828 | ·003866 | 28 | 9·996134 | 22 |
| 39 | 9·124248 | 1569 | ·875752 | 9·128130 | 1597 | 10·871870 | ·003883 | 28 | 9·996117 | 21 |
| 40 | 9·125187 | 1566 | ·874813 | 9·129087 | 1594 | 10·870913 | ·003900 | 28 | 9·996100 | 20 |
| 41 | 9·126125 | 1562 | ·873875 | 9·130041 | 1591 | 10·869959 | ·003917 | 28 | 9·996083 | 19 |
| 42 | 9·127060 | 1559 | ·872940 | 9·130994 | 1587 | 10·869006 | ·003934 | 28 | 9·996066 | 18 |
| 43 | 9·127993 | 1556 | ·872007 | 9·131944 | 1584 | 10·868056 | ·003951 | 28 | 9·996049 | 17 |
| 44 | 9·128925 | 1552 | ·871075 | 9·132893 | 1581 | 10·867107 | ·003968 | 28 | 9·996032 | 16 |
| 45 | 9·129854 | 1549 | ·870146 | 9·133839 | 1577 | 10·866161 | ·003985 | 28 | 9·996015 | 15 |
| 46 | 9·130781 | 1545 | ·869219 | 9·134784 | 1574 | 10·865216 | ·004002 | 28 | 9·995998 | 14 |
| 47 | 9·131706 | 1542 | ·868294 | 9·135726 | 1571 | 10·864274 | ·004020 | 30 | 9·995980 | 13 |
| 48 | 9·132630 | 1539 | ·867370 | 9·136667 | 1567 | 10·863333 | ·004037 | 28 | 9·995963 | 12 |
| 49 | 9·133551 | 1535 | ·866449 | 9·137605 | 1564 | 10·862395 | ·004054 | 28 | 9·995946 | 11 |
| 50 | 9·134470 | 1532 | ·865530 | 9·138542 | 1561 | 10·861458 | ·004072 | 30 | 9·995928 | 10 |
| 51 | 9·135387 | 1529 | ·864613 | 9·139476 | 1558 | 10·860524 | ·004089 | 28 | 9·995911 | 9 |
| 52 | 9·136303 | 1525 | ·863697 | 9·140409 | 1555 | 10·859591 | ·004106 | 28 | 9·995894 | 8 |
| 53 | 9·137216 | 1522 | ·862784 | 9·141340 | 1551 | 10·858660 | ·004124 | 30 | 9·995876 | 7 |
| 54 | 9·138128 | 1519 | ·861872 | 9·142269 | 1548 | 10·857731 | ·004141 | 28 | 9·995859 | 6 |
| 55 | 9·139037 | 1516 | ·860963 | 9·143196 | 1545 | 10·856804 | ·004159 | 30 | 9·995841 | 5 |
| 56 | 9·139944 | 1512 | ·860056 | 9·144121 | 1542 | 10·855879 | ·004177 | 30 | 9·995823 | 4 |
| 57 | 9·140850 | 1509 | ·859150 | 9·145044 | 1539 | 10·854956 | ·004194 | 28 | 9·995806 | 3 |
| 58 | 9·141754 | 1506 | ·858246 | 9·145966 | 1535 | 10·854034 | ·004212 | 30 | 9·995788 | 2 |
| 59 | 9·142655 | 1503 | ·857345 | 9·146885 | 1532 | 10·853115 | ·004229 | 28 | 9·995771 | 1 |
| 60 | 9·143555 | 1500 | ·856445 | 9·147803 | 1529 | 10·852197 | ·004247 | 30 | 9·995753 | 0 |
| ' | Cosine. | | Secant. | Cotangent. | | Tangent. | Cosecant. | | Sine. | ' |

82 DEG.

8 DEG.

| ′ | Sine. | Diff. 100″ | Cosecant. | Tangent. | Diff. 100″ | Cotangent. | Secant. | Diff. 100″ | Cosine. | ′ |
|---|---|---|---|---|---|---|---|---|---|---|
| 0 | 9·143555 | | ·856445 | 9·147803 | | 10·852197 | ·004247 | | 9·995753 | 60 |
| 1 | 9·144453 | 1496 | ·855547 | 9·148718 | 1526 | 10·851282 | ·004265 | 30 | 9·995735 | 59 |
| 2 | 9·145349 | 1493 | ·854651 | 9·149632 | 1523 | 10·850368 | ·004283 | 30 | 9·995717 | 58 |
| 3 | 9·146243 | 1490 | ·853757 | 9·150544 | 1520 | 10·849456 | ·004301 | 30 | 9·995699 | 57 |
| 4 | 9·147136 | 1487 | ·852864 | 9·151454 | 1517 | 10·848546 | ·004319 | 30 | 9·995681 | 56 |
| 5 | 9·148026 | 1484 | ·851974 | 9·152363 | 1514 | 10·847637 | ·004336 | 28 | 9·995664 | 55 |
| 6 | 9·148915 | 1481 | ·851085 | 9·153269 | 1511 | 10·846731 | ·004354 | 30 | 9·995646 | 54 |
| 7 | 9·149802 | 1478 | ·850198 | 9·154174 | 1508 | 10·845826 | ·004372 | 30 | 9·995628 | 53 |
| 8 | 9·150686 | 1475 | ·849314 | 9·155077 | 1505 | 10·844923 | ·004390 | 30 | 9·995610 | 52 |
| 9 | 9·151569 | 1472 | ·848431 | 9·155978 | 1502 | 10·844022 | ·004409 | 32 | 9·995591 | 51 |
| 10 | 9·152451 | 1469 | ·847549 | 9·156877 | 1499 | 10·843123 | ·004427 | 30 | 9·995573 | 50 |
| 11 | 9·153330 | 1466 | ·846670 | 9·157775 | 1496 | 10·842225 | ·004445 | 30 | 9·995555 | 49 |
| 12 | 9·154208 | 1462 | ·845792 | 9·158671 | 1493 | 10·841329 | ·004463 | 30 | 9·995537 | 48 |
| 13 | 9·155083 | 1460 | ·844917 | 9·159565 | 1490 | 10·840435 | ·004481 | 30 | 9·995519 | 47 |
| 14 | 9·155957 | 1457 | ·844043 | 9·160457 | 1487 | 10·839543 | ·004499 | 30 | 9·995501 | 46 |
| 15 | 9·156830 | 1454 | ·843170 | 9·161347 | 1484 | 10·838653 | ·004518 | 32 | 9·995482 | 45 |
| 16 | 9·157700 | 1451 | ·842300 | 9·162236 | 1481 | 10·837764 | ·004536 | 30 | 9·995464 | 44 |
| 17 | 9·158569 | 1448 | ·841431 | 9·163123 | 1478 | 10·836877 | ·004554 | 30 | 9·995446 | 43 |
| 18 | 9·159435 | 1445 | ·840565 | 9·164008 | 1475 | 10·835992 | ·004573 | 32 | 9·995427 | 42 |
| 19 | 9·160301 | 1442 | ·839699 | 9·164892 | 1473 | 10·835108 | ·004591 | 30 | 9·995409 | 41 |
| 20 | 9·161164 | 1439 | ·838836 | 9·165774 | 1470 | 10·834226 | ·004610 | 32 | 9·995390 | 40 |
| 21 | 9·162025 | 1436 | ·837975 | 9·166654 | 1467 | 10·833346 | ·004628 | 30 | 9·995372 | 39 |
| 22 | 9·162885 | 1433 | ·837115 | 9·167532 | 1464 | 10·832468 | ·004647 | 32 | 9·995353 | 38 |
| 23 | 9·163743 | 1430 | ·836257 | 9·168409 | 1461 | 10·831591 | ·004666 | 32 | 9·995334 | 37 |
| 24 | 9·164600 | 1427 | ·835400 | 9·169284 | 1458 | 10·830716 | ·004684 | 30 | 9·995316 | 36 |
| 25 | 9·165454 | 1424 | ·834546 | 9·170157 | 1455 | 10·829843 | ·004703 | 32 | 9·995297 | 35 |
| 26 | 9·166307 | 1422 | ·833693 | 9·171029 | 1453 | 10·828971 | ·004722 | 32 | 9·995278 | 34 |
| 27 | 9·167159 | 1419 | ·832841 | 9·171899 | 1450 | 10·828101 | ·004740 | 30 | 9·995260 | 33 |
| 28 | 9·168008 | 1416 | ·831992 | 9·172767 | 1447 | 10·827233 | ·004759 | 32 | 9·995241 | 32 |
| 29 | 9·168856 | 1413 | ·831144 | 9·173634 | 1444 | 10·826366 | ·004778 | 32 | 9·995222 | 31 |
| 30 | 9·169702 | 1410 | ·830298 | 9·174499 | 1442 | 10·825501 | ·004797 | 32 | 9·995203 | 30 |
| 31 | 9·170547 | 1407 | ·829453 | 9·175362 | 1439 | 10·824638 | ·004816 | 32 | 9·995184 | 29 |
| 32 | 9·171389 | 1405 | ·828611 | 9·176224 | 1436 | 10·823776 | ·004835 | 32 | 9·995165 | 28 |
| 33 | 9·172230 | 1402 | ·827770 | 9·177084 | 1433 | 10·822916 | ·004854 | 32 | 9·995146 | 27 |
| 34 | 9·173070 | 1399 | ·826930 | 9·177942 | 1431 | 10·822058 | ·004873 | 32 | 9·995127 | 26 |
| 35 | 9·173908 | 1396 | ·826092 | 9·178799 | 1428 | 10·821201 | ·004892 | 32 | 9·995108 | 25 |
| 36 | 9·174744 | 1394 | ·825256 | 9·179655 | 1425 | 10·820345 | ·004911 | 32 | 9·995089 | 24 |
| 37 | 9·175578 | 1391 | ·824422 | 9·180508 | 1423 | 10·819492 | ·004930 | 32 | 9·995070 | 23 |
| 38 | 9·176411 | 1388 | ·823589 | 9·181360 | 1420 | 10·818640 | ·004949 | 32 | 9·995051 | 22 |
| 39 | 9·177242 | 1385 | ·822758 | 9·182211 | 1417 | 10·817789 | ·004968 | 32 | 9·995032 | 21 |
| 40 | 9·178072 | 1383 | ·821928 | 9·183059 | 1415 | 10·816941 | ·004987 | 32 | 9·995013 | 20 |
| 41 | 9·178900 | 1380 | ·821100 | 9·183907 | 1412 | 10·816093 | ·005007 | 33 | 9·994993 | 19 |
| 42 | 9·179726 | 1377 | ·820274 | 9·184752 | 1409 | 10·815248 | ·005026 | 32 | 9·994974 | 18 |
| 43 | 9·180551 | 1374 | ·819449 | 9·185597 | 1407 | 10·814403 | ·005045 | 32 | 9·994955 | 17 |
| 44 | 9·181374 | 1372 | ·818626 | 9·186439 | 1404 | 10·813561 | ·005065 | 33 | 9·994935 | 16 |
| 45 | 9·182196 | 1369 | ·817804 | 9·187280 | 1402 | 10·812720 | ·005084 | 32 | 9·994916 | 15 |
| 46 | 9·183016 | 1367 | ·816984 | 9·188120 | 1399 | 10·811880 | ·005104 | 33 | 9·994896 | 14 |
| 47 | 9·183834 | 1364 | ·816166 | 9·188958 | 1396 | 10·811042 | ·005123 | 32 | 9·994877 | 13 |
| 48 | 9·184651 | 1361 | ·815349 | 9·189794 | 1394 | 10·810206 | ·005143 | 33 | 9·994857 | 12 |
| 49 | 9·185466 | 1359 | ·814534 | 9·190629 | 1391 | 10·809371 | ·005162 | 32 | 9·994838 | 11 |
| 50 | 9·186280 | 1356 | ·813720 | 9·191462 | 1389 | 10·808538 | ·005182 | 33 | 9·994818 | 10 |
| 51 | 9·187092 | 1353 | ·812908 | 9·192294 | 1386 | 10·807706 | ·005202 | 33 | 9·994798 | 9 |
| 52 | 9·187903 | 1351 | ·812097 | 9·193124 | 1384 | 10·806876 | ·005221 | 32 | 9·994779 | 8 |
| 53 | 9·188712 | 1348 | ·811288 | 9·193953 | 1381 | 10·806047 | ·005241 | 33 | 9·994759 | 7 |
| 54 | 9·189519 | 1346 | ·810481 | 9·194780 | 1379 | 10·805220 | ·005261 | 33 | 9·994739 | 6 |
| 55 | 9·190325 | 1343 | ·809675 | 9·195606 | 1376 | 10·804394 | ·005281 | 32 | 9·994720 | 5 |
| 56 | 9·191130 | 1341 | ·808870 | 9·196430 | 1374 | 10·803570 | ·005300 | 33 | 9·994700 | 4 |
| 57 | 9·191933 | 1338 | ·808067 | 9·197253 | 1371 | 10·802747 | ·005320 | 33 | 9·994680 | 3 |
| 58 | 9·192734 | 1336 | ·807266 | 9·198074 | 1369 | 10·801926 | ·005340 | 33 | 9·994660 | 2 |
| 59 | 9·193534 | 1333 | ·806466 | 9·198894 | 1366 | 10·801106 | ·005360 | 33 | 9·994640 | 1 |
| 60 | 9·194332 | 1330 | ·805668 | 9·199713 | 1364 | 10·800287 | ·005380 | 33 | 9·994620 | 0 |
| ′ | Cosine. | | Secant. | Cotangent. | | Tangent. | Cosecant. | | Sine. | ′ |

81 DEG.

9 DEG.

| ′ | Sine. | Diff. 100″ | Cosecant. | Tangent. | Diff. 100″ | Cotangent. | Secant. | Diff. 100″ | Cosine. | ′ |
|---|---|---|---|---|---|---|---|---|---|---|
| 0 | 9·194332 | | ·805668 | 9·199713 | | 10·800287 | ·005380 | | 9·994620 | 60 |
| 1 | 9·195129 | 1328 | ·804871 | 9·200529 | 1361 | 10·799471 | ·005400 | 33 | 9·994600 | 59 |
| 2 | 9·195925 | 1326 | ·804075 | 9·201345 | 1359 | 10·798655 | ·005420 | 33 | 9·994580 | 58 |
| 3 | 9·196719 | 1323 | ·803281 | 9·202159 | 1356 | 10·797841 | ·005440 | 33 | 9·994560 | 57 |
| 4 | 9·197511 | 1321 | ·802489 | 9·202971 | 1354 | 10·797029 | ·005460 | 33 | 9·994540 | 56 |
| 5 | 9·198302 | 1318 | ·801698 | 9·203782 | 1352 | 10·796218 | ·005481 | 35 | 9·994519 | 55 |
| 6 | 9·199091 | 1316 | ·800909 | 9·204592 | 1349 | 10·795408 | ·005501 | 33 | 9·994499 | 54 |
| 7 | 9·199879 | 1313 | ·800121 | 9·205400 | 1347 | 10·794600 | ·005521 | 33 | 9·994479 | 53 |
| 8 | 9·200666 | 1311 | ·799334 | 9·206207 | 1345 | 10·793793 | ·005541 | 33 | 9·994459 | 52 |
| 9 | 9·201451 | 1308 | ·798549 | 9·207013 | 1342 | 10·792987 | ·005562 | 35 | 9·994438 | 51 |
| 10 | 9·202234 | 1306 | ·797766 | 9·207817 | 1340 | 10·792183 | ·005582 | 33 | 9·994418 | 50 |
| 11 | 9·203017 | 1304 | ·796983 | 9·208619 | 1338 | 10·791381 | ·005602 | 33 | 9·994398 | 49 |
| 12 | 9·203797 | 1301 | ·796203 | 9·209420 | 1335 | 10·790580 | ·005623 | 35 | 9·994377 | 48 |
| 13 | 9·204577 | 1299 | ·795423 | 9·210220 | 1333 | 10·789780 | ·005643 | 33 | 9·994357 | 47 |
| 14 | 9·205354 | 1296 | ·794646 | 9·211018 | 1331 | 10·788982 | ·005664 | 35 | 9·994336 | 46 |
| 15 | 9·206131 | 1294 | ·793869 | 9·211815 | 1328 | 10·788185 | ·005684 | 33 | 9·994316 | 45 |
| 16 | 9·206906 | 1292 | ·793094 | 9·212611 | 1326 | 10·787389 | ·005705 | 35 | 9·994295 | 44 |
| 17 | 9·207679 | 1289 | ·792321 | 9·213405 | 1324 | 10·786595 | ·005726 | 35 | 9·994274 | 43 |
| 18 | 9·208452 | 1287 | ·791548 | 9·214198 | 1321 | 10·785802 | ·005746 | 33 | 9·994254 | 42 |
| 19 | 9·209222 | 1285 | ·790778 | 9·214989 | 1319 | 10·785011 | ·005767 | 35 | 9·994233 | 41 |
| 20 | 9·209992 | 1282 | ·790008 | 9·215780 | 1317 | 10·784220 | ·005788 | 35 | 9·994212 | 40 |
| 21 | 9·210760 | 1280 | ·789240 | 9·216568 | 1315 | 10·783432 | ·005809 | 35 | 9·994191 | 39 |
| 22 | 9·211526 | 1278 | ·788474 | 9·217356 | 1312 | 10·782644 | ·005829 | 33 | 9·994171 | 38 |
| 23 | 9·212291 | 1275 | ·787709 | 9·218142 | 1310 | 10·781858 | ·005850 | 35 | 9·994150 | 37 |
| 24 | 9·213055 | 1273 | ·786945 | 9·218926 | 1308 | 10·781074 | ·005871 | 35 | 9·994129 | 36 |
| 25 | 9·213818 | 1271 | ·786182 | 9·219710 | 1305 | 10·780290 | ·005892 | 35 | 9·994108 | 35 |
| 26 | 9·214579 | 1268 | ·785421 | 9·220492 | 1303 | 10·779508 | ·005913 | 35 | 9·994087 | 34 |
| 27 | 9·215338 | 1266 | ·784662 | 9·221272 | 1301 | 10·778728 | ·005934 | 35 | 9·994066 | 33 |
| 28 | 9·216097 | 1264 | ·783903 | 9·222052 | 1299 | 10·777948 | ·005955 | 35 | 9·994045 | 32 |
| 29 | 9·216854 | 1261 | ·783146 | 9·222830 | 1297 | 10·777170 | ·005976 | 35 | 9·994024 | 31 |
| 30 | 9·217609 | 1259 | ·782391 | 9·223607 | 1294 | 10·776393 | ·005997 | 35 | 9·994003 | 30 |
| 31 | 9·218363 | 1257 | ·781637 | 9·224382 | 1292 | 10·775618 | ·006018 | 35 | 9·993982 | 29 |
| 32 | 9·219116 | 1255 | ·780884 | 9·225156 | 1290 | 10·774844 | ·006040 | 37 | 9·993960 | 28 |
| 33 | 9·219868 | 1253 | ·780132 | 9·225929 | 1288 | 10·774071 | ·006061 | 35 | 9·993939 | 27 |
| 34 | 9·220618 | 1250 | ·779382 | 9·226700 | 1286 | 10·773300 | ·006082 | 35 | 9·993918 | 26 |
| 35 | 9·221367 | 1248 | ·778633 | 9·227471 | 1284 | 10·772529 | ·006103 | 35 | 9·993897 | 25 |
| 36 | 9·222115 | 1246 | ·777885 | 9·228239 | 1281 | 10·771761 | ·006125 | 37 | 9·993875 | 24 |
| 37 | 9·222861 | 1244 | ·777139 | 9·229007 | 1279 | 10·770993 | ·006146 | 35 | 9·993854 | 23 |
| 38 | 9·223606 | 1242 | ·776394 | 9·229773 | 1277 | 10·770227 | ·006168 | 37 | 9·993832 | 22 |
| 39 | 9·224349 | 1239 | ·775651 | 9·230539 | 1275 | 10·769461 | ·006189 | 35 | 9·993811 | 21 |
| 40 | 9·225092 | 1237 | ·774908 | 9·231302 | 1273 | 10·768698 | ·006211 | 37 | 9·993789 | 20 |
| 41 | 9·225833 | 1235 | ·774167 | 9·232065 | 1271 | 10·767935 | ·006232 | 35 | 9·993768 | 19 |
| 42 | 9·226573 | 1233 | ·773427 | 9·232826 | 1269 | 10·767174 | ·006254 | 37 | 9·993746 | 18 |
| 43 | 9·227311 | 1231 | ·772689 | 9·233586 | 1267 | 10·766414 | ·006275 | 35 | 9·993725 | 17 |
| 44 | 9·228048 | 1228 | ·771952 | 9·234345 | 1265 | 10·765655 | ·006297 | 37 | 9·993703 | 16 |
| 45 | 9·228784 | 1226 | ·771216 | 9·235103 | 1262 | 10·764897 | ·006319 | 37 | 9·993681 | 15 |
| 46 | 9·229518 | 1224 | ·770482 | 9·235859 | 1260 | 10·764141 | ·006340 | 35 | 9·993660 | 14 |
| 47 | 9·230252 | 1222 | ·769748 | 9·236614 | 1258 | 10·763386 | ·006362 | 37 | 9·993638 | 13 |
| 48 | 9·230984 | 1220 | ·769016 | 9·237368 | 1256 | 10·762632 | ·006384 | 37 | 9·993616 | 12 |
| 49 | 9·231715 | 1218 | ·768285 | 9·238120 | 1254 | 10·761880 | ·006406 | 37 | 9·993594 | 11 |
| 50 | 9·232444 | 1216 | ·767556 | 9·238872 | 1252 | 10·761128 | ·006428 | 37 | 9·993572 | 10 |
| 51 | 9·233172 | 1214 | ·766828 | 9·239622 | 1250 | 10·760378 | ·006450 | 37 | 9·993550 | 9 |
| 52 | 9·233899 | 1212 | ·766101 | 9·240371 | 1248 | 10·759629 | ·006472 | 37 | 9·993528 | 8 |
| 53 | 9·234625 | 1209 | ·765375 | 9·241118 | 1246 | 10·758882 | ·006494 | 37 | 9·993506 | 7 |
| 54 | 9·235349 | 1207 | ·764651 | 9·241865 | 1244 | 10·758135 | ·006516 | 37 | 9·993484 | 6 |
| 55 | 9·236073 | 1205 | ·763927 | 9·242610 | 1242 | 10·757390 | ·006538 | 37 | 9·993462 | 5 |
| 56 | 9·236795 | 1203 | ·763205 | 9·243354 | 1240 | 10·756646 | ·006560 | 37 | 9·993440 | 4 |
| 57 | 9·237515 | 1201 | ·762485 | 9·244097 | 1238 | 10·755903 | ·006582 | 37 | 9·993418 | 3 |
| 58 | 9·238235 | 1199 | ·761765 | 9·244839 | 1236 | 10·755161 | ·006604 | 37 | 9·993396 | 2 |
| 59 | 9·238953 | 1197 | ·761047 | 9·245579 | 1234 | 10·754421 | ·006626 | 37 | 9·993374 | 1 |
| 60 | 9·239670 | 1195 | ·760330 | 9·246319 | 1232 | 10·753681 | ·006649 | 38 | 9·993351 | 0 |
| ′ | Cosine. | | Secant. | Cotangent. | | Tangent. | Cosecant. | | Sine. | ′ |

80 DEG.

10 DEG.

| ′ | Sine. | Diff. 100″ | Cosecant. | Tangent. | Diff. 100″ | Cotangent. | Secant. | Diff. 100″ | Cosine. | ′ |
|---|---|---|---|---|---|---|---|---|---|---|
| 0 | 9·239670 | | ·760330 | 9·246319 | | 10·753681 | ·006649 | | 9·993351 | 60 |
| 1 | 9·240386 | 1193 | ·759614 | 9·247057 | 1230 | 10·752943 | ·006671 | 37 | 9·993329 | 59 |
| 2 | 9·241101 | 1191 | ·758899 | 9·247794 | 1228 | 10·752206 | ·006693 | 37 | 9·993307 | 58 |
| 3 | 9·241814 | 1189 | ·758181 | 9·248530 | 1226 | 10·751470 | ·006715 | 38 | 9·993285 | 57 |
| 4 | 9·242526 | 1187 | ·757474 | 9·249264 | 1224 | 10·750736 | ·006738 | 37 | 9·993262 | 56 |
| 5 | 9·243237 | 1185 | ·756763 | 9·249998 | 1222 | 10·750002 | ·006760 | 37 | 9·993240 | 55 |
| 6 | 9·243947 | 1183 | ·756053 | 9·250730 | 1220 | 10·749270 | ·006783 | 38 | 9·993217 | 54 |
| 7 | 9·244656 | 1181 | ·755344 | 9·251461 | 1218 | 10·748539 | ·006805 | 37 | 9·993195 | 53 |
| 8 | 9·245363 | 1179 | ·754637 | 9·252191 | 1217 | 10·747809 | ·006828 | 38 | 9·993172 | 52 |
| 9 | 9·246069 | 1177 | ·753931 | 9·252920 | 1215 | 10·747080 | ·006851 | 38 | 9·993149 | 51 |
| 10 | 9·246775 | 1175 | ·753225 | 9·253648 | 1213 | 10·746352 | ·006873 | 37 | 9·993127 | 50 |
| 11 | 9·247478 | 1173 | ·752522 | 9·254374 | 1211 | 10·745626 | ·006896 | 38 | 9·993104 | 49 |
| 12 | 9·248181 | 1171 | ·751819 | 9·255100 | 1209 | 10·744900 | ·006919 | 38 | 9·993081 | 48 |
| 13 | 9·248883 | 1169 | ·751117 | 9·255824 | 1207 | 10·744176 | ·006941 | 37 | 9·993059 | 47 |
| 14 | 9·249583 | 1167 | ·750417 | 9·256547 | 1205 | 10·743453 | ·006964 | 38 | 9·993036 | 46 |
| 15 | 9·250282 | 1165 | ·749718 | 9·257269 | 1203 | 10·742731 | ·006987 | 38 | 9·993013 | 45 |
| 16 | 9·250980 | 1163 | ·749020 | 9·257990 | 1201 | 10·742010 | ·007010 | 38 | 9·992990 | 44 |
| 17 | 9·251677 | 1161 | ·748323 | 9·258710 | 1200 | 10·741290 | ·007033 | 38 | 9·992967 | 43 |
| 18 | 9·252373 | 1159 | ·747627 | 9·259429 | 1198 | 10·740571 | ·007056 | 38 | 9·992944 | 42 |
| 19 | 9·253067 | 1158 | ·746933 | 9·260146 | 1196 | 10·739854 | ·007079 | 38 | 9·992921 | 41 |
| 20 | 9·253761 | 1156 | ·746239 | 9·260863 | 1194 | 10·739137 | ·007102 | 38 | 9·992898 | 40 |
| 21 | 9·254453 | 1154 | ·745547 | 9·261578 | 1192 | 10·738422 | ·007125 | 38 | 9·992875 | 39 |
| 22 | 9·255144 | 1152 | ·744856 | 9·262292 | 1190 | 10·737708 | ·007148 | 38 | 9·992852 | 38 |
| 23 | 9·255834 | 1150 | ·744166 | 9·263005 | 1189 | 10·736995 | ·007171 | 38 | 9·992829 | 37 |
| 24 | 9·256523 | 1148 | ·743477 | 9·263717 | 1187 | 10·736283 | ·007194 | 38 | 9·992806 | 36 |
| 25 | 9·257211 | 1146 | ·742789 | 9·264428 | 1185 | 10·735572 | ·007217 | 38 | 9·992783 | 35 |
| 26 | 9·257898 | 1144 | ·742102 | 9·265138 | 1183 | 10·734862 | ·007241 | 40 | 9·992759 | 34 |
| 27 | 9·258583 | 1142 | ·741417 | 9·265847 | 1181 | 10·734153 | ·007264 | 38 | 9·992736 | 33 |
| 28 | 9·259268 | 1141 | ·740732 | 9·266555 | 1179 | 10·733445 | ·007287 | 38 | 9·992713 | 32 |
| 29 | 9·259951 | 1139 | ·740049 | 9·267261 | 1178 | 10·732739 | ·007311 | 38 | 9·992690 | 31 |
| 30 | 9·260633 | 1137 | ·739367 | 9·267967 | 1176 | 10·732033 | ·007334 | 40 | 9·992666 | 30 |
| 31 | 9·261314 | 1135 | ·738686 | 9·268671 | 1174 | 10·731329 | ·007357 | 38 | 9·992643 | 29 |
| 32 | 9·261994 | 1133 | ·738006 | 9·269375 | 1172 | 10·730625 | ·007381 | 40 | 9·992619 | 28 |
| 33 | 9·262673 | 1131 | ·737327 | 9·270077 | 1170 | 10·729923 | ·007404 | 38 | 9·992596 | 27 |
| 34 | 9·263351 | 1130 | ·736649 | 9·270779 | 1169 | 10·729221 | ·007428 | 40 | 9·992572 | 26 |
| 35 | 9·264027 | 1128 | ·735973 | 9·271479 | 1167 | 10·728521 | ·007451 | 38 | 9·992549 | 25 |
| 36 | 9·264703 | 1126 | ·735297 | 9·272178 | 1165 | 10·727822 | ·007475 | 40 | 9·992525 | 24 |
| 37 | 9·265377 | 1124 | ·734623 | 9·272876 | 1164 | 10·727124 | ·007499 | 40 | 9·992501 | 23 |
| 38 | 9·266051 | 1122 | ·733949 | 9·273573 | 1162 | 10·726427 | ·007522 | 38 | 9·992478 | 22 |
| 39 | 9·266723 | 1120 | ·733277 | 9·274260 | 1160 | 10·725731 | ·007546 | 40 | 9·992454 | 21 |
| 40 | 9·267395 | 1119 | ·732605 | 9·274964 | 1158 | 10·725036 | ·007570 | 40 | 9·992430 | 20 |
| 41 | 9·268065 | 1117 | ·731935 | 9·275658 | 1157 | 10·724342 | ·007594 | 40 | 9·992406 | 19 |
| 42 | 9·268734 | 1115 | ·731266 | 9·276351 | 1155 | 10·723649 | ·007618 | 40 | 9·992382 | 18 |
| 43 | 9·269402 | 1113 | ·730598 | 9·277043 | 1153 | 10·722957 | ·007642 | 38 | 9·992358 | 17 |
| 44 | 9.270069 | 1111 | ·729931 | 9·277734 | 1151 | 10·722266 | ·007665 | 40 | 9·992335 | 16 |
| 45 | 9·270735 | 1110 | ·729265 | 9·278424 | 1150 | 10·721576 | ·007689 | 40 | 9·992311 | 15 |
| 46 | 9·271400 | 1108 | ·728600 | 9·279113 | 1148 | 10·720887 | ·007713 | 40 | 9·992287 | 14 |
| 47 | 9·272064 | 1106 | ·727936 | 9·279801 | 1146 | 10·720199 | ·007737 | 40 | 9·992263 | 13 |
| 48 | 9·272726 | 1105 | ·727274 | 9·280488 | 1145 | 10·719512 | ·007761 | 40 | 9·992239 | 12 |
| 49 | 9·273388 | 1103 | ·726612 | 9·281174 | 1143 | 10·718826 | ·007786 | 42 | 9·992214 | 11 |
| 50 | 9·274049 | 1101 | ·725951 | 9·281858 | 1141 | 10·718142 | ·007810 | 40 | 9·992190 | 10 |
| 51 | 9·274708 | 1099 | ·725292 | 9·282542 | 1140 | 10·717458 | ·007834 | 40 | 9·992166 | 9 |
| 52 | 9·275367 | 1098 | ·724633 | 9·283225 | 1138 | 10·716775 | ·007858 | 40 | 9·992142 | 8 |
| 53 | 9·276025 | 1096 | ·723975 | 9·283907 | 1136 | 10·716093 | ·007882 | 40 | 9·992118 | 7 |
| 54 | 9·276681 | 1094 | ·723319 | 9·284588 | 1135 | 10·715412 | ·007907 | 42 | 9·992093 | 6 |
| 55 | 9·277337 | 1092 | ·722663 | 9·285268 | 1133 | 10·714732 | ·007931 | 40 | 9·992069 | 5 |
| 56 | 9·277991 | 1091 | ·722009 | 9·285947 | 1131 | 10·714053 | ·007956 | 42 | 9·992044 | 4 |
| 57 | 9·278645 | 1089 | ·721355 | 9·286624 | 1130 | 10·713376 | ·007980 | 40 | 9·992020 | 3 |
| 58 | 9·279297 | 1087 | ·720703 | 9·287301 | 1128 | 10·712699 | ·008004 | 40 | 9·991996 | 2 |
| 59 | 9·279948 | 1086 | ·720052 | 9·287977 | 1126 | 10·712023 | ·008029 | 42 | 9·991971 | 1 |
| 60 | 9·280599 | 1084 | ·719401 | 9·288652 | 1125 | 10·711348 | ·008053 | 40 | 9·991947 | 0 |
| ′ | Cosine. | | Secant. | Cotangent. | | Tangent. | Cosecant. | | Sine. | ′ |



79 DEG.

11 DEG.

| ′ | Sine. | Diff. 100″ | Cosecant. | Tangent. | Diff. 100″ | Cotangent. | Secant. | Diff. 100″ | Cosine. | ′ |
|---|---|---|---|---|---|---|---|---|---|---|
| 0 | 9·280599 | | ·719401 | 9·288652 | | 10·711348 | ·008053 | | 9·991947 | 60 |
| 1 | 9·281248 | 1082 | ·718752 | 9·289326 | 1123 | 10·710674 | ·008078 | 42 | 9·991922 | 59 |
| 2 | 9·281897 | 1081 | ·718103 | 9·289999 | 1122 | 10·710001 | ·008103 | 42 | 9·991897 | 58 |
| 3 | 9·282544 | 1079 | ·717456 | 9·290671 | 1120 | 10·709329 | ·008127 | 40 | 9·991873 | 57 |
| 4 | 9·283190 | 1077 | ·716810 | 9·291342 | 1118 | 10·708658 | ·008152 | 42 | 9·991848 | 56 |
| 5 | 9·283836 | 1076 | ·716164 | 9·292013 | 1117 | 10·707987 | ·008177 | 42 | 9·991823 | 55 |
| 6 | 9·284480 | 1074 | ·715520 | 9·292682 | 1115 | 10·707318 | ·008201 | 40 | 9·991799 | 54 |
| 7 | 9·285124 | 1072 | ·714876 | 9·293350 | 1114 | 10·706650 | ·008226 | 42 | 9·991774 | 53 |
| 8 | 9·285766 | 1071 | ·714234 | 9·294017 | 1112 | 10·705983 | ·008251 | 42 | 9·991749 | 52 |
| 9 | 9·286408 | 1069 | ·713592 | 9·294684 | 1111 | 10·705316 | ·008276 | 42 | 9·991724 | 51 |
| 10 | 9·287048 | 1067 | ·712952 | 9·295349 | 1109 | 10·704651 | ·008301 | 42 | 9·991699 | 50 |
| 11 | 9·287688 | 1066 | ·712312 | 9·296013 | 1107 | 10·703987 | ·008326 | 42 | 9·991674 | 49 |
| 12 | 9·288326 | 1064 | ·711674 | 9·296677 | 1106 | 10·703323 | ·008351 | 42 | 9·991649 | 48 |
| 13 | 9·288964 | 1063 | ·711036 | 9·297339 | 1104 | 10·702661 | ·008376 | 42 | 9·991624 | 47 |
| 14 | 9·289600 | 1061 | ·710400 | 9·298001 | 1103 | 10·701999 | ·008401 | 42 | 9·991599 | 46 |
| 15 | 9·290236 | 1059 | ·709764 | 9·298662 | 1101 | 10·701338 | ·008426 | 42 | 9·991574 | 45 |
| 16 | 9·290870 | 1058 | ·709130 | 9·299322 | 1100 | 10·700678 | ·008451 | 42 | 9·991549 | 44 |
| 17 | 9·291504 | 1056 | ·708496 | 9·299980 | 1098 | 10·700020 | ·008476 | 42 | 9·991524 | 43 |
| 18 | 9·292137 | 1054 | ·707863 | 9·300638 | 1096 | 10·699362 | ·008502 | 43 | 9·991498 | 42 |
| 19 | 9·292768 | 1053 | ·707232 | 9·301295 | 1095 | 10·698705 | ·008527 | 42 | 9·991473 | 41 |
| 20 | 9·293399 | 1051 | ·706601 | 9·301951 | 1093 | 10·698049 | ·008552 | 42 | 9·991448 | 40 |
| 21 | 9·294029 | 1050 | ·705971 | 9·302607 | 1092 | 10·697393 | ·008578 | 43 | 9·991422 | 39 |
| 22 | 9·294658 | 1048 | ·705342 | 9·303261 | 1090 | 10·696739 | ·008603 | 42 | 9·991397 | 38 |
| 23 | 9·295286 | 1046 | ·704714 | 9·303914 | 1089 | 10·696086 | ·008628 | 42 | 9·991372 | 37 |
| 24 | 9·295913 | 1045 | ·704087 | 9·304567 | 1087 | 10·695433 | ·008654 | 43 | 9·991346 | 36 |
| 25 | 9·296539 | 1043 | ·703461 | 9·305218 | 1086 | 10·694782 | ·008679 | 42 | 9·991321 | 35 |
| 26 | 9·297164 | 1042 | ·702836 | 9·305869 | 1084 | 10·694131 | ·008705 | 43 | 9·991295 | 34 |
| 27 | 9·297788 | 1040 | ·702212 | 9·306519 | 1083 | 10·693481 | ·008730 | 42 | 9·991270 | 33 |
| 28 | 9·298412 | 1039 | ·701588 | 9·307168 | 1081 | 10·692832 | ·008756 | 43 | 9·991244 | 32 |
| 29 | 9·299034 | 1037 | ·700966 | 9·307815 | 1080 | 10·692185 | ·008782 | 43 | 9·991218 | 31 |
| 30 | 9·299655 | 1036 | ·700345 | 9·308463 | 1078 | 10·691537 | ·008807 | 42 | 9·991193 | 30 |
| 31 | 9·300276 | 1034 | ·699724 | 9·309109 | 1077 | 10·690891 | ·008833 | 43 | 9·991167 | 29 |
| 32 | 9·300895 | 1032 | ·699105 | 9·309754 | 1075 | 10·690246 | ·008859 | 43 | 9·991141 | 28 |
| 33 | 9·301514 | 1031 | ·698486 | 9·310398 | 1074 | 10·689602 | ·008885 | 43 | 9·991115 | 27 |
| 34 | 9·302132 | 1029 | ·697868 | 9·311042 | 1073 | 10·688958 | ·008910 | 42 | 9·991090 | 26 |
| 35 | 9·302748 | 1028 | ·697252 | 9·311685 | 1071 | 10·688315 | ·008936 | 43 | 9·991064 | 25 |
| 36 | 9·303364 | 1026 | ·696636 | 9·312327 | 1070 | 10·687673 | ·008962 | 43 | 9·991038 | 24 |
| 37 | 9·303979 | 1025 | ·696021 | 9·312967 | 1068 | 10·687033 | ·008988 | 43 | 9·991012 | 23 |
| 38 | 9·304593 | 1023 | ·695407 | 9·313608 | 1067 | 10·686392 | ·009014 | 43 | 9·990986 | 22 |
| 39 | 9·305207 | 1022 | ·694793 | 9·314247 | 1065 | 10·685753 | ·009040 | 43 | 9·990960 | 21 |
| 40 | 9·305819 | 1020 | ·694181 | 9·314885 | 1064 | 10·685115 | ·009066 | 43 | 9·990934 | 20 |
| 41 | 9·306430 | 1019 | ·693570 | 9·315523 | 1062 | 10·684477 | ·009092 | 43 | 9·990908 | 19 |
| 42 | 9·307041 | 1017 | ·692959 | 9·316159 | 1061 | 10·683841 | ·009118 | 43 | 9·990882 | 18 |
| 43 | 9·307650 | 1016 | ·692350 | 9·316795 | 1060 | 10·683205 | ·009145 | 45 | 9·990855 | 17 |
| 44 | 9·308259 | 1014 | ·691741 | 9·317430 | 1058 | 10·682570 | ·009171 | 43 | 9·990829 | 16 |
| 45 | 9·308867 | 1013 | ·691133 | 9·318064 | 1057 | 10·681936 | ·009197 | 43 | 9·990803 | 15 |
| 46 | 9·309474 | 1011 | ·690526 | 9·318697 | 1055 | 10·681303 | ·009223 | 43 | 9·990777 | 14 |
| 47 | 9·310080 | 1010 | ·689920 | 9·319329 | 1054 | 10·680671 | ·009250 | 45 | 9·990750 | 13 |
| 48 | 9·310685 | 1008 | ·689315 | 9·319961 | 1053 | 10·680039 | ·009276 | 43 | 9·990724 | 12 |
| 49 | 9·311289 | 1007 | ·688711 | 9·320592 | 1051 | 10·679408 | ·009303 | 45 | 9·990697 | 11 |
| 50 | 9·311893 | 1006 | ·688107 | 9·321222 | 1050 | 10·678778 | ·009329 | 43 | 9·990671 | 10 |
| 51 | 9·312495 | 1004 | ·687505 | 9·321851 | 1048 | 10·678149 | ·009355 | 43 | 9·990645 | 9 |
| 52 | 9·313097 | 1003 | ·686903 | 9·322479 | 1047 | 10·677521 | ·009382 | 45 | 9·990618 | 8 |
| 53 | 9·313698 | 1001 | ·686302 | 9·323106 | 1045 | 10·676894 | ·009409 | 45 | 9·990591 | 7 |
| 54 | 9·314297 | 1000 | ·685703 | 9·323733 | 1044 | 10·676267 | ·009435 | 43 | 9·990565 | 6 |
| 55 | 9·314897 | 998 | ·685103 | 9·324358 | 1043 | 10·675642 | ·009462 | 45 | 9·990538 | 5 |
| 56 | 9·315495 | 997 | ·684505 | 9·324983 | 1041 | 10·675017 | ·009489 | 45 | 9·990511 | 4 |
| 57 | 9·316092 | 996 | ·683908 | 9·325607 | 1040 | 10·674393 | ·009515 | 43 | 9·990485 | 3 |
| 58 | 9·316689 | 994 | ·683311 | 9·326231 | 1039 | 10·673769 | ·009542 | 45 | 9·990458 | 2 |
| 59 | 9·317284 | 993 | ·682716 | 9·326853 | 1037 | 10·673147 | ·009569 | 45 | 9·990431 | 1 |
| 60 | 9·317879 | 991 | ·682121 | 9·327475 | 1036 | 10·672525 | ·009596 | 45 | 9·990404 | 0 |
| ′ | Cosine. | | Secant. | Cotangent. | | Tangent. | Cosecant. | | Sine. | ′ |

78 DEG.

12 DEG.

| ′ | Sine. | Diff. 100″ | Cosecant. | Tangent. | Diff. 100′ | Cotangent. | Secant. | Diff. 100″ | Cosine. | ′ |
|---|---|---|---|---|---|---|---|---|---|---|
| 0 | 9·317879 | | ·682121 | 9·327474 | | 10·672526 | ·009596 | | 9·990404 | 60 |
| 1 | 9·318473 | 990 | ·681527 | 9·328095 | 1035 | 10·671905 | ·009622 | 43 | 9·990378 | 59 |
| 2 | 9·319066 | 988 | ·680934 | 9·328715 | 1033 | 10·671285 | ·009649 | 45 | 9·990351 | 58 |
| 3 | 9·319658 | 987 | ·680342 | 9·329334 | 1032 | 10·670666 | ·009676 | 45 | 9·990324 | 57 |
| 4 | 9·320249 | 986 | ·679751 | 9·329953 | 1030 | 10·670047 | ·009703 | 45 | 9·990297 | 56 |
| 5 | 9·320840 | 984 | ·679160 | 9·330570 | 1029 | 10·669430 | ·009730 | 45 | 9·990270 | 55 |
| 6 | 9·321430 | 983 | ·678570 | 9·331187 | 1028 | 10·668813 | ·009757 | 45 | 9·990243 | 54 |
| 7 | 9·322019 | 982 | ·677981 | 9·331803 | 1026 | 10·668197 | ·009785 | 47 | 9·990215 | 53 |
| 8 | 9·322607 | 980 | ·677393 | 9·332418 | 1025 | 10·667582 | ·009812 | 45 | 9·990188 | 52 |
| 9 | 9·323194 | 979 | ·676806 | 9·333033 | 1024 | 10·666967 | ·009839 | 45 | 9·990161 | 51 |
| 10 | 9·323780 | 977 | ·676220 | 9·333646 | 1023 | 10·666354 | ·009866 | 45 | 9·990134 | 50 |
| 11 | 9·324366 | 976 | ·675634 | 9·334259 | 1021 | 10·665741 | ·009893 | 45 | 9·990107 | 49 |
| 12 | 9·324950 | 975 | ·675050 | 9·334871 | 1020 | 10·665129 | ·009921 | 47 | 9·990079 | 48 |
| 13 | 9·325534 | 973 | ·674466 | 9·335482 | 1019 | 10·664518 | ·009948 | 45 | 9·990052 | 47 |
| 14 | 9·326117 | 972 | ·673883 | 9·336093 | 1017 | 10·663907 | ·009975 | 45 | 9·990025 | 46 |
| 15 | 9·326700 | 970 | ·673300 | 9·336702 | 1016 | 10·663298 | ·010003 | 47 | 9·989997 | 45 |
| 16 | 9·327281 | 969 | ·672719 | 9·337311 | 1015 | 10·662689 | ·010030 | 45 | 9·989970 | 44 |
| 17 | 9·327862 | 968 | ·672138 | 9·337919 | 1013 | 10·662081 | ·010058 | 47 | 9·989942 | 43 |
| 18 | 9·328442 | 966 | ·671558 | 9·338527 | 1012 | 10·661473 | ·010085 | 45 | 9·989915 | 42 |
| 19 | 9·329021 | 965 | ·670979 | 9·339133 | 1011 | 10·660867 | ·010113 | 47 | 9·989887 | 41 |
| 20 | 9·329599 | 964 | ·670401 | 9·339739 | 1010 | 10·660261 | ·010140 | 45 | 9·989860 | 40 |
| 21 | 9·330176 | 962 | ·669824 | 9·340344 | 1008 | 10·659656 | ·010168 | 47 | 9·989832 | 39 |
| 22 | 9·330753 | 961 | ·669247 | 9·340948 | 1007 | 10·659052 | ·010196 | 47 | 9·989804 | 38 |
| 23 | 9·331329 | 960 | ·668671 | 9·341552 | 1006 | 10·658448 | ·010223 | 45 | 9·989777 | 37 |
| 24 | 9·331903 | 958 | ·668097 | 9·342155 | 1004 | 10·657845 | ·010251 | 47 | 9·989749 | 36 |
| 25 | 9·332478 | 957 | ·667522 | 9·342757 | 1003 | 10·657243 | ·010279 | 47 | 9·989721 | 35 |
| 26 | 9·333051 | 956 | ·666949 | 9·343358 | 1002 | 10·656642 | ·010307 | 47 | 9·989693 | 34 |
| 27 | 9·333624 | 954 | ·666376 | 9·343958 | 1001 | 10·656042 | ·010335 | 47 | 9·989665 | 33 |
| 28 | 9·334195 | 953 | ·665805 | 9·344558 | 999 | 10·655442 | ·010363 | 47 | 9·989637 | 32 |
| 29 | 9·334767 | 952 | ·665233 | 9·345157 | 998 | 10·654843 | ·010390 | 45 | 9·989610 | 31 |
| 30 | 9·335337 | 950 | ·664663 | 9·345755 | 997 | 10·654245 | ·010418 | 47 | 9·989582 | 30 |
| 31 | 9·335906 | 949 | ·664094 | 9·346353 | 996 | 10·653647 | ·010447 | 48 | 9·989553 | 29 |
| 32 | 9·336475 | 948 | ·663525 | 9·346949 | 994 | 10·653051 | ·010475 | 47 | 9·989525 | 28 |
| 33 | 9·337043 | 946 | ·662957 | 9·347545 | 993 | 10·652455 | ·010503 | 47 | 9·989497 | 27 |
| 34 | 9·337610 | 945 | ·662390 | 9·348141 | 992 | 10·651859 | ·010531 | 47 | 9·989469 | 26 |
| 35 | 9·338176 | 944 | ·661824 | 9·348735 | 991 | 10·651265 | ·010559 | 47 | 9·989441 | 25 |
| 36 | 9·338742 | 943 | ·661258 | 9·349329 | 990 | 10·650671 | ·010587 | 47 | 9·989413 | 24 |
| 37 | 9·339307 | 941 | ·660693 | 9·349922 | 988 | 10·650078 | ·010615 | 47 | 9·989385 | 23 |
| 38 | 9·339871 | 940 | ·660129 | 9·350514 | 987 | 10·649486 | ·010644 | 48 | 9·989356 | 22 |
| 39 | 9·340434 | 939 | ·659566 | 9·351106 | 986 | 10·648894 | ·010672 | 47 | 9·989328 | 21 |
| 40 | 9·340996 | 937 | ·659004 | 9·351697 | 985 | 10·648303 | ·010700 | 47 | 9·989300 | 20 |
| 41 | 9·341558 | 936 | ·658442 | 9·352287 | 983 | 10·647713 | ·010729 | 48 | 9·989271 | 19 |
| 42 | 9·342119 | 935 | ·657881 | 9·352876 | 982 | 10·647124 | ·010757 | 47 | 9·989243 | 18 |
| 43 | 9·342679 | 934 | ·657321 | 9·353465 | 981 | 10·646535 | ·010786 | 48 | 9·989214 | 17 |
| 44 | 9·343239 | 932 | ·656761 | 9·354053 | 980 | 10·645947 | ·010814 | 47 | 9·989186 | 16 |
| 45 | 9·343797 | 931 | ·656203 | 9·354640 | 979 | 10·645360 | ·010843 | 48 | 9·989157 | 15 |
| 46 | 9·344355 | 930 | ·655645 | 9·355227 | 977 | 10·644773 | ·010872 | 48 | 9·989128 | 14 |
| 47 | 9·344912 | 929 | ·655088 | 9·355813 | 976 | 10·644187 | ·010900 | 47 | 9·989100 | 13 |
| 48 | 9·345469 | 927 | ·654531 | 9·356398 | 975 | 10·643602 | ·010929 | 48 | 9·989071 | 12 |
| 49 | 9·346024 | 926 | ·653976 | 9·356982 | 974 | 10·643018 | ·010958 | 48 | 9·989042 | 11 |
| 50 | 9·346579 | 925 | ·653421 | 9·357566 | 973 | 10·642434 | ·010986 | 47 | 9·989014 | 10 |
| 51 | 9·347134 | 924 | ·652866 | 9·358149 | 971 | 10·641851 | ·011015 | 48 | 9·988985 | 9 |
| 52 | 9·347687 | 922 | ·652313 | 9·358731 | 970 | 10·641269 | ·011044 | 48 | 9·988956 | 8 |
| 53 | 9·348240 | 921 | ·651760 | 9·359313 | 969 | 10·640687 | ·011073 | 48 | 9·988927 | 7 |
| 54 | 9·348792 | 920 | ·651208 | 9·359893 | 968 | 10·640107 | ·011102 | 48 | 9·988898 | 6 |
| 55 | 9·349343 | 919 | ·650657 | 9·360474 | 967 | 10·639526 | ·011131 | 48 | 9·988869 | 5 |
| 56 | 9·349893 | 917 | ·650107 | 9·361053 | 966 | 10·638947 | ·011160 | 48 | 9·988840 | 4 |
| 57 | 9·350443 | 916 | ·649557 | 9·361632 | 965 | 10·638368 | ·011189 | 48 | 9·988811 | 3 |
| 58 | 9·350992 | 915 | ·649008 | 9·362210 | 963 | 10·637790 | ·011218 | 48 | 9·988782 | 2 |
| 59 | 9·351540 | 914 | ·648460 | 9·362787 | 962 | 10·637213 | ·011247 | 48 | 9·988753 | 1 |
| 60 | 9·352088 | 913 | ·647912 | 9·363364 | 961 | 10·636636 | ·011276 | 48 | 9·988724 | 0 |
| ′ | Cosine. | | Secant. | Cotangent. | | Tangent. | Cosecant. | | Sine. | ′ |

77 DEG.

13 DEG.

| ′ | Sine. | Diff. 100″ | Cosecant. | Tangent. | Diff. 100″ | Cotangent. | Secant. | Diff 100″ | Cosine. | ′ |
|---|---|---|---|---|---|---|---|---|---|---|
| 0 | 9·352088 | | ·647912 | 9·363364 | | 10·636636 | ·011276 | | 9·988724 | 60 |
| 1 | 9·352635 | 911 | ·647365 | 9·363940 | 960 | 10·636060 | ·011305 | 48 | 9·988695 | 59 |
| 2 | 9·353181 | 910 | ·646819 | 9·364515 | 959 | 10·635485 | ·011334 | 48 | 9·988666 | 58 |
| 3 | 9·353726 | 909 | ·646274 | 9·365090 | 958 | 10·634910 | ·011364 | 50 | 9·988636 | 57 |
| 4 | 9·354271 | 908 | ·645729 | 9·365664 | 957 | 10·634336 | ·011393 | 48 | 9·988607 | 56 |
| 5 | 9·354815 | 907 | ·645185 | 9·366237 | 955 | 10·633763 | ·011422 | 48 | 9·988578 | 55 |
| 6 | 9·355358 | 905 | ·644642 | 9·366810 | 954 | 10·633190 | ·011452 | 50 | 9·988548 | 54 |
| 7 | 9·355901 | 904 | ·644099 | 9·367382 | 953 | 10·632618 | ·011481 | 48 | 9·988519 | 53 |
| 8 | 9·356443 | 903 | ·643557 | 9·367953 | 952 | 10·632047 | ·011511 | 50 | 9·988489 | 52 |
| 9 | 9·356984 | 902 | ·643016 | 9·368524 | 951 | 10·631476 | ·011540 | 48 | 9·988460 | 51 |
| 10 | 9·357524 | 901 | ·642476 | 9·369094 | 950 | 10·630906 | ·011570 | 50 | 9·988430 | 50 |
| 11 | 9·358064 | 899 | ·641936 | 9·369663 | 949 | 10·630337 | ·011599 | 48 | 9·988401 | 49 |
| 12 | 9·358603 | 898 | ·641397 | 9·370232 | 948 | 10·629768 | ·011629 | 50 | 9·988371 | 48 |
| 13 | 9·359141 | 897 | ·640859 | 9·370799 | 946 | 10·629201 | ·011658 | 48 | 9·988342 | 47 |
| 14 | 9·359678 | 896 | ·640322 | 9·371367 | 945 | 10·628633 | ·011688 | 50 | 9·988312 | 46 |
| 15 | 9·360215 | 895 | ·639785 | 9·371933 | 944 | 10·628067 | ·011718 | 50 | 9·988282 | 45 |
| 16 | 9·360752 | 893 | ·639248 | 9·372499 | 943 | 10·627501 | ·011748 | 50 | 9·988252 | 44 |
| 17 | 9·361287 | 892 | ·638713 | 9·373064 | 942 | 10·626936 | ·011777 | 48 | 9·988223 | 43 |
| 18 | 9·361822 | 891 | ·638178 | 9·373629 | 941 | 10·626371 | ·011807 | 50 | 9·988193 | 42 |
| 19 | 9·362356 | 890 | ·637644 | 9·374193 | 940 | 10·625807 | ·011837 | 50 | 9·988163 | 41 |
| 20 | 9·362889 | 889 | ·637111 | 9·374756 | 939 | 10·625244 | ·011867 | 50 | 9·988133 | 40 |
| 21 | 9·363422 | 888 | ·636578 | 9·375319 | 938 | 10·624681 | ·011897 | 50 | 9·988103 | 39 |
| 22 | 9·363954 | 887 | ·636046 | 9·375881 | 937 | 10·624119 | ·011927 | 50 | 9·988073 | 38 |
| 23 | 9·364485 | 885 | ·635515 | 9·376442 | 935 | 10·623558 | ·011957 | 50 | 9·988043 | 37 |
| 24 | 9·365016 | 884 | ·634984 | 9·377003 | 934 | 10·622997 | ·011987 | 50 | 9·988013 | 36 |
| 25 | 9·365546 | 883 | ·634454 | 9·377563 | 933 | 10·622437 | ·012017 | 50 | 9·987983 | 35 |
| 26 | 9·366075 | 882 | ·633925 | 9·378122 | 932 | 10·621878 | ·012047 | 50 | 9·987953 | 34 |
| 27 | 9·366604 | 881 | ·633396 | 9·378681 | 931 | 10·621319 | ·012078 | 52 | 9·987922 | 33 |
| 28 | 9·367131 | 880 | ·632869 | 9·379239 | 930 | 10·620761 | ·012108 | 50 | 9·987892 | 32 |
| 29 | 9·367659 | 879 | ·632341 | 9·379797 | 929 | 10·620203 | ·012138 | 50 | 9·987862 | 31 |
| 30 | 9·368185 | 878 | ·631815 | 9·380354 | 928 | 10·619646 | ·012168 | 50 | 9·987832 | 30 |
| 31 | 9·368711 | 876 | ·631289 | 9·380910 | 927 | 10·619090 | ·012199 | 52 | 9·987801 | 29 |
| 32 | 9·369236 | 875 | ·630764 | 9·381466 | 926 | 10·618534 | ·012229 | 50 | 9·987771 | 28 |
| 33 | 9·369761 | 874 | ·630239 | 9·382020 | 925 | 10·617980 | ·012260 | 52 | 9·987740 | 27 |
| 34 | 9·370285 | 873 | ·629715 | 9·382575 | 924 | 10·617425 | ·012290 | 50 | 9·987710 | 26 |
| 35 | 9·370808 | 872 | ·629192 | 9·383129 | 923 | 10·616871 | ·012321 | 52 | 9·987679 | 25 |
| 36 | 9·371330 | 871 | ·628670 | 9·383682 | 922 | 10·616318 | ·012351 | 50 | 9·987649 | 24 |
| 37 | 9·371852 | 870 | ·628148 | 9·384234 | 921 | 10·615766 | ·012382 | 52 | 9·987618 | 23 |
| 38 | 9·372373 | 869 | ·627627 | 9·384786 | 920 | 10·615214 | ·012412 | 50 | 9·987588 | 22 |
| 39 | 9·372894 | 867 | ·627106 | 9·385337 | 919 | 10·614663 | ·012443 | 52 | 9·987557 | 21 |
| 40 | 9·373414 | 866 | ·626586 | 9·385888 | 918 | 10·614112 | ·012474 | 52 | 9·987526 | 20 |
| 41 | 9·373933 | 865 | ·626067 | 9·386438 | 917 | 10·613562 | ·012504 | 50 | 9·987496 | 19 |
| 42 | 9·374452 | 864 | ·625548 | 9·386987 | 916 | 10·613013 | ·012535 | 52 | 9·987465 | 18 |
| 43 | 9·374970 | 863 | ·625030 | 9·387536 | 914 | 10·612464 | ·012566 | 52 | 9·987434 | 17 |
| 44 | 9·375487 | 862 | ·624513 | 9·388084 | 913 | 10·611916 | ·012597 | 52 | 9·987403 | 16 |
| 45 | 9·376003 | 861 | ·623997 | 9·388631 | 912 | 10·611369 | ·012628 | 52 | 9·987372 | 15 |
| 46 | 9·376519 | 860 | ·623481 | 9·389178 | 911 | 10·610822 | ·012659 | 52 | 9·987341 | 14 |
| 47 | 9·377035 | 859 | ·622965 | 9·389724 | 910 | 10·610276 | ·012690 | 52 | 9·987310 | 13 |
| 48 | 9·377549 | 858 | ·622451 | 9·390270 | 909 | 10·609730 | ·012721 | 52 | 9·987279 | 12 |
| 49 | 9·378063 | 857 | ·621937 | 9·390815 | 908 | 10·609185 | ·012752 | 52 | 9·987248 | 11 |
| 50 | 9·378577 | 856 | ·621423 | 9·391360 | 907 | 10·608640 | ·012783 | 52 | 9·987217 | 10 |
| 51 | 9·379089 | 854 | ·620911 | 9·391903 | 906 | 10·608097 | ·012814 | 52 | 9·987186 | 9 |
| 52 | 9·379601 | 853 | ·620399 | 9·392447 | 905 | 10·607553 | ·012845 | 52 | 9·987155 | 8 |
| 53 | 9·380113 | 852 | ·619887 | 9·392989 | 904 | 10·607011 | ·012876 | 52 | 9·987124 | 7 |
| 54 | 9·380624 | 851 | ·619376 | 9·393531 | 903 | 10·606469 | ·012908 | 53 | 9·987092 | 6 |
| 55 | 9·381134 | 850 | ·618866 | 9·394073 | 902 | 10·605927 | ·012939 | 52 | 9·987061 | 5 |
| 56 | 9·381643 | 849 | ·618357 | 9·394614 | 901 | 10·605386 | ·012970 | 52 | 9·987030 | 4 |
| 57 | 9·382152 | 848 | ·617848 | 9·395154 | 900 | 10·604846 | ·013002 | 53 | 9·986998 | 3 |
| 58 | 9·382661 | 847 | ·617339 | 9·395694 | 899 | 10·604306 | ·013033 | 52 | 9·986967 | 2 |
| 59 | 9·383168 | 846 | ·616832 | 9·396233 | 898 | 10·603767 | ·013064 | 52 | 9·986936 | 1 |
| 60 | 9·383675 | 845 | ·616325 | 9·396771 | 897 | 10·603229 | ·013096 | 53 | 9·986904 | 0 |
| ′ | Cosine. | | Secant. | Cotangent. | | Tangent. | Cosecant. | | Sine. | ′ |

76 DEG.

14 DEG.

| ′ | Sine. | Diff. 100″ | Cosecant. | Tangent. | Diff. 100″ | Cotangent. | Secant. | Diff. 100″ | Cosine. | ′ |
|---|---|---|---|---|---|---|---|---|---|---|
| 0 | 9·383675 | | ·616325 | 9·396771 | | 10·603229 | ·013096 | | 9·986904 | 60 |
| 1 | 9·384182 | 844 | ·615818 | 9·397309 | 896 | 10·602691 | ·013127 | 52 | 9·986873 | 59 |
| 2 | 9·384687 | 843 | ·615313 | 9·397846 | 896 | 10·602154 | ·013159 | 53 | 9·986841 | 58 |
| 3 | 9·385192 | 842 | ·614808 | 9·398383 | 895 | 10·601617 | ·013191 | 53 | 9·986809 | 57 |
| 4 | 9·385697 | 841 | ·614303 | 9·398919 | 894 | 10·601081 | ·013222 | 52 | 9·986778 | 56 |
| 5 | 9·386201 | 840 | ·613799 | 9·399455 | 893 | 10·600545 | ·013254 | 53 | 9·986746 | 55 |
| 6 | 9·386704 | 839 | ·613296 | 9·399990 | 892 | 10·600010 | ·013286 | 53 | 9·986714 | 54 |
| 7 | 9·387207 | 838 | ·612793 | 9·400524 | 891 | 10·599476 | ·013317 | 52 | 9·986683 | 53 |
| 8 | 9·387709 | 837 | ·612291 | 9·401058 | 890 | 10·598942 | ·013349 | 53 | 9·986651 | 52 |
| 9 | 9·388210 | 836 | ·611790 | 9·401591 | 889 | 10·598409 | ·013381 | 53 | 9·986619 | 51 |
| 10 | 9·388711 | 835 | ·611289 | 9·402124 | 888 | 10·597876 | ·013413 | 53 | 9·986587 | 50 |
| 11 | 9·389211 | 834 | ·610789 | 9·402656 | 887 | 10·597344 | ·013445 | 53 | 9·986555 | 49 |
| 12 | 9·389711 | 833 | ·610289 | 9·403187 | 886 | 10·596813 | ·013477 | 53 | 9·986523 | 48 |
| 13 | 9·390210 | 832 | ·609790 | 9·403718 | 885 | 10·596282 | ·013509 | 53 | 9·986491 | 47 |
| 14 | 9·390708 | 831 | ·609292 | 9·404249 | 884 | 10·595751 | ·013541 | 53 | 9·986459 | 46 |
| 15 | 9·391206 | 830 | ·608794 | 9·404778 | 883 | 10·595222 | ·013573 | 53 | 9·986427 | 45 |
| 16 | 9·391703 | 828 | ·608297 | 9·405308 | 882 | 10·594692 | ·013605 | 53 | 9·986395 | 44 |
| 17 | 9·392199 | 827 | ·607801 | 9·405836 | 881 | 10·594164 | ·013637 | 53 | 9·986363 | 43 |
| 18 | 9·392695 | 826 | ·607305 | 9·406364 | 880 | 10·593636 | ·013669 | 53 | 9·986331 | 42 |
| 19 | 9·393191 | 825 | ·606810 | 9·406892 | 879 | 10·593108 | ·013701 | 53 | 9·986299 | 41 |
| 20 | 9·393685 | 824 | ·606315 | 9·407419 | 878 | 10·592581 | ·013734 | 55 | 9·986266 | 40 |
| 21 | 9·394179 | 823 | ·605821 | 9·407945 | 877 | 10·592055 | ·013766 | 53 | 9·986234 | 39 |
| 22 | 9·394673 | 822 | ·605327 | 9·408471 | 876 | 10·591529 | ·013798 | 53 | 9·986202 | 38 |
| 23 | 9·395166 | 821 | ·604834 | 9·408997 | 875 | 10·591003 | ·013831 | 55 | 9·986169 | 37 |
| 24 | 9·395658 | 820 | ·604342 | 9·409521 | 874 | 10·590479 | ·013863 | 53 | 9·986137 | 36 |
| 25 | 9·396150 | 819 | ·603850 | 9·410045 | 874 | 10·589955 | ·013896 | 55 | 9·986104 | 35 |
| 26 | 9·396641 | 818 | ·603359 | 9·410569 | 873 | 10·589431 | ·013928 | 53 | 9·986072 | 34 |
| 27 | 9·397132 | 817 | ·602868 | 9·411092 | 872 | 10·588908 | ·013961 | 55 | 9·986039 | 33 |
| 28 | 9·397621 | 817 | ·602379 | 9·411615 | 871 | 10·588385 | ·013993 | 53 | 9·986007 | 32 |
| 29 | 9·398111 | 816 | ·601889 | 9·412137 | 870 | 10·587863 | ·014026 | 55 | 9·985974 | 31 |
| 30 | 9·398600 | 815 | ·601400 | 9·412658 | 869 | 10·587342 | ·014058 | 53 | 9·985942 | 30 |
| 31 | 9·399088 | 814 | ·600912 | 9·413179 | 868 | 10·586821 | ·014091 | 55 | 9·985909 | 29 |
| 32 | 9·399575 | 813 | ·600425 | 9·413699 | 867 | 10·586301 | ·014124 | 55 | 9·985876 | 28 |
| 33 | 9·400062 | 812 | ·599938 | 9·414219 | 866 | 10·585781 | ·014157 | 55 | 9·985843 | 27 |
| 34 | 9·400549 | 811 | ·599451 | 9·414738 | 865 | 10·585262 | ·014189 | 53 | 9·985811 | 26 |
| 35 | 9·401035 | 810 | ·598965 | 9·415257 | 864 | 10·584743 | ·014222 | 55 | 9·985778 | 25 |
| 36 | 9·401520 | 809 | ·598480 | 9·415775 | 864 | 10·584225 | ·014255 | 55 | 9·985745 | 24 |
| 37 | 9·402005 | 808 | ·597995 | 9·416293 | 863 | 10·583707 | ·014288 | 55 | 9·985712 | 23 |
| 38 | 9·402489 | 807 | ·597511 | 9·416810 | 862 | 10·583190 | ·014321 | 55 | 9·985679 | 22 |
| 39 | 9·402972 | 806 | ·597028 | 9·417326 | 861 | 10·582674 | ·014354 | 55 | 9·985646 | 21 |
| 40 | 9·403455 | 805 | ·596545 | 9·417842 | 860 | 10·582158 | ·014387 | 55 | 9·985613 | 20 |
| 41 | 9·403938 | 804 | ·596062 | 9·418358 | 859 | 10·581642 | ·014420 | 55 | 9·985580 | 19 |
| 42 | 9·404420 | 803 | ·595580 | 9·418873 | 858 | 10·581127 | ·014453 | 55 | 9·985547 | 18 |
| 43 | 9·404901 | 802 | ·595099 | 9·419387 | 857 | 10·580613 | ·014486 | 55 | 9·985514 | 17 |
| 44 | 9·405382 | 801 | ·594618 | 9·419901 | 856 | 10·580099 | ·014520 | 57 | 9·985480 | 16 |
| 45 | 9·405862 | 800 | ·594138 | 9·420415 | 855 | 10·579585 | ·014553 | 55 | 9·985447 | 15 |
| 46 | 9·406341 | 799 | ·593659 | 9·420927 | 855 | 10·579073 | ·014586 | 55 | 9·985414 | 14 |
| 47 | 9·406820 | 798 | ·593180 | 9·421440 | 854 | 10·578560 | ·014619 | 55 | 9·985381 | 13 |
| 48 | 9·407299 | 797 | ·592701 | 9·421952 | 853 | 10·578048 | ·014653 | 57 | 9·985347 | 12 |
| 49 | 9·407777 | 796 | ·592223 | 9·422463 | 852 | 10·577537 | ·014686 | 55 | 9·985314 | 11 |
| 50 | 9·408254 | 795 | ·591746 | 9·422974 | 851 | 10·577026 | ·014720 | 57 | 9·985280 | 10 |
| 51 | 9·408731 | 794 | ·591269 | 9·423484 | 850 | 10·576516 | ·014753 | 55 | 9·985247 | 9 |
| 52 | 9·409207 | 794 | ·590793 | 9·423993 | 849 | 10·576007 | ·014787 | 57 | 9·985213 | 8 |
| 53 | 9·409682 | 793 | ·590318 | 9·424503 | 848 | 10·575497 | ·014820 | 55 | 9·985180 | 7 |
| 54 | 9·410157 | 792 | ·589843 | 9·425011 | 848 | 10·574989 | ·014854 | 57 | 9·985146 | 6 |
| 55 | 9·410632 | 791 | ·589368 | 9·425519 | 847 | 10·574481 | ·014887 | 55 | 9·985113 | 5 |
| 56 | 9·411106 | 790 | ·588894 | 9·426027 | 846 | 10·573973 | ·014921 | 57 | 9·985079 | 4 |
| 57 | 9·411579 | 789 | ·588421 | 9·426534 | 845 | 10·573466 | ·014955 | 57 | 9·985045 | 3 |
| 58 | 9·412052 | 788 | ·587948 | 9·427041 | 844 | 10·572959 | ·014989 | 57 | 9·985011 | 2 |
| 59 | 9·412524 | 787 | ·587476 | 9·427547 | 843 | 10·572453 | ·015022 | 55 | 9·984978 | 1 |
| 60 | 9·412996 | 786 | ·587004 | 9·428052 | 843 | 10·571948 | ·015056 | 57 | 9·984944 | 0 |
| ′ | Cosine. | | Secant. | Cotangent. | | Tangent. | Cosecant. | | Sine. | ′ |

75 DEG.

15 DEG.

| ′ | Sine. | Diff. 100″ | Cosecant. | Tangent. | Diff. 100″ | Cotangent. | Secant. | Diff. 100″ | Cosine. | ′ |
|---|---|---|---|---|---|---|---|---|---|---|
| 0 | 9·412996 | | ·587004 | 9·428052 | | 10·571948 | ·015056 | | 9·984944 | 60 |
| 1 | 9·413467 | 785 | ·586533 | 9·428557 | 842 | 10·571443 | ·015090 | 57 | 9·984910 | 59 |
| 2 | 9·413938 | 784 | ·586062 | 9·429062 | 841 | 10·570938 | ·015124 | 57 | 9·984876 | 58 |
| 3 | 9·414408 | 783 | ·585592 | 9·429566 | 840 | 10·570434 | ·015158 | 57 | 9·984842 | 57 |
| 4 | 9·414878 | 783 | ·585122 | 9·430070 | 839 | 10·569930 | ·015192 | 57 | 9·984808 | 56 |
| 5 | 9·415347 | 782 | ·584653 | 9·430573 | 838 | 10·569427 | ·015226 | 57 | 9·984774 | 55 |
| 6 | 9·415815 | 781 | ·584185 | 9·431075 | 838 | 10·568925 | ·015260 | 57 | 9·984740 | 54 |
| 7 | 9·416283 | 780 | ·583717 | 9·431577 | 837 | 10·568423 | ·015294 | 57 | 9·984706 | 53 |
| 8 | 9·416751 | 779 | ·583249 | 9·432079 | 836 | 10·567921 | ·015328 | 57 | 9·984672 | 52 |
| 9 | 9·417217 | 778 | ·582783 | 9·432580 | 865 | 10·567420 | ·015362 | 57 | 9·984638 | 51 |
| 10 | 9·417684 | 777 | ·582316 | 9·433080 | 834 | 10·566920 | ·015397 | 58 | 9·984603 | 50 |
| 11 | 9·418150 | 776 | ·581850 | 9·433580 | 833 | 10·566420 | ·015431 | 57 | 9·984569 | 49 |
| 12 | 9·418615 | 775 | ·581385 | 9·434080 | 832 | 10·565920 | ·015465 | 57 | 9·984535 | 48 |
| 13 | 9·419079 | 774 | ·580921 | 9·434579 | 832 | 10·565421 | ·015500 | 58 | 9·984500 | 47 |
| 14 | 9·419544 | 773 | ·580456 | 9·435078 | 831 | 10·564922 | ·015534 | 57 | 9·984466 | 46 |
| 15 | 9·420007 | 773 | ·579993 | 9·435576 | 830 | 10·564424 | ·015568 | 57 | 9·984432 | 45 |
| 16 | 9·420470 | 772 | ·579530 | 9·436073 | 829 | 10·563927 | ·015603 | 58 | 9·984397 | 44 |
| 17 | 9·420983 | 771 | ·579067 | 9·436570 | 828 | 10·563430 | ·015637 | 57 | 9·984363 | 43 |
| 18 | 9·421395 | 770 | ·578605 | 9·437067 | 828 | 10·562933 | ·015672 | 58 | 9·984328 | 42 |
| 19 | 9·421857 | 769 | ·578143 | 9·437563 | 827 | 10·562437 | ·015706 | 58 | 9·984294 | 41 |
| 20 | 9·422318 | 768 | ·577682 | 9·438059 | 826 | 10·561941 | ·015741 | 57 | 9·984259 | 40 |
| 21 | 9·422778 | 767 | ·577222 | 9·438554 | 825 | 10·561446 | ·015776 | 58 | 9·984224 | 39 |
| 22 | 9·423238 | 767 | ·576762 | 9·439048 | 824 | 10·560952 | ·015810 | 57 | 9·984190 | 38 |
| 23 | 9·423697 | 766 | ·576303 | 9·439543 | 823 | 10·560457 | ·015845 | 58 | 9·984155 | 37 |
| 24 | 9·424156 | 765 | ·575844 | 9·440036 | 823 | 10·559964 | ·015880 | 58 | 9·984120 | 36 |
| 25 | 9·424615 | 764 | ·575385 | 9·440529 | 822 | 10·559471 | ·015915 | 58 | 9·984085 | 35 |
| 26 | 9·425073 | 763 | ·574927 | 9·441022 | 821 | 10·558978 | ·015950 | 58 | 9·984050 | 34 |
| 27 | 9·425530 | 762 | ·574470 | 9·441514 | 820 | 10·558486 | ·015985 | 58 | 9·984015 | 33 |
| 28 | 9·425987 | 761 | ·574013 | 9·442006 | 819 | 10·557994 | ·016019 | 57 | 9·983981 | 32 |
| 29 | 9·426443 | 760 | ·573557 | 9·442497 | 819 | 10·557503 | ·016054 | 58 | 9·983946 | 31 |
| 30 | 9·426899 | 760 | ·573101 | 9·442988 | 818 | 10·557012 | ·016089 | 58 | 9·983911 | 30 |
| 31 | 9·427354 | 759 | ·572646 | 9·443479 | 817 | 10·556521 | ·016125 | 60 | 9·983875 | 29 |
| 32 | 9·427809 | 758 | ·572191 | 9·443968 | 816 | 10·556032 | ·016160 | 58 | 9·983840 | 28 |
| 33 | 9·428263 | 757 | ·571737 | 9·444458 | 816 | 10·555542 | ·016195 | 58 | 9·983805 | 27 |
| 34 | 9·428717 | 756 | ·571283 | 9·444947 | 815 | 10·555053 | ·016230 | 58 | 9·983770 | 26 |
| 35 | 9·429170 | 755 | ·570830 | 9·445435 | 814 | 10·554565 | ·016265 | 58 | 9·983735 | 25 |
| 36 | 9·429623 | 754 | ·570377 | 9·445923 | 813 | 10·554077 | ·016300 | 58 | 9·983700 | 24 |
| 37 | 9·430075 | 753 | ·569925 | 9·446411 | 812 | 10·553589 | ·016336 | 60 | 9·983664 | 23 |
| 38 | 9·430527 | 752 | ·569473 | 9·446898 | 812 | 10·553102 | ·016371 | 58 | 9·983629 | 22 |
| 39 | 9·430978 | 752 | ·569022 | 9·447384 | 811 | 10·552616 | ·016406 | 58 | 9·983594 | 21 |
| 40 | 9·431429 | 751 | ·568571 | 9·447870 | 810 | 10·552130 | ·016442 | 60 | 9·983558 | 20 |
| 41 | 9·431879 | 750 | ·568121 | 9·448356 | 809 | 10·551644 | ·016477 | 58 | 9·983523 | 19 |
| 42 | 9·432329 | 749 | ·567671 | 9·448841 | 809 | 10·551159 | ·016513 | 60 | 9·983487 | 18 |
| 43 | 9·432778 | 749 | ·567222 | 9·449326 | 808 | 10·550674 | ·016548 | 58 | 9·983452 | 17 |
| 44 | 9·433226 | 748 | ·566774 | 9·449810 | 807 | 10·550190 | ·016584 | 60 | 9·983416 | 16 |
| 45 | 9·433675 | 747 | ·566325 | 9·450294 | 806 | 10·549706 | ·016619 | 58 | 9·983381 | 15 |
| 46 | 9·434122 | 746 | ·565878 | 9·450777 | 806 | 10·549223 | ·016655 | 60 | 9·983345 | 14 |
| 47 | 9·434569 | 745 | ·565431 | 9·451260 | 805 | 10·548740 | ·016691 | 60 | 9·983309 | 13 |
| 48 | 9·435016 | 744 | ·564984 | 9·451743 | 804 | 10·548257 | ·016727 | 60 | 9·983273 | 12 |
| 49 | 9·435462 | 744 | ·564538 | 9·452225 | 803 | 10·547775 | ·016762 | 58 | 9·983238 | 11 |
| 50 | 9·435908 | 743 | ·564092 | 9·452706 | 802 | 10·547294 | ·016798 | 60 | 9·983202 | 10 |
| 51 | 9·436353 | 742 | ·563647 | 9·453187 | 802 | 10·546813 | ·016834 | 60 | 9·983166 | 9 |
| 52 | 9·436798 | 741 | ·563202 | 9·453668 | 801 | 10·546332 | ·016870 | 60 | 9·983130 | 8 |
| 53 | 9·437242 | 740 | ·562758 | 9·454148 | 800 | 10·545852 | ·016906 | 60 | 9·983094 | 7 |
| 54 | 9·437686 | 740 | ·562314 | 9·454628 | 799 | 10·545372 | ·016942 | 60 | 9·983058 | 6 |
| 55 | 9·438129 | 739 | ·561871 | 9·455107 | 799 | 10·544893 | ·016978 | 60 | 9·983022 | 5 |
| 56 | 9·438572 | 738 | ·561428 | 9·455586 | 798 | 10·544414 | ·017014 | 60 | 9·982986 | 4 |
| 57 | 9·439014 | 737 | ·560986 | 9·456064 | 797 | 10·543936 | ·017050 | 60 | 9·982950 | 3 |
| 58 | 9·439456 | 736 | ·560544 | 9·456542 | 796 | 10·543458 | ·017086 | 60 | 9·982914 | 2 |
| 59 | 9·439897 | 736 | ·560103 | 9·457019 | 796 | 10·542981 | ·017122 | 60 | 9·982878 | 1 |
| 60 | 9·440338 | 735 | ·559662 | 9·457496 | 795 | 10·542504 | ·017158 | 60 | 9·982842 | 0 |
| ′ | Cosine. | | Secant. | Cotangent. | | Tangent. | Cosecant. | | Sine. | ′ |

74 DEG.

16 DEG.

| ′ | Sine. | Diff. 100″ | Cosecant. | Tangent. | Diff. 100″ | Cotangent. | Secant. | Diff. 100″ | Cosine. | ′ |
|---|---|---|---|---|---|---|---|---|---|---|
| 0 | 9·440338 | | ·559662 | 9·457496 | | 10·542504 | ·017158 | | 9·982842 | 60 |
| 1 | 9·440778 | 734 | ·559222 | 9·457973 | 794 | 10·542027 | ·017195 | 62 | 9·982805 | 59 |
| 2 | 9·441218 | 733 | ·558782 | 9·458449 | 793 | 10·541551 | ·017231 | 60 | 9·982769 | 58 |
| 3 | 9·441658 | 732 | ·558342 | 9·458925 | 793 | 10·541075 | ·017267 | 60 | 9·982733 | 57 |
| 4 | 9·442096 | 731 | ·557904 | 9·459400 | 792 | 10·540600 | ·017304 | 62 | 9·982696 | 56 |
| 5 | 9·442535 | 731 | ·557465 | 9·459875 | 791 | 10·540125 | ·017340 | 60 | 9·982660 | 55 |
| 6 | 9·442973 | 730 | ·557027 | 9·460349 | 790 | 10·539651 | ·017376 | 60 | 9·982624 | 54 |
| 7 | 9·443410 | 729 | ·556590 | 9·460823 | 790 | 10·539177 | ·017413 | 62 | 9·982587 | 53 |
| 8 | 9·443847 | 728 | ·556153 | 9·461297 | 789 | 10·538703 | ·017449 | 60 | 9·982551 | 52 |
| 9 | 9·444284 | 727 | ·555716 | 9·461770 | 788 | 10·538230 | ·017486 | 62 | 9·982514 | 51 |
| 10 | 9·444720 | 727 | ·555280 | 9·462242 | 788 | 10·537758 | ·017523 | 62 | 9·982477 | 50 |
| 11 | 9·445155 | 726 | ·554845 | 9·462714 | 787 | 10·537286 | ·017559 | 60 | 9·982441 | 49 |
| 12 | 9·445590 | 725 | ·554410 | 9·463186 | 786 | 10·536814 | ·017596 | 62 | 9·982404 | 48 |
| 13 | 9·446025 | 724 | ·553975 | 9·463658 | 785 | 10·536342 | ·017633 | 62 | 9·982367 | 47 |
| 14 | 9·446459 | 723 | ·553541 | 9·464128 | 785 | 10·535872 | ·017669 | 60 | 9·982331 | 46 |
| 15 | 9·446893 | 723 | ·553107 | 9·464599 | 784 | 10·535401 | ·017706 | 62 | 9·982294 | 45 |
| 16 | 9·447326 | 722 | ·552674 | 9·465069 | 783 | 10·534931 | ·017743 | 62 | 9·982257 | 44 |
| 17 | 9·447759 | 721 | ·552241 | 9·465539 | 783 | 10·534461 | ·017780 | 62 | 9·982220 | 43 |
| 18 | 9·448191 | 720 | ·551809 | 9·466008 | 782 | 10·533992 | ·017817 | 62 | 9·982183 | 42 |
| 19 | 9·448623 | 720 | ·551377 | 9·466476 | 781 | 10·533524 | ·017854 | 62 | 9·982146 | 41 |
| 20 | 9·449054 | 719 | ·550946 | 9·466945 | 780 | 10·533055 | ·017891 | 62 | 9·982109 | 40 |
| 21 | 9·449485 | 718 | ·550515 | 9·467413 | 780 | 10·532587 | ·017928 | 62 | 9·982072 | 39 |
| 22 | 9·449915 | 717 | ·550085 | 9·467880 | 779 | 10·532120 | ·017965 | 62 | 9·982035 | 38 |
| 23 | 9·450345 | 716 | ·549655 | 9·468347 | 778 | 10·531653 | ·018002 | 62 | 9·981998 | 37 |
| 24 | 9·450775 | 716 | ·549225 | 9·468814 | 778 | 10·531186 | ·018039 | 62 | 9·981961 | 36 |
| 25 | 9·451204 | 715 | ·548796 | 9·469280 | 777 | 10·530720 | ·018076 | 62 | 9·981924 | 35 |
| 26 | 9·451632 | 714 | ·548368 | 9·469746 | 776 | 10·530254 | ·018114 | 63 | 9·981886 | 34 |
| 27 | 9·452060 | 713 | ·547940 | 9·470211 | 775 | 10·529789 | ·018151 | 62 | 9·981849 | 33 |
| 28 | 9·452488 | 713 | ·547512 | 9·470676 | 775 | 10·529324 | ·018188 | 62 | 9·981812 | 32 |
| 29 | 9·452915 | 712 | ·547085 | 9·471141 | 774 | 10·528859 | ·018226 | 63 | 9·981774 | 31 |
| 30 | 9·453342 | 711 | ·546658 | 9·471605 | 773 | 10·528395 | ·018263 | 62 | 9·981737 | 30 |
| 31 | 9·453768 | 710 | ·546232 | 9·472068 | 773 | 10·527932 | ·018300 | 62 | 9·981700 | 29 |
| 32 | 9·454194 | 710 | ·545806 | 9·472532 | 772 | 10·527468 | ·018338 | 63 | 9·981662 | 28 |
| 33 | 9·454619 | 709 | ·545381 | 9·472995 | 771 | 10·527005 | ·018375 | 62 | 9·981625 | 27 |
| 34 | 9·455044 | 708 | ·544956 | 9·473457 | 771 | 10·526543 | ·018413 | 63 | 9·981587 | 26 |
| 35 | 9·455469 | 707 | ·544531 | 9·473919 | 770 | 10·526081 | ·018451 | 63 | 9·981549 | 25 |
| 36 | 9·455893 | 707 | ·544107 | 9·474381 | 769 | 10·525619 | ·018488 | 62 | 9·981512 | 24 |
| 37 | 9·456316 | 706 | ·543684 | 9·474842 | 769 | 10·525158 | ·018526 | 63 | 9·981474 | 23 |
| 38 | 9·456739 | 705 | ·543261 | 9·475303 | 768 | 10·524697 | ·018564 | 63 | 9·981436 | 22 |
| 39 | 9·457162 | 704 | ·542838 | 9·475763 | 767 | 10·524237 | ·018601 | 62 | 9·981399 | 21 |
| 40 | 9·457584 | 704 | ·542416 | 9·476223 | 767 | 10·523777 | ·018639 | 63 | 9·981361 | 20 |
| 41 | 9·458006 | 703 | ·541994 | 9·476683 | 766 | 10·523317 | ·018677 | 63 | 9·981323 | 19 |
| 42 | 9·458427 | 702 | ·541573 | 9·477142 | 765 | 10·522858 | ·018715 | 63 | 9·981285 | 18 |
| 43 | 9·458848 | 701 | ·541152 | 9·477601 | 765 | 10·522399 | ·018753 | 63 | 9·981247 | 17 |
| 44 | 9·459268 | 701 | ·540732 | 9·478059 | 764 | 10·521941 | ·018791 | 63 | 9·981209 | 16 |
| 45 | 9·459688 | 700 | ·540312 | 9·478517 | 763 | 10·521483 | ·018829 | 63 | 9·981171 | 15 |
| 46 | 9·460108 | 699 | ·539892 | 9·478975 | 763 | 10·521025 | ·018867 | 63 | 9·981133 | 14 |
| 47 | 9·460527 | 698 | ·539473 | 9·479432 | 762 | 10·520568 | ·018905 | 63 | 9·981095 | 13 |
| 48 | 9·460946 | 698 | ·539054 | 9·479889 | 761 | 10·520111 | ·018943 | 63 | 9·981057 | 12 |
| 49 | 9·461364 | 697 | ·538636 | 9·480345 | 761 | 10·519655 | ·018981 | 63 | 9·981019 | 11 |
| 50 | 9·461782 | 696 | ·538218 | 9·480801 | 760 | 10·519199 | ·019019 | 63 | 9·980981 | 10 |
| 51 | 9·462199 | 695 | ·537801 | 9·481257 | 759 | 10·518743 | ·019058 | 65 | 9·980942 | 9 |
| 52 | 9·462616 | 695 | ·537384 | 9·481712 | 759 | 10·518288 | ·019096 | 63 | 9·980904 | 8 |
| 53 | 9·463032 | 694 | ·536968 | 9·482167 | 758 | 10·517833 | ·019134 | 63 | 9·980866 | 7 |
| 54 | 9·463448 | 693 | ·536552 | 9 482621 | 757 | 10·517379 | ·019173 | 65 | 9·980827 | 6 |
| 55 | 9·463864 | 693 | ·536136 | 9·483075 | 757 | 10·516925 | ·019211 | 63 | 9·980789 | 5 |
| 56 | 9·464279 | 692 | ·535721 | 9·483529 | 756 | 10·516471 | ·019250 | 65 | 9·980750 | 4 |
| 57 | 9·464694 | 691 | ·535306 | 9·483982 | 755 | 10·516018 | ·019288 | 63 | 9·980712 | 3 |
| 58 | 9·465108 | 690 | ·534892 | 9·484435 | 755 | 10·515565 | ·019327 | 65 | 9·980673 | 2 |
| 59 | 9·465522 | 690 | ·534478 | 9·484887 | 754 | 10·515113 | ·019365 | 63 | 9·980635 | 1 |
| 60 | 9·465935 | 689 | ·534065 | 9·485339 | 753 | 10·514661 | ·019404 | 65 | 9·980596 | 0 |
| ′ | Cosine. | | Secant. | Cotangent. | | Tangent. | Cosecant. | | Sine. | ′ |

73 DEG.

17 DEG.

| ′ | Sine. | Diff. 100″ | Cosecant. | Tangent. | Diff. 100″ | Cotangent. | Secant. | Diff. 100″ | Cosine. | ′ |
|---|---|---|---|---|---|---|---|---|---|---|
| 0 | 9·465935 | | ·534065 | 9·485339 | | 10·514661 | ·019404 | | 9·980596 | 60 |
| 1 | 9·466348 | 688 | ·533652 | 9·485791 | 753 | 10·514209 | ·019442 | 63 | 9·980558 | 59 |
| 2 | 9·466761 | 688 | ·533239 | 9·486242 | 752 | 10·513758 | ·019481 | 65 | 9·980519 | 58 |
| 3 | 9·467173 | 687 | ·532827 | 9·486693 | 751 | 10·513307 | ·019520 | 65 | 9·980480 | 57 |
| 4 | 9·467585 | 686 | ·532415 | 9·487143 | 751 | 10·512857 | ·019558 | 63 | 9·980442 | 56 |
| 5 | 9·467996 | 685 | ·532004 | 9·487593 | 750 | 10·512407 | ·019597 | 65 | 9·980403 | 55 |
| 6 | 9·468407 | 685 | ·531593 | 9·488043 | 749 | 10·511957 | ·019636 | 65 | 9·980364 | 54 |
| 7 | 9·468817 | 684 | ·531183 | 9·488492 | 749 | 10·511508 | ·019675 | 65 | 9·980325 | 53 |
| 8 | 9·469227 | 683 | ·530773 | 9·488941 | 748 | 10·511059 | ·019714 | 65 | 9·980286 | 52 |
| 9 | 9·469637 | 683 | ·530363 | 9·489390 | 747 | 10·510610 | ·019753 | 65 | 9·980247 | 51 |
| 10 | 9·470046 | 682 | ·529954 | 9·489838 | 747 | 10·510162 | ·019792 | 65 | 9·980208 | 50 |
| 11 | 9·470455 | 681 | ·529545 | 9·490286 | 746 | 10·509714 | ·019831 | 65 | 9·980169 | 49 |
| 12 | 9·470863 | 680 | ·529137 | 9·490733 | 746 | 10·509267 | ·019870 | 65 | 9·980130 | 48 |
| 13 | 9·471271 | 680 | ·528729 | 9·491180 | 745 | 10·508820 | ·019909 | 65 | 9·980091 | 47 |
| 14 | 9·471679 | 679 | ·528321 | 9·491627 | 744 | 10·508373 | ·019948 | 65 | 9·980052 | 46 |
| 15 | 9·472086 | 678 | ·527914 | 9·492073 | 744 | 10·507927 | ·019988 | 67 | 9·980012 | 45 |
| 16 | 9·472492 | 678 | ·527508 | 9·492519 | 743 | 10·507481 | ·020027 | 65 | 9·979973 | 44 |
| 17 | 9·472898 | 677 | ·527102 | 9·492965 | 743 | 10·507035 | ·020066 | 65 | 9·979934 | 43 |
| 18 | 9·473304 | 676 | ·526696 | 9·493410 | 742 | 10·506590 | ·020105 | 65 | 9·979895 | 42 |
| 19 | 9·473710 | 676 | ·526290 | 9·493854 | 741 | 10·506146 | ·020145 | 67 | 9·979855 | 41 |
| 20 | 9·474115 | 675 | ·525885 | 9·494299 | 740 | 10·505701 | ·020184 | 65 | 9·979816 | 40 |
| 21 | 9·474519 | 674 | ·525481 | 9·494743 | 740 | 10·505257 | ·020224 | 67 | 9·979776 | 39 |
| 22 | 9·474923 | 674 | ·525077 | 9·495186 | 739 | 10·504814 | ·020263 | 65 | 9·979737 | 38 |
| 23 | 9·475327 | 673 | ·524673 | 9·495630 | 739 | 10·504370 | ·020303 | 67 | 9·979697 | 37 |
| 24 | 9·475730 | 672 | ·524270 | 9·496073 | 738 | 10·503927 | ·020342 | 65 | 9·979658 | 36 |
| 25 | 9·476133 | 672 | ·523867 | 9·496515 | 737 | 10·503485 | ·020382 | 67 | 9·979618 | 35 |
| 26 | 9·476536 | 671 | ·523464 | 9·496957 | 737 | 10·503043 | ·020421 | 65 | 9·979579 | 34 |
| 27 | 9·476938 | 670 | ·523062 | 9·497399 | 736 | 10·502601 | ·020461 | 67 | 9·979539 | 33 |
| 28 | 9·477340 | 669 | ·522660 | 9·497841 | 736 | 10·502159 | ·020501 | 67 | 9·979499 | 32 |
| 29 | 9·477741 | 669 | ·522259 | 9·498282 | 735 | 10·501718 | ·020541 | 67 | 9·979459 | 31 |
| 30 | 9·478142 | 668 | ·521858 | 9·498722 | 734 | 10·501278 | ·020580 | 65 | 9·979420 | 30 |
| 31 | 9·478542 | 667 | ·521458 | 9·499163 | 734 | 10·500837 | ·020620 | 67 | 9·979380 | 29 |
| 32 | 9·478942 | 667 | ·521058 | 9·499603 | 733 | 10·500397 | ·020660 | 67 | 9·979340 | 28 |
| 33 | 9·479342 | 666 | ·520658 | 9·500042 | 733 | 10·499958 | ·020700 | 67 | 9·979300 | 27 |
| 34 | 9·479741 | 665 | ·520259 | 9·500481 | 732 | 10·499519 | ·020740 | 67 | 9·979260 | 26 |
| 35 | 9·480140 | 665 | ·519860 | 9·500920 | 731 | 10·499080 | ·020780 | 67 | 9·979220 | 25 |
| 36 | 9·480539 | 664 | ·519461 | 9·501359 | 731 | 10·498641 | ·020820 | 67 | 9·979180 | 24 |
| 37 | 9·480937 | 663 | ·519063 | 9·501797 | 730 | 10·498203 | ·020860 | 67 | 9·979140 | 23 |
| 38 | 9·481334 | 663 | ·518666 | 9·502235 | 730 | 10·497765 | ·020900 | 67 | 9·979100 | 22 |
| 39 | 9·481731 | 662 | ·518269 | 9·502672 | 729 | 10·497328 | ·020941 | 68 | 9·979059 | 21 |
| 40 | 9·482128 | 661 | ·517872 | 9·503109 | 728 | 10·496891 | ·020981 | 67 | 9·979019 | 20 |
| 41 | 9·482525 | 661 | ·517475 | 9·503546 | 728 | 10·496454 | ·021021 | 67 | 9·978979 | 19 |
| 42 | 9·482921 | 660 | ·517079 | 9·503982 | 727 | 10·496018 | ·021061 | 67 | 9·978939 | 18 |
| 43 | 9·483316 | 659 | ·516684 | 9·504418 | 727 | 10·495582 | ·021102 | 68 | 9·978898 | 17 |
| 44 | 9·483712 | 659 | ·516288 | 9·504854 | 726 | 10·495146 | ·021142 | 67 | 9·978858 | 16 |
| 45 | 9·484107 | 658 | ·515893 | 9·505289 | 725 | 10·494711 | ·021183 | 68 | 9·978817 | 15 |
| 46 | 9·484501 | 657 | ·515499 | 9·505724 | 725 | 10·494276 | ·021223 | 67 | 9·978777 | 14 |
| 47 | 9·484895 | 657 | ·515105 | 9·506159 | 724 | 10·493841 | ·021263 | 67 | 9·978737 | 13 |
| 48 | 9·485289 | 656 | ·514711 | 9·506593 | 724 | 10·493407 | ·021304 | 68 | 9·978696 | 12 |
| 49 | 9·485682 | 655 | ·514318 | 9·507027 | 723 | 10·492973 | ·021345 | 68 | 9·978655 | 11 |
| 50 | 9·486075 | 655 | ·513925 | 9·507460 | 722 | 10·492540 | ·021385 | 67 | 9·978615 | 10 |
| 51 | 9·486467 | 654 | ·513533 | 9·507893 | 722 | 10·492107 | ·021426 | 68 | 9·978574 | 9 |
| 52 | 9·486860 | 653 | ·513140 | 9·508326 | 721 | 10·491674 | ·021467 | 68 | 9·978533 | 8 |
| 53 | 9·487251 | 653 | ·512749 | 9·508759 | 721 | 10·491241 | ·021507 | 67 | 9·978493 | 7 |
| 54 | 9·487643 | 652 | ·512357 | 9·509191 | 720 | 10·490809 | ·021548 | 68 | 9·978452 | 6 |
| 55 | 9·488034 | 651 | ·511966 | 9·509622 | 719 | 10·490378 | ·021589 | 68 | 9·978411 | 5 |
| 56 | 9·488424 | 651 | ·511576 | 9·510054 | 719 | 10·489946 | ·021630 | 68 | 9·978370 | 4 |
| 57 | 9·488814 | 650 | ·511186 | 9·510485 | 718 | 10·489515 | ·021671 | 68 | 9·978329 | 3 |
| 58 | 9·489204 | 650 | ·510796 | 9·510916 | 718 | 10·489084 | ·021712 | 68 | 9·978288 | 2 |
| 59 | 9·489593 | 649 | ·510407 | 9·511346 | 717 | 10·488654 | ·021753 | 68 | 9·978247 | 1 |
| 60 | 9·489982 | 648 | ·510018 | 9·511776 | 717 | 10·488224 | ·021794 | 68 | 9·978206 | 0 |
| ′ | Cosine. | | Secant. | Cotangent. | | Tangent. | Cosecant. | | Sine. | ′ |

72 DEG.

18 DEG.

| ′ | Sine. | Diff. 100″ | Cosecant. | Tangent. | Diff. 100″ | Cotangent. | Secant. | Diff. 100″ | Cosine. | ′ |
|---|---|---|---|---|---|---|---|---|---|---|
| 0 | 9·489982 | | ·510018 | 9·511776 | | 10·488224 | ·021794 | | 9·978206 | 60 |
| 1 | 9·490371 | 648 | ·509629 | 9·512206 | 716 | 10·487794 | ·021835 | 68 | 9·978165 | 59 |
| 2 | 9·490759 | 648 | ·509241 | 9·512635 | 716 | 10·487365 | ·021876 | 68 | 9·978124 | 58 |
| 3 | 9·491147 | 647 | ·508853 | 9·513064 | 715 | 10·486936 | ·021917 | 68 | 9·978083 | 57 |
| 4 | 9·491535 | 646 | ·508465 | 9·513493 | 714 | 10·486507 | ·021958 | 69 | 9·978042 | 56 |
| 5 | 9·491922 | 646 | ·508078 | 9·513921 | 714 | 10·486079 | ·021999 | 69 | 9·978001 | 55 |
| 6 | 9·492308 | 645 | ·507692 | 9·514349 | 713 | 10·485651 | ·022041 | 69 | 9·977959 | 54 |
| 7 | 9·492695 | 644 | ·507305 | 9·514777 | 713 | 10·485223 | ·022082 | 69 | 9·977918 | 53 |
| 8 | 9·493081 | 644 | ·506919 | 9·515204 | 712 | 10·484796 | ·022123 | 69 | 9·977877 | 52 |
| 9 | 9·493466 | 643 | ·506534 | 9·515631 | 712 | 10·484369 | ·022165 | 69 | 9·977835 | 51 |
| 10 | 9·493851 | 642 | ·506149 | 9·516057 | 711 | 10·483943 | ·022206 | 69 | 9·977794 | 50 |
| 11 | 9·494236 | 642 | ·505764 | 9·516484 | 710 | 10·483516 | ·022248 | 69 | 9·977752 | 49 |
| 12 | 9·494621 | 641 | ·505379 | 9·516910 | 710 | 10·483090 | ·022289 | 69 | 9·977711 | 48 |
| 13 | 9·495005 | 641 | ·504995 | 9·517335 | 709 | 10·482665 | ·022331 | 69 | 9·977669 | 47 |
| 14 | 9·495388 | 640 | ·504612 | 9·517761 | 709 | 10·482239 | ·022372 | 69 | 9·977628 | 46 |
| 15 | 9·495772 | 639 | ·504228 | 9·518185 | 708 | 10·481815 | ·022414 | 69 | 9·977586 | 45 |
| 16 | 9·496154 | 639 | ·503846 | 9·518610 | 708 | 10·481390 | ·022456 | 69 | 9·977544 | 44 |
| 17 | 9·496537 | 638 | ·503463 | 9·519034 | 707 | 10·480966 | ·022497 | 70 | 9·977503 | 43 |
| 18 | 9·496919 | 637 | ·503081 | 9·519458 | 706 | 10·480542 | ·022539 | 70 | 9·977461 | 42 |
| 19 | 9·497301 | 637 | ·502699 | 9·519882 | 706 | 10·480118 | ·022581 | 70 | 9·977419 | 41 |
| 20 | 9·497682 | 636 | ·502318 | 9·520305 | 705 | 10·479695 | ·022623 | 70 | 9·977377 | 40 |
| 21 | 9·498064 | 636 | ·501936 | 9·520728 | 705 | 10·479272 | ·022665 | 70 | 9·977335 | 39 |
| 22 | 9·498444 | 635 | ·501556 | 9·521151 | 704 | 10·478849 | ·022707 | 70 | 9·977293 | 38 |
| 23 | 9·498825 | 634 | ·501175 | 9·521573 | 704 | 10·478427 | ·022749 | 70 | 9·977251 | 37 |
| 24 | 9·499204 | 634 | ·500796 | 9·521995 | 703 | 10·478005 | ·022791 | 70 | 9·977209 | 36 |
| 25 | 9·499584 | 633 | ·500416 | 9·522417 | 703 | 10·477583 | ·022833 | 70 | 9·977167 | 35 |
| 26 | 9·499963 | 632 | ·500037 | 9·522838 | 702 | 10·477162 | ·022875 | 70 | 9·977125 | 34 |
| 27 | 9·500342 | 632 | ·499658 | 9·523259 | 702 | 10·476741 | ·022917 | 70 | 9·977083 | 33 |
| 28 | 9·500721 | 631 | ·499279 | 9·523680 | 701 | 10·476320 | ·022959 | 70 | 9·977041 | 32 |
| 29 | 9·501099 | 631 | ·498901 | 9·524100 | 701 | 10·475900 | ·023001 | 70 | 9·976999 | 31 |
| 30 | 9·501476 | 630 | ·498524 | 9·524520 | 700 | 10·475480 | ·023043 | 70 | 9·976957 | 30 |
| 31 | 9·501854 | 629 | ·498146 | 9·524939 | 699 | 10·475061 | ·023086 | 70 | 9·976914 | 29 |
| 32 | 9·502231 | 629 | ·497769 | 9·525359 | 699 | 10·474641 | ·023128 | 70 | 9·976872 | 28 |
| 33 | 9·502607 | 628 | ·497393 | 9·525778 | 698 | 10·474222 | ·023170 | 70 | 9·976830 | 27 |
| 34 | 9·502984 | 628 | ·497016 | 9·526197 | 698 | 10·473803 | ·023213 | 71 | 9·976787 | 26 |
| 35 | 9·503360 | 627 | ·496640 | 9·526615 | 697 | 10·473385 | ·023255 | 71 | 9·976745 | 25 |
| 36 | 9·503735 | 626 | ·496265 | 9·527033 | 697 | 10·472967 | ·023298 | 71 | 9·976702 | 24 |
| 37 | 9·504110 | 626 | ·495890 | 9·527451 | 696 | 10·472549 | ·023340 | 71 | 9·976660 | 23 |
| 38 | 9·504485 | 625 | ·495515 | 9·527868 | 696 | 10·472132 | ·023383 | 71 | 9·976617 | 22 |
| 39 | 9·504860 | 625 | ·495140 | 9·528285 | 695 | 10·471715 | ·023426 | 71 | 9·976574 | 21 |
| 40 | 9·505234 | 024 | ·494766 | 9·528702 | 695 | 10·471298 | ·023468 | 71 | 9·976532 | 20 |
| 41 | 9·505608 | 623 | ·494392 | 9·529119 | 694 | 10·470881 | ·023511 | 71 | 9·976489 | 19 |
| 42 | 9·505981 | 623 | ·494019 | 9·529535 | 694 | 10·470465 | ·023554 | 71 | 9·976446 | 18 |
| 43 | 9·506354 | 622 | ·493646 | 9·529950 | 693 | 10·470050 | ·023596 | 71 | 9·976404 | 17 |
| 44 | 9·506727 | 622 | ·493273 | 9·530366 | 693 | 10·469634 | ·023639 | 71 | 9·976361 | 16 |
| 45 | 9·507099 | 621 | ·492901 | 9·530781 | 692 | 10·469219 | ·023682 | 71 | 9·976318 | 15 |
| 46 | 9·507471 | 620 | ·492529 | 9·531196 | 691 | 10·468804 | ·023725 | 71 | 9·976275 | 14 |
| 47 | 9·507843 | 620 | ·492157 | 9·531611 | 691 | 10·468389 | ·023768 | 71 | 9·976232 | 13 |
| 48 | 9·508214 | 619 | ·491786 | 9·532025 | 690 | 10·467975 | ·023811 | 72 | 9·976189 | 12 |
| 49 | 9·508585 | 619 | ·491415 | 9·532439 | 690 | 10·467561 | ·023854 | 72 | 9·976146 | 11 |
| 50 | 9·508956 | 618 | ·491044 | 9·532853 | 689 | 10·467147 | ·023897 | 72 | 9·976103 | 10 |
| 51 | 9·509326 | 618 | ·490674 | 9·533266 | 689 | 10 466734 | ·023940 | 72 | 9·976060 | 9 |
| 52 | 9·509696 | 617 | ·490304 | 9·533679 | 688 | 10·466321 | ·023983 | 72 | 9·976017 | 8 |
| 53 | 9·510065 | 616 | ·489935 | 9·534092 | 688 | 10·465908 | ·024026 | 72 | 9·975974 | 7 |
| 54 | 9·510434 | 616 | ·489566 | 9·534504 | 687 | 10·465496 | ·024070 | 72 | 9·975930 | 6 |
| 55 | 9·510803 | 615 | ·489197 | 9·534916 | 687 | 10·465084 | ·024113 | 72 | 9·975887 | 5 |
| 56 | 9·511172 | 615 | ·488828 | 9·535328 | 686 | 10·464672 | ·024156 | 72 | 9·975844 | 4 |
| 57 | 9·511540 | 614 | ·488460 | 9·535739 | 686 | 10·464261 | ·024200 | 72 | 9·975800 | 3 |
| 58 | 9·511907 | 613 | ·488093 | 9·536150 | 685 | 10·463850 | ·024243 | 72 | 9·975757 | 2 |
| 59 | 9·512275 | 613 | ·487725 | 9·536561 | 685 | 10·463439 | ·024286 | 72 | 9·975714 | 1 |
| 60 | 9·512642 | 612 | ·487358 | 9·536972 | 684 | 10·463028 | ·024330 | 72 | 9·975670 | 0 |
| ′ | Cosine. | | Secant. | Cotangent. | | Tangent. | Cosecant. | | Sine. | ′ |

71 DEG.

19 DEG.

| ′ | Sine. | Diff. 100″ | Cosecant. | Tangent. | Diff. 100″ | Cotangent. | Secant. | Diff. 100″ | Cosine. | ′ |
|---|---|---|---|---|---|---|---|---|---|---|
| 0 | 9·512642 | | ·487358 | 9·536972 | | 10·463028 | ·024330 | | 9·975670 | 60 |
| 1 | 9·513009 | 612 | ·486991 | 9·537382 | 684 | 10·462618 | ·024373 | 72 | 9·975627 | 59 |
| 2 | 9·513375 | 611 | ·486625 | 9·537792 | 683 | 10·462208 | ·024417 | 73 | 9·975583 | 58 |
| 3 | 9·513741 | 611 | ·486259 | 9·538202 | 683 | 10·461798 | ·024461 | 73 | 9·975539 | 57 |
| 4 | 9·514107 | 610 | ·485893 | 9·538611 | 682 | 10·461389 | ·024504 | 73 | 9·975496 | 56 |
| 5 | 9·514472 | 609 | ·485528 | 9·539020 | 682 | 10·460980 | ·024548 | 73 | 9·975452 | 55 |
| 6 | 9·514837 | 609 | ·485163 | 9·539429 | 681 | 10·460571 | ·024592 | 73 | 9·975408 | 54 |
| 7 | 9·515202 | 608 | ·484798 | 9·539837 | 681 | 10·460163 | ·024635 | 73 | 9·975365 | 53 |
| 8 | 9·515566 | 608 | ·484434 | 9·540245 | 680 | 10·459755 | ·024679 | 73 | 9·975321 | 52 |
| 9 | 9·515930 | 607 | ·484070 | 9·540653 | 680 | 10·459347 | ·024723 | 73 | 9·975277 | 51 |
| 10 | 9·516294 | 607 | ·483706 | 9·541061 | 679 | 10·458939 | ·024767 | 73 | 9·975233 | 50 |
| 11 | 9·516657 | 606 | ·483343 | 9·541468 | 679 | 10·458532 | ·024811 | 73 | 9·975189 | 49 |
| 12 | 9·517020 | 605 | ·482980 | 9·541875 | 678 | 10·458125 | ·024855 | 73 | 9·975145 | 48 |
| 13 | 9·517382 | 605 | ·482618 | 9·542281 | 678 | 10·457719 | ·024899 | 73 | 9·975101 | 47 |
| 14 | 9·517745 | 604 | ·482255 | 9·542688 | 677 | 10·457312 | ·024943 | 73 | 9·975057 | 46 |
| 15 | 9·518107 | 604 | ·481893 | 9·543094 | 677 | 10·456906 | ·024987 | 73 | 9·975013 | 45 |
| 16 | 9·518468 | 603 | ·481532 | 9·543499 | 676 | 10·456501 | ·025031 | 73 | 9·974969 | 44 |
| 17 | 9·518829 | 603 | ·481171 | 9·543905 | 676 | 10·456095 | ·025075 | 74 | 9·974925 | 43 |
| 18 | 9·519190 | 602 | ·480810 | 9·544310 | 675 | 10·455690 | ·025120 | 74 | 9·974880 | 42 |
| 19 | 9·519551 | 601 | ·480449 | 9·544715 | 675 | 10·455285 | ·025164 | 74 | 9·974836 | 41 |
| 20 | 9·519911 | 601 | ·480089 | 9·545119 | 674 | 10·454881 | ·025208 | 74 | 9·974792 | 40 |
| 21 | 9·520271 | 600 | ·479729 | 9·545524 | 674 | 10·454476 | ·025252 | 74 | 9·974748 | 39 |
| 22 | 9·520631 | 600 | ·479369 | 9·545928 | 673 | 10·454072 | ·025297 | 74 | 9·974703 | 38 |
| 23 | 9·520990 | 599 | ·479010 | 9·546331 | 673 | 10·453669 | ·025341 | 74 | 9·974659 | 37 |
| 24 | 9·521349 | 599 | ·478651 | 9·546735 | 672 | 10·453265 | ·025386 | 74 | 9·974614 | 36 |
| 25 | 9·521707 | 598 | ·478293 | 9·547138 | 672 | 10·452862 | ·025430 | 74 | 9·974570 | 35 |
| 26 | 9·522066 | 598 | ·477934 | 9·547540 | 671 | 10·452460 | ·025475 | 74 | 9·974525 | 34 |
| 27 | 9·522424 | 597 | ·477576 | 9·547943 | 671 | 10·452057 | ·025519 | 74 | 9·974481 | 33 |
| 28 | 9·522781 | 596 | ·477219 | 9·548345 | 670 | 10·451655 | ·025564 | 74 | 9·974436 | 32 |
| 29 | 9·523138 | 596 | ·476862 | 9·548747 | 670 | 10·451253 | ·025609 | 74 | 9·974391 | 31 |
| 30 | 9·523495 | 595 | ·476505 | 9·549149 | 669 | 10·450851 | ·025653 | 74 | 9·974347 | 30 |
| 31 | 9·523852 | 595 | ·476148 | 9·549550 | 669 | 10·450450 | ·025698 | 75 | 9·974302 | 29 |
| 32 | 9·524208 | ·594 | ·475792 | 9·549951 | 668 | 10·450049 | ·025743 | 75 | 9·974257 | 28 |
| 33 | 9·524564 | 594 | ·475436 | 9·550352 | 668 | 10·449648 | ·025788 | 75 | 9·974212 | 27 |
| 34 | 9·524920 | 593 | ·475080 | 9·550752 | 667 | 10·449248 | ·025833 | 75 | 9·974167 | 26 |
| 35 | 9·525275 | 593 | ·474725 | 9·551152 | 667 | 10·448848 | ·025878 | 75 | 9·974122 | 25 |
| 36 | 9·525630 | 592 | ·474370 | 9·551552 | 666 | 10·448448 | ·025923 | 75 | 9·974077 | 24 |
| 37 | 9·525984 | 591 | ·474016 | 9·551952 | 666 | 10·448048 | ·025968 | 75 | 9·974032 | 23 |
| 38 | 9·526339 | 591 | ·473661 | 9·552351 | 665 | 10·447649 | ·026013 | 75 | 9·973987 | 22 |
| 39 | 9·526693 | 590 | ·473307 | 9·552750 | 665 | 10·447250 | ·026058 | 75 | 9·973942 | 21 |
| 40 | 9·527046 | 590 | ·472954 | 9·553149 | 665 | 10·446851 | ·026103 | 75 | 9·973897 | 20 |
| 41 | 9·527400 | 589 | ·472600 | 9·553548 | 664 | 10·446452 | ·026148 | 75 | 9·973852 | 19 |
| 42 | 9·527753 | 589 | ·472247 | 9·553946 | 664 | 10·446054 | ·026193 | 75 | 9·973807 | 18 |
| 43 | 9·528105 | 588 | ·471895 | 9·554344 | 663 | 10·445656 | ·026239 | 75 | 9·973761 | 17 |
| 44 | 9·528458 | 588 | ·471542 | 9·554741 | 663 | 10·445259 | ·026284 | 75 | 9·973716 | 16 |
| 45 | 9·528810 | 587 | ·471190 | 9·555139 | 662 | 10·444861 | ·026329 | 76 | 9·973671 | 15 |
| 46 | 9·529161 | 587 | ·470839 | 9·555536 | 662 | 10·444464 | ·026375 | 76 | 9·973625 | 14 |
| 47 | 9·529513 | 586 | ·470487 | 9·555933 | 661 | 10·444067 | ·026420 | 76 | 9·973580 | 13 |
| 48 | 9·529864 | 586 | ·470136 | 9·556329 | 661 | 10·443671 | ·026465 | 76 | 9·973535 | 12 |
| 49 | 9·530215 | 585 | ·469785 | 9·556725 | 660 | 10·443275 | ·026511 | 76 | 9·973489 | 11 |
| 50 | 9·530565 | 585 | ·469435 | 9·557121 | 660 | 10·442879 | ·026556 | 76 | 9·973444 | 10 |
| 51 | 9·530915 | 584 | ·469085 | 9·557517 | 659 | 10·442483 | ·026602 | 76 | 9·973398 | 9 |
| 52 | 9·531265 | 584 | ·468735 | 9·557913 | 659 | 10·442087 | ·026648 | 76 | 9·973352 | 8 |
| 53 | 9·531614 | 583 | ·468386 | 9·558308 | 659 | 10·441692 | ·026693 | 76 | 9·973307 | 7 |
| 54 | 9·531963 | 582 | ·468037 | 9·558702 | 658 | 10·441298 | ·026739 | 76 | 9·973261 | 6 |
| 55 | 9·532312 | 582 | ·467688 | 9·559097 | 658 | 10·440903 | ·026785 | 76 | 9·973215 | 5 |
| 56 | 9·532661 | 581 | ·467339 | 9·559491 | 657 | 10·440509 | ·026831 | 76 | 9·973169 | 4 |
| 57 | 9·533009 | 581 | ·466991 | 9·559885 | 657 | 10·440115 | ·026876 | 76 | 9·973124 | 3 |
| 58 | 9·533357 | 580 | ·466643 | 9·560279 | 656 | 10·439721 | ·026922 | 76 | 9·973078 | 2 |
| 59 | 9·533704 | 580 | ·466296 | 9·560673 | 656 | 10·439327 | ·026968 | 76 | 9·973032 | 1 |
| 60 | 9·534052 | 579 | ·465948 | 9·561066 | 655 | 10·438934 | ·027014 | 76 | 9·972986 | 0 |
| ′ | Cosine. | | Secant. | Cotangent. | | Tangent. | Cosecant. | | Sine. | ′ |

70 DEG.

20 DEG.

| ′ | Sine. | Diff. 100″ | Cosecant. | Tangent. | Diff. 100″ | Cotangent. | Secant. | Diff. 100″ | Cosine. | ′ |
|---|---|---|---|---|---|---|---|---|---|---|
| 0 | 9·534052 | | ·465948 | 9·561066 | | 10·438934 | ·027014 | | 9·972986 | 60 |
| 1 | 9·534399 | 578 | ·465601 | 9·561459 | 655 | 10·438541 | ·027060 | 77 | 9·972940 | 59 |
| 2 | 9·534745 | 577 | ·465255 | 9·561851 | 654 | 10·438149 | ·027106 | 77 | 9·972894 | 58 |
| 3 | 9·535092 | 577 | ·464908 | 9·562244 | 654 | 10·437756 | ·027152 | 77 | 9·972848 | 57 |
| 4 | 9·535438 | 577 | ·464562 | 9·562636 | 653 | 10·437364 | ·027198 | 77 | 9·972802 | 56 |
| 5 | 9·535783 | 576 | ·464217 | 9·563028 | 653 | 10·436972 | ·027245 | 78 | 9·972755 | 55 |
| 6 | 9·536129 | 576 | ·463871 | 9·563419 | 653 | 10·436581 | ·027291 | 77 | 9·972709 | 54 |
| 7 | 9·536474 | 575 | ·463526 | 9·563811 | 652 | 10·436189 | ·027337 | 77 | 9·972663 | 53 |
| 8 | 9·536818 | 574 | ·463182 | 9·564202 | 652 | 10·435798 | ·027383 | 77 | 9·972617 | 52 |
| 9 | 9·537163 | 574 | ·462837 | 9·564592 | 651 | 10·435408 | ·027430 | 78 | 9·972570 | 51 |
| 10 | 9·537507 | 573 | ·462493 | 9·564983 | 651 | 10·435017 | ·027476 | 77 | 9·972524 | 50 |
| 11 | 9·537851 | 573 | ·462149 | 9·565373 | 650 | 10·434627 | ·027522 | 77 | 9·972478 | 49 |
| 12 | 9·538194 | 572 | ·461806 | 9·565763 | 650 | 10·434237 | ·027569 | 78 | 9·972431 | 48 |
| 13 | 9·538538 | 572 | ·461462 | 9·566153 | 649 | 10·433847 | ·027615 | 77 | 9·972385 | 47 |
| 14 | 9·538880 | 571 | ·461120 | 9·566542 | 649 | 10·433458 | ·027662 | 78 | 9·972338 | 46 |
| 15 | 9·539223 | 571 | ·460777 | 9·566932 | 649 | 10·433068 | ·027709 | 78 | 9·972291 | 45 |
| 16 | 9·539565 | 570 | ·460435 | 9·567320 | 648 | 10·432680 | ·027755 | 77 | 9·972245 | 44 |
| 17 | 9·539907 | 570 | ·460093 | 9·567709 | 648 | 10·432291 | ·027802 | 78 | 9·972198 | 43 |
| 18 | 9·540249 | 569 | ·459751 | 9·568098 | 647 | 10·431902 | ·027849 | 78 | 9·972151 | 42 |
| 19 | 9·540590 | 569 | ·459410 | 9·568486 | 647 | 10·431514 | ·027895 | 77 | 9·972105 | 41 |
| 20 | 9·540931 | 568 | ·459069 | 9·568873 | 646 | 10·431127 | ·027942 | 78 | 9·972058 | 40 |
| 21 | 9·541272 | 568 | ·458728 | 9·569261 | 646 | 10·430739 | ·027989 | 78 | 9·972011 | 39 |
| 22 | 9·541613 | 567 | ·458387 | 9·569648 | 645 | 10·430352 | ·028036 | 78 | 9·971964 | 38 |
| 23 | 9·541953 | 567 | ·458047 | 9·570035 | 645 | 10·429965 | ·028083 | 78 | 9·971917 | 37 |
| 24 | 9·542293 | 566 | ·457707 | 9·570422 | 645 | 10·429578 | ·028130 | 78 | 9·971870 | 36 |
| 25 | 9·542632 | 566 | ·457368 | 9·570809 | 644 | 10·429191 | ·028177 | 78 | 9·971823 | 35 |
| 26 | 9·542971 | 565 | ·457029 | 9·571195 | 644 | 10·428805 | ·028224 | 78 | 9·971776 | 34 |
| 27 | 9·543310 | 565 | ·456690 | 9·571581 | 643 | 10·428419 | ·028271 | 78 | 9·971729 | 33 |
| 28 | 9·543649 | 564 | ·456351 | 9·571967 | 643 | 10·428033 | ·028318 | 78 | 9·971682 | 32 |
| 29 | 9·543987 | 564 | ·456013 | 9·572352 | 642 | 10·427648 | ·028365 | 78 | 9·971635 | 31 |
| 30 | 9·544325 | 563 | ·455675 | 9·572738 | 642 | 10·427262 | ·028412 | 78 | 9·971588 | 30 |
| 31 | 9·544663 | 563 | ·455337 | 9·573123 | 642 | 10·426877 | ·028460 | 80 | 9·971540 | 29 |
| 32 | 9·545000 | 562 | ·455000 | 9·573507 | 641 | 10·426493 | ·028507 | 78 | 9·971493 | 28 |
| 33 | 9·545338 | 562 | ·454662 | 9·573892 | 641 | 10·426108 | ·028554 | 78 | 9·971446 | 27 |
| 34 | 9·545674 | 561 | ·454326 | 9·574276 | 640 | 10·425724 | ·028602 | 80 | 9·971398 | 26 |
| 35 | 9·546011 | 561 | ·453989 | 9·574660 | 640 | 10·425340 | ·028649 | 78 | 9·971351 | 25 |
| 36 | 9·546347 | 560 | ·453653 | 9·575044 | 639 | 10·424956 | ·028697 | 80 | 9·971303 | 24 |
| 37 | 9·546683 | 560 | ·453317 | 9·575427 | 639 | 10·424573 | ·028744 | 78 | 9·971256 | 23 |
| 38 | 9·547019 | 559 | ·452981 | 9·575810 | 639 | 10·424190 | ·028792 | 80 | 9·971208 | 22 |
| 39 | 9·547354 | 559 | ·452646 | 9·576193 | 638 | 10·423807 | ·028839 | 78 | 9·971161 | 21 |
| 40 | 9·547689 | 538 | ·452311 | 9·576576 | 638 | 10·423424 | ·028887 | 80 | 9·971113 | 20 |
| 41 | 9·548024 | 558 | ·451976 | 9·576958 | 637 | 10·423042 | ·028934 | 78 | 9·971066 | 19 |
| 42 | 9·548359 | 557 | ·451641 | 9·577341 | 637 | 10·422659 | ·028982 | 80 | 9·971018 | 18 |
| 43 | 9·548693 | 557 | ·451307 | 9·577723 | 636 | 10·422277 | ·029030 | 80 | 9·970970 | 17 |
| 44 | 9·549027 | 556 | ·450973 | 9·578104 | 636 | 10·421896 | ·029078 | 80 | 9·970922 | 16 |
| 45 | 9·549360 | 556 | ·450640 | 9·578486 | 636 | 10·421514 | ·029126 | 80 | 9·970874 | 15 |
| 46 | 9·549693 | 555 | ·450307 | 9·578867 | 635 | 10·421133 | ·029173 | 78 | 9·970827 | 14 |
| 47 | 9·550026 | 555 | ·449974 | 9·579248 | 635 | 10·420752 | ·029221 | 80 | 9·970779 | 13 |
| 48 | 9·550359 | 554 | ·449641 | 9·579629 | 634 | 10·420371 | ·029269 | 80 | 9·970731 | 12 |
| 49 | 9·550692 | 554 | ·449308 | 9·580009 | 634 | 10·419991 | ·029317 | 80 | 9·970683 | 11 |
| 50 | 9·551024 | 553 | ·448976 | 9·580389 | 634 | 10·419611 | ·029365 | 80 | 9·970635 | 10 |
| 51 | 9·551356 | 553 | ·448644 | 9·580769 | 633 | 10·419231 | ·029414 | 82 | 9·970586 | 9 |
| 52 | 9·551687 | 552 | ·448313 | 9·581149 | 633 | 10·418851 | ·029462 | 80 | 9·970538 | 8 |
| 53 | 9·552018 | 552 | ·447982 | 9·581528 | 632 | 10·418472 | ·029510 | 80 | 9·970490 | 7 |
| 54 | 9·552349 | 552 | ·447651 | 9·581907 | 632 | 10·418093 | ·029558 | 80 | 9·970442 | 6 |
| 55 | 9·552680 | 551 | ·447320 | 9·582286 | 632 | 10·417714 | ·029606 | 80 | 9·970394 | 5 |
| 56 | 9·553010 | 551 | ·446990 | 9·582665 | 631 | 10·417335 | ·029655 | 82 | 9·970345 | 4 |
| 57 | 9·553341 | 550 | ·446659 | 9·583043 | 631 | 10·416957 | ·029703 | 80 | 9·970297 | 3 |
| 58 | 9·553670 | 550 | ·446330 | 9·583422 | 630 | 10·416578 | ·029751 | 80 | 9·970249 | 2 |
| 59 | 9·554000 | 549 | ·446000 | 9·583800 | 630 | 10·416200 | ·029800 | 82 | 9·970200 | 1 |
| 60 | 9·554329 | 549 | ·445671 | 9·584177 | 629 | 10·415823 | ·029848 | 80 | 9·970152 | 0 |
| ′ | Cosine. | | Secant. | Cotangent. | | Tangent. | Cosecant. | | Sine. | ′ |

69 DEG.

21 DEG.

| ′ | Sine. | Diff. 100″ | Cosecant. | Tangent. | Diff. 100″ | Cotangent. | Secant. | Diff. 100″ | Cosine. | ′ |
|---|---|---|---|---|---|---|---|---|---|---|
| 0 | 9·554329 | | ·445671 | 9·584177 | | 10·415823 | ·029848 | | 9·970152 | 60 |
| 1 | 9·554658 | 548 | ·445342 | 9·584555 | 629 | 10·415445 | ·029897 | 81 | 9·970103 | 59 |
| 2 | 9·554987 | 548 | ·445013 | 9·584932 | 629 | 10·415068 | ·029945 | 81 | 9·970055 | 58 |
| 3 | 9·555315 | 547 | ·444685 | 9·585309 | 628 | 10·414691 | ·029994 | 81 | 9·970006 | 57 |
| 4 | 9·555643 | 547 | ·444357 | 9·585686 | 628 | 10·414314 | ·030043 | 81 | 9·969957 | 56 |
| 5 | 9·555971 | 546 | ·444029 | 9·586062 | 627 | 10·413938 | ·030091 | 81 | 9·969909 | 55 |
| 6 | 9·556299 | 546 | ·443701 | 9·586439 | 627 | 10·413561 | ·030140 | 81 | 9·969860 | 54 |
| 7 | 9·556626 | 545 | ·443374 | 9·586815 | 627 | 10·413185 | ·030189 | 81 | 9·969811 | 53 |
| 8 | 9·556953 | 545 | ·443047 | 9·587190 | 626 | 10·412810 | ·030238 | 81 | 9·969762 | 52 |
| 9 | 9·557280 | 544 | ·442720 | 9·587566 | 626 | 10·412434 | ·030286 | 81 | 9·969714 | 51 |
| 10 | 9·557606 | 544 | ·442394 | 9·587941 | 625 | 10·412059 | ·030335 | 81 | 9·969665 | 50 |
| 11 | 9·557932 | 543 | ·442068 | 9·588316 | 625 | 10·411684 | ·030384 | 81 | 9·969616 | 49 |
| 12 | 9·558258 | 543 | ·441742 | 9·588691 | 625 | 10·411309 | ·030433 | 82 | 9·969567 | 48 |
| 13 | 9·558583 | 543 | ·441417 | 9·589066 | 624 | 10·410934 | ·030482 | 82 | 9·969518 | 47 |
| 14 | 9·558909 | 542 | ·441091 | 9·589440 | 624 | 10·410560 | ·030531 | 82 | 9·969469 | 46 |
| 15 | 9·559234 | 542 | ·440766 | 9·589814 | 623 | 10·410186 | ·030580 | 82 | 9·969420 | 45 |
| 16 | 9·559558 | 541 | ·440442 | 9·590188 | 623 | 10·409812 | ·030630 | 82 | 9·969370 | 44 |
| 17 | 9·559883 | 541 | ·440117 | 9·590562 | 623 | 10·409438 | ·030679 | 82 | 9·969321 | 43 |
| 18 | 9·560207 | 540 | ·439793 | 9·590935 | 622 | 10·409065 | ·030728 | 82 | 9·969272 | 42 |
| 19 | 9·560531 | 540 | ·439469 | 9·591308 | 622 | 10·408692 | ·030777 | 82 | 9·969223 | 41 |
| 20 | 9·560855 | 539 | ·439145 | 9·591681 | 622 | 10·408319 | ·030827 | 82 | 9·969173 | 40 |
| 21 | 9·561178 | 539 | ·438822 | 9·592054 | 621 | 10·407946 | ·030876 | 82 | 9·969124 | 39 |
| 22 | 9·561501 | 538 | ·438499 | 9·592426 | 621 | 10·407574 | ·030925 | 82 | 9·969075 | 38 |
| 23 | 9·561824 | 538 | ·438176 | 9·592798 | 620 | 10·407202 | ·030975 | 82 | 9·969025 | 37 |
| 24 | 9·562146 | 537 | ·437854 | 9·593171 | 620 | 10·406829 | ·031024 | 82 | 9·968976 | 36 |
| 25 | 9·562468 | 537 | ·437532 | 9·593542 | 620 | 10·406458 | ·031074 | 82 | 9·968926 | 35 |
| 26 | 9·562790 | 536 | ·437210 | 9·593914 | 619 | 10·406086 | ·031123 | 83 | 9·968877 | 34 |
| 27 | 9·563112 | 536 | ·436888 | 9·594285 | 619 | 10·405715 | ·031173 | 83 | 9·968827 | 33 |
| 28 | 9·563433 | 536 | ·436567 | 9·594656 | 618 | 10·405344 | ·031223 | 83 | 9·968777 | 32 |
| 29 | 9·563755 | 535 | ·436245 | 9·595027 | 618 | 10·404973 | ·031272 | 83 | 9·968728 | 31 |
| 30 | 9·564075 | 535 | ·435925 | 9·595398 | 618 | 10·404602 | ·031322 | 83 | 9·968678 | 30 |
| 31 | 9·564396 | 534 | ·435604 | 9·595768 | 617 | 10·404232 | ·031372 | 83 | 9·968628 | 29 |
| 32 | 9·564716 | 534 | ·435284 | 9·596138 | 617 | 10·403862 | ·031422 | 83 | 9·968578 | 28 |
| 33 | 9·565036 | 533 | ·434964 | 9·596508 | 616 | 10·403492 | ·031472 | 83 | 9·968528 | 27 |
| 34 | 9·565356 | 533 | ·434644 | 9·596878 | 616 | 10·403122 | ·031521 | 83 | 9·968479 | 26 |
| 35 | 9·565676 | 532 | ·434324 | 9·597247 | 616 | 10·402753 | ·031571 | 83 | 9·968429 | 25 |
| 36 | 9·565995 | 532 | ·434005 | 9·597616 | 615 | 10·402384 | ·031621 | 83 | 9·968379 | 24 |
| 37 | 9·566314 | 531 | ·433686 | 9·597985 | 615 | 10·402015 | ·031671 | 83 | 9·968329 | 23 |
| 38 | 9·566632 | 531 | ·433368 | 9·598354 | 615 | 10·401646 | ·031722 | 83 | 9·968278 | 22 |
| 39 | 9·566951 | 531 | ·433049 | 9·598722 | 614 | 10·401278 | ·031772 | 83 | 9·968228 | 21 |
| 40 | 9·567269 | 530 | ·432731 | 9·599091 | 614 | 10·400909 | ·031822 | 84 | 9·968178 | 20 |
| 41 | 9·567587 | 530 | ·432413 | 9·599459 | 613 | 10·400541 | ·031872 | 84 | 9·968128 | 19 |
| 42 | 9·567904 | 529 | ·432096 | 9·599827 | 613 | 10·400173 | ·031922 | 84 | 9·968078 | 18 |
| 43 | 9·568222 | 529 | ·431778 | 9·600194 | 613 | 10·399806 | ·031973 | 84 | 9·968027 | 17 |
| 44 | 9·568539 | 528 | ·431461 | 9·600562 | 612 | 10·399438 | ·032023 | 84 | 9·967977 | 16 |
| 45 | 9·568856 | 528 | ·431144 | 9·600929 | 612 | 10·399071 | ·032073 | 84 | 9·967927 | 15 |
| 46 | 9·569172 | 528 | ·430828 | 9·601296 | 611 | 10·398704 | ·032124 | 84 | 9·967876 | 14 |
| 47 | 9·569488 | 527 | ·430512 | 9·601662 | 611 | 10·398338 | ·032174 | 84 | 9·967826 | 13 |
| 48 | 9·569804 | 527 | ·430196 | 9·602029 | 611 | 10·397971 | ·032225 | 84 | 9·967775 | 12 |
| 49 | 9·570120 | 526 | ·429880 | 9·602395 | 610 | 10·397605 | ·032275 | 84 | 9·967725 | 11 |
| 50 | 9·570435 | 526 | ·429565 | 9·602761 | 610 | 10·397239 | ·032326 | 84 | 9·967674 | 10 |
| 51 | 9·570751 | 525 | ·429249 | 9·603127 | 610 | 10·396873 | ·032376 | 84 | 9·967624 | 9 |
| 52 | 9·571066 | 525 | ·428934 | 9·603493 | 609 | 10·396507 | ·032427 | 84 | 9·967573 | 8 |
| 53 | 9·571380 | 524 | ·428620 | 9·603858 | 609 | 10·396142 | ·032478 | 84 | 9·967522 | 7 |
| 54 | 9·571695 | 524 | ·428305 | 9·604223 | 609 | 10·395777 | ·032529 | 85 | 9·967471 | 6 |
| 55 | 9·572009 | 523 | ·427991 | 9·604588 | 608 | 10·395412 | ·032579 | 85 | 9·967421 | 5 |
| 56 | 9·572323 | 523 | ·427677 | 9·604953 | 608 | 10·395047 | ·032630 | 85 | 9·967370 | 4 |
| 57 | 9·572636 | 523 | ·427364 | 9·605317 | 607 | 10·394683 | ·032681 | 85 | 9·967319 | 3 |
| 58 | 9·572950 | 522 | ·427050 | 9·605682 | 607 | 10·394318 | ·032732 | 85 | 9·967268 | 2 |
| 59 | 9·573263 | 522 | ·426737 | 9·606046 | 607 | 10·393954 | ·032783 | 85 | 9·967217 | 1 |
| 60 | 9·573575 | 521 | ·426425 | 9·606410 | 606 | 10·393590 | ·032834 | 85 | 9·967166 | 0 |
| ′ | Cosine. | | Secant. | Cotangent. | | Tangent. | Cosecant. | | Sine. | ′ |

68 DEG.

22 DEG.

| ′ | Sine. | Diff. 100″ | Cosecant. | Tangent. | Diff. 100″ | Cotangent. | Secant. | Diff. 100″ | Cosine. | ′ |
|---|---|---|---|---|---|---|---|---|---|---|
| 0 | 9·573575 | | ·426425 | 9·606410 | | 10·393590 | ·032834 | | 9·967166 | 60 |
| 1 | 9·573888 | 521 | ·426112 | 9·606773 | 606 | 10·393227 | ·032885 | 85 | 9·967115 | 59 |
| 2 | 9·574200 | 520 | ·425800 | 9·607137 | 606 | 10·392863 | ·032936 | 85 | 9·967064 | 58 |
| 3 | 9·574512 | 520 | ·425488 | 9·607500 | 605 | 10·392500 | ·032987 | 85 | 9·967013 | 57 |
| 4 | 9·574824 | 519 | ·425176 | 9·607863 | 605 | 10·392137 | ·033039 | 85 | 9·966961 | 56 |
| 5 | 9·575136 | 519 | ·424864 | 9·608225 | 604 | 10·391775 | ·033090 | 85 | 9·966910 | 55 |
| 6 | 9·575447 | 519 | ·424553 | 9·608588 | 604 | 10·391412 | ·033141 | 85 | 9·966859 | 54 |
| 7 | 9·575758 | 518 | ·424242 | 9·608950 | 604 | 10·391050 | ·033192 | 85 | 9·966808 | 53 |
| 8 | 9·576069 | 518 | ·423931 | 9·609312 | 603 | 10·390688 | ·033244 | 85 | 9·966756 | 52 |
| 9 | 9·576379 | 517 | ·423621 | 9·609674 | 603 | 10·390326 | ·033295 | 86 | 9·966705 | 51 |
| 10 | 9·576689 | 517 | ·423311 | 9·610036 | 603 | 10·389964 | ·033347 | 86 | 9·966653 | 50 |
| 11 | 9·576999 | 516 | ·423001 | 9·610397 | 602 | 10·389603 | ·033398 | 86 | 9·966602 | 49 |
| 12 | 9·577309 | 516 | ·422691 | 9·610759 | 602 | 10·389211 | ·033450 | 86 | 9·966550 | 48 |
| 13 | 9·577618 | 516 | ·422382 | 9·611120 | 602 | 10·388880 | ·033501 | 86 | 9·966499 | 47 |
| 14 | 9·577927 | 515 | ·422073 | 9·611480 | 601 | 10·388520 | ·033553 | 86 | 9·966447 | 46 |
| 15 | 9·578236 | 515 | ·421764 | 9·611841 | 601 | 10·388159 | ·033605 | 86 | 9·966395 | 45 |
| 16 | 9·578545 | 514 | ·421455 | 9·612201 | 601 | 10·387799 | ·033656 | 86 | 9·966344 | 44 |
| 17 | 9·578853 | 514 | ·421147 | 9·612561 | 600 | 10·387439 | ·033708 | 86 | 9·966292 | 43 |
| 18 | 9·579162 | 513 | ·420838 | 9·612921 | 600 | 10·387079 | ·033760 | 86 | 9·966240 | 42 |
| 19 | 9·579470 | 513 | ·420530 | 9·613281 | 600 | 10·386719 | ·033812 | 86 | 9·966188 | 41 |
| 20 | 9·579777 | 513 | ·420223 | 9·613641 | 599 | 10·386359 | ·033864 | 86 | 9·966136 | 40 |
| 21 | 9·580085 | 512 | ·419915 | 9·614000 | 599 | 10·386000 | ·033915 | 86 | 9·966085 | 39 |
| 22 | 9·580392 | 512 | ·419608 | 9·614359 | 598 | 10·385641 | ·033967 | 87 | 9·966033 | 38 |
| 23 | 9·580699 | 511 | ·419301 | 9·614718 | 598 | 10·385282 | ·034019 | 87 | 9·965981 | 37 |
| 24 | 9·581005 | 511 | ·418995 | 9·615077 | 598 | 10·384923 | ·034071 | 87 | 9·965929 | 36 |
| 25 | 9·581312 | 511 | ·418688 | 9·615435 | 597 | 10·384565 | ·034124 | 87 | 9·965876 | 35 |
| 26 | 9·581618 | 510 | ·418382 | 9·615793 | 597 | 10·384207 | ·034176 | 87 | 9·965824 | 34 |
| 27 | 9·581924 | 510 | ·418076 | 9·616151 | 597 | 10·383849 | ·034228 | 87 | 9·965772 | 33 |
| 28 | 9·582229 | 509 | ·417771 | 9·616509 | 596 | 10·383491 | ·034280 | 87 | 9·965720 | 32 |
| 29 | 9·582535 | 509 | ·417465 | 9·616867 | 596 | 10·383133 | ·034332 | 87 | 9·965668 | 31 |
| 30 | 9·582840 | 509 | ·417160 | 9·617224 | 596 | 10·382776 | ·034385 | 87 | 9·965615 | 30 |
| 31 | 9·583145 | 508 | ·416855 | 9·617582 | 595 | 10·382418 | ·034437 | 87 | 9·965563 | 29 |
| 32 | 9·583449 | 508 | ·416551 | 9·617939 | 595 | 10·382061 | ·034489 | 87 | 9·965511 | 28 |
| 33 | 9·583754 | 507 | ·416246 | 9·618295 | 595 | 10·381705 | ·034542 | 87 | 9·965458 | 27 |
| 34 | 9·584058 | 507 | ·415942 | 9·618652 | 594 | 10·381348 | ·034594 | 87 | 9·965406 | 26 |
| 35 | 9·584361 | 506 | ·415639 | 9·619008 | 594 | 10·380992 | ·034647 | 87 | 9·965353 | 25 |
| 36 | 9·584665 | 506 | ·415335 | 9·619364 | 594 | 10·380636 | ·034699 | 88 | 9·965301 | 24 |
| 37 | 9·584968 | 506 | ·415032 | 9·619721 | 593 | 10·380279 | ·034752 | 88 | 9·965248 | 23 |
| 38 | 9·585272 | 505 | ·414728 | 9·620076 | 593 | 10·379924 | ·034805 | 88 | 9·965195 | 22 |
| 39 | 9·585574 | 505 | ·414426 | 9·620432 | 593 | 10·379568 | ·034857 | 88 | 9·965143 | 21 |
| 40 | 9·585877 | 504 | ·414123 | 9·620787 | 592 | 10·379213 | ·034910 | 88 | 9·965090 | 20 |
| 41 | 9·586179 | 504 | ·413821 | 9·621142 | 592 | 10·378858 | ·034963 | 88 | 9·965037 | 19 |
| 42 | 9·586482 | 503 | ·413518 | 9·621497 | 592 | 10·378503 | ·035016 | 88 | 9·964984 | 18 |
| 43 | 9·586783 | 503 | ·413217 | 9·621852 | 591 | 10·378148 | ·035069 | 88 | 9·964931 | 17 |
| 44 | 9·587085 | 503 | ·412915 | 9·622207 | 591 | 10·377793 | ·035121 | 88 | 9·964879 | 16 |
| 45 | 9·587386 | 502 | ·412614 | 9·622561 | 590 | 10·377439 | ·035174 | 88 | 9·964826 | 15 |
| 46 | 9·587688 | 502 | ·412312 | 9·622915 | 590 | 10·377085 | ·035227 | 88 | 9·964773 | 14 |
| 47 | 9·587989 | 501 | ·412011 | 9·623269 | 590 | 10·376731 | ·035280 | 88 | 9·964720 | 13 |
| 48 | 9·588289 | 501 | ·411711 | 9·623623 | 589 | 10·376377 | ·035334 | 88 | 9·964666 | 12 |
| 49 | 9·588590 | 501 | ·411410 | 9·623976 | 589 | 10·376024 | ·035387 | 89 | 9·964613 | 11 |
| 50 | 9·588890 | 500 | ·411110 | 9·624330 | 589 | 10·375670 | ·035440 | 89 | 9·964560 | 10 |
| 51 | 9·589190 | 500 | ·410810 | 9·624683 | 588 | 10·375317 | ·035493 | 89 | 9·964507 | 9 |
| 52 | 9·589489 | 499 | ·410511 | 9·625036 | 588 | 10·374964 | ·035546 | 89 | 9·964454 | 8 |
| 53 | 9·589789 | 499 | ·410211 | 9·625388 | 588 | 10·374612 | ·035600 | 89 | 9·964400 | 7 |
| 54 | 9·590088 | 499 | ·409912 | 9·625741 | 587 | 10·374259 | ·035653 | 89 | 9·964347 | 6 |
| 55 | 9·590387 | 498 | ·409613 | 9·626093 | 587 | 10·373907 | ·035706 | 89 | 9·964294 | 5 |
| 56 | 9·590686 | 498 | ·409314 | 9·626445 | 587 | 10·373555 | ·035760 | 89 | 9·964240 | 4 |
| 57 | 9·590984 | 497 | ·409016 | 9·626797 | 586 | 10·373203 | ·035813 | 89 | 9·964187 | 3 |
| 58 | 9·591282 | 497 | ·408718 | 9·627149 | 586 | 10·372851 | ·035867 | 89 | 9·964133 | 2 |
| 59 | 9·591580 | 497 | ·408420 | 9·627501 | 586 | 10·372499 | ·035920 | 89 | 9·964080 | 1 |
| 60 | 9·591878 | 496 | ·408122 | 9·627852 | 585 | 10·372148 | ·035974 | 89 | 9·964026 | 0 |
| ′ | Cosine. | | Secant. | Cotangent. | | Tangent. | Cosecant. | | Sine. | ′ |

67 DEG.

23 DEG.

| ′ | Sine. | Diff. 100″ | Cosecant. | Tangent. | Diff. 100″ | Cotangent. | Secant. | Diff. 100″ | Cosine. | ′ |
|---|---|---|---|---|---|---|---|---|---|---|
| 0 | 9·591878 | | ·408122 | 9·627852 | | 10·372148 | ·035974 | | 9·964026 | 60 |
| 1 | 9·592176 | 496 | ·407824 | 9·628203 | 585 | 10·371797 | ·036028 | 89 | 9·963972 | 59 |
| 2 | 9·592473 | 495 | ·407527 | 9·628554 | 585 | 10·371446 | ·036081 | 89 | 9·963919 | 58 |
| 3 | 9·592770 | 495 | ·407230 | 9·628905 | 585 | 10·371095 | ·036135 | 89 | 9·963865 | 57 |
| 4 | 9·593067 | 495 | ·406933 | 9·629255 | 584 | 10·370745 | ·036189 | 90 | 9·963811 | 56 |
| 5 | 9·593363 | 494 | ·406637 | 9·629606 | 584 | 10·370394 | ·036243 | 90 | 9·963757 | 55 |
| 6 | 9·593659 | 494 | ·406341 | 9·629956 | 583 | 10·370044 | ·036296 | 90 | 9·963704 | 54 |
| 7 | 9·593955 | 493 | ·406045 | 9·630306 | 583 | 10·369694 | ·036350 | 90 | 9·963650 | 53 |
| 8 | 9·594251 | 493 | ·405749 | 9·630656 | 583 | 10·369344 | ·036404 | 90 | 9·963596 | 52 |
| 9 | 9·594547 | 493 | ·405453 | 9·631005 | 583 | 10·368995 | ·036458 | 90 | 9·963542 | 51 |
| 10 | 9·594842 | 492 | ·405158 | 9·631355 | 582 | 10·368645 | ·036512 | 90 | 9·963488 | 50 |
| 11 | 9·595137 | 492 | ·404863 | 9·631704 | 582 | 10·368296 | ·036566 | 90 | 9·963434 | 49 |
| 12 | 9·595432 | 491 | ·404568 | 9·632053 | 582 | 10·367947 | ·036621 | 90 | 9·963379 | 48 |
| 13 | 9·595727 | 491 | ·404273 | 9·632401 | 581 | 10·367599 | ·036675 | 90 | 9·963325 | 47 |
| 14 | 9·596021 | 491 | ·403979 | 9·632750 | 581 | 10·367250 | ·036729 | 90 | 9·963271 | 46 |
| 15 | 9·596315 | 490 | ·403685 | 9·633098 | 581 | 10·366902 | ·036783 | 90 | 9·963217 | 45 |
| 16 | 9·596609 | 490 | ·403391 | 9·633447 | 580 | 10·366553 | ·036837 | 90 | 9·963163 | 44 |
| 17 | 9·596903 | 489 | ·403097 | 9·633795 | 580 | 10·366205 | ·036892 | 90 | 9·963108 | 43 |
| 18 | 9·597196 | 489 | ·402804 | 9·634143 | 580 | 10·365857 | ·036946 | 91 | 9·963054 | 42 |
| 19 | 9·597490 | 489 | ·402510 | 9·634490 | 579 | 10·365510 | ·037001 | 91 | 9·962999 | 41 |
| 20 | 9·597783 | 488 | ·402217 | 9·634838 | 579 | 10·365162 | ·037055 | 91 | 9·962945 | 40 |
| 21 | 9·598075 | 488 | ·401925 | 9·635185 | 579 | 10·364815 | ·037110 | 91 | 9·962890 | 39 |
| 22 | 9·598368 | 487 | ·401632 | 9·635532 | 578 | 10·364468 | ·037164 | 91 | 9·962836 | 38 |
| 23 | 9·598660 | 487 | ·401340 | 9·635879 | 578 | 10·364121 | ·037219 | 91 | 9·962781 | 37 |
| 24 | 9·598952 | 487 | ·401048 | 9·636226 | 578 | 10·363774 | ·037273 | 91 | 9·962727 | 36 |
| 25 | 9·599244 | 486 | ·400756 | 9·636572 | 577 | 10·363428 | ·037328 | 91 | 9·962672 | 35 |
| 26 | 9·599536 | 486 | ·400464 | 9·636919 | 577 | 10·363081 | ·037383 | 91 | 9·962617 | 34 |
| 27 | 9·599827 | 485 | ·400173 | 9·637265 | 577 | 10·362735 | ·037438 | 91 | 9·962562 | 33 |
| 28 | 9·600118 | 485 | ·399882 | 9·637611 | 577 | 10·362389 | ·037492 | 91 | 9·962508 | 32 |
| 29 | 9·600409 | 485 | ·399591 | 9·637956 | 576 | 10·362044 | ·037547 | 91 | 9·962453 | 31 |
| 30 | 9·600700 | 484 | ·399300 | 9·638302 | 576 | 10·361698 | ·037602 | 91 | 9·962398 | 30 |
| 31 | 9·600990 | 484 | ·399010 | 9·638647 | 576 | 10·361353 | ·037657 | 92 | 9·962343 | 29 |
| 32 | 9·601280 | 484 | ·398720 | 9·638992 | 575 | 10·361008 | ·037712 | 92 | 9·962288 | 28 |
| 33 | 9·601570 | 483 | ·398430 | 9·639337 | 575 | 10·360663 | ·037767 | 92 | 9·962233 | 27 |
| 34 | 9·601860 | 483 | ·398140 | 9·639682 | 575 | 10·360318 | ·037822 | 92 | 9·962178 | 26 |
| 35 | 9·602150 | 482 | ·397850 | 9·640027 | 574 | 10·359973 | ·037877 | 92 | 9·962123 | 25 |
| 36 | 9·602439 | 482 | ·397561 | 9·640371 | 574 | 10·359629 | ·037933 | 92 | 9·962067 | 24 |
| 37 | 9·602728 | 482 | ·397272 | 9·640716 | 574 | 10·359284 | ·037988 | 92 | 9·962012 | 23 |
| 38 | 9·603017 | 481 | ·396983 | 9·641060 | 573 | 10·358940 | ·038043 | 92 | 9·961957 | 22 |
| 39 | 9·603305 | 481 | ·396695 | 9·641404 | 573 | 10·358596 | ·038098 | 92 | 9·961902 | 21 |
| 40 | 9·603594 | 481 | ·396406 | 9·641747 | 573 | 10·358253 | ·038154 | 92 | 9·961846 | 20 |
| 41 | 9·603882 | 480 | ·396118 | 9·642091 | 572 | 10·357909 | ·038209 | 92 | 9·961791 | 19 |
| 42 | 9·604170 | 480 | ·395830 | 9·642434 | 572 | 10·357566 | ·038265 | 92 | 9·961735 | 18 |
| 43 | 9·604457 | 479 | ·395543 | 9·642777 | 572 | 10·357223 | ·038320 | 92 | 9·961680 | 17 |
| 44 | 9·604745 | 479 | ·395255 | 9·643120 | 572 | 10·356880 | ·038376 | 92 | 9·961624 | 16 |
| 45 | 9·605032 | 479 | ·394968 | 9·643463 | 571 | 10·356537 | ·038431 | 93 | 9·961569 | 15 |
| 46 | 9·605319 | 478 | ·394681 | 9·643806 | 571 | 10·356194 | ·038487 | 93 | 9·961513 | 14 |
| 47 | 9·605606 | 478 | ·394394 | 9·644148 | 571 | 10·355852 | ·038542 | 93 | 9·961458 | 13 |
| 48 | 9·605892 | 478 | ·394108 | 9·644490 | 570 | 10·355510 | ·038598 | 93 | 9·961402 | 12 |
| 49 | 9·606179 | 477 | ·393821 | 9·644832 | 570 | 10·355168 | ·038654 | 93 | 9·961346 | 11 |
| 50 | 9·606465 | 477 | ·393535 | 9·645174 | 570 | 10·354826 | ·038710 | 93 | 9·961290 | 10 |
| 51 | 9·606751 | 476 | ·393249 | 9·645516 | 570 | 10·354484 | ·038765 | 93 | 9·961235 | 9 |
| 52 | 9·607036 | 476 | ·392964 | 9·645857 | 569 | 10·354143 | ·038821 | 93 | 9·961179 | 8 |
| 53 | 9·607322 | 476 | ·392678 | 9·646199 | 569 | 10·353801 | ·038877 | 93 | 9·961123 | 7 |
| 54 | 9·607607 | 475 | ·392393 | 9·646540 | 569 | 10·353460 | ·038933 | 93 | 9·961067 | 6 |
| 55 | 9·607892 | 475 | ·392108 | 9·646881 | 568 | 10·353119 | ·038989 | 93 | 9·961011 | 5 |
| 56 | 9·608177 | 474 | ·391823 | 9·647222 | 568 | 10·352778 | ·039045 | 93 | 9·960955 | 4 |
| 57 | 9·608461 | 474 | ·391539 | 9·647562 | 568 | 10·352438 | ·039101 | 93 | 9·960899 | 3 |
| 58 | 9·608745 | 474 | ·391255 | 9·647903 | 567 | 10·352097 | ·039157 | 93 | 9·960843 | 2 |
| 59 | 9·609029 | 473 | ·390971 | 9·648243 | 567 | 10·351757 | ·039214 | 94 | 9·960786 | 1 |
| 60 | 9·609313 | 473 | ·390687 | 9·648583 | 567 | 10·351417 | ·039270 | 94 | 9·960730 | 0 |
| ′ | Cosine. | | Secant. | Cotangent. | | Tangent. | Cosecant. | | Sine. | ′ |

66 DEG.

24 DEG.

| ′ | Sine. | Diff. 100″ | Cosecant. | Tangent. | Diff. 100″ | Cotangent. | Secant. | Diff 100″ | Cosine. | ′ |
|---|---|---|---|---|---|---|---|---|---|---|
| 0 | 9·609313 | | ·390687 | 9·648583 | | 10·351417 | ·039270 | | 9·960730 | 60 |
| 1 | 9·609597 | 473 | ·390403 | 9·648923 | 566 | 10·351077 | ·039326 | 94 | 9·960674 | 59 |
| 2 | 9·609880 | 472 | ·390120 | 9·649263 | 566 | 10·350737 | ·039382 | 94 | 9·960618 | 58 |
| 3 | 9·610164 | 472 | ·389836 | 9·649602 | 566 | 10·350398 | ·039439 | 94 | 9·960561 | 57 |
| 4 | 9·610447 | 472 | ·389553 | 9·649942 | 566 | 10·350058 | ·039495 | 94 | 9·960505 | 56 |
| 5 | 9·610729 | 471 | ·389271 | 9·650281 | 565 | 10·349719 | ·039552 | 94 | 9·960448 | 55 |
| 6 | 9·611012 | 471 | ·388988 | 9·650620 | 565 | 10·349380 | ·039608 | 94 | 9·960392 | 54 |
| 7 | 9·611294 | 470 | ·388706 | 9·650959 | 565 | 10·349041 | ·039665 | 94 | 9·960335 | 53 |
| 8 | 9·611576 | 470 | ·388424 | 9·651297 | 564 | 10·348703 | ·039721 | 94 | 9·960279 | 52 |
| 9 | 9·611858 | 470 | ·388142 | 9·651636 | 564 | 10·348364 | ·039778 | 94 | 9·960222 | 51 |
| 10 | 9·612140 | 469 | ·387860 | 9·651974 | 564 | 10·348026 | ·039835 | 94 | 9·960165 | 50 |
| 11 | 9·612421 | 469 | ·387579 | 9·652312 | 563 | 10·347688 | ·039891 | 94 | 9·960109 | 49 |
| 12 | 9·612702 | 469 | ·387298 | 9·652650 | 563 | 10·347350 | ·039948 | 95 | 9·960052 | 48 |
| 13 | 9·612983 | 468 | ·387017 | 9·652988 | 563 | 10·347012 | ·040005 | 95 | 9·959995 | 47 |
| 14 | 9·613264 | 468 | ·386736 | 9·653326 | 563 | 10·346674 | ·040062 | 95 | 9·959938 | 46 |
| 15 | 9·613545 | 467 | ·386455 | 9·653663 | 562 | 10·346337 | ·040118 | 95 | 9·959882 | 45 |
| 16 | 9·613825 | 467 | ·386175 | 9·654000 | 562 | 10·346000 | ·040175 | 95 | 9·959825 | 44 |
| 17 | 9·614105 | 467 | ·385895 | 9·654337 | 562 | 10·345663 | ·040232 | 95 | 9·959768 | 43 |
| 18 | 9·614385 | 466 | ·385615 | 9·654674 | 561 | 10·345326 | ·040289 | 95 | 9·959711 | 42 |
| 19 | 9·614665 | 466 | ·385335 | 9·655011 | 561 | 10·344989 | ·040346 | 95 | 9·959654 | 41 |
| 20 | 9·614944 | 466 | ·385056 | 9·655348 | 561 | 10·344652 | ·040404 | 95 | 9·959596 | 40 |
| 21 | 9·615223 | 465 | ·384777 | 9·655684 | 561 | 10·344316 | ·040461 | 95 | 9·959539 | 39 |
| 22 | 9·615502 | 465 | ·384498 | 9·656020 | 560 | 10·343980 | ·040518 | 95 | 9·959482 | 38 |
| 23 | 9·615781 | 465 | ·384219 | 9·656356 | 560 | 10·343644 | ·040575 | 95 | 9·959425 | 37 |
| 24 | 9·616060 | 464 | ·383940 | 9·656692 | 560 | 10·343308 | ·040632 | 95 | 9·959368 | 36 |
| 25 | 9·616338 | 464 | ·383662 | 9·657028 | 559 | 10·342972 | ·040690 | 95 | 9·959310 | 35 |
| 26 | 9·616616 | 464 | ·383384 | 9·657364 | 559 | 10·342636 | ·040747 | 96 | 9·959253 | 34 |
| 27 | 9·616894 | 463 | ·383106 | 9·657699 | 559 | 10·342301 | ·040805 | 96 | 9·959195 | 33 |
| 28 | 9·617172 | 463 | ·382828 | 9·658034 | 559 | 10·341966 | ·040862 | 96 | 9·959138 | 32 |
| 29 | 9·617450 | 462 | ·382550 | 9·658369 | 558 | 10·341631 | ·040919 | 96 | 9·959080 | 31 |
| 30 | 9·617727 | 462 | ·382273 | 9·658704 | 558 | 10·341296 | ·040977 | 96 | 9·959023 | 30 |
| 31 | 9·618004 | 462 | ·381996 | 9·659039 | 558 | 10·340961 | ·041035 | 96 | 9·958965 | 29 |
| 32 | 9·618281 | 461 | ·381719 | 9·659373 | 558 | 10·340627 | ·041092 | 96 | 9·958908 | 28 |
| 33 | 9·618558 | 461 | ·381442 | 9·659708 | 557 | 10·340292 | ·041150 | 96 | 9·958850 | 27 |
| 34 | 9·618834 | 461 | ·381166 | 9·660042 | 557 | 10·339958 | ·041208 | 96 | 9·958792 | 26 |
| 35 | 9·619110 | 460 | ·380890 | 9·660376 | 557 | 10·339624 | ·041266 | 96 | 9·958734 | 25 |
| 36 | 9·619386 | 460 | ·380614 | 9·660710 | 556 | 10·339290 | ·041323 | 96 | 9·958677 | 24 |
| 37 | 9·619662 | 460 | ·380338 | 9·661043 | 556 | 10·338957 | ·041381 | 96 | 9·958619 | 23 |
| 38 | 9·619938 | 459 | ·380062 | 9·661377 | 556 | 10·338623 | ·041439 | 96 | 9·958561 | 22 |
| 39 | 9·620213 | 459 | ·379787 | 9·661710 | 556 | 10·338290 | ·041497 | 96 | 9·958503 | 21 |
| 40 | 9·620188 | 459 | ·379512 | 9·662043 | 555 | 10·337957 | ·041555 | 97 | 9·958445 | 20 |
| 41 | 9·620763 | 458 | ·379237 | 9·662376 | 555 | 10·337624 | ·041613 | 97 | 9·958387 | 19 |
| 42 | 9·621038 | 458 | ·378962 | 9·662709 | 555 | 10·337291 | ·041671 | 97 | 9·958329 | 18 |
| 43 | 9·621313 | 457 | ·378687 | 9·663042 | 554 | 10·336958 | ·041729 | 97 | 9·958271 | 17 |
| 44 | 9·621587 | 457 | ·378413 | 9·663375 | 554 | 10·336625 | ·041787 | 97 | 9·958213 | 16 |
| 45 | 9·621861 | 457 | ·378139 | 9·663707 | 554 | 10·336293 | ·041846 | 97 | 9·958154 | 15 |
| 46 | 9·622135 | 456 | ·377865 | 9·664039 | 554 | 10·335961 | ·041904 | 97 | 9·958096 | 14 |
| 47 | 9·622409 | 456 | ·377591 | 9·664371 | 553 | 10·335629 | ·041962 | 97 | 9·958038 | 13 |
| 48 | 9·622682 | 456 | ·377318 | 9·664703 | 553 | 10·335297 | ·042021 | 97 | 9·957979 | 12 |
| 49 | 9·622956 | 455 | ·377044 | 9·665035 | 553 | 10·334965 | ·042079 | 97 | 9·957921 | 11 |
| 50 | 9·623229 | 455 | ·376771 | 9·665366 | 553 | 10·334634 | ·042137 | 97 | 9·957863 | 10 |
| 51 | 9·623502 | 455 | ·376498 | 9·665697 | 552 | 10·334303 | ·042196 | 97 | 9·957804 | 9 |
| 52 | 9·623774 | 454 | ·376226 | 9·666029 | 552 | 10·333971 | ·042254 | 97 | 9·957746 | 8 |
| 53 | 9·624047 | 454 | ·375953 | 9·666360 | 552 | 10·333640 | ·042313 | 98 | 9·957687 | 7 |
| 54 | 9·624319 | 454 | ·375681 | 9·666691 | 551 | 10·333309 | ·042372 | 98 | 9·957628 | 6 |
| 55 | 9·624591 | 453 | ·375409 | 9·667021 | 551 | 10·332979 | ·042430 | 98 | 9·957570 | 5 |
| 56 | 9·624863 | 453 | ·375137 | 9·667352 | 551 | 10·332648 | ·042489 | 98 | 9·957511 | 4 |
| 57 | 9·625135 | 453 | ·374865 | 9·667682 | 551 | 10·332318 | ·042548 | 98 | 9·957452 | 3 |
| 58 | 9·625406 | 452 | ·374594 | 9·668013 | 550 | 10·331987 | ·042607 | 98 | 9·957393 | 2 |
| 59 | 9·625677 | 452 | ·374323 | 9·668343 | 550 | 10·331657 | ·042665 | 98 | 9·957335 | 1 |
| 60 | 9·625948 | 452 | ·374052 | 9·668672 | 550 | 10·331328 | ·042724 | 98 | 9·957276 | 0 |
| ′ | Cosine. | | Secant. | Cotangent. | | Tangent. | Cosecant. | | Sine. | ′ |

65 DEG.

25 DEG.

| ′ | Sine. | Diff. 100″ | Cosecant. | Tangent. | Diff. 100″ | Cotangent. | Secant. | Diff. 100″ | Cosine. | ′ |
|---|---|---|---|---|---|---|---|---|---|---|
| 0 | 9·625948 | | ·374052 | 9·668673 | | 10·331327 | ·042724 | | 9·957276 | 60 |
| 1 | 9·626219 | 451 | ·373781 | 9·669002 | 550 | 10·330998 | ·042783 | 98 | 9·957217 | 59 |
| 2 | 9·626490 | 451 | ·373510 | 9·669332 | 549 | 10·330668 | ·042842 | 98 | 9·957158 | 58 |
| 3 | 9·626760 | 451 | ·373240 | 9·669661 | 549 | 10·330339 | ·042901 | 98 | 9·957099 | 57 |
| 4 | 9·627030 | 450 | ·372970 | 9·669991 | 549 | 10·330009 | ·042960 | 98 | 9·957040 | 56 |
| 5 | 9·627300 | 450 | ·372700 | 9·670320 | 548 | 10·329680 | ·043019 | 98 | 9·956981 | 55 |
| 6 | 9·627570 | 450 | ·372430 | 9·670649 | 548 | 10·329351 | ·043079 | 98 | 9·956921 | 54 |
| 7 | 9·627840 | 449 | ·372160 | 9·670977 | 548 | 10·329023 | ·043138 | 99 | 9·956862 | 53 |
| 8 | 9·628109 | 449 | ·371891 | 9·671306 | 548 | 10·328694 | ·043197 | 99 | 9·956803 | 52 |
| 9 | 9·628348 | 449 | ·371622 | 9·671634 | 547 | 10·328366 | ·043256 | 99 | 9·956744 | 51 |
| 10 | 9·628647 | 448 | ·371353 | 9·671963 | 547 | 10·328037 | ·043316 | 99 | 9·956684 | 50 |
| 11 | 9·628916 | 448 | ·371084 | 9·672291 | 547 | 10·327709 | ·043375 | 99 | 9·956625 | 49 |
| 12 | 9·629185 | 447 | ·370815 | 9·672619 | 547 | 10·327381 | ·043434 | 99 | 9·956566 | 48 |
| 13 | 9·629453 | 447 | ·370547 | 9·672947 | 546 | 10·327053 | ·043494 | 99 | 9·956506 | 47 |
| 14 | 9·629721 | 447 | ·370279 | 9·673274 | 546 | 10·326726 | ·043553 | 99 | 9·956447 | 46 |
| 15 | 9·629989 | 446 | ·370011 | 9·673602 | 546 | 10·326398 | ·043613 | 99 | 9·956387 | 45 |
| 16 | 9·630257 | 446 | ·369743 | 9·673929 | 546 | 10·326071 | ·043673 | 99 | 9·956327 | 44 |
| 17 | 9·630524 | 446 | ·369476 | 9·674257 | 545 | 10·325743 | ·043732 | 99 | 9·956268 | 43 |
| 18 | 9·630792 | 446 | ·369208 | 9·674584 | 545 | 10·325416 | ·043792 | 99 | 9·956208 | 42 |
| 19 | 9·631059 | 445 | ·368941 | 9·674910 | 545 | 10·325090 | ·043852 | 100 | 9·956148 | 41 |
| 20 | 9·631326 | 445 | ·368674 | 9·675237 | 544 | 10·324763 | ·043911 | 100 | 9·956089 | 40 |
| 21 | 9·631593 | 445 | ·368407 | 9·675564 | 544 | 10·324436 | ·043971 | 100 | 9·956029 | 39 |
| 22 | 9·631859 | 444 | ·368141 | 9·675890 | 544 | 10·324110 | ·044031 | 100 | 9·955969 | 38 |
| 23 | 9·632125 | 444 | ·367875 | 9·676217 | 544 | 10·323783 | ·044091 | 100 | 9·955909 | 37 |
| 24 | 9·632392 | 444 | ·367608 | 9·676543 | 543 | 10·323457 | ·044151 | 100 | 9·955849 | 36 |
| 25 | 9·632658 | 443 | ·367342 | 9·676869 | 543 | 10·323131 | ·044211 | 100 | 9·955789 | 35 |
| 26 | 9·632923 | 443 | ·367077 | 9·677194 | 543 | 10·322806 | ·044271 | 100 | 9·955729 | 34 |
| 27 | 9·633189 | 443 | ·366811 | 9·677520 | 543 | 10·322480 | ·044331 | 100 | 9·955669 | 33 |
| 28 | 9·633454 | 442 | ·366546 | 9·677846 | 542 | 10·322154 | ·044391 | 100 | 9·955609 | 32 |
| 29 | 9·633719 | 442 | ·366281 | 9·678170 | 542 | 10·321829 | ·044452 | 100 | 9·955548 | 31 |
| 30 | 9·633984 | 442 | ·366016 | 9·678496 | 542 | 10·321504 | ·044512 | 100 | 9·955488 | 30 |
| 31 | 9·634249 | 441 | ·365751 | 9·678821 | 542 | 10·321179 | ·044572 | 100 | 9·955428 | 29 |
| 32 | 9·634514 | 441 | ·365486 | 9·679146 | 541 | 10·320854 | ·044632 | 101 | 9·955368 | 28 |
| 33 | 9·634778 | 440 | ·365222 | 9·679471 | 541 | 10·320529 | ·044693 | 101 | 9·955307 | 27 |
| 34 | 9·635042 | 440 | ·364958 | 9·679795 | 541 | 10·320205 | ·044753 | 101 | 9·955247 | 26 |
| 35 | 9·635306 | 440 | ·364694 | 9·680120 | 541 | 10·319880 | ·044814 | 101 | 9·955186 | 25 |
| 36 | 9·635570 | 439 | ·364430 | 9·680444 | 540 | 10·319556 | ·044874 | 101 | 9·955126 | 24 |
| 37 | 9·635834 | 439 | ·364166 | 9·680768 | 540 | 10·319232 | ·044935 | 101 | 9·955065 | 23 |
| 38 | 9·636097 | 439 | ·363903 | 9·681092 | 540 | 10·318908 | ·044995 | 101 | 9·955005 | 22 |
| 39 | 9·636360 | 438 | ·363640 | 9·681416 | 540 | 10·318584 | ·045056 | 101 | 9·954944 | 21 |
| 40 | 9·636623 | 438 | ·363377 | 9·681740 | 539 | 10·318260 | ·045117 | 101 | 9·954883 | 20 |
| 41 | 9·636886 | 438 | ·363114 | 9·682063 | 539 | 10·317937 | ·045177 | 101 | 9·954823 | 19 |
| 42 | 9·637148 | 437 | ·362852 | 9·682387 | 539 | 10·317613 | ·045238 | 101 | 9·954762 | 18 |
| 43 | 9·637411 | 437 | ·362589 | 9·682710 | 539 | 10·317290 | ·045299 | 101 | 9·954701 | 17 |
| 44 | 9·637673 | 437 | ·362327 | 9·683033 | 538 | 10·316967 | ·045360 | 101 | 9·954640 | 16 |
| 45 | 9·637935 | 437 | ·362065 | 9·683356 | 538 | 10·316644 | ·045421 | 101 | 9·954579 | 15 |
| 46 | 9·638197 | 436 | ·361803 | 9·683679 | 538 | 10·316321 | ·045482 | 101 | 9·954518 | 14 |
| 47 | 9·638458 | 436 | ·361542 | 9·684001 | 538 | 10·315999 | ·045543 | 102 | 9·954457 | 13 |
| 48 | 9·638720 | 436 | ·361280 | 9·684324 | 537 | 10·315676 | ·045604 | 102 | 9·954396 | 12 |
| 49 | 9·638981 | 435 | ·361019 | 9·684646 | 537 | 10·315354 | ·045665 | 102 | 9·954335 | 11 |
| 50 | 9·639242 | 435 | ·360758 | 9·684968 | 537 | 10·315032 | ·045726 | 102 | 9·954274 | 10 |
| 51 | 9·639503 | 435 | ·360497 | 9·685290 | 537 | 10·314710 | ·045787 | 102 | 9·954213 | 9 |
| 52 | 9·639764 | 434 | ·360236 | 9·685612 | 536 | 10·314388 | ·045848 | 102 | 9·954152 | 8 |
| 53 | 9·640024 | 434 | ·359976 | 9·685934 | 536 | 10·314066 | ·045910 | 102 | 9·954090 | 7 |
| 54 | 9·640284 | 434 | ·359716 | 9·686255 | 536 | 10·313745 | ·045971 | 102 | 9·954029 | 6 |
| 55 | 9·640544 | 433 | ·359456 | 9·686577 | 536 | 10·313423 | ·046032 | 102 | 9·953968 | 5 |
| 56 | 9·640804 | 433 | ·359196 | 9·686898 | 535 | 10·313102 | ·046094 | 102 | 9·953906 | 4 |
| 57 | 9·641064 | 433 | ·358936 | 9·687219 | 535 | 10·312781 | ·046155 | 102 | 9·953845 | 3 |
| 58 | 9·641324 | 432 | ·358676 | 9·687540 | 535 | 10·312460 | ·046217 | 102 | 9·953783 | 2 |
| 59 | 9·641583 | 432 | ·358417 | 9·687861 | 535 | 10·312139 | ·046278 | 102 | 9·953722 | 1 |
| 60 | 9·641842 | 432 | ·358158 | 9·688182 | 534 | 10·311818 | ·046340 | 103 | 9·953660 | 0 |
| ′ | Cosine. | | Secant. | Cotangent. | | Tangent. | Cosecant. | | Sine. | ′ |

64 DEG.

26 DEG.

| ′ | Sine. | Diff. 100″ | Cosecant. | Tangent. | Diff. 100″ | Cotangent. | Secant. | Diff. 100″ | Cosine. | ′ |
|---|---|---|---|---|---|---|---|---|---|---|
| 0 | 9·641842 | | ·358158 | 9·688182 | | 10·311818 | ·046340 | | 9·953660 | 60 |
| 1 | 9·642101 | 431 | ·357899 | 9·688502 | 534 | 10·311498 | ·046401 | 103 | 9·953599 | 59 |
| 2 | 9·642360 | 431 | ·357640 | 9·688823 | 534 | 10·311177 | ·046463 | 103 | 9·953537 | 58 |
| 3 | 9·642618 | 431 | ·357382 | 9·689143 | 534 | 10·310857 | ·046525 | 103 | 9·953475 | 57 |
| 4 | 9·642877 | 430 | ·357123 | 9·689463 | 533 | 10·310537 | ·046587 | 103 | 9·953413 | 56 |
| 5 | 9·643135 | 430 | ·356865 | 9·689783 | 533 | 10·310217 | ·046648 | 103 | 9·953352 | 55 |
| 6 | 9·643393 | 430 | ·356607 | 9·690103 | 533 | 10·309897 | ·046710 | 103 | 9·953290 | 54 |
| 7 | 9·643650 | 430 | ·356350 | 9·690423 | 533 | 10·309577 | ·046772 | 103 | 9·953228 | 53 |
| 8 | 9·643908 | 429 | ·356092 | 9·690742 | 533 | 10·309258 | ·046834 | 103 | 9·953166 | 52 |
| 9 | 9·644165 | 429 | ·355835 | 9·691062 | 532 | 10·308938 | ·046896 | 103 | 9·953104 | 51 |
| 10 | 9·644423 | 429 | ·355577 | 9·691381 | 532 | 10·308619 | ·046958 | 103 | 9·953042 | 50 |
| 11 | 9·644680 | 428 | ·355320 | 9·691700 | 532 | 10·308300 | ·047020 | 103 | 9·952980 | 49 |
| 12 | 9·644936 | 428 | ·355064 | 9·692019 | 531 | 10·307981 | ·047082 | 104 | 9·952918 | 48 |
| 13 | 9·645193 | 428 | ·354807 | 9·692338 | 531 | 10·307662 | ·047145 | 104 | 9·952855 | 47 |
| 14 | 9·645450 | 427 | ·354550 | 9·692656 | 531 | 10·307344 | ·047207 | 104 | 9·952793 | 46 |
| 15 | 9·645706 | 427 | ·354294 | 9·692975 | 531 | 10·307025 | ·047269 | 104 | 9·952731 | 45 |
| 16 | 9·645962 | 427 | ·354038 | 9·693293 | 531 | 10·306707 | ·047331 | 104 | 9·952669 | 44 |
| 17 | 9·646218 | 426 | ·353782 | 9·693612 | 530 | 10·306388 | ·047394 | 104 | 9·952606 | 43 |
| 18 | 9·646474 | 426 | ·353526 | 9·693930 | 530 | 10·306070 | ·047456 | 104 | 9·952544 | 42 |
| 19 | 9·646729 | 426 | ·353271 | 9·694248 | 530 | 10·305752 | ·047519 | 104 | 9·952481 | 41 |
| 20 | 9·646984 | 425 | ·353016 | 9·694566 | 530 | 10·305434 | ·047581 | 104 | 9·952419 | 40 |
| 21 | 9·647240 | 425 | ·352760 | 9·694883 | 529 | 10·305117 | ·047644 | 104 | 9·952356 | 39 |
| 22 | 9·647494 | 425 | ·352506 | 9·695201 | 529 | 10·304799 | ·047706 | 104 | 9·952294 | 38 |
| 23 | 9·647749 | 424 | ·352251 | 9·695518 | 529 | 10·304482 | ·047769 | 104 | 9·952231 | 37 |
| 24 | 9·648004 | 424 | ·351996 | 9·695836 | 529 | 10·304164 | ·047832 | 104 | 9·952168 | 36 |
| 25 | 9·648258 | 424 | ·351742 | 9·696153 | 529 | 10·303847 | ·047894 | 105 | 9·952106 | 35 |
| 26 | 9·648512 | 424 | ·351488 | 9·696470 | 528 | 10·303530 | ·047957 | 105 | 9·952043 | 34 |
| 27 | 9·648766 | 423 | ·351234 | 9·696787 | 528 | 10·303213 | ·048020 | 105 | 9·951980 | 33 |
| 28 | 9·649020 | 423 | ·350980 | 9·697103 | 528 | 10·302897 | ·048083 | 105 | 9·951917 | 32 |
| 29 | 9·649274 | 423 | ·350726 | 9·697420 | 528 | 10·302580 | ·048146 | 105 | 9·951854 | 31 |
| 30 | 9·649527 | 422 | ·350473 | 9·697736 | 527 | 10·302264 | ·048209 | 105 | 9·951791 | 30 |
| 31 | 9·649781 | 422 | ·350219 | 9·698053 | 527 | 10·301947 | ·048272 | 105 | 9·951728 | 29 |
| 32 | 9·650034 | 422 | ·349966 | 9·698369 | 527 | 10·301631 | ·048335 | 105 | 9·951665 | 28 |
| 33 | 9·650287 | 422 | ·349713 | 9·698685 | 527 | 10·301315 | ·048398 | 105 | 9·951602 | 27 |
| 34 | 9·650539 | 421 | ·349461 | 9·699001 | 526 | 10·300999 | ·048461 | 105 | 9·951539 | 26 |
| 35 | 9·650792 | 421 | ·349208 | 9·699316 | 526 | 10·300684 | ·048524 | 105 | 9·951476 | 25 |
| 36 | 9·651044 | 421 | ·348956 | 9·699632 | 526 | 10·300368 | ·048588 | 105 | 9·951412 | 24 |
| 37 | 9·651297 | 420 | ·348703 | 9·699947 | 526 | 10·300053 | ·048651 | 105 | 9·951349 | 23 |
| 38 | 9·651549 | 420 | ·348451 | 9·700263 | 526 | 10·299737 | ·048714 | 106 | 9·951286 | 22 |
| 39 | 9·651800 | 420 | ·348200 | 9·700578 | 525 | 10·299422 | ·048778 | 106 | 9·951222 | 21 |
| 40 | 9·652052 | 419 | ·347948 | 9·700893 | 525 | 10·299107 | ·048841 | 106 | 9·951159 | 20 |
| 41 | 9·652304 | 419 | ·347696 | 9·701208 | 525 | 10·298792 | ·048904 | 106 | 9·951096 | 19 |
| 42 | 9·652555 | 419 | ·347445 | 9·701523 | 525 | 10·298477 | ·048968 | 106 | 9·951032 | 18 |
| 43 | 9·652806 | 418 | ·347194 | 9·701837 | 524 | 10·298163 | ·049032 | 106 | 9·950968 | 17 |
| 44 | 9·653057 | 418 | ·346943 | 9·702152 | 524 | 10·297848 | ·049095 | 106 | 9·950905 | 16 |
| 45 | 9·653308 | 418 | ·346692 | 9·702466 | 524 | 10·297534 | ·049159 | 106 | 9·950841 | 15 |
| 46 | 9·653558 | 418 | ·346442 | 9·702780 | 524 | 10·297220 | ·049222 | 106 | 9·950778 | 14 |
| 47 | 9·653808 | 417 | ·346192 | 9·703095 | 523 | 10·296905 | ·049286 | 106 | 9·950714 | 13 |
| 48 | 9·654059 | 417 | ·345941 | 9·703409 | 523 | 10·296591 | ·049350 | 106 | 9·950650 | 12 |
| 49 | 9·654309 | 417 | ·345691 | 9·703723 | 523 | 10·296277 | ·049414 | 106 | 9·950586 | 11 |
| 50 | 9·654558 | 416 | ·345442 | 9·704036 | 523 | 10·295964 | ·049478 | 106 | 9·950522 | 10 |
| 51 | 9·654808 | 416 | ·345192 | 9·704350 | 523 | 10·295650 | ·049542 | 107 | 9·950458 | 9 |
| 52 | 9·655058 | 416 | ·344942 | 9·704663 | 522 | 10·295337 | ·049606 | 107 | 9·950394 | 8 |
| 53 | 9·655307 | 415 | ·344693 | 9·704977 | 522 | 10·295023 | ·049670 | 107 | 9·950330 | 7 |
| 54 | 9·655556 | 415 | ·344444 | 9·705290 | 522 | 10·294710 | ·049734 | 107 | 9·950266 | 6 |
| 55 | 9·655805 | 415 | ·344195 | 9·705603 | 522 | 10·294397 | ·049798 | 107 | 9·950202 | 5 |
| 56 | 9·656054 | 415 | ·343946 | 9·705916 | 521 | 10·294084 | ·049862 | 107 | 9·950138 | 4 |
| 57 | 9·656302 | 414 | ·343698 | 9·706228 | 521 | 10·293772 | ·049926 | 107 | 9·950074 | 3 |
| 58 | 9·656551 | 414 | ·343449 | 9·706541 | 521 | 10·293459 | ·049990 | 107 | 9·950010 | 2 |
| 59 | 9·656799 | 414 | ·343201 | 9·706854 | 521 | 10·293146 | ·050055 | 107 | 9·949945 | 1 |
| 60 | 9·657047 | 413 | ·342953 | 9·707166 | 521 | 10·292834 | ·050119 | 107 | 9·949881 | 0 |
| ′ | Cosine. | | Secant. | Cotangent. | | Tangent. | Cosecant. | | Sine. | ′ |



63 DEG.

27 DEG.

| ′ | Sine. | Diff. 100″ | Cosecant. | Tangent. | Diff. 100″ | Cotangent. | Secant. | Diff. 100″ | Cosine. | ′ |
|---|---|---|---|---|---|---|---|---|---|---|
| 0 | 9·657047 | | ·342953 | 9·707166 | | 10·292834 | ·050119 | | 9·949881 | 60 |
| 1 | 9·657295 | 413 | ·342705 | 9·707478 | 520 | 10·292522 | ·050184 | 107 | 9·949816 | 59 |
| 2 | 9·657542 | 413 | ·342458 | 9·707790 | 520 | 10·292210 | ·050248 | 107 | 9·949752 | 58 |
| 3 | 9·657790 | 412 | ·342210 | 9·708102 | 520 | 10·291898 | ·050312 | 107 | 9·949688 | 57 |
| 4 | 9·658037 | 412 | ·341963 | 9·708414 | 520 | 10·291586 | ·050377 | 108 | 9·949623 | 56 |
| 5 | 9·658284 | 412 | ·341716 | 9·708726 | 519 | 10·291274 | ·050442 | 108 | 9·949558 | 55 |
| 6 | 9·658531 | 412 | ·341469 | 9·709037 | 519 | 10·290963 | ·050506 | 108 | 9·949494 | 54 |
| 7 | 9·658778 | 411 | ·341222 | 9·709349 | 519 | 10·290651 | ·050571 | 108 | 9·949429 | 53 |
| 8 | 9·659025 | 411 | ·340975 | 9·709660 | 519 | 10·290340 | ·050636 | 108 | 9·949364 | 52 |
| 9 | 9·659271 | 411 | ·340729 | 9·709971 | 519 | 10·290029 | ·050700 | 108 | 9·949300 | 51 |
| 10 | 9·659517 | 410 | ·340483 | 9·710282 | 518 | 10·289718 | ·050765 | 108 | 9·949235 | 50 |
| 11 | 9·659763 | 410 | ·340237 | 9·710593 | 518 | 10·289407 | ·050830 | 108 | 9·949170 | 49 |
| 12 | 9·660009 | 410 | ·339991 | 9·710904 | 518 | 10·289096 | ·050895 | 108 | 9·949105 | 48 |
| 13 | 9·660255 | 409 | ·339745 | 9·711215 | 518 | 10·288785 | ·050960 | 108 | 9·949040 | 47 |
| 14 | 9·660501 | 409 | ·339499 | 9·711525 | 518 | 10·288475 | ·051025 | 108 | 9·948975 | 46 |
| 15 | 9·660746 | 409 | ·339254 | 9·711836 | 517 | 10·288164 | ·051090 | 108 | 9·948910 | 45 |
| 16 | 9·660991 | 409 | ·339009 | 9·712146 | 517 | 10·287854 | ·051155 | 108 | 9·948845 | 44 |
| 17 | 9·661236 | 408 | ·338764 | 9·712456 | 517 | 10·287544 | ·051220 | 108 | 9·948780 | 43 |
| 18 | 9·661481 | 408 | ·338519 | 9·712766 | 517 | 10·287234 | ·051285 | 109 | 9·948715 | 42 |
| 19 | 9·661726 | 408 | ·338274 | 9·713076 | 516 | 10·286924 | ·051350 | 109 | 9·948650 | 41 |
| 20 | 9·661970 | 407 | ·338030 | 9·713386 | 516 | 10·286614 | ·051416 | 109 | 9·948584 | 40 |
| 21 | 9·662214 | 407 | ·337786 | 9·713696 | 516 | 10·286304 | ·051481 | 109 | 9·948519 | 39 |
| 22 | 9·662459 | 407 | ·337541 | 9·714005 | 516 | 10·285995 | ·051546 | 109 | 9·948454 | 38 |
| 23 | 9·662703 | 407 | ·337297 | 9·714314 | 516 | 10·285686 | ·051612 | 109 | 9·948388 | 37 |
| 24 | 9·662946 | 406 | ·337054 | 9·714624 | 515 | 10·285376 | ·051677 | 109 | 9·948323 | 36 |
| 25 | 9·663190 | 406 | ·336810 | 9·714933 | 515 | 10·285067 | ·051743 | 109 | 9·948257 | 35 |
| 26 | 9·663433 | 406 | ·336567 | 9·715242 | 515 | 10·284758 | ·051808 | 109 | 9·948192 | 34 |
| 27 | 9·663677 | 405 | ·336323 | 9·715551 | 515 | 10·284449 | ·051874 | 109 | 9·948126 | 33 |
| 28 | 9·663920 | 405 | ·336080 | 9·715860 | 514 | 10·284140 | ·051940 | 109 | 9·948060 | 32 |
| 29 | 9·664163 | 405 | ·335837 | 9·716168 | 514 | 10·283832 | ·052005 | 109 | 9·947995 | 31 |
| 30 | 9·664406 | 405 | ·335594 | 9·716477 | 514 | 10·283523 | ·052071 | 110 | 9·947929 | 30 |
| 31 | 9·664648 | 404 | ·335352 | 9·716785 | 514 | 10·283215 | ·052137 | 110 | 9·947863 | 29 |
| 32 | 9·664891 | 404 | ·335109 | 9·717093 | 514 | 10·282907 | ·052203 | 110 | 9·947797 | 28 |
| 33 | 9·665133 | 404 | ·334867 | 9·717401 | 513 | 10·282599 | ·052269 | 110 | 9·947731 | 27 |
| 34 | 9·665375 | 403 | ·334625 | 9·717709 | 513 | 10·282291 | ·052335 | 110 | 9·947665 | 26 |
| 35 | 9·665617 | 403 | ·334383 | 9·718017 | 513 | 10·281983 | ·052400 | 110 | 9·947600 | 25 |
| 36 | 9·665859 | 403 | ·334141 | 9·718325 | 513 | 10·281675 | ·052467 | 110 | 9·947533 | 24 |
| 37 | 9·666100 | 402 | ·333900 | 9·718633 | 513 | 10·281367 | ·052533 | 110 | 9·947467 | 23 |
| 38 | 9·666342 | 402 | ·333658 | 9·718940 | 512 | 10·281060 | ·052599 | 110 | 9·947401 | 22 |
| 39 | 9·666583 | 402 | ·333417 | 9·719248 | 512 | 10·280752 | ·052665 | 110 | 9·947335 | 21 |
| 40 | 9·666824 | 402 | ·333176 | 9·719555 | 512 | 10·280445 | ·052731 | 110 | 9·947269 | 20 |
| 41 | 9·667065 | 401 | ·332935 | 9·719862 | 512 | 10·280138 | ·052797 | 110 | 9·947203 | 19 |
| 42 | 9·667305 | 401 | ·332695 | 9·720169 | 512 | 10·279831 | ·052864 | 110 | 9·947136 | 18 |
| 43 | 9·667546 | 401 | ·332454 | 9·720476 | 511 | 10·279524 | ·052930 | 111 | 9·947070 | 17 |
| 44 | 9·667786 | 401 | ·332214 | 9·720783 | 511 | 10·279217 | ·052996 | 111 | 9·947004 | 16 |
| 45 | 9·668027 | 400 | ·331973 | 9·721089 | 511 | 10·278911 | ·053063 | 111 | 9·946937 | 15 |
| 46 | 9·668267 | 400 | ·331733 | 9·721396 | 511 | 10·278604 | ·053129 | 111 | 9·946871 | 14 |
| 47 | 9·668506 | 400 | ·331494 | 9·721702 | 511 | 10·278298 | ·053196 | 111 | 9·946804 | 13 |
| 48 | 9·668746 | 399 | ·331254 | 9·722009 | 510 | 10·277991 | ·053262 | 111 | 9·946738 | 12 |
| 49 | 9·668986 | 399 | ·331014 | 9·722315 | 510 | 10·277685 | ·053329 | 111 | 9·946671 | 11 |
| 50 | 9·669225 | 399 | ·330775 | 9·722621 | 510 | 10·277379 | ·053396 | 111 | 9·946604 | 10 |
| 51 | 9·669464 | 399 | ·330536 | 9·722927 | 510 | 10·277073 | ·053462 | 111 | 9·946538 | 9 |
| 52 | 9·669703 | 398 | ·330297 | 9·723232 | 510 | 10·276768 | ·053529 | 111 | 9·946471 | 8 |
| 53 | 9·669942 | 398 | ·330058 | 9·723538 | 509 | 10·276462 | ·053596 | 111 | 9·946404 | 7 |
| 54 | 9·670181 | 398 | ·329819 | 9·723844 | 509 | 10·276156 | ·053663 | 111 | 9·946337 | 6 |
| 55 | 9·670419 | 397 | ·329581 | 9·724149 | 509 | 10·275851 | ·053730 | 111 | 9·946270 | 5 |
| 56 | 9·670658 | 397 | ·329342 | 9·724454 | 509 | 10·275546 | ·053797 | 112 | 9·946203 | 4 |
| 57 | 9·670896 | 397 | ·329104 | 9·724759 | 509 | 10·275241 | ·053864 | 112 | 9·946136 | 3 |
| 58 | 9·671134 | 397 | ·328866 | 9·725065 | 508 | 10·274935 | ·053931 | 112 | 9·946069 | 2 |
| 59 | 9·671372 | 396 | ·328628 | 9·725369 | 508 | 10·274631 | ·053998 | 112 | 9·946002 | 1 |
| 60 | 9·671609 | 396 | ·328391 | 9·725674 | 508 | 10·274326 | ·054065 | 112 | 9·945935 | 0 |
| ′ | Cosine. | | Secant. | Cotangent. | | Tangent. | Cosecant. | | Sine. | ′ |

62 DEG.

28 DEG.

| ′ | Sine. | Diff. 100″ | Cosecant. | Tangent. | Diff. 100″ | Cotangent. | Secant. | Diff. 100″ | Cosine. | ′ |
|---|---|---|---|---|---|---|---|---|---|---|
| 0 | 9·671609 | | ·328391 | 9·725674 | | 10·274326 | ·054065 | | 9·945935 | 60 |
| 1 | 9·671847 | 396 | ·328153 | 9·725979 | 508 | 10·274021 | ·054132 | 112 | 9·945868 | 59 |
| 2 | 9·672084 | 395 | ·327916 | 9·726284 | 508 | 10·273716 | ·054200 | 112 | 9·945800 | 58 |
| 3 | 9·672321 | 395 | ·327679 | 9·726588 | 507 | 10·273412 | ·054267 | 112 | 9·945733 | 57 |
| 4 | 9·672558 | 395 | ·327442 | 9·726892 | 507 | 10·273108 | ·054334 | 112 | 9·945666 | 56 |
| 5 | 9·672795 | 395 | ·327205 | 9·727197 | 507 | 10·272803 | ·054402 | 112 | 9·945598 | 55 |
| 6 | 9·673032 | 394 | ·326968 | 9·727501 | 507 | 10·272499 | ·054469 | 112 | 9·945531 | 54 |
| 7 | 9·673268 | 394 | ·326732 | 9·727805 | 507 | 10·272195 | ·054536 | 112 | 9·945464 | 53 |
| 8 | 9·673505 | 394 | ·326495 | 9·728109 | 506 | 10·271891 | ·054604 | 113 | 9·945396 | 52 |
| 9 | 9·673741 | 394 | ·326259 | 9·728412 | 506 | 10·271588 | ·054672 | 113 | 9·945328 | 51 |
| 10 | 9·673977 | 393 | ·326023 | 9·728716 | 506 | 10·271284 | ·054739 | 113 | 9·945261 | 50 |
| 11 | 9·674213 | 393 | ·325787 | 9·729020 | 506 | 10·270980 | ·054807 | 113 | 9·945193 | 49 |
| 12 | 9·674448 | 393 | ·325552 | 9·729323 | 506 | 10·270677 | ·054875 | 113 | 9·945125 | 48 |
| 13 | 9·674684 | 392 | ·325316 | 9·729626 | 505 | 10·270374 | ·054942 | 113 | 9·945058 | 47 |
| 14 | 9·674919 | 392 | ·325081 | 9·729929 | 505 | 10·270071 | ·055010 | 113 | 9·944990 | 46 |
| 15 | 9·675155 | 392 | ·324845 | 9·730233 | 505 | 10·269767 | ·055078 | 113 | 9·944922 | 45 |
| 16 | 9·675390 | 392 | ·324610 | 9·730535 | 505 | 10·269465 | ·055146 | 113 | 9·944854 | 44 |
| 17 | 9·675624 | 391 | ·324376 | 9·730838 | 505 | 10·269162 | ·055214 | 113 | 9·944786 | 43 |
| 18 | 9·675859 | 391 | ·324141 | 9·731141 | 504 | 10·268859 | ·055282 | 113 | 9·944718 | 42 |
| 19 | 9·676094 | 391 | ·323906 | 9·731444 | 504 | 10·268556 | ·055350 | 113 | 9·944650 | 41 |
| 20 | 9·676328 | 391 | ·323672 | 9·731746 | 504 | 10·268254 | ·055418 | 113 | 9·944582 | 40 |
| 21 | 9·676562 | 390 | ·323438 | 9·732048 | 504 | 10·267952 | ·055486 | 114 | 9·944514 | 39 |
| 22 | 9·676796 | 390 | ·323204 | 9·732351 | 504 | 10·267649 | ·055554 | 114 | 9·944446 | 38 |
| 23 | 9·677030 | 390 | ·322970 | 9·732653 | 503 | 10·267347 | ·055623 | 114 | 9·944377 | 37 |
| 24 | 9·677264 | 390 | ·322736 | 9·732955 | 503 | 10·267045 | ·055691 | 114 | 9·944309 | 36 |
| 25 | 9·677498 | 389 | ·322502 | 9·733257 | 503 | 10·266743 | ·055759 | 114 | 9·944241 | 35 |
| 26 | 9·677731 | 389 | ·322269 | 9·733558 | 503 | 10·266442 | ·055828 | 114 | 9·944172 | 34 |
| 27 | 9·677964 | 389 | ·322036 | 9·733860 | 503 | 10·266140 | ·055896 | 114 | 9·944104 | 33 |
| 28 | 9·678197 | 388 | ·321803 | 9·734162 | 503 | 10·265838 | ·055964 | 114 | 9·944036 | 32 |
| 29 | 9·678430 | 388 | ·321570 | 9·734463 | 502 | 10·265537 | ·056033 | 114 | 9·943967 | 31 |
| 30 | 9·678663 | 388 | ·321337 | 9·734764 | 502 | 10·265236 | ·056102 | 114 | 9·943899 | 30 |
| 31 | 9·678895 | 388 | ·321105 | 9·735066 | 502 | 10·264934 | ·056170 | 114 | 9·943830 | 29 |
| 32 | 9·679128 | 387 | ·320872 | 9·735367 | 502 | 10·264633 | ·056239 | 114 | 9·943761 | 28 |
| 33 | 9·679360 | 387 | ·320640 | 9·735668 | 502 | 10·264332 | ·056307 | 114 | 9·943693 | 27 |
| 34 | 9·679592 | 387 | ·320408 | 9·735969 | 501 | 10·264031 | ·056376 | 115 | 9·943624 | 26 |
| 35 | 9·679824 | 387 | ·320176 | 9·736269 | 501 | 10·263731 | ·056445 | 115 | 9·943555 | 25 |
| 36 | 9·680056 | 386 | ·319944 | 9·736570 | 501 | 10·263430 | ·056514 | 115 | 9·943486 | 24 |
| 37 | 9·680288 | 386 | ·319712 | 9·736871 | 501 | 10·263129 | ·056583 | 115 | 9·943417 | 23 |
| 38 | 9·680519 | 386 | ·319481 | 9·737171 | 501 | 10·262829 | ·056652 | 115 | 9·943348 | 22 |
| 39 | 9·680750 | 385 | ·319250 | 9·737471 | 500 | 10·262529 | ·056721 | 115 | 9·943279 | 21 |
| 40 | 9·680982 | 385 | ·319018 | 9·737771 | 500 | 10·262229 | ·056790 | 115 | 9·943210 | 20 |
| 41 | 9·681213 | 385 | ·318787 | 9·738071 | 500 | 10·261929 | ·056859 | 115 | 9·943141 | 19 |
| 42 | 9·681443 | 385 | ·318557 | 9·738371 | 500 | 10·261629 | ·056928 | 115 | 9·943072 | 18 |
| 43 | 9·681674 | 384 | ·318326 | 9·738671 | 500 | 10·261329 | ·056997 | 115 | 9·943003 | 17 |
| 44 | 9·681905 | 384 | ·318095 | 9·738971 | 500 | 10·261029 | ·057066 | 115 | 9·942934 | 16 |
| 45 | 9·682135 | 384 | ·317865 | 9·739271 | 499 | 10·260729 | ·057136 | 115 | 9·942864 | 15 |
| 46 | 9·682365 | 384 | ·317635 | 9·739570 | 499 | 10·260430 | ·057205 | 115 | 9·942795 | 14 |
| 47 | 9·682595 | 383 | ·317405 | 9·739870 | 499 | 10·260130 | ·057274 | 116 | 9·942726 | 13 |
| 48 | 9·682825 | 383 | ·317175 | 9·740169 | 499 | 10·259831 | ·057344 | 116 | 9·942656 | 12 |
| 49 | 9·683055 | 383 | ·316945 | 9·740468 | 499 | 10·259532 | ·057413 | 116 | 9·942587 | 11 |
| 50 | 9·683284 | 383 | ·316716 | 9·740767 | 498 | 10·259233 | ·057483 | 116 | 9·942517 | 10 |
| 51 | 9·683514 | 382 | ·316486 | 9·741066 | 498 | 10·258934 | ·057552 | 116 | 9·942448 | 9 |
| 52 | 9·683743 | 382 | ·316257 | 9·741365 | 498 | 10·258635 | ·057622 | 116 | 9·942378 | 8 |
| 53 | 9·683972 | 382 | ·316028 | 9·741664 | 498 | 10·258336 | ·057692 | 116 | 9·942308 | 7 |
| 54 | 9·684201 | 382 | ·315799 | 9·741962 | 498 | 10·258038 | ·057761 | 116 | 9·942239 | 6 |
| 55 | 9·684430 | 381 | ·315570 | 9·742261 | 498 | 10·257739 | ·057831 | 116 | 9·942169 | 5 |
| 56 | 9·684658 | 381 | ·315342 | 9·742559 | 497 | 10·257441 | ·057901 | 116 | 9·942099 | 4 |
| 57 | 9·684887 | 381 | ·315113 | 9·742858 | 497 | 10·257142 | ·057971 | 116 | 9·942029 | 3 |
| 58 | 9·685115 | 380 | ·314885 | 9·743156 | 497 | 10·256844 | ·058041 | 116 | 9·941959 | 2 |
| 59 | 9·685343 | 380 | ·314657 | 9·743454 | 497 | 10·256546 | ·058111 | 116 | 9·941889 | 1 |
| 60 | 9·685571 | 380 | ·314429 | 9·743752 | 497 | 10·256248 | ·058181 | 117 | 9·941819 | 0 |
| ′ | Cosine. | | Secant. | Cotangent. | | Tangent. | Cosecant. | | Sine. | ′ |

61 DEG.

29 DEG.

| ′ | Sine. | Diff. 100″ | Cosecant. | Tangent. | Diff. 100″ | Cotangent. | Secant. | Diff 100″ | Cosine. | ′ |
|---|---|---|---|---|---|---|---|---|---|---|
| 0 | 9·685571 | | ·314429 | 9·743752 | | 10·256248 | ·058181 | | 9·941819 | 60 |
| 1 | 9·685799 | 380 | ·314201 | 9·744050 | 496 | 10·255950 | ·058251 | 117 | 9·941749 | 59 |
| 2 | 9·686027 | 379 | ·313973 | 9·744348 | 496 | 10·255652 | ·058321 | 117 | 9·941679 | 58 |
| 3 | 9·686254 | 379 | ·313746 | 9·744645 | 496 | 10·255355 | ·058391 | 117 | 9·941609 | 57 |
| 4 | 9·686482 | 379 | ·313518 | 9·744943 | 496 | 10·255057 | ·058461 | 117 | 9·941539 | 56 |
| 5 | 9·686709 | 379 | ·313291 | 9·745240 | 496 | 10·254760 | ·058531 | 117 | 9·941469 | 55 |
| 6 | 9·686936 | 378 | ·313064 | 9·745538 | 496 | 10·254462 | ·058602 | 117 | 9·941398 | 54 |
| 7 | 9·687163 | 378 | ·312837 | 9·745835 | 495 | 10·254165 | ·058672 | 117 | 9·941328 | 53 |
| 8 | 9·687389 | 378 | ·312611 | 9·746132 | 495 | 10·253868 | ·058742 | 117 | 9·941258 | 52 |
| 9 | 9·687616 | 378 | ·312384 | 9·746429 | 495 | 10·253571 | ·058813 | 117 | 9·941187 | 51 |
| 10 | 9·687843 | 377 | ·312157 | 9·746726 | 495 | 10·253274 | ·058883 | 117 | 9·941117 | 50 |
| 11 | 9·688069 | 377 | ·311931 | 9·747023 | 495 | 10·252977 | ·058954 | 117 | 9·941046 | 49 |
| 12 | 9·688295 | 377 | ·311705 | 9·747319 | 494 | 10·252681 | ·059025 | 118 | 9·940975 | 48 |
| 13 | 9·688521 | 377 | ·311479 | 9·747616 | 494 | 10·252384 | ·059095 | 118 | 9·940905 | 47 |
| 14 | 9·688747 | 376 | ·311253 | 9·747913 | 494 | 10·252087 | ·059166 | 118 | 9·940834 | 46 |
| 15 | 9·688972 | 376 | ·311028 | 9·748209 | 494 | 10·251791 | ·059237 | 118 | 9·940763 | 45 |
| 16 | 9·689198 | 376 | ·310802 | 9·748505 | 494 | 10·251495 | ·059367 | 118 | 9·940693 | 44 |
| 17 | 9·689423 | 376 | ·310577 | 9·748801 | 494 | 10·251199 | ·059378 | 118 | 9·940622 | 43 |
| 18 | 9·689648 | 375 | ·310352 | 9·749097 | 493 | 10·250903 | ·059449 | 118 | 9·940551 | 42 |
| 19 | 9·689873 | 375 | ·310127 | 9·749393 | 493 | 10·250607 | ·059520 | 118 | 9·940480 | 41 |
| 20 | 9·690098 | 375 | ·309902 | 9·749689 | 493 | 10·250311 | ·059591 | 118 | 9·940409 | 40 |
| 21 | 9·690323 | 375 | ·309677 | 9·749985 | 493 | 10·250015 | ·059662 | 118 | 9·940338 | 39 |
| 22 | 9·690548 | 374 | ·309452 | 9·750281 | 493 | 10·249719 | ·059733 | 118 | 9·940267 | 38 |
| 23 | 9·690772 | 374 | ·309228 | 9·750576 | 493 | 10·249424 | ·059804 | 118 | 9·940196 | 37 |
| 24 | 9·690996 | 374 | ·309004 | 9·750872 | 492 | 10·249128 | ·059875 | 118 | 9·940125 | 36 |
| 25 | 9·691220 | 374 | ·308780 | 9·751167 | 492 | 10·248833 | ·059946 | 119 | 9·940054 | 35 |
| 26 | 9·691444 | 373 | ·308556 | 9·751462 | 492 | 10·248538 | ·060018 | 119 | 9·939982 | 34 |
| 27 | 9·691668 | 373 | ·308332 | 9·751757 | 492 | 10·248243 | ·060089 | 119 | 9·939911 | 33 |
| 28 | 9·691892 | 373 | ·308108 | 9·752052 | 492 | 10·247948 | ·060160 | 119 | 9·939840 | 32 |
| 29 | 9·692115 | 373 | ·307885 | 9·752347 | 491 | 10·247653 | ·060232 | 119 | 9·939768 | 31 |
| 30 | 9·692339 | 372 | ·307661 | 9·752642 | 491 | 10·247358 | ·060303 | 119 | 9·939697 | 30 |
| 31 | 9·692562 | 372 | ·307438 | 9·752937 | 491 | 10·247063 | ·060375 | 119 | 9·939625 | 29 |
| 32 | 9·692785 | 372 | ·307215 | 9·753231 | 491 | 10·246769 | ·060446 | 119 | 9·939554 | 28 |
| 33 | 9·693008 | 371 | ·306992 | 9·753526 | 491 | 10·246474 | ·060518 | 119 | 9·939482 | 27 |
| 34 | 9·693231 | 371 | ·306769 | 9·753820 | 491 | 10·246180 | ·060590 | 119 | 9·939410 | 26 |
| 35 | 9·693453 | 371 | ·306547 | 9·754115 | 490 | 10·245885 | ·060661 | 119 | 9·939339 | 25 |
| 36 | 9·693676 | 371 | ·306324 | 9·754409 | 490 | 10·245591 | ·060733 | 119 | 9·939267 | 24 |
| 37 | 9·693898 | 370 | ·306102 | 9·754703 | 490 | 10·245297 | ·060805 | 120 | 9·939195 | 23 |
| 38 | 9·694120 | 370 | ·305880 | 9·754997 | 490 | 10·245003 | ·060877 | 120 | 9·939123 | 22 |
| 39 | 9·694342 | 370 | ·305658 | 9·755291 | 490 | 10·244709 | ·060948 | 120 | 9·939052 | 21 |
| 40 | 9·694564 | 370 | ·305436 | 9·755585 | 490 | 10·244415 | ·061020 | 120 | 9·938980 | 20 |
| 41 | 9·694786 | 369 | ·305214 | 9·755878 | 489 | 10·244122 | ·061092 | 120 | 9·938908 | 19 |
| 42 | 9·695007 | 369 | ·304993 | 9·756172 | 489 | 10·243828 | ·061164 | 120 | 9·938836 | 18 |
| 43 | 9·695229 | 369 | ·304771 | 9·756465 | 489 | 10·243535 | ·061237 | 120 | 9·938763 | 17 |
| 44 | 9·695450 | 369 | ·304550 | 9·756759 | 489 | 10·243241 | ·061309 | 120 | 9·938691 | 16 |
| 45 | 9·695671 | 368 | ·304329 | 9·757052 | 489 | 10·242948 | ·061381 | 120 | 9·938619 | 15 |
| 46 | 9·695892 | 368 | ·304108 | 9·757345 | 489 | 10·242655 | ·061453 | 120 | 9·938547 | 14 |
| 47 | 9·696113 | 368 | ·303887 | 9·757638 | 488 | 10·242362 | ·061525 | 120 | 9·938475 | 13 |
| 48 | 9·696334 | 368 | ·303666 | 9·757931 | 488 | 10·242069 | ·061598 | 120 | 9·938402 | 12 |
| 49 | 9·696554 | 367 | ·303446 | 9·758224 | 488 | 10·241776 | ·061670 | 121 | 9·938330 | 11 |
| 50 | 9·696775 | 367 | ·303225 | 9·758517 | 488 | 10·241483 | ·061742 | 121 | 9·938258 | 10 |
| 51 | 9·696995 | 367 | ·303005 | 9·758810 | 488 | 10·241190 | ·061815 | 121 | 9·938185 | 9 |
| 52 | 9·697215 | 367 | ·302785 | 9·759102 | 488 | 10·240898 | ·061887 | 121 | 9·938113 | 8 |
| 53 | 9·697435 | 366 | ·302565 | 9·759395 | 487 | 10·240605 | ·061960 | 121 | 9·938040 | 7 |
| 54 | 9·697654 | 366 | ·302346 | 9·759687 | 487 | 10·240313 | ·062033 | 121 | 9·937967 | 6 |
| 55 | 9·697874 | 366 | ·302126 | 9·759979 | 487 | 10·240021 | ·062105 | 121 | 9·937895 | 5 |
| 56 | 9·698094 | 366 | ·301906 | 9·760272 | 487 | 10·239728 | ·062178 | 121 | 9·937822 | 4 |
| 57 | 9·698313 | 365 | ·301687 | 9·760564 | 487 | 10·239436 | ·062251 | 121 | 9·937749 | 3 |
| 58 | 9·698532 | 365 | ·301468 | 9·760856 | 487 | 10·239144 | ·062324 | 121 | 9·937676 | 2 |
| 59 | 9·698751 | 365 | ·301249 | 9·761148 | 486 | 10·238852 | ·062396 | 121 | 9·937604 | 1 |
| 60 | 9·698970 | 365 | ·301030 | 9·761439 | 486 | 10·238561 | ·062469 | 121 | 9·937531 | 0 |
| ′ | Cosine. | | Secant. | Cotangent. | | Tangent. | Cosecant. | | Sine. | ′ |

60 DEG.

30 DEG.

| ′ | Sine. | Diff. 100″ | Cosecant. | Tangent. | Diff. 100″ | Cotangent. | Secant. | Diff. 100″ | Cosine. | ′ |
|---|---|---|---|---|---|---|---|---|---|---|
| 0 | 9·698970 | | ·301030 | 9·761439 | | 10·238561 | ·062469 | | 9·937531 | 60 |
| 1 | 9·699189 | 364 | ·300811 | 9·761731 | 486 | 10·238269 | ·062542 | 121 | 9·937458 | 59 |
| 2 | 9·699407 | 364 | ·300593 | 9·762023 | 486 | 10·237977 | ·062615 | 122 | 9·937385 | 58 |
| 3 | 9·699626 | 364 | ·300374 | 9·762314 | 486 | 10·237686 | ·062688 | 122 | 9·937312 | 57 |
| 4 | 9·699844 | 364 | ·300156 | 9·762606 | 486 | 10·237394 | ·062762 | 122 | 9·937238 | 56 |
| 5 | 9·700062 | 363 | ·299938 | 9·762897 | 485 | 10·237103 | ·062835 | 122 | 9·937165 | 55 |
| 6 | 9·700280 | 363 | ·299520 | 9·763188 | 485 | 10·236812 | ·062908 | 122 | 9·937092 | 54 |
| 7 | 9·700498 | 363 | ·299702 | 9·763479 | 485 | 10·236521 | ·062981 | 122 | 9·937019 | 53 |
| 8 | 9·700716 | 363 | ·299284 | 9·763770 | 485 | 10·236230 | ·063054 | 122 | 9·936946 | 52 |
| 9 | 9·700933 | 363 | ·299067 | 9·764061 | 485 | 10·235939 | ·063128 | 122 | 9·936872 | 51 |
| 10 | 9·701151 | 362 | ·298849 | 9·764352 | 485 | 10·235648 | ·063201 | 122 | 9·936799 | 50 |
| 11 | 9·701368 | 362 | ·298632 | 9·764643 | 485 | 10·235357 | ·063275 | 122 | 9·936725 | 49 |
| 12 | 9·701585 | 362 | ·298415 | 9·764933 | 484 | 10·235067 | ·063348 | 122 | 9·936652 | 48 |
| 13 | 9·701802 | 362 | ·298198 | 9·765224 | 484 | 10·234776 | ·063422 | 123 | 9·936578 | 47 |
| 14 | 9·702019 | 361 | ·297981 | 9·765514 | 484 | 10·234486 | ·063495 | 123 | 9·936505 | 46 |
| 15 | 9·702236 | 361 | ·297764 | 9·765805 | 484 | 10·234195 | ·063569 | 123 | 9·936431 | 45 |
| 16 | 9·702452 | 361 | ·297548 | 9·766095 | 484 | 10·233905 | ·063643 | 123 | 9·936357 | 44 |
| 17 | 9·702669 | 361 | ·297331 | 9·766385 | 484 | 10·233615 | ·063716 | 123 | 9·936284 | 43 |
| 18 | 9·702885 | 360 | ·297115 | 9·766675 | 483 | 10·233325 | ·063790 | 123 | 9·936210 | 42 |
| 19 | 9·703101 | 360 | ·296899 | 9·766965 | 483 | 10·233035 | ·063864 | 123 | 9·936136 | 41 |
| 20 | 9·703317 | 360 | ·296683 | 9·767255 | 483 | 10·232745 | ·063938 | 123 | 9·936062 | 40 |
| 21 | 9·703533 | 360 | ·296467 | 9·767545 | 483 | 10·232455 | ·064012 | 123 | 9·935988 | 39 |
| 22 | 9·703749 | 359 | ·296251 | 9·767834 | 483 | 10·232166 | ·064086 | 123 | 9·935914 | 38 |
| 23 | 9·703964 | 359 | ·296036 | 9·768124 | 483 | 10·231876 | ·064160 | 123 | 9·935840 | 37 |
| 24 | 9·704179 | 359 | ·295821 | 9·768414 | 482 | 10·231586 | ·064234 | 123 | 9·935766 | 36 |
| 25 | 9·704395 | 359 | ·295605 | 9·768703 | 482 | 10·231297 | ·064308 | 124 | 9·935692 | 35 |
| 26 | 9·704610 | 359 | ·295390 | 9·768992 | 482 | 10·231008 | ·064382 | 124 | 9·935618 | 34 |
| 27 | 9·704825 | 358 | ·295175 | 9·769281 | 482 | 10·230719 | ·064457 | 124 | 9·935543 | 33 |
| 28 | 9·705040 | 358 | ·294960 | 9·769570 | 482 | 10·230430 | ·064531 | 124 | 9·935469 | 32 |
| 29 | 9·705254 | 358 | ·294746 | 9·769860 | 482 | 10·230140 | ·064605 | 124 | 9·935395 | 31 |
| 30 | 9·705469 | 358 | ·294531 | 9·770148 | 481 | 10·229852 | ·064680 | 124 | 9·935320 | 30 |
| 31 | 9·705683 | 357 | ·294317 | 9·770437 | 481 | 10·229563 | ·064754 | 124 | 9·935246 | 29 |
| 32 | 9·705898 | 357 | ·294102 | 9·770726 | 481 | 10·229274 | ·064829 | 124 | 9·935171 | 28 |
| 33 | 9·706112 | 357 | ·293888 | 9·771015 | 481 | 10·228985 | ·064903 | 124 | 9·935097 | 27 |
| 34 | 9·706326 | 357 | ·293674 | 9·771303 | 481 | 10·228697 | ·064978 | 124 | 9·935022 | 26 |
| 35 | 9·706539 | 356 | ·293461 | 9·771592 | 481 | 10·228408 | ·065052 | 124 | 9·934948 | 25 |
| 36 | 9·706753 | 356 | ·293247 | 9·771880 | 481 | 10·228120 | ·065127 | 124 | 9·934873 | 24 |
| 37 | 9·706967 | 356 | ·293033 | 9·772168 | 480 | 10·227832 | ·065202 | 124 | 9·934798 | 23 |
| 38 | 9·707180 | 356 | ·292820 | 9·772457 | 480 | 10·227543 | ·065277 | 125 | 9·934723 | 22 |
| 39 | 9·707393 | 355 | ·292607 | 9·772745 | 480 | 10·227255 | ·065351 | 125 | 9·934649 | 21 |
| 40 | 9·707606 | 355 | ·292394 | 9·773033 | 480 | 10·226967 | ·065426 | 125 | 9·934574 | 20 |
| 41 | 9·707819 | 355 | ·292181 | 9·773321 | 480 | 10·226679 | ·065501 | 125 | 9·934499 | 19 |
| 42 | 9·708032 | 355 | ·291968 | 9·773608 | 480 | 10·226392 | ·065576 | 125 | 9·934424 | 18 |
| 43 | 9·708245 | 354 | ·291755 | 9·773896 | 480 | 10·226104 | ·065651 | 125 | 9·934349 | 17 |
| 44 | 9·708458 | 354 | ·291542 | 9·774184 | 479 | 10·225816 | ·065726 | 125 | 9·934274 | 16 |
| 45 | 9·708670 | 354 | ·291330 | 9·774471 | 479 | 10·225529 | ·065801 | 125 | 9·934199 | 15 |
| 46 | 9·708882 | 354 | ·291118 | 9·774759 | 479 | 10·225241 | ·065877 | 125 | 9·934123 | 14 |
| 47 | 9·709094 | 353 | ·290906 | 9·775046 | 479 | 10·224954 | ·065952 | 125 | 9·934048 | 13 |
| 48 | 9·709306 | 353 | ·290694 | 9·775333 | 479 | 10·224667 | ·066027 | 125 | 9·933973 | 12 |
| 49 | 9·709518 | 353 | ·290482 | 9·775621 | 479 | 10·224379 | ·066102 | 125 | 9·933898 | 11 |
| 50 | 9·709730 | 353 | ·290270 | 9·775908 | 478 | 10·224092 | ·066178 | 126 | 9·933822 | 10 |
| 51 | 9·709941 | 353 | ·290059 | 9·776195 | 478 | 10·223805 | ·066253 | 126 | 9·933747 | 9 |
| 52 | 9·710153 | 352 | ·289847 | 9·776482 | 478 | 10·223518 | ·066329 | 126 | 9·933671 | 8 |
| 53 | 9·710364 | 352 | ·289636 | 9·776769 | 478 | 10·223231 | ·066404 | 126 | 9·933596 | 7 |
| 54 | 9·710575 | 352 | ·289425 | 9·777055 | 478 | 10·222945 | ·066480 | 126 | 9·933520 | 6 |
| 55 | 9·710786 | 352 | ·289214 | 9·777342 | 478 | 10·222658 | ·066555 | 126 | 9·933445 | 5 |
| 56 | 9·710997 | 351 | ·289003 | 9·777628 | 478 | 10·222372 | ·066631 | 126 | 9·933369 | 4 |
| 57 | 9·711208 | 351 | ·288792 | 9·777915 | 477 | 10·222085 | ·066707 | 126 | 9·933293 | 3 |
| 58 | 9·711419 | 351 | ·288581 | 9·778201 | 477 | 10·221799 | ·066783 | 126 | 9·933217 | 2 |
| 59 | 9·711629 | 351 | ·288371 | 9·778487 | 477 | 10·221513 | ·066859 | 126 | 9·933141 | 1 |
| 60 | 9·711839 | 350 | ·288161 | 9·778774 | 477 | 10·221226 | ·066934 | 126 | 9·933066 | 0 |
| ′ | Cosine. | | Secant. | Cotangent. | | Tangent. | Cosecant. | | Sine. | ′ |



59 DEG.

31 DEG.

| ′ | Sine. | Diff. 100″ | Cosecant. | Tangent. | Diff. 100″ | Cotangent. | Secant. | Diff. 100″ | Cosine. | ′ |
|---|---|---|---|---|---|---|---|---|---|---|
| 0 | 9·711839 | | ·288161 | 9·778774 | | 10·221226 | ·066934 | | 9·933066 | 60 |
| 1 | 9·712050 | 350 | ·287950 | 9·779060 | 477 | 10·220940 | ·067010 | 126 | 9·932990 | 59 |
| 2 | 9·712260 | 350 | ·287740 | 9·779346 | 477 | 10·220654 | ·067086 | 127 | 9·932914 | 58 |
| 3 | 9·712469 | 350 | ·287531 | 9·779632 | 477 | 10·220368 | ·067162 | 127 | 9·932838 | 57 |
| 4 | 9·712679 | 349 | ·287321 | 9·779918 | 476 | 10·220082 | ·067238 | 127 | 9·932762 | 56 |
| 5 | 9·712889 | 349 | ·287111 | 9·780203 | 476 | 10·219797 | ·067315 | 127 | 9·932685 | 55 |
| 6 | 9·713098 | 349 | ·286902 | 9·780489 | 476 | 10·219511 | ·067391 | 127 | 9·932609 | 54 |
| 7 | 9·713308 | 349 | ·286692 | 9·780775 | 476 | 10·219225 | ·067467 | 127 | 9·932533 | 53 |
| 8 | 9·713517 | 349 | ·286483 | 9·781060 | 476 | 10·218940 | ·067543 | 127 | 9·932457 | 52 |
| 9 | 9·713726 | 348 | ·286274 | 9·781346 | 476 | 10·218654 | ·067620 | 127 | 9·932380 | 51 |
| 10 | 9·713935 | 348 | ·286065 | 9·781631 | 476 | 10·218369 | ·067696 | 127 | 9·932304 | 50 |
| 11 | 9·714144 | 348 | ·285856 | 9·781916 | 475 | 10·218084 | ·067772 | 127 | 9·932228 | 49 |
| 12 | 9·714352 | 348 | ·285648 | 9·782201 | 475 | 10·217799 | ·067849 | 127 | 9·932151 | 48 |
| 13 | 9·714561 | 347 | ·285439 | 9·782486 | 475 | 10·217514 | ·067925 | 127 | 9·932075 | 47 |
| 14 | 9·714769 | 347 | ·285231 | 9·782771 | 475 | 10·217229 | ·068002 | 128 | 9·931998 | 46 |
| 15 | 9·714978 | 347 | ·285022 | 9·783056 | 475 | 10·216944 | ·068079 | 128 | 9·931921 | 45 |
| 16 | 9·715186 | 347 | ·284814 | 9·783341 | 475 | 10·216659 | ·068155 | 128 | 9·931845 | 44 |
| 17 | 9·715394 | 347 | ·284606 | 9·783626 | 475 | 10·216374 | ·068232 | 128 | 9·931768 | 43 |
| 18 | 9·715602 | 346 | ·284398 | 9·783910 | 474 | 10·216090 | ·068309 | 128 | 9·931691 | 42 |
| 19 | 9·715809 | 346 | ·284191 | 9·784195 | 474 | 10·215805 | ·068386 | 128 | 9·931614 | 41 |
| 20 | 9·716017 | 346 | ·283983 | 9·784479 | 474 | 10·215521 | ·068463 | 128 | 9·931537 | 40 |
| 21 | 9·716224 | 346 | ·283776 | 9·784764 | 474 | 10·215236 | ·068540 | 128 | 9·931460 | 39 |
| 22 | 9·716432 | 345 | ·283568 | 9·785048 | 474 | 10·214952 | ·068617 | 128 | 9·931383 | 38 |
| 23 | 9·716639 | 345 | ·283361 | 9·785332 | 474 | 10·214668 | ·068694 | 128 | 9·931306 | 37 |
| 24 | 9·716846 | 345 | ·283154 | 9·785616 | 474 | 10·214384 | ·068771 | 128 | 9·931229 | 36 |
| 25 | 9·717053 | 345 | ·282947 | 9·785900 | 473 | 10·214100 | ·068848 | 129 | 9·931152 | 35 |
| 26 | 9·717259 | 345 | ·282741 | 9·786184 | 473 | 10·213816 | ·068925 | 129 | 9·931075 | 34 |
| 27 | 9·717466 | 344 | ·282534 | 9·786468 | 473 | 10·213532 | ·069002 | 129 | 9·930998 | 33 |
| 28 | 9·717673 | 344 | ·282327 | 9·786752 | 473 | 10·213248 | ·069079 | 129 | 9·930921 | 32 |
| 29 | 9·717879 | 344 | ·282121 | 9·787036 | 473 | 10·212964 | ·069157 | 129 | 9·930843 | 31 |
| 30 | 9·718085 | 344 | ·281915 | 9·787319 | 473 | 10·212681 | ·069234 | 129 | 9·930766 | 30 |
| 31 | 9·718291 | 343 | ·281709 | 9·787603 | 473 | 10·212397 | ·069312 | 129 | 9·930688 | 29 |
| 32 | 9·718497 | 343 | ·281503 | 9·787886 | 472 | 10·212114 | ·069389 | 129 | 9·930611 | 28 |
| 33 | 9·718703 | 343 | ·281297 | 9·788170 | 472 | 10·211830 | ·069467 | 129 | 9·930533 | 27 |
| 34 | 9·718909 | 343 | ·281091 | 9·788453 | 472 | 10·211547 | ·069544 | 129 | 9·930456 | 26 |
| 35 | 9·719114 | 343 | ·280886 | 9·788736 | 472 | 10·211264 | ·069622 | 129 | 9·930378 | 25 |
| 36 | 9·719320 | 342 | ·280680 | 9·789019 | 472 | 10·210981 | ·069700 | 129 | 9·930300 | 24 |
| 37 | 9·719525 | 342 | ·280475 | 9·789302 | 472 | 10·210698 | ·069777 | 130 | 9·930223 | 23 |
| 38 | 9·719730 | 342 | ·280270 | 9·789585 | 472 | 10·210415 | ·069855 | 130 | 9·930145 | 22 |
| 39 | 9·719935 | 342 | ·280065 | 9·789868 | 471 | 10·210132 | ·069933 | 130 | 9·930067 | 21 |
| 40 | 9·720140 | 341 | ·279860 | 9·790151 | 471 | 10·209849 | ·070011 | 130 | 9·929989 | 20 |
| 41 | 9·720345 | 341 | ·279655 | 9·790433 | 471 | 10·209567 | ·070089 | 130 | 9·929911 | 19 |
| 42 | 9·720549 | 341 | ·279451 | 9·790716 | 471 | 10·209284 | ·070167 | 130 | 9·929833 | 18 |
| 43 | 9·720754 | 341 | ·279246 | 9·790999 | 471 | 10·209001 | ·070245 | 130 | 9·929755 | 17 |
| 44 | 9·720958 | 340 | ·279042 | 9·791281 | 471 | 10·208719 | ·070323 | 130 | 9·929677 | 16 |
| 45 | 9·721162 | 340 | ·278838 | 9·791563 | 471 | 10·208437 | ·070401 | 130 | 9·929599 | 15 |
| 46 | 9·721366 | 340 | ·278634 | 9·791846 | 470 | 10·208154 | ·070479 | 130 | 9·929521 | 14 |
| 47 | 9·721570 | 340 | ·278430 | 9·792128 | 470 | 10·207872 | ·070558 | 130 | 9·929442 | 13 |
| 48 | 9·721774 | 340 | ·278226 | 9·792410 | 470 | 10·207590 | ·070636 | 130 | 9·929364 | 12 |
| 49 | 9·721978 | 339 | ·278022 | 9·792692 | 470 | 10·207308 | ·070714 | 131 | 9·929286 | 11 |
| 50 | 9·722181 | 339 | ·277819 | 9·792974 | 470 | 10·207026 | ·070793 | 131 | 9·929207 | 10 |
| 51 | 9·722385 | 339 | ·277615 | 9·793256 | 470 | 10·206744 | ·070871 | 131 | 9·929129 | 9 |
| 52 | 9·722588 | 339 | ·277412 | 9·793538 | 470 | 10·206462 | ·070950 | 131 | 9·929050 | 8 |
| 53 | 9·722791 | 339 | ·277209 | 9·793819 | 469 | 10·206181 | ·071028 | 131 | 9·928972 | 7 |
| 54 | 9·722994 | 338 | ·277006 | 9·794101 | 469 | 10·205899 | ·071107 | 131 | 9·928893 | 6 |
| 55 | 9·723197 | 338 | ·276803 | 9·794383 | 469 | 10·205617 | ·071185 | 131 | 9·928815 | 5 |
| 56 | 9·723400 | 338 | ·276600 | 9·794664 | 469 | 10·205336 | ·071264 | 131 | 9·928736 | 4 |
| 57 | 9·723603 | 338 | ·276397 | 9·794945 | 469 | 10·205055 | ·071343 | 131 | 9·928657 | 3 |
| 58 | 9·723805 | 337 | ·276195 | 9·795227 | 469 | 10·204773 | ·071422 | 131 | 9·928578 | 2 |
| 59 | 9·724007 | 337 | ·275993 | 9·795508 | 469 | 10·204492 | ·071501 | 131 | 9·928499 | 1 |
| 60 | 9·724210 | 337 | ·275790 | 9·795789 | 468 | 10·204211 | ·071580 | 131 | 9·928420 | 0 |
| ′ | Cosine. | | Secant. | Cotangent. | | Tangent. | Cosecant. | | Sine. | ′ |

58 DEG.

32 DEG.

| ′ | Sine. | Diff. 100″ | Cosecant. | Tangent. | Diff. 100″ | Cotangent. | Secant. | Diff. 100″ | Cosine. | ′ |
|---|---|---|---|---|---|---|---|---|---|---|
| 0 | 9·724210 | | ·275790 | 9·795789 | | 10·204211 | ·071580 | | 9·928420 | 60 |
| 1 | 9·724412 | 337 | ·275588 | 9·796070 | 468 | 10·203930 | ·071658 | 132 | 9·928342 | 59 |
| 2 | 9·724614 | 337 | ·275386 | 9·796351 | 468 | 10·203649 | ·071737 | 132 | 9·928263 | 58 |
| 3 | 9·724816 | 336 | ·275184 | 9·796632 | 468 | 10·203368 | ·071817 | 132 | 9·928183 | 57 |
| 4 | 9·725017 | 336 | ·274983 | 9·796913 | 468 | 10·203087 | ·071896 | 132 | 9·928104 | 56 |
| 5 | 9·725219 | 336 | ·274781 | 9·797194 | 468 | 10·202806 | ·071975 | 132 | 9·928025 | 55 |
| 6 | 9·725420 | 336 | ·274580 | 9·797475 | 468 | 10·202525 | ·072054 | 132 | 9·927946 | 54 |
| 7 | 9·725622 | 335 | ·274378 | 9·797755 | 468 | 10·202245 | ·072133 | 132 | 9·927867 | 53 |
| 8 | 9·725823 | 335 | ·274177 | 9·798036 | 467 | 10·201964 | ·072213 | 132 | 9·927787 | 52 |
| 9 | 9·726024 | 335 | ·273976 | 9·798316 | 467 | 10·201684 | ·072292 | 132 | 9·927708 | 51 |
| 10 | 9·726225 | 335 | ·273775 | 9·798596 | 467 | 10·201404 | ·072371 | 132 | 9·927629 | 50 |
| 11 | 9·726426 | 335 | ·273574 | 9·798877 | 467 | 10·201123 | ·072451 | 132 | 9·927549 | 49 |
| 12 | 9·726626 | 334 | ·273374 | 9·799157 | 467 | 10·200843 | ·072530 | 132 | 9·927470 | 48 |
| 13 | 9·726827 | 334 | ·273173 | 9·799437 | 467 | 10·200563 | ·072610 | 133 | 9·927390 | 47 |
| 14 | 9·727027 | 334 | ·272973 | 9·799717 | 467 | 10·200283 | ·072690 | 133 | 9·927310 | 46 |
| 15 | 9·727228 | 334 | ·272772 | 9·799997 | 467 | 10·200003 | ·072769 | 133 | 9·927231 | 45 |
| 16 | 9·727428 | 334 | ·272572 | 9·800277 | 466 | 10·199723 | ·072849 | 133 | 9·927151 | 44 |
| 17 | 9·727628 | 333 | ·272372 | 9·800557 | 466 | 10·199443 | ·072929 | 133 | 9·927071 | 43 |
| 18 | 9·727828 | 333 | ·272172 | 9·800836 | 466 | 10·199164 | ·073009 | 133 | 9·926991 | 42 |
| 19 | 9·728027 | 333 | ·271973 | 9·801116 | 466 | 10·198884 | ·073089 | 133 | 9·926911 | 41 |
| 20 | 9·728227 | 333 | ·271773 | 9·801396 | 466 | 10·198604 | ·073169 | 133 | 9·926831 | 40 |
| 21 | 9·728427 | 333 | ·271573 | 9·801675 | 466 | 10·198325 | ·073249 | 133 | 9·926751 | 39 |
| 22 | 9·728626 | 332 | ·271374 | 9·801955 | 466 | 10·198045 | ·073329 | 133 | 9·926671 | 38 |
| 23 | 9·728825 | 332 | ·271175 | 9·802234 | 466 | 10·197766 | ·073409 | 133 | 9·926591 | 37 |
| 24 | 9·729024 | 332 | ·270976 | 9·802513 | 465 | 10·197487 | ·073489 | 133 | 9·926511 | 36 |
| 25 | 9·729223 | 332 | ·270777 | 9·802792 | 465 | 10·197208 | ·073569 | 134 | 9·926431 | 35 |
| 26 | 9·729422 | 331 | ·270578 | 9·803072 | 465 | 10·196928 | ·073649 | 134 | 9·926351 | 34 |
| 27 | 9·729621 | 331 | ·270379 | 9·803351 | 465 | 10·196649 | ·073730 | 134 | 9·926270 | 33 |
| 28 | 9·729820 | 331 | ·270180 | 9·803630 | 465 | 10·196370 | ·073810 | 134 | 9·926190 | 32 |
| 29 | 9·730018 | 331 | ·269982 | 9·803908 | 465 | 10·196092 | ·073890 | 134 | 9·926110 | 31 |
| 30 | 9·730217 | 330 | ·269783 | 9·804187 | 465 | 10·195813 | ·073971 | 134 | 9·926029 | 30 |
| 31 | 9·730415 | 330 | ·269585 | 9·804466 | 465 | 10·195534 | ·074051 | 134 | 9·925949 | 29 |
| 32 | 9·730613 | 330 | ·269387 | 9·804745 | 464 | 10·195255 | ·074132 | 134 | 9·925868 | 28 |
| 33 | 9·730811 | 330 | ·269189 | 9·805023 | 464 | 10·194977 | ·074212 | 134 | 9·925788 | 27 |
| 34 | 9·731009 | 330 | ·268991 | 9·805302 | 464 | 10·194698 | ·074293 | 134 | 9·925707 | 26 |
| 35 | 9·731206 | 329 | ·268794 | 9·805580 | 464 | 10·194420 | ·074374 | 134 | 9·925626 | 25 |
| 36 | 9·731404 | 329 | ·268596 | 9·805859 | 464 | 10·194141 | ·074455 | 134 | 9·925545 | 24 |
| 37 | 9·731602 | 329 | ·268398 | 9·806137 | 464 | 10·193863 | ·074535 | 135 | 9·925465 | 23 |
| 38 | 9·731799 | 329 | ·268201 | 9·806415 | 464 | 10·193585 | ·074616 | 135 | 9·925384 | 22 |
| 39 | 9·731996 | 329 | ·268004 | 9·806693 | 464 | 10·193307 | ·074697 | 135 | 9·925303 | 21 |
| 40 | 9·732193 | 328 | ·267807 | 9·806971 | 463 | 10·193029 | ·074778 | 135 | 9·925222 | 20 |
| 41 | 9·732390 | 328 | ·267610 | 9·807249 | 463 | 10·192751 | ·074859 | 135 | 9·925141 | 19 |
| 42 | 9·732587 | 328 | ·267413 | 9·807527 | 463 | 10·192473 | ·074940 | 135 | 9·925060 | 18 |
| 43 | 9·732784 | 328 | ·267216 | 9·807805 | 463 | 10·192195 | ·075021 | 135 | 9·924979 | 17 |
| 44 | 9·732980 | 328 | ·267020 | 9·808083 | 463 | 10·191917 | ·075103 | 135 | 9·924897 | 16 |
| 45 | 9·733177 | 327 | ·266823 | 9·808361 | 463 | 10·191639 | ·075184 | 135 | 9·924816 | 15 |
| 46 | 9·733373 | 327 | ·266627 | 9·808638 | 463 | 10·191362 | ·075265 | 135 | 9·924735 | 14 |
| 47 | 9·733569 | 327 | ·266431 | 9·808916 | 463 | 10·191084 | ·075346 | 136 | 9·924654 | 13 |
| 48 | 9·733765 | 327 | ·266235 | 9·809193 | 462 | 10·190807 | ·075428 | 136 | 9·924572 | 12 |
| 49 | 9·733961 | 327 | ·266039 | 9·809471 | 462 | 10·190529 | ·075509 | 136 | 9·924491 | 11 |
| 50 | 9·734157 | 326 | ·265843 | 9·809748 | 462 | 10·190252 | ·075591 | 136 | 9·924409 | 10 |
| 51 | 9·734353 | 326 | ·265647 | 9·810025 | 462 | 10·189975 | ·075672 | 136 | 9·924328 | 9 |
| 52 | 9·734549 | 326 | ·265451 | 9·810302 | 462 | 10·189698 | ·075754 | 136 | 9·924246 | 8 |
| 53 | 9·734744 | 326 | ·265256 | 9·810580 | 462 | 10·189420 | ·075836 | 136 | 9·924164 | 7 |
| 54 | 9·734939 | 325 | ·265061 | 9·810857 | 462 | 10·189143 | ·075917 | 136 | 9·924083 | 6 |
| 55 | 9·735135 | 325 | ·264865 | 9·811134 | 462 | 10·188866 | ·075999 | 136 | 9·924001 | 5 |
| 56 | 9·735330 | 325 | ·264670 | 9·811410 | 461 | 10·188590 | ·076081 | 136 | 9·923919 | 4 |
| 57 | 9·735525 | 325 | ·264475 | 9·811687 | 461 | 10·188313 | ·076163 | 136 | 9·923837 | 3 |
| 58 | 9·735719 | 325 | ·264281 | 9·811964 | 461 | 10·188036 | ·076245 | 136 | 9·923755 | 2 |
| 59 | 9·735914 | 324 | ·264086 | 9·812241 | 461 | 10·187759 | ·076327 | 137 | 9·923673 | 1 |
| 60 | 9·736109 | 324 | ·263891 | 9·812517 | 461 | 10·187483 | ·076409 | 137 | 9·923591 | 0 |
| ′ | Cosine. | | Secant. | Cotangent. | | Tangent. | Cosecant. | | Sine. | ′ |

57 DEG.

33 DEG.

| ′ | Sine. | Diff. 100″ | Cosecant. | Tangent. | Diff. 100″ | Cotangent. | Secant. | Diff. 100″ | Cosine. | ′ |
|---|---|---|---|---|---|---|---|---|---|---|
| 0 | 9·736109 | | ·263891 | 9·812517 | | 10·187483 | ·076409 | | 9·923591 | 60 |
| 1 | 9·736303 | 324 | ·263697 | 9·812794 | 461 | 10·187206 | ·076491 | 137 | 9·923509 | 59 |
| 2 | 9·736498 | 324 | ·263502 | 9·813070 | 461 | 10·186930 | ·076573 | 137 | 9·923427 | 58 |
| 3 | 9·736692 | 324 | ·263308 | 9·813347 | 461 | 10·186653 | ·076655 | 137 | 9·923345 | 57 |
| 4 | 9·736886 | 323 | ·263114 | 9·813623 | 460 | 10·186377 | ·076737 | 137 | 9·923263 | 56 |
| 5 | 9·737080 | 323 | ·262920 | 9·813899 | 460 | 10·186101 | ·076819 | 137 | 9·923181 | 55 |
| 6 | 9·737274 | 323 | ·262726 | 9·814175 | 460 | 10·185825 | ·076902 | 137 | 9·923098 | 54 |
| 7 | 9·737467 | 323 | ·262533 | 9·814452 | 460 | 10·185548 | ·076984 | 137 | 9·923016 | 53 |
| 8 | 9·737661 | 323 | ·262339 | 9·814728 | 460 | 10·185272 | ·077067 | 137 | 9·922933 | 52 |
| 9 | 9·737855 | 322 | ·262145 | 9·815004 | 460 | 10·184996 | ·077149 | 137 | 9·922851 | 51 |
| 10 | 9·738048 | 322 | ·261952 | 9·815279 | 460 | 10·184721 | ·077232 | 137 | 9·922768 | 50 |
| 11 | 9·738241 | 322 | ·261759 | 9·815555 | 460 | 10·184445 | ·077314 | 138 | 9·922686 | 49 |
| 12 | 9·738434 | 322 | ·261566 | 9·815831 | 460 | 10·184169 | ·077397 | 138 | 9·922603 | 48 |
| 13 | 9·738627 | 322 | ·261373 | 9·816107 | 459 | 10·183893 | ·077480 | 138 | 9·922520 | 47 |
| 14 | 9·738820 | 321 | ·261180 | 9·816382 | 459 | 10·183618 | ·077562 | 138 | 9·922438 | 46 |
| 15 | 9·739013 | 321 | ·260987 | 9·816658 | 459 | 10·183342 | ·077645 | 138 | 9·922355 | 45 |
| 16 | 9·739206 | 321 | ·260794 | 9·816933 | 459 | 10·183067 | ·077728 | 138 | 9·922272 | 44 |
| 17 | 9·739398 | 321 | ·260602 | 9·817209 | 459 | 10·182791 | ·077811 | 138 | 9·922189 | 43 |
| 18 | 9·739590 | 321 | ·260410 | 9·817484 | 459 | 10·182516 | ·077894 | 138 | 9·922106 | 42 |
| 19 | 9·739783 | 320 | ·260217 | 9·817759 | 459 | 10·182241 | ·077977 | 138 | 9·922023 | 41 |
| 20 | 9·739975 | 320 | ·260025 | 9·818035 | 459 | 10·181965 | ·078060 | 138 | 9·921940 | 40 |
| 21 | 9·740167 | 320 | ·259833 | 9·818310 | 459 | 10·181690 | ·078143 | 138 | 9·921857 | 39 |
| 22 | 9·740359 | 320 | ·259641 | 9·818585 | 458 | 10·181415 | ·078226 | 139 | 9·921774 | 38 |
| 23 | 9·740550 | 320 | ·259450 | 9·818860 | 458 | 10·181140 | ·078309 | 139 | 9·921691 | 37 |
| 24 | 9·740742 | 319 | ·259258 | 9·819135 | 458 | 10·180865 | ·078393 | 139 | 9·921607 | 36 |
| 25 | 9·740934 | 319 | ·259066 | 9·819410 | 458 | 10·180590 | ·078476 | 139 | 9·921524 | 35 |
| 26 | 9·741125 | 319 | ·258875 | 9·819684 | 458 | 10·180316 | ·078559 | 139 | 9·921441 | 34 |
| 27 | 9·741316 | 319 | ·258684 | 9·819959 | 458 | 10·180041 | ·078643 | 139 | 9·921357 | 33 |
| 28 | 9·741508 | 319 | ·258492 | 9·820234 | 458 | 10·179766 | ·078726 | 139 | 9·921274 | 32 |
| 29 | 9·741699 | 318 | ·258301 | 9·820508 | 458 | 10·179492 | ·078810 | 139 | 9·921190 | 31 |
| 30 | 9·741889 | 318 | ·258111 | 9·820783 | 458 | 10·179217 | ·078893 | 139 | 9·921107 | 30 |
| 31 | 9·742080 | 318 | ·257920 | 9·821057 | 457 | 10·178943 | ·078977 | 139 | 9·921023 | 29 |
| 32 | 9·742271 | 318 | ·257729 | 9·821332 | 457 | 10·178668 | ·079061 | 139 | 9·920939 | 28 |
| 33 | 9·742462 | 318 | ·257538 | 9·821606 | 457 | 10·178394 | ·079144 | 139 | 9·920856 | 27 |
| 34 | 9·742652 | 317 | ·257348 | 9·821880 | 457 | 10·178120 | ·079228 | 140 | 9·920772 | 26 |
| 35 | 9·742842 | 317 | ·257158 | 9·822154 | 457 | 10·177846 | ·079312 | 140 | 9·920688 | 25 |
| 36 | 9·743033 | 317 | ·256967 | 9·822429 | 457 | 10·177571 | ·079396 | 140 | 9·920604 | 24 |
| 37 | 9·743223 | 317 | ·256777 | 9·822703 | 457 | 10·177297 | ·079480 | 140 | 9·920520 | 23 |
| 38 | 9·743413 | 317 | ·256587 | 9·822977 | 457 | 10·177023 | ·079564 | 140 | 9·920436 | 22 |
| 39 | 9·743602 | 316 | ·256398 | 9·823250 | 457 | 10·176750 | ·079648 | 140 | 9·920352 | 21 |
| 40 | 9·743792 | 316 | ·256208 | 9·823524 | 456 | 10·176476 | ·079732 | 140 | 9·920268 | 20 |
| 41 | 9·743982 | 316 | ·256018 | 9·823798 | 456 | 10·176202 | ·079816 | 140 | 9·920184 | 19 |
| 42 | 9·744171 | 316 | ·255829 | 9·824072 | 456 | 10·175928 | ·079901 | 140 | 9·920099 | 18 |
| 43 | 9·744361 | 316 | ·255639 | 9·824345 | 456 | 10·175655 | ·079985 | 140 | 9·920015 | 17 |
| 44 | 9·744550 | 315 | ·255450 | 9·824619 | 456 | 10·175381 | ·080069 | 140 | 9·919931 | 16 |
| 45 | 9·744739 | 315 | ·255261 | 9·824893 | 456 | 10·175107 | ·080154 | 141 | 9·919846 | 15 |
| 46 | 9·744928 | 315 | ·255072 | 9·825166 | 456 | 10·174834 | ·080238 | 141 | 9·919762 | 14 |
| 47 | 9·745117 | 315 | ·254883 | 9·825439 | 456 | 10·174561 | ·080323 | 141 | 9·919677 | 13 |
| 48 | 9·745306 | 315 | ·254694 | 9·825713 | 456 | 10·174287 | ·080407 | 141 | 9·919593 | 12 |
| 49 | 9·745494 | 314 | ·254506 | 9·825986 | 455 | 10·174014 | ·080492 | 141 | 9·919508 | 11 |
| 50 | 9·745683 | 314 | ·254317 | 9·826259 | 455 | 10·173741 | ·080576 | 141 | 9·919424 | 10 |
| 51 | 9·745871 | 314 | ·254129 | 9·826532 | 455 | 10·173468 | ·080661 | 141 | 9·919339 | 9 |
| 52 | 9·746060 | 314 | ·253940 | 9·826805 | 455 | 10·173195 | ·080746 | 141 | 9·919254 | 8 |
| 53 | 9·746248 | 314 | ·253752 | 9·827078 | 455 | 10·172922 | ·080831 | 141 | 9·919169 | 7 |
| 54 | 9·746436 | 313 | ·253564 | 9·827351 | 455 | 10·172649 | ·080915 | 141 | 9·919085 | 6 |
| 55 | 9·746624 | 313 | ·253376 | 9·827624 | 455 | 10·172376 | ·081000 | 141 | 9·919000 | 5 |
| 56 | 9·746812 | 313 | ·253188 | 9·827897 | 455 | 10·172103 | ·081085 | 142 | 9·918915 | 4 |
| 57 | 9·746999 | 313 | ·253001 | 9·828170 | 455 | 10·171830 | 81170 | 142 | 9·918830 | 3 |
| 58 | 9·747187 | 313 | ·252813 | 9·828442 | 454 | 10·171558 | ·081255 | 142 | 9·918745 | 2 |
| 59 | 9·747374 | 312 | ·252626 | 9·828715 | 454 | 10·171285 | ·081341 | 142 | 9·918659 | 1 |
| 60 | 9·747562 | 312 | ·252438 | 9·828987 | 454 | 10·171013 | ·081426 | 142 | 9·918574 | 0 |
| ′ | Cosine. | | Secant. | Cotangent. | | Tangent. | Cosecant. | | Sine. | ′ |

56 DEG.

34 DEG.

| ′ | Sine. | Diff. 100″ | Cosecant. | Tangent. | Diff. 100″ | Cotangent. | Secant. | Diff. 100′ | Cosine. | ′ |
|---|---|---|---|---|---|---|---|---|---|---|
| 0 | 9·747562 | | ·252438 | 9·828987 | | 10·171013 | ·081426 | | 9·918574 | 60 |
| 1 | 9·747749 | 312 | ·252251 | 9·829260 | 454 | 10·170740 | ·081511 | 142 | 9·918489 | 59 |
| 2 | 9·747936 | 312 | ·252064 | 9·829532 | 454 | 10·170468 | ·081596 | 142 | 9·918404 | 58 |
| 3 | 9·748123 | 312 | ·251877 | 9·829805 | 454 | 10·170195 | ·081682 | 142 | 9·918318 | 57 |
| 4 | 9·748310 | 311 | ·251690 | 9·830077 | 454 | 10·169923 | ·081767 | 142 | 9·918233 | 56 |
| 5 | 9·748497 | 311 | ·251503 | 9·830349 | 454 | 10·169651 | ·081853 | 142 | 9·918147 | 55 |
| 6 | 9·748683 | 311 | ·251317 | 9·830621 | 454 | 10·169379 | ·081938 | 142 | 9·918062 | 54 |
| 7 | 9·748870 | 311 | ·251130 | 9·830893 | 453 | 10·169107 | ·082024 | 143 | 9·917976 | 53 |
| 8 | 9·749056 | 311 | ·250944 | 9·831165 | 453 | 10·168835 | ·082109 | 143 | 9·917891 | 52 |
| 9 | 9·749243 | 310 | ·250757 | 9·831437 | 453 | 10·168563 | ·082195 | 143 | 9·917805 | 51 |
| 10 | 9·749429 | 310 | ·250571 | 9·831709 | 453 | 10·168291 | ·082281 | 143 | 9·917719 | 50 |
| 11 | 9·749615 | 310 | ·250385 | 9·831981 | 453 | 10·168019 | ·082366 | 143 | 9·917634 | 49 |
| 12 | 9·749801 | 310 | ·250199 | 9·832253 | 453 | 10·167747 | ·082452 | 143 | 9·917548 | 48 |
| 13 | 9·749987 | 310 | ·250013 | 9·832525 | 453 | 10·167475 | ·082538 | 143 | 9·917462 | 47 |
| 14 | 9·750172 | 309 | ·249828 | 9·832796 | 453 | 10·167204 | ·082624 | 143 | 9·917376 | 46 |
| 15 | 9·750358 | 309 | ·249642 | 9·833068 | 453 | 10·166932 | ·082710 | 143 | 9·917290 | 45 |
| 16 | 9·750543 | 309 | ·249457 | 9·833339 | 453 | 10·166661 | ·082796 | 143 | 9·917204 | 44 |
| 17 | 9·750729 | 309 | ·249271 | 9·833611 | 452 | 10·166389 | ·082882 | 143 | 9·917118 | 43 |
| 18 | 9·750914 | 309 | ·249086 | 9·833882 | 452 | 10·166118 | ·082968 | 144 | 9·917032 | 42 |
| 19 | 9·751099 | 308 | ·248901 | 9·834154 | 452 | 10·165846 | ·083054 | 144 | 9·916946 | 41 |
| 20 | 9·751284 | 308 | ·248716 | 9·834425 | 452 | 10·165575 | ·083141 | 144 | 9·916859 | 40 |
| 21 | 9·751469 | 308 | ·248531 | 9·834696 | 452 | 10·165304 | ·083227 | 144 | 9·916773 | 39 |
| 22 | 9·751654 | 308 | ·248346 | 9·834967 | 452 | 10·165033 | ·083313 | 144 | 9·916687 | 38 |
| 23 | 9·751839 | 308 | ·248161 | 9·835238 | 452 | 10·164762 | ·083400 | 144 | 9·916600 | 37 |
| 24 | 9·752023 | 308 | ·247977 | 9·835509 | 452 | 10·164491 | ·083486 | 144 | 9·916514 | 36 |
| 25 | 9·752208 | 307 | ·247792 | 9·835780 | 452 | 10·164220 | ·083573 | 144 | 9·916427 | 35 |
| 26 | 9·752392 | 307 | ·247608 | 9·836051 | 452 | 10·163949 | ·083659 | 144 | 9·916341 | 34 |
| 27 | 9·752576 | 307 | ·247424 | 9·836322 | 451 | 10·163678 | ·083746 | 144 | 9·916254 | 33 |
| 28 | 9·752760 | 307 | ·247240 | 9·836593 | 451 | 10·163407 | ·083833 | 144 | 9·916167 | 32 |
| 29 | 9·752944 | 307 | ·247056 | 9·836864 | 451 | 10·163136 | ·083919 | 145 | 9·916081 | 31 |
| 30 | 9·753128 | 306 | ·246872 | 9·837134 | 451 | 10·162866 | ·084006 | 145 | 9·915994 | 30 |
| 31 | 9·753312 | 306 | ·246688 | 9·837405 | 451 | 10·162595 | ·084093 | 145 | 9·915907 | 29 |
| 32 | 9·753495 | 306 | ·246505 | 9·837675 | 451 | 10·162325 | ·084180 | 145 | 9·915820 | 28 |
| 33 | 9·753679 | 306 | ·246321 | 9·837946 | 451 | 10·162054 | ·084267 | 145 | 9·915733 | 27 |
| 34 | 9·753862 | 306 | ·246138 | 9·838216 | 451 | 10·161784 | ·084354 | 145 | 9·915646 | 26 |
| 35 | 9·754046 | 305 | ·245954 | 9·838487 | 451 | 10·161513 | ·084441 | 145 | 9·915559 | 25 |
| 36 | 9·754229 | 305 | ·245771 | 9·838757 | 451 | 10·161243 | ·084528 | 145 | 9·915472 | 24 |
| 37 | 9·754412 | 305 | ·245588 | 9·839027 | 450 | 10·160973 | ·084615 | 145 | 9·915385 | 23 |
| 38 | 9·754595 | 305 | ·245405 | 9·839297 | 450 | 10·160703 | ·084703 | 145 | 9·915297 | 22 |
| 39 | 9·754778 | 305 | ·245222 | 9·839568 | 450 | 10·160432 | ·084790 | 145 | 9·915210 | 21 |
| 40 | 9·754960 | 304 | ·245040 | 9·839838 | 450 | 10·160162 | ·084877 | 145 | 9·915123 | 20 |
| 41 | 9·755143 | 304 | ·244857 | 9·840108 | 450 | 10·159892 | ·084965 | 146 | 9·915035 | 19 |
| 42 | 9·755326 | 304 | ·244674 | 9·840378 | 450 | 10·159622 | ·085052 | 146 | 9·914948 | 18 |
| 43 | 9·755508 | 304 | ·244492 | 9·840647 | 450 | 10·159353 | ·085140 | 146 | 9·914860 | 17 |
| 44 | 9·755690 | 304 | ·244310 | 9·840917 | 450 | 10·159083 | ·085227 | 146 | 9·914773 | 16 |
| 45 | 9·755872 | 304 | ·244128 | 9·841187 | 450 | 10·158813 | ·085315 | 146 | 9·914685 | 15 |
| 46 | 9·756054 | 303 | ·243946 | 9·841457 | 450 | 10·158543 | ·085402 | 146 | 9·914598 | 14 |
| 47 | 9·756236 | 303 | ·243764 | 9·841726 | 449 | 10·158274 | ·085490 | 146 | 9·914510 | 13 |
| 48 | 9·756418 | 303 | ·243582 | 9·841996 | 449 | 10·158004 | ·085578 | 146 | 9·914422 | 12 |
| 49 | 9·756600 | 303 | ·243400 | 9·842266 | 449 | 10·157734 | ·085666 | 146 | 9·914334 | 11 |
| 50 | 9·756782 | 303 | ·243218 | 9·842535 | 449 | 10·157465 | ·085754 | 146 | 9·914246 | 10 |
| 51 | 9·756963 | 302 | ·243037 | 9·842805 | 449 | 10·157195 | ·085842 | 147 | 9·914158 | 9 |
| 52 | 9·757144 | 302 | ·242856 | 9·843074 | 449 | 10·156926 | ·085930 | 147 | 9·914070 | 8 |
| 53 | 9·757326 | 302 | ·242674 | 9·843343 | 449 | 10·156657 | ·086018 | 147 | 9·913982 | 7 |
| 54 | 9·757507 | 302 | ·242493 | 9·843612 | 449 | 10·156388 | ·086106 | 147 | 9·913894 | 6 |
| 55 | 9·757688 | 302 | ·242312 | 9·843882 | 449 | 10·156118 | ·086194 | 147 | 9·913806 | 5 |
| 56 | 9·757869 | 301 | ·242131 | 9·844151 | 449 | 10·155849 | ·086282 | 147 | 9·913718 | 4 |
| 57 | 9·758050 | 301 | ·241950 | 9·844420 | 448 | 10·155580 | ·086370 | 147 | 9·913630 | 3 |
| 58 | 9·758230 | 301 | ·241770 | 9·844689 | 448 | 10·155311 | ·086459 | 147 | 9·913541 | 2 |
| 59 | 9·758411 | 301 | ·241589 | 9·844958 | 448 | 10·155042 | ·086547 | 147 | 9·913453 | 1 |
| 60 | 9·758591 | 301 | ·241409 | 9·845227 | 448 | 10·154773 | ·086635 | 147 | 9·913365 | 0 |
| ′ | Cosine. | | Secant. | Cotangent. | | Tangent. | Cosecant. | | Sine. | ′ |

55 DEG.

35 DEG.

| ′ | Sine. | Diff. 100″ | Cosecant. | Tangent. | Diff. 100″ | Cotangent. | Secant. | Diff. 100″ | Cosine. | ′ |
|---|---|---|---|---|---|---|---|---|---|---|
| 0 | 9·758591 | | ·241409 | 9·845227 | | 10·154773 | ·086635 | | 9·913365 | 60 |
| 1 | 9·758772 | 301 | ·241228 | 9·845496 | 448 | 10·154504 | ·086724 | 147 | 9·913276 | 59 |
| 2 | 9·758952 | 300 | ·241048 | 9·845764 | 448 | 10·154236 | ·086813 | 147 | 9·913187 | 58 |
| 3 | 9·759132 | 300 | ·240868 | 9·846033 | 448 | 10·153967 | ·086901 | 148 | 9·913099 | 57 |
| 4 | 9·759312 | 300 | ·240688 | 9·846302 | 448 | 10·153698 | ·086990 | 148 | 9·913010 | 56 |
| 5 | 9·759492 | 300 | ·240508 | 9·846570 | 448 | 10·153430 | ·087078 | 148 | 9·912922 | 55 |
| 6 | 9·759672 | 300 | ·240328 | 9·846839 | 448 | 10·153161 | ·087167 | 148 | 9·912833 | 54 |
| 7 | 9·759852 | 299 | ·240148 | 9·847107 | 448 | 10·152893 | ·087256 | 148 | 9·912744 | 53 |
| 8 | 9·760031 | 299 | ·239969 | 9·847376 | 447 | 10·152624 | ·087345 | 148 | 9·912655 | 52 |
| 9 | 9·760211 | 299 | ·239789 | 9·847644 | 447 | 10·152356 | ·087434 | 148 | 9·912566 | 51 |
| 10 | 9·760390 | 299 | ·239610 | 9·847913 | 447 | 10·152087 | ·087523 | 148 | 9·912477 | 50 |
| 11 | 9·760569 | 299 | ·239431 | 9·848181 | 447 | 10·151819 | ·087612 | 148 | 9·912388 | 49 |
| 12 | 9·760748 | 298 | ·239252 | 9·848449 | 447 | 10·151551 | ·087701 | 148 | 9·912299 | 48 |
| 13 | 9·760927 | 298 | ·239073 | 9·848717 | 447 | 10·151283 | ·087790 | 149 | 9·912210 | 47 |
| 14 | 9·761106 | 298 | ·238894 | 9·848986 | 447 | 10·151014 | ·087879 | 149 | 9·912121 | 46 |
| 15 | 9·761285 | 298 | ·238715 | 9·849254 | 447 | 10·150746 | ·087969 | 149 | 9·912031 | 45 |
| 16 | 9·761464 | 298 | ·238536 | 9·849522 | 447 | 10·150478 | ·088058 | 149 | 9·911942 | 44 |
| 17 | 9·761642 | 298 | ·238358 | 9·849790 | 447 | 10·150210 | ·088147 | 149 | 9·911853 | 43 |
| 18 | 9·761821 | 297 | ·238179 | 9·850058 | 446 | 10·149942 | ·088237 | 149 | 9·911763 | 42 |
| 19 | 9·761999 | 297 | ·238001 | 9·850325 | 446 | 10·149675 | ·088326 | 149 | 9·911674 | 41 |
| 20 | 9·762177 | 297 | ·237823 | 9·850593 | 446 | 10·149407 | ·088416 | 149 | 9·911584 | 40 |
| 21 | 9·762356 | 297 | ·237644 | 9·850861 | 446 | 10·149139 | ·088505 | 149 | 9·911495 | 39 |
| 22 | 9·762534 | 297 | ·237466 | 9·851129 | 446 | 10·148871 | ·088595 | 149 | 9·911405 | 38 |
| 23 | 9·762712 | 296 | ·237288 | 9·851396 | 446 | 10·148604 | ·088685 | 149 | 9·911315 | 37 |
| 24 | 9·762889 | 296 | ·237111 | 9·851664 | 446 | 10·148336 | ·088774 | 150 | 9·911226 | 36 |
| 25 | 9·763067 | 296 | ·236933 | 9·851931 | 446 | 10·148069 | ·088864 | 150 | 9·911136 | 35 |
| 26 | 9·763245 | 296 | ·236755 | 9·852199 | 446 | 10·147801 | ·088954 | 150 | 9·911046 | 34 |
| 27 | 9·763422 | 296 | ·236578 | 9·852466 | 446 | 10·147534 | ·089044 | 150 | 9·910956 | 33 |
| 28 | 9·763600 | 296 | ·236400 | 9·852733 | 446 | 10·147267 | ·089134 | 150 | 9·910866 | 32 |
| 29 | 9·763777 | 295 | ·236223 | 9·853001 | 445 | 10·146999 | ·089224 | 150 | 9·910776 | 31 |
| 30 | 9·763954 | 295 | ·236046 | 9·853268 | 445 | 10·146732 | ·089314 | 150 | 9·910686 | 30 |
| 31 | 9·764131 | 295 | ·235869 | 9·853535 | 445 | 10·146465 | ·089404 | 150 | 9·910596 | 29 |
| 32 | 9·764308 | 295 | ·235692 | 9·853802 | 445 | 10·146198 | ·089494 | 150 | 9·910506 | 28 |
| 33 | 9·764485 | 295 | ·235515 | 9·854069 | 445 | 10·145931 | ·089585 | 150 | 9·910415 | 27 |
| 34 | 9·764662 | 294 | ·235338 | 9·854336 | 445 | 10·145664 | ·089675 | 150 | 9·910325 | 26 |
| 35 | 9·764838 | 294 | ·235162 | 9·854603 | 445 | 10·145397 | ·089765 | 151 | 9·910235 | 25 |
| 36 | 9·765015 | 294 | ·234985 | 9·854870 | 445 | 10·145130 | ·089856 | 151 | 9·910144 | 24 |
| 37 | 9·765191 | 294 | ·234809 | 9·855137 | 445 | 10·144863 | ·089946 | 151 | 9·910054 | 23 |
| 38 | 9·765367 | 294 | ·234633 | 9·855404 | 445 | 10·144596 | ·090037 | 151 | 9·909963 | 22 |
| 39 | 9·765544 | 294 | ·234456 | 9·855671 | 445 | 10·144329 | ·090127 | 151 | 9·909873 | 21 |
| 40 | 9·765720 | 293 | ·234280 | 9·855938 | 444 | 10·144062 | ·090218 | 151 | 9·909782 | 20 |
| 41 | 9·765896 | 293 | ·234104 | 9·856204 | 444 | 10·143796 | ·090309 | 151 | 9·909691 | 19 |
| 42 | 9·766072 | 293 | ·233928 | 9·856471 | 444 | 10·143529 | ·090399 | 151 | 9·909601 | 18 |
| 43 | 9·766247 | 293 | ·233753 | 9·856737 | 444 | 10·143263 | ·090490 | 151 | 9·909510 | 17 |
| 44 | 9·766423 | 293 | ·233577 | 9·857004 | 444 | 10·142996 | ·090581 | 151 | 9·909419 | 16 |
| 45 | 9·766598 | 293 | ·233402 | 9·857270 | 444 | 10·142730 | ·090672 | 151 | 9·909328 | 15 |
| 46 | 9·766774 | 292 | ·233226 | 9·857537 | 444 | 10·142463 | ·090763 | 152 | 9·909237 | 14 |
| 47 | 9·766949 | 292 | ·233051 | 9·857803 | 444 | 10·142197 | ·090854 | 152 | 9·909146 | 13 |
| 48 | 9·767124 | 292 | ·232876 | 9·858069 | 444 | 10·141931 | ·090945 | 152 | 9·909055 | 12 |
| 49 | 9·767300 | 292 | ·232700 | 9·858336 | 444 | 10·141664 | ·091036 | 152 | 9·908964 | 11 |
| 50 | 9·767475 | 292 | ·232525 | 9·858602 | 444 | 10·141398 | ·091127 | 152 | 9·908873 | 10 |
| 51 | 9·767649 | 291 | ·232351 | 9·858868 | 444 | 10·141132 | ·091219 | 152 | 9·908781 | 9 |
| 52 | 9·767824 | 291 | ·232176 | 9·859134 | 443 | 10·140866 | ·091310 | 152 | 9·908690 | 8 |
| 53 | 9·767999 | 291 | ·232001 | 9·859400 | 443 | 10·140600 | ·091401 | 152 | 9·908599 | 7 |
| 54 | 9·768173 | 291 | ·231827 | 9·859666 | 443 | 10·140334 | ·091493 | 152 | 9·908507 | 6 |
| 55 | 9·768348 | 291 | ·231652 | 9·859932 | 443 | 10·140068 | ·091584 | 152 | 9·908416 | 5 |
| 56 | 9·768522 | 290 | ·231478 | 9·860198 | 443 | 10·139802 | ·091676 | 153 | 9·908324 | 4 |
| 57 | 9·768697 | 290 | ·231303 | 9·860464 | 443 | 10·139536 | ·091767 | 153 | 9·908233 | 3 |
| 58 | 9·768871 | 290 | ·231129 | 9·860730 | 443 | 10·139270 | ·091859 | 153 | 9·908141 | 2 |
| 59 | 9·769045 | 290 | ·230955 | 9·860995 | 443 | 10·139005 | ·091951 | 153 | 9·908049 | 1 |
| 60 | 9·769219 | 290 | ·230781 | 9·861261 | 443 | 10·138739 | ·092042 | 153 | 9·907958 | 0 |
| ′ | Cosine. | | Secant. | Cotangent. | | Tangent. | Cosecant. | | Sine. | ′ |

54 DEG.

36 DEG.

| ′ | Sine. | Diff. 100″ | Cosecant. | Tangent. | Diff. 100″ | Cotangent. | Secant. | Diff. 100″ | Cosine. | ′ |
|---|---|---|---|---|---|---|---|---|---|---|
| 0 | 9·769219 | | ·230781 | 9·861261 | | 10·138739 | ·092042 | | 9·907958 | 60 |
| 1 | 9·769393 | 290 | ·230607 | 9·861527 | 443 | 10·138473 | ·092134 | 153 | 9·907866 | 59 |
| 2 | 9·769566 | 289 | ·230434 | 9·861792 | 443 | 10·138208 | ·092226 | 153 | 9·907774 | 58 |
| 3 | 9·769740 | 289 | ·230260 | 9·862058 | 443 | 10·137942 | ·092318 | 153 | 9·907682 | 57 |
| 4 | 9·769913 | 289 | ·230087 | 9·862323 | 442 | 10·137677 | ·092410 | 153 | 9·907590 | 56 |
| 5 | 9·770087 | 289 | ·229913 | 9·862589 | 442 | 10·137411 | ·092502 | 153 | 9·907498 | 55 |
| 6 | 9·770260 | 289 | ·229740 | 9·862854 | 442 | 10·137146 | ·092594 | 153 | 9·907406 | 54 |
| 7 | 9·770433 | 288 | ·229567 | 9·863119 | 442 | 10·136881 | ·092686 | 153 | 9·907314 | 53 |
| 8 | 9·770606 | 288 | ·229394 | 9·863385 | 442 | 10·136615 | ·092778 | 154 | 9·907222 | 52 |
| 9 | 9·770779 | 288 | ·229221 | 9·863650 | 442 | 10·136350 | ·092871 | 154 | 9·907129 | 51 |
| 10 | 9·770952 | 288 | ·229048 | 9·863915 | 442 | 10·136085 | ·092963 | 154 | 9·907037 | 50 |
| 11 | 9·771125 | 288 | ·228875 | 9·864180 | 442 | 10·135820 | ·093055 | 154 | 9·906945 | 49 |
| 12 | 9·771298 | 288 | ·228702 | 9·864445 | 442 | 10·135555 | ·093148 | 154 | 9·906852 | 48 |
| 13 | 9·771470 | 287 | ·228530 | 9·864710 | 442 | 10·135290 | ·093240 | 154 | 9·906760 | 47 |
| 14 | 9·771643 | 287 | ·228357 | 9·864975 | 442 | 10·135025 | ·093333 | 154 | 9·906667 | 46 |
| 15 | 9·771815 | 287 | ·228185 | 9·865240 | 442 | 10·134760 | ·093425 | 154 | 9·906575 | 45 |
| 16 | 9·771987 | 287 | ·228013 | 9·865505 | 441 | 10·134495 | ·093518 | 154 | 9·906482 | 44 |
| 17 | 9·772159 | 287 | ·227841 | 9·865770 | 441 | 10·134230 | ·093611 | 154 | 9·906389 | 43 |
| 18 | 9·772331 | 287 | ·227669 | 9·866035 | 441 | 10·133965 | ·093704 | 155 | 9·906296 | 42 |
| 19 | 9·772503 | 286 | ·227497 | 9·866300 | 441 | 10·133700 | ·093796 | 155 | 9·906204 | 41 |
| 20 | 9·772675 | 286 | ·227325 | 9·866564 | 441 | 10·133436 | ·093889 | 155 | 9·906111 | 40 |
| 21 | 9·772847 | 286 | ·227153 | 9·866829 | 441 | 10·133171 | ·093982 | 155 | 9·906018 | 39 |
| 22 | 9·773018 | 286 | ·226982 | 9·867094 | 441 | 10·132906 | ·094075 | 155 | 9·905925 | 38 |
| 23 | 9·773190 | 286 | ·226810 | 9·867358 | 441 | 10·132642 | ·094168 | 155 | 9·905832 | 37 |
| 24 | 9·773361 | 286 | ·226639 | 9·867623 | 441 | 10·132377 | ·094261 | 155 | 9·905739 | 36 |
| 25 | 9·773533 | 285 | ·226467 | 9·867887 | 441 | 10·132113 | ·094355 | 155 | 9·905645 | 35 |
| 26 | 9·773704 | 285 | ·226296 | 9·868152 | 441 | 10·131848 | ·094448 | 155 | 9·905552 | 34 |
| 27 | 9·773875 | 285 | ·226125 | 9·868416 | 441 | 10·131584 | ·094541 | 155 | 9·905459 | 33 |
| 28 | 9·774046 | 285 | ·225954 | 9·868680 | 441 | 10·131320 | ·094634 | 155 | 9·905366 | 32 |
| 29 | 9·774217 | 285 | ·225783 | 9·868945 | 440 | 10·131055 | ·094728 | 156 | 9·905272 | 31 |
| 30 | 9·774388 | 285 | ·225612 | 9·869209 | 440 | 10·130791 | ·094821 | 156 | 9·905179 | 30 |
| 31 | 9·774558 | 284 | ·225442 | 9·869473 | 440 | 10·130527 | ·094915 | 156 | 9·905085 | 29 |
| 32 | 9·774729 | 284 | ·225271 | 9·869737 | 440 | 10·130263 | ·095008 | 156 | 9·904992 | 28 |
| 33 | 9·774899 | 284 | ·225101 | 9·870001 | 440 | 10·129999 | ·095102 | 156 | 9·904898 | 27 |
| 34 | 9·775070 | 284 | ·224930 | 9·870265 | 440 | 10·129735 | ·095196 | 156 | 9·904804 | 26 |
| 35 | 9·775240 | 284 | ·224760 | 9·870529 | 440 | 10·129471 | ·095289 | 156 | 9·904711 | 25 |
| 36 | 9·775410 | 284 | ·224590 | 9·870793 | 440 | 10·129207 | ·095383 | 156 | 9·904617 | 24 |
| 37 | 9·775580 | 283 | ·224420 | 9·871057 | 440 | 10·128943 | ·095477 | 156 | 9·904523 | 23 |
| 38 | 9·775750 | 283 | ·224250 | 9·871321 | 440 | 10·128679 | ·095571 | 156 | 9·904429 | 22 |
| 39 | 9·775920 | 283 | ·224080 | 9·871585 | 440 | 10·128415 | ·095665 | 157 | 9·904335 | 21 |
| 40 | 9·776090 | 283 | ·223910 | 9·871849 | 440 | 10·128151 | ·095759 | 157 | 9·904241 | 20 |
| 41 | 9·776259 | 283 | ·223741 | 9·872112 | 440 | 10·127888 | ·095853 | 157 | 9·904147 | 19 |
| 42 | 9·776429 | 283 | ·223571 | 9·872376 | 439 | 10·127624 | ·095947 | 157 | 9·904053 | 18 |
| 43 | 9·776598 | 282 | ·223402 | 9·872640 | 439 | 10·127360 | ·096041 | 157 | 9·903959 | 17 |
| 44 | 9·776768 | 282 | ·223232 | 9·872903 | 439 | 10·127097 | ·096136 | 157 | 9·903864 | 16 |
| 45 | 9·776937 | 282 | ·223063 | 9·873167 | 439 | 10·126833 | ·096230 | 157 | 9·903770 | 15 |
| 46 | 9·777106 | 282 | ·222894 | 9·873430 | 439 | 10·126570 | ·096324 | 157 | 9·903676 | 14 |
| 47 | 9·777275 | 282 | ·222725 | 9·873694 | 439 | 10·126306 | ·096419 | 157 | 9·903581 | 13 |
| 48 | 9·777444 | 281 | ·222556 | 9·873957 | 439 | 10·126043 | ·096513 | 157 | 9·903487 | 12 |
| 49 | 9·777613 | 281 | ·222387 | 9·874220 | 439 | 10·125780 | ·096608 | 157 | 9·903392 | 11 |
| 50 | 9·777781 | 281 | ·222219 | 9·874484 | 439 | 10·125516 | ·096702 | 158 | 9·903298 | 10 |
| 51 | 9·777950 | 281 | ·222050 | 9·874747 | 439 | 10·125253 | ·096797 | 158 | 9·903203 | 9 |
| 52 | 9·778119 | 281 | ·221881 | 9·875010 | 439 | 10·124990 | ·096892 | 158 | 9·903108 | 8 |
| 53 | 9·778287 | 281 | ·221713 | 9·875273 | 439 | 10·124727 | ·096986 | 158 | 9·903014 | 7 |
| 54 | 9·778455 | 280 | ·221545 | 9·875536 | 439 | 10·124464 | ·097081 | 158 | 9·902919 | 6 |
| 55 | 9·778624 | 280 | ·221376 | 9·875800 | 439 | 10·124200 | ·097176 | 158 | 9·902824 | 5 |
| 56 | 9·778792 | 280 | ·221208 | 9·876063 | 438 | 10·123937 | ·097271 | 158 | 9·902729 | 4 |
| 57 | 9·778960 | 280 | ·221040 | 9·876326 | 438 | 10·123674 | ·097366 | 158 | 9·902634 | 3 |
| 58 | 9·779128 | 280 | ·220872 | 9·876589 | 438 | 10·123411 | ·097461 | 158 | 9·902539 | 2 |
| 59 | 9·779295 | 280 | ·220705 | 9·876851 | 438 | 10·123149 | ·097556 | 159 | 9·902444 | 1 |
| 60 | 9·779463 | 279 | ·220537 | 9·877114 | 438 | 10·122886 | ·097651 | 159 | 9·902349 | 0 |
| ′ | Cosine. | | Secant. | Cotangent. | | Tangent. | Cosecant. | | Sine. | ′ |

53 DEG.

37 DEG.

| ′ | Sine. | Diff. 100″ | Cosecant. | Tangent. | Diff. 100″ | Cotangent. | Secant. | Diff. 100″ | Cosine. | ′ |
|---|---|---|---|---|---|---|---|---|---|---|
| 0 | 9·779463 | | ·220537 | 9·877114 | | 10·122886 | ·097651 | | 9·902349 | 60 |
| 1 | 9·779631 | 279 | ·220369 | 9·877377 | 438 | 10·122623 | ·097747 | 159 | 9·902253 | 59 |
| 2 | 9·779798 | 279 | ·220202 | 9·877640 | 438 | 10·122360 | ·097842 | 159 | 9·902158 | 58 |
| 3 | 9·779966 | 279 | ·220034 | 9·877903 | 438 | 10·122097 | ·097937 | 159 | 9·902063 | 57 |
| 4 | 9·780133 | 279 | ·219867 | 9·878165 | 438 | 10·121835 | ·098033 | 159 | 9·901967 | 56 |
| 5 | 9·780300 | 279 | ·219700 | 9·878428 | 438 | 10·121572 | ·098128 | 159 | 9·901872 | 55 |
| 6 | 9·780467 | 278 | ·219533 | 9·878691 | 438 | 10·121309 | ·098224 | 159 | 9·901776 | 54 |
| 7 | 9·780634 | 278 | ·219366 | 9·878953 | 438 | 10·121047 | ·098319 | 159 | 9·901681 | 53 |
| 8 | 9·780801 | 278 | ·219199 | 9·879216 | 437 | 10·120784 | ·098415 | 159 | 9·901585 | 52 |
| 9 | 9·780968 | 278 | ·219032 | 9·879478 | 437 | 10·120522 | ·098510 | 159 | 9·901490 | 51 |
| 10 | 9·781134 | 278 | ·218866 | 9·879741 | 437 | 10·120259 | ·098606 | 159 | 9·901394 | 50 |
| 11 | 9·781301 | 278 | ·218699 | 9·880003 | 437 | 10·119997 | ·098702 | 160 | 9·901298 | 49 |
| 12 | 9·781468 | 277 | ·218532 | 9·880265 | 437 | 10·119735 | ·098798 | 160 | 9·901202 | 48 |
| 13 | 9·781634 | 277 | ·218366 | 9·880528 | 437 | 10·119472 | ·098894 | 160 | 9·901106 | 47 |
| 14 | 9·781800 | 277 | ·218200 | 9·880790 | 437 | 10·119210 | ·098990 | 160 | 9·901010 | 46 |
| 15 | 9·781966 | 277 | ·218034 | 9·881052 | 437 | 10·118948 | ·099086 | 160 | 9·900914 | 45 |
| 16 | 9·782132 | 277 | ·217868 | 9·881314 | 437 | 10·118686 | ·099182 | 160 | 9·900818 | 44 |
| 17 | 9·782298 | 277 | ·217702 | 9·881576 | 437 | 10·118424 | ·099278 | 160 | 9·900722 | 43 |
| 18 | 9·782464 | 276 | ·217536 | 9·881839 | 437 | 10·118161 | ·099374 | 160 | 9·900626 | 42 |
| 19 | 9·782630 | 276 | ·217370 | 9·882101 | 437 | 10·117899 | ·099471 | 160 | 9·900529 | 41 |
| 20 | 9·782796 | 276 | ·217204 | 9·882363 | 437 | 10·117637 | ·099567 | 160 | 9·900433 | 40 |
| 21 | 9·782961 | 276 | ·217039 | 9·882625 | 437 | 10·117375 | ·099663 | 161 | 9·900337 | 39 |
| 22 | 9·783127 | 276 | ·216873 | 9·882887 | 436 | 10·117113 | ·099760 | 161 | 9·900240 | 38 |
| 23 | 9·783292 | 276 | ·216708 | 9·883148 | 436 | 10·116852 | ·099856 | 161 | 9·900144 | 37 |
| 24 | 9·783458 | 275 | ·216542 | 9·883410 | 436 | 10·116590 | ·099953 | 161 | 9·900047 | 36 |
| 25 | 9·783623 | 275 | ·216377 | 9·883672 | 436 | 10·116328 | ·100049 | 161 | 9·899951 | 35 |
| 26 | 9·783788 | 275 | ·216212 | 9·883934 | 436 | 10·116066 | ·100146 | 161 | 9·899854 | 34 |
| 27 | 9·783953 | 275 | ·216047 | 9·884196 | 436 | 10·115804 | ·100243 | 161 | 9·899757 | 33 |
| 28 | 9·784118 | 275 | ·215882 | 9·884457 | 436 | 10·115543 | ·100340 | 161 | 9·899660 | 32 |
| 29 | 9·784282 | 275 | ·215718 | 9·884719 | 436 | 10·115281 | ·100436 | 161 | 9·899564 | 31 |
| 30 | 9·784447 | 274 | ·215553 | 9·884980 | 436 | 10·115020 | ·100533 | 161 | 9·899467 | 30 |
| 31 | 9·784612 | 274 | ·215388 | 9·885242 | 436 | 10·114758 | ·100630 | 162 | 9·899370 | 29 |
| 32 | 9·784776 | 274 | ·215224 | 9·885503 | 436 | 10·114497 | ·100727 | 162 | 9·899273 | 28 |
| 33 | 9·784941 | 274 | ·215059 | 9·885765 | 436 | 10·114235 | ·100824 | 162 | 9·899176 | 27 |
| 34 | 9·785105 | 274 | ·214895 | 9·886026 | 436 | 10·113974 | ·100922 | 162 | 9·899078 | 26 |
| 35 | 9·785269 | 274 | ·214731 | 9·886288 | 436 | 10·113712 | ·101019 | 162 | 9·898981 | 25 |
| 36 | 9·785433 | 273 | ·214567 | 9·886549 | 436 | 10·113451 | ·101116 | 162 | 9·898884 | 24 |
| 37 | 9·785597 | 273 | ·214403 | 9·886810 | 436 | 10·113190 | ·101213 | 162 | 9·898787 | 23 |
| 38 | 9·785761 | 273 | ·214239 | 9·887072 | 435 | 10·112928 | ·101311 | 162 | 9·898689 | 22 |
| 39 | 9·785925 | 273 | ·214075 | 9·887333 | 435 | 10·112667 | ·101408 | 162 | 9·898592 | 21 |
| 40 | 9·786089 | 273 | ·213911 | 9·887594 | 435 | 10·112406 | ·101506 | 162 | 9·898494 | 20 |
| 41 | 9·786252 | 273 | ·213748 | 9·887855 | 435 | 10·112145 | ·101603 | 163 | 9·898397 | 19 |
| 42 | 9·786416 | 272 | ·213584 | 9·888116 | 435 | 10·111884 | ·101701 | 163 | 9·898299 | 18 |
| 43 | 9·786579 | 272 | ·213421 | 9·888377 | 435 | 10·111623 | ·101798 | 163 | 9·898202 | 17 |
| 44 | 9·786742 | 272 | ·213258 | 9·888639 | 435 | 10·111361 | ·101896 | 163 | 9·898104 | 16 |
| 45 | 9·786906 | 272 | ·213094 | 9·888900 | 435 | 10·111100 | ·101994 | 163 | 9·898006 | 15 |
| 46 | 9·787069 | 272 | ·212931 | 9·889160 | 435 | 10·110840 | ·102092 | 163 | 9·897908 | 14 |
| 47 | 9·787232 | 272 | ·212768 | 9·889421 | 435 | 10·110579 | ·102190 | 163 | 9·897810 | 13 |
| 48 | 9·787395 | 271 | ·212605 | 9·889682 | 435 | 10·110318 | ·102288 | 163 | 9·897712 | 12 |
| 49 | 9·787557 | 271 | ·212443 | 9·889943 | 435 | 10·110057 | ·102386 | 163 | 9·897614 | 11 |
| 50 | 9·787720 | 271 | ·212280 | 9·890204 | 435 | 10·109796 | ·102484 | 163 | 9·897516 | 10 |
| 51 | 9·787883 | 271 | ·212117 | 9·890465 | 435 | 10·109535 | ·102582 | 163 | 9·897418 | 9 |
| 52 | 9·788045 | 271 | ·211955 | 9·890725 | 435 | 10·109275 | ·102680 | 164 | 9·897320 | 8 |
| 53 | 9·788208 | 271 | ·211792 | 9·890986 | 434 | 10·109014 | ·102778 | 164 | 9·897222 | 7 |
| 54 | 9·788370 | 271 | ·211630 | 9·891247 | 434 | 10·108753 | ·102877 | 164 | 9·897123 | 6 |
| 55 | 9·788532 | 270 | ·211468 | 9·891507 | 434 | 10·108493 | ·102975 | 164 | 9·897025 | 5 |
| 56 | 9·788694 | 270 | ·211306 | 9·891768 | 434 | 10·108232 | ·103074 | 164 | 9·896926 | 4 |
| 57 | 9·788856 | 270 | ·211144 | 9·892028 | 434 | 10·107972 | ·103172 | 164 | 9·896828 | 3 |
| 58 | 9·789018 | 270 | ·210982 | 9·892289 | 434 | 10·107711 | ·103271 | 164 | 9·896729 | 2 |
| 59 | 9·789180 | 270 | ·210820 | 9·892549 | 434 | 10·107451 | ·103369 | 164 | 9·896631 | 1 |
| 60 | 9·789342 | 270 | ·210658 | 9·892810 | 434 | 10·107190 | ·103468 | 164 | 9·896532 | 0 |
| ′ | Cosine. | | Secant. | Cotangent. | | Tangent. | Cosecant. | | Sine. | ′ |

52 DEG.

38 DEG.

| ′ | Sine. | Diff. 100″ | Cosecant. | Tangent. | Diff. 100″ | Cotangent. | Secant. | Diff. 100″ | Cosine. | ′ |
|---|---|---|---|---|---|---|---|---|---|---|
| 0 | 9·789342 | | ·210658 | 9·892810 | | 10·107190 | ·103468 | | 9·896532 | 60 |
| 1 | 9·789504 | 269 | ·210496 | 9·893070 | 434 | 10·106930 | ·103567 | 164 | 9·896433 | 59 |
| 2 | 9·789665 | 269 | ·210335 | 9·893331 | 434 | 10·106669 | ·103665 | 165 | 9·896335 | 58 |
| 3 | 9·789827 | 269 | ·210173 | 9·893591 | 434 | 10·106409 | ·103764 | 165 | 9·896236 | 57 |
| 4 | 9·789988 | 269 | ·210012 | 9·893851 | 434 | 10·106149 | ·103863 | 165 | 9·896137 | 56 |
| 5 | 9·790149 | 269 | ·209851 | 9·894111 | 434 | 10·105889 | ·103962 | 165 | 9·896038 | 55 |
| 6 | 9·790310 | 269 | ·209690 | 9·894371 | 434 | 10·105629 | ·104061 | 165 | 9·895939 | 54 |
| 7 | 9·790471 | 268 | ·209529 | 9·894632 | 434 | 10·105368 | ·104160 | 165 | 9·895840 | 53 |
| 8 | 9·790632 | 268 | ·209368 | 9·894892 | 434 | 10·105108 | ·104259 | 165 | 9·895741 | 52 |
| 9 | 9·790793 | 268 | ·209207 | 9·895152 | 433 | 10·104848 | ·104359 | 165 | 9·895641 | 51 |
| 10 | 9·790954 | 268 | ·209046 | 9·895412 | 433 | 10·104588 | ·104458 | 165 | 9·895542 | 50 |
| 11 | 9·791115 | 268 | ·208885 | 9·895672 | 433 | 10·104328 | ·104557 | 165 | 9·895443 | 49 |
| 12 | 9·791275 | 268 | ·208725 | 9·895932 | 433 | 10·104068 | ·104657 | 166 | 9·895343 | 48 |
| 13 | 9·791436 | 267 | ·208564 | 9·896192 | 433 | 10·103808 | ·104756 | 166 | 9·895244 | 47 |
| 14 | 9·791596 | 267 | ·208404 | 9·896452 | 433 | 10·103548 | ·104855 | 166 | 9·895145 | 46 |
| 15 | 9·791757 | 267 | ·208243 | 9·896712 | 433 | 10·103288 | ·104955 | 166 | 9·895045 | 45 |
| 16 | 9·791917 | 267 | ·208083 | 9·896971 | 433 | 10·103029 | ·105055 | 166 | 9·894945 | 44 |
| 17 | 9·792077 | 267 | ·207923 | 9·897231 | 433 | 10·102769 | ·105154 | 166 | 9·894846 | 43 |
| 18 | 9·792237 | 267 | ·207763 | 9·897491 | 433 | 10·102509 | ·105254 | 166 | 9·894746 | 42 |
| 19 | 9·792397 | 266 | ·207603 | 9·897751 | 433 | 10·102249 | ·105354 | 166 | 9·894646 | 41 |
| 20 | 9·792557 | 266 | ·207443 | 9·898010 | 433 | 10·101990 | ·105454 | 166 | 9·894546 | 40 |
| 21 | 9·792716 | 266 | ·207284 | 9·898270 | 433 | 10·101730 | ·105554 | 166 | 9·894446 | 39 |
| 22 | 9·792876 | 266 | ·207124 | 9·898530 | 433 | 10·101470 | ·105654 | 167 | 9·894346 | 38 |
| 23 | 9·793035 | 266 | ·206965 | 9·898789 | 433 | 10·101211 | ·105754 | 167 | 9·894246 | 37 |
| 24 | 9·793195 | 266 | ·206805 | 9·899049 | 433 | 10·100951 | ·105854 | 167 | 9·894146 | 36 |
| 25 | 9·793354 | 265 | ·206646 | 9·899308 | 432 | 10·100692 | ·105954 | 167 | 9·894046 | 35 |
| 26 | 9·793514 | 265 | ·206486 | 9·899568 | 432 | 10·100432 | ·106054 | 167 | 9·893946 | 34 |
| 27 | 9·793673 | 265 | ·206327 | 9·899827 | 432 | 10·100173 | ·106154 | 167 | 9·893846 | 33 |
| 28 | 9·793832 | 265 | ·206168 | 9·900086 | 432 | 10·099914 | ·106255 | 167 | 9·893745 | 32 |
| 29 | 9·793991 | 265 | ·206009 | 9·900346 | 432 | 10·099654 | ·106355 | 167 | 9·893645 | 31 |
| 30 | 9·794150 | 265 | ·205850 | 9·900605 | 432 | 10·099395 | ·106456 | 167 | 9·893544 | 30 |
| 31 | 9·794308 | 264 | ·205692 | 9·900864 | 432 | 10·099136 | ·106556 | 167 | 9·893444 | 29 |
| 32 | 9·794467 | 264 | ·205533 | 9·901124 | 432 | 10·098876 | ·106657 | 168 | 9·893343 | 28 |
| 33 | 9·794626 | 264 | ·205374 | 9·901383 | 432 | 10·098617 | ·106757 | 168 | 9·893243 | 27 |
| 34 | 9·794784 | 264 | ·205216 | 9·901642 | 432 | 10·098358 | ·106858 | 168 | 9·893142 | 26 |
| 35 | 9·794942 | 264 | ·205058 | 9·901901 | 432 | 10·098099 | ·106959 | 168 | 9·893041 | 25 |
| 36 | 9·795101 | 264 | ·204899 | 9·902160 | 432 | 10·097840 | ·107060 | 168 | 9·892940 | 24 |
| 37 | 9·795259 | 264 | ·204741 | 9·902419 | 432 | 10·097581 | ·107161 | 168 | 9·892839 | 23 |
| 38 | 9·795417 | 263 | ·204583 | 9·902679 | 432 | 10·097321 | ·107261 | 168 | 9·892739 | 22 |
| 39 | 9·795575 | 263 | ·204425 | 9·902938 | 432 | 10·097062 | ·107362 | 168 | 9·892638 | 21 |
| 40 | 9·795733 | 263 | ·204267 | 9·903197 | 432 | 10·096803 | ·107464 | 168 | 9·892536 | 20 |
| 41 | 9·795891 | 263 | ·204109 | 9·903455 | 432 | 10·096545 | ·107565 | 168 | 9·892435 | 19 |
| 42 | 9·796049 | 263 | ·203951 | 9·903714 | 431 | 10·096286 | ·107666 | 169 | 9·892334 | 18 |
| 43 | 9·796206 | 263 | ·203794 | 9·903973 | 431 | 10·096027 | ·107767 | 169 | 9·892233 | 17 |
| 44 | 9·796364 | 263 | ·203636 | 9·904232 | 431 | 10·095768 | ·107868 | 169 | 9·892132 | 16 |
| 45 | 9·796521 | 262 | ·203479 | 9·904491 | 431 | 10·095509 | ·107970 | 169 | 9·892030 | 15 |
| 46 | 9·796679 | 262 | ·203321 | 9·904750 | 431 | 10·095250 | ·108071 | 169 | 9·891929 | 14 |
| 47 | 9·796836 | 262 | ·203164 | 9·905008 | 431 | 10·094992 | ·108173 | 169 | 9·891827 | 13 |
| 48 | 9·796993 | 262 | ·203007 | 9·905267 | 431 | 10·094733 | ·108274 | 169 | 9·891726 | 12 |
| 49 | 9·797150 | 262 | ·202850 | 9·905526 | 431 | 10·094474 | ·108376 | 169 | 9·891624 | 11 |
| 50 | 9·797307 | 262 | ·202693 | 9·905784 | 431 | 10·094216 | ·108477 | 169 | 9·891523 | 10 |
| 51 | 9·797464 | 261 | ·202536 | 9·906043 | 431 | 10·093957 | ·108579 | 169 | 9·891421 | 9 |
| 52 | 9·797621 | 261 | ·202379 | 9·906302 | 431 | 10·093698 | ·108681 | 170 | 9·891319 | 8 |
| 53 | 9·797777 | 261 | ·202223 | 9·906560 | 431 | 10·093440 | ·108783 | 170 | 9·891217 | 7 |
| 54 | 9·797934 | 261 | ·202066 | 9·906819 | 431 | 10·093181 | ·108885 | 170 | 9·891115 | 6 |
| 55 | 9·798091 | 261 | ·201909 | 9·907077 | 431 | 10·092923 | ·108987 | 170 | 9·891013 | 5 |
| 56 | 9·798247 | 261 | ·201753 | 9·907336 | 431 | 10·092664 | ·109089 | 170 | 9·890911 | 4 |
| 57 | 9·798403 | 261 | ·201597 | 9·907594 | 431 | 10·092406 | ·109191 | 170 | 9·890809 | 3 |
| 58 | 9·798560 | 260 | ·201440 | 9·907852 | 431 | 10·092148 | ·109293 | 170 | 9·890707 | 2 |
| 59 | 9·798716 | 260 | ·201284 | 9·908111 | 431 | 10·091889 | ·109395 | 170 | 9·890605 | 1 |
| 60 | 9·798872 | 260 | ·201128 | 9·908369 | 431 | 10·091631 | ·109497 | 170 | 9·890503 | 0 |
| ′ | Cosine. | | Secant. | Cotangent. | | Tangent. | Cosecant. | | Sine. | ′ |

51 DEG.

39 DEG.

| ′ | Sine. | Diff. 100″ | Cosecant. | Tangent. | Diff. 100″ | Cotangent. | Secant. | Diff. 100 | Cosine. | ′ |
|---|---|---|---|---|---|---|---|---|---|---|
| 0 | 9·798872 | | ·201128 | 9·908369 | | 10·091631 | ·109497 | | 9·890503 | 60 |
| 1 | 9·799028 | 260 | ·200972 | 9·908628 | 430 | 10·091372 | ·109600 | 170 | 9·890400 | 59 |
| 2 | 9·799184 | 260 | ·200816 | 9·908886 | 430 | 10·091114 | ·109702 | 171 | 9·890298 | 58 |
| 3 | 9·799339 | 260 | ·200661 | 9·909144 | 430 | 10·090856 | ·109805 | 171 | 9·890195 | 57 |
| 4 | 9·799495 | 259 | ·200505 | 9·909402 | 430 | 10·090598 | ·109907 | 171 | 9·890093 | 56 |
| 5 | 9·799651 | 259 | ·200349 | 9·909660 | 430 | 10·090340 | ·110010 | 171 | 9·889990 | 55 |
| 6 | 9·799806 | 259 | ·200194 | 9·909918 | 430 | 10·090082 | ·110112 | 171 | 9·889888 | 54 |
| 7 | 9·799962 | 259 | ·200038 | 9·910177 | 430 | 10·089823 | ·110215 | 171 | 9·889785 | 53 |
| 8 | 9·800117 | 259 | ·199883 | 9·910435 | 430 | 10·089565 | ·110318 | 171 | 9·889682 | 52 |
| 9 | 9·800272 | 259 | ·199728 | 9·910693 | 430 | 10·089307 | ·110421 | 171 | 9·889579 | 51 |
| 10 | 9·800427 | 258 | ·199573 | 9·910951 | 430 | 10·089049 | ·110523 | 171 | 9·889477 | 50 |
| 11 | 9·800582 | 258 | ·199418 | 9·911209 | 430 | 10·088791 | ·110626 | 171 | 9·889374 | 49 |
| 12 | 9·800737 | 258 | ·199263 | 9·911467 | 430 | 10·088533 | ·110729 | 172 | 9·889271 | 48 |
| 13 | 9·800892 | 258 | ·199108 | 9·911724 | 430 | 10·088276 | ·110832 | 172 | 9·889168 | 47 |
| 14 | 9·801047 | 258 | ·198953 | 9·911982 | 430 | 10·088018 | ·110936 | 172 | 9·889064 | 46 |
| 15 | 9·801201 | 258 | ·198799 | 9·912240 | 430 | 10·087760 | ·111039 | 172 | 9·888961 | 45 |
| 16 | 9·801356 | 258 | ·198644 | 9·912498 | 430 | 10·087502 | ·111142 | 172 | 9·888858 | 44 |
| 17 | 9·801511 | 257 | ·198489 | 9·912756 | 430 | 10·087244 | ·111245 | 172 | 9·888755 | 43 |
| 18 | 9·801665 | 257 | ·198335 | 9·913014 | 430 | 10·086986 | ·111349 | 172 | 9·888651 | 42 |
| 19 | 9·801819 | 257 | ·198181 | 9·913271 | 430 | 10·086729 | ·111452 | 172 | 9·888548 | 41 |
| 20 | 9·801973 | 257 | ·198027 | 9·913529 | 429 | 10·086471 | ·111556 | 172 | 9·888444 | 40 |
| 21 | 9·802128 | 257 | ·197872 | 9·913787 | 429 | 10·086213 | ·111659 | 173 | 9·888341 | 39 |
| 22 | 9·802282 | 257 | ·197718 | 9·914044 | 429 | 10·085956 | ·111763 | 173 | 9·888237 | 38 |
| 23 | 9·802436 | 256 | ·197564 | 9·914302 | 429 | 10·085698 | ·111866 | 173 | 9·888134 | 37 |
| 24 | 9·802589 | 256 | ·197411 | 9·914560 | 429 | 10·085440 | ·111970 | 173 | 9·888030 | 36 |
| 25 | 9·802743 | 256 | ·197257 | 9·914817 | 429 | 10·085183 | ·112074 | 173 | 9·887926 | 35 |
| 26 | 9·802897 | 256 | ·197103 | 9·915075 | 429 | 10·084925 | ·112178 | 173 | 9·887822 | 34 |
| 27 | 9·803050 | 256 | ·196950 | 9·915332 | 429 | 10·084668 | ·112282 | 173 | 9·887718 | 33 |
| 28 | 9·803204 | 256 | ·196796 | 9·915590 | 429 | 10·084410 | ·112386 | 173 | 9·887614 | 32 |
| 29 | 9·803357 | 256 | ·196643 | 9·915847 | 429 | 10·084153 | ·112490 | 173 | 9·887510 | 31 |
| 30 | 9·803511 | 255 | ·196489 | 9·916104 | 429 | 10·083896 | ·112594 | 173 | 9·887406 | 30 |
| 31 | 9·803664 | 255 | ·196336 | 9·916362 | 429 | 10·083638 | ·112698 | 174 | 9·887302 | 29 |
| 32 | 9·803817 | 255 | ·196183 | 9·916619 | 429 | 10·083381 | ·112802 | 174 | 9·887198 | 28 |
| 33 | 9·803970 | 255 | ·196030 | 9·916877 | 429 | 10·083123 | ·112907 | 174 | 9·887093 | 27 |
| 34 | 9·804123 | 255 | ·195877 | 9·917134 | 429 | 10·082866 | ·113011 | 174 | 9·886989 | 26 |
| 35 | 9·804276 | 255 | ·195724 | 9·917391 | 429 | 10·082609 | ·113115 | 174 | 9·886885 | 25 |
| 36 | 9·804428 | 254 | ·195572 | 9·917648 | 429 | 10·082352 | ·113220 | 174 | 9·886780 | 24 |
| 37 | 9·804581 | 254 | ·195419 | 9·917905 | 429 | 10·082095 | ·113324 | 174 | 9·886676 | 23 |
| 38 | 9·804734 | 254 | ·195266 | 9·918163 | 429 | 10·081837 | ·113429 | 174 | 9·886571 | 22 |
| 39 | 9·804886 | 254 | ·195114 | 9·918420 | 429 | 10·081580 | ·113534 | 174 | 9·886466 | 21 |
| 40 | 9·805039 | 254 | ·194961 | 9·918677 | 429 | 10·081323 | ·113638 | 174 | 9·886362 | 20 |
| 41 | 9·805191 | 254 | ·194809 | 9·918934 | 429 | 10·081066 | ·113743 | 175 | 9·886257 | 19 |
| 42 | 9·805343 | 254 | ·194657 | 9·919191 | 428 | 10·080809 | ·113848 | 175 | 9·886152 | 18 |
| 43 | 9·805495 | 253 | ·194505 | 9·919448 | 428 | 10·080552 | ·113953 | 175 | 9·886047 | 17 |
| 44 | 9·805647 | 253 | ·194353 | 9·919705 | 428 | 10·080295 | ·114058 | 175 | 9·885942 | 16 |
| 45 | 9·805799 | 253 | ·194201 | 9·919962 | 428 | 10·080038 | ·114163 | 175 | 9·885837 | 15 |
| 46 | 9·805951 | 253 | ·194049 | 9·920219 | 428 | 10·079781 | ·114268 | 175 | 9·885732 | 14 |
| 47 | 9·806103 | 253 | ·193897 | 9·920476 | 428 | 10·079524 | ·114373 | 175 | 9·885627 | 13 |
| 48 | 9·806254 | 253 | ·193746 | 9·920733 | 428 | 10·079267 | ·114478 | 175 | 9·885522 | 12 |
| 49 | 9·806406 | 253 | ·193594 | 9·920990 | 428 | 10·079010 | ·114584 | 175 | 9·885416 | 11 |
| 50 | 9·806557 | 252 | ·193443 | 9·921247 | 428 | 10·078753 | ·114689 | 175 | 9·885311 | 10 |
| 51 | 9·806709 | 252 | ·193291 | 9·921503 | 428 | 10·078497 | ·114795 | 176 | 9·885205 | 9 |
| 52 | 9·806860 | 252 | ·193140 | 9·921760 | 428 | 10·078240 | ·114900 | 176 | 9·885100 | 8 |
| 53 | 9·807011 | 252 | ·192989 | 9·922017 | 428 | 10·077983 | ·115006 | 176 | 9·884994 | 7 |
| 54 | 9·807163 | 252 | ·192837 | 9·922274 | 428 | 10·077726 | ·115111 | 176 | 9·884889 | 6 |
| 55 | 9·807314 | 252 | ·192686 | 9·922530 | 428 | 10·077470 | ·115217 | 176 | 9·884783 | 5 |
| 56 | 9·807465 | 252 | ·192535 | 9·922787 | 428 | 10·077213 | ·115323 | 176 | 9·884677 | 4 |
| 57 | 9·807615 | 251 | ·192385 | 9·923044 | 428 | 10·076956 | ·115428 | 176 | 9·884572 | 3 |
| 58 | 9·807766 | 251 | ·192234 | 9·923300 | 428 | 10·076700 | ·115534 | 176 | 9·884466 | 2 |
| 59 | 9·807917 | 251 | ·192083 | 9·923557 | 428 | 10·076443 | ·115640 | 176 | 9·884360 | 1 |
| 60 | 9·808067 | 251 | ·191933 | 9·923813 | 428 | 10·076187 | ·115746 | 176 | 9·884254 | 0 |
| ′ | Cosine. | | Secant. | Cotangent. | | Tangent. | Cosecant. | | Sine. | ′ |

50 DEG.

40 DEG.

| ′ | Sine. | Diff. 100″ | Cosecant. | Tangent. | Diff. 100″ | Cotangent. | Secant. | Diff. 100″ | Cosine. | ′ |
|---|---|---|---|---|---|---|---|---|---|---|
| 0 | 9·808067 | | ·191933 | 9·923813 | | 10·076187 | ·115746 | | 9·884254 | 60 |
| 1 | 9·808218 | 251 | ·191782 | 9·924070 | 428 | 10·075930 | ·115852 | 177 | 9·884148 | 59 |
| 2 | 9·808368 | 251 | ·191632 | 9·924327 | 428 | 10·075673 | ·115958 | 177 | 9·884042 | 58 |
| 3 | 9·808519 | 251 | ·191481 | 9·924583 | 428 | 10·075417 | ·116064 | 177 | 9·883936 | 57 |
| 4 | 9·808669 | 250 | ·191331 | 9·924840 | 427 | 10·075160 | ·116171 | 177 | 9·883829 | 56 |
| 5 | 9·808819 | 250 | ·191181 | 9·925096 | 427 | 10·074904 | ·116277 | 177 | 9·883723 | 55 |
| 6 | 9·808969 | 250 | ·191031 | 9·925352 | 427 | 10·074648 | ·116383 | 177 | 9·883617 | 54 |
| 7 | 9·809119 | 250 | ·190881 | 9·925609 | 427 | 10·074391 | ·116490 | 177 | 9·883510 | 53 |
| 8 | 9·809269 | 250 | ·190731 | 9·925865 | 427 | 10·074135 | ·116596 | 177 | 9·883404 | 52 |
| 9 | 9·809419 | 250 | ·190581 | 9·926122 | 427 | 10·073878 | ·116703 | 177 | 9·883297 | 51 |
| 10 | 9·809569 | 249 | ·190431 | 9·926378 | 427 | 10·073622 | ·116809 | 178 | 9·883191 | 50 |
| 11 | 9·809718 | 249 | ·190282 | 9·926634 | 427 | 10·073366 | ·116916 | 178 | 9·883084 | 49 |
| 12 | 9·809868 | 249 | ·190132 | 9·926890 | 427 | 10·073110 | ·117023 | 178 | 9·882977 | 48 |
| 13 | 9·810017 | 249 | ·189983 | 9·927147 | 427 | 10·072853 | ·117129 | 178 | 9·882871 | 47 |
| 14 | 9·810167 | 249 | ·189833 | 9·927403 | 427 | 10·072597 | ·117236 | 178 | 9·882764 | 46 |
| 15 | 9·810316 | 249 | ·189684 | 9·927659 | 427 | 10·072341 | ·117343 | 178 | 9·882657 | 45 |
| 16 | 9·810465 | 248 | ·189535 | 9·927915 | 427 | 10·072085 | ·117450 | 178 | 9·882550 | 44 |
| 17 | 9·810614 | 248 | ·189386 | 9·928171 | 427 | 10·071829 | ·117557 | 178 | 9·882443 | 43 |
| 18 | 9·810763 | 248 | ·189237 | 9·928427 | 427 | 10·071573 | ·117664 | 178 | 9·882336 | 42 |
| 19 | 9·810912 | 248 | ·189088 | 9·928683 | 427 | 10·071317 | ·117771 | 179 | 9·882229 | 41 |
| 20 | 9·811061 | 248 | ·188939 | 9·928940 | 427 | 10·071060 | ·117879 | 179 | 9·882121 | 40 |
| 21 | 9·811210 | 248 | ·188790 | 9·929196 | 427 | 10·070804 | ·117986 | 179 | 9·882014 | 39 |
| 22 | 9·811358 | 248 | ·188642 | 9·929452 | 427 | 10·070548 | ·118093 | 179 | 9·881907 | 38 |
| 23 | 9·811507 | 247 | ·188493 | 9·929708 | 427 | 10·070292 | ·118201 | 179 | 9·881799 | 37 |
| 24 | 9·811655 | 247 | ·188345 | 9·929964 | 427 | 10·070036 | ·118308 | 179 | 9·881692 | 36 |
| 25 | 9·811804 | 247 | ·188196 | 9·930220 | 427 | 10·069780 | ·118416 | 179 | 9·881584 | 35 |
| 26 | 9·811952 | 247 | ·188048 | 9·930475 | 427 | 10·069525 | ·118523 | 179 | 9·881477 | 34 |
| 27 | 9·812100 | 247 | ·187900 | 9·930731 | 427 | 10·069269 | ·118631 | 179 | 9·881369 | 33 |
| 28 | 9·812248 | 247 | ·187752 | 9·930987 | 426 | 10·069013 | ·118739 | 179 | 9·881261 | 32 |
| 29 | 9·812396 | 247 | ·187604 | 9·931243 | 426 | 10·068757 | ·118847 | 180 | 9·881153 | 31 |
| 30 | 9·812544 | 246 | ·187456 | 9·931499 | 426 | 10·068501 | ·118954 | 180 | 9·881046 | 30 |
| 31 | 9·812692 | 246 | ·187308 | 9·931755 | 426 | 10·068245 | ·119062 | 180 | 9·880938 | 29 |
| 32 | 9·812840 | 246 | ·187160 | 9·932010 | 426 | 10·067990 | ·119170 | 180 | 9·880830 | 28 |
| 33 | 9·812988 | 246 | ·187012 | 9·932266 | 426 | 10·067734 | ·119178 | 180 | 9·880722 | 27 |
| 34 | 9·813135 | 246 | ·186865 | 9·932522 | 426 | 10·067478 | ·119387 | 180 | 9·880613 | 26 |
| 35 | 9·813283 | 246 | ·186717 | 9·932778 | 426 | 10·067222 | ·119495 | 180 | 9·880505 | 25 |
| 36 | 9·813430 | 246 | ·186570 | 9·933033 | 426 | 10·066967 | ·119603 | 180 | 9·880397 | 24 |
| 37 | 9·813578 | 245 | ·186422 | 9·933289 | 426 | 10·066711 | ·119711 | 180 | 9·880289 | 23 |
| 38 | 9·813725 | 245 | ·186275 | 9·933545 | 426 | 10·066455 | ·119820 | 181 | 9·880180 | 22 |
| 39 | 9·813872 | 245 | ·186128 | 9·933800 | 426 | 10·066200 | ·119928 | 181 | 9·880072 | 21 |
| 40 | 9·814019 | 245 | ·185981 | 9·934056 | 426 | 10·065944 | ·120037 | 181 | 9·879963 | 20 |
| 41 | 9·814166 | 245 | ·185834 | 9·934311 | 426 | 10·065689 | ·120145 | 181 | 9·879855 | 19 |
| 42 | 9·814313 | 245 | ·185687 | 9·934567 | 426 | 10·065433 | ·120254 | 181 | 9·879746 | 18 |
| 43 | 9·814460 | 245 | ·185540 | 9·934823 | 426 | 10·065177 | ·120363 | 181 | 9·879637 | 17 |
| 44 | 9·814607 | 244 | ·185393 | 9·935078 | 426 | 10·064922 | ·120471 | 181 | 9·879529 | 16 |
| 45 | 9·814753 | 244 | ·185247 | 9·935333 | 426 | 10·064667 | ·120580 | 181 | 9·879420 | 15 |
| 46 | 9·814900 | 244 | ·185100 | 9·935589 | 426 | 10·064411 | ·120689 | 181 | 9·879311 | 14 |
| 47 | 9·815046 | 244 | ·184954 | 9·935844 | 426 | 10·064156 | ·120798 | 181 | 9·879202 | 13 |
| 48 | 9·815193 | 244 | ·184807 | 9·936100 | 426 | 10·063900 | ·120907 | 182 | 9·879093 | 12 |
| 49 | 9·815339 | 244 | ·184661 | 9·936355 | 426 | 10·063645 | ·121016 | 182 | 9·878984 | 11 |
| 50 | 9·815485 | 244 | ·184515 | 9·936610 | 426 | 10·063390 | ·121125 | 182 | 9·878875 | 10 |
| 51 | 9·815632 | 243 | ·184368 | 9·936866 | 426 | 10·063134 | ·121234 | 182 | 9·878766 | 9 |
| 52 | 9·815778 | 243 | ·184222 | 9·937121 | 426 | 10·062879 | ·121344 | 182 | 9·878656 | 8 |
| 53 | 9·815924 | 243 | ·184076 | 9·937376 | 426 | 10·062624 | ·121453 | 182 | 9·878547 | 7 |
| 54 | 9·816069 | 243 | ·183931 | 9·937632 | 425 | 10·062368 | ·121562 | 182 | 9·878438 | 6 |
| 55 | 9·816215 | 243 | ·183785 | 9·937887 | 425 | 10·062113 | ·121672 | 182 | 9·878328 | 5 |
| 56 | 9·816361 | 243 | ·183639 | 9·938142 | 425 | 10·061858 | ·121781 | 182 | 9·878219 | 4 |
| 57 | 9·816507 | 243 | ·183493 | 9·938398 | 425 | 10·061602 | ·121891 | 183 | 9·878109 | 3 |
| 58 | 9·816652 | 242 | ·183348 | 9·938653 | 425 | 10·061347 | ·122001 | 183 | 9·877999 | 2 |
| 59 | 9·816798 | 242 | ·183202 | 9·938908 | 425 | 10·061092 | ·122110 | 183 | 9·877890 | 1 |
| 60 | 9·816943 | 242 | ·183057 | 9·939163 | 425 | 10·060837 | ·122220 | 183 | 9·877780 | 0 |
| ′ | Cosine. | | Secant. | Cotangent. | | Tangent. | Cosecant. | | Sine. | ′ |

49 DEG.

41 DEG.

| ′ | Sine. | Diff. 100″ | Cosecant. | Tangent. | Diff. 100″ | Cotangent. | Secant. | Diff. 100″ | Cosine. | ′ |
|---|---|---|---|---|---|---|---|---|---|---|
| 0 | 9·816943 | | ·183057 | 9·939163 | | 10·060837 | ·122220 | | 9·877780 | 60 |
| 1 | 9·817088 | 242 | ·182912 | 9·939418 | 425 | 10·060582 | ·122330 | 183 | 9·877670 | 59 |
| 2 | 9·817233 | 242 | ·182767 | 9·939673 | 425 | 10·060327 | ·122440 | 183 | 9·877560 | 58 |
| 3 | 9·817379 | 242 | ·182621 | 9·939928 | 425 | 10·060072 | ·122550 | 183 | 9·877450 | 57 |
| 4 | 9·817524 | 242 | ·182476 | 9·940183 | 425 | 10·059817 | ·122660 | 183 | 9·877340 | 56 |
| 5 | 9·817668 | 241 | ·182332 | 9·940438 | 425 | 10·059562 | ·122770 | 183 | 9·877230 | 55 |
| 6 | 9·817813 | 241 | ·182187 | 9·940694 | 425 | 10·059306 | ·122880 | 184 | 9·877120 | 54 |
| 7 | 9·817958 | 241 | ·182042 | 9·940949 | 425 | 10·059051 | ·122990 | 184 | 9·877010 | 53 |
| 8 | 9·818103 | 241 | ·181897 | 9·941204 | 425 | 10·058796 | ·123101 | 184 | 9·876899 | 52 |
| 9 | 9·818247 | 241 | ·181753 | 9·941458 | 425 | 10·058542 | ·123211 | 184 | 9·876789 | 51 |
| 10 | 9·818392 | 241 | ·181608 | 9·941714 | 425 | 10·058286 | ·123322 | 184 | 9·876678 | 50 |
| 11 | 9·818536 | 241 | ·181464 | 9·941968 | 425 | 10·058032 | ·123432 | 184 | 9·876568 | 49 |
| 12 | 9·818681 | 240 | ·181319 | 9·942223 | 425 | 10·057777 | ·123543 | 184 | 9·876457 | 48 |
| 13 | 9·818825 | 240 | ·181175 | 9·942478 | 425 | 10·057522 | ·123653 | 184 | 9·876347 | 47 |
| 14 | 9·818969 | 240 | ·181031 | 9·942733 | 425 | 10·057267 | ·123764 | 184 | 9·876236 | 46 |
| 15 | 9·819113 | 240 | ·180887 | 9·942988 | 425 | 10·057012 | ·123875 | 185 | 9·876125 | 45 |
| 16 | 9·819257 | 240 | ·180743 | 9·943243 | 425 | 10·056757 | ·123986 | 185 | 9·876014 | 44 |
| 17 | 9·819401 | 240 | ·180599 | 9·943498 | 425 | 10·056502 | ·124096 | 185 | 9·875904 | 43 |
| 18 | 9·819545 | 240 | ·180455 | 9·943752 | 425 | 10·056248 | ·124207 | 185 | 9·875793 | 42 |
| 19 | 9·819689 | 239 | ·180311 | 9·944007 | 425 | 10·055993 | ·124318 | 185 | 9·875682 | 41 |
| 20 | 9·819832 | 239 | ·180168 | 9·944262 | 425 | 10·055738 | ·124429 | 185 | 9·875571 | 40 |
| 21 | 9·819976 | 239 | ·180024 | 9·944517 | 425 | 10·055483 | ·124541 | 185 | 9·875459 | 39 |
| 22 | 9·820120 | 239 | ·179880 | 9·944771 | 425 | 10·055229 | ·124652 | 185 | 9·875348 | 38 |
| 23 | 9·820263 | 239 | ·179737 | 9·945026 | 425 | 10·054974 | ·124763 | 185 | 9·875237 | 37 |
| 24 | 9·820406 | 239 | ·179594 | 9·945281 | 425 | 10·054719 | ·124874 | 185 | 9·875126 | 36 |
| 25 | 9·820550 | 239 | ·179450 | 9·945535 | 425 | 10·054465 | ·124986 | 186 | 9·875014 | 35 |
| 26 | 9·820693 | 238 | ·179307 | 9·945790 | 425 | 10·054210 | ·125097 | 186 | 9·874903 | 34 |
| 27 | 9·820836 | 238 | ·179164 | 9·946045 | 425 | 10·053955 | ·125209 | 186 | 9·874791 | 33 |
| 28 | 9·820979 | 238 | ·179021 | 9·946299 | 425 | 10·053701 | ·125320 | 186 | 9·874680 | 32 |
| 29 | 9·821122 | 238 | ·178878 | 9·946554 | 425 | 10·053446 | ·125432 | 186 | 9·874568 | 31 |
| 30 | 9·821265 | 238 | ·178735 | 9·946808 | 425 | 10·053192 | ·125544 | 186 | 9·874456 | 30 |
| 31 | 9·821407 | 238 | ·178593 | 9·947063 | 425 | 10·052937 | ·125656 | 186 | 9·874344 | 29 |
| 32 | 9·821550 | 238 | ·178450 | 9·947318 | 424 | 10·052682 | ·125768 | 186 | 9·874232 | 28 |
| 33 | 9·821693 | 238 | ·178307 | 9·947572 | 424 | 10·052428 | ·125879 | 186 | 9·874121 | 27 |
| 34 | 9·821835 | 237 | ·178165 | 9·947826 | 424 | 10·052174 | ·125991 | 187 | 9·874009 | 26 |
| 35 | 9·821977 | 237 | ·178023 | 9·948081 | 424 | 10·051919 | ·126104 | 187 | 9·873896 | 25 |
| 36 | 9·822120 | 237 | ·177880 | 9·948336 | 424 | 10·051664 | ·126216 | 187 | 9·873784 | 24 |
| 37 | 9·822262 | 237 | ·177738 | 9·948590 | 424 | 10·051410 | ·126328 | 187 | 9·873672 | 23 |
| 38 | 9·822404 | 237 | ·177596 | 9·948844 | 424 | 10·051156 | ·126440 | 187 | 9·873560 | 22 |
| 39 | 9·822546 | 237 | ·177454 | 9·949099 | 424 | 10·050901 | ·126552 | 187 | 9·873448 | 21 |
| 40 | 9·822688 | 237 | ·177312 | 9·949353 | 424 | 10·050647 | ·126665 | 187 | 9·873335 | 20 |
| 41 | 9·822830 | 236 | ·177170 | 9·949607 | 424 | 10·050393 | ·126777 | 187 | 9·873223 | 19 |
| 42 | 9·822972 | 236 | ·177028 | 9·949862 | 424 | 10·050138 | ·126890 | 187 | 9·873110 | 18 |
| 43 | 9·823114 | 236 | ·176886 | 9·950116 | 424 | 10·049884 | ·127002 | 188 | 9·872998 | 17 |
| 44 | 9·823255 | 236 | ·176745 | 9·950370 | 424 | 10·049630 | ·127115 | 188 | 9·872885 | 16 |
| 45 | 9·823397 | 236 | ·176603 | 9·950625 | 424 | 10·049375 | ·127228 | 188 | 9·872772 | 15 |
| 46 | 9·823539 | 236 | ·176461 | 9·950879 | 424 | 10·049121 | ·127341 | 188 | 9·872659 | 14 |
| 47 | 9·823680 | 236 | ·176320 | 9·951133 | 424 | 10·048867 | ·127453 | 188 | 9·872547 | 13 |
| 48 | 9·823821 | 235 | ·176179 | 9·951388 | 424 | 10·048612 | ·127566 | 188 | 9·872434 | 12 |
| 49 | 9·823963 | 235 | ·176037 | 9·951642 | 424 | 10·048358 | ·127679 | 188 | 9·872321 | 11 |
| 50 | 9·824104 | 235 | ·175896 | 9·951896 | 424 | 10·048104 | ·127792 | 188 | 9·872208 | 10 |
| 51 | 9·824245 | 235 | ·175755 | 9·952150 | 424 | 10·047850 | ·127905 | 188 | 9·872095 | 9 |
| 52 | 9·824386 | 235 | ·175614 | 9·952405 | 424 | 10·047595 | ·128019 | 189 | 9·871981 | 8 |
| 53 | 9·824527 | 235 | ·175473 | 9·952659 | 424 | 10·047341 | ·128132 | 189 | 9·871868 | 7 |
| 54 | 9·824668 | 235 | ·175332 | 9·952913 | 424 | 10·047087 | ·128245 | 189 | 9·871755 | 6 |
| 55 | 9·824808 | 234 | ·175192 | 9·953167 | 424 | 10·046833 | ·128359 | 189 | 9·871641 | 5 |
| 56 | 9·824949 | 234 | ·175051 | 9·953421 | 423 | 10·046579 | ·128472 | 189 | 9·871528 | 4 |
| 57 | 9·825090 | 234 | ·174910 | 9·953675 | 423 | 10·046325 | ·128586 | 189 | 9·871414 | 3 |
| 58 | 9·825230 | 234 | ·174770 | 9·953929 | 423 | 10·046071 | ·128699 | 189 | 9·871301 | 2 |
| 59 | 9·825371 | 234 | ·174629 | 9·954183 | 423 | 10·045817 | ·128813 | 189 | 9·871187 | 1 |
| 60 | 9·825511 | 234 | ·174489 | 9·954437 | 423 | 10·045563 | ·128927 | 189 | 9·871073 | 0 |
| ′ | Cosine. | | Secant. | Cotangent. | | Tangent. | Cosecant. | | Sine. | ′ |

48 DEG.

42 DEG.

| ′ | Sine. | Diff. 100″ | Cosecant. | Tangent. | Diff. 100″ | Cotangent. | Secant. | Diff. 100″ | Cosine. | ′ |
|---|---|---|---|---|---|---|---|---|---|---|
| 0 | 9·825511 | | ·174489 | 9·954437 | | 10·045563 | ·128927 | | 9·871073 | 60 |
| 1 | 9·825651 | 234 | ·174349 | 9·954691 | 423 | 10·045309 | ·129040 | 190 | 9·870960 | 59 |
| 2 | 9·825791 | 233 | ·174209 | 9·954945 | 423 | 10·045055 | ·129154 | 190 | 9·870846 | 58 |
| 3 | 9·825931 | 233 | ·174069 | 9·955200 | 423 | 10·044800 | ·129268 | 190 | 9·870732 | 57 |
| 4 | 9·826071 | 233 | ·173929 | 9·955454 | 423 | 10·044546 | ·129382 | 190 | 9·870618 | 56 |
| 5 | 9·826211 | 233 | ·173789 | 9·955707 | 423 | 10·044293 | ·129496 | 190 | 9·870504 | 55 |
| 6 | 9·826351 | 233 | ·173649 | 9·955961 | 423 | 10·044039 | ·129610 | 190 | 9·870390 | 54 |
| 7 | 9·826491 | 233 | ·173509 | 9·956215 | 423 | 10·043785 | ·129724 | 190 | 9·870276 | 53 |
| 8 | 9·826631 | 233 | ·173369 | 9·956469 | 423 | 10·043531 | ·129839 | 190 | 9·870161 | 52 |
| 9 | 9·826770 | 233 | ·173230 | 9·956723 | 423 | 10·043277 | ·129953 | 190 | 9·870047 | 51 |
| 10 | 9·826910 | 232 | ·173090 | 9·956977 | 423 | 10·043023 | ·130067 | 191 | 9·869933 | 50 |
| 11 | 9·827049 | 232 | ·172951 | 9·957231 | 423 | 10·042769 | ·130182 | 191 | 9·869818 | 49 |
| 12 | 9·827189 | 232 | ·172811 | 9·957485 | 423 | 10·042515 | ·130296 | 191 | 9·869704 | 48 |
| 13 | 9·827328 | 232 | ·172672 | 9·957739 | 423 | 10·042261 | ·130411 | 191 | 9·869589 | 47 |
| 14 | 9·827467 | 232 | ·172533 | 9·957993 | 423 | 10·042007 | ·130526 | 191 | 9·869474 | 46 |
| 15 | 9·827606 | 232 | ·172394 | 9·958246 | 423 | 10·041754 | ·130640 | 191 | 9·869360 | 45 |
| 16 | 9·827745 | 232 | ·172255 | 9·958500 | 423 | 10·041500 | ·130755 | 191 | 9·869245 | 44 |
| 17 | 9·827884 | 232 | ·172116 | 9·958754 | 423 | 10·041246 | ·130870 | 191 | 9·869130 | 43 |
| 18 | 9·828023 | 231 | ·171977 | 9·959008 | 423 | 10·040992 | ·130985 | 191 | 9·869015 | 42 |
| 19 | 9·828162 | 231 | ·171838 | 9·959262 | 423 | 10·040738 | ·131100 | 192 | 9·868900 | 41 |
| 20 | 9·828301 | 231 | ·171699 | 9·959516 | 423 | 10·040484 | ·131215 | 192 | 9·868785 | 40 |
| 21 | 9·828439 | 231 | ·171561 | 9·959769 | 423 | 10·040231 | ·131330 | 192 | 9·868670 | 39 |
| 22 | 9·828578 | 231 | ·171422 | 9·960023 | 423 | 10·039977 | ·131445 | 192 | 9·868555 | 38 |
| 23 | 9·828716 | 231 | ·171284 | 9·960277 | 423 | 10·039723 | ·131560 | 192 | 9·868440 | 37 |
| 24 | 9·828855 | 231 | ·171145 | 9·960531 | 423 | 10·039469 | ·131676 | 192 | 9·868324 | 36 |
| 25 | 9·828993 | 230 | ·171007 | 9·960784 | 423 | 10·039216 | ·131791 | 192 | 9·868209 | 35 |
| 26 | 9·829131 | 230 | ·170869 | 9·961038 | 423 | 10·038962 | ·131907 | 192 | 9·868093 | 34 |
| 27 | 9·829269 | 230 | ·170731 | 9·961291 | 423 | 10·038709 | ·132022 | 192 | 9·867978 | 33 |
| 28 | 9·829407 | 230 | ·170593 | 9·961545 | 423 | 10·038455 | ·132138 | 193 | 9·867862 | 32 |
| 29 | 9·829545 | 230 | ·170455 | 9·961799 | 423 | 10·038201 | ·132253 | 193 | 9·867747 | 31 |
| 30 | 9·829683 | 230 | ·170317 | 9·962052 | 423 | 10·037948 | ·132369 | 193 | 9·867631 | 30 |
| 31 | 9·829821 | 230 | ·170179 | 9·962306 | 423 | 10·037694 | ·132485 | 193 | 9·867515 | 29 |
| 32 | 9·829959 | 229 | ·170041 | 9·962560 | 423 | 10·037440 | ·132601 | 193 | 9·867399 | 28 |
| 33 | 9·830097 | 229 | ·169903 | 9·962813 | 423 | 10·037187 | ·132717 | 193 | 9·867283 | 27 |
| 34 | 9·830234 | 229 | ·169766 | 9·963067 | 423 | 10·036933 | ·132833 | 193 | 9·867167 | 26 |
| 35 | 9·830372 | 229 | ·169628 | 9·963320 | 423 | 10·036680 | ·132949 | 193 | 9·867051 | 25 |
| 36 | 9·830509 | 229 | ·169491 | 9·963574 | 423 | 10·036426 | ·133065 | 193 | 9·866935 | 24 |
| 37 | 9·830646 | 229 | ·169354 | 9·963827 | 423 | 10·036173 | ·133181 | 194 | 9·866819 | 23 |
| 38 | 9·830784 | 229 | ·169216 | 9·964081 | 423 | 10·035919 | ·133297 | 194 | 9·866703 | 22 |
| 39 | 9·830921 | 229 | ·169079 | 9·964335 | 423 | 10·035665 | ·133414 | 194 | 9·866586 | 21 |
| 40 | 9·831058 | 228 | ·168942 | 9·964588 | 423 | 10·035412 | ·133530 | 194 | 9·866470 | 20 |
| 41 | 9·831195 | 228 | ·168805 | 9·964842 | 423 | 10·035158 | ·133647 | 194 | 9·866353 | 19 |
| 42 | 9·831332 | 228 | ·168668 | 9·965095 | 422 | 10·034905 | ·133763 | 194 | 9·866237 | 18 |
| 43 | 9·831469 | 228 | ·168531 | 9·965349 | 422 | 10·034651 | ·133880 | 194 | 9·866120 | 17 |
| 44 | 9·831606 | 228 | ·168394 | 9·965602 | 422 | 10·034398 | ·133996 | 194 | 9·866004 | 16 |
| 45 | 9·831742 | 228 | ·168258 | 9·965855 | 422 | 10·034145 | ·134113 | 195 | 9·865887 | 15 |
| 46 | 9·831879 | 228 | ·168121 | 9·966109 | 422 | 10·033891 | ·134230 | 195 | 9·865770 | 14 |
| 47 | 9·832015 | 228 | ·167985 | 9·966362 | 422 | 10·033638 | ·134347 | 195 | 9·865653 | 13 |
| 48 | 9·832152 | 227 | ·167848 | 9·966616 | 422 | 10·033384 | ·134464 | 195 | 9·865536 | 12 |
| 49 | 9·832288 | 227 | ·167712 | 9·966869 | 422 | 10·033131 | ·134581 | 195 | 9·865419 | 11 |
| 50 | 9·832425 | 227 | ·167575 | 9·967123 | 422 | 10·032877 | ·134698 | 195 | 9·865302 | 10 |
| 51 | 9·832561 | 227 | ·167439 | 9·967376 | 422 | 10·032624 | ·134815 | 195 | 9·865185 | 9 |
| 52 | 9·832697 | 227 | ·167303 | 9·967629 | 422 | 10·032371 | ·134932 | 195 | 9·865068 | 8 |
| 53 | 9·832833 | 227 | ·167167 | 9·967883 | 422 | 10·032117 | ·135050 | 195 | 9·864950 | 7 |
| 54 | 9·832969 | 227 | ·167031 | 9·968136 | 422 | 10·031864 | ·135167 | 195 | 9·864833 | 6 |
| 55 | 9·833105 | 226 | ·166895 | 9·968389 | 422 | 10·031611 | ·135284 | 196 | 9·864716 | 5 |
| 56 | 9·833241 | 226 | ·166759 | 9·968643 | 422 | 10·031357 | ·135402 | 196 | 9·864598 | 4 |
| 57 | 9·833377 | 226 | ·166623 | 9·968896 | 422 | 10·031104 | ·135519 | 196 | 9·864481 | 3 |
| 58 | 9·833512 | 226 | ·166488 | 9·969149 | 422 | 10·030851 | ·135637 | 196 | 9·864363 | 2 |
| 59 | 9·833648 | 226 | ·166352 | 9·969403 | 422 | 10·030597 | ·135755 | 196 | 9·864245 | 1 |
| 60 | 9·833783 | 226 | ·166217 | 9·969656 | 422 | 10·030344 | ·135873 | 196 | 9·864127 | 0 |
| ′ | Cosine. | | Secant. | Cotangent. | | Tangent. | Cosecant. | | Sine. | ′ |



47 DEG.

43 DEG.

| ′ | Sine. | Diff. 100″ | Cosecant. | Tangent. | Diff. 100″ | Cotangent. | Secant. | Diff. 100″ | Cosine. | ′ |
|---|---|---|---|---|---|---|---|---|---|---|
| 0 | 9·833783 | | ·166217 | 9·969656 | | 10·030344 | ·135873 | | 9·864127 | 60 |
| 1 | 9·833919 | 226 | ·166081 | 9·969909 | 422 | 10·030091 | ·135990 | 196 | 9·864010 | 59 |
| 2 | 9·834054 | 225 | ·165946 | 9·970162 | 422 | 10·029838 | ·136108 | 196 | 9·863892 | 58 |
| 3 | 9·834189 | 225 | ·165811 | 9·970416 | 422 | 10·029584 | ·136226 | 197 | 9·863774 | 57 |
| 4 | 9·834325 | 225 | ·165675 | 9·970669 | 422 | 10·029331 | ·136344 | 197 | 9·863656 | 56 |
| 5 | 9·834460 | 225 | ·165540 | 9·970922 | 422 | 10·029078 | ·136462 | 197 | 9·863538 | 55 |
| 6 | 9·834595 | 225 | ·165405 | 9·971175 | 422 | 10·028825 | ·136581 | 197 | 9·863419 | 54 |
| 7 | 9·834730 | 225 | ·165270 | 9·971429 | 422 | 10·028571 | ·136699 | 197 | 9·863301 | 53 |
| 8 | 9·834865 | 225 | ·165135 | 9·971682 | 422 | 10·028318 | ·136817 | 197 | 9·863183 | 52 |
| 9 | 9·834999 | 225 | ·165001 | 9·971935 | 422 | 10·028065 | ·136936 | 197 | 9·863064 | 51 |
| 10 | 9·835134 | 224 | ·164866 | 9·972188 | 422 | 10·027812 | ·137054 | 197 | 9·862946 | 50 |
| 11 | 9·835269 | 224 | ·164731 | 9·972441 | 422 | 10·027559 | ·137173 | 198 | 9·862827 | 49 |
| 12 | 9·835403 | 224 | ·164597 | 9·972694 | 422 | 10·027306 | ·137291 | 198 | 9·862709 | 48 |
| 13 | 9·835538 | 224 | ·164462 | 9·972948 | 422 | 10·027052 | ·137410 | 198 | 9·862590 | 47 |
| 14 | 9·835672 | 224 | ·164328 | 9·973201 | 422 | 10·026799 | ·137529 | 198 | 9·862471 | 46 |
| 15 | 9·835807 | 224 | ·164193 | 9·973454 | 422 | 10·026546 | ·137647 | 198 | 9·862353 | 45 |
| 16 | 9·835941 | 224 | ·164059 | 9·973707 | 422 | 10·026293 | ·137766 | 198 | 9·862234 | 44 |
| 17 | 9·836075 | 224 | ·163925 | 9·973960 | 422 | 10·026040 | ·137885 | 198 | 9·862115 | 43 |
| 18 | 9·836209 | 223 | ·163791 | 9·974213 | 422 | 10·025787 | ·138004 | 198 | 9·861996 | 42 |
| 19 | 9·836343 | 223 | ·163657 | 9·974466 | 422 | 10·025534 | ·138123 | 198 | 9·861877 | 41 |
| 20 | 9·836477 | 223 | ·163523 | 9·974719 | 422 | 10·025281 | ·138242 | 198 | 9·861758 | 40 |
| 21 | 9·836611 | 223 | ·163389 | 9·974973 | 422 | 10·025027 | ·138362 | 199 | 9·861638 | 39 |
| 22 | 9·836745 | 223 | ·163255 | 9·975226 | 422 | 10·024774 | ·138481 | 199 | 9·861519 | 38 |
| 23 | 9·836878 | 223 | ·163122 | 9·975479 | 422 | 10·024521 | ·138600 | 199 | 9·861400 | 37 |
| 24 | 9·837012 | 223 | ·162988 | 9·975732 | 422 | 10·024268 | ·138720 | 199 | 9·861280 | 36 |
| 25 | 9·837146 | 222 | ·162854 | 9·975985 | 422 | 10·024015 | ·138839 | 199 | 9·861161 | 35 |
| 26 | 9·837279 | 222 | ·162721 | 9·976238 | 422 | 10·023762 | ·138959 | 199 | 9·861041 | 34 |
| 27 | 9·837412 | 222 | ·162588 | 9·976491 | 422 | 10·023509 | ·139078 | 199 | 9·860922 | 33 |
| 28 | 9·837546 | 222 | ·162454 | 9·976744 | 422 | 10·023256 | ·139198 | 199 | 9·860802 | 32 |
| 29 | 9·837679 | 222 | ·162321 | 9·976997 | 422 | 10·023003 | ·139318 | 199 | 9·860682 | 31 |
| 30 | 9·837812 | 222 | ·162188 | 9·977250 | 422 | 10·022750 | ·139438 | 200 | 9·860562 | 30 |
| 31 | 9·837945 | 222 | ·162055 | 9·977503 | 422 | 10·022497 | ·139558 | 200 | 9·860442 | 29 |
| 32 | 9·838078 | 222 | ·161922 | 9·977756 | 422 | 10·022244 | ·139678 | 200 | 9·860322 | 28 |
| 33 | 9·838211 | 221 | ·161789 | 9·978009 | 422 | 10·021991 | ·139798 | 200 | 9·860202 | 27 |
| 34 | 9·838344 | 221 | ·161656 | 9·978262 | 422 | 10·021738 | ·139918 | 200 | 9·860082 | 26 |
| 35 | 9·838477 | 221 | ·161523 | 9·978515 | 422 | 10·021485 | ·140038 | 200 | 9·859962 | 25 |
| 36 | 9·838610 | 221 | ·161390 | 9·978768 | 422 | 10·021232 | ·140158 | 200 | 9·859842 | 24 |
| 37 | 9·838742 | 221 | ·161258 | 9·979021 | 422 | 10·020979 | ·140279 | 200 | 9·859721 | 23 |
| 38 | 9·838875 | 221 | ·161125 | 9·979274 | 422 | 10·020726 | ·140399 | 201 | 9·859601 | 22 |
| 39 | 9·839007 | 221 | ·160993 | 9·979527 | 422 | 10·020473 | ·140520 | 201 | 9·859480 | 21 |
| 40 | 9·839140 | 221 | ·160860 | 9·979780 | 422 | 10·020220 | ·140640 | 201 | 9·859360 | 20 |
| 41 | 9·839272 | 220 | ·160728 | 9·980033 | 422 | 10·019967 | ·140761 | 201 | 9·859239 | 19 |
| 42 | 9·839404 | 220 | ·160596 | 9·980286 | 422 | 10·019714 | ·140881 | 201 | 9·859119 | 18 |
| 43 | 9·839536 | 220 | ·160464 | 9·980538 | 422 | 10·019462 | ·141002 | 201 | 9·858998 | 17 |
| 44 | 9·839668 | 220 | ·160332 | 9·980791 | 422 | 10·019209 | ·141123 | 201 | 9·858877 | 16 |
| 45 | 9·839800 | 220 | ·160200 | 9·981044 | 422 | 10·018956 | ·141244 | 201 | 9·858756 | 15 |
| 46 | 9·839932 | 220 | ·160068 | 9·981297 | 422 | 10·018703 | ·141365 | 202 | 9·858635 | 14 |
| 47 | 9·840064 | 220 | ·159936 | 9·981550 | 422 | 10·018450 | ·141486 | 202 | 9·858514 | 13 |
| 48 | 9·840196 | 219 | ·159804 | 9·981803 | 422 | 10·018197 | ·141607 | 202 | 9·858393 | 12 |
| 49 | 9·840328 | 219 | ·159672 | 9·982056 | 422 | 10·017944 | ·141728 | 202 | 9·858272 | 11 |
| 50 | 9·840459 | 219 | ·159541 | 9·982309 | 422 | 10·017691 | ·141849 | 202 | 9·858151 | 10 |
| 51 | 9·840591 | 219 | ·159409 | 9·982562 | 421 | 10·017438 | ·141971 | 202 | 9·858029 | 9 |
| 52 | 9·840722 | 219 | ·159278 | 9·982814 | 421 | 10·017186 | ·142092 | 202 | 9·857908 | 8 |
| 53 | 9·840854 | 219 | ·159146 | 9·983067 | 421 | 10·016933 | ·142214 | 202 | 9·857786 | 7 |
| 54 | 9·840985 | 219 | ·159015 | 9·983320 | 421 | 10·016680 | ·142335 | 202 | 9·857665 | 6 |
| 55 | 9·841116 | 219 | ·158884 | 9·983573 | 421 | 10·016427 | ·142457 | 203 | 9·857543 | 5 |
| 56 | 9·841247 | 218 | ·158753 | 9·983826 | 421 | 10·016174 | ·142578 | 203 | 9·857422 | 4 |
| 57 | 9·841378 | 218 | ·158622 | 9·984079 | 421 | 10·015921 | ·142700 | 203 | 9·857300 | 3 |
| 58 | 9·841509 | 218 | ·158491 | 9·984331 | 421 | 10·015669 | ·142822 | 203 | 9·857178 | 2 |
| 59 | 9·841640 | 218 | ·158360 | 9·984584 | 421 | 10·015416 | ·142944 | 203 | 9·857056 | 1 |
| 60 | 9·841771 | 218 | ·158229 | 9·984837 | 421 | 10·015163 | ·143066 | 203 | 9·856934 | 0 |
| ′ | Cosine. | | Secant. | Cotangent. | | Tangent. | Cosecant. | | Sine. | ′ |

46 DEG.

44 DEG.

| ′ | Sine. | Diff. 100″ | Cosecant. | Tangent. | Diff. 100″ | Cotangent. | Secant. | Diff. 100″ | Cosine. | ′ |
|---|---|---|---|---|---|---|---|---|---|---|
| 0 | 9·841771 | | ·158229 | 9·984837 | | 10·015163 | ·143066 | | 9·856934 | 60 |
| 1 | 9·841902 | 218 | ·158098 | 9·985090 | 421 | 10·014910 | ·143188 | 203 | 9 856812 | 59 |
| 2 | 9·842033 | 218 | ·157967 | 9·985343 | 421 | 10·014657 | ·143310 | 203 | 9·856690 | 58 |
| 3 | 9·842163 | 218 | ·157837 | 9·985596 | 421 | 10·014404 | ·143432 | 204 | 9·856568 | 57 |
| 4 | 9·842294 | 217 | ·157706 | 9·985848 | 421 | 10·014152 | ·143554 | 204 | 9·856446 | 56 |
| 5 | 9·842424 | 217 | ·157576 | 9·986101 | 421 | 10·013899 | ·143677 | 204 | 9·856323 | 55 |
| 6 | 9·842555 | 217 | ·157445 | 9·986354 | 421 | 10·013646 | ·143799 | 204 | 9·856201 | 54 |
| 7 | 9·842685 | 217 | ·157315 | 9·986607 | 421 | 10·013393 | ·143922 | 204 | 9·856078 | 53 |
| 8 | 9·842815 | 217 | ·157185 | 9·986860 | 421 | 10·013140 | ·144044 | 204 | 9·855956 | 52 |
| 9 | 9·842946 | 217 | ·157054 | 9·987112 | 421 | 10·012888 | ·144167 | 204 | 9·855833 | 51 |
| 10 | 9·843076 | 217 | ·156924 | 9·987365 | 421 | 10·012635 | ·144289 | 204 | 9·855711 | 50 |
| 11 | 9·843206 | 217 | ·156794 | 9·987618 | 421 | 10·012382 | ·144412 | 205 | 9·855588 | 49 |
| 12 | 9·843336 | 216 | ·156664 | 9·987871 | 421 | 10·012129 | ·144535 | 205 | 9·855465 | 48 |
| 13 | 9·843466 | 216 | ·156534 | 9·988123 | 421 | 10·011877 | ·144658 | 205 | 9·855342 | 47 |
| 14 | 9·843595 | 216 | ·156405 | 9·988376 | 421 | 10·011624 | ·144781 | 205 | 9·855212 | 46 |
| 15 | 9·843725 | 216 | ·156275 | 9·988629 | 421 | 10·011371 | ·144904 | 205 | 9·855096 | 45 |
| 16 | 9·843855 | 216 | ·156145 | 9·988882 | 421 | 10·011118 | ·145027 | 205 | 9·854973 | 44 |
| 17 | 9·843984 | 216 | ·156016 | 9·989134 | 421 | 10·010866 | ·145150 | 205 | 9·854850 | 43 |
| 18 | 9·844114 | 216 | ·155886 | 9·989387 | 421 | 10·010613 | ·145273 | 205 | 9·854727 | 42 |
| 19 | 9·844243 | 216 | ·155757 | 9·989640 | 421 | 10·010360 | ·145397 | 206 | 9·854603 | 41 |
| 20 | 9·844372 | 215 | ·155628 | 9·989893 | 421 | 10·010107 | ·145520 | 206 | 9·854480 | 40 |
| 21 | 9·844502 | 215 | ·155498 | 9·990145 | 421 | 10·009855 | ·145644 | 206 | 9·854356 | 39 |
| 22 | 9·844631 | 215 | ·155369 | 9·990398 | 421 | 10·009602 | ·145767 | 206 | 9·854233 | 38 |
| 23 | 9·844760 | 215 | ·155240 | 9·990651 | 421 | 10·009349 | ·145891 | 206 | 9·854109 | 37 |
| 24 | 9·844889 | 215 | ·155111 | 9·990903 | 421 | 10·009097 | ·146014 | 206 | 9·853986 | 36 |
| 25 | 9·845018 | 215 | ·154982 | 9·991156 | 421 | 10·008844 | ·146138 | 206 | 9·853862 | 35 |
| 26 | 9·845147 | 215 | ·154853 | 9·991409 | 421 | 10·008591 | ·146262 | 206 | 9·853738 | 34 |
| 27 | 9·845276 | 215 | ·154724 | 9·991662 | 421 | 10·008338 | ·146386 | 206 | 9·853614 | 33 |
| 28 | 9·845405 | 214 | ·154595 | 9·991914 | 421 | 10·008086 | ·146510 | 207 | 9·853490 | 32 |
| 29 | 9·845533 | 214 | ·154467 | 9·992167 | 421 | 10·007833 | ·146634 | 207 | 9·853366 | 31 |
| 30 | 9·845662 | 214 | ·154338 | 9·992420 | 421 | 10·007580 | ·146758 | 207 | 9·853242 | 30 |
| 31 | 9·845790 | 214 | ·154210 | 9·992672 | 421 | 10·007328 | ·146882 | 207 | 9·853118 | 29 |
| 32 | 9·845919 | 214 | ·154081 | 9·992925 | 421 | 10·007075 | ·147006 | 207 | 9·852994 | 28 |
| 33 | 9·846047 | 214 | ·153953 | 9·993178 | 421 | 10·006822 | ·147131 | 207 | 9·852869 | 27 |
| 34 | 9·846175 | 214 | ·153825 | 9·993430 | 421 | 10·006570 | ·147255 | 207 | 9·852745 | 26 |
| 35 | 9·846304 | 214 | ·153696 | 9·993683 | 421 | 10·006317 | ·147380 | 207 | 9·852620 | 25 |
| 36 | 9·846432 | 214 | ·153568 | 9·993936 | 421 | 10·006064 | ·147504 | 207 | 9·852496 | 24 |
| 37 | 9·846560 | 213 | ·153440 | 9·994189 | 421 | 10·005811 | ·147629 | 208 | 9·852371 | 23 |
| 38 | 9·846688 | 213 | ·153312 | 9·994441 | 421 | 10·005559 | ·147753 | 208 | 9·852247 | 22 |
| 39 | 9·846816 | 213 | ·153184 | 9·994694 | 421 | 10·005306 | ·147878 | 208 | 9·852122 | 21 |
| 40 | 9·846944 | 213 | ·153056 | 9·994947 | 421 | 10·005053 | ·148003 | 208 | 9·851997 | 20 |
| 41 | 9·847071 | 213 | ·152929 | 9·995199 | 421 | 10·004801 | ·148128 | 208 | 9·851872 | 19 |
| 42 | 9·847199 | 213 | ·152801 | 9·995452 | 421 | 10·004548 | ·148253 | 208 | 9·851747 | 18 |
| 43 | 9·847327 | 213 | ·152673 | 9·995705 | 421 | 10·004295 | ·148378 | 208 | 9·851622 | 17 |
| 44 | 9·847454 | 213 | ·152546 | 9·995957 | 421 | 10·004043 | ·148503 | 208 | 9·851497 | 16 |
| 45 | 9·847582 | 212 | ·152418 | 9·996210 | 421 | 10·003790 | ·148628 | 209 | 9·851372 | 15 |
| 46 | 9·847709 | 212 | ·152291 | 9·996463 | 421 | 10·003537 | ·148754 | 209 | 9·851246 | 14 |
| 47 | 9·847836 | 212 | ·152164 | 9·996715 | 421 | 10·003285 | ·148879 | 209 | 9·851121 | 13 |
| 48 | 9·847964 | 212 | ·152036 | 9·996968 | 421 | 10·003032 | ·149004 | 209 | 9·850996 | 12 |
| 49 | 9·848091 | 212 | ·151909 | 9·997221 | 421 | 10·002779 | ·149130 | 209 | 9·850870 | 11 |
| 50 | 9·848218 | 212 | ·151782 | 9·997473 | 421 | 10·002527 | ·149255 | 209 | 9·850745 | 10 |
| 51 | 9·848345 | 212 | ·151655 | 9·997726 | 421 | 10·002274 | ·149381 | 209 | 9·850619 | 9 |
| 52 | 9·848472 | 212 | ·151528 | 9·997979 | 421 | 10·002021 | ·149507 | 209 | 9·850493 | 8 |
| 53 | 9·848599 | 211 | ·151401 | 9·998231 | 421 | 10·001769 | ·149632 | 210 | 9·850368 | 7 |
| 54 | 9·848726 | 211 | ·151274 | 9·998484 | 421 | 10·001516 | ·149758 | 210 | 9·850242 | 6 |
| 55 | 9·848852 | 211 | ·151148 | 9·998737 | 421 | 10·001263 | ·149884 | 210 | 9·850116 | 5 |
| 56 | 9·848979 | 211 | ·151021 | 9·998989 | 421 | 10·001011 | ·150010 | 210 | 9·849990 | 4 |
| 57 | 9·849106 | 211 | ·150894 | 9·999242 | 421 | 10·000758 | ·150136 | 210 | 9·849864 | 3 |
| 58 | 9·849232 | 211 | ·150768 | 9·999495 | 421 | 10·000505 | ·150262 | 210 | 9·849738 | 2 |
| 59 | 9·849359 | 211 | ·150641 | 9·999747 | 421 | 10·000253 | ·150389 | 210 | 9·849611 | 1 |
| 60 | 9·849485 | 211 | ·150515 | 10·000000 | 421 | 10·000000 | ·150515 | 210 | 9·849485 | 0 |
| ′ | Cosine. | | Secant. | Cotangent. | | Tangent. | Cosecant. | | Sine. | ′ |

45 DEG.

# INDEX.

THE END.

## THE PRACTICAL COTTON-SPINNER AND MANUFACTURER; OR, THE MANAGER'S AND OVERLOOKER'S COMPANION.

This work contains a Comprehensive System of Calculations for Mill Gearing and Machinery, from the first moving power through the different processes of Carding, Drawing, Slabbing, Roving, Spinning, and Weaving, adapted to American Machinery, Practice, and Usages. Compendious Tables of Yarns and Reeds are added. Illustrated by large Working-Drawings of the most approved American Cotton Machinery. Complete in One Volume, octavo..................$3.50

This edition of Scott's Cotton-Spinner, by OLIVER BYRNE, is designed for the American Operative. It will be found intensely practical, and will be of the greatest possible value to the Manager, Overseer, and Workman.

---

## THE PRACTICAL METAL-WORKER'S ASSISTANT;

For Tin-Plate Workers, Brasiers, Coppersmiths, Zinc-Plate Ornamenters and Workers, Wire Workers, Whitesmiths, Blacksmiths, Bell Hangers, Jewellers, Silver and Gold Smiths, Electrotypers, and all other Workers in Alloys and Metals. By CHARLES HOLTZAPPFEL. Edited, with important additions, by OLIVER BYRNE. Complete in One Volume, octavo.................................$4.00

It will treat of Casting, Founding, and Forging; of Tongs and other Tools; Degrees of Heat and Management of Fires; Welding; of Heading and Swage Tools; of Punches and Anvils; of Hardening and Tempering; of Malleable Iron Castings, Case Hardening, Wrought and Cast Iron. The management and manipulation of Metals and Alloys, Melting and Mixing. The management of Furnaces, Casting and Founding with Metallic Moulds, Joining and Working Sheet Metal. Peculiarities of the different Tools employed. Processes dependent on the ductility of Metals. Wire Drawing, Drawing Metal Tubes, Soldering. The use of the Blowpipe, and every other known Metal-Worker's Tool. To the works of Holtzappfel, OLIVER BYRNE has added all that is useful and peculiar to the American Metal-Worker.

---

## A COMPLETE TREATISE ON TANNING, CURRYING, AND EVERY BRANCH OF LEATHER-DRESSING.

From the French and from original sources. By CAMPBELL MORFIT, one of the Editors of the "Encyclopedia of Chemistry," Author of "Chemistry Applied to the Manufacture of Soap and Candles," and other Scientific Treatises. Illustrated with several hundred Engravings. Complete in One Volume, royal 8vo. (In press.)

This important treatise will be issued from the press at as early a day as the duties of the editor will permit, and it is believed that in no other branch of applied science could more signal service be rendered to American Manufacturers.

The publisher is not aware that in any other work heretofore issued in this country, more space has been devoted to this subject than a single chapter; and in offering this volume to so large and intelligent a class as American Tanners and Leather Dressers, he feels confident of their substantial support and encouragement.

---

## THE MANUFACTURE OF IRON IN ALL ITS VARIOUS BRANCHES:

To which is added an Essay on the Manufacture of Steel, by FREDERICK OVERMAN, Mining Engineer, with one hundred and fifty Wood Engravings. A new edition. In One Volume, octavo, five hundred pages............$5.00

We have now to announce the appearance of another valuable work on the subject which, in our humble opinion, supplies any deficiency which late improvements and discoveries may have caused, from the lapse of time since the date of "Mushet" and "Schrivenor." It is the production of one of our transatlantic brethren, Mr. Frederick Overman, Mining Engineer: and we do not hesitate to set it down as a work of great importance to all connected with the iron interest; one which, while it is sufficiently technological fully to explain chemical analysis, and the various phenomena of iron under different circumstances, to the satisfaction of the most fastidious, is written in that clear and comprehensive style as to be available to the capacity of the humblest mind, and consequently will be of much advantage to those works where the proprietors may see the desirability of placing it in the hands of their operatives.—*London Morning Journal.*

# PRACTICAL SERIES.

THE volumes in this Series are published in duodecimo form, and the design is to furnish to Artisans, for a moderate sum, Hand-books of the different Arts and Manufactures, in order that they may be enabled to keep pace with the improvements of the age.

There have already appeared—

THE AMERICAN MILLER AND MILLWRIGHT'S ASSISTANT. $1.
THE TURNER'S COMPANION. 75 cts.
THE PAINTER, GILDER, AND VARNISHER'S COMPANION. 75 cts.
THE DYER AND COLOUR-MAKER'S COMPANION. 75 cts.
THE BUILDER'S COMPANION. $1.
THE CABINET-MAKER'S COMPANION. 75 cts.

The following, among others, are in preparation:—

A TREATISE ON A BOX OF INSTRUMENTS. By THOMAS KENTISH.
THE PAPER-HANGER'S COMPANION. By J. ARROWSMITH.
THE ASSAYER'S GUIDE. By OSCAR M. LIEBER.

## THE AMERICAN MILLER AND MILLWRIGHT'S ASSISTANT:

By WILLIAM CARTER HUGHES, Editor of "The American Miller," (newspaper), Buffalo, N. Y. Illustrated by Drawings of the most approved Machinery. In One Volume, 12mo....$1

The author offers it as a substantial reference, instead of speculative theories, which belong only to those not immediately attached to the business. Special notice is also given of most of the essential improvements which have of late been introduced for the benefit of the Miller.—*Savannah Republican.*

The whole business of making flour is most thoroughly treated by him.—*Bulletin.*

A very comprehensive view of the Millwright's business.—*Southern Literary Messenger.*

## THE TURNER'S COMPANION:

Containing Instructions in Concentric, Elliptic, and Eccentric Turning. Also, various Plates of Chucks, Tools, and Instruments, and Directions for using the Eccentric Cutter, Drill, Vertical Cutter, and Circular Rest; with Patterns and Instructions for working them. Illustrated by numerous Engravings. In One Volume, 12mo....75 cts.

The object of the Turner's Companion is to explain in a clear, concise, and intelligible manner, the rudiments of this beautiful art.—*Savannah Republican.*

There is no description of turning or lathe-work that this elegant little treatise does not describe and illustrate.—*Western Lit. Messenger.*

## THE PAPER-HANGER'S COMPANION:

In which the Practical Operations of the Trade are systematically laid down; with copious Directions Preparatory to Papering; Preventions against the effect of Damp in Walls; the various Cements and Pastes adapted to the several purposes of the Trade; Observations and Directions for the Panelling and Ornamenting of Rooms, &c., &c. By JAMES ARROWSMITH. In One Volume, 12mo.

## THE PAINTER, GILDER, AND VARNISHER'S COMPANION:

Containing Rules and Regulations for every thing relating to the arts of Painting, Gilding, Varnishing, and Glass Staining; numerous useful and valuable Receipts; Tests for the detection of adulterations in Oils, Colours, &c., and a Statement of the Diseases and Accidents to which Painters, Gilders, and Varnishers are particularly liable; with the simplest methods of Prevention and Remedy. Second Edition. In One Volume, 12mo, cloth....................75 cts.

Rejecting all that appeared foreign to the subject, the compiler has omitted nothing of real practical worth.—*Hunt's Merchants' Magazine.*

An excellent *practical work*, and one which the practical man cannot afford to be without.—*Farmer and Mechanic.*

It contains every thing that is of interest to persons engaged in this trade.—*Bulletin.*

This book will prove valuable to all whose business is in any way connected with painting.—*Scott's Weekly.*

Cannot fail to be useful.—*N. Y. Commercial.*

---

## THE DYER AND COLOUR-MAKER'S COMPANION:

Containing upwards of two hundred Receipts for making Colours, on the most approved principles, for all the various styles and fabrics now in existence; with the Scouring Process, and plain Directions for Preparing, Washing-off, and Finishing the Goods. Second Edition. In One Volume, 12mo, cloth..........................................................................................75 cts.

This is another of that most excellent class of practical books, which the publisher is giving to the public. Indeed, we believe there is not, for manufacturers, a more valuable work, having been prepared for, and expressly adapted to their business.—*Farmer and Mechanic.*

It is a valuable book.—*Otsego Republican.*

We have shown it to some practical men, who all pronounced it the completest thing of the kind they had seen.—*N. Y. Nation.*

---

## THE BUILDER'S POCKET COMPANION:

Containing the Elements of Building, Surveying, and Architecture; with Practical Rules and Instructions connected with the subject. By A. C. SMEATON, Civil Engineer, &c. Second Edition. In One Volume, 12mo.......$1

CONTENTS.—The Builder, Carpenter, Joiner, Mason, Plasterer, Plumber, Painter, Smith, Practical Geometry, Surveyor, Cohesive Strength of Bodies, Architect.

---

## THE ASSAYER'S GUIDE;

Or, Practical Directions to Assayers, Miners, and Smelters, for the Tests and Assays by Heat and by Wet Processes of the Ores of all the principal Metals, and of Gold and Silver Coins and Alloys. By OSCAR M. LIEBER, Mining Engineer, Geologist to the State of Mississippi, etc. 12mo. (In press.)

---

## A TREATISE ON A BOX OF INSTRUMENTS,

And the SLIDE RULE, with the Theory of Trigonometry and Logarithms, including Practical Geometry, Surveying, Measuring of Timber, Cask and Malt Gauging, Heights and Distances. By THOMAS KENTISH. In One Volume, 12mo.

## THE CABINET-MAKER AND UPHOLSTERER'S COMPANION:

Comprising the Rudiments and Principles of Cabinet-making and Upholstery, with familiar Instructions, illustrated by Examples, for attaining a proficiency in the Art of Drawing, as applicable to Cabinet-Work; the processes of Veneering, Inlaying, and Buhl Work; the art of Dyeing and Staining Wood, Bone, Tortoise-shell, &c. Directions for Lackering, Japanning, and Varnishing; to make French Polish; to prepare the best Glues, Cements, and Compositions, and a number of Receipts particularly useful for Workmen generally, with Explanatory and Illustrative Engravings. By J. STOKES. In One Volume, 12mo, with Illustrations........................................................................75 cts.

A large amount of practical information, of great service to all concerned in those branches of business. —*Ohio State Journal.*

---

## PROPELLERS AND STEAM NAVIGATION:

With Biographical Sketches of Early Inventors. By ROBERT MACFARLANE, C. E., Editor of the "Scientific American." In One Volume, 12mo. Illustrated by over Eighty Wood Engravings..........................................75 cts.

The object of this "History of Propellers and Steam Navigation" is twofold. One is the arrangement and description of many devices which have been invented to propel vessels, in order to prevent many ingenious men from wasting their time, talents, and money on such projects. The immense amount of time, study, and money thrown away on such contrivances is beyond calculation. In this respect, it is hoped that it will be the means of doing some good.—*Preface.*

---

## TABLES OF LOGARITHMS FOR ENGINEERS AND MACHINISTS:

Containing the Logarithms of the Natural Numbers, from 1 to 100000, by the help of Proportional Differences. And Logarithmic Sines, Cosines, Tangents, Co-tangents, Secants, and Co-secants, for every Degree and Minute in the Quadrant. To which are added, Differences for every 100 Seconds. By OLIVER BYRNE, Civil, Military, and Mechanical Engineer. In One Vol. 8vo. cl.... $1

---

## THE FRUIT, FLOWER, AND KITCHEN GARDEN.

By PATRICK NEILL, L. L. D., F. R. S. E., Secretary to the Royal Caledonian Horticultural Society. Adapted to the United States, from the Fourth Edition, revised and improved by the Author. Illustrated by fifty Wood Engravings of Hothouses, &c. &c. In One Volume, 12mo.........................$1.25

This volume supplies a desideratum much felt, and gives within a moderate compass all the horticultural information necessary for practical use.—*Newark Mercury.*

A valuable addition to the horticulturist's library.—*Baltimore Patriot.*

This work is the production of a most celebrated British horticulturist, Dr. NEILL, of Scotland, for upwards of thirty years the Secretary of the Caledonian Horticultural Society, and in every way qualified to make a standard book upon the subject it discusses. The careful adaptation of the work to the peculiar circumstances and necessities of our own people, is a subject of congratulation, since good books upon horticulture cannot be too much multiplied. We are pleased with the comprehensiveness of Dr. NEILL'S treatise.—*Southern Literary Gazette.*

---

## ELEMENTARY PRINCIPLES OF CARPENTRY.

By THOMAS TREDGOLD. In One Volume, quarto, with numerous Illustrations...................................................................................$2.50

## THE ENCYCLOPEDIA OF CHEMISTRY, PRACTICAL AND THEORETICAL:

Embracing its Application to the Arts, Metallurgy, Mineralogy, Geology, Medicine, and Pharmacy. By JAMES C. BOOTH, Melter and Refiner in the United States Mint, Professor of Applied Chemistry in the Franklin Institute, &c.; assisted by CAMPBELL MORFIT, Author of "Chemical Manipulations," &c. Complete in One Volume, royal octavo, 978 pages, with numerous Woodcuts and other Illustrations. Second Edition. Full bound ....................................$5

It covers the whole field of Chemistry as applied to Arts and Sciences. * * * As no library is complete without a common dictionary, it is also our opinion that none can be without this Encyclopedia of Chemistry.—*Scientific American.*

A work of time and labour, and a treasury of chemical information.—*North American.*

By far the best manual of the kind which has been presented to the American public.—*Boston Courier.*

An invaluable work for the dissemination of sound practical knowledge.—*Ledger.*

A treasury of chemical information, including all the latest and most important discoveries.—*Baltimore American.*

At the first glance at this massive volume, one is amazed at the amount of reading furnished in its compact double pages, about one thousand in number. A further examination shows that every page is richly stored with information, and that while the labours of the authors have covered a wide field, they have neglected or slighted nothing. Every chemical term, substance, and process is elaborately, but intelligibly, described. The whole science of Chemistry is placed before the reader as fully as is practicable with a science continually progressing. * * * Unlike most American works of this class, the authors have not depended upon any one European work for their materials. They have gathered theirs from works on Chemistry in all languages, and in all parts of Europe and America; their own experience, as practical chemists, being ever ready to settle doubts or reconcile conflicting authorities. The fruit of so much toil is a work that must ever be an honour to American science.—*Evening Bulletin.*

---

## SYLLABUS OF A COMPLETE COURSE OF LECTURES ON CHEMISTRY:

Including its Application to the Arts, Agriculture, and Mining, prepared for the use of the Gentlemen Cadets at the Hon. E. I. Co.'s Military Seminary, Addiscombe. By Professor E. SOLLY, Lecturer on Chemistry in the Hon. E. I. Co.'s Military Seminary. Revised by the Author of "Chemical Manipulations." In One Volume, octavo, cloth.............................................$1.25

The present work is designed to occupy a vacant place in the libraries of Chemical text-books. It is admirably adapted to the wants of both TEACHER and PUPIL; and will be found especially convenient to the latter, either as a companion in the class-room, or as a remembrancer in the study. It gives, at a glance, under appropriate headings, a classified view of the whole science, which is at the same time compendious and minutely accurate; and its wide margins afford sufficient blank space for such manuscript notes as the student may wish to add during lectures or recitations.

The almost indispensable advantages of such an impressive aid to memory are evident to every student who has used one in other branches of study. Therefore, as there is now no Chemical Syllabus, we have been induced by the excellences of this work to recommend its republication in this country; confident that an examination of the contents will produce full conviction of its intrinsic worth and usefulness.—*Editor's Preface.*

---

## AN ELEMENTARY COURSE OF INSTRUCTION ON ORDNANCE AND GUNNERY.

Prepared for the use of the Midshipmen at the Naval School. By JAMES H. WARD, U. S. N. In One Volume, octavo............................$1.50

---

## STEAM FOR THE MILLION.

An Elementary Outline Treatise on the Nature and Management of Steam, and the Principles and Arrangement of the Engine. Adapted for Popular Instruction, for Apprentices, and for the use of the Navigator. With an Appendix containing Notes on Expansive Steam, &c. In One Volume, 8vo...37½ cts.

## HOUSEHOLD SURGERY; OR, HINTS ON EMERGENCIES.

By J. F. South, one of the Surgeons of St. Thomas's Hospital. In One Volume, 12mo. Illustrated by nearly fifty Engravings............$1.25

CONTENTS:

*The Doctor's Shop.*—Poultices, Fomentations, Lotions, Liniments, Ointments, Plasters. *Surgery.*—Blood-letting, Blistering, Vaccination, Tooth-drawing, How to put on a Roller, Lancing the Gums, Swollen Veins, Bruises, Wounds, Torn or Cut Achilles Tendon, What is to be done in cases of sudden Bleeding from various causes, Scalds and Burns, Frost-bite, Chilblains, Sprains, Broken Bones, Bent Bones, Dislocations, Ruptures, Piles, Protruding Bowels, Wetting the Bed, Whitlow, Boils, Black-heads, Ingrowing Nails, Bunions, Corns, Sty in the Eye, Blight in the Eye, Tumours in the Eyelids, Inflammation on the Surface of the Eye, Pustules on the Eye, Milk Abscesses, Sore Nipples, Irritable Breast, Breathing, Stifling, Choking, Things in the Eye, On Dress, Exercise and Diet of Children, Bathing, Infections, Observations on Ventilation.

---

## HOUSEHOLD MEDICINE.

By D. Francis Condie, M. D. In One Volume, 12mo. Uniform with, and a companion to, the above. (In immediate preparation.)

---

## ELWOOD'S GRAIN TABLES:

Showing the value of Bushels and Pounds of different kinds of Grain, calculated in Federal Money, so arranged as to exhibit upon a single page the value at a given price from *ten cents to two dollars* per bushel, of any quantity from *one pound to ten thousand bushels.* By J. L. Elwood. A new Edition. In One Volume, 12mo..........................................................................$1

To Millers and Produce Dealers this work is pronounced by all who have it in use, to be superior in arrangement to any work of the kind published—and *unerring accuracy in every calculation may be relied upon in every instance.*

☞ A reward of Twenty-five Dollars is offered for an error of one cent found in the work.

---

## PERFUMERY; ITS MANUFACTURE AND USE:

With Instructions in every branch of the Art, and Receipts for all the Fashionable Preparations; the whole forming a valuable aid to the Perfumer, Druggist, and Soap Manufacturer. Illustrated by numerous Woodcuts. From the French of Celnart, and other late authorities. With Additions and Improvements, by Campbell Morfit, one of the Editors of the "Encyclopedia of Chemistry." In One Volume, 12mo, cloth..............................................$1

---

## ELECTROTYPE MANIPULATION:

Being the Theory and Plain Instructions in the Art of Working in Metals, by Precipitating them from their Solutions, through the agency of Galvanic or Voltaic Electricity. By Charles V. Walker, Hon. Secretary to the London Electrical Society, &c. Illustrated by Woodcuts. From the Thirteenth London Edition. In One Volume, 24mo, cloth........................................62 cts.

## PHOTOGENIC MANIPULATION:

Containing the Theory and Plain Instructions in the Art of Photography, or the Production of Pictures through the Agency of Light; including Calotype, Chrysotype, Cyanotype, Chromatype, Energiatype, Anthotype, Amphitype, Daguerreotype, Thermography, Electrical and Galvanic Impressions. By George Thomas Fisher, Jr., Assistant in the Laboratory of the London Institution. Illustrated by Wood-cuts. In One Volume, 24mo, cloth...........62 cts.

---

## MATHEMATICS FOR PRACTICAL MEN:

Being a Common-Place Book of Principles, Theorems, Rules, and Tables, in various Departments of Pure and Mixed Mathematics, with their Applications, especially to the pursuits of Surveyors, Architects, Mechanics, and Civil Engineers. With numerous Engravings. By Olinthus Gregory, L. L. D., F. R. A. S.........................................................................................$1.50

Only let men awake, and fix their eye, one while on the nature of things, another while on the application of them to the use and service of mankind.—*Lord Bacon.*

---

## SHEEP HUSBANDRY IN THE SOUTH:

Comprising a Treatise on the Acclimation of Sheep in the Southern States, and an Account of the different Breeds. Also, a Complete Manual of Breeding, Summer and Winter Management, and of the Treatment of Diseases. With Portraits and other Illustrations. By Henry S. Randall. In One Volume, octavo.........................................................................$1.25

---

## MISS LESLIE'S COMPLETE COOKERY.

Directions for Cookery, in its Various Branches. By Miss Leslie. Forty-first Edition. Thoroughly Revised, with the Addition of New Receipts. In One Volume, 12mo, half bound, or in sheep.............................$1

In preparing a new and carefully revised edition of this my first work on cookery, I have introduced improvements, corrected errors, and added new receipts, that I trust will on trial be found satisfactory. The success of the book (proved by its immense and increasing circulation) affords conclusive evidence that it has obtained the approbation of a large number of my countrywomen; many of whom have informed me that it has made practical housewives of young ladies who have entered into married life with no other acquirements than a few showy accomplishments. Gentlemen, also, have told me of great improvements in the family table, after presenting their wives with this manual of domestic cookery, and that, after a morning devoted to the fatigues of business, they no longer find themselves subjected to the annoyance of an ill-dressed dinner.—*Preface.*

---

## MISS LESLIE'S TWO HUNDRED RECEIPTS IN FRENCH COOKERY.

A new Edition, in cloth. ....................................................................25 cts.

---

## TWO HUNDRED DESIGNS FOR COTTAGES AND VILLAS, &c. &c.,

Original and Selected. By Thomas U. Walter, Architect of Girard College, and John Jay Smith, Librarian of the Philadelphia Library. In Four Parts, quarto..................................................................................$10

# STANDARD ILLUSTRATED POETRY.

## THE TALES AND POEMS OF LORD BYRON:

Illustrated by Henry Warren. In One Volume, royal 8vo, with 10 Plates, scarlet cloth, gilt edges........$5

Morocco extra........$7

It is illustrated by several elegant engravings, from original designs by Warren, and is a most splendid work for the parlour or study.—*Boston Evening Gazette.*

## CHILDE HAROLD; A ROMAUNT BY LORD BYRON:

Illustrated by 12 Splendid Plates, by Warren and others. In One Volume, royal 8vo, cloth extra, gilt edges........$5

Morocco extra........$7

Printed in elegant style, with splendid pictures, far superior to any thing of the sort usually found in books of this kind.—*N. Y. Courier.*

## SPECIMENS OF THE BRITISH POETS.

From the time of Chaucer to the end of the Eighteenth Century. By Thomas Campbell. In One Volume, royal 8vo. (In press.)

## THE FEMALE POETS OF AMERICA.

By Rufus W. Griswold. A new Edition. In One Volume, royal 8vo. Cloth, gilt........$2.50

Cloth extra, gilt edges........$3

Morocco super extra........$4.50

The best production which has yet come from the pen of Dr. Griswold, and the most valuable contribution which he has ever made to the literary celebrity of the country.—*N. Y. Tribune.*

## THE LADY OF THE LAKE:

By Sir Walter Scott. Illustrated with 10 Plates, by Corbould and Meadows. In One Volume, royal 8vo. Bound in cloth extra, gilt edges........$5

Turkey morocco super extra........$7

This is one of the most truly beautiful books which has ever issued from the American press.

## LALLA ROOKH; A ROMANCE BY THOMAS MOORE:

Illustrated by 13 Plates, from Designs by Corbould, Meadows, and Stephanoff. In One Volume, royal 8vo. Bound in cloth extra, gilt edges...$5

Turkey morocco super extra........$7

This is published in a style uniform with the "Lady of the Lake."

## THE POETICAL WORKS OF THOMAS GRAY:

With Illustrations by C. W. Radcliffe. Edited with a Memoir, by Henry Reed, Professor of English Literature in the University of Pennsylvania. In One Volume, 8vo. Bound in cloth extra, gilt edges..............$3.50

Turkey morocco super extra..........................................................$5.50

It is many a day since we have seen issued from the press of our country a volume so complete and truly elegant in every respect. The typography is faultless, the illustrations superior, and the binding superb.—*Troy Whig.*

We have not seen a specimen of typographical luxury from the American press which can surpass this volume in choice elegance.—*Boston Courier.*

It is eminently calculated to consecrate among American readers (if they have not been consecrated already in their hearts) the pure, the elegant, the refined, and, in many respects, the sublime imaginings of Thomas Gray.—*Richmond Whig.*

---

## THE POETICAL WORKS OF HENRY WADSWORTH LONGFELLOW:

Illustrated by 10 Plates, after Designs by D. Huntingdon, with a Portrait. Ninth Edition. In One Volume, royal 8vo. Bound in cloth extra, gilt edges..........................................................................$5

Morocco super extra..........................................................$7

This is the very luxury of literature—Longfellow's charming poems presented in a form of unsurpassed beauty.—*Neal's Gazette.*

---

## POETS AND POETRY OF ENGLAND IN THE NINETEENTH CENTURY:

By Rufus W. Griswold. Illustrated. In One Volume, royal 8vo. Bound in cloth..........................................................$3

Cloth extra, gilt edges..........................................................$3.50

Morocco super extra..........................................................$5

Such is the critical acumen discovered in these selections, that scarcely a page is to be found but is redolent with beauties, and the volume itself may be regarded as a galaxy of literary pearls.—*Democratic Review.*

---

## THE POETS AND POETRY OF THE ANCIENTS:

By William Peter, A. M. Comprising Translations and Specimens of the Poets of Greece and Rome, with an elegant engraved View of the Coliseum at Rome. Bound in cloth..........................................................$3

Cloth extra, gilt edges..........................................................$3.50

Turkey morocco super extra..........................................................$5

---

## THE FEMALE POETS OF GREAT BRITAIN.

With Copious Selections and Critical Remarks. By Frederic Rowton. With Additions by an American Editor, and finely engraved Illustrations by celebrated Artists. In One Volume, royal 8vo. Bound in cloth extra, gilt edges..........................................................$5

Turkey morocco..........................................................$7

Mr. Rowton has presented us with admirably selected specimens of nearly one hundred of the most celebrated female poets of Great Britain, from the time of Lady Juliana Bernes, the first of whom there is any record, to the Mitfords, the Hewitts, the Cooks, the Barretts, and others of the present day.—*Hunt's Merchants' Magazine.*

### THE TASK, AND OTHER POEMS.

By WILLIAM COWPER. Illustrated by 10 Steel Engravings. In One Volume, 12mo. Cloth extra, gilt edges........$2
Morocco extra........$3

---

### THE POETICAL WORKS OF NATHANIEL P. WILLIS.

Illustrated by 16 Plates, after Designs by E. LEUTZE. In One Volume, royal 8vo. A new Edition. Bound in cloth extra, gilt edges........$5
Turkey morocco super extra........$7

This is one of the most beautiful works ever published in this country.—*Courier and Inquirer.*

---

# MISCELLANEOUS.

---

### ADVENTURES OF CAPTAIN SIMON SUGGS;

And other Sketches. By JOHNSON J. HOOPER. With Illustrations. 12mo, paper........50 cts.
Cloth........62 cts.

### AUNT PATTY'S SCRAP-BAG.

By Mrs. CAROLINE LEE HENTZ, Author of "Linda." 12mo. Paper covers........50 cts.
Cloth........62 cts.

### BIG BEAR OF ARKANSAS;

And other Western Sketches. Edited by W. T. PORTER. In One Volume, 12mo, paper........50 cts.
Cloth........62 cts.

### COMIC BLACKSTONE.

By GILBERT ABBOT A' BECKET. Illustrated. Complete in One Volume. Cloth........75 cts

### GHOST STORIES.

Illustrated by Designs by DARLEY. In One Volume, 12mo, paper covers........50 cts.

### MODERN CHIVALRY; OR, THE ADVENTURES OF CAPTAIN FARRAGO AND TEAGUE O'REGAN.

By H. H. BRACKENRIDGE. Second Edition since the Author's death. With a Biographical Notice, a Critical Disquisition on the Work, and Explanatory Notes. With Illustrations, from Original Designs, by DARLEY. Two Volumes, paper covers........$1.00
Cloth or sheep........$1.25

## THE COMPLETE WORKS OF LORD BOLINGBROKE:

With a Life, prepared expressly for this Edition, containing Additional Information relative to his Personal and Public Character, selected from the best authorities. In Four Volumes, 8vo. Bound in cloth............$6.00
In sheep......................................................................$7.50

## FAMILY ENCYCLOPEDIA

Of Useful Knowledge and General Literature; containing about Four Thousand Articles upon Scientific and Popular Subjects. With Plates. By JOHN L. BLAKE, D. D. In One Volume, 8vo, full bound..........................$5

## CHRONICLES OF PINEVILLE.

By the Author of "Major Jones's Courtship." Illustrated by DARLEY. 12mo, paper...........................................................50 cts.
Cloth......................................................................62 cts.

## GILBERT GURNEY.

By THEODORE HOOK. With Illustrations. In One Volume, 8vo, paper......................................................................50 cts.

## MEMOIRS OF THE GENERALS, COMMODORES, AND OTHER COMMANDERS,

Who distinguished themselves in the American Army and Navy, during the War of the Revolution, the War with France, that with Tripoli, and the War of 1812, and who were presented with Medals, by Congress, for their gallant services. By THOMAS WYATT, A. M., Author of "History of the Kings of France." Illustrated with Eighty-two Engravings from the Medals. 8vo, cloth gilt ......................................................................$2.00
Half morocco......................................................................$2.50

## VISITS TO REMARKABLE PLACES:

Old Halls, Battle Fields, and Scenes Illustrative of striking passages in English History and Poetry. By WILLIAM HOWITT. In Two Volumes, 8vo, cloth......................................................................$3.50

## THE MISCELLANEOUS WORKS OF WILLIAM HAZLITT;

Including Table-talk; Opinions of Books, Men and Things; Lectures on Dramatic Literature of the Age of Elizabeth; Lectures on the English Comic Writers; The Spirit of the Age, or Contemporary Portraits. Five Volumes, 12mo, cloth ......................................................................$5.00
Half calf......................................................................$6.25

## FLORAL OFFERING:

A Token of Friendship. Edited by FRANCES S. OSGOOD. Illustrated by 10 beautiful Bouquets of Flowers. In One Volume, 4to, muslin, gilt edges......................................................................$3.50
Turkey morocco super extra......................................................................$5.50

## THE HISTORICAL ESSAYS,

Published under the title of "Dix Ans D'Etude Historique," and Narratives of the Merovingian Era; or, Scenes in the Sixth Century. With an Autobiographical Preface. By AUGUSTUS THIERRY, Author of the "History of the Conquest of England by the Normans." 8vo, paper......................75 cts.
Cloth......................................................................$1.00

www.ingramcontent.com/pod-product-compliance
Lightning Source LLC
LaVergne TN
LVHW021220110826
845150LV00002B/207

* 9 7 8 1 4 2 5 5 6 4 6 5 0 *